Springer Monographs in Mathematics

More information about series at www.springer.com/series/3733

Jean-Pierre Tignol • Adrian R. Wadsworth

Value Functions on Simple Algebras, and Associated Graded Rings

Springer

Jean-Pierre Tignol
ICTEAM Institute
Université Catholique de Louvain
Louvain-la-Neuve
Belgium

Adrian R. Wadsworth
Department of Mathematics
University of California, San Diego
La Jolla, California
USA

ISSN 1439-7382
ISSN 2196-9922 (electronic)
Springer Monographs in Mathematics
ISBN 978-3-319-16359-8
ISBN 978-3-319-16360-4 (eBook)
DOI 10.1007/978-3-319-16360-4

Library of Congress Control Number: 2015937371

Mathematics Subject Classification: 16W60, 16W50, 16K20, 16K50, 16-03

Springer Cham Heidelberg New York Dordrecht London

Printed on acid-free paper

Springer is part of Springer Science+Business Media (www.springer.com)

Preface

The theory of finite-dimensional division algebras witnessed several breakthroughs in the latter decades of the twentieth century. Important advances, such as Amitsur's construction of noncrossed product division algebras and Platonov's solution of the Tannaka–Artin problem, relied on an inventive use of valuation theory, applied in the context of noncommutative rings. The subsequent development of valuation theory for finite-dimensional division algebras led to significant simplifications of the initial results and to a host of new constructions of division algebras satisfying various conditions, which shed much light on the structure of these algebras. In this research area, valuation theory has become a standard tool, for which this book is intended to provide a useful reference.

The theory of valuations and valuation rings has been extended to division rings in several different ways. We treat here only the most stringent of these extensions, which is the one that has turned out to be most useful in applications. Thus, our valuations on division algebras are defined by the same axioms as the (Krull) valuations on fields; hence they restrict to a valuation in the classical sense on the center of the division ring. Yet, noncommutative valuation theory has some significant features that give it a different flavor from the commutative theory. Notably, there are many fewer valuations on division algebras than on fields: A valuation always extends from a field F to any field containing F; often there are many such extensions. But if D is a division algebra with center F and finite-dimensional over F, a valuation on F extends to D if and only if it has a *unique* extension to every field between F and D. Thus, very often it has no extension to D at all. But if it does extend, then the extension is unique. Consequently, the presence of a valuation on a division algebra D is a rather special phenomenon. When this occurs, it often gives a great deal of information about D and its subalgebras that can be virtually inaccessible for most division algebras. For this reason, valuation theory has had some of its greatest success in the construction of examples, such as noncrossed product algebras and division algebras with nontrivial reduced Whitehead group SK_1.

Henselian valuations on fields and Henselizations play a role for general valuations analogous to that of complete valuations and completions for rank 1 valuations. Henselian valuations on the center are even more important in the noncommutative theory because a Henselian valuation on a field F has a unique extension to each field algebraic over F. Consequently, it extends (uniquely) to each division algebra finite-dimensional over F. Much of the work on valued division algebra has thus focussed on algebras over Henselian fields. Also, notable results on arbitrary valued division algebras have been obtained by first proving the Henselian case (e.g., "Ostrowski's Theorem" on the defect of valued division algebras).

Another distinctive feature of valuation theory on division algebras is a greater complexity of the residue structure, and some notable interaction between the residue algebra and the value group: There is a canonical action of the value group of a valued division algebra on the center of its residue division algebra. This provides an important piece of information even in the most classical cases studied by Hasse in the 1930s, as it is related to the local invariant of division algebras over local fields.

When division algebras are being investigated, simple algebras with zero divisors frequently arise, e.g., as tensor products or scalar extensions of division algebras. Therefore, it has been a drawback for noncommutative valuation theory that valuations make sense only for division algebras: The basic axiom that

$$v(ab) = v(a) + v(b) \qquad (*)$$

breaks down if $ab = 0$ for nonzero a and b. A few years ago the authors found a way to address this difficulty by defining a more general notion of value function that we call a *gauge*, which can exist on a (finite-dimensional) semisimple algebra A over a field F, with respect to a valuation on F. For a function α on A to be a gauge, we replace the multiplicative condition $(*)$ for a valuation with the following *surmultiplicativity* condition:

$$\alpha(ab) \geq \alpha(a) + \alpha(b) \quad \text{for all } a, b \in A.$$

The filtration of A induced by α yields an associated graded algebra $\mathsf{gr}(A)$, and gauges are distinguished among surmultiplicative value functions by a condition on $\mathsf{gr}(A)$: This graded algebra must be graded semisimple, which means that it has no nilpotent homogeneous ideals.

Gauges work remarkably well. They show good behavior with respect to tensor products of algebras and scalar extensions. Moreover, there are natural constructions of gauges on many symbol algebras, cyclic algebras, and crossed product algebras. Additionally, over a Henselian base field, the gauges on the endomorphism algebra of a vector space are exactly the operator norms that are familiar in functional analysis.

Even for valuations on division algebras associated graded structures prove particularly useful. They encapsulate all the information about the residue algebra, the value group, and the canonical action of the value group on the center of the residue algebra. They should be regarded as a substantially enhanced analogue of the residue algebra. Usually, when one passes from a ring to an associated graded ring one obtains a simplified structure, but at the price of significant loss of information about the original ring. With valued division algebras, the graded ring is definitely simpler and easier to work with than the original valued algebra, with surprisingly little lost in the transition to the graded setting. Indeed, if the valuation on the center is Henselian, we will see that under mild tameness conditions (which hold automatically whenever the characteristic of the residue field is 0 or prime to the degree of the division algebra) the graded algebra $\mathsf{gr}(D)$ associated to a division algebra D determines D up to isomorphism; moreover, the graded subalgebras of $\mathsf{gr}(D)$ then classify the subalgebras of D.

Graded structures are thus central to our approach of valuation theory. Our general strategy is to prove results first in the graded setting, where the arguments are often easier and more transparent. With gauges at our disposal, the passage to the corresponding results for valued division algebras is often very quick. To take full advantage of this method, we build a solid foundation on graded algebras with grade set lying in a torsion-free abelian group. It is worth pointing out that, in contrast with the classical theory, which mostly deals with valuations with value group $\mathbb{Z}$, our valuations take their values in arbitrary totally ordered abelian groups. Valuations of higher rank (i.e., with value group not embeddable in $\mathbb{R}$) allow a greater richness in the possible structure of the value group and of the residue algebra. Moreover, new phenomena occur, such as totally ramified division algebras and algebras with noncyclic center of the residue—these have been particularly important in the construction of significant examples.

The material in this book can be roughly divided into three parts, which we briefly outline below, referring to the introduction of each chapter for additional information.

The first part consists of Chapters 1–4. They lay the groundwork for the theory of valuations on finite-dimensional division algebras and its extension to the theory of gauges on finite-dimensional semisimple algebras. The first chapter introduces the fundamental notions associated with valuations on division algebras and provides assorted examples. We view a valuation on the algebra as an extension of a known valuation on its center. In Chapter 2, the focus shifts to graded structures with a torsion-free abelian grade group. Graded rings in which the nonzero homogeneous elements are invertible are called graded division rings, because they display properties that are strikingly similar to those of the usual division rings. We are thus led to introduce graded vector spaces, and we develop a graded analogue of the Wedderburn and Noether theory of simple algebras. In Chapter 3, we return to the theme of valuations, which we extend to vector spaces and algebras over valued fields in order to define gauges on semisimple algebras. This first part of the book culminates in Chapter 4 with a determination of the necessary and sufficient condition for the existence of gauges. This condition involves the division algebras Brauer-equivalent to the simple components of the semisimple algebra after scalar extension to a Henselization of the base field: These division algebras must each be defectless, which means that their dimension over their center must be the product of the residue degree and the ramification index. In particular, gauges always exist when the residue characteristic is zero.

The second part, comprising Chapters 5–7, addresses various topics related to the Brauer group of valued fields. We first discuss graded field extensions in Chapter 5, and review properties of valued field extensions from the perspective of their associated graded field extensions. Brauer groups of graded fields and of valued fields form the subject of Chapter 6. Valuation-theoretic properties define an ascending sequence of three subgroups of the Brauer group $Br(F)$ of a valued field: the inertial part $Br_{in}(F)$, the inertially split part $Br_{is}(F)$, and the tamely ramified part $Br_{tr}(F)$. We use gauges to relate these subgroups to corresponding subgroups of the Brauer group $\mathsf{Br}(\mathsf{gr}(F))$ of the associated graded field $\mathsf{gr}(F)$. The main result of this part of the book yields for a Henselian field F a canonical index-preserving isomorphism $Br_{tr}(F) \xrightarrow{\sim} \mathsf{Br}(\mathsf{gr}(F))$ mapping the Brauer class of a tame division algebra D to the Brauer class of $\mathsf{gr}(D)$. We can then easily read off information about the pieces of $Br_{tr}(F)$ from the corresponding data about $\mathsf{Br}(\mathsf{gr}(F))$. The inertial, or unramified part of the Brauer group is canonically isomorphic to the Brauer group of the residue field: $Br_{in}(F) \cong Br(\overline{F})$. The inertially split part $Br_{is}(F)$ consists of the classes of division algebras split by the maximal inertial (= unramified) extension field of F. We give a generalization of Witt's classical description of the Brauer group of a complete discretely-valued field, in the form of a "ramification" isomorphism from the quotient $Br_{is}(F)/\,Br_{in}(F)$ to a group of characters of the absolute Galois group of the residue field $\overline{F}$. The next quotient $Br_{tr}(F)/\,Br_{is}(F)$ is described in Chapter 7, where division algebras totally ramified over their centers are thoroughly investigated. When the base field is Henselian, the properties of such algebras can be read off from the extension of value groups, with the help of a canonical alternating pairing with values in the group of roots of unity of the residue field. Since totally ramified division algebras arise only when the value group has rank at least 2, such algebras have been relatively less studied in the literature; yet their structure is very simple and explicit.

In the third part of the book, Chapters 8–12, we apply the preceding results to investigate the structure of division algebras over Henselian fields, and we present several applications. Following the same methodology as in previous chapters, in Chapter 8 we first consider the structure of graded division algebras; we then derive corresponding structure theorems for division algebras over Henselian fields by relating the algebra to its associated graded algebra. We thereby recover easily several results that have been previously established by much more complicated methods. Historically, a primary application of valuation theory has been in the construction of significant examples. Our last four chapters are devoted to the presentation of such examples. In Chapter 9 we obtain information on the maximal subfields and splitting fields of valued division algebras, and construct noncyclic division algebras with pure maximal subfields, noncyclic p-algebras, and noncrossed product algebras. Examples of division algebras that do not decompose into tensor products of proper subalgebras are given in Chapter 10, and Chapter 11 discusses reduced Whitehead group computations: We show that if D is a division algebra tamely ramified over a Henselian field then $SK_1(D) \cong SK_1(\mathsf{gr}(D))$. This leads to quick proofs of many formulas for $SK_1(D)$. Finally, we give in Chapter 12 a modified version of recent results of Merkurjev and Baek–Merkurjev using valuation theory to obtain lower bounds on the essential dimension of central simple algebras of given degree and exponent.

The assumed background for this book is acquaintance with the classical theory of central simple algebras, together with a basic knowledge of the valuation theory of fields, as given for example in Bourbaki, *Algèbre Commutative*, Ch. VI. For the convenience of the reader, we have included an appendix covering some of the more technical facts we need in commutative valuation theory, especially concerning Henselian valuations and Henselizations. The theoretical aspects developed throughout the book are illustrated by many examples, which are listed by chapter in another appendix.

We thank Maurício Ferreira for his collaboration on the material in §4.3.4. In addition, we are grateful to Cécile Coyette, Maurício Ferreira, Timo Hanke, and Mélanie Raczek for reading drafts of parts of the book and making many valuable comments. A significant part of the book was written while the first author was a Senior Fellow of the Zukunftskolleg of the Universität Konstanz (Germany) between April 2010 and January 2012. He gratefully acknowledges the excellent working conditions and stimulating atmosphere enjoyed there, and the hospitality of Karim-Johannes Becher and the staff of the Zukunftskolleg. He also acknowledges support from the *Fonds de la Recherche Scientifique–FNRS* under grants n° 1.5181.08, 1.5009.11, and 1.5054.12.

A note on notation

As pointed out above, we compare throughout most of the book algebras over valued fields and graded algebras. As a visual aid to help the reader determine whether a given statement lies in the context of graded algebras, we use sans serif letters (A, F, V, ...) to designate graded structures and associated constructions. Thus, for instance $\mathsf{End}_{\mathsf{D}}(\mathsf{V})$ denotes the graded algebra of endomorphisms of the graded vector space V over the graded division algebra D, and we write $\mathsf{gr}(D)$ for the graded algebra associated to the valued division algebra D, and $\mathsf{Br}(\mathsf{F})$ for the Brauer group of a graded field F.

The blackboard bold symbols $\mathbb{C}$, $\mathbb{F}_q$, $\mathbb{N}$, $\mathbb{Q}$, $\mathbb{Q}_p$, $\mathbb{R}$, and $\mathbb{Z}$ have their customary meanings: the complex numbers, the finite field of cardinality q, the nonnegative integers, the rational numbers, the p-adic completion of $\mathbb{Q}$, the real numbers, and the integers.

Louvain-la-Neuve, La Jolla
June 2014

Jean-Pierre Tignol
Adrian Wadsworth

Contents

Chapter 1
Valuations on Division Rings

In this chapter we introduce the central object of study in this book: valuations on division algebras D finite-dimensional over their centers. In §1.1 we define valuations (not assuming finite-dimensionality of D) and describe the associated structures familiar from commutative valuation theory: the valuation ring $\mathcal{O}_D$, its unique maximal left and maximal right ideal $\mathfrak{m}_D$, the residue division algebra $\overline{D}$, and the value group Γ_D. While the residue field $\overline{Z(D)}$ of the center of D lies in the center $Z(\overline{D})$ of $\overline{D}$, the inclusion is often strict. We describe an important and distinctively noncommutative feature, namely a canonical homomorphism θ_D from Γ_D to the automorphism group $Aut\big(Z(\overline{D})/\overline{Z(D)}\big)$; θ_D is induced by conjugation by elements of $D^\times$.

In §1.2 we focus on a division algebra D finite-dimensional over its center F. We prove the "Fundamental Inequality" for valued division algebras. We then look at valuations on D from the perspective of F. We show that a valuation on F has at most one extension to D, and prove a criterion for when such an extension exists. When this occurs, we show that $Z(\overline{D})$ is a finite-dimensional normal field extension of $\overline{F}$ and that θ_D is surjective. We also describe the technical adjustments needed to apply the classical method of "composition" of valuations to division algebras.

The filtration on D induced by a valuation leads to an associated graded ring $\mathsf{gr}(D)$, which we describe in §1.3. Throughout the book we emphasize use of $\mathsf{gr}(D)$ to help understand the valuation on D. We give many examples of division algebras with valuations throughout the chapter.

1.1 Basic definitions and examples

The fundamental structures associated to valuations on division rings are defined in §1.1.1, and illustrated in §1.1.2–§1.1.4 for several examples obtained by various kinds of series constructions.

J.-P. Tignol, A.R. Wadsworth, *Value Functions on Simple Algebras, and Associated Graded Rings*, Springer Monographs in Mathematics,
DOI 10.1007/978-3-319-16360-4_1

1.1.1 Valuations and associated structures

The valuations we consider on a division ring D are functions

$$v\colon D \longrightarrow \Gamma \cup \{\infty\}$$

where Γ is a totally ordered additive abelian group and ∞ is a symbol such that $\gamma < \infty$ and $\gamma + \infty = \infty + \infty = \infty$ for all $\gamma \in \Gamma$, subject to the following conditions: for all $x,\ y \in D$,

(i) $v(x) = \infty$ if and only if $x = 0$;
(ii) $v(x+y) \geq min\big(v(x), v(y)\big)$;
(iii) $v(xy) \ = \ v(x) + v(y)$.

Thus, the restriction of v to the multiplicative group of units $D^\times$ is a group homomorphism $D^\times \to \Gamma$. It readily follows that

$$v(1) \ = \ 0 \qquad \text{and} \qquad v(x^{-1}) \ = \ -v(x) \quad \text{for all } x \in D^\times.$$

Also, since Γ has no torsion, we have

$$v(-1) \ = \ 0 \qquad \text{hence} \qquad v(-x) \ = \ v(x) \quad \text{for all } x \in D.$$

By writing $x = (x+y)-y$, it follows from (ii) that $v(x) \geq min\big(v(x+y), v(y)\big)$; hence, $v(x+y) = v(x)$ if $v(y) > v(x)$. Therefore, we have for all $x,\ y \in D$

$$v(x+y) \ = \ min\big(v(x), v(y)\big) \qquad \text{if } v(x) \neq v(y).$$

Associated to the valuation v there are the following structures:

- $\Gamma_D = v(D^\times)$, the *value group* of v, which is a subgroup of Γ;
- $\mathcal{O}_D = \{x \in D \mid v(x) \geq 0\}$, which is a subring of D called the *valuation ring* of D;
- $\mathfrak{m}_D = \{x \in D \mid v(x) > 0\}$, which is a two-sided ideal that is the unique maximal left and maximal right ideal of $\mathcal{O}_D$ since the group of units in $\mathcal{O}_D$ is $\mathcal{O}_D^\times = \{x \in D \mid v(x) = 0\} = \mathcal{O}_D \setminus \mathfrak{m}_D$;
- $\overline{D} = \mathcal{O}_D/\mathfrak{m}_D$, the residue division ring.

If we need to specify the valuation on D, we will write $\Gamma_{D,v}$, $\mathcal{O}_{D,v}$, $\mathfrak{m}_{D,v}$, and $\overline{D}^{v}$. But most of the time we will be considering only one valuation on D, and the simpler notation will suffice.

For $x \in \mathcal{O}_D$ we let $\overline{x}$ be the image of x in $\overline{D}$,

$$\overline{x} \ = \ x + \mathfrak{m}_D \ \in \ \overline{D}.$$

A distinctive feature of noncommutative valuation theory is the interaction between the value group and the residue division ring, which takes the following form: any valuation v on a division ring D restricts to a valuation on its center $Z(D)$, and we may consider $\overline{Z(D)} \subseteq \overline{D}$. Clearly, we have

$$\overline{Z(D)} \ \subseteq \ Z(\overline{D}).$$

Certain automorphisms of this field extension are associated to elements in the value group: any $d \in D^\times$ determines the inner automorphism $int(d)\colon D \to D$ given by $x \mapsto dxd^{-1}$. Since $v\big(int(d)(x)\big) = v(x)$, we have

$$int(d)(\mathcal{O}_D) = \mathcal{O}_D \qquad \text{and} \qquad int(d)(\mathfrak{m}_D) = \mathfrak{m}_D,$$

so $int(d)$ induces an automorphism $\overline{int(d)}$ of $\overline{D}$, hence by restriction an automorphism of $Z(\overline{D})$ fixing every element of $\overline{Z(D)}$. If $v(d) = 0$, then $\overline{int(d)} = int(\overline{d})$, and the restriction of $int(\overline{d})$ to $Z(\overline{D})$ is the identity. Thus, conjugation induces a group homomorphism mapping $D^\times/\mathcal{O}_D^\times$ to $Aut\big(Z(\overline{D})/\,\overline{Z(D)}\,\big)$. On the other hand, v induces an isomorphism $D^\times/\mathcal{O}_D^\times \xrightarrow{\sim} \Gamma_D$, so there is a well-defined group homomorphism

$$\theta_D\colon \Gamma_D \longrightarrow Aut\big(Z(\overline{D})/\,\overline{Z(D)}\,\big), \tag{1.1}$$

which can be described as follows: for any $\gamma \in \Gamma_D$ and any $d \in D^\times$ with $v(d) = \gamma$ and any $x \in \mathcal{O}_D$ with $\overline{x} \in Z(\overline{D})$,

$$\theta_D(\gamma)(\overline{x}) = \overline{dxd^{-1}}.$$

We call θ_D the *canonical homomorphism* of the valuation v on D. It is clear that $\Gamma_{Z(D)} \subseteq \ker\theta_D$, so we may also consider the induced homomorphism

$$\overline{\theta}_D\colon \Gamma_D/\Gamma_{Z(D)} \longrightarrow Aut\big(Z(\overline{D})/\,\overline{Z(D)}\,\big).$$

1.1.2 Examples: twisted Laurent series

Let A be a division ring, and let σ be an automorphism of A. The *twisted Laurent series ring* $A((x;\sigma))$ is defined as the set of formal series

$$\sum_{i=k}^{\infty} a_i x^i, \qquad \text{where } k \in \mathbb{Z} \text{ and } a_i \in A \text{ for all } i.$$

The addition is defined as usual, and multiplication is given by

$$\sum_i a_i x^i \cdot \sum_j b_j x^j = \sum_{i,j} a_i \sigma^i(b_j) x^{i+j} \qquad \text{for } a_i,\, b_j \in A.$$

Let $D = A((x;\sigma))$. For $d = \sum_{i=k}^{\infty} a_i x^i \in D$, let

$$supp(d) = \{i \in \mathbb{Z} \mid a_i \neq 0\} \quad \text{and} \quad v_x(d) = min\big(supp(d)\big) \ \ (\text{so } v_x(0) = \infty).$$

If $v_x(d) > 0$, then the element $1 + d + d^2 + \ldots$ is defined in D, and

$$(1-d)(1+d+d^2+\ldots) = 1 = (1+d+d^2+\ldots)(1-d),$$

so $1-d$ is invertible. It is then easy to see that D is a division ring: for an arbitrary nonzero element $d = \sum_{i=k}^{\infty} a_i x^i$ with $a_k \neq 0$, we have

$$\sigma^{-k}(a_k^{-1})x^{-k}d = x^{-k}a_k^{-1}d = 1 - d_0$$

and

$$d\sigma^{-k}(a_k^{-1})x^{-k} \;=\; dx^{-k}a_k^{-1} \;=\; 1-d_1$$

for some $d_0,\ d_1 \in D$ with $v_x(d_0),\ v_x(d_1) > 0$. As $1-d_0$ and $1-d_1$ are invertible, it follows that d is invertible. Therefore, D is a division ring. It is easy to check that v_x is a valuation on D with $\Gamma_D = \mathbb{Z}$. It is known as the *x-adic valuation* on D. Clearly $\overline{D} = A$.

A series $\sum_{i=k}^{\infty} a_i x^i \in D$ lies in the center of D if and only if it commutes with x and centralizes A. Therefore,

$$Z(D) \;=\; \big\{\sum_{i=k}^{\infty} a_i x^i \mid \sigma(a_i) = a_i \text{ and } ba_i = a_i\sigma^i(b) \text{ for all } i \text{ and all } b \in A\big\}.$$

In particular, the residue field of $Z(D)$ is the subfield of $Z(A)$ fixed under σ,

$$\overline{Z(D)} \;=\; Z(A)^{\sigma}.$$

The homomorphism $\theta_D\colon \mathbb{Z} \to Aut(Z(A)/Z(A)^{\sigma})$ maps $1 \in \mathbb{Z}$ to $\sigma|_{Z(A)}$.

1.1.3 Examples: iterated Laurent series

The construction above can of course be iterated: if τ is an automorphism of $A((x;\sigma))$ we may consider the division ring $A((x;\sigma))((y;\tau))$. This division ring carries the y-adic valuation v_y with value group $\mathbb{Z}$, but it also has a composite valuation $v_x * v_y$ with value group $\mathbb{Z}^2$, as we will see shortly. We will make use of the following result:

Proposition 1.1. *Every automorphism τ of $A((x;\sigma))$ preserves the x-adic valuation v_x, i.e., $v_x \circ \tau = v_x$.*

Proof. We need to show that $v_x\big(\tau(d)\big) = v_x(d)$ for all $d \in A((x;\sigma))$. We proceed in four steps:

Step 1: If $v_x(d) > 0$, then $v_x\big(\tau(d)\big) \geq 0$. Let n be any positive integer prime to the characteristic of A. By substituting d for the variable X in the Taylor expansion of the function $\sqrt[n]{1+X}$, we obtain a series $s \in A((x;\sigma))$ such that $s^n = 1+d$, hence also $\tau(s)^n = 1+\tau(d)$. If $v_x\big(\tau(d)\big) < 0$, this equality shows that $v_x\big(\tau(d)\big) = n\, v_x\big(\tau(s)\big) \in n\mathbb{Z}$. This relation cannot hold for infinitely many integers n, hence it is impossible that $v_x\big(\tau(d)\big) < 0$.

Step 2: If $v_x(d) = 0$, then $v_x\big(\tau(d)\big) = 0$. Suppose instead that $v_x\big(\tau(d)\big) \neq 0$. We may then find an integer $z \in \mathbb{Z}$ such that $z\, v_x\big(\tau(d)\big) < -v_x\big(\tau(x)\big)$. Then $v_x\big(\tau(d^z x)\big) = z\, v_x\big(\tau(d)\big) + v_x\big(\tau(x)\big) < 0$ while $v_x(d^z x) = 1$, a contradiction to step 1.

Step 3: If $d \neq 0$, then $v_x\big(\tau(d)\big) = v_x(d) v_x\big(\tau(x)\big)$. Let $z = v_x(d)$. We have $v_x(d^{-1}x^z) = 0$; hence step 2 yields $v_x\big(\tau(d^{-1}x^z)\big) = 0$, and it follows that $v_x\big(\tau(d)\big) = z\, v_x\big(\tau(x)\big)$.

Step 4: $v_x(\tau(x)) = 1$. Step 1 shows that $v_x(\tau(x)) \geq 0$, and step 3 shows that $v_x(\tau(x))$ divides $v_x(\tau(d))$ for all nonzero $d \in A((x;\sigma))$. Since τ is onto, $v_x(\tau(x))$ divides every integer, so $v_x(\tau(x)) = 1$.

The proposition follows from steps 3 and 4. □

One way to build an automorphism on $A((x,\sigma))$ is by extending an automorphism of A:

Lemma 1.2. *Suppose ρ is an automorphism of A, and suppose there is a $b \in A^\times$ with $\mathrm{int}(b)\sigma\rho = \rho\sigma$. Then, there is an automorphism $\widehat{\rho}$ of $A((x;\sigma))$ with $\widehat{\rho}|_A = \rho$ and $\widehat{\rho}(x) = bx$.*

Proof. Define $\widehat{\rho}$ by $\widehat{\rho}\big(\sum_{i=k}^{\infty} a_i x^i\big) = \sum_{i=k}^{\infty} \rho(a_i)(bx)^i$. The hypothesis on b implies that $(bx)\rho(c) = \rho\sigma(c)(bx)$ and $(bx)^{-1}\rho(c) = \rho\sigma^{-1}(c)(bx)^{-1}$. Hence, by upward and downward induction on i,

$$(bx)^i\rho(c) \;=\; \rho\sigma^i(c)\,(bx)^i \qquad \text{for all } i \in \mathbb{Z},\ c \in A. \tag{1.2}$$

Clearly $\widehat{\rho}(1) = 1$ and $\widehat{\rho}(s+t) = \widehat{\rho}(s) + \widehat{\rho}(t)$ for all $s, t \in A((x;\sigma))$. To verify that $\widehat{\rho}(st) = \widehat{\rho}(s)\widehat{\rho}(t)$, it suffices to check this for monomials. Say $s = ax^i$ and $t = cx^j$ with $a, c \in A$, $i, j \in \mathbb{Z}$. Then, using (1.2),

$$\begin{aligned}\widehat{\rho}(st) &= \widehat{\rho}\big(a\sigma^i(c)x^{i+j}\big) \;=\; \rho(a)\rho\sigma^i(c)(bx)^{i+j}\\ &= \big[\rho(a)(bx)^i\big]\big[\rho(c)(bx)^j\big] \;=\; \widehat{\rho}(s)\,\widehat{\rho}(t),\end{aligned}$$

as desired. □

Now, fix some automorphism τ of $A((x;\sigma))$ and consider the division ring $E = A((x;\sigma))((y;\tau))$. Let v_x be the x-adic valuation on $A((x;\sigma))$ and v_y the y-adic valuation on E. For any nonzero series $s = \sum_{i=k}^{\infty} d_i y^i \in E$ with $d_i \in A((x;\sigma))$ define

$$v_{x,y}(s) \;=\; (v_x(d_k), k) \in \mathbb{Z}^2 \qquad \text{if } d_k \neq 0.$$

Let also $v_{x,y}(0) = \infty$. If $s = d_k y^k + \sum_{i>k} d_i y^i$ and $s' = d'_\ell y^\ell + \sum_{j>\ell} d'_j y^j$, then

$$ss' \;=\; d_k\tau^k(d'_\ell)y^{k+\ell} + \sum_{i+j>k+\ell} d_i\tau^i(d'_j)y^{i+j}.$$

Since $v_x\big(d_k\tau^k(d'_\ell)\big) = v_x(d_k) + v_x(d'_\ell)$ by Prop. 1.1, we have

$$v_{x,y}(ss') \;=\; v_{x,y}(s) + v_{x,y}(s').$$

It is also easy to see that $v_{x,y}(s+s') \geq min\big(v_{x,y}(s), v_{x,y}(s')\big)$ when $\mathbb{Z}^2$ is given the right-to-left lexicographic ordering: indeed we have $v_{x,y}(x+y) = (1,0)$, hence $(1,0) < (0,1)$. The map $v_{x,y}$ is thus a valuation on E with value group $\mathbb{Z}^2$. It is called the (x,y)-*adic valuation*, which is the composite $v_x * v_y$ of the y-adic valuation v_y on E and the x-adic valuation v_x on the residue division ring $\overline{E}^{\,v_y} = A((x;\sigma))$. See Exercise 1.2 for a general construction of composite valuations on division rings of Laurent series.

With a view toward the general construction of Mal'cev–Neumann series below, observe that the nonzero elements in E are formal series $s = \sum_{i,j} a_{i,j} x^i y^j$ (with $a_{i,j} \in A$) for which the set $S = \{(i,j) \in \mathbb{Z}^2 \mid a_{i,j} \neq 0\}$, called the *support* of s, has a characteristic property with respect to the right-to-left lexicographic ordering on $\mathbb{Z}^2$: it has a minimum, which is $v_{x,y}(s)$, but also every subset of S has a minimum because for $(i,j) \in S$ there are only finitely many $k \in \mathbb{Z}$ such that $k \leq i$ and $(k,j) \in S$. This property is expressed by saying that the set S is *well-ordered* for the right-to-left lexicographic ordering. Conversely, if $S \subset \mathbb{Z}^2$ is any nonempty well-ordered subset, then the series $\sum_{(i,j)\in S} a_{i,j} x^i y^j$ lies in E for any choice of the coefficients $a_{i,j} \in A$.

This construction can be used inductively to obtain iterated twisted Laurent series rings with a valuation with value group $\mathbb{Z}^n$, for any integer $n \geq 1$. As a typical example, consider a field M with a family $\boldsymbol{\sigma} = (\sigma_i)_{i=1}^n$ of pairwise commuting automorphisms, and fix a collection $\boldsymbol{u} = (u_{i,j})_{i,j=1}^n$ of elements in $M^\times$ subject to the following conditions:

$$u_{i,i} = 1, \quad u_{i,j} u_{j,i} = 1, \quad u_{i,j} u_{j,k} u_{k,i} = \sigma_k(u_{i,j}) \sigma_i(u_{j,k}) \sigma_j(u_{k,i}) \tag{1.3}$$

for all i, j, $k = 1, \ldots, n$. Consider the division ring of iterated Laurent series

$$\mathcal{L}((M; \boldsymbol{\sigma}, \boldsymbol{u})) = M((x_1; \sigma_1))((x_2; \widehat{\sigma_2})) \ldots ((x_n; \widehat{\sigma_n})), \tag{1.4}$$

where the automorphism $\widehat{\sigma_i}$ of $M((x_1; \sigma_1))((x_2; \widehat{\sigma_2})) \ldots ((x_{i-1}; \widehat{\sigma_{i-1}}))$ is defined for $i = 2, \ldots, n$ by

$$\widehat{\sigma_i}\big(\textstyle\sum m_{k_1 \ldots k_{i-1}} x_1^{k_1} \ldots x_{i-1}^{k_{i-1}}\big) = \sum \sigma_i(m_{k_1 \ldots k_{i-1}}) (u_{i,1} x_1)^{k_1} \ldots (u_{i,i-1} x_{i-1})^{k_{i-1}}.$$

To see that $\widehat{\sigma_i}$ is an automorphism, assume inductively that $\widehat{\sigma_2}, \ldots, \widehat{\sigma_{i-1}}$ are automorphisms, and let

$$A_0 = M \quad \text{and} \quad A_j = M((x_1; \sigma_1))((x_2; \widehat{\sigma_2})) \ldots ((x_j; \widehat{\sigma_j}))$$

for $1 \leq j \leq i-1$, and let $\sigma_{i,j} = \widehat{\sigma_i}|_{A_j}$. Then, $\sigma_{i,0}$ is the automorphism σ_i on M. For ℓ with $1 \leq \ell \leq i-1$, assume $\sigma_{i,\ell-1}$ is an automorphism of $A_{\ell-1}$. We apply Lemma 1.2 to $A_\ell = A_{\ell-1}((x_\ell; \widehat{\sigma_\ell}))$ with $\rho = \sigma_{i,\ell-1}$ and $b = u_{i,\ell}$ to see that $\widehat{\rho} = \sigma_{i,\ell}$ is an automorphism of A_ℓ. For this we need

$$\mathit{int}(u_{i,\ell})\, \widehat{\sigma_\ell}\, \sigma_{i,\ell-1}(a) = \sigma_{i,\ell-1}\, \widehat{\sigma_\ell}(a) \qquad \text{for all } a \in A_{\ell-1}. \tag{1.5}$$

It suffices to check this equality for $a \in M$ and $a = x_j$ for $1 \leq j \leq \ell - 1$. It holds for $a \in M$ as M is commutative, $\widehat{\sigma_\ell}|_M = \sigma_\ell$, $\sigma_{i,\ell-1}|_M = \sigma_i$, and $\sigma_i \sigma_\ell = \sigma_\ell \sigma_i$. For $a = x_j$, (1.5) becomes

$$u_{i,\ell}\, \sigma_\ell(u_{i,j,})\, u_{\ell,j}\, x_j\, u_{i,\ell}^{-1} = \sigma_i(u_{\ell,j})\, u_{i,j}\, x_j,$$

which holds by (1.3) as $x_j u_{i,\ell}^{-1} = \sigma_j(u_{i,\ell}^{-1}) x_j$ and M is commutative. Thus, by induction on ℓ, $\widehat{\sigma_i} = \sigma_{i,i-1}$ is an automorphism of A_{i-1}. Note that the fundamental relations in $\mathcal{L}((M; \boldsymbol{\sigma}, \boldsymbol{u}))$ are

$$x_i m = \sigma_i(m) x_i \text{ and } x_i x_j = u_{i,j} x_j x_i \quad \text{for all } m \in M \text{ and } i,\ j = 1, \ldots, n.$$

For simplicity, let $D = \mathcal{L}((M; \boldsymbol{\sigma}, \boldsymbol{u}))$. The ring D carries the $(x_1, \dots, x_n)$-adic valuation $v_{x_1,\dots,x_n}$ with value group $\mathbb{Z}^n$ given the right-to-left lexicographic ordering. This is the total ordering in which

$$(r_1, \dots, r_n) < (s_1, \dots, s_n) \quad \text{just when} \quad \begin{cases} \text{there is an } \ell \text{ with } r_\ell < s_\ell \\ \text{and } r_j = s_j \text{ for } \ell + 1 \le j \le n. \end{cases}$$

Since $\mathbb{Z}^n$ with this ordering is discrete of rank n as an ordered abelian group, the valuation $v_{x_1,\dots,x_n}$ is discrete of rank n. (See the review of ranks for valuations in §A.4 of Appendix A, and Remark A.34 on discrete valuations.) The residue ring of D is $\overline{D} = M$ $(= Z(\overline{D}))$. A series in $\mathcal{O}_D \cap Z(D)$ has constant term fixed under $\sigma_1, \dots, \sigma_n$, hence $\overline{Z(D)}$ is the subfield of M fixed under $\sigma_1, \dots, \sigma_n$. The map $\theta_D \colon \mathbb{Z}^n \to Aut(Z(\overline{D})/\overline{Z(D)})$ carries $(i_1, \dots, i_n) \in \mathbb{Z}^n$ to $\sigma_1^{i_1} \dots \sigma_n^{i_n}$.

1.1.4 Examples: Mal'cev–Neumann series

All the examples above are particular cases of the following general construction due to Mal'cev and Neumann (see Cohn [55, §2.4] or Lam [121, §14]): let Γ be an arbitrary totally ordered abelian group and let D be an arbitrary division ring. Let $f \colon \Gamma \times \Gamma \to D^\times$ and $\omega \colon \Gamma \to Aut(D)$ be maps satisfying the following properties:

$$\omega_\gamma\big(f(\delta, \varepsilon)\big) f(\gamma, \delta + \varepsilon) = f(\gamma, \delta) f(\gamma + \delta, \varepsilon) \qquad \text{for all } \gamma, \delta, \varepsilon \in \Gamma, \tag{1.6}$$

$$\omega_\gamma \circ \omega_\delta(d) = f(\gamma, \delta) \omega_{\gamma+\delta}(d) f(\gamma, \delta)^{-1} \qquad \text{for all } \gamma, \delta \in \Gamma \text{ and } d \in D, \tag{1.7}$$

and moreover

$$\omega_0 = id_D, \qquad f(0, \gamma) = f(\gamma, 0) = 1 \qquad \text{for all } \gamma \in \Gamma. \tag{1.8}$$

Define the *support* of a map $\varphi \colon \Gamma \to D$ by

$$supp(\varphi) = \{\gamma \in \Gamma \mid \varphi(\gamma) \neq 0\}.$$

In the additive group $\mathcal{F}(\Gamma, D)$ of all maps $\Gamma \to D$, the set $\mathcal{F}_{\mathrm{wo}}(\Gamma, D)$ of maps with well-ordered support is a subgroup. The following modified convolution product is well-defined for φ, $\psi \in \mathcal{F}_{\mathrm{wo}}(\Gamma, D)$:

$$(\varphi * \psi)(\gamma) = \sum_{\delta \in \Gamma} \varphi(\delta) \omega_\delta\big(\psi(\gamma - \delta)\big) f(\delta, \gamma - \delta) \qquad \text{for } \gamma \in \Gamma,$$

because the sum on the right has only finitely many nonzero terms. For if the set $supp(\varphi) \cap (\gamma - supp(\psi))$ were infinite, it would contain a strictly increasing infinite sequence $\delta_1 < \delta_2 < \dots$, as $supp(\varphi)$ is well-ordered; but then $supp(\psi)$ would contain the strictly descending infinite sequence $\gamma - \delta_1 > \gamma - \delta_2 > \dots$, contradicting the well-ordering of $supp(\psi)$. Moreover, see Cohn [55, p. 75] or Lam [121, p. 243], $\varphi * \psi \in \mathcal{F}_{\mathrm{wo}}(\Gamma, D)$, and the sum and convolution product define a ring structure on $\mathcal{F}_{\text{wo}}(\Gamma, D)$; we use the notation $D((\Gamma; \omega, f))$ for this ring to emphasize the dependence on the maps ω and f. Mapping each $d \in D$ to the map φ_d such that $\varphi_d(0) = d$ and $\varphi_d(\gamma) = 0$ for $\gamma \neq 0$ yields an identification of D with a subring of $D((\Gamma; \omega, f))$.

To make the definition of the multiplication in $D((\Gamma; \omega, f))$ more transparent, it is useful to change notation. For each $\gamma \in \Gamma$ we let z^γ denote an indeterminate. Each $\varphi \in \mathcal{F}(\Gamma, D)$ is identified with a formal series as follows:

$$\varphi = \sum_{\gamma \in \Gamma} \varphi(\gamma) z^\gamma \quad \Big(= \sum_{\gamma \in supp(\varphi)} \varphi(\gamma) z^\gamma\Big).$$

The multiplication in $D((\Gamma; \omega, f))$ is then

$$\Big(\sum_{\gamma \in \Gamma} \varphi(\gamma) z^\gamma\Big) \cdot \Big(\sum_{\delta \in \Gamma} \psi(\delta) z^\delta\Big) = \sum_{\gamma, \delta \in \Gamma} \varphi(\gamma) \omega_\gamma\big(\psi(\delta)\big) f(\gamma, \delta) z^{\gamma+\delta}.$$

We have $z^0 = 1$ in $D((\Gamma; \omega, f))$, and each element $d \in D$ is identified with $dz^0 \in D((\Gamma; \omega, f))$. Thus, we have

$$z^\gamma \cdot d = \omega_\gamma(d) \cdot z^\gamma \quad \text{and} \quad z^\gamma \cdot z^\delta = f(\gamma, \delta) z^{\gamma+\delta} \quad \text{for } \gamma,\ \delta \in \Gamma \text{ and } d \in D.$$

Define a map $v\colon D((\Gamma; \omega, f)) \to \Gamma \cup \{\infty\}$ by

$$v(\varphi) = min(supp\, \varphi) \qquad \text{for } \varphi \in D((\Gamma; \omega, f))$$

(and $v(0) = min(\varnothing) = \infty$). It is easy to see that this map satisfies all the conditions that define a valuation, so $D((\Gamma; \omega, f))$ has no zero divisors. If $\varphi \in D((\Gamma; \omega, f))$ satisfies $v(\varphi) > 0$, then with some effort one can prove that the set $\bigcup_{n \in \mathbb{N}} supp(\varphi^n)$ is well-ordered (see Cohn [55, Lemma 2.4.3, Th. 2.4.4, pp. 73–75] or Lam [121, Lemma 14.22(1), p. 244]); hence, $\sum_{n \in \mathbb{N}} \varphi^n \in D((\Gamma; \omega, f))$, and it is easy to check that $\sum_{n=0}^{\infty} \varphi^n = (1 - \varphi)^{-1}$. Arguing as for twisted Laurent series at the beginning of this subsection, one can then see that $D((\Gamma; \omega, f))$ is a division ring, and v is a valuation on $D((\Gamma; \omega, f))$. Its value group is Γ and the residue division algebra is D. The map $\theta_{D((\Gamma; \omega, f))}$ carries each $\gamma \in \Gamma$ to $\omega_\gamma|_{Z(D)}$.

Let D be any division ring and let Γ be a totally ordered abelian group. Note that if we let $\omega\colon \Gamma \to Aut(D)$ be the trivial homomorphism, and define $\mathbf{1}\colon \mathbb{Z} \times \mathbb{Z} \to D^\times$ by $\mathbf{1}(\gamma, \delta) = 1$ for all $\gamma,\ \delta \in \Gamma$, then ω and $\mathbf{1}$ satisfy conditions (1.6)–(1.8). The resulting Mal'cev–Neumann ring $D((\Gamma, \omega, \mathbf{1}))$ has value group Γ. Thus, every totally ordered group Γ is the value group of some division algebra. If we take $D = F$, a field, and $\Gamma = \mathbb{Z}^n$ with right-to-left lexicographic ordering, then $F((\mathbb{Z}^n, \omega, \mathbf{1}))$ can be identified with the n-fold iterated Laurent series field:

$$F((\mathbb{Z}^n, \omega, \mathbf{1})) = F((x_1)) \dots ((x_n)).$$

If τ is an automorphism of D and $\Gamma = \mathbb{Z}$, we define $\omega\colon \mathbb{Z} \to Aut(D)$ and $\mathbf{1}\colon \mathbb{Z} \times \mathbb{Z} \to D^\times$ by $\omega_\gamma = \tau^\gamma$ and $\mathbf{1}(\gamma, \delta) = 1$ for all $\gamma,\ \delta \in \mathbb{Z}$. With this choice,

ω and $\mathbf{1}$ satisfy the conditions (1.6)–(1.8), and the Mal'cev–Neumann ring $D((\mathbb{Z};\omega,\mathbf{1}))$ can be identified with a division ring of Laurent series:

$$D((\mathbb{Z};\omega,\mathbf{1})) = D((z^1;\tau)).$$

Similarly, rings of n-fold iterated twisted Laurent series can be identified with Mal'cev–Neumann rings with $\Gamma = \mathbb{Z}^n$.

1.2 Valuations on finite-dimensional division algebras

After proving in §1.2.1 a fundamental inequality relating the dimension of an extension of valued division rings to the product of the ramification index and residue degree, we focus in this section on division rings D that are finite-dimensional algebras over a field F, viewed as a subfield of the center $Z(D)$. Any valuation w on D restricts to a valuation v on F, and we may consider w as an extension of v. Thus, we may try to define valuations on D by extending a given valuation on F. A necessary and sufficient condition for the existence of such an extension is given in §1.2.2: see Th. 1.4. This condition features one of the most striking differences between commutative and noncommutative valuation theory: recall that for any extension field L of F, the valuation v extends to a valuation on L; often there are many such extensions. By contrast, Th. 1.4 below shows that a valuation on the center of a division algebra D has at most one extension to D, but may have none at all. Likewise, a composition of valuations cannot be defined without an added condition, which is given in §1.2.3. We conclude this section with various examples in §1.2.4–§1.2.9.

1.2.1 The fundamental inequality

Let D be a division ring with valuation v, and let E be any sub-division ring of D. Then the restriction $v|_E$ of v to E is clearly a valuation on E, and we have

$$\Gamma_E \subseteq \Gamma_D, \qquad \mathcal{O}_E = \mathcal{O}_D \cap E, \quad \text{and} \quad \mathfrak{m}_E = \mathfrak{m}_D \cap E;$$

hence, there is a canonical injection $\overline{E} \hookrightarrow \overline{D}$, which we will treat as an inclusion. We write $[D{:}E]_\ell$ (resp. $[D{:}E]_r$) for the dimension of D as a left (resp. right) E-vector space.

Proposition 1.3 (Fundamental Inequality).

$$[\overline{D}{:}\overline{E}]_\ell\,|\Gamma_D{:}\Gamma_E| \le [D{:}E]_\ell \quad \textit{and} \quad [\overline{D}{:}\overline{E}]_r\,|\Gamma_D{:}\Gamma_E| \le [D{:}E]_r.$$

Proof. We prove only the left inequality. The proof of the right one is analogous. Pick $\{d_i\}_{i\in I} \subseteq \mathcal{O}_D^\times$ such that the images $\{\overline{d_i}\}_{i\in I} \subseteq \overline{D}^\times$ form a base of

$\overline{D}$ as a left $\overline{E}$-vector space. Also pick a set of coset representatives $\{\delta_j\}_{j\in J}$ of Γ_E in Γ_D, and for each j choose some $c_j \in D$ with $v(c_j) = \delta_j$. To verify the inequality of the proposition, we show that $\{d_i c_j\}_{i\in I\, j\in J}$ is left E-linearly independent in D.

For this, suppose $\{i_1, \ldots, i_r\} \subseteq I$ and $\{j_1, \ldots, j_s\} \subseteq J$ are finite subsets and $\{a_{k\ell} \mid 1 \le k \le r,\ 1 \le \ell \le s\}$ is a set of elements in E. We want to show

$$\sum_{k=1}^{r}\sum_{\ell=1}^{s} a_{k\ell} d_{i_k} c_{j_\ell} \;=\; 0 \quad \text{implies} \quad a_{k\ell} \;=\; 0 \text{ for all } k,\ \ell.$$

For $\ell = 1, \ldots, s$, let

$$b_\ell = \sum_{k=1}^{r} a_{k\ell} d_{i_k} \quad \text{and} \quad \gamma_\ell = \mathit{min}\{v(a_{k\ell}) \mid 1 \le k \le r\} \in \Gamma_E \cup \{\infty\}.$$

We claim that $v(b_\ell) = \gamma_\ell$ for all ℓ. Assuming this equation holds, we have

$$v(b_\ell c_{j_\ell}) \;=\; \gamma_\ell + \delta_{j_\ell} \;\equiv\; \delta_{j_\ell} \ (\mathit{mod}\ \Gamma_E) \quad \text{if } b_\ell \neq 0.$$

Because the $v(b_\ell c_{j_\ell})$ are thus distinct for those ℓ with $b_\ell \neq 0$, we have

$$v\Big(\sum_{k,\ell} a_{k\ell} d_{i_k} c_{j_\ell}\Big) \;=\; v\Big(\sum_{\ell} b_\ell c_{j_\ell}\Big) \;=\; \min_{1\le\ell\le s} v(b_\ell c_{j_\ell}) \;=\; \min_{1\le\ell\le s} (\gamma_\ell + \delta_\ell).$$

Therefore, if $\sum a_{k\ell} d_{i_k} c_{j_\ell} = 0$ we must have $\gamma_\ell = \infty$ for all ℓ, i.e., $a_{k\ell} = 0$ for all k, ℓ, and the proof is complete.

To prove the claim, suppose $\gamma_\ell \neq \infty$. By reordering $i_1, \ldots, i_r$, we may assume $v(a_{1\ell}) = \gamma_\ell$. The $\overline{E}$-independence of the $\overline{d_i}$ shows $\sum_{k=1}^{r} \overline{a_{1\ell}^{-1} a_{k\ell}}\, \overline{d_{i_k}} \neq 0$; hence, $v\big(a_{1\ell}^{-1} \sum_{k=1}^{r} a_{k\ell} d_{i_k}\big) = 0$ and therefore

$$v(b_\ell) \;=\; v\Big(\sum_{k=1}^{r} a_{k\ell} d_{i_k}\Big) \;=\; v(a_{1\ell}) \;=\; \gamma_\ell,$$

proving the claim. □

1.2.2 *Extension of a valuation from the center*

Our focus henceforward will be on division rings that are finite-dimensional algebras over a field. Let $v\colon F \to \Gamma \cup \{\infty\}$ be a valuation on a field F. By replacing Γ by its divisible hull $\Gamma \otimes_{\mathbb{Z}} \mathbb{Q}$ (to which the ordering on Γ extends uniquely compatibly with the group structure), we may assume at the outset that Γ is divisible. Let D be a finite-dimensional division F-algebra. If v extends to a valuation w on D, then the fundamental inequality (Prop. 1.3) shows that $|w(D^\times){:}v(F^\times)| < \infty$. Since Γ is divisible and torsion-free, the inclusion $\Gamma_F = v(F^\times) \hookrightarrow \Gamma$ therefore extends uniquely to a monomorphism $\iota\colon w(D^\times) \hookrightarrow \Gamma$. Moreover, ι is order-preserving, since the ordering on $w(D^\times)$ is determined by the ordering on Γ_F. Therefore, we may assume without loss of generality that all valuations extending v to a finite-dimensional division algebra over F take their values in the divisible group Γ.

Throughout this section, we fix a valuation $v\colon F \to \Gamma \cup \{\infty\}$ where Γ is a divisible totally ordered abelian group, and we refer to the pair (F, v) as a *valued field.* The division algebras with center F are called *central* division algebras over F; they are always assumed to be finite-dimensional over F.

Theorem 1.4. *Let (F, v) be a valued field and let D be a (finite-dimensional) central division algebra over F. The valuation v extends to a valuation w on D if and only if v has a unique extension to each field L with $F \subseteq L \subseteq D$. When this condition holds, the valuation w is unique and is given in terms of the reduced norm Nrd and the index $\operatorname{ind}(D) = \sqrt{[D{:}F]}$ by the formula*

$$w(x) = \tfrac{1}{\operatorname{ind}(D)}\, v\big(\operatorname{Nrd}(x)\big) \qquad \text{for } x \in D^\times. \tag{1.9}$$

Moreover, $\mathcal{O}_D = \{x \in D \mid x \text{ is integral over } \mathcal{O}_F\}$.

The main technical tool for the proof is Wedderburn's Factorization Theorem on the minimal polynomial of elements in D. Before proving Th. 1.4 we gather its main consequences in the following proposition:

Proposition 1.5. *Let D be a central division F-algebra. Suppose w is a valuation on D extending v and let $P \subseteq Z(\overline{D})$ be the purely inseparable closure of $\overline{F}$ in $Z(\overline{D})$.*

(i) *For all $a \in D^\times$ we have $v\big(\operatorname{Nrd}(a)\big) = \operatorname{ind}(D)\, w(a)$.*
(ii) *Every element in $\mathcal{O}_D$ is integral over $\mathcal{O}_F$.*
(iii) *The field extension $Z(\overline{D})/\overline{F}$ is normal.*
(iv) *$P = Z(\overline{D})^{\theta_D(\Gamma_D)}$, the subfield fixed under $\theta_D(\Gamma_D)$ (where θ_D is defined in (1.1)).*
(v) *$Z(\overline{D})/P$ is Galois with abelian Galois group $\theta_D(\Gamma_D)$.*

Proof. Let $a \in D^\times$ and let $f(X) = X^n + \alpha_{n-1}X^{n-1} + \ldots + \alpha_0 \in F[X]$ be the minimal polynomial of a over F. Thus, $n = [F(a){:}F]$ and $\operatorname{Nrd}(a) = (-1)^{\operatorname{ind}(D)}\alpha_0^{\operatorname{ind}(D)/n}$, so $v\big(\operatorname{Nrd}(a)\big) = \frac{\operatorname{ind}(D)}{n} v(\alpha_0)$. By Wedderburn's Factorization Theorem (see Lam [121, Th. 16.9, p. 265]), we may find conjugates of a,

$$a_1 = d_1 a d_1^{-1}, \quad \ldots, \quad a_n = d_n a d_n^{-1}$$

such that $f(X) = (X - a_1)\ldots(X - a_n)$. In particular, $\alpha_0 = (-1)^n a_1 \ldots a_n$. Since $w(a_i) = w(a)$ for all i, it follows that $v(\alpha_0) = n\, w(a)$; hence, $v\big(\operatorname{Nrd}(a)\big) = \operatorname{ind}(D) w(a)$, proving (i).

If $a \in \mathcal{O}_D$, then $a_i \in \mathcal{O}_D$ for all i; hence, $f(X) \in \mathcal{O}_D[X] \cap F[X] = \mathcal{O}_F[X]$, and (ii) follows.

Now, suppose $a \in \mathcal{O}_D$ and $\overline{a} \in Z(\overline{D})$. By definition of θ_D we have

$$\overline{a_i} = \theta_D\big(w(d_i)\big)(\overline{a}) \in Z(\overline{D});$$

hence, $\overline{f}(X)$ splits in $Z(\overline{D})[X]$. The minimal polynomial of $\overline{a}$ over $\overline{F}$ also splits over $Z(\overline{D})$ since it divides $\overline{f}$, hence (iii) is proved.

Assuming $\overline{a} \in Z(\overline{D})^{\theta_D(\Gamma_D)}$, we have $\overline{a_i} = \overline{a}$ for all i, hence $\overline{f}(X) = (X - \overline{a})^n$ and $\overline{a}$ is purely inseparable over $\overline{F}$. Thus, $Z(\overline{D})^{\theta_D(\Gamma_D)} \subseteq P$. Statements (iv) and (v) then readily follow by Galois theory. □

For the proof of Th. 1.4, we also need the following lemma:

Lemma 1.6. *Let L be a finite-degree field extension of F. If v has a unique extension to a valuation w on L, then for each $x \in L$ we have*

$$v(N_{L/F}(x)) = [L{:}F]\, w(x),$$

where $N_{L/F}$ denotes the norm from L to F.

Proof. Let K be a normal closure of L over F, and let v' be any extension of v to K. We have $N_{L/F}(x) = x_1 \ldots x_n$ where $n = [L{:}F]$ and each x_i is a conjugate of x in K. For each i, there is an F-automorphism σ_i of K with $\sigma_i(x) = x_i$. Since $v' \circ \sigma_i|_L$ is a valuation of L extending v, we have $v' \circ \sigma_i|_L = w$. Hence, $v'(x_i) = v'(\sigma_i(x)) = w(x)$. Thus,

$$v(N_{L/F}(x)) = v'(N_{L/F}(x)) = v'(x_1) \ldots v'(x_n) = [L{:}F]\, w(x). \qquad \square$$

Proof of Th. 1.4. Suppose v has a unique extension to each field L with $F \subseteq L \subseteq D$. We show that the formula (1.9) defines a valuation on D. Clearly, $w(x) = \infty$ if and only if $x = 0$, and $w(xy) = w(x) + w(y)$ for x, $y \in D$ since $Nrd(xy) = Nrd(x)\, Nrd(y)$. To prove $w(x+y) \geq min\big(w(x), w(y)\big)$ for x, $y \in D$, we may of course assume $y \neq 0$. Since $w(x+y) = w(xy^{-1}+1) + w(y)$ and $w(x) = w(xy^{-1}) + w(y)$, it suffices to show

$$w(xy^{-1} + 1) \geq min(w(xy^{-1}), 0). \tag{1.10}$$

Let $L \subseteq D$ be any maximal subfield containing xy^{-1}. Since $N_{L/F} = Nrd|_L$, Lemma 1.6 shows that $w|_L$ is the unique valuation on L extending v. Therefore, (1.10) holds and w is a valuation on D.

Now, suppose v extends to a valuation w on D. Proposition 1.5(i) yields the formula (1.9), which shows that w is unique. If $a \in D$ is integral over $\mathcal{O}_F$, then we have

$$a^n = \alpha_{n-1} a^{n-1} + \ldots + \alpha_0 \ \text{ for some } n \geq 1 \text{ and some } \alpha_{n-1}, \ldots, \alpha_0 \in \mathcal{O}_F.$$

If $w(a) < 0$, then $w(a^n) < w(\alpha_i a^i)$ for $i = 0, \ldots, n-1$, and the equality above is impossible. Therefore, $a \in \mathcal{O}_D$. By Prop. 1.5(ii), it follows that $\mathcal{O}_D$ consists of the elements in D that are integral over $\mathcal{O}_F$. Likewise, for any field L with $F \subseteq L \subseteq D$ the valuation ring $\mathcal{O}_L$ of $w|_L$ is integral over $\mathcal{O}_F$. But in any finite degree field extension of F the integral closure of $\mathcal{O}_F$ is the irredundant intersection of the valuation rings of all the extensions of v to L, see Engler–Prestel [73, Cor. 3.1.4, p. 60; Lemma 3.2.8, p. 64]. The integrality of $\mathcal{O}_L$ over $\mathcal{O}_F$ therefore implies that $w|_L$ is the only extension of v to L. □

Theorem 1.4 indicates the importance of Henselian valuations in the valuation theory of division algebras. Recall that the valuation v on F is Henselian[1] if and only if v has a unique extension to each field L algebraic over F; the extension of v to L is also clearly Henselian. Thus we have the following corollary to Th. 1.4:

Corollary 1.7. *Let D be a division algebra finite-dimensional over F. If v is a Henselian valuation on F, then v has a unique extension to a valuation on D.*

Proof. Let L be the center of D. Since $[L{:}F] < \infty$, v has a unique extension to a valuation v_L of L, and v_L is Henselian. Therefore, by Th. 1.4, v_L has a unique extension to a valuation v_D on D. Then, v_D is the unique extension of v to D, since any valuation on D restricts to a valuation on L. □

1.2.3 Composite valuations

When F is a field, we can "compose" a valuation v on F with any valuation u on $\overline{F}^v$ to obtain a valuation $u * v$ on F that is a refinement of v. This is the valuation whose ring is $\pi^{-1}(\mathcal{O}_{\overline{F}^v,u})$, where $\pi\colon \mathcal{O}_{F,v} \to \overline{F}^v$ is the canonical surjection; see Engler–Prestel [73, p. 45]. For valuations on a division algebra, such a composition is possible only with an added condition:

Proposition 1.8. *Let D be a central division algebra over F and let w be a valuation on D. Let u be a valuation on $\overline{D}^w$. Then, the composite valuation $u|_{\overline{F}^w} * w|_F$ on F extends to a valuation on D if and only if u has a unique extension from $\overline{F}^w$ to $Z(\overline{D}^w)$. When this extension exists, its valuation ring is $\pi^{-1}(\mathcal{O}_{\overline{D}^w,u})$, where $\pi\colon \mathcal{O}_{D,w} \to \overline{D}^w$ is the canonical surjection.*

Here is why the added condition is needed: While the ring $V = \pi^{-1}(\mathcal{O}_{\overline{D}^w,u})$ is always a total valuation ring of D, i.e., d or d^{-1} lies in V for all $d \in D^\times$, this V need not have the further property of being stable under inner automorphisms, which is required for V to be the valuation ring of a valuation on D. The added condition in Prop. 1.8 is equivalent to: $dVd^{-1} = V$ for all $d \in D^\times$.

Proof of Prop. 1.8. We write $\overline{D}$ for $\overline{D}^w$ and $\overline{F}$ for $\overline{F}^w$ throughout the proof. Let z be the valuation $u|_{\overline{F}^w} * w|_F$ on F. Suppose z extends to a valuation v on D. Take any element of $\overline{D}$, and write it as $\overline{d}$ for some $d \in \mathcal{O}_{D,w}$. Let L be the field $F(d)$, and set $\overline{L} = \overline{L}^w$. If u' is any valuation on $\overline{L}$ with $u'|_{\overline{F}} = u|_{\overline{F}}$, then $u' * w|_L$ and $u|_{\overline{L}} * w|_L$ are each valuations on L extending $v|_F = z$. Since

[1] Properties of Henselian valuations and Henselizations are reviewed in Appendix A.

z on F extends to v on D, by Th. 1.4 z has a unique extension to L. Hence, $u' * w|_L = v|_L = u|_{\overline{L}} * w|_L$; therefore, u' must coincide with $u|_{\overline{L}}$, since they each have the valuation ring $\pi(\mathcal{O}_{L,v})$. Thus, $u|_{\overline{L}}$ is the unique extension of the valuation $u|_{\overline{F}}$ on $\overline{F}$ to $\overline{L}$. Therefore, $u|_{\overline{F}}$ extends uniquely to $\overline{F}(\overline{d}) \subseteq \overline{L}$. It follows that $u|_{\overline{F}}$ must extend uniquely to $Z(\overline{D})$, since different extensions to $Z(\overline{D})$ would restrict to different extensions to $\overline{F}(\overline{d})$ for some $\overline{d} \in Z(\overline{D})$. The valuation ring $\mathcal{O}_{D,v}$ is the union of its restrictions, $\mathcal{O}_{D,v} = \bigcup_{L \in \mathcal{L}} \mathcal{O}_{L,v}$, where $\mathcal{L} = \{\text{fields } F(d) \mid d \in D\}$. But we saw above that $v|_L = u|_{\overline{L}} * w|_L$ for all $L \in \mathcal{L}$; hence, $\mathcal{O}_{L,v} = \pi^{-1}(\mathcal{O}_{\overline{L},u}) \cap L$. Thus,

$$\mathcal{O}_{D,v} = \bigcup_{L \in \mathcal{L}} \big(\pi^{-1}(\mathcal{O}_{\overline{L},u}) \cap L\big) = \pi^{-1}(\mathcal{O}_{\overline{D},u}).$$

Conversely, suppose $u|_{Z(\overline{D})}$ is the unique extension of $u|_{\overline{F}}$ to $Z(\overline{D})$. Let K be any subfield of $\overline{D}$. By Th. 1.4, $u|_{KZ(\overline{D})}$ is the unique extension of the valuation $u|_{Z(\overline{D})}$ from $Z(\overline{D})$ to the subfield $KZ(\overline{D})$ of $\overline{D}$. Hence, $u|_K$ must be the unique extension of $u|_{\overline{F}}$ from $\overline{F}$ to K. Now, let M be any field with $F \subseteq M \subseteq D$, and let y be a valuation on M extending z $(= u|_{\overline{F}} * w|_F)$ on F. Since $w|_F$ is a coarsening of z, there is a prime ideal $\mathfrak{p}$ of $\mathcal{O}_{M,y}$ with $\mathfrak{p} \cap \mathcal{O}_{F,z} = \mathfrak{m}_{F,w}$, which is a prime ideal of $\mathcal{O}_{F,z}$. Then, $y = \widetilde{u} * \widetilde{w}$, where $\widetilde{w}$ is the valuation on M with valuation ring the localization $(\mathcal{O}_{M,y})_{\mathfrak{p}}$ of $\mathcal{O}_{M,y}$ at $\mathfrak{p}$ and $\widetilde{u}$ is the valuation on $\overline{M}^{\widetilde{w}}$ with ring $\mathcal{O}_{M,y}/\mathfrak{p}$. Since $\mathcal{O}_{M,\widetilde{w}} \cap F = \mathcal{O}_{F,w}$, this $\widetilde{w}$ is an extension of $w|_F$ to M. Hence, as $w|_F$ extends to D, Th. 1.4 shows that $\widetilde{w} = w|_M$, so $\overline{M}^{\widetilde{w}} = \overline{M}^{w}$. Thus, $\widetilde{u}$ is an extension of $u|_{\overline{F}}$ from $\overline{F}$ to $\overline{M}^{w}$. We saw above that the only such extension is $u|_{\overline{M}^w}$. Hence, $y = u|_{\overline{M}^w} * w|_M$, which shows that z has a unique extension from F to M. Since this is true for each subfield M of D, Th. 1.4 shows that z extends to a valuation on D. □

Remark 1.9. The composite valuation v of Prop. 1.8 is denoted by $u * w$. Note that $\overline{D}^{u*w} = \overline{\overline{D}^{w}}^{u}$ and we have a short exact sequence of ordered abelian groups

$$0 \longrightarrow \Gamma_{\overline{D}^{w},u} \longrightarrow \Gamma_{D,u*w} \longrightarrow \Gamma_{D,w} \longrightarrow 0. \tag{1.11}$$

If there is an ordered group homomorphism $\Gamma_{D,w} \to \Gamma_{D,u*w}$ that is a splitting map for (1.11), then we have an ordered group isomorphism

$$\Gamma_{D,u*w} \cong \Gamma_{\overline{D}^{w},u} \times \Gamma_{D,w},$$

where the direct product is given the right-to-left lexicographic ordering (defined by $(\gamma, \delta) \leq (\gamma', \delta')$ if and only if either $\delta < \delta'$, or $\delta = \delta'$ and $\gamma \leq \gamma'$). We saw a case of this when we discussed iterated Laurent series in §1.1.3. This construction will also be used in Ex. 7.77.

If v is a valuation on D and w is any coarser valuation, i.e., $\mathcal{O}_{D,w} \supseteq \mathcal{O}_{D,v}$, then there is a "residue valuation" on $\overline{D}^{w}$ with associated valuation ring $\mathcal{O}_{D,v}/\mathfrak{m}_{D,w}$. This ring is invariant under inner automorphisms of $\overline{D}^{w}$ since all such automorphisms are induced by inner automorphisms of D. We denote the residue valuation by v/w. Clearly $v = (v/w) * w$.

1.2.4 Examples: rational functions over a division algebra

Let D be a division ring finite-dimensional over its center F, and let x be an indeterminate commuting with the elements of D. The center of the polynomial ring $D[x]$ is $F[x]$. The quotient field of $F[x]$ is the rational function field $F(x)$, and the division ring of quotients of $D[x]$ is $D(x) = D[x] \otimes_{F[x]} F(x) = D \otimes_F F(x)$, which is finite-dimensional over its center $F(x)$. Take any irreducible $f \in F[x]$. Let v_f be the f-adic valuation on $F(x)$, which is defined by $v_f(f^j g/h) = j$ for all $j \in \mathbb{Z}$ and $g, h \in F[x]$ with $f \nmid g$ and $f \nmid h$. So, $\Gamma_{F(x),v_f} = \mathbb{Z}$ and $\overline{F(x)}^{v_f} \cong F[x]/fF[x]$.

Proposition 1.10. *The valuation v_f on $F(x)$ extends to a valuation on $D(x)$ if and only if $D \otimes_F \overline{F(x)}^{v_f}$ is a division ring.*

Proof. Let $L = F[x]/fF[x] \cong \overline{F(x)}^{v_f}$. Then, $D \otimes_F L \cong D[x]/fD[x]$. Suppose this is a division ring. We first extend v_f on $F[x]$ to a map w_f on $D[x]$: any $g \in D[x] \setminus \{0\}$ is expressible uniquely as $g = f^n g'$ for some integer $n \geq 0$ and $g' \in D[x]$ with $f \nmid g'$; set $w_f(g) = n$. (Since f is central in $D[x]$, there is no difference between left and right divisibility by f.) Also, set $w_f(0) = \infty$. Clearly $w_f(g+h) \geq min\big(w_f(g), w_f(h)\big)$ for all $g, h \in D[x]$. Take any nonzero $g, h \in D[x]$, and write $g = f^n g'$ and $h = f^m h'$ with $f \nmid g'$, $f \nmid h'$ in $D[x]$. Since the division ring $D[x]/fD[x]$ has no zero divisors, we have $f \nmid g'h'$ in $D[x]$. Hence,

$$w_f(gh) = w_f(g) + w_f(h). \tag{1.12}$$

We now extend w_f to $D(x)$ by defining $w_f(g/k) = w_f(g) - w_f(k)$ for all $g \in D[x]$ and $k \in F[x] \setminus \{0\}$. Formula (1.12) assures that w_f is well-defined, and it is easy to check that w_f is a valuation on $D(x)$ with $w_f|_{F(x)} = v_f$.

Conversely, suppose v_f extends to a valuation w_f on $D(x)$. Since $[D:F] < \infty$, the trivial valuation on D is the unique extension to D of the trivial valuation $v_f|_F$ on F. Hence, $D \subseteq \mathcal{O}_{D(x)}$; so D maps into $\overline{D(x)}$ via the canonical surjection $\mathcal{O}_{D(x)} \to \overline{D(x)}$. Also, $L \cong \overline{F(x)}$ so L maps into $\overline{D(x)}$. Its image is central, so it centralizes the image of D. Hence, there is an F-algebra homomorphism $\rho\colon D \otimes_F L \to \overline{D(x)}$. This ρ must be injective, as its domain is simple. Therefore, $D \otimes_F L$ has no zero divisors, since its isomorphic copy $\rho(D \otimes_F L)$ is a subring of a division ring. Because $D \otimes_F L$ is also finite-dimensional over F, it must be a division ring. □

It is easy to see that when v_f extends to $D(x)$, we have $\Gamma_{D(x)} = \Gamma_{F(x)} = \mathbb{Z}$ and $\overline{D(x)} \cong D \otimes_F \overline{F(x)}$. Assuming $D \neq F$ (hence $|F| = \infty$ by Wedderburn's

Theorem), it follows that there are infinitely many different valuations v_f that extend to $D(x)$; for example, take $f = x - d$ for $d \in D$. But there are also infinitely many v_f that do not extend. To see this, take a maximal subfield N of D with N separable over F. There are infinitely many choices of $a \in N$ with $N = F(a)$. By taking f to be the minimal polynomial a over F, we obtain a v_f that does not extend to $D(x)$ by Prop. 1.10, as $\overline{F(x)}^{v_f} \cong N$ which splits D.

Note that for irreducible polynomials $f, g \in F[x]$, $v_f = v_g$ if and only if $fF[x] = gF[x]$. Moreover, the set of all nontrivial valuations on $F(x)$ that are trivial on F consists of all the v_f together with the $1/x$-adic valuation for the ring $F[1/x]$; call the latter valuation v_∞. Since $\overline{F(x)}^{v_\infty} = F$, Prop. 1.10, applied with $1/x$ replacing x, shows that v_∞ extends to $D(x)$.

1.2.5 Examples: Laurent series

Let A be a division ring, let σ be an automorphism of A, and let $D = A((x;\sigma))$ be the division ring of twisted Laurent series defined in §1.1.2. We determine when D is finite-dimensional over its center $Z(D)$.

Recall from §1.1.2 that

$$Z(D) = \Big\{ \sum_{i=k}^{\infty} a_i x^i \mid \sigma(a_i) = a_i \text{ and } ba_i = a_i \sigma^i(b) \text{ for all } i \text{ and all } b \in A \Big\}. \tag{1.13}$$

If no power of σ with nonzero exponent is an inner automorphism of A, it is clear from the description of $Z(D)$ above that $Z(D) = \overline{Z(D)} = Z(A)^\sigma$; D is then infinite-dimensional over its center. Similarly, if A is infinite-dimensional over $Z(A)$, then $\overline{D} = A$ is infinite-dimensional over $\overline{Z(D)} = Z(A)^\sigma$, hence the fundamental inequality (Prop. 1.3) shows that D is infinite-dimensional over its center. We now consider the case where A is finite-dimensional over its center.

Lemma 1.11. *Suppose $[A{:}Z(A)]$ is finite. The following conditions on an integer $r \geq 1$ are equivalent:*

(a) *σ has order r modulo inner automorphisms of A;*
(b) *$\sigma|_{Z(A)}$ has order r in $Aut(Z(A))$;*
(c) *$[Z(A){:}Z(A)^\sigma] = r$.*

If these equivalent conditions hold, then there exists $c \in A^\times$ such that

$$\sigma^r = int(c) \qquad and \qquad \sigma(c) = c.$$

Proof. (a) and (b) are equivalent because the Skolem–Noether Theorem shows that the kernel of the restriction map $Aut(A) \to Aut(Z(A))$ is the group of inner automorphisms of A. The equivalence of (b) and (c) is clear by Galois theory.

Suppose that (a), (b), and (c) hold and let $\sigma^r = int(a)$ for some $a \in A^\times$. By applying σ to the equation $\sigma^r(b) = aba^{-1}$ for $b \in A$, we find that $\sigma^r = int\big(\sigma(a)\big)$, hence

$$\sigma(a) = za \qquad \text{for some } z \in Z(A)^\times.$$

Then $\sigma^i(a) = z\sigma(z)\ldots\sigma^{i-1}(z)a$ for all $i \geq 1$. Since $\sigma^r(a) = int(a)(a) = a$, it follows that

$$\prod_{i=0}^{r-1} \sigma^i(z) = 1. \tag{1.14}$$

Now, the left term in (1.14) is the norm $N_{Z(A)/Z(A)^\sigma}(z)$; hence, Hilbert's Theorem 90 yields an element $z' \in Z(A)^\times$ such that $z = z'/\sigma(z')$. Then $c = z'a$ satisfies the required conditions. □

Proposition 1.12. *Let A be a division ring and let $\sigma \in Aut(A)$. The division ring $D = A((x;\sigma))$ is finite-dimensional over its center if and only if $[A{:}Z(A)]$ is finite and $\sigma|_{Z(A)}$ has finite order. If these conditions hold and $\sigma|_{Z(A)}$ has order r, then*

$$[D{:}Z(D)] = [A{:}Z(A)]\, r^2 = [\overline{D}{:}\overline{Z(D)}]\, |\Gamma_D{:}\Gamma_{Z(D)}|.$$

Proof. If $[A{:}Z(A)]$ is infinite or if σ has infinite order modulo inner automorphisms in A, it was already observed in the discussion preceding Lemma 1.11 that $[D{:}Z(D)]$ is infinite. For the rest of the proof, assume $[A{:}Z(A)]$ is finite and $\sigma|_{Z(A)}$ has order r. Lemma 1.11 yields an element $c \in A^\times$ such that $\sigma^r = int(c)$ and $\sigma(c) = c$. Then from the description of $Z(D)$ in (1.13) it follows that $c^{-1}x^r \in Z(D)$, and that every element in $Z(D)$ can be written as a Laurent series over $Z(A)^\sigma$ in $c^{-1}x^r$,

$$Z(D) = Z(A)^\sigma((c^{-1}x^r)). \tag{1.15}$$

If $(a_i)_{i\in I}$ is a $Z(A)^\sigma$-vector space base of A, then $(a_i x^j \mid i \in I,\, 0 \leq j \leq r-1)$ is a $Z(D)$-base of D; hence,

$$[D{:}Z(D)] = [A{:}Z(A)^\sigma]\, r = [A{:}Z(A)]\, r^2.$$

We have $\overline{D} = A$, $\overline{Z(D)} = Z(A)^\sigma$, and $\Gamma_{Z(D)} = r\mathbb{Z}$; hence,

$$[\overline{D}{:}\overline{Z(D)}]\, |\Gamma_D{:}\Gamma_{Z(D)}| = [A{:}Z(A)^\sigma]\, r.$$

The proof is thus complete. □

1.2.6 Examples: iterated Laurent series

Proposition 1.12 may be applied inductively to determine when a division ring of iterated Laurent series is finite-dimensional over its center. We discuss here a couple of special cases.

First, let M/F be a finite Galois extension of fields with abelian Galois group G. Fix a decomposition of G into a direct sum of cyclic subgroups $G = G_1 \oplus \ldots \oplus G_n$, let $\sigma_i \in G$ be a generator of G_i for $i = 1, \ldots, n$, and let $\boldsymbol{\sigma} = (\sigma_i)_{i=1}^n$. Consider also a collection $\boldsymbol{u} = (u_{i,j})_{i,j=1}^n$ of elements of $M^\times$ subject to the conditions in (1.3) in §1.1.3, and let $D = \mathcal{L}((M; \boldsymbol{\sigma}, \boldsymbol{u}))$ be the division ring of iterated Laurent series defined in (1.4), in which the multiplication satisfies

$$x_i m = \sigma_i(m) x_i \quad \text{and} \quad x_i x_j = u_{i,j} x_j x_i \quad \text{for } m \in M \text{ and } i, j = 1, \ldots, n.$$

As observed in §1.1.3, we have $\overline{D} = M$ and $\overline{Z(D)} = F$ for the $(x_1, \ldots, x_n)$-adic valuation.

For $i = 1, \ldots, n$, let $N_i : M \to M$ be the map defined by $N_i(x) = \prod_{\sigma \in G_i} \sigma(x)$.

Proposition 1.13. *If $N_i \circ N_j(u_{i,j}) = 1$ for all $i, j = 1, \ldots, n$, then $|\Gamma_D : \Gamma_{Z(D)}| = [M:F]$ and*

$$[D:Z(D)] = [M:F]^2 = [\overline{D}:\overline{Z(D)}] \cdot |\Gamma_D:\Gamma_{Z(D)}|.$$

In proving the proposition we will use the following lemma, which is a version of Hilbert's Theorem 90 for the field extension M/F:

Lemma 1.14. *If $(a_i)_{i=1}^n$ is a collection of elements of $M^\times$ satisfying*

$$N_i(a_i) = 1 \quad \text{and} \quad \sigma_i(a_j)a_j^{-1} = \sigma_j(a_i)a_i^{-1} \quad \text{for all } i, j = 1, \ldots, n, \tag{1.16}$$

then there exists a $b \in M^\times$ such that $a_i = \sigma_i(b)b^{-1}$ for all $i = 1, \ldots, n$.

Proof. We apply induction on n. If $n = 1$, then $G = G_1$ is cyclic, and the lemma is the classical version of Hilbert's Theorem 90 (see Draxl [63, p. 35] or Pierce [178, p. 312]). Now, assume $n > 1$ and let $L = M^{G_n}$ be the subfield of M fixed under σ_n. The Galois group of L/F may be identified with $G_1 \oplus \ldots \oplus G_{n-1}$. Since $N_n(a_n) = 1$, Hilbert's Theorem 90 yields an element $c \in M^\times$ such that $a_n = \sigma_n(c)c^{-1}$. By (1.16), it follows that for all $i = 1, \ldots, n-1$,

$$[\sigma_i\sigma_n(c)\sigma_i(c)^{-1}]\,[\sigma_n(c)^{-1}c] = \sigma_n(a_i)a_i^{-1};$$

hence,

$$a_i\sigma_i(c)^{-1}c = \sigma_n(a_i\sigma_i(c)^{-1}c).$$

Let $a_i' = a_i\sigma_i(c)^{-1}c$ for $i = 1, \ldots, n-1$. The preceding equation shows that $a_i' \in L^\times$. For $i, j = 1, \ldots, n$ we have $N_i(a_i') = N_i(a_i)$ and

$$\sigma_i(a_j')a_j'^{-1} = [\sigma_i(a_j)a_j^{-1}]\,[\sigma_i\sigma_j(c)^{-1}\sigma_i(c)]\,[\sigma_j(c)c^{-1}].$$

Therefore, the conditions (1.16) for $(a_i)_{i=1}^n$ imply analogous conditions for $(a_i')_{i=1}^{n-1}$. By induction, there exists a $b' \in L^\times$ such that $a_i' = \sigma_i(b')b'^{-1}$ for $i = 1, \ldots, n-1$. Letting $b = b'c$ we then have $a_i = \sigma_i(b)b^{-1}$ for $i = 1, \ldots, n$.

□

Proof of Prop. 1.13. Suppose that $N_i \circ N_j(u_{i,j}) = 1$ for all $i, j = 1, \ldots, n$. Note that the conditions (1.3) on the $u_{i,j}$ then imply that for each fixed $k = 1, \ldots, n$ the family $(a_i)_{i=1}^n$ with $a_i = N_k(u_{i,k})$ satisfies (1.16). Therefore, by Lemma 1.14 there is an element $b_k \in M^\times$ such that

$$N_k(u_{i,k}) = \sigma_i(b_k)b_k^{-1} \qquad \text{for all } i = 1, \ldots, n.$$

For $i = 1, \ldots, n$, let e_i be the order of σ_i in G. Then,

$$x_i x_k^{e_k} = N_k(u_{i,k})x_k^{e_k}x_i = \sigma_i(b_k)b_k^{-1}x_k^{e_k}x_i \qquad \text{for all } i, k = 1, \ldots, n;$$

hence, $b_k^{-1}x_k^{e_k}$ centralizes $x_1, \ldots, x_n$ for $k = 1, \ldots, n$. On the other hand, a monomial $x_1^{\alpha_1} \ldots x_n^{\alpha_n}$ centralizes M if and only if α_i is a multiple of e_i for $i = 1, \ldots, n$. Therefore, the center $Z(D)$ is

$$Z(D) = \Big\{ \sum_{\alpha_1, \ldots, \alpha_n} \lambda_{\alpha_1, \ldots, \alpha_n} (b_1^{-1}x_1^{e_1})^{\alpha_1} \ldots (b_n^{-1}x_n^{e_n})^{\alpha_n} \mid \lambda_{\alpha_1, \ldots, \alpha_n} \in F \Big\}. \quad (1.17)$$

It can be identified with a field of iterated Laurent series,

$$Z(D) = F((b_1^{-1}x_1^{e_1})) \ldots ((b_n^{-1}x_n^{e_n})).$$

It is then clear that $\Gamma_{Z(D)} = e_1\mathbb{Z} \times \ldots \times e_n\mathbb{Z} \subseteq \mathbb{Z}^n = \Gamma_D$; hence,

$$|\Gamma_D{:}\Gamma_{Z(D)}| = e_1 \ldots e_n = [M{:}F].$$

If $(m_i)_{i\in I}$ is an F-base of M, then the monomials $m_i x_1^{\alpha_1} \ldots x_n^{\alpha_n}$ with $i \in I$ and $0 \leq \alpha_j < e_j$ for $j = 1, \ldots, n$ form a base of D over $Z(D)$. Proposition 1.13 is thus proved. □

For another special case of the iterated Laurent series construction, consider on an arbitrary field M the collection of automorphisms $\boldsymbol{I} = (\sigma_i)_{i=1}^n$ where $\sigma_i = id_M$ for all $i = 1, \ldots, n$, and let $\boldsymbol{u} = (u_{i,j})_{i,j=1}^n$ be a collection of elements of M satisfying the conditions in (1.3), which in the present case simplify to

$$u_{i,i} = 1 \quad \text{and} \quad u_{i,j}u_{j,i} = 1 \qquad \text{for all } i, j = 1, \ldots, n.$$

Let $E = \mathcal{L}((M; \boldsymbol{I}, \boldsymbol{u}))$.

Proposition 1.15. *The division ring E is finite-dimensional over its center $Z(E)$ if and only if every $u_{i,j}$ is a root of unity. Then $\overline{E} = \overline{Z(E)}$ and $[E{:}Z(E)] = |\Gamma_E{:}\Gamma_{Z(E)}|$.*

Proof. Suppose x_i is algebraic of degree m over $Z(E)$, say

$$x_i^m + \lambda_{m-1}x_i^{m-1} + \ldots + \lambda_0 = 0 \qquad \text{with} \quad \lambda_0, \ldots, \lambda_{m-1} \in Z(E).$$

By conjugating this equation by x_j, we obtain

$$u_{j,i}^m x_i^m + \lambda_{m-1}u_{j,i}^{m-1}x_i^{m-1} + \ldots + \lambda_0 = 0.$$

Therefore, for any k such that $\lambda_k \neq 0$ we must have $u_{i,j}^{k-m} = 1$. This shows that if E is finite-dimensional over $Z(E)$, then each $u_{i,j}$ is a root of unity.

For the converse, we define a pairing $\langle \cdot, \cdot \rangle$ on $\mathbb{Z}^n$ $(=\Gamma_E)$ with values in $M^\times$ as follows: for $\boldsymbol{\alpha} = (\alpha_1, \ldots, \alpha_n)$ and $\boldsymbol{\beta} = (\beta_1, \ldots, \beta_n)$ in $\mathbb{Z}^n$ we let $x^{\boldsymbol{\alpha}} = x_1^{\alpha_1} \ldots x_n^{\alpha_n}$, $x^{\boldsymbol{\beta}} = x_1^{\beta_1} \ldots x_n^{\beta_n}$, and

$$\langle \boldsymbol{\alpha}, \boldsymbol{\beta} \rangle \;=\; x^{\boldsymbol{\alpha}} x^{\boldsymbol{\beta}} (x^{\boldsymbol{\alpha}})^{-1} (x^{\boldsymbol{\beta}})^{-1} \;=\; \prod_{i,j=1}^{n} (u_{i,j})^{\alpha_i \beta_j} \in M^\times.$$

The last equality holds by commutator identities since each x_k commutes with every $u_{i,j}$. It is readily seen that this pairing is bilinear and alternating, i.e., for $\boldsymbol{\alpha}$, $\boldsymbol{\beta}$, $\boldsymbol{\gamma} \in \mathbb{Z}^n$ we have

$$\langle \boldsymbol{\alpha}, \boldsymbol{\beta} + \boldsymbol{\gamma} \rangle \;=\; \langle \boldsymbol{\alpha}, \boldsymbol{\beta} \rangle \cdot \langle \boldsymbol{\alpha}, \boldsymbol{\gamma} \rangle \qquad \text{and} \qquad \langle \boldsymbol{\alpha}, \boldsymbol{\alpha} \rangle \;=\; 1.$$

Let Λ be the radical of this pairing, i.e.,

$$\Lambda \;=\; \{\boldsymbol{\alpha} \in \mathbb{Z}^n \mid \langle \boldsymbol{\alpha}, \boldsymbol{\beta} \rangle = 1 \text{ for all } \boldsymbol{\beta} \in \mathbb{Z}^n\}.$$

If each $u_{i,j}$ is a root of unity in M, then Λ is a subgroup of finite index in $\mathbb{Z}^n$, and we have

$$Z(E) \;=\; \big\{ \sum_{\boldsymbol{\alpha} \in \mathbb{Z}^n} \lambda_{\boldsymbol{\alpha}} x^{\boldsymbol{\alpha}} \mid \lambda_{\boldsymbol{\alpha}} \in M, \; \boldsymbol{\alpha} \in \Lambda \big\}.$$

Therefore, $\Gamma_{Z(E)} \;=\; \Lambda$ and $[E{:}Z(E)] \;=\; |\mathbb{Z}^n{:}\Lambda|$; any collection of monomials $x^{\boldsymbol{\alpha}}$ where $\boldsymbol{\alpha}$ runs over a set of coset representatives of $\mathbb{Z}^n$ modulo Λ is a $Z(E)$-base of E. □

1.2.7 Examples: discrete rank 1 *valuations*

We will prove later (see Th. 4.1) that if (F, v) is a valued field with Henselization (F_h, v_h) and D is a central division F-algebra, then v extends to D if and only if $D \otimes_F F_h$ is a division ring. This will provide a very useful approach to proving results about arbitrary valued division algebras, by first proving them in the Henselian case. For convenience, we now prove a special case of this with the Henselization replaced by the completion.

A valuation v on a field F is said to be *discrete of rank* 1 if $\Gamma_F \cong \mathbb{Z}$. (See §A.4 for the terminology on ranks for ordered groups and valuations.)

Proposition 1.16 (Cohn). *Let F be a field with a discrete rank 1 valuation v. Let $(\widehat{F}, \widehat{v})$ be the completion of (F, v) and let D be a central division algebra over F. The valuation v extends to a valuation on D if and only if $D \otimes_F \widehat{F}$ is a division ring.*

Proof. $\Leftarrow$ Suppose $D \otimes_F \widehat{F}$ is a division ring. The valuation $\widehat{v}$ on $\widehat{F}$ is discrete of rank 1 (since v is) and complete, hence $\widehat{v}$ is Henselian (see Engler–Prestel [73, Th. 1.3.1, p. 20]). Therefore, as $D \otimes_F \widehat{F}$ is a division ring, Cor. 1.7 shows that $\widehat{v}$ has an extension to a valuation v' of $D \otimes_F \widehat{F}$. We may view $D \subseteq D \otimes_F \widehat{F}$. Then $v'|_D$ is a valuation on D that extends v.

$\Rightarrow$ Suppose v extends to a valuation w on D. Since $|\Gamma_D{:}\Gamma_F| \le [D{:}F] < \infty$, we have $\Gamma_D \simeq \mathbb{Z}$. Therefore, $\bigcap_{i=1}^{\infty} \mathfrak{m}_D^i = \{0\}$. Thus, we can form the completion $\widehat{D}$ of D with respect to the filtration of ideals $\{\mathfrak{m}_D^i \mid i = 1, 2, \ldots\}$. Just as in the commutative case, Bourbaki [30, Ch. VI, §5.3, Prop. 5] shows that $\widehat{D}$ is a division ring and w extends continuously to a valuation $\widehat{w}$ on $\widehat{D}$. The closure $\check{F}$ of the image of F in $\widehat{D}$ is isomorphic to the completion $\widehat{F}$ of F with respect to v. Since D centralizes F, the image of D in $\widehat{D}$ centralizes $\check{F}$. Therefore we have an F-homomorphism $D \otimes_F \widehat{F} \to \widehat{D}$ built from the maps $D \to \widehat{D}$ and $\widehat{F} \xrightarrow{\sim} \check{F} \subseteq \widehat{D}$. Because $D \otimes_F \widehat{F}$ is a central simple $\widehat{F}$-algebra, this homomorphism must be injective. Hence, $D \otimes_F \widehat{F}$ has no zero divisors, so this central simple algebra must be a division ring. □

Example 1.17. *Quaternion algebras.* Let F be a field with a discrete rank 1 valuation v such that $char\,\overline{F} \neq 2$, and let $Q = \big(a, b/F\big)$ be a quaternion algebra over F. Thus, a, $b \in F^\times$ and Q is the 4-dimensional central simple F-algebra with F-vector space base $(1, i, j, k)$ satisfying $i^2 = a$, $j^2 = b$, and $k = ij = -ji$. Recall that Q is a division algebra if and only if $a \notin F^{\times 2}$ and $b \notin im\,N_{F(\sqrt{a})/F}$, where $N_{F(\sqrt{a})/F}$ is the norm from the field $F(\sqrt{a})$ to F (see, e.g., Lam [122, Th. III.2.7, p. 58]). Assume this holds. We determine when v extends to a valuation on Q. We will use the completion criterion of Prop. 1.16, but will end up with readily checkable conditions on a and b within F. In each case, the condition given for v to extend to Q also implies that Q is a division algebra.

Recall that $Q \cong \big(ar^2, bs^2/F\big)$ for any $r, s \in F^\times$. Thus, by altering a or b if necessary, we may assume that $v(a), v(b) \in \{0, 1\}$. This leads to four possible cases. Let $\widehat{Q} = Q \otimes_F \widehat{F} = \big(a, b/\widehat{F}\,\big)$, where $(\widehat{F}, \widehat{v})$ is the completion of (F, v).

Case 1. $v(a) = v(b) = 0$. The reduced norm for $\widehat{Q}$ is given by

$$Nrd_{\widehat{Q}}(r + si + tj + uk) \;=\; r^2 - as^2 - bt^2 + abu^2 \qquad \text{for all } r, s, t, u \in \widehat{F}.$$

Recall that $\widehat{Q}$ is a division algebra if and only if there is no nonzero $q \in \widehat{Q}$ with $Nrd_{\widehat{Q}}(q) = 0$, i.e., $Nrd_{\widehat{Q}}$ viewed as a quadratic form on $\widehat{Q}$ over $\widehat{F}$ is anisotropic. Since $v(1) = v(-a) = v(-b) = v(ab) = 0$, Springer's Theorem on quadratic forms over a field with nondyadic complete discrete rank 1 valuation (see Lam [122, Prop. VI.1.9(1), p. 148]) says that $Nrd_{\widehat{Q}}$ is anisotropic if and only if its first residue quadratic form $\partial_1 Nrd_{\widehat{Q}}$ is anisotropic, where

$$\partial_1 Nrd_{\widehat{Q}}(x, y, z, w) \;=\; x^2 + \overline{-a}\, y^2 + \overline{-b}\, z^2 + \overline{ab}\, w^2 \quad \text{for all } x, y, z, w \in \overline{\widehat{F}} = \overline{F}.$$

That is, $\partial_1 Nrd_{\widehat{Q}}$ is isometric to the reduced norm form Nrd_B, where $B = \big(\overline{a}, \overline{b}/\,\overline{F}\,\big)$. Thus, Prop. 1.16 yields that v extends to a valuation on Q if and only if $\big(\overline{a}, \overline{b}/\,\overline{F}\,\big)$ is a division algebra. When this occurs, $\overline{Q} = \big(\overline{a}, \overline{b}/\,\overline{F}\,\big)$ and $\Gamma_Q = \Gamma_F$.

Case 2. $v(a) = 0$ and $v(b) = 1$. The quaternion algebra $\widehat{Q}$ is a division algebra if and only if $a \notin \widehat{F}^{\times 2}$ and $b \notin im\,N_{\widehat{Q}(\sqrt{a})/\widehat{Q}}$. By Hensel's

Lemma applied to the polynomial $X^2 - a \in \mathcal{O}_F[X]$ (whose residue polynomial $X^2 - \overline{a}$ is separable), $a \notin \widehat{F}^{\times 2}$ if and only if $\overline{a} \notin \overline{F}^{\times 2}$. Assume these conditions hold. Let $L = \widehat{F}(\sqrt{a})$, and let v_L be an extension of $\widehat{v}$ to L. Since $\sqrt{\overline{a}} = \overline{\sqrt{a}} \in \overline{L} \setminus \overline{F}$, we have $[\overline{L}:\overline{F}] \geq 2$. Because $[L:\widehat{F}] = 2$, the Fundamental Inequality (see Th. A.2) shows that v_L is the only extension of $\widehat{v}$ to L, and that $[\overline{L}:\overline{F}] = 2$ and $\Gamma_L = \Gamma_{\widehat{F}}$. Let τ be the nonidentity $\widehat{F}$-automorphism of L. The uniqueness of v_L implies that τ must preserve v_L. Hence, for any $t \in L^\times$,

$$\widehat{v}\big(N_{L/\widehat{F}}(t)\big) \ = \ v_L(t\,\tau(t)) \ = \ 2v_L(t) \ \in 2\Gamma_L \ = \ 2\Gamma_{\widehat{F}}.$$

Since $\widehat{v}(b) = 1 \notin 2\Gamma_{\widehat{F}}$, this shows that b in $\widehat{F}$ is not a norm from L. Thus, Prop. 1.16 shows that v extends to a valuation on Q if and only if $\overline{a}$ is not a square in $\overline{F}$. When this occurs, $\overline{Q} \cong \overline{F}(\sqrt{\overline{a}})$ and $\Gamma_Q = \frac{1}{2}\Gamma_F$.

Case 3. $v(a) = 1$ and $v(b) = 0$. By Case 2 with a and b reversed, v extends to Q if and only if $\overline{b} \notin \overline{F}^{\times 2}$.

Case 4. $v(a) = v(b) = 1$. Since $(b^{-1}k)^2 = -ab^{-1}$ and $(b^{-1}k)j = -j(b^{-1}k)$, we have $Q \cong \big(-ab^{-1}, b/F\big)$, which is covered by Case 2, as $v(-ab^{-1}) = 0$. Thus, v extends to Q if and only if $\overline{-ab^{-1}} \notin \overline{F}^{\times 2}$.

Remark 1.18. Proposition 1.16 still holds, with the same proof, if we drop the discreteness condition and assume only that v has rank 1. For it is still true that $\bigcap_{i=1}^{\infty} \mathfrak{m}_D^i = \{0\}$, the results in Bourbaki apply, and that $\widehat{v}$ on $\widehat{F}$ is Henselian. However, if $rk(v) \geq 2$, then the result is no longer true, and the completion is problematical, as the following example illustrates.

Example 1.19. Let k be any field, and let $F = k(x, y)$ with x and y algebraically independent over k. Let w be the y-adic valuation on F, i.e., the restriction to F of the y-adic valuation on $k(x)((y))$, and let v be the (x, y)-adic valuation on F, which is the restriction of the (x, y)-adic valuation on $k((x))((y))$. We have $\Gamma_{F,w} \cong \mathbb{Z}$ and $\Gamma_{F,v} \cong \mathbb{Z}^2$ ordered lexicographically right-to-left. For $\gamma \in \Gamma_{F,v}$ set

$$F^v_{>\gamma} \ = \ \{c \in F \mid v(c) > \gamma\},$$

and define $F^w_{>\delta}$ analogously for $\delta \in \Gamma_{F,w}$. In the topology on F determined by v (resp. by w) the set $\mathcal{T}_v = \{F^v_{>\gamma} \mid \gamma \in \Gamma_{F,v}\}$ (resp. $\mathcal{T}_w$) is a fundamental system of neighborhoods of 0. Since for any $n, m \in \mathbb{Z}$,

$$F^v_{>(n,m)} \ \supseteq \ F^w_{>m} \ \supseteq \ F^v_{>(0,m+1)}$$

the sets $\mathcal{T}_v$ and $\mathcal{T}_w$ are cofinal; so, v and w determine the same topology on F. Hence, the completions (defined by (generalized) Cauchy sequences as in Engler–Prestel [73, p. 50] or equivalently by Cauchy filters as in Bourbaki [32, Ch. II, §3, Ch. III, §3]) $\widehat{F}^v$ and $\widehat{F}^w$ coincide! Indeed,

$$\widehat{F}^v \ = \ \widehat{F}^w \ = \ k(x)((y)),$$

with $\widehat{w}$ the y-adic valuation and $\widehat{v}$ the restriction to $\widehat{F}^v$ of the (x,y)-adic valuation on $k((x))((y))$. Assume *char* $k \neq 2$ and let D be the quaternion algebra $\big(1+x, y/F\big)$; then

$$D \otimes_F \widehat{F}^w \cong \big(1+x, y\,/\,k(x)((y))\big),$$

which is a division algebra by the preceding example, Case 2, since

$$\overline{1+x} \notin \big(\overline{\widehat{F}}^{\widehat{w}}\big)^{\times 2} = k(x)^{\times 2}$$

and $w(y) = 1$. Hence, the discrete rank 1 valuation w extends to D. On the other hand, for the rank 2 valuation v, we have $D \otimes_F \widehat{F}^v = D \otimes_F \widehat{F}^w$ is division algebra, but v does not extend to D by Th. 1.4, since v has two different extensions to the subfield $F(i) \cong F(\sqrt{1+x}\,)$ of D. (In one extension $\overline{i} = 1$ and in the other, $\overline{i} = -1$.)

1.2.8 Examples: local fields

Let F be a local field, i.e., a field with a complete valuation v such that $\overline{F}$ is finite and $\Gamma_F = \mathbb{Z}$, and let D be a central division F-algebra. Let $n = ind(D)$. By Cor. 1.7, the valuation v has a unique extension to a valuation on D, which we denote again by v. We proceed to give an explicit description of this extension (and of the division algebra D).

Since $\overline{D}$ is a finite division ring, it is commutative hence $Z(\overline{D}) = \overline{D}$ and $\overline{D}/\overline{F}$ is a cyclic field extension. By Prop. 1.5, the canonical homomorphism θ_D induces a surjective map

$$\overline{\theta}_D \colon \Gamma_D/\Gamma_F \longrightarrow \mathcal{G}(\overline{D}/\overline{F}),$$

so $|\Gamma_D{:}\Gamma_F| \geq [\overline{D}{:}\overline{F}]$. It is known (see Serre [229, Prop. XII.2.2, p. 183] or Prop. 8.75 below) that D contains a maximal subfield L that is an inertial extension of F, i.e., $[\overline{L}{:}\overline{F}] = [L{:}F]$ and $\overline{L}$ is separable over $\overline{F}$. Since $\overline{L} \subseteq \overline{D}$, we have

$$[\overline{D}{:}\overline{F}] \geq [\overline{L}{:}\overline{F}] = [L{:}F] = n.$$

On the other hand, the fundamental inequality (Prop. 1.3) yields

$$n^2 \geq [\overline{D}{:}\overline{F}]\,|\Gamma_D{:}\Gamma_F|,$$

so the inequalities above must be equalities:

$$[\overline{D}{:}\overline{F}] = n = |\Gamma_D{:}\Gamma_F|,$$

hence $\Gamma_D = \frac{1}{n}\mathbb{Z}$, $\overline{D} = \overline{L}$, and θ_D is an isomorphism. Since L/F is inertial, there is a canonical isomorphism $\mathcal{G}(L/F) \simeq \mathcal{G}(\overline{L}/\overline{F})$, and there is in $\mathcal{G}(L/F)$

a distinguished generator, the *Frobenius automorphism* $\varphi\colon L \to L$ whose induced automorphism $\overline{\varphi}$ of $\overline{L}$ is given by

$$\overline{\varphi}(\overline{x}) = \overline{\varphi(x)} = \overline{x}^{|\overline{F}|} \qquad \text{for } x \in O_L.$$

The Skolem–Noether Theorem yields an element $z \in D^\times$ such that

$$zx = \varphi(x)z \qquad \text{for all } x \in L. \tag{1.18}$$

Then $\theta_D\big(v(z)\big) = \overline{\varphi}$, so the image of $v(z)$ in Γ_D/Γ_F is a generator. Therefore, there is an integer j prime to n and an element $\alpha \in F^\times$ such that

$$\tfrac{1}{n} = j\,v(z) + v(\alpha).$$

Let $\pi_D = \alpha z^j \in D^\times$. We have $v(\pi_D) = 1/n$, so π_D is a uniformizing parameter of v on D, and

$$\pi_D\, x = \varphi^j(x)\pi_D \qquad \text{for } x \in L. \tag{1.19}$$

It follows that π_D^n centralizes L, hence $\pi_D^n \in L^\times$. Since moreover π_D^n commutes with π_D, we have $\pi_D^n \in F^\times$. Let $\pi_F = \pi_D^n \in F^\times$. This element is a uniformizing parameter of v on F, and D is a cyclic algebra:

$$D = L \oplus L\pi_D \oplus \ldots \oplus L\pi_D^{n-1}$$

where the multiplication is defined by (1.19) and $\pi_D^n = \pi_F$. Since $v(1)$, $v(\pi_D)$, ..., $v(\pi_D^{n-1})$ lie in different cosets of Γ_D modulo Γ_F, and since $\Gamma_L = \Gamma_F$, we have

$$v\big(\textstyle\sum_{i=0}^{n-1} x_i\pi_D^i\big) = \min\limits_{0\le i\le n-1}\big(v(x_i) + \tfrac{i}{n}\big) \qquad \text{for } x_0, \ldots, x_{n-1} \in L.$$

Local class field theory yields a description of the Brauer group $Br(F)$ by means of a canonical isomorphism

$$inv\colon\ Br(F) \xrightarrow{\sim} \mathbb{Q}/\mathbb{Z}$$

known as the (*local*) *invariant map*.The Brauer class of the division algebra D above is mapped to $v(z) + \mathbb{Z}$, for any z as in (1.18).

As an explicit example, consider $D = \big(-1,-1/\mathbb{Q}_2\big)$, the Hamilton quaternion division algebra over the dyadic local field $\mathbb{Q}_2$. Let $(1,i,j,k)$ be the standard base of D, where $i^2 = j^2 = -1$ and $k = ij = -ji$. Let $u = \frac{1}{2}(-1+i+j+k)$ and $\pi_D = i-j$. Then $u^2+u+1 = 0$, $\pi_D u\pi_D^{-1} = -u-1$, and $\pi_D^2 = -2$. The field $L = \mathbb{Q}_2(u)$ is the unique quadratic inertial extension of $\mathbb{Q}_2$. The invariant of D is $\frac{1}{2} + \mathbb{Z} \in \mathbb{Q}/\mathbb{Z}$.

1.2.9 Example: the field of rational numbers

Let $F = \mathbb{Q}$ and let $\mathcal{P}$ be the set of prime numbers. For any $p \in \mathcal{P}$, let $v_p\colon \mathbb{Q} \to \mathbb{Z} \cup \{\infty\}$ be the p-adic valuation on $\mathbb{Q}$ defined by $v_p(p^k m/n) = k$

for all k, m, $n \in \mathbb{Z}$, with $p \nmid m$, $p \nmid n$. Ostrowski's Theorem (see Engler–Prestel [73, Th. 2.1.4, p. 30]) says that the v_p are all the nonarchimedean valuations on $\mathbb{Q}$. Let $\mathbb{Q}_p$ be the field of p-adic numbers, which is the completion of $\mathbb{Q}$ with respect to v_p. Take any central division $\mathbb{Q}$-algebra D. By Prop. 1.16, v_p extends to a valuation on D if and only if $D \otimes_{\mathbb{Q}} \mathbb{Q}_p$ is a division ring. This occurs for at most finitely many p, since global class field theory yields an injection

$$Br(\mathbb{Q}) \hookrightarrow \big(\bigoplus_{p \in \mathcal{P}} Br(\mathbb{Q}_p)\big) \oplus Br(\mathbb{R}),$$

which maps any Brauer class $[A] \in Br(\mathbb{Q})$ to $\big(([A \otimes \mathbb{Q}_p])_{p \in \mathcal{P}}, [A \otimes \mathbb{R}]\big)$. Class field theory also shows that $ind(D)$ is the order of the Brauer class $[D] \in Br(\mathbb{Q})$. Therefore, if $ind(D) = q^r$ for some prime number q and some $r \geq 1$, then there exists a $p \in \mathcal{P}$ such that $[D \otimes \mathbb{Q}_p] \in Br(\mathbb{Q}_p)$ has order q^r. From the description of $Br(\mathbb{Q}_p)$ by the local invariant in the previous example, it follows that $D \otimes \mathbb{Q}_p$ is a division algebra, hence v_p extends to D. However, it is easy to construct a D of composite index such that all the local indices $ind(D \otimes_{\mathbb{Q}} \mathbb{Q}_p)$ are strictly smaller than $ind(D)$; when (and only when) this occurs, there is no valuation on D.

1.3 The graded ring associated to a valued division algebra

In this section, we introduce a very convenient tool, which encodes all the residue information on a valued division ring D, as well as the value group and the canonical homomorphism θ_D: it is the graded ring $\mathsf{gr}(D)$ associated to the filtration of D induced by the valuation. This graded ring is often easier to work with than D itself, and the various structures we will consider over $\mathsf{gr}(D)$ also have a natural grading that yields much extra information. Throughout this book we will emphasize the use of associated graded rings. This approach has led to substantial simplifications in many proofs that were originally given without using associated graded rings. In Chapter 6 below, we will also see that in an important case (when D is finite-dimensional and tame over its center $Z(D)$, and $v|_{Z(D)}$ is Henselian), $\mathsf{gr}(D)$ actually determines D up to isomorphism.

1.3.1 Definition

Let D be an arbitrary division ring. Any valuation $v\colon D \to \Gamma \cup \{\infty\}$ defines a filtration on D, and we consider along with the residue algebra $\overline{D}$ the other slices of this filtration: for each $\gamma \in \Gamma$, set

$$D_{\geq\gamma} = \{x \in D \mid v(x) \geq \gamma\} \quad \text{and} \quad D_{>\gamma} = \{x \in D \mid v(x) > \gamma\}.$$

Then $D_{\geq\gamma}$ is a sub-$\mathcal{O}_D$-bimodule of D, and $D_{>\gamma}$ is a sub-$\mathcal{O}_D$-bimodule of $D_{\geq\gamma}$ containing $\mathfrak{m}_D \cdot D_{\geq\gamma}$ and $D_{\geq\gamma} \cdot \mathfrak{m}_D$. Of course, $D_{>\gamma} = D_{\geq\gamma}$ if and only if $\gamma \notin \Gamma_D$. On the other hand, if $x \in D^\times$ satisfies $v(x) = \gamma$, then

$$D_{\geq\gamma} = x\mathcal{O}_D = \mathcal{O}_D x \qquad \text{and} \qquad D_{>\gamma} = x\mathfrak{m}_D = \mathfrak{m}_D x.$$

Set

$$D_\gamma = D_{\geq\gamma}/D_{>\gamma},$$

a left and right $\overline{D}$-vector space of dimension 1 (if $\gamma \in \Gamma_D$) or 0 (if $\gamma \notin \Gamma_D$). For any $\gamma,\ \delta \in \Gamma$ we have

$$D_{\geq\gamma}D_{\geq\delta} \subseteq D_{\geq\gamma+\delta}, \quad D_{>\gamma}D_{\geq\delta} \subseteq D_{>\gamma+\delta}, \quad \text{and} \quad D_{\geq\gamma}D_{>\delta} \subseteq D_{>\gamma+\delta},$$

hence there is a well-defined multiplication operation

$$D_\gamma \times D_\delta \longrightarrow D_{\gamma+\delta}$$

given by

$$(x + D_{>\gamma}) \cdot (y + D_{>\delta}) = (xy) + D_{>(\gamma+\delta)} \qquad \text{for all } x \in D_{\geq\gamma},\ y \in D_{\geq\delta}.$$

The *associated graded ring* of v on D is defined to be

$$\mathsf{gr}(D) = \bigoplus_{\gamma\in\Gamma} D_\gamma.$$

(See §2.1.1 for the general definition of a graded ring.) We will use the notation $\mathsf{gr}_v(D)$ when it is useful to keep in mind that this construction depends on v. The addition on $\mathsf{gr}(D)$ arises from the addition on each component D_γ. The multiplication is defined by extending the multiplication on the components bilinearly to all of $\mathsf{gr}(D)$. Routine calculations show that with these operations $\mathsf{gr}(D)$ is a ring. Moreover, since $D_\gamma \cdot D_\delta \subseteq D_{\gamma+\delta}$ for all $\gamma,\ \delta \in \Gamma$, the ring $\mathsf{gr}(D)$ is graded by the abelian group Γ. The *grade group* of $\mathsf{gr}(D)$ is defined to be

$$\Gamma_{\mathsf{gr}(D)} = \{\gamma \in \Gamma \mid D_\gamma \neq 0\};$$

clearly $\Gamma_{\mathsf{gr}(D)} = \Gamma_D$. Note that

$$\mathcal{O}_D = D_{\geq 0}, \qquad \mathfrak{m}_D = D_{>0}, \quad \text{and} \quad \overline{D} = D_0.$$

Thus, two of the most basic objects associated to the valuation, $\overline{D}$ and Γ_D, show up naturally in the structure of $\mathsf{gr}(D)$. The canonical homomorphism θ_D can also be defined in terms of the associated graded ring, as we proceed to show.

For $x \in D^\times$, we write $\widetilde{x}$ for the image $x + D_{>v(x)}$ of x in $D_{v(x)}$,

$$\widetilde{x} = x + D_{>v(x)} \in D_{v(x)}.$$

We also set $\widetilde{0} = 0$. Thus, $\widetilde{x} = \overline{x}$ if $v(x) = 0$.

The *homogeneous elements* of $\mathsf{gr}(D)$ are the elements of $\bigcup_{\gamma\in\Gamma} D_\gamma$. Clearly the nonzero homogeneous elements are just the elements $\widetilde{x}$ as x ranges over $D^\times$. Note that the multiplicative property (iii) for the valuation shows that

$$\widetilde{x} \cdot \widetilde{y} = \widetilde{xy} \qquad \text{for all } x,\, y \in D^\times.$$

It follows that $\mathsf{gr}(D)$ is a *graded division ring*, i.e., every nonzero homogeneous element of $\mathsf{gr}(D)$ is a unit; indeed, $(\widetilde{x})^{-1} = \widetilde{(x^{-1})}$ for all $x \in D^\times$. (But $\mathsf{gr}(D)$ is *not* a division ring if $|\Gamma_D| > 1$.)

We now define a canonical homomorphism $\theta_{\mathsf{gr}(D)} \colon \Gamma_D \to Aut\big(Z(D_0)\big)$ as follows: for $\gamma \in \Gamma$, pick an arbitrary nonzero homogeneous element $\xi \in D_\gamma$, and for $z \in Z(D_0)$ let

$$\theta_{\mathsf{gr}(D)}(\gamma) \colon z \mapsto \xi z \xi^{-1}.$$

Since D_γ is a 1-dimensional right vector space over D_0, it is clear that $\xi z \xi^{-1}$ depends only on γ, and not on the choice of ξ in D_γ. It is also easy to see that $\xi z \xi^{-1} \in Z(D_0)$ if $z \in Z(D_0)$. Under the identifications $\Gamma_{\mathsf{gr}(D)} = \Gamma_D$ and $D_0 = \overline{D}$, we clearly have $\theta_{\mathsf{gr}(D)} = \theta_D$.

1.3.2 Examples

The graded ring construction transforms Laurent series into Laurent polynomials: suppose σ is an automorphism of an arbitrary division ring A and consider the x-adic valuation v_x on the division ring $D := A((x;\sigma))$ of twisted Laurent series (see §1.1.2). For all $\gamma \in \mathbb{Z}$ $(= \Gamma_D = \Gamma_{\mathsf{gr}(D)})$ we have

$$D_\gamma = A\widetilde{x}^\gamma = \widetilde{x}^\gamma A,$$

hence $\mathsf{gr}(D) = \bigoplus_{\gamma \in \mathbb{Z}} A\widetilde{x}^\gamma$. The multiplication in this ring satisfies

$$\widetilde{x}a = \sigma(a)\widetilde{x} \qquad \text{for } a \in A.$$

Thus, $\mathsf{gr}(D)$ is the ring of twisted Laurent polynomials in the indeterminate $\widetilde{x}$ with coefficients in A:

$$\mathsf{gr}(D) = A[\widetilde{x}, \widetilde{x}^{-1}; \sigma].$$

The grading on $\mathsf{gr}(D)$ is by degree in $\widetilde{x}$.

The same observation applies to division rings of iterated Laurent series: consider for example an arbitrary field M, an element $w \in M^\times$, and the division ring $E = M((x_1))((x_2;\tau))$ where $\tau\big(\sum m_i x_1^i\big) = \sum m_i w^{-i} x_1^i$; i.e., $E = \mathcal{L}((M; \boldsymbol{I}, \boldsymbol{u}))$ with the notation of §1.1.3, where $\boldsymbol{I} = (id_M, id_M)$ and $\boldsymbol{u} = (u_{i,j})_{i,j=1}^2$ is given by $u_{1,1} = u_{2,2} = 1$ and $u_{1,2} = w = u_{2,1}^{-1}$. Then $\mathsf{gr}(E)$ is a Laurent polynomial ring in two noncommuting indeterminates

$$\mathsf{gr}(E) = M[\widetilde{x}_1, \widetilde{x}_1^{-1}, \widetilde{x}_2, \widetilde{x}_2^{-1}]$$

where the multiplication is determined by $\widetilde{x}_1\widetilde{x}_2 = w\widetilde{x}_2\widetilde{x}_1$ and $\widetilde{x}_i m = m\widetilde{x}_i$ for all $m \in M$ and $i = 1, 2$. The grade group is $\mathbb{Z}^2$ and the grading is by degree in $\widetilde{x}_1$ and $\widetilde{x}_2$. Note that the graded ring $\mathsf{gr}(E)$ "forgets" the ordering on $\mathbb{Z}^2$: for $E' = M((x_2))((x_1;\sigma))$ where $\sigma(\sum m_i x_2^i) = \sum m_i w^i x_2^i$, with the (x_2, x_1)-adic valuation we have

$$\mathsf{gr}(E') = \mathsf{gr}(E);$$

but of course $E' \neq E$: for example $\sum_{i=0}^{\infty} x_1^{-i} x_2^i$ lies in E but not in E'.

As an example where the characteristic of $\mathsf{gr}(D)$ is different from the characteristic of D, consider the quaternion division algebra $D = \big(-1, -1/\mathbb{Q}_2\big)$ with the (unique) extension v of the 2-adic valuation. Let $(1, i, j, k)$ be the standard base of D. As observed in §1.2.8, the element $\pi_D = i - j$ satisfies $\pi_D^2 = -2$, hence $v(\pi_D) = \frac{1}{2}$. We have $\overline{D} = \mathbb{F}_4$ and $\Gamma_D = \frac{1}{2}\mathbb{Z}$. The graded field $\mathsf{gr}(\mathbb{Q}_2)$ is the Laurent polynomial ring $\mathbb{F}_2[t, t^{-1}]$, where $t = \widetilde{2}$ has degree 1. The graded division ring $\mathsf{gr}(D)$ is the twisted Laurent polynomial ring $\mathbb{F}_4[\widetilde{\pi_D}, \widetilde{\pi_D}^{-1}; \varphi]$, where φ is the Frobenius automorphism of $\mathbb{F}_4$ and $\widetilde{\pi_D}$ has degree $\frac{1}{2}$ with $\widetilde{\pi_D}^2 = t$.

Exercises

Exercise 1.1. A total valuation ring not the ring of a valuation. A subring T of a division algebra D is called a *total valuation ring* of D if for each $d \in D^\times$ we have $d \in T$ or $d^{-1} \in T$. If D were commutative it is well-known that such a T is the valuation ring of a valuation in D. But, when D is not commutative, T is the ring of a valuation if and only if in addition it is closed under conjugation. This exercise gives an example of a total valuation ring that is not the ring of a valuation.

Let k be any field with $char(k) \neq 2$. Let x be transcendental over k and let $F = k(x)((y))$, the Laurent power series ring over $k(x)$. Let D be the quaternion algebra $\big(1+x, y/F\big)$, with its standard generators i and j satisfying $i^2 = 1+x$, $j^2 = y$, and $ij = -ji$. Let $w\colon F \to \mathbb{Z} \cup \{\infty\}$ be the complete discrete rank 1 (so Henselian) y-adic valuation on F, with $\overline{F}^w = k(x)$ and $\Gamma_{F,w} = \mathbb{Z}$. Since $1+x \notin (\overline{F}^w)^{\times 2}$ and $w(y) = 1 \notin 2\Gamma_{F,w}$, Ex. 1.17 Case 2 shows that D is a division algebra and w extends to a valuation on D with $\overline{D}^w = k(x)(\sqrt{1+x})$ and $\Gamma_{D,w} = \frac{1}{2}\mathbb{Z}$. (See also Ex. 1.19.)

(i) Let u be the x-adic valuation on $k(x)$, which is the restriction to $k(x)$ of the x-adic valuation on $k((x))$. Since $u(1+x) = 0$ and $\overline{1+x} = \overline{1}$, which is a square in $\overline{k(x)}^u$, the valuation u has two different extensions u_1 and u_2 to $k(x)(\sqrt{1+x}) = \overline{D}^w$, say with $\overline{\sqrt{1+x}}^{u_1} = 1$ and $\overline{\sqrt{1+x}}^{u_2} = -1$. Let $\pi\colon \mathcal{O}_{D,w} \to \overline{D}^w$ be the residue map and let $T_\ell = \pi^{-1}(\mathcal{O}_{\overline{D}^w, u_\ell})$, $\ell = 1, 2$. Prove that T_1 and T_2 are total valuation rings of D. Prove also that the inner automorphism $int(j)$ of D given by conjugation by j permutes T_1 and T_2. Hence, the T_ℓ are not invariant under inner automorphisms. Prove also that every inner automorphism of D either preserves or interchanges T_1 and T_2.

(ii) Let the valuation v on F be the restriction of the (x,y)-adic valuation on $k((x))((y))$, so v is the composition $u{*}w$. Since $\overline{1+x}^{v} = 1 \in (\overline{F}^{v})^{\times 2}$, Ex. 1.17 Case 2 shows that v does not extend to a valuation on D. Hence, there can be no total valuation ring V of D invariant under inner automorphisms with $V \cap F = \mathcal{O}_{F,v}$. Prove that in fact T_1 and T_2 are the only total valuation rings of D whose intersection with F is $\mathcal{O}_{F,v}$.

Exercise 1.2. Composite valuations on division rings of Laurent series. Suppose that D is a division ring with valuation u and that τ is an automorphism of D preserving u, in the sense that $u(\tau(d)) = u(d)$ for all $d \in D$. Let $E = D((y;\tau))$ with its y-adic valuation v_y. Build a composite valuation $w = u * v_y$ as follows: give $\Gamma_{D,u} \times \mathbb{Z}$ the right-to-left lexicographic ordering in which

$$(\gamma, i) \le (\delta, j) \quad \text{if and only if} \quad i < j \ \text{ or } (i = j \text{ and } \gamma \le \delta).$$

Then define $w\colon E \to (\Gamma_{D,u} \times \mathbb{Z}) \cup \{\infty\}$ by

$$w\Big(\sum_{i=k}^{\infty} d_i y^i\Big) \;=\; (u(d_j), j) \quad \text{where } j \text{ is minimal with } d_j \neq 0, \text{ if some } d_i \neq 0,$$

and $w(0) = \infty$. Prove that w is a valuation on E with $\overline{E}^{w} = \overline{D}^{u}$ and that $\Gamma_{E,w} = \Gamma_{D,u} \times \mathbb{Z}$. Prove also that the canonical homomorphism $\theta_{E,w}\colon \Gamma_{D,u} \times \mathbb{Z} \to Aut(Z(\overline{D}^{u}))$ is given by

$$(u(d), j) \;\mapsto\; \theta_{D,u}(u(d)) \circ (\overline{\tau}|_{Z(\overline{D}^{u})})^{j},$$

where $\overline{\tau}$ is the automorphism of $\overline{D}$ induced by τ.

Exercise 1.3. The Gaussian valuation on $D(x)$. Let D be any division ring, let $D[x]$ be the polynomial ring over D (with x commuting with the elements of D). Since $D[x]$ is a right Ore domain, it has a right ring of quotients

$$D(x) \;=\; \{fg^{-1} \mid f, g \in D[x],\ g \neq 0\}.$$

Suppose v is a valuation on D.

(i) Extend v first to a map on $D[x]$ by defining

$$v\Big(\sum_{i=0}^{n} d_i x^i\Big) \;=\; \min_{0 \le i \le n} v(d_i).$$

Prove that this v on $D[x]$ satisfies the axioms for a valuation, including the multiplicative property $v(fg) = v(f) + v(g)$ for all $f,\, g \in D[x]$.

(ii) Extend v to $D(x)$ by defining $v(fg^{-1}) = v(f) - v(g)$ for all $f, g \in D[x]$, $g \neq 0$. Prove that v is well-defined and is a valuation on $D(x)$. This valuation is called the *Gaussian valuation* on $D(x)$ obtained from v on D.

(iii) Prove that $\Gamma_{D(x)} = \Gamma_D$ and $\overline{D(x)} = \overline{D}(\overline{x})$, which is the right ring of quotients of the polynomial ring $\overline{D}[\overline{x}]$.

Notes

The early history of commutative valuation theory—up to the work of Krull—is ably recounted by Roquette in [207]. To our knowledge the earliest use of valuations on noncommutative division algebras was by Hasse in [98] in his analysis of maximal orders in finite-dimensional division algebras over local fields. Schilling in [226] and [227] was the first to consider valuations on arbitrary division algebras. He showed in [227, Th. 9, Th. 10, pp. 53–54] that a Henselian valuation (which he called a 'relatively complete" valuation, see [227, Def. 17, p. 52]) on the center of a finite-dimensional division algebra extends uniquely to a valuation on the division algebra.

Invariant valuation rings. Let w be a valuation on an division ring D. For the associated valuation ring $\mathcal{O}_D$ observe that for every $d \in D^\times$ we have $d \in \mathcal{O}_D$ or $d^{-1} \in \mathcal{O}_D$. Furthermore, for every $d \in D^\times$ we have $d\mathcal{O}_D d^{-1} = \mathcal{O}_D$. Because of this invariance under conjugation, $\mathcal{O}_D$ is called an *invariant valuation ring.* The invariance is equivalent to $d\mathcal{O}_D = \mathcal{O}_D d$ for all $d \in D^\times$; hence, $\mathcal{O}_D$ is a duo ring, i.e., every left ideal is a right ideal, and vice versa. Furthermore, the left=right=two-sided ideals of $\mathcal{O}_D$ are linearly ordered by inclusion. We have

$$\mathcal{O}_D^\times = \{d \in D \mid w(d) = 0\} = \mathit{ker}(w|_{D^\times}).$$

Thus, w induces an isomorphism

$$D^\times/\mathcal{O}_D^\times \xrightarrow{\sim} \Gamma_D.$$

Since in addition $w(d) \geq w(e)$ if and only if $d\mathcal{O}_D \subseteq e\mathcal{O}_D$, it follows that the ring $\mathcal{O}_D$ fully determines the valuation w (up to an ordered group isomorphism of Γ_D). Indeed, if R is any subring of D such that for each $d \in D$, (i) d or $d^{-1} \in R$, and (ii) $dRd^{-1} = R$, and $D^\times/R^\times$ is abelian, then the relation $\leq$ on $D^\times/R^\times$ given by

$$dR^\times \leq eR^\times \quad \text{if and only if} \quad dR \supseteq eR$$

makes $D^\times/R^\times$ into a totally ordered abelian group; moreover the map $y\colon D \to D^\times/R^\times \cup \{\infty\}$ given by $y(d) = dR^\times$ if $d \neq 0$ and $y(0) = \infty$ is a valuation on D with $\mathcal{O}_D = R$.

Other valuation rings. For fields, there are three basic structures in valuation theory: (i) valuations; (ii) valuation rings; and (iii) places. These are equivalent in that a valuation v on F determines its valuation ring $\mathcal{O}_F$, and conversely $\mathcal{O}_F$ determines v by the inclusion ordering on its principal ideals. The associated place of v is the map $\pi_v\colon F \to \overline{F} \cup \{\infty\}$ given by $c \mapsto \overline{c}$

if $c \in \mathcal{O}_F$ and $c \mapsto \infty$ otherwise. Thus, $\mathcal{O}_F$ determines the place π_v, but also π_v determines $\mathcal{O}_F$, since $\mathcal{O}_F = \pi_v^{-1}(\overline{F})$. (The terminology "composition" of valuations comes from the composition as functions of the associated places.) People working with division algebras and central simple algebras have found generalizations of each of these structures, but they are no longer equivalent. Generalizing (i) we have the notion of a valuation on a division algebra as defined in §1.1, for which the associated rings are the *invariant valuation rings* described above. Another natural generalization of valuation rings on fields, in the spirit of (ii), is given by *total valuation rings*: a subring T of a division algebra D is called a total valuation ring if $d \in T$ or $d^{-1} \in T$ for every $d \in D^\times$. Total valuation rings have an associated "value set" of right ideals, which is totally ordered by inclusion, but this set is typically not a group. Every invariant valuation ring of D is a total valuation ring, but the converse is not true, as Exercise 1.1 illustrates. For more on total valuation rings on division algebras see Brungs–Gräter [34] and [35], Mathiak [134], [135], [136], Wadsworth [255]. It is known, for example, see [255, Th. G] that if D is a division algebra finite-dimensional over its center F and V is a valuation ring of F (i.e., with quotient field F), then there are at most finitely many total valuation rings $T_1, \ldots, T_m$ of D with $T_i \cap F = V$. (But there may well be no such T_i at all.) When the T_i exist, they are all conjugate in D, and the number m of T_i equals the matrix size of $D \otimes_F F_h$, where F_h is the Henselization of F with respect to v. Also, if $m > 1$, then V has Krull dimension at least 2.

Invariant and total valuation rings exist only in division algebras. For matrix rings over division algebras, Dubrovin has defined and studied in depth in [68] and [69] a family of rings based on the idea of places, (iii) above, for the category of simple Artinian rings; such rings are now called *Dubrovin valuation rings*. Invariant valuation rings and total valuation rings are Dubrovin valuation rings, as are matrix rings over such rings. Moreover, Dubrovin's rings have good extension behavior from the center: If A is a central simple algebra over a field F and V is a valuation ring of F, then Dubrovin showed in [69] that there is a Dubrovin valuation ring B of A with $B \cap F = V$; moreover, Brungs and Gräter in [35] (for V of finite rank) and Wadsworth in [255] (in general) showed that such a B is unique up to conjugacy in A. See the Notes to Ch. 3 and to Ch. 4 for the definition and more properties of Dubrovin valuation rings and connections between these rings and the rings associated with gauges on central simple algebras. A good reference for the substantial theory of Dubrovin valuation rings is the book [132] by Marubayashi et al.

This book is about the most restrictive of these three kinds of noncommutative valuation ring. Invariant valuation rings and their associated valuations have proved the most fruitful in applications. Such a valuation ring may not exist in a given division algebra D, as Th. 1.4 demonstrates; but the presence of such a valuation ring in D is often a source of considerable insight into properties of D.

§1.1: *Value groups.* In our definition of a valuation on a division ring D, the target group Γ is assumed to be abelian. This might seem an unnatural restriction since D itself is typically noncommutative. If we drop from the definition of a valuation the requirement that the totally ordered group Γ be abelian, there is still an associated valuation ring with unique maximal left and maximal right ideal and residue division ring and value group Γ_D, and associated graded ring. The Mal'cev–Neumann construction of §1.1.4 still works with nonabelian Γ, and shows that any totally ordered group can be the value group of a valued division algebra (see Cohn [55, §2.4]). However, if D is finite-dimensional over its center, it turns out (cf. Wadsworth [253, p. 21]) that Γ_D must be abelian, even if that were not initially assumed. Since the division algebras considered in this book are essentially always finite-dimensional over their centers, there is no loss in assuming at the outset that Γ is abelian.

Mal'cev–Neumann series algebras were originally defined independently by Mal'cev in [131] and Neumann in [172].

§1.2: Theorem 1.4 was proved by Ershov in [76] and rediscovered independently by Wadsworth in [253]. It had earlier been proved by Cohn in [54] for rank 1 valuations. The theorem generalizes readily to the case where the division algebra D is infinite-dimensional over its center F but locally finite, i.e., every finite subset of D lies in some finite-dimensional F-subalgebra of D. A possible further generalization was proved by Mahdavi-Hezavehi in [130]: Let $D^{\times\prime}$ be the multiplicative commutator subgroup of $D^\times$. He proved that if D is algebraic over F, then a valuation v on F extends to D if and only if (i) v has a unique extension to each subfield L of D containing F, and (ii) $D^{\times\prime} \cap F^\times \subseteq \mathcal{O}_F^\times$. Whether this result is actually more general is uncertain because it is still unknown whether there exist algebraic division algebras that are not locally finite.

Proposition 1.16 is due to Cohn [54, Th. 1].

§1.3: Early contributions to the study of graded rings associated to filtrations defined by valuations are due to Krasner [117] and Dedecker [58].

Chapter 2
Graded Algebra

Since our approach to structures over valued fields relies in a fundamental way on filtrations and associated graded structures, our arguments often require information that is specific to graded modules. We collect in this chapter the basic definitions and results on graded algebras and modules that will be of constant use in subsequent chapters.

Because of the intended applications to valuation theory, we are interested only in graded structures where the grade group is abelian and torsion-free. With this restriction, the structure theory of graded algebras closely parallels the classical Wedderburn theory of algebras, provided that homogeneity conditions are imposed. We emphasize the analogy by using suggestive terminology: A commutative graded ring is said to be a *graded field* if its nonzero homogeneous elements are invertible (even when the ring is not a field), and graded modules over graded fields are called *graded vector spaces*. In §2.1 we lay the groundwork for linear algebra over graded fields, discussing graded vector spaces over graded division rings, their homomorphisms, and their tensor products. We next develop in §2.2 the graded analogue of the Wedderburn structure theory of semisimple algebras, showing that semisimple graded algebras are direct products of endomorphism algebras of graded vector spaces over graded division rings. We also establish in that section graded versions of the Double Centralizer Theorem (Th. 2.35) and of the Skolem–Noether Theorem (Th. 2.37). Finally, in §2.3 special attention is directed toward the degree zero component of simple graded algebras, which is a key part of their structure. This component is semisimple but usually not simple, and it carries a canonical action of the grade group. In the structures associated to algebras over valued fields, it plays the role of the residue algebra.

None of the results in this chapter is particularly deep. A major tool, introduced in §2.2.1, is the central quotients construction, which often allows us to reduce statements about graded algebras to corresponding statements on ungraded algebras. A deeper study of graded structures will unfold in subsequent chapters, to match our needs for the investigation of algebras

J.-P. Tignol, A.R. Wadsworth, *Value Functions on Simple Algebras, and Associated Graded Rings*, Springer Monographs in Mathematics,
DOI 10.1007/978-3-319-16360-4_2

over valued fields: See Ch. 5 for the ramification theory and Galois theory of graded field extensions, Ch. 6 for the Brauer group of graded fields, Ch. 7 for the description of totally ramified graded division algebras, Ch. 8 for further discussion of graded division algebras, and Ch. 11 for the calculation of their SK_1 groups.

2.1 Graded linear algebra

This section gives the definitions and basic results concerning linear algebra over graded rings. We successively discuss graded rings (§2.1.1), graded modules (§2.1.2), homomorphisms (§2.1.3), and tensor products (§2.1.4). Throughout this section (and actually throughout the chapter), we let Γ denote a torsion-free abelian group, written additively, which will contain the set of degrees of all the graded objects we consider. The restriction to torsion-free abelian grade groups entails significant simplifications, which make the graded linear algebra a close analogue of the theory of vector spaces over division rings.

2.1.1 Graded rings

Let R be a Γ-graded ring, i.e., a ring with direct sum decomposition $\mathsf{R} = \bigoplus_{\gamma\in\Gamma} \mathsf{R}_\gamma$ where each R_γ is an additive abelian group and $\mathsf{R}_\gamma \cdot \mathsf{R}_\delta \subseteq \mathsf{R}_{\gamma+\delta}$ for all γ, $\delta \in \Gamma$. Thus, every $r \in \mathsf{R}$ is uniquely expressible as $r = \sum_{\gamma\in\Gamma} r_\gamma$ with each $r_\gamma \in \mathsf{R}_\gamma$ and at most finitely many r_γ nonzero. The r_γ are called the *homogeneous components* of r. The *grade set* of R is the subset of Γ defined by

$$\Gamma_{\mathsf{R}} = \{\gamma \in \Gamma \mid \mathsf{R}_\gamma \neq \{0\}\}.$$

Note that Γ_{R} need not be a subgroup of the torsion-free abelian group Γ. The *homogeneous elements* of R are the elements of $\bigcup_{\gamma\in\Gamma} \mathsf{R}_\gamma$. If $a \in \mathsf{R}_\gamma$ and $a \neq 0$, we say that γ is the *degree* of a and write $\gamma = \deg a$. We assume always that a ring has a 1. For the graded ring R we have R_0 is a subring of R, as $\mathsf{R}_0 \cdot \mathsf{R}_0 \subseteq \mathsf{R}_0$, and it is easy to check that $1 \in \mathsf{R}_0$. Let $\mathsf{R}^\times$ denote the group of units of R. Define

$$\Gamma_{\mathsf{R}}^\times = \{\gamma \in \Gamma_{\mathsf{R}} \mid \mathsf{R}_\gamma \cap \mathsf{R}^\times \neq \varnothing\}. \tag{2.1}$$

Thus, $\Gamma_{\mathsf{R}}^\times$ is the set of degrees of homogeneous units of R. Clearly, $\Gamma_{\mathsf{R}}^\times$ is a subgroup of Γ (but it may not be the largest subgroup of Γ lying in Γ_{R}; see Ex. 2.46 below).

The ring R with the grading forgotten is denoted by $\mathsf{R}^\natural$. In other words, $\mathsf{R}^\natural$ is the underlying ungraded ring of R.

A graded ring D is called a *graded division ring* if $1 \neq 0$ in D and every nonzero homogeneous element of D is a unit. Commutative graded division rings are called *graded fields*. For example, if D is a graded division ring, then its center $Z(\mathsf{D})$ is a graded field. The ungraded ring $\mathsf{D}^\natural$ of a graded division ring D is never a division algebra unless $\Gamma_{\mathsf{D}} = \{0\}$, by Prop. 2.3(iii) below.

Example 2.1. Let D_0 be an arbitrary division ring and let $\varepsilon\colon \Gamma \to Aut(D_0)$ be a group homomorphism. The twisted group ring $D_0(\Gamma;\varepsilon)$ consists of the finite formal sums $\sum_{\gamma\in\Gamma} a_\gamma\gamma$ with $a_\gamma \in D_0$ for all γ (and $a_\gamma \neq 0$ for finitely many γ's only). Multiplication in $D_0(\Gamma;\varepsilon)$ is defined by

$$a\gamma \cdot b\delta = a\varepsilon(\gamma)(b)\,(\gamma+\delta) \qquad \text{for } a,\, b \in D_0 \text{ and } \gamma,\, \delta \in \Gamma,$$

and a Γ-grading is defined by $D_0(\Gamma;\varepsilon)_\gamma = D_0\gamma$ for $\gamma \in \Gamma$. More general examples can be obtained by considering factor sets of Γ in D_0. See Exercise 5.1 for an example of a graded division ring that is *not* obtained by a group ring construction.

Remark 2.2. Although we do not need to fix an ordering on Γ in this chapter, note that the torsion-free abelian group Γ carries total orderings making it into an ordered abelian group: Let $\mathbb{H}(\Gamma)$ denote the divisible hull of Γ,

$$\mathbb{H}(\Gamma) = \Gamma \otimes_{\mathbb{Z}} \mathbb{Q} = \varinjlim_n \tfrac{1}{n}\Gamma.$$

We may choose a total ordering on a $\mathbb{Q}$-base of $\mathbb{H}(\Gamma)$ and use it to build a lexicographical ordering on $\mathbb{H}(\Gamma)$. The group Γ then inherits an ordering from its embedding into $\mathbb{H}(\Gamma)$.

Basic properties of graded division rings are collected in the following proposition:

Proposition 2.3. *Let $\mathsf{D} = \bigoplus_{\gamma\in\Gamma} \mathsf{D}_\gamma$ be a graded division ring.*

(i) *D has no zero divisors.*
(ii) *For $a,\, b \in \mathsf{D}\setminus\{0\}$, if ab is homogeneous, then a and b are homogeneous.*
(iii) *$\mathsf{D}^\times$ consists of the nonzero homogeneous elements of D, and the degree map*

$$deg\colon \mathsf{D}^\times \to \Gamma$$

is a group homomorphism.
(iv) *$\Gamma_{\mathsf{D}} = \Gamma_{\mathsf{D}}^\times$, which is a group.*
(v) *D_0 is a division ring.*
(vi) *Each nonzero D_γ is a 1-dimensional left and right D_0-vector space.*

Proof. (i) and (ii): Choose a total ordering on Γ compatible with the addition (see Remark 2.2), and take nonzero $a,\, b \in \mathsf{D}$. Write $a = a_1 + \ldots + a_k$ with $a_1, \ldots, a_k$ homogeneous of increasing degrees; likewise write $b = b_1 + \ldots + b_\ell$. Then a_1b_1 is the homogeneous component of degree $deg(a_1) + deg(b_1)$ of ab, and $a_1b_1 \neq 0$ since a_1 and b_1 are homogeneous, hence invertible. Therefore, $ab \neq 0$, showing that D has no zero divisors. Likewise, the highest degree homogeneous component of ab is $a_k b_\ell$. If ab is homogeneous, we must have $deg(a_1) + deg(b_1) = deg(a_k) + deg(b_\ell)$, hence $deg(a_1) = deg(a_k)$ and $deg(b_1) = deg(b_\ell)$, which means that a and b are homogeneous.

(iii): This is immediate from (ii).

(iv): This is clear since every nonzero homogeneous element of D is a unit.

(v): We noted above that D_0 is a ring since D is a graded ring. The nonzero elements in D_0 form the kernel of $deg\colon \mathsf{D}^\times \to \Gamma$, which is a group. Therefore, D_0 is a division ring.

(vi): If $\mathsf{D}_\gamma \neq \{0\}$, take any nonzero $a \in \mathsf{D}_\gamma$. Then $\mathsf{D}_\gamma a^{-1}$ is a nonzero left D_0-subspace of D_0, hence $\mathsf{D}_\gamma a^{-1} = \mathsf{D}_0$. Thus, $\mathsf{D}_\gamma = \mathsf{D}_0 a$, which is a 1-dimensional left D_0-vector space. Likewise, $\mathsf{D}_\gamma = a\mathsf{D}_0$. □

From this we obtain a convenient characterization of graded division rings:

Proposition 2.4. *Let* R *be a graded ring with* $1 \neq 0$. *Then,* R *is a graded division ring if and only if* R_0 *is a division ring and* $\Gamma_{\mathsf{R}}^\times = \Gamma_{\mathsf{R}}$.

Proof. Suppose R_0 is a division ring and $\Gamma_{\mathsf{R}}^\times = \Gamma_{\mathsf{R}}$. Take any $\delta \in \Gamma_{\mathsf{R}} = \Gamma_{\mathsf{R}}^\times$ and any $b \in \mathsf{R}_\delta \cap \mathsf{R}^\times$. Write $b^{-1} = \sum_{\gamma \in \Gamma_{\mathsf{R}}} c_\gamma$ where each $c_\gamma \in \mathsf{R}_\gamma$. The equation $1 = b^{-1} b = \sum_{\gamma \in \Gamma_{\mathsf{R}}} c_\gamma b$ with each $c_\gamma b \in \mathsf{R}_{\gamma+\delta}$ and $1 \in \mathsf{R}_0$ implies that $c_\gamma b = 0$ for $\gamma \neq -\delta$. Hence, $c_{-\delta} b = 1$; likewise $b c_{-\delta} = 1$, showing that $b^{-1} = c_{-\delta} \in \mathsf{R}_{-\delta}$. Therefore, $b^{-1} R_\delta \subseteq \mathsf{R}_{-\delta} \cdot \mathsf{R}_\delta \subseteq \mathsf{R}_0$, so that $\mathsf{R}_\delta = b b^{-1} \mathsf{R}_\delta \subseteq b\mathsf{R}_0$. Thus,

$$\mathsf{R}_\delta \setminus \{0\} \subseteq b(\mathsf{R}_0 \setminus \{0\}) \subseteq \mathsf{R}^\times,$$

as R_0 is a division ring. Therefore, every nonzero homogeneous element of R is a unit; so, R is a graded division ring. This proves one implication of the proposition, and the converse is given in Prop. 2.3(iv) and (v). □

A *graded subring* of a graded ring R is a subring $\mathsf{S} \subseteq \mathsf{R}$ such that $\mathsf{S} = \bigoplus_{\gamma \in \Gamma} (\mathsf{R}_\gamma \cap \mathsf{S})$. This decomposition defines a grading on S. As a special case, note that the center $Z(\mathsf{R})$ of R is a graded subring. Its 0-component $Z(\mathsf{R})_0$ satisfies

$$Z(\mathsf{R})_0 = Z(\mathsf{R}) \cap \mathsf{R}_0 \subseteq Z(\mathsf{R}_0).$$

A graded subring that is also a graded division ring (resp. a graded field) is called a *graded sub-division ring* (resp. *graded subfield*).

2.1.2 Graded modules

Let R be a graded ring. A right module M over $\mathsf{R}^\natural$ equipped with a decomposition $\mathsf{M} = \bigoplus_{\gamma \in \Gamma} \mathsf{M}_\gamma$, where each M_γ is an additive subgroup of M and $\mathsf{M}_\gamma \cdot \mathsf{R}_\delta \subseteq \mathsf{M}_{\gamma+\delta}$ for γ, $\delta \in \Gamma$, is called a *right graded* R*-module*. Left graded modules are defined likewise. The *grade set* Γ_{M} is defined by

$$\Gamma_{\mathsf{M}} = \{\gamma \in \Gamma \mid \mathsf{M}_\gamma \neq \{0\}\}.$$

A *graded submodule* of a graded R-module M is an $R^\natural$-submodule $N \subseteq M$ such that $N = \bigoplus_{\gamma\in\Gamma}(M_\gamma \cap N)$. This decomposition defines a graded R-module structure on N. Just as for graded rings, we let $M^\natural$ denote the underlying ungraded $R^\natural$-module of M.

For the rest of this subsection, we focus on the case where the base graded ring is a graded *division* ring D. Graded modules over a graded division ring are called *graded vector spaces*, because they are free modules: see Prop. 2.5 below. Note that if V is a graded (left or right) D-vector space, then each homogeneous component V_γ is a D_0-vector space. The grade set Γ_V need not be a group, but it is a union of cosets of the group Γ_D. We denote by $|\Gamma_V{:}\Gamma_D|$ the number of these cosets (which may be infinite). Let $\Gamma_V = \coprod_{i\in I}\Gamma_i$ be the decomposition of Γ_V into disjoint cosets of Γ_D; there is a corresponding direct sum decomposition of V into graded subspaces:

$$V = \bigoplus_{i\in I} V_i \qquad \text{where} \quad V_i = \bigoplus_{\gamma\in\Gamma_i} V_\gamma \quad \text{for } i \in I. \tag{2.2}$$

We call (2.2) the *canonical decomposition* of V.

Proposition 2.5. *Every graded vector space over a graded division ring is a free module with a homogeneous base. More precisely, let V be a right graded vector space over the graded division ring D, with canonical decomposition (2.2). For $i \in I$, fix some $\gamma_i \in \Gamma_i$ and some D_0-base $(e_{ij})_{j\in J_i}$ of V_{γ_i}. Then $(e_{ij})_{j\in J_i}$ is a homogeneous D-base of V_i, and $(e_{ij})_{i\in I,\, j\in J_i}$ is a homogeneous D-base of V. Moreover, every homogeneous D-base of V has cardinality $\sum_{i\in I} dim_{D_0} V_{\gamma_i}$.*

Proof. The decomposition $V_{\gamma_i} = \bigoplus_{j\in J_i} e_{ij}D_0$ yields $V_{\gamma_i+\delta} = \bigoplus_{j\in J_i} e_{ij}D_\delta$ for all $\delta \in \Gamma$, hence

$$V_i = \bigoplus_{j\in J_i} e_{ij}D,$$

proving that $(e_{ij})_{j\in J_i}$ is a D-base of V_i. In view of the canonical decomposition, it follows that $(e_{ij})_{i\in I,\, j\in J_i}$ is a D-base of V.

Assume now that V is a right graded D-vector space. Let $(b_k)_{k\in K}$ be any homogeneous D-base of V. For $i \in I$, let $K_i = \{k \in K \mid deg(b_k) \in \Gamma_i\}$. Since $\Gamma_i = \gamma_i + \Gamma_D$, for each $k \in K_i$ there is a nonzero $d_k \in D_{\gamma_i - deg(b_k)}$; so, $b_k d_k \in V_{\gamma_i}$. Then, $(b_k d_k)_{k\in K}$ is a homogeneous D-base of V. Hence, for each $i \in I$, $(b_k d_k)_{k\in K_i}$ must be a homogeneous D-base of V_i, and hence also a D_0-base of V_{γ_i}. Thus, $|K| = \sum_{i\in I}|K_i| = \sum_{i\in I} dim_{D_0} V_{\gamma_i}$. □

Corollary 2.6. *Every graded subspace of a graded vector space has a complementary graded subspace.*

Proof. Let V be a right graded vector space over a graded division ring D and let $U \subseteq V$ be a graded subspace. We claim that there is a graded subspace $W \subseteq V$ such that $V = U \oplus W$. Consider the canonical decomposition of V as in (2.2) above: let $\Gamma_V = \coprod_{i\in I}\Gamma_i$ and $V = \bigoplus_{i\in I} V_i$ where $V_i = \bigoplus_{\gamma\in\Gamma_i} V_\gamma$

for $i \in I$. Since $\mathsf{U} \subseteq \mathsf{V}$ is a graded subspace, we also have a canonical decomposition of U: there is a subset $I_{\mathsf{U}} \subseteq I$ such that $\Gamma_{\mathsf{U}} = \coprod_{i \in I_{\mathsf{U}}} \Gamma_i$ and, for each $i \in I_{\mathsf{U}}$, there is a D_0-subspace $\mathsf{U}_\gamma \subseteq \mathsf{V}_\gamma$ such that

$$\mathsf{U} = \bigoplus_{i \in I_{\mathsf{U}}} \mathsf{U}_i \qquad \text{where} \quad \mathsf{U}_i = \bigoplus_{\gamma \in \Gamma_i} \mathsf{U}_\gamma \qquad \text{for } i \in I_{\mathsf{U}}.$$

For $i \in I_{\mathsf{U}}$, fix some $\gamma_i \in \Gamma_i$ and some D_0-subspace $\mathsf{U}'_{\gamma_i} \subseteq \mathsf{V}_{\gamma_i}$ such that

$$\mathsf{V}_{\gamma_i} = \mathsf{U}_{\gamma_i} \oplus \mathsf{U}'_{\gamma_i}.$$

By combining a D_0-base $\mathcal{B}$ of U_{γ_i} with a D_0-base $\mathcal{B}'$ of U'_{γ_i}, we obtain a D_0-base of V_{γ_i}, which is a D-base of V_i by Prop. 2.5. Likewise, $\mathcal{B}$ is a D-base of U_i. Therefore, letting U'_i be the D-span of $\mathcal{B}'$, we obtain

$$\mathsf{V}_i = \mathsf{U}_i \oplus \mathsf{U}'_i \qquad \text{for } i \in I_{\mathsf{U}}.$$

Therefore, the following is a complementary subspace of U in V:

$$\mathsf{W} = \Big(\bigoplus_{i \in I_{\mathsf{U}}} \mathsf{U}'_i\Big) \oplus \Big(\bigoplus_{i \in I \setminus I_{\mathsf{U}}} \mathsf{V}_i\Big). \qquad \square$$

The rank of a graded vector space V over a graded division ring D, which is the number of elements in any homogeneous base, is also called its *dimension*, and is denoted by $dim_{\mathsf{D}} \mathsf{V}$ or $[\mathsf{V}{:}\mathsf{D}]$. Proposition 2.5 shows that

$$dim_{\mathsf{D}} \mathsf{V} = \sum_{i \in I} dim_{\mathsf{D}} \mathsf{V}_i = \sum_{i \in I} dim_{\mathsf{D}_0} \mathsf{V}_{\gamma_i}, \tag{2.3}$$

where I is a set with $|\Gamma_{\mathsf{V}}{:}\Gamma_{\mathsf{D}}|$ elements and $\{\gamma_i\}_{i \in I}$ is a set of coset representatives of Γ_{V} modulo Γ_{D}. In particular, if $dim_{\mathsf{D}} \mathsf{V} = d < \infty$, then $|\Gamma_{\mathsf{V}}{:}\Gamma_{\mathsf{D}}| \le d < \infty$ and $dim_{\mathsf{D}_0} \mathsf{V}_\gamma \le d < \infty$ for every $\gamma \in \Gamma$.

In one important case the dimensions $dim_{\mathsf{D}_0} \mathsf{V}_{\gamma_i}$ are all equal, and the sum above can therefore be rewritten as a product: suppose D is a graded sub-division ring of another graded division ring $\mathsf{E} = \bigoplus_{\gamma \in \Gamma} \mathsf{E}_\gamma$. We may then consider E as a left (resp. right) graded D-vector space. We write $[\mathsf{E}{:}\mathsf{D}]_\ell$ (resp. $[\mathsf{E}{:}\mathsf{D}]_r$) for its dimension.[1] Clearly D_0 is a subring of E_0 and Γ_{D} is a subgroup of Γ_{E}.

Corollary 2.7. *With the notation above,*

$$[\mathsf{E}{:}\mathsf{D}]_\ell = [\mathsf{E}_0{:}\mathsf{D}_0]_\ell \, |\Gamma_{\mathsf{E}}{:}\Gamma_{\mathsf{D}}| \quad \textit{and} \quad [\mathsf{E}{:}\mathsf{D}]_r = [\mathsf{E}_0{:}\mathsf{D}_0]_r \, |\Gamma_{\mathsf{E}}{:}\Gamma_{\mathsf{D}}|.$$

Proof. Consider E as a left graded D-vector space. For all $\gamma \in \Gamma_{\mathsf{E}}$, the homogeneous component E_γ is a 1-dimensional E_0-vector space; hence,

$$dim_{\mathsf{D}_0}(\mathsf{E}_\gamma) = [\mathsf{E}_0{:}\mathsf{D}_0]_\ell.$$

By Prop. 2.5, it follows that $dim_{\mathsf{D}}(\mathsf{E}) = [\mathsf{E}_0{:}\mathsf{D}_0]_\ell \, |\Gamma_{\mathsf{E}}{:}\Gamma_{\mathsf{D}}|$, proving the left equality. The proof of the right equation is analogous. $\square$

---

[1] When the dimensions as left and right vector spaces coincide—e.g., when D lies in the center of E—we write simply $[\mathsf{E}{:}\mathsf{D}]$.

2.1.3 Homomorphisms

Let $\mathsf{M} = \bigoplus_{\gamma\in\Gamma} \mathsf{M}_\gamma$ and $\mathsf{N} = \bigoplus_{\gamma\in\Gamma} \mathsf{N}_\gamma$ be right graded modules over a graded ring R, and let $Hom_{\mathsf{R}^\natural}(\mathsf{M}^\natural, \mathsf{N}^\natural)$ be the group of $\mathsf{R}^\natural$-linear maps (i.e., $\mathsf{R}^\natural$-module homomorphisms) from the ungraded module $\mathsf{M}^\natural$ to $\mathsf{N}^\natural$; so $Hom_{\mathsf{R}^\natural}(\mathsf{M}^\natural, \mathsf{N}^\natural)$ is a module over $Z(\mathsf{R}^\natural)$. For $\gamma \in \Gamma$, we denote by $\mathsf{Hom}_\mathsf{R}(\mathsf{M}, \mathsf{N})_\gamma$ the group of R-linear maps that shift the degree by γ,

$$\mathsf{Hom}_\mathsf{R}(\mathsf{M}, \mathsf{N})_\gamma = \{f \in Hom_{\mathsf{R}^\natural}(\mathsf{M}^\natural, \mathsf{N}^\natural) \mid f(\mathsf{M}_\delta) \subseteq \mathsf{N}_{\delta+\gamma} \text{ for all } \delta \in \Gamma\}.$$

Thus, $\mathsf{Hom}_\mathsf{R}(\mathsf{M}, \mathsf{N})_0$ consists of the (degree-preserving) graded homomorphisms from M to N, which are the morphisms in the category of graded R-modules. If there is an isomorphism in $\mathsf{Hom}_\mathsf{R}(\mathsf{M}, \mathsf{N})_0$, we write $\mathsf{M} \cong_g \mathsf{N}$ to emphasize that M and N are isomorphic as graded modules (i.e., by a degree-preserving isomorphism), not simply as $\mathsf{R}^\natural$-modules.

For $\gamma \neq 0$, homomorphisms in $\mathsf{Hom}_\mathsf{R}(\mathsf{M}, \mathsf{N})_\gamma$ can also be viewed as degree-preserving maps after changing the grading on M (or N, or both). For any right graded R-module M and any $\gamma \in \Gamma$, we define the *shifted*[2] graded module $\mathsf{M}(\gamma)$: it is the $\mathsf{R}^\natural$-module $\mathsf{M}^\natural$ with a grading defined by shifting the grading of M, so that for $\delta \in \Gamma$,

$$\mathsf{M}(\gamma)_\delta = \mathsf{M}_{\gamma+\delta}.$$

Thus, $\Gamma_{\mathsf{M}(\gamma)} = -\gamma + \Gamma_\mathsf{M}$, and we may identify

$$\mathsf{Hom}_\mathsf{R}(\mathsf{M}, \mathsf{N})_\gamma = \mathsf{Hom}_\mathsf{R}(\mathsf{M}(-\gamma), \mathsf{N})_0 = \mathsf{Hom}_\mathsf{R}\big(\mathsf{M}, \mathsf{N}(\gamma)\big)_0$$
$$\big(= \mathsf{Hom}_\mathsf{R}\big(\mathsf{M}(\varepsilon), \mathsf{N}(\gamma+\varepsilon)\big)_0 \text{ for all } \varepsilon \in \Gamma\big).$$

More generally, for all γ, δ, $\varepsilon \in \Gamma$ we may identify

$$\mathsf{Hom}_\mathsf{R}\big(\mathsf{M}(\gamma), \mathsf{N}(\delta)\big)_\varepsilon = \mathsf{Hom}_\mathsf{R}(\mathsf{M}, \mathsf{N})_{\delta-\gamma+\varepsilon}. \tag{2.4}$$

Proposition 2.8. *If $\mathsf{M}^\natural$ is a finitely generated $\mathsf{R}^\natural$-module, then*

$$Hom_{\mathsf{R}^\natural}(\mathsf{M}^\natural, \mathsf{N}^\natural) = \bigoplus_{\gamma\in\Gamma} \mathsf{Hom}_\mathsf{R}(\mathsf{M}, \mathsf{N})_\gamma.$$

With this decomposition, $Hom_{\mathsf{R}^\natural}(\mathsf{M}^\natural, \mathsf{N}^\natural)$ is a graded $Z(\mathsf{R})$-module denoted by $\mathsf{Hom}_\mathsf{R}(\mathsf{M}, \mathsf{N})$. For all γ, $\delta \in \Gamma$ we have an identification of graded $Z(\mathsf{R})$-modules

$$\mathsf{Hom}_\mathsf{R}\big(\mathsf{M}(\gamma), \mathsf{N}(\delta)\big) = \mathsf{Hom}_\mathsf{R}(\mathsf{M}, \mathsf{N})(\delta - \gamma). \tag{2.5}$$

Proof. The sum $\sum_{\gamma\in\Gamma} \mathsf{Hom}_\mathsf{R}(\mathsf{M}, \mathsf{N})_\gamma$ is clearly direct, so it suffices to show that this sum is all of $Hom_{\mathsf{R}^\natural}(\mathsf{M}^\natural, \mathsf{N}^\natural)$ to establish the first part. By decomposing generators of $\mathsf{M}^\natural$ into homogeneous components, we may find a generating set $\{m_i\}_{i=1}^n$ consisting of homogeneous elements. Let $\delta_i = deg\, m_i$

[2] The shift construction is also classically known as the *twist* construction. We will avoid this terminology because it may be confusing in a noncommutative context.

and let $f \in \mathit{Hom}_{\mathsf{R}^\natural}(\mathsf{M}^\natural, \mathsf{N}^\natural)$. For $i = 1, \ldots, n$, let $y_{i,\gamma}$ be the homogeneous component of degree γ of $f(m_i)$, so

$$f(m_i) = \sum_{\gamma\in\Gamma} y_{i,\gamma}.$$

We claim that for each $\gamma \in \Gamma$ there exists an $f_\gamma \in \mathit{Hom}_{\mathsf{R}^\natural}(\mathsf{M}^\natural, \mathsf{N}^\natural)$ such that

$$f_\gamma(m_i) = y_{i,\gamma+\delta_i} \qquad \text{for } i = 1, \ldots, n.$$

It is clear that $f_\gamma \in \mathsf{Hom}_{\mathsf{R}}(\mathsf{M}, \mathsf{N})_\gamma$ and $f = \sum_{\gamma\in\Gamma} f_\gamma$, hence the first part of the proposition follows.

To prove the claim, it suffices to show that if $a_1, \ldots, a_n \in \mathsf{R}$ satisfy $\sum_{i=1}^n m_i a_i = 0$, then $\sum_{i=1}^n y_{i,\gamma+\delta_i} a_i = 0$ for all $\gamma \in \Gamma$; for then the map f_γ is well-defined by

$$f_\gamma\Big(\sum_{i=1}^n m_i a_i\Big) = \sum_{i=1}^n y_{i,\gamma+\delta_i} a_i \qquad \text{for } a_1, \ldots, a_n \in \mathsf{R}.$$

Suppose $\sum_{i=1}^n m_i a_i = 0$, and let $a_{i,\gamma}$ denote the homogeneous component of degree γ of a_i. For all $\delta \in \Gamma$, the homogeneous component of degree δ of $m_i a_i$ is $m_i a_{i,\delta-\delta_i}$, so we have $\sum_i m_i a_{i,\delta-\delta_i} = 0$. By applying f, we obtain $\sum_i f(m_i) a_{i,\delta-\delta_i} = 0$ for all $\delta \in \Gamma$. Now, for any $\gamma \in \Gamma$, the homogeneous component of degree $\gamma+\delta$ of $f(m_i)a_{i,\delta-\delta_i}$ is $y_{i,\gamma+\delta_i} a_{i,\delta-\delta_i}$, so $\sum_i y_{i,\gamma+\delta_i} a_{i,\delta-\delta_i} = 0$ for all γ, $\delta \in \Gamma$. Summing over δ, we obtain $\sum_i y_{i,\gamma+\delta_i} a_i = 0$ for all $\gamma \in \Gamma$, proving the claim.

If $f \in \mathsf{Hom}_{\mathsf{R}}(\mathsf{M}, \mathsf{N})_\gamma$ and $z \in Z(\mathsf{R})_\delta$, then clearly $zf \in \mathsf{Hom}_{\mathsf{R}}(\mathsf{M}, \mathsf{N})_{\gamma+\delta}$, hence $\mathsf{Hom}_{\mathsf{R}}(\mathsf{M}, \mathsf{N})$ is a graded $Z(\mathsf{R})$-module. Equation (2.5) follows immediately from (2.4). □

If $\mathsf{M}^\natural$ is a finitely generated $\mathsf{R}^\natural$-module, we set $\mathsf{End}_{\mathsf{R}}\,\mathsf{M} = \mathsf{Hom}_{\mathsf{R}}(\mathsf{M}, \mathsf{M})$. The grading on $\mathsf{End}_{\mathsf{R}}\,\mathsf{M}$ is compatible with the composition of maps, so $\mathsf{End}_{\mathsf{R}}\,\mathsf{M}$ has a natural graded ring structure. We next represent this graded ring as a graded matrix ring, assuming M is a free graded module.

For any positive integer n, the $n \times n$ matrix ring $M_n(\mathsf{R})$ has an obvious grading, where for each $\gamma \in \Gamma$,

$$M_n(\mathsf{R})_\gamma = \begin{pmatrix} \mathsf{R}_\gamma & \ldots & \mathsf{R}_\gamma \\ \vdots & \ddots & \vdots \\ \mathsf{R}_\gamma & \ldots & \mathsf{R}_\gamma \end{pmatrix}. \tag{2.6}$$

But other gradings, as follows, arise naturally: Take any $\delta_1, \ldots, \delta_n \in \Gamma$. Let $\mathsf{M}_n(\mathsf{R})(\delta_1, \ldots, \delta_n)$ denote $M_n(\mathsf{R}^\natural)$ as a ring, but graded so that

$$\mathsf{M}_n(\mathsf{R})(\delta_1, \ldots, \delta_n) = \begin{pmatrix} \mathsf{R} & \mathsf{R}(\delta_1-\delta_2) & \ldots & \mathsf{R}(\delta_1-\delta_n) \\ \mathsf{R}(\delta_2-\delta_1) & \mathsf{R} & \ldots & \mathsf{R}(\delta_2-\delta_n) \\ \vdots & \vdots & \ddots & \vdots \\ \mathsf{R}(\delta_n-\delta_1) & \mathsf{R}(\delta_n-\delta_2) & \ldots & \mathsf{R} \end{pmatrix}. \tag{2.7}$$

Thus, for any $\gamma \in \Gamma$ the component of degree γ is

$$\mathsf{M}_n(\mathsf{R})(\delta_1,\ldots,\delta_n)_\gamma = \begin{pmatrix} \mathsf{R}_\gamma & \mathsf{R}_{\gamma+\delta_1-\delta_2} & \cdots & \mathsf{R}_{\gamma+\delta_1-\delta_n} \\ \mathsf{R}_{\gamma+\delta_2-\delta_1} & \mathsf{R}_\gamma & \cdots & \mathsf{R}_{\gamma+\delta_2-\delta_n} \\ \vdots & \vdots & \ddots & \vdots \\ \mathsf{R}_{\gamma+\delta_n-\delta_1} & \mathsf{R}_{\gamma+\delta_n-\delta_2} & \cdots & \mathsf{R}_\gamma \end{pmatrix}.$$

Let $\mathsf{T} = \mathsf{M}_n(\mathsf{R})(\delta_1,\ldots,\delta_n)$. Since additively T is a direct sum of copies of R and for any $\delta \in \Gamma$ we have $\mathsf{R} = \bigoplus_{\gamma\in\Gamma} \mathsf{R}(\delta)_\gamma$, clearly $\mathsf{T} = \bigoplus_{\gamma\in\Gamma} \mathsf{T}_\gamma$. Moreover, if $A = (a_{ij}) \in \mathsf{T}_\gamma$ and $B = (b_{ij}) \in \mathsf{T}_\varepsilon$, then each $a_{ij} \in \mathsf{R}_{\gamma+\delta_i-\delta_j}$ and $b_{jk} \in \mathsf{R}_{\varepsilon+\delta_j-\delta_k}$, so $a_{ij}b_{jk} \in \mathsf{R}_{\gamma+\delta_i+\varepsilon-\delta_k} = \mathsf{R}(\delta_i-\delta_k)_{\gamma+\varepsilon}$. This yields that $AB \in \mathsf{T}_{\gamma+\varepsilon}$, verifying that T is a graded ring.

Proposition 2.9. *Let* R *be a graded ring, let* N *be a finitely generated right graded* R*-module, and let* $\mathsf{E} = \mathsf{End}_\mathsf{R}\,\mathsf{N}$. *For any* $\delta_1,\ldots,\delta_n \in \Gamma$,

$$\mathsf{End}_\mathsf{R}\big(\mathsf{N}(\delta_1)\oplus\ldots\oplus\mathsf{N}(\delta_n)\big) \cong_g \mathsf{M}_n(\mathsf{E})(\delta_1,\ldots,\delta_n),$$

a graded ring isomorphism.

Proof. For $i = 1, \ldots, n$, consider the canonical maps

$$q_i\colon \mathsf{N}(\delta_i) \longrightarrow \mathsf{N}(\delta_1)\oplus\ldots\oplus\mathsf{N}(\delta_n) \quad\text{and}\quad p_i\colon \mathsf{N}(\delta_1)\oplus\ldots\oplus\mathsf{N}(\delta_n) \longrightarrow \mathsf{N}(\delta_i).$$

To each endomorphism f of $\mathsf{N}(\delta_1)\oplus\ldots\oplus\mathsf{N}(\delta_n)$ we associate the matrix $(f_{ij})_{i,j=1}^n$ where

$$f_{ij} = p_i \circ f \circ q_j\colon \mathsf{N}(\delta_j) \longrightarrow \mathsf{N}(\delta_i).$$

Thus, we have

$$\mathsf{End}_\mathsf{R}\big(\mathsf{N}(\delta_1)\oplus\ldots\oplus\mathsf{N}(\delta_n)\big) \cong_g$$
$$\begin{pmatrix} \mathsf{Hom}_\mathsf{R}\big(\mathsf{N}(\delta_1),\mathsf{N}(\delta_1)\big) & \mathsf{Hom}_\mathsf{R}\big(\mathsf{N}(\delta_2),\mathsf{N}(\delta_1)\big) & \cdots & \mathsf{Hom}_\mathsf{R}\big(\mathsf{N}(\delta_n),\mathsf{N}(\delta_1)\big) \\ \mathsf{Hom}_\mathsf{R}\big(\mathsf{N}(\delta_1),\mathsf{N}(\delta_2)\big) & \mathsf{Hom}_\mathsf{R}\big(\mathsf{N}(\delta_2),\mathsf{N}(\delta_2)\big) & \cdots & \mathsf{Hom}_\mathsf{R}\big(\mathsf{N}(\delta_n),\mathsf{N}(\delta_2)\big) \\ \vdots & \vdots & \ddots & \vdots \\ \mathsf{Hom}_\mathsf{R}\big(\mathsf{N}(\delta_1),\mathsf{N}(\delta_n)\big) & \mathsf{Hom}_\mathsf{R}\big(\mathsf{N}(\delta_2),\mathsf{N}(\delta_n)\big) & \cdots & \mathsf{Hom}_\mathsf{R}\big(\mathsf{N}(\delta_n),\mathsf{N}(\delta_n)\big) \end{pmatrix}.$$

This is a ring isomorphism, which is compatible with the gradings on each side. For the ij-entry on the right, Prop. 2.8 yields

$$\mathsf{Hom}_\mathsf{R}\big(\mathsf{N}(\delta_j),\mathsf{N}(\delta_i)\big) \cong_g (\mathsf{End}_\mathsf{R}\,\mathsf{N})(\delta_i-\delta_j),$$

which is the ij-entry of $\mathsf{M}_n(\mathsf{E})(\delta_1,\ldots,\delta_n)$. The result follows. □

The "left" version of Prop. 2.9 says that if M is a finitely generated left graded R-module and $\mathsf{B} = \mathsf{End}_\mathsf{R}\,\mathsf{M}$, then

$$\mathsf{End}_\mathsf{R}\big(\mathsf{M}(\gamma_1)\oplus\ldots\oplus\mathsf{M}(\gamma_n)\big) \cong_g \mathsf{M}_n(\mathsf{B})(-\gamma_1,\ldots,-\gamma_n).$$

Indeed, endomorphisms of left modules act on the right of their argument, hence their composition is in the reverse order as compared to the previous case. Thus, the matrix representation of an endomorphism f of $\mathsf{M}(\gamma_1) \oplus \ldots \oplus \mathsf{M}(\gamma_n)$ is $(f_{ij})_{i,j=1}^n$ where

$$f_{ij} = q_i \circ f \circ p_j \colon \mathsf{M}(\gamma_i) \longrightarrow \mathsf{M}(\gamma_j).$$

To see that Prop. 2.9 applies in particular to endomorphism rings of graded vector spaces, we rephrase Prop. 2.5 as follows:

Corollary 2.10. *Every graded vector space over a graded division ring* D *is isomorphic to a direct sum of shifted* 1*-dimensional spaces* $\mathsf{D}(\gamma)$.

Proof. In the notation of Prop. 2.5 and its proof, let $\mathsf{D}^{(J_i)}$ for $i \in I$ denote the graded vector space of $|J_i|$-tuples $(d_j)_{j\in J_i}$ such that $d_j \neq 0$ for only a finite number of $j \in J_i$. This graded vector space is a direct sum of copies of D and the map $(d_j)_{j\in J_i} \mapsto \sum_{j\in J_i} e_{ij} d_j$ is a graded vector space isomorphism $\mathsf{D}^{(J_i)}(-\gamma_i) \cong_g \mathsf{V}_i$. Therefore,

$$\mathsf{V} \cong_g \bigoplus_{i\in I} \mathsf{D}(-\gamma_i)^{(J_i)}. \qquad \square$$

Let V, W be finite-dimensional right graded vector spaces over a graded division ring D. Consider decompositions of V and W as direct sums of shifted 1-dimensional graded vector spaces as in Cor. 2.10:

$$\mathsf{V} \cong_g \mathsf{D}(\gamma_1) \oplus \ldots \oplus \mathsf{D}(\gamma_n), \quad \mathsf{W} \cong_g \mathsf{D}(\delta_1) \oplus \ldots \oplus \mathsf{D}(\delta_m)$$

for some $\gamma_1, \ldots, \delta_m \in \Gamma$.

We have $\mathsf{End}_{\mathsf{D}} \mathsf{D} = \mathsf{D}$. Therefore, arguing as in Prop. 2.9, we obtain

$$\mathsf{Hom}_{\mathsf{D}}(\mathsf{V},\mathsf{W}) \cong_g$$

$$\begin{pmatrix} \mathsf{Hom}_{\mathsf{D}}\big(\mathsf{D}(\gamma_1),\mathsf{D}(\delta_1)\big) & \mathsf{Hom}_{\mathsf{D}}\big(\mathsf{D}(\gamma_2),\mathsf{D}(\delta_1)\big) & \ldots & \mathsf{Hom}_{\mathsf{D}}\big(\mathsf{D}(\gamma_n),\mathsf{D}(\delta_1)\big) \\ \mathsf{Hom}_{\mathsf{D}}\big(\mathsf{D}(\gamma_1),\mathsf{D}(\delta_2)\big) & \mathsf{Hom}_{\mathsf{D}}\big(\mathsf{D}(\gamma_2),\mathsf{D}(\delta_2)\big) & \ldots & \mathsf{Hom}_{\mathsf{D}}\big(\mathsf{D}(\gamma_n),\mathsf{D}(\delta_2)\big) \\ \vdots & \vdots & \ddots & \vdots \\ \mathsf{Hom}_{\mathsf{D}}\big(\mathsf{D}(\gamma_1),\mathsf{D}(\delta_m)\big) & \mathsf{Hom}_{\mathsf{D}}\big(\mathsf{D}(\gamma_2),\mathsf{D}(\delta_m)\big) & \ldots & \mathsf{Hom}_{\mathsf{D}}\big(\mathsf{D}(\gamma_n),\mathsf{D}(\delta_m)\big) \end{pmatrix}$$

$$\cong_g \begin{pmatrix} \mathsf{D}(\delta_1-\gamma_1) & \mathsf{D}(\delta_1-\gamma_2) & \ldots & \mathsf{D}(\delta_1-\gamma_n) \\ \mathsf{D}(\delta_2-\gamma_1) & \mathsf{D}(\delta_2-\gamma_2) & \ldots & \mathsf{D}(\delta_2-\gamma_n) \\ \vdots & \vdots & \ddots & \vdots \\ \mathsf{D}(\delta_m-\gamma_1) & \mathsf{D}(\delta_m-\gamma_2) & \ldots & \mathsf{D}(\delta_m-\gamma_n) \end{pmatrix}.$$

Consequently, we have

$$\Gamma_{\mathsf{Hom}_{\mathsf{D}}(\mathsf{V},\mathsf{W})} = \bigcup_{i,j} (\delta_i - \gamma_j + \Gamma_{\mathsf{D}}) = \{\delta - \gamma \mid \delta \in \Gamma_{\mathsf{W}},\ \gamma \in \Gamma_{\mathsf{V}}\}.$$

In particular, when $\mathsf{V} = \mathsf{W}$ we obtain:

Corollary 2.11. *Let* V *be a finite-dimensional graded vector space over a graded division ring* D. *If* $\mathsf{V} \cong_g \mathsf{D}(\gamma_1) \oplus \ldots \oplus \mathsf{D}(\gamma_n)$ *for some* $\gamma_1, \ldots, \gamma_n$, *then*

$$\mathsf{End}_{\mathsf{D}} \mathsf{V} \cong_g \mathsf{M}_n(\mathsf{D})(\gamma_1, \ldots, \gamma_n) \quad \textit{and} \quad \Gamma_{\mathsf{End}_{\mathsf{D}} \mathsf{V}} = \bigcup_{i,j} (\gamma_i - \gamma_j + \Gamma_{\mathsf{D}}).$$

2.1.4 Tensor products

Let $\mathsf{M} = \bigoplus_{\gamma \in \Gamma} \mathsf{M}_\gamma$ be a right graded module and let $\mathsf{N} = \bigoplus_{\gamma \in \Gamma} \mathsf{N}_\gamma$ be a left graded module over the graded ring R. Letting $\mathsf{P}_\gamma = \sum_{\delta+\varepsilon=\gamma} \mathsf{M}_\delta \otimes_{\mathbb{Z}} \mathsf{N}_\varepsilon$ for $\gamma \in \Gamma$, we have a grading on the $\mathbb{Z}$-module $\mathsf{M} \otimes_{\mathbb{Z}} \mathsf{N}$:

$$\mathsf{M} \otimes_{\mathbb{Z}} \mathsf{N} = \bigoplus_{\gamma \in \Gamma} \mathsf{P}_\gamma.$$

The tensor product $\mathsf{M} \otimes_{\mathsf{R}} \mathsf{N}$ is the quotient of $\mathsf{M} \otimes_{\mathbb{Z}} \mathsf{N}$ by the submodule Q spanned by the elements of the form $(mr) \otimes n - m \otimes (rn)$ for $m \in \mathsf{M}$, $n \in \mathsf{N}$ and $r \in \mathsf{R}$. By decomposing m, n, and r into homogeneous components, we see that Q is also spanned by elements of this form with m, n, and r homogeneous. Therefore, Q is a graded submodule of $\mathsf{M} \otimes_{\mathbb{Z}} \mathsf{N}$, and it follows that $\mathsf{M} \otimes_{\mathsf{R}} \mathsf{N}$ inherits the grading of $\mathsf{M} \otimes_{\mathbb{Z}} \mathsf{N}$. For $\gamma \in \Gamma$, the homogeneous component $(\mathsf{M} \otimes_{\mathsf{R}} \mathsf{N})_\gamma$ is the additive subgroup generated by the products $m_\delta \otimes n_\varepsilon$ with $m_\delta \in \mathsf{M}_\delta$, $n_\varepsilon \in \mathsf{N}_\varepsilon$, and $\delta + \varepsilon = \gamma$. Therefore, $\Gamma_{\mathsf{M} \otimes_{\mathsf{R}} \mathsf{N}} \subseteq \Gamma_{\mathsf{M}} + \Gamma_{\mathsf{N}}$. Note that this grading provides $\mathsf{M} \otimes_{\mathsf{R}} \mathsf{N}$ with a graded $Z(\mathsf{R})$-module structure.

The graded tensor product construction will be used mostly in the case where R is a graded division ring D and M, N are graded D-vector spaces V, W. For δ, $\varepsilon \in \Gamma$ we have a canonical map $\mathsf{V}_\delta \otimes_{\mathsf{D}_0} \mathsf{W}_\varepsilon \to \mathsf{V} \otimes_{\mathsf{D}} \mathsf{W}$. This map is injective since any D_0-base of V_δ (resp. W_ε) is D-linearly independent in V (resp. W), see Prop. 2.5. Therefore, we may identify $\mathsf{V}_\delta \otimes_{\mathsf{D}_0} \mathsf{W}_\varepsilon$ with a subgroup of $\mathsf{V} \otimes_{\mathsf{D}} \mathsf{W}$, and we have for $\gamma \in \Gamma$

$$(\mathsf{V} \otimes_{\mathsf{D}} \mathsf{W})_\gamma = \sum_{\delta+\varepsilon=\gamma} (\mathsf{V}_\delta \otimes_{\mathsf{D}_0} \mathsf{W}_\varepsilon). \tag{2.8}$$

Note however that for $v_\delta \in \mathsf{V}_\delta$, $w_\varepsilon \in \mathsf{W}_\varepsilon$, and $d_\lambda \in \mathsf{D}_\lambda$, the elements

$$(v_\delta d_\lambda) \otimes w_\varepsilon \in \mathsf{V}_{\delta+\lambda} \otimes_{\mathsf{D}_0} \mathsf{W}_\varepsilon \quad \text{and} \quad v_\delta \otimes (d_\lambda w_\varepsilon) \in \mathsf{V}_\delta \otimes_{\mathsf{D}_0} \mathsf{W}_{\lambda+\varepsilon}$$

are identified in $\mathsf{V} \otimes_{\mathsf{D}} \mathsf{W}$. Thus, the sum on the right in (2.8) is not a direct sum.

Proposition 2.12. *Let* V *be a right graded vector space and* W *a left graded vector space over a graded division ring* D. *Let* $(\gamma_i)_{i \in I}$ *be a collection of representatives in* Γ *of the cosets of* $\Gamma_{\mathsf{V}} \cap \Gamma_{\mathsf{W}}$ *modulo* Γ_{D}, *so that*

$$\Gamma_{\mathsf{V}} \cap \Gamma_{\mathsf{W}} = \coprod_{i \in I} (\gamma_i + \Gamma_{\mathsf{D}}).$$

Then $\Gamma_{\mathsf{V}\otimes_{\mathsf{D}}\mathsf{W}} = \Gamma_{\mathsf{V}} + \Gamma_{\mathsf{W}}$ *and, for* $\delta \in \Gamma_{\mathsf{V}}$, $\varepsilon \in \Gamma_{\mathsf{W}}$,

$$(\mathsf{V} \otimes_{\mathsf{D}} \mathsf{W})_{\delta+\varepsilon} = \bigoplus_{i\in I} (\mathsf{V}_{\delta+\gamma_i} \otimes_{\mathsf{D}_0} \mathsf{W}_{\varepsilon-\gamma_i}).$$

Proof. The equality $\Gamma_{\mathsf{V}\otimes\mathsf{W}} = \Gamma_{\mathsf{V}} + \Gamma_{\mathsf{W}}$ is clear from (2.8). Now, fix $\delta \in \Gamma_{\mathsf{V}}$ and $\varepsilon \in \Gamma_{\mathsf{W}}$. If $\lambda \in \Gamma_{\mathsf{V}}$ and $\mu \in \Gamma_{\mathsf{W}}$ satisfy $\lambda + \mu = \delta + \varepsilon$, then $\lambda - \delta = \varepsilon - \mu \in \Gamma_{\mathsf{V}} \cap \Gamma_{\mathsf{W}}$; hence, we may find $i \in I$ and $d \in \mathsf{D}^\times$ such that

$$\lambda - \delta = \varepsilon - \mu = \gamma_i + deg(d).$$

For $v_\lambda \in \mathsf{V}_\lambda$ and $w_\mu \in \mathsf{W}_\mu$ we have in $\mathsf{V} \otimes_{\mathsf{D}} \mathsf{W}$

$$v_\lambda \otimes w_\mu = v_\lambda d^{-1} \otimes dw_\mu \in \mathsf{V}_{\delta+\gamma_i} \otimes_{\mathsf{D}_0} \mathsf{W}_{\varepsilon-\gamma_i}.$$

Thus, for all λ, $\mu \in \Gamma$ such that $\lambda + \mu = \delta + \varepsilon$, the image of $\mathsf{V}_\lambda \otimes_{\mathsf{D}_0} \mathsf{W}_\mu$ in $\mathsf{V} \otimes_{\mathsf{D}} \mathsf{W}$ lies in the image of $\sum_{i\in I} (\mathsf{V}_{\delta+\gamma_i} \otimes_{\mathsf{D}_0} \mathsf{W}_{\varepsilon-\gamma_i})$.

The latter sum is direct because for $i \in I$ the $\mathsf{V}_{\delta+\gamma_i}$'s lie in different summands of the canonical decomposition (2.2) of V. More explicitly, for every coset $\Lambda = \gamma + \Gamma_{\mathsf{D}}$ in Γ_{V} consider the projection

$$p_\Lambda \colon \mathsf{V} \longrightarrow \bigoplus_{\lambda\in\Lambda} \mathsf{V}_\lambda.$$

This map is D-linear; it yields a $Z(\mathsf{D})$-linear map

$$p_\Lambda \otimes id_{\mathsf{W}} \colon \mathsf{V} \otimes_{\mathsf{D}} \mathsf{W} \longrightarrow \big(\bigoplus_{\lambda\in\Lambda} \mathsf{V}_\lambda\big) \otimes_{\mathsf{D}} \mathsf{W}.$$

The restriction of this map to $\mathsf{V}_{\delta+\gamma_i} \otimes_{\mathsf{D}_0} \mathsf{W}_{\varepsilon-\gamma_i}$ is the identity if $\delta + \gamma_i \in \Lambda$; it is the 0 map otherwise. Therefore, if $\sum_{i\in I} x_i = 0$ with $x_i \in \mathsf{V}_{\delta+\gamma_i} \otimes_{\mathsf{D}_0} \mathsf{W}_{\varepsilon-\gamma_i}$ for $i \in I$, then by applying $p_{\delta+\gamma_j+\Gamma_{\mathsf{D}}} \otimes id_{\mathsf{W}}$ we obtain $x_j = 0$. □

2.2 Wedderburn structure theory

In this section, we develop the analogue for graded rings of the Wedderburn structure theory of semisimple algebras, with a view toward its application to the definition of gauges in the next chapter. It turns out that, under the appropriate restrictions to homogeneous ideals and graded modules, the whole theory can be carried out in the graded case. The arguments we use should make this point clear; they are slight variations on the most classical proofs (dealing with finite-dimensional algebras). The only special feature of the graded case is that the classification of simple graded algebras up to isomorphism involves a collection of elements in the grade group Γ; see Prop. 2.27.

Throughout the section, we fix a base graded field F with torsion-free abelian grade group Γ. The group Γ contains the degrees of all the graded structures we consider in this section. We let $\mathsf{F}^\natural$ denote F with its grading forgotten, i.e., $\mathsf{F}^\natural$ is the underlying ungraded ring of F. Proposition 2.3 shows that $\mathsf{F}^\natural$ is an integral domain; hence, we may consider its quotient field, for which we use the notation $q(\mathsf{F})$:

$$q(\mathsf{F}) = Quot(\mathsf{F}^\natural).$$

We start with the basic definitions, then we show in §2.2.2 that semisimple graded algebras are direct products of simple graded algebras. In §2.2.3, we show that simple graded algebras are algebras of endomorphisms of graded vector spaces over graded division rings, and we discuss tensor products of graded algebras and analogues of the Double Centralizer Theorem and the Skolem–Noether Theorem in §2.2.4.

2.2.1 Semisimple graded algebras and central quotients

Mimicking the classical (ungraded) theory, we define a *graded* F-*algebra* as a graded ring A with 1 that is also a graded F-vector space, in which the multiplication and the scalar multiplication are related by

$$(\lambda a)b = \lambda(ab) = a(\lambda b) \qquad \text{for } \lambda \in \mathsf{F} \text{ and } a,\, b \in \mathsf{A}.$$

We identify F with a graded subring of A by mapping $\lambda \in \mathsf{F}$ to $\lambda 1 \in \mathsf{A}$ and we say that the graded F-algebra A is *central* if its center $Z(\mathsf{A})$ is F. If A is also a graded division ring, we call it a *graded division algebra.* If A and B are graded F-algebras, we define a grading on the direct product $\mathsf{A} \times \mathsf{B}$ by

$$\mathsf{A} \times \mathsf{B} = \bigoplus_{\gamma \in \Gamma} (\mathsf{A}_\gamma \times \mathsf{B}_\gamma).$$

Homogeneous left (resp. right) ideals in a graded algebra A are simply the graded submodules of A for its structure of left (resp. right) module. A finite-dimensional graded algebra A is said to be *semisimple*[3] if $\{0\}$ is its only homogeneous two-sided nilpotent ideal. Then $\{0\}$ is also the only homogeneous left nilpotent ideal and the only homogeneous right nilpotent ideal because for any left (resp. right) ideal $\mathsf{J} \subseteq \mathsf{A}$ the two-sided ideal JA (resp. AJ) satisfies for every integer $n \geq 1$,

$$(\mathsf{JA})^n = \mathsf{J}(\mathsf{AJ})^{n-1}\mathsf{A} = \mathsf{J}^n\mathsf{A} \qquad (\text{resp. } (\mathsf{AJ})^n = \mathsf{A}(\mathsf{JA})^{n-1}\mathsf{J} = \mathsf{AJ}^n).$$

Clearly, a finite direct product of semisimple graded algebras is again graded semisimple.

When A has enough homogeneous units, we can verify the graded simplicity of A by examining A_0. For this, let

$$\Gamma_{\mathsf{A}}^{\times} = \{deg(a) \mid a \in \mathsf{A}^\times \text{ and } a \text{ is homogeneous}\} \subseteq \Gamma_{\mathsf{A}}.$$

[3] We sometimes call such a graded algebra *graded semisimple*, for emphasis.

Lemma 2.13. *Let* A *be a graded algebra over a graded field* F*. If* A_0 *is semisimple and* $\Gamma_{\mathsf{A}}^{\times} = \Gamma_{\mathsf{A}}$*, then* A *is graded semisimple.*

Proof. Suppose A is not graded semisimple. Then, there is a nonzero nilpotent homogeneous ideal N of A. Take $\gamma \in \Gamma_A$ with $\mathsf{N}_\gamma \neq \{0\}$. By hypothesis, there is a $u \in \mathsf{A}^\times \cap \mathsf{A}_\gamma$. Then, $\{0\} \subsetneq u^{-1}\mathsf{N}_\gamma \subseteq \mathsf{N}_0$. Hence, N_0 is a nonzero nilpotent ideal of A_0. This cannot occur since A_0 is assumed semisimple. Hence, A must be graded semisimple. □

The converse of Lemma 2.13 also holds, without the hypothesis on $\Gamma_{\mathsf{A}}^{\times}$. See Cor. 2.42 below.

A *simple graded* F*-algebra* is a finite-dimensional graded F-algebra $\mathsf{A} \neq \{0\}$ in which $\{0\}$ and A are the only homogeneous two-sided ideals. To prove that a graded algebra is simple, it suffices to extend scalars to the quotient field $q(\mathsf{F})$, as the next lemma shows:

Lemma 2.14. *Let* A *be a graded algebra over a graded field* F *and let*

$$q_{\mathsf{F}}(\mathsf{A}) = \mathsf{A}^\natural \otimes_{\mathsf{F}^\natural} q(\mathsf{F}),$$

which we consider as an algebra over $q(\mathsf{F})$*. We have*

$$[q_{\mathsf{F}}(\mathsf{A}){:}q(\mathsf{F})] = [\mathsf{A}{:}\mathsf{F}] \quad \textit{and} \quad Z\big(q_{\mathsf{F}}(\mathsf{A})\big) = Z(\mathsf{A})^\natural \otimes_{\mathsf{F}^\natural} q(\mathsf{F}) \quad \big(=: q_{\mathsf{F}}\big(Z(\mathsf{A})\big)\big).$$

Moreover, if $q_{\mathsf{F}}(\mathsf{A})$ *is simple, then* A *is graded simple. If* A *is a finite-dimensional graded division algebra, then* $q_{\mathsf{F}}(\mathsf{A})$ *is a division algebra.*

Note that the converse of these last statements also holds: see Prop. 2.28 below.

Proof. Every element of $q_{\mathsf{F}}(\mathsf{A})$ has the form $\sum_i a_i \otimes \lambda_i^{-1}$ for some $a_i \in \mathsf{A}$ and some nonzero $\lambda_i \in \mathsf{F}$. Reducing to the common denominator $\lambda = \prod_i \lambda_i$, we may rewrite the element in the form $a \otimes \lambda^{-1}$, where $a = \sum_i (a_i \prod_{j \neq i} \lambda_j)$. Therefore, it is easy to see that any F-base of A is a $q(\mathsf{F})$-base of $q_{\mathsf{F}}(\mathsf{A})$, and that the center of $q_{\mathsf{F}}(\mathsf{A})$ is $q_{\mathsf{F}}\big(Z(\mathsf{A})\big)$.

Now, let $\mathsf{I} \subseteq \mathsf{A}$ be a nonzero homogeneous two-sided ideal. By Cor. 2.6, there is a graded F-subspace $\mathsf{W} \subseteq \mathsf{A}$ such that $\mathsf{A} = \mathsf{I} \oplus \mathsf{W}$. The tensor product $\mathsf{I}^\natural \otimes_{\mathsf{F}^\natural} q(\mathsf{F})$ is a nonzero two-sided ideal of $q_{\mathsf{F}}(\mathsf{A})$, and we have

$$q_{\mathsf{F}}(\mathsf{A}) = \big(\mathsf{I}^\natural \otimes_{\mathsf{F}^\natural} q(\mathsf{F})\big) \oplus \big(\mathsf{W}^\natural \otimes_{\mathsf{F}^\natural} q(\mathsf{F})\big).$$

Therefore, if $q_{\mathsf{F}}(\mathsf{A})$ is simple we must have $\mathsf{I}^\natural \otimes_{\mathsf{F}^\natural} q(\mathsf{F}) = q_{\mathsf{F}}(\mathsf{A})$, hence $\mathsf{W}^\natural \otimes_{\mathsf{F}^\natural} q(\mathsf{F}) = \{0\}$, and therefore $W = \{0\}$. It follows that $\mathsf{I} = \mathsf{A}$. Thus, A is graded simple.

If A is a finite-dimensional graded division algebra, then Prop. 2.3 shows that $\mathsf{A}^\natural$ has no zero divisors. Then $q_{\mathsf{F}}(\mathsf{A})$ is a finite-dimensional $q(\mathsf{F})$-algebra without zero divisors; hence, it is a division algebra. □

The $q(\mathsf{F})$-algebra $q_{\mathsf{F}}(\mathsf{A}) = \mathsf{A}^\natural \otimes_{F^\natural} q(F)$ is less dependent on F than its definition suggests, as we see by relating it to the ring of central quotients of $\mathsf{A}^\natural$. Recall that if R is a commutative ring and $\mathcal{N}$ is its set of non-zero-divisors, then the *total quotient ring* of R is its localization

$$q(R) = R_{\mathcal{N}} = \{rs^{-1} \mid r \in R,\ s \in \mathcal{N}\}.$$

The set $\mathcal{N}$ is the largest multiplicatively closed subset S of $R \setminus \{0\}$ such that R embeds in the localization R_S. If B is a noncommutative ring, then the *ring of central quotients* of B is

$$q(B) = B \otimes_{Z(B)} q(Z(B)),$$

where $Z(B)$ is the center of B. The canonical map $B \to q(B)$, $b \mapsto b \otimes 1$, need not be injective, however, since a non-zero-divisor of $Z(B)$ could become a zero-divisor in B. (See Exercise 2.3 below for an example.) For a graded F-algebra A, we write $q(\mathsf{A})$ for $q(\mathsf{A}^\natural)$. The next lemma shows that for the graded F-algebras usually considered in this book, $q_{\mathsf{F}}(\mathsf{A})$ and $q(\mathsf{A})$ coincide.

Lemma 2.15. *Let* A *be a graded algebra over a graded field* F*. Then,* $\mathsf{A}^\natural \subseteq q_{\mathsf{F}}(\mathsf{A})$*. If* $Z(\mathsf{A})^\natural$ *is integral over* $\mathsf{F}^\natural$ *(e.g., if* $[\mathsf{A}{:}\mathsf{F}] < \infty$*), then*

$$q_{\mathsf{F}}(\mathsf{A}) = q(\mathsf{A}).$$

Proof. Let $\varphi\colon \mathsf{A}^\natural \to q_{\mathsf{F}}(\mathsf{A})$ be the canonical map given by $a \mapsto a \otimes 1$. If there is an $a \in \ker\varphi$ with $a \neq 0$, choose $s \in \mathsf{F} \setminus \{0\}$ with $sa = 0$. Let $\leq$ be some total ordering on the torsion-free abelian group Γ, and let s_γ (resp. a_δ) be the nonzero homogeneous component of s (resp. a) of lowest degree. Then $s_\gamma a_\delta$ must vanish, since it is a homogeneous component of sa. But $s_\gamma \in \mathsf{F}^\times$, so $a_\delta = 0$, a contradiction. Hence, φ is injective, and we identify $\mathsf{A}^\natural$ with its image $\varphi(\mathsf{A}^\natural)$ in $q_{\mathsf{F}}(\mathsf{A})$.

Let $\mathcal{N}$ be the set of non-zero-divisors of $Z(\mathsf{A})^\natural$. We have just shown that $\mathsf{F} \setminus \{0\} \subseteq \mathcal{N}$. Hence, there is a canonical monomorphism

$$\psi\colon Z(\mathsf{A})^\natural \otimes_{\mathsf{F}^\natural} q(\mathsf{F}) \longrightarrow q(Z(\mathsf{A}))$$

given by $z \otimes (bt^{-1}) \mapsto (zb)t^{-1}$ for all $z \in Z(\mathsf{A})$, $b \in \mathsf{F}$, $t \in \mathsf{F} \setminus \{0\}$. Suppose that $Z(\mathsf{A})^\natural$ is integral over $\mathsf{F}^\natural$. To see that ψ is surjective, take any $t \in \mathcal{N}$; let $f = c_n X^n + \ldots + c_0 \in \mathsf{F}[X]$ be a nonzero polynomial of minimal degree such that $f(t) = 0$. Then, $c_0 \neq 0$ as $c_n t^{n-1} + \ldots + c_1 \neq 0$ and t is not a zero divisor. Hence, $t^{-1} = -c_0^{-1}(c_n t^{n-1} + \ldots + c_1) \in \operatorname{im}\psi$. Since $\{t^{-1} \mid t \in \mathcal{N}\}$ generates the ring $q(Z(\mathsf{A}))$ over $Z(\mathsf{A})$, the map ψ is an isomorphism. Thus, $Z(\mathsf{A})^\natural \otimes_{\mathsf{F}^\natural} q(\mathsf{F}) \cong q(Z(\mathsf{A}))$, and hence,

$$\begin{aligned} q_{\mathsf{F}}(\mathsf{A}) = \mathsf{A}^\natural \otimes_{\mathsf{F}^\natural} q(\mathsf{F}) &\cong \mathsf{A}^\natural \otimes_{Z(\mathsf{A})^\natural} \big(Z(\mathsf{A})^\natural \otimes_{\mathsf{F}^\natural} q(\mathsf{F})\big) \\ &\cong \mathsf{A}^\natural \otimes_{Z(\mathsf{A})^\natural} q(Z(\mathsf{A})) = q(\mathsf{A}). \end{aligned}$$

We use these canonical isomorphisms to identify $q_{\mathsf{F}}(A)$ with $q(\mathsf{A})$. □

We conclude this subsection with examples of simple graded algebras. The easiest ones are obtained by scalar extension from F_0. For any F_0-algebra A, the F-algebra $A \otimes_{\mathsf{F}_0} \mathsf{F}$ has an obvious grading, for which $A \otimes_{\mathsf{F}_0} \mathsf{F}_\gamma$ is the homogeneous component of degree γ, for any $\gamma \in \Gamma$.

Proposition 2.16. *For any finite-dimensional simple algebra A over F_0, the graded F-algebra $A \otimes_{\mathsf{F}_0} \mathsf{F}$ is graded simple, and $Z(A \otimes_{\mathsf{F}_0} \mathsf{F}) = Z(A) \otimes_{\mathsf{F}_0} \mathsf{F}$. If A is a division algebra, then $A \otimes_{\mathsf{F}_0} \mathsf{F}$ is a graded division algebra.*

Proof. Suppose first A is a division algebra. Let $\gamma \in \Gamma$. The homogeneous component of degree γ in $A \otimes_{\mathsf{F}_0} \mathsf{F}$ is $A \otimes \lambda$ for any $\lambda \in \mathsf{F}^\times$ homogeneous of degree γ. It is clear that each nonzero element in this homogeneous component is invertible, hence $A \otimes_{\mathsf{F}_0} \mathsf{F}$ is a graded division algebra.

When A is assumed only to be simple (and finite-dimensional), the preceding argument applies to $Z(A)$, and shows that $Z(A) \otimes_{\mathsf{F}_0} \mathsf{F}$ is a graded field, hence $q(Z(A) \otimes_{\mathsf{F}_0} \mathsf{F})$ is a field. The equality $Z(A \otimes_{\mathsf{F}_0} \mathsf{F}) = Z(A) \otimes_{\mathsf{F}_0} \mathsf{F}$ is clear. Since $[A \otimes_{\mathsf{F}_0} \mathsf{F} : \mathsf{F}] = [A{:}\mathsf{F}_0] < \infty$, we have $q_{\mathsf{F}}(A \otimes_{\mathsf{F}_0} \mathsf{F}) = q(A \otimes_{\mathsf{F}_0} \mathsf{F})$ by Lemma 2.15, hence

$$q_{\mathsf{F}}(A \otimes_{\mathsf{F}_0} \mathsf{F}) = (A \otimes_{\mathsf{F}_0} \mathsf{F}) \otimes_{Z(A) \otimes_{\mathsf{F}_0} \mathsf{F}} q(Z(A) \otimes_{\mathsf{F}_0} \mathsf{F}) = A \otimes_{Z(A)} q(Z(A) \otimes_{\mathsf{F}_0} \mathsf{F}).$$

It follows that $q_{\mathsf{F}}(A \otimes_{\mathsf{F}_0} \mathsf{F})$ is simple (see Draxl [63, Cor. 3, p. 30] or Pierce [178, Prop. b, p. 226]). Therefore, Lemma 2.14 shows that $A \otimes_{\mathsf{F}_0} \mathsf{F}$ is graded simple. □

The following examples turn out to be typical of all simple graded F-algebras by Th. 2.26 below.

Proposition 2.17. *Let D be a (finite-dimensional) graded division algebra over F and let V be a finite-dimensional right graded D-vector space. The graded F-algebra $\mathsf{End}_{\mathsf{D}} \mathsf{V}$ is graded simple. Its center consists of scalar multiplications by elements in the center of D, so $Z(\mathsf{End}_{\mathsf{D}} \mathsf{V}) = Z(\mathsf{D})$.*

Proof. Using a homogeneous base of V, we may identify $\mathsf{End}_{\mathsf{D}} \mathsf{V}$ with a graded matrix ring (see Cor. 2.11), so $(\mathsf{End}_{\mathsf{D}} \mathsf{V})^\natural \cong M_n(\mathsf{D}^\natural)$ for some integer n. It follows that $q_{\mathsf{F}}(\mathsf{End}_{\mathsf{D}} \mathsf{V}) \cong M_n\big(q_{\mathsf{F}}(\mathsf{D})\big)$. The latter is a simple $q(\mathsf{F})$-algebra, since Lemma 2.14 shows that $q_{\mathsf{F}}(\mathsf{D})$ is a division algebra over $q(\mathsf{F})$. Therefore, it follows from Lemma 2.14 that $\mathsf{End}_{\mathsf{D}} \mathsf{V}$ is a simple graded algebra. The equality $Z(\mathsf{End}_{\mathsf{D}} \mathsf{V}) = Z(\mathsf{D})$ is easily proved using a matrix representation of $\mathsf{End}_{\mathsf{D}} \mathsf{V}$. □

The graded symbol algebras that we define next are explicit examples of central simple graded algebras; such symbol algebras will occur frequently in the sequel (see for example §7.2.2).

Definition 2.18. Let F be a graded field and $n \geq 2$ an integer such that F_0 contains a primitive n-th root of unity ω (in particular, *char* $\mathsf{F}_0 \nmid n$). For any homogeneous elements a, $b \in \mathsf{F}^\times$, consider the $q(\mathsf{F})$-algebra A generated by two elements i, j subject to the following relations:

$$i^n = a, \qquad j^n = b, \qquad ij = \omega ji.$$

This algebra is a *symbol algebra* as defined in Draxl[4] [63, §11] or Gille–Szamuely [84, p. 36]; it is a central simple algebra of degree n (i.e., dimension n^2) over $q(\mathsf{F})$, and we have

$$A = \bigoplus_{k,\ell=0}^{n-1} q(\mathsf{F})i^k j^\ell.$$

For this algebra we use the symbol algebra notation

$$A = \big(a, b/q(F)\big)_n \text{ or } \big(a, b/q(F)\big)_{\omega,n}.$$

Since i^n, $j^n \in \mathsf{F}$, the $\mathsf{F}^\natural$-submodule of A spanned by $(i^k j^\ell)_{k,\ell=0}^{n-1}$ is an $\mathsf{F}^\natural$-algebra. We provide it with a grading extending the grading on F by declaring i and j to be homogeneous of degree $\frac{1}{n}\, deg\, a$ and $\frac{1}{n}\, deg\, b$ respectively. Thus, for any $\gamma \in \Gamma$, we let

$$\mathsf{S}_\gamma = \textstyle\bigoplus_{k,\ell=0}^{n-1} \mathsf{F}_{\varepsilon(\gamma,k,\ell)} i^k j^\ell \quad \text{where } \varepsilon(\gamma,k,\ell) = \gamma - \frac{k}{n}\, deg\, a - \frac{\ell}{n}\, deg\, b,$$

and we define

$$\mathsf{S} = \bigoplus_{k,\ell=0}^{n-1} \mathsf{F} i^k j^\ell = \bigoplus_{\gamma\in\Gamma} \mathsf{S}_\gamma.$$

Easy calculations show that $\mathsf{S}_\gamma \cdot \mathsf{S}_\delta \subseteq \mathsf{S}_{\gamma+\delta}$ for all γ, $\delta \in \Gamma$; hence, S is a graded F-algebra in which S_γ is the homogeneous component of degree γ. Note that since the powers of i and j are units of S, we have

$$\Gamma_{\mathsf{S}} = \Gamma_{\mathsf{S}}^\times = \big\langle \tfrac{1}{n}\, deg\, a, \tfrac{1}{n}\, deg\, b \big\rangle + \Gamma_{\mathsf{F}}. \tag{2.9}$$

The graded algebra S is said to be a *graded symbol algebra* of degree n over F; we use the following notation:

$$\mathsf{S} = \big(a, b/\mathsf{F}\big)_n \text{ or } \big(a, b/\mathsf{F}\big)_{\omega,n}. \tag{2.10}$$

Graded symbol algebras of degree 2 are also called *graded quaternion algebras.*

Proposition 2.19. *Every graded symbol algebra of degree n over F is a central simple graded F-algebra of dimension n^2.*

Proof. It is clear from the construction of graded symbol algebras that extending scalars to the quotient field yields an (ungraded) symbol algebra over $q(\mathsf{F})$:

$$q_{\mathsf{F}}\big(\big(a, b/\mathsf{F}\big)_n\big) = \big(a, b/q(\mathsf{F})\big)_n.$$

Symbol algebras over $q(\mathsf{F})$ are known to be central simple: see Draxl [63, Th. 1, p. 78] or Gille–Szamuely [84, Cor. 2.5.5]. Therefore, the proposition follows from Lemma 2.14. □

As further examples of graded simple algebras, graded analogues of cyclic and crossed product algebras will be defined in §6.1.2.

[4] Draxl uses the term "power norm residue algebra."

2.2.2 The structure of semisimple graded algebras

Our goal in this subsection is to show that every semisimple graded algebra is a direct product of simple graded algebras; see Th. 2.23. Our arguments also yield information on graded modules over semisimple graded algebras. Defining a *simple graded module* over a graded algebra to be a graded module $\mathsf{M} \neq \{0\}$ in which $\{0\}$ and M are the only graded submodules, we show in Prop. 2.24 that every finitely generated graded module over a semisimple graded algebra is a direct sum of simple graded submodules.

Lemma 2.20. *Every minimal homogeneous left ideal* J *in a semisimple graded* F*-algebra* A *is generated by a homogeneous idempotent.*

Proof. Since $\mathsf{J}^2 \neq \{0\}$, there exists an $a \in \mathsf{J}$ such that $\mathsf{J}a \neq \{0\}$, and we may choose a to be homogeneous. Then $\mathsf{J}a$ is a homogeneous left ideal contained in J; hence $\mathsf{J}a = \mathsf{J}$. Therefore, there exists an $e \in \mathsf{J}$ such that $ea = a$. By substituting for e its homogeneous component of degree 0, we may assume that e is homogeneous of degree 0. By multiplying by e we obtain $e^2a = ea$; hence

$$e^2 - e \in \{x \in \mathsf{J} \mid xa = 0\}.$$

The set on the right is a homogeneous left ideal contained in J; it is not J since $\mathsf{J}a \neq \{0\}$, hence it is $\{0\}$ and therefore $e^2 - e = 0$; i.e., e is an idempotent. Since $e \in \mathsf{J}$ we have $\mathsf{J} = \mathsf{A}e$. □

Note that in a graded algebra every nonzero homogeneous idempotent has degree 0.

Proposition 2.21. *Let* J *be a nonzero homogeneous left ideal in a semisimple graded* F*-algebra* A*. There are nonzero homogeneous idempotents* $e_1, \dots, e_r$ *such that*

- $e_1, \dots, e_r$ *are pairwise orthogonal, i.e.,* $e_ie_j = 0$ *for* $i \neq j$*;*
- $\mathsf{A}e_i$ *is a minimal homogeneous left ideal for all* $i = 1, \dots, r$*, and*
- $\mathsf{J} = \mathsf{A}e_1 \oplus \dots \oplus \mathsf{A}e_r = \mathsf{A}(e_1 + \dots + e_r)$*.*

In particular, $e_1 + \dots + e_r$ *is an idempotent that generates* J*.*

Proof. We argue by induction on $dim_{\mathsf{F}}\,\mathsf{J}$. If J is minimal, the proposition readily follows from the lemma. Otherwise, let $\mathsf{J}_1 \subset \mathsf{J}$ be a minimal homogeneous left ideal. Lemma 2.20 yields a homogeneous idempotent e_1 such that $\mathsf{J}_1 = \mathsf{A}e_1$. We have $e_1 = e_1^2 \in \mathsf{J}e_1$; hence $\mathsf{J}e_1$ is a nonzero homogeneous left ideal contained in J_1. Therefore, $\mathsf{J}e_1 = \mathsf{J}_1$. We have

$$\mathsf{J} = \mathsf{J}e_1 \oplus \mathsf{J}(1 - e_1) = \mathsf{A}e_1 \oplus \mathsf{J}(1 - e_1) \tag{2.11}$$

and $\mathsf{J}(1 - e_1)$ is a homogeneous left ideal since $1 - e_1$ is homogeneous. By induction, there are pairwise orthogonal nonzero homogeneous idempotents

$e_2', \ldots, e_r'$ such that $\mathsf{A}e_i'$ is a minimal homogeneous left ideal for $i = 2, \ldots, r$, and

$$\mathsf{J}(1 - e_1) = \mathsf{A}e_2' \oplus \ldots \oplus \mathsf{A}e_r'. \tag{2.12}$$

Since $e_i' \in \mathsf{J}(1 - e_1)$, we have

$$e_i' e_1 = 0 \qquad \text{for } i = 2, \ldots, r. \tag{2.13}$$

For $i = 2, \ldots, r$, let $e_i = (1 - e_1)e_i'$. Then $e_i \in \mathsf{A}e_i'$ and, by (2.13)

$$e_i' e_i = e_i'(1 - e_1)e_i' = e_i'^2 = e_i' \qquad \text{for } i = 2, \ldots, r.$$

Hence $e_i' \in \mathsf{A}e_i$ and it follows that $\mathsf{A}e_i = \mathsf{A}e_i'$ for $i = 2, \ldots, r$. Therefore, by (2.11) and (2.12), we have

$$\mathsf{J} = \mathsf{A}e_1 \oplus \mathsf{A}e_2 \oplus \ldots \oplus \mathsf{A}e_r.$$

Moreover, for $i = 2, \ldots, r$ we have $e_i e_1 = 0$ by (2.13), and

$$e_1 e_i = e_1(1 - e_1)e_i' = 0.$$

Using (2.13) again, we have also for $i, j \in \{2, \ldots, r\}$ with $i \neq j$

$$e_i e_j = (1 - e_1)e_i'(1 - e_1)e_j' = (1 - e_1)e_i' e_j' = 0.$$

Finally, e_i is an idempotent for $i = 2, \ldots, r$ since

$$e_i^2 = (1 - e_1)e_i'(1 - e_1)e_i' = (1 - e_1)e_i'^2 = (1 - e_1)e_i'.$$

Thus, $e_1, \ldots, e_r$ satisfy all the requirements, and it remains only to show that J is generated by $e_1 + \ldots + e_r$.

To complete the proof, observe that

$$e_i = e_i(e_1 + \ldots + e_r) \in \mathsf{A}(e_1 + \ldots + e_r) \qquad \text{for } i = 1, \ldots, r,$$

hence

$$\mathsf{A}e_1 \oplus \ldots \oplus \mathsf{A}e_r \subseteq \mathsf{A}(e_1 + \ldots + e_r).$$

The reverse inclusion is obvious, and it follows that $\mathsf{J} = \mathsf{A}(e_1 + \ldots + e_r)$. □

Corollary 2.22. *Every two-sided homogeneous ideal* J *in a semisimple graded algebra* A *is generated by a homogeneous central idempotent.*

Proof. By Prop. 2.21 we have $\mathsf{J} = \mathsf{A}e$ for some homogeneous idempotent $e \in \mathsf{J}$. Since J is two-sided, we have $ex \in \mathsf{J}$ for all $x \in \mathsf{A}$, hence $ex = exe$. It follows that $(1 - e)\mathsf{A}e$ is a homogeneous right ideal of A. This ideal satisfies $((1 - e)\mathsf{A}e)^2 = \{0\}$ since e is idempotent, hence $(1 - e)\mathsf{A}e = \{0\}$ since A is graded semisimple. Therefore, we have $(1 - e)xe = 0$ for all $x \in \mathsf{A}$, hence $xe = exe$. Thus, for all $x \in \mathsf{A}$ we have

$$xe = exe = ex,$$

which shows that e is a central idempotent. □

Theorem 2.23. *Every semisimple graded algebra* A *is a direct product of simple graded algebras, which are uniquely determined. Its center* $Z(\mathsf{A})$ *is a direct product of graded fields.*

Proof. The center of a simple graded algebra is a graded field, since any nonzero central homogeneous element that is not invertible generates a nontrivial two-sided ideal. Therefore, the theorem is clear if A is graded simple. If it is not, Cor. 2.22 yields a homogeneous central idempotent $e \neq 0, 1$. Then

$$\mathsf{A} = \mathsf{A}e \oplus \mathsf{A}(1-e) \cong_g \mathsf{A}e \times \mathsf{A}(1-e),$$

and $\mathsf{A}e$, $\mathsf{A}(1-e)$ are semisimple graded algebras. Arguing by induction on dimension, we may assume that each of these algebras is a direct product of simple graded algebras, hence A also is such a product. If

$$\mathsf{A} \cong_g \mathsf{A}_1 \times \ldots \times \mathsf{A}_n$$

with $\mathsf{A}_1, \ldots, \mathsf{A}_n$ simple graded algebras, then the center $Z(\mathsf{A})$ satisfies

$$Z(\mathsf{A}) \cong_g Z(\mathsf{A}_1) \times \ldots \times Z(\mathsf{A}_n).$$

Since each $Z(\mathsf{A}_i)$ is a graded field, it follows that $Z(\mathsf{A})$ is a direct product of graded fields. Moreover, the simple graded components $\mathsf{A}_1, \ldots, \mathsf{A}_n$ are isomorphic to $\mathsf{A}e_1, \ldots, \mathsf{A}e_n$ for $e_1, \ldots, e_n$ the primitive idempotents of $Z(\mathsf{A})$, hence they are uniquely determined. □

By applying Prop. 2.21 to the left ideal $\mathsf{J} = \mathsf{A}$, we obtain a decomposition of A into a direct sum of minimal homogeneous left ideals

$$\mathsf{A} = \mathsf{A}e_1 \oplus \ldots \oplus \mathsf{A}e_r. \tag{2.14}$$

There is a similar decomposition for every finitely generated left graded A-module, as the next proposition shows.

Proposition 2.24. *Every finitely generated left graded module over a semisimple graded algebra* A *is a direct sum of simple left graded* A*-submodules.*

Proof. Let $g_1, \ldots, g_n$ be homogeneous generators of some left graded A-module M. From (2.14) it follows that

$$\mathsf{M} = \sum_{i=1}^{r} \sum_{j=1}^{n} \mathsf{A}e_i g_j.$$

Each summand $\mathsf{A}e_i g_j$ is either $\{0\}$ or a simple graded submodule of M since for any graded submodule $\mathsf{N} \subset \mathsf{A}e_i g_j$ the set $\{x \in \mathsf{A}e_i \mid x g_j \in \mathsf{N}\}$ is a homogeneous left ideal contained in the minimal homogeneous left ideal $\mathsf{A}e_i$. Therefore, M is a sum of finitely many simple graded submodules $\mathsf{M} = \sum_{i \in I} \mathsf{M}_i$. Pick a maximal subset $J \subseteq I$ such that the sum $\sum_{j \in J} \mathsf{M}_j$ is direct. For each $i \in I \setminus J$ we have

$$\mathsf{M}_i \cap \Big(\bigoplus_{j \in J} \mathsf{M}_j\Big) \neq \{0\},$$

otherwise J would not be maximal. But this intersection is a graded submodule of the simple graded module M_i, hence it equals M_i; therefore, $\mathsf{M}_i \subseteq \bigoplus_{j\in J} \mathsf{M}_j$. Thus, it follows that $\sum_{i\in I} \mathsf{M}_i = \bigoplus_{j\in J} \mathsf{M}_j$, hence M is a direct sum of simple graded submodules. □

2.2.3 The Wedderburn Theorem for simple graded algebras

We start our discussion of the graded analogue of Wedderburn's Theorem with the graded version of Schur's Lemma:

Lemma 2.25. *If* M *and* N *are simple graded modules over a graded algebra* A*, then every graded homomorphism* $f\colon \mathsf{M} \to \mathsf{N}$ *is either* 0 *or an isomorphism. In particular,* $\mathsf{End}_{\mathsf{A}} \mathsf{M}$ *is a graded division ring for the grading induced by the grading on* M *(see Prop. 2.8).*

Proof. This readily follows from the observation that the kernel and the image of any graded A-module homomorphism $\mathsf{M} \to \mathsf{N}$ are graded submodules of M and N respectively. □

Now, let A be a simple graded algebra over F. Since $[\mathsf{A}{:}\mathsf{F}]$ is finite, there is a minimal nonzero homogeneous left ideal $\mathsf{J} \subset \mathsf{A}$. Let $\mathsf{D} = \mathsf{End}_{\mathsf{A}} \mathsf{J}$ (acting on J on the right). Since J is a finite-dimensional graded F-vector space, the F-algebra D is graded and finite-dimensional. Moreover, J carries a right D-module structure, and multiplication on the left by elements of A defines a canonical graded F-algebra homomorphism

$$\rho\colon \mathsf{A} \longrightarrow \mathsf{End}_{\mathsf{D}} \mathsf{J}.$$

Theorem 2.26. *The* F*-algebra* D *is a graded division* F*-algebra and the map* ρ *is an isomorphism of graded* F*-algebras. So,*

$$\mathsf{A} \cong_g \mathsf{End}_{\mathsf{D}} \mathsf{J}.$$

Moreover, every simple left graded A*-module* S *is isomorphic to some shift* $\mathsf{J}(\gamma)$ *of* J *for some* $\gamma \in \Gamma$*, and* $\mathsf{End}_{\mathsf{A}} \mathsf{S} \cong_g \mathsf{D}$ *as graded* F*-algebras.*

As a consequence, the graded division algebra D is uniquely determined by A up to graded isomorphism, independent of the choice of J. (This is also shown in Prop. 2.27 below.) This D is called the *graded division algebra associated to* A.

Proof. Since J is a simple graded module, D is a graded division algebra by the graded analogue of Schur's Lemma, see Lemma 2.25. Therefore, J is a graded D-vector space. If $a \in \mathsf{A}$, $x \in \mathsf{J}$, and $f \in \mathsf{D}$ are homogeneous, then by the definition of $deg\, f$ we have whenever $ax^f \neq 0$,

$$deg(ax^f) = deg\, a + deg\, x + deg\, f.$$

Thus, the map ρ is a homomorphism of graded algebras. It is therefore injective since A is graded simple. To show that ρ is surjective, it suffices to show that its image is a left ideal in $\mathsf{End_D\,J}$, since this image contains the identity. Since A is graded simple and $\mathsf{J}\cdot\mathsf{A}$ is a two-sided homogeneous ideal in A, we have $\mathsf{J}\cdot\mathsf{A}=\mathsf{A}$; hence $\rho(\mathsf{J})\cdot\rho(\mathsf{A})=\rho(\mathsf{A})$ and therefore

$$(\mathsf{End_D\,J})\cdot\rho(\mathsf{A}) \;=\; (\mathsf{End_D\,J})\cdot\rho(\mathsf{J})\cdot\rho(\mathsf{A}).$$

We claim that $(\mathsf{End_D\,J})\cdot\rho(\mathsf{J})\subset\rho(\mathsf{J})$; it then follows that

$$(\mathsf{End_D\,J})\cdot\rho(\mathsf{A}) \;\subseteq\; \rho(\mathsf{J})\cdot\rho(\mathsf{A}) \;\subseteq\; \rho(\mathsf{A}),$$

which shows that $\rho(\mathsf{A})$ is a left ideal in $\mathsf{End_D\,J}$. Thus, to prove the claim it suffices to show that ρ is an isomorphism. Note that for x, $y\in\mathsf{J}$ we have $xy\in\mathsf{J}$, hence multiplication on the right by y defines an element in $\mathsf{D}=\mathsf{End_A\,J}$. Therefore, for $g\in\mathsf{End_D\,J}$ we have $g(xy)=g(x)y$, which means that

$$g\circ\rho(x) \;=\; \rho\big(g(x)\big)\in\rho(\mathsf{J}).$$

The claim is thus proved.

Now, let S be a simple left graded A-module. The set $\{x\in\mathsf{A}\mid x\mathsf{S}=\{0\}\}$ is a two-sided homogeneous ideal of A, hence it is $\{0\}$. Therefore, $\mathsf{J}\cdot\mathsf{S}\neq\{0\}$ and we may find a homogeneous element $s\in\mathsf{S}$ such that $\mathsf{J}\cdot s\neq\{0\}$. The map $x\mapsto xs$ is a nonzero homomorphism of left graded A-modules $\mathsf{J}\to\mathsf{S}$. Since J and S are graded simple, this homomorphism is bijective, and it identifies S with the shift $\mathsf{J}(-\mathit{deg}\,s)$. Since $\mathsf{End_A}\big(\mathsf{J}(\gamma)\big)=\mathsf{End_A\,J}$ for all $\gamma\in\Gamma$, we have $\mathsf{End_A\,S}\cong_g\mathsf{D}$. □

Another way to view a central simple graded algebra A is as a matrix ring with a shifted grading. Using the notation of Th. 2.26, we fix a decomposition of the right graded D-vector space J into 1-dimensional vector spaces as in Cor. 2.10:

$$\mathsf{J}\cong_g\mathsf{D}(\delta_1)\oplus\ldots\oplus\mathsf{D}(\delta_n). \tag{2.15}$$

By Th. 2.26 and Prop. 2.9, we have a matrix representation of A as in Cor. 2.11:

$$\mathsf{A}\cong_g\mathsf{End_D\,J}\cong_g\mathsf{M}_n(\mathsf{D})(\delta_1,\ldots,\delta_n). \tag{2.16}$$

If we let $\mathsf{A}^\natural$, $\mathsf{D}^\natural$ denote the F-algebras A and D with their grading forgotten, we thus have $\mathsf{A}^\natural\cong M_n(\mathsf{D}^\natural)$.

With the notation above, we have:

Proposition 2.27. *Let* V *be a finite-dimensional right graded vector space over some graded division* F*-algebra* E *and let* A *be as in* (2.16)*. There is a graded algebra isomorphism* $\mathsf{A}\cong_g\mathsf{End_E\,V}$ *if and only if the following conditions hold:*

(i) $\mathsf{E}\cong_g\mathsf{D}$, *and*
(ii) *there exists a* $\gamma\in\Gamma$ *such that* $\mathsf{V}\cong_g\mathsf{E}(\gamma+\delta_1)\oplus\ldots\oplus\mathsf{E}(\gamma+\delta_n)$.

Proof. Suppose there is an isomorphism $\mathsf{A} \cong_g \mathsf{End}_\mathsf{E}\,\mathsf{V}$. We use it as an identification. The grading on V gives it a left graded A-module structure since by definition of the grading on $\mathsf{End}_\mathsf{E}\,\mathsf{V}$ we have for all homogeneous elements $a \in \mathsf{A}$, $v \in \mathsf{V}$ such that $a(v) \neq 0$

$$deg\big(a(v)\big) \;=\; deg\, a + deg\, v.$$

It is clear that V is a simple graded A-module since for every nonzero homogeneous element $v \in \mathsf{V}$ we have $\mathsf{A} \cdot v = \mathsf{V}$. Therefore, Th. 2.26 shows that there is an isomorphism of graded A-modules $\mathsf{V} \cong_g \mathsf{J}(\gamma)$ for some $\gamma \in \Gamma$, and an isomorphism of graded F-algebras $\mathsf{D} \cong_g \mathsf{End}_\mathsf{A}\,\mathsf{V}$. On the other hand, we claim that $\mathsf{End}_\mathsf{A}\,\mathsf{V}$ consists of scalar multiplications by elements in E, so $\mathsf{D} \cong_g \mathsf{E}$. To prove the claim, choose a homogeneous E-base $(v_i)_{i=1}^n$ of V, and for i, $j = 1, \ldots, n$ let $\varepsilon_{ij} \in \mathsf{A}$ be the linear transformation of V satisfying

$$\varepsilon_{ij}(v_k) \;=\; \begin{cases} v_i & \text{if } k = j, \\ 0 & \text{if } k \neq j. \end{cases}$$

Suppose $g \in \mathsf{End}_\mathsf{A}\,\mathsf{V}$ and let

$$v_j^g \;=\; \sum_{i=1}^n v_i e_{ij} \qquad \text{for } j = 1, \ldots, n, \text{ with } e_{ij} \in \mathsf{E}.$$

For k, $\ell = 1, \ldots, n$ we have

$$\varepsilon_{k\ell}(v_j^g) \;=\; v_k e_{\ell j} \qquad \text{and} \qquad \varepsilon_{k\ell}(v_j)^g \;=\; \begin{cases} v_k^g & \text{if } \ell = j, \\ 0 & \text{if } \ell \neq j. \end{cases}$$

But $\varepsilon_{k\ell}(v_j^g) = \varepsilon_{k\ell}(v_j)^g$ since g is A-linear, hence

$$e_{\ell j} \;=\; 0 \text{ if } \ell \neq j \quad \text{and} \quad v_k^g \;=\; v_k e_{jj} \;\; \text{for all } j,\, k = 1, \ldots, n.$$

Therefore, $e_{11} = \ldots = e_{nn}$, and g is scalar multiplication by e_{11}, which proves the claim that $\mathsf{End}_\mathsf{A}\,\mathsf{V} = \mathsf{E}$, hence $\mathsf{D} \cong_g \mathsf{E}$.

By counting dimensions over F, we see that $dim_\mathsf{E}\,\mathsf{V} = n$, and condition (ii) follows from the isomorphism $\mathsf{V} \cong_g \mathsf{J}(\gamma)$. Thus, conditions (i) and (ii) hold.

Conversely, suppose (i) and (ii) hold. Since $(\gamma + \delta_i) - (\gamma + \delta_j) = \delta_i - \delta_j$, by Prop. 2.9 the matrix representation of $\mathsf{End}_\mathsf{E}\,\mathsf{V}$ is

$$\mathsf{End}_\mathsf{E}\,\mathsf{V} \;\cong_g\; \mathsf{M}_n(\mathsf{E})(\delta_1, \ldots, \delta_n).$$

In view of (2.16), it follows that any isomorphism $\mathsf{D} \cong_g \mathsf{E}$ induces an isomorphism $\mathsf{A} \cong_g \mathsf{End}_\mathsf{E}\,\mathsf{V}$. □

As a consequence of Th. 2.26, we can now prove the converse of the last statement of Lemma 2.14, relating properties of a finite-dimensional graded F-algebra A to properties of its associated $q(\mathsf{F})$-algebra $q(\mathsf{A}) = q_\mathsf{F}(\mathsf{A}) = \mathsf{A}^\natural \otimes_{\mathsf{F}^\natural} q(\mathsf{F})$ (see Lemma 2.15).

Proposition 2.28. *Let* A *be a finite-dimensional graded* F*-algebra.*

(i) A *is a graded division algebra if and only if* $q(\mathsf{A})$ *is a division algebra.*
(ii) A *is a simple graded algebra if and only if* $q(\mathsf{A})$ *is a simple algebra.*
(iii) A *is a semisimple graded algebra if and only if* $q(\mathsf{A})$ *is a semisimple algebra.*

Proof. Lemma 2.14 shows that $q(\mathsf{A})$ is finite-dimensional; it also establishes the "only if" part of (i) and the "if" part of (ii). If $q(\mathsf{A})$ is a division algebra, the "if" part of (ii) shows that A is graded simple. Moreover, $\mathsf{A}^\natural$ has no zero divisors, so in a representation of A as in (2.16) above we must have $n = 1$, hence A is a graded division algebra. This completes the proof of (i).

For (ii), it remains only to prove the "only if" part. Assume A is a central simple graded algebra, and fix a representation as in (2.16) above:

$$\mathsf{A} \cong_g \mathsf{M}_n(\mathsf{D})(\delta_1, \ldots, \delta_n)$$

where D is a graded division algebra. Then,

$$q(\mathsf{A}) \cong M_n\big(q(\mathsf{D})\big).$$

By (i), we know that $q(\mathsf{D})$ is a division algebra; hence, $q(\mathsf{A})$ is a simple algebra.

For (iii), first assume that A is semisimple. Then, by Th. 2.23,

$$\mathsf{A} = \mathsf{B}_1 \times \ldots \times \mathsf{B}_n$$

for some simple graded algebras B_i. Hence,

$$q(\mathsf{A}) \cong q(\mathsf{B}_1) \times \ldots \times q(\mathsf{B}_n),$$

and each $q(\mathsf{B}_i)$ is simple by (ii). Thus, $q(\mathsf{A})$ is semisimple. On the other hand, if A is not semisimple, then it has a nonzero nilpotent homogeneous ideal N. Then, $\mathsf{N}^\natural \otimes_{\mathsf{F}^\natural} q(\mathsf{F})$ is a nonzero nilpotent ideal of $q(\mathsf{A})$. Hence, $q(\mathsf{A})$ is not semisimple. □

Corollary 2.29. *If* A *is a central simple graded* F*-algebra, then* $[\mathsf{A}{:}\mathsf{F}] = d^2$ *for some integer* d.

Proof. By Lemma 2.14 and Prop. 2.28(ii), the $q(\mathsf{F})$-algebra $q(\mathsf{A}) = q_{\mathsf{F}}(\mathsf{A})$ is central simple, hence there is an integer d such that $[q(\mathsf{A}){:}q(\mathsf{F})] = d^2$. (The integer d is the degree of the central simple $q(\mathsf{F})$-algebra $q(\mathsf{A})$.) By Lemma 2.14, we also have $d^2 = [\mathsf{A}{:}\mathsf{F}]$. □

The integer $d \geq 1$ as in Cor. 2.29 is called the *degree* of the central simple graded algebra A. We use the notation *deg* A for the degree of A; thus

$$deg\, \mathsf{A} = deg\, q(\mathsf{A}).$$

Note that this definition is consistent with the definition of a graded symbol algebra of degree n, see Prop. 2.19. We also define the (*Schur*) *index* of a central simple graded algebra A as the degree of the division algebra D associated to A by Wedderburn's Theorem. Thus (see Th. 2.26),

$$ind\, \mathsf{A} = deg\, \mathsf{D} \quad (= deg\, q(\mathsf{D}) = ind\, q(\mathsf{A})).$$

2.2.4 Centralizers and simple subalgebras

For arbitrary graded F-algebras A, B the tensor product $\mathsf{A} \otimes_{\mathsf{F}} \mathsf{B}$ carries a grading defined in §2.1.4, which is compatible with the multiplication. Thus, $\mathsf{A} \otimes_{\mathsf{F}} \mathsf{B}$ is a graded F-algebra. It contains as graded subalgebras $\mathsf{A} \otimes 1$ and $1 \otimes \mathsf{B}$, which we identify with A and B. In this subsection, we determine when a tensor product of semisimple graded algebras is graded semisimple. We then prove graded analogues of two essential tools for the study of subalgebras of central simple algebras: the Double Centralizer Theorem and the Skolem–Noether Theorem. We start with some general remarks on centralizers.

If $\mathsf{A}' \subseteq \mathsf{A}$ is a graded subalgebra, we let $C_{\mathsf{A}}(\mathsf{A}')$ denote the centralizer of A' in A, i.e.,

$$C_{\mathsf{A}}(\mathsf{A}') = \{a \in \mathsf{A} \mid aa' = a'a \text{ for all } a' \in \mathsf{A}'\}.$$

It is a graded subalgebra of A.

Proposition 2.30. *If $\mathsf{A}' \subseteq \mathsf{A}$ and $\mathsf{B}' \subseteq \mathsf{B}$ are graded subalgebras of arbitrary graded F-algebras A, B, then*

$$C_{\mathsf{A}\otimes_{\mathsf{F}}\mathsf{B}}(\mathsf{A}' \otimes_{\mathsf{F}} \mathsf{B}') = C_{\mathsf{A}}(\mathsf{A}') \otimes_{\mathsf{F}} C_{\mathsf{B}}(\mathsf{B}').$$

Proof. Let $x \in C_{\mathsf{A}\otimes_{\mathsf{F}}\mathsf{B}}(\mathsf{A}' \otimes_{\mathsf{F}} \mathsf{B}')$. For some F-base $(a_i)_{i\in I}$ of A, write $x = \sum_{i\in I} a_i \otimes x_i$ with each $x_i \in \mathsf{B}$. Since x centralizes $1 \otimes \mathsf{B}'$, we must have $x_i \in C_{\mathsf{B}}(\mathsf{B}')$ for all $i \in I$. Rewriting $x = \sum_{j\in J} u_j \otimes b_j$ where $(b_j)_{j\in J}$ is an F-base of $C_{\mathsf{B}}(\mathsf{B}')$, we have $u_j \in C_{\mathsf{A}}(\mathsf{A}')$ for all $j \in J$ since x centralizes $\mathsf{A}' \otimes 1$. Therefore, x lies in $C_{\mathsf{A}}(\mathsf{A}') \otimes_{\mathsf{F}} C_{\mathsf{B}}(\mathsf{B}')$, and we have proved $C_{\mathsf{A}\otimes_{\mathsf{F}}\mathsf{B}}(\mathsf{A}' \otimes_{\mathsf{F}} \mathsf{B}') \subseteq C_{\mathsf{A}}(\mathsf{A}') \otimes_{\mathsf{F}} C_{\mathsf{B}}(\mathsf{B}')$. The reverse inclusion is obvious. □

The particular case where $\mathsf{A}' = \mathsf{A}$ and $\mathsf{B}' = \mathsf{B}$ determines the center of $\mathsf{A} \otimes_{\mathsf{F}} \mathsf{B}$:

Corollary 2.31. $Z(\mathsf{A} \otimes_{\mathsf{F}} \mathsf{B}) = Z(\mathsf{A}) \otimes_{\mathsf{F}} Z(\mathsf{B})$.

We now study the semisimplicity of tensor products, in the special case where one of the factors is graded simple with center F.

Proposition 2.32. *Let A be a central simple graded F-algebra, and let B be a (possibly infinite-dimensional) graded F-algebra with $Z(\mathsf{B})$ integral over F. The graded $Z(\mathsf{B})$-algebra $\mathsf{A} \otimes_{\mathsf{F}} \mathsf{B}$ is graded simple (resp. graded semisimple) if and only if B is a graded simple (resp. graded semisimple) $Z(\mathsf{B})$-algebra.*

Proof. If B is not graded simple, then any nontrivial two-sided homogeneous ideal $\mathsf{J} \subseteq \mathsf{B}$ yields a nontrivial two-sided homogeneous ideal $\mathsf{A} \otimes_{\mathsf{F}} \mathsf{J} \subseteq \mathsf{A} \otimes_{\mathsf{F}} \mathsf{B}$. Similarly, if B is not graded semisimple, we may take for J a nontrivial homogeneous nilpotent ideal in B and obtain a nontrivial homogeneous nilpotent ideal in $\mathsf{A} \otimes_{\mathsf{F}} \mathsf{B}$. The "only if" part of the proposition is thus clear.

Now, suppose B is graded simple, so finite-dimensional over the graded field $Z(\mathsf{B})$. Then $q(\mathsf{B})$ is simple by Prop. 2.28; similarly, $q(\mathsf{A})$ is central simple over $q(\mathsf{F})$. Hence, $q(\mathsf{A}) \otimes_{q(\mathsf{F})} q(\mathsf{B})$ is a simple algebra (see Draxl [63, Cor. 3, p. 30] or Pierce [178, Prop. b, p. 226]). Now, $Z(\mathsf{A} \otimes_{\mathsf{F}} \mathsf{B}) = \mathsf{F} \otimes_{\mathsf{F}} Z(\mathsf{B}) = Z(\mathsf{B})$, which is integral over F by hypothesis. Hence, by Lemma 2.15,

$$q(\mathsf{A} \otimes_{\mathsf{F}} \mathsf{B}) = q_{\mathsf{F}}(\mathsf{A} \otimes_{\mathsf{F}} \mathsf{B}) = q_{\mathsf{F}}(A) \otimes_{q(\mathsf{F})} q_{\mathsf{F}}(\mathsf{B}) = q(\mathsf{A}) \otimes_{q(\mathsf{F})} q(\mathsf{B}).$$

It therefore follows from Prop. 2.28 (or Lemma 2.14) that $\mathsf{A} \otimes_{\mathsf{F}} \mathsf{B}$ is graded simple.

If B is graded semisimple, we have $\mathsf{B} \cong_g \mathsf{B}_1 \times \ldots \times \mathsf{B}_r$ for some simple graded F-algebras $\mathsf{B}_1, \ldots, \mathsf{B}_r$ by Th. 2.23. Then,

$$\mathsf{A} \otimes_{\mathsf{F}} \mathsf{B} \cong_g (\mathsf{A} \otimes_{\mathsf{F}} \mathsf{B}_1) \times \ldots \times (\mathsf{A} \otimes_{\mathsf{F}} \mathsf{B}_r).$$

Each direct factor is graded simple, hence $\mathsf{A} \otimes_{\mathsf{F}} \mathsf{B}$ is graded semisimple. □

As a corollary, we obtain a criterion for the graded semisimplicity of the tensor product of semisimple graded algebras:

Corollary 2.33. *Let A, B be semisimple graded F-algebras. The graded algebra $\mathsf{A} \otimes_{\mathsf{F}} \mathsf{B}$ is graded semisimple if and only if $Z(\mathsf{A}) \otimes_{\mathsf{F}} Z(\mathsf{B})$ is graded semisimple.*

Proof. Suppose first A and B are graded simple. Since

$$\mathsf{A} \otimes_{\mathsf{F}} \mathsf{B} = \mathsf{A} \otimes_{Z(\mathsf{A})} (Z(\mathsf{A}) \otimes_{\mathsf{F}} \mathsf{B}),$$

Prop. 2.32 shows that $\mathsf{A} \otimes_{\mathsf{F}} \mathsf{B}$ is graded semisimple if and only if $Z(\mathsf{A}) \otimes_{\mathsf{F}} \mathsf{B}$ is semisimple. But

$$Z(\mathsf{A}) \otimes_{\mathsf{F}} \mathsf{B} = \big(Z(\mathsf{A}) \otimes_{\mathsf{F}} Z(\mathsf{B})\big) \otimes_{Z(\mathsf{B})} \mathsf{B};$$

so, one more application of Prop. 2.32 shows that this condition holds if and only if $Z(\mathsf{A}) \otimes_{\mathsf{F}} Z(\mathsf{B})$ is graded semisimple. The corollary is thus proved if A and B are simple.

In the general case of semisimple graded algebras A, B, we use Th. 2.23 to decompose A and B into direct products of simple algebras. Let

$$\mathsf{A} = \mathsf{A}_1 \times \ldots \times \mathsf{A}_n \qquad \text{and} \qquad \mathsf{B} = \mathsf{B}_1 \times \ldots \times \mathsf{B}_m$$

where $\mathsf{A}_1, \ldots, \mathsf{A}_n$ and $\mathsf{B}_1, \ldots, \mathsf{B}_m$ are simple graded F-algebras. We have

$$\mathsf{A} \otimes_{\mathsf{F}} \mathsf{B} = \prod_{i,j} \mathsf{A}_i \otimes_{\mathsf{F}} \mathsf{B}_j \quad \text{and} \quad Z(\mathsf{A}) \otimes_{\mathsf{F}} Z(\mathsf{B}) = \prod_{i,j} Z(\mathsf{A}_i) \otimes_{\mathsf{F}} Z(\mathsf{B}_j).$$

Since a direct product of graded algebras is graded semisimple if and only if each factor is graded semisimple, $\mathsf{A} \otimes_{\mathsf{F}} \mathsf{B}$ is graded semisimple if and only if $\mathsf{A}_i \otimes_{\mathsf{F}} \mathsf{B}_j$ is graded semisimple for all i, j. Likewise, $Z(\mathsf{A}) \otimes_{\mathsf{F}} Z(\mathsf{B})$ is graded semisimple if and only if $Z(\mathsf{A}_i) \otimes_{\mathsf{F}} Z(\mathsf{B}_j)$ is graded semisimple for all i, j. The special case of simple graded algebras considered above shows that $\mathsf{A}_i \otimes_{\mathsf{F}} \mathsf{B}_j$ is graded semisimple if and only if $Z(\mathsf{A}_i) \otimes_{\mathsf{F}} Z(\mathsf{B}_j)$ is graded semisimple. The corollary follows. □

Since a semisimple graded F-algebra is a direct product of simple graded F-algebras and the center of a simple graded F-algebra is a graded field, Cor. 2.33 reduces the question of semisimplicity of $\mathsf{A}\otimes_{\mathsf{F}}\mathsf{B}$ to consideration of tensor products of graded fields. We will see below in Prop. 5.21 that if L is a finite-degree separable (= étale) graded field extension of F and K is any graded field extension of F, then $\mathsf{L}\otimes_{\mathsf{F}}\mathsf{K}$ is a semisimple K-algebra.

Proposition 2.34. *Let* A *be a simple graded* F*-algebra with center* Z*. Let* N *be a finitely generated left graded* A*-module. Let* $\mathsf{E}=\mathsf{End}_{\mathsf{A}}\mathsf{N}$ *(acting on* N *on the right) Then,* E *is graded simple with* $Z(\mathsf{E})=\mathsf{Z}$ *and*

$$[\mathsf{A}{:}\mathsf{F}]\cdot[\mathsf{E}{:}\mathsf{F}] \;=\; (dim_{\mathsf{F}}\,\mathsf{N})^2. \tag{2.17}$$

Moreover, $\mathsf{End}_{\mathsf{E}}\mathsf{N}\cong_g\mathsf{A}$ *and* $\mathsf{A}\otimes_{\mathsf{Z}}\mathsf{E}^{\mathrm{op}}\cong_g\mathsf{End}_{\mathsf{Z}}\mathsf{N}$.

Proof. Let J be a minimal nonzero homogeneous left ideal of A, and let $\mathsf{D}=\mathsf{End}_{\mathsf{A}}\mathsf{J}$. As in §2.2.3, we write $\mathsf{J}\cong_g\mathsf{D}(\delta_1)\oplus\ldots\oplus\mathsf{D}(\delta_n)$; then, as in (2.16),

$$\mathsf{A}\;\cong_g\;\mathsf{M}_n(\mathsf{D})(\delta_1,\ldots,\delta_n),$$

so $Z(\mathsf{A})=Z(\mathsf{D})$. By Prop. 2.24 and Th. 2.26 we have $\mathsf{N}\cong_g\bigoplus_{i=1}^k\mathsf{J}(\lambda_i)$ for suitable λ_i in Γ. Hence, by the "left version" of Prop. 2.9,

$$\mathsf{E}\;\cong_g\;\mathsf{M}_k(\mathsf{D})(-\lambda_1,\ldots,-\lambda_k).$$

Let $\mathsf{P}=\bigoplus_{i=1}^k\mathsf{D}(-\lambda_i)$ viewed as a right graded D-vector space. By Prop. 2.9,

$$\mathsf{End}_{\mathsf{D}}\mathsf{P}\;\cong_g\;\mathsf{M}_k(\mathsf{D})(-\lambda_1,\ldots,-\lambda_k)\;\cong_g\;\mathsf{E}.$$

Hence, by Prop. 2.17, E is graded simple and $Z(\mathsf{E})=Z(\mathsf{D})=Z(\mathsf{A})$. Let $d=[\mathsf{D}{:}\mathsf{F}]$. Then, $[\mathsf{A}{:}\mathsf{F}]=n^2d$, $[\mathsf{E}{:}\mathsf{F}]=k^2d$, and $dim_{\mathsf{F}}\,\mathsf{N}=k\,dim_{\mathsf{F}}\,\mathsf{J}=knd$. These equalities yield formula (2.17).

Let $\mathsf{B}=\mathsf{End}_{\mathsf{E}}\mathsf{N}$ (acting on N on the left). The map $\rho\colon\mathsf{A}\to\mathsf{B}$ given by $\rho(a)(x)=ax$ for $a\in\mathsf{A}$, $x\in\mathsf{N}$ is a graded F-algebra homomorphism, which is injective, as A is graded simple. Because E is graded simple, formula (2.17) applies with E (resp. B) replacing A (resp. E), showing that

$$[\mathsf{E}{:}\mathsf{F}]\cdot[\mathsf{B}{:}\mathsf{F}]\;=\;(dim_{\mathsf{F}}\,\mathsf{N})^2\;=\;[\mathsf{A}{:}\mathsf{F}]\cdot[\mathsf{E}{:}\mathsf{F}].$$

Therefore, $[\mathsf{B}{:}\mathsf{F}]=[\mathsf{A}{:}\mathsf{F}]$, so ρ is a graded isomorphism. The last isomorphism of the proposition is proved similarly: the module actions of A and E on N give graded F-algebra homomorphisms $\mathsf{A}\to\mathsf{End}_{\mathsf{Z}}\mathsf{N}$ and $\mathsf{E}^{\mathrm{op}}\to\mathsf{End}_{\mathsf{Z}}\mathsf{N}$ whose images commute. Moreover, the elements of Z in A have the same action on N as their images in E. Hence, there is a graded F-algebra homomorphism $\tau\colon\mathsf{A}\otimes_{\mathsf{Z}}\mathsf{E}^{\mathrm{op}}\to\mathsf{End}_{\mathsf{Z}}\mathsf{N}$. This τ is injective, as $\mathsf{A}\otimes_{\mathsf{Z}}\mathsf{E}^{\mathrm{op}}$ is graded simple by Prop. 2.32. Then τ is also surjective, since by (2.17),

$$\begin{aligned}[\mathsf{A}\otimes_{\mathsf{Z}}\mathsf{E}^{\mathrm{op}}:\mathsf{Z}]\;&=\;[\mathsf{A}{:}\mathsf{Z}]\cdot[\mathsf{E}{:}\mathsf{Z}]\;=\;[\mathsf{A}{:}\mathsf{F}]\cdot[\mathsf{E}{:}\mathsf{F}]/[\mathsf{Z}{:}\mathsf{F}]^2\\&=\;(dim_{\mathsf{F}}\,\mathsf{N})^2/[\mathsf{Z}{:}\mathsf{F}]^2\;=\;[\mathsf{End}_{\mathsf{Z}}\mathsf{N}:\mathsf{Z}].\end{aligned}$$

□

Theorem 2.35 (Double Centralizer Theorem). *Let* A *be a central simple graded* F*-algebra, let* B *be a simple graded* F*-subalgebra of* A*, and let* $\mathsf{C} = C_{\mathsf{A}}(\mathsf{B})$*. Then,*

(i) C *is a simple graded* F*-algebra with* $Z(\mathsf{C}) = Z(\mathsf{B})$*, and* $C_{\mathsf{A}}(\mathsf{C}) = \mathsf{B}$.
(ii) $[\mathsf{C}{:}\mathsf{F}] \cdot [\mathsf{B}{:}\mathsf{F}] = [\mathsf{A}{:}\mathsf{F}]$.
(iii) $\mathsf{B} \otimes_{Z(\mathsf{B})} \mathsf{C} \cong_g C_{\mathsf{A}}(Z(\mathsf{B}))$*. In particular, if* $Z(\mathsf{B}) = \mathsf{F}$*, then* $\mathsf{B} \otimes_{\mathsf{F}} \mathsf{C} \cong_g \mathsf{A}$.

Proof. By using the multiplication in A, we may view A as a left graded B-module and a right graded A-module. Since the multiplication is associative, A is a graded B-A-bimodule, or, equivalently, a right graded $\mathsf{B}^{\mathrm{op}} \otimes_{\mathsf{F}} \mathsf{A}$-module. The module action is given by

$$t \cdot \left(\sum b_i^{\mathrm{op}} \otimes a_i\right) = \sum b_i t a_i \quad \text{for } t \in \mathsf{A},\ b_i \in \mathsf{B},\ a_i \in \mathsf{A}.$$

Let $\mathsf{T} = \mathsf{B}^{\mathrm{op}} \otimes_{\mathsf{F}} \mathsf{A}$. Since A is graded central simple over F and B^{op} is graded simple, Prop. 2.32 shows that T is graded simple with $Z(\mathsf{T}) = Z(\mathsf{B}^{\mathrm{op}}) = Z(\mathsf{B})$. Let $\mathsf{End}_{\mathsf{A}}\,\mathsf{A}$ denote the graded endomorphism ring of A as a right graded A-module. With $\mathsf{End}_{\mathsf{A}}\,\mathsf{A}$ acting on A on the left, we have the graded ring isomorphism $\eta\colon \mathsf{End}_{\mathsf{A}}\,\mathsf{A} \to \mathsf{A}$ given by $\eta(f) = f(1)$. The inverse map is given by $\eta^{-1}(a)(y) = ay$ for all $a, y \in \mathsf{A}$. Now, $\mathsf{End}_{\mathsf{T}}\,\mathsf{A}$ is the graded subring of $\mathsf{End}_{\mathsf{A}}\,\mathsf{A}$ consisting of those A-endomorphisms of A that also commute with the action of B on A. That is,

$$\eta(\mathsf{End}_{\mathsf{T}}\,\mathsf{A}) = \{a \in \mathsf{A} \mid aby = bay \text{ for all } b \in B,\ y \in A\} = \mathsf{C}.$$

Thus, $\mathsf{C} \cong_g \mathsf{End}_{\mathsf{T}}\,\mathsf{A}$. By Prop. 2.34, $\mathsf{End}_{\mathsf{T}}\,\mathsf{A}$, hence also C, is graded simple, with center $Z(\mathsf{C}) = Z(\mathsf{End}_{\mathsf{T}}\,\mathsf{A}) = Z(\mathsf{T}) = Z(\mathsf{B})$. Furthermore, formula (2.17) gives

$$[\mathsf{A}{:}\mathsf{F}]^2 = [\mathsf{T}{:}\mathsf{F}] \cdot [\mathsf{End}_{\mathsf{T}}\,\mathsf{A}{:}\mathsf{F}] = [\mathsf{B}{:}\mathsf{F}] \cdot [\mathsf{A}{:}\mathsf{F}] \cdot [\mathsf{C}{:}\mathsf{F}],$$

which yields (ii). Clearly $\mathsf{B} \subseteq C_{\mathsf{A}}(\mathsf{C})$. But since C is graded simple, part (ii) applies with C replacing B. Thus,

$$[C_{\mathsf{A}}(\mathsf{C}){:}\mathsf{F}] \cdot [\mathsf{C}{:}\mathsf{F}] = [\mathsf{A}{:}\mathsf{F}] = [\mathsf{C}{:}\mathsf{F}] \cdot [\mathsf{B}{:}\mathsf{F}],$$

which yields $C_{\mathsf{A}}(\mathsf{C}) = \mathsf{B}$ by dimension count. This proves (i).

For (iii), the inclusions $\mathsf{B} \hookrightarrow \mathsf{A}$ and $\mathsf{C} \hookrightarrow \mathsf{A}$ with images that commute yield a graded ring homomorphism $\sigma\colon \mathsf{B} \otimes_{Z(\mathsf{B})} \mathsf{C} \to \mathsf{A}$ given by $\sum b_i \otimes c_i \mapsto \sum b_i c_i$. Let $\mathsf{B} \cdot \mathsf{C}$ denote $im(\sigma)$, which is the graded subalgebra of A generated by B and C. So, using (i),

$$C_{\mathsf{A}}(\mathsf{B} \cdot \mathsf{C}) = C_{\mathsf{A}}(\mathsf{B}) \cap C_{\mathsf{A}}(\mathsf{C}) = C_{\mathsf{A}}(\mathsf{B}) \cap \mathsf{B} = Z(\mathsf{B}).$$

Since $\mathsf{B} \otimes_{Z(\mathsf{B})} \mathsf{C}$ is graded simple by Prop. 2.32, σ is injective, so

$$\mathsf{B} \cdot \mathsf{C} \cong_g \mathsf{B} \otimes_{Z(\mathsf{B})} \mathsf{C}.$$

This isomorphism shows that $\mathsf{B} \cdot \mathsf{C}$ is graded simple, and hence, by (i),

$$\mathsf{B} \cdot \mathsf{C} = C_{\mathsf{A}}(C_{\mathsf{A}}(\mathsf{B} \cdot \mathsf{C})) = C_{\mathsf{A}}(Z(\mathsf{B})).$$

Thus, $\mathsf{B} \otimes_{Z(\mathsf{B})} \mathsf{C} \cong_g \mathsf{B} \cdot \mathsf{C} = C_{\mathsf{A}}(Z(\mathsf{B}))$, which is the first isomorphism of (iii). The second isomorphism of (iii) is a special case of the first, as $C_{\mathsf{A}}(\mathsf{F}) = \mathsf{A}$. □

The graded Double Centralizer Theorem gives an upper bound on the dimensions of graded subfields of a graded simple algebra:

Corollary 2.36. *Let* A *be a central simple graded* F*-algebra, and let* K *be a graded subfield of* A *with* $\mathsf{F} \subseteq \mathsf{K}$*. Then,*

(i) $[\mathsf{K}:\mathsf{F}] \mid deg\, \mathsf{A}$;
(ii) *If* A *is a graded division algebra, then* K *is a maximal graded subfield of* A *if and only if* $[\mathsf{K}:\mathsf{F}] = deg\, \mathsf{A}$.

Proof. Let $\mathsf{C} = C_{\mathsf{A}}(\mathsf{K}) \supseteq \mathsf{K}$. By Th. 2.35(ii),

$$[\mathsf{K}:\mathsf{F}]^2 \mid [\mathsf{K}:\mathsf{F}] \cdot [\mathsf{C}:\mathsf{F}] \;=\; [\mathsf{A}:\mathsf{F}] \;=\; (deg\, A)^2. \tag{2.18}$$

This yields (i), and it shows that if $[\mathsf{K}:\mathsf{F}] = deg\, \mathsf{A}$, then K is a maximal subfield of A. (This could also be proved by passing to $q(\mathsf{F})$ and invoking Lemma 2.14.) For the other implication in (ii), assume A is a graded division algebra and suppose $[\mathsf{K}:\mathsf{F}] < deg\, \mathsf{A}$. Then, (2.18) shows that $[\mathsf{C}:\mathsf{F}] > deg\, \mathsf{A} > [\mathsf{K}:\mathsf{F}]$; hence, $\mathsf{C} \supsetneqq \mathsf{K}$. Therefore, there is a nonzero homogeneous $c \in \mathsf{C} \setminus \mathsf{K}$. The commutative graded subalgebra $\mathsf{K}[c]$ of A has no zero divisors, since A has none, and $\mathsf{K}[c]$ is finite-dimensional over F. Hence, $\mathsf{K}[c]$ is a graded subfield of A strictly containing K, showing that K is not a maximal graded subfield of A. □

The graded version of the Skolem–Noether Theorem is more delicate than its ungraded counterpart. Graded isomorphisms of simple graded subalgebras of a central simple graded algebra A will be shown always to be induced by an inner automorphism of A. But what is really desired is a *graded* inner automorphism of A, which is only assured if one conjugates by a homogeneous unit of A. This is not always possible, but we will clarify exactly when it can be done. Thus, part (i) of the next theorem is general, but of little value. Parts (ii) and (iii) are the useful analogues of the Skolem–Noether Theorem in the graded setting.

Theorem 2.37. *Let* A *be a central simple graded* F*-algebra. Let* B *be a simple graded subalgebra of* A*, let* $\mathsf{C} = C_{\mathsf{A}}(\mathsf{B})$ *and* $\mathsf{Z} = Z(\mathsf{B})$*. Let* $f\colon \mathsf{B} \to \mathsf{A}$ *be a graded* F*-algebra homomorphism. Then,*

(i) *There is a* $t \in \mathsf{A}^\times$ *with* $f(b) = tbt^{-1}$ *for all* $b \in \mathsf{B}$.
(ii) *If* C *is a graded division ring, then the* t *of part* (i) *can be chosen to be homogeneous in* A.
(iii) *The* t *in part* (i) *can be chosen to be homogeneous if and only if there is a graded homomorphism* $g\colon \mathsf{C} \to \mathsf{A}$ *such that* $g|_{\mathsf{Z}} = f|_{\mathsf{Z}}$ *and* $g(\mathsf{C})$ *centralizes* $f(\mathsf{B})$ *in* A.

Proof. Let $\mathsf{T} = \mathsf{B} \otimes_{\mathsf{F}} \mathsf{A}^{\mathrm{op}}$, which is a simple graded algebra by Prop. 2.32. We have two ways to view A as a graded B-A-bimodule, hence a left graded T-module. The first action is given, as usual, by $\big(\sum b_i \otimes a_i^{\mathrm{op}}\big) \cdot y = \sum b_i y a_i$. The second is given by $\big(\sum b_i \otimes a_i^{\mathrm{op}}\big) \cdot y = \sum f(b_i) y a_i$. To distinguish these two module structures, we let A denote A with the first T-action and write A' for A with the second T-action. Let S be a simple left graded T-module. By Prop. 2.24 and Th. 2.26,

$$\mathsf{A} \cong_g \bigoplus_{i=1}^{k} \mathsf{S}(\gamma_i) \qquad \text{and} \qquad \mathsf{A}' \cong_g \bigoplus_{j=1}^{\ell} \mathsf{S}(\delta_j)$$

for some $\gamma_1, \ldots, \gamma_k, \delta_1, \ldots \delta_\ell \in \Gamma$. Clearly $\ell = k$ by dimension count. As usual, we write $\mathsf{T}^\natural$ for T with its grading ignored, and likewise $\mathsf{A}^\natural, \mathsf{A}'^\natural, \mathsf{S}^\natural$ for $\mathsf{A}, \mathsf{A}', \mathsf{S}$ with gradings suppressed. Then, as $\mathsf{T}^\natural$-modules we have $\mathsf{A}^\natural \cong \bigoplus_{i=1}^{k} \mathsf{S}^\natural \cong \mathsf{A}'^\natural$. Let $h\colon \mathsf{A}^\natural \to \mathsf{A}'^\natural$ be a $\mathsf{T}^\natural$-module isomorphism. This means

$$h(bya) \;=\; f(b)h(y)a \qquad \text{for all } b \in \mathsf{B},\ y, a \in \mathsf{A}. \tag{2.19}$$

Let $t = h(1)$ and $s = h^{-1}(1)$. By setting $b = a = 1$ in (2.19), we obtain $h(a) = ta$ for all $a \in \mathsf{A}$. Using this, then setting $a = y = 1$ in (2.19), we have

$$tb \;=\; h(b) \;=\; h(b \cdot 1 \cdot 1) \;=\; f(b)t \qquad \text{for all } b \in \mathsf{B}. \tag{2.20}$$

Now, $1 = h(s) = st$. Also, $h(ts - 1) = tst - t = t(st - 1) = 0$. Since h is injective, this shows $ts = 1$. Thus, $t \in \mathsf{A}^\times$ and $s = t^{-1}$. Formula (2.20) yields $f(b) = tbt^{-1}$ for all $b \in \mathsf{B}$, proving (i).

For (ii), suppose C is a graded division ring. Since $s = t^{-1}$, we have

$$f(b) \;=\; tbt^{-1} \;=\; s^{-1}bs$$

for $b \in \mathsf{B}$. By combining this with (2.20), we see that

$$tb = f(b)t \quad \text{and} \quad bs \;=\; sf(b) \qquad \text{for all } b \in \mathsf{B}. \tag{2.21}$$

Since $st = 1 \neq 0$ there must be homogeneous components s_δ of s and t_γ of t with $s_\delta t_\gamma \neq 0$. Let $\delta = deg\, s_\delta$ and $\gamma = deg\, t_\gamma$. For any $\varepsilon \in \Gamma_{\mathsf{B}}$, take any $b_\varepsilon \in \mathsf{B}_\varepsilon$. Then $f(b_\varepsilon) \in \mathsf{B}_\varepsilon$ as f is a graded homomorphism. The $(\gamma + \varepsilon)$-component in the equation $tb_\varepsilon = f(b_\varepsilon)t$ yields $t_\gamma b_\varepsilon = f(b_\varepsilon)t_\gamma$. Since every $b \in \mathsf{B}$ is the sum of its homogeneous components, we obtain

$$t_\gamma b \;=\; f(b)t_\gamma \quad \text{and, likewise,} \quad bs_\delta \;=\; s_\delta f(b) \qquad \text{for all } b \in \mathsf{B}. \tag{2.22}$$

Thus, $s_\delta t_\gamma b = s_\delta f(b) t_\gamma = b s_\delta t_\gamma$ for all $b \in \mathsf{B}$; hence, $s_\delta t_\gamma \in C_{\mathsf{A}}(\mathsf{B}) = \mathsf{C}$. Since $s_\delta t_\gamma$ is homogeneous in the graded division ring C, we have $s_\delta t_\gamma \in \mathsf{C}^\times \subseteq \mathsf{A}^\times$. Therefore, $t_\gamma \in \mathsf{A}^\times$, and (2.22) shows $f(b) = t_\gamma b t_\gamma^{-1}$ for all $b \in \mathsf{B}$, proving (ii).

(iii): Suppose there is a homogeneous unit $t \in \mathsf{A}^\times$ with $f(b) = tbt^{-1}$ for all $b \in \mathsf{B}$. Define $g\colon \mathsf{C} \to \mathsf{A}$ by $g(c) = tct^{-1}$. Because t is homogeneous, g is a graded homomorphism. For all $b \in \mathsf{B}$, $c \in \mathsf{C}$, and $z \in \mathsf{Z}$, we have

$$f(b)g(c) \;=\; tbct^{-1} \;=\; tcbt^{-1} \;=\; g(c)f(b)$$

and $g(z) = f(z)$. Thus, g has the desired properties.

Conversely, suppose there is a $g\colon \mathsf{C} \to \mathsf{A}$ as described in (iii). Then, there is a well-defined graded homomorphism $\mu\colon \mathsf{B} \otimes_{\mathsf{Z}} \mathsf{C} \to \mathsf{A}$ given by $\mu(\sum b_i \otimes c_i) = \sum f(b_i)g(c_i)$. By the graded Double Centralizer Theorem 2.35 there is a graded isomorphism

$$\eta\colon \mathsf{B} \otimes_{\mathsf{Z}} \mathsf{C} \longrightarrow C_{\mathsf{A}}(\mathsf{Z}), \qquad \textstyle\sum b_i \otimes c_i \mapsto \sum b_i c_i.$$

Let $f' = \mu \circ \eta^{-1}\colon C_{\mathsf{A}}(\mathsf{Z}) \to \mathsf{A}$, a graded homomorphism. Since B is graded simple, its center Z is a graded field, so also graded simple. Therefore, the graded Double Centralizer Theorem 2.35 applied to Z shows that $C_{\mathsf{A}}(\mathsf{Z})$ is graded simple and $C_{\mathsf{A}}(C_{\mathsf{A}}(\mathsf{Z})) = \mathsf{Z}$. Therefore, by parts (i) and (ii), there is a homogeneous unit $t \in \mathsf{A}^\times$ with $f'(y) = tyt^{-1}$ for all $y \in C_{\mathsf{A}}(\mathsf{Z})$. Then, $f(b) = tbt^{-1}$ for all $b \in \mathsf{B}$, as $f'|_{\mathsf{B}} = f$. □

Corollary 2.38. *Let* A *be a central simple graded* F*-algebra. Then every graded* F*-automorphism of* A *has the form* $\mathrm{int}(t)$ *for some homogeneous unit* $t \in \mathsf{A}^\times$.

Proof. This is immediate from Th. 2.37(ii) since $C_{\mathsf{A}}(\mathsf{A}) = \mathsf{F}$, which is a graded field. □

The following example illustrates the need for the added conditions in (ii) and (iii) of Th. 2.37 in order to assure that one can conjugate by a homogeneous unit.

Example 2.39. Let F be a graded field with $\Gamma_{\mathsf{F}} = \mathbb{Z}$. We have the graded F-vector spaces $\mathsf{V} = \mathsf{F}(0) \oplus \mathsf{F}(\frac{1}{2})$, $\mathsf{W} = \mathsf{V}$, and $\mathsf{W}' = \mathsf{F}(0) \oplus \mathsf{F}(0)$. It is immediate from the definitions that $\mathsf{F}(\gamma) \otimes_{\mathsf{F}} \mathsf{F}(\delta) \cong_g \mathsf{F}(\gamma + \delta)$ for any $\gamma, \delta \in \Gamma$. Hence,

$$\mathsf{V} \otimes_{\mathsf{F}} \mathsf{W} \cong_g \mathsf{F}(0) \oplus \mathsf{F}(\tfrac{1}{2}) \oplus \mathsf{F}(\tfrac{1}{2}) \oplus \mathsf{F}(1) \cong_g \mathsf{F}(0) \oplus \mathsf{F}(\tfrac{1}{2}) \oplus \mathsf{F}(0) \oplus \mathsf{F}(\tfrac{1}{2}) \cong_g \mathsf{V} \otimes_{\mathsf{F}} \mathsf{W}'.$$

Using Prop. 2.9, let

$$\begin{aligned}
\mathsf{A} &= M_4(\mathsf{F})(0, \tfrac{1}{2}, 0, \tfrac{1}{2}) \cong_g \mathrm{End}_{\mathsf{F}}(\mathsf{V} \otimes_{\mathsf{F}} \mathsf{W}) \cong_g \mathrm{End}_{\mathsf{F}}(\mathsf{V} \otimes_{\mathsf{F}} \mathsf{W}'),\\
\mathsf{B} &= M_2(\mathsf{F})(0, \tfrac{1}{2}) \cong_g \mathrm{End}_{\mathsf{F}} \mathsf{V},\\
\mathsf{C} &= \mathsf{B} \cong_g \mathrm{End}_{\mathsf{F}} \mathsf{W},\\
\mathsf{C}' &= M_2(\mathsf{F})(0, 0) \cong_g \mathrm{End}_{\mathsf{F}} \mathsf{W}'.
\end{aligned}$$

Then, $\mathsf{C} \not\cong_g \mathsf{C}'$, since $\Gamma_{\mathsf{C}} = \frac{1}{2}\mathbb{Z}$ while $\Gamma'_{\mathsf{C}} = \mathbb{Z}$. Note that the F-bilinear map $\mathrm{End}_{\mathsf{F}} \mathsf{V} \times \mathrm{End}_{\mathsf{F}} \mathsf{W} \to \mathrm{End}_{\mathsf{F}}(\mathsf{V} \otimes_{\mathsf{F}} \mathsf{W})$ given by $(g, h) \mapsto g \otimes h$ induces a well-defined graded F-algebra homomorphism

$$\psi\colon \mathrm{End}_{\mathsf{F}} \mathsf{V} \otimes_{\mathsf{F}} \mathrm{End}_{\mathsf{F}} \mathsf{W} \longrightarrow \mathrm{End}_{\mathsf{F}}(\mathsf{V} \otimes_{\mathsf{F}} \mathsf{W}).$$

This map is injective since its domain is graded simple by Prop. 2.17 and Prop. 2.32, and it is surjective by dimension count; hence, ψ is a graded isomorphism. Thus, there are graded isomorphisms $\mu\colon \mathsf{B}\otimes_{\mathsf{F}}\mathsf{C} \xrightarrow{\sim} \mathsf{A}$ and $\eta\colon \mathsf{B}\otimes_{\mathsf{F}}\mathsf{C}' \xrightarrow{\sim} \mathsf{A}$ which are the compositions

$$\mathsf{B}\otimes_{\mathsf{F}}\mathsf{C} \xrightarrow{\sim} \mathsf{End}_{\mathsf{F}}\mathsf{V}\otimes_{\mathsf{F}}\mathsf{End}_{\mathsf{F}}\mathsf{W} \xrightarrow{\sim} \mathsf{End}_{\mathsf{F}}(\mathsf{V}\otimes_{\mathsf{F}}\mathsf{W}) \xrightarrow{\sim} \mathsf{A}, \quad \text{and}$$
$$\mathsf{B}\otimes_{\mathsf{F}}\mathsf{C}' \xrightarrow{\sim} \mathsf{End}_{\mathsf{F}}\mathsf{V}\otimes_{\mathsf{F}}\mathsf{End}_{\mathsf{F}}\mathsf{W}' \xrightarrow{\sim} \mathsf{End}_{\mathsf{F}}(\mathsf{V}\otimes_{\mathsf{F}}\mathsf{W}') \xrightarrow{\sim} \mathsf{A}.$$

Now, B is graded simple by Prop. 2.17, so $\mu(\mathsf{B})$ is a simple graded subalgebra of A. The graded homomorphism $f=\eta\circ\mu^{-1}|_{\mu(\mathsf{B})}\colon \mu(\mathsf{B})\to\mathsf{A}$ has image $\eta(\mathsf{B})$. Since $C_{\mathsf{B}\otimes_{\mathsf{F}}\mathsf{C}}(\mathsf{B})=\mathsf{C}$ and $C_{\mathsf{B}\otimes_{\mathsf{F}}\mathsf{C}'}(\mathsf{B})=\mathsf{C}'$ by Prop. 2.30, the isomorphisms μ and η yield $C_{\mathsf{A}}\big(\mu(\mathsf{B})\big)=\mu(\mathsf{C})$ and $C_{\mathsf{A}}\big(\eta(\mathsf{B})\big)=\eta(\mathsf{C}')$. By Th. 2.37(i) there is a $t\in\mathsf{A}^\times$ with $f(y)=tyt^{-1}$ for all $y\in\mu(\mathsf{B})$. But there can be no such t that is homogeneous. For, if t were homogeneous, $int(t)$ would be a graded automorphism of A, which would yield a graded isomorphism of centralizers, $\mu(\mathsf{C})\cong_g\eta(\mathsf{C}')$. This cannot occur, as $\mu(\mathsf{C})\cong_g\mathsf{C}\not\cong_g\mathsf{C}'\cong_g\eta(\mathsf{C}')$.

2.3 Degree zero elements in simple graded algebras

The focus in this section is on the degree zero component of a central simple graded algebra, which is described explicitly in §2.3.1; see Prop. 2.41. We also give an explicit description of the group of degrees of invertible homogeneous elements, and of its action on the center of the degree zero component; see Prop. 2.44. As a result, we characterize in Prop. 2.45 the grade group of the associated graded division algebra. In §2.3.2, we consider in particular graded algebras obtained by scalar extension from their zero component. These algebras, which are said to be *inertial*, have a particularly simple behavior under tensor products.

2.3.1 The grade group action

For a central simple graded algebra A, the grade set Γ_{A} and the homogeneous component A_0 of degree zero have special connections, which are made explicit in this subsection. We start with the case of graded division rings, which need not be finite-dimensional over any subfield.

For a graded division ring D, any $x\in\mathsf{D}^\times$ is homogeneous, hence the inner automorphism $int(x)$ preserves the grading and induces an automorphism $int(x)_0$ of D_0. This automorphism restricts to an automorphism of $Z(\mathsf{D}_0)$ fixing $Z(\mathsf{D})_0$. The restriction is the identity when $x\in\mathsf{D}_0^\times$. Since $\mathsf{D}_0^\times$ is the kernel of the degree homomorphism $deg\colon\mathsf{D}^\times\to\Gamma_{\mathsf{D}}$, we have a well-defined canonical homomorphism

$$\theta_{\mathsf{D}}\colon\Gamma_{\mathsf{D}}\longrightarrow Aut\big(Z(\mathsf{D}_0)/Z(\mathsf{D})_0\big) \tag{2.23}$$

such that

$$\theta_{\mathsf{D}}\big(deg(x)\big) = int(x)_0|_{Z(\mathsf{D}_0)} \qquad \text{for } x\in\mathsf{D}^\times.$$

The analogy between this canonical homomorphism and the canonical homomorphism (1.1) of a valuation is illustrated in the following proposition (cf. Prop. 1.5(v)).

Proposition 2.40. *Let* D *be a graded division ring with center* F. *The fixed ring of* $Z(\mathsf{D}_0)$ *under the automorphisms in* $\theta_{\mathsf{D}}(\Gamma_{\mathsf{D}})$ *is* F_0. *Hence, if* $[\mathsf{D}_0{:}\mathsf{F}_0] < \infty$, *then* $Z(\mathsf{D}_0)$ *is abelian Galois over* F_0, *and* $im(\theta_{\mathsf{D}})$ *is the entire Galois group* $\mathcal{G}(Z(\mathsf{D}_0)/\mathsf{F}_0)$.

Proof. By the definition of θ_{D}, every element b of $Z(\mathsf{D}_0)$ lying in the fixed ring of $im(\theta_{\mathsf{D}})$ commutes with all nonzero homogeneous elements of D. Since the homogeneous elements additively generate D, we have $b \in Z(\mathsf{D})_0 = \mathsf{F}_0$. If $[\mathsf{D}_0{:}\mathsf{F}_0] < \infty$ then $[Z(\mathsf{D}_0){:}\mathsf{F}_0] < \infty$, and it follows by Galois theory that $Z(\mathsf{D}_0)$ is Galois over F_0 and $im(\theta_{\mathsf{D}}) = \mathcal{G}(Z(\mathsf{D}_0)/\mathsf{F}_0)$. This Galois group is abelian since it is a homomorphic image of the abelian group Γ_{D}. □

Now, let A be a (finite-dimensional) simple graded F-algebra, let J be a minimal nonzero homogeneous left ideal of A, and let $\mathsf{D} = \mathsf{End}_{\mathsf{A}}\,\mathsf{J}$, the graded division algebra associated to A. In view of Th. 2.26, we identify A with $\mathsf{End}_{\mathsf{D}}\,\mathsf{J}$. For convenience, assume A is central, i.e., $\mathsf{F} = Z(\mathsf{A}) = Z(\mathsf{D})$. We let $n = dim_{\mathsf{D}}\,\mathsf{J}$.

While A is graded simple, A_0 need not be simple. The structure of A_0 is determined by the canonical decomposition of J as a right graded D-vector space, as follows: let $\Gamma_1, \ldots, \Gamma_k$ be the distinct cosets of Γ_{D} in Γ_{J}, and for each Γ_i choose a coset representative γ_i, so

$$\Gamma_{\mathsf{J}} = \Gamma_1 \cup \ldots \cup \Gamma_k \qquad \text{with} \qquad \Gamma_i = \gamma_i + \Gamma_{\mathsf{D}}.$$

Take the canonical decomposition of J as in (2.2),

$$\mathsf{J} = \mathsf{J}_1 \oplus \ldots \oplus \mathsf{J}_k, \quad \text{where} \quad \mathsf{J}_i = \bigoplus_{\gamma \in \Gamma_i} \mathsf{J}_\gamma.$$

For $i = 1, \ldots, k$, let

$$r_i = dim_{\mathsf{D}}\,\mathsf{J}_i, \qquad \text{so} \quad \mathsf{J}_i \cong_g \mathsf{D}(\gamma_i)^{r_i} \ \text{and} \ \sum_{i=1}^{k} r_i = n. \tag{2.24}$$

The decomposition $\mathsf{J} = \bigoplus_{i=1}^k \mathsf{J}_i \cong_g \bigoplus_{i=1}^k \mathsf{D}(\gamma_i)^{r_i}$ yields a matrix representation of $\mathsf{A} = \mathsf{End}_{\mathsf{D}}\,\mathsf{J}$:

$$\mathsf{A} \cong_g M_n(\mathsf{D})(\underbrace{\gamma_1, \ldots, \gamma_1}_{r_1}, \underbrace{\gamma_2, \ldots, \gamma_2}_{r_2}, \ldots, \underbrace{\gamma_k, \ldots, \gamma_k}_{r_k}). \tag{2.25}$$

In $\mathsf{A} = \mathsf{End}_{\mathsf{D}}\,\mathsf{J}$, let e_i be the projection on J_i parallel to $\bigoplus_{j \neq i} \mathsf{J}_j$, so

$$e_i|_{\mathsf{J}_i} = id_{\mathsf{J}_i} \qquad \text{and} \qquad e_i|_{\mathsf{J}_j} = 0 \text{ for } j \neq i.$$

Each e_i is homogeneous of degree 0 and idempotent. Set

$$\begin{aligned} B_i &= e_i \mathsf{A}_0 e_i = \{f \in \mathsf{A}_0 \mid f(\mathsf{J}_i) \subseteq \mathsf{J}_i, \text{ and } f(\mathsf{J}_j) = 0 \text{ for } j \neq i\} \\ &\cong (\mathsf{End}_{\mathsf{D}}\, \mathsf{J}_i)_0. \end{aligned} \tag{2.26}$$

Proposition 2.41. *The idempotents $e_1, \ldots, e_k$ are the primitive central idempotents of A_0, and*

$$\mathsf{A}_0 = B_1 \oplus \ldots \oplus B_k \cong M_{r_1}(\mathsf{D}_0) \times \ldots \times M_{r_k}(\mathsf{D}_0).$$

Therefore,

$$Z(\mathsf{A}_0) = Z(\mathsf{A}_0)e_1 \oplus \ldots \oplus Z(\mathsf{A}_0)e_k \cong Z(\mathsf{D}_0) \times \ldots \times Z(\mathsf{D}_0).$$

Proof. Note that A_0 consists of the grade-preserving endomorphisms of J. For $i \neq j$ we have $\Gamma_{\mathsf{J}_i} \cap \Gamma_{\mathsf{J}_j} = \Gamma_i \cap \Gamma_j = \varnothing$. So, any $f \in \mathsf{A}_0$ must map each J_i to itself, and hence $\mathsf{A}_0 = \bigoplus_{i=1}^k B_i$. To see that $e_1, \ldots, e_k$ are the primitive central idempotents in A_0, it now suffices to prove that B_i is simple for each i. Let $(b_{ij})_{j=1}^{r_i}$ be a D_0-base of J_{γ_i}; it is also a D-base of J_i by Prop. 2.5. Thus, we may identify J_i with $\mathsf{J}_{\gamma_i} \otimes_{\mathsf{D}_0} \mathsf{D}$. Any map in $\mathsf{End}_{\mathsf{D}}(\mathsf{J}_i)_0$ is degree-preserving, and hence maps J_{γ_i} to itself. Thus, there is a ring homomorphism $(\mathsf{End}_{\mathsf{D}}\, \mathsf{J}_i)_0 \to End_{\mathsf{D}_0}(\mathsf{J}_{\gamma_i})$, given by $g \mapsto g|_{\mathsf{J}_{\gamma_i}}$. This has an inverse given by the map sending $h \in End_{\mathsf{D}_0}(\mathsf{J}_{\gamma_i})$ to $h \otimes id_{\mathsf{D}} \in \mathsf{End}_{\mathsf{D}}(\mathsf{J}_{\gamma_i} \otimes_{\mathsf{D}_0} \mathsf{D})$. Thus,

$$(\mathsf{End}_{\mathsf{D}}\, \mathsf{J}_i)_0 \cong End_{\mathsf{D}_0}(\mathsf{J}_{\gamma_i}) \cong M_{r_i}(\mathsf{D}_0),$$

which shows that B_i is simple and yields the isomorphisms of the proposition. (In the matrix representation (2.25), A_0 is realized in block diagonal form, with i-th block $M_{r_i}(\mathsf{D}_0)$, because whenever $i \neq j$ we have $\mathsf{D}(\gamma_i - \gamma_j)_0 = \mathsf{D}_{\gamma_i - \gamma_j} = \{0\}$. The idempotent e_i is identified with the identity matrix of the i-th block.) □

Corollary 2.42. *Let S be a semisimple graded F-algebra. Then, S_0 is semisimple.*

Proof. Since S is graded semisimple, by Th. 2.23 $\mathsf{S} = \mathsf{A}_1 \times \ldots \times \mathsf{A}_k$, where each A_i is graded simple. Then S_0 is the direct product of the degree-0 parts of the A_i, each of which is semisimple, by Prop. 2.41. Hence, S_0 is semisimple. □

Proposition 2.41 also yields a convenient criterion for when A is a graded division algebra:

Corollary 2.43. *A finite-dimensional simple graded F-algebra A is a graded division algebra if and only if A_0 is a division ring.*

Proof. If A is a graded division algebra then A_0 is a division ring, as noted in Prop. 2.3(v). Conversely, suppose A_0 is a division ring. Then, in the notation of Prop. 2.41, $k = 1$ and r_1; so $n = r_1 = 1$, and $\mathsf{A} \cong_g M_1(\mathsf{D}) \cong_g \mathsf{D}$, where D is a graded division ring. □

We have seen in Prop. 2.40 that there is an action of Γ_{D} on $Z(\mathsf{D}_0)$ given by the epimorphism $\theta_{\mathsf{D}}\colon \Gamma_{\mathsf{D}} \to \mathcal{G}(Z(\mathsf{D}_0)/\mathsf{F}_0)$. We now describe the corresponding map for the simple graded algebra A. It is more complicated because the homogeneous elements of A need not all be units and A_0 need not be simple. We continue with the notation preceding Prop. 2.41. Since $\Gamma_{\mathsf{J}} = \bigcup_{i=1}^{k}(\gamma_i + \Gamma_{\mathsf{D}})$, we have as in Cor. 2.11,

$$\Gamma_{\mathsf{A}} = \bigcup_{i,j} (\gamma_i - \gamma_j + \Gamma_{\mathsf{D}}), \tag{2.27}$$

which need not be a group. Recall from (2.1) that

$$\Gamma_{\mathsf{A}}^{\times} = \{deg(a) \mid a \in \mathsf{A}^{\times} \text{ and } a \text{ is homogeneous}\}. \tag{2.28}$$

Clearly, $\Gamma_{\mathsf{A}}^{\times}$ is a subgroup of Γ with

$$\Gamma_{\mathsf{D}} = \Gamma_{\mathsf{D}}^{\times} \subseteq \Gamma_{\mathsf{A}}^{\times} \subseteq \Gamma_{\mathsf{A}}, \qquad \text{and} \qquad \left|\Gamma_{\mathsf{A}}^{\times}/\Gamma_{\mathsf{D}}\right| \leq k^2 < \infty.$$

For each homogeneous $a \in \mathsf{A}^{\times}$, its inner automorphism $int(a)$ preserves the homogeneous components of A, so is a graded automorphism of A; thus, $int(a)$ restricts to an automorphism of A_0 and of $Z(\mathsf{A}_0)$. Define

$$\theta_{\mathsf{A}}\colon \Gamma_{\mathsf{A}}^{\times} \longrightarrow Aut(Z(\mathsf{A}_0)/\mathsf{F}_0) \tag{2.29}$$

by

$$\theta_{\mathsf{A}}(\gamma) = int(a)|_{Z(\mathsf{A}_0)} \text{ for any homogeneous } a \in \mathsf{A}^{\times} \text{ with } deg(a) = \gamma. \tag{2.30}$$

This θ_{A} is a well-defined group homomorphism since homogeneous units of degree 0 act trivially on $Z(\mathsf{A}_0)$ by conjugation. To help understand $\Gamma_{\mathsf{A}}^{\times}$ and θ_{A}, we partition Γ_{J} according to the r_i of (2.24): let $\ell_1, \ldots, \ell_m$ be the distinct values in $\{r_1, \ldots, r_k\}$. For $t = 1, \ldots, m$, let

$$S_t = \{i \in \{1, \ldots, k\} \mid r_i = \ell_t\} \qquad \text{and} \qquad \Sigma_t = \bigcup_{i \in S_t} \Gamma_i.$$

Thus,

$$\{1, \ldots, k\} = S_1 \cup \ldots \cup S_m \qquad \text{and} \qquad \Gamma_{\mathsf{J}} = \Sigma_1 \cup \ldots \cup \Sigma_m.$$

Proposition 2.44.

(i) *We have*

$$\Gamma_{\mathsf{A}}^{\times} = \{\varepsilon \in \Gamma \mid \varepsilon + \Sigma_t = \Sigma_t \text{ for } 1 \leq t \leq m\}. \tag{2.31}$$

Hence, $\Gamma_{\mathsf{A}}^{\times}$ is the largest subgroup of Γ such that each Σ_t is a union of cosets of $\Gamma_{\mathsf{A}}^{\times}$. Also, Γ_{A} is a union of cosets of $\Gamma_{\mathsf{A}}^{\times}$ (though $\Gamma_{\mathsf{A}}^{\times}$ need not be the largest group with this property; see Ex. 2.46 below).

(ii) *$\Gamma_{\mathsf{A}}^{\times}$ acts by translation on the set of cosets $\Gamma_{\mathsf{J}}/\Gamma_{\mathsf{D}} = \{\Gamma_1, \ldots, \Gamma_k\}$, sending each $\Sigma_t/\Gamma_{\mathsf{D}}$ to itself. Each coset Γ_i has stabilizer Γ_{D} under this action, so its orbit has size $\left|\Gamma_{\mathsf{A}}^{\times}/\Gamma_{\mathsf{D}}\right|$. The action of $\Gamma_{\mathsf{A}}^{\times}$ on the central* *primitive idempotents* $e_1, \ldots, e_k$ *of* $Z(\mathsf{A}_0)$ *via* θ_{A} *corresponds to its action on* $\Gamma_1, \ldots, \Gamma_k$*: for* $\varepsilon \in \Gamma_{\mathsf{A}}^{\times}$ *and* $i, j \in \{1, \ldots, k\}$*, we have*

$$\varepsilon + \Gamma_i = \Gamma_j \qquad \textit{if and only if} \qquad \theta_{\mathsf{A}}(\varepsilon)(e_i) = e_j.$$

Proof. Take any $\varepsilon \in \Gamma_{\mathsf{A}}^{\times}$ and any $f \in \mathsf{A}^{\times} \cap \mathsf{A}_{\varepsilon}$; recall that $\mathsf{A}_{\varepsilon} = \mathsf{End}_{\mathsf{D}}(\mathsf{J})(\varepsilon)$. Take any canonical component J_i of J and any nonzero homogeneous $z \in \mathsf{J}_i$. Then $f(z)$ is homogeneous and nonzero as f is bijective, say $f(z) \in \mathsf{J}_j$. Since $\varepsilon + deg\, z = deg\, f(z) \in \Gamma_j$, we have $\varepsilon + \Gamma_i = \Gamma_j$. Thus f, which shifts all degrees by ε, maps every homogeneous element of J_i to J_j; hence, $f(\mathsf{J}_i) \subseteq \mathsf{J}_j$. As f is injective, this yields

$$r_i = dim_{\mathsf{D}}\, \mathsf{J}_i = dim_{\mathsf{D}}\, f(\mathsf{J}_i) \leq dim_{\mathsf{D}}\, \mathsf{J}_j = r_j.$$

Since $f^{-1} \in \mathsf{A}^{\times} \cap \mathsf{A}_{-\varepsilon}$ and $-\varepsilon + \Gamma_j = \Gamma_i$, the same argument shows that $f^{-1}(\mathsf{J}_j) \subseteq \mathsf{J}_i$ and $r_j \leq r_i$. Therefore, $r_j = r_i$, which shows that J_i and J_j lie in the same piece Σ_t of Γ_{J}.

Let $\tau_\varepsilon \colon \Gamma/\Gamma_{\mathsf{D}} \to \Gamma/\Gamma_{\mathsf{D}}$ be the translation-by-ε map given by $\gamma + \Gamma_{\mathsf{D}} \mapsto \varepsilon + \gamma + \Gamma_{\mathsf{D}}$. Clearly τ_ε is bijective, and we have just shown that τ_ε maps each $\Sigma_t/\Gamma_{\mathsf{D}}$ into itself; necessarily $\tau_\varepsilon(\Sigma_t) = \Sigma_t$ as $\left|\Sigma_t/\Gamma_{\mathsf{D}}\right| < \infty$. Thus, for any $\varepsilon \in \Gamma_{\mathsf{A}}^{\times}$, we have $\varepsilon + \Sigma_t = \Sigma_t$ for $t = 1, \ldots, m$.

Conversely, take any $\rho \in \Gamma$ with $\rho + \Sigma_t = \Sigma_t$ for all t. For each t and $i \in S_t$, there is some $\pi(i) \in S_t$ such that $\rho + \Gamma_i = \Gamma_{\pi(i)}$. Since i and $\pi(i)$ lie in the same S_t, we have $r_i = r_{\pi(i)}$, hence $dim_{\mathsf{D}}\, \mathsf{J}_i = dim_{\mathsf{D}}\, \mathsf{J}_{\pi(i)}$. For the ρ-shift $\mathsf{J}_i(\rho)$ we have

$$\Gamma_{\mathsf{J}_i(\rho)} = \rho + \Gamma_i = \Gamma_{\pi(i)} \qquad \text{and} \qquad dim_{\mathsf{D}}\, \mathsf{J}_i(\rho) = dim_{\mathsf{D}}\, \mathsf{J}_i = dim_{\mathsf{D}}\, \mathsf{J}_{\pi(i)}.$$

Therefore, there is a graded D-vector space isomorphism $g_i \colon \mathsf{J}_i(\rho) \to \mathsf{J}_{\pi(i)}$. Thus, $g_i \in \mathsf{Hom}_{\mathsf{D}}(\mathsf{J}_i, \mathsf{J}_{\pi(i)})(\rho)$ and g_i is bijective. Choose such a g_i for each J_i, and let

$$g = (g_1, \ldots, g_k) \in \mathsf{Hom}_{\mathsf{D}}\big(\bigoplus_{i=1}^{k} \mathsf{J}_i\,,\ \bigoplus_{i=1}^{k} \mathsf{J}_{\pi(i)}\big)(\rho).$$

Since the injective translation-by-ρ map τ_ρ sends $\Gamma_{\mathsf{J}}/\Gamma_{\mathsf{D}} = \bigcup_{t=1}^{m} \Sigma_t/\Gamma_{\mathsf{D}}$ to itself, π is a permutation of $\{1, \ldots, k\}$; hence,

$$\bigoplus_{i=1}^{k} \mathsf{J}_{\pi(i)} = \mathsf{J} = \bigoplus_{i=1}^{k} \mathsf{J}_i.$$

Therefore, $g \in \mathsf{End}_{\mathsf{D}}(\mathsf{J})(\rho) = \mathsf{A}_\rho$. Moreover, g is bijective since each g_i is an isomorphism, hence $g \in \mathsf{A}^{\times}$. Thus, $\rho \in \Gamma_{\mathsf{A}}^{\times}$. This proves (2.31), from which it is immediate that each Σ_t is a union of cosets of $\Gamma_{\mathsf{A}}^{\times}$, and $\Gamma_{\mathsf{A}}^{\times}$ is the largest group with this property.

For any $\gamma \in \Gamma_{\mathsf{A}}$ and $\varepsilon \in \Gamma_{\mathsf{A}}^{\times}$, choose a nonzero $h \in \mathsf{A}_\gamma$ and $g \in \mathsf{A}^{\times} \cap \mathsf{A}_\varepsilon$. Then $0 \neq hg \in \mathsf{A}_{\gamma+\varepsilon}$, so $\gamma + \varepsilon \in \Gamma_{\mathsf{A}}$. This shows that Γ_{A} is a union of cosets of $\Gamma_{\mathsf{A}}^{\times}$.

(ii) For each $\varepsilon \in \Gamma_{\mathsf{A}}^{\times}$ we have seen that the translation-by-ε map τ_ε sends each $\Sigma_t/\Gamma_{\mathsf{D}}$ to itself. Clearly, for any $\gamma \in \Gamma$, $\tau_\varepsilon(\gamma + \Gamma_{\mathsf{D}}) = \gamma + \Gamma_{\mathsf{D}}$ if and only if $\varepsilon \in \Gamma_{\mathsf{D}}$. Thus, for the group action of $\Gamma_{\mathsf{A}}^{\times}$ on $\Gamma_{\mathsf{J}}/\Gamma_{\mathsf{D}}$ by translation, each coset Γ_i has stabilizer group Γ_{D}, so it has orbit size $\left|\Gamma_{\mathsf{A}}^{\times}/\Gamma_{\mathsf{D}}\right|$.

Take any $\varepsilon \in \Gamma_{\mathsf{A}}^{\times}$ and any $h \in \mathsf{A}^{\times} \cap \mathsf{A}_{\varepsilon}$. We have seen that h permutes $\mathsf{J}_1, \ldots, \mathsf{J}_k$: for $i = 1, \ldots, k$ there is a $j \in \{1, \ldots, k\}$ such that $h(\mathsf{J}_i) = \mathsf{J}_j$, where $\Gamma_j = \varepsilon + \Gamma_i$. It follows that $he_i = e_jh$, hence $\theta_{\mathsf{A}}(\varepsilon)(e_i) = e_j$ when $\tau_{\varepsilon}(\Gamma_i) = \Gamma_j$, proving (ii). □

By using the action of $\Gamma_{\mathsf{A}}^{\times}$ on the idempotents of $Z(\mathsf{A}_0)$, we can also recover Γ_{D} as a subgroup of $\Gamma_{\mathsf{A}}^{\times}$. For the following statement, fix an isomorphism of F_0-algebras

$$\varphi\colon Z(\mathsf{A}_0) \xrightarrow{\sim} Z(\mathsf{D}_0) \times \ldots \times Z(\mathsf{D}_0).$$

We use φ to associate to every $\sigma \in \mathcal{G}(Z(\mathsf{D}_0)/\mathsf{F}_0)$ the automorphism $d(\sigma)$ of $Z(\mathsf{A}_0)$ defined by

$$d(\sigma)\big(\varphi^{-1}(z_1, \ldots, z_k)\big) = \varphi^{-1}\big(\sigma(z_1), \ldots, \sigma(z_k)\big) \qquad \text{for } z_1, \ldots, z_k \in Z(\mathsf{A}_0).$$

Since $\mathcal{G}(Z(\mathsf{D}_0)/\mathsf{F}_0)$ is abelian by Prop. 2.40, the automorphism $d(\sigma)$ does not depend on the choice of φ: this is because for any other isomorphism φ' we may find a permutation π of $\{1, \ldots, k\}$ and $\tau_1, \ldots, \tau_k \in \mathcal{G}(Z(\mathsf{D}_0)/\mathsf{F}_0)$ such that

$$\varphi'^{-1}(z_1, \ldots, z_k) = \varphi^{-1}\big(\tau_1(z_{\pi(1)}), \ldots, \tau_k(z_{\pi(k)})\big) \qquad \text{for } z_1, \ldots, z_k \in Z(\mathsf{A}_0).$$

Then

$$\begin{aligned} d(\sigma)\big(\varphi'^{-1}(z_1, \ldots, z_k)\big) &= \varphi^{-1}\big(\tau_1\sigma(z_{\pi(1)}), \ldots, \tau_k\sigma(z_{\pi(k)})\big) \\ &= \varphi'^{-1}\big(\sigma(z_1), \ldots, \sigma(z_k)\big). \end{aligned}$$

Thus, the map

$$d\colon \mathcal{G}(Z(\mathsf{D}_0)/\mathsf{F}_0) \longrightarrow Aut(Z(\mathsf{A}_0)/\mathsf{F}_0), \qquad \sigma \mapsto d(\sigma) \tag{2.32}$$

is a group homomorphism that does not depend on the choice of φ.

Proposition 2.45. *With the same notation as in Prop. 2.41 and 2.44,*

(i) $\Gamma_{\mathsf{D}} = \{\gamma \in \Gamma_{\mathsf{A}}^{\times} \mid \theta_{\mathsf{A}}(\gamma)(e_1) = e_1\} = \ldots = \{\gamma \in \Gamma_{\mathsf{A}}^{\times} \mid \theta_{\mathsf{A}}(\gamma)(e_k) = e_k\}$.

(ii) *There is a commutative diagram*

$$\begin{array}{ccc} \Gamma_{\mathsf{D}} & \overset{i}{\hookrightarrow} & \Gamma_{\mathsf{A}}^{\times} \\ {\scriptstyle \theta_{\mathsf{D}}}\downarrow & & \downarrow{\scriptstyle \theta_{\mathsf{A}}} \\ \mathcal{G}(Z(\mathsf{D}_0)/\mathsf{F}_0) & \overset{d}{\hookrightarrow} & Aut(Z(\mathsf{A}_0)/\mathsf{F}_0), \end{array}$$

where $i\colon \Gamma_{\mathsf{D}} \to \Gamma_{\mathsf{A}}^{\times}$ *is the inclusion map and* d *is the map* (2.32).

Proof. Take any $\varepsilon \in \Gamma_{\mathsf{A}}^{\times}$. From Prop. 2.44(ii), we know that for any $i \in \{1, \ldots, k\}$ the equation $\theta_{\mathsf{A}}(\varepsilon)(e_i) = e_i$ holds if and only if $\tau_{\varepsilon}(\Gamma_i) = \Gamma_i$, if and only if $\varepsilon \in \Gamma_{\mathsf{D}}$. This proves (i).

(ii) Take any homogeneous D-vector space base $(b_1, \ldots, b_n)$ of J. For $c \in \mathsf{D}$, define $f_c \in A$ by

$$f_c\big(\sum_{i=1}^{n} b_i d_i\big) = \sum_{i=1}^{n} b_i c d_i$$

for $d_i \in \mathsf{D}$. The map $c \mapsto f_c$ is an injective graded ring homomorphism $\mathsf{D} \to \mathsf{A}$. We identify $Z(\mathsf{D}_0)$ with $Z(B_i)$ by $z \mapsto e_i f_z$. For any $\delta \in \Gamma_{\mathsf{D}}$, choose a nonzero $c \in \mathsf{D}_\delta$; so $f_c \in \mathsf{A}^\times \cap \mathsf{A}_\delta$. Then, for $z \in Z(\mathsf{D}_0)$, we have $\theta_{\mathsf{D}}(\delta)(z) = czc^{-1}$ while $\theta_{\mathsf{A}}(\delta) = int(f_c)|_{Z(\mathsf{A}_0)}$. Since $\delta \in \Gamma_{\mathsf{D}}$, we have seen in (i) that $\theta_{\mathsf{A}}(\delta)(e_i) = e_i$ for each i. So,

$$\theta_{\mathsf{A}}(\delta)(e_i f_z) = e_i f_c f_z f_c^{-1} = e_i f_{czc^{-1}} = e_i f_{\theta_{\mathsf{D}}(\delta)(z)}.$$

Thus, the diagram in (ii) is commutative. □

Example 2.46. Let D be a graded division algebra with $\Gamma_{\mathsf{D}} = \mathbb{Z}$. Consider the graded D-vector space $\mathsf{J} = \mathsf{D} \oplus \mathsf{D}(\frac{1}{3})$. We have $\Gamma_{\mathsf{J}} = \mathbb{Z} \cup (\frac{1}{3} + \mathbb{Z})$ and J has canonical components $\mathsf{J}_1 \cong_g \mathsf{D}$ and $\mathsf{J}_2 \cong_g \mathsf{D}(\frac{1}{3})$; so, $\Gamma_1 = \mathbb{Z}$, $\Gamma_2 = \frac{1}{3} + \mathbb{Z}$, $r_1 = r_2 = 1$, and $\Sigma_1 = \Gamma_{\mathsf{J}}$. For $\mathsf{A} = \mathsf{End}_{\mathsf{D}}\,\mathsf{J}$ we have $\Gamma_{\mathsf{A}}^\times = \Gamma_{\mathsf{D}} = \mathbb{Z}$, since no larger subgroup of Γ translates Γ_{J} to itself. But

$$\Gamma_{\mathsf{A}} = \mathbb{Z} \cup (\tfrac{1}{3} + \mathbb{Z}) \cup (-\tfrac{1}{3} + \mathbb{Z}) = \tfrac{1}{3}\mathbb{Z},$$

which is a group strictly containing $\Gamma_{\mathsf{A}}^\times$.

When A_0 is simple, the structure is less complicated:

Proposition 2.47. *For any central simple graded* F*-algebra* A*, the following conditions are equivalent:*

(a) A_0 *is simple.*
(b) $\Gamma_{\mathsf{A}} = \Gamma_{\mathsf{D}}$.
(c) $dim_{\mathsf{D}_0} \mathsf{A}_0 = dim_{\mathsf{D}} \mathsf{A}$.
(d) $Z(\mathsf{A}_0)$ *is a field.*

When these conditions hold, $\mathsf{A} \cong_g M_n(\mathsf{D})$ *with the standard grading on* $M_n(\mathsf{D})$ *(as in* (2.6)*), and*

$$\mathsf{A}_0 \cong M_n(\mathsf{D}_0), \qquad \Gamma_{\mathsf{A}}^\times = \Gamma_{\mathsf{A}} = \Gamma_{\mathsf{D}}, \qquad \textit{and} \qquad \theta_{\mathsf{A}} = \theta_{\mathsf{D}}.$$

Proof. Suppose Γ_{J} is a single coset $\gamma + \Gamma_{\mathsf{D}}$ of Γ_{D}. Then, the canonical decomposition of J has only one component, J itself, and conditions (a)–(d) hold by Prop. 2.41 and (2.27). Furthermore, by Prop. 2.9,

$$\mathsf{A} \cong_g \mathsf{M}_n(\mathsf{D})(\gamma, \ldots, \gamma) \cong_g \mathsf{M}_n(\mathsf{D})(0, \ldots, 0),$$

which is $M_n(\mathsf{D})$ with its standard grading as in (2.6). Since $\Gamma_{\mathsf{D}} \subseteq \Gamma_{\mathsf{A}}^\times \subseteq \Gamma_{\mathsf{A}}$, we must then have $\Gamma_{\mathsf{A}}^\times = \Gamma_{\mathsf{A}}$, and $\theta_{\mathsf{A}} = \theta_{\mathsf{D}}$ by Prop. 2.45. On the other hand, if $|\Gamma_{\mathsf{J}}/\Gamma_{\mathsf{D}}| > 1$, then by Prop. 2.41 and (2.27) none of (a)–(d) holds. □

We can also identify when $\Gamma_{\mathsf{A}}^{\times} = \Gamma_{\mathsf{A}}$:

Proposition 2.48. *For any central simple graded* F*-algebra* A*, the following conditions are equivalent:*

(a) $\Gamma_{\mathsf{A}}^{\times} = \Gamma_{\mathsf{A}}$.
(b) $\Gamma_{\mathsf{A}}^{\times}$ *acts transitively on the primitive central idempotents of* A_0.
(c) $Z(\mathsf{A}_0)$ *is a Galois étale* F_0*-algebra with group* $im(\theta_{\mathsf{A}})$.
(d) Γ_{J} *is a coset of a group and all the* r_i *are equal. (The group is then* $\Gamma_{\mathsf{A}}^{\times}$*.)*

When these equivalent conditions hold,

$$[\mathsf{A}:\mathsf{F}] \;=\; [\mathsf{A}_0:\mathsf{F}_0]\cdot|\Gamma_{\mathsf{A}}:\Gamma_{\mathsf{F}}|. \tag{2.33}$$

Proof. We will repeatedly use Prop. 2.44 and 2.45 without specific mention.

(a) ⇒ (b) Suppose $\Gamma_{\mathsf{A}}^{\times} = \Gamma_{\mathsf{A}}$. For any two cosets $\gamma_i + \Gamma_{\mathsf{D}}$ and $\gamma_j + \Gamma_{\mathsf{D}}$ in Γ_J/Γ_D, let $\varepsilon = \gamma_j - \gamma_i \in \Gamma_{\mathsf{A}}$ (see (2.27)). Then, $\varepsilon \in \Gamma_{\mathsf{A}}^{\times}$ by hypothesis, and $\varepsilon + (\gamma_i + \Gamma_{\mathsf{D}}) = \gamma_j + \Gamma_{\mathsf{D}}$. Hence, $\Gamma_{\mathsf{A}}^{\times}$ acts transitively by translation on $\Gamma_{\mathsf{J}}/\Gamma_{\mathsf{D}}$, so its equivalent action on the simple components of $Z(\mathsf{A}_0)$ is also transitive.

(b) ⇒ (c) By using a diagonal embedding of D in A as in the proof of Prop. 2.45(ii), we have

$$Z(\mathsf{A}_0) \;=\; Z(\mathsf{A}_0)e_1 \oplus \ldots \oplus Z(\mathsf{A}_0)e_k$$

with each $Z(\mathsf{A}_0)e_i \cong Z(\mathsf{D}_0)$. Recall from Prop. 2.40 that the field $Z(\mathsf{D}_0)$ is Galois over F_0 with Galois group $im(\theta_{\mathsf{D}})$. Hence, $Z(\mathsf{A}_0)$ is étale over F_0. Let

$$G \;=\; im(\theta_{\mathsf{A}}) \;\subseteq\; Aut_{\mathsf{F}_0}(Z(\mathsf{A}_0)).$$

For $Z(\mathsf{A}_0)$ to be G-Galois over F_0 we need that $|G| = dim_{\mathsf{F}_0} Z(\mathsf{A}_0)$ and the fixed ring $Z(\mathsf{A}_0)^G$ is F_0 (cf. Knus et al. [115, Def. (18.15)]). We have $|\,im(\theta_{\mathsf{D}})| = [Z(\mathsf{D}_0):\mathsf{F}_0]$. Since $\Gamma_{\mathsf{A}}^{\times}$ acts transitively on $\{e_1, \ldots, e_k\}$, and the stabilizer of each e_i is Γ_{D}, we have $|\Gamma_{\mathsf{A}}^{\times}:\Gamma_{\mathsf{D}}| = k$ and $ker(\theta_{\mathsf{A}}) \subseteq \Gamma_{\mathsf{D}}$. The diagonal action of Γ_{D} on $Z(\mathsf{A}_0)$ via θ_{A} then shows that $ker(\theta_{\mathsf{A}}) = ker(\theta_{\mathsf{D}})$. Thus,

$$|G| \;=\; |\,im(\theta_{\mathsf{A}})| \;=\; |\Gamma_{\mathsf{A}}^{\times}:\Gamma_{\mathsf{D}}|\cdot|\,im(\theta_{\mathsf{D}})| \;=\; k\,[Z(\mathsf{D}_0):\mathsf{F}_0] \;=\; dim_{\mathsf{F}_0} Z(\mathsf{A}_0).$$

Now take any $e_1z_1 + \ldots + e_kz_k \in Z(\mathsf{A}_0)^G$ with each $z_i \in Z(\mathsf{D}_0)$. From the diagonal action of Γ_{D} on $Z(\mathsf{A}_0)$ via θ_{A}, each z_i lies in $Z(\mathsf{D}_0)^{im(\theta_{\mathsf{D}})} = \mathsf{F}_0$. Because of the transitive F_0-linear action of G on $\{e_1, \ldots, e_k\}$, we have $z_1 = \ldots = z_k$, so

$$e_1z_1 + \ldots + e_kz_k \;=\; (e_1 + \ldots + e_k)z_1 \;=\; z_1 \;\in\; \mathsf{F}_0.$$

Thus, $Z(\mathsf{A}_0)^G = \mathsf{F}_0$ and $Z(\mathsf{A}_0)$ is G-Galois over F_0.

(c) ⇒ (b) Suppose $Z(\mathsf{A}_0)$ is Galois over F_0 with group $im(\theta_{\mathsf{A}})$. If the orbit of e_1 under the action of $\Gamma_{\mathsf{A}}^{\times}$ is $\{e_{i_1}, \ldots, e_{i_q}\}$, then the nonzero idempotent $e_{i_1} + \ldots + e_{i_q}$ of $Z(\mathsf{A}_0)$ is $\Gamma_{\mathsf{A}}^{\times}$-stable so lies in the field $Z(\mathsf{A}_0)^{im(\theta_{\mathsf{A}})}$. Hence, $e_{i_1} + \ldots + e_{i_q} = 1$, so $\{i_1, \ldots, i_q\} = \{1, \ldots, k\}$. This shows that $\Gamma_{\mathsf{A}}^{\times}$ acts transitively on the e_i.

(b) $\Rightarrow$ (d) If $\Gamma_{\mathsf{A}}^{\times}$ acts transitively on the e_i, then its corresponding translation action on the cosets $\Gamma_1, \ldots, \Gamma_k$ of Γ_{D} is transitive. So, $\Gamma_{\mathsf{J}} = \Gamma_1 \cup \ldots \cup \Gamma_k$ must be a single coset of $\Gamma_{\mathsf{A}}^{\times}$. Since $\Gamma_{\mathsf{J}} = \coprod_{t=1}^{m} \Sigma_t$ and each Σ_t is a union of cosets of $\Gamma_{\mathsf{A}}^{\times}$, there can be only one Σ_t. Hence $r_1 = \ldots = r_k$.

(d) $\Rightarrow$ (a) Suppose Γ_{J} is a coset of a group Ω and $r_1 = \ldots = r_k$. The equality of the r_i implies that there is only one Σ_t, which must then be all of Γ_{J}. Hence,

$$\Gamma_{\mathsf{A}}^{\times} = \{\varepsilon \in \Gamma \mid \varepsilon + \Gamma_{\mathsf{J}} = \Gamma_{\mathsf{J}}\} = \Omega.$$

So, Γ_{J} is a coset of $\Gamma_{\mathsf{A}}^{\times}$, which shows that $\Gamma_{\mathsf{A}}^{\times} = \{\gamma - \delta \mid \gamma, \delta \in \Gamma_{\mathsf{J}}\} = \Gamma_{\mathsf{A}}$ (see (2.27)).

Suppose conditions (a)–(d) hold. Then, for each $\gamma \in \Gamma_{\mathsf{A}} = \Gamma_{\mathsf{A}}^{\times}$ there is a $c_\gamma \in \mathsf{A}^{\times} \cap \mathsf{A}_\gamma$. Hence, the map $\mathsf{A}_0 \to \mathsf{A}_\gamma$ given by $a \mapsto c_\gamma a$ is an F_0-vector space isomorphism. If $\Gamma_{\mathsf{A}} = \bigcup_{i=1}^{k} (\gamma_i + \Gamma_{\mathsf{F}})$ (disjoint union), then by Prop. 2.5,

$$[\mathsf{A}:\mathsf{F}] = \sum_{i=1}^{k} dim_{\mathsf{F}_0} \mathsf{A}_{\gamma_i} = k\,[\mathsf{A}_0:\mathsf{F}_0] = |\Gamma_{\mathsf{A}}:\Gamma_{\mathsf{F}}| \cdot [\mathsf{A}_0:\mathsf{F}_0]. \qquad \square$$

Note that the equality (2.33) fails to hold in Ex. 2.46 where $\Gamma_{\mathsf{A}}^{\times} \neq \Gamma_{\mathsf{A}}$, $[\mathsf{A}:\mathsf{F}] = 4$, $[\mathsf{A}_0:\mathsf{F}_0] = 2$, and $|\Gamma_{\mathsf{A}}:\Gamma_{\mathsf{F}}| = 3$.

2.3.2 Inertial graded algebras

One easy way to build a graded algebra over a graded field F is by scalar extension from algebras over F_0; we already met this construction in Prop. 2.16 and consider here an important special case.

Lemma 2.49. *For a finite-dimensional graded* F*-algebra* A*, the following conditions are equivalent:*

(a) $\mathsf{A} = \mathsf{A}_0 \otimes_{\mathsf{F}_0} \mathsf{F}$ (*with each* $\mathsf{A}_\gamma = \mathsf{A}_0 \cdot \mathsf{F}_\gamma$).
(b) $[\mathsf{A}:\mathsf{F}] = [\mathsf{A}_0:\mathsf{F}_0]$.
(c) $\Gamma_{\mathsf{A}} = \Gamma_{\mathsf{F}}$.

When they hold, we have $\Gamma_{\mathsf{A}}^{\times} = \Gamma_{\mathsf{A}} = \Gamma_{\mathsf{F}}$.

Proof. (a)$\Rightarrow$(b) is clear. (b)$\Rightarrow$(a) Let $\mathsf{B} = \mathsf{A}_0 \cdot \mathsf{F} = \mathsf{A}_0 \otimes_{\mathsf{F}_0} \mathsf{F}$, which is a graded F-subalgebra of A with each $\mathsf{B}_\gamma = \mathsf{A}_0 \otimes_{\mathsf{F}_0} \mathsf{F}_\gamma$ and $[\mathsf{B}:\mathsf{F}] = [\mathsf{A}_0:\mathsf{F}_0]$. If (b) holds, then $[\mathsf{B}:\mathsf{F}] = [\mathsf{A}:\mathsf{F}] < \infty$, which shows that $\mathsf{A} = \mathsf{B} = \mathsf{A}_0 \otimes_{\mathsf{F}_0} \mathsf{F}$.

(b)$\Leftrightarrow$(c) Write Γ_{A} as a disjoint union of Γ_{F}-cosets, $\Gamma_{\mathsf{A}} = \bigcup_{i \in I} (\gamma_i + \Gamma_{\mathsf{F}})$ with, say, $\gamma_j = 0$. By Prop. 2.5,

$$[\mathsf{A}:\mathsf{F}] = [\mathsf{A}_0:\mathsf{F}_0] + \sum_{i \neq j} dim_{\mathsf{F}_0} \mathsf{A}_{\gamma_i}\,,$$

with each $dim_{\mathsf{F}_0} A_{\gamma_i} \geq 1$. Hence, $[\mathsf{A}:\mathsf{F}] = [\mathsf{A}_0:\mathsf{F}_0]$ if and only if Γ_{A} contains no other cosets of Γ_{F} but Γ_{F} itself.

When (c) holds, we have $\Gamma_{\mathsf{F}} \subseteq \Gamma_{\mathsf{A}}^{\times} \subseteq \Gamma_{\mathsf{A}} = \Gamma_{\mathsf{F}}$, hence equality holds throughout. $\square$

Definition 2.50. Let F be a graded field. A (finite-dimensional) semisimple graded F-algebra A is said to be *inertial* (*over* F) if the equivalent conditions of Lemma 2.49 hold and $Z(\mathsf{A}_0)$ is a separable (= étale) algebra over the field F_0; i.e., $Z(\mathsf{A}_0)$ is a direct product of separable field extensions of F_0. This is an analogue for graded algebras of the notion of an inertial extension in valuation theory, and it encompasses the finite-dimensional case of inertial graded field extensions that will be considered in §5.1.3. Corollary 2.42 shows that A_0 is a semisimple F_0-algebra when A is inertial over F. Then $Z(\mathsf{A}_0) \cong L_1 \times \ldots \times L_k$, where each L_i is a field finite-dimensional over F_0. The condition that $Z(\mathsf{A}_0)$ be separable over F_0 is equivalent to: each field L_i is separable over F_0.

Conversely, if A is a finite-dimensional graded F-algebra for which the conditions in Lemma 2.49 hold, then Lemma 2.13 shows that A is semisimple, hence inertial over F, if A_0 is semisimple and $Z(\mathsf{A}_0)$ is separable over F_0.

Remark 2.51. If A is an inertial graded F-algebra, then the graded subalgebras B of A all have the form $\mathsf{B} = \mathsf{B}_0 \otimes_{\mathsf{F}_0} \mathsf{F}$, where B_0 is an F_0-subalgebra of A_0. Note that $Z(\mathsf{A}) = Z(\mathsf{A}_0) \otimes_{\mathsf{F}_0} \mathsf{F}$. Also, A is graded simple if and only if A_0 is simple, if and only if $Z(\mathsf{A}_0)$ is a field: see Prop. 2.16 and 2.47.

Proposition 2.52. *Let* A *be a simple graded* F*-algebra with associated graded division algebra* D*. Then* A *is inertial over* F *if and only if* D *is inertial over* F *and* $\mathsf{A} \cong_g M_n(\mathsf{D})$ *for some integer* n*, with the standard grading. When this occurs, we have*

$$\mathsf{A}_0 \cong M_n(\mathsf{D}_0), \quad \Gamma_{\mathsf{A}} = \Gamma_{\mathsf{A}}^{\times} = \Gamma_{\mathsf{D}} = \Gamma_{\mathsf{F}}, \quad \textit{and} \quad \theta_{\mathsf{A}} = \theta_{\mathsf{D}}.$$

Proof. By the graded version of Wedderburn's Theorem (Th. 2.26), we have $\mathsf{A} \cong_g \mathsf{End}_{\mathsf{D}}(\mathsf{J})$ for some right graded D-vector space J. Let $n = dim_{\mathsf{D}}\, \mathsf{J}$. If A is inertial over F, then

$$\Gamma_{\mathsf{F}} \subseteq \Gamma_{\mathsf{D}} \subseteq \Gamma_{\mathsf{A}}^{\times} \subseteq \Gamma_{\mathsf{A}} = \Gamma_{\mathsf{F}},$$

hence $\Gamma_{\mathsf{A}} = \Gamma_{\mathsf{A}}^{\times} = \Gamma_{\mathsf{D}} = \Gamma_{\mathsf{F}}$. Proposition 2.47 then shows that $\mathsf{A} \cong_g M_n(\mathsf{D})$, $\mathsf{A}_0 \cong M_n(\mathsf{D}_0)$, and $\theta_{\mathsf{A}} = \theta_{\mathsf{D}}$. It follows that $Z(\mathsf{D}_0) \cong Z(\mathsf{A}_0)$, which is separable over F_0, hence D is inertial over F. Conversely, if D is inertial over F, then for $\mathsf{A} = M_n(\mathsf{D})$ we have $\Gamma_{\mathsf{A}} = \Gamma_{\mathsf{D}} = \Gamma_{\mathsf{F}}$ and $Z(\mathsf{A}_0) \cong Z(\mathsf{D}_0)$, so A is inertial over F. □

Example 2.53. Let n be a positive integer, and let F be a graded field such that F_0 contains a primitive n-th root of unity ω (so *char* F_0 is 0 or prime to n), and let $a, b \in \mathsf{F}_0^{\times}$. Let S be the symbol algebra $(a, b/\mathsf{F}_0)_n$, a central simple F_0-algebra of degree n. Then, $S \otimes_{\mathsf{F}_0} \mathsf{F}$ is the graded symbol algebra $(a, b/\mathsf{F})_n$ as in Def. 2.18 above. Thus, $(a, b/\mathsf{F})_n$ is an inertial graded F-algebra with degree 0 component S.

Inertial graded algebras are well-behaved with respect to tensor products and scalar extensions:

Proposition 2.54. *If* A *and* B *are inertial graded* F*-algebras, then* $\mathsf{A} \otimes_{\mathsf{F}} \mathsf{B}$ *is also inertial over* F *with* $(\mathsf{A} \otimes_{\mathsf{F}} \mathsf{B})_0 = \mathsf{A}_0 \otimes_{\mathsf{F}_0} \mathsf{B}_0$.

Proof. We have

$$\mathsf{A} \otimes_{\mathsf{F}} \mathsf{B} \;=\; (\mathsf{A}_0 \otimes_{\mathsf{F}_0} \mathsf{F}) \otimes_{\mathsf{F}} (\mathsf{B}_0 \otimes_{\mathsf{F}_0} \mathsf{F}) \;=\; (\mathsf{A}_0 \otimes_{\mathsf{F}_0} \mathsf{B}_0) \otimes_{\mathsf{F}_0} \mathsf{F}. \tag{2.34}$$

So, $(\mathsf{A} \otimes_{\mathsf{F}} \mathsf{B})_0 = \mathsf{A}_0 \otimes_{\mathsf{F}_0} \mathsf{B}_0$. Let

$$Z \;=\; Z((\mathsf{A} \otimes_{\mathsf{F}} \mathsf{B})_0) \;=\; Z(\mathsf{A}_0 \otimes_{\mathsf{F}_0} \mathsf{B}_0) \;=\; Z(\mathsf{A}_0) \otimes_{\mathsf{F}_0} Z(\mathsf{B}_0).$$

Then, Z is a separable F_0-algebra since $Z(\mathsf{A}_0)$ and $Z(\mathsf{B}_0)$ are each separable F_0-algebras. Say $Z \cong L_1 \times \ldots \times L_k$, for fields $L_1, \ldots, L_k$, each separable over F_0. Each L_i is an algebra over $Z(\mathsf{A}_0)$ and over $Z(\mathsf{B}_0)$, and $\mathsf{A}_0 \otimes_{Z(\mathsf{A}_0)} L_i \otimes_{Z(\mathsf{B}_0)} \mathsf{B}_0$ is a central simple L_i-algebra. So, as

$$\begin{aligned} \mathsf{A}_0 \otimes_{\mathsf{F}_0} \mathsf{B}_0 \;&=\; \mathsf{A}_0 \otimes_{Z(\mathsf{A}_0)} \big(Z(\mathsf{A}_0) \otimes_{\mathsf{F}_0} Z(\mathsf{B}_0)\big) \otimes_{Z(\mathsf{B}_0)} \mathsf{B}_0 \\ &\cong\; \mathsf{A}_0 \otimes_{Z(\mathsf{A}_0)} (L_1 \times \ldots \times L_k) \otimes_{Z(\mathsf{B}_0)} \mathsf{B}_0 \\ &\cong\; \prod_{i=1}^{k} \mathsf{A}_0 \otimes_{Z(\mathsf{A}_0)} L_i \otimes_{Z(\mathsf{B}_0)} \mathsf{B}_0, \end{aligned}$$

$\mathsf{A}_0 \otimes_{\mathsf{F}_0} \mathsf{B}_0$ is semisimple. Moreover, (2.34) yields $\Gamma_{\mathsf{A}\otimes\mathsf{B}} = \Gamma_{\mathsf{F}} = \Gamma^{\times}_{\mathsf{A}\otimes\mathsf{B}}$. Hence, by (2.34) and Lemma 2.49, $\mathsf{A} \otimes_{\mathsf{F}} \mathsf{B}$ is graded semisimple. Thus, $\mathsf{A} \otimes_{\mathsf{F}} \mathsf{B}$ is inertial over F. □

Proposition 2.55. *Let* A *be a semisimple graded* F*-algebra which is inertial over* F*, and let* K *be any graded field extension of* F*. Then,* $\mathsf{A} \otimes_{\mathsf{F}} \mathsf{K}$ *is an inertial* K*-algebra with* $(\mathsf{A} \otimes_{\mathsf{F}} \mathsf{K})_0 = \mathsf{A}_0 \otimes_{\mathsf{F}_0} \mathsf{K}_0$.

Proof. Since A is inertial, we have

$$\mathsf{A} \otimes_{\mathsf{F}} \mathsf{K} \;=\; \mathsf{A}_0 \otimes_{\mathsf{F}_0} \mathsf{K} \;=\; (\mathsf{A}_0 \otimes_{\mathsf{F}_0} \mathsf{K}_0) \otimes_{\mathsf{K}_0} \mathsf{K} \quad \text{and} \quad \Gamma_{\mathsf{A}\otimes_{\mathsf{F}}\mathsf{K}} \;=\; \Gamma_{\mathsf{K}}.$$

Therefore,

$$(\mathsf{A} \otimes_{\mathsf{F}} \mathsf{K})_0 \;=\; \mathsf{A}_0 \otimes_{\mathsf{F}_0} \mathsf{K}_0 \qquad \text{and} \qquad Z\big((\mathsf{A} \otimes_{\mathsf{F}} \mathsf{K})_0\big) \;=\; Z(\mathsf{A}_0) \otimes_{\mathsf{F}_0} \mathsf{K}_0.$$

Since $Z(\mathsf{A}_0)$ is separable over F_0, it follows that $Z\big((\mathsf{A} \otimes_{\mathsf{F}} \mathsf{K})_0\big)$ is separable over K_0. Moreover, A_0 is semisimple by Cor. 2.42, hence $\mathsf{A}_0 \otimes_{\mathsf{F}_0} \mathsf{K}_0$ is semisimple. By Lemma 2.13, it follows that $\mathsf{A} \otimes_{\mathsf{F}} \mathsf{K}$ is semisimple, because $\Gamma^{\times}_{\mathsf{A}\otimes_{\mathsf{F}}\mathsf{K}} = \Gamma_{\mathsf{A}\otimes_{\mathsf{F}}\mathsf{K}}$ as $\Gamma_{\mathsf{K}} \subseteq \Gamma^{\times}_{\mathsf{A}\otimes_{\mathsf{F}}\mathsf{K}} \subseteq \Gamma_{\mathsf{A}\otimes_{\mathsf{F}}\mathsf{K}} = \Gamma_{\mathsf{K}}$. Therefore, $\mathsf{A} \otimes_{\mathsf{F}} \mathsf{K}$ is inertial over K. □

Proposition 2.56. *Let* A *be a central simple graded* F*-algebra which is inertial over* F*, and let* B *be any simple graded* F*-algebra. Let* $\mathsf{C} = \mathsf{A} \otimes_{\mathsf{F}} \mathsf{B}$*, which is a simple graded* F*-algebra. Then,*

$$\mathsf{C}_0 \;=\; \mathsf{A}_0 \otimes_{\mathsf{F}_0} \mathsf{B}_0, \quad \Gamma_{\mathsf{C}} \;=\; \Gamma_{\mathsf{B}}, \quad Z(\mathsf{C}_0) \;=\; Z(\mathsf{B}_0),$$
$$\Gamma^{\times}_{\mathsf{C}} \;=\; \Gamma^{\times}_{\mathsf{B}}, \quad \textit{and} \quad \theta_{\mathsf{C}} \;=\; \theta_{\mathsf{B}}.$$

Proof. The graded simplicity of C is given by Prop. 2.32. Since

$$\mathsf{C} = \mathsf{A} \otimes_{\mathsf{F}} \mathsf{B} = \mathsf{A}_0 \otimes_{\mathsf{F}_0} \mathsf{F} \otimes_{\mathsf{F}} \mathsf{B} = \mathsf{A}_0 \otimes_{\mathsf{F}_0} \mathsf{B},$$

we have $\mathsf{C}_\gamma = \mathsf{A}_0 \otimes_{\mathsf{F}_0} \mathsf{B}_\gamma$, for each $\gamma \in \Gamma$. Hence, $\mathsf{C}_0 = \mathsf{A}_0 \otimes_{\mathsf{F}_0} \mathsf{B}_0$ and $\Gamma_{\mathsf{C}} = \Gamma_{\mathsf{B}}$. Since $\mathsf{F} = Z(\mathsf{A}) = Z(\mathsf{A}_0) \otimes_{\mathsf{F}_0} \mathsf{F}$, we must have $Z(\mathsf{A}_0) = \mathsf{F}_0$, and hence

$$Z(\mathsf{C}_0) = Z(\mathsf{A}_0) \otimes_{\mathsf{F}_0} Z(\mathsf{B}_0) = Z(\mathsf{B}_0).$$

If $\gamma \in \Gamma_{\mathsf{B}}^\times$, then there is some $b \in \mathsf{B}_\gamma \cap \mathsf{B}^\times$. Then $1 \otimes b \in \mathsf{C}_\gamma \cap \mathsf{C}^\times$, so $\gamma \in \Gamma_{\mathsf{C}}^\times$ and

$$\theta_{\mathsf{C}}(\gamma) = int(1 \otimes b)|_{Z(\mathsf{C}_0)} = int(b)|_{Z(\mathsf{B}_0)} = \theta_{\mathsf{B}}(\gamma).$$

Hence, $\Gamma_{\mathsf{B}}^\times \subseteq \Gamma_{\mathsf{C}}^\times$ and $\theta_{\mathsf{C}}|_{\Gamma_{\mathsf{B}}^\times} = \theta_{\mathsf{B}}$.

To prove that $\Gamma_{\mathsf{B}}^\times = \Gamma_{\mathsf{C}}^\times$, it suffices to show that $\Gamma_{\mathsf{C}}^\times \subseteq \Gamma_{\mathsf{B}}^\times$. Let $\gamma \in \Gamma_{\mathsf{C}}^\times$. The homogeneous component $\mathsf{C}_\gamma = \mathsf{A}_0 \otimes_{\mathsf{F}_0} \mathsf{B}_\gamma$ then contains an element u that is invertible in C. Let $m = [\mathsf{A}{:}\mathsf{F}] = [\mathsf{A}_0{:}\mathsf{F}_0]$. Note that A_0 is central simple over F_0, by Remark 2.51. Therefore, we have F_0-algebra isomorphisms

$$\mathsf{A}_0^{\mathrm{op}} \otimes_{\mathsf{F}_0} \mathsf{A}_0 \cong End_{\mathsf{F}_0} \mathsf{A}_0^{\mathrm{op}} \cong M_m(\mathsf{F}_0).$$

(Compare Lemma 6.3 for the graded analogue.) By tensoring with $\mathsf{A}_0^{\mathrm{op}}$, we obtain isomorphisms

$$\mathsf{A}_0^{\mathrm{op}} \otimes_{\mathsf{F}_0} \mathsf{C} = \mathsf{A}_0^{\mathrm{op}} \otimes_{\mathsf{F}_0} \mathsf{A}_0 \otimes_{\mathsf{F}_0} \mathsf{B} \cong_g M_m(\mathsf{F}_0) \otimes_{\mathsf{F}_0} \mathsf{B} = M_m(\mathsf{B}),$$

for the standard grading on $M_m(\mathsf{B})$. Now, $1 \otimes u \in \mathsf{A}_0^{\mathrm{op}} \otimes_{\mathsf{F}_0} \mathsf{C}_\gamma \cong M_m(\mathsf{B})_\gamma$. This element is invertible in $M_m(\mathsf{B})$ since u is invertible in C, hence $\gamma \in \Gamma_{M_m(\mathsf{B})}^\times$. Suppose $\mathsf{B} \cong_g \mathsf{End}_{\mathsf{E}} \mathsf{J}$ for some graded division algebra E and some right graded E-vector space J. Then $M_m(\mathsf{B}) \cong_g \mathsf{End}_{\mathsf{E}}(\mathsf{J}^m)$. Clearly, $\Gamma_{\mathsf{J}^m} = \Gamma_{\mathsf{J}}$ and if $\mathsf{J} = \bigoplus_{i=1}^k \mathsf{J}_i$ is the canonical decomposition of J determined by the cosets of Γ_{E} in Γ_{J} as in (2.2), then the canonical decomposition of J^m is $\bigoplus_{i=1}^k \mathsf{J}_i^{\,m}$. Since each $\Gamma_{\mathsf{J}_i^{\,m}} = \Gamma_{\mathsf{J}_i}$ and $dim_{\mathsf{E}}\, \mathsf{J}_i^{\,m} = m\, dim_{\mathsf{E}}\, \mathsf{J}_i$, Γ_{J^m} has the same pieces Σ_t as Γ_{J} in the partition of Γ_{J} by the dimensions of the J_i as in Prop. 2.44. Hence, $\Gamma_{M_m(\mathsf{B})}^\times = \Gamma_{\mathsf{B}}^\times$ by the characterization in Prop. 2.44(i), and it follows that $\gamma \in \Gamma_{\mathsf{B}}^\times$. Thus, $\Gamma_{\mathsf{C}}^\times \subseteq \Gamma_{\mathsf{B}}^\times$. When combined with the first paragraph results, this yields $\Gamma_{\mathsf{C}}^\times = \Gamma_{\mathsf{B}}^\times$ and $\theta_{\mathsf{C}} = \theta_{\mathsf{B}}$. □

We now consider scalar extension of central simple graded algebras by inertial graded field extensions.

Proposition 2.57. *Let* L *be an inertial graded field extension of a graded field* F*, and let* A *be a central simple graded* F*-algebra. Then,*

$$(\mathsf{A} \otimes_{\mathsf{F}} \mathsf{L})_0 = \mathsf{A}_0 \otimes_{\mathsf{F}_0} \mathsf{L}_0, \qquad Z\big((\mathsf{A} \otimes_{\mathsf{F}} \mathsf{L})_0\big) = Z(\mathsf{A}_0) \otimes_{\mathsf{F}_0} \mathsf{L}_0,$$

$$\Gamma_{\mathsf{A} \otimes_{\mathsf{F}} \mathsf{L}} = \Gamma_{\mathsf{A}}, \qquad \Gamma_{\mathsf{A} \otimes_{\mathsf{F}} \mathsf{L}}^\times = \Gamma_{\mathsf{A}}^\times.$$

Moreover, the following diagram commutes:

$$\begin{array}{ccc} \Gamma_{\mathsf{A}}^{\times} & = & \Gamma_{\mathsf{A}\otimes\mathsf{L}}^{\times} \\ {\scriptstyle \theta_{\mathsf{A}}}\downarrow & & \downarrow{\scriptstyle \theta_{\mathsf{A}\otimes\mathsf{L}}} \\ Aut(Z(\mathsf{A}_0)/\mathsf{F}_0) & \overset{\iota}{\hookrightarrow} & Aut(Z(\mathsf{A}\otimes_{\mathsf{F}}\mathsf{L})_0/\mathsf{L}_0) \end{array}$$

where

$$\iota(\tau) \;=\; \tau\otimes id_{\mathsf{L}_0} \qquad \textit{for } \tau\in Aut(Z(\mathsf{A}_0)/\mathsf{F}_0).$$

Proof. Let $\mathsf{T} = \mathsf{A}\otimes_{\mathsf{F}}\mathsf{L} = \mathsf{A}\otimes_{\mathsf{F}_0}\mathsf{L}_0$. It is clear that $\mathsf{T}_0 = \mathsf{A}_0\otimes_{\mathsf{F}_0}\mathsf{L}_0$ and hence $Z(\mathsf{T}_0) = Z(\mathsf{A}_0)\otimes_{\mathsf{F}_0}\mathsf{L}_0$. Since A is a graded subalgebra of T, we have

$$\Gamma_{\mathsf{A}} \subseteq \Gamma_{\mathsf{T}} \subseteq \Gamma_{\mathsf{A}}+\Gamma_{\mathsf{L}} \;=\; \Gamma_{\mathsf{A}}+\Gamma_{\mathsf{F}} \;=\; \Gamma_{\mathsf{A}},$$

so $\Gamma_{\mathsf{T}} = \Gamma_{\mathsf{A}}$, and $\Gamma_{\mathsf{A}}^{\times} \subseteq \Gamma_{\mathsf{T}}^{\times}$. To get equality in the last inclusion, let $\mathsf{B} = \mathsf{End}_{\mathsf{F}}(\mathsf{L})$, and view L as a graded F-subalgebra of B by the left regular representation. Since L is inertial over F, we have $\mathsf{L}\cong_g \mathsf{F}^n$ as graded F-vector spaces where $n = [\mathsf{L}{:}\mathsf{F}]$. Hence, B is an inertial central simple graded F-algebra. Therefore, Prop. 2.56 shows that $\Gamma_{\mathsf{A}\otimes_{\mathsf{F}}\mathsf{B}}^{\times} = \Gamma_{\mathsf{A}}^{\times}$. But the inclusions of graded algebras $\mathsf{A}\subseteq\mathsf{T}\subseteq\mathsf{B}\otimes_{\mathsf{F}}\mathsf{A}$ yield $\Gamma_{\mathsf{A}}^{\times}\subseteq\Gamma_{\mathsf{T}}^{\times}\subseteq\Gamma_{\mathsf{A}\otimes_{\mathsf{F}}\mathsf{B}}^{\times}$. Hence, $\Gamma_{\mathsf{T}}^{\times} = \Gamma_{\mathsf{A}}^{\times}$. So, for any $\gamma\in\Gamma_{\mathsf{T}}^{\times}$ there exists an $a\in\mathsf{A}_\gamma\cap\mathsf{A}^{\times}$. Then $a\otimes 1\in\mathsf{T}_\gamma\cap\mathsf{T}^{\times}$, so as $\theta_{\mathsf{A}}(\gamma) = int_{\mathsf{A}}(a)|_{Z(\mathsf{A}_0)}$, we have

$$\theta_{\mathsf{T}}(\gamma) \;=\; int_{\mathsf{T}}(a\otimes 1)|_{Z(\mathsf{T}_0)} \;=\; (int_{\mathsf{A}}(a)|_{Z(\mathsf{A}_0)})\otimes id_{\mathsf{L}_0} \;=\; \iota(\theta_{\mathsf{A}}(\gamma)).$$

Thus, $\theta_{\mathsf{T}} = \iota\circ\theta_{\mathsf{A}}$. □

Our final result in this section demonstrates how the action of the grade group of a graded division algebra on the homogeneous component of degree zero behaves under an inertial graded field extension.

We let D be a central graded division algebra over a graded field F and consider an inertial graded field extension L of F (possibly of infinite degree). Recall from Prop. 2.40 that $Z(\mathsf{D}_0)$ is a Galois field extension of F_0 with abelian Galois group. Let $G = \mathcal{G}(Z(\mathsf{D}_0)/\mathsf{F}_0)$. With G acting on the left factor, the étale L_0-algebra $Z(\mathsf{D}_0)\otimes_{\mathsf{F}_0}\mathsf{L}_0$ is G-Galois. Since G is abelian, any element that fixes any one of the primitive idempotents of $Z(\mathsf{D}_0)\otimes_{\mathsf{F}_0}\mathsf{L}_0$ fixes each of the idempotents. Let $H\subseteq G$ be the subgroup of all such stabilizing elements. There is a field extension K of L_0 such that

$$Z(\mathsf{D}_0)\otimes_{\mathsf{F}_0}\mathsf{L}_0 \;\cong\; K\times\ldots\times K. \tag{2.35}$$

The extension K/L_0 is Galois with Galois group isomorphic to H. The isomorphism $H\xrightarrow{\sim}\mathcal{G}(K/\mathsf{L}_0)$ is obtained via the inclusions

$$H\subseteq G\subseteq Aut\big((Z(\mathsf{D}_0)\otimes_{\mathsf{F}_0}\mathsf{L}_0)/\mathsf{L}_0\big)$$

and the isomorphism (2.35), but it does not depend on the choice of this isomorphism: see the discussion preceding Prop. 2.45. The isomorphism (2.35) yields F_0-embeddings of $Z(\mathsf{D}_0)$ in K. Since $Z(\mathsf{D}_0)$ is Galois over F_0, all these embeddings have the same image, and K can be viewed as the field compositum of L_0 and $Z(\mathsf{D}_0)$. Since moreover G is abelian, the intersection $Z(\mathsf{D}_0)\cap\mathsf{L}_0$ in K is independent of the choice of embedding $Z(\mathsf{D}_0)\hookrightarrow K$, and we have by Galois theory

$$H \cong \mathcal{G}(K/\mathsf{L}_0) \cong \mathcal{G}\big(Z(\mathsf{D}_0)/(Z(\mathsf{D}_0)\cap \mathsf{L}_0)\big).$$

Proposition 2.58. *Use the notation above, and let* E *be the graded division algebra associated to* $\mathsf{D}\otimes_{\mathsf{F}}\mathsf{L}$. *Then,*

(i) E_0 *is the associated division algebra of* $\mathsf{D}_0\otimes_{Z(\mathsf{D}_0)}K$.
(ii) $\Gamma_{\mathsf{E}}=\theta_{\mathsf{D}}^{-1}(H)\subseteq\Gamma_{\mathsf{D}}$, *and there is a commutative diagram:*

$$\begin{array}{ccc}
\Gamma_{\mathsf{E}} & \hookrightarrow & \Gamma_{\mathsf{D}} \\
\downarrow{\scriptstyle \theta_{\mathsf{E}}} & & \downarrow{\scriptstyle \theta_{\mathsf{D}}} \\
\mathcal{G}(K/\mathsf{L}_0) & \hookrightarrow & G
\end{array}$$

where the upper horizontal map is the inclusion and the lower horizontal map is the composition $\mathcal{G}(K/\mathsf{L}_0)\xrightarrow{\sim} H\hookrightarrow G$.

Proof. Proposition 2.41 shows that E_0 is the division algebra associated to each simple factor of the semisimple algebra $(\mathsf{D}\otimes_{\mathsf{F}}\mathsf{L})_0$. By Prop. 2.57, we have

$$(\mathsf{D}\otimes_{\mathsf{F}}\mathsf{L})_0 \;=\; \mathsf{D}_0\otimes_{\mathsf{F}_0}\mathsf{L}_0.$$

By decomposing $\mathsf{D}_0\otimes_{\mathsf{F}_0}\mathsf{L}_0=\mathsf{D}_0\otimes_{Z(\mathsf{D}_0)}(Z(\mathsf{D}_0)\otimes_{\mathsf{F}_0}\mathsf{L}_0)$ and using (2.35), we obtain

$$(\mathsf{D}\otimes_{\mathsf{F}}\mathsf{L})_0 \cong (\mathsf{D}_0\otimes_{Z(\mathsf{D}_0)}K)\times\ldots\times(\mathsf{D}_0\otimes_{Z(\mathsf{D}_0)}K).$$

Therefore, the simple factors of $(\mathsf{D}\otimes_{\mathsf{F}}\mathsf{L})_0$ are isomorphic to $\mathsf{D}_0\otimes_{Z(\mathsf{D}_0)}K$. Assertion (i) follows.

To prove (ii), note that by Prop. 2.57 we have $Z\big((\mathsf{D}\otimes_{\mathsf{F}}\mathsf{L})_0\big)=Z(\mathsf{D}_0)\otimes_{\mathsf{F}_0}\mathsf{L}_0$ and $\Gamma^{\times}_{\mathsf{D}\otimes\mathsf{L}}=\Gamma^{\times}_{\mathsf{D}}=\Gamma_{\mathsf{D}}$, and the following diagram commutes:

$$\begin{array}{ccc}
\Gamma_{\mathsf{D}} & = & \Gamma^{\times}_{\mathsf{D}\otimes\mathsf{L}} \\
\downarrow{\scriptstyle \theta_{\mathsf{D}}} & & \downarrow{\scriptstyle \theta_{\mathsf{D}\otimes\mathsf{L}}} \\
G & \xrightarrow{\iota} & Aut((Z(\mathsf{D}_0)\otimes_{\mathsf{F}_0}\mathsf{L}_0)/\mathsf{L}_0)
\end{array} \qquad (2.36)$$

Now, Prop. 2.45 shows that Γ_{E} consists of the $\gamma\in\Gamma^{\times}_{\mathsf{D}\otimes\mathsf{L}}$ such that $\theta_{\mathsf{D}\otimes\mathsf{L}}(\gamma)$ fixes the primitive idempotents of $Z(\mathsf{D}_0)\otimes_{\mathsf{F}_0}\mathsf{L}_0$. Since diagram (2.36) commutes, these γ lie in Γ_{D} and can also be characterized by the condition that $\theta_{\mathsf{D}}(\gamma)\in H$. Moreover, Prop. 2.45 also yields a commutative diagram

$$\begin{array}{ccc} \Gamma_{\mathsf{E}} & \hookrightarrow & \Gamma^{\times}_{\mathsf{D}\otimes\mathsf{L}} \\ {\scriptstyle \theta_{\mathsf{E}}}\downarrow & & \downarrow{\scriptstyle \theta_{\mathsf{D}\otimes\mathsf{L}}} \\ \mathcal{G}(K/\mathsf{L}_0) & \xrightarrow{\ d\ } & Aut((Z(\mathsf{D}_0)\otimes_{\mathsf{F}_0}\mathsf{L}_0)/\mathsf{L}_0) \end{array}$$

Then (ii) follows by combining this diagram with (2.36). □

Exercises

Exercise 2.1. Let D be a graded division ring.

(i) Let $\gamma,\ \delta \in \Gamma$. Show that the shifted right graded D-vector spaces $\mathsf{D}(\gamma)$ and $\mathsf{D}(\delta)$ are isomorphic as graded D-modules if and only if $\gamma \equiv \delta\ (mod\ \Gamma_{\mathsf{D}})$.

(ii) Let V be a right graded D-vector space and W a left graded D-vector space. Establish a canonical isomorphism of graded $Z(\mathsf{D})$-vector spaces $\mathsf{V}(\gamma)\otimes_{\mathsf{D}}\mathsf{W}(\delta)\cong_g(\mathsf{V}\otimes_{\mathsf{D}}\mathsf{W})(\gamma+\delta)$.

Exercise 2.2. Let D be a graded division ring and let $\delta_1,\ldots,\delta_n,\varepsilon_1,\ldots,\varepsilon_n\in\Gamma$. Show that $M_n(\mathsf{D})(\delta_1,\ldots,\delta_n)\cong_g M_n(\mathsf{D})(\varepsilon_1,\ldots,\varepsilon_n)$ if and only if there is a permutation σ of $\{1,\ldots,n\}$ and a $\gamma\in\Gamma$ such that $\delta_i-\varepsilon_{\sigma(i)}\in\gamma+\Gamma_{\mathsf{D}}$ for all $i=1,\ \ldots,\ n$.

Exercise 2.3. Let F be a field, let t be a commuting indeterminate over F, and let $A=F[t]\{x,y\}$, the free algebra in noncommuting indeterminates x and y over the polynomial ring $F[t]$. Thus, A has a base as a free $F[t]$-module consisting of all words of finite length in x and y. Clearly $Z(A)=F[t]$. Let I be the two-sided ideal of A generated by tx, and let $B=A/I$, which is generated as an F-algebra by the images $\overline{t},\overline{x},\overline{y}$ of t,x,y. Show that $Z(B)=F[\overline{t}]\cong F[t]$, and $\overline{x}\neq 0$, but $\overline{t}\,\overline{x}=0$. Thus, $q(Z(B))\cong F(t)$, a rational function field over F, but B does not embed in its ring of central quotients $q(B)=B\otimes_{Z(B)}q(Z(B))\cong B\otimes_{F[t]}F(t)$. In fact, $q(B)\cong F(\overline{t})[\overline{y}]$, a commutative polynomial ring in $\overline{y}$ over the field $F(t)$. (Hint: There is a $\mathbb{Z}\times\mathbb{Z}$-grading on A given by degree in t and total degree in x and y. Since I is a homogeneous ideal of A with respect to this grading, there is an induced grading on B. From this one can determine an F-vector space base of B.)

Exercise 2.4. This exercise gives the graded version of a standard identity for symbol algebras, cf. Draxl [63, Lemma 7, p. 81]. Let $n=n_1n_2$ for some relatively prime integers $n_1,\ n_2\geq 2$, and suppose F is a graded field such that F_0 contains a primitive n-th root of unity ω. If $m_1,m_2,n_1,n_2\in\mathbb{Z}$ satisfy $m_1n_1+m_2n_2=1$, show that for all homogeneous elements $a,\ b\in\mathsf{F}^{\times}$

$$\big(a,b/\mathsf{F}\big)_{\omega,n}\cong_g\big(a^{m_2},b/\mathsf{F}\big)_{\omega^{n_2},n_1}\otimes_{\mathsf{F}}\big(a^{m_1},b/\mathsf{F}\big)_{\omega^{n_1},n_2}.$$

Exercise 2.5. Use the same notation as in the preceding exercise, and let ω_1, $\omega_2 \in \mathsf{F}_0$ be primitive roots of unity of order n_1 and n_2 respectively. Show that for all homogeneous elements $a_1, b_1, a_2, b_2 \in \mathsf{F}^\times$

$$\left(a_1, b_1/\mathsf{F}\right)_{\omega_1, n_1} \otimes_{\mathsf{F}} \left(a_2, b_2/\mathsf{F}\right)_{\omega_2, n_2} \cong_g \left(a_1^{n_2} a_2^{n_1}, b_1^{n_2} b_2^{n_1}/\mathsf{F}\right)_{\omega_1\omega_2, n}.$$

Exercise 2.6. Let F be a graded field, let A be a central simple graded F-algebra, and let B be an arbitrary graded F-algebra. Show that there is a bijection between two-sided homogeneous ideals of $\mathsf{A} \otimes_{\mathsf{F}} \mathsf{B}$ and two-sided homogeneous ideals of B, which maps $\mathsf{J} \subseteq \mathsf{A} \otimes \mathsf{B}$ to $\mathsf{J} \cap (1 \otimes \mathsf{B}) \subseteq \mathsf{B}$ and $\mathsf{K} \subseteq \mathsf{B}$ to $\mathsf{A} \otimes_{\mathsf{F}} \mathsf{K} \subseteq \mathsf{A} \otimes_{\mathsf{F}} \mathsf{B}$. (This result yields an alternative proof of Prop. 2.32.)

Exercise 2.7. Show that if A and B are finite-dimensional graded algebras over a graded field F, then $\mathsf{A} \times \mathsf{B}$ is inertial over F if and only if A and B are each inertial over F.

Exercise 2.8. Let A and B be (finite-dimensional) central graded division algebras over a graded field F. Assume that $\Gamma_{\mathsf{A}} \cap \Gamma_{\mathsf{B}} = \Gamma_{\mathsf{F}}$.

(i) Prove that $(\mathsf{A} \otimes_{\mathsf{F}} \mathsf{B})_0 = \mathsf{A}_0 \otimes_{\mathsf{F}_0} \mathsf{B}_0$, $\Gamma_{\mathsf{A}\otimes_{\mathsf{F}}\mathsf{B}} = \Gamma^\times_{\mathsf{A}\otimes_{\mathsf{F}}\mathsf{B}} = \Gamma_{\mathsf{A}} + \Gamma_{\mathsf{B}}$ and that $\theta_{\mathsf{A}\otimes_{\mathsf{F}}\mathsf{B}} \colon \Gamma_{\mathsf{A}} + \Gamma_{\mathsf{B}} \to \mathit{Aut}\left(Z(\mathsf{A})_0 \otimes_{\mathsf{F}_0} Z(\mathsf{B})_0 / \mathsf{F}_0\right)$ is given by $\gamma + \delta \mapsto \theta_{\mathsf{A}}(\gamma) \otimes \theta_{\mathsf{B}}(\delta)$, for all $\gamma \in \Gamma_{\mathsf{A}}$, $\delta \in \Gamma_{\mathsf{B}}$.

Let S be a separable closure of F_0. Since $Z(\mathsf{A})_0$ (resp. $Z(\mathsf{B})_0$) is Galois over F_0, we may identify it with its unique F_0-isomorphic copy in S. Then the compositum $Z(\mathsf{A})_0 \cdot Z(\mathsf{B})_0$ and the intersection $Z(\mathsf{A})_0 \cap Z(\mathsf{B})_0$ are well-defined subfields of S. Moreover, as $Z(\mathsf{A})_0$ is abelian Galois over F_0, the identification of $\mathcal{G}(Z(\mathsf{A})_0/\mathsf{F}_0)$ with the Galois group of the image of $Z(\mathsf{A})_0$ in S is independent of the choice of F_0-homomorphism $Z(\mathsf{A})_0 \hookrightarrow S$. Likewise for $Z(\mathsf{B})_0$. See the remarks preceding Prop. 2.45.

(ii) Let D be the graded division algebra associated to $\mathsf{A} \otimes_{\mathsf{F}} \mathsf{B}$. Prove that $Z(\mathsf{D}_0) \cong Z(\mathsf{A})_0 \cdot Z(\mathsf{B})_0$ and that D_0 is the division algebra associated to

$$\mathsf{A}_0 \otimes_{Z(\mathsf{A})_0} \left(Z(\mathsf{A})_0 \cdot Z(\mathsf{B})_0\right) \otimes_{Z(\mathsf{B})_0} \mathsf{B}_0.$$

Let $Z = Z(\mathsf{A})_0 \cap Z(\mathsf{B})_0 \subseteq S$. Recall (see Pierce [178, Lemma b, p. 256]) that since Z is Galois over F_0, we have

$$Z \otimes_{\mathsf{F}_0} Z \cong \prod_{\sigma \in \mathcal{G}(Z/\mathsf{F}_0)} e_\sigma Z,$$

where the primitive idempotents $\{e_\sigma \mid \sigma \in \mathcal{G}(Z/\mathsf{F}_0)\}$ of $Z \otimes_{\mathsf{F}_0} Z$ are characterized by the condition that $e_\sigma(c \otimes 1) = e_\sigma(1 \otimes \sigma(c))$ for all $c \in Z$.

(iii) Prove that $\Gamma_{\mathsf{D}} = \{\gamma + \delta \mid \gamma \in \Gamma_{\mathsf{A}},\ \delta \in \Gamma_{\mathsf{B}}, \text{ and } \theta_{\mathsf{A}}(\gamma)|_Z = \theta_{\mathsf{B}}(\delta)|_Z\}$. Hence $|\Gamma_{\mathsf{A}} + \Gamma_{\mathsf{B}} : \Gamma_{\mathsf{D}}| = [Z{:}\mathsf{F}_0]$.

Notes

Graded rings and modules are a classical topic, which is well-documented in the literature; see for example Bourbaki [31, § II.11, § III.3]. The idea to consider a graded ring in which nonzero homogeneous elements are invertible as a "graded field" can be traced back to Năstăsescu [169] in the special case where the grade group is $\mathbb{Z}$. Elaborating on this idea, it is natural to develop for semisimple graded algebras the analogue of the Wedderburn theory of semisimple algebras; this was done by Năstăsescu–Van Oystaeyen [170, § II.9] (for $\Gamma = \mathbb{Z}$); see also Boulagouaz [25] and Hwang–Wadsworth [103, §1]. The graded version of the Skolem–Noether Theorem (Th. 2.37) is due to Hwang–Wadsworth [103, Prop. 1.6]. The description of the zero-component of a simple graded algebra in §2.3 comes from Tignol–Wadsworth [246, §2].

Exercise 2.8 is a graded version of Morandi–Wadsworth [163, Cor. 3.12], which is for valued division algebras.

Chapter 3
Value Functions

Value functions on vector spaces over valued division algebras are the natural extensions of the notion of a valuation to vector spaces. In §3.1 we discuss the properties of value functions in relation to various linear algebra constructions: subspaces, homomorphisms, tensor products. In §3.2 we turn our attention to algebras and define gauges on semisimple algebras, which will be our main object of study in subsequent chapters. Like Ch. 2, the present chapter is thus mostly foundational in nature; its main purpose is to introduce basic concepts, develop tools, and lay the groundwork for more substantial later results. One of the highlights is a theorem of Morandi (Th. 3.43), which yields a very useful criterion for proving that a tensor product of valued division algebras is a division algebra.

Each value function on a vector space defines a filtration indexed by the set of values. Throughout the chapter, we emphasize the use of the graded structures associated to this filtration. When dealing with algebras, we will restrict to value functions with the property that the algebra multiplication is compatible with the filtration, so the associated graded structure is a graded algebra. Among these value functions, we define gauges by the conditions that the associated graded algebra has the right dimension and is semisimple.

3.1 Value functions on vector spaces

On a vector space over a valued division ring, value functions are defined by the same axioms as valuations, substituting scalar multiplication for the product. The most versatile value functions are those that can be computed as $\alpha(\sum e_i x_i) = min\{\alpha(e_i x_i)\}$ for some base (e_i). These value functions, which we call *norms*, are characterized by the property that the graded vector space associated to the filtration defined by the value function has the same dimension as the vector space itself. This property is reminiscent of the defectless condition for the extension of valuations (see Def. 3.31); we show in §3.1.2

J.-P. Tignol, A.R. Wadsworth, *Value Functions on Simple Algebras, and Associated Graded Rings*, Springer Monographs in Mathematics,
DOI 10.1007/978-3-319-16360-4_3

that over a maximally complete field all value functions are norms. We then show that norms restrict to norms on subspaces, and construct norms on spaces of homomorphisms and on tensor products.

Throughout this section, D will be a division ring with a valuation $v\colon D \to \Gamma \cup \{\infty\}$, where the totally ordered abelian group Γ can be chosen arbitrarily large. We consider the associated graded division ring $\mathsf{gr}(D)$ defined in §1.3.1.

3.1.1 Value functions and associated graded vector spaces

Definition 3.1. Let V be a right D-vector space (possibly infinite-dimensional). A map $\alpha\colon V \to \Gamma \cup \{\infty\}$ is called a *v-value function* if for all x, $y \in V$ and $d \in D$,

(i) $\alpha(x) = \infty$ if and only if $x = 0$;
(ii) $\alpha(x + y) \geq min(\alpha(x), \alpha(y))$;
(iii) $\alpha(xd) = \alpha(x) + v(d)$.

It clearly follows from the axioms that $\alpha(-x) = \alpha(x)$ and

$$\alpha(x + y) = min(\alpha(x), \alpha(y)) \qquad \text{whenever } \alpha(x) \neq \alpha(y).$$

Associated to α we have its *value set*

$$\Gamma_V = \{\alpha(x) \mid x \in V,\, x \neq 0\} \subseteq \Gamma.$$

This set Γ_V need not be a group, but it is a union of cosets of Γ_D, and we use the notation $|\Gamma_V{:}\Gamma_D|$ for the number of cosets.

The value function also yields an *associated graded vector space* $\mathsf{gr}(V)$, defined on the same model as the graded division ring $\mathsf{gr}(D)$: for $\gamma \in \Gamma$, we set

$$V_{\geq\gamma} = \{x \in V \mid \alpha(x) \geq \gamma\},\; V_{>\gamma} = \{x \in V \mid \alpha(x) > \gamma\},\; V_\gamma = V_{\geq\gamma}/V_{>\gamma},$$

and

$$\mathsf{gr}(V) = \bigoplus_{\gamma\in\Gamma} V_\gamma.$$

The D-vector space structure on V induces a well-defined $\mathsf{gr}(D)$-module action on $\mathsf{gr}(V)$, so $\mathsf{gr}(V)$ is a graded $\mathsf{gr}(D)$-vector space.

When we need to specify the value function, we write $\Gamma_{V,\alpha}$ for Γ_V and $\mathsf{gr}_\alpha(V)$ for $\mathsf{gr}(V)$.

Consistently with the notation for $\mathsf{gr}(D)$, for $x \in V$, $x \neq 0$ we write $\widetilde{x}$ for the image $x + V_{>\alpha(x)}$ of x in $V_{\alpha(x)}$. We also set $\widetilde{0} = 0$. By definition of the scalar multiplication in $\mathsf{gr}(V)$, we then have

$$\widetilde{xd} = \widetilde{x}\widetilde{d} \qquad \text{for all } x \in V,\, d \in D.$$

The following easy observations, whose proofs are omitted, are often useful:

Lemma 3.2. *For nonzero $x,\ y \in V$,*

$$\widetilde{x+y} \;=\; \begin{cases} \widetilde{x} & \text{if } \alpha(x) < \alpha(y); \\ \widetilde{y} & \text{if } \alpha(x) > \alpha(y). \end{cases}$$

Moreover, if $\alpha(x) = \alpha(y)$ and $x + y \neq 0$, then

$$\widetilde{x+y} = \widetilde{x} + \widetilde{y} \iff \alpha(x+y) = \alpha(x) = \alpha(y) \iff \widetilde{x} + \widetilde{y} \neq 0.$$

Example 3.3. Let V be a finite-dimensional D-vector space. Take any base $(e_i)_{i=1}^n$ of V and any sequence $(\gamma_i)_{i=1}^n$ in Γ, and define $\alpha\colon V \to \Gamma \cup \{\infty\}$ by

$$\alpha\big(\sum_{i=1}^{n} e_i d_i\big) \;=\; \min_{1\le i\le n} \big(\gamma_i + v(d_i)\big) \qquad \text{for } d_1,\ \ldots,\ d_n \in D. \tag{3.1}$$

It is easy to check that α satisfies the axioms for a v-value function on V.

This class of examples is particularly tractable and important; we single it out by giving it a special name:

Definition 3.4. A v-value function α on a finite-dimensional D-vector space V is a *v-norm* if there is a base $(e_i)_{i=1}^n$ of V for which formula (3.1) holds. Every such base is called a *splitting base* of V for α. More generally, a family of vectors $(x_i)_{i=1}^m$ is called a *splitting family* for a value function α if

$$\alpha\big(\sum_{i=1}^{m} x_i d_i\big) \;=\; \min_{1\le i\le m} \big(\alpha(x_i) + v(d_i)\big) \qquad \text{for all } d_1,\ \ldots,\ d_m \in D.$$

We will show in §3.1.2 below that if $[D{:}Z(D)] < \infty$ and $v|_{Z(D)}$ is maximally complete, then each v-value function on a finite-dimensional D-vector space is a norm. This holds for instance if $v|_{Z(D)}$ is a complete discrete (rank one) valuation, or if $Z(D)$ is an iterated Laurent series field and $v|_{Z(D)}$ is the usual valuation (see Ex. 3.11 below).

It is a key fact that norms and splitting bases are detected by properties of the associated graded vector space:

Proposition 3.5. *Let α be a v-value function on an arbitrary D-vector space V, and let $x_1,\ \ldots,\ x_m \in V$ be nonzero vectors. Then, $(\widetilde{x}_i)_{i=1}^m$ is $\mathsf{gr}(D)$-linearly independent in $\mathsf{gr}(V)$ if and only if $(x_i)_{i=1}^m$ is a splitting family. When this condition holds, $(x_i)_{i=1}^m$ is D-linearly independent in V.*

Proof. We prove the contrapositive. Suppose $\widetilde{x}_1, \ldots, \widetilde{x}_m$ are $\mathsf{gr}(D)$-dependent, say

$$\sum_{i=1}^{m} \widetilde{x}_i a_i \;=\; 0 \qquad \text{for some } a_1,\ \ldots,\ a_m \in \mathsf{gr}(D) \text{ not all zero.} \tag{3.2}$$

Suppose $a_k \neq 0$ and choose $\delta \in \Gamma$ such that a_k has nonzero δ-component. For $\varepsilon = \delta + \alpha(x_k)$ the ε-component of equation (3.2) reads

$$\sum_{i=1}^{m} \widetilde{x}_i b_i \; = \; 0,$$

where b_i is the homogeneous component of a_i of degree $\varepsilon - \alpha(x_i)$. Whenever $b_i \neq 0$, choose $d_i \in D^\times$ with $\widetilde{d_i} = b_i$, so $v(d_i) = \varepsilon - \alpha(x_i)$; whenever $b_i = 0$, set $d_i = 0$. The equation $\sum_i \widetilde{x}_i \widetilde{d_i} = 0$ in V_ε shows that

$$\alpha\big(\sum_{i=1}^{n} x_i d_i\big) \; > \; \varepsilon \; = \; \min_{1 \le i \le n} \big(\alpha(x_i) + v(d_i)\big).$$

Thus, $(x_i)_{i=1}^m$ is not a splitting family.

Conversely, suppose that for some $d_1, \ldots, d_m \in D$ we have

$$\alpha\big(\sum_{i=1}^{m} x_i d_i\big) \; > \; \min_{1 \le i \le m} \big(\alpha(x_i) + v(d_i)\big).$$

Let γ be the minimum quantity on the right, and let $I = \{i \mid \alpha(x_i) + v(d_i) = \gamma\}$. Then, $x_i d_i \in V_{\ge\gamma}$ for all $i = 1, \ldots, m$, and $x_i d_i \in V_{>\gamma}$ if and only if $i \notin I$. Since $\sum_{i=1}^m x_i d_i \in V_{>\gamma}$, we have in V_γ

$$0 \; = \; \sum_{i=1}^{m} x_i d_i + V_{>\gamma} \; = \; \sum_{i \in I} x_i d_i + V_{>\gamma} \; = \; \sum_{i \in I} \widetilde{x}_i \widetilde{d}_i.$$

Since $d_i \neq 0$ for $i \in I$, it follows that $\widetilde{x}_1, \ldots, \widetilde{x}_m$ are $\mathsf{gr}(D)$-linearly dependent in $\mathsf{gr}(V)$. The argument applies in particular if $\sum_{i=1}^m x_i d_i = 0$ with $d_1, \ldots, d_m$ not all zero, showing that $(\widetilde{x}_i)_{i=1}^m$ is $\mathsf{gr}(D)$-linearly dependent in $\mathsf{gr}(V)$ if $(x_i)_{i=1}^m$ is D-linearly dependent in V. □

The following are easy consequences of the proposition:

Corollary 3.6. *Let α be a v-value function on a finite-dimensional D-vector space V. Then,*

(i) *$\dim_{\mathsf{gr}(D)} \mathsf{gr}(V) \le \dim_D V$, and equality holds if and only if α is a norm.*
(ii) *Suppose α is a norm. A family of vectors $(x_i)_{i=1}^n$ is a splitting base of V for α if and only if $(\widetilde{x}_i)_{i=1}^n$ is a homogeneous base of $\mathsf{gr}(V)$.*

Example 3.7. Let $F \subseteq K$ be a field extension of finite degree, let w be a valuation of K, and let $v = w|_F$. When we view K as an F-vector space, w is clearly a v-value function on K. By Cor. 2.7 and the Fundamental Inequality (Prop. 1.3), we have

$$[\mathsf{gr}(K) \colon \mathsf{gr}(F)] \; = \; [\overline{K} \colon \overline{F}] \cdot |\Gamma_K \colon \Gamma_F| \; \le \; [K \colon F]. \tag{3.3}$$

Thus, Cor. 3.6 shows that w is a v-norm on K if and only if equality holds in (3.3). It follows from the Fundamental Inequality (see Th. A.2 in Appendix A) that if w is not the only extension of v to K, then the inequality in (3.3) is strict, so w is not a v-norm. See §3.2.2 and §4.3.3 below for further considerations of building v-norms on K from extensions of v to K.

3.1.2 Value functions over maximally complete fields

A valued field (F, v) is said to be *maximally complete* if it has no proper immediate extension, i.e., there is no valued field extension (K, v_K) of (F, v) with $K \neq F$ but $\Gamma_K = \Gamma_F$ and $\overline{K} = \overline{F}$. We will show that maximally complete fields are precisely those for which every value function on a finite-dimensional vector space is a norm. The proof will use characterizations of maximally complete fields in terms of pseudo-Cauchy sequences and compatible systems of congruences.

Let F be a field with valuation v. A *pseudo-Cauchy sequence* (also called a pseudo-convergent sequence) in F is a sequence $(a_\sigma)_{\sigma\in\Sigma}$ of elements of F, where Σ is a well-ordered set with no largest element, and for all $\sigma, \tau, \rho \in \Sigma$,

$$v(a_\sigma - a_\tau) < v(a_\tau - a_\rho) \qquad \text{whenever} \qquad \sigma < \tau < \rho.$$

For $\sigma \in \Sigma$, let $\sigma + 1$ denote the least $\rho \in \Sigma$ with $\sigma < \rho$, which exists as Σ is well-ordered with no largest element. Note that for all $\sigma, \tau \in \Sigma$,

$$v(a_\sigma - a_\tau) = v(a_\sigma - a_{\sigma+1}) \qquad \text{whenever} \qquad \sigma < \tau. \tag{3.4}$$

This is clear if $\tau = \sigma+1$. Otherwise, $\tau > \sigma+1$, so $v(a_\sigma - a_{\sigma+1}) < v(a_{\sigma+1} - a_\tau)$; then,

$$\begin{aligned} v(a_\sigma - a_\tau) &= v\big((a_\sigma - a_{\sigma+1}) + (a_{\sigma+1} - a_\tau)\big) \\ &= \min\big(v(a_\sigma - a_{\sigma+1}), v(a_{\sigma+1} - a_\tau)\big) = v(a_\sigma - a_{\sigma+1}). \end{aligned}$$

An element $b \in F$ is called a *pseudolimit* of the pseudo-Cauchy sequence $(a_\sigma)_{\sigma\in\Sigma}$ if

$$v(b - a_\sigma) = v(a_\sigma - a_{\sigma+1}) \quad \text{for each } \sigma \in \Sigma.$$

A *compatible system of congruences* in F is a family of congruence conditions

$$x \equiv c_\sigma \pmod{I_\sigma} \quad \text{for each } \sigma \in \Sigma,$$

where Σ is a well-ordered set with no largest element, each $c_\sigma \in F$ and each I_σ is an $\mathcal{O}_F$-submodule of F, such that whenever $\sigma \leq \tau$ we have $I_\tau \subseteq I_\sigma$ and $c_\tau \equiv c_\sigma \pmod{I_\sigma}$. A solution of the system is a $d \in F$ with $d \equiv c_\sigma \pmod{I_\sigma}$ for each $\sigma \in \Sigma$.

Proposition 3.8. *For any valued field (F, v), the following conditions are equivalent:*

(a) *(F, v) is maximally complete.*
(b) *Every v-value function on a finite-dimensional F-vector space is a norm.*
(c) *Every pseudo-Cauchy sequence in F has a pseudolimit.*
(d) *Every compatible system of congruences in F has a solution.*

Proof. (a) $\Leftrightarrow$ (c) is a theorem of Kaplansky. See Schilling [227, pp. 47–51] for a proof.

(b) $\Rightarrow$ (a) Let (K, v_K) be any valued field extension of (F, v) with $K \neq F$. Take any F-subspace V of K with $1 < dim_F V < \infty$ and $F \subseteq V$; let $\alpha = v_K|_V$, which is a v-value function on V. By (b) α is a v-norm, so $dim_{\mathsf{gr}(F)} \mathsf{gr}(V) = dim_F V > 1$. Hence, by Prop. 2.5, $\Gamma_V \supsetneqq \Gamma_F$ or $V_0 \supsetneqq F_0 = \overline{F}$. Then since $\Gamma_K \supseteq \Gamma_V$ and $\overline{K} \supseteq V_0$, v_K is not an immediate extension of v. Thus, (F, v) is maximally complete.

(d) $\Rightarrow$ (c) Let $(a_\sigma)_{\sigma \in \Sigma}$ be a pseudo-Cauchy sequence in F. For $\sigma \in \Sigma$, let $I_\sigma = F_{\geq v(a_\sigma - a_{\sigma+1})}$. For any $\tau > \sigma$, (3.4) yields

$$v(a_\tau - a_{\tau+1}) > v(a_\sigma - a_\tau) = v(a_\sigma - a_{\sigma+1});$$

hence, $I_\tau \subseteq I_\sigma$. Also, (3.4) shows that $a_\tau \equiv a_\sigma (mod\, I_\sigma)$. Thus, $x \equiv a_\sigma (mod\, I_\sigma)$ is a compatible system of congruences. By (d) there is a solution b for this system. So, for any $\sigma \in \Sigma$ we have

$$v(b - a_{\sigma+1}) \geq v(a_{\sigma+1} - a_{(\sigma+1)+1}) > v(a_\sigma - a_{\sigma+1});$$

hence,

$$v(b - a_\sigma) = min\big(v(b - a_{\sigma+1}), v(a_\sigma - a_{\sigma+1})\big) = v(a_\sigma - a_{\sigma+1}).$$

Thus, b is a pseudolimit of the pseudo-Cauchy sequence $(a_\sigma)_{\sigma \in \Sigma}$.

To prove (c) $\Rightarrow$ (d) we will use the following lemma:

Lemma 3.9. *Let $x \equiv a_\sigma$ (mod I_σ) for $\sigma \in \Sigma$ be a compatible system of congruences. Suppose no a_ρ is a solution of the system. Then there is a cofinal subset Σ_0 of Σ such that $(a_\tau)_{\tau \in \Sigma_0}$ is a pseudo-Cauchy sequence.*

Proof of Lemma 3.9. Take any $\sigma \in \Sigma$. Since a_σ is assumed not to be a solution to the system of congruences, there is a $\sigma' \in \Sigma$ with $a_\sigma \not\equiv a_{\sigma'}$ (mod $I_{\sigma'}$). The compatibility of the system forces $\sigma' > \sigma$. For each $\sigma \in \Sigma$ choose and fix such a σ'.

Let Σ_0 be a nonempty subset of Σ which is maximal with respect to the property that for any $\sigma, \tau \in \Sigma_0$, if $\sigma < \tau$ then $\sigma' \leq \tau$. Such a Σ_0 exists by Zorn's Lemma, and Σ_0 is well-ordered since Σ is well-ordered. Furthermore, Σ_0 must be cofinal in Σ. For otherwise there is a $\theta \in \Sigma$ with $\sigma < \theta$ for all $\sigma \in \Sigma_0$. If Σ_0 has no largest element, fix any such θ. Then for each $\tau \in \Sigma_0$ there is a $\rho \in \Sigma_0$ with $\tau < \rho$; so, $\tau' \leq \rho < \theta$. If Σ_0 has a largest element ω, choose $\theta = \omega'$. Then, for $\tau \in \Sigma_0$ with $\tau \neq \omega$ we have $\tau < \omega$, so $\tau' \leq \omega < \omega' = \theta$. In either case, $\Sigma_0 \cup \{\theta\}$ satisfies the same condition as Σ_0, contradicting the maximality of Σ_0. This yields the asserted cofinality. Then Σ_0 has no largest element, since this is true for Σ.

We check that $(a_\tau)_{\tau \in \Sigma_0}$ is a pseudo-Cauchy sequence. Take any $\sigma, \tau, \rho \in \Sigma_0$ with $\sigma < \tau < \rho$. So, $\sigma < \sigma' \leq \tau < \rho$. By compatibility, $a_\tau - a_\rho \in I_\tau \subseteq I_{\sigma'}$; so, $F_{\geq v(a_\tau - a_\rho)} \subseteq I_{\sigma'}$. But $a_\sigma - a_{\sigma'} \notin I_{\sigma'}$ by the definition of σ', while by compatibility $a_{\sigma'} - a_\tau \in I_{\sigma'}$; together these show that $a_\sigma - a_\tau \notin I_{\sigma'}$. Hence, $v(a_\sigma - a_\tau) < v(a_\tau - a_\rho)$, proving that $(a_\tau)_{\tau \in \Sigma_0}$ is pseudo-Cauchy. □

Proof of Prop. 3.8 (continued). (c) $\Rightarrow$ (d) Suppose every pseudo-Cauchy sequence in F has a pseudolimit. Take any compatible system of congruences $x \equiv a_\sigma \ (mod\ I_\sigma)$ for $\sigma \in \Sigma$. If some a_τ is a solution of the system, we are done. If not, the lemma yields a cofinal subset Σ_0 of Σ with $(a_\tau)_{\tau\in\Sigma_0}$ a pseudo-Cauchy sequence. By hypothesis there is a pseudolimit b of this sequence. For each $\sigma \in \Sigma$ there is a $\tau \in \Sigma_0$ with $\sigma < \tau$, so $a_\sigma - a_\tau \in I_\sigma$. For the successor $\tau + 1$ of τ in Σ_0 we have $v(b - a_\tau) = v(a_\tau - a_{\tau+1})$; so as $a_\tau - a_{\tau+1} \in I_\tau$ we obtain $b - a_\tau \in I_\tau \subseteq I_\sigma$. Thus, $b - a_\sigma = (b - a_\tau) - (a_\sigma - a_\tau) \in I_\sigma$, showing that b is a solution of the original system of compatible congruences.

(d) $\Rightarrow$ (b) Let α be a v-value function on a finite-dimensional F-vector space V. Suppose α is not a norm on V. Then, $dim_{\mathsf{gr}(F)}\,\mathsf{gr}(V) < dim_F V$ by Cor. 3.6(i). Pick any $b_1, \ldots, b_k \in V \setminus \{0\}$ such that $\{\widetilde{b}_1, \ldots, \widetilde{b}_k\}$ is a homogeneous $\mathsf{gr}(F)$-base of $\mathsf{gr}(V)$, and let $W = F\text{-span}\{b_1, \ldots, b_k\}$, a proper subspace of V. Our α restricts to a v-value function on W, and $\mathsf{gr}(W)$ embeds canonically into $\mathsf{gr}(V)$, so we view $\mathsf{gr}(W)$ as a graded $\mathsf{gr}(F)$-subspace of $\mathsf{gr}(V)$. Since $\widetilde{b}_1, \ldots, \widetilde{b}_k$ are $\mathsf{gr}(F)$-independent and lie in $\mathsf{gr}(W)$, we have

$$k \le dim_{\mathsf{gr}(F)}\,\mathsf{gr}(W) \le dim_{\mathsf{gr}(F)}\,\mathsf{gr}(V) = k;$$

hence, $\mathsf{gr}(W) = \mathsf{gr}(V)$, and $(b_1, \ldots, b_k)$ is a splitting base of W for the norm $\alpha|_W$, by Cor. 3.6.

Take any $t \in V \setminus W$, and let $\Sigma = \{\gamma \in \Gamma \mid \gamma = \alpha(t - y) \text{ for some } y \in W\}$. Then, Σ has no largest element. This can be seen as follows: if $\gamma = \alpha(t - y)$ with $y \in W$, choose $s \in W$ with $\widetilde{s} = \widetilde{t - y}$ in $\mathsf{gr}(W) = \mathsf{gr}(V)$. Then, $s + y \in W$ and

$$\alpha(t - (s + y)) = \alpha((t - y) - s) > \alpha(t - y) = \gamma.$$

Let Σ_1 be a well-ordered subset of Σ with Σ_1 cofinal in Σ. So, Σ_1 has no largest element. For each $\sigma \in \Sigma_1$ choose $y_\sigma \in W$ with $\alpha(t - y_\sigma) = \sigma$. Then write $y_\sigma = \sum_{i=1}^k b_i c_{i,\sigma}$ with each $c_{i,\sigma} \in F$. Let $\delta_i = \alpha(b_i)$, $1 \le i \le k$. Then, for $\sigma, \rho \in \Sigma_1$ with $\sigma < \rho$,

$$\alpha(y_\sigma - y_\rho) = \alpha\Big(\sum_{i=1}^k b_i(c_{i,\sigma} - c_{i,\rho})\Big) = \min_{1\le i\le k}\big(v(c_{i,\sigma} - c_{i,\rho}) + \delta_i\big).$$

But also, as $\alpha(t - y_\sigma) = \sigma < \rho = \alpha(t - y_\rho)$,

$$\alpha(y_\sigma - y_\rho) = \min\big(\alpha(t - y_\sigma), \alpha(t - y_\rho)\big) = \sigma.$$

These equations yield, for each i,

$$v(c_{i,\sigma} - c_{i,\rho}) \ge \sigma - \delta_i \qquad \text{for all } \sigma, \rho \in \Sigma_1 \text{ with } \sigma < \rho. \tag{3.5}$$

Fix i for the moment. For $\sigma \in \Sigma_1$, let $I_\sigma = \{a \in F \mid v(a) \ge \sigma - \delta_i\}$, which is an $\mathcal{O}_F$-submodule of F. Clearly, $I_\rho \subseteq I_\sigma$ whenever $\sigma < \rho$. Equation (3.5) shows that $c_{i,\sigma} \equiv c_{i,\rho} \ (mod\ I_\sigma)$ whenever $\sigma < \rho$ in Σ_1. Thus, $x \equiv c_{i,\sigma} \ (mod\ I_\sigma)$ for $\sigma \in \Sigma_1$ is a compatible system of congruences in F. By hypothesis (d) there is an $x_i \in F$ which is a solution to this system. Given such an x_i for each i, let $z = b_1x_1 + \ldots + b_kx_k$. For each $\sigma \in \Sigma_1$ we have $v(x_i - c_{i,\sigma}) \geq \sigma - \delta_i$; hence,

$$\begin{aligned}\alpha(y_\sigma - z) &= \alpha\big(\sum_{i=1}^{k} b_i(c_{i,\sigma} - x_i)\big) = \min_{1\leq i\leq k} \big(v(c_{i,\sigma} - x_i) + \delta_i\big) \\ &\geq \min_{1\leq i\leq k} \big((\sigma - \delta_i) + \delta_i\big) = \sigma.\end{aligned}$$

Therefore, $\alpha(t - z) \geq \min\big(\alpha(t - y_\sigma), \alpha(y_\sigma - z)\big) = \sigma$; this is true for every $\sigma \in \Sigma_1$. Since $\alpha(t - z)$ is thus an upper bound for Σ_1 and Σ has no largest element, we have a contradiction to the cofinality of Σ_1 in Σ. Thus, α must be a norm on V, proving (b). □

Corollary 3.10. *Let (F, v) be a maximally complete valued field, let D be a finite-dimensional division algebra over F, and let v_D be the extension of v to D. Then, any v_D-value function on a finite-dimensional D-vector space is a norm.*

Proof. Note that v must be Henselian, since the Henselization of (F, v) is an immediate extension by Cor. A.28. So, by Cor. 1.7 the valuation v_D exists, and is the unique extension of v to D. Take a v_D-value function α on a finite-dimensional D-vector space V. Then, V is finite-dimensional as an F-vector space and α is a v-value function on V. By Prop. 3.8 v_D and α are each v-norms. Thus, by Cor. 3.6(i),

$$\begin{aligned}[D{:}F]\ dim_D V &= dim_F V = dim_{\mathsf{gr}(F)} \mathsf{gr}(V) \\ &= [\mathsf{gr}(D){:}\, \mathsf{gr}(F)]\ dim_{\mathsf{gr}(D)} \mathsf{gr}(V) = [D{:}F]\ dim_{\mathsf{gr}(D)} \mathsf{gr}(V).\end{aligned}$$

Hence, $dim_{\mathsf{gr}(D)} \mathsf{gr}(V) = dim_D(V)$, so by Cor. 3.6(i), α is a v_D-norm. □

Example 3.11. (i) A field F with a complete discrete (rank 1) valuation is maximally complete. This is immediate from (c) $\Rightarrow$ (a) of Prop. 3.8 since when $\Gamma_F = \mathbb{Z}$ every pseudo-Cauchy sequence is a Cauchy sequence, and the limit of the sequence is a pseudo-limit.

(ii) Let (F, w) be a valued field, let u be a valuation on the residue field $\overline{F}^{w}$, and let $v = u * w$, the composite valuation on F. It is easy to check that if (F, w) and $(\overline{F}^{w}, u)$ are maximally complete, then (F, v) is also maximally complete. It follows by induction using (i) that for any field k, the iterated Laurent series field $k((x_1))\ldots((x_n))$ with its usual valuation is maximally complete.

3.1.3 Subspaces

Let U be any subspace of an arbitrary D-vector space V. Clearly, every v-value function α on V restricts to a v-value function $\alpha|_U$ on U, with $\Gamma_U \subseteq \Gamma_V$. Moreover, for each $\gamma \in \Gamma$ we have $U_{\geq\gamma} = V_{\geq\gamma} \cap U$ and $U_{>\gamma} = V_{>\gamma} \cap U$, hence there is a natural injection $U_\gamma \hookrightarrow V_\gamma$, which we view as an inclusion. Then $\mathsf{gr}(U)$ is a graded $\mathsf{gr}(D)$-subspace of $\mathsf{gr}(V)$.

Definition 3.12. A subspace $W \subseteq V$ is called a *splitting complement* of U for α if $V = U \oplus W$ and

$$\alpha(u+w) \;=\; min\big(\alpha(u), \alpha(w)\big) \qquad \text{for all } u \in U,\ w \in W.$$

We will prove that if α is a norm on V, then $\alpha|_U$ is a norm on U and U has a splitting complement.

Lemma 3.13. *If $(x_i)_{i=1}^m$ is a splitting family in V, then*

$$\mathsf{gr}\big(\sum_{i=1}^m x_i D\big) \;=\; \sum_{i=1}^m \widetilde{x}_i\,\mathsf{gr}(D).$$

Proof. The inclusion $\mathsf{gr}(\sum_{i=1}^m x_i D) \supseteq \sum_{i=1}^m \widetilde{x}_i\,\mathsf{gr}(D)$ is clear. To prove the reverse inclusion, it suffices to show that every homogeneous element in the left set also lies in the right set. For this, take any nonzero $\sum_i x_i d_i$, and let

$$\gamma \;=\; \alpha\big(\sum_i x_i d_i\big) \;=\; \min_i\big(\alpha(x_i) + v(d_i)\big) \in \Gamma.$$

Then, letting $I = \{i \mid \alpha(x_i) + v(d_i) = \gamma\}$, we have in V_γ

$$\big(\sum_{i=1}^m x_i d_i\big)^{\sim} \;=\; \sum_{i\in I} \widetilde{x}_i \widetilde{d}_i \;\in\; \sum_{i=1}^m \widetilde{x}_i\,\mathsf{gr}(D). \qquad \square$$

Proposition 3.14. *Suppose V is finite-dimensional and α is a v-norm on V. Then for any subspace $U \subseteq V$ the restriction $\alpha|_U$ is a norm, and U has a splitting complement in V. If W is a subspace of V, then W is a splitting complement of U if and only if*

$$\mathsf{gr}(V) \;=\; \mathsf{gr}(U) \oplus \mathsf{gr}(W).$$

Proof. By Cor. 3.6(i), we have $dim_{\mathsf{gr}(D)}\,\mathsf{gr}(U) \le dim_D\,U$. We will prove the dimensions are equal. Let $m = dim_{\mathsf{gr}(D)}\,\mathsf{gr}(U)$. Choose a homogeneous base $(\widetilde{x}_i)_{i=1}^m$ of $\mathsf{gr}(U)$ with $x_1, \ldots, x_m \in U$, and extend it to a homogeneous base $(\widetilde{x}_i)_{i=1}^n$ of $\mathsf{gr}(V)$. Then $(x_i)_{i=1}^n$ is a splitting base of V for α, by Cor. 3.6(ii). Let

$$W \;=\; \sum_{i=m+1}^n x_i D \subseteq V$$

be the span of $x_{m+1}, \ldots, x_n$. Since $(x_i)_{i=m+1}^n$ is a splitting family, Lemma 3.13 yields

$$\mathsf{gr}(W) \;=\; \sum_{i=m+1}^n \widetilde{x}_i\,\mathsf{gr}(D),$$

hence, $\mathsf{gr}(U) \cap \mathsf{gr}(W) = \{0\}$. Note that $m = dim_D V - dim_D W$. If $dim_D U > m$, then there is a nonzero $y \in U \cap W$, and we have $\widetilde{y} \in \mathsf{gr}(U) \cap \mathsf{gr}(W)$, a contradiction since $\widetilde{y} \ne 0$. Therefore, $dim_D\,U = m$, showing that $\alpha|_U$ is a norm by Cor. 3.6(i). Moreover, Cor. 3.6(ii) shows that $(x_i)_{i=1}^m$ is a splitting base of U. Since $(x_i)_{i=m+1}^n$ is a splitting base of W and $(x_i)_{i=1}^n$ is a splitting base of V, it follows that W is a splitting complement of U.

The preceding argument shows that if W is any subspace of V satisfying $\mathsf{gr}(U) \oplus \mathsf{gr}(W) = \mathsf{gr}(V)$, then W is a splitting complement of U in V. Conversely, suppose W is an arbitrary splitting complement of U in V. If $(u_i)_{i=1}^m$ is a splitting base of U and $(u_i)_{i=m+1}^n$ a splitting base of W, then $(u_i)_{i=1}^n$ is a splitting base of V. By Cor. 3.6(ii), $(\widetilde{u}_i)_{i=1}^m$ is a $\mathsf{gr}(D)$-base of $\mathsf{gr}(U)$, $(\widetilde{u}_i)_{i=m+1}^n$ a $\mathsf{gr}(D)$-base of $\mathsf{gr}(W)$, and $(\widetilde{u}_i)_{i=1}^n$ a $\mathsf{gr}(D)$-base of $\mathsf{gr}(V)$. Therefore, $\mathsf{gr}(V) = \mathsf{gr}(U) \oplus \mathsf{gr}(W)$. □

Remark 3.15. With the notation as in Prop. 3.14, suppose W is a splitting complement of U. The map $\mathsf{gr}(U) \oplus \mathsf{gr}(W) \to \mathsf{gr}(V)$ arises from the canonical inclusions $\mathsf{gr}(U) \hookrightarrow \mathsf{gr}(V)$ and $\mathsf{gr}(W) \hookrightarrow \mathsf{gr}(V)$. The inverse map is given as follows: for any $u \in U$, $w \in W$,

$$\widetilde{u+w} \;\mapsto\; \begin{cases} (\widetilde{u}, \widetilde{w}) & \text{if } \alpha(u) = \alpha(w), \\ (\widetilde{u}, 0) & \text{if } \alpha(u) < \alpha(w), \\ (0, \widetilde{w}) & \text{if } \alpha(u) > \alpha(w). \end{cases}$$

3.1.4 Homomorphisms

Let V and W be right D-vector spaces with respective v-value functions α and β.

Lemma 3.16. *Let $f\colon V \to W$ be a D-linear map. If for some $\gamma \in \Gamma$ we have*

$$\beta(f(x)) \geq \alpha(x) + \gamma \qquad \textit{for all } x \in V, \tag{3.6}$$

then f induces a map $\widehat{f}^\gamma \in \mathsf{Hom}_{\mathsf{gr}(D)}\big(\mathsf{gr}(V), \mathsf{gr}(W)\big)_\gamma$ determined by

$$\widehat{f}^\gamma(\widetilde{x}) \;=\; f(x) + W_{>\alpha(x)+\gamma} \;\in\; W_{\alpha(x)+\gamma} \qquad \textit{for all } x \in V \setminus \{0\}.$$

The map $\widehat{f}^\gamma$ is injective if and only if equality holds in (3.6) for all $x \in V$, while $\widehat{f}^\gamma = 0$ if and only if the inequality in (3.6) is strict for all $x \in V$.

Proof. For any $\delta \in \Gamma$, (3.6) shows that $f(V_{\geq\delta}) \subseteq W_{\geq\delta+\gamma}$ and $f(V_{>\delta}) \subseteq W_{>\delta+\gamma}$; hence, f induces a map $(\widehat{f}^\gamma)_\delta\colon V_\delta \to W_{\delta+\gamma}$ given by

$$(\widehat{f}^\gamma)_\delta(x + V_{>\delta}) \;=\; f(x) + W_{>\delta+\gamma}.$$

Then set $\widehat{f}^\gamma = \bigoplus_{\delta\in\Gamma}(\widehat{f}^\gamma)_\delta$. Clearly, $\widehat{f}^\gamma$ lies in $\mathsf{Hom}_{\mathsf{gr}(D)}\big(\mathsf{gr}(V), \mathsf{gr}(W)\big)_\gamma$. Its kernel is therefore spanned by homogeneous elements, and we have $\widehat{f}^\gamma(\widetilde{x}) = 0$ if and only if $\beta(f(x)) > \alpha(x) + \gamma$. Thus, the stated conditions for injectivity of $\widehat{f}^\gamma$ and for $\widehat{f}^\gamma = 0$ hold. □

If V is infinite-dimensional, it is not difficult to construct examples where (3.6) does not hold for any $\gamma \in \Gamma$. However, if V is finite-dimensional and α is a norm, then the following lemma shows that there is an optimal choice for γ:

Lemma 3.17. *Suppose* $\dim_D V < \infty$ *and* α *is a* v*-norm on* V. *Let* $(e_i)_{i=1}^n$ *be a splitting base of* V *and let* $f \in \mathit{Hom}_D(V, W)$. *Then* (3.6) *holds with*

$$\gamma = \min_{1\leq i\leq n} \big(\beta(f(e_i)) - \alpha(e_i)\big) \in \Gamma \cup \{\infty\}.$$

Proof. For $x = \sum_{i=1}^n e_i d_i \in V$, we have $f(x) = \sum_{i=1}^n f(e_i) d_i$, hence

$$\beta(f(x)) \geq \min_{1\leq i\leq n} \big(\beta(f(e_i)) + v(d_i)\big) = \min_{1\leq i\leq n} \big(\beta(f(e_i)) - \alpha(e_i) + \alpha(e_i) + v(d_i)\big).$$

On the other hand, since $(e_i)_{i=1}^n$ is a splitting base, we have

$$\alpha(x) = \min_{1\leq i\leq n} \big(\alpha(e_i) + v(d_i)\big).$$

Since

$$\min_{1\leq i\leq n} \big(\beta(f(e_i)) - \alpha(e_i) + \alpha(e_i) + v(d_i)\big) \geq \min_{1\leq i\leq n} \big(\beta(f(e_i)) - \alpha(e_i)\big) + \min_{1\leq i\leq n} \big(\alpha(e_i) + v(d_i)\big),$$

we have $\beta(f(x)) \geq \gamma + \alpha(x)$, and the lemma follows. □

Lemma 3.17 shows that for $f \neq 0$ in $\mathit{Hom}_D(V, W)$ there is a minimum in Γ for the set $\{\beta(f(x)) - \alpha(x) \mid x \in V,\ x \neq 0\}$: this minimum is in fact $\min_{1\leq i\leq n}\big(\beta(f(e_i)) - \alpha(e_i)\big)$ for any splitting base $(e_i)_{i=1}^n$ of V for α. For $f \in \mathit{Hom}_D(V, W)$, we set

$$\mathit{Hom}(\alpha, \beta)(f) = \min\{\beta(f(x)) - \alpha(x) \mid x \in V,\, x \neq 0\} \in \Gamma \cup \{\infty\}. \tag{3.7}$$

If $f \neq 0$, then $\gamma = \mathit{Hom}(\alpha, \beta)(f)$ exists and lies in Γ. We thus have the map $\widehat{f}^\gamma \in \mathsf{Hom}_{\mathsf{gr}(D)}\big(\mathsf{gr}(V), \mathsf{gr}(W)\big)_\gamma$ of Lemma 3.16.

Proposition 3.18. *Suppose* $\dim_D V < \infty$ *and* α *is a* v*-norm. Then, the map* $\mathit{Hom}(\alpha, \beta)$ *defined in* (3.7) *is a* $v|_{Z(D)}$*-value function on* $\mathit{Hom}_D(V, W)$ *(viewed as a* $Z(D)$*-vector space). Moreover, there is a canonical (degree-preserving) isomorphism of graded* $\mathsf{gr}(Z(D))$*-vector spaces*

$$\Phi\colon \mathsf{gr}_{\mathit{Hom}(\alpha,\beta)}\big(\mathit{Hom}_D(V, W)\big) \xrightarrow{\sim} \mathsf{Hom}_{\mathsf{gr}(D)}\big(\mathsf{gr}_\alpha(V), \mathsf{gr}_\beta(W)\big)$$

mapping $f + \mathit{Hom}_D(V, W)_{>\gamma}$ *to* $\widehat{f}^\gamma$ *for* $f \neq 0$ *and* $\gamma = \mathit{Hom}(\alpha, \beta)(f)$.

If, in addition, $\dim_D W < \infty$, β *is a* v*-norm,* $[D{:}Z(D)] < \infty$, *and* v *is a* $v|_{Z(D)}$*-norm on* D, *then* $\mathit{Hom}(\alpha, \beta)$ *is a* $v|_{Z(D)}$*-norm.*

Proof. The axioms for a $v|_{Z(D)}$-value function are easily verified. It is clear from the definitions that Φ is a well-defined degree-preserving $\mathsf{gr}(Z(D))$-linear map of graded vector spaces. Therefore, it suffices to check injectivity and surjectivity on homogeneous elements. If $f \neq 0$ in $\mathit{Hom}_D(V, W)$

and $\gamma = \mathit{Hom}(\alpha,\beta)(f)$, then there exists a nonzero $x \in V$ such that $\beta(f(x)) = \alpha(x)+\gamma$, hence $\widehat{f}^{\gamma}(\widetilde{x}) = \widetilde{f(x)} \neq 0$. Therefore, $\widehat{f}^{\gamma} \neq 0$ and Φ is injective.

To prove surjectivity, fix a splitting base $(e_i)_{i=1}^n$ of V for α. Then $(\widetilde{e}_i)_{i=1}^n$ is a $\mathsf{gr}(D)$-base of $\mathsf{gr}_\alpha(V)$, by Cor. 3.6(ii). Let $\varphi \in \mathsf{Hom}_{\mathsf{gr}(D)}\big(\mathsf{gr}_\alpha(V),\mathsf{gr}_\beta(W)\big)_\gamma$, for some $\gamma \in \Gamma$. Choose $w_1, \ldots, w_n \in W$ satisfying

$$\varphi(\widetilde{e}_i) = w_i + W_{>\alpha(e_i)+\gamma} \in W_{\alpha(e_i)+\gamma} \qquad \text{for } i = 1, \ldots, n.$$

Then define $f \in \mathit{Hom}_D(V,W)$ by

$$f(e_i) = w_i \qquad \text{for } i = 1, \ldots, n.$$

We have $\beta(w_i) \geq \alpha(e_i)+\gamma$ for all i, with equality for at least one i if $\varphi \neq 0$. In this case we have $\mathit{Hom}(\alpha,\beta)(f) = \gamma$ and $\widehat{f}^{\gamma} = \varphi$, proving that Φ is surjective.

For the rest of the proof, suppose that $\mathit{dim}_D W < \infty$, $[D{:}Z(D)] < \infty$, β is a v-norm, and v is a $v|_{Z(D)}$-norm. Since Φ is an isomorphism we have

$$\mathit{dim}_{\mathsf{gr}(Z(D))}\big(\mathsf{gr}_{\mathit{Hom}(\alpha,\beta)}\big(\mathit{Hom}_D(V,W)\big)\big) = \\ [\mathsf{gr}(D){:}\, \mathsf{gr}(Z(D))]\ \mathit{dim}_{\mathsf{gr}(D)}\big(\mathsf{gr}_\alpha(V)\big)\ \mathit{dim}_{\mathsf{gr}(D)}\big(\mathsf{gr}_\beta(W)\big).$$

On the other hand, since v, α, and β are norms we have by Cor. 3.6(i)

$$[\mathsf{gr}(D){:}\, \mathsf{gr}(Z(D))] = [D{:}Z(D)], \qquad \mathit{dim}_{\mathsf{gr}(D)}\big(\mathsf{gr}_\alpha(V)\big) = \mathit{dim}_D V,$$

and

$$\mathit{dim}_{\mathsf{gr}(D)}\big(\mathsf{gr}_\beta(W)\big) = \mathit{dim}_D W.$$

Therefore,

$$\mathit{dim}_{\mathsf{gr}(Z(D))}\big(\mathsf{gr}_{\mathit{Hom}(\alpha,\beta)}\big(\mathit{Hom}_D(V,W)\big)\big) = \mathit{dim}_{Z(D)}\, \mathit{Hom}_D(V,W),$$

and it follows from Cor. 3.6 that $\mathit{Hom}(\alpha,\beta)$ is a norm. □

In view of Prop. 3.18, it is consistent with our general notation to write simply $\widetilde{f}$ for $\widehat{f}^{\gamma}$ if $\gamma = \mathit{Hom}(\alpha,\beta)(f)$, using Φ as an identification (assuming that α is a norm, but not necessarily that $\mathit{dim}_D W < \infty$). Thus, by definition we have for $x \in V$ and $f \in \mathit{Hom}_D(V,W)$:

$$\widetilde{f}(\widetilde{x}) = \begin{cases} \widetilde{f(x)} & \text{if } \beta(f(x)) = \mathit{Hom}(\alpha,\beta)(f)+\alpha(x), \\ 0 & \text{if } \beta(f(x)) > \mathit{Hom}(\alpha,\beta)(f)+\alpha(x). \end{cases} \tag{3.8}$$

Since $\widetilde{f}$ is homogeneous, $\mathit{ker}\,\widetilde{f}$ and $\mathit{im}\,\widetilde{f}$ are graded subspaces of $\mathsf{gr}(V)$ and $\mathsf{gr}(W)$ respectively. We have obvious inclusions

$$\mathsf{gr}(\mathit{ker}\, f) \subseteq \mathit{ker}\,\widetilde{f} \qquad \text{and} \qquad \mathit{im}\,\widetilde{f} \subseteq \mathsf{gr}(\mathit{im}\, f). \tag{3.9}$$

Proposition 3.19. *Suppose α and β are v-norms on V and W. The following conditions on $f \in \mathit{Hom}_D(V,W)$ are equivalent:*

(a) $\mathsf{gr}(\mathit{ker}\, f) = \mathit{ker}\,\widetilde{f}$;
(b) $\mathit{im}\,\widetilde{f} = \mathsf{gr}(\mathit{im}\, f)$;

(c) $dim_{\mathsf{gr}(D)}\, \mathsf{gr}(ker\, f) = dim_{\mathsf{gr}(D)}\, ker\, \widetilde{f}$;
(d) $dim_{\mathsf{gr}(D)}\, im\, \widetilde{f} = dim_{\mathsf{gr}(D)}\, \mathsf{gr}(im\, f)$;
(e) *for every* $y \in im\, f$ *there exists* $x \in V$ *such that* $f(x) = y$ *and* $\beta(y) = \alpha(x) + Hom(\alpha,\beta)(f)$.

Proof. The equivalences (a) $\Leftrightarrow$ (c) and (b) $\Leftrightarrow$ (d) are clear in view of the inclusions (3.9). We have

$$dim_D\, V \;=\; dim_D\, ker\, f + dim_D\, im\, f$$

and

$$dim_{\mathsf{gr}(D)}\, \mathsf{gr}(V) \;=\; dim_{\mathsf{gr}(D)}\, ker\, \widetilde{f} + dim_{\mathsf{gr}(D)}\, im\, \widetilde{f}.$$

Moreover, $dim_D\, V = dim_{\mathsf{gr}(D)}\, \mathsf{gr}(V)$ and $dim_D\, ker\, f = dim_{\mathsf{gr}(D)}\, \mathsf{gr}(ker\, f)$ since α is a norm, and $dim_D\, im\, f = dim_{\mathsf{gr}(D)}\, \mathsf{gr}(im\, f)$ since β is a norm. Therefore,

$$dim_{\mathsf{gr}(D)}\, \mathsf{gr}(ker\, f) + dim_{\mathsf{gr}(D)}\, \mathsf{gr}(im\, f) \;=\; dim_{\mathsf{gr}(D)}\, ker\, \widetilde{f} + dim_{\mathsf{gr}(D)}\, im\, \widetilde{f},$$

and (c) $\Leftrightarrow$ (d) follows.

If $x \in V$ satisfies $\beta\big(f(x)\big) = \alpha(x) + Hom(\alpha,\beta)(f)$ and $x \neq 0$, then $\widetilde{f}(\widetilde{x}) = \widetilde{f(x)}$. Therefore, (e)$\Rightarrow$(b). To complete the proof, we show (a)$\Rightarrow$(e). Let $W \subseteq V$ be a splitting complement of $ker\, f$ in V. For each $y \in im\, f$ we can find $x \in W$ such that $f(x) = y$. If $\beta\big(f(x)\big) > \alpha(x) + Hom(\alpha,\beta)(f)$, then $\widetilde{f}(\widetilde{x}) = 0$ so $\widetilde{x} \in ker\, \widetilde{f} \cap \mathsf{gr}(W)$. This shows $ker\, \widetilde{f} \neq \mathsf{gr}(ker\, f)$ since $\mathsf{gr}(ker\, f) \cap \mathsf{gr}(W) = \{0\}$. □

The following proposition clarifies how the value function $Hom(\alpha,\beta)$ depends on α and β:

Proposition 3.20. *Suppose* α, α' *are* v*-norms on* V *and* β, β' *are* v*-value functions on* W. *We have* $Hom(\alpha,\beta) = Hom(\alpha',\beta')$ *if and only if there exists a* $\gamma \in \Gamma$ *such that*

$$\alpha'(x) \;=\; \alpha(x) + \gamma \quad \textit{and} \quad \beta'(y) \;=\; \beta(y) + \gamma \quad \textit{for all } x \in V \textit{ and } y \in W.$$

Proof. If there exists a γ in Γ with the stated property, then it is clear from the definitions that $Hom(\alpha,\beta) = Hom(\alpha',\beta')$. To prove the converse, we show that for any $x \in V \setminus \{0\}$ and $y \in W$ the difference $\beta(y) - \alpha(x)$ can be determined in terms of $Hom(\alpha,\beta)$: let

$$H(x,y) \;=\; \{f \in Hom_D(V,W) \mid f(x) = y\};$$

we show that

$$\beta(y) - \alpha(x) \;=\; \max_{f \in H(x,y)} \big(Hom(\alpha,\beta)(f)\big). \tag{3.10}$$

Consequently, if $Hom(\alpha,\beta) = Hom(\alpha',\beta')$ then for all $x \in V \setminus \{0\}$ and $y \in W$,

$$\beta(y) - \alpha(x) \;=\; \beta'(y) - \alpha'(x).$$

The proposition follows.

To prove (3.10), observe that by definition of $H(x,y)$ and of $Hom(\alpha,\beta)$ we have

$$\beta(y) - \alpha(x) \geq Hom(\alpha,\beta)(f) \qquad \text{for } f \in H(x,y).$$

On the other hand, Prop. 3.14 shows that there is a splitting base $(e_i)_{i=1}^n$ of V for α with $e_1 = x$. If $f \in Hom_D(V,W)$ is defined by

$$f(e_i) = \begin{cases} y & \text{if } i = 1, \\ 0 & \text{if } i > 1, \end{cases}$$

then $f \in H(x,y)$ and $Hom(\alpha,\beta)(f) = \beta(y) - \alpha(x)$. This proves (3.10). □

3.1.5 Tensor products

In this subsection, V is a right D-vector space with v-value function α and W is a left D-vector space with v-value function β. We may then consider the tensor product $V \otimes_D W$, which is a $Z(D)$-vector space.

Theorem 3.21. *Suppose α or β is a norm. Then there is a unique $v|_{Z(D)}$-value function τ on $V \otimes_D W$ such that there is a (degree-preserving) isomorphism of graded* $\mathsf{gr}(Z(D))$*-vector spaces*

$$\Psi_\tau \colon \mathsf{gr}_\tau(V \otimes_D W) \xrightarrow{\sim} \mathsf{gr}_\alpha(V) \otimes_{\mathsf{gr}(D)} \mathsf{gr}_\beta(W)$$

sending $\widetilde{x \otimes y}$ to $\widetilde{x} \otimes \widetilde{y}$ for all $x \in V$, $y \in W$. Moreover, τ satisfies

(i) *$\tau(x \otimes y) = \alpha(x) + \beta(y)$ for all $x \in V$ and $y \in W$.*
(ii) *If α is a norm and $(e_i)_{i=1}^n$ is a splitting base of V for α, then*

$$\tau\Big(\sum_{i=1}^n e_i \otimes z_i\Big) = \min_{1 \leq i \leq n} \big(\alpha(e_i) + \beta(z_i)\big) \qquad \text{for all } z_1, \ldots, z_n \in W.$$

(iii) *For any $z \in V \otimes_D W$,*

$$\tau(z) = \max\Big\{ \min_{i \in I}\big(\alpha(x_i) + \beta(y_i)\big) \mid z = \sum_{i \in I} x_i \otimes y_i \Big\},$$

where the maximum is over the set of all representations of z in the form $z = \sum_{i \in I} x_i \otimes y_i$.
(iv) *If α and β are v-norms, and if v is a $v|_{Z(D)}$-norm on D, then τ is a $v|_{Z(D)}$-norm.*

The unique value function τ on $V \otimes_D W$ described in the proposition will be denoted $\alpha \otimes \beta$, and the canonical isomorphism of graded vector spaces Ψ_τ will be viewed as an identification.

Proof. Suppose α is a norm on V, and fix some splitting base $(e_i)_{i=1}^n$ of V for α. Define τ by the formula in (ii) above. It is clear that τ is a $v|_{Z(D)}$-value function on $V \otimes_D W$. To prove (i), take any $x \in V$ and $y \in W$, and write $x = \sum_{i=1}^n e_i d_i$ with $d_1, \ldots, d_n \in D$. Since $(e_i)_{i=1}^n$ is a splitting base of V, we have $\alpha(x) = min_{1\le i\le n}\big(\alpha(e_i) + v(d_i)\big)$. On the other hand, since $x \otimes y = \sum_{i=1}^n e_i \otimes d_i y$, we have

$$\tau(x\otimes y) = \min_{1\le i\le n}\big(\alpha(e_i)+\beta(d_i y)\big) = \min_{1\le i\le n}\big(\alpha(e_i)+v(d_i)+\beta(y)\big) = \alpha(x)+\beta(y),$$

which proves (i).

To obtain the graded vector space isomorphism, note that the definition of τ shows that for all $\gamma \in \Gamma$

$$(V\otimes_D W)_{\ge\gamma} = \sum_{i=1}^n e_i\otimes W_{\ge\gamma-\alpha(e_i)} \quad\text{and}\quad (V\otimes_D W)_{>\gamma} = \sum_{i=1}^n e_i\otimes W_{>\gamma-\alpha(e_i)}.$$

Also, since $(\widetilde{e}_i)_{i=1}^n$ is a homogeneous $\mathsf{gr}(D)$-base of $\mathsf{gr}(V)$, we have by the definition of the grading on $\mathsf{gr}(V) \otimes_{\mathsf{gr}(D)} \mathsf{gr}(W)$

$$\big(\mathsf{gr}(V) \otimes_{\mathsf{gr}(D)} \mathsf{gr}(W)\big)_\gamma = \bigoplus_{i=1}^n \widetilde{e}_i \otimes W_{\gamma-\alpha(e_i)}.$$

For any $\gamma \in \Gamma$, define $\psi_\gamma\colon (V \otimes_D W)_{\ge\gamma} \longrightarrow \big(\mathsf{gr}(V) \otimes_{\mathsf{gr}(D)} \mathsf{gr}(W)\big)_\gamma$ by

$$\psi_\gamma\big(\sum_{i=1}^n e_i \otimes z_i\big) = \sum_{i=1}^n \widetilde{e}_i \otimes \big(z_i + W_{>\gamma-\alpha(e_i)}\big) \qquad \text{for } z_i \in W_{\ge\gamma-\alpha(e_i)}.$$

Clearly, ψ_γ is a surjective group homomorphism with $ker(\psi_\gamma) = (V\otimes_D W)_{>\gamma}$. Thus, ψ_γ induces a group isomorphism

$$\overline{\psi_\gamma}\colon (V \otimes_D W)_\gamma \xrightarrow{\sim} \big(\mathsf{gr}(V) \otimes_{\mathsf{gr}(D)} \mathsf{gr}(W)\big)_\gamma;$$

moreover,

$$\Psi_\tau = \bigoplus_{i=1}^n \overline{\psi_\gamma}\colon \mathsf{gr}_\tau(V \otimes_D W) \longrightarrow \mathsf{gr}_\alpha(V) \otimes_{\mathsf{gr}(D)} \mathsf{gr}_\beta(W)$$

is a graded $\mathsf{gr}(Z(D))$-vector space isomorphism. To compute $\Psi_\tau(\widetilde{x \otimes y})$ for $x \in V$ and $y \in W$, write $x = \sum_{i=1}^n e_i d_i$ as above. We have $\alpha(x) = min_{1\le i\le n}\big(\alpha(e_i) + v(d_i)\big)$ since $(e_i)_{i=1}^n$ is a splitting base of V. Consider $I = \{i \in \{1, \ldots, n\} \mid \alpha(e_i) + v(d_i) = \alpha(x)\}$. Then in $V_{\alpha(x)}$ we have

$$\widetilde{x} = \sum_{i\in I} \widetilde{e_i d_i} = \sum_{i\in I} \widetilde{e}_i \widetilde{d}_i.$$

Let $\delta = \tau(x \otimes y)$. By (i) we have $\delta = \alpha(x) + \beta(y)$ and

$$\tau(e_i \otimes d_i y) = \alpha(e_i) + v(d_i) + \beta(y) \ge \delta \qquad \text{for } i = 1, \ldots, n,$$

with equality if and only if $i \in I$. Therefore,

$$\Psi_\tau(\widetilde{x \otimes y}) = \psi_\delta\big(\sum_{i=1}^n e_i \otimes d_i y\big) = \sum_{i\in I} \widetilde{e}_i \otimes \widetilde{d_i y} = \sum_{i\in I} \widetilde{e}_i \widetilde{d}_i \otimes \widetilde{y} = \widetilde{x} \otimes \widetilde{y}.$$

Thus, Ψ_τ has the required property.

For the uniqueness of τ, suppose $\sigma\colon V\otimes_D W\to\Gamma\cup\{\infty\}$ is another $v|_{Z(D)}$-value function such that there is a graded $Z(\mathsf{gr}(D))$-isomorphism

$$\Psi_\sigma\colon\ \mathsf{gr}_\sigma(V\otimes_D W)\ \longrightarrow\ \mathsf{gr}_\alpha(V)\otimes\mathsf{gr}_\beta(W)$$

with

$$\Psi_\sigma\big(\widetilde{x\otimes y}^{\sigma}\big)\ =\ \widetilde{x}\otimes\widetilde{y}\ =\ \Psi_\tau\big(\widetilde{x\otimes y}^{\tau}\big)\qquad\text{for all } x\in V,\ y\in W.$$

For any x and y, since $\widetilde{x}\otimes\widetilde{y}$ has degree $\alpha(x)+\beta(y)$, it follows that $\sigma(x\otimes y)=\alpha(x)+\beta(y)$. Thus, using the splitting base $(e_i)_{i=1}^n$ for V, we have for all $z_1,\ldots,z_n\in W$,

$$\sigma\big(\sum_{i=1}^n e_i\otimes z_i\big)\ \geq\ \min_{1\leq i\leq n}\big(\sigma(e_i\otimes z_i)\big)\ =\ \min_{1\leq i\leq n}\big(\alpha(e_i)+\beta(z_i)\big)\ =\ \tau\big(\sum_{i=1}^n e_i\otimes z_i\big).$$

By Lemma 3.16, the identity map $id\colon V\otimes_D W\to V\otimes_D W$ induces a degree-preserving homomorphism of graded $Z(D)$-vector spaces

$$\widetilde{id}^{\,0}\colon\ \mathsf{gr}_\tau(V\otimes_D W)\ \longrightarrow\ \mathsf{gr}_\sigma(V\otimes_D W)$$

such that $\widetilde{id}^{\,0}(\widetilde{x\otimes y}^{\tau})=\widetilde{x\otimes y}^{\sigma}$ for all $x\in V$, $y\in W$. Then $\Psi_\sigma\circ\widetilde{id}^{\,0}=\Psi_\tau$, since these maps agree on a generating set of $\mathsf{gr}_\tau(V\otimes_D W)$. Because Ψ_τ and Ψ_σ are isomorphisms, $\widetilde{id}^{\,0}$ is injective. Then, by Lemma 3.16, we have $\tau(z)=\sigma(z)$ for all $z\in V\otimes_D W$.

The uniqueness of τ just proved shows that the formula in (ii) is valid for any choice of splitting base of V for α. Also, it is clear from the formula that if β is also a v-norm and v is a $v|_{Z(D)}$-norm, then for any splitting bases $(e_i)_{i=1}^n$ of V for α, $(f_j)_{j=1}^m$ of W for β, and $(d_k)_{k=1}^r$ of D for v, the family $(e_id_k\otimes f_j\mid 1\leq i\leq n, 1\leq j\leq m, 1\leq k\leq r)$ is a splitting base for τ, so τ is a norm.

It remains to verify the base-independent formula for τ in (iii). For this, take any $z\in V\otimes_D W$ and any representation $z=\sum_{i\in I}x_i\otimes y_i$, and choose a splitting base $(e_j)_{j=1}^n$ of V for α. Let $x_i=\sum_{j=1}^n e_jd_{ji}$ with $d_{ji}\in D$. Then $\alpha(x_i)=\min_{1\leq j\leq n}\big(\alpha(e_j)+v(d_{ji})\big)$, so

$$\min_{i\in I}\big(\alpha(x_i)+\beta(y_i)\big)\ =\ \min\big\{\alpha(e_j)+v(d_{ji})+\beta(y_i)\mid 1\leq j\leq n,\ i\in I\big\}.$$

On the other hand, we have $z=\sum_{j=1}^n e_j\otimes d_{ji}y_i$, so the formula in (ii) yields

$$\begin{aligned}\tau(z)\ &=\ \min_{1\leq j\leq n}\big(\alpha(e_j)+\beta\big(\sum_{i\in I} d_{ji}y_i\big)\big)\\ &\geq\ \min\big\{\alpha(e_j)+v(d_{ji})+\beta(y_i)\mid 1\leq j\leq n,\ i\in I\big\}.\end{aligned}$$

Therefore, for any representation $z=\sum_{i\in I}x_i\otimes y_i$ we have

$$\tau(z)\ \geq\ \min_{i\in I}\big(\alpha(x_i)+\beta(y_i)\big).$$

But when we use a representation where the x_i form a splitting base of V, this inequality becomes an equality by (ii). Thus, $\tau(z)$ is the maximum value of the set $\{\, min_{i\in I}(\alpha(x_i)+\beta(y_i)) \mid z=\sum_{i\in I} x_i\otimes y_i\}$. □

3.2 Value functions on algebras

Just as with vector spaces, value functions on algebras give rise to associated graded vector spaces. We consider in this section value functions for which the associated graded vector space is actually a graded algebra, with a multiplication induced by the algebra multiplication. These value functions are said to be *surmultiplicative.* Valuations on division algebras are examples of surmultiplicative value functions; they can be characterized by the property that the associated graded algebra is a graded division ring; see Cor. 3.30. Another type of particularly useful surmultiplicative value functions are the *gauges* defined in Def. 3.31. In contrast to valuations, gauges can be carried by semisimple algebras with zero divisors. A large part of this section is devoted to the construction of examples; we study in detail the case of symbol algebras in §3.2.3, and consider gauges that are tensor products of gauges in §3.2.4. We also reverse the construction of the associated graded algebra of a surmultiplicative value function: In §3.2.5 we show how to associate to a graded algebra A a surmultiplicative norm on the algebra of central quotients $q(\mathsf{A})$ so that $\mathsf{gr}_\alpha(q(\mathsf{A}))=\mathsf{A}$.

Another feature that distinguishes gauges from valuations is that gauges are *not* completely defined by their gauge ring, as seen in Ex. 3.50. However, when the 0-component of the associated graded ring is simple the gauge is determined by its gauge ring; see Prop. 3.48.

Throughout this section, F is an arbitrary field and $v\colon F\to\Gamma\cup\{\infty\}$ is a valuation on F. All the v-value functions we consider on F-vector spaces have values in Γ, which may be chosen large enough from the outset. In particular, we will assume in §3.2.7 that Γ is divisible. (For most practical purposes, it suffices to take for Γ the divisible hull $\mathbb{H}(\Gamma_F)$ of the value group Γ_F.)

3.2.1 Surmultiplicativity

Definition 3.22. Let A be an F-algebra and let $\alpha\colon A\to\Gamma\cup\{\infty\}$ be a v-value function on A (viewed as an F-vector space). The v-value function α is said to be *surmultiplicative* if

$$\alpha(1)=0 \qquad \text{and} \qquad \alpha(xy)\geq\alpha(x)+\alpha(y) \quad \text{for all } x,\, y\in A.$$

Just as for valuations, we define

$$\mathcal{O}_A = \{x \in A \mid \alpha(x) \geq 0\} \qquad \text{and} \qquad \mathfrak{m}_A = \{x \in A \mid \alpha(x) > 0\}.$$

By surmultiplicativity, $\mathcal{O}_A$ is a subring of A (and an $\mathcal{O}_F$-algebra) and $\mathfrak{m}_A$ is a two-sided ideal of $\mathcal{O}_A$. Moreover, the multiplication in A induces a multiplication on the graded $\mathsf{gr}(F)$-vector space $\mathsf{gr}_\alpha(A)$ such that for x, $y \in A$,

$$\widetilde{x} \cdot \widetilde{y} = xy + A_{>\alpha(x)+\alpha(y)},$$

so

$$\widetilde{x} \cdot \widetilde{y} = \begin{cases} \widetilde{xy} & \text{if } \alpha(xy) = \alpha(x) + \alpha(y), \\ 0 & \text{if } \alpha(xy) > \alpha(x) + \alpha(y). \end{cases} \tag{3.11}$$

Thus, $\mathsf{gr}_\alpha(A)$ is a graded $\mathsf{gr}(F)$-algebra.

Of course, if A is a division ring, every valuation on A extending v is a surmultiplicative v-value function. To give other examples, we use the following lemma, which is a very useful tool for verifying when a v-norm is surmultiplicative.

Lemma 3.23. *Suppose α is a v-norm such that $\alpha(1) = 0$ on an F-algebra A. Let $(a_i)_{i=1}^n$ be a splitting base of A for α. If $\alpha(a_i a_j) \geq \alpha(a_i) + \alpha(a_j)$ for all i, $j = 1, \ldots, n$, then α is surmultiplicative.*

Proof. Let $x = \sum_{i=1}^n a_i x_i$ and $y = \sum_{i=1}^n a_i y_i$ in A with x_i, $y_i \in F$ for all i. We have

$$\alpha(x) = \min_{1 \leq i \leq n} \big(\alpha(a_i) + v(x_i)\big) \qquad \text{and} \qquad \alpha(y) = \min_{1 \leq i \leq n} \big(\alpha(a_i) + v(y_i)\big),$$

while

$$\begin{aligned} \alpha(xy) = \alpha\big(\sum_{i,j} a_i a_j x_i y_j\big) &\geq \min_{i,j}\big(\alpha(a_i a_j x_i y_j)\big) \\ &\geq \min_{i,j}\big(\alpha(a_i) + v(x_i) + \alpha(a_j) + v(y_j)\big). \end{aligned}$$

Therefore, $\alpha(xy) \geq \alpha(x) + \alpha(y)$. □

Example 3.24. Let A be a quadratic algebra, $A = F \oplus Fu$ where $u^2 = a \in F$. For $\gamma \in \Gamma$, define a norm α_γ on A by

$$\alpha_\gamma(x + yu) = \min\big(v(x), v(y) + \gamma\big) \qquad \text{for } x, y \in F.$$

Then 1, u is a splitting base of A for α_γ. Lemma 3.23 shows that α_γ is surmultiplicative if and only if $\gamma \leq \frac{1}{2}v(a)$. We have $\Gamma_A = \Gamma_F \cup (\gamma + \Gamma_F)$ and $\mathsf{gr}_\alpha(A) = \mathsf{gr}(F) \oplus \mathsf{gr}(F)\widetilde{u}$ with $\widetilde{u} \in A_\gamma$ and

$$(\widetilde{u})^2 = \begin{cases} \widetilde{a} & \text{if } \gamma = \frac{1}{2}v(a), \\ 0 & \text{if } \gamma < \frac{1}{2}v(a). \end{cases}$$

Example 3.25. Let $A = M_2(F)$, the algebra of 2×2 matrices over F. For γ, $\delta \in \Gamma$, define a norm $\alpha_{\gamma,\delta}$ on A by

$$\alpha_{\gamma,\delta}\left(\begin{smallmatrix} a_{11} & a_{12} \\ a_{21} & a_{22} \end{smallmatrix}\right) \;=\; \mathit{min}\big(v(a_{11}), v(a_{12}) + \gamma, v(a_{21}) + \delta, v(a_{22})\big),$$

so the standard matrix base is a splitting base of A for $\alpha_{\gamma,\delta}$. By Lemma 3.23, it is easy to check that $\alpha_{\gamma,\delta}$ is surmultiplicative if and only if $\gamma + \delta \leq 0$. Clearly, $\Gamma_A = \Gamma_F \cup (\gamma + \Gamma_F) \cup (\delta + \Gamma_F)$ and for all $\varepsilon \in \Gamma$ we have

$$\mathsf{gr}(A)_\varepsilon \;=\; \begin{pmatrix} \mathsf{gr}(F)_\varepsilon & \mathsf{gr}(F)_{\varepsilon-\gamma} \\ \mathsf{gr}(F)_{\varepsilon-\delta} & \mathsf{gr}(F)_\varepsilon \end{pmatrix}.$$

In particular,

$$\mathit{deg}\left(\begin{smallmatrix} 0 & \widetilde{1} \\ 0 & 0 \end{smallmatrix}\right) \;=\; \gamma \qquad \text{and} \qquad \mathit{deg}\left(\begin{smallmatrix} 0 & 0 \\ \widetilde{1} & 0 \end{smallmatrix}\right) \;=\; \delta,$$

so

$$\left(\begin{smallmatrix} 0 & \widetilde{1} \\ 0 & 0 \end{smallmatrix}\right) \cdot \left(\begin{smallmatrix} 0 & 0 \\ \widetilde{1} & 0 \end{smallmatrix}\right) \;=\; \left(\begin{smallmatrix} 0 & 0 \\ \widetilde{1} & 0 \end{smallmatrix}\right) \cdot \left(\begin{smallmatrix} 0 & \widetilde{1} \\ 0 & 0 \end{smallmatrix}\right) \;=\; 0 \quad \text{if } \gamma + \delta < 0.$$

In this case, $\left(\begin{smallmatrix} 0 & \mathsf{gr}(F) \\ 0 & 0 \end{smallmatrix}\right)$ is a two-sided homogeneous nilpotent ideal in $\mathsf{gr}_{\alpha_{\gamma,\delta}}(A)$, hence this graded algebra is not graded semisimple. By contrast, if $\gamma + \delta = 0$, then using the notation of (2.7) we have

$$\mathsf{gr}_{\alpha_{\gamma,\delta}}(A) \;=\; M_2\big(\mathsf{gr}(F)\big)(0,\gamma) \;=\; M_2\big(\mathsf{gr}(F)\big)(\delta, 0),$$

hence $\mathsf{gr}_{\alpha_{\gamma,\delta}}(A)$ is graded simple by Prop. 2.9 and Prop. 2.17.

Further examples can be obtained by the tensor product construction, as will be shown in Prop. 3.27 below.

Lemma 3.26. *Let A and B be F-algebras with respective surmultiplicative v-value functions α and β. Suppose there is an F-algebra homomorphism $f\colon A \to B$ such that*

$$\beta\big(f(a)\big) \;\geq\; \alpha(a) \qquad \textit{for all } a \in A.$$

Then the graded $\mathsf{gr}(F)$-vector space homomorphism

$$\widehat{f}^0\colon\; \mathsf{gr}_\alpha(A) \;\longrightarrow\; \mathsf{gr}_\beta(B)$$

of Lemma 3.16 is a $\mathsf{gr}(F)$-algebra homomorphism.

Proof. Let $\widehat{f} = \widehat{f}^0$. Recall that $\widehat{f}$ is degree-preserving and is defined on homogeneous elements by

$$\widehat{f}(\widetilde{x}) \;=\; f(x) + B_{>\alpha(x)} \;\in\; \mathsf{gr}(B)_{\alpha(x)}, \qquad \text{for all } x \in A.$$

So, for $x \neq 0$, we have $\widehat{f}(\widetilde{x}) \neq 0$ if and only if $\beta\big(f(x)\big) = \alpha(x)$. To verify that the $\mathsf{gr}(F)$-linear map $\widehat{f}$ is compatible with multiplication, it suffices to check this for homogeneous elements, i.e., for any nonzero x, $y \in A$ we need

$$\widehat{f}(\widetilde{x}\,\widetilde{y}) = \widehat{f}(\widetilde{x})\widehat{f}(\widetilde{y}). \tag{3.12}$$

When the left expression in (3.12) is nonzero, it equals $\widetilde{f(xy)}$, and when the right expression is nonzero it equals $\widetilde{f(x)}\widetilde{f(y)} = \widetilde{f(x)f(y)} = \widetilde{f(xy)}$. So equality indeed holds in (3.12) when each side is nonzero. Now, we have

$$\beta\big(f(xy)\big) \geq \alpha(xy) \geq \alpha(x) + \alpha(y) \tag{3.13}$$

and

$$\beta\big(f(xy)\big) = \beta\big(f(x)f(y)\big) \geq \beta\big(f(x)\big) + \beta\big(f(y)\big) \geq \alpha(x) + \alpha(y). \tag{3.14}$$

The left expression in (3.12) is nonzero if and only if $\alpha(xy) = \alpha(x)+\alpha(y)$ (so $\widetilde{x}\,\widetilde{y} = \widetilde{xy} \neq 0$) and $\beta\big(f(xy)\big) = \alpha(xy)$, i.e., equality holds through (3.13). The right expression in (3.12) is nonzero if and only if $\beta\big(f(x)\big) = \alpha(x)$ and $\beta\big(f(y)\big) = \alpha(y)$, and $\beta\big(f(x)f(y)\big) = \beta\big(f(x)\big) + \beta\big(f(y)\big)$, i.e., equality holds throughout (3.14). Each of these conditions holds if and only if $\beta\big(f(xy)\big) = \alpha(x) + \alpha(y)$. Thus, we have equality in (3.12) in all cases. □

Proposition 3.27. *Let A and B be F-algebras with respective surmultiplicative v-value functions α and β. If α is a norm (which entails $[A{:}F] < \infty$), then the v-value function $\alpha \otimes \beta$ on $A \otimes_F B$ is surmultiplicative and the graded $\mathsf{gr}(F)$-vector space isomorphism of Th. 3.21*

$$\Psi\colon \mathsf{gr}_{\alpha\otimes\beta}(A \otimes_F B) \xrightarrow{\sim} \mathsf{gr}_\alpha(A) \otimes_{\mathsf{gr}(F)} \mathsf{gr}_\beta(B)$$

is a $\mathsf{gr}(F)$-algebra isomorphism. Moreover, using Ψ as an identification, we have

$$Z\big(\mathsf{gr}_{\alpha\otimes\beta}(A \otimes_F B)\big) = Z\big(\mathsf{gr}_\alpha(A)\big) \otimes_{\mathsf{gr}(F)} Z\big(\mathsf{gr}_\beta(B)\big).$$

Proof. Let $(a_i)_{i=1}^n$ be a splitting base of A for α and let x, $y \in A \otimes_F B$. We may write

$$x = \sum_{i=1}^n a_i \otimes x_i \quad \text{and} \quad y = \sum_{i=1}^n a_i \otimes y_i \qquad \text{for some } x_1, \ldots, y_n \in B.$$

By Th. 3.21, we have

$$(\alpha{\otimes}\beta)(x) = \min_{1\leq i\leq n}\big(\alpha(a_i){+}\beta(x_i)\big) \quad \text{and} \quad (\alpha{\otimes}\beta)(y) = \min_{1\leq i\leq n}\big(\alpha(a_i){+}\beta(y_i)\big).$$

On the other hand, since $(\alpha \otimes \beta)(a \otimes b) = \alpha(a) + \beta(b)$ for $a \in A$ and $b \in B$,

$$(\alpha \otimes \beta)(xy) = (\alpha \otimes \beta)\Big(\sum_{i,j=1}^n a_i a_j \otimes x_i y_j\Big) \geq \min_{1\leq i,j\leq n}\big(\alpha(a_i a_j) + \beta(x_i y_j)\big).$$

Since α and β are surmultiplicative, we have $\alpha(a_i a_j) \geq \alpha(a_i) + \alpha(a_j)$ and $\beta(x_i y_j) \geq \beta(x_i) + \beta(y_j)$, hence the preceding inequality yields

$$(\alpha \otimes \beta)(xy) \geq \min_{1\leq i,j\leq n}\big(\alpha(a_i) + \alpha(a_j) + \beta(x_i) + \beta(y_j)\big).$$

Therefore, $(\alpha \otimes \beta)(xy) \geq (\alpha \otimes \beta)(x) + (\alpha \otimes \beta)(y)$, proving that $\alpha \otimes \beta$ is surmultiplicative.

To see that the graded $\mathsf{gr}(F)$-vector space isomorphism Ψ is compatible with multiplication, we prove this for Ψ^{-1}. For the canonical mapping $i_A\colon A \to A \otimes_F B$ given by $a \mapsto a \otimes 1$, we have

$$(\alpha \otimes \beta)\big(i_A(a)\big) \;=\; (\alpha \otimes \beta)(a \otimes 1) \;=\; \alpha(a).$$

Therefore, by Lemma 3.26, the induced map $\widehat{\imath}_A^0\colon \mathsf{gr}(A) \to \mathsf{gr}(A \otimes_F B)$ which sends $\widetilde{a}$ to $\widetilde{a \otimes 1}$ is a graded $\mathsf{gr}(F)$-algebra homomorphism. Likewise, the canonical map $i_B\colon B \to A \otimes_F B$ induces a graded $\mathsf{gr}(F)$-algebra homomorphism $\widehat{\imath}_B^0\colon \mathsf{gr}(B) \to \mathsf{gr}(A \otimes_F B)$. Note that for any $a \in A$, $b \in B$ we have by Th. 3.21(i)

$$\begin{aligned}(\alpha \otimes \beta)\big((a \otimes 1)\cdot(1 \otimes b)\big) \;&=\; (\alpha \otimes \beta)(a \otimes b)\\ &=\; \alpha(a) + \beta(b) \;=\; (\alpha \otimes \beta)(a \otimes 1) + (\alpha \otimes \beta)(1 \otimes b).\end{aligned}$$

Hence $\widetilde{a \otimes 1}\cdot\widetilde{1 \otimes b} = \big((a \otimes 1)(1 \otimes b)\big)^{\sim} = \widetilde{a \otimes b}$ in $\mathsf{gr}(A \otimes_F B)$, and likewise $\widetilde{1 \otimes b}\cdot\widetilde{a \otimes 1} = \widetilde{a \otimes b}$. This shows that the images of $\widehat{\imath}_A^0$ and $\widehat{\imath}_B^0$ in $\mathsf{gr}(A \otimes_F B)$ commute. Therefore, there is an induced $\mathsf{gr}(F)$-algebra homomorphism

$$\Psi'\colon\; \mathsf{gr}(A) \otimes_{\mathsf{gr}(F)} \mathsf{gr}(B) \longrightarrow \mathsf{gr}(A \otimes_F B)$$

which maps $\widetilde{a} \otimes \widetilde{b}\ \big(= (\widetilde{a} \otimes \widetilde{1})\cdot(\widetilde{1} \otimes \widetilde{b})\big)$ to $\widehat{\imath}_A^0(\widetilde{a})\cdot\widehat{\imath}_B^0(\widetilde{b}) = \widetilde{a \otimes 1}\cdot\widetilde{1 \otimes b} = \widetilde{a \otimes b}$. Clearly, $\Psi' = \Psi^{-1}$, so Ψ is also an algebra homomorphism.

The restriction of Ψ to the center is a $\mathsf{gr}(F)$-algebra isomorphism

$$Z\big(\mathsf{gr}_{\alpha\otimes\beta}(A \otimes B)\big) \;\xrightarrow{\sim}\; Z\big(\mathsf{gr}_\alpha(A) \otimes_{\mathsf{gr}(F)} \mathsf{gr}_\beta(B)\big).$$

By Cor. 2.31, we have

$$Z\big(\mathsf{gr}_\alpha(A) \otimes_{\mathsf{gr}(F)} \mathsf{gr}_\beta(B)\big) \;=\; Z\big(\mathsf{gr}_\alpha(A)\big) \otimes_{\mathsf{gr}(F)} Z\big(\mathsf{gr}_\beta(B)\big),$$

and the proof is complete. □

Corollary 3.28. *Let A be a finite-dimensional F-algebra with a surmultiplicative v-norm α and let L/F be an arbitrary field extension. Let also $w\colon L \to \Gamma \cup \{\infty\}$ be a valuation extending v. Then $\alpha \otimes w$ is a surmultiplicative w-norm on the L-algebra $A \otimes_F L$, and there is a canonical isomorphism of $\mathsf{gr}_w(L)$-algebras, which will be treated as an identification,*

$$\mathsf{gr}_{\alpha\otimes w}(A \otimes_F L) \;=\; \mathsf{gr}_\alpha(A) \otimes_{\mathsf{gr}(F)} \mathsf{gr}_w(L).$$

Proof. Let $(a_i)_{i=1}^n$ be a splitting base of A for α. For any $\ell_1, \ldots, \ell_n \in L$ we have

$$(\alpha \otimes w)\Big(\sum_{i=1}^n a_i \otimes \ell_i\Big) \;=\; \min_{1\le i\le n}\big(\alpha(a_i) + w(\ell_i)\big).$$

From this formula, it is clear that $\alpha \otimes w$ is a w-norm on $A \otimes_F L$ with splitting base $(a_i \otimes 1)_{i=1}^n$. The other statements readily follow from Prop. 3.27. □

The notion of surmultiplicative value function is somewhat too general for our purposes, and we often impose additional conditions, which will be spelled out in the next subsection. But no added conditions are needed for the following result characterizing the invertible homogeneous elements in the associated graded algebra.

Proposition 3.29. *Let α be a surmultiplicative v-value function on a finite-dimensional F-algebra A. For any nonzero $a \in A$, the following conditions are equivalent:*

(a) $\widetilde{a} \in \mathsf{gr}_\alpha(A)^\times$, *the group of units of* $\mathsf{gr}_\alpha(A)$;
(b) $\alpha(ax) = \alpha(a) + \alpha(x)$ *for all* $x \in A$;
(c) $\alpha(xa) = \alpha(x) + \alpha(a)$ *for all* $x \in A$;
(d) $a \in A^\times$ *and* $\alpha(a) + \alpha(a^{-1}) = 0$.

Proof. Since $[A{:}F]$ is finite, Cor. 3.6 shows that $[\mathsf{gr}(A){:}\,\mathsf{gr}(F)]$ is finite. Therefore, $\widetilde{a}$ is invertible if and only if it is not a left zero-divisor, if and only if it is not a right zero-divisor. In view of (3.11), it follows that (a), (b), and (c) are equivalent.

(b) $\Rightarrow$ (d): Condition (b) shows that a is not a left zero divisor in the finite-dimensional algebra A, hence $a \in A^\times$. The formula in (d) follows by setting $x = a^{-1}$ in (b).

(d) $\Rightarrow$ (b): For any $x \in A$, we have $\alpha(x) = \alpha(a^{-1}ax) \geq \alpha(a^{-1}) + \alpha(ax)$. Therefore, (d) yields

$$\alpha(ax) \;\leq\; \alpha(x) - \alpha(a^{-1}) \;=\; \alpha(a) + \alpha(x).$$

The reverse inequality holds by surmultiplicativity, so (b) holds. □

The values of the elements described in Prop. 3.29 form a significant subset of the value set Γ_A. Define

$$\Gamma_A^\times \;=\; \{\alpha(a) \mid a \in A^\times \text{ and } \alpha(a^{-1}) = -\alpha(a)\}. \tag{3.15}$$

Thus, $\Gamma_A^\times = \Gamma_{\mathsf{gr}(A)}^\times$, the group of degrees of homogeneous units of $\mathsf{gr}(A)$ (see (2.28)). So, $\Gamma_A^\times$ is a subgroup of Γ lying in Γ_A (which need not be a group), though $\Gamma^\times$ need not be the largest group in Γ_A. The group $\Gamma_A^\times$ is the appropriate domain for the generalization of the map θ_D for valued division algebras D. Define $\theta_A\colon \Gamma_A^\times \to Aut_{F_0}(Z(A_0))$ as follows: for $\gamma \in \Gamma_A^\times$ and any $a \in A^\times$ with $\alpha(a) = \gamma = -\alpha(a^{-1})$,

$$\theta_A(\gamma) \;=\; int(\widetilde{a})|_{Z(A_0)}. \tag{3.16}$$

Then, θ_A is a well-defined group homomorphism, and clearly $\theta_A = \theta_{\mathsf{gr}(A)}$ (see (2.29) and (2.30)).

Proposition 3.29 yields a convenient way to check whether a surmultiplicative value function on an algebra is a valuation.

Corollary 3.30. *Let A be a finite-dimensional algebra over F, and let α be a surmultiplicative v-value function on A. Then, $\mathsf{gr}_\alpha(A)$ is a graded division ring if and only if A is a division ring and α is a valuation.*

Proof. It was already observed in §1.3 that the graded ring associated with a valuation on a division algebra is a graded division ring. Conversely, if $\mathsf{gr}_\alpha(A)$ is a graded division ring, then for each nonzero $a \in A$, condition (a) of Prop. 3.29 holds, hence condition (b) also holds. This shows that the finite-dimensional F-algebra A has no zero divisors, and hence A is a division ring. Since condition (b) of Prop. 3.29 holds for every nonzero element a of A, the surmultiplicative value function α is actually a valuation. □

3.2.2 Gauges

The notion of semisimple graded algebra (see §2.2.1) is crucial to characterize the most useful type of value functions:

Definition 3.31. A surmultiplicative v-value function α on a finite-dimensional F-algebra A is called a *v-gauge* if it is a norm on A (which means that $[\mathsf{gr}_\alpha(A){:}\,\mathsf{gr}(F)] = [A{:}F]$) and $\mathsf{gr}_\alpha(A)$ is a semisimple graded $\mathsf{gr}(F)$-algebra. Note that if A has a v-gauge then it must itself be semisimple: if A has a nonzero ideal N with $N^2 = \{0\}$, then $\mathsf{gr}(N)$ is a nonzero homogeneous ideal of $\mathsf{gr}(A)$ with $\mathsf{gr}(N)^2 = \{0\}$, hence $\mathsf{gr}(A)$ is not semisimple. If A and A' are semisimple F-algebras with respective v-gauges α and α', we say that these v-gauges are *isomorphic* if there is an F-algebra isomorphism $\xi\colon A \xrightarrow{\sim} A'$ such that $\alpha'(\xi(a)) = \alpha(a)$ for all $a \in A$.

If α is a v-gauge on A, the associated ring

$$O_{A,\alpha} = \{a \in A \mid \alpha(a) \geq 0\}$$

is called the *gauge ring* of α.

It is clear from Cor. 3.30 that a valuation on a division algebra D over F is a v-gauge if and only if it is a v-norm, and Cor. 2.7 shows that the norm condition holds if and only if

$$[D{:}F] = [\overline{D}{:}\overline{F}] \cdot |\Gamma_D{:}\Gamma_F|.$$

When this equality holds, the valuation is said to be *defectless*[1] *over* F; it is then the only v-gauge on D, see Prop. 4.33 below. On the other hand, §3.2.3 below yields examples of gauges on division algebras that are not valuations. Of course, gauges also exist on algebras that are not division algebras. For example, the v-value function α_γ of Ex. 3.24 is a gauge if and only if $\gamma = \frac{1}{2}v(a)$, and the v-value function $\alpha_{\gamma,\delta}$ of Ex. 3.25 is a gauge if and only if $\gamma + \delta = 0$. The gauge ring of the gauge $\alpha_{\gamma,-\gamma}$ of that example has a simple residue ring if and only if $\gamma \in \Gamma_F$. For further examples see Prop. 3.34 and §3.2.3 below.

---

[1] See §4.2 for a detailed discussion of the defect.

Example 3.32. If K/F is a finite-degree field extension and $v_1, \ldots, v_r$ is the collection of all the distinct extensions of v to K, the formula

$$\alpha(x) = min(v_1(x), \ldots, v_r(x)) \qquad \text{for } x \in K \tag{3.17}$$

defines a surmultiplicative v-value function on K. We show in Prop. 4.34 below that

$$\mathsf{gr}_\alpha(K) \cong_g \mathsf{gr}_{v_1}(K) \times \ldots \times \mathsf{gr}_{v_r}(K).$$

Therefore, $\mathsf{gr}_\alpha(K)$ is graded semisimple, and α is a v-gauge if and only if $[K{:}F] = [\mathsf{gr}_\alpha(K){:}\, \mathsf{gr}(F)]$, i.e.,

$$[K{:}F] = \sum_{i=1}^{r} [\mathsf{gr}_{v_i}(K){:}\, \mathsf{gr}(F)].$$

For $i = 1, \ldots, r$, let $e_i = |\Gamma_{K,v_i}{:}\Gamma_F|$ be the ramification index and $f_i = [\overline{K}^{v_i}{:}\overline{F}]$ be the residue degree, so $[\mathsf{gr}_{v_i}(K){:}\, \mathsf{gr}(F)] = e_i f_i$. We say that v is *defectless in* K if equality holds in the Fundamental Inequality, i.e.,

$$[K{:}F] = \sum_{i=1}^{r} e_i f_i. \tag{3.18}$$

Thus, formula (3.17) defines a v-gauge on K if and only if v is defectless in K. We will show in §4.3.3 below that α is the unique v-gauge on K in this case, and that there is no v-gauge on K if the fundamental equality (3.18) does not hold, see Cor. 4.37.

More generally, we will show in Th. 4.50 that a suitable absence of defect is all that is needed for a semisimple F-algebra to have a v-gauge.

Remark 3.33. Let A be a (finite-dimensional) semisimple algebra over a field F with valuation v. If α is a v-gauge on A, then $\alpha|_{Z(A)}$ is a v-gauge on $Z(A)$. For by Prop. 3.14, $\alpha|_{Z(A)}$ is a v-norm on $Z(A)$, which is clearly surmultiplicative. Moreover, a nilpotent homogeneous element z of $\mathsf{gr}(Z(A))$ is central in the graded semisimple algebra $\mathsf{gr}(A)$; hence $z = 0$. Therefore, the commutative graded $\mathsf{gr}(F)$-algebra $\mathsf{gr}(Z(A))$ is graded semisimple, showing that $\alpha|_{Z(A)}$ is a v-gauge.

We turn next to a fundamental example of gauges on endomorphism algebras of vector spaces. The gauges defined in the next proposition will be called *End*-gauges.

Proposition 3.34. *Let D be a finite-dimensional division F-algebra with a valuation $v_D\colon D \to \Gamma \cup \{\infty\}$ extending v, and let V be a finite-dimensional D-vector space with a v_D-norm $\alpha\colon V \to \Gamma \cup \{\infty\}$. Suppose that v_D is defectless over F. Then, the v-value function $End(\alpha) = Hom(\alpha, \alpha)$ on $\mathrm{End}_D V$ of Prop. 3.18 is a v-gauge, and there is a canonical identification of graded $\mathsf{gr}(F)$-algebras*

$$\mathsf{gr}_{End(\alpha)}(\mathrm{End}_D V) = \mathsf{End}_{\mathsf{gr}(D)}(\mathsf{gr}_\alpha(V)).$$

Proof. Let $\varepsilon = End(\alpha)\colon End_D V \to \Gamma \cup \{\infty\}$. Proposition 3.18 shows that ε is a v-norm and yields a canonical isomorphism of graded $\mathsf{gr}(F)$-vector spaces

$$\Phi\colon \mathsf{gr}_\varepsilon(End_D V) \xrightarrow{\sim} \mathsf{End}_{\mathsf{gr}(D)}\big(\mathsf{gr}_\alpha(V)\big).$$

Recall that for any $f \in End_D V$ we have $\Phi(\widetilde{f}) = \widehat{f}^{\,\varepsilon(f)}$, where

$$\widehat{f}^{\,\varepsilon(f)}(x + V_{>\gamma}) = f(x) + V_{>\varepsilon(f)+\gamma} \qquad \text{for } \gamma \in \Gamma \text{ and } x \in V_{\geq\gamma}.$$

For any f, $g \in End_D V$ and any $x \in V$ we have

$$\alpha\big(f(g(x))\big) \geq \varepsilon(f) + \alpha\big(g(x)\big) \geq \varepsilon(f) + \varepsilon(g) + \alpha(x).$$

Thus, if $x \neq 0$, we have $\alpha\big((f \circ g)(x)\big) - \alpha(x) \geq \varepsilon(f) + \varepsilon(g)$. This shows that $\varepsilon(f \circ g) \geq \varepsilon(f) + \varepsilon(g)$; hence, ε is surmultiplicative.

To see that Φ is a multiplicative homomorphism, take any $\gamma \in \Gamma$ and $x \in V_{\geq\gamma}$. Then

$$\begin{aligned}\big(\Phi(\widetilde{f}) \circ \Phi(\widetilde{g})\big)(x + V_{>\gamma}) &= \widehat{f}^{\,\varepsilon(f)}\big(\widehat{g}^{\,\varepsilon(g)}(x + V_{>\gamma})\big) \\ &= \widehat{f}^{\,\varepsilon(f)}\big(g(x) + V_{>\varepsilon(g)+\gamma}\big) = f\big(g(x)\big) + V_{>\varepsilon(f)+\varepsilon(g)+\gamma}.\end{aligned} \tag{3.19}$$

If $\varepsilon(f \circ g) = \varepsilon(f) + \varepsilon(g)$, then $\widetilde{f} \cdot \widetilde{g} = \widetilde{f \circ g}$, and (3.19) shows that

$$\Phi(\widetilde{f}) \circ \Phi(\widetilde{g}) = \Phi(\widetilde{f \circ g}) = \Phi(\widetilde{f} \cdot \widetilde{g}).$$

If $\varepsilon(f \circ g) > \varepsilon(f) + \varepsilon(g)$, then $\widetilde{f} \cdot \widetilde{g} = 0$, and

$$(f \circ g)(V_{\geq\gamma}) \subseteq V_{\geq\varepsilon(f\circ g)+\gamma} \subseteq V_{>\varepsilon(f)+\varepsilon(g)+\gamma};$$

then (3.19) shows that

$$\Phi(\widetilde{f}) \circ \Phi(\widetilde{g}) = 0 = \Phi(\widetilde{f} \cdot \widetilde{g}).$$

Thus, in all cases $\Phi(\widetilde{f}) \circ \Phi(\widetilde{g}) = \Phi(\widetilde{f} \cdot \widetilde{g})$; hence, the $\mathsf{gr}(F)$-vector space isomorphism Φ is actually a $\mathsf{gr}(F)$-algebra isomorphism.

To complete the proof, it suffices to observe that $\mathsf{gr}_\varepsilon(End_D V)$ is a semisimple graded algebra (indeed, a simple graded algebra) by Prop. 2.17, since it is isomorphic under Φ to the simple graded algebra $\mathsf{End}_{\mathsf{gr}(D)}\big(\mathsf{gr}_\alpha(V)\big)$. □

For example, let V be a two-dimensional F-vector space with base (e_1, e_2). For any $\gamma \in \Gamma$, let α be the v-norm on V given by

$$\alpha(ae_1 + be_2) = \min\big(v(a) + \gamma, v(b)\big).$$

Using the base (e_1, e_2), we define an isomorphism $End_F V \cong M_2(F)$; the gauge $End(\alpha)$ on $End_F V$ corresponds under this isomorphism to the gauge $\alpha_{\gamma,-\gamma}$ of Ex. 3.25 on $M_2(F)$.

Corollary 3.35. *Let A be a finite-dimensional simple F-algebra, and let D be its associated division algebra. If D carries a valuation v_D extending v and defectless over F, then A carries a v-gauge with $\Gamma_A = \Gamma_D$.*

Proof. Fix an identification $A = End_D V$ for some finite-dimensional right D-vector space V. Define a v_D-norm α on V by picking an arbitrary base $(e_i)_{i=1}^n$ of V and letting

$$\alpha\Big(\sum_{i=1}^n e_i d_i\Big) = \min_{1\le i\le n}\big(v_D(d_i)\big) \qquad \text{for } d_1, \dots, d_n \in D.$$

Then $End(\alpha)$ is a v-gauge on A by Prop. 3.34, and we have $\Gamma_A = \Gamma_D$. □

We next show that gauges on direct products arise from gauges on the direct factors. For many purposes, this reduces the study of gauges on semisimple algebras to the case of simple algebras.

Proposition 3.36. *Let $A_1, \dots, A_n$ be finite-dimensional semisimple F-algebras with respective surmultiplicative v-value functions $\alpha_1, \dots, \alpha_n$, and let $A = A_1 \times \dots \times A_n$. The map $\alpha\colon A \to \Gamma \cup \{\infty\}$ defined by*

$$\alpha(x_1, \dots, x_n) = \min_{1\le i\le n}\big(\alpha_i(x_i)\big) \qquad \textit{for } x_1 \in A_1, \dots, x_n \in A_n$$

is a surmultiplicative v-value function on A, and for $\gamma = \alpha(x_1, \dots, x_n)$ the map

$$(x_1, \dots, x_n) + A_{>\gamma} \mapsto (x_1 + (A_1)_{>\gamma}, \dots, x_n + (A_n)_{>\gamma})$$

defines a canonical identification of graded $\mathsf{gr}(F)$-algebras

$$\mathsf{gr}_\alpha(A) = \mathsf{gr}_{\alpha_1}(A_1) \times \dots \times \mathsf{gr}_{\alpha_n}(A_n).$$

The map α is a norm if and only if $\alpha_1, \dots, \alpha_n$ are norms; it is a gauge if and only if $\alpha_1, \dots, \alpha_n$ are gauges.

Moreover, every v-gauge β on A is of the type above: more precisely, if $\beta\colon A \to \Gamma \cup \{\infty\}$ is an arbitrary surmultiplicative v-value function (not necessarily a norm) such that $\mathsf{gr}_\beta(A)$ is semisimple, then for each $i = 1, \dots, n$ the map $\beta_i\colon A_i \to \Gamma \cup \{\infty\}$ defined by

$$\beta_i(x) = \beta(0, \dots, 0, x, 0, \dots, 0) \qquad (x \textit{ in the } i\textit{-th position})$$

is a surmultiplicative v-value function on A_i, and

$$\beta(x_1, \dots, x_n) = \min_{1\le i\le n}\big(\beta_i(x_i)\big) \qquad \textit{for } x_1 \in A_1, \dots, x_n \in A_n. \tag{3.20}$$

Proof. It suffices by induction to consider the case $n = 2$. It is easily checked that α is a surmultiplicative v-value function on A. For $\gamma \in \Gamma$ and $x_1 \in A_1$, $x_2 \in A_2$ we have

$$\alpha(x_1, x_2) \ge \gamma \quad \text{if and only if} \quad \alpha_1(x_1) \ge \gamma \text{ and } \alpha_2(x_2) \ge \gamma,$$

hence $A_{\ge\gamma} = (A_1)_{\ge\gamma} \times (A_2)_{\ge\gamma}$. Similarly, $A_{>\gamma} = (A_1)_{>\gamma} \times (A_2)_{>\gamma}$. Therefore,

$$\mathsf{gr}_\alpha(A) = \mathsf{gr}_{\alpha_1}(A_1) \times \mathsf{gr}_{\alpha_2}(A_2).$$

Hence,

$$[\mathsf{gr}_\alpha(A) : \mathsf{gr}(F)] = [\mathsf{gr}_{\alpha_1}(A_1) : \mathsf{gr}(F)] + [\mathsf{gr}_{\alpha_2}(A_2) : \mathsf{gr}(F)].$$

Since $[\mathsf{gr}_{\alpha_i}(A_i):\mathsf{gr}(F)] \leq [A_i:F]$ for $i = 1, 2$, it follows that

$$[\mathsf{gr}_\alpha(A):\mathsf{gr}(F)] = [A:F] \text{ if and only if } [\mathsf{gr}_{\alpha_i}(A_i):\mathsf{gr}(F)] = [A_i:F] \text{ for } i=1,2,$$

hence α is a norm if and only if α_1 and α_2 are norms. Finally, the projections of a homogeneous nilpotent two-sided ideal of $\mathsf{gr}_\alpha(A)$ are homogeneous nilpotent two-sided ideals of $\mathsf{gr}_{\alpha_1}(A_1)$ and $\mathsf{gr}_{\alpha_2}(A_2)$, hence $\mathsf{gr}_\alpha(A)$ is semisimple if $\mathsf{gr}_{\alpha_1}(A_1)$ and $\mathsf{gr}_{\alpha_2}(A_2)$ are semisimple. On the other hand, if I_i is a homogeneous nilpotent two-sided ideal of $\mathsf{gr}_{\alpha_i}(A_i)$ for $i = 1, 2$, then $\mathsf{I}_1 \times \mathsf{I}_2$ is a homogeneous nilpotent two-sided ideal of $\mathsf{gr}_\alpha(A)$, hence $\mathsf{gr}_{\alpha_1}(A_1)$ and $\mathsf{gr}_{\alpha_2}(A_2)$ are semisimple if $\mathsf{gr}_\alpha(A)$ is semisimple. It follows that α is a gauge if and only if α_1 and α_2 are gauges.

Now, suppose β is an arbitrary surmultiplicative v-value function on A such that $\mathsf{gr}_\beta(A)$ is semisimple. Consider the central idempotents

$$e_1 = (1,0) \in A \qquad \text{and} \qquad e_2 = (0,1) \in A.$$

Since e_1 is central, its image $\widetilde{e}_1 \in \mathsf{gr}_\beta(A)$ is central. By surmultiplicativity we have

$$\beta(e_1) = \beta(e_1^2) \geq 2\beta(e_1),$$

so $\beta(e_1) \leq 0$. If $\beta(e_1) < 0$, then the inequality above is strict, hence $(\widetilde{e}_1)^2 = 0$. But then $\widetilde{e}_1\,\mathsf{gr}_\beta(A)$ is a nonzero homogeneous nilpotent ideal of $\mathsf{gr}_\beta(A)$ (as $\widetilde{e}_1 \in Z(\mathsf{gr}_\beta(A))$), contradicting the semisimplicity of $\mathsf{gr}_\beta(A)$. Therefore, we must have $\beta(e_1) = 0$, hence

$$\beta(e_1 x) \geq \beta(e_1) + \beta(x) = \beta(x) \qquad \text{for } x \in A.$$

Likewise, $\beta(e_2x) \geq \beta(x)$, hence $\beta(x) \leq \min\big(\beta(e_1x), \beta(e_2x)\big)$. The reverse inequality holds because $x = e_1x + e_2x$; hence,

$$\beta(x) = \min\big(\beta(e_1x), \beta(e_2x)\big) \qquad \text{for } x \in A.$$

For $x = (x_1, x_2)$ with $x_1 \in A_1$ and $x_2 \in A_2$ we have by definition of β_1 and β_2

$$\beta(e_1x) = \beta(x_1, 0) = \beta_1(x_1) \qquad \text{and} \qquad \beta(e_2x) = \beta(0, x_2) = \beta_2(x_2);$$

hence, (3.20) is proved. The maps β_1 and β_2 are surmultiplicative v-value functions because β is a surmultiplicative v-value function. □

3.2.3 Gauges on symbol algebras

Symbol algebras provide diverse examples of gauges, which we discuss in this subsection.

Let F be a field with valuation v such that F contains a primitive n-th root of unity ω for some $n \geq 2$. So, $v(\omega) = 0$. Let $\overline{\omega}$ be the image of ω in $\overline{F}$; then, $\overline{\omega}^n = 1$. Take any $a, b \in F^\times$, and let S be the symbol algebra of degree n,

$$S = \big(a, b/F\big)_n,$$

with its standard generators i, j satisfying

$$i^n = a, \qquad j^n = b, \qquad \text{and} \qquad ij = \omega ji.$$

Define a v-norm α on S by

$$\alpha\Big(\sum_{k,\ell=0}^{n-1} c_{k\ell}\, i^k j^\ell\Big) = \min_{k,\ell}\big(v(c_{k\ell}) + \tfrac{k}{n}v(a) + \tfrac{\ell}{n}v(b)\big) \qquad \text{where each } c_{k\ell} \in F. \tag{3.21}$$

Thus, the standard F-base $\big(i^k j^\ell\big)_{k,\ell=0}^{n-1}$ for S is a splitting base for α. Since $\alpha(i) = \frac{1}{n}v(a)$ and $\alpha(j) = \frac{1}{n}v(b)$ and $v(\omega) = 0$, it is easy to check that

$$\alpha(i^k j^\ell) = k\alpha(i) + \ell\alpha(j) = \tfrac{k}{n}v(a) + \tfrac{\ell}{n}v(b) \qquad \text{for all } k, \ell \in \mathbb{Z}. \tag{3.22}$$

Hence, $\alpha\big((i^k j^\ell)(i^{k'} j^{\ell'})\big) = \alpha(i^k j^\ell) + \alpha(i^{k'} j^{\ell'})$, for all $k, \ell, k', \ell' \in \mathbb{Z}$, which implies that α is a surmultiplicative v-norm on S by Lemma 3.23. Formula (3.22) also implies that $\widetilde{i^k j^\ell} = \widetilde{i}^{\,k}\,\widetilde{j}^{\,\ell}$ for all k, ℓ. The associated graded algebra $\mathsf{gr}_\alpha(S)$ has the homogeneous $\mathsf{gr}(F)$-base $\big(\widetilde{i}^{\,k}\,\widetilde{j}^{\,\ell}\big)_{k,\ell=0}^{n-1}$. Moreover,

$$\widetilde{i}^{\,n} = \widetilde{a}, \qquad \widetilde{j}^{\,n} = \widetilde{b}, \qquad \text{and} \qquad \widetilde{i}\,\widetilde{j} = \widetilde{\omega}\,\widetilde{j}\,\widetilde{i}. \tag{3.23}$$

Clearly,

$$\Gamma_{\mathsf{gr}_\alpha(S)} = \Gamma_S = \langle \tfrac{1}{n}v(a), \tfrac{1}{n}v(b)\rangle + \Gamma_F.$$

Note also that since $\widetilde{i}^{\,k}\,\widetilde{j}^{\,\ell} \in \mathsf{gr}_\alpha(S)^\times$ and $deg(\widetilde{i}^{\,k}\,\widetilde{j}^{\,\ell}) = \frac{k}{n}v(a) + \frac{\ell}{n}v(b)$, we have

$$\Gamma^\times_{\mathsf{gr}_\alpha(S)} = \Gamma_{\mathsf{gr}_\alpha(S)}. \tag{3.24}$$

From this we obtain:

Proposition 3.37. *The following are equivalent:*

(a) *α is a v-gauge;*
(b) *$\mathsf{gr}_\alpha(S)$ is graded semisimple;*
(c) *the homogeneous component S_0 is semisimple.*

Proof. (a) $\Leftrightarrow$ (b) follows from the definition of a gauge. (b) $\Rightarrow$ (c) is clear from the Wedderburn structure theory of graded semisimple algebras (see Prop. 2.41). (c) $\Rightarrow$ (b) follows by Lemma 2.13, as $\Gamma_S^\times = \Gamma_S$. □

The conditions for α to be a valuation are more stringent:

Proposition 3.38. *The following are equivalent:*

(a) *S is a division algebra and α is a valuation on S;*
(b) *$\mathsf{gr}_\alpha(S)$ is a graded division algebra;*
(c) *the homogeneous component S_0 is a division algebra.*

Proof. (a) $\Leftrightarrow$ (b) is given in Cor. 3.30. Also, (b) $\Rightarrow$ (c) is clear and (c) $\Rightarrow$ (b) follows easily from (3.24). □

When α is a valuation, we will see by Prop. 4.33(i) below that α is the only v-gauge on S. But when it is not a valuation, there can be other v-gauges on S besides α. This will be illustrated for a particular quaternion algebra in Ex. 3.50 below.

For the symbol algebra S, we now obtain a presentation of S_0: let $\mathbb{Z}^2 = \mathbb{Z} \times \mathbb{Z}$, and let $\eta\colon \mathbb{Z}^2 \to (\frac{1}{n}\Gamma_F)/\Gamma_F$ be the homomorphism given by $(k,\ell) \mapsto \big(\frac{k}{n}v(a) + \frac{\ell}{n}v(b)\big) + \Gamma_F$. Let

$$H \;=\; \ker \eta \;=\; \{(k,\ell) \in \mathbb{Z}^2 \mid \widetilde{i}^{\,k}\widetilde{j}^{\,\ell}\,\mathsf{gr}(F) \cap S_0 \neq \{0\}\}. \tag{3.25}$$

So, $(n\mathbb{Z})^2 \subseteq H \subseteq \mathbb{Z}^2$. For $(r,s) \in H$ let $\overline{(r,s)} = (r,s) + (n\mathbb{Z})^2 \in H/(n\mathbb{Z})^2$. Note that

$$\begin{aligned}[S_0{:}F_0] &= [\mathsf{gr}(S){:}\,\mathsf{gr}(F)]/|\Gamma_{\mathsf{gr}_\alpha(S)}{:}\Gamma_{\mathsf{gr}(F)}| \\ &\qquad\qquad = |\mathbb{Z}/n\mathbb{Z}|^2/|\operatorname{im}\eta| \;=\; \big|\ker\eta/(n\mathbb{Z}^2)\big|.\end{aligned}$$

The first equality here follows by the argument for (2.33), as $\Gamma_{\mathsf{gr}_\alpha(S)} = \Gamma^{\times}_{\mathsf{gr}_\alpha(S)}$. The finite abelian group $H/(n\mathbb{Z})^2$ is a direct product of at most two cyclic groups, since it is a subgroup of the bicyclic group $(\mathbb{Z}/n\mathbb{Z})^2$. We consider the different possible cases for this group:

Case I. Suppose $H = (n\mathbb{Z})^2$, i.e., $|\Gamma_S{:}\Gamma_F| = n^2$. This occurs if and only if the images of $v(a)$ and $v(b)$ generate a subgroup of $\Gamma_F/n\Gamma_F$ of order n^2. Then, $S_0 = F_0$. By Prop. 3.38, α is a valuation (so also a v-gauge) on S, which must be a division algebra. Clearly, the valuation α is totally ramified over v.

Case II. Suppose $H/(n\mathbb{Z})^2$ is a cyclic group, say of order q, with a generator $\overline{(r,s)}$ for some $(r,s) \in H$. Then there is a $c \in \mathsf{gr}(F)^\times$ with $deg(c) = \frac{r}{n}v(a) + \frac{s}{n}v(b)$. Let $u = c^{-1}\widetilde{i}^{\,r}\widetilde{j}^{\,s} \in S_0^\times$, and let $t = u^q$. Since $(qr, qs) \in (n\mathbb{Z})^2$, we have $t \in \mathsf{gr}(F)^\times \cap \mathsf{S}_0 = F_0^\times$. We claim that

$$S_0 \;=\; F_0[u] \;\cong\; F_0[X]/(X^q - t).$$

For every power u^m of u is expressible as $u^m = c_m\,\widetilde{i}^{\,mr}\,\widetilde{j}^{\,ms} = d_m\,\widetilde{i}^{\,k_m}\widetilde{j}^{\,\ell_m}$ with $c_m, d_m \in \mathsf{gr}(F)^\times$ and $k_m, \ell_m \in \{0,1,2,\ldots,n-1\}$ satisfying

$$(k_m, \ell_m) \;\equiv\; (mr, ms)\,(mod\;(n\mathbb{Z})^2).$$

Since the (mr, ms) are distinct mod $(n\mathbb{Z})^2$ for $m = 0, 1, \ldots, q-1$ the corresponding ordered pairs (k_m, ℓ_m) are distinct; hence, the powers $u^0, u^1, \ldots, u^{q-1}$ are $\mathsf{gr}(F)$-independent, so F_0-independent. Thus, the epimorphism $F_0[X]/(X^q - t) \to F_0[u]$ is an isomorphism by dimension count. Therefore, $[F_0[u]{:}F_0] = q = [S_0{:}F_0]$, showing that $S_0 = F_0[u]$ as claimed. Here, S_0 is commutative, but need not be a field. Indeed, S_0 need not even be semisimple when $char(\overline{F}) \mid q$.

Case III. Suppose $H/(n\mathbb{Z})^2$ is not cyclic. It is then bicyclic. Write

$$H/(n\mathbb{Z})^2 \;=\; \big\langle\, \overline{(r,s)}\,\big\rangle \times \big\langle\, \overline{(r',s')}\,\big\rangle,$$

and let $q = |\langle \overline{(r,s)} \rangle|$ and $q' = |\langle \overline{(r',s')} \rangle|$. So, $qq' = |H/(n\mathbb{Z})^2| = [S_0{:}F_0]$. There exist $c, c' \in \mathsf{gr}(F)^\times$ with

$$deg(c) = \tfrac{r}{n}v(a) + \tfrac{s}{n}v(b) \quad \text{and} \quad deg(c') = \tfrac{r'}{n}v(a) + \tfrac{s'}{n}v(b).$$

Let $u = c^{-1}\widetilde{i}^{\,r}\widetilde{j}^{\,s} \in S_0^\times$ and $u' = c'^{-1}\widetilde{i}^{\,r'}\widetilde{j}^{\,s'} \in S_0^\times$, and let $t = u^q$ and $t' = u'^{q'}$. Since $(qr, qs), (q'r', q's') \in (n\mathbb{Z})^2$, we have $t, t' \in \mathsf{gr}(F) \cap S_0^\times = F_0^\times$. Thus,

$$u^q = t, \quad u'^{q'} = t', \quad \text{and} \quad uu' = \xi u'u, \quad \text{where } \xi = \overline{\omega}^{\,rs'-r's}. \tag{3.26}$$

We claim that $S_0 = F_0[u, u']$, the (in general noncommutative) F_0-algebra generated by u and u', and that the relations (3.26) give a presentation of S_0. Indeed, arguing as in Case II, since the elements $(mr + m'r', ms + m's')$ of $\mathbb{Z}^2$ are distinct modulo $(n\mathbb{Z})^2$ for (m, m') in $\{0, 1, \ldots, q-1\} \times \{0, 1, \ldots, q'-1\}$, the corresponding $u^m u'^{m'}$ are $\mathsf{gr}(F)$-independent, and hence F_0-independent in S_0. Thus, $[F_0[u, u']{:}F_0] \geq qq' = [S_0{:}F_0]$, showing that $S_0 = F_0[u, u']$. But the relations in (3.26) imply that $[F_0[u, u']{:}F_0] \leq qq'$. Therefore, those relations generate all the relations on u and u' in S_0.

The structure of $\mathsf{gr}_\alpha(S)$ depends on $char(\overline{F})$: we consider successively the cases where $char(\overline{F}) \nmid n$, and where $char(\overline{F}) = \overline{p} \neq 0$ and n is a power of $\overline{p}$.

Suppose first $char(\overline{F}) \nmid n$. Then, $\overline{\omega}$ is a primitive n-th root of unity in $\overline{F}$. For the polynomial equation $X^n - 1 = \prod_{k=0}^{n-1}(X - \omega^k)$ in $\mathcal{O}_F[X]$ maps in $\overline{F}[X]$ to $X^n - \overline{1} = \prod_{k=0}^{n-1}(X - \overline{\omega^k}) = \prod_{k=0}^{n-1}(X - \overline{\omega}^k)$. Since $X^n - \overline{1}$ has no repeated roots as $char(\overline{F}) \nmid n$, the powers $\overline{1}, \overline{\omega}, \overline{\omega}^2, \ldots, \overline{\omega}^{n-1}$ are all distinct, showing that $\overline{\omega}$ has order n in $\overline{F}^\times$. The relations in (3.23) satisfied by $\widetilde{i}$ and $\widetilde{j}$ show that $\mathsf{gr}_\alpha(S)$ is the graded symbol algebra of degree n

$$\mathsf{gr}_\alpha(S) \cong_g \big(\widetilde{a}, \widetilde{b}/\,\mathsf{gr}(F)\big)_n, \tag{3.27}$$

as in Def. 2.18. Since $\mathsf{gr}(S)$ is therefore a simple graded algebra by Prop. 2.19, α is a v-gauge on S.

At the other extreme, suppose $char(\overline{F}) = \overline{p} \neq 0$, and suppose n is a power of $\overline{p}$. Then $\overline{\omega} = 1$, so $\mathsf{gr}_\alpha(S)$ is commutative. Indeed,

$$\mathsf{gr}_\alpha(S) \cong_g \mathsf{gr}(F)[X, Y]/(X^n - \widetilde{a}, Y^n - \widetilde{b}). \tag{3.28}$$

Let $K = q(\mathsf{gr}(F))$, which is a field of characteristic $\overline{p}$. Then,

$$q(\mathsf{gr}_\alpha(S)) \cong K \otimes_{\mathsf{gr}(F)} \mathsf{gr}_\alpha(S) \cong K[X, Y]/(X^n - \widetilde{a}, Y^n - \widetilde{b}).$$

Note that the commutative ring $q(\mathsf{gr}_\alpha(S))$ is local, as n is a power of $char(K)$. Hence, either $q(\mathsf{gr}_\alpha(S))$ is a field purely inseparable over K or it is not semisimple. By combining Prop. 3.37, Prop. 3.38, and Prop. 2.28, we obtain:

Proposition 3.39. *If $char(\overline{F}) = \overline{p} \neq 0$ and n is a power of $\overline{p}$, the following conditions are all equivalent:*

(a) *α is a v-gauge;*
(b) *$\mathsf{gr}_\alpha(S)$ is graded semisimple;*

(c) $q(\mathsf{gr}_\alpha(S))$ *is semisimple;*
(d) $q(\mathsf{gr}_\alpha(S))$ *is a field;*
(e) α *is a valuation on* S*;*
(f) $[K(\sqrt[n]{\widetilde{a}}, \sqrt[n]{\widetilde{b}}) : K] = n^2$*;*
(g) $[K(\sqrt[\overline{p}]{\widetilde{a}}, \sqrt[\overline{p}]{\widetilde{b}}) : K] = \overline{p}^2$.

The construction of value functions given here for symbol algebras will be generalized to tensor products of symbol algebras in §7.2.3 below.

We give a specific example illustrating some of the different possibilities described above:

Example 3.40. Let S be the quaternion algebra

$$S = \big(-3, 30/\mathbb{Q}\big),$$

with generators i and j satisfying

$$i^2 = -3, \qquad j^2 = 30, \quad \text{and} \quad ij = -ji.$$

For each prime number p, let $v_p\colon \mathbb{Q} \to \mathbb{Z} \cup \{\infty\}$ be the p-adic discrete valuation on $\mathbb{Q}$. We have $\Gamma_{\mathbb{Q},v_p} = \mathbb{Z}$, $\overline{\mathbb{Q}}^{v_p} = \mathbb{F}_p$, and

$$\mathsf{gr}_{v_p}(\mathbb{Q}) = \mathbb{F}_p[\widetilde{p}, \widetilde{p}^{-1}] \cong_g \mathbb{F}_p[X, X^{-1}].$$

For each p, we consider the surmultiplicative v_p-norm α_p on S described at the beginning of this subsection. For every $p \neq 2$, α_p is a v_p-gauge on S with associated graded ring the graded quaternion algebra

$$\mathsf{gr}_{\alpha_p}(S) = \big(\widetilde{-3}, \widetilde{30}/\,\mathsf{gr}_{v_p}(\mathbb{Q})\big).$$

Let H_p denote the group $H = \mathit{ker}\,\eta$ of (3.25) for α_p.

(i) $p = 3$. We have $\alpha_3(i) = \frac{1}{2}$, $\alpha_3(j) = \frac{1}{2}$, and $\Gamma_{S,\alpha_3} = \frac{1}{2}\mathbb{Z}$. Here,

$$H_3 = \{(k, \ell) \in \mathbb{Z}^2 \mid k \equiv \ell \ (\mathit{mod}\ 2)\},$$

so $H_3/(2\mathbb{Z})^2$ is a cyclic group with generator $\overline{(-1, 1)}$, and we are in Case II above. We can take $u = \widetilde{i}^{-1}\widetilde{j} \in \mathsf{gr}_{\alpha_3}(S)_0$ and $t = u^2 = \overline{30} = \overline{1} \in \mathbb{F}_3$. So,

$$\mathsf{gr}_{\alpha_3}(S)_0 = \mathbb{F}_3[u] \cong \mathbb{F}_3[X]/(X^2 - \overline{1}) \cong \mathbb{F}_3 \times \mathbb{F}_3,$$

which is semisimple, but not a field. So, the gauge α_3 on S is not a valuation.

(ii) $p = 5$. We have $\alpha_5(i) = 0$, $\alpha_5(j) = \frac{1}{2}$, and $\Gamma_{S,\alpha_5} = \frac{1}{2}\mathbb{Z}$. Also, $H_5 = \mathbb{Z} \times 2\mathbb{Z}$, so $H/(2\mathbb{Z}^2)^2$ is cyclic with generator $\overline{(1, 0)}$, and we are in Case II. We can take $u = \widetilde{i} \in \mathsf{gr}_{\alpha_5}(S)_0$, so $t = \widetilde{i}^{\,2} = \widetilde{i^2} = \overline{-3} \in \mathbb{F}_5$, and

$$\mathsf{gr}_{\alpha_5}(S)_0 = \mathbb{F}_5[u] \cong \mathbb{F}_5[X]/\big(X^2 - (\overline{-3})\big) \cong \mathbb{F}_5\big(\sqrt{\overline{-3}}\big),$$

which is the field with 25 elements. Hence, α_5 is a v_5-gauge and a valuation on S.

(iii) $p > 5$. We have $\alpha_p(i) = 0$, $\alpha_p(j) = 0$, and $\Gamma_{S,\alpha_p} = \mathbb{Z}$. So, $H_p = \mathbb{Z}^2$, and the bicyclic group $H_p/(2\mathbb{Z})^2$ has independent generators $\overline{(1,0)}$ and $\overline{(0,1)}$, and we are in Case III. We can take $u = \widetilde{i}$ and $u' = \widetilde{j}$, so $t = u^2 = \overline{-3}$, $t' = u'^2 = \overline{30}$ and

$$\mathsf{gr}_{\alpha_p}(S)_0 \cong \big(\overline{-3}, \overline{30} \,/\, \mathbb{F}_p\big).$$

This is a quaternion algebra over $\mathbb{F}_p$, which is necessarily split as $\mathbb{F}_p$ is finite. Hence, α_p is a v_p-gauge on S, but is not a valuation.

(iv) $p = 2$. We have $\alpha_2(i) = 0$, $\alpha_2(j) = \frac{1}{2}$, and $\Gamma_{S,\alpha_2} = \frac{1}{2}\mathbb{Z}$. So, $H_2 = \mathbb{Z} \times 2\mathbb{Z}$, and $H_2/(2\mathbb{Z})^2$ is cyclic with generator $\overline{(1,0)}$, and we are in Case II. We may take $u = \widetilde{i} \in \mathsf{gr}_{\alpha_2}(S)_0$ and $t = u^2 = \overline{-3}$ in $\mathbb{F}_2$. Thus,

$$\mathsf{gr}_{\alpha_2}(S)_0 \;=\; \mathbb{F}_2[u] \;\cong\; \mathbb{F}_2[X]/\big(X^2 - (\overline{-3})\big) \;\cong\; \mathbb{F}_2[X]/(X - \overline{1})^2,$$

which is not semisimple. Hence, $\mathsf{gr}_{\alpha_2}(S)$ is not graded simple, so the surmultiplicative v_2-norm α_2 is not a v_2-gauge on S. Indeed,

$$\begin{aligned}\mathsf{gr}_{\alpha_2}(S) &\cong_g \mathsf{gr}_{v_2}(\mathbb{Q})[X,Y]/(X^2 - (\widetilde{-3}), Y^2 - \widetilde{30})\\ &\cong_g \mathsf{gr}_{v_2}(\mathbb{Q})[X,Y]/\big((X - \widetilde{1})^2, Y^2 - \widetilde{2}\big).\end{aligned}$$

(To see the non-semisimplicity of $\mathsf{gr}_{\alpha_2}(S)$ explicitly, note that in S we have $\alpha_2(1+i) = 0$ but $\alpha_2\big((1+i)^2\big) = \alpha_2(-2+2i) = 1$. Hence, $\widetilde{1+i}$ is a nonzero homogeneous nilpotent element of the commutative ring $\mathsf{gr}_{\alpha_2}(S)$.) Setting $K = q(\mathsf{gr}_{v_2}(\mathbb{Q})) \cong \mathbb{F}_2(\widetilde{2})$, where $\widetilde{2}$ is transcendental over $\mathbb{F}_2$, we have

$$q(\mathsf{gr}_{\alpha_2}(S)) \;\cong\; K[X,Y]/\big((X - \widetilde{1})^2, Y^2 - \widetilde{2}\big),$$

which is a commutative and local but not semisimple ring.

We take a closer look at the case $p = 2$. The construction above did not produce a v_2-gauge on S but did not prove that no such gauge exists. In fact, Th. 4.50 below says that S must have a v_2-gauge, since $\Gamma_{\mathbb{Q},v_2} = \mathbb{Z}$. Moreover, we can see that S must have a valuation extending v_2. Indeed, $S \otimes_{\mathbb{Q}} \mathbb{R} \cong \big(-3, 30/\mathbb{R}\big) \cong M_2(\mathbb{R})$, and for $p = 3$ and all primes $p > 5$, $S \otimes_{\mathbb{Q}} \mathbb{Q}_p$ is split, where $\mathbb{Q}_p$ is the p-adic completion of $\mathbb{Q}$. This can be seen from Prop. 4.33 below which implies that since S has a v_p-gauge which is not a valuation, S has no valuation extending v_p on $\mathbb{Q}$. Hence, the prime-degree algebra $S \otimes_{\mathbb{Q}} \mathbb{Q}_p$ is not a division algebra by Prop. 1.16, so it is split. But $S \otimes_{\mathbb{Q}} \mathbb{Q}_5$ is nonsplit, since α_5 is a valuation on S extending v_5 on $\mathbb{Q}$. Hilbert's Reciprocity Law (see Lam [122, p. 180]) says that $S \otimes_{\mathbb{Q}} \mathbb{Q}_p$ must be nonsplit for an even number of $\mathbb{Q}_p$ (when we include $\mathbb{R}$ as $\mathbb{Q}_\infty$). Therefore, $S \otimes_{\mathbb{Q}} \mathbb{Q}_2$ must be nonsplit, and its valuation extending the complete discrete 2-adic valuation on $\mathbb{Q}_2$ must restrict to a valuation on S extending v_2 on $\mathbb{Q}$. (This valuation is then the unique v_2-gauge on S, by Prop. 4.33(i) below.)

We can describe this v_2-gauge and valuation on S explicitly as follows: Let $\zeta = \frac{1}{2} + \frac{1}{2}i \in S$, and let β be the v_2-value function on S defined by

$$\beta(a+b\zeta+cj+d\zeta j) = \min\big(v_2(a), v_2(b), \tfrac{1}{2}v_2(c), \tfrac{1}{2}v_2(d)\big) \quad \text{for all } a,b,c,d \in \mathbb{Q}.$$

So β is a v_2-norm on S with splitting base $(1, \zeta, j, \zeta j)$. Since $\zeta^2 = \zeta - 1$, so $\beta(\zeta^2) = 0$, and $j\zeta j^{-1} = 1 - \zeta$ with $\beta(1-\zeta) = 0$, it is easy to check that β is surmultiplicative for products of elements of the splitting base. Hence, β is surmultiplicative by Lemma 3.23. Because $\zeta^2 = 1 - \zeta$, so $\beta(\zeta^2) = \beta(1-\zeta) = 0 = \beta(\zeta)$, we have $\widetilde{\zeta}^2 = \widetilde{1-\zeta} = \widetilde{1} - \widetilde{\zeta}$ in $\mathsf{gr}_\beta(S)$. Hence,

$$\mathsf{gr}_\beta(S)_0 = \mathsf{gr}_{v_2}(\mathbb{Q})_0 \oplus \mathsf{gr}_{v_2}(\mathbb{Q})_0\,\widetilde{\zeta} = \mathbb{F}_2 \oplus \mathbb{F}_2\widetilde{\zeta} \cong \mathbb{F}_2[X]/(X^2 - X + 1),$$

which is the field with 4 elements. Since $\beta(j^2) = v_2(30) = 1 = 2\beta(j)$, we have $\widetilde{j}^{\,2} = \widetilde{j^2} = \widetilde{30} \in \mathsf{gr}_{v_2}(\mathbb{Q})^\times$; hence, $\widetilde{j} \in \mathsf{gr}_\beta(S)^\times$ which yields that $\Gamma^\times_{S,\beta} = \frac{1}{2}\mathbb{Z} = \Gamma_{S,\beta}$. Since $\mathsf{gr}_\beta(S)_0$ is semisimple, Lemma 2.13 thus shows that $\mathsf{gr}_\beta(S)$ is graded semisimple, so β is a v_2-gauge. Moreover, $\mathsf{gr}_\beta(S)$ is a graded division ring since $\mathsf{gr}_\beta(S)_0$ is a field and $\Gamma^\times_{S,\beta} = \Gamma_{S,\beta}$; hence by Cor. 3.30; β is a valuation on S extending v_2 on $\mathbb{Q}$.

3.2.4 Gauges on tensor products

In this subsection, we use tensor products to obtain further examples of gauges. In particular, scalar extensions are discussed in Prop. 3.42. As a first illustration of the utility of gauges, we will prove a theorem of Morandi (Th. 3.43) on tensor products of division algebras.

Proposition 3.41. *Let A and B be finite-dimensional semisimple F-algebras with respective v-gauges α and β. The following conditions are equivalent:*

(a) *$\alpha \otimes \beta$ is a v-gauge on $A \otimes_F B$;*
(b) *$\mathsf{gr}_\alpha(A) \otimes_{\mathsf{gr}(F)} \mathsf{gr}_\beta(B)$ is a semisimple graded $\mathsf{gr}(F)$-algebra;*
(c) *$Z(\mathsf{gr}_\alpha(A)) \otimes_{\mathsf{gr}(F)} Z(\mathsf{gr}_\beta(B))$ is graded semisimple, i.e., a direct product of graded fields.*

These conditions hold in particular whenever $Z(\mathsf{gr}_\alpha(A)) = \mathsf{gr}(F)$ or $Z(\mathsf{gr}_\beta(B)) = \mathsf{gr}(F)$.

Proof. Theorem 3.21(iv) shows that $\alpha \otimes \beta$ is a v-norm, and Prop. 3.27 shows that this norm is surmultiplicative and yields the canonical identification

$$\mathsf{gr}_{\alpha\otimes\beta}(A \otimes_F B) = \mathsf{gr}_\alpha(A) \otimes_{\mathsf{gr}(F)} \mathsf{gr}_\beta(B).$$

Therefore, $\alpha \otimes \beta$ is a gauge if and only if the right expression is graded semisimple, proving the equivalence of (a) and (b). The equivalence of (b) and (c) follows from Cor. 2.33. □

Proposition 3.42. *Let A be a finite-dimensional semisimple F-algebra with a v-gauge α and let L/F be an arbitrary field extension. Let $w\colon L \to \Gamma \cup \{\infty\}$ be a valuation extending v. Then $\alpha\otimes w$ is a w-gauge on the L-algebra $A\otimes_F L$ if* *and only if* $Z(\mathsf{gr}_\alpha(A))\otimes_{\mathsf{gr}(F)}\mathsf{gr}_w(L)$ *is a semisimple graded* $\mathsf{gr}(L)$*-algebra, i.e., a direct product of graded fields. This condition holds in particular whenever* $Z(\mathsf{gr}_\alpha(A)) = \mathsf{gr}(F)$*, and also when the extension* L/F *is immediate.*

Proof. Corollary 3.28 shows that $\alpha \otimes w$ is a surmultiplicative w-norm and yields the canonical identification

$$\mathsf{gr}_{\alpha\otimes w}(A \otimes_F L) \;=\; \mathsf{gr}_\alpha(A) \otimes_{\mathsf{gr}(F)} \mathsf{gr}_w(L).$$

Therefore, $\alpha \otimes w$ is a w-gauge if and only if $\mathsf{gr}_\alpha(A) \otimes_{\mathsf{gr}(F)} \mathsf{gr}_w(L)$ is a semisimple graded algebra. By Cor. 2.33, this condition holds if and only if $Z(\mathsf{gr}_\alpha(A)) \otimes_{\mathsf{gr}(F)} \mathsf{gr}_w(L)$ is a semisimple graded $\mathsf{gr}(L)$-algebra. The proposition follows. □

The following theorem is a very useful tool to show that certain tensor products of division algebras are division algebras.

Theorem 3.43 (Morandi). *Let* D *and* E *be division* F*-algebras with respective valuations* v_D*,* v_E *extending* v*. Suppose* D *is finite-dimensional and* $[D{:}F] = [\overline{D}{:}\overline{F}]|\Gamma_D{:}\Gamma_F|$*; but* E *may be infinite-dimensional. If* $\overline{D} \otimes_{\overline{F}} \overline{E}$ *is a division ring and* $\Gamma_D \cap \Gamma_E = \Gamma_F$*, then* $D \otimes_F E$ *is a division ring and* $v_D \otimes v_E$ *is a valuation on* $D \otimes_F E$ *such that*

$$\overline{D \otimes_F E} \;=\; \overline{D} \otimes_{\overline{F}} \overline{E} \qquad \textit{and} \qquad \Gamma_{D\otimes_F E} \;=\; \Gamma_D + \Gamma_E.$$

Proof. The hypotheses imply that v_D is a v-gauge, hence $v_D \otimes v_E$ is a surmultiplicative v-value function by Prop. 3.27. Its value set is the group $\Gamma_D + \Gamma_E$ by Prop. 2.12. Since $\Gamma_D \cap \Gamma_E = \Gamma_F$, it follows from Prop. 2.12 that for all $\gamma \in \Gamma_D$ and $\delta \in \Gamma_E$ we have

$$(D \otimes_F E)_{\gamma+\delta} \;=\; D_\gamma \otimes_{\overline{F}} E_\delta. \tag{3.29}$$

Pick $d \in D^\times$ with $v_D(d) = \gamma$ and $e \in E^\times$ with $v_E(e) = \delta$. Then

$$D_\gamma \otimes_{\overline{F}} E_\delta \;=\; (\widetilde{d \otimes e}) \cdot (\overline{D} \otimes_{\overline{F}} \overline{E}) \;=\; (\overline{D} \otimes_{\overline{F}} \overline{E}) \cdot (\widetilde{d \otimes e}).$$

Since $\widetilde{d \otimes e}$ is invertible and $\overline{D} \otimes_{\overline{F}} \overline{E}$ is a division ring, it follows that all the nonzero elements in $(D \otimes_F E)_{\gamma+\delta}$ are invertible, hence $\mathsf{gr}_{v_D\otimes v_E}(D \otimes_F E)$ is a graded division ring. If E is finite-dimensional, then the theorem follows from Cor. 3.30; but it turns out that this hypothesis is not needed: for any nonzero $x, y \in D \otimes_F E$, the homogeneous elements $\widetilde{x}$, $\widetilde{y}$ are invertible, hence $\widetilde{x} \cdot \widetilde{y} \neq 0$ and therefore

$$(v_D \otimes v_E)(xy) \;=\; (v_D \otimes v_E)(x) + (v_D \otimes v_E)(y). \tag{3.30}$$

It follows that $D \otimes_F E$ has no zero-divisors, hence it is a division ring since it is a finite-dimensional E-vector space. By (3.30), the v-value function $v_D \otimes v_E$ is a valuation. The equality $\overline{D \otimes_F E} = \overline{D} \otimes_{\overline{F}} \overline{E}$ is the particular case of (3.29) where $\gamma = \delta = 0$. □

3.2.5 Gauges from graded algebras

Let F be a graded field, and let A be a finite-dimensional graded F-algebra; so $\Gamma_{\mathsf{F}} \subseteq \Gamma_{\mathsf{A}} \subseteq \Gamma$. In this subsection we show how to use an ordering on Γ to build a valuation v on $q(\mathsf{F})$ and a surmultiplicative v-norm α on $q(\mathsf{A})$ so that $\mathsf{gr}_\alpha(q(\mathsf{A})) \cong_g \mathsf{A}$. Such an α is a v-gauge if and only if A is graded semisimple.

Because the ambient abelian group Γ is torsion-free, there are total orderings on Γ making it into a totally ordered abelian group, see Remark 2.2. Choose one such ordering $\leq$ on Γ, and fix it throughout this subsection.

Any s in the graded field F is expressible uniquely as $s = \sum_{\gamma\in\Gamma} s_\gamma$ with each $s_\gamma \in \mathsf{F}_\gamma$. Let

$$supp(s) \ = \ \{\gamma \in \Gamma \mid s_\gamma \neq 0\},$$

a finite set. If $s \neq 0$, set

$$\lambda(s) \ = \ s_\delta, \quad \text{where } \delta \text{ is minimal in } supp(s).$$

Thus, $\lambda(s)$ is the homogeneous component of s of least degree, relative to the chosen ordering on Γ. The map $s \mapsto \lambda(s)$ defines a function $\lambda\colon \mathsf{F}\setminus\{0\} \to \mathsf{F}^\times$. For $s, t \in \mathsf{F}\setminus\{0\}$, we have

$$\begin{aligned} st \ &= \ \big(\lambda(s) + \text{higher degree terms}\big)\cdot\big(\lambda(t) + \text{higher degree terms}\big) \\ &= \lambda(s)\lambda(t) + \text{higher degree terms.} \end{aligned}$$

Since the graded field F has no zero divisors, this shows that

$$\lambda(st) \ = \ \lambda(s)\cdot\lambda(t), \qquad \text{for all } s,\, t \in \mathsf{F}\setminus\{0\}. \tag{3.31}$$

This equality allows us to extend λ to the multiplicative group $q(\mathsf{F})^\times$ of the quotient field of $\mathsf{F}^\natural$ by defining

$$\lambda(st^{-1}) \ = \ \lambda(s)\lambda(t)^{-1} \qquad \text{for all } s,\, t \in \mathsf{F}\setminus\{0\}.$$

It follows easily from (3.31) that λ on $q(\mathsf{F})^\times$ is well-defined, and that

$$\lambda(cd) = \lambda(c)\cdot\lambda(d) \qquad \text{for all } c,\, d, \in q(\mathsf{F})^\times. \tag{3.32}$$

Equation (3.32) shows that $\lambda\colon q(\mathsf{F})^\times \to \mathsf{F}^\times$ is a group homomorphism; it yields a split short exact sequence:

$$1 \longrightarrow \mathsf{F}^\times \underset{\lambda}{\leftrightarrows} q(\mathsf{F})^\times \longrightarrow q(\mathsf{F})^\times/\mathsf{F}^\times \longrightarrow 1 \tag{3.33}$$

We use λ to build a valuation on $q(\mathsf{F})$: Define a function $v\colon q(\mathsf{F}) \to \Gamma\cup\{\infty\}$ by

$$v(c) \ = \ \begin{cases} deg(\lambda(c)) & \text{if } c \neq 0, \\ \infty & \text{if } c = 0. \end{cases}$$

Thus, $v(s) = min(supp(s))$ for $s \in \mathsf{F} \setminus \{0\}$. We adopt the usual convention that $\gamma < \infty$ for all $\gamma \in \Gamma$ and $\gamma + \infty = \infty + \gamma = \infty + \infty = \infty$. Since $deg\colon \mathsf{F}^\times \to \Gamma$ is a group homomorphism, (3.32) yields

$$v(cd) = v(c) + v(d) \qquad \text{for all } c,\, d \in q(\mathsf{F}). \tag{3.34}$$

Clearly $v(c) = \infty$ if and only if $c = 0$. Furthermore, we claim that

$$v(c+d) \geq min\big(v(c), v(d)\big) \qquad \text{for all } c,\, d \in q(\mathsf{F}). \tag{3.35}$$

This is clear if $c = 0$ or $d = 0$ or $c + d = 0$. It also holds if $c,\, d \in \mathsf{F} \setminus \{0\}$ and $c + d \neq 0$. For then,

$$\begin{aligned} v(c+d) &= min(supp(c+d)) \geq min\big(supp(c) \cup supp(d)\big) \\ &= min\big(min(supp(c)), min(supp(d))\big) = min\big(v(c), v(d)\big). \end{aligned}$$

For the general case, write $c = st^{-1}$, $d = s't'^{-1}$ where $s,\, t,\, s',\, t' \in \mathsf{F}$ with $tt' \neq 0$. Then by using (3.34) and the cases already covered, we have

$$\begin{aligned} v(c+d) &= v(st' + s't) - v(tt') \geq min\big(v(st'), v(s't)\big) - v(t) - v(t') \\ &= min\big(v(s) - v(t), v(s') - v(t')\big) = min\big(v(c), v(d)\big), \end{aligned}$$

proving the claim. We have now shown that v is a valuation on $q(\mathsf{F})$. Clearly $\Gamma_{q(\mathsf{F}),v} = \Gamma_{\mathsf{F}}$, and we will see in Prop. 3.45 below that $\overline{q(\mathsf{F})} = \mathsf{F}_0$. Note that v is completely determined by the grading on F and the choice of ordering on Γ_{F}.

In verifying (3.35) we proved the result for $c,\, d \in q(\mathsf{F})$ by reducing to the previously proved case where $c,\, d \in \mathsf{F}$. We call this process "clearing denominators."

Now let A be a finite-dimensional graded F-algebra with $\Gamma_{\mathsf{A}} \subseteq \Gamma$. We continue to use the chosen total ordering on Γ. Recall from Lemma 2.15 that the ring of central quotients of A is

$$q(\mathsf{A}) = q_{\mathsf{F}}(\mathsf{A}) = \mathsf{A}^\natural \otimes_{\mathsf{F}^\natural} q(\mathsf{F}) = \big\{as^{-1} \mid a \in \mathsf{A},\ s \in \mathsf{F} \setminus \{0\}\big\}.$$

We define analogues for $q(\mathsf{A})$ to the functions λ and v for $q(\mathsf{F})$: As above, for $a = \sum_{\gamma\in\Gamma} a_\gamma$ with each $a_\gamma \in \mathsf{A}_\gamma$, define

$$supp(a) = \{\gamma \in \Gamma \mid a_\gamma \neq 0\},$$

which is a finite subset of Γ_{A}. If $a \neq 0$, define

$$\mu(a) = a_\delta, \quad \text{where } \delta \text{ is minimal in } supp(a).$$

Since no element of $\mathsf{F} \setminus \{0\}$ is a zero divisor on A, analogous to (3.31) we have

$$\mu(as) = \mu(a) \cdot \lambda(s) \qquad \text{for all } a \in \mathsf{A} \setminus \{0\},\ s \in \mathsf{F} \setminus \{0\}. \tag{3.36}$$

We then extend the domain of μ by defining for $r \in q(\mathsf{A}) \setminus \{0\}$

$$\mu(r) = \mu(a)\lambda(s)^{-1} \ \text{ if } r = as^{-1} \text{ with } a \in \mathsf{A} \setminus \{0\} \text{ and } s \in \mathsf{F} \setminus \{0\}.$$

It follows from (3.36) by clearing denominators that μ is a well-defined function from $q(\mathsf{A}) \setminus \{0\}$ to $\mathsf{A}^h \setminus \{0\}$, where A^h denotes the set of homogeneous elements of A. Moreover, (3.36) and (3.31) yield by clearing denominators (since F is central in A)

$$\mu(rc) = \mu(r) \cdot \lambda(c) \qquad \text{for all } r \in q(\mathsf{A}) \setminus \{0\},\ c \in q(\mathsf{F}) \setminus \{0\}. \tag{3.37}$$

We now define a function $\alpha\colon q(\mathsf{A}) \to \Gamma \cup \{\infty\}$. Set

$$\alpha(r) = \begin{cases} deg(\mu(r)) & \text{if } r \neq 0, \\ \infty & \text{if } r = 0, \end{cases} \qquad \text{for all } r \in q(\mathsf{A}).$$

Note that

$$\alpha(a) = min(supp(a)) \qquad \text{for all } a \in \mathsf{A} \setminus \{0\}. \tag{3.38}$$

We show that α is a surmultiplicative v-value function on $q(\mathsf{A})$. It is clear from the definition that $\alpha(r) = \infty$ if and only if $r = 0$. Also, formula (3.37) yields

$$\alpha(rc) = \alpha(r) + v(c) \qquad \text{for all } r \in q(\mathsf{A}),\ c \in q(\mathsf{F}). \tag{3.39}$$

Furthermore,

$$\alpha(r + q) \geq min\big(\alpha(r), \alpha(q)\big) \qquad \text{for all } r,\ q \in q(\mathsf{A}).$$

This is clear if $r = 0$ or $q = 0$ or if $r+q = 0$, and it follows easily from (3.38) if $r, q \in \mathsf{A}\setminus\{0\}$. It then follows in general by clearing denominators using (3.39), as in the proof of (3.35). Moreover, $\alpha(1) = 0$ and

$$\alpha(rq) \geq \alpha(r) + \alpha(q) \qquad \text{for all } r,\ q \in q(\mathsf{A}).$$

This is clear if $r = 0$ or $q = 0$ and follows easily from (3.38) if $r, q \in \mathsf{A} \setminus \{0\}$. It follows in general by clearing denominators using (3.39). So, α is surmultiplicative.

Lemma 3.44. *α is a v-norm on $q(\mathsf{A})$. Indeed, every homogeneous F-base of A is a splitting base of $q(\mathsf{A})$ for α.*

Proof. Let $(b_1, \ldots, b_n)$ be a homogeneous F-base of A. Take any $a \in \mathsf{A}$ and write $a = b_1 s_1 + \ldots + b_n s_n$ with the s_i in F. We claim that

$$\alpha(a) = min\big(\alpha(b_1 s_1), \ldots, \alpha(b_n s_n)\big). \tag{3.40}$$

For this, let $\delta = min\big(\alpha(b_1 s_1), \ldots, \alpha(b_n s_n)\big)$. It is clear from (3.38) that $\alpha(a) \geq \delta$. For the reverse inequality, renumber the b_i so that $\alpha(b_1 s_1) = \ldots = \alpha(b_k s_k) = \delta$ and $\alpha(b_j s_j) > \delta$ for $j > k$. Then $s_i \neq 0$ for $i = 1, \ldots, k$, and the δ-component of $b_i s_i$ is $\mu(b_i s_i) = \mu(b_i) \cdot \lambda(s_i) = b_i \lambda(s_i)$; but for $j > k$, the δ-component of $b_j s_j$ is 0. So, the δ-component of a is $b_1\lambda(s_1) + \ldots + b_k\lambda(s_k)$, which is nonzero since the $\lambda(s_i) \neq 0$ and the b_i are F-independent. Hence, $\alpha(a) \leq \delta$, so with the earlier inequality we have $\alpha(a) = \delta$, proving (3.40). It follows by clearing denominators that for any $r \in q(\mathsf{A})$, when we write $r = b_1c_1 + \ldots + b_nc_n$ with the $c_i \in q(\mathsf{F})$, then

$$\alpha(r) = \mathit{min}\big(\alpha(b_1c_1), \ldots, \alpha(b_nc_n)\big) = \mathit{min}\big(\alpha(b_1)+v(c_1), \ldots, \alpha(b_n)+v(c_n)\big).$$

Thus, $(b_1, \ldots, b_n)$ is a splitting base of $q(\mathsf{A})$ for the v-value function α, which shows that α is a v-norm. $\square$

Proposition 3.45. *We have*

$$\mathsf{gr}_v(q(\mathsf{F})) \cong_g \mathsf{F} \qquad \textit{and} \qquad \mathsf{gr}_\alpha(q(\mathsf{A})) \cong_g \mathsf{A}.$$

Hence,

$$\overline{q(\mathsf{A})} \cong \mathsf{A}_0, \qquad \Gamma_{q(\mathsf{A})} = \Gamma_{\mathsf{A}}, \qquad \Gamma^{\times}_{q(\mathsf{A})} = \Gamma^{\times}_{\mathsf{A}}, \quad \textit{and} \quad \theta_{q(\mathsf{A})} = \theta_{\mathsf{A}}.$$

Moreover, the surmultiplicative v-norm α on $q(\mathsf{A})$ is a v-gauge if and only if A is graded semisimple.

Proof. We prove the second graded isomorphism. The first is proved the same way.

For any $\gamma \in \Gamma$, define a map $\eta_\gamma \colon q(\mathsf{A})_{\geq\gamma} \to \mathsf{A}_\gamma$ by

$$\eta_\gamma(r) = \begin{cases} \mu(r) & \text{if } \alpha(r) = \gamma, \\ 0 & \text{if } \alpha(r) > \gamma. \end{cases}$$

So, if $r \in q(\mathsf{A})_{\geq\gamma} \cap \mathsf{A}$, then $\eta_\gamma(r)$ is the γ-component of r. Note that for any $r \in q(\mathsf{A})_{\geq\gamma}$ and $c \in q(\mathsf{F})^\times$,

$$\eta_{\gamma+v(c)}(rc) = \eta_\gamma(r) \cdot \lambda(c). \tag{3.41}$$

This follows easily from (3.37) and (3.39). We next show that

$$\eta_\gamma(r+q) = \eta_\gamma(r) + \eta_\gamma(q) \qquad \text{for all } r,\, q \in q(\mathsf{A})_{\geq\gamma}.$$

This is clear when $r,\, q \in q(\mathsf{A})_{\geq\gamma} \cap \mathsf{A}$, since the sum of the γ-components of r and q is the γ-component of $r + q$. The formula then follows for any $r,\, q \in q(\mathsf{A})_{\geq\gamma}$ by clearing denominators using (3.41). Thus, η_γ is an additive group homomorphism. It is surjective since $\eta_\gamma(a) = \mu(a) = a$ for all $a \in \mathsf{A}_\gamma \setminus \{0\}$. Since evidently $\mathit{ker}\,\eta_\gamma = q(\mathsf{A})_{>\gamma}$, the map η_γ induces an additive group isomorphism $\overline{\eta}_\gamma \colon q(\mathsf{A})_\gamma \xrightarrow{\sim} \mathsf{A}_\gamma$. The direct sum of these maps yields a graded group isomorphism

$$\overline{\eta} = \oplus\overline{\eta}_\gamma \colon \mathsf{gr}_\alpha(q(\mathsf{A})) = \bigoplus_{\gamma\in\Gamma} q(\mathsf{A})_\gamma \xrightarrow{\sim} \bigoplus_{\gamma\in\Gamma} \mathsf{A}_\gamma = \mathsf{A}.$$

It remains to check that $\overline{\eta}$ is compatible with multiplication. For this note first that if $a \in \mathsf{A}_\gamma \setminus \{0\}$, then $\alpha(a) = \mathit{deg}\, a = \gamma$, so $\widetilde{a} \in q(\mathsf{A})_\gamma$ and $\overline{\eta}(\widetilde{a}) = \eta_\gamma(a) = a$. Since each $\overline{\eta}_\gamma$ is an isomorphism, it follows that

$\{\widetilde{a} \mid a \in \mathsf{A}^h \setminus \{0\}\}$ generates the ring $\mathsf{gr}_\alpha(q(\mathsf{A}))$. Take any $a,\, b \in \mathsf{A}^h \setminus \{0\}$. We compute $\overline{\eta}(\widetilde{a}\,\widetilde{b})$. If $ab \neq 0$, then

$$\alpha(ab) \;=\; deg(ab) \;=\; deg(a) + deg(b) \;=\; \alpha(a) + \alpha(b).$$

So, $\widetilde{a}\,\widetilde{b} = \widetilde{ab}$ and

$$\overline{\eta}(\widetilde{a}\,\widetilde{b}) \;=\; \overline{\eta}(\,\widetilde{ab}\,) \;=\; ab \;=\; \overline{\eta}(\widetilde{a})\,\overline{\eta}(\widetilde{b}).$$

On the other hand, if $ab = 0$, then $\alpha(ab) = \infty > \alpha(a) + \alpha(b)$, so $\widetilde{a}\,\widetilde{b} = 0$ and

$$\overline{\eta}(\widetilde{a}\,\widetilde{b}) \;=\; \overline{\eta}(0) \;=\; 0 \;=\; ab \;=\; \overline{\eta}(\widetilde{a})\,\overline{\eta}(\widetilde{b}).$$

Thus $\overline{\eta}$ is multiplicative on a generating set of $\mathsf{gr}_\alpha(q(\mathsf{A}))$; hence, $\overline{\eta}$ is a graded ring isomorphism $\mathsf{gr}_\alpha(q(\mathsf{A})) \xrightarrow{\sim} \mathsf{A}$.

The surmultiplicative v-norm α is by definition a v-gauge on $q(\mathsf{A})$ if and only if $\mathsf{gr}_\alpha(q(\mathsf{A}))$ is graded semisimple. By what we have just proved, this occurs if and only if A is graded semisimple.

It is clear from the definition of α that $\Gamma_{q(\mathsf{A})} = \Gamma_{\mathsf{A}}$. The equalities $\overline{q(\mathsf{A})} = \mathsf{A}_0$, $\Gamma^\times_{q(\mathsf{A})} = \Gamma^\times_{\mathsf{A}}$, and $\theta_{q(\mathsf{A})} = \theta_{\mathsf{A}}$ follow from the isomorphism $\mathsf{gr}_\alpha(q(\mathsf{A})) \cong_g \mathsf{A}$. For the last two equalities, see the comments following the proof of Prop. 3.29. □

Note that $q(\mathsf{A})$ is a division algebra if and only if A is a graded division algebra. When this occurs, α is a valuation on $q(\mathsf{A})$ extending v on $q(\mathsf{F})$, by Cor. 3.30 and Prop. 3.45.

The proposition shows that every semisimple graded algebra A finite-dimensional over its center is the associated graded algebra of some gauge. In fact, every such A is the associated graded algebra of a gauge for a Henselian valuation. This follows from Prop. 3.42 below by passing from the valued field $(q(\mathsf{F}), v)$ to its Henselization $(q(\mathsf{F})_h, v_h)$.

Proposition 3.45 indicates the close link between algebras over graded fields and algebras over valued fields. We will see much more of this in subsequent chapters. Notably, the theory of algebraic extensions of graded fields (see §5.1 below) is very similar to that of algebraic extensions of Henselian valued fields. Indeed, some of the graded field theory is deducible from the Henselian field theory by using Prop. 3.45. Likewise, we will see in Ch. 6 that the theory of division algebras and the Brauer group for a graded field is strongly analogous to the theory of tame division algebras and the tame subgroup of the Brauer group of a Henselian field.

When we pass from a semisimple graded algebra A to an algebra with gauge over a valued field, then to the associated graded algebra of the gauge, Prop. 3.45 shows that we recover the graded algebra A we began with. But notice that if we start with an algebra B with a gauge over a valued field (F, w) and pass to its associated graded algebra $\mathsf{gr}(B)$, then back to the algebra $q(\mathsf{gr}(B))$ over the valued field $q(\mathsf{gr}(F))$ (using the ordering on Γ associated with w) by the process described above, in general we do not recover the algebra we started with. For example, let $B = F = \mathbb{Q}_p$ with its complete discrete p-adic valuation, for any prime p. Then $\mathsf{gr}(F) \cong_g \mathbb{F}_p[X, X^{-1}]$, with its usual grading; the ordering on $\Gamma_{\mathsf{gr}(F)}$ inherited from the ordering on Γ_F is the usual ordering on $\mathbb{Z}$. So, $q(\mathsf{gr}(F)) \cong \mathbb{F}_p(X)$ and the valuation on it given by Prop. 3.45 is the X-adic valuation. Thus, F is uncountable of characteristic 0 while $q(\mathsf{gr}(F))$ is countable of characteristic p.

In general, the category of semisimple graded algebras is less complicated than that of algebras over a valued field (even when the valuation is Henselian). It is quite remarkable how much of the valued theory, especially in the Henselian case, is reflected in the graded theory.

3.2.6 Residually simple gauges

One of the familiar facts in commutative valuation theory is that a valuation on a field is completely determined by its associated valuation ring. This property does not hold for gauges in general (see Ex. 3.50 below), but we now describe a significant class of gauges for which it does hold.

Definition 3.46. Let (F, v) be a field with valuation, and let A be a (finite-dimensional) simple F-algebra. A v-gauge α on A is said to be *residually simple* if A_0 is a simple F-algebra. When this holds, $\mathsf{gr}(A)$ satisfies the special properties given in Prop. 2.47 above.

Example 3.47. (i) Clearly, if D is a division algebra over F and v_D is a valuation on D extending v on F, then v_D is a residually simple v-gauge whenever it is a v-gauge. This holds whenever v_D is defectless over F.

(ii) It follows from the proof of Prop. 4.57 below that if v on F has rank 1, then every simple F-algebra A which has a v-gauge has a residually simple v-gauge. But not every v-gauge on A need be residually simple.

(iii) Suppose $A = End_D(M)$, where D is a division algebra over F with a valuation v_D and M is a finite-dimensional right D-vector space. Suppose v_D is a v-gauge on D. Let η be a v_D-norm on M, and let $\alpha = End(\eta)$ be the v-gauge on A determined by η. Then, α is a v-gauge by Prop. 3.34, and α is residually simple if and only if Γ_M is a single coset of Γ_D by Prop. 3.34 and Prop. 2.41. Thus, whenever $dim_D\, M > 1$ and $\Gamma_D \subsetneq \Gamma$, there are choices of η for which α is residually simple and other choices of η for which α is not residually simple.

Proposition 3.48. *Let α be a residually simple v-gauge on a simple F-algebra A. Then, α is determined by its gauge ring $O_{A,\alpha}$.*

We will use the following notation in proving the Proposition:

$$\begin{aligned} R &= \mathcal{O}_{A,\alpha} = \{a \in A \mid \alpha(a) \geq 0\}; \\ J &= \mathfrak{m}_{A,\alpha} = \{a \in A \mid \alpha(a) > 0\}; \\ \mathcal{S} &= \{s \in A^\times \mid sRs^{-1} = R\}; \\ \mathcal{T} &= \{t \in A^\times \mid \alpha(t^{-1}) = -\alpha(t)\} = \{t \in A^\times \mid \widetilde{t} \in \mathsf{gr}(A)^\times\} \\ &= \{t \in A^\times \mid \alpha(ta) = \alpha(t) + \alpha(a) = \alpha(at) \text{ for all } a \in A\}. \end{aligned}$$

The equalities for $\mathcal{T}$ follow from Prop. 3.29. Also, clearly $\mathcal{S}$ and $\mathcal{T}$ are subgroups of $A^\times$ with $R^\times \subseteq \mathcal{T} \subseteq \mathcal{S}$.

Lemma 3.49.

(i) $1 + J \subseteq R^\times$.
(ii) $\mathcal{S} \cap (R \setminus J) = R^\times$.
(iii) $\alpha(\mathcal{T}) = \Gamma_A$.
(iv) $\mathcal{T} = \mathcal{S}$.

Proof. (i) Let $a \in 1 + J$. Then $\widetilde{a} = \widetilde{1} \in \mathsf{gr}_\alpha(A)^\times$. By Prop. 3.29, $a \in A^\times$ and $\alpha(a^{-1}) = -\alpha(a) = 0$. Hence, $a \in R^\times$. (Since $R/J = A_0$, which is semisimple, it follows that J is the Jacobson radical $J(R)$.)

(ii) Take any $s \in \mathcal{S} \cap (R \setminus J)$. For $r \in R$, write $\overline{r}$ for the image $r + J$ of r in A_0. Since $\overline{s} \neq 0$ and A_0 is simple, there are $a_i, b_i \in R$ with $\overline{1} = \sum_i \overline{a_i}\,\overline{s}\,\overline{b_i} = \overline{\sum_i a_i s b_i}$. Hence, $\sum_i a_i s b_i \in 1 + J \subseteq R^\times$, by (i). Therefore, $RsR = R$. But $sR = Rs$, as $s \in \mathcal{S}$. So, $R = RsR = sR$, which shows that $s \in R^\times$.

(iii) By Prop. 2.47, as A_0 is simple we have $\Gamma_{\mathsf{gr}(A)} = \Gamma^\times_{\mathsf{gr}(A)}$, where

$$\Gamma^\times_{\mathsf{gr}(A)} = \{\gamma \in \Gamma_A \mid A_\gamma \cap \mathsf{gr}(A)^\times \neq \varnothing\} = \{\alpha(t) \mid t \in \mathcal{T}\}.$$

Hence, $\alpha(\mathcal{T}) = \Gamma_{\mathsf{gr}(A)} = \Gamma_A$.

(iv) Take any $s \in \mathcal{S}$. By (iii), there is a $t \in \mathcal{T}$ with $\alpha(t) = \alpha(s)$. Then, $st^{-1} \in \mathcal{S}$, as $\mathcal{T} \subseteq \mathcal{S}$. But $\alpha(st^{-1}) = \alpha(s) - \alpha(t) = 0$, as $t \in \mathcal{T}$. Hence, $st^{-1} \in \mathcal{S} \cap (R \setminus J)$. Therefore by (ii), $st^{-1} \in R^\times \subseteq \mathcal{T}$. Hence, $s \in \mathcal{T}$. □

Proof of Prop. 3.48. The restriction $\alpha|_{\mathcal{S}}$ of α to $\mathcal{S}$ maps $\mathcal{S}$ onto Γ_A by (iii) and (iv) of Lemma 3.49, and $\alpha|_{\mathcal{S}}$ is a group homomorphism by (iv) of the lemma, since it is multiplicative on elements of $\mathcal{T}$. Furthermore, $ker(\alpha|_{\mathcal{S}}) = \mathcal{S} \cap (R \setminus J) = R^\times$ by (ii) of the lemma. Thus, $\Gamma_A \cong \mathcal{S}/R^\times$. Moreover, for $s, s' \in \mathcal{S} = \mathcal{T}$, we have $\alpha(s) \leq \alpha(s')$ if and only if $\alpha(s^{-1}s') \geq 0$, if and only if $s^{-1}s' \in R$, if and only if $s'R \subseteq sR$. Thus, the ordering on Γ_A is determined by inclusions of principal fractional right ideals of R. Now, for $a \in A \setminus \{0\}$, choose any $s \in \mathcal{S}$ with $\alpha(s) = \alpha(a)$. Then, $\alpha(s^{-1}a) = 0$, as $s \in \mathcal{T}$. The argument for (ii) of the lemma shows that $Rs^{-1}aR = R$. So, $sR = sRs^{-1}aR = RaR$. The map α corresponds to the map $A \setminus \{0\} \to \mathcal{S}/R^\times$

given by $a \mapsto sR^\times$ for any $s \in \mathcal{S}$ such that $RaR = sR = Rs$. This map is well-defined since if $s, s' \in \mathcal{S}$ with $s'R = sR$, then $s^{-1}s'R = R$, showing that $s \equiv s' \ (mod\ R^\times)$. Since $\mathcal{S}$ is defined in terms of R, this shows that α is completely determined by R. □

Here is an example of a simple algebra with multiple non-isomorphic gauges all having the same gauge ring.

Example 3.50. Let L be a field with $char(L) \neq 2$, and let $F = L(x, y)$ with x and y algebraically independent over L. Let v be the valuation on F obtained by restriction from the standard rank 2 Henselian valuation on $L((x))((y))$. Thus, $\overline{F}^{v} = L$, $\Gamma_{F,v} = \mathbb{Z} \times \mathbb{Z}$ with the right-to-left lexicographic ordering, $v(x) = (1, 0)$, $v(y) = (0, 1)$, and $\mathsf{gr}_v(F) = L[X, X^{-1}, Y, Y^{-1}]$, a twice iterated Laurent polynomial ring, where $X = \widetilde{x}$ and $Y = \widetilde{y}$. Let w be the rank 1 coarsening of v, which is the discrete valuation on F obtained by restriction from the complete discrete y-adic valuation on $L(x)((y))$. So, $\overline{F}^{w} = L(x)$, $\Gamma_{F,w} = \mathbb{Z}$, and the epimorphism $\varepsilon\colon \Gamma_{F,v} \to \Gamma_{F,w}$ given by $v(c) \mapsto w(c)$ for $c \in F^\times$ is the projection $(\ell, m) \mapsto m$. Since we will be working primarily with v, we write $\overline{F}$ for $\overline{F}^{v}$ and Γ_F for $\Gamma_{F,v}$. Let

$$D = (1 + x,\ y/F),$$

a quaternion division algebra over F with its standard F-base $(1, i, j, k)$, where $i^2 = 1 + x$, $j^2 = y$, and $k = ij = -ji$.

Let $K = F(t)$, where $t^2 = 1 + x$. Thus, K is a quadratic extension field of F, and since $\overline{1 + x} = \overline{1}$ in $\overline{F}$, the valuation v has two extensions to K distinguished by whether $\overline{t} = \overline{1}$ or $\overline{-1}$ in $\overline{K}$. Let v' denote the extension of v to K with $\overline{t} = 1$. Then, $\overline{K} = \overline{F} = L$ and $\Gamma_K = \Gamma_F = \mathbb{Z} \times \mathbb{Z}$, so $\mathsf{gr}(K) = \mathsf{gr}(F)$. Note that as $x = (t - 1)(t + 1)$ and $v'(t + 1) = 0$, we have $v'(t-1) = v(x) = (1, 0)$. The rank 1 coarsening of v' is the unique, unramified, extension w' of w to K, with $\overline{K}^{w'} = L(x)(\sqrt{1 + x}\,)$ and $\Gamma_{K,w'} = \mathbb{Z}$. Also, K is a splitting field of D, as $K \cong F(i)$, which is a maximal subfield of D. Explicitly, let $S = M_2(K)$, and view D as an F-subalgebra of S by identifying

$$1 = \left(\begin{smallmatrix} 1 & 0 \\ 0 & 1 \end{smallmatrix}\right), \quad i = \left(\begin{smallmatrix} t & 0 \\ 0 & -t \end{smallmatrix}\right), \quad j = \left(\begin{smallmatrix} 0 & y \\ 1 & 0 \end{smallmatrix}\right), \quad k = \left(\begin{smallmatrix} 0 & ty \\ -t & 0 \end{smallmatrix}\right).$$

Give $\mathbb{Q} \times \mathbb{Q}$ the right-to-left lexicographic ordering. Fix any $\gamma \in \mathbb{Q}$ with $0 < \gamma < \frac{1}{2}$, and let $\delta = (\gamma, \frac{1}{2}) \in \mathbb{Q} \times \mathbb{Q}$. Let α' be the v'-gauge on S given by

$$\alpha'\left(\begin{smallmatrix} p & q \\ r & s \end{smallmatrix}\right) = min\left(v'(p),\ v'(q) - \delta,\ v'(r) + \delta,\ v'(s)\right).$$

Indeed, let $M = K\text{-}span\{\left(\begin{smallmatrix} 1 \\ 0 \end{smallmatrix}\right), \left(\begin{smallmatrix} 0 \\ 1 \end{smallmatrix}\right)\}$, and identify $S = End_K(M)$. Then α' is the v'-gauge $End(\eta)$, where $\eta\colon M \to \mathbb{Q} \times \mathbb{Q} \cup \{\infty\}$ is the v'-norm on M given by $\eta\left(\begin{smallmatrix} p \\ q \end{smallmatrix}\right) = min\left(v'(p), v'(q) + \delta\right)$. Thus,

$$\mathsf{gr}_{\alpha'}(S) = \mathsf{End}_{\mathsf{gr}(K)}(\mathsf{gr}(M)) \cong_g M_2(\mathsf{gr}(K))(0, \delta).$$

Let $\alpha = \alpha'|_D$, which is a surmultiplicative v-value function on D. While α' is a v'-gauge, we must still verify that α is a v-gauge. For this, note that for any $z = a + bi + cj + dk \in D$ with $a, b, c, d \in F$, we have $z = \left(\begin{smallmatrix} a+bt & (c+dt)y \\ c-dt & a-bt \end{smallmatrix}\right)$ in S. Thus, as $v(y) = (0, 1)$,

$$\begin{aligned} \alpha(z) &= \min\big(v'(a+bt),\ v'((c+dt)y) - \delta,\ v'(c-dt) + \delta,\ v'(a-bt)\big) \\ &= \min\big(v'(a+bt), v'(a-bt), v'(c+dt) + (-\gamma, \tfrac{1}{2}), v'(c-dt) + (\gamma, \tfrac{1}{2})\big). \end{aligned} \tag{3.42}$$

So, $\alpha(1) = \alpha(i) = 0$ and

$$\alpha(j) = \alpha(k) = \min\big((-\gamma, \tfrac{1}{2}), (\gamma, \tfrac{1}{2})\big) = (-\gamma, \tfrac{1}{2}).$$

Since $v'(1-t) = (1, 0)$, we have

$$\alpha(j-k) = \min\big((-\gamma, \tfrac{1}{2}) + (1, 0), (\gamma, \tfrac{1}{2})\big) = (\gamma, \tfrac{1}{2}),$$

as $\gamma < \frac{1}{2}$. So, in $\mathsf{gr}_\alpha(D) \subseteq \mathsf{gr}_{\alpha'}(S)$,

$$\widetilde{1+i} = \left(\begin{smallmatrix} 2 & 0 \\ 0 & 0 \end{smallmatrix}\right) \in D_0, \qquad \widetilde{1-i} = \left(\begin{smallmatrix} 0 & 0 \\ 0 & 2 \end{smallmatrix}\right) \in D_0,$$
$$\widetilde{j} = \widetilde{k} = \left(\begin{smallmatrix} 0 & \widetilde{y} \\ 0 & 0 \end{smallmatrix}\right) \in D_{(-\gamma, \frac{1}{2})}, \qquad \widetilde{j-k} = \left(\begin{smallmatrix} 0 & 0 \\ 2 & 0 \end{smallmatrix}\right) \in D_{(\gamma, \frac{1}{2})}.$$

Since $1+i$, $1-i$, j, $j-k$ have images in $\mathsf{gr}_\alpha(D)$ which are clearly $\mathsf{gr}(F)$-independent, they comprise a splitting base of α as a v-value function, by Prop. 3.5; this shows that α is a v-norm on D. Moreover,

$$[\mathsf{gr}_\alpha(D) \colon \mathsf{gr}(F)] = 4 = [\mathsf{gr}_{\alpha'}(S) \colon \mathsf{gr}(K)] = [\mathsf{gr}_{\alpha'}(S) \colon \mathsf{gr}(F)],$$

since $\mathsf{gr}(K) = \mathsf{gr}(F)$. Hence, $\mathsf{gr}_\alpha(D) = \mathsf{gr}_{\alpha'}(S)$, which is graded simple. Thus, α is a v-gauge on D. Note that

$$\Gamma_{D,\alpha} = \Gamma_{S,\alpha'} = \mathbb{Z}^2 \cup (\delta + \mathbb{Z}^2) \cup (-\delta + \mathbb{Z}^2). \tag{3.43}$$

Also, α is not residually simple, as $D_0 = S_0 = L \times L$.

From (3.42), we have

$$\begin{aligned} O_{D,\alpha} = \{a + bi + cj + dk \in D \mid\ & v'(a+bt) \geq 0,\ v'(a-bt) \geq 0, \\ & v'(c+dt) \geq (\gamma, -\tfrac{1}{2}),\ v'(c-dt) \geq (-\gamma, -\tfrac{1}{2})\}. \end{aligned}$$

Let $v'(c+dt) = (\ell, m) \in \Gamma_K = \mathbb{Z} \times \mathbb{Z}$. Then, $v'(c+dt) \geq (\gamma, -\frac{1}{2})$ if and only if $m \geq 0$, i.e., $w'(c+dt) \geq 0$ where w' is the rank 1 coarsening of v'. Likewise, $v'(c-dt) \geq (-\gamma, -\frac{1}{2})$ if and only if $w'(c-dt) \geq 0$. Therefore, each of the infinitely many choices of γ yields the same gauge ring for the associated v-gauge α. But different choices of γ yield different gauges since the gauges have different value sets (see (3.43)). Thus, the gauge ring $O_{D,\alpha}$ does not determine α.

Since $char(\overline{F}) \neq 2$, the description of $O_{D,\alpha}$ simplifies to

$$O_{D,\alpha} = \{a + bi + cj + dk \mid a, b \in O_{F,v},\ c, d \in O_{F,w}\}.$$

For this example, one can show that D has exactly two Dubrovin valuation rings B_1, B_2 with center $\mathcal{O}_{F,v}$. (See the Notes at the end of this chapter for properties of Dubrovin valuation rings.) These rings are

$$B_\ell = \{a + bi + cj + dk \mid a + (-1)^\ell bt \in \mathcal{O}_{K,v'},\ c,\, d \in \mathcal{O}_{F,w}\}, \quad \ell = 1,\, 2.$$

Each B_ℓ is actually a "total valuation ring," i.e., for each $z \in D^\times$ we have $z \in B_\ell$ or $z^{-1} \in B_\ell$. Also, $\mathcal{O}_{D,\alpha} = B_1 \cap B_2$.

3.2.7 Composition of value functions and gauges

The construction discussed in this subsection is an analogue for gauges of the composition of valuations (see §1.2.3). The rather technical Prop. 3.54 will be a crucial tool to establish the existence theorem of gauges (Th. 4.50).

Let $v\colon F \to \Gamma \cup \{\infty\}$ be a valuation on a field F with a *divisible* totally ordered group Γ, and let $\Delta \subset \Gamma$ be a convex subgroup, i.e., if $0 \le \gamma \le \delta$ with $\gamma \in \Gamma$ and $\delta \in \Delta$, then $\gamma \in \Delta$. Let $\Lambda = \Gamma/\Delta$, and let $\varepsilon\colon \Gamma \to \Lambda$ be the canonical map. The ordering on Γ induces a total ordering on Λ such that for $\gamma_1, \gamma_2 \in \Gamma$, if $\gamma_1 \le \gamma_2$, then $\varepsilon(\gamma_1) \le \varepsilon(\gamma_2)$. Consequently,

$$\text{if } \ \varepsilon(\gamma_2) < \varepsilon(\gamma_1), \quad \text{then } \ \gamma_2 < \gamma_1. \tag{3.44}$$

Because Γ is assumed to be divisible, Δ and Λ are also divisible. By composing v with ε, we obtain a coarser valuation on F,

$$w = \varepsilon \circ v\colon F \longrightarrow \Lambda \cup \{\infty\}.$$

Let $\overline{F}^{v}$ (resp. $\overline{F}^{w}$) denote the residue field of F for the valuation v (resp. w). The valuation v induces a valuation

$$u\colon \overline{F}^{w} \longrightarrow \Delta \cup \{\infty\},$$

with residue field

$$\overline{\overline{F}^{w}}^{u} = \overline{F}^{v},$$

see Engler–Prestel [73, pp. 44–45].

Now, let V be an F-vector space and let $\alpha\colon V \to \Gamma \cup \{\infty\}$ be a v-value function. Composition with ε yields a w-value function

$$\beta = \varepsilon \circ \alpha\colon V \longrightarrow \Lambda \cup \{\infty\}.$$

Each $\lambda \in \Lambda = \Gamma/\Delta$ is a coset of Δ, and may therefore be viewed as a subset of Γ. For $x \in V$, we have by definition

$$\beta(x) = \lambda \in \Lambda \quad \text{if and only if} \quad \alpha(x) \in \lambda \subseteq \Gamma.$$

For $\lambda \in \Lambda$, let

$$V^{\beta}_{\geq\lambda} = \{x \in V \mid \beta(x) \geq \lambda\}, \quad V^{\beta}_{>\lambda} = \{x \in V \mid \beta(x) > \lambda\}, \quad V^{\beta}_{\lambda} = V^{\beta}_{\geq\lambda}/V^{\beta}_{>\lambda}.$$

The group V^{β}_{λ} is an $\overline{F}^{w}$-vector space.

Lemma 3.51. *If $x,\, y \in V^{\beta}_{\geq\lambda}$ satisfy $x \equiv y \not\equiv 0 \pmod{V^{\beta}_{>\lambda}}$, then $\alpha(x) = \alpha(y)$.*

Proof. We have $\beta(x-y) > \lambda = \beta(y)$. Since $\beta = \varepsilon \circ \alpha$, (3.44) shows that $\alpha(x-y) > \alpha(y)$. Hence, $\alpha(x) = \min\big(\alpha(x-y), \alpha(y)\big) = \alpha(y)$. □

In view of this lemma, we may define

$$\alpha_\lambda \colon V^{\beta}_{\lambda} \longrightarrow \lambda \cup \{\infty\} \quad \text{by} \quad x + V^{\beta}_{>\lambda} \mapsto \begin{cases} \alpha(x) & \text{if } \beta(x) = \lambda, \\ \infty & \text{if } \beta(x) > \lambda. \end{cases}$$

Clearly, α_λ is a u-value function on V^{β}_{λ}. For $\gamma \in \lambda$ we have

$$(V^{\beta}_{\lambda})^{\alpha_\lambda}_{\gamma} = V^{\alpha}_{\gamma}.$$

Therefore,

$$\mathsf{gr}_{\alpha}(V) = \bigoplus_{\lambda\in\Lambda} \mathsf{gr}_{\alpha_\lambda}(V^{\beta}_{\lambda}) \quad \text{where} \quad \mathsf{gr}_{\alpha_\lambda}(V^{\beta}_{\lambda}) = \bigoplus_{\gamma\in\lambda} V^{\alpha}_{\gamma}$$

while

$$\mathsf{gr}_{\beta}(V) = \bigoplus_{\lambda\in\Lambda} V^{\beta}_{\lambda}.$$

Now, let

$$\Gamma_F = v(F^\times) \subseteq \Gamma, \quad \Delta_F = \Delta \cap \Gamma_F \subseteq \Delta, \quad \Lambda_F = w(F^\times) = \Gamma_F/\Delta_F \subseteq \Lambda.$$

These groups are the value groups of, respectively, v, u, and w. Similarly, let

$$\Gamma_V = \alpha(V \setminus \{0\}) \subseteq \Gamma \quad \text{and} \quad \Lambda_V = \beta(V \setminus \{0\}) \subseteq \Lambda.$$

Further, for each $\lambda \in \Lambda_V$, let

$$\lambda_V = \alpha_\lambda(V^{\beta}_{\lambda} \setminus \{0\}) \subseteq \lambda.$$

Clearly, $\lambda_V = \lambda \cap \Gamma_V$. Note that Γ_F (resp. Δ_F, resp. Λ_F) is a subgroup of Γ (resp. Δ, resp. Λ), while Γ_V (resp. Λ_V, resp. λ_V for $\lambda \in \Lambda_V$) is a union of cosets of Γ_F (resp. Λ_F, resp. Δ_F). We denote by $|\Gamma_V{:}\Gamma_F|$ the cardinality of the set of cosets of Γ_F in Γ_V, and define likewise $|\Lambda_V{:}\Lambda_F|$ and $|\lambda_V{:}\Delta_F|$ for $\lambda \in \Lambda_V$.

Lemma 3.52. *If $\dim_F V$ is finite, then $|\Gamma_V{:}\Gamma_F|$, $|\Lambda_V{:}\Lambda_F|$, and $|\lambda_V{:}\Delta_F|$ for $\lambda \in \Lambda_V$ are finite. If $\lambda_1, \ldots, \lambda_r \in \Lambda_V$ are representatives of the distinct cosets of Λ_V modulo Λ_F, then*

$$|\Gamma_V{:}\Gamma_F| = \sum_{i=1}^{r} |(\lambda_i)_V{:}\Delta_F|.$$

Proof. By Cor. 3.6 and Prop. 2.5 we have

$$|\Gamma_V{:}\Gamma_F| \leq dim_F\, V, \qquad |\Lambda_V{:}\Lambda_F| \leq dim_F\, V,$$

and also

$$|\lambda_V{:}\Delta_F| \leq dim_{\overline{F}^w}\, V_\lambda \leq dim_F\, V \qquad \text{for } \lambda \in \Lambda_V.$$

For $i = 1, \ldots, r$, let $\gamma_{i1}, \ldots, \gamma_{is_i} \in (\lambda_i)_V \subseteq \Gamma$ be representatives of the distinct cosets of $(\lambda_i)_V$ modulo Δ_F. Thus,

$$(\lambda_i)_V = \coprod_{j=1}^{s_i} (\gamma_{ij} + \Delta_F),$$

where $\coprod$ denotes the disjoint union. For $\gamma \in \Gamma_V$, we have $\varepsilon(\gamma) \in \Lambda_V$, hence

$$\varepsilon(\gamma) = \lambda_i + w(a) \qquad \text{for some } i \in \{1, \ldots, r\} \text{ and some } a \in F^\times.$$

It follows that $\gamma - v(a) \in (\lambda_i)_V$; hence,

$$\gamma - v(a) = \gamma_{ij} + v(b) \qquad \text{for some } j \in \{1, \ldots, s_i\} \text{ and some } b \in F^\times.$$

This shows that $\gamma \equiv \gamma_{ij}\ (mod\ \Gamma_F)$, hence

$$\Gamma_V = \bigcup_{i=1}^{r} \bigcup_{j=1}^{s_i} (\gamma_{ij} + \Gamma_F). \tag{3.45}$$

To complete the proof, it suffices to show the union is disjoint. If $\gamma_{ij} \equiv \gamma_{k\ell}\ (mod\ \Gamma_F)$ for some i, j, k, ℓ, then $\varepsilon(\gamma_{ij}) \equiv \varepsilon(\gamma_{k\ell})\ (mod\ \Lambda_F)$, hence $i = k$ since $\varepsilon(\gamma_{ij}) = \lambda_i$ and $\varepsilon(\gamma_{k\ell}) = \lambda_k$. Moreover, from $\varepsilon(\gamma_{ij}) = \varepsilon(\gamma_{k\ell})$ it follows that $\gamma_{ij} - \gamma_{k\ell} \in \Delta$; hence, $\gamma_{ij} \equiv \gamma_{k\ell}\ (mod\ \Gamma_F)$ implies that $\gamma_{ij} \equiv \gamma_{k\ell}\ (mod\ \Delta_F)$, hence also that $j = \ell$. □

Proposition 3.53. *Suppose $dim_F\, V$ is finite, and let $\lambda_1, \ldots, \lambda_r \in \Lambda_V$ be representatives of the distinct cosets of Λ_V modulo Λ_F. The following conditions are equivalent:*

(a) *α is a v-norm;*
(b) *β is a w-norm and α_λ is a u-norm for all $\lambda \in \Lambda_V$;*
(c) *β is a w-norm and α_{λ_i} is a u-norm for $i = 1, \ldots, r$.*

Proof. Use the same notation as in the lemma. For simplicity, denote $\alpha_i = \alpha_{\lambda_i}$ and $V_i = V^\beta_{\lambda_i}$ for $i = 1, \ldots, r$, and $V_{ij} = V^\alpha_{\gamma_{ij}}$ for $i = 1, \ldots, r$ and $j = 1, \ldots, s_i$, and use the notation $[V{:}F]$ for $dim_F\, V$. From (3.45) it follows that

$$[\mathsf{gr}_\alpha(V){:}\ \mathsf{gr}_v(F)] = \sum_{i=1}^{r} \sum_{j=1}^{s_i} [V_{ij}{:}\overline{F}^v]. \tag{3.46}$$

Likewise,

$$[\mathsf{gr}_\beta(V){:}\ \mathsf{gr}_w(F)] = \sum_{i=1}^{r} [V_i{:}\overline{F}^w] \tag{3.47}$$

and

$$[\mathsf{gr}_{\alpha_i}(V_i){:}\ \mathsf{gr}_u(\overline{F}^w)] = \sum_{j=1}^{s_i} [V_{ij}{:}\overline{F}^v] \qquad \text{for } i = 1, \ldots, r. \tag{3.48}$$

If α_k is not a u-norm for some $k \in \{1, \ldots, r\}$, then

$$[V_k : \overline{F}^w] > [\mathsf{gr}_{\alpha_k}(V_k) : \mathsf{gr}_u(\overline{F}^w)].$$

On the other hand, we have

$$[V_i : \overline{F}^w] \geq [\mathsf{gr}_{\alpha_i}(V_i) : \mathsf{gr}_u(\overline{F}^w)] \quad \text{for all } i;$$

hence, by (3.47) and (3.48),

$$[\mathsf{gr}_\beta(V) : \mathsf{gr}_w(F)] > \sum_{i=1}^r [\mathsf{gr}_{\alpha_i}(V_i) : \mathsf{gr}_u(\overline{F}^w)] = \sum_{i=1}^r \sum_{j=1}^{s_i} [V_{ij} : \overline{F}^v].$$

In view of (3.46), it follows that $[\mathsf{gr}_\beta(V) : \mathsf{gr}_w(F)] > [\mathsf{gr}_\alpha(V) : \mathsf{gr}_v(F)]$. Since $[V:F] \geq [\mathsf{gr}_\beta(V) : \mathsf{gr}_w(F)]$, we have $[V:F] > [\mathsf{gr}_\alpha(V) : \mathsf{gr}_v(F)]$; hence, α is not a v-norm.

If each α_i is a u-norm, then $[V_i : \overline{F}^w] = [\mathsf{gr}_{\alpha_i}(V_i) : \mathsf{gr}_u(F)]$ for $i = 1, \ldots, r$; hence, (3.47), (3.48), and (3.46) yield

$$[\mathsf{gr}_\beta(V) : \mathsf{gr}_w(F)] = \sum_{i=1}^r \sum_{j=1}^{s_i} [V_{ij} : \overline{F}^v] = [\mathsf{gr}_\alpha(V) : \mathsf{gr}_v(F)].$$

It follows that α is a v-norm if and only if β is a w-norm. We have thus proved (a) $\Leftrightarrow$ (c). Since any $\lambda \in \Lambda$ can be chosen as a representative of its coset, the arguments above also show (a) $\Rightarrow$ (b). Since (b) $\Rightarrow$ (c) is clear, the proof is complete. □

To put Prop. 3.53 in perspective, we relate the graded vector spaces $\mathsf{gr}_\alpha(V)$ and $\mathsf{gr}_\beta(V)$ by means of a value-function-like map

$$\alpha_* : \mathsf{gr}_\beta(V) \longrightarrow \Gamma \cup \{\infty\}$$

defined as follows: for $\xi \in \mathsf{gr}_\beta(V)$, $\xi \neq 0$, let $\ell(\xi)$ be the homogeneous component of ξ of least degree, and let $\lambda = deg\big(\ell(\xi)\big)$, so $\ell(\xi) \in V_\lambda^\beta$; then let

$$\alpha_*(\xi) = \alpha_\lambda\big(\ell(\xi)\big) \in \lambda \subseteq \Gamma.$$

Let also $\alpha_*(0) = \infty$. For $x \in V$ we thus have

$$\alpha_*(\widetilde{x}^\beta) = \alpha(x), \tag{3.49}$$

where $\widetilde{x}^\beta$ denotes the image of x in $\mathsf{gr}_\beta(V)$.

A similar construction applies to the valuation v, and yields a map

$$v_* : \mathsf{gr}_w(F) \longrightarrow \Gamma \cup \{\infty\},$$

which satisfies the same properties as a valuation, and such that the image $v_*(\rho)$ of any nonzero $\rho \in \mathsf{gr}_w(F)$ depends only on its homogeneous component of least degree. The map α_* deserves the name of a *graded v_*-value function* since it satisfies the following properties:

(i) $\alpha_*(\xi) = \infty$ if and only if $\xi = 0$; if $\xi \neq 0$, then $\alpha_*(\xi) = \alpha_*(\ell(\xi))$ and $\varepsilon \circ \alpha_*(\xi) = deg\, \ell(\xi)$;
(ii) $\alpha_*(\xi + \eta) \geq min(\alpha_*(\xi), \alpha_*(\eta))$ for ξ, $\eta \in \mathsf{gr}_\beta(V)$;
(iii) $\alpha_*(\xi\rho) = \alpha_*(\xi) + v_*(\rho)$ for $\xi \in \mathsf{gr}_\beta(V)$ and $\rho \in \mathsf{gr}_w(F)$.

We may thus consider the associated graded structure $\mathsf{gr}_{\alpha_*}(\mathsf{gr}_\beta(V))$. If $x \in V$ satisfies $\beta(x) = \lambda$ and $\alpha(x) = \gamma$, we may identify

$$(x + V^\beta_{>\lambda}) + \mathsf{gr}_\beta(V)^{\alpha_*}_{>\gamma} \;=\; x + V^\alpha_{>\gamma};$$

thus

$$\mathsf{gr}_{\alpha_*}(\mathsf{gr}_\beta(V)) \;=\; \mathsf{gr}_\alpha(V). \tag{3.50}$$

We define α_* to be a *graded v_*-norm* if $[\mathsf{gr}_{\alpha_*}(\mathsf{gr}_\beta(V))\colon \mathsf{gr}_{v_*}(\mathsf{gr}_w(F))] = [\mathsf{gr}_\beta(V)\colon \mathsf{gr}_w(F)]$. It is easy to check that this holds if and only if each α_λ is a u-norm. By an argument analogous to the one in Prop. 3.14 for ungraded norms, one can check that if α_* is a graded norm, then for any graded subspace W of $\mathsf{gr}_\beta(V)$, $\alpha_*|_{\mathsf{W}}$ is a graded norm on W. Consequently, by dimension count, the functor $\mathsf{gr}_{\alpha_*}(\cdot)$ preserves strict inclusions of graded subspaces of $\mathsf{gr}_\beta(V)$. Proposition 3.53 may be rephrased as follows: α is a v-norm if and only if β is a w-norm and α_* is a graded v_*-norm. Indeed, if $(e_i)_{i=1}^n$ is a splitting base of V for α, then it is also a splitting base for β, and $(\widetilde{e}_i^{\,\beta})_{i=1}^n$ is a splitting base of $\mathsf{gr}_\beta(V)$ for α_*.

We now apply this construction to a finite-dimensional F-algebra A. If $\alpha\colon A \to \Gamma \cup \{\infty\}$ is a surmultiplicative v-value function, then the coarser w-value function $\beta = \varepsilon \circ \alpha$ is clearly surmultiplicative, and the map α_* is also surmultiplicative, by an easy calculation using (3.49). The notion of gauge for graded norms is defined analogously to the ungraded case: a *graded gauge* is a surmultiplicative graded norm whose associated graded ring is graded semisimple.

Proposition 3.54. *The map α is a v-gauge if and only if β is a w-gauge and α_* is a graded v_*-gauge. When these conditions hold, we also have*

(i) $\mathsf{gr}_\alpha(Z(A)) \;=\; Z(\mathsf{gr}_\alpha(A))$ *if and only if* $\mathsf{gr}_\beta(Z(A)) \;=\; Z(\mathsf{gr}_\beta(A))$ *and* $\mathsf{gr}_{\alpha*}(Z(\mathsf{gr}_\beta(A))) = Z(\mathsf{gr}_{\alpha*}(\mathsf{gr}_\beta(A)))$.
(ii) α_0 *is a u-gauge on* A_0^β.

Proof. Proposition 3.53 already shows that α is a v-norm if and only if β is a w-norm and α_* is a graded v_*-norm. We noted above that α is surmultiplicative if and only if β and α_* are surmultiplicative.

Suppose α is a v-gauge. Since $\mathsf{gr}_{\alpha_*}(\mathsf{gr}_\beta(A)) = \mathsf{gr}_\alpha(A)$ and $\mathsf{gr}_\alpha(A)$ is semisimple, it follows that $\mathsf{gr}_\beta(A)$ is semisimple. For, if I is a nontrivial nilpotent homogeneous left ideal of $\mathsf{gr}_\beta(A)$, then $\mathsf{gr}_{\alpha_*}(\mathsf{I})$ is a nontrivial nilpotent homogeneous left ideal of $\mathsf{gr}_{\alpha_*}(\mathsf{gr}_\beta(A))$. Thus, β is a w-gauge. Also, $\mathsf{gr}_{\alpha_*}(\mathsf{gr}_\beta(A))$ is semisimple by hypothesis; hence, α_* is a graded v_*-gauge. Conversely, if β is a w-gauge and α_* is a graded v_*-gauge, then α is a v-gauge since $\mathsf{gr}_\alpha(A) = \mathsf{gr}_{\alpha_*}(\mathsf{gr}_\beta(A))$.

Assume now that α is a v-gauge. We prove (i) and (ii).

(i) For the centers we have the obvious inclusions

$$\begin{aligned}\mathsf{gr}_\alpha\big(Z(A)\big) &= \mathsf{gr}_{\alpha_*}\big(\mathsf{gr}_\beta(Z(A))\big) \subseteq \mathsf{gr}_{\alpha_*}\big(Z(\mathsf{gr}_\beta(A))\big)\\ &\subseteq Z\big(\mathsf{gr}_{\alpha_*}(\mathsf{gr}_\beta(A))\big) = Z\big(\mathsf{gr}_\alpha(A)\big).\end{aligned} \tag{3.51}$$

Thus, $Z\big(\mathsf{gr}_\alpha(A)\big) = \mathsf{gr}_\alpha\big(Z(A)\big)$ if and only if we have equalities throughout (3.51); since $\mathsf{gr}_{\alpha_*}(\_)$ preserves strict inclusions, this holds if and only if

$$\mathsf{gr}_\beta\big(Z(A)\big) = Z\big(\mathsf{gr}_\beta(A)\big) \quad\text{and}\quad \mathsf{gr}_{\alpha_*}\big(Z(\mathsf{gr}_\beta(A))\big) = Z\big(\mathsf{gr}_{\alpha_*}(\mathsf{gr}_\beta(A))\big).$$

(ii) Consider the u-value function α_0 on $A_0^\beta = A_{\geq 0}^\beta / A_{>0}^\beta$. Recall that for $x \in A_{\geq 0}^\beta$, we have $\beta(x) = 0$ if and only if $\alpha(x) \in \Lambda$, and

$$\alpha_0(x + A_{>0}^\beta) = \begin{cases}\alpha(x) & \text{if } \beta(x) = 0,\\ \infty & \text{if } \beta(x) > 0.\end{cases}$$

By Prop. 3.53, α_0 is a u-norm on A_0^β; also α_0 is surmultiplicative, as α is surmultiplicative. To see that α_0 is a u-gauge, it remains only to verify that $\mathsf{gr}_{\alpha_0}(A_0^\beta)$ is graded semisimple. For this, let $\mathsf{E} = \mathsf{gr}_\alpha(A)$. Then,

$$\mathsf{gr}_{\alpha_0}(A_0^\beta) = \bigoplus_{\lambda\in\Lambda} A_\lambda = \bigoplus_{\lambda\in\Lambda} \mathsf{E}_\lambda = \mathsf{E}_\Lambda,$$

which is the Λ-part of E. Since E_Λ is the product of the Λ-parts of the simple components of the semisimple graded algebra E, we may assume that E is graded simple. By the graded Wedderburn Theorem (Th. 2.26), we then have $\mathsf{E} = \mathsf{End}_{\mathsf{D}}(\mathsf{V})$ for some finite-dimensional graded right vector space V over a graded division algebra D. Let $(b_1, \ldots, b_n)$ be a homogeneous D-vector-space base of V, and partition $\{b_1, \ldots, b_n\}$ via cosets of $\Gamma_{\mathsf{D}} + \Lambda$. That is, let $\varepsilon_1, \ldots, \varepsilon_k$ be representatives of the distinct cosets of $\Gamma_{\mathsf{D}} + \Lambda$ in the union $\bigcup_{j=1}^n \big(deg(b_j) + \Gamma_{\mathsf{D}} + \Lambda\big)$, and let $S_i = \{b_j \mid deg(b_j) \in \varepsilon_i + \Gamma_{\mathsf{D}} + \Lambda\}$ for $i = 1, 2, \ldots, k$. Then, $\{b_1, \ldots, b_n\} = \bigcup_{i=1}^k S_i$, a disjoint union. Re-index the b_j so that each $S_i = \{b_{i1}, \ldots, b_{im_i}\}$, and let $\gamma_{i\ell} = deg(b_{i\ell})$ for all i, ℓ. Since each $\gamma_{i\ell} - \gamma_{i1} \in \Gamma_{\mathsf{D}} + \Lambda$, there exist $c_{il} \in D^\times$ such that $c_{i1} = 1$ and $\gamma_{i\ell} + deg(c_{i\ell}) - \gamma_{i1} \in \Lambda$ for all i, ℓ. Replace each $b_{i\ell}$ by $b_{i\ell}c_{i\ell}$. We then have

$$\gamma_{i\ell} - \gamma_{i\ell'} = \big(\gamma_{i\ell} - \gamma_{i1}\big) - \big(\gamma_{i\ell'} - \gamma_{i1}\big) \in \Lambda \qquad \text{for all } i, \ell, \ell'.$$

Let $\mathsf{D}_\Lambda = \bigoplus_{\lambda\in\Lambda} \mathsf{D}_\lambda$, the Λ-part of D, which is a graded division algebra. For $i = 1, 2, \ldots, k$, let W_i be the graded D_Λ-subspace of V spanned by S_i. We claim that

$$\mathsf{E}_\Lambda = \mathsf{End}_{\mathsf{D}_\Lambda}(\mathsf{W}_1) \times \ldots \times \mathsf{End}_{\mathsf{D}_\Lambda}(\mathsf{W}_k). \tag{3.52}$$

This will yield the desired graded semisimplicity of $\mathsf{gr}_{\alpha_0}(A_0^\beta) = \mathsf{E}_\Lambda$. To verify the claim, note first that by Cor. 2.11

$$\Gamma_{\mathsf{End}_{\mathsf{D}_\Lambda}(\mathsf{W}_i)} = \bigcup_{\ell,\ell'} \big(\gamma_{i\ell} - \gamma_{i\ell'} + \Gamma_{\mathsf{D}_\Lambda}\big).$$

This lies in Λ, as $\Gamma_{\mathsf{D}_\Lambda} \subseteq \Lambda$ and each $\gamma_{i\ell} - \gamma_{i\ell'} \in \Lambda$. Hence,

$$\prod_{i=1}^{k} \mathsf{End}_{\mathsf{D}_\Lambda}(\mathsf{W}_i) \subseteq \mathsf{E}_\Lambda .$$

For the reverse inclusion, take any homogeneous $f \in \mathsf{E}_\Lambda$, and write

$$f(b_{i\ell}) = \sum_{i',\ell'} b_{i'\ell'} d_{i\ell i'\ell'},$$

with all $d_{i\ell i'\ell'}$ homogeneous in D. Let $\lambda = deg(f) \in \Lambda$. Then, for any nonzero coefficient $d_{i\ell i'\ell'}$ we have

$$\lambda = deg\big(f(b_{i\ell})\big) - deg(b_{i\ell}) = \gamma_{i'\ell'} + deg(d_{i\ell i'\ell'}) - \gamma_{i\ell}.$$

Hence, $\gamma_{i'\ell'} - \gamma_{i\ell} = \lambda - deg(d_{i\ell i'\ell'}) \in \Lambda + \Gamma_{\mathsf{D}}$. When $i' \neq i$, $\gamma_{i\ell}$ and $\gamma_{i'\ell'}$ lie in different cosets of $\Gamma_{\mathsf{D}} + \Lambda$. Hence, whenever $d_{i\ell i'\ell'} \neq 0$ we must have $i' = i$ and

$$deg(d_{i\ell i\ell'}) = \lambda + \gamma_{i\ell} - \gamma_{i\ell'} \in \Lambda;$$

so $d_{i\ell i\ell'} \in \mathsf{D}_\Lambda^\times$. This shows that each $f(b_{i\ell}) \in \mathsf{W}_i$. Hence, $f \in \prod_{i=1}^{k} \mathsf{End}_{\mathsf{D}_\Lambda}(\mathsf{W}_i)$. Since E_Λ is additively generated by homogeneous elements, this proves claim (3.52). Thus, $\mathsf{gr}_{\alpha_0}(A_0^\beta)$ is graded semisimple, completing the proof that α_0 is a u-gauge on A_0^β. □

Exercises

Exercise 3.1. Let D be a division ring with valuation v, and let M be a free right module of finite rank over the valuation ring $\mathcal{O}_D$. Define $V = M \otimes_{\mathcal{O}_D} D$, so V is a finite-dimensional right D-vector space. Identify M with its image in V. Show that for a v-norm α on V, the following conditions are equivalent:

(a) $M = \{x \in V \mid \alpha(x) \geq 0\}$ and $\Gamma_{V,\alpha} = \Gamma_D$;
(b) $M = \{x \in V \mid \alpha(x) \geq 0\}$ and $M \otimes_{\mathcal{O}_D} \mathfrak{m}_D = \{x \in V \mid \alpha(x) > 0\}$;
(c) $M = \{x \in V \mid \alpha(x) \geq 0\}$ and $M \otimes_{\mathcal{O}_D} \overline{D} = \mathsf{gr}_\alpha(V)_0$.

Show that there is a unique v-norm α for which these conditions hold.

Exercise 3.2. Let D be a division ring with valuation v, and let V be an infinite-dimensional right D-vector space with a v-value function α. Define α to be a *v-norm* on V if V has a splitting base for α. Suppose for this exercise that α is such a norm.

(i) Show that for every finite-dimensional subspace W of V, the restriction $\alpha|_W$ is a norm on W.
(ii) Now show that for every countably infinite-dimensional subspace W of V, the restriction $\alpha|_W$ is a norm on W.

(iii) Suppose that Γ_V is nontrivial and lies in the divisible hull of Γ_D. Show that there exist maximal splitting families in v that are not splitting bases. [Hint: If $(e_i)_{i\in\mathbb{N}}$ is a splitting base of V, consider $(e_i\lambda_i + e_{i+1})_{i\in\mathbb{N}}$ where $\lambda_i \in D$ is chosen such that $\alpha(e_i\lambda_i) < \alpha(e_{i+1})$.]

(iv) With the hypotheses of (iii), deduce that V has a subspace W such that $\alpha|_W$ is a norm but W has no splitting complement in V.

Exercise 3.3. Let U, W be subspaces of a vector space V over a division ring D, and let α be a v-norm on V. Show that

$$\mathsf{gr}(U \cap W) \subseteq \mathsf{gr}(U) \cap \mathsf{gr}(W) \qquad \text{and} \qquad \mathsf{gr}(U + W) \supseteq \mathsf{gr}(U) + \mathsf{gr}(W).$$

If U and W are finite-dimensional, show that $\mathsf{gr}(U \cap W) = \mathsf{gr}(U) \cap \mathsf{gr}(W)$ if and only if $\mathsf{gr}(U + W) = \mathsf{gr}(U) + \mathsf{gr}(W)$.

Exercise 3.4. Let α be a v-norm on a finite-dimensional D-vector space V, and let $U \subseteq V$ be a subspace. Show that a v-norm β can be defined on the quotient space V/U by the formula

$$\beta(x + U) = \max\{\alpha(x + u) \mid u \in U\} \qquad \text{for } x \in V.$$

If W is a splitting complement of U in V, establish an isomorphism of normed vector spaces $(W, \alpha|_W) \cong (V/U, \beta)$.

Exercise 3.5. Let α be a v-value function on a finite-dimensional D-vector space V. One can define the *norm defect* of α to be

$$\partial(\alpha) = \dim_D(V) - \dim_{\mathsf{gr}(D)}(\mathsf{gr}(V)).$$

By Cor. 3.6(i), $\partial(\alpha) \geq 0$ and $\partial(\alpha) = 0$ if and only if α is a norm. Show that if W is any D-subspace of V, then $\partial(\alpha|_W) \leq \partial(\alpha)$.

Exercise 3.6. Let (D, w) be a valued division ring, let α, α' be v-norms on a right D-vector space V and let β, β' be v-value functions on a left D-vector space W. Show that $\alpha \otimes \beta = \alpha' \otimes \beta'$ on $V \otimes_D W$ if and only if there is a constant $\gamma \in \Gamma$ such that $\alpha(x) - \alpha'(x) = \gamma = \beta'(y) - \beta(y)$ for all $x \in V$ and $y \in W$.

Exercise 3.7. Let D be a finite-dimensional division algebra over a field F with a valuation v, and let g be a v-gauge on D. Define a notion of g-norm on finite-dimensional D-vector spaces and show that if α is a g-norm on a finite-dimensional D-vector space V, then $End(\alpha)$ is a v-gauge on $End_D V$.

Exercise 3.8. Let $(F, v) \subseteq (K, w)$ be an extension of valued fields. Assume that $k \in K$ and $\gamma \in \Gamma_K$ satisfy (i) $\gamma > w(\lambda - k)$ for all $\lambda \in F$ and (ii) for every $\lambda \in F$ there exists a $\lambda' \in F$ such that $w(\lambda' - k) > w(\lambda - k)$. Consider the F-vector space $V = F^2$. Define a w-norm α on $K \otimes_F V = K^2$ by

$$\alpha\big(x(1,0) + y(k,1)\big) = \min\big(w(x), w(y) + \gamma\big) \qquad \text{for } x,\, y \in K.$$

Show that $\alpha|_V$ is not a v-norm on V. Conclude that α is not a v-norm on $V \otimes_F K$.

(To obtain a specific example, consider $F = \mathbb{Q}((t))$, $K = \mathbb{Q}(i)((t))$, with v the composite of the 5-adic valuation on $\mathbb{Q}$ with the t-adic valuation, and w either of the two extensions of v to K; let $k = i$ and $\gamma = v(t)$.)

Exercise 3.9. Show that, under the hypotheses of Prop. 3.48, the ideal $\mathfrak{m}_A$ lies in the Jacobson radical $J(O_A)$. Deduce that if α is a gauge then $\mathfrak{m}_A = J(O_A)$.

Notes

The extension of the notion of valuation to vector spaces dates back at least to Monna [155] and Cohen [53]. Our exposition owes much to Bruhat–Tits [33, §1], where tensor products of norms and norms on endomorphism algebras are defined. The idea to view properties of norms through the graded structure associated to the induced filtration was first systematically explored in Renard et al. [200, §2], which also contains some additional information. In particular, the result that any two norms on a vector space share a common splitting base, which is crucial for the Bruhat–Tits theory [33, Prop. 1.26, App.], is proved in full generality (i.e., without restriction to valuations of rank 1) in [200, Th. 2.8]. The results on the composition of gauges (§3.2.7) first appeared in Tignol–Wadsworth [247, §4].

§3.2.6: *Dubrovin valuation rings and residually simple gauges.* Let A be a (finite-dimensional) central simple F-algebra. A subring B of A is called a *Dubrovin valuation ring* of A if there is an ideal J of B such that B/J is a simple ring and for every $a \in A \backslash B$ there are $s, t \in B$ such that $sa, at \in B \setminus J$. There is an extensive theory of Dubrovin valuation rings—see the thorough treatment in the book [132] by Marubayashi et al. or the original papers by Dubrovin [67], [68], [69], Brungs–Gräter [35], Gräter [88], [89], Morandi [157], [163], [158], and Wadsworth [255]. See also surveys of the theory in Brungs–Gräter [36], Gräter [90], and Wadsworth [256, §10], and the brief account in Marubayashi–Van Oystaeyen [133, §1.6]. If B is a Dubrovin valuation ring of A, then every finitely-generated right or left ideal of B is principal and projective as a B-module. The two-sided ideals of B are linearly ordered by inclusion. If P is any prime ideal of B, then B/P is a Dubrovin valuation ring. Every ring C with $B \subseteq C \subseteq A$ is also a Dubrovin valuation ring. Moreover, B is a prime ring with $B \cdot F = A$, and $Z(B) = B \cap F$, which is a valuation ring of (i.e., with quotient field) F. If V is any valuation ring of F, a more difficult result is that there is a Dubrovin valuation ring B of A with $Z(B) = V$, and B is unique up to conjugacy; call such a B a V-Dubrovin valuation ring of A. Also, if e is any nonzero idempotent of B, then eBe is a V-Dubrovin valuation ring of eAe; in addition, the matrix ring $M_m(B)$ is a V-Dubrovin valuation ring of $M_m(A)$ for every positive integer m.

In general, a V-Dubrovin valuation ring B need not be integral over V. But whenever B is integral over V, Morandi proved in [157] that there is an associated value function v_B on A such that $B = \mathcal{O}_{A,v_B} = \{a \in A \mid v_B(a) \geq 0\}$. Whenever v_B is defectless over F, then v_B is actually a residually simple gauge on A (see Tignol–Wadsworth [246, Prop. 2.5]). Conversely, if α is a residually simple gauge on A, it is easy to see using Lemma 3.49(iii) that $\mathcal{O}_{A,\alpha}$ is a Dubrovin valuation ring that is integral over its center (see Lemma 4.24 below for the integrality), and in fact α is the Morandi value function associated to $\mathcal{O}_{A,\alpha}$. Let $v = \alpha|_F$. It follows from the Dubrovin theory that if α' is another residually simple v-gauge, then $\mathcal{O}_{A,\alpha'}$ and $\mathcal{O}_{A,\alpha}$ are conjugate in A, and hence by Prop. 3.48 that α and α' are related by $\alpha = \alpha' \circ \mathit{int}(u)$ for any $u \in A^\times$ such that $\mathcal{O}_{A,\alpha'} = u\mathcal{O}_{A,\alpha}u^{-1}$. Furthermore, if $A \cong M_n(D)$ for a division algebra D, then D has a residually simple v-gauge β and $\mathcal{O}_{A,\alpha} \cong M_n(\mathcal{O}_{D,\beta})$. These facts do not seem to have short direct proofs without invoking the Dubrovin theory. See the Notes for the next chapter for further connections between gauges and Dubrovin valuation rings.

Chapter 4
Existence and Fundamental Properties of Gauges

This chapter is crucial for the study of gauges: We give in Th. 4.50 a necessary and sufficient condition for the existence of a gauge on a semisimple algebra over an arbitrary valued field, and provide a complete description of gauges on semisimple algebras over Henselian fields (Th. 4.26, Th. 4.28). We also obtain information on gauges over arbitrary valued fields by scalar extension to a Henselization.

First, we discuss in §4.1 the influence of Henselization. The main result of that section is a theorem of Morandi asserting that a valuation extends to a central division algebra if and only if scalar extension to a Henselization does not reduce its index. This criterion allows us to carry over to arbitrary valued division algebras results that are proved when the valuation on the center is Henselian. This theme is illustrated in §4.2, where we derive a general "Ostrowski Theorem" for arbitrary valued division algebras by reduction to the Henselian case. This section also introduces the notion of defect for the extension of a valuation to a division algebra, which is the main ingredient in the criterion for the existence of gauges: Theorem 4.50 roughly states that gauges exist only in defectless situations. This result is the main topic of §4.4, while §4.3 yields various structural results on gauges, including the description of gauges on semisimple algebras over Henselian fields and on commutative semisimple algebras over arbitrary valued fields. As a result, we will see in Prop. 4.33 that when a division algebra D carries a valuation v_D extending a valuation v on its center, then v_D is the only possible v-gauge on D, and that v_D is a v-gauge if and only if it is defectless.

4.1 The Henselization of a division algebra

In this section we prove the following criterion (already stated in §1.2.7) for the extension of a valuation v on a field F to a central division algebra D

J.-P. Tignol, A.R. Wadsworth, *Value Functions on Simple Algebras, and Associated Graded Rings*, Springer Monographs in Mathematics,
DOI 10.1007/978-3-319-16360-4_4

over F: the valuation v extends to D if and only if D remains a division algebra after scalar extension to a Henselization F_h of F with respect to v.

Throughout this section, we let (F_h, v_h) denote a Henselization of a valued field (F, v). Occasionally, when F carries several valuations of interest, we will use the more explicit notation $F_{h,v}$ for F_h, to emphasize the dependence on v.

Recall that the rank of v, denoted $rk(v)$, is the Krull dimension of its valuation ring $\mathcal{O}_{F,v}$ or, equivalently, the rank of its value group $rk(\Gamma_{F,v})$. See (A.13)–(A.16) in Appendix A.

Theorem 4.1 (Morandi). *Let v be an arbitrary valuation on a field F and let D be a central division F-algebra. The valuation v extends to a valuation v_D on D if and only if $D \otimes_F F_h$ is a division algebra. When this occurs, v_D is the restriction to D of the valuation on $D \otimes_F F_h$ extending v_h, and we have*

$$\overline{D \otimes_F F_h} = \overline{D} \quad \text{and} \quad \Gamma_{D \otimes F_h} = \Gamma_D,$$

and

$$\mathsf{gr}(D \otimes_F F_h) \cong_g \mathsf{gr}(D).$$

Note that the extension of v to D is unique when it exists, by Th. 1.4.

Proof. If $D \otimes_F F_h$ is a division algebra, then v_h extends to a valuation on $D \otimes_F F_h$ by Cor. 1.7, and it is clear that the restriction of this valuation to D is a valuation on D. Thus, we only have to prove the "only if" part and the equalities of residue algebras and value groups. This is very easy when the valuation v_D is defectless, see Remark 4.2 below. In the general case, we first prove the result when the valuation v has finite rank, by induction on the rank.

If $rk(v) = 0$, i.e., v is the trivial valuation, then $F_h = F$ and there is nothing to prove. Suppose next that $rk(v) = 1$; then Γ_F may be identified with a subgroup of $\mathbb{R}$; hence, Γ_D may also be viewed as a subgroup of $\mathbb{R}$, as Γ_D/Γ_F is finite. We may then argue as in Prop. 1.16: consider the completion $\widehat{D}$ of D for the topology induced by the valuation v_D (see Pierce [178, Chap. 17.4] or Schilling [227, Chap. 2]). The algebra $\widehat{D}$ is a division algebra. It contains the completion $\widehat{F}$ of F for the topology of v, which centralizes D; hence, $D \otimes_F \widehat{F}$ embeds in $\widehat{D}$. Since $\widehat{F}$ is Henselian (see, e.g., Engler–Prestel [73, Th. 1.3.1]), F_h embeds in $\widehat{F}$, hence we have $D \otimes_F F_h \subseteq \widehat{D}$. It is now clear that the finite-dimensional F_h-algebra $D \otimes_F F_h$ is a division algebra, as $\widehat{D}$ is a division algebra. Moreover, we have $\overline{\widehat{D}} = \overline{D}$ and $\Gamma_{\widehat{D}} = \Gamma_D$; hence, the inclusions

$$\overline{D} \subseteq \overline{D \otimes_F F_h} \subseteq \overline{\widehat{D}} \quad \text{and} \quad \Gamma_D \subseteq \Gamma_{D \otimes F_h} \subseteq \Gamma_{\widehat{D}}$$

yield $\overline{D \otimes_F F_h} = \overline{D}$ and $\Gamma_{D \otimes F_h} = \Gamma_D$. The theorem is thus proved if $rk(v) = 1$.

Now, assume $rk(v) > 1$ and $rk(v)$ is finite, and consider an arbitrary nontrivial convex subgroup $\Delta_F \subsetneqq \Gamma_{F,v}$. The ordering on $\Gamma_{F,v}$ induces an ordering on the factor group $\Lambda_F = \Gamma_{F,v}/\Delta_F$. As in §3.2.7, let $\varepsilon\colon \Gamma_{F,v} \to \Lambda_F$ be the canonical epimorphism. The composition of v and ε defines a valuation w coarser than v:

$$w = \varepsilon \circ v\colon F \longrightarrow \Lambda_F \cup \{\infty\}.$$

If $x \in F$ satisfies $w(x) = 0$, then $v(x) \in \Delta_F$. Moreover, if x, $x' \in F$ satisfy $w(x) = w(x') = 0$ and $w(x - x') > 0$, then $v(x - x') > v(x')$, hence

$$v(x) = v\big((x - x') + x'\big) = v(x').$$

Therefore, we have a valuation $u\colon \overline{F}^w \to \Delta_F \cup \{\infty\}$ given by

$$u(\overline{x}) = v(x) \qquad \text{for } x \in F \text{ such that } w(x) = 0. \tag{4.1}$$

We sometimes write v/w for this valuation u. The valuation v is thus the composite valuation $v = u * w = (v/w) * w$, and we have

$$\Gamma_{F,w} = \Lambda_F, \qquad \Gamma_{\overline{F}^w,u} = \Delta_F, \qquad \overline{\overline{F}^w}^u = \overline{F}^v$$

and $rk(v) = rk(u) + rk(w)$.

Let

$$w' = \varepsilon \circ v_h\colon F_{h,v} \longrightarrow \Lambda \cup \{\infty\},$$

which is a valuation on $F_{h,v}$ and a coarsening of v_h. By Prop. A.15, w' is a Henselian valuation on $F_{h,v}$ extending w; hence there is a unique embedding of the Henselization $F_{h,w}$ into $F_{h,v}$, which we regard as an inclusion

$$(F_{h,w}, w_h) \subseteq (F_{h,v}, w').$$

Thus, w' is the unique extension of w_h to $F_{h,v}$. By Prop. A.31, the extension $(F_{h,v}, w')$ of $(F_{h,w}, w_h)$ is inertial, as defined in Def. A.4, and the residue field $\overline{F_{h,v}}^{w'}$, with the valuation induced by v_h, is a Henselization of $\overline{F_{h,w}}^{w_h} = \overline{F}^w$:

$$\overline{F_{h,v}}^{w'} = (\overline{F}^w)_{h,u}. \tag{4.2}$$

We therefore let u_h denote the valuation induced by v_h on $\overline{F_{h,v}}^{w'}$, and we have $v_h = u_h * w'$.

The construction above may also be applied to D, to obtain a coarsening w_D of v_D: let $\Delta_D \subset \Gamma_{D,v_D}$ be the convex hull of Δ_F in Γ_{D,v_D}, let $\Lambda_D = \Gamma_{D,v_D}/\Delta_D$, and write again ε for the canonical epimorphism $\Gamma_{D,v_D} \to \Lambda_D$, which extends $\varepsilon\colon \Gamma_{F,v} \to \Lambda_F$. Let

$$w_D = \varepsilon \circ v_D\colon D \longrightarrow \Lambda_D \cup \{\infty\}.$$

This is a valuation on D that extends w, and there is an induced valuation

$$u_D\colon \overline{D}^{w_D} \longrightarrow \Delta_D \cup \{\infty\},$$

which extends u. Just as for v and v_h above, we have $v_D = u_D * w_D$. By Prop. 1.8, it follows that u has a unique extension from $\overline{F}^w$ to $Z(\overline{D}^{w_D})$;

hence, $Z(\overline{D}^{w_D})$ and $(\overline{F}^{w})_{h,u}$ are linearly disjoint extensions of $\overline{F}^{w}$, and a Henselization of $Z(\overline{D}^{w_D})$ is given by

$$Z(\overline{D}^{w_D})_{h,u} = Z(\overline{D}^{w_D}) \otimes_{\overline{F}^w} (\overline{F}^{w})_{h,u}.$$

Since u_D extends u and $rk(u) < rk(v)$, the induction hypothesis shows that $\overline{D}^{w_D} \otimes_{Z(\overline{D}^{w_D})} Z(\overline{D}^{w_D})_{h,u}$ is a division algebra. As

$$\overline{D}^{w_D} \otimes_{Z(\overline{D}^{w_D})} Z(\overline{D}^{w_D})_{h,u} = \overline{D}^{w_D} \otimes_{\overline{F}^w} (\overline{F}^{w})_{h,u}, \tag{4.3}$$

it follows that $\overline{D}^{w_D} \otimes_{\overline{F}^w} (\overline{F}^{w})_{h,u}$ is a division algebra.

On the other hand, we also have $rk(w) < rk(v)$ and w_D extends w, so the induction hypothesis shows that $D \otimes_F F_{h,w}$ is a division algebra, and that

$$\overline{D \otimes_F F_{h,w}}^{w_D} = \overline{D}^{w_D} \qquad \text{and} \qquad \Gamma_{D\otimes F_{h,w}, w_D} = \Gamma_{D,w_D}.$$

To show that D remains a division algebra when scalars are further extended from $F_{h,w}$ to $F_{h,v}$, it suffices to consider finite-degree extensions (M, w_M) of $(F_{h,w}, w_h)$ in $(F_{h,v}, w')$ since $F_{h,v}$ is the union of such extensions. To see that $(D \otimes_F F_{h,w}) \otimes_{F_{h,w}} M$ is a division algebra, we apply Th. 3.43: since $(F_{h,v}, w')$ is an inertial extension of $(F_{h,w}, w_h)$, we know that M is an inertial extension of finite degree of $F_{h,w}$, hence it is defectless and $\Gamma_{M,w_M} = \Gamma_{F_{h,w},w_h}$. Therefore, Th. 3.43 shows that $(D \otimes_F F_{h,w}) \otimes_{F_{h,w}} M$ is a division algebra provided that $\overline{D \otimes_F F_{h,w}}^{w_D} \otimes_{\overline{F_{h,w}}^{w_h}} \overline{M}^{w_M}$ is a division algebra. We have

$$\overline{D \otimes_F F_{h,w}}^{w_D} \otimes_{\overline{F_{h,w}}^{w_h}} \overline{M}^{w_M} = \overline{D}^{w_D} \otimes_{\overline{F}^w} \overline{M}^{w_M} \quad \text{and} \quad \overline{M}^{w_M} \subseteq \overline{F_{h,v}}^{w'}. \tag{4.4}$$

Since $\overline{F_{h,v}}^{w'} = (\overline{F}^{w})_{h,u}$ by (4.2) and since $\overline{D}^{w_D} \otimes_{\overline{F}^w} (\overline{F}^{w})_{h,u}$ is a division algebra, it follows that $\overline{D}^{w_D} \otimes_{\overline{F}^w} \overline{M}^{w_M}$ is a division algebra; hence, $(D \otimes_F F_{h,w}) \otimes_{F_{h,w}} M$ is a division algebra. Since this holds for every subfield $M \subseteq F_{h,v}$ of finite degree over $F_{h,w}$, the algebra $D \otimes_F F_{h,v}$ is a division algebra. Theorem 3.43 and the equation in (4.4) also yield

$$\overline{(D \otimes_F F_{h,w}) \otimes_{F_{h,w}} M}^{w_D} = \overline{D}^{w_D} \otimes_{\overline{F}^w} \overline{M}^{w_M}$$

and

$$\Gamma_{(D\otimes_F F_{h,w})\otimes_{F_{h,w}} M, w_D} = \Gamma_{D\otimes_F F_{h,w}, w_D} = \Gamma_{D,w_D}.$$

Hence, by taking the union over all M in $F_{h,v}$ we obtain

$$\overline{D \otimes_F F_{h,v}}^{w_D} = \overline{D}^{w_D} \otimes_{\overline{F}^w} \overline{F_{h,v}}^{w'} \quad \text{and} \quad \Gamma_{D\otimes_F F_{h,v}, w_D} = \Gamma_{D,w_D}. \tag{4.5}$$

By (4.2) and (4.3), we have

$$\overline{D}^{w_D} \otimes_{\overline{F}^w} \overline{F_{h,v}}^{w'} = \overline{D}^{w_D} \otimes_{Z(\overline{D}^{w_D})} Z(\overline{D}^{w_D})_{h,u}. \tag{4.6}$$

Since $rk(u) < rk(v)$, the induction hypothesis yields

$$\overline{\overline{D}^{w_D} \otimes_{Z(\overline{D}^{w_D})} Z(\overline{D}^{w_D})_{h,u}}^{u_D} = \overline{\overline{D}^{w_D}}^{u_D}. \tag{4.7}$$

Therefore, by (4.5), (4.6), and (4.7),

$$\overline{\overline{D \otimes_F F_{h,v}}^{w_D}}^{u_D} = \overline{\overline{D}^{w_D}}^{u_D},$$

hence $\overline{D \otimes_F F_{h,v}}^{v_D} = \overline{D}^{v_D}$.

For the value groups, we have the following commutative diagram with exact rows (see Remark 1.9), where the vertical maps are inclusions:

$$\begin{array}{ccccccccc}
0 & \longrightarrow & \Gamma_{\overline{D}^{w_D},u_D} & \longrightarrow & \Gamma_{D,v_D} & \longrightarrow & \Gamma_{D,w_D} & \longrightarrow & 0 \\
 & & \downarrow & & \downarrow & & \downarrow & & \\
0 & \longrightarrow & \Gamma_{\overline{D \otimes_F F_{h,v}}^{w_D},u_D} & \longrightarrow & \Gamma_{D \otimes_F F_{h,v},v_D} & \longrightarrow & \Gamma_{D \otimes_F F_{h,v},w_D} & \longrightarrow & 0
\end{array} \tag{4.8}$$

By (4.5), the right vertical map is an equality. By (4.5) and (4.6), we have $\overline{D \otimes_F F_{h,v}}^{w_D} = \overline{D}^{w_D} \otimes_{Z(\overline{D}^{w_D})} Z(\overline{D}^{w_D})_{h,u}$. Since $rk(u) < rk(v)$, the induction hypothesis yields

$$\Gamma_{\overline{D}^{w_D} \otimes_{Z(\overline{D}^{w_D})} Z(\overline{D}^{w_D})_{h,u},u_D} = \Gamma_{\overline{D}^{w_D},u_D}.$$

Therefore, the left vertical map in diagram (4.8) also is an equality. The middle vertical map then also is an equality. Now, from the equality of value groups and residue rings, it is clear that the canonical inclusion $\mathsf{gr}(D) \hookrightarrow \mathsf{gr}(D \otimes_F F_{h,v})$ is an equality, and the proof is complete when $rk(v)$ is finite.

For arbitrary $rk(v)$, consider an F-base $(d_i)_{i=1}^n$ of D and the corresponding structure constants $\lambda_{ijk} \in F$, defined by the equation

$$d_i d_j = \sum_{k=1}^{n} \lambda_{ijk} d_k \qquad \text{for } i, j = 1, \ldots, n.$$

Let F_0 be the prime subfield of F and let $F_1 = F_0(\lambda_{ijk} \mid 1 \leq i, j, k \leq n)$. Then, $rk(v_{F_0}) \leq 1$. The field F is the union of its subfields K that are finitely generated over F_1 (hence also over F_0). For any such subfield K, let $D(K)$ be the K-span of the base $(d_i)_{i=1}^n$. We have $D(K) \otimes_K F = D$, hence $D(K)$ is a central division K-algebra, and $D = \bigcup_K D(K)$. Since F_h is a Henselian extension of K, there is a unique Henselization K_h of K in F_h, by Endler [72, 17.11]. The union $\bigcup_K K_h$ is Henselian and contains F, hence $\bigcup_K K_h = F_h$ and $\bigcup_K (D(K) \otimes_K K_h) = D \otimes_F F_h$. By rank formulas (A.15), (A.17), and (A.21) and the transcendence degree inequality, Th. A.35,

$$\begin{aligned} rk(v|_K) &= rk(\Gamma_K) = rk(\Gamma_{F_0}) + rk(\Gamma_K/\Gamma_{F_0}) \\ &\leq 1 + rr(\Gamma_K/\Gamma_{F_0}) \leq 1 + trdeg\, K/F_0 < \infty. \end{aligned}$$

Hence, the finite rank case of the proof shows that $D(K) \otimes_K K_h$ is a division algebra and

$$\overline{D(K) \otimes_K K_h} = \overline{D(K)} \qquad \text{and} \qquad \Gamma_{D(K)\otimes_K K_h} = \Gamma_{D(K)}.$$

By taking the union over all the finitely generated extensions K of F_1 in F, we see that $D \otimes_F F_h$ is a division algebra, and $\overline{D \otimes_F F_h} = \overline{D}$ and $\Gamma_{D\otimes_F F_h} = \Gamma_D$. Therefore, the inclusion $\mathsf{gr}(D) \hookrightarrow \mathsf{gr}(D \otimes_F F_h)$ is an equality and the proof is complete. □

Remark 4.2. In the notation of Th. 4.1, suppose the valuation v_D is defectless over F, i.e., $[D:F] = [\overline{D}:\overline{F}] \cdot |\Gamma_D:\Gamma_F|$ (see §3.2.2); then v_D is a v-gauge on D. Therefore, since $\mathsf{gr}(F_h) = \mathsf{gr}(F)$, Prop. 3.42 shows that $v_D \otimes v_h$ is a v_h-gauge on $D \otimes_F F_h$, and we have

$$\mathsf{gr}(D \otimes_F F_h) \cong_g \mathsf{gr}(D) \otimes_{\mathsf{gr}(F)} \mathsf{gr}(F_h) = \mathsf{gr}(D).$$

It follows that $\mathsf{gr}(D \otimes_F F_h)$ is a graded division ring, hence by Cor. 3.30 $D \otimes_F F_h$ is a division algebra (and $v_D \otimes v_h$ is the valuation on $D \otimes_F F_h$ extending v_h). This gives a substantially easier proof of the nontrivial implication of Th. 4.1 for the special case where v_D is defectless over F.

4.2 Defect

If D is a division algebra over the field F and D has a valuation v_D extending a valuation v on F, then the *defect* $\partial_{D/F}$ measures the failure of equality in the Fundamental Inequality for v_D over v. This is relevant for gauges since v_D is a v-gauge just when v is defectless in D, i.e., $\partial_{D/F} = 1$. In §4.2.1 we prove an Ostrowski-type result, Th. 4.3, for $\partial_{D/F}$. In §4.2.2 we extend the notion of defectlessness of v from division algebras to arbitrary (finite-dimensional) semisimple F-algebras A, by passage to the Henselization. Defectlessness turns out to be a crucial feature for gauges: we will show in Th. 4.50 below that A has a v-gauge if and only if v is defectless in A.

4.2.1 Defect of valued division algebras

Let D be a finite-dimensional division algebra over a field F. A valuation v_D on D is said to be *defectless over* F if

$$[D:F] = [\overline{D}:\overline{F}] \cdot |\Gamma_D:\Gamma_F|,$$

where $\overline{F}$ (resp. Γ_F) is the residue field (resp. the value group) of the restriction $v_D|_F$. In that case, v_D is also defectless over $Z(D)$, and $Z(D)$ is defectless over F, because of the fundamental inequalities (see Prop. 1.3)

$$[D:Z(D)] \geq [\overline{D}:\overline{Z(D)}]\cdot|\Gamma_D:\Gamma_{Z(D)}| \;\text{ and }\; [Z(D):F] \geq [\overline{Z(D)}:\overline{F}]\cdot|\Gamma_{Z(D)}:\Gamma_F|$$

and of the transitivity of dimensions and indices

$$[D{:}F] = [D{:}Z(D)] \cdot [Z(D){:}F], \qquad [\overline{D}{:}\overline{F}] = [\overline{D}{:}\overline{Z(D)}] \cdot [\overline{Z(D)}{:}\overline{F}],$$

and

$$|\Gamma_D{:}\Gamma_F| = |\Gamma_D{:}\Gamma_{Z(D)}| \cdot |\Gamma_{Z(D)}{:}\Gamma_F|.$$

In particular, if v_D is defectless over F, then $v_D|_F$ extends uniquely to $Z(D)$ (hence also to D, by Th. 1.4) because the fundamental equality (3.18) holds for $Z(D)/F$.

If $v_D|_F$ extends uniquely to $Z(D)$, we define the *defect* $\partial_{D/F}$ of D over F by

$$\partial_{D/F} = [D{:}F] \big/ \big([\overline{D}{:}\overline{F}] \cdot |\Gamma_D{:}\Gamma_F|\big).$$

In particular, $\partial_{D/Z(D)}$ is always defined; we call it simply the *defect of* D and use the simpler notation ∂_D for $\partial_{D/Z(D)}$. When $\partial_{D/F}$ is defined we have

$$\partial_{D/F} = \partial_D \cdot \partial_{Z(D)/F} \tag{4.9}$$

by the transitivity of dimensions and indices.

It is clear from the Fundamental Inequality (Prop. 1.3) that $\partial_{D/F} \geq 1$, but not at all clear at this point that the defect is an integer. The proof of this fact will occupy most of this section.

Theorem 4.3 (Morandi). *Let D be a finite-dimensional division algebra over a field F. Suppose v_D is a valuation on D such that $v_D|_F$ extends uniquely to $Z(D)$, and let $\overline{p} = \operatorname{char} \overline{F}$. Then $\partial_{D/F} = \overline{p}^{\ell}$ for some integer $\ell \geq 0$ if $\overline{p} \neq 0$, and $\partial_{D/F} = 1$ if $\overline{p} = 0$.*

An example where $\partial_{D/F} \neq 1$ is given in Ex. 4.15. Note that $\partial_{D/F} = 1$ when v_D is discrete of rank 1 and $F = Z(D)$, see Prop. 4.16. We may however have $\partial_{D/F} \neq 1$ with $rk(v_D) = 1$, see Ex. 4.18.

The commutative case of Th. 4.3 is known as *Ostrowski's Theorem*:

Proposition 4.4 (Ostrowski's Theorem). *Let L/F be a finite degree field extension, and let v be a valuation on F that extends uniquely to L. Let $\overline{p} = \operatorname{char} \overline{F}$. Then $\partial_{L/F} = \overline{p}^{\ell}$ for some integer $\ell \geq 0$ if $\overline{p} \neq 0$, and $\partial_{L/F} = 1$ if $\overline{p} = 0$.*

Proof. See Th. A.12. □

As a result of Prop. 4.4, it suffices to prove Th. 4.3 for ∂_D ($= \partial_{D/Z(D)}$), because it then follows for $\partial_{D/F}$ by (4.9) and Ostrowski's Theorem for $\partial_{Z(D)/F}$.

The main part of the proof of Th. 4.3 consists in establishing the result under the additional hypothesis that F is Henselian. The general case is derived from the Henselian case by scalar extension to a Henselization. Thus, the following is the crucial special case:

Theorem 4.5 (Draxl). *Let F be a field with Henselian valuation v. Let D be a central division F-algebra, and let $\overline{p} = \operatorname{char} \overline{F}$. Then, for the unique valuation v_D on D extending v, we have $\partial_D = \overline{p}^{\ell}$ for some integer $\ell \geq 0$ if $\overline{p} \neq 0$, and $\partial_D = 1$ if $\overline{p} = 0$.*

The next lemma will be used in the proof of Th. 4.5, and will have further use subsequently.

Suppose we have a valued division algebra D over a field F and a subfield S of $\overline{D}$ which is separable over $\overline{F}$. An *inertial lift* of S over F in D is a field L with $F \subseteq L \subseteq D$, $[L{:}F] = [S{:}\overline{F}] < \infty$, and $\overline{L} = S$ in $\overline{D}$, cf. Prop. A.17.

Lemma 4.6. *Let F be a field with a Henselian valuation v, and let D be a central division algebra over F. Let T be a maximal subfield of $\overline{D}$ such that T is separable over $Z(\overline{D})$, and let S be the separable closure of $\overline{F}$ in T. Then, there is an inertial lift L of S over F in D. For the centralizer $C = C_D(L)$, we have*

$$\overline{C} = T, \qquad \Gamma_C = \mathit{ker}(\theta_D), \qquad \textit{and} \qquad \partial_C = \partial_D,$$

where $\theta_D \colon \Gamma_D \to \mathit{Aut}(Z(\overline{D})/\overline{F})$ is the canonical homomorphism of (1.1).

Proof. Since S is separable over $\overline{F}$, we have $S = \overline{F}(s)$ for some s in S. Choose any $a \in \mathcal{O}_D$ with $\overline{a} = s$, and let $M = F(a)$, a subfield of D. Then, $S = \overline{F}(\overline{a}) \subseteq \overline{M}$. The valuation on M is Henselian, as $[M{:}F] < \infty$. Therefore, by Prop. A.17, there is an inertial lift L of S over F with $L \subseteq M$.

Let $Z = Z(\overline{D})$. Since S is the separable closure of $\overline{F}$ in T, it follows that $S \cap Z$ is the separable closure of $\overline{F}$ in Z. The field $S \cap Z$ is also a normal extension of $\overline{F}$ by Prop. 1.5. Hence, it is a Galois extension and its Galois group $\mathcal{G}(S \cap Z/\overline{F})$ consists of the restrictions to $S \cap Z$ of the automorphisms of $Z/\overline{F}$:

$$\mathit{Aut}(Z/\overline{F}) \xrightarrow{\sim} \mathcal{G}(S \cap Z/\overline{F}). \tag{4.10}$$

Recall from Prop. 1.5 that $\mathit{im}\,\theta_D = \mathit{Aut}(Z/\overline{F})$. Therefore,

$$|\mathit{im}\,\theta_D| = |\mathit{Aut}(Z/\overline{F})| = [S \cap Z : \overline{F}]. \tag{4.11}$$

Note that T is purely inseparable over S and separable over Z, hence it is purely inseparable and separable over $S{\cdot}Z$, and therefore $T = S{\cdot}Z$. Also, S is separable over $S \cap Z$ and Z is purely inseparable over $S \cap Z$; hence, S and Z are linearly disjoint over $S \cap Z$ and it follows that

$$[T{:}Z] = [S{\cdot}Z : Z] = [S : S \cap Z]. \tag{4.12}$$

Now, consider the centralizer $C = C_D(L)$. For any $c \in C^\times$, the automorphism of $\overline{D}$ induced by $\mathit{int}(c)$ must be the identity on $\overline{L} = S$, hence it is the identity on Z by (4.10). Therefore, $v(c) \in \mathit{ker}\,\theta_D$ and $\Gamma_C \subseteq \mathit{ker}\,\theta_D$. By (4.11) we have

$$|\Gamma_D{:}\Gamma_C| \geq |\Gamma_D{:}\, \mathit{ker}\,\theta_D| = [S \cap Z : \overline{F}]. \tag{4.13}$$

If $c \in \mathcal{O}_C^\times$, then the automorphism of $\overline{D}$ induced by $\mathit{int}(c)$ is $\mathit{int}(\overline{c})$. We have just seen that it is the identity on S and Z, hence also on $T = S{\cdot}Z$. Since T is a maximal subfield of $\overline{D}$, it follows that $\overline{C} \subseteq T$. Therefore, by (4.12),

$$[\overline{D}{:}\overline{C}] \geq [\overline{D}{:}T] = [T{:}Z] = [S : S \cap Z]. \tag{4.14}$$

By combining (4.13) and (4.14) with the Fundamental Inequality, we obtain

$$[D{:}C] \geq [\overline{D}{:}\overline{C}] \cdot |\Gamma_D{:}\Gamma_C| \geq [S : S \cap Z] \cdot [S \cap Z : \overline{F}] = [S{:}\overline{F}]. \qquad (4.15)$$

The Double Centralizer Theorem yields $[D{:}C] = [L{:}F]$; since $[L{:}F] = [S{:}\overline{F}]$, equality must hold throughout (4.15), hence also in (4.13) and (4.14). Therefore, $\Gamma_C = \mathit{ker}\,\theta_D$ and $\overline{C} = T$. Now, compute the defect:

$$\partial_D = \frac{[D{:}F]}{[\overline{D}{:}\overline{F}] \cdot |\Gamma_D{:}\Gamma_F|} = \frac{[D{:}C] \cdot [C{:}L] \cdot [L{:}F]}{[\overline{D}{:}\overline{C}] \cdot [\overline{C}{:}\overline{L}] \cdot [\overline{L}{:}\overline{F}] \cdot |\Gamma_D{:}\Gamma_C| \cdot |\Gamma_C{:}\Gamma_F|}.$$

Since $[L{:}F] = [\overline{L}{:}\overline{F}]$ and since equality holds throughout (4.15), this reduces to

$$\partial_D = [C{:}L] \big/ \big([\overline{C}{:}\overline{L}] \cdot |\Gamma_C{:}\Gamma_F|\big).$$

But $\Gamma_L = \Gamma_F$ since $[L{:}F] = [\overline{L}{:}\overline{F}]$, hence $\partial_D = \partial_C$. □

Corollary 4.7. *Let F be a field with a Henselian valuation v, and let D be a central division algebra over F. For any inertial field extension M/F (possibly of infinite degree), we have $\partial_D = \partial_{D_M}$, where D_M is the central division algebra over M associated to $D \otimes_F M$.*

Proof. Let F_i be the maximal inertial extension of F in a separable closure of M. So, $M \subseteq F_i$. We have $D_{F_i} = (D_M)_{F_i}$ since $D \otimes_F F_i = (D \otimes_F M) \otimes_M F_i$. Hence, if we show $\partial_D = \partial_{D_{F_i}}$, then by substituting D_M for D we also have $\partial_{D_M} = \partial_{(D_M)_{F_i}} = \partial_{D_{F_i}}$, hence $\partial_D = \partial_{D_M}$. Thus, it suffices to prove the corollary for $M = F_i$.

Let T, S, L, C be as in Lemma 4.6. We may assume $L \subseteq F_i$, hence $D_L = C$. We claim that $C \otimes_L F_i$ is a division algebra, hence $D_{F_i} = C \otimes_L F_i$. To prove the claim, we consider F_i as the union $\bigcup K$ of all the finite-degree extensions K/L with $K \subseteq F_i$. Each of these extensions is defectless and for the unique extension of v to such a field K we have

$$\Gamma_C \cap \Gamma_K = \Gamma_L = \Gamma_F$$

because K/F and L/F are inertial; moreover $\overline{C} \otimes_{\overline{L}} \overline{K} = T \otimes_S \overline{K}$, and this tensor product is a field since T/S is purely inseparable and $\overline{K}/S$ is separable. Therefore, Th. 3.43 shows that $C \otimes_L K$ is a division algebra and moreover, for the unique extension of v to $C \otimes_L K$ we have

$$\overline{C \otimes_L K} = T \otimes_S \overline{K} \qquad \text{and} \qquad \Gamma_{C \otimes K} = \Gamma_C + \Gamma_K = \Gamma_C.$$

By taking the union over all finite extensions K/L in F_i, we see that $C \otimes_L F_i$ is a division algebra, proving the claim that $D_{F_i} = C \otimes_L F_i$, and moreover

$$\overline{D_{F_i}} = T \otimes_S \overline{F_i} \qquad \text{and} \qquad \Gamma_{D_{F_i}} = \Gamma_C.$$

It follows that $[\overline{D_{F_i}}:\overline{F_i}] = [T:S] = [\overline{C}:\overline{L}]$ and $|\Gamma_{D_{F_i}}:\Gamma_{F_i}| = |\Gamma_C:\Gamma_L|$; hence,

$$\begin{aligned}\partial_{D_{F_i}} &= [D_{F_i}:F_i] \big/ \big([\overline{D_{F_i}}:\overline{F_i}] \cdot |\Gamma_{D_{F_i}}:\Gamma_{F_i}|\big) \\ &= [C:L] \big/ \big([\overline{C}:\overline{L}] \cdot |\Gamma_C:\Gamma_L|\big) = \partial_C.\end{aligned}$$

The corollary follows, since $\partial_C = \partial_D$ by Lemma 4.6. □

As a first step in the proof of Th. 4.5, we observe an easy consequence of Ostrowski's Theorem for fields:

Lemma 4.8. *Let F be a field with a Henselian valuation v, and let D be a central division F-algebra. Every prime divisor of $[\overline{D}:\overline{F}]$ or $|\Gamma_D:\Gamma_F|$ divides $[D:F]$.*

Proof. Let q be a prime divisor of $[\overline{D}:\overline{F}]$. There is an element $a \in \mathcal{O}_D$ such that $q \mid [\overline{F}(\overline{a}):\overline{F}]$. Let $L = F(a)$. Then $\overline{F}(\overline{a}) \subseteq \overline{L}$, so $q \mid [\overline{L}:\overline{F}]$. By Ostrowski's Theorem (Prop. 4.4), q also divides $[L:F]$ since $\partial_{L/F}$ is an integer; hence, $q \mid [D:F]$. Similarly, if q is a prime divisor of $|\Gamma_D:\Gamma_F|$, then there is an element $d \in D$ such that $v_D(d)$ has order q modulo Γ_F. For $M = F(d)$ we have $v_D(d) \in \Gamma_M$, hence $q \mid |\Gamma_M:\Gamma_F|$. By Ostrowski's Theorem, q also divides $[M:F]$ since $\partial_{M/F}$ is an integer; hence, $q \mid [D:F]$. □

We next prove an important special case of Th. 4.5:

Proposition 4.9. *Let F be a field with a Henselian valuation v, and let D be a central division algebra over F. Let $\overline{p} = \operatorname{char} \overline{F}$. If $\overline{p} \nmid [D:F]$ (in particular, if $\overline{p} = 0$), then $\partial_D = 1$.*

Proof. We argue by induction on $[D:F]$. Since $\partial_{F/F} = 1$, we may assume that $[D:F] > 1$. Let T be a maximal subfield of $\overline{D}$. By Lemma 4.8, $\overline{p} \nmid [\overline{D}:\overline{F}]$, hence T is separable over $\overline{F}$. Lemma 4.6 yields a subfield $L \subseteq D$ such that $\overline{L} = T$ and the centralizer $C = C_D(L)$ satisfies $\partial_C = \partial_D$. If $[C:L] < [D:F]$, then we are done by induction. Thus, we may assume that $[C:L] = [D:F]$, hence $L = F$ and therefore $\overline{D} = \overline{F}$. In particular,

$$\partial_D = [D:F] / |\Gamma_D:\Gamma_F|.$$

Take any prime divisor q of $[D:F]$. By hypothesis, we have $q \neq \overline{p}$. We adjust D to have a q-th root of unity in the center: let ω be a primitive q-th root of unity in the separable closure of F, and let $F' = F(\omega)$. The Henselian valuation v extends to a unique valuation on F'. The minimal polynomial f of ω over F has coefficients in $\mathcal{O}_F$, and its image $\overline{f}$ in $\overline{F}[X]$ is separable, since it divides $X^q - 1$ and $q \neq \overline{p}$. By Hensel's Lemma it follows that $\overline{f}$ is irreducible, hence F' is an inertial extension of F: we have $[F':F] = [\overline{F'}:\overline{F}]$ and $\Gamma_{F'} = \Gamma_F$ (and $\overline{F'}/\overline{F}$ is separable). Let $D' = D \otimes_F F'$. Since $\overline{D} = \overline{F}$ and $\Gamma_{F'} = \Gamma_F$, we have

$$\overline{D} \otimes_{\overline{F}} \overline{F'} = \overline{F'} \qquad \text{and} \qquad \Gamma_D \cap \Gamma_{F'} = \Gamma_F.$$

Moreover, $[F':F] = [\overline{F'}:\overline{F}] \cdot |\Gamma_{F'}:\Gamma_F|$. Therefore, Th. 3.43 shows that D' is a division algebra, and $v_D \otimes v$ is a valuation on D' with

$$\overline{D'} = \overline{D} \otimes_{\overline{F}} \overline{F'} = \overline{F'} \qquad \text{and} \qquad \Gamma_{D'} = \Gamma_D + \Gamma_{F'} = \Gamma_D.$$

The center of D' is F' and we have

$$\partial_{D'} = [D':F'] / ([\overline{D'}:\overline{F'}] \cdot |\Gamma_{D'}:\Gamma_{F'}|) = [D:F] / |\Gamma_D:\Gamma_F| = \partial_D.$$

Therefore, by substituting D' for D, we may assume for the rest of the proof that F contains a primitive q-th root of unity ω. We have $\overline{\omega} \neq 1$ since the equation $\sum_{i=0}^{q-1} \omega^i = 0$ yields $q = 0$ in $\overline{F}$ if $\overline{\omega} = 1$, contradicting the assumption that $q \neq \overline{p}$.

Let $M \subseteq D$ be a maximal subfield. Then, $\overline{p} \nmid [M:F]$ since $\overline{p} \nmid [D:F]$; hence, Ostrowski's Theorem (Prop. 4.4) yields $\partial_{M/F} = 1$. We have $\overline{F} \subseteq \overline{M} \subseteq \overline{D} = \overline{F}$, hence $\overline{M} = \overline{F}$. Therefore,

$$|\Gamma_M:\Gamma_F| = [M:F] = \sqrt{[D:F]}.$$

Since $q \mid [D:F]$, we may find an element $t_0 \in M$ such that $v_D(t_0)$ has order q in the factor group Γ_M/Γ_F. Let $a_0 \in F^\times$ satisfy $v_D(t_0^q) = v(a_0)$. Since $\overline{M} = \overline{F}$, we may find an $a_1 \in \mathcal{O}_F^\times$ such that $\overline{t_0^q a_0^{-1}} = \overline{a_1}$; hence, letting $a = a_0 a_1 \in F^\times$,

$$t_0^q = a(1+m) \qquad \text{for some } m \in \mathfrak{m}_M.$$

The polynomial $X^q - (1+m) \in \mathcal{O}_M[X]$ has image $X^q - 1$ in $\overline{M}[X]$. Since 1 is a simple root of $X^q - 1$ and M is Henselian, we may find a $t_1 \in M^\times$ such that $t_1^q = 1 + m$, hence $v_D(t_1) = 0$. For $t = t_0 t_1^{-1} \in M$, we have

$$t^q = a \in F^\times$$

and $v_D(t) = v_D(t_0)$ has order q in Γ_M/Γ_F. Therefore, for $N = F(t)$ we have

$$[N:F] = |\Gamma_N:\Gamma_F| = q,$$

and N carries an automorphism τ such that $\tau(t) = \omega t$.

Consider the centralizer $E = C_D(N)$. By the Double Centralizer Theorem, E has center N and $[D:F] = q^2[E:N]$. By induction, we have $\partial_E = 1$, hence $[E:N] = |\Gamma_E:\Gamma_N|$. Therefore,

$$\partial_D = \frac{[D:F]}{|\Gamma_D:\Gamma_F|} = \frac{q^2\,[E:N]}{|\Gamma_D:\Gamma_E| \cdot |\Gamma_E:\Gamma_N| \cdot |\Gamma_N:\Gamma_F|} = \frac{q}{|\Gamma_D:\Gamma_E|}.$$

Since $\partial_D \geq 1$ by the Fundamental Inequality, it now suffices to show that $|\Gamma_D:\Gamma_E| \geq q$ to complete the proof.

By the Skolem–Noether Theorem, the automorphism τ of N extends to an inner automorphism of D. We may thus find $s \in D^\times$ such that $sts^{-1} = \omega t$.

If $i \in \mathbb{Z}$ satisfies $v_D(s^i) \in \Gamma_E$, choose $e \in E^\times$ so that $v_D(s^i) = v_D(e)$. Since $\overline{D} = \overline{F}$, we may find $u \in \mathcal{O}_F^\times$ such that $\overline{s^i e^{-1}} = \overline{u}$, hence

$$s^i e^{-1} = u(1+m') \qquad \text{for some } m' \in \mathfrak{m}_D.$$

Since $ts^it^{-1} = \omega^{-i}s^i$ and t commutes with e, we have

$$ts^ie^{-1}t^{-1} = \omega^{-i}s^ie^{-1}.$$

By substituting $u(1+m')$ for s^ie^{-1}, we obtain

$$u(1+tm't^{-1}) = \omega^{-i}u(1+m').$$

By comparing the images of each side in the residue field, we find that

$$\overline{u} = \overline{\omega}^{-i}\,\overline{u},$$

which implies that i is a multiple of q since $\overline{\omega}$ is a primitive q-th root of unity in $\overline{F}$. Therefore, $v_D(s)$ has order at least q in the factor group Γ_D/Γ_E, and it follows that $|\Gamma_D{:}\Gamma_E| \geq q$, completing the proof. □

Proof of Th. 4.5. If $\overline{p} = 0$, the theorem follows from Prop. 4.9. If $\overline{p} \neq 0$, the primary decomposition of D yields a decomposition $D \cong P \otimes_F Q$ where P and Q are subalgebras of D with $[P{:}F]$ a power of $\overline{p}$ and $[Q{:}F]$ prime to $\overline{p}$. By Prop. 4.9, $[Q{:}F] = [\overline{Q}{:}\overline{F}] \cdot |\Gamma_Q{:}\Gamma_F|$, so $v_D|_Q$ is a gauge. On the other hand, Lemma 4.8 shows that $[\overline{P}{:}\overline{F}]$ and $|\Gamma_P{:}\Gamma_F|$ are powers of $\overline{p}$, so $\Gamma_P \cap \Gamma_Q = \Gamma_F$ and $\overline{P} \otimes_{\overline{F}} \overline{Q}$ is a division algebra. Therefore, Th. 3.43 applies to show that $v_D = v_D|_P \otimes v_D|_Q$, and that

$$\overline{D} = \overline{P} \otimes_{\overline{F}} \overline{Q} \qquad \text{and} \qquad \Gamma_D = \Gamma_P + \Gamma_Q.$$

It follows that

$$[\overline{D}{:}\overline{F}] = [\overline{P}{:}\overline{F}] \cdot [\overline{Q}{:}\overline{F}] \qquad \text{and} \qquad |\Gamma_D{:}\Gamma_F| = |\Gamma_P{:}\Gamma_F| \cdot |\Gamma_Q{:}\Gamma_F|,$$

hence $\partial_D = \partial_P \cdot \partial_Q$. Now, $\partial_Q = 1$ by Prop. 4.9, and $[P{:}F]$, $[\overline{P}{:}\overline{F}]$, $|\Gamma_P{:}\Gamma_F|$ are all powers of $\overline{p}$, hence

$$\partial_D = \partial_P = \overline{p}^\ell \qquad \text{for some } \ell \in \mathbb{Z}.$$

The Fundamental Inequality shows that $\ell \geq 0$, which completes the proof. □

We now give the proof in the general (non-Henselian) case:

Proof of Th. 4.3. As observed just after Prop. 4.4, we may assume that $F = Z(D)$. Let (F_h, v_h) be a Henselization of $(F, v_D|_F)$ and let $D_h = D \otimes_F F_h$. Theorem 4.1 shows that D_h is a division algebra, and that $\overline{D_h} = \overline{D}$ and $\Gamma_{D_h} = \Gamma_D$, hence

$$[\overline{D_h}{:}\overline{F_h}] = [\overline{D}{:}\overline{F}] \qquad \text{and} \qquad |\Gamma_{D_h}{:}\Gamma_{F_h}| = |\Gamma_D{:}\Gamma_F|.$$

Since moreover $[D_h{:}F_h] = [D{:}F]$, it is clear that $\partial_{D_h} = \partial_D$. Therefore, Th. 4.3 follows readily from Th. 4.5. □

Corollary 4.10. *Let D be a central division algebra over a field F. For any valuation on D, the residue degree $[\overline{D}{:}\overline{F}]$ and the ramification index $|\Gamma_D{:}\Gamma_F|$ each divide $[D{:}F]$.*

Proof. This follows at once from the fact that $\partial_D \in \mathbb{Z}$, as shown by Th. 4.3. □

Corollary 4.11. *Let D be a central division algebra over a field F. Suppose $D = A \otimes_F B$ where A, B are central division F-algebras of relatively prime degrees. Then for any valuation v_D on D,*

$$\overline{D} = \overline{A} \otimes_{\overline{F}} \overline{B}, \ \ \Gamma_D/\Gamma_F = (\Gamma_A/\Gamma_F) \times (\Gamma_B/\Gamma_F), \ \ \mathsf{gr}(D) = \mathsf{gr}(A) \otimes_{\mathsf{gr}(F)} \mathsf{gr}(B),$$

and

$$\partial_D = \partial_A \cdot \partial_B.$$

Proof. Since $\deg A$ and $\deg B$ are relatively prime, they cannot be both divisible by $\operatorname{char} \overline{F}$, hence at least one of A, B is defectless by Th. 4.3. Corollary 4.10 shows that $|\Gamma_A{:}\Gamma_F|$ and $|\Gamma_B{:}\Gamma_F|$ are relatively prime, so $\Gamma_A \cap \Gamma_B = \Gamma_F$, and also that $[\overline{A}{:}\overline{F}]$ and $[\overline{B}{:}\overline{F}]$ are relatively prime, so $\overline{A} \otimes_{\overline{F}} \overline{B}$ is a division ring. By Th. 3.43, the valuation v_D on D is the tensor product $v_D = v_D|_A \otimes v_D|_B$, hence $\mathsf{gr}(D) = \mathsf{gr}(A) \otimes_{\mathsf{gr}(F)} \mathsf{gr}(B)$. Theorem 3.43 also shows that $\overline{D} = \overline{A} \otimes_{\overline{F}} \overline{B}$ and $\Gamma_D = \Gamma_A + \Gamma_B$. The corollary follows. □

By applying Cor. 4.11 to the primary decomposition of D, we obtain:

Corollary 4.12. *Let D be a central division algebra over a field F. Let $p_1, \ldots, p_n$ be the distinct prime factors of $\deg D$, and let*

$$D = D_1 \otimes_F \ldots \otimes_F D_n$$

where D_i is a division subalgebra of p_i-power degree for $i = 1, \ldots, n$. Then for any valuation on D,

$$\overline{D} = \overline{D_1} \otimes_{\overline{F}} \ldots \otimes_{\overline{F}} \overline{D_n}, \qquad \Gamma_D/\Gamma_F = (\Gamma_{D_1}/\Gamma_F) \times \ldots \times (\Gamma_{D_n}/\Gamma_F),$$

$$\mathsf{gr}(D) = \mathsf{gr}(D_1) \otimes_{\mathsf{gr}(F)} \cdots \otimes_{\mathsf{gr}(F)} \mathsf{gr}(D_n), \quad \textit{and} \quad \partial_D = \partial_{D_1} \cdot \ldots \cdot \partial_{D_n}.$$

In Ex. 4.15, we will construct a valuation with nontrivial defect by composing a defectless valuation on a division algebra with commutative residue algebra and a defective valuation on the residue field. We first show that the defect is multiplicative for the composition of valuations.

Let D be a central division algebra over a field F, and let v be a valuation on D that is a composite $v = u * w$ of a valuation w on D and a valuation u on $\overline{D}^{w}$, see §1.2.3. Recall from Prop. 1.8 that in this case the valuation u has a unique extension from $\overline{F}^{w}$ to $Z(\overline{D}^{w})$, so the defect $\partial_{\overline{D}^{w}/\overline{F}^{w},u}$ is defined.

Proposition 4.13. $\partial_{D,v} = \partial_{\overline{D}^{w}/\overline{F}^{w},u} \cdot \partial_{D,w}$.

Proof. By definition of the defect, we have

$$[\overline{D}^{w}{:}\overline{F}^{w}] \;=\; [\overline{\overline{D}^{w}}^{u}{:}\overline{\overline{F}^{w}}^{u}]\cdot|\Gamma_{\overline{D}^{w},u}{:}\Gamma_{\overline{F}^{w},u}|\cdot\partial_{\overline{D}^{w}/\overline{F}^{w},u}$$

and

$$[D{:}F] \;=\; [\overline{D}^{w}{:}\overline{F}^{w}]\cdot|\Gamma_{D,w}{:}\Gamma_{F,w}|\cdot\partial_{D,w}.$$

Since $\overline{\overline{D}^{w}}^{u}=\overline{D}^{v}$ and $\overline{\overline{F}^{w}}^{u}=\overline{F}^{v}$, substituting the expression for $[\overline{D}^{w}{:}\overline{F}^{w}]$ in the second equation yields

$$[D{:}F] \;=\; [\overline{D}^{v}{:}\overline{F}^{v}]\cdot|\Gamma_{\overline{D}^{w},u}{:}\Gamma_{\overline{F}^{w},u}|\cdot|\Gamma_{D,w}{:}\Gamma_{F,w}|\cdot\partial_{\overline{D}^{w}/\overline{F}^{w},u}\cdot\partial_{D,w}. \tag{4.16}$$

The exact sequence (1.11) for $\Gamma_{D,v}$ yields an exact sequence

$$0 \longrightarrow \Gamma_{\overline{D}^{w},u}/\Gamma_{\overline{F}^{w},u} \longrightarrow \Gamma_{D,v}/\Gamma_{F,v} \longrightarrow \Gamma_{D,w}/\Gamma_{F,w} \longrightarrow 0.$$

It follows that $|\Gamma_{D,v}{:}\Gamma_{F,v}| = |\Gamma_{\overline{D}^{w},u}{:}\Gamma_{\overline{F}^{w},u}|\cdot|\Gamma_{D,w}{:}\Gamma_{F,w}|$, hence the proposition follows from (4.16). □

To give in Ex. 4.15 an example of a valued division algebra with nontrivial defect, we first recall an example of a field extension with nontrivial defect.

Example 4.14. (*A field extension with nontrivial defect.*) Let k be a field with *char* $k = p > 0$. In the Laurent series field $k((x))$, choose any element y that is transcendental over $k(x)$. For example, we could take $y = \sum_{i=1}^{\infty} x^{i!}$. Then y^p is also transcendental over $k(x)$. Let $K = k(x, y^p)$ and $M = k(x, y)$. We have $k(x) \subseteq K \subseteq M \subseteq k((x))$ and M is purely inseparable over K with $[M{:}K] = p$. Let v_x be the usual x-adic discrete valuation on $k((x))$. With respect to v_x, we have

$$k \;=\; \overline{k(x)} \;\subseteq\; \overline{K} \;\subseteq\; \overline{M} \;\subseteq\; \overline{k((x))} \;=\; k$$

and

$$\mathbb{Z} = \Gamma_{k(x)} \;\subseteq\; \Gamma_K \;\subseteq\; \Gamma_M \;\subseteq\; \Gamma_{k((x))} \;=\; \mathbb{Z},$$

hence $\overline{M} = \overline{K}$ and $\Gamma_M = \Gamma_K$. Because M is purely inseparable over K, $v_x|_M$ is the unique extension of $v_x|_K$ to M. Thus, $\partial_{M/K} = p$.

Example 4.15. *A division algebra with nontrivial defect.* Let $K \subseteq M$ be fields of characteristic $p > 0$ with M purely inseparable of degree p over K, say $M = K(a^{1/p})$. Let $F = K((t))$, and let v_t be the usual discrete t-adic valuation on F. Let $L = F(s)$, where $s^p - s = t^{-1}$; so L is an Artin–Schreier cyclic Galois extension of F with Galois group $\mathcal{G}(L/F)$ generated by $\sigma\colon L \to L$ given by $\sigma(s) = s + 1$. Let D be the cyclic algebra

$$D \;=\; (L/F, \sigma, a) \;=\; \bigoplus_{i=0}^{p-1} Lz^i$$

with $z^p = a$ and $z\ell z^{-1} = \sigma(\ell)$ for all $\ell \in L$. Thus, $(s^i z^j)_{i,j=0}^{p-1}$ is an F-base of D, and multiplication is defined by

$$s^p - s \;=\; t^{-1}, \qquad z^p \;=\; a, \qquad \text{and} \quad zs \;=\; (s+1)z,$$

and we have

$$s^i z^j \cdot s^k z^\ell \;=\; s^i(s+j)^k z^{j+\ell} \;=\; s^i\Big(\sum_{m=0}^{k} \tbinom{k}{m} s^m j^{k-m}\Big) z^{j+\ell}. \tag{4.17}$$

Define a v_t-norm α on D with splitting F-base $(s^i z^j)_{i,j=0}^{p-1}$ by setting

$$\alpha(s^i z^j) \;=\; -i/p \in \tfrac{1}{p}\mathbb{Z}.$$

Since $min_{0\leq m\leq k}(-(i+m)/p) = -(i+k)/p$, it follows from (4.17) that

$$\alpha(s^i z^j \cdot s^k z^\ell) \;=\; -(i+k)/p \;=\; \alpha(s^i z^j) + \alpha(s^k z^\ell).$$

Therefore, Lemma 3.23 shows that α is surmultiplicative. Observe that $\mathsf{gr}_{v_t}(F) = K[\widetilde{t}, \widetilde{t}^{-1}]$ and $\mathsf{gr}_\alpha(D) = \mathsf{gr}_{v_t}(F)[\widetilde{s}, \widetilde{z}]$, where

$$\widetilde{s}^p \;=\; \widetilde{(s^p - s)} \;=\; \widetilde{t}^{-1}, \qquad \widetilde{z}^p \;=\; a, \qquad \text{and} \quad \widetilde{z}\,\widetilde{s} \;=\; \widetilde{(s+1)}\widetilde{z} \;=\; \widetilde{s}\,\widetilde{z}.$$

Thus, $\mathsf{gr}_\alpha(D) \cong_g M[\widetilde{s}, \widetilde{s}^{-1}]$, which is a graded field. Since $\mathsf{gr}_\alpha(D)$ has no zero-divisors, D must be a division ring and α is a valuation on D. We have $\alpha|_F = v_t$, $\overline{D}^{\alpha} \cong M$, and $\Gamma_{D,\alpha} = \frac{1}{p}\mathbb{Z}$, while $\overline{F}^{v_t} = K$ and $\Gamma_{F,v_t} = \mathbb{Z}$. So, $\partial_{D,\alpha} = 1$.

Now, assume u is a valuation on M with $\partial_{M/K} = p$, as in Ex. 4.14. Since $Z(\overline{D}^{\alpha}) = \overline{D}^{\alpha} \cong M$ and the valuation u on $\overline{F}^{v_t} = K$ has a unique extension to $Z(\overline{D}^{\alpha})$ since M/K is a purely inseparable field extension, it follows from Prop. 1.8 that there is a composite valuation $v = u * \alpha$ on D. By Prop. 4.13, this valuation satisfies

$$\partial_{D,v} \;=\; \partial_{M/K,u} \cdot \partial_{D,\alpha} \;=\; p.$$

We next show that discrete valuations of rank 1 are defectless.

Proposition 4.16. *Let D be a finite-dimensional central division algebra over F. Suppose v_D is a valuation on D with $\Gamma_D \cong \mathbb{Z}$. Then v_D is defectless over F.*

Proof. Consider the completion $\widehat{D}$ of D for the topology induced by the valuation, and the continuous extension of v_D to a valuation $v_{\widehat{D}}$ of $\widehat{D}$. Then, $\widehat{D}$ contains the completion $\widehat{F}$ of F, and $v_{\widehat{F}} = v_{\widehat{D}}|_{\widehat{F}}$ is the continuous extension of $v_D|_F$ to $\widehat{F}$. Let $E = D \otimes_F \widehat{F}$, which is a central simple $\widehat{F}$-algebra. Because $\widehat{F}$ centralizes D in $\widehat{D}$, there is a ring homomorphism $E \to \widehat{D}$, which must be injective as E is simple. Therefore, we may view E as a subalgebra of $\widehat{D}$. With respect to the valuation $v_E = v_{\widehat{D}}|_E$ on E we have $\overline{D} \subseteq \overline{E} \subseteq \overline{\widehat{D}} = \overline{D}$ and $\Gamma_D \subseteq \Gamma_E \subseteq \Gamma_{\widehat{D}} = \Gamma_D$, so all these inclusions are equalities. Since also $\overline{\widehat{F}} = \overline{F}$, $\Gamma_{\widehat{F}} = \Gamma_F$, and $[E{:}\widehat{F}] = [D{:}F]$, we have $\partial_{E/\widehat{F}} = \partial_{D/F}$. But as $\Gamma_{\widehat{F}} = \Gamma_F \cong \mathbb{Z}$, $v_{\widehat{F}}$ is a complete discrete valuation, and hence maximally complete (see Ex. 3.11(i)). Therefore v_E is a $v_{\widehat{F}}$-norm by Prop. 3.8. Hence E is defectless over $\widehat{F}$, so D is defectless over F. □

We next give an example of a Galois extension K of a field F with rank 1 (nondiscrete) Henselian valuation with $\partial_{K/F} > 1$. We will use it to construct a division algebra with (nondiscrete) valuation of rank 1 with defect.

Example 4.17. (*Ostrowski*) Let $\mathbb{Q}_2$ be the 2-adic completion of $\mathbb{Q}$ with its complete discrete rank 1 (so Henselian) valuation v. So, $\overline{\mathbb{Q}_2} = \mathbb{F}_2$ and $\Gamma_{\mathbb{Q}_2} = \mathbb{Z}$. In an algebraic closure of $\mathbb{Q}_2$, choose $a_1, a_2, \ldots, a_n, \ldots$ with $a_1^2 = \frac{1}{2}$, $a_2^2 = a_1, \ldots, a_n^2 = a_{n-1}, \ldots$, and let

$$F = \bigcup_{n=1}^{\infty} \mathbb{Q}_2(a_n) = \bigcup_{n=1}^{\infty} \mathbb{Q}_2(\sqrt[2^n]{2}) \qquad \text{and} \qquad K = F(\sqrt{3}).$$

Let v denote also the unique extension of v to K. For each n, let $F_n = \mathbb{Q}_2(a_n)$ and $K_n = F_n(\sqrt{3})$. Since $a_n^{2^n} = \frac{1}{2}$ we have $v(a_n) = -1/2^n$, so F_n is totally ramified of degree 2^n over $\mathbb{Q}_2$ with $\Gamma_{F_n} = 2^{-n}\mathbb{Z}$ and $\overline{F_n} = \overline{\mathbb{Q}_2} = \mathbb{F}_2$. Therefore, as $F = \bigcup_{n=1}^{\infty} F_n$, we have $\Gamma_F = \bigcup_{n=1}^{\infty} 2^{-n}\mathbb{Z} = \mathbb{Z}[\frac{1}{2}]$ and $\overline{F} = \mathbb{F}_2$. Let

$$s_n = a_1 + a_2 + \ldots + a_n \in F_n, \quad \text{so} \quad v(s_n) = v(a_1) = -\tfrac{1}{2}.$$

Since $a_j^2 = a_{j-1}$ for $j > 1$, we have

$$s_n^2 = \tfrac{1}{2} + s_n + t, \quad \text{where} \quad t = -a_n + \sum_{i=1}^{n-1} \sum_{j=i+1}^{n} 2a_i a_j.$$

Then, $v(t) = v(-a_n) = -2^{-n}$, as $v(2a_ia_j) \geq 0$ for all i, j. Let

$$b = (1 + \sqrt{3})/2 \in \mathbb{Q}_2(\sqrt{3}), \quad \text{so} \quad b^2 = b + \tfrac{1}{2}.$$

Thus,

$$t = (s_n^2 - s_n) - (b^2 - b) = (s_n - b)(s_n + b - 1) = (s_n - b)((s_n - b) + \sqrt{3}).$$

If $v(s_n - b) \geq 0$, then as $v(\sqrt{3}) = 0$ we would have $v(t) \geq 0$. But $v(t) < 0$. Hence $v(s_n - b) < 0$, so $v(s_n - b + \sqrt{3}) = v(s_n - b)$, and

$$-2^{-n} = v(t) = v(s_n - b) + v(s_n - b + \sqrt{3}) = 2v(s_n - b).$$

Since $s_n - b \in K_n$ and $v(s_n - b) = -2^{-(n+1)} \notin \Gamma_{F_n}$, the field K_n strictly contains F_n, and K_n is totally ramified over F_n, with $\Gamma_{K_n} = 2^{-(n+1)}\mathbb{Z}$ and $\overline{K_n} = \overline{F_n} = \mathbb{F}_2$. Also, $\sqrt{3} \notin F$, as $\sqrt{3} \notin F_n$ for each n. So, $[K{:}F] = 2$ and, since $K = \bigcup_{n=1}^{\infty} K_n$, we have $\Gamma_K = \bigcup_{n=1}^{\infty} \Gamma_{K_n} = \mathbb{Z}[\frac{1}{2}] = \Gamma_F$ and $\overline{K} = \mathbb{F}_2 = \overline{F}$. Thus, K is an immediate extension of F and $\partial_{K/F} = [K{:}F] = 2$. Note that

$$\mathsf{gr}(K) = \mathsf{gr}(F) = \bigcup_{n=1}^{\infty} \mathbb{F}_2[T^{1/2^n}, T^{-1/2^n}] \quad \text{where} \quad T = \widetilde{2} \text{ and } \deg T = 1.$$

Example 4.18. *A valuation of rank 1 on a division algebra with nontrivial defect.* Let F_0, K_0, and v be the F, K, and v of the preceding example. Thus, $F_0 = \bigcup_{n=1}^{\infty} \mathbb{Q}_2(2^{1/n})$, $K_0 = F_0(i)$ where $i^2 = 3$, and v is a rank 1 Henselian valuation with $\overline{F_0} = \overline{K_0} = \mathbb{F}_2$ and $\Gamma_{F_0} = \Gamma_{K_0} = \mathbb{Z}[\frac{1}{2}]$. Let x be an indeterminate over K_0, and let $K = K_0(x)$ and $F = F_0(x)$. Extend v to the Gaussian valuation v on K determined by

$$v(a_0 + a_1 x + \ldots + a_n x^n) = \mathit{min}\big(v(a_0), \ldots, v(a_n)\big) \quad \text{for } a_0, \ldots, a_n \in K_0.$$

Then $\overline{K} = \mathbb{F}_2(\overline{x})$ with $\overline{x}$ transcendental over $\mathbb{F}_2$, $\Gamma_K = \Gamma_{K_0} = \mathbb{Z}[\frac{1}{2}]$,

$$\mathsf{gr}(K) = \bigcup_{n=1}^{\infty} \mathbb{F}_2(\overline{x})[T^{1/2^n}, T^{-1/2^n}], \quad \text{where } T = \widetilde{2} \text{ and } \mathit{deg}\, T = 1,$$

and K is an immediate extension of F.

Consider the quaternion algebra

$$D = \big(3, x/F\big) = K \oplus Kj,$$

with multiplication given by $j^2 = x$ and $ji = -ij$. Define the v-norm w on the left K-vector space D by

$$w(a + bj) = \mathit{min}\big(v(a), v(b)\big) \quad \text{for all } a, b \in K.$$

Then, $\mathsf{gr}(D)$ is a 2-dimensional graded $\mathsf{gr}(K)$-vector space with base $\big(\widetilde{1}, \widetilde{j}\big)$. Moreover, as $w(j^2) = 0 = 2w(j)$, w is surmultiplicative for products of elements in the splitting K-base $\big(1, j\big)$ of D. Since in addition $w(jaj^{-1}) = w(a)$ for all $a \in K$, the argument for Lemma 3.23 shows that w is surmultiplicative. Therefore, $\mathsf{gr}(D)$ is a graded ring, which is commutative as $\widetilde{j}$ centralizes $\mathsf{gr}(K)$ (since $v \circ \mathit{int}\, j = v$ and $\mathit{int}\, j$ induces the identity on $\overline{K}$). Because $\overline{x} \notin \mathsf{gr}(K)^2$ and $\widetilde{j}^2 = \overline{x}$, the ring $\mathsf{gr}(K)\big[\sqrt{\overline{x}}\big]$ is a graded field and there is a graded ring epimorphism $\mathsf{gr}(K)\big[\sqrt{\overline{x}}\big] \to \mathsf{gr}(D)$. This map is also injective since its domain is a graded field. Thus, $\mathsf{gr}(D)$ is a graded field; hence, by Cor. 3.30 D is a division ring and w is a valuation on D extending v on K and F. Moreover, as $\mathsf{gr}(F) = \mathsf{gr}(K)$, we have $\partial_D = [D{:}F]/[\mathsf{gr}(D){:}\,\mathsf{gr}(F)] = 2$.

Note that

$$D = \big(3, x\,/F\big) = \big(3, x\,/\mathbb{Q}_2(x)\big) \otimes_{\mathbb{Q}_2(x)} F,$$

but the valuation on D is *not* the tensor product of its restrictions to $D' = \big(-1, x/\mathbb{Q}_2(x)\big)$ and to F: this tensor product is a surmultiplicative norm by Prop. 3.27 but it is not a gauge since the associated graded algebra is not semisimple. Indeed, let $r = 1 + i \in D'$. Then, $v(r) = \frac{1}{2}$ as $r^2 = 2r + 2$. Thus, $\mathsf{gr}(D') = \mathbb{F}_2(\overline{x})[\widetilde{r}, \widetilde{r}^{-1}]\big[\sqrt{\overline{x}}\big]$, and in the commutative graded ring $\mathsf{gr}_{w|_{D'} \otimes v|_F}(D) = \mathsf{gr}(D') \otimes_{\mathsf{gr}(\mathbb{Q}_2(x))} \mathsf{gr}(F)$ we have

$$\big(\widetilde{r} \otimes 1 - 1 \otimes \widetilde{\sqrt{2}}\big)^2 = (\widetilde{r} \otimes 1)^2 - \big(1 \otimes \widetilde{\sqrt{2}}\big)^2 = \widetilde{2} - \widetilde{2} = 0.$$

4.2.2 Defectlessness of valuations in semisimple algebras

Throughout this subsection, let F be a field with valuation v. We now develop the notion of defectlessness of v in a (finite-dimensional) semisimple F-algebra A. This condition is significant in the theory of gauges, because we will see in Th. 4.50 below that A has a v-gauge if and only if v is defectless in A.

Definition 4.19. Let A be a (finite-dimensional) semisimple F-algebra, let (F_h, v_h) be a Henselization of (F, v), and let $A_h = A \otimes_F F_h$. Since F_h is a separable extension of F, the F_h-algebra A_h is semisimple, hence there is a decomposition into simple components

$$A_h \cong M_{n_1}(D_1) \times \ldots \times M_{n_r}(D_r)$$

for some integers $n_1, \ldots, n_r$ and some division algebras $D_1, \ldots, D_r$ over F_h. We say *v is defectless in A* if for each $i = 1, \ldots, r$ the unique valuation on D_i extending v_h is defectless over F, i.e., $\partial_{D_i/F_h} = 1$ for $i = 1, \ldots, r$.

It is clear from the definition that v is defectless in A if and only if v is defectless in each simple component of A, and that this condition holds if and only if v_h is defectless in A_h. We single out two particular cases:

- If K is a finite-degree field extension of F, then v is defectless in K if and only if equality holds in the Fundamental Inequality for extensions of v to K (see (A.1)). This follows readily from Th. A.32.
- If A is a central simple F-algebra, then v is defectless in A if and only if the valuation on the division algebra associated to A_h is defectless over F_h. In particular, if v is defectless in A, it is also defectless in every algebra that is Brauer-equivalent to A.

In simple algebras that are not central, we have the following reduction to the central case:

Proposition 4.20. *Let A be a simple F-algebra, and let $v_1, \ldots, v_r$ be the valuations on $Z(A)$ extending v. The following conditions are equivalent:*

(a) *v is defectless in A;*
(b) *v is defectless in $Z(A)$ and $v_1, \ldots, v_r$ are defectless in A.*

Proof. For each $i = 1, \ldots, r$, let $(Z_i, v_{i,h})$ be a Henselization of $(Z(A), v_i)$. Recall from Th. A.32 that

$$Z(A)_h \cong Z_1 \times \ldots \times Z_r.$$

Since $Z(A_h) = Z(A)_h$, the number of simple components of A_h is r, and we have division algebras $D_1, \ldots, D_r$ with centers $Z_1, \ldots, Z_r$ respectively such that

$$A_h \cong M_{n_1}(D_1) \times \ldots \times M_{n_r}(D_r)$$

for some integers $n_1, \ldots, n_r$. Now, we have $\partial_{D_i/F_h} = \partial_{D_i} \cdot \partial_{Z_i/F_h}$ (cf. (4.9)), hence $\partial_{D_i/F_h} = 1$ if and only if $\partial_{D_i} = 1$ and $\partial_{Z_i/F_h} = 1$. The equivalence of (a) and (b) follows. □

The following result lists several cases where defectlessness is automatic:

Proposition 4.21.

(i) *If* $\operatorname{char} \overline{F} = 0$, *then* v *is defectless in every semisimple* F*-algebra.*
(ii) *If* $\operatorname{char} \overline{F} = \overline{p} \neq 0$, *then* v *is defectless in every central simple* F*-algebra* A *whose index* $\operatorname{ind}(A)$ *is not divisible by* $\overline{p}$.
(iii) *If* $\Gamma_F \cong \mathbb{Z}$, *then* v *is defectless in any étale* F*-algebra and in any central simple* F*-algebra.*
(iv) *If* $\Gamma_F \cong \mathbb{Z}$ *and* A *is a semisimple* F*-algebra, then* v *is defectless in* A *if and only if it is defectless in* $Z(A)$.

Proof. Since v is defectless in an algebra A if and only if v_h is defectless in A_h, we may substitute (F_h, v_h) for (F, v) and thus assume (F, v) is Henselian. Then (i) readily follows from Ostrowski's Theorem, Prop. 4.4, and Draxl's Theorem, Th. 4.5. Likewise, (ii) follows from Prop. 4.9. For the rest of the proof, assume v is a discrete (rank 1) valuation. Such a valuation is defectless in any separable field extension (see Engler–Prestel [73, Th. 3.3.5, p. 76]); hence v is also defectless in any étale F-algebra E, as E is a direct product of fields separable over F. It is also defectless in any central division algebra by Prop. 4.16, hence also in any central simple algebra, which proves (iii). Finally, in case (iv) we may restrict to simple components and thus assume A is a simple F-algebra. The extension w of v to $Z(A)$ is also a discrete valuation, hence w is defectless in A by (iii). Therefore, (iv) follows from Prop. 4.20. □

Lemma 4.22. *Let* w *be any coarsening of the valuation* v *on* F, *and let* K *be a finite-degree field extension of* F. *If* v *is defectless in* K, *then* w *is defectless in* K.

Proof. Let $w_1, \ldots, w_\ell$ be the extensions of w to K. For $j = 1, \ldots, \ell$, let $v_{1j}, v_{2j}, \ldots, v_{k_j j}$ be the extensions of v to K that are refinements of w_j. Thus, the v_{ij} are all the extensions of v to K. Let v/w denote the valuation on $\overline{F}^{w}$ induced by v and likewise v_{ij}/w_j the valuation on $\overline{K}^{w_j}$ induced by v_{ij}. Note that $v_{1j}/w_j, v_{2j}/w_j, \ldots, v_{k_j j}/w_j$ are all the extensions of v/w to $\overline{K}^{w_j}$. For each i, j there is a commutative diagram of value groups with exact rows:

$$\begin{array}{ccccccccc}
0 & \longrightarrow & \Gamma_{\overline{F}^{w}, v/w} & \longrightarrow & \Gamma_{F,v} & \longrightarrow & \Gamma_{F,w} & \longrightarrow & 0 \\
 & & \downarrow & & \downarrow & & \downarrow & & \\
0 & \longrightarrow & \Gamma_{\overline{K}^{w_j}, v_{ij}/w_j} & \longrightarrow & \Gamma_{K,v_{ij}} & \longrightarrow & \Gamma_{K,w_j} & \longrightarrow & 0
\end{array}$$

Because the map $\Gamma_{F,w} \to \Gamma_{K,w_j}$ is injective, the Five-Lemma yields a short exact sequence of cokernels of the columns:

$$0 \longrightarrow \Gamma_{\overline{K}^{w_j},v_{ij}/w_j}/\Gamma_{\overline{F}^{w},v/w} \longrightarrow \Gamma_{K,v_{ij}}/\Gamma_{F,v} \longrightarrow \Gamma_{K,w_j}/\Gamma_{F,w} \longrightarrow 0.$$

Hence,

$$\big|\Gamma_{K,v_{ij}} : \Gamma_{F,v}\big| = \big|\Gamma_{\overline{K}^{w_j},v_{ij}/w_j} : \Gamma_{\overline{F}^{w},v/w}\big|\,\big|\Gamma_{K,w_j} : \Gamma_{F,w}\big|. \tag{4.18}$$

Since v is defectless in K, we have

$$[K{:}F] = \sum_{j=1}^{\ell}\sum_{i=1}^{k_j}\big[\overline{K}^{v_{ij}} : \overline{F}^{v}\big]\,\big|\Gamma_{K,v_{ij}} : \Gamma_{F,v}\big|.$$

Equation (4.18) together with the Fundamental Inequality (Th. A.2) for each $\overline{K}^{w_j}$ over $\overline{F}^{w}$ and for K over F then yield

$$\begin{aligned}
[K{:}F] &= \sum_{j=1}^{\ell}\sum_{i=1}^{k_j}\big[\overline{K}^{v_{ij}} : \overline{F}^{v}\big]\,\big|\Gamma_{\overline{K}^{w_j},v_{ij}/w_j} : \Gamma_{\overline{F}^{w},v/w}\big|\,\big|\Gamma_{K,w_j} : \Gamma_{F,w}\big| \\
&= \sum_{j=1}^{\ell}\Big(\sum_{i=1}^{k_j}\big[\overline{\overline{K}^{w_j}}^{\,v_{ij}/w_j} : \overline{\overline{F}^{w}}^{\,v/w}\big]\,\big|\Gamma_{\overline{K}^{w_j},v_{ij}/w_j} : \Gamma_{\overline{F}^{w},v/w}\big|\Big)\big|\Gamma_{K,w_j} : \Gamma_{F,w}\big| \\
&\le \sum_{j=1}^{\ell}\big[\overline{K}^{w_j} : \overline{F}^{w}\big]\,\big|\Gamma_{K,w_j} : \Gamma_{F,w}\big| \le [K{:}F].
\end{aligned}$$

The last inequality must therefore be an equality, showing that w is defectless in K. □

Proposition 4.23. *Let w be any coarsening of the valuation v on F. Let A be a semisimple F-algebra. If v is defectless in A, then w is defectless in A.*

Proof. Assume first that A is central simple over F. We write $(F_{h,v}, v_h)$ for a Henselization of (F,v). Let w' be the valuation on $F_{h,v}$ with ring $\mathcal{O}_{F,w}\cdot\mathcal{O}_{F_{h,v},v_h}$. So, w' is the extension of w which is a coarsening of v_h. Since v_h is Henselian, its coarsening w' is also Henselian by Prop. A.15, so $(F_{h,v}, w')$ contains a Henselization $(F_{h,w}, w_h)$ of (F,w). Recall from Prop. A.31 that w' is inertial over w_h. Let $D_{h,v}$ (resp. $D_{h,w}$) be the central division algebra over $F_{h,v}$ (resp. $F_{h,w}$) associated to $A\otimes_F F_{h,v}$ (resp. $A\otimes_F F_{h,w}$). Since v is defectless in A, $D_{h,v}$ is defectless for v_h; it is then also defectless for the coarser valuation w' by Prop. 4.13. Then, by Cor. 4.7 applied to the inertial extension $(F_{h,v}, w')$ of $(F_{h,w}, w_h)$ with $D = D_{h,w}$ (so $D_M = D_{h,v}$), $D_{h,w}$ is defectless for w_h. Hence, w is defectless in A, as desired.

Now assume only that A is simple. Let $K = Z(A)$, and let $v_1, \ldots, v_r$ be the extensions of v to K, and $w_1, \ldots, w_\ell$ the extensions of w to K. For $i \in \{1,2,\ldots,r\}$, let $j(i) \in \{1,2,\ldots,\ell\}$ be the index such that $w_{j(i)}$ is the w-coarsening of v_i. Since v is defectless in A, v is defectless in K and each v_i is defectless in A. Then, w is defectless in K by Lemma 4.22, and each $w_{j(i)}$ is defectless in A by the central simple case just considered. Since $w_{j(1)}, \ldots, w_{j(r)}$ are all the extensions of w to K, it follows by definition that

w is defectless in A. This completes the proof for A simple, and the general case for A semisimple follows easily by considering simple components of A. □

4.3 The structure of gauges

In this section we prove some key theorems on the structure of gauges. When a valuation v on a field F is Henselian we show in Th. 4.26 that all v-gauges on simple F-algebras are *End*-gauges, i.e., gauges on endomorphism algebras determined by a norm on a vector space over a valued division algebra (see Prop. 3.34). This enables us to see that even if v is not Henselian, any gauge on a central simple F-algebra A is closely related to the valuation on the associated division algebra of the scalar extension of A to the Henselization of F; see Th. 4.30. These are pivotal theorems in our general approach of using gauges on simple algebras to study valuations on division algebras. In addition, we will characterize the v-gauges on a finite-degree field extension K of F as "*min*-gauges" with respect to the extensions of v to K. We show likewise that v-gauges on simple but not central simple algebras are *min*-gauges with respect to gauges for the extensions of v to the center of the algebra.

4.3.1 Gauges on algebras over Henselian fields

Our goal in this subsection is to relate gauges on semisimple algebras over a Henselian field to valuations on the division algebras associated to the simple factors.

Our first result does not use the Henselian hypothesis. Let (F, v) be a valued field and let A be a finite-dimensional F-algebra. Let $\alpha\colon A \to \Gamma \cup \{\infty\}$ be a surmultiplicative v-value function on A and let O_A be the corresponding ring (as in §3.2.1):

$$O_A = \{a \in A \mid \alpha(a) \geq 0\}.$$

Lemma 4.24. *If α is a v-norm, then the ring O_A is integral over the valuation ring O_F.*

Proof. Let $B = End_F A$ and let $\ell_\bullet\colon A \hookrightarrow B$ be the regular representation, which maps every $a \in A$ to $\ell_a\colon x \mapsto ax$. Since α is a v-norm on the F-vector space A, we may consider the v-gauge $\beta = End(\alpha)$ on B, see Prop. 3.34. Since α is surmultiplicative, we have

$$\alpha(\ell_a(x)) - \alpha(x) \geq \alpha(a) \qquad \text{for all } a, x \in A,$$

hence $\beta(\ell_a) \geq \alpha(a)$. Therefore, the monomorphism $\ell_\bullet$ maps $\mathcal{O}_A$ into $\mathcal{O}_B$. To prove the lemma, it suffices to show that $\mathcal{O}_B$ is integral over $\mathcal{O}_F$.

Let $(a_i)_{i=1}^n$ be an F-base of A that splits the v-norm α. For $f \in B$ with matrix $(f_{ij})_{i,j=1}^n$ relative to the base $(a_i)_{i=1}^n$, i.e., $f(a_j) = \sum_{i=1}^n a_i f_{ij}$ for $j = 1, \ldots, n$, we have

$$\beta(f) = \min_{1\leq j\leq n}\big(\alpha(f(a_j)) - \alpha(a_j)\big) = \min_{1\leq i,j\leq n}\big(\alpha(a_i) + v(f_{ij}) - \alpha(a_j)\big).$$

Therefore, $f \in \mathcal{O}_B$ if and only if $v(f_{ij}) \geq \alpha(a_j) - \alpha(a_i)$ for all $i, j = 1, \ldots, n$. To complete the proof, it suffices to show that this condition implies that the coefficients of the characteristic polynomial P of f lie in $\mathcal{O}_F$. To see this, adjoin an indeterminate X to F and extend the valuation v to a valuation on $F(X)$ defined by

$$v(c_0 + c_1X + \ldots + c_nX^n) = \min\big(v(c_0), \ldots, v(c_n)\big) \qquad \text{for } c_0, \ldots, c_n \in F.$$

For $i, j = 1, \ldots, n$, let

$$f'_{ij} = \begin{cases} X - f_{ii} & \text{if } i = j, \\ -f_{ij} & \text{if } i \neq j, \end{cases}$$

so the characteristic polynomial of f is $P = \det(f'_{ij})$. Observe that $v(f'_{ij}) \geq \alpha(a_j) - \alpha(a_i)$ for all i, j; hence, for any permutation π of $\{1, \ldots, n\}$ we have

$$v(f'_{1\pi(1)} \cdots f'_{n\pi(n)}) \geq \sum_{j=1}^n \alpha(a_{\pi(j)}) - \sum_{i=1}^n \alpha(a_i) = 0.$$

Since $\det(f'_{ij}) = \sum_\pi \operatorname{sgn}(\pi) f'_{1\pi(1)} \cdots f'_{n\pi(n)}$, it follows that $v(P) \geq 0$, which means that all the coefficients of P lie in $\mathcal{O}_F$. □

Until the end of this subsection, we assume the valuation v on F is Henselian. This hypothesis is mainly invoked to allow use of the following idempotent-lifting result:

Proposition 4.25. *Suppose (F, v) is Henselian, and let A be a finite-dimensional F-algebra with a surmultiplicative v-value function α that is a norm. Let $J \subset \mathcal{O}_A$ be a two-sided ideal. For any $a \in \mathcal{O}_A$ such that $a^2 - a \in J$, there exists an idempotent $e \in \mathcal{O}_A$ such that $e - a \in J$.*

Proof. Let $R = \mathcal{O}_F[a] \subseteq \mathcal{O}_A$; it is a commutative ring that is finitely generated as an $\mathcal{O}_F$-module, since a is integral over $\mathcal{O}_F$ by Lemma 4.24. Since $\mathcal{O}_F$ is a Henselian local ring, by Th. 43.15 in Nagata's book [168, p. 185] (see also Raynaud [194, Prop. 5, p. 2]) R is a direct product

$$R = S_1 \times \ldots \times S_k$$

where each S_i is a local ring. Let $J \cap R = K_1 \times \ldots \times K_k$, so

$$R/(J \cap R) = (S_1/K_1) \times \ldots \times (S_k/K_k).$$

Since each quotient S_i/K_i is a local ring, it has no nontrivial idempotent, hence the image of a in $R/(J \cap R)$ is a k-tuple whose entries are 0 or 1. Therefore, there is an idempotent $e \in R$ such that $e \equiv a \pmod{J \cap R}$. □

The main result of this section is next. We show that every gauge on a simple algebra over a Henselian field is an *End*-gauge. This is a very strong classification result. It yields a close connection between the structure of the gauge and that of the valuation on the associated division algebra. Recall from Th. 1.4 that the Henselian valuation v has a unique extension to each finite-dimensional division algebra over F.

Theorem 4.26. *Let F be a field with a Henselian valuation v. Let A be a simple F-algebra with a v-gauge α. Let D be the associated division algebra of A, and let v_D be the valuation on D extending v. Then $\mathsf{gr}_\alpha(A)$ is graded simple and D is defectless over F. Furthermore, given an identification $A = \operatorname{End}_D M$ for some D-vector space M, there is a v_D-norm α_M on M such that*

$$\alpha = End(\alpha_M) \qquad \text{and} \qquad \mathsf{gr}_\alpha(A) = \mathsf{End}_{\mathsf{gr}(D)} \mathsf{gr}_{\alpha_M}(M).$$

The proof of Th. 4.26 will invoke the following easy but very useful observation:

Lemma 4.27. *Let F be a field with a valuation v, and let A be an F-algebra with a surmultiplicative v-value function α. If $e,\ f \in A$ are idempotents such that $\alpha(e) = \alpha(f) = 0$, then*

$$\mathsf{gr}(eAf) = \widetilde{e}\,\mathsf{gr}(A)\widetilde{f}. \tag{4.19}$$

Proof. For any $x \in V$ we have

$$\widetilde{e}\widetilde{x}\widetilde{f} = \begin{cases} \widetilde{exf} & \text{if } \alpha(exf) = \alpha(e) + \alpha(x) + \alpha(f), \\ 0 & \text{if } \alpha(exf) > \alpha(e) + \alpha(x) + \alpha(f), \end{cases}$$

hence $\widetilde{e}\,\mathsf{gr}(A)\widetilde{f} \subseteq \mathsf{gr}(eAf)$. On the other hand, since e and f are idempotents and $\alpha(e) = \alpha(f) = 0$, we have

$$\alpha(exf) = \alpha\big(e(exf)f\big) = \alpha(e) + \alpha(exf) + \alpha(f).$$

Therefore,

$$[e(exf)f]^{\sim} = \widetilde{e}\,\widetilde{(exf)}\widetilde{f}.$$

It follows that $\widetilde{(exf)} \in \widetilde{e}\,\mathsf{gr}(A)\widetilde{f}$, hence $\mathsf{gr}(eAf) \subseteq \widetilde{e}\,\mathsf{gr}(A)\widetilde{f}$. □

Proof of Th. 4.26. We first show that $\mathsf{gr}_\alpha(A)$ is graded simple. If it is not, then it contains a nontrivial central homogeneous idempotent e_0, since it is graded semisimple by the hypothesis that α is a gauge. Since $e_0^2 = e_0$, this idempotent necessarily has degree 0, so $e_0 \in \mathsf{gr}_\alpha(A)_0 = O_A/\mathfrak{m}_A$. By Prop. 4.25, there is an idempotent $e \in A$ with $\widetilde{e} = e_0$ (hence $\alpha(e) = 0$). For the complementary idempotent $1-e$, we have $\widetilde{1-e} = 1-\widetilde{e}$ since $\alpha(e)=0$ and $\widetilde{e}\neq 1$. So, $\alpha(1-e)=0$, and by Lemma 4.27,

$$\mathsf{gr}_\alpha\big(eA(1-e)\big) \;=\; \widetilde{e}\,\mathsf{gr}_\alpha(A)\widetilde{1-e} \;=\; e_0\,\mathsf{gr}_\alpha(A)(1-e_0).$$

The right expression equals $\{0\}$ since e_0 is central in $\mathsf{gr}_\alpha(A)$ and $e_0(1-e_0)=0$. It follows that $eA(1-e)=\{0\}$ and, likewise, $(1-e)Ae=\{0\}$. Hence, we have

$$A \;\cong\; eAe\oplus(1-e)A(1-e) \;\cong\; eAe\times(1-e)A(1-e),$$

contradicting the simplicity of A. Thus, $\mathsf{gr}_\alpha(A)$ must be simple.

Now, let $e_1\in\mathsf{gr}_\alpha(A)$ be any primitive homogeneous idempotent. Just as for e_0 above, we may find an idempotent $e\in A$ with $\alpha(e)=0$ and $\widetilde{e}=e_1$. The restriction $\alpha|_{eAe}$ is a surmultiplicative v-value function on the finite-dimensional F-algebra eAe, and we have $\mathsf{gr}(eAe)=\widetilde{e}\,\mathsf{gr}(A)\widetilde{e}$ by Lemma 4.27. The latter is a graded division algebra since $\widetilde{e}=e_1$ is a primitive homogeneous idempotent. By Cor. 3.30, it follows that eAe is a division algebra and $\alpha|_{eAe}$ is a valuation. Hence, e is a primitive idempotent of A, and we may identify eAe with D, and $\alpha|_{eAe}$ with v_D. Since the gauge α is a v-norm on A, its restriction to D also is a v-norm by Prop. 3.14, so D is defectless over F.

If M is a D-vector space such that $A\cong \mathit{End}_D\,M$, we may identify M with Ae so the isomorphism is given by the map

$$\rho\colon A\longrightarrow \mathit{End}_D\,M \quad\text{defined by}\quad \rho(a)\colon m\mapsto am.$$

Let $\alpha_M=\alpha|_{Ae}$, which is a v-norm on M by Prop. 3.14. By Lemma 4.27, we have $\mathsf{gr}_{\alpha_M}(M)=\mathsf{gr}_\alpha(A)\widetilde{e}$, which is a (unital) graded right module over $\widetilde{e}\,\mathsf{gr}_\alpha(A)\widetilde{e}=\mathsf{gr}_{v_D}(D)$. For $m\in M$ and $d\in D$ nonzero, we have $\widetilde{m}\widetilde{d}\neq 0$ since $\widetilde{d}$ is invertible; hence,

$$\alpha_M(md) \;=\; \alpha_M(m)+\alpha(d) \;=\; \alpha_M(m)+v_D(d).$$

This shows that α_M is actually a v_D-value function on M. Furthermore, we have

$$[\mathsf{gr}(M)\colon\mathsf{gr}(D)] = |\mathsf{gr}(M)\colon\mathsf{gr}(F)|/[\mathsf{gr}(D)\colon\mathsf{gr}(F)] = [M{:}F]/[D{:}F] = [M{:}D];$$

hence, α_M is a v_D-norm on M.

We next prove that the gauge α corresponds to $\mathit{End}(\alpha_M)$ under the isomorphism ρ. Let $\beta=\mathit{End}(\alpha_M)$. For $a\in A$ we have

$$\beta\big(\rho(a)\big) \;=\; \mathit{min}\{\alpha_M(am)-\alpha_M(m)\mid m\in M,\ m\neq 0\},$$

hence $\beta\big(\rho(a)\big)\geq\alpha(a)$ since $\alpha(am)\geq\alpha(a)+\alpha(m)$ for all m. If $a\neq 0$ and $\beta\big(\rho(a)\big)>\alpha(a)$, then $\alpha(am)>\alpha(a)+\alpha(m)$ for all $m\in M$, hence $\widetilde{a}\cdot\widetilde{m}=0$ in $\mathsf{gr}(M)$. Since $\mathsf{gr}(M)$ is spanned by its homogeneous elements, it follows that $\widetilde{a}\cdot\mathsf{gr}(M)=\{0\}$. Now, $\mathsf{gr}(M)=\mathsf{gr}(A)\cdot\widetilde{e}$, and $\mathsf{gr}(A)\cdot\widetilde{e}\cdot\mathsf{gr}(A)=\mathsf{gr}(A)$ since $\mathsf{gr}(A)$ is graded simple. Therefore, $\widetilde{a}\cdot\mathsf{gr}(M)=\{0\}$ implies

$$\widetilde{a}\cdot\mathsf{gr}(A) \;=\; \widetilde{a}\cdot\mathsf{gr}(A)\cdot\widetilde{e}\cdot\mathsf{gr}(A) \;=\; \{0\},$$

which shows $\widetilde{a} = 0$, contradicting $a \neq 0$. Thus, $\beta\big(\rho(a)\big) = \alpha(a)$ for all $a \in A$, and α corresponds to $End(\alpha_M)$ under the isomorphism ρ. By Prop. 3.34, ρ induces a graded isomorphism

$$\mathsf{gr}_\alpha(A) \cong_g \mathsf{End}_{\mathsf{gr}(D)}\big(\mathsf{gr}_{\alpha_M}(M)\big).$$

The proof of Th. 4.26 is thus complete. □

Because of the compatibility of gauges with direct products (see Prop. 3.36), Th. 4.26 extends easily to semisimple algebras. In the next statement, we use the notation $ms(A)$ to denote the *matrix size* of a simple (or simple graded) algebra A, i.e., when $A \cong End_D\, M$ for some division algebra D and some D-vector space M,

$$ms(A) = dim_D\, M; \tag{4.20}$$

so, $A \cong M_{ms(A)}(D)$.

Theorem 4.28. *Let F be a field with a Henselian valuation v, and let A be a semisimple F-algebra with a v-gauge α. Let $A_1, \ldots, A_n$ be the simple components of A,*

$$A = A_1 \times \ldots \times A_n.$$

For $i = 1, \ldots, n$, the restriction $\alpha_i = \alpha|_{A_i}$ is a v-gauge on A_i, the graded algebra $\mathsf{gr}_{\alpha_i}(A_i)$ is graded simple with $\mathrm{ms}(A_i) = \mathrm{ms}\big(\mathsf{gr}_{\alpha_i}(A_i)\big)$, and $\mathsf{gr}(A_1), \ldots, \mathsf{gr}(A_n)$ are the graded simple components of $\mathsf{gr}(A)$,

$$\mathsf{gr}_\alpha(A) = \mathsf{gr}_{\alpha_1}(A_1) \times \ldots \times \mathsf{gr}_{\alpha_n}(A_n).$$

Moreover, for $a = (a_1, \ldots, a_n) \in A_1 \times \ldots \times A_n$, we have

$$\alpha(a) = \min_{1 \le i \le n}\big(\alpha_i(a_i)\big).$$

Proof. This is immediate from Th. 4.26 and Prop. 3.36. □

Corollary 4.29. *Let F be a field with a Henselian valuation v. For any semisimple F-algebra A, the following conditions are equivalent:*

(a) *A carries a v-gauge;*
(b) *A carries a v-gauge with $\Gamma_A \subseteq \mathbb{H}(\Gamma_F)$;*
(c) *v is defectless in A, i.e., the associated division algebras of the simple components of A are each defectless over F.*

In particular, all these conditions hold if $\operatorname{char} \overline{F} = 0$ or if both $\Gamma_F \cong \mathbb{Z}$ and v is defectless in $Z(A)$.

Proof. If A carries a v-gauge, then by Th. 4.28 each of its simple components carries a v-gauge, and the corresponding division algebras are defectless by Th. 4.26, hence (a) ⇒ (c). Conversely, if the division algebras are defectless, their valuations extending v are v-gauges with value group lying in $\mathbb{H}(\Gamma_F)$, as formula (1.9) shows. Therefore, the simple components of A carry a v-gauge by Prop. 3.34, and the value sets of these gauges can be assumed to be the value groups of the corresponding division algebras by Cor. 3.35. Then A also carries a v-gauge with $\Gamma_A \subseteq \mathbb{H}(\Gamma_F)$ by Prop. 3.36, proving (c) $\Rightarrow$ (b). Since (b) $\Rightarrow$ (a) is obvious, we have the equivalence of (a) – (c). Proposition 4.21(i) and (iv) show that if $char\,\overline{F} = 0$ or $\Gamma_F \cong \mathbb{Z}$ and v is defectless in $Z(A)$, then condition (c) holds. □

We will prove in Th. 4.50 below that the preceding corollary holds without the assumption that v is Henselian.

Examples of gauges with $\Gamma_A \not\subseteq \mathbb{H}(\Gamma_F)$ are easy to construct using Prop. 3.34. For example, the norm $\alpha_{\gamma,-\gamma}$ on $A = M_2(F)$ in Ex. 3.25 has $\Gamma_A \not\subseteq \mathbb{H}(\Gamma_F)$ if $\gamma \notin \mathbb{H}(\Gamma_F)$.

4.3.2 Gauges on central simple algebras

Throughout this subsection, let (F, v) be a valued field, and let (F_h, v_h) be its Henselization. If α is a gauge on a central simple F-algebra A, then by Prop. 3.42 $\alpha_h = \alpha \otimes v_h$ is a v_h-gauge on $A \otimes_F F_h$, and, as $\mathsf{gr}_{v_h}(F_h) = \mathsf{gr}_v(F)$,

$$\mathsf{gr}_{\alpha_h}(A \otimes_F F_h) \cong_g \mathsf{gr}_\alpha(A) \otimes_{\mathsf{gr}_v(F)} \mathsf{gr}_{v_h}(F_h) \cong_g \mathsf{gr}_\alpha(A). \tag{4.21}$$

When this is combined with Th. 4.26, we can see that the structure of $\mathsf{gr}_\alpha(A)$ strongly reflects the structure of the valuation on the associated division algebra of $A \otimes_F F_h$. In addition, this yields a criterion for when v extends to a valuation on the associated division algebra of A; this criterion can be applied using *any* v-gauge on A.

Theorem 4.30. *Let A be a central simple F-algebra with a v-gauge α. Let D be the associated division algebra of A, let E be the associated division algebra of $A \otimes_F F_h$, and let w be the valuation on E extending v_h on F_h. Then,*

(i) $\mathsf{gr}_\alpha(A)$ *is a simple graded algebra, and* $\mathsf{gr}_w(E)$ *is its associated graded division algebra. Hence,* $Z(\mathsf{gr}_\alpha(A)) \cong_g Z(\mathsf{gr}_w(E))$ *and* $\overline{E}$ *is the associated division algebra of each simple component of* A_0*. Also,* $\Gamma_E \subseteq \Gamma_A$*, with equality holding if and only if* A_0 *is simple.*

(ii) *We have*

$$ms(\mathsf{gr}_\alpha(A)) = ms(A \otimes_F F_h) \geq ms(A).$$

Moreover, v extends to a valuation on D if and only if $ms(\mathsf{gr}_\alpha(A)) = ms(A)$.

Proof. (i) Since the extension $\alpha_h = \alpha \otimes v_h$ of α to $A \otimes_F F_h$ is a v_h-gauge with v_h Henselian, Th. 4.26 shows that $\mathsf{gr}_{\alpha_h}(A \otimes_F F_h)$ is graded simple with associated graded division algebra $\mathsf{gr}_w(E)$. This is likewise true with $\mathsf{gr}_\alpha(A)$ replacing $\mathsf{gr}_{\alpha_h}(A \otimes_F F_h)$, by (4.21). Hence, $Z(\mathsf{gr}_\alpha(A)) \cong_g Z(\mathsf{gr}_w(D))$ (see Prop. 2.17). Moreover, every simple component of A_0 has associated division algebra E_0 by Prop. 2.41, and $E_0 \cong \overline{E}$. We have $\Gamma_A = \Gamma_{\mathsf{gr}(A)} \supseteq \Gamma_{\mathsf{gr}(E)} = \Gamma_E$

by (2.27), and Prop. 2.47 shows that the inclusion is an equality if and only if A_0 is simple.

(ii) Let $A_h = A \otimes_F F_h$. We have $ms(\mathsf{gr}_\alpha(A)) = ms(\mathsf{gr}_{\alpha_h}(A_h))$ by (4.21); $ms(\mathsf{gr}_{\alpha_h}(A_h)) = ms(A_h)$ by Th. 4.26; and $ms(A_h) \geq ms(A)$ as A is central simple over F. This yields the displayed formula in (ii). By Th. 4.1 v extends to D if and only if $ms(D \otimes_F F_h) = 1 = ms(D)$. Since A (resp. A_h) is Brauer-equivalent to D (resp. $D \otimes_F F_h$) we have

$$ms(A_h)/\, ms(D \otimes_F F_h) \;=\; ms(A)/\, ms(D).$$

Hence, $ms(D \otimes_F F_h) = ms(D)$ if and only if $ms(A_h) = ms(A)$, which we have just seen holds if and only if $ms(\mathsf{gr}_\alpha(A)) = ms(A)$. This yields the last assertion of the theorem. □

For any v-gauge α on a central simple F-algebra A it is clear that $\mathsf{gr}(F) = \mathsf{gr}(Z(A)) \subseteq Z(\mathsf{gr}(A))$. However, the last inclusion is sometimes strict, as in Ex. 6.36(ii) below. We will show however, by passage to the Henselian case, that $Z(\mathsf{gr}(A))$ is always purely inseparable over $\mathsf{gr}(F)$. A graded field K is said to be *purely inseparable* over a graded subfield F if $q(\mathsf{K})$ is algebraic over $q(\mathsf{F})$ and every homogeneous element of K is the unique root of its minimal polynomial over $q(\mathsf{F})$. Further characterizations of pure inseparability are given in Def. 5.25 below.

Lemma 4.31. *Let D be a central division algebra over F with a valuation v_D extending v. Then, the graded field $Z(\mathsf{gr}(D))$ is purely inseparable over $\mathsf{gr}(F)$.*

Proof. Take $c \in D^\times$ with $\widetilde{c} \in Z(\mathsf{gr}(D))$, and let

$$f = X^n + a_{n-1}X^{n-1} + \ldots + a_0 \in F[X]$$

be the minimal polynomial of c over F. By Wedderburn's Factorization Theorem (see Lam [121, (16.9), pp. 251–252]), there are conjugates $c = c_1, \ldots, c_n$ of c in D with $f = (X - c_1)\ldots(X - c_n) \in D[X]$. Say $c_i = d_i c d_i^{-1}$. Then, $v_D(c_i) = v_D(c)$ for each i, and in $\mathsf{gr}(D)$ we have

$$\widetilde{c_i} \;=\; \widetilde{d_i c d_i^{-1}} \;=\; \widetilde{d_i}\,\widetilde{c}\,\widetilde{d_i}^{-1} = \widetilde{c}$$

as $\widetilde{c}$ is central. Let $g = (x - \widetilde{c})^n = (X - \widetilde{c_1})\ldots(X - \widetilde{c_n}) \in \mathsf{gr}(D)[X]$, and write $g = X^n + b_{n-1}X^{n-1} + \ldots + b_0$. Each b_i is a sum of monomials

$$(-1)^{n-i}\, \widetilde{c}_{j_1} \ldots \widetilde{c}_{j_{n-i}} \;=\; \big((-1)^{n-i} c_{j_1} \ldots c_{j_{n-i}}\big)^{\sim} \in D_{(n-i)v_D(c)}.$$

The corresponding coefficient a_i of f is the sum of the corresponding monomials $(-1)^{n-i} c_{j_1} \ldots c_{j_{n-i}}$. Because these monomials in D all have the same value, we have $b_i = \widetilde{a}_i$ if $v_D(a_i) = (n-i)v_D(c)$, while $b_i = 0$ if $v_D(a_i) > (n-i)v_D(c)$. In either case $b_i \in \mathsf{gr}(F)$ for each i, so $g \in \mathsf{gr}(F)[X]$. Since $\widetilde{c}$ is the only root of g, $\widetilde{c}$ must be purely inseparable over $\mathsf{gr}(F)$. Thus, every homogeneous element of $Z(\mathsf{gr}(D))$ is purely inseparable over $\mathsf{gr}(F)$, which implies that $Z(\mathsf{gr}(D))$ is purely inseparable over $\mathsf{gr}(F)$. □

Corollary 4.32. *Let A be a central simple F-algebra with a v-gauge α. Then, $Z(\mathsf{gr}_\alpha(A))$ is a graded field purely inseparable over $\mathsf{gr}_v(F)$.*

Proof. As in Th. 4.30, let E be the associated division algebra of $A \otimes_F F_h$; so, E has a valuation extending v_h. Let $\mathsf{K} = Z(\mathsf{gr}(E))$. By the preceding lemma, applied to the central division algebra E over F_h, K is a graded field purely inseparable over $\mathsf{gr}_{v_h}(F_h) = \mathsf{gr}_v(F)$. Theorem 4.30(i) shows that $Z(\mathsf{gr}(A)) \cong_g \mathsf{K}$. □

The same approach yields the uniqueness of gauges on valued division algebras:

Proposition 4.33. *Let D be an F-central division algebra with a valuation v_D extending v on F. Then,*

(i) *v_D is the only possible v-gauge on D; it is a v-gauge if and only if D is defectless over F.*
(ii) *Let $B = M_n(D)$ for some n, and suppose B has a v-gauge β. Then, v_D is a v-gauge, and $\mathsf{gr}_{v_D}(D)$ is the associated graded division algebra of the simple graded algebra $\mathsf{gr}_\beta(B)$.*

Proof. (i) By Th. 4.1, $D \otimes_F F_h$ is a division ring, call it D_h. Let w be the valuation on D_h extending the Henselian valuation v_h on F_h. Let α be a v-gauge on D, and let $\alpha_h = \alpha \otimes v_h$, which is a v_h-gauge on D_h. By (4.21) and Th. 4.26

$$\mathsf{gr}_\alpha(D) \cong_g \mathsf{gr}_{\alpha_h}(D_h) \cong_g \mathsf{gr}_w(D_h), \tag{4.22}$$

which is a graded division ring. By Cor. 3.30, α is a valuation on D extending v on F. But v_D is the unique extension of v to D by Th. 1.4, so $\alpha = v_D$. As observed in §3.2.2, v_D is a gauge if and only if it is a norm, which means by Cor. 2.7 that D is defectless over F.

(ii) Let β be a v-gauge on $B = M_n(D)$, let $B_h = B \otimes_F F_h = M_n(D_h)$, and let $\beta_h = \beta \otimes v_h$, which is a v_h-gauge on B_h with $\mathsf{gr}_{\beta_h}(B_h) \cong_g \mathsf{gr}_\beta(B)$ by (4.21). Since B_h has a v_h-gauge, D_h is defectless over F_h by Th. 4.26, which also shows that $\mathsf{gr}_w(D_h)$ is the associated graded division algebra of $\mathsf{gr}_{\beta_h}(B_h)$. Then, D is defectless over F, as

$$[D{:}F] = [D_h{:}F_h] = [\mathsf{gr}_w(D_h){:}\,\mathsf{gr}_{v_h}(F_h)] = [\mathsf{gr}_{v_D}(D){:}\,\mathsf{gr}(F)].$$

So by (i) v_D is a v-gauge. Since $\mathsf{gr}_{v_D}(D) \cong_g \mathsf{gr}_w(D_h)$ by (4.22), $\mathsf{gr}_{v_D}(D)$ is the associated graded division algebra of $\mathsf{gr}_\beta(B)$. □

4.3.3 Gauges on commutative algebras

For a commutative finite-dimensional algebra A over a field F, semisimplicity is equivalent to the absence of nonzero elements $x \in A$ such that $x^2 = 0$. A similar observation holds for graded algebras: a commutative graded algebra is semisimple if and only if it does not contain any homogeneous nilpotent element. Thus, if F has a valuation v and A has a surmultiplicative v-value function α, the following conditions are equivalent:

(a) $\mathsf{gr}_\alpha(A)$ is semisimple;
(b) $(\widetilde{x})^2 \neq 0$ for all nonzero $x \in A$;
(c) $(\widetilde{x})^n \neq 0$ for all nonzero $x \in A$ and for every positive integer n.

In view of (3.11), these conditions are also equivalent to:

(d) $\alpha(x^2) = 2\alpha(x)$ for all $x \in A$;
(e) $\alpha(x^n) = n\alpha(x)$ for all $x \in A$.

In analyzing gauges on commutative F-algebras A, we first consider the case where A is a field. The following proposition yields an important example of surmultiplicative value functions.

Proposition 4.34. *Let K/F be a finite-degree field extension and let $v_1,\ldots,v_n$ be valuations on K extending a valuation v on F. Define*

$$\alpha = min(v_1, \ldots, v_n)\colon K \to \Gamma \cup \{\infty\}$$

by

$$\alpha(x) = \min_{1 \leq i \leq n} \big(v_i(x)\big) \qquad \text{for } x \in K.$$

Then α is a surmultiplicative v-value function and there is a natural degree-preserving isomorphism of graded $\mathsf{gr}(F)$-algebras

$$\mathsf{gr}_\alpha(K) \xrightarrow{\sim} \mathsf{gr}_{v_1}(K) \times \ldots \times \mathsf{gr}_{v_n}(K) \tag{4.23}$$

such that for $x \in K^\times$

$$x + K^\alpha_{>\alpha(x)} \mapsto \big(x + K^{v_1}_{>\alpha(x)}, \ldots, x + K^{v_n}_{>\alpha(x)}\big).$$

Proof. It is clear from the definitions that α is a v-value function. It is surmultiplicative because $v_i(1) = 0$ for all i and

$$\min_{1\leq i\leq n} \big(v_i(x) + v_i(y)\big) \geq \min_{1\leq i\leq n} \big(v_i(x)\big) + \min_{1\leq i\leq n} \big(v_i(y)\big) \qquad \text{for } x,\, y \in K.$$

For $i = 1, \ldots, n$ we have $\alpha(x) \leq v_i(x)$ for all $x \in K$, hence the map (4.23) is a well-defined homomorphism of graded $\mathsf{gr}(F)$-algebras. This map is injective since for every $x \in K^\times$ there is some index i such that $\alpha(x) = v_i(x)$. To prove surjectivity, we will apply an approximation lemma, for which we will use the following notation: for each pair i, $j \in \{1, 2, \ldots, n\}$, let v_{ij} be the finest common coarsening of v_i and v_j (i.e., v_{ij} is the valuation on K with ring $\mathcal{O}_{K,v_{ij}} = \mathcal{O}_{K,v_i} \cdot \mathcal{O}_{K,v_j}$). Let Δ_{ij} be the convex subgroup of the divisible hull Γ of $\Gamma_{F,v}$ associated to v_{ij}; so v_{ij} is a map $K \to \Gamma/\Delta_{ij} \cup \{\infty\}$. For any $\delta_1, \ldots, \delta_n \in \Gamma$ we say that the n-tuple $(\delta_1, \ldots, \delta_n)$ is *compatible* in Γ^n if for all i, j we have $\delta_i - \delta_j \in \Delta_{ij}$.

Lemma 4.35. *With the notation just defined, fix some* $k \in \{1, 2, \ldots, n\}$, *and let* $(\delta_1, \ldots, \delta_n)$ *be a compatible* n*-tuple in* Γ^n *with* $\delta_k = 0$. *Then, there is* $c \in K^\times$ *with* $v_i(c) > \delta_i$ *for each* $i \neq k$, $v_k(c) = 0$, *and* $v_k(c-1) > 0$.

Proof. For each pair of indices i, j, we have

$$0 \leq \big||\delta_i| - |\delta_j|\big| \leq |\delta_i - \delta_j|.$$

Since $\delta_i - \delta_j \in \Delta_{ij}$ and Δ_{ij} is convex, it follows that $|\delta_i| - |\delta_j| \in \Delta_{ij}$. Furthermore, as $\Gamma/\Gamma_{F,v}$ is a torsion group, there is an $m \in \mathbb{N}$ such that each $m|\delta_i|$ lies in $\Gamma_{F,v}$. Hence, each $m|\delta_i| \in \Gamma_{K,v_i}$ and $m|\delta_i| - m|\delta_j| \in \Delta_{ij}$. These are the conditions needed for $(m|\delta_1|, \ldots, m|\delta_n|)$ to be compatible in $\Gamma_{K,v_1} \times \ldots \times \Gamma_{K,v_n}$ in the terminology of Ribenboim's paper [202]. Since this compatibility holds, the general approximation theorem for incomparable valuations on K [202, Th. 5] says that there exists a $d \in K^\times$ with

$$v_i(d) = m\,|\delta_i|, \qquad \text{for all } i.$$

In particular, $v_k(d) = m\,|\delta_k| = 0$. Now, let $T = \bigcap_{i=1}^n \mathcal{O}_{K,v_i} \subseteq K$. A weaker approximation theorem for incomparable valuations on K (see Engler–Prestel [73, Th. 3.2.7(3), p. 64]) says that the canonical map $\rho\colon T \to \overline{K}^{v_1} \times \ldots \times \overline{K}^{v_n}$ is surjective. Therefore, there is a $t \in T$ with $\rho(t) = (\overline{0}, \overline{0}, \ldots, \overline{d}^{\,-1}, \ldots, \overline{0})$, i.e., $v_i(t) > 0$ for $i \neq k$ and $v_k(t) = 0$ with $\overline{t} = \overline{d}^{\,-1}$ in $\overline{K}^{v_k}$. Let $c = td$. Then for $i \neq k$ we have

$$v_i(c) = v_i(t) + v_i(d) > 0 + m|\delta_i| \geq \delta_i,$$

hence $v_i(c) > \delta_i$. Also, $v_k(c) = v_k(t) + v_k(d) = 0$, and in $\overline{K}^{v_k}$

$$\overline{c} = \overline{t} \cdot \overline{d} = \overline{d}^{\,-1}\overline{d} = \overline{1}.$$

Thus, c has all the required properties. □

Proof of Prop. 4.34, completed. It remains only to prove the surjectivity of the map in (4.23). Fix any $k \in \{1, 2, \ldots, n\}$ and take any $b \in K^\times$. Let $\delta_i = v_k(b) - v_i(b)$ for $i = 1, 2, \ldots, n$. Then, for all i, j, as $v_i(b)$ and $v_j(b)$ each have image $v_{ij}(b)$ in Γ/Δ_{ij},

$$\delta_i - \delta_j = v_j(b) - v_i(b) \in \Delta_{ij}.$$

Hence, $(\delta_1, \ldots, \delta_n)$ is compatible in Γ^n. Also, $\delta_k = 0$. Lemma 4.35 yields $c \in K^\times$ with $v_i(c) > \delta_i$ for all $i \neq k$ and $v_k(c) = 0$ with $\overline{c} = \overline{1}$ in $\overline{K}^{v_k}$. Let $a = cb$. Then, for $i \neq k$,

$$v_i(a) = v_i(c) + v_i(b) > \delta_i + v_i(b) = v_k(b),$$

so $v_i(a) > v_k(b)$. Also, $v_k(a) = 0 + v_k(b) = v_k(b)$ and

$$v_k(a-b) = v_k(c-1) + v_k(b) > v_k(b).$$

Hence, for our α described in Prop. 4.34, we have

$$\alpha(a) = min\big(v_1(a), \ldots, v_n(a)\big) = v_k(b),$$

and the image of $\widetilde{a} = a + K^{\alpha}_{>\alpha(a)}$ under the map (4.23) is $(0, \ldots, 0, \widetilde{b}, 0, \ldots, 0)$ in $\mathsf{gr}_{v_1}(K) \times \ldots \times \mathsf{gr}_{v_n}(K)$. Since such n-tuples, for all choices of k and b, generate $\prod_{i=1}^n \mathsf{gr}_{v_i}(K)$, the map (4.23) is surjective. This completes the proof of Prop. 4.34. □

From the isomorphism (4.23), it is clear that the graded algebra $\mathsf{gr}_\alpha(K)$ is semisimple. It turns out that every surmultiplicative value function on K whose associated graded algebra is semisimple is of the above type:

Theorem 4.36. *Let K/F be a finite-degree field extension. Suppose that $\alpha\colon K \to \Gamma \cup \{\infty\}$ is a surmultiplicative v-value function such that $\mathsf{gr}_\alpha(K)$ is semisimple. Then, there exist valuations v_1, ..., v_n on K extending v such that $\alpha = min(v_1, \ldots, v_n)$.*

Proof. Since $\mathsf{gr}_\alpha(K)$ is assumed to be semisimple, we have $\alpha(x^n) = n\alpha(x)$ for all $x \in K$ (see condition (e) above), hence the ideal $\mathfrak{m}_K$ is radical. We may therefore find a set of prime ideals $P_\lambda \subseteq \mathcal{O}_K$ (indexed by some set Λ) such that $\mathfrak{m}_K = \bigcap_{\lambda\in\Lambda} P_\lambda$. By Chevalley's Extension Theorem (see Engler–Prestel [73, Th. 3.1.1]), we may find for each $\lambda \in \Lambda$ a valuation ring $\mathcal{O}_\lambda$ of K with maximal ideal $\mathfrak{m}_\lambda$ such that $\mathcal{O}_K \subseteq \mathcal{O}_\lambda$ and $P_\lambda = \mathcal{O}_K \cap \mathfrak{m}_\lambda$.

Claim: The valuation v_λ corresponding to $\mathcal{O}_\lambda$ extends v. Since $\mathcal{O}_F \subseteq \mathcal{O}_K \subseteq \mathcal{O}_\lambda$, we have $\mathcal{O}_F \subseteq \mathcal{O}_\lambda \cap F$. Similarly, $\mathfrak{m}_F \subseteq \mathfrak{m}_K \subseteq P_\lambda \subseteq \mathfrak{m}_\lambda$, so $\mathfrak{m}_F \subseteq \mathfrak{m}_\lambda \cap F$. Since $\mathcal{O}_\lambda \cap F$ is a valuation ring of F with maximal ideal $\mathfrak{m}_\lambda \cap F$, the inclusions $\mathcal{O}_F \subseteq \mathcal{O}_\lambda \cap F$ and $\mathfrak{m}_F \subseteq \mathfrak{m}_\lambda \cap F$ imply that $\mathcal{O}_F = \mathcal{O}_\lambda \cap F$, proving the claim. Hence, each value group Γ_{K,v_i} embeds canonically into the divisible group Γ.

Since there are only finitely many extensions of v to K, and since for λ, $\lambda' \in \Lambda$ the equality $\mathcal{O}_\lambda = \mathcal{O}_{\lambda'}$ implies $\mathfrak{m}_\lambda = \mathfrak{m}_{\lambda'}$, hence $P_\lambda = P_{\lambda'}$, it follows that Λ is a finite set. Let $\Lambda = \{1, \ldots, n\}$ and, for $x \in K$, let

$$w(x) = \min_{1\le i\le n}\big(v_i(x)\big).$$

Since $\mathcal{O}_K \subseteq \bigcap_{i=1}^n \mathcal{O}_i$, we have

$$\alpha(x) \ge 0 \Rightarrow w(x) \ge 0 \qquad \text{for } x \in K.$$

Similarly, since $\mathfrak{m}_K = \bigcap_{i=1}^n P_i = \mathcal{O}_K \cap \big(\bigcap_{i=1}^n \mathfrak{m}_i\big)$, we have

$$\alpha(x) > 0 \iff \big(\alpha(x) \ge 0 \text{ and } w(x) > 0\big).$$

It follows that

$$\alpha(x) = 0 \Rightarrow w(x) = 0 \qquad \text{for } x \in K^\times. \tag{4.24}$$

Now, fix some $x \in K^\times$. Since $[K{:}F]$ is finite, it follows from Cor. 3.6(i) and Prop. 2.5 that $|\Gamma_K{:}\Gamma_F|$ is finite, hence Γ_K/Γ_F is a torsion group. Therefore, there is an integer $m > 0$ and an element $u \in F^\times$ such that $m\alpha(x) = v(u)$, hence $\alpha(x^m u^{-1}) = 0$. By (4.24), we then have $w(x^m u^{-1}) = 0$. Because each v_i extends v, we have $w(x^m u^{-1}) = w(x^m) - v(u)$. Hence,

$$mw(x) = w(x^m) = v(u) = m\alpha(x).$$

Since Γ has no torsion, it follows that $w(x) = \alpha(x)$, which proves that $\alpha = min(v_1, \ldots, v_n)$. □

Let $v_1, \ldots, v_r$ be all the extensions of v to K. For $i = 1, \ldots, r$, let $e_i = |\Gamma_{K,v_i} : \Gamma_F|$ be the ramification index and let $f_i = [\overline{K}^{v_i} : \overline{F}]$ be the residue degree. Recall from Def. 4.19 that we say v is defectless in K if the Fundamental Equality holds for K/F, i.e.,

$$[K:F] = \sum_{i=1}^{r} e_i f_i.$$

Corollary 4.37. *There is a v-gauge on the field K if and only if v is defectless in K. When that condition holds, the v-gauge on K is unique and is defined by*

$$\alpha = \min_{1 \le i \le r} (v_i) \tag{4.25}$$

where $v_1, \ldots, v_r$ are all the extensions of v to K.

Proof. Suppose α is a v-gauge on K. By Th. 4.36, we may find some extensions $v_1, \ldots, v_n$ of v to K such that $\mathsf{gr}_\alpha(K) \cong_g \mathsf{gr}_{v_1}(K) \times \ldots \times \mathsf{gr}_{v_n}(K)$. Now, $[\mathsf{gr}_{v_i}(K) : \mathsf{gr}(F)] = e_i f_i$ by Cor. 2.10, and $[\mathsf{gr}_\alpha(K) : \mathsf{gr}(F)] = [K:F]$ since α is a v-norm, so

$$[K:F] = \sum_{i=1}^{n} e_i f_i.$$

This implies that $v_1, \ldots, v_n$ are *all* extensions of v to K by Engler–Prestel [73, Th. 3.3.4] and v is defectless in K. Conversely, if v is defectless in K, then formula (4.25) defines a v-gauge on K. □

The following special case is noteworthy:

Corollary 4.38. *Suppose char $\overline{F} \nmid [K:F]$. If v extends uniquely to K, this extension is a v-gauge on K.*

Proof. In view of Cor. 4.37, it suffices to show that

$$[K:F] = [\overline{K}:\overline{F}]\,|\Gamma_K : \Gamma_F|.$$

But this is immediate from Ostrowski's Theorem, Th. A.12, since the defect $\partial_{K/F}$ is both a power of *char* $\overline{F}$ and a divisor of $[K:F]$. □

We can now easily treat the general type of commutative semisimple algebras.

Proposition 4.39. *Let $K_1, \ldots, K_m$ be finite-degree field extensions of F and let $A = K_1 \times \ldots \times K_m$. There is a v-gauge on A if and only if v is defectless* *in A, i.e., equality holds in the Fundamental Inequality for each K_i/F. When that condition holds, there is a unique v-gauge α on A, defined by*

$$\alpha(a) = \min_{1\leq i\leq m}(\alpha_i(a_i))$$

where each α_i is the unique v-gauge on K_i as in Cor. 4.37.

Proof. This follows readily from Prop. 3.36 and Cor. 4.37. □

The results in this section on gauges in field extensions yield a necessary condition for the existence of a v-gauge on an arbitrary semisimple F-algebra. We will see in Th. 4.50 below that this necessary condition is also sufficient.

Proposition 4.40. *Let A be any semisimple (finite-dimensional) F-algebra. If A has a v-gauge, then v is defectless in A.*

Proof. A has a v-gauge if and only if each simple component of A has a v-gauge, by Prop. 3.36. Also, by definition, v is defectless in A if and only if v is defectless in each simple component of A. Thus, we may assume that A is simple. Let $K = Z(A)$, and let $v_1, \ldots, v_r$ be the extensions of v to the field K. Let (F_h, v_h) be a Henselization of (F, v), and let $(K_{h,v_i}, v_{i,h})$ be a Henselization of (K, v_i). Let α be a v-gauge on A. Then its restriction $\alpha|_K$ is a surmultiplicative v-norm on K with $\mathsf{gr}_{\alpha|_K}(K)$ graded semisimple, since it is a central graded subalgebra of the graded semisimple algebra $\mathsf{gr}_\alpha(A)$. Hence $\alpha|_K$ is a v-gauge on K, so v is defectless in K by Cor. 4.37. Moreover, $\alpha \otimes v_h$ is a v_h-gauge on $A \otimes_F F_h$ and

$$A \otimes_F F_h \cong (A \otimes_K K_{h,v_1}) \times \ldots \times (A \otimes_K K_{h,v_r}).$$

By Prop. 3.36 each $A \otimes_K K_{h,v_i}$ carries a gauge, hence the corresponding central division algebra over K_{h,v_i} is defectless by Th. 4.26(i). □

4.3.4 Gauges on simple and semisimple algebras

Throughout this section, let v be an arbitrary valuation on a field F, and let A be a (finite-dimensional) simple F-algebra. Let K be the center of A, so K is a finite-degree field extension of F. Let $v_1, \ldots, v_r$ be all the extensions of v to K. Our goal is to characterize v-gauges on A in terms of v_i-gauges for $i = 1, \ldots, r$.

Proposition 4.41. *Let L be a field with $F \subseteq L \subseteq K$, and suppose v has a unique extension to a valuation v_L of L. If A has a v-gauge α, then $\alpha|_L = v_L$, which is a v-gauge on L, and α is a v_L-gauge. Thus, whenever v_L is a v-gauge on L, v-gauges on A are the same as v_L-gauges on A.*

Proof. The restriction $\alpha|_L$ of α to L is clearly surmultiplicative, and is a v-norm on L by Prop. 3.14. Moreover, as $L \subseteq Z(A)$, we have $\mathsf{gr}_{\alpha|_L}(L) \subseteq Z(\mathsf{gr}_\alpha(A))$.

Because $\mathsf{gr}_\alpha(A)$ is graded semisimple, it contains no nonzero central homogeneous nilpotent elements. Therefore, the commutative $\mathsf{gr}(F)$-algebra $\mathsf{gr}_{\alpha|_L}(L)$ is semisimple, and hence $\alpha|_L$ is a v-gauge on L. Because v_L is the unique extension of v to L, Cor. 4.37 shows that $\alpha|_L = v_L$. Hence, for $c \in L^\times$, we have

$$\alpha(c^{-1}) = v_L(c^{-1}) = -v_L(c) = -\alpha(c).$$

So, by Prop. 3.29, $\alpha(ca) = \alpha(c) + \alpha(a)$ for all $a \in A$. This shows that the v-value function α on A is actually a v_L-value function. Since α and v_L are v-norms, we have

$$\begin{aligned}[L{:}F]\,[\mathsf{gr}_\alpha(A){:}\,\mathsf{gr}_{v_L}(L)] &= [\mathsf{gr}_{v_L}(L){:}\,\mathsf{gr}_v(F)]\,[\mathsf{gr}_\alpha(A){:}\,\mathsf{gr}_{v_L}(L)]\\ &= [\mathsf{gr}_\alpha(A){:}\,\mathsf{gr}_v(F)] = [A{:}F] = [L{:}F]\,[A{:}L].\end{aligned}$$

Hence, $[\mathsf{gr}_\alpha(A){:}\,\mathsf{gr}_{v_L}(L)] = [A{:}L]$, showing that α is a v_L-norm. The other conditions needed for α to be a v_L-gauge hold because α is a v-gauge. Conversely, whenever v_L is a v-norm and β is a v_L-gauge on A, then β is also a v-norm and hence a v-gauge on A. □

Theorem 4.42. *Let α be a v-gauge on A. Then there exist v_i-gauges α_i on A for $i = 1, \ldots, r$ such that*

$$\alpha(a) = \min\big(\alpha_1(a), \ldots, \alpha_r(a)\big) \quad \text{for all } a \in A.$$

Furthermore,

$$\mathsf{gr}_\alpha(A) \cong_g \mathsf{gr}_{\alpha_1}(A) \times \ldots \times \mathsf{gr}_{\alpha_r}(A).$$

Hence, the semisimple $\overline{F}$-algebra $\mathsf{gr}_\alpha(A)_0$ has at least r simple components. Moreover, the $\mathsf{gr}_{\alpha_i}(A)$ are the graded simple components of $\mathsf{gr}_\alpha(A)$ and

$$[\mathsf{gr}_{\alpha_i}(A){:}\,\mathsf{gr}(F)] = [A{:}K]\ [\mathsf{gr}_{v_i}(K){:}\,\mathsf{gr}(F)].$$

We call the α_i of the theorem the v_i-*component* of α, for $i = 1, \ldots, r$.

Proof. Let (F_h, v_h) be the Henselization of (F, v), and let

$$B = A \otimes_F F_h \qquad \text{and} \qquad L = Z(B) = K \otimes_F F_h.$$

Then, L is a direct product of finitely many fields. Let $e_1, \ldots, e_r$ be the primitive idempotents of L, so

$$L = L_1 \times \ldots \times L_r \qquad \text{where} \qquad L_i = e_i L,$$

and each L_i is a field. The L_i are indexed by the v_i, as we will recall below. Correspondingly, B is a product of simple algebras

$$B = B_1 \times \ldots \times B_r \qquad \text{where} \qquad B_i = e_i B.$$

So, each B_i is a central simple L_i-algebra. We identify K, A, F_h with their isomorphic copies $K\otimes 1$, $A\otimes 1$, $1\otimes F_h$ in B. But we do *not* identify them with their isomorphic copies e_iK, e_iA, e_iF_h in B_i. For each i we have canonical inclusions

$$p_i\colon A \hookrightarrow B_i, \quad a \mapsto e_i(a \otimes 1) \qquad \text{and} \qquad q_i\colon F_h \hookrightarrow B_i, \quad c \mapsto e_i(1 \otimes c).$$

Thus, B_i has subalgebras $p_i(A)$, $p_i(K)$, and L_i, with

$$B_i \;=\; p_i(A) \otimes_{p_i(K)} L_i \;\cong\; A \otimes_K L_i, \qquad \text{hence} \qquad [B_i{:}L_i] \;=\; [A{:}K].$$

Each field L_i is a compositum of fields, $L_i = p_i(K) \cdot q_i(F_h)$. The Henselian valuation v_h on F_h has an isomorphic (Henselian) valuation $v_h \circ q_i^{-1}$ on $q_i(F_h)$, which extends uniquely to a Henselian valuation w_i on L_i. This pulls back to a valuation $w_i \circ p_i$ on K which extends v on F. It is known (see Th. A.32) that the valuations $w_1 \circ p_1, \ldots, w_r \circ p_r$ are all distinct and are all the extensions of v to K. Thus, after renumbering the e_i if necessary, we can assume $w_i \circ p_i = v_i$ for $i = 1, \ldots, r$. That is,

$$v_i(d) \;=\; w_i(e_i(d \otimes 1)) \qquad \text{for all } d \in K. \tag{4.26}$$

From Th. A.32, we have also that (L_i, w_i) is a Henselization of (K, v_i).

Let $\beta = \alpha \otimes v_h$, which is a v_h-gauge on B with

$$\mathsf{gr}_\beta(B) \;\cong_g\; \mathsf{gr}_\alpha(A) \otimes_{\mathsf{gr}(F)} \mathsf{gr}(F_h) \;\cong_g\; \mathsf{gr}_\alpha(A) \tag{4.27}$$

by Prop. 3.42. Let $\beta_i = \beta|_{B_i}$, which is a v_h-gauge on B_i via the embedding $q_i\colon F_h \to B_i$. By Prop. 3.36,

$$\beta(b) \;=\; \min_{1 \le i \le r} \big(\beta_i(e_i b)\big) \qquad \text{for all} \;\; b \in B, \tag{4.28}$$

and

$$\mathsf{gr}_\beta(B) \;\cong_g\; \prod_{i=1}^{r} \mathsf{gr}_{\beta_i}(B_i). \tag{4.29}$$

Since w_i is the unique extension of the Henselian valuation v_h to L_i, Prop. 4.41 above shows that each β_i is a w_i-gauge.

Define v-value functions $\alpha_1, \ldots, \alpha_r$ on A by

$$\alpha_i(a) \;=\; \beta(p_i(a)) \;=\; \beta_i(e_i(a \otimes 1)).$$

Then, each α_i is surmultiplicative as β_i is surmultiplicative, and for all $a \in A$,

$$\alpha(a) \;=\; \beta(a \otimes 1) \;=\; \min_{1 \le i \le r} \big(\beta_i(e_i(a \otimes 1))\big) \;=\; \min_{1 \le i \le r} \big(\alpha_i(a)\big). \tag{4.30}$$

Furthermore, as β_i is a w_i-value function, for all $c \in K$ and $a \in A$, we have

$$\begin{aligned} \alpha_i(ca) \;&=\; \beta_i(e_i(ca \otimes 1)) \;=\; \beta_i\big(e_i(c \otimes 1) \cdot e_i(a \otimes 1)\big) \\ &=\; w_i(e_i(c \otimes 1)) \;+\; \beta_i(e_i(a \otimes 1)) \;=\; v_i(c) + \alpha_i(a). \end{aligned}$$

Thus, α_i is a v_i-value function on A. The following diagram shows the algebras related to B_i and the associated value functions being considered here.

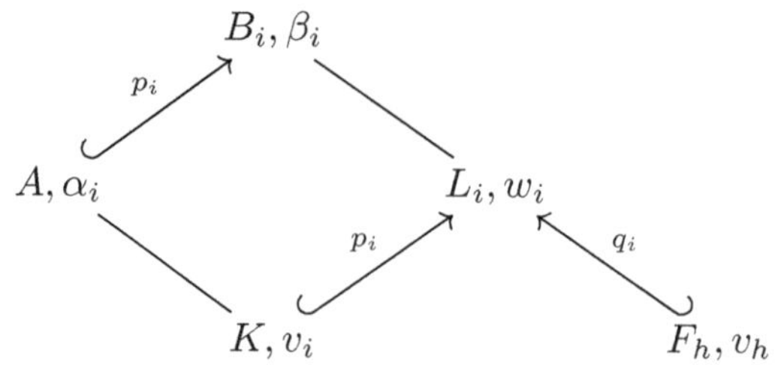

(4.31)

Now, for all $\gamma \in \Gamma$, the definition of the β_i and (4.28) and (4.30) above show that we have a commutative diagram for each i:

$$\begin{array}{ccc} A^{\alpha}_{\geq\gamma} & \xrightarrow{id} & A^{\alpha_i}_{\geq\gamma} \\ \downarrow & & \downarrow \\ B^{\beta}_{\geq\gamma} & \xrightarrow{e_i\cdot} & B^{\beta_i}_{i,\geq\gamma} \end{array}$$

where the right vertical map is $a \mapsto e_i(a \otimes 1)$. There is a corresponding commutative diagram with $> \gamma$ replacing $\geq \gamma$, and these together yield a commutative diagram of graded $\mathsf{gr}(F)$-algebra homomorphisms:

$$\begin{array}{ccc} \mathsf{gr}_{\alpha}(A) & \longrightarrow & \prod_{i=1}^{r} \mathsf{gr}_{\alpha_i}(A) \\ \cong\downarrow & & \downarrow \\ \mathsf{gr}_{\beta}(B) & \xrightarrow{\cong} & \prod_{i=1}^{r} \mathsf{gr}_{\beta_i}(B_i) \end{array} \tag{4.32}$$

Here, the top map is injective by (4.30); the left map is the isomorphism in (4.27); the right map is injective since for each i, the definition of α_i shows that $\mathsf{gr}_{\alpha_i}(A) \to \mathsf{gr}_{\beta_i}(B)$ is injective; and the bottom map is the isomorphism of (4.29). Therefore, all the maps in this diagram must be isomorphisms. Hence, for each i, $\mathsf{gr}_{\alpha_i}(A) \cong_g \mathsf{gr}_{\beta_i}(B_i)$, which is semisimple as β_i is a gauge. Since (L_i, w_i) (resp. (F_h, v_h)) is an immediate extension of (K, v_i) (resp. (F, v)), we have

$$[L_i{:}F_h] \;\geq\; [\mathsf{gr}_{w_i}(L_i){:}\, \mathsf{gr}_{v_h}(F_h)] \;=\; [\mathsf{gr}_{v_i}(K){:}\, \mathsf{gr}(F)].$$

So, as β_i is a v_h-norm,

$$\begin{aligned} &[\mathsf{gr}_{\alpha_i}(A){:}\, \mathsf{gr}_{v_i}(K)]\; [\mathsf{gr}_{v_i}(K){:}\, \mathsf{gr}(F)] \\ &\quad= [\mathsf{gr}_{\alpha_i}(A){:}\, \mathsf{gr}(F)] \;=\; [\mathsf{gr}_{\beta_i}(B_i){:}\, \mathsf{gr}_{v_h}(F_h)] \\ &\quad= [B_i{:}F_h] \;=\; [B_i{:}L_i]\,[L_i{:}F_h] \;=\; [A{:}K]\,[L_i{:}F_h] \\ &\quad\geq [A{:}K]\; [\mathsf{gr}_{v_i}(K){:}\, \mathsf{gr}(F)], \end{aligned} \tag{4.33}$$

and hence $[\mathsf{gr}_{\alpha_i}(A){:}\, \mathsf{gr}_{v_i}(K)] \geq [A{:}K]$. Since the reverse inequality holds for any v_i-value function, we have $[\mathsf{gr}_{\alpha_i}(A){:}\, \mathsf{gr}_{v_i}(K)] = [A{:}K]$. Hence, α_i is a

v_i-norm on A; with the semisimplicity noted above, this yields that α_i is a v_i-gauge. Furthermore, equality holds in (4.33), yielding

$$[\mathsf{gr}_{\alpha_i}(A): \mathsf{gr}(F)] = [A:K]\,[\mathsf{gr}_{v_i}(K): \mathsf{gr}(F)].$$

The isomorphism for $\mathsf{gr}_\alpha(A)$ in the theorem is the top isomorphism in the commutative diagram (4.32). Since each $\mathsf{gr}_{\alpha_i}(A)$ is graded simple by Th. 4.30(i) the $\mathsf{gr}_{\alpha_i}(A)$ are the graded simple components of $\mathsf{gr}_\alpha(A)$. Because $\mathsf{gr}_\alpha(A)$ has r graded simple components, its degree-0 part must have at least r simple components. □

Let v be a valuation on some division algebra D, and let α, β, η_1, …, η_r be value functions on a finite-dimensional D-vector space V. We write $\alpha \leq \beta$ if $\alpha(z) \leq \beta(z)$ for all $z \in V$. Likewise, we write $\alpha = min\,(\eta_1, \ldots, \eta_r)$ if $\alpha(z) = min\,(\eta_1(z), \ldots, \eta_r(z))$ for all $z \in V$. It is easy to construct examples of v-norms α, β on V with $\alpha \leq \beta$ and $\alpha \neq \beta$. However, we will see that for gauges on semisimple algebras, if $\alpha \leq \beta$ then $\alpha = \beta$. This will be proved by showing a minimality property characterizing the components of a gauge on a simple algebra.

Theorem 4.43. *Let α be any v-gauge on the simple F-algebra A, and, as in Th. 4.42 above, let α_i be the v_i-component of α for $i = 1, 2, \ldots, r$. Suppose η is a v_k-gauge on A, for some k. If $\alpha \leq \eta$, then $\eta = \alpha_k$.*

Proof. Pick a homogeneous base $(b_1, \ldots, b_n)$ of $\mathsf{gr}_{\alpha_k}(A)$ as a graded $\mathsf{gr}_{v_k}(K)$-vector space, and let $\gamma_j = deg(b_j)$. Then, pick $a_1, \ldots, a_n \in A$ with each $\widetilde{a_j}^{\alpha} \mapsto (0, \ldots, 0, b_j, 0, \ldots, 0)$ (b_j in the k-th position) under the isomorphism $\mathsf{gr}_\alpha(A) \cong_g \prod_{i=1}^r \mathsf{gr}_{\alpha_j}(A)$ of Th. 4.42. This means, for each j,

$$\alpha_i(a_j) > \gamma_j \text{ for } i \neq k, \ \ \alpha(a_j) = \alpha_k(a_j) = \gamma_j, \text{ and } \widetilde{a_j}^{\alpha_k} = b_j \text{ in } \mathsf{gr}_{\alpha_k}(A).$$

Since the a_j map to a graded vector space base of $\mathsf{gr}_{\alpha_k}(A)$, by Cor. 3.6(ii) they form a splitting base of the K-vector space A for α_k. Furthermore, $\alpha_k(a_j) = \alpha(a_j) \leq \eta(a_j)$, for all j. Therefore, for any $a \in A$, writing $a = \sum_{j=1}^n a_j c_j$ with $c_j \in K$, we have

$$\begin{aligned}\alpha_k(a) &= \min_{1\leq j\leq n} \big(\alpha_k(a_j) + v_k(c_j)\big) \leq \min_{1\leq j\leq n} \big(\eta(a_j) + v_k(c_j)\big)\\ &\leq \eta\big(\sum_{j=1}^{n} a_j c_j\big) = \eta(a).\end{aligned}$$

Thus, $\alpha_k \leq \eta$ as v_k-gauges on A.

Since $\alpha_k \leq \eta$, Lemma 3.26 (with $f = id_A$) shows that there is a well-defined graded $\mathsf{gr}_{v_k}(K)$-algebra homomorphism $\varphi\colon \mathsf{gr}_{\alpha_k}(A) \to \mathsf{gr}_\eta(A)$ given by $\widetilde{a}^{\alpha_k} \mapsto \widetilde{a}^{\eta}$ for all $a \in A$. But, as α_k is a v_k-gauge on the central simple K-algebra A, Th. 4.30(i) shows that $\mathsf{gr}_{\alpha_k}(A)$ is a simple graded algebra. Hence φ must be injective. Therefore, by Lemma 3.16 we have $\alpha_k = \eta$. □

Corollary 4.44. *Let α be any v-gauge on the simple F-algebra A, and, as in Th. 4.42 above, let α_i be the v_i-component of α for $i = 1, 2, \ldots, r$. Suppose $\alpha = \min(\eta_1, \ldots, \eta_r)$ for some v_i-gauges η_i. Then each $\eta_i = \alpha_i$.*

Proof. For each i, we have $\alpha \leq \eta_i$. Hence, $\eta_i = \alpha_i$ by the preceding theorem. □

Corollary 4.45. *Let α and η be v-gauges on a semisimple F-algebra C. If $\alpha \leq \eta$, then $\alpha = \eta$.*

Proof. It suffices to check this for the restrictions of α and η on the simple components of C. Therefore, we may assume that C is a simple F-algebra. Then, let $v_1, \ldots, v_r$ be the extensions of v to $Z(C)$, and let α_i (resp. η_i) be the v_i-component of α (resp. η). Since for each i we have $\alpha \leq \eta \leq \eta_i$, Th. 4.43 shows that $\alpha_i = \eta_i$. Hence,

$$\alpha = \min(\alpha_1, \ldots, \alpha_r) = \min(\eta_1, \ldots, \eta_r) = \eta. \qquad \square$$

We next determine when given v_i-gauges $\eta_1, \ldots, \eta_r$ yield a v-gauge as $\min(\eta_1, \ldots, \eta_r)$.

Let α be any v-value function on A, and let Δ be any convex subgroup of Γ. We have seen in §3.2.7 that there is a value function β on A determined from α and Δ as $\beta = \varepsilon \circ \alpha$, where $\varepsilon \colon \Gamma \to \Gamma/\Delta$ is the canonical epimorphism. We call such a β a *coarsening* of α. Let $w = \beta|_F$, a valuation on F which is a coarsening of v. If $\Gamma_{A,\alpha}$ lies in the divisible hull $\mathbb{H}(\Gamma_{F,v})$ of $\Gamma_{F,v}$, then the distinct coarsenings of α are in one-to-one correspondence with the distinct convex subgroups of $\mathbb{H}(\Gamma_{F,v})$, which are in one-to-one correspondence with the coarsenings of v. When this occurs, Δ is determined by $w = \beta|_F$, and we call β the *w-coarsening* of α.

Corollary 4.46. *Let α be a v-gauge on the simple F-algebra A, with v_i-components α_i, for $i = 1, \ldots, r$, and suppose $\Gamma_{A,\alpha}$ lies in the divisible hull of $\Gamma_{F,v}$. Let w be any valuation on F which is a coarsening of v, and let $w_1, \ldots, w_\ell$ be the extensions of w to K. Let β be the w-coarsening of α, and let β_j be the w_j-component of β for $j = 1, \ldots, \ell$. For $i \in \{1, \ldots, r\}$, let $j(i) \in \{1, \ldots, \ell\}$ be the index such that $w_{j(i)}$ is the w-coarsening of v_i (i.e., $O_{K,w_{j(i)}} = O_{F,w} \cdot O_{K,v_i}$). Then, $\beta_{j(i)}$ is the $w_{j(i)}$-coarsening of α_i.*

Proof. Let $\alpha_{i,w}$ denote the $w_{j(i)}$-coarsening of α_i. Let Δ be the convex subgroup of Γ associated to w. Recall that $\beta = \varepsilon \circ \alpha$, where $\varepsilon \colon \Gamma \to \Gamma/\Delta$ is the canonical surjection. This ε is compatible with the orderings on Γ and Γ/Δ. Likewise, $w_{j(i)} = \varepsilon \circ v_i$ and $\alpha_{i,w} = \varepsilon \circ \alpha_i$. Now fix any i. Since $\alpha \leq \alpha_i$,

$$\beta = \varepsilon \circ \alpha \leq \varepsilon \circ \alpha_i = \alpha_{i,w},$$

i.e., $\beta \leq \alpha_{i,w}$. Since $\alpha_{i,w}$ is a $w_{j(i)}$-gauge by Prop. 3.54, Th. 4.43 shows that $\alpha_{i,w} = \beta_{j(i)}$. □

In the next theorem we use the notation defined at the beginning of this subsection. In addition, for $i, j \in \{1, 2, \dots, r\}$ let v_{ij} denote the finest common coarsening of the K-valuations v_i and v_j. That is, v_{ij} is the valuation on K with $\mathcal{O}_{K,v_{ij}} = \mathcal{O}_{K,v_i} \cdot \mathcal{O}_{K,v_j}$.

Theorem 4.47. *Suppose v is defectless in the simple F-algebra A. For $i = 1, \dots, r$, let η_i be a v_i-gauge on A with Γ_{A,η_i} in the divisible hull of $\Gamma_{F,v}$. Let $\alpha = \min(\eta_1, \dots, \eta_r)$. Then, α is a v-gauge on A if and only if η_i and η_j have the same v_{ij}-coarsening for all pairs i, j. When this occurs, each η_i is the v_i-component of α.*

Proof. $\Rightarrow$ Suppose our α is a v-gauge on A. By Cor. 4.44, each η_i is the v_i-component of α. Fix indices i, j, let $w = v_{ij}|_F$, and let β be the w-coarsening of α. The v_{ij}-coarsenings of η_i and η_j must be the same, since by Cor. 4.46 they each coincide with the v_{ij}-component of β.

$\Leftarrow$ The proof here is very similar to the surjectivity proof for Prop. 4.34 above. Suppose each η_i and η_j have the same v_{ij}-coarsening. Now, $\alpha = \min(\eta_1, \dots, \eta_r)$ is clearly a surmultiplicative v-value function on A. Consider the graded $\mathsf{gr}(F)$-algebra homomorphism

$$\Psi\colon \mathsf{gr}_\alpha(A) \to \prod_{i=1}^{r} \mathsf{gr}_{\eta_i}(A) \text{ given by } a + A^{\alpha}_{>\alpha(a)} \mapsto \big(a + A^{\eta_1}_{>\alpha(a)}, \dots, a + A^{\eta_r}_{>\alpha(a)}\big).$$

Then, Ψ is well-defined and injective because $\alpha = \min(\eta_1, \dots, \eta_r)$. We will show that Ψ is an isomorphism. It then follows that $\mathsf{gr}_\alpha(A)$ is semisimple, as each $\mathsf{gr}_{\eta_i}(A)$ is semisimple. Furthermore, as v is defectless in K since it is defectless in A, we then have

$$\begin{aligned}
[\mathsf{gr}_\alpha(A)\colon \mathsf{gr}(F)] &= \sum_{i=1}^{r} [\mathsf{gr}_{\eta_i}(A)\colon \mathsf{gr}(F)] \\
&= \sum_{i=1}^{r} [\mathsf{gr}_{\eta_i}(A)\colon \mathsf{gr}_{v_i}(K)]\,[\mathsf{gr}_{v_i}(K)\colon \mathsf{gr}(F)] \\
&= \sum_{i=1}^{r} [A{:}K]\,[\mathsf{gr}_{v_i}(K)\colon \mathsf{gr}(F)] \;=\; [A{:}K]\,[K{:}F] \;=\; [A{:}F].
\end{aligned}$$

Hence, α is v-gauge.

It remains to prove that Ψ is surjective. For this, fix arbitrary $k \in \{1, \dots, r\}$, and fix any $b \in A \setminus \{0\}$. Let

$$\delta_i \;=\; \eta_k(b) - \eta_i(b) \qquad \text{for } i = 1, 2, \dots, r.$$

For each pair of indices i, j, let Δ_{ij} be the convex subgroup of $\mathbb{H}(\Gamma_F)$ associated to the finest common coarsening v_{ij} of v_i and v_j on K. Then, since η_i and η_j are assumed to have the same v_{ij}-coarsening, $\eta_i(b)$ and $\eta_j(b)$ have the same image in Γ/Δ_{ij}; hence,

$$\delta_i - \delta_j \;=\; \eta_j(b) - \eta_i(b) \;\in\; \Delta_{ij}.$$

Thus, $(\delta_1, \ldots, \delta_r)$ is a compatible r-tuple in $\mathbb{H}(\Gamma_F)^r$, as defined preceding Lemma 4.35. Since also $\delta_k = 0$, that lemma yields $c \in K^\times$ with $v_i(c) > \delta_i$ for all $i \neq k$ and $v_k(c) = 0$ with $\overline{c} = \overline{1}$ in $\overline{K}^{v_k}$. Let $a = cb \in A$. Then, for $i \neq k$,

$$\eta_i(a) \;=\; v_i(c) + \eta_i(b) \;>\; \delta_i + \eta_i(b) \;=\; \eta_k(b),$$

so $\eta_i(a) > \eta_k(b)$. But

$$\eta_k(a) \;=\; v_k(c) + \eta_k(b) \;=\; \eta_k(b).$$

Hence, $\alpha(a) = \mathit{min}\big(\eta_1(a), \ldots, \eta_r(a)\big) = \eta_k(b)$. Moreover,

$$\eta_k(a-b) \;=\; \eta_k(cb-b) \;=\; v_k(c-1) + \eta_k(b) \;>\; \eta_k(b).$$

Thus, for $\widetilde{a} = a + A^{\alpha}_{>\alpha(a)} \in \mathsf{gr}_{\alpha}(A)$ and $\widetilde{b} = b + A^{\eta_k}_{>\eta_k(b)} \in \mathsf{gr}_{\eta_k}(A)$, we have

$$\Psi(\widetilde{a}) = \big(a + A^{\eta_1}_{>\alpha(a)}, \ldots, a + A^{\eta_r}_{>\alpha(a)}\big) = \big(0, \ldots, 0, \underset{k}{\widetilde{b}}, 0, \ldots, 0\big) \in \prod_{i=1}^{r} \mathsf{gr}_{\eta_i}(A).$$

Since for arbitrary k and b such elements generate $\prod_{i=1}^{r} \mathsf{gr}_{\eta_i}(A)$, the map Ψ is surjective. This completes the proof. □

Corollary 4.48. *Suppose v is defectless in A, and suppose the extensions $v_1, \ldots, v_r$ of v to K are pairwise independent. Take any v_i-gauges η_i on A, $i = 1, 2, \ldots, r$. Then $\mathit{min}\big(\eta_1, \ldots, \eta_r\big)$ is a v-gauge on A with components $\eta_1, \ldots \eta_r$. In particular, this holds whenever v has rank 1.*

Proof. This is immediate from the preceding theorem. For, when v_i and v_j are independent valuations, their finest common coarsening v_{ij} is the trivial valuation, so the compatibility condition on η_i and η_j holds automatically. □

Example 4.49. (*A quaternion algebra with many gauges for a rank 2 valuation on a subfield of the center.*) Let L be a field with $\mathit{char}\, L \neq 2$, and let $F = L(x, y)$ with x and y algebraically independent over L. Let $s = \sqrt{1+x}$ and let $K = F(s) = L(s, y)$, a quadratic extension of F. Consider the quaternion division algebra

$$D \;=\; \big(s, 1 + y/K\big) \;=\; K \oplus Ki \oplus Kj \oplus Kk$$

where

$$i^2 = s, \quad j^2 = 1 + y, \quad k = ij = -ji.$$

Let v be the valuation on F obtained by restriction from the rank 2 Henselian (x, y)-adic valuation on $L((x))((y))$. So, $\Gamma_{F,v} \;=\; \mathbb{Z} \times \mathbb{Z}$ with the right-to-left lexicographic ordering, $v(x) \;=\; (1, 0)$, $v(y) \;=\; (0, 1)$, $\overline{F}^{v} \;=\; L$, and $\mathsf{gr}_v(F) = L[X, X^{-1}, Y, Y^{-1}]$ where $X = \widetilde{x}$ and $Y = \widetilde{y}$. We construct v-gauges on D.

Let w be the rank 1 coarsening of v; so, w is the restriction to F of the complete discrete y-adic valuation on $L(x)((y))$, with $\Gamma_{F,w} \;=\; \mathbb{Z}$,

$w(x) = 0$, $w(y) = 1$, $\overline{F}^w = L(x)$, and $\mathsf{gr}_w(F) = L(x)[Y,Y^{-1}]$. The map $\varepsilon\colon \Gamma_{F,v} \to \Gamma_{F,w}$ associated with the coarsening from v to w is given by $(\ell,m) \mapsto m$. This w has a unique, inertial, extension to K, again denoted w, with $\Gamma_{K,w} = \Gamma_{F,w}$, $w(s) = 0$, $\overline{K}^w = L(s)$, and $\mathsf{gr}_w(K) = L(s)[Y,Y^{-1}]$. However, since $v(1+x) = 0$ and $\overline{1+x} = \overline{1} \in (\overline{F}^v)^{\times 2}$, this v has two extensions to K, say v_1 and v_2, with each v_ℓ an immediate extension of v with $v_\ell(s) = 0$. If we let η_1 be the $(s-1)$-adic discrete valuation on $\overline{K}^w = L(s)$ and η_2 the $(s+1)$-adic valuation on $\overline{K}^w$ (each of which restricts to the x-adic valuation on $\overline{F}^v$), then the v_ℓ are the composite valuations, say $v_1 = \eta_1 * w$ and $v_2 = \eta_2 * w$. Thus, $v_1(s-1) = (1,0)$, so $\overline{s} = 1$ in $\overline{K}^{v_1}$, while $v_2(s+1) = (1,0)$, so $\overline{s} = -1$ in $\overline{K}^{v_2}$.

Note that $\overline{1+y} = \overline{1}$ in $\overline{K}^w$. Hence $D \otimes_K K_h$ is split for any Henselization (K_h, w_h) of (K,w), as $1+y \in K_h^{\times 2}$. Hence, w cannot extend to a valuation on D, by Th. 4.1. Moreover, by Th. 4.30(i), for any w-gauge β on D, we have $\mathsf{gr}_\beta(D) \cong_g M_2(\mathsf{gr}_w(K))(0,\delta)$ for some δ in a totally ordered group containing $\Gamma_{K,w}$. Likewise, the valuations v_ℓ cannot extend to valuations on D, and for any v_ℓ-gauge α_ℓ on D, the associated graded ring $\mathsf{gr}_{\alpha_\ell}(D)$ has the form $M_2(\mathsf{gr}_v(K))(0,\gamma_\ell)$.

For any valuation u on K with *char* $\overline{K}^u \neq 2$, we can use the quaternion base $\mathcal{B} = (1,i,j,k)$ of D to build a u-gauge[1] on D: define a u-norm φ on D with $\mathcal{B}$ a splitting base, and set

$$\varphi(1) = \tfrac{1}{2}u(1) = 0, \qquad \varphi(i) = \tfrac{1}{2}u(i^2) = \tfrac{1}{2}u(s),$$
$$\varphi(j) = \tfrac{1}{2}u(j^2) = \tfrac{1}{2}u(1+y), \quad \varphi(k) = \tfrac{1}{2}u(k^2) = \tfrac{1}{2}u(-s(1+y)).$$

Then, for all a, b, c, $d \in K$, set

$$\varphi(a+bi+cj+dk) = \mathit{min}\big(u(a), u(b)+\varphi(i), u(c)+\varphi(j), u(d)+\varphi(k)\big).$$

It is easy to check that for all $p, q \in \mathcal{B}$, we have

$$\varphi(pq) = \tfrac{1}{2}u((pq)^2) = \varphi(p)+\varphi(q);$$

hence φ is surmultiplicative, by Lemma 3.23. Furthermore, in $\mathsf{gr}_\varphi(D)$ we have $\widetilde{i}^2 = \widetilde{s}$, $\widetilde{j}^2 = \widetilde{1+y}$, and $\widetilde{i}\,\widetilde{j} = -\widetilde{j}\,\widetilde{i}$, which (as *char* $\overline{K}^u \neq 2$) shows that $\mathsf{gr}_\varphi(D)$ is the central simple graded quaternion algebra $\big(\widetilde{s}, \widetilde{1+y} \,/\, \mathsf{gr}_u(K)\big)$. Thus, φ is a u-gauge on D. Let α_ℓ be the v_ℓ-gauge φ obtained this way by taking $u = v_\ell$, for $\ell = 1,2$. Then,

$$\alpha_\ell(a+bi+cj+dk) = \mathit{min}\big(v_\ell(a), v_\ell(b), v_\ell(c), v_\ell(d)\big) \text{ for all } a,b,c,d \in K,$$

so $\Gamma_{D,\alpha_\ell} = \Gamma_{K,v} = \mathbb{Z}\times\mathbb{Z}$ and

$$\mathsf{gr}_{\alpha_\ell}(D) \cong_g \big((-1)^{\ell-1}, 1 / \mathsf{gr}_{v_\ell}(K)\big) \cong_g M_2\big(K[X,X^{-1},Y,Y^{-1}]\big)(0,0).$$

---

[1] This construction is an instance of an *armature gauge*, see §7.2.3.

Let β be the w-gauge φ on D obtained analogously by taking $u = w$. So,

$$\mathsf{gr}_\beta(D) \cong_g \big(s, 1/\,\mathsf{gr}_w(K)\big) \cong_g M_2\big(L(s)[Y, Y^{-1}]\big)(0,0).$$

It is clear from the construction that β is the w-coarsening of α_1 and also of α_2.

Consider the v-value function $\alpha = min(\alpha_1, \alpha_2)$ on D. Since w is the finest common coarsening of v_1 and v_2 on K, and α_1 and α_2 have the same w-coarsening, Th. 4.47 shows that α is a v-gauge on D. The gauge is given explicitly as follows: let $\psi = min(v_1, v_2)$, which is the unique v-gauge on K by Cor. 4.37. Then,

$$\alpha(a + bi + cj + dk) = min\big(\psi(a), \psi(b), \psi(c), \psi(d)\big) \quad \text{for all} \quad a, b, c, d \in K$$

and

$$\mathsf{gr}_\alpha(D) \cong_g \mathsf{gr}_{\alpha_1}(D) \times \mathsf{gr}_{\alpha_2}(D).$$

We can build other v_ℓ- and w-gauges on D from gauges on scalar extensions of D. For example, let $t = \sqrt{1+y}$ and set $N = K(t) = L(s,t)$, so $[N{:}K] = 2$. Let

$$A = D \otimes_K N = \big(s, t^2/N\big) \cong M_2(N).$$

Since $1+y$ is a square in the Henselization K_h of K re w, the field N embeds in K_h. We may view N as a small but sufficient step from K toward K_h—sufficient because N already splits D. There are two immediate extensions of w from K to N which differ by whether $\overline{t} = 1$ or $\overline{t} = -1$. We fix the extension with $\overline{t} = 1$ and call it w'. Likewise, each v_ℓ has two immediate extensions from K to N, and we fix and call v'_ℓ the extension satisfying $\overline{t} = 1$; thus, v'_ℓ has rank 1 coarsening w'. Let $A = M_2(N)$. In A, let

$$i = \left(\begin{smallmatrix} 0 & s \\ 1 & 0 \end{smallmatrix}\right), \quad j = \left(\begin{smallmatrix} t & 0 \\ 0 & -t \end{smallmatrix}\right), \quad \text{and} \quad k = ij = \left(\begin{smallmatrix} 0 & -st \\ t & 0 \end{smallmatrix}\right).$$

Since $i^2 = \left(\begin{smallmatrix} s & 0 \\ 0 & s \end{smallmatrix}\right)$, $j^2 = \left(\begin{smallmatrix} t^2 & 0 \\ 0 & t^2 \end{smallmatrix}\right) = \left(\begin{smallmatrix} 1+y & 0 \\ 0 & 1+y \end{smallmatrix}\right)$, and $ji = -ij$, we can identify

$$D = K + Ki + Kj + Kk \subseteq A = M_2(N).$$

For any valuation u on N and any γ in a totally ordered abelian group containing $\Gamma_{N,u}$, there is a u-gauge $\rho = \rho_{u,\gamma}$ on A given by

$$\rho\left(\begin{smallmatrix} c_{11} & c_{12} \\ c_{21} & c_{22} \end{smallmatrix}\right) = min\big(u(c_{11}),\, u(c_{12}) + \gamma,\, u(c_{21}) - \gamma,\, u(c_{11})\big),$$

with

$$\Gamma_{A,\rho} = \Gamma_{N,u} \cup (\gamma + \Gamma_{N,u}) \cup (-\gamma + \Gamma_{N,u}) \text{ and } \mathsf{gr}_\rho(A) \cong_g M_2(\mathsf{gr}_u(N))(0,\gamma).$$

Then, $\rho|_D$ is a surmultiplicative $u|_K$-value function on D. Since

$$a + bi + cj + dk = \left(\begin{smallmatrix} a+ct & (b-dt)s \\ b+dt & a-ct \end{smallmatrix}\right) \quad \text{for } a,\, b,\, c,\, d \in K,$$

we have

$$\rho(a+bi+cj+dk) \;=\; \min\big(u(a+ct), u((b-dt)s)+\gamma, u(b+dt)-\gamma, u(a-ct)\big).$$

Specifically, choose $u = v_1'$ and let γ be some $\gamma_1 \in \mathbb{Q}\times\mathbb{Q} = \mathbb{H}(\Gamma_{N,v_1'})$; let α_1' be the restriction to D of ρ_{v_1',γ_1} on A. If $\gamma_1 = 0$, then α_1' is the gauge α_1 constructed earlier. But now choose $\gamma_1 \notin \mathbb{Z}\times\mathbb{Z} = \Gamma_{K,v_1}$, and for convenience assume $\gamma_1 > 0$. Then, $(1,i,j,k)$ is no longer a splitting base for α_1' on D; indeed, $\widetilde{i} = \left(\begin{smallmatrix} 0 & 0 \\ 1 & 0 \end{smallmatrix}\right)^{\sim} = \widetilde{k}$. Nonetheless, we can obtain a splitting base as follows: We have $v_1((1-t)^n) = (0,n)$ while $v_1((1+t)^n) = 0$ for any $n \in \mathbb{Z}$. Choose a positive integer n sufficiently large that $(0,n) > 2\gamma_1$ in $\mathbb{Q}\times\mathbb{Q}$. By separating even and odd degree monomials in $(1-t)^n$, write

$$(1-t)^n \;=\; f(t^2) + g(t^2)t \;=\; b + dt,$$

where $b = f(t^2) \in K$ and $d = g(t^2) \in K$. Then, $b - dt = (1+t)^n$. So,

$$(bi+dk)^{\sim} \;=\; \left(\begin{smallmatrix} 0 & (1+t)^n s \\ (1-t)^n & 0 \end{smallmatrix}\right)^{\sim} \;=\; \left(\begin{smallmatrix} 0 & 2^n \\ 0 & 0 \end{smallmatrix}\right)^{\sim} \quad \text{in } \mathsf{gr}_{\alpha_1'}(D).$$

Using this, one can check that $(1,i,j,bi+dk)$ is a v_1-splitting base for α_1 on D, and $\mathsf{gr}_{\alpha_1'}(D) \cong_g M_2(\mathsf{gr}_{v_1}(K))(0,\gamma_1)$, which is graded simple. Hence, α_1' is a v_1-gauge on D.

Likewise, by choosing any $\gamma_2 \in \mathbb{Q}\times\mathbb{Q}$ and using $u = v_2'$, we obtain a v_2'-gauge ρ_{v_2',γ_2} on A whose restriction to D, call it α_2', is a v_2-gauge on D with $\mathsf{gr}_{\alpha_2'}(D) \cong_g M_2(\mathsf{gr}_{v_2}(K))(0,\gamma_2)$. Let $\varepsilon\colon \mathbb{Q}\times\mathbb{Q} \to \mathbb{Q}$ be projection onto the second component. Then, the w-coarsening of α_ℓ' coincides with the restriction to D of the w'-gauge $\rho_{w',\varepsilon(\gamma_\ell)}$ on A. Thus, α_1' and α_2' have the same w-coarsening if and only if $\varepsilon(\gamma_1) = \varepsilon(\gamma_2)$. When this occurs (and only then), $\alpha' = \min(\alpha_1',\alpha_2')$ is a v-gauge by Th. 4.47 and $\mathsf{gr}_{\alpha'}(D) \cong_g \mathsf{gr}_{\alpha_1'}(D) \times \mathsf{gr}_{\alpha_2'}(D)$. If we choose γ_2 so that $\varepsilon(\gamma_2) = \varepsilon(\gamma_1)$ but $\gamma_2 \notin (\mathbb{Z}\times\mathbb{Z}) \cup (\gamma_1 + \mathbb{Z}\times\mathbb{Z}) \cup (-\gamma_1 + \mathbb{Z}\times\mathbb{Z})$, then $\Gamma_{D,\alpha_1'} \neq \Gamma_{D,\alpha_2'}$; thus, the graded simple components of $\mathsf{gr}_{\alpha'}(D)$ are not graded isomorphic even though they are Brauer-equivalent in $\mathsf{Br}(\mathsf{gr}_v(F))$.

4.4 Existence of gauges on semisimple algebras

Let v be an arbitrary valuation on a field F and let (F_h, v_h) be a Henselization of (F,v). Let A be a (finite-dimensional) semisimple F-algebra. The goal of this section is to prove the following existence criterion for gauges on A:

Theorem 4.50. *The following conditions are equivalent:*

(a) *A carries a v-gauge;*
(b) *A carries a v-gauge with $\Gamma_A \subseteq \mathbb{H}(\Gamma_F)$;*
(c) *$A \otimes_F F_h$ carries a v_h-gauge;*
(d) *v is defectless in A (see Def. 4.19).*

In particular, these conditions hold for all A whenever char $\overline{F} = 0$ *or both* $\Gamma_F \cong \mathbb{Z}$ *and v is defectless in $Z(A)$.*

Recall that when v is Henselian, this theorem was proved earlier in Cor. 4.29. For arbitrary v, some implications in this theorem follow easily from earlier results: (a) $\Rightarrow$ (c) holds since if α is a v-gauge on A, then $\alpha \otimes v_h$ is a v_h-gauge on $A \otimes_F F_h$ by Prop. 3.42. (Note that $\mathsf{gr}(F_h) = \mathsf{gr}(F)$ since F_h is an immediate extension of F.) Also, (c) $\Leftrightarrow$ (d) holds since $A \otimes_F F_h$ has a v_h-gauge if and only if v_h is defectless in $A \otimes_F F_h$ by Cor. 4.29, and this holds if and only if v is defectless in A by definition. (b) $\Rightarrow$ (a) is obvious. All of this section will be aimed toward proving (c) $\Rightarrow$ (b). The cycle of implications will then be complete. If *char* $\overline{F} = 0$ or $\Gamma_F \cong \mathbb{Z}$ and v is defectless in $Z(A)$, then we have seen in Prop. 4.21(i), (iv) that condition (d) holds. In view of the compatibility of gauges with direct product decompositions of algebras (see Prop. 3.36), it suffices to prove (c) $\Rightarrow$ (b) when A is a simple F-algebra.

Let $A_h = A \otimes_F F_h$. If A_h carries a v_h-gauge, then by Cor. 4.29 A_h has a v_h-gauge also satisfying $\Gamma_{A_h} \subseteq \mathbb{H}(\Gamma_F)$. Our goal is to show that A then has a v-gauge with $\Gamma_A \subseteq \mathbb{H}(\Gamma_F)$. The main difficulty for this is that in general the restriction to A of a v_h-gauge on A_h may not be a v-norm, hence not a v-gauge. This restriction property does hold when $rk(v) = 1$, see Prop. 4.54 and Ex. 4.58. As a key step toward completing the proof of the theorem when $rk(v) > 1$, we consider in the next subsection the problem of restriction of norms in a general context. We will obtain an inductive criterion for the restriction of a v_h-norm to be a v-norm when $rk(v)$ is finite, see Prop. 4.56.

4.4.1 Descent of norms

Throughout this subsection, we use the following notation: V is a finite-dimensional vector space over a field F with a valuation v, and (F_h, v_h) is a Henselization of (F, v). Let (K, v_K) be an arbitrary valued field extension of (F, v), and let $\alpha\colon V \otimes_F K \to \Gamma \cup \{\infty\}$ be a v_K-norm. We identify V with its canonical image in $V \otimes_F K$. For any $x \in V$ and $c \in K$ we have

$$\alpha(x \otimes c) \;=\; \alpha((x \otimes 1) \cdot c) \;=\; \alpha|_V(x) + v_K(c).$$

Therefore, for any $\gamma \in \Gamma_{V,\alpha|_V}$ and $\delta \in \Gamma_K$ the usual F-bilinear map $V \times K \to V \otimes_F K$ sends $V_{\geq\gamma} \times K_{\geq\delta}$ into $(V \otimes_F K)_{\geq\gamma+\delta}$. Likewise, $V_{>\gamma} \times K_{\geq\delta}$ and $V_{\geq\gamma} \times K_{>\delta}$ map into $(V \otimes_F K)_{>\gamma+\delta}$. Consequently, there is a well-defined induced map $V_\gamma \times K_\delta \to (V \otimes_F K)_{\gamma+\delta}$ given by $(\widetilde{x}, \widetilde{c}) \mapsto \widetilde{x \otimes c}$. The direct sum of these maps over all such γ, δ yields a map

$$\mathsf{gr}_{\alpha|_V}(V) \times \mathsf{gr}_{v_K}(K) \longrightarrow \mathsf{gr}_\alpha(V \otimes_F K),$$

which is clearly $\mathsf{gr}_v(F)$-bilinear; hence, there is a canonical map

$$\chi\colon \mathsf{gr}_{\alpha|_V}(V)\otimes_{\mathsf{gr}_v(F)}\mathsf{gr}_{v_K}(K)\longrightarrow \mathsf{gr}_\alpha(V\otimes_F K)$$

which maps $\widetilde{x}\otimes\widetilde{c}$ to $\widetilde{x\otimes c}$ for $x\in V$ and $c\in K$.

On the other hand, recall from Th. 3.21 that if $\alpha|_V$ is a v-norm on V, then there is a canonical isomorphism of $\mathsf{gr}_{v_K}(K)$-vector spaces

$$\Psi\colon \mathsf{gr}_{\alpha|_V\otimes v_K}(V\otimes_F K)\xrightarrow{\sim}\mathsf{gr}_{\alpha|_V}(V)\otimes_{\mathsf{gr}_v(F)}\mathsf{gr}_{v_K}(K)$$

which maps $\widetilde{x\otimes c}$ to $\widetilde{x}\otimes\widetilde{c}$ for $x\in V$ and $c\in K$.

Lemma 4.51. *The following conditions are equivalent:*

(a) $\alpha|_V$ *is a v-norm on V and* $\alpha=\alpha|_V\otimes v_K$.
(b) V *contains a K-splitting base of the norm α on* $V\otimes_F K$.
(c) $\alpha|_V$ *is a v-norm on V and the canonical map χ is injective.*

When these conditions hold, the map χ is a graded isomorphism, which is the inverse of Ψ, and $\Gamma_{V\otimes_F K,\alpha}=\Gamma_{V,\alpha|_V}+\Gamma_{K,v_K}$.

Proof. (b) $\Rightarrow$ (a): Every splitting base $(e_i)_{i=1}^n$ of $V\otimes_F K$ for α which lies in V is clearly also a splitting base of V for $\alpha|_V$, so $\alpha|_V$ is a v-norm. Furthermore, by the definition of $\alpha|_V\otimes v_K$, we have for any $k_1,\ldots,k_n\in K$,

$$\begin{aligned}(\alpha|_V\otimes v_K)\big(\sum_{i=1}^n e_i\otimes k_i\big) &= \min_{1\le i\le n}\big(\alpha|_V(e_i)+v_K(k_i)\big)\\ &= \min_{1\le i\le n}\big(\alpha(e_i\otimes 1)+v_K(k_i)\big)\\ &= \alpha\big(\sum_{i=1}^n (e_i\otimes 1)\cdot k_i\big) = \alpha\big(\sum_{i=1}^n e_i\otimes k_i\big),\end{aligned}$$

showing that $\alpha|_V\otimes v_K=\alpha$.

(a) $\Rightarrow$ (c): When (a) holds, $\alpha|_V$ is a norm, and χ is clearly the inverse of Ψ, so χ is injective.

(c) $\Rightarrow$ (b): Suppose (c) holds. Let $(e_i)_{i=1}^n$ be an F-splitting base for $\alpha|_V$ on V, where $n=[V{:}K]$. Then, by Prop. 3.5 $\widetilde{e_1},\ldots,\widetilde{e_n}$ are $\mathsf{gr}_v(F)$-linearly independent in $\mathsf{gr}_{\alpha|_V}V$. Hence, $\widetilde{e_1}\otimes\widetilde{1},\ldots,\widetilde{e_n}\otimes\widetilde{1}$ are $\mathsf{gr}_{v_K}(K)$-linearly independent in $\mathsf{gr}_{\alpha|_V}(V)\otimes_{\mathsf{gr}_v(F)}\mathsf{gr}_{v_K}(K)$. By the injectivity of χ the $\chi(\widetilde{e_i}\otimes\widetilde{1})=\widetilde{e_i\otimes 1}$ are $\mathsf{gr}_{v_K}(K)$-linearly independent in $\mathsf{gr}_\alpha(V\otimes_F K)$. But, since α is a v_K-norm, we have

$$[\mathsf{gr}_\alpha(V\otimes_F K):\mathsf{gr}_{v_K}(K)] = [(V\otimes_F K){:}K] = [V{:}F] = n.$$

So, $\big(\widetilde{e_i\otimes 1}\big)_{i=1}^n$ is a homogeneous $\mathsf{gr}_{v_K}(K)$-vector space base of $\mathsf{gr}_\alpha(V\otimes_F K)$; hence, $(e_i\otimes 1)_{i=1}^n$ is a K-splitting base for α on $V\otimes_F K$ by Cor. 3.6.

When the conditions (a)–(c) hold, we have

$$\Gamma_{V\otimes_F K,\alpha} = \Gamma_{V\otimes_F K,\alpha|_V\otimes v_K} = \Gamma_{V,\alpha|_V}+\Gamma_{K,v_K}$$

and the map χ is the inverse of Ψ, so χ is an isomorphism. □

Note that under the hypotheses of Lemma 4.51 if $\alpha|_V$ is a norm then $\alpha \geq \alpha|_V \otimes v_K$. For, if $(e_i)_{i=1}^n$ is a v-splitting base for $\alpha|_V$ on V, then for any $k_1, \ldots, k_n \in K$,

$$\begin{aligned} \alpha\big(\sum_{i=1}^n e_i \otimes k_i\big) &\geq \min_{1\leq i\leq n} \big(\alpha(e_i \otimes k_i)\big) \\ &= \min_{1\leq i\leq n} \big(\alpha(e_i) + v_K(k_i)\big) = (\alpha|_V \otimes v_K)\big(\sum_{i=1}^n e_i \otimes k_i\big). \end{aligned}$$

We next show that the inequality $\alpha \geq \alpha|_V \otimes v_K$ is actually an equality when K is immediate over F, though not in general.

Corollary 4.52. *Let (K, v_K) be an immediate valued field extension of (F, v) and let $\alpha\colon V \otimes_F K \to \Gamma \cup \{\infty\}$ be a v_K-norm. If $\alpha|_V$ is a norm, then $\alpha = \alpha|_V \otimes v_K$. So, the canonical map $\mathsf{gr}_{\alpha|_V}(V) \to \mathsf{gr}_\alpha(V \otimes_F K)$ is a graded isomorphism $\mathsf{gr}_{\alpha|_V}(V) \cong_g \mathsf{gr}_\alpha(V \otimes_F K)$, and $\Gamma_V = \Gamma_{V\otimes_F K}$.*

Proof. Since v_K is immediate over v, we have $\mathsf{gr}_{v_K}(K) = \mathsf{gr}_v(F)$, so the canonical map χ of Lemma 4.51(c) is the injection $\mathsf{gr}_{\alpha|_V} V \hookrightarrow \mathsf{gr}_\alpha(V \otimes_F K)$ arising from the canonical inclusion $V \hookrightarrow V \otimes_F K$. Thus, the corollary follows from Lemma 4.51, using $\Gamma_K = \Gamma_F$ for the last assertion. □

Example 4.53. *A norm α on $V \otimes_F K$ with $\alpha|_V$ a norm but $\alpha|_V \otimes v_K < \alpha$.* Let (K, v_K) be an extension of (F, v) with $\overline{F} \subsetneqq \overline{K}$. Let $\xi \in K$ satisfy $v_K(\xi) = 0$ and $\overline{\xi} \notin \overline{F}$, and let V be a 2-dimensional F-vector space with base (e_1, e_2). Let $f = e_1 \otimes 1 + e_2 \otimes \xi \in V \otimes_F K$, and consider the v_K-norm α on $V \otimes_F K$ with splitting base $(e_1 \otimes 1, f)$ such that

$$\alpha(e_1 \otimes 1) = 0 \qquad \text{and} \qquad \alpha(f) > 0.$$

Then, as $e_2 = (f - e_1)\xi^{-1}$, we have for $c_1, c_2 \in F$

$$\begin{aligned} \alpha|_V(e_1c_1 + e_2c_2) &= \alpha\big((e_1(c_1 - \xi^{-1}c_2) + fc_2\big) \\ &= \min\big(\alpha(e_1) + v_K(c_1 - \xi^{-1}c_2), \alpha(f) + v_K(c_2)\big) \\ &= \min\big(v(c_1), v(c_2)\big). \end{aligned}$$

Hence, (e_1, e_2) is a v-splitting base of V for $\alpha|_V$, showing that $\alpha|_V$ is a v-norm on V. However, $\alpha|_V \otimes v_K < \alpha$ since

$$(\alpha|_V \otimes v_K)(f) = \min\big(v_K(1), v_K(\xi)\big) = 0 < \alpha(f).$$

Thus, the first condition in Lemma 4.51(a) holds, but not the second. The lemma shows that V does not contain any splitting base for the v_K-norm α on $V \otimes_F K$. Also, the second condition in part (c) of the lemma fails, since the canonical map χ satisfies

$$\chi(\widetilde{e_1} \otimes 1 + \widetilde{e_2} \otimes \widetilde{\xi}) = \widetilde{e_1 \otimes 1 + e_2 \otimes \xi} = 0.$$

We now turn to the descent problem for $(K, v_K) = (F_h, v_h)$ a Henselization of (F, v). The rank one case is easy:

Proposition 4.54. *Let $\alpha\colon V\otimes_F F_h \to \Gamma\cup\{\infty\}$ be a v_h-norm. If $rk(\Gamma_F)=1$ and $\Gamma_V \subseteq \gamma+\mathbb{H}(\Gamma_F)$ for some $\gamma\in\Gamma$, then $\alpha|_V$ is a v-norm and $\alpha=\alpha|_V\otimes v_h$.*

Proof. Let $(e_i)_{i=1}^n$ be an arbitrary F-base of V and let $x\in V\otimes_F F_h$; write

$$x = \sum_{i=1}^n e_i\otimes k_i \qquad \text{for some } k_i\in F_h.$$

Since $rk(\Gamma_F)=1$, the field F is dense in F_h for the topology of the valuation v_h, because F is dense in its completion $\widehat{F}$ and F_h embeds in $\widehat{F}$ by Endler [72, Th. 17.18]. Furthermore, Γ_F is dense in its divisible hull $\Gamma_F\otimes_{\mathbb{Z}}\mathbb{Q}$. For each $i=1,\ldots,n$, since $\alpha(x)-\alpha(e_i\otimes 1)\in\Gamma_F\otimes_{\mathbb{Z}}\mathbb{Q}$, we may therefore find an element $f_i\in F$ such that

$$v_h(k_i-f_i) > \alpha(x)-\alpha(e_i\otimes 1).$$

Let $y=\sum_{i=1}^n e_i\otimes f_i=\sum_{i=1}^n e_if_i\otimes 1\in V$. Then,

$$\begin{aligned}\alpha(x-y) &= \alpha\Big(\sum_{i=1}^n e_i\otimes(k_i-f_i)\Big) \ge \min_{1\le i\le n}\big(\alpha(e_i\otimes(k_i-f_i))\big)\\ &= \min_{1\le i\le n}\big(\alpha(e_i)+v_h(k_i-f_i)\big) > \alpha(x).\end{aligned}$$

Hence,

$$\widetilde{x} = \widetilde{y} \in \mathsf{gr}_{\alpha|_V}(V).$$

This proves that the monomorphism $\mathsf{gr}_{\alpha|_V}(V)\hookrightarrow\mathsf{gr}_\alpha(V\otimes_F F_h)$ is an isomorphism. Hence, as α is a norm,

$$[\mathsf{gr}_{\alpha|_V}(V):\mathsf{gr}_v(F)] = [\mathsf{gr}_\alpha(V\otimes_F F_h):\mathsf{gr}_{v_h}(F_h)] = [(V\otimes_F F_h){:}F_h] = [V{:}F],$$

which shows that $\alpha|_V$ is a v-norm. □

Remark 4.55. The proof holds without change when the completion $(\widehat{F},\widehat{v})$ is substituted for the Henselization (F_h,v_h) in Prop. 4.54.

Now, suppose $\Gamma=\mathbb{H}(\Gamma_F)$, with $rk(\Gamma)>1$, and let Δ be a nontrivial convex subgroup of Γ. As in §3.2.7 and §4.1, we consider the canonical map $\varepsilon\colon\Gamma\to\Gamma/\Delta=\Lambda$ and the coarser valuation

$$w = \varepsilon\circ v\colon F \longrightarrow \Lambda\cup\{\infty\}.$$

Let $(F_{h,v},v_h)$ be a Henselization of (F,v) and $(F_{h,w},w_h)$ a Henselization of (F,w). We also consider the coarsening of v_h, call it w',

$$w' = \varepsilon\circ v_h\colon F_{h,v} \longrightarrow \Lambda\cup\{\infty\}.$$

Since v_h is Henselian, its coarsening w' is also Henselian, see Prop. A.15. Moreover, as w' is an extension of w, we may assume that $(F_{h,w},w_h)\subseteq(F_{h,v},w')$, cf. Prop. A.31.

Let $\alpha\colon V\otimes_F F_{h,v}\to\Gamma\cup\{\infty\}$ be a v_h-norm, and let

$$\beta = \varepsilon\circ\alpha\colon V\otimes_F F_{h,v} \longrightarrow \Lambda\cup\{\infty\}.$$

By Prop. 3.53, the map β is a w'-norm.

Proposition 4.56. *Suppose Δ has rank 1, i.e., the residue valuation v/w has rank 1. If $\beta|_V$ is a w-norm and $\beta = \beta|_V \otimes w'$, then $\alpha|_V$ is a v-norm on V and $\alpha = \alpha|_V \otimes v_h$.*

Proof. To simplify notation, let

$$u\colon \overline{F}^{w} \longrightarrow \Delta \cup \{\infty\}$$

denote the valuation v/w on the residue field $\overline{F}^{w}$ induced by v, as in (4.1). Note that the value group of u is $\Delta_F = \Gamma_F \cap \Delta$ and, as Δ is divisible and torsion-free,

$$\Delta_F \otimes_{\mathbb{Z}} \mathbb{Q} \;=\; \big(\Gamma_F \otimes_{\mathbb{Z}} \mathbb{Q}\big) \cap \big(\Delta \otimes_{\mathbb{Z}} \mathbb{Q}\big) \;=\; \Gamma \cap \Delta \;=\; \Delta.$$

Let $\Lambda_V = \beta|_V(V \setminus \{0\}) \subseteq \Lambda$. Clearly, $\beta|_V = \varepsilon \circ (\alpha|_V)$. In order to show that $\alpha|_V$ is a norm, it therefore suffices, by Prop. 3.53, to show that each map

$$(\alpha|_V)_\lambda\colon V_\lambda^{\beta|_V} \longrightarrow \lambda \cup \{\infty\}, \qquad \text{for } \lambda \in \Lambda_V, \tag{4.34}$$

is a u-norm. To simplify notation, we write V_λ for $V_\lambda^{\beta|_V}$. Note that the canonical inclusion $V \hookrightarrow V \otimes_F F_{h,v}$ is compatible with the respective value functions $\beta|_V$ so β yields an injection $V_\lambda \hookrightarrow (V \otimes_F F_{h,v})_\lambda^\beta$; let V'_λ denote the image of V_λ. Then clearly $\alpha_\lambda|_{V'_\lambda} \cong (\alpha|_V)_\lambda$.

Let $u_h\colon \overline{F_{h,v}}^{w'} \to \Delta \cup \{\infty\}$ be the valuation v_h/w' induced by v_h. By Prop. A.31, $(\overline{F_{h,v}}^{w'}, u_h)$ is a Henselization of $(\overline{F}^{w}, u)$. Since α is a v_h-norm, Prop. 3.53 shows that

$$\alpha_\lambda\colon (V \otimes_F F_{h,v})_\lambda^\beta \longrightarrow \lambda \cup \{\infty\}$$

is a u_h-norm for every $\lambda \in \Lambda_V$. Since $(F_{h,v}, w')$ is an inertial extension of $(F_{h,w}, w_h)$, we have

$$\mathsf{gr}_{w'}(F_{h,v}) \cong_g \mathsf{gr}_{w_h}(F_{h,w}) \otimes_{\overline{F_{h,w}}^{w_h}} \overline{F_{h,v}}^{w'} \;=\; \mathsf{gr}_w(F) \otimes_{\overline{F}^w} \overline{F_{h,v}}^{w'}.$$

Because $\beta = \beta|_V \otimes w'$, this yields graded isomorphisms

$$\mathsf{gr}_\beta(V \otimes_F F_{h,v}) \cong_g \mathsf{gr}_{\beta|_V}(V) \otimes_{\mathsf{gr}_w(F)} \mathsf{gr}_{w'}(F_{h,v}) \cong_g \mathsf{gr}_{\beta|_V}(V) \otimes_{\overline{F}^w} \overline{F_{h,v}}^{w'}.$$

For any $\lambda \in \Lambda_F$, when we restrict these graded isomorphisms to the λ-component we obtain the $\overline{F_{h,v}}^{w'}$-vector space isomorphism

$$\psi\colon (V \otimes_F F_{h,v})_\lambda^\beta \xrightarrow{\sim} V_\lambda \otimes_{\overline{F}^w} \overline{F_{h,v}}^{w'}.$$

Let $\widehat{\alpha} = \alpha_\lambda \circ \psi^{-1}\colon V_\lambda \otimes_{\overline{F}^w} \overline{F_{h,v}}^{w'} \to \lambda \cup \{\infty\}$, which is the u_h-value function on $im(\psi)$ corresponding to α_λ on the domain of ψ. Since α_λ is a u_h-norm, so is $\widehat{\alpha}$. Because $(\overline{F_{h,v}}^{w'}, u_h)$ is a Henselization of $(\overline{F}^{w}, u)$ and λ is a coset of $\Delta = \Delta_F \otimes_{\mathbb{Z}} \mathbb{Q}$, which has rank 1, with Δ_F the value group of u, Prop. 4.54 applies to $\widehat{\alpha}$, and shows that $\widehat{\alpha}|_{V_\lambda}$ is a u-norm. Note that ψ maps the V'_λ

defined above after (4.34) to the copy of V_λ in $im(\psi)$. So, $\alpha_\lambda|_{V'_\lambda} \cong \widehat{\alpha}|_{V_\lambda}$. But we saw above that $(\alpha|_V)_\lambda \cong \alpha_\lambda|_{V'_\lambda}$. Since $\widehat{\alpha}|_{V_\lambda}$ is a u-norm, these isomorphisms show that $(\alpha|_V)_\lambda$ is also a u-norm. Thus, by Prop. 3.53 $\alpha|_V$ is a v-norm; then $\alpha = \alpha|_V \otimes v_h$ by Cor. 4.52. □

4.4.2 Existence of gauges

We use the notation introduced at the beginning of §4.4: (F, v) is an arbitrary valued field, with Henselization (F_h, v_h), and A is a semisimple F-algebra. Let $A_h = A \otimes_F F_h$. A part of the proof of Th. 4.50 was given in the comments after the statement of the theorem. To complete the proof, we must show that A carries a v-gauge with $\Gamma_A \subseteq \mathbb{H}(\Gamma_F)$ if v is defectless in A, or, equivalently, if A_h carries a v_h-gauge with $\Gamma_{A_h} \subseteq \mathbb{H}(\Gamma_F)$. If $rk(v) = 1$, Th. 4.50 follows from this proposition:

Proposition 4.57. *Let α be a v_h-gauge on A_h. If $\Gamma_{A_h} \subseteq \mathbb{H}(\Gamma_F)$ and $rk(v) = 1$, then $\alpha|_A$ is a v-gauge on A with* $\mathsf{gr}_{\alpha|_A}(A) = \mathsf{gr}_\alpha(A_h)$.

Proof. Proposition 4.54 shows that $\alpha|_A$ is a v-norm and $\alpha = \alpha|_A \otimes v_h$. Therefore,

$$\mathsf{gr}_\alpha(A_h) = \mathsf{gr}_{\alpha|_A}(A) \otimes_{\mathsf{gr}(F)} \mathsf{gr}_{v_h}(F_h).$$

Since F_h/F is an immediate extension, we have $\mathsf{gr}_{v_h}(F_h) = \mathsf{gr}(F)$, hence $\mathsf{gr}_\alpha(A_h) = \mathsf{gr}_{\alpha|_A}(A)$. Since α is a v_h-gauge, the graded algebra $\mathsf{gr}_\alpha(A_h)$ is graded semisimple, hence $\mathsf{gr}_{\alpha|_A}(A)$ is also graded semisimple, and $\alpha|_A$ is a v-gauge on A. □

The following example shows that the property in Prop. 4.57 is specific to valuations of rank 1.

Example 4.58. (*A gauge on A_h not restricting to a norm on A.*) Let k be any field with *char* $k \neq 2$, and let $F = k(x, y)$ with x and y algebraically independent over k. Let v be the valuation on F obtained by restriction from the canonical Henselian valuation on $k((x))((y))$, so $\Gamma_F = \mathbb{Z} \times \mathbb{Z}$ and $\overline{F} = k$. Let (F_h, v_h) be a Henselization of (F, v). Let $A = (1 + x,\ y/F)$, a quaternion division algebra over F, and let $A_h = A \otimes_F F_h$. The algebra A_h is split since $1 + x \in F_h^{\times 2}$. Therefore, we may find a v_h-gauge α on A_h such that $\Gamma_{A_h,\alpha} = \Gamma_F$. We claim that $\alpha|_A$ is not a v-norm on A.

Suppose the contrary. Then $\mathsf{gr}_{\alpha|_A}(A) = \mathsf{gr}_\alpha(A_h)$ by Cor. 4.52, so $\alpha|_A$ is a v-gauge. Consider the convex subgroup $\Delta = \mathbb{Z} \times \{0\} \subseteq \Gamma_F$ and the canonical epimorphism

$$\varepsilon \colon \Gamma_F \to \Gamma_F/\Delta = \mathbb{Z}.$$

Let $w = \varepsilon \circ v \colon F \to \mathbb{Z} \cup \{\infty\}$, which is the y-adic valuation on F, and let $\beta = \varepsilon \circ \alpha \colon A_h \to \mathbb{Z} \cup \{\infty\}$. Proposition 3.54 shows that $\beta|_A$ is a w-gauge on A.

However, the y-adic valuation w extends to A, so by Prop. 4.33(i) $\beta|_A$ is the (unique) valuation on A that extends w. In particular, if $j \in A$ satisfies $j^2 = y$ we must have $\beta|_A(j) = \frac{1}{2}$. This is a contradiction since $\Gamma_{A_h,\beta} = \mathbb{Z}$.

This example shows that we have to take a different approach to prove Th. 4.50 when $rk(v) > 1$. Just as for Th. 4.1, we argue by induction on $rk(v)$ if $rk(v)$ is finite, and we reduce the general case to the finite rank case by regarding F as the union of finitely generated subfields over the prime subfield.

The key to proving Th. 4.50 for v of finite rank is the following inductive lemma:

Lemma 4.59. *Let v be a valuation of rank n on F with $1 \le n < \infty$, and let w be the coarsening of v of rank $n-1$. Let A be a central simple F-algebra. Suppose that v is defectless in A and A has a w-gauge β with $\Gamma_{A,\beta} \subseteq \mathbb{H}(\Gamma_{F,w})$. Then, A has a v-gauge α such that β is the w-coarsening of α and $\Gamma_{A,\alpha} \subseteq \mathbb{H}(\Gamma_{F,v})$.*

Proof. Let Δ be the convex subgroup of Γ associated to w, let $\Lambda = \Gamma/\Delta$, and let $\varepsilon\colon \Gamma \to \Lambda$ be the canonical epimorphism. Thus, $w = \varepsilon \circ v\colon F \to \Lambda \cup \{\infty\}$, and $rk(\Delta) = rk(v) - rk(w) = 1$. We write $F_{h,v}$ for F_h. Let $w' = \varepsilon \circ v_h$, which is Henselian as v_h is Henselian; hence, $(F_{h,v}, w')$ contains a Henselization $(F_{h,w}, w_h)$ of (F, w). Note that

$$\Gamma_{F_h,w'} = \varepsilon(\Gamma_{F_{h,v},v_h}) = \varepsilon(\Gamma_{F,v}) = \Gamma_{F,w}.$$

Let D_h denote the central division algebra over $F_{h,v}$ associated to $A \otimes_F F_{h,v}$. Then, D_h is defectless for v_h, as v is defectless in A.

Since β is a w-gauge, by Prop. 3.27 we have a surmultiplicative w'-value function on A_h:

$$\beta \otimes w'\colon A_h = A \otimes_F F_{h,v} \longrightarrow \Lambda \cup \{\infty\}$$

such that

$$\mathsf{gr}_{\beta\otimes w'}(A_h) = \mathsf{gr}_\beta(A) \otimes_{\mathsf{gr}_w(F)} \mathsf{gr}_{w'}(F_{h,v}). \tag{4.35}$$

Now, $\mathsf{gr}_w(F) = \mathsf{gr}_{w_h}(F_{h,w})$ as w_h is immediate over w, and, since $(F_{h,v}, w')$ is inertial over $(F_{h,w}, w_h)$ by Prop. A.31,

$$\mathsf{gr}_{w'}(F_{h,v}) \cong_g \mathsf{gr}_{w_h}(F_{h,w}) \otimes_{\overline{F}^w} \overline{F_{h,v}}^{w'}.$$

Equation (4.35) thus yields

$$\mathsf{gr}_{\beta\otimes w'}(A_h) = \mathsf{gr}_\beta(A) \otimes_{\overline{F}^w} \overline{F_{h,v}}^{w'}.$$

Hence, for the quotient algebras,

$$q(\mathsf{gr}_{\beta\otimes w'}(A_h)) = q(\mathsf{gr}_\beta(A)) \otimes_{\overline{F}^w} \overline{F_{h,v}}^{w'}.$$

Since $\mathsf{gr}_\beta(A)$ is graded semisimple, by Prop. 2.28(iii) $q(\mathsf{gr}_\beta(A))$ is semisimple. Hence, $q(\mathsf{gr}_\beta(A)) \otimes_{\overline{F}^w} \overline{F_{h,v}}^{w'}$ is semisimple, as the field $\overline{F_{h,v}}^{w'}$ is separable over $\overline{F}^w$. It follows from the preceding equation and Prop. 2.28(iii) again that $\mathsf{gr}_{\beta\otimes w'}(A_h)$ is graded semisimple; hence, $\beta \otimes w'$ is a w'-gauge on A_h. By Th. 4.26, we may find a right D_h-vector space V and a w'-norm $\beta_V : V \to \Lambda \cup \{\infty\}$ with identifications

$$A_h = End_{D_h} V \qquad \text{and} \qquad \beta \otimes w' = End(\beta_V).$$

Moreover, the proof of Th. 4.26 shows that we can take V to be a minimal left ideal of A_h, and $\beta_V = (\beta \otimes w')|_V$. Hence,

$$\Gamma_{V,\beta_V} \subseteq \Gamma_{A_h,\beta\otimes w'} \subseteq \Gamma_{A,\beta} + \Gamma_{F_h,w'} \subseteq \mathbb{H}(\Gamma_{F,w}).$$

Choose a splitting base $(e_i)_{i=1}^n$ of V for β_V, and elements $\gamma_1, \ldots, \gamma_n \in \mathbb{H}(\Gamma_{F,v})$ such that

$$\varepsilon(\gamma_i) = \beta_V(e_i) \qquad \text{for } i = 1, \ldots, n.$$

Define a v_h-norm $\alpha_V : V \to \Gamma \cup \{\infty\}$ by the condition that $(e_i)_{i=1}^n$ is a splitting base of V and

$$\alpha_V(e_i) = \gamma_i \qquad \text{for } i = 1, \ldots, n.$$

Thus, $\varepsilon \circ \alpha_V = \beta_V$. Let also $\alpha = End(\alpha_V)$, which is a v_h-gauge on A_h by Prop. 3.34, since D_h is defectless for v_h. Moreover,

$$\Gamma_{A_h,\alpha} \subseteq \mathbb{H}(\Gamma_{V,\alpha_V}) \subseteq \mathbb{H}(\Gamma_{F,v}).$$

From the defining formula for an *End* gauge in (3.7), it is clear that $End(\beta_V) = \varepsilon \circ End(\alpha_V)$, i.e.,

$$\beta \otimes w' = \varepsilon \circ \alpha.$$

By Prop. 4.56, it follows that $\alpha|_A$ is a v-norm and $\alpha = \alpha|_A \otimes v_h$. Therefore, using the fact that $F_{h,v}/F$ is immediate, we have

$$\mathsf{gr}_\alpha(A_h) = \mathsf{gr}_{\alpha|_A}(A) \otimes_{\mathsf{gr}(F)} \mathsf{gr}_{v_h}(F_h) = \mathsf{gr}_{\alpha|_A}(A).$$

Since α is a v_h-gauge, the graded algebra $\mathsf{gr}_\alpha(A_h)$ is graded semisimple (actually graded simple by Th. 4.26), hence $\mathsf{gr}_{\alpha|_A}(A)$ is graded semisimple, and $\alpha|_A$ is a v-gauge on A. Furthermore, since $\beta \otimes w' = \varepsilon \circ \alpha$, we have $\alpha|_A = \varepsilon \circ (\beta \otimes w')|_A = \varepsilon \circ \beta$, i.e., β is the w-coarsening of α. □

We can now prove Th. 4.50 for finite rank valuations. The infinite rank case will be proved after Lemma 4.61 below.

Proposition 4.60. *Theorem 4.50 is true if* $rk(v) < \infty$.

Proof. It suffices to prove the theorem for the simple components of a semisimple F-algebra. So, assume A is simple. In light of the comments just after the theorem, it remains only to show (c)⇒(b). Let $n = rk(v)$, with $0 \le n < \infty$. We argue by induction on n. If $n = 0$, there is nothing to prove. The case $n = 1$ was proved in Prop. 4.57. Now assume $n > 1$, and let w be the coarsening of v with $rk(w) = n - 1$. Let $K = Z(A)$, and let

$v_1, \ldots, v_r$ be the extensions of v to K, and $w_1, \ldots, w_\ell$ the extensions of w to K. Assume condition (c): $A \otimes_F F_h$ has a v_h-gauge. Since (c)$\Leftrightarrow$(d) has been proved, v is defectless in A. Therefore, w is defectless in A by Prop. 4.23. Since the theorem holds for w by induction, (d)$\Rightarrow$(b) says that there is a w-gauge β on A with $\Gamma_{A,\beta} \subseteq \mathbb{H}(\Gamma_{F,w})$. Let β_j be the w_j-component of β for $j = 1, \ldots, \ell$. The construction of the β_j in Th. 4.42 shows that each $\Gamma_{A,\beta_j} \subseteq \Gamma_{A,\beta} \subseteq \mathbb{H}(\Gamma_{F,w})$. By Th. 4.47, β_j and β_k have the same w_{jk}-coarsening for all $j, k \in \{1, 2, \ldots, \ell\}$.

For each $i \in \{1, 2, \ldots, r\}$, let $j(i) \in \{1, 2, \ldots, \ell\}$ be the index such that $w_{j(i)}$ is the rank $n-1$ coarsening of v_i. For each i we apply Lemma 4.59 for the K-valuations v_i and $w_{j(i)}$ on the central simple K-algebra A. Note that v_i is defectless in A by definition, since v is defectless in A. The lemma shows that for each i there is a v_i-gauge η_i on A such that the w_j-coarsening of η_i is the $j(i)$-component $\beta_{j(i)}$ of β and $\Gamma_{A,\eta_i} \subseteq \mathbb{H}(\Gamma_{F,v})$. Let $\alpha = min\big(\eta_1, \ldots, \eta_r\big)$. To see that α is a v-gauge we must show that the η_i satisfy the compatibility condition in Th. 4.47. For this, take any distinct $i, k \in \{1, 2, \ldots, r\}$. Consider first the case where $j(i) = j(k)$. Then, $w_{j(i)}$ is the finest common coarsening v_{ik} of v_i and v_k. By construction, $\beta_{j(i)}$ is the v_{ik}-coarsening of both η_i and η_k. Now, consider the other case, where $j(i) \neq j(k)$. Then, v_{ik} must be coarser than $w_{j(i)}$ and coarser than $w_{j(k)}$, so v_{ik} must coincide with $w_{j(i),j(k)}$. Thus, the v_{ik}-coarsening of η_i coincides with the v_{ik}-coarsening of the $w_{j(i)}$-coarsening of η_i; this is the $w_{j(i),j(k)}$-coarsening of $\beta_{j(i)}$. Likewise, the v_{ik}-coarsening of η_k is the $w_{j(i),j(k)}$-coarsening of $\beta_{j(k)}$. These coarsenings of $\beta_{j(i)}$ and $\beta_{j(k)}$ agree by Th. 4.47. So, η_i and η_k have the same v_{ik}-coarsening. Since this is true in all possible cases, Th. 4.47 shows that α is a v-gauge. Furthermore,

$$\Gamma_{A,\alpha} \subseteq \bigcup_{i=1}^{r} \Gamma_{A,\eta_i} \subseteq \mathbb{H}(\Gamma_{F,v}),$$

proving (b). □

For the infinite rank case of Th. 4.50, we will need the following lemma.

Lemma 4.61. *Let $(F, v) \subseteq (L, v_L)$ be valued fields. Let A be a semisimple F-algebra with v-gauges α and α'. Then, $\alpha \otimes v_L$ is a v_L-gauge on $A \otimes_F L$ if and only if $\alpha' \otimes v_L$ is a v_L-gauge on $A \otimes_F L$.*

Proof. If A has simple components $A_1, \ldots, A_k$, then $\alpha|_{A_i}$ is a v-gauge for each i, by Prop. 3.36. Moreover, $\alpha \otimes v_L$ is a v_L-gauge on $A \otimes_F L$ if and only if each $\alpha|_{A_i} \otimes v_L$ is a v_L-gauge on $A_i \otimes_F L$. Hence, the lemma is true for the gauges on A if it is true for their restrictions to each A_i. It thus suffices to prove the lemma for simple algebras.

Assume that A is simple. Proposition 3.42 tells us that

$$\alpha \otimes v_L \text{ is a } v_L\text{-gauge} \quad \Leftrightarrow \quad Z(\mathsf{gr}_\alpha(A)) \otimes_{\mathsf{gr}(F)} \mathsf{gr}_{v_L}(L) \text{ is semisimple.} \tag{4.36}$$

Let $K = Z(A)$, let $v_1, \ldots, v_r$ be the extensions of v to K, and let α_i (resp. α_i') be the v_i-component of α (resp. α') as in Th. 4.42 for $i = 1, 2, \ldots, r$. By that theorem,

$$\mathsf{gr}_\alpha(A) \cong_g \prod_{i=1}^r \mathsf{gr}_{\alpha_i}(A), \qquad \text{so} \qquad Z(\mathsf{gr}_\alpha(A)) \cong_g \prod_{i=1}^r Z(\mathsf{gr}_{\alpha_i}(A)).$$

Likewise for α'. Let $(K_{h,v_i}, v_{i,h})$ be a Henselization of (K, v_i), let D_i be the associated division algebra of $A \otimes_K K_{h,v_i}$, and let u_i be the valuation on D_i extending $v_{i,h}$ on its center. By Th. 4.30(i), $Z(\mathsf{gr}_{\alpha_i}(A)) \cong_g Z(\mathsf{gr}_{u_i}(D_i))$. Likewise, $Z(\mathsf{gr}_{\alpha_i'}(A)) \cong_g Z(\mathsf{gr}_{u_i}(D_i))$. Hence,

$$\begin{aligned} Z(\mathsf{gr}_\alpha(A)) &\cong_g \prod_{i=1}^r Z(\mathsf{gr}_{\alpha_i}(A)) \cong_g \prod_{i=1}^r Z(\mathsf{gr}_{u_i}(D_i)) \\ &\cong_g \prod_{i=1}^r Z(\mathsf{gr}_{\alpha_i'}(A)) \cong_g Z(\mathsf{gr}_{\alpha'}(A)). \end{aligned}$$

Thus,

$$Z(\mathsf{gr}_\alpha(A)) \otimes_{\mathsf{gr}(F)} \mathsf{gr}_{v_L}(L) \cong_g Z(\mathsf{gr}_{\alpha'}(A)) \otimes_{\mathsf{gr}(F)} \mathsf{gr}_{v_L}(L).$$

Therefore, the criterion in (4.36) for $\alpha \otimes v_L$ to be a v_L-gauge is satisfied if and only if this criterion is satisfied for $\alpha' \otimes v_L$. □

We can now prove Th. 4.50 for valuations of arbitrary rank.

Proof of Th. 4.50, completed. Let (F_v, v_h) be a Henselization of (F, v), and let $A_h = A \otimes_F F_h$. In view of the comments after the statement of the theorem, it remains only to show that (c) A_h has a v_h-gauge implies (b) A has a v-gauge with $\Gamma_A \subseteq \mathbb{H}(\Gamma_F)$. Let α be a v_h-gauge on A_h. Fix an F-base $(a_i)_{i=1}^m$ of A and let $(\lambda_{ijk})_{i,j,k=1}^m$ be the corresponding structure constants, defined by the equation

$$a_i a_j = \sum_{k=1}^m \lambda_{ijk} a_k \qquad \text{for } i,\ j = 1,\ \ldots,\ m.$$

Let also $(\mu_{ij})_{i,j=1}^m$ be a set of elements in F_h such that $\left(\sum_{i=1}^m a_i \otimes \mu_{ij}\right)_{j=1}^m$ is a splitting F_h-base of A_h for α. Let $F_0 \subseteq F$ be the prime subfield, and let $F_1 = F_0(\lambda_{ijk} \mid 1 \le i, j, k \le m) \subseteq F$. The field F is the union of its subfields L that are finitely generated over F_1 (hence over F_0). Each such field L has a unique Henselization L_h for $v|_L$ lying in F_h, and the union $\bigcup_L L_h$ is Henselian and contains F, hence $F_h = \bigcup_L L_h$. Therefore, we may find a field $L \subseteq F$ finitely generated over F_1 such that $\mu_{ij} \in L_h$ for all $i,\ j = 1,\ \ldots,\ m$. Let $A(L) \subseteq A$ be the L-span of $(a_i)_{i=1}^m$, and let

$$\widehat{A} = A(L) \otimes_L L_h, \qquad \widehat{v} = v|_{L_h}, \qquad \text{and} \qquad \widehat{\alpha} = \alpha|_{\widehat{A}}.$$

So, $\widehat{A} \otimes_{L_h} F_h = A(L) \otimes_L F_h = A_h$. Since A_h is semisimple, $A(L)$ and $\widehat{A}$ must be semisimple. For a nontrivial nilpotent two-sided ideal of $A(L)$ (resp. $\widehat{A}$) would tensor up to a nontrivial nilpotent two-sided ideal of A_h. Since $\widehat{A}$ contains the splitting base $\left(\sum_{i=1}^{m} a_i \otimes \mu_{ij}\right)_{j=1}^{m}$ for α, Lemma 4.51 shows that $\widehat{\alpha}$ is a $\widehat{v}$-norm and $\alpha = \widehat{\alpha} \otimes v_h$. In particular,

$$\mathsf{gr}_\alpha(A_h) = \mathsf{gr}_{\widehat{\alpha}}(\widehat{A}) \otimes_{\mathsf{gr}(L_h)} \mathsf{gr}(F_h).$$

Since $\mathsf{gr}_\alpha(A_h)$ is graded semisimple, it follows that $\mathsf{gr}_{\widehat{\alpha}}(\widehat{A})$ also is graded semisimple, because any two-sided homogeneous nilpotent ideal J of $\mathsf{gr}_{\widehat{\alpha}}(\widehat{A})$ would yield a two-sided homogeneous nilpotent ideal $\mathsf{J} \otimes \mathsf{gr}(F_h)$ of $\mathsf{gr}_\alpha(A_h)$. Therefore, $\widehat{\alpha}$ must be a $\widehat{v}$-gauge on $\widehat{A}$.

Since L is finitely generated over its prime field F_0, inequality (A.21) and the Transcendence Degree Inequality (Th. A.35) show that $rk(v|_L) - 1$ is bounded above by the finite transcendence degree of L over F_0; so, $rk(v|_L) < \infty$. Note that $v_h|_L = v|_L$, and $(\widehat{L}, \widehat{v})$ is a Henselization of $(L, v|_L)$. Since $\widehat{A}$ $(= A(L) \otimes_L L_h)$ has a $\widehat{v}$-gauge, the finite rank case of Th. 4.50 proved in Prop. 4.60 above, applied to $(L, v|_L)$, shows that $A(L)$ carries a $v|_L$-gauge α' with $\Gamma_{A(L),\alpha'} \subseteq \mathbb{H}(\Gamma_{L,v|_L})$. Since the extension L_h/L is immediate, the tensor product $\alpha' \otimes \widehat{v}$ is a $\widehat{v}$-gauge on $\widehat{A}$ by Prop. 3.42, and we have

$$\mathsf{gr}_{\alpha'}(A(L)) \cong_g \mathsf{gr}_{\alpha' \otimes \widehat{v}}(\widehat{A}).$$

We now have two $v|_{L_h}$-gauges on $\widehat{A}$, namely $\widehat{\alpha}$ and $\alpha' \otimes \widehat{v}$. Since $\widehat{\alpha} \otimes v_h$ is our original gauge α on $\widehat{A} \otimes_{L_h} F_h = A_h$, Lemma 4.61 applied to $\widehat{A}$ and the valued field extension $(L_h, \widehat{v}) \subseteq (F_h, v_h)$ shows that $(\alpha' \otimes \widehat{v}) \otimes v_h$ must also be a v_h-gauge on A_h.

Consider now the v-value function $\alpha' \otimes v$ on $A(L) \otimes_L F = A$. This is our candidate for a v-gauge on A. Note that on the F-algebra

$$(A(L) \otimes_L F) \otimes_F F_h = (A(L) \otimes_L \widehat{L}) \otimes_{\widehat{L}} F_h$$

we have $(\alpha' \otimes v) \otimes v_h = (\alpha' \otimes \widehat{v}) \otimes v_h$ which is a v_h-gauge. But we must still work back to $\alpha' \otimes v$: By Prop. 3.42 for the extension $(L, v|_L) \subseteq (F, v)$, $\alpha' \otimes v$ is a v-gauge if and only if $Z(\mathsf{gr}_{\alpha'}(A(L))) \otimes_{\mathsf{gr}_{v|_L}(L)} \mathsf{gr}_v(F)$ is graded semisimple. For this, we have the following sequence of graded isomorphisms, which will be justified below.

$$\begin{aligned}
&Z(\mathsf{gr}_{\alpha'}(A(L)) \otimes_{\mathsf{gr}_{v|_L}(L)} \mathsf{gr}_v(F) \\
&\qquad \cong_g Z(\mathsf{gr}_{\alpha'}(A(L))) \otimes_{\mathsf{gr}_{v|_L}(L)} \mathsf{gr}_{v_h}(F_h) \\
&\qquad \cong_g \left[Z(\mathsf{gr}_{\alpha'}(A(L))) \otimes_{\mathsf{gr}_{v|_L}(L)} \mathsf{gr}_{\widehat{v}}(L_h)\right] \otimes_{\mathsf{gr}_{\widehat{v}}(L_h)} \mathsf{gr}_{v_h}(F_h) \\
&\qquad \cong_g Z\big(\mathsf{gr}_{\alpha' \otimes \widehat{v}}(\widehat{A})\big) \otimes_{\mathsf{gr}_{\widehat{v}}(L_h)} \mathsf{gr}_{v_h}(F_h) \\
&\qquad \cong_g Z\big(\mathsf{gr}_{(\alpha' \otimes \widehat{v}) \otimes v_h}(\widehat{A} \otimes_{L_h} F_h)\big).
\end{aligned} \tag{4.37}$$

The first isomorphism holds because $\mathsf{gr}_v(F) \cong_g \mathsf{gr}_{v_h}(F_h)$, as (F_h, v_h) is an immediate extension of (F, v). The second isomorphism follows from the inclusions of graded fields $\mathsf{gr}_{v|_L}(L) \subseteq \mathsf{gr}_{\widehat{v}}(L_h) \subseteq \mathsf{gr}_{v_h}(F_h)$. The third isomorphism is the isomorphism of centers (cf. Cor. 2.31) following from the isomorphism

$$\mathsf{gr}_{\alpha'\otimes\widehat{v}}(\widehat{A}) \cong_g \mathsf{gr}_{\alpha'}(A(L)) \otimes_{\mathsf{gr}_{v|_L}(L)} \mathsf{gr}_{\widehat{v}}(L_h)$$

given by Cor. 3.28 for the immediate extension $(L, v|_L) \subseteq (L_h, \widehat{v})$. This uses the fact that the $v|_L$-gauge α' is a $v|_L$-norm. The fourth isomorphism in (4.37) is another isomorphism of centers following from the isomorphism

$$\mathsf{gr}_{\alpha'\otimes\widehat{v}}(\widehat{A}) \otimes_{\mathsf{gr}_{\widehat{v}}(L_h)} \mathsf{gr}_{v_h}(F_h) \cong_g \mathsf{gr}_{(\alpha'\otimes\widehat{v})\otimes v_h}(\widehat{A} \otimes_{L_h} F_h)$$

given by Cor. 3.28 for the extension $(L_h, \widehat{v}) \subseteq (F_h, v_h)$. This uses the fact that the $\widehat{v}$-gauge $\alpha' \otimes \widehat{v}$ is a $\widehat{v}$-norm. The last term in (4.37) is graded semisimple since we have seen that $(\alpha' \otimes \widehat{v}) \otimes v_h$ is a v_h-gauge. Therefore, the first term in (4.37) is graded semisimple. As noted above, Prop. 3.42 thus implies that $\alpha' \otimes v$ is a v-gauge on A. Moreover,

$$\Gamma_{A,\alpha'\otimes v} \subseteq \Gamma_{A(L),\alpha'} + \Gamma_{F,v} \subseteq \mathbb{H}(\Gamma_{L,v|_L}) + \Gamma_{F,v} \subseteq \mathbb{H}(\Gamma_{F,v}).$$

Thus (b) holds for v and A, assuming (c). This completes the proof of Th. 4.50. □

Exercises

Exercise 4.1. Let v be a valuation on a field F, let w be any coarsening of v, and let K be a finite-degree field extension of F. Prove that v is defectless in K if and only if w is defectless in K and for each extension w_j of w to K, the residue valuation v/w on $\overline{F}^w$ is defectless in $\overline{K}^{w_j}$.

Exercise 4.2. Let p be a prime number and let G be a finite group. Assuming the order of G is prime to p, show that the group algebra $\mathbb{Q}_p[G]$ over the field of p-adic numbers is a direct product of matrix algebras over unramified field extensions of $\mathbb{Q}_p$. [Hint: Let v be the p-adic valuation on $\mathbb{Q}_p$. Define a v-gauge α on $\mathbb{Q}_p[G]$ by $\alpha(\sum_g a_g g) = min_g(v(a_g))$, and use Th. 4.28.]

Exercise 4.3. Let (F, v) be a valued field, and let A be a finite-dimensional simple F-algebra with a v-gauge α such that $\Gamma_{A,\alpha}$ lies in the divisible hull $\mathbb{H}(\Gamma_{F,v})$. Let $K = Z(A)$, and let v_k be an extension of v to K, and α_k the v_k-component of α. Prove that for any $a \in A \setminus \{0\}$,

$$\alpha_k(a) = max\{\alpha(ca) \mid c \in K^\times,\ v_k(c) = 0\}.$$

Notes

§4.1: Morandi's Henselization Theorem (Th. 4.1) was proved in Morandi [156, Th. 2]. Earlier, Ershov had asserted this theorem in [76], however his proof had an error. He later gave a valid proof in [77, Prop. 3]. Yet another proof, more ring-theoretic, can be found in Marubayashi et al. [132, Cor. 8.5, p. 43]. A slightly different version for valuations of rank 1, with the completion of the base field substituted for the Henselization, was proved earlier by Cohn [54, Th. 1]; see Prop. 1.16.

§4.2: Ostrowski proved the defect formula for field extensions, Prop. 4.4, for rank 1 Henselian valuations in [175, II, §55]. The defect formula for valued division algebras was first proved in the Henselian case (Th. 4.5) in Draxl's posthumous paper [64, Th. 2], then established in general (Th. 4.3) by Morandi [156, Th. 3], by using his Henselization theorem (Th. 4.1). Defectlessness for discrete valuations of rank 1 (Prop. 4.16) was observed by Tomchin–Yanchevskiĭ [249, Prop. 2.2] and by Morandi [159, p. 359]. The same authors constructed division algebras with defect, which can be any power of the residue characteristic dividing the dimension of the algebra: For any prime p and any integers n, ℓ, q with $2n \geq \ell \geq 0$ and $p \nmid q$ there exist division algebras of degree $p^n q$ and defect p^ℓ; see [249, Th. 4.2], [159, Th. 5, Cor. 7]. On the other hand, Tomchin–Yanchevskiĭ [249, Th. 3.1] also give a criterion for a division algebra of prime degree to be defectless: Let F be a Henselian field with $\mathit{char}\,\overline{F} = p > 0$. Every division algebra D of degree p with $Z(D)$ finite-dimensional over F is defectless if and only if every finite separable field extension of F is defectless. The latter condition was studied by Kuhlmann in [118], [119]: a valued field (F, v) is said to be *defectless* (resp. *separably defectless*) if the fundamental inequality (A.1) is an equality for every finite (resp. finite separable) field extension of F. By [119, Th. 2.3], a valued field is defectless (resp. separably defectless) if and only if its Henselization has the same property. Moreover, if F is a finitely generated field extension of a subfield F_0, and if v is a valuation on F extending a valuation v_0 on F_0 so that the Transcendence Degree Inequality of Th. A.35 is an equality, then (F, v) is defectless if (F_0, v_0) is defectless [118, Th. 1.1]. This applies in particular when the valuation v_0 is trivial, and yields many examples of defectless fields.

Example 4.17 is essentially due to Ostrowski [175, II, §39]. (His ground field was $\mathbb{Q}$ instead of $\mathbb{Q}_2$.) It is described in Efrat [70, Ex. 17.2.3, pp. 155–157] and in Ribenboim [203, p. 246]. Our account of it is based on Efrat's.

§4.3: Most of the material in §4.3.1–§4.3.3 first appeared in Tignol–Wadsworth [246, §1, §3]. Note that Th. 4.26 yields the following characterization of *End*-gauges: Let D be a division algebra over a field F with a Henselian valuation v, let v_D be the valuation on D extending v, and let M be a finite-dimensional right D-vector space. Assume that D is defectless over F. A surmultiplicative v-norm α on $\mathit{End}_D\, M$ has the form $\alpha = \mathit{End}(\alpha_M)$ for some v_D-norm α_M on M if and only if the graded algebra $\mathsf{gr}_\alpha(\mathit{End}_D\, M)$ is graded semisimple. Indeed, for multiplicative v-norms of the form $\alpha = \mathit{End}(\alpha_M)$ the algebra $\mathsf{gr}_\alpha(\mathit{End}_D\, M)$ is graded simple by Prop. 3.34. Conversely, if α is a surmultiplicative v-norm such that $\mathsf{gr}_\alpha(\mathit{End}_D\, M)$ is graded simple, then α is a v-gauge, so Th. 4.26 applies to show $\alpha = End(\alpha_M)$ for some v_D-norm α_M on M. This observation points to an important motivation for the use of the associated graded structures in the study of value functions.

The material in §4.3.4 is from Ferreira–Wadsworth [79].

§4.4: The results on the descent of norms (§4.4.1) are taken from Tignol–Wadsworth [247, §5], where gauges are used to define for central simple algebras with involution a notion that plays a role analogous to valuations for division algebras. Here, an involution on a central simple algebra is meant to be a ring anti-automorphism of order 2. An involution σ on a central simple F-algebra A is said to be *anisotropic* if $\sigma(a)a = 0$ holds for $a \in A$ only when $a = 0$. Let $F_0 \subseteq F$ be the subfield fixed under σ. The involution σ is said to be *orthogonal* if $F_0 = F$ and σ becomes isomorphic to the adjoint involution of a symmetric nonalternating bilinear form after scalar extension to a splitting field of A. (Involutions that are not orthogonal fall into two other types: symplectic and unitary; see Knus et al. [115] for background information on involutions on central simple algebras.) Tignol–Wadsworth assume in [247] that F_0 carries a valuation v, and when $char\,\overline{F_0} = 2$ they avoid the case where σ is orthogonal. In all the other cases, they show in [247, Th. 2.2] that if F_0 is Henselian and σ is anisotropic, then there exists a unique v-value function α on A such that $\alpha(\sigma(a)a) = 2\alpha(a)$ for all $a \in A$. This value function is actually a v-gauge, and it is the unique v-gauge such that $\alpha \circ \sigma = \alpha$. They also prove in [247, Th. 6.1] an analogue of Morandi's Henselization Theorem: Unless σ is orthogonal and $char\,\overline{F_0} = 2$, there exists a v-gauge α on A such that $\alpha(\sigma(a)a) = 2\alpha(a)$ for all $a \in A$ if and only if σ remains anisotropic after scalar extension to a Henselization of F_0.

Minimal gauges and intersections of Dubrovin valuation rings. Let (F, v) be a valued field, let $V = \mathcal{O}_{F,v}$, and let A be a central simple F-algebra. As mentioned in the notes to Ch. 3, there exists a Dubrovin valuation ring B of A with $Z(B) = B \cap F = V$, and B is unique up to conjugacy. Suppose v is defectless in A. Then $B = \mathcal{O}_{A,\alpha}$ for some residually simple v-gauge α on A if and only if B is integral over V; moreover, each residually simple v-gauge α' of A is determined by $\mathcal{O}_{A,\alpha'}$ (see Prop. 3.48), which is a conjugate of B. Of course, if B is not integral over V, then B cannot be a gauge ring since such rings are always integral over their centers (see Lemma 4.24). In that case, there can be no residually simple v-gauge on A. A recent paper by Ferreira–Wadsworth [79] describes what happens then:

While Dubrovin valuation rings are in general not integral over their centers, Gräter found in [88] special families of V-Dubrovin valuation rings $B_1, \ldots, B_k$ of A such that $S = B_1 \cap \ldots \cap B_k$ is integral over V and the B_i and their overrings coincide with the localizations of S at its prime ideals. Such rings S have other very nice properties (e.g., they are semilocal left and right Bézout rings, and Morandi proved in [158] that there is an approximation theorem for $B_1, \ldots, B_k$). Gräter also showed that S is determined uniquely up to conjugacy by V and A. The number k of B_i in the intersection equals the number of maximal ideals of S, and is thus an invariant of A and v. Let $\xi_{v,A}$ denote this number k, which Gräter called the *extension number* of v in A. (The same number had appeared earlier as n_B/t_B in an Ostrowski-type theorem for Dubrovin valuation rings in Wadsworth [255].) For example, if $rk(v) = 1$, then it is known that $\xi_{v,A} = 1$, which is equivalent to the condition that the V-Dubrovin valuation rings of A are integral over V. If $rk(v) = 2$, let w be the rank 1 coarsening of v, let $(F_{h,w}, w_h)$ be the Henselization of (F, w), and D_w the associated division algebra of $A \otimes_F F_{h,w}$; then, $\xi_{v,A}$ equals the number of valuations of $Z(\overline{D_w}^{w_h})$ extending the residue valuation v/w on $\overline{F}^{w}$. The book by Marubayashi et al. [132] is a good reference for all of these results.

If α is a v-gauge on A, let $\omega(\alpha)$ denote the number of simple components of the semisimple $\overline{F}$-algebra $A_0 = \mathcal{O}_{A,\alpha}/\mathfrak{m}_{A,\alpha}$. It is proved by Wadsworth–Ferreira [79] that

$$\omega(\alpha) \geq \xi_{v,A}.$$

The v-gauge α is said to be *minimal* if $\omega(\alpha) = \xi_{v,A}$. It is shown in [79] that whenever v is defectless in A (so A has a v-gauge by Th. 4.50) then in fact A has a minimal gauge. The proof is substantially different from the one given here for Th. 4.50. It is shown further that if α is a minimal gauge of A, then its gauge ring $\mathcal{O}_{A,\alpha}$ is one of Gräter's special intersections of V-Dubrovin valuation rings. Conversely, if v is defectless in A, then any of Gräter's special intersections is the gauge ring $\mathcal{O}_{A,\alpha}$ of some minimal v-gauge α of A, though α is not uniquely determined by $\mathcal{O}_{A,\alpha}$. These results on minimal gauges generalize from central simple F-algebras A to arbitrary (finite-dimensional) semisimple F-algebras. See [79, §4].

Chapter 5
Graded and Valued Field Extensions

We pursue in this chapter the investigation of valuations through the graded structures associated to the valuation filtration. The graded field of a valued field is an enhanced version of the residue field, inasmuch as it encapsulates information about the value group in addition to the residue field. It thus captures much of the structure of the field, particularly in the Henselian case. This point is made clear in §5.2, where we show that—when the ramification is tame—Galois groups and their inertia subgroups of Galois extensions of valued fields can be determined from the corresponding extension of graded fields. Henselian fields are shown to satisfy a tame lifting property from graded field extensions, generalizing the inertial lifting property. In §5.1, we lay the groundwork for the subsequent developments by an independent study of graded fields, their algebraic extensions and their Galois theory.

As in Chapter 2, we fix throughout this chapter a torsion-free abelian group Γ, whose operation is additive. This group is to contain the grade sets of all the graded algebras we consider; therefore, to allow ramified extensions we assume Γ is divisible.

5.1 Graded field extensions

In many ways, graded fields are close to fields. Through the quotient field construction laid out in §5.1.1, the category of algebraic graded field extensions of a given graded field F is embedded as a full subcategory in the category of algebraic field extensions of the quotient field $q(\mathsf{F})$, see §5.1.2. Yet, the grading imposes some restrictions, so that in nonzero characteristic it may occur that some algebraic extensions of $q(\mathsf{F})$ are not quotient fields of algebraic graded extensions of F. The grading also brings to the theory of graded fields an abelian group aspect, so that each graded field F appears as a hybrid of its degree 0 component F_0, which is a field, and its grade set Γ_{F}, which

J.-P. Tignol, A.R. Wadsworth, *Value Functions on Simple Algebras, and Associated Graded Rings*, Springer Monographs in Mathematics,
DOI 10.1007/978-3-319-16360-4_5

is a torsion-free abelian group. Ramification theory, discussed in §5.1.3 is an important tool in the analysis of graded field extensions: its role is to separate the field-theoretic aspects related to extensions of the degree 0 component from the abelian group aspects reflecting value group extensions. In the final §5.1.4, we develop the Galois theory of algebraic graded field extensions, which reduces to the classical theory for ungraded fields through the quotient field construction. Ramification theory enters the picture since the grading leads to the definition of the inertia subgroup, a canonical subgroup of the Galois group.

The analogy between graded fields and Henselian-valued fields should be kept in mind throughout this section. It will be made explicit in §5.2.

Throughout this section, we let $\mathsf{F} = \bigoplus_{\gamma\in\Gamma} \mathsf{F}_\gamma$ be an arbitrary graded field, as defined in §2.1.1.

5.1.1 The quotient field

Let $\mathsf{F}^\natural$ be the underlying ungraded ring of F. Recall from Prop. 2.3 that $\mathsf{F}^\natural$ has no zero-divisors; we may thus consider its quotient field, for which we use the notation $q(\mathsf{F})$, as in §2.2:

$$q(\mathsf{F}) = Quot(\mathsf{F}^\natural).$$

This field contains F_0 as a subfield.

Recall also from Prop. 2.3 that the invertible elements in F are the nonzero homogeneous elements, and that the degree map fits into an exact sequence of abelian groups

$$1 \longrightarrow \mathsf{F}_0^\times \longrightarrow \mathsf{F}^\times \xrightarrow{deg} \Gamma_\mathsf{F} \longrightarrow 0,$$

where $\Gamma_\mathsf{F} = \{\gamma \in \Gamma \mid \mathsf{F}_\gamma \neq \{0\}\}$ is the grade group of F. If this exact sequence splits, we can find a system of homogeneous elements $(x^\gamma)_{\gamma\in\Gamma_\mathsf{F}}$ with $\mathsf{F}_\gamma = \mathsf{F}_0 \cdot x^\gamma$ for each $\gamma \in \Gamma_\mathsf{F}$ and $x^\gamma \cdot x^\delta = x^{\gamma+\delta}$ for γ, $\delta \in \Gamma_\mathsf{F}$. Thus, F is identified with a group ring:

$$\mathsf{F} = \mathsf{F}_0[\Gamma_\mathsf{F}].$$

In this case, F is said to be *of group-ring type*. This is always the case when Γ_F is finitely generated, for then it is a free abelian group. See Exercise 5.1 for an example of a graded field that is *not* of group-ring type.

Lemma 5.1. *Let* F *be a graded field. Every finitely generated subring* $R \subseteq \mathsf{F}^\natural$ *is contained in a subring isomorphic to a Laurent polynomial ring in finitely many indeterminates* $\mathsf{F}_0[t_1, t_1^{-1}, \ldots, t_r, t_r^{-1}]$.

Proof. Suppose the subring $R \subseteq \mathsf{F}^\natural$ is generated by a finite number of elements $g_1, \ldots, g_s$. Let $\Delta \subseteq \Gamma_\mathsf{F}$ be the subgroup generated by the degrees of the homogeneous components of $g_1, \ldots, g_s$, and let $\mathsf{E} = \bigoplus_{\delta \in \Delta} \mathsf{F}_\delta$, which is a graded subfield of F. Thus, $R \subseteq \mathsf{E}^\natural$ and Δ is a finitely generated torsion-free abelian group, hence $\Delta \cong \mathbb{Z}^r$ for some integer $r \geq 0$. Any splitting of the degree map $deg\colon \mathsf{E}^\times \to \Delta$ yields an isomorphism

$$\mathsf{E} \cong \mathsf{F}_0[\Delta] \cong \mathsf{F}_0[t_1, t_1^{-1}, \ldots, t_r, t_r^{-1}]. \qquad \square$$

Proposition 5.2. *For any graded field* F*, the integral domain* $\mathsf{F}^\natural$ *is integrally closed, and* F_0 *is algebraically closed in* $q(\mathsf{F})$.

Proof. Suppose a, $b \in \mathsf{F}$ are such that $b \neq 0$ and $ab^{-1} \in q(\mathsf{F})$ is integral over $\mathsf{F}^\natural$. Fix some polynomial $P = X^n + a_{n-1}X^{n-1} + \ldots + a_0 \in \mathsf{F}^\natural[X]$ such that $P(ab^{-1}) = 0$. Lemma 5.1 shows that a, b, a_{n-1}, $\ldots$, a_0 are all contained in a subring $R \subseteq \mathsf{F}^\natural$ isomorphic to a Laurent polynomial ring in finitely many indeterminates over F_0. The ring R is integrally closed as it is a unique factorization domain; hence, $ab^{-1} \in R \subseteq \mathsf{F}^\natural$. Likewise, if ab^{-1} is algebraic over F_0, then $ab^{-1} \in \mathsf{F}_0$ since F_0 is algebraically closed in R. $\square$

5.1.2 Algebraic extensions

A graded field K is called a *graded field extension* of a graded field F if F is a graded subfield of K.

Proposition 5.3. *For a graded field extension* K *of* F*, the following conditions are equivalent:*

(a) *every element in* K *is a root of some nonzero polynomial in* $\mathsf{F}[X]$*;*
(b) *every homogeneous element in* K *is a root of some nonzero polynomial in* $\mathsf{F}[X]$ *with homogeneous coefficients;*
(c) $\mathsf{K}^\natural$ *is integral over* $\mathsf{F}^\natural$*;*
(d) $q(\mathsf{K})$ *is algebraic over* $q(\mathsf{F})$.

When these equivalent conditions hold, every element of $q(\mathsf{K})$ *has the form* ab^{-1} *for some* $a \in \mathsf{K}$ *and some* $b \in \mathsf{F}$, $b \neq 0$. *Moreover, we have*

$$\mathsf{F}^\natural = \mathsf{K}^\natural \cap q(\mathsf{F}), \qquad q(\mathsf{K}) = \mathsf{K}^\natural \otimes_{\mathsf{F}^\natural} q(\mathsf{F}), \qquad [\mathsf{K}{:}\mathsf{F}] = [q(\mathsf{K}){:}q(\mathsf{F})],$$

and $\mathsf{K}^\natural$ *is the integral closure of* $\mathsf{F}^\natural$ *in* $q(\mathsf{K})$.

Proof. (a) $\Rightarrow$ (b) Let $u \in \mathsf{K}$ be a nonzero homogeneous element, and suppose

$$a_n u^n + \ldots + a_1 u + a_0 = 0 \qquad \text{for some } a_0, \ldots, a_n \in \mathsf{F} \text{ with } a_n \neq 0. \quad (5.1)$$

Let $a_n' \in \mathsf{F}$ be a nonzero homogeneous component of a_n and let $\gamma = deg\, a_n'$. The homogeneous component of degree $\gamma + n\, deg\, u$ in (5.1) is

$$a_n' u^n + \ldots + a_1' u + a_0' = 0,$$

where a_i' is the homogeneous component of degree $\gamma + (n-i)\,deg\, u$ of a_i, for $i = 0, \ldots, n$. Therefore, u is a root of a nonzero polynomial in $\mathsf{F}[X]$ with homogeneous coefficients.

(b) $\Rightarrow$ (c) Since nonzero homogeneous elements in F are invertible, it follows from (b) that every homogeneous element in K is integral over $\mathsf{F}^\natural$. Therefore, we have (c) since every element in K is a sum of finitely many homogeneous elements.

(c) $\Rightarrow$ (d) The field $q(\mathsf{K})$ is generated by the elements of $\mathsf{K}^\natural$, which are integral over $\mathsf{F}^\natural$, hence algebraic over $q(\mathsf{F})$. Therefore $q(\mathsf{K})$ is algebraic over $q(\mathsf{F})$.

(d) $\Rightarrow$ (a) Let $u \in \mathsf{K}$. Condition (d) shows that $P(u) = 0$ for some nonzero polynomial $P \in q(\mathsf{F})[X]$. Clearing denominators in P yields (a).

Now, assume (a)–(d) hold and let $u \in q(\mathsf{K})$. By (d), we may find $a_{n-1}, \ldots, a_0, b \in \mathsf{F}$ such that $b \neq 0$ and

$$u^n + b^{-1}a_{n-1}u^{n-1} + \ldots + b^{-1}a_0 \;=\; 0.$$

By multiplying by b^n, we obtain

$$(bu)^n + a_{n-1}(bu)^{n-1} + \ldots + b^{n-1}a_0 \;=\; 0,$$

hence bu is integral over $\mathsf{F}^\natural$. Therefore $bu \in \mathsf{K}^\natural$ since $\mathsf{K}^\natural$ is integrally closed by Prop. 5.2. Letting $a = bu \in \mathsf{K}$, we have $u = ab^{-1}$ with $b \in \mathsf{F}$, $b \neq 0$. Since $\mathsf{K}^\natural \otimes_{\mathsf{F}^\natural} q(\mathsf{F})$ may be identified with the set of fractions in $q(\mathsf{K})$ whose denominator lies in F, we thus have $q(\mathsf{K}) = \mathsf{K}^\natural \otimes_{\mathsf{F}^\natural} q(\mathsf{F})$. Since $\mathsf{K}^\natural$ is integral over $\mathsf{F}^\natural$ by (c) and is also integrally closed, it follows that $\mathsf{K}^\natural$ is the integral closure of $\mathsf{F}^\natural$ in $q(\mathsf{K})$. Moreover, since $\mathsf{F}^\natural$ is integrally closed, we have $\mathsf{K}^\natural \cap q(\mathsf{F}) = \mathsf{F}^\natural$. To complete the proof, observe that any base $(u_i)_{i\in I}$ of K over F is also a $q(\mathsf{F})$-base of $q(\mathsf{K})$; hence, $[\mathsf{K}{:}\mathsf{F}] = [q(\mathsf{K}){:}q(\mathsf{F})]$. □

When the equivalent conditions of Prop. 5.3 hold, we say that K is an *algebraic graded field extension* of F. Examples of algebraic graded field extensions are obtained as in the ungraded case as residue rings of polynomial rings modulo the ideal generated by an irreducible polynomial. However, this polynomial has to satisfy certain conditions, which are spelled out next.

Definition 5.4. Let F be a graded field and let $\delta \in \Gamma$, not necessarily in Γ_{F}. A polynomial $f = \sum_{i=0}^{n} a_i X^i \in \mathsf{F}[X]$ is said to be *δ-homogenizable* if each a_i is homogeneous in F and

$$deg\, a_j - deg\, a_i \;=\; \delta(i-j) \qquad \text{for all nonzero } a_i,\, a_j.$$

If $a_n \neq 0$, this condition can be rephrased as

$$deg\, a_i \;=\; \delta(n-i) + deg\, a_n \qquad \text{for all nonzero } a_i.$$

We say that f is *homogenizable* if it is δ-homogenizable for some δ, which is then uniquely determined unless f is a monomial (or 0). The condition on f

(and its name) can be understood by considering a grading on $\mathsf{F}[X]$: let

$$\mathsf{F}[X]^{(\delta)}_\gamma = \bigoplus_{i=0}^{\infty} \mathsf{F}_{\gamma - i\delta} \cdot X^i \qquad \text{for } \gamma \in \Gamma. \tag{5.2}$$

So, we have a direct sum decomposition

$$\mathsf{F}[X] = \bigoplus_{\gamma\in\Gamma} \mathsf{F}[X]^{(\delta)}_\gamma,$$

which turns $\mathsf{F}[X]$ into a graded F-algebra in which the indeterminate X is homogeneous of degree δ. Let $\mathsf{F}[X]^{(\delta)}$ denote $\mathsf{F}[X]$ with the grading given in (5.2). Then, a polynomial f is δ-homogenizable just when f is homogeneous in $\mathsf{F}[X]^{(\delta)}$.

Note that for nonzero $f, g \in F[X]$, if fg is δ-homogenizable, then f and g are each δ-homogenizable. This follows by the same argument as for Prop. 2.3(ii), as $\mathsf{F}[X]^{(\delta)}$ has no zero-divisors and $\Gamma_{\mathsf{F}[X]^{(\delta)}}$ is torsion-free.

The divisibility theory for $\mathsf{F}[X]^{(\delta)}$ in the graded setting is strongly analogous to that for a polynomial ring over a field in the ungraded setting. Note specifically that the division algorithm can be used to divide by any homogenizable polynomial, since the leading coefficient is invertible. As in the ungraded case, we may then show:

Proposition 5.5. *For any $\delta \in \Gamma$, the graded ring $\mathsf{F}[X]^{(\delta)}$ is a graded principal ideal domain and a graded unique factorization domain, i.e., it is a graded integral domain in which every homogeneous ideal is generated by a single homogeneous element, and every nonconstant homogeneous polynomial can be written as a product of (homogeneous) irreducibles, and this expression is unique up to units and to the order of factors. Also, if $h \in \mathsf{F}[X]^{(\delta)}$ is homogeneous and irreducible, then h is a prime element and $\mathsf{F}[X]^{(\delta)}/(h)$ is a graded field.*

Proof (sketch). Take any nonzero homogeneous $f, g \in \mathsf{F}[X]^{(\delta)}$ with $\mathit{deg}\, f = \gamma$ and $\mathit{deg}\, g = \varepsilon$. Observe that when we apply the division algorithm to write

$$g = qf + r \qquad \text{with} \qquad q, r \in \mathsf{F}[X] \text{ and } \mathit{deg}_X\, r < \mathit{deg}_X\, f \text{ or } r = 0,$$

then q and r are homogeneous of respective degrees $\varepsilon - \gamma$ and ε. Thus, the usual argument shows that any nonzero homogeneous ideal of $\mathsf{F}[X]^{(\delta)}$ is a principal ideal generated by any nonzero homogeneous element of minimal degree in X.

Now take an irreducible homogeneous polynomial h of $\mathsf{F}[X]^{(\delta)}$. Because homogeneous ideals are principal and h is irreducible, (h) must be a maximal (proper) homogeneous ideal of $\mathsf{F}[X]^{(\delta)}$. Therefore, with respect to the grading inherited from $F[X]^{(\delta)}$, the factor ring $\mathsf{F}[X]^{(\delta)}/(h)$ is a graded simple commutative finite-dimensional graded F-algebra. Hence, $\mathsf{F}[X]^{(\delta)}/(h)$ is a graded field. This implies that $\mathsf{F}[X]/(h)^\natural$ is an integral domain, and hence h is a prime element of $\mathsf{F}[X]^\natural$. Because the homogeneous irreducibles of $\mathsf{F}[X]^{(\delta)}$

are all prime elements, the usual argument yields the unique factorization of homogeneous elements into products of irreducibles. □

The following observation will be particularly useful:

Proposition 5.6. *Let $f \in \mathsf{F}[X]$ be a nonzero δ-homogenizable polynomial. If g, $h \in q(\mathsf{F})[X]$ satisfy $f = gh$ and g is monic, then g, $h \in \mathsf{F}[X]$ and g and h are δ-homogenizable. In particular, f is irreducible in $q(\mathsf{F})[X]$ if and only if f is irreducible in $\mathsf{F}[X]$.*

Proof. Since the leading coefficient of f is homogeneous in F, hence a unit, we may assume that f is monic. So, every root of g in a field containing $q(\mathsf{F})$ is integral over F, since it is also a root of f. The coefficients of g, which are, up to sign, the elementary symmetric polynomials of the roots of g, are thus also integral over F. Since $\mathsf{F}^\natural$ is integrally closed by Prop. 5.2, these coefficients lie in F. So, $g \in \mathsf{F}[X]$, and the division algorithm shows that $h \in \mathsf{F}[X]$. Then, g and h are δ-homogenizable since f is δ-homogenizable, as noted after Def. 5.4. □

The following proposition shows that the minimal polynomials of homogeneous elements in an algebraic graded field extension are homogenizable. It also highlights a very significant fact: the grading on an algebraic graded field extension K of F is actually determined by the grading on F and the ring structure of $\mathsf{K}^\natural$.

Proposition 5.7. *Let K be an algebraic graded field extension of a graded field F, let $u \in \mathsf{K} \setminus \{0\}$, and let $P = X^n + a_{n-1}X^{n-1} + \ldots + a_0 \in q(\mathsf{F})[X]$ be the minimal polynomial of u over $q(\mathsf{F})$. Then $a_{n-1}, \ldots, a_0 \in \mathsf{F}$. Moreover,*

(i) *if u is homogeneous of degree δ, then P is δ-homogenizable;*
(ii) *if P is δ-homogenizable, then u is homogeneous of degree δ;*
(iii) *if u is homogeneous of degree δ, then $\mathsf{F}[u]$ is a graded field with $[\mathsf{F}[u]{:}\mathsf{F}] = n$ and*

$$\mathsf{F}[u] \cong_g \mathsf{F}[X]^{(\delta)}/(P).$$

Proof. In an algebraic closure of $q(\mathsf{F})$, we may find $u_1, \ldots, u_n$ such that

$$P = \prod_{i=1}^{n}(X - u_i).$$

Then $a_0, \ldots, a_{n-1}$ are (up to sign) the elementary symmetric functions in $u_1, \ldots, u_n$. Since u is integral over $\mathsf{F}^\natural$ by Prop. 5.3, there is a monic polynomial $Q \in \mathsf{F}[X]$ such that $Q(u) = 0$, hence P divides Q in $q(\mathsf{F})[X]$. It follows that $u_1, \ldots, u_n$ are roots of Q; hence, they are integral over $\mathsf{F}^\natural$, and therefore $a_0, \ldots, a_{n-1}$ also are integral over $\mathsf{F}^\natural$. Since $\mathsf{F}^\natural$ is integrally closed by Prop. 5.2, it follows that $a_0, \ldots, a_{n-1} \in \mathsf{F}$.

If u is homogeneous, then the proof of (a) $\Rightarrow$ (b) in Prop. 5.3 shows that $a_0, \ldots, a_{n-1}$ are homogeneous and satisfy $deg\, a_i = (n-i)\, deg\, u$, proving (i). On the other hand, suppose P is δ-homogenizable. Since $X - u$ divides P in $\mathsf{K}[X]$, it must also be δ-homogenizable, by Prop. 5.6; hence u is homogeneous with

$$deg\, u \;=\; deg\, X \;=\; \delta,$$

proving (ii).

For (iii), suppose u is homogeneous of degree δ. The evaluation homomorphism $\mathrm{ev}_u \colon F[X]^{(\delta)} \to \mathsf{F}[u]$ given by $f \mapsto f(u)$ is a graded ring epimorphism with kernel $(P \cdot q(\mathsf{F})[X]) \cap \mathsf{F}[X]^{(\delta)}$. This intersection equals $P \cdot \mathsf{F}[X]^{(\delta)}$, as P is monic. Thus, ev_u induces a graded isomorphism

$$\mathsf{F}[u] \;\cong_g\; \mathsf{F}[X]^{(\delta)}/(P),$$

where $\mathsf{F}[X]^{(\delta)}/(P)$ has the grading induced by the grading on $\mathsf{F}[X]^{(\delta)}$. Since P is δ-homogenizable in $\mathsf{F}[X]$ and irreducible in $q(\mathsf{F})[X]$, it is irreducible in $\mathsf{F}[X]$ by Prop. 5.6. Hence, Prop. 5.5 and the isomorphism above show that $\mathsf{F}[u]$ is a graded field, with

$$[\mathsf{F}[u]{:}\mathsf{F}] \;=\; [\mathsf{F}[X]^{(\delta)}/(P){:}\mathsf{F}] \;=\; deg_X\, P \;=\; n. \qquad \square$$

Corollary 5.8. *For a graded field extension* K *of a graded field* F*, the following conditions are equivalent:*

(a) K *is an algebraic graded field extension of* F*;*
(b) K_0 *is an algebraic extension of* F_0 *and the factor group* $\Gamma_{\mathsf{K}}/\Gamma_{\mathsf{F}}$ *is torsion.*

Proof. (a) $\Rightarrow$ (b) By (a), every element $u \in \mathsf{K}_0$ is algebraic over $q(\mathsf{F})$. Proposition 5.7(i) shows that the coefficients of its minimal polynomial lie in F_0, hence $\mathsf{K}_0/\mathsf{F}_0$ is an algebraic extension. Proposition 5.7(i) also shows that if $u \in \mathsf{K}$ is homogeneous and its minimal polynomial over $q(\mathsf{F})$ has degree n, then $n\, deg\, u \in \Gamma_{\mathsf{F}}$. Therefore, $\Gamma_{\mathsf{K}}/\Gamma_{\mathsf{F}}$ is a torsion group.

(b) $\Rightarrow$ (a) Assuming (b), we claim that every homogeneous element $u \in \mathsf{K}$ is a root of some nonzero polynomial in $\mathsf{F}[X]$, which suffices to prove (a). Since $\Gamma_{\mathsf{K}}/\Gamma_{\mathsf{F}}$ is a torsion group, there is an integer $n > 0$ such that $n\, deg\, u \in \Gamma_{\mathsf{F}}$. We may therefore find a nonzero homogeneous element $a \in \mathsf{F}$ such that $deg(au^n) = 0$. Since the extension $\mathsf{K}_0/\mathsf{F}_0$ is algebraic, it follows that au^n is a root of some nonzero polynomial in $\mathsf{F}_0[X]$, proving the claim. $\square$

We now change the viewpoint and use homogenizable polynomials to construct graded field extensions.

Lemma 5.9. *Let* F *be a graded field, let* E *be a field containing* $\mathsf{F}^\natural$ *and let* $u \in E \setminus \{0\}$ *be a root of some homogenizable monic polynomial* $P \in \mathsf{F}[X]$*. Then, there is a uniquely determined grading on the subring* $\mathsf{F}^\natural[u] \subseteq E$ *that turns it into a graded field extension of* F*. With respect to this grading,* u *is homogeneous.*

Proof. Since P is homogenizable, it is δ-homogenizable for some $\delta \in \Gamma$. Let $f \in q(\mathsf{F})[X]$ be the (monic) minimal polynomial of u over $q(\mathsf{F})$. Since $P(u) = 0$, we have $f \mid P$ in $q(\mathsf{F})[X]$. By Prop. 5.6 above, $f \in \mathsf{F}[X]$ and f is also δ-homogenizable and irreducible in $\mathsf{F}[X]$, since it is irreducible in $q(F)[X]$. By Prop. 5.5, $\mathsf{F}[X]^{(\delta)}/(f)$ is a graded field with the grading inherited from $\mathsf{F}[X]^{(\delta)}$. Since f is the minimal polynomial of u over $q(\mathsf{F})$, and $(f \cdot q(F)[X]) \cap \mathsf{F}[X] = f \cdot \mathsf{F}[X]$ as f is monic, evaluation of polynomials at u induces an isomorphism

$$\mathsf{F}[X]/(f) \xrightarrow{\sim} \mathsf{F}^\natural[u] \subseteq E,$$

which can be used to transfer the graded field structure to $\mathsf{F}^\natural[u]$. Since X is homogeneous in $F[X]^{(\delta)}$, its image u is homogeneous in $\mathsf{F}^\natural[u]$. Uniqueness of the grading on $\mathsf{F}^\natural[u]$ follows from Prop. 5.7, which shows that any nonzero $t \in \mathsf{F}^\natural[u]$ is homogeneous if and only if its minimal polynomial P_t over $q(\mathsf{F})$ is homogenizable in $\mathsf{F}[X]$, and when t is homogeneous, P_t determines $deg\, t$. For, when P_t is homogenizable and not a monomial, it is clear from the definition that P_t is δ-homogenizable for a unique $\delta \in \Gamma$. □

Proposition 5.10. *Let* F *be a graded field and let* $q(\mathsf{F})_{\mathrm{alg}}$ *be an algebraic closure of its quotient field. There is a unique algebraic graded field extension* $\mathsf{F}_{\mathrm{alg}}$ *of* F *with the following properties:*

(i) $(\mathsf{F}_{\mathrm{alg}})^\natural \subseteq q(\mathsf{F})_{\mathrm{alg}}$;
(ii) *every graded field extension* K *of* F *such that* $\mathsf{K}^\natural \subseteq q(\mathsf{F})_{\mathrm{alg}}$ *is a graded subfield of* $\mathsf{F}_{\mathrm{alg}}$.

The graded field $\mathsf{F}_{\mathrm{alg}}$ *has the following additional properties:*

(iii) *every algebraic graded field extension of* F *is isomorphic to a graded subfield of* $\mathsf{F}_{\mathrm{alg}}$;
(iv) $(\mathsf{F}_{\mathrm{alg}})_0$ *is the algebraic closure of* F_0 *in* $q(\mathsf{F})_{\mathrm{alg}}$;
(v) $\Gamma_{\mathsf{F}_{\mathrm{alg}}} = \mathbb{H}(\Gamma_{\mathsf{F}})$, *the divisible hull of* Γ_{F};
(vi) $q(\mathsf{F}_{\mathrm{alg}})$ *is a normal extension of* $q(\mathsf{F})$.

The graded field $\mathsf{F}_{\mathrm{alg}}$ is said to be a *graded algebraic closure* of the graded field F.

Proof. Uniqueness of F_{alg} is clear, since if F_1, F_2 are two algebraic graded field extensions of F satisfying (i) and (ii), then we must have $\mathsf{F}_1 \subseteq \mathsf{F}_2$ because F_1 satisfies (i) and F_2 satisfies (ii), and vice versa. To prove the existence of F_{alg}, consider the union $\bigcup_{\mathsf{K}} \mathsf{K}^\natural$, where K runs over the graded field extensions of F such that $\mathsf{K}^\natural \subseteq q(\mathsf{F})_{alg}$. We claim that this union is the underlying (ungraded) ring of a graded field, which satisfies (i) and (ii) and is necessarily an algebraic graded field extension of F in view of (i).

To see that the union is a ring, it suffices to prove that it is closed under sums and products. If $u_1, u_2 \in \bigcup_{\mathsf{K}} \mathsf{K}^\natural$, let K_1 and K_2 be graded field extensions of F with $\mathsf{K}_1^\natural$, $\mathsf{K}_2^\natural \subseteq q(\mathsf{F})_{alg}$ and $u_1 \in \mathsf{K}_1$, $u_2 \in \mathsf{K}_2$. Let $u_{2,\gamma}$ be any nonzero homogeneous component of u_2. Proposition 5.7 shows that $u_{2,\gamma}$ is a root of some homogenizable polynomial in $\mathsf{F}[X]$, hence it follows from Lemma 5.9 that $\mathsf{K}_1[u_{2,\gamma}]$ is a graded field extension of K_1, hence also of F. By applying this procedure inductively for all the homogeneous components of u_2, we obtain after a finite number of steps a graded field extension L of K_1 such that $\mathsf{L}^\natural \subseteq q(\mathsf{F})_{alg}$ and $u_2 \in \mathsf{L}$. Then $u_1 + u_2$ and $u_1 u_2$ lie in L, hence also in $\bigcup_{\mathsf{K}} \mathsf{K}^\natural$. Therefore, this union is a ring. Proposition 5.7 shows that the grading on F extends uniquely to a grading on each $\mathsf{K}^\natural$, hence also to $\bigcup_{\mathsf{K}} \mathsf{K}^\natural$. Every homogeneous element in this union lies in some K, hence it is invertible. Therefore, there is a graded field F_{alg} such that $(\mathsf{F}_{alg})^\natural = \bigcup_{\mathsf{K}} \mathsf{K}^\natural$; this graded field satisfies (i) and (ii). To see that it also satisfies (iii), it suffices to observe that every algebraic graded field extension L of F is isomorphic to a graded field K with $\mathsf{K}^\natural \subseteq q(\mathsf{F})_{alg}$; this is because $q(\mathsf{L})$ is isomorphic to a subfield of $q(\mathsf{F})_{alg}$. To prove (iv), consider the algebraic closure $(\mathsf{F}_0)_{alg}$ of F_0 in $q(\mathsf{F})_{alg}$. Since $(\mathsf{F}_0)_{alg} \otimes_{\mathsf{F}_0} \mathsf{F}$ is an algebraic graded field extension of F, we must have $(\mathsf{F}_0)_{alg} \otimes_{\mathsf{F}_0} \mathsf{F} \subseteq \mathsf{F}_{alg}$, hence $(\mathsf{F}_0)_{alg} \subseteq (\mathsf{F}_{alg})_0$. The reverse inclusion readily follows from Prop. 5.7(i), which shows that $(\mathsf{F}_{alg})_0$ is algebraic over F_0. For (v), we have $\Gamma_{\mathsf{F}} \subseteq \Gamma_{\mathsf{F}_{alg}} \subseteq \mathbb{H}(\Gamma_{\mathsf{F}})$ since $\Gamma_{\mathsf{F}_{alg}}/\Gamma_{\mathsf{F}}$ is a torsion group by Cor. 5.8. Suppose $\Gamma_{\mathsf{F}_{alg}}$ is not divisible. Then, there is some $\gamma \in \Gamma_{\mathsf{F}_{alg}}$ and $n \in \mathbb{N}$ with $\frac{1}{n}\gamma \notin \Gamma_{\mathsf{F}_{alg}}$. Take any nonzero $b \in \mathsf{F}_{alg,\gamma}$, let $f = X^n - b \in \mathsf{F}_{alg}[X]$, and let u be a root of f in $q(F)_{alg}$. Since f is homogenizable, Lemma 5.9 shows that $\mathsf{F}_{alg}[u]$ is a graded field lying in $q(\mathsf{F})_{alg}$ and u is homogeneous in $\mathsf{F}_{alg}[u]$. So, $u \in \mathsf{F}_{alg}[u] \subseteq \mathsf{F}_{alg}$ by (ii); but $deg\, u = \frac{1}{n}\, deg\, b = \frac{1}{n}\gamma \notin \Gamma_{\mathsf{F}_{alg}}$. This contradiction shows that $\Gamma_{\mathsf{F}_{alg}}$ must be divisible; hence, $\Gamma_{\mathsf{F}_{alg}} = \mathbb{H}(\Gamma_{\mathsf{F}})$.

To complete the proof, we show (vi). Since every element in $q(\mathsf{F}_{alg})$ has the form ab^{-1} with $a \in \mathsf{F}_{alg}$ and $b \in \mathsf{F}$ by Prop. 5.3, it suffices to show that the minimal polynomial over $q(\mathsf{F})$ of every element $a \in \mathsf{F}_{alg}$ splits over $q(\mathsf{F}_{alg})$. Suppose first that a is homogeneous and let $P \in q(\mathsf{F})[X]$ be its minimal polynomial over $q(\mathsf{F})$. Proposition 5.7 shows that P is a homogenizable polynomial in $\mathsf{F}[X]$. By Lemma 5.9, it follows that for every other root v of P in $q(\mathsf{F})_{alg}$ a grading can be defined on the ring $\mathsf{F}^\natural[v] \subseteq q(\mathsf{F})_{alg}$, turning it into a graded field extension $\mathsf{F}[v]$ of F. Therefore, property (ii) shows that $\mathsf{F}[v] \subseteq \mathsf{F}_{alg}$, hence $v \in \mathsf{F}_{alg}$. Thus, P splits in $q(\mathsf{F}_{alg})$. We have thus shown that $q(\mathsf{F}_{alg})$ contains the splitting field of the minimal polynomial of each homogeneous element. If $a \in \mathsf{F}_{alg}$ is not homogeneous, it lies in the splitting field of the product of the minimal polynomials of its homogeneous components, which is a normal extension of $q(\mathsf{F})$ contained in $q(\mathsf{F}_{alg})$; hence, the minimal polynomial of a splits over $q(\mathsf{F}_{alg})$. □

Example 5.11. Fix an integer $r > 0$. Let F_0 be an arbitrary field and let $\mathsf{F} = \mathsf{F}_0[t_1, t_1^{-1}, \ldots, t_r, t_r^{-1}]$ be the Laurent polynomial ring in r indeterminates over F_0, with the $\mathbb{Z}^r$-grading defined by

$$\mathsf{F}_{(\gamma_1,\ldots,\gamma_r)} = \mathsf{F}_0 \cdot t_1^{\gamma_1} \ldots t_r^{\gamma_r} \qquad \text{for } (\gamma_1, \ldots, \gamma_r) \in \mathbb{Z}^r.$$

Thus, F is a graded field. To describe F_{alg}, we fix an algebraic closure $q(\mathsf{F})_{alg}$ of $q(\mathsf{F})$ and consider the algebraic closure $(\mathsf{F}_0)_{alg}$ of F_0 in $q(\mathsf{F})_{alg}$. For each integer $n > 0$ we define the graded field

$$\mathsf{K}(n) = (\mathsf{F}_0)_{alg}[t_1^{1/n}, t_1^{-1/n}, \ldots, t_r^{1/n}, t_r^{-1/n}]$$

with the $(\frac{1}{n}\mathbb{Z})^r$-grading defined by

$$\mathsf{K}(n)_{(\gamma_1,\ldots,\gamma_r)} = (\mathsf{F}_0)_{alg} \cdot t_1^{\gamma_1} \ldots t_r^{\gamma_r} \qquad \text{for } (\gamma_1, \ldots, \gamma_r) \in (\tfrac{1}{n}\mathbb{Z})^r.$$

Each $\mathsf{K}(n)$ is an algebraic graded field extension of F, and we identify $\mathsf{K}(n)^\natural$ with a subring of $q(\mathsf{F})_{alg}$. Thus, $\mathsf{K}(n) \subseteq \mathsf{K}(m)$ if $n \mid m$, and we may consider

$$\mathsf{K}(\infty) = \bigcup_n \mathsf{K}(n) \subseteq q(\mathsf{F})_{alg}.$$

We claim that $\mathsf{F}_{alg} = \mathsf{K}(\infty)$. The inclusion $\mathsf{K}(\infty) \subseteq \mathsf{F}_{alg}$ readily follows from property (iii) in Prop. 5.10. We have $K(\infty)_0 = (\mathsf{F}_0)_{alg} = (\mathsf{F}_{alg})_0$ and

$$\Gamma_{K(\infty)} = \bigcup_{n\in\mathbb{N}} (\tfrac{1}{n}\mathbb{Z})^r = (\mathbb{Q}/\mathbb{Z})^r = \mathbb{H}(\Gamma_\mathsf{F}) = \Gamma_{\mathsf{F}_{alg}}$$

(see Prop. 5.10(iv) and (v)). Hence, $K(\infty) = \mathsf{F}_{alg}$.

Observe that $q(\mathsf{K}(n)) = (\mathsf{F}_0)_{alg}(t_1^{1/n}, \ldots, t_r^{1/n})$, so

$$q(\mathsf{F}_{alg}) = \bigcup_n (\mathsf{F}_0)_{alg}(t_1^{1/n}, \ldots, t_r^{1/n}).$$

We have $q(\mathsf{F}_{alg}) \neq q(\mathsf{F})_{alg}$, as there are examples of algebraic extensions of $q(\mathsf{F})$ that do not lie in any field of the form $q(\mathsf{L})$, for L an algebraic graded field extension of F: see Exercise 5.2. Alternatively, note that $q(\mathsf{F}_{alg})$ is abelian Galois over $(F_0)_{alg}(t_1, \ldots, t_r)$ if $char\, \mathsf{F} = 0$ or is the compositum of an abelian Galois extension and a purely inseparable extension if $char(\mathsf{F}) \neq 0$. In either case, any nonabelian Galois extension of $(F_0)_{alg}(t_1, \ldots, t_r)$ is algebraic over $q(\mathsf{F})$ but does not lie in $q(\mathsf{F}_{alg})$.

There are norms and traces for graded field extensions just as in the ungraded case. Recall first the corresponding functions for algebras over an arbitrary commutative ring R: Let S be an associative R-algebra with 1 such that S is a free R-module of finite rank. Let $\ell_\bullet \colon S \to End_R\, S$ be the left regular representation, which sends each $s \in S$ to $\ell_s : t \mapsto st$. Since $End_R\, S \cong M_n(R)$ where $n = rk_R\, S$, the trace and determinant are well-defined on $End_R\, S$, and the *trace* and *norm* from S to R are defined by[1]

$$tr_{S/R}(s) = tr(\ell_s) \qquad \text{and} \qquad N_{S/R}(s) = det(\ell_s), \qquad \text{for all } s \in S. \tag{5.3}$$

Define also the *characteristic polynomial for* S/R of any $s \in S$ by

$$\begin{aligned} Pchar_{S/R,s} &= det(id_S\, X - \ell_s) \\ &= X^n - tr_{S/R}(s)X^{n-1} + \ldots + (-1)^n N_{S/R}(s) \in R[X]. \end{aligned} \tag{5.4}$$

Elementary properties of matrices yield:

$$tr_{S/R} \colon S \to R \ \text{ is } R\text{-linear} \quad \text{and} \quad tr_{S/R}(r) = nr \quad \text{for all } r \in R; \tag{5.5}$$

and, for all s, $t \in S$ and $r \in R$,

$$N_{S/R}(st) = N_{S/R}(s) \cdot N_{S/R}(t), \qquad \text{and} \qquad N_{S/R}(r) = r^n. \tag{5.6}$$

[1] We could use the right regular representation of S instead of the left in defining these functions. For all the cases considered in this book, either representation gives the same trace, norm, and characteristic polynomial. But this is not true in general, e.g. when S is a ring of triangular matrices over R.

Furthermore, these functions have good behavior with respect to scalar extensions: If R' is any ring containing R as a subring, and $S' = S \otimes_R R'$, then S' is an R'-algebra and a free R'-module with $rk_{R'}\, S' = rk_R\, S$, and a matrix representing $\ell_s \in End_R\, S$ also represents $\ell_{s\otimes 1} \in End_{R'}\, S'$. Hence, when we identify S with its image $S \otimes 1 \subseteq S'$, for any $s \in S$,

$$tr_{S'/R'}(s) = tr_{S/R}(s),\ N_{S'/R'} = N_{S/R}(s),\ \text{and}\ Pchar_{S'/R',s} = Pchar_{S/R,s}. \tag{5.7}$$

Now consider the case of graded field extensions.

Proposition 5.12. *Let* K/F *be any finite-degree extension of graded fields with* $[\mathsf{K}{:}\mathsf{F}] = n$.

(i) *For any* $b \in \mathsf{K}$, *we have* $Pchar_{\mathsf{K}/\mathsf{F},b} = Pchar_{q(\mathsf{K})/q(\mathsf{F}),b}$, *and*

$$tr_{\mathsf{K}/\mathsf{F}}(b) = tr_{q(\mathsf{K})/q(\mathsf{F})}(b) \qquad and \qquad N_{\mathsf{K}/\mathsf{F}}(b) = N_{q(\mathsf{K})/q(\mathsf{F})}(b).$$

(ii) *For* $b \in \mathsf{K}$, *let*

$$p_b = X^d + c_{d-1}X^{d-1} + \ldots + c_0 \in \mathsf{F}[X]$$

be the minimal polynomial of b *over* $q(\mathsf{F})$. *Then,* $Pchar_{\mathsf{K}/\mathsf{F},b} = (p_b)^{n/d}$ *and*

$$tr_{\mathsf{K}/\mathsf{F}}(b) = -\tfrac{n}{d}c_{d-1} \qquad and \qquad N_{\mathsf{K}/\mathsf{F}}(b) = (-1)^n c_0^{n/d}.$$

(iii) *If* a *is homogeneous of degree* δ *in* K, *then* $Pchar_{\mathsf{K}/\mathsf{F},a}$ *is a* δ*-homogenizable polynomial in* $\mathsf{F}[X]$. *Hence,* $tr_{\mathsf{K}/\mathsf{F}}(a) \in \mathsf{F}_\delta$ *and* $N_{\mathsf{K}/\mathsf{F}}(a) \in \mathsf{F}_{n\delta}$.

(iv) *Suppose* $a \in \mathsf{K}_0$. *Then* $p_a \in \mathsf{F}_0[X]$. *Hence,* $Pchar_{\mathsf{K}/\mathsf{F},a} = (Pchar_{\mathsf{K}_0/\mathsf{F}_0,a})^e$, *where* $e = [\mathsf{K}{:}\mathsf{F}]/[\mathsf{K}_0{:}\mathsf{F}_0]$, *and*

$$tr_{\mathsf{K}/\mathsf{F}}(a) = e\, tr_{\mathsf{K}_0/\mathsf{F}_0}(a) \qquad and \qquad N_{\mathsf{K}/\mathsf{F}}(a) = N_{\mathsf{K}_0/\mathsf{F}_0}(a)^e.$$

Proof. Since $q(\mathsf{K}) = \mathsf{K}^\natural \otimes_{\mathsf{F}^\natural} q(\mathsf{K})$ (see Lemma 2.15) (i) is a special case of (5.7). Note that $p_b \in \mathsf{F}[X]$ for any $b \in \mathsf{K}$ since b is integral over $\mathsf{F}^\natural$, so the coefficients of p_b in $q(\mathsf{F})$ are integral over $\mathsf{F}^\natural$, which is integrally closed by Prop. 5.2. (ii) then follows from (i) and the corresponding result for ungraded field extensions.

(iii) Suppose $a \in \mathsf{K}_\delta$. Let $(b_1, \ldots, b_n)$ be a homogeneous base of K as a graded F-vector space, and let $\gamma_i = deg\, b_i$ for all i. Let (c_{ij}) in $M_n(\mathsf{F})$ be the matrix of ℓ_a relative to this base. The equation

$$a\, b_j = \sum_{i=1}^{n} c_{ij} b_i \in \mathsf{K}_{\delta+\gamma_j},$$

implies that each $c_{ij} \in \mathsf{F}_{\delta+\gamma_j-\gamma_i}$. For if c'_{ij} is the $(\delta + \gamma_j - \gamma_i)$-component of c_{ij}, then the equation for ab_j holds with c'_{ij} replacing c_{ij}. The uniqueness of the c_{ij} then implies that $c_{ij} = c'_{ij} \in \mathsf{F}_{\delta+\gamma_j-\gamma_i}$. In $\mathsf{K}[X]$, let

$$t_{ij} = \begin{cases} X - c_{ij} & \text{if } j = i, \\ -c_{ij} & \text{if } j \neq i. \end{cases}$$

In the graded ring $\mathsf{K}[X]^{(\delta)}$ of (5.2), in which X has degree δ, each t_{ij} lies in the $\delta + \gamma_j - \gamma_i$-component. Then,

$$Pchar_{\mathsf{K}/\mathsf{F},a} \;=\; det\,(t_{ij}) \;=\; \sum_{\sigma\in S_n} sgn(\sigma)t_{1\sigma(1)}\dots t_{n\sigma(n)} \;\in\; \mathsf{F}[X]^{(\delta)}_{n\delta}$$

since $deg\,\big(sgn(\sigma)t_{1\sigma(1)}\dots t_{n\sigma(n)}\big) \;=\; \sum_{i=1}^{n}\big(\delta + \gamma_{\sigma(i)} - \gamma_i\big) \;=\; n\delta$ for each $\sigma \in S_n$. Because $Pchar_{\mathsf{K}/\mathsf{F},a}$ is homogeneous in $\mathsf{F}[X]^{(\delta)}$, it is a δ-homogenizable polynomial. Since its coefficient of X^n lies in F_0, it follows that the coefficient of X^{n-1} lies in F_δ and the constant term lies in $\mathsf{F}_{n\delta}$. This yields the degree formulas for $tr_{\mathsf{K}/\mathsf{F}}(a)$ and $N_{\mathsf{K}/\mathsf{F}}(a)$.

For (iv) suppose $a \in \mathsf{K}_0$. Then p_a is a monic 0-homogenizable polynomial in $\mathsf{F}[X]$ by (iii), i.e., $p_a \in \mathsf{F}_0[X]$. Hence, p_a is the minimal polynomial of a over F_0. Let $d = deg\,p_a$. It follows by (ii) and its ungraded counterpart that

$$Pchar_{\mathsf{K}/\mathsf{F},a} \;=\; p_a^{\,[\mathsf{K}:\mathsf{F}]/d} \;=\; \big(p_a^{\,[\mathsf{K}_0:\mathsf{F}_0]/d}\big)^e \;=\; \big(Pchar_{\mathsf{K}_0/\mathsf{F}_0,a}\big)^e.$$

The trace and norm formulas in (iv) follow at once. □

To conclude this subsection, we give a categorical interpretation of Prop. 5.3 and Prop. 5.7. For a given graded field F, consider the category $\mathfrak{grFields}_{\mathsf{F}}$, whose objects are the graded field extensions of F and whose morphisms are the graded F-algebra homomorphisms; consider also the full subcategory $\mathfrak{algrFields}_{\mathsf{F}}$ of algebraic graded field extensions of F. We compare these categories to the category $\mathfrak{Fields}_{q(\mathsf{F})}$ of field extensions of $q(\mathsf{F})$.

Proposition 5.13. *The quotient field construction mapping each algebraic graded field extension* K *of* F *to the field* $q(\mathsf{K})$ *defines a faithful functor*

$$q\colon\ \mathfrak{grFields}_{\mathsf{F}} \;\to\; \mathfrak{Fields}_{q(\mathsf{F})}\,.$$

The restriction of this functor to $\mathfrak{algrFields}_{\mathsf{F}}$ *is full. Indeed, if* K *and* L *are algebraic graded field extensions of* F*, any* $q(\mathsf{F})$*-algebra homomorphism* $\psi\colon q(\mathsf{K})\to q(\mathsf{L})$ *restricts to a homomorphism of graded* F*-algebras* $\psi|_{\mathsf{K}}\colon \mathsf{K}\to\mathsf{L}$*. Thus, the functor* q *identifies* $\mathfrak{algrFields}_{\mathsf{F}}$ *with a full subcategory of* $\mathfrak{Fields}_{q(\mathsf{F})}$*.*

Proof. To see that q is a functor, note that every graded F-algebra homomorphism $\varphi\colon \mathsf{K}\to\mathsf{L}$ of graded field extensions of F is injective, since its kernel is a homogeneous ideal and every nonzero homogeneous element in K is invertible. Therefore, φ induces a $q(\mathsf{F})$-algebra homomorphism $q(\varphi)\colon q(\mathsf{K})\to q(\mathsf{L})$. The functor q is faithful because φ is the restriction of $q(\varphi)$ to $\mathsf{K}^\natural \subseteq q(\mathsf{K})$. To prove that the restriction of q to $\mathfrak{algrFields}_{\mathsf{F}}$ is also full, consider two algebraic graded field extensions K, L of F and a $q(\mathsf{F})$-algebra homomorphism $\psi\colon q(\mathsf{K})\to q(\mathsf{L})$. Proposition 5.3 shows that $\mathsf{K}^\natural$ and $\mathsf{L}^\natural$ are the integral closures of $\mathsf{F}^\natural$ in $q(\mathsf{K})$ and $q(\mathsf{L})$ respectively, hence $\psi(\mathsf{K}^\natural)\subseteq\mathsf{L}^\natural$. If $u\in\mathsf{K}$ is a nonzero homogeneous element whose minimal polynomial P over $q(\mathsf{F})$ has degree n and constant term a_0, then P is also the minimal polynomial of $\psi(u)$ over $q(\mathsf{F})$; hence, Prop. 5.7 shows that $\psi(u)$ is homogeneous of degree $\frac{1}{n}\,deg\,a_0 = deg\,u$.

Therefore, the restriction of ψ to $\mathsf{K}^\natural$ preserves the grading and thus defines a homomorphism of graded F-algebras $\psi|_{\mathsf{K}}\colon \mathsf{K} \to \mathsf{L}$ such that $q(\psi|_{\mathsf{K}}) = \psi$. □

Corollary 5.14. *For a given algebraic graded field extension* M/F*, the functor* q *defines an injection*

$$\left\{\begin{matrix}\text{graded fields } \mathsf{L}\\ \text{with } \mathsf{F}\subseteq\mathsf{L}\subseteq\mathsf{M}\end{matrix}\right\} \hookrightarrow \left\{\begin{matrix}\text{fields } E \text{ with}\\ q(\mathsf{F})\subseteq E\subseteq q(\mathsf{M})\end{matrix}\right\} \tag{5.8}$$

Proof. If K and L are field extensions of F in M, then the proposition above with ψ the identity map shows that $q(\mathsf{K}) = q(\mathsf{L})$ implies $\mathsf{K} = \mathsf{L}$. □

The following example shows that the injection (5.8) is not onto. This kind of example can occur only when *char* $\mathsf{F}_0 \neq 0$: see Prop. 5.35 below.

Example 5.15. Let $\mathsf{F}_0 = k(x,y)$ be the rational function field in two indeterminates x, y over an arbitrary field k of characteristic $p \neq 0$. Let $\mathsf{K}_0 = \mathsf{F}_0(\sqrt[p]{x}, \sqrt[p]{y})$ and consider the rings of Laurent polynomials in a new indeterminate t:

$$\mathsf{F} = \mathsf{F}_0[t,t^{-1}], \qquad \mathsf{K} = \mathsf{K}_0[t,t^{-1}]$$

with the $\mathbb{Z}$-grading by powers of t. Clearly, K is an algebraic graded field extension of F, and

$$q(\mathsf{F}) = \mathsf{F}_0(t) = k(x,y,t), \qquad q(\mathsf{K}) = \mathsf{K}_0(t) = q(\mathsf{F})(\sqrt[p]{x}, \sqrt[p]{y}).$$

Now, let $u = \sqrt[p]{x} + \sqrt[p]{y}\,t \in \mathsf{K}$ and let $E = q(\mathsf{F})(u) \subseteq q(\mathsf{K})$. Since $u^p = x + y\,t^p \in q(\mathsf{F})$, we have $[E{:}q(\mathsf{F})] = p$. If L is a graded field extension of F with $E = q(\mathsf{L})$, then we must have $[\mathsf{L}{:}\mathsf{F}] = p$ by Prop. 5.3. Moreover, we must have $u \in \mathsf{L}$ since u is integral over F, hence its homogeneous components $\sqrt[p]{x}$ and $\sqrt[p]{y}\,t$ must lie in L. It follows that L_0 contains $\sqrt[p]{x}$ and $\sqrt[p]{y}$, hence $[\mathsf{L}{:}\mathsf{F}] > p$, a contradiction. We have thus shown that E is not of the form $q(\mathsf{L})$ for any graded field extension L of F.

Remark 5.16. (i) For a given graded field F, let $\mathfrak{Alg}_{\mathsf{F}^\natural}$ be the category of $\mathsf{F}^\natural$-algebras. The forgetful functor

$$\natural\colon \mathfrak{algrFields}_{\mathsf{F}} \to \mathfrak{Alg}_{\mathsf{F}^\natural}$$

is clearly faithful but not full, whereas $q\colon \mathfrak{algrFields}_{\mathsf{F}} \to \mathfrak{Fields}_{q(\mathsf{F})}$ is faithful and full, by Prop. 5.13. This observation illustrates the fact that the grading makes graded fields closer in nature to fields than to rings.

(ii) Without the restriction to *algebraic* graded field extensions, the functor q is in general not full. As an example, take for F a field with the trivial grading (so $\Gamma_{\mathsf{F}} = \{0\}$) and let $\mathsf{K} = \mathsf{F}[t,t^{-1}]$ be the ring of Laurent polynomials in one indeterminate t, with the $\mathbb{Z}$-grading by powers of t, so $\mathsf{K}_\gamma = \mathsf{F}t^\gamma$ for $\gamma \in \mathbb{Z}$. We have $q(\mathsf{F}) = \mathsf{F}$, $q(\mathsf{K}) = \mathsf{F}(t)$, and for any nonconstant rational fraction $f \in \mathsf{F}(t)$ there is a $q(\mathsf{F})$-algebra endomorphism $\psi_f\colon q(\mathsf{K}) \to q(\mathsf{K})$ that maps t to f. There is no graded F-algebra endomorphism $\varphi\colon \mathsf{K} \to \mathsf{K}$ such that $\psi_f = q(\varphi)$, unless $f = at$ for some $a \in \mathsf{F}^\times$. In particular, note that the map

$\psi_{t^{-1}}$ restricts to an $\mathsf{F}^\natural$-algebra isomorphism $\mathsf{K}^\natural \to \mathsf{K}^\natural$ that preserves all the homogeneous components of K; yet it is not an isomorphism of graded fields since $deg\, \psi_{t^{-1}}(u) = -\,deg\, u$ for each nonzero homogeneous element $u \in \mathsf{K}$.

5.1.3 Ramification in graded field extensions

The basic definitions for the ramification theory of algebraic graded field extensions are the following:

Definition 5.17. Let K be an algebraic graded field extension of some graded field F. The extension K/F is said to be

- *unramified* if $\Gamma_\mathsf{K} = \Gamma_\mathsf{F}$;
- *totally ramified* if $\mathsf{K}_0 = \mathsf{F}_0$;
- *tame* (or *tamely ramified*) if the extension $\mathsf{K}_0/\mathsf{F}_0$ is separable and moreover either $char\, \mathsf{F}_0 = 0$, or $char\, \mathsf{F}_0 = p > 0$ and there is no p-torsion element in $\Gamma_\mathsf{K}/\Gamma_\mathsf{F}$;
- *inertial* if it is unramified and the extension $\mathsf{K}_0/\mathsf{F}_0$ is separable.

Thus, the inertial extensions are exactly those that are tame and unramified.

All these properties are clearly transitive: given a tower of algebraic graded field extensions $\mathsf{F} \subseteq \mathsf{K} \subseteq \mathsf{L}$, we have L/F unramified (resp. totally ramified, resp. tame, resp. inertial) if and only if K/F and L/K have the same property.

Note that Cor. 2.7 yields

$$[\mathsf{K}{:}\mathsf{F}] = [\mathsf{K}_0{:}\mathsf{F}_0] \cdot |\Gamma_\mathsf{K}{:}\Gamma_\mathsf{F}|.$$

Therefore, if $[\mathsf{K}{:}\mathsf{F}]$ is finite, the extension K/F is unramified if and only if $[\mathsf{K}{:}\mathsf{F}] = [\mathsf{K}_0{:}\mathsf{F}_0]$, and it is totally ramified if and only if $[\mathsf{K}{:}\mathsf{F}] = |\Gamma_\mathsf{K}{:}\Gamma_\mathsf{F}|$.

For an arbitrary algebraic graded field extension K/F, we may identify the tensor product $\mathsf{K}_0 \otimes_{\mathsf{F}_0} \mathsf{F}$ with the graded subfield $\bigoplus_{\gamma\in\Gamma_\mathsf{F}} \mathsf{K}_\gamma \subseteq \mathsf{K}$, since $\mathsf{K}_\gamma = \mathsf{K}_0 \otimes_{\mathsf{F}_0} \mathsf{F}_\gamma$ for each $\gamma \in \Gamma_\mathsf{F}$ as K_γ is 1-dimensional over K_0. Thus, $\mathsf{K}_0 \otimes_{\mathsf{F}_0} \mathsf{F}$ is an unramified graded field extension of F contained in K. Since $\bigoplus_{\gamma\in\Gamma_\mathsf{F}} \mathsf{K}_\gamma$ contains every unramified graded field extension of F in K, we call $\mathsf{K}_0 \otimes_{\mathsf{F}_0} \mathsf{F}$ the *unramified closure* of F in K. Likewise, if $S \subseteq \mathsf{K}_0$ is the separable closure of F_0 in K_0, then $S \otimes_{\mathsf{F}_0} \mathsf{F}$ is an inertial graded field extension of F in K, which contains every inertial graded field extension of F in K. This graded field is called the *inertial closure* of F in K. It is clear from the definitions that K is a totally ramified graded field extension of $\mathsf{K}_0 \otimes_{\mathsf{F}_0} \mathsf{F}$. Thus, for an arbitrary algebraic graded field extension K/F there is a canonically defined intermediate graded field $\mathsf{L} = \mathsf{K}_0 \otimes_{\mathsf{F}_0} \mathsf{F}$ with

$$\mathsf{F} \subseteq \mathsf{L} \subseteq \mathsf{K}, \quad \mathsf{L}/\mathsf{F} \text{ unramified}, \quad \text{and} \quad \mathsf{K}/\mathsf{L} \text{ totally ramified.}$$

Of course, $\mathsf{L} = \mathsf{K}$ if and only if K/F is unramified, and $\mathsf{L} = \mathsf{F}$ if and only if K/F is totally ramified. Note that by Prop. 5.3 we have

$$q(\mathsf{K}_0 \otimes_{\mathsf{F}_0} \mathsf{F}) = (\mathsf{K}_0 \otimes_{\mathsf{F}_0} \mathsf{F}) \otimes_\mathsf{F} q(\mathsf{F}) = \mathsf{K}_0 \otimes_{\mathsf{F}_0} q(\mathsf{F}). \tag{5.9}$$

The observations above show that unramified graded field extensions are obtained by extending the field of homogeneous elements of degree 0. Totally ramified extensions are described in the next proposition.

Proposition 5.18. *Let* K *be an algebraic graded field extension of some graded field* F. *If* K/F *is totally ramified, then the minimal polynomial over* $q(\mathsf{F})$ *of every nonzero homogeneous element* $u \in \mathsf{K}$ *is a binomial: it has the form* $X^n - a$ *for some nonzero homogeneous element* $a \in \mathsf{F}$. *The degree of this polynomial is the order of the image of* $\deg u$ *in* $\Gamma_{\mathsf{K}}/\Gamma_{\mathsf{F}}$.

Proof. By Cor. 5.8, $\Gamma_{\mathsf{K}}/\Gamma_{\mathsf{F}}$ is a torsion group. Let n be the order of the image of $\deg u$ in this factor group, so $\deg(u^n) \in \Gamma_{\mathsf{F}}$. Since K/F is totally ramified, we have $\mathsf{K}_{\deg(u^n)} = \mathsf{F}_{\deg(u^n)}$, hence $u^n = a$ for some nonzero homogeneous element $a \in \mathsf{F}$. Proposition 5.7 shows that u is not a root of a polynomial over $q(\mathsf{F})$ of degree $< n$, hence $X^n - a$ is the minimal polynomial of u over $q(\mathsf{F})$. □

We may now characterize the tame graded field extensions as follows:

Proposition 5.19. *Let* K *be an algebraic graded field extension of a graded field* F. *The following conditions are equivalent:*

(a) K *is a tame extension of* F;
(b) $q(\mathsf{K})$ *is a separable field extension of* $q(\mathsf{F})$.

Proof. Let $\mathsf{L} = \mathsf{K}_0 \otimes_{\mathsf{F}_0} \mathsf{F}$ be the unramified closure of F in K, so $q(\mathsf{L}) = \mathsf{K}_0 \otimes_{\mathsf{F}_0} q(\mathsf{F})$ by (5.9), and let S be the separable closure of F_0 in K_0. Then K_0 is purely inseparable over S and S is separable over F_0, hence $q(\mathsf{L})$ is purely inseparable over $S \otimes_{\mathsf{F}_0} q(\mathsf{F})$ and $S \otimes_{\mathsf{F}_0} q(\mathsf{F})$ is separable over $q(\mathsf{F})$. Thus, $\mathsf{K}_0/\mathsf{F}_0$ is separable if and only if $\mathsf{K}_0 = S$, and this holds if and only if $q(\mathsf{L})/q(\mathsf{F})$ is separable. Therefore, L/F is tame if and only if $q(\mathsf{L})/q(\mathsf{F})$ is a separable field extension. On the other hand, the field $q(\mathsf{K})$ is generated as an extension of $q(\mathsf{L})$ by the homogeneous elements in K, and Prop. 5.18 shows that any such element $u \in \mathsf{K}$ is separable over $q(\mathsf{F})$ if and only if the order of the image of $\deg u$ in $\Gamma_{\mathsf{K}}/\Gamma_{\mathsf{F}}$ is not divisible by $\operatorname{char} \mathsf{F}_0$. Hence, K/L is tame if and only if $q(\mathsf{K})/q(\mathsf{L})$ is a separable field extension. By transitivity of tameness and separability, we thus have (a) $\iff$ (b). □

Remark 5.20. Use the same notation as in Prop. 5.19, and assume $[\mathsf{K}{:}\mathsf{F}]$ is finite. Then conditions (a) and (b) are also equivalent to

(c) $\mathsf{K}^\natural$ is a separable algebra over $\mathsf{F}^\natural$.

Indeed, since $q(\mathsf{K}) = \mathsf{K}^\natural \otimes_{\mathsf{F}^\natural} q(\mathsf{F})$ by Prop. 5.3, it follows that (c) $\Rightarrow$ (b) because separability is preserved under scalar extension (see Knus–Ojanguren [116, Prop. 2.1, p. 77]) (and because the finite-degree field extension $q(\mathsf{K})/q(\mathsf{F})$ is separable when $q(\mathsf{K})$ is a separable $q(\mathsf{F})$-algebra). Similarly, if L/F is tame (i.e., $\mathsf{K}_0/\mathsf{F}_0$ is a separable field extension, hence K_0 is a separable F_0-algebra), then $\mathsf{L}^\natural$ is a separable $\mathsf{F}^\natural$-algebra since $\mathsf{L}^\natural = \mathsf{K}_0 \otimes_{\mathsf{F}_0} \mathsf{F}$. By transitivity of separability (Knus–Ojanguren [116, Prop. 2.4, p. 78]), it now suffices to prove (a) ⇒ (c) for totally ramified extensions. Arguing by induction on $[\mathsf{K}{:}\mathsf{F}]$ and using Prop. 5.18, we may reduce to the case where $\mathsf{K} = \mathsf{F}[u]$ where the minimal polynomial of u over $q(\mathsf{F})$ is $X^n - a$ for some integer n not divisible by *char* F_0 and some nonzero homogeneous element $a \in \mathsf{F}$. In that case, $\mathsf{K}^\natural$ is a free $\mathsf{F}^\natural$-module with base $1, u, \ldots, u^{n-1}$, and the trace map $T\colon \mathsf{K}^\natural \to \mathsf{F}^\natural$ satisfies $T(u^i) = 0$ if $i \not\equiv 0 \ (mod\ n)$, while $T(u^i)$ is invertible if $i \equiv 0 \ (mod\ n)$ since *char* $\mathsf{F}_0 \nmid n$. Therefore, the separability criterion involving the discriminant (see Knus–Ojanguren [116, Th. 4.7, p. 89]) shows that $\mathsf{K}^\natural$ is a separable $\mathsf{F}^\natural$-algebra.

Note that (c) cannot hold if $[\mathsf{K}{:}\mathsf{F}]$ is infinite, since $\mathsf{K}^\natural$ is a free module over $\mathsf{F}^\natural$; see Knus–Ojanguren [116, Prop. 3.2, p. 82].

We show next that tame graded field extensions behave like separable field extensions with respect to tensor products.

Proposition 5.21. *Let* L *be a tame finite-degree graded field extension of a graded field* F. *Let* K *be any graded field extension of* F. *Then,* $\mathsf{L} \otimes_{\mathsf{F}} \mathsf{K}$ *is a finite direct product of graded fields each tame over* K.

Proof. This is deducible from the preservation of separability under scalar extensions of rings, cf. Knus–Ojanguren [116, Prop. 2.1, p. 77]. But here is a more elementary argument. Let $\mathsf{A} = \mathsf{L} \otimes_{\mathsf{F}} \mathsf{K}$, which is a finite-dimensional commutative graded K-algebra, with

$$\begin{aligned} q(\mathsf{A}) &= \mathsf{A}^\natural \otimes_{\mathsf{K}^\natural} q(\mathsf{K}) = (\mathsf{L}^\natural \otimes_{\mathsf{F}^\natural} \mathsf{K}^\natural) \otimes_{\mathsf{K}^\natural} q(\mathsf{K}) = \mathsf{L}^\natural \otimes_{\mathsf{F}^\natural} q(\mathsf{K}) \\ &= (\mathsf{L}^\natural \otimes_{\mathsf{F}^\natural} q(\mathsf{F})) \otimes_{q(\mathsf{F})} q(\mathsf{K}) = q(\mathsf{L}) \otimes_{q(\mathsf{F})} q(\mathsf{K}). \end{aligned}$$

Since L is tame over F, by Prop. 5.19 $q(\mathsf{L})$ is a finite-degree separable field extension of $q(\mathsf{F})$. Hence, as $q(\mathsf{K})$ is a field, $q(\mathsf{L}) \otimes_{q(\mathsf{F})} q(\mathsf{K})$ is a finite direct product of fields, each finite-dimensional and separable over $q(\mathsf{K})$ (see Knus–Ojanguren [116, Prop. 4.1, p. 85]). Since $q(\mathsf{A})$ is thus semisimple, by Prop. 2.28(iii) A is graded semisimple. Hence, by Th. 2.23 $\mathsf{A} = \mathsf{B}_1 \times \ldots \times \mathsf{B}_n$, where each B_i is graded simple. Since A is commutative, each B_i must be a graded field. Moreover,

$$q(\mathsf{A}) \cong q(\mathsf{B}_1) \times \ldots \times q(\mathsf{B}_n).$$

Hence the fields $q(\mathsf{B}_i)$ are the simple components of $q(\mathsf{A})$, each of which is separable over $q(\mathsf{K})$. By Prop. 5.19 each graded field B_i is tame over K. □

There is also a graded analogue to the separable closure of a field in a larger field:

Proposition 5.22. *Let* K *be an algebraic graded field extension of a graded field* F, *and let* L *be a graded field extension of* F *in* K. *The following conditions are equivalent:*

(a) L/F *is a maximal tame extension of* F *in* K*;*
(b) *either char* $\mathsf{F}_0 = 0$ *and* $\mathsf{L} = \mathsf{K}$*, or char* $\mathsf{F}_0 = p > 0$ *and the following conditions hold:* L_0 *is the separable closure of* F_0 *in* K_0 *and* Γ_L *is the "prime-to-p closure" of* Γ_F *in* Γ_K *in the following sense: the factor group* $\Gamma_\mathsf{L}/\Gamma_\mathsf{F}$ *has no p-torsion and every element in* $\Gamma_\mathsf{K}/\Gamma_\mathsf{L}$ *has order a power of p;*
(c) $q(\mathsf{L})$ *is the separable closure of* $q(\mathsf{F})$ *in* $q(\mathsf{K})$*;*
(d) *every graded field* M *with* $\mathsf{F} \subseteq \mathsf{M} \subseteq \mathsf{K}$ *and* M/F *tame satisfies* $\mathsf{M} \subseteq \mathsf{L}$.

Moreover, there is a unique graded field extension L *of* F *in* K *satisfying (a)–(d).*

The graded field L satisfying conditions (a)–(d) is called the *tame closure* of F in K.

Proof. If *char* $\mathsf{F}_0 = 0$, then every algebraic graded field extension of F is tame and every algebraic extension of $q(\mathsf{F})$ is separable, hence conditions (a), (b), and (d) hold only for $\mathsf{L} = \mathsf{K}$. Corollary 5.14 shows that condition (c) also holds only for $\mathsf{L} = \mathsf{K}$. For the rest of the proof, we may thus assume *char* $\mathsf{F}_0 = p > 0$.

Suppose L satisfies (a). Since L/F is tame, Prop. 5.19 shows that L_0 is separable over F_0 and $\Gamma_\mathsf{L}/\Gamma_\mathsf{F}$ has no p-torsion. If K_0 contains an element $u \notin \mathsf{L}_0$ separable over L_0, then $\mathsf{L}[u]$ is a proper tame extension of L in K. This cannot occur since L satisfies (a), hence L_0 is the separable closure of F_0 in K_0. To complete the proof of (a) $\Rightarrow$ (b), it only remains to see that every element in $\Gamma_\mathsf{K}/\Gamma_\mathsf{L}$ has p-power order. Let $v \in \mathsf{K}$ be a nonzero homogeneous element, and let d be the order of *deg* v in $\Gamma_\mathsf{K}/\Gamma_\mathsf{L}$. Write $d = p^r e$ where $r \geq 0$ and p does not divide e. Let also $\ell \in \mathsf{L}$ be a nonzero homogeneous element with $\deg \ell = p^r e \deg v$, so $v^{p^r e}\ell^{-1} \in \mathsf{K}_0$. Since $\mathsf{K}_0/\mathsf{L}_0$ is purely inseparable, we have

$$(v^{p^r e}\ell^{-1})^{p^m} = \ell' \qquad \text{for some integer } m \text{ and for some } \ell' \in \mathsf{L}_0.$$

Therefore, $v^{p^{r+m}}$ is a root of $X^e - \ell^{p^m}\ell' \in \mathsf{L}[X]$, and $\mathsf{L}[v^{p^{r+m}}]$ is a tame extension of L since the order of $\deg v^{p^{r+m}}$ in $\Gamma_\mathsf{K}/\Gamma_\mathsf{L}$ is e. Since L is maximal among the tame extensions of F in K, we must have $e = 1$, hence L satisfies (b).

Now, assume L satisfies (b). Then L is tame over F, hence $q(\mathsf{L})$ is separable over $q(\mathsf{F})$ by Prop. 5.19. To prove that $q(\mathsf{L})$ is the separable closure of $q(\mathsf{F})$ in $q(\mathsf{K})$, we show that every element $u \in q(\mathsf{K})$ is purely inseparable over $q(\mathsf{L})$, i.e., there exists an integer n such that $u^{p^n} \in q(\mathsf{L})$. By Prop. 5.3, we have $u = ab^{-1}$ for some $a \in \mathsf{K}$ and $b \in \mathsf{F}$, $b \neq 0$. Clearly, $u^{p^n} \in q(\mathsf{L})$ if and only if $a^{p^n} \in q(\mathsf{L})$, hence it suffices to find n such that $a^{p^n} \in q(\mathsf{L})$. It even suffices to prove that for every *homogeneous* element $v \in \mathsf{K}$ there exists a suitable power of p such that $v^{p^n} \in q(\mathsf{L})$, since then we may find $n \in \mathbb{N}$ large enough so that $a_\gamma^{p^n} \in q(\mathsf{L})$ for each homogeneous component of a, which implies that $a^{p^n} \in q(\mathsf{L})$. So, let $v \in \mathsf{K}$ be a nonzero homogeneous element. Our hypothesis on L implies that $p^r \deg v \in \Gamma_L$ for some $r \geq 0$. Let $\ell \in \mathsf{L}$ be a nonzero homogeneous element such that $p^r \, deg\, v = deg\, \ell$, hence $v^{p^r}\ell^{-1} \in \mathsf{K}_0$. Since $\mathsf{K}_0/\mathsf{L}_0$ is purely inseparable by hypothesis, we have $(v^{p^r}\ell^{-1})^{p^m} \in \mathsf{L}_0$ for some $m \geq 0$, hence $v^{p^{r+m}} \in \mathsf{L}$. This proves that L satisfies (c).

To prove (c) $\Rightarrow$ (d), we observe that for every graded subfield M with $\mathsf{F} \subseteq \mathsf{M} \subseteq \mathsf{K}$ we have $\mathsf{M}^\natural = q(\mathsf{M}) \cap \mathsf{K}^\natural$ by Prop. 5.3, since K is an algebraic graded field extension of M. Now, if M/F is tame, then $q(\mathsf{M})$ is separable over $q(\mathsf{F})$, hence $q(\mathsf{M}) \subseteq q(\mathsf{L})$ if L satisfies (c). Therefore, we then have

$$\mathsf{M}^\natural = q(\mathsf{M}) \cap \mathsf{K}^\natural \subseteq q(\mathsf{L}) \cap \mathsf{K}^\natural = \mathsf{L}^\natural.$$

Since Prop. 5.7 shows that the gradings on M and L are uniquely determined by the ring structure, it follows that $\mathsf{M} \subseteq \mathsf{L}$; hence, L satisfies (d). Since (d) $\Rightarrow$ (a) is obvious, we have thus shown that the conditions (a)–(d) are all equivalent. Zorn's Lemma readily yields a graded field L satisfying (a), and this graded field is unique in view of (d). □

Definition 5.23. Given a graded field F, we may apply the construction of the inertial, the unramified, and the tame closure to a graded algebraic closure F_{alg}, as defined in Prop. 5.10. The inertial (resp. unramified, resp. tame) closure of F in F_{alg} is said to be an *inertial* (resp. *unramified*, resp. *tame* or *tamely ramified*) *closure* of F and is denoted by F_{in} (resp. F_{nr}, resp. F_{tr}). Thus, if $(\mathsf{F}_0)_{sep}$ and $(\mathsf{F}_0)_{alg}$ are the separable and algebraic closures of F_0, viewed as subfields of $q(\mathsf{F}_{alg})$, we have

$$\mathsf{F}_{in} = (\mathsf{F}_0)_{sep} \otimes_{\mathsf{F}_0} \mathsf{F}, \qquad \mathsf{F}_{nr} = (\mathsf{F}_0)_{alg} \otimes_{\mathsf{F}_0} \mathsf{F},$$

hence also, as in (5.9),

$$q(\mathsf{F}_{in}) = (\mathsf{F}_0)_{sep} \otimes_{\mathsf{F}_0} q(\mathsf{F}) \quad \text{and} \quad q(\mathsf{F}_{nr}) = (\mathsf{F}_0)_{alg} \otimes_{\mathsf{F}_0} q(\mathsf{F}).$$

For F_{tr}, Prop. 5.22 shows that $(\mathsf{F}_{tr})_0 = (\mathsf{F}_0)_{sep}$, and

$$\Gamma_{\mathsf{F}_{tr}} = \begin{cases} \mathbb{H}(\Gamma_{\mathsf{F}}), & \text{if } char(\mathsf{F}) = 0, \\ \text{the "prime-to-}p\text{ closure" of } \Gamma_{\mathsf{F}} \text{ in } \mathbb{H}(\Gamma_{\mathsf{F}}), & \text{if } char(\mathsf{F}) = p \neq 0. \end{cases}$$

Moreover, $q(\mathsf{F}_{tr})$ is the separable closure of $q(\mathsf{F})$ in $q(\mathsf{F}_{alg})$.

The graded field F_{in} contains as a graded subfield an isomorphic copy of every inertial graded field extension of F. Likewise, F_{nr} contains an isomorphic copy of every unramified graded field extension of F, and F_{tr} contains an isomorphic copy of every tame graded field extension of F.

Example 5.24. As in Ex. 5.11, let

$$\mathsf{F} = \mathsf{F}_0[t_1, t_1^{-1}, \ldots, t_r, t_r^{-1}]$$

be the graded field of Laurent polynomials in r indeterminates over an arbitrary field F_0, with the usual $\mathbb{Z}^r$-grading. As proved there, we have

$$\mathsf{F}_{alg} = \bigcup_n (\mathsf{F}_0)_{alg}[t_1^{1/n}, t_1^{-1/n}, \ldots, t_r^{1/n}, t_r^{-1/n}],$$

where n runs over the nonzero integers. Let $\mathbb{N}' \subseteq \mathbb{N}$ denote the set of integers that are not divisible by $char\, \mathsf{F}_0$ (so $\mathbb{N}' = \mathbb{N} \setminus \{0\}$ if $char\, \mathsf{F}_0 = 0$). Then

$$\mathsf{F}_{in} = (\mathsf{F}_0)_{sep}[t_1, t_1^{-1}, \ldots, t_r, t_r^{-1}], \qquad \mathsf{F}_{nr} = (\mathsf{F}_0)_{alg}[t_1, t_1^{-1}, \ldots, t_r, t_r^{-1}],$$

and

$$\mathsf{F}_{tr} = \bigcup_{n \in \mathbb{N}'} (\mathsf{F}_0)_{sep}[t_1^{1/n}, t_1^{-1/n}, \ldots, t_r^{1/n}, t_r^{-1/n}].$$

The last equality follows from the characterization of F_{tr} in Prop. 5.22(b).

While we will be primarily concerned with tame graded field extensions, other types of extensions will arise occasionally.

Definition 5.25. Let $\mathsf{F} \subseteq \mathsf{K}$ be graded fields with K algebraic over F. We say that $a \in \mathsf{K}$ is *purely inseparable over* F if a is the only root of the minimal polynomial of a over $q(\mathsf{F})$. So, if $char\, \mathsf{F} = 0$, then a is purely inseparable over F if and only if $a \in \mathsf{F}$. But, if $char\, \mathsf{F} = p \neq 0$, then a is purely inseparable over F if and only if $a^{p^n} \in \mathsf{F}$ for some positive integer n. We say that K is *purely inseparable over* F if every homogeneous element of K is purely inseparable over F. So, if $char\, \mathsf{F} = 0$, then K is purely inseparable over F if and only if $\mathsf{K} = \mathsf{F}$. But if $char\, \mathsf{F} = p \neq 0$ it is easy to check that the following conditions are equivalent: (a) K is purely inseparable over F; (b) K_0 is purely inseparable over F_0 and $\Gamma_{\mathsf{K}}/\Gamma_{\mathsf{F}}$ is a p-primary torsion group; (c) $q(\mathsf{K})$ is purely inseparable over $q(\mathsf{F})$.

Example 5.26. Suppose (K, w) is an extension of the valued field (F, v) and K is purely inseparable over F. Then $\mathsf{gr}_w(K)$ is purely inseparable over $\mathsf{gr}_v(F)$ because any homogeneous element a of $\mathsf{gr}(K)$ has the form $a = \widetilde{c}$ for some $c \in K$. If $char\, F = p \neq 0$, then $c^{p^n} \in F$ for some integer n; then $a^{p^n} = \widetilde{c}^{p^n} = \widetilde{c^{p^n}} \in \mathsf{gr}(F)$.

Definition 5.27. Let $\mathsf{F} \subseteq \mathsf{K}$ be graded fields with K algebraic over F. We say that K is *normal* over F if every irreducible homogenizable polynomial in $\mathsf{F}[X]$ which has a root in K factors into a product of polynomials of degree 1 in $\mathsf{K}[X]$.

Proposition 5.28. *Let $\mathsf{F} \subseteq \mathsf{K}$ be graded fields with K algebraic over F. Then, the following conditions are equivalent:*

(a) *K is normal over F.*
(b) *For every homogeneous $b \in \mathsf{K}$, the minimal polynomial P_b of b over $q(\mathsf{F})$ (which lies in $\mathsf{F}[X]$) factors into a product of polynomials of degree 1 in $\mathsf{K}[X]$.*
(c) *$q(\mathsf{K})$ is normal over $q(\mathsf{F})$.*

Proof. (a) $\Rightarrow$ (b) Suppose K is normal over F, and take any homogeneous $b \in \mathsf{K}$. Its minimal polynomial P_b over $q(\mathsf{F})$ lies in $\mathsf{F}[X]$ and is homogenizable by Prop. 5.7. Since P_b is homogenizable in $\mathsf{F}[X]$ and irreducible in $q(\mathsf{F})[X]$, it is irreducible in $\mathsf{F}[X]$, by Prop. 5.6. The normality implies that P_b factors into a product of polynomials of degree 1 in $\mathsf{K}[X]$. So, condition (b) holds.

(b) $\Rightarrow$ (a) Take any irreducible homogenizable $f \in \mathsf{F}[X]$ with a root, say d, in K. Let c be the leading coefficient of f. Since f is homogenizable, c is homogeneous in F, so $c \in \mathsf{F}^\times$. Hence, letting $g = c^{-1}f$, we have g is monic, homogenizable, and irreducible in $\mathsf{F}[X]$; so, g is irreducible in $q(\mathsf{F})[X]$ by Prop. 5.6. Hence, $g = P_d$, the minimal polynomial of d over $q(F)$. By condition (b), P_d factors into a product of polynomials of degree 1 in $\mathsf{K}[X]$; so, $f = cP_d$ also has such a factorization.

(b) $\Rightarrow$ (c) Since $q(\mathsf{K})$ is generated over $q(\mathsf{F})$ by $\mathsf{K}^\times$, condition (b) implies that $q(\mathsf{K})$ is a splitting field over $q(\mathsf{F})$ of the set of minimal polynomials $\{P_b \mid b \in \mathsf{K}^\times\}$. Hence, $q(\mathsf{K})$ is normal over $q(\mathsf{F})$.

(c) $\Rightarrow$ (b) Suppose $q(\mathsf{K})$ is normal over $q(\mathsf{F})$. For any homogeneous b in K, this normality implies that P_b splits over $q(\mathsf{K})$. Since P_b lies in $\mathsf{F}[X]$ by Prop. 5.7, its roots in $q(\mathsf{K})$ are integral over F; so, they lie in $\mathsf{K}^\natural$, which is the integral closure of $\mathsf{F}^\natural$ in $q(\mathsf{K})$, by Prop. 5.3. Therefore, the monic degree 1 factors of P_b in $q(\mathsf{K})[X]$ lie in $\mathsf{F}[X]$. So, condition (b) holds. □

5.1.4 Galois theory of graded fields

The definition of a Galois extension of graded fields closely follows the classical definition for fields:

Definition 5.29. Let K be an algebraic graded field extension of some graded field F, and let $Aut(\mathsf{K}/\mathsf{F})$ denote the group of graded F-algebra automorphisms of K. For any subgroup $G \subseteq Aut(\mathsf{K}/\mathsf{F})$, let K^G denote the set of fixed elements

$$\mathsf{K}^G = \{x \in \mathsf{K} \mid \sigma(x) = x \text{ for all } \sigma \in G\}.$$

This set is a graded subfield of K containing F. The extension K/F is defined to be a *Galois extension* if $\mathsf{K}^{Aut(\mathsf{K}/\mathsf{F})} = \mathsf{F}$. When this condition holds, we use the notation $\mathcal{G}(\mathsf{K}/\mathsf{F})$ for $Aut(\mathsf{K}/\mathsf{F})$, and call this group the *Galois group* of the Galois extension K/F.

The Krull topology on $Aut(\mathsf{K}/\mathsf{F})$ is defined analogously to the ungraded case: It is the topology that makes $Aut(\mathsf{K}/\mathsf{F})$ into a topological group with a fundamental system of open neighborhoods of id_{K} given by the subgroups $Aut(\mathsf{K}/\mathsf{L})$ for the graded fields L with $\mathsf{F} \subseteq \mathsf{L} \subseteq \mathsf{K}$ and $[\mathsf{L}{:}\mathsf{F}] < \infty$.

Proposition 5.30. *For an algebraic graded field extension* K/F*, the following conditions are equivalent:*

(a) K *is a Galois extension of* F*;*
(b) $q(\mathsf{K})$ *is a Galois field extension of* $q(\mathsf{F})$*.*

When they hold, the functorial map $\mathcal{G}(\mathsf{K}/\mathsf{F}) \to \mathcal{G}\big(q(\mathsf{K})/q(\mathsf{F})\big)$ *is an isomorphism of topological groups, whose inverse map is given by restriction to* K.

Proof. By Prop. 5.13, the functor $q\colon \mathfrak{algrFields}_{\mathsf{F}} \to \mathfrak{Fields}_{q(\mathsf{F})}$ defines a bijective group homomorphism $\psi\colon Aut(\mathsf{K}/\mathsf{F}) \to Aut\big(q(\mathsf{K})/q(\mathsf{F})\big)$; the inverse of this bijection is given by restriction to K. Therefore,

$$\mathsf{K}^{Aut(\mathsf{K}/\mathsf{F})} = \mathsf{K} \cap q(\mathsf{K})^{Aut(q(\mathsf{K})/q(\mathsf{F}))}.$$

Since $\mathsf{K} \cap q(\mathsf{F}) = \mathsf{F}$ by Prop. 5.3, it follows that (b) $\Rightarrow$ (a). For the converse, note that every element of $q(\mathsf{K})$ can be written in the form ab^{-1} with $a \in \mathsf{K}$ and $b \in \mathsf{F}$, by Prop. 5.3. If this ab^{-1} is fixed under $Aut\big(q(\mathsf{K})/q(\mathsf{F})\big)$, then $a \in \mathsf{K}^{Aut(\mathsf{K}/\mathsf{F})}$. Therefore, (a) $\Rightarrow$ (b). It remains only to prove that the group isomorphism ψ is also a homeomorphism. This will be done during the proof of Prop. 5.32 below. □

Remark 5.31. Use the same notation as in Prop. 5.30, and assume $[\mathsf{K}{:}\mathsf{F}]$ is finite. Then conditions (a) and (b) are also equivalent to

(c) $\mathsf{K}^\natural$ is a Galois ring extension of $\mathsf{F}^\natural$.

When conditions (a)–(c) hold, the functorial map $\mathcal{G}(\mathsf{K}/\mathsf{F}) \to \mathcal{G}(\mathsf{K}^\natural/\mathsf{F}^\natural)$ is an isomorphism. Indeed, since $q(\mathsf{K}) = \mathsf{K}^\natural \otimes_{\mathsf{F}^\natural} q(\mathsf{F})$ by Prop. 5.3, we have (c) $\Rightarrow$ (b) by scalar extension (see Greither [91, Lemma 1.11, p. 5]). According to the characterization of Galois ring extensions in Greither [91, Th. 1.6, p. 2] or Chase et al. [50, Th. 1.3], to prove (a) $\Rightarrow$ (c) it suffices to show that for any $\sigma \in \mathcal{G}(\mathsf{K}/\mathsf{F}) \setminus \{id\}$ and any proper ideal $I \subsetneq \mathsf{K}^\natural$, there exists an element $x \in \mathsf{K}^\natural$ such that $\sigma(x) - x \notin I$. Taking for x a homogeneous element such that $\sigma(x) \neq x$, we readily have $\sigma(x) - x \notin I$ for every proper ideal I, since $\sigma(x) - x$ is homogeneous and nonzero, hence invertible. (Alternatively, (a) $\Rightarrow$ (c) follows from Prop. 5.19 and the characterization of Galois ring extensions in terms of separability, see [91, Th. 2.1, p. 7] or [50, Th. 1.3].) Note that when (c) holds the Galois group $\mathcal{G}(\mathsf{K}^\natural/\mathsf{F}^\natural)$ is uniquely determined (and $\mathcal{G}(\mathsf{K}^\natural/\mathsf{F}^\natural) = Aut(\mathsf{K}^\natural/\mathsf{F}^\natural)$) since the ring $\mathsf{F}^\natural$ is connected (i.e., 0 and 1 are the only idempotents): see [91, Prop. 7.3] or [50, Cor. 3.3].

Proposition 5.32. *Suppose* K *is a Galois graded field extension of* F. *Let* $G = \mathcal{G}(\mathsf{K}/\mathsf{F})$ *and* $\mathcal{G} = \mathcal{G}(q(\mathsf{K})/q(\mathsf{F}))$.

(i) *Let* L *be a graded field with* $\mathsf{F} \subseteq \mathsf{L} \subseteq \mathsf{K}$. *Then* K *is Galois over* L *and* $\mathcal{G}(\mathsf{K}/\mathsf{L})$ *is closed in* G. *If* $[\mathsf{L}{:}\mathsf{F}] < \infty$ *then* $|G{:}\,\mathcal{G}(\mathsf{K}/\mathsf{L})| = [\mathsf{L}{:}\mathsf{F}]$.
(ii) *Let* H *be a subgroup of* G. *Then* $\mathcal{G}(\mathsf{K}/\mathsf{K}^H) = \overline{H}$, *the closure of* H *in the Krull topology.*
(iii) *The map* $\mathsf{L} \mapsto \mathcal{G}(\mathsf{K}/\mathsf{L})$ *gives a one-to-one inclusion-reversing correspondence between the graded fields* L *with* $\mathsf{F} \subseteq \mathsf{L} \subseteq \mathsf{K}$ *and the closed subgroups of* G. *Furthermore, L is Galois over* F *if and only if* $\mathcal{G}(\mathsf{K}/\mathsf{L})$ *is a normal subgroup of* G. *When this occurs,* $\mathcal{G}(\mathsf{L}/\mathsf{F}) \cong G/\,\mathcal{G}(\mathsf{K}/\mathsf{L})$, *a topological group isomorphism, where the factor group is given the quotient topology.*
(iv) *If* $[\mathsf{K}{:}\mathsf{F}] < \infty$ *then* $|G| = [\mathsf{K}{:}\mathsf{F}]$.

Proof. We first prove (i) and show that the canonical isomorphism $\psi\colon G \to \mathcal{G}$ is a homeomorphism. Take a graded field L as in (i). So, $q(\mathsf{F}) \subseteq q(\mathsf{L}) \subseteq q(\mathsf{K})$. Since K is Galois over F, by Prop. 5.30 $q(\mathsf{K})$ is Galois over $q(\mathsf{F})$, so $q(\mathsf{K})$ is Galois over $q(\mathsf{L})$; hence K is Galois over L by Prop. 5.30 again. The map ψ clearly restricts to the corresponding map for $\mathcal{G}(\mathsf{K}/\mathsf{L})$; hence, $\psi(\mathcal{G}(\mathsf{K}/\mathsf{L})) = \mathcal{G}(q(\mathsf{K})/q(\mathsf{L}))$. Therefore, if $[\mathsf{L}{:}\mathsf{F}] < \infty$, then

$$|G\colon \mathcal{G}(\mathsf{K}/\mathsf{L})| \;=\; |\mathcal{G}\colon \mathcal{G}(q(\mathsf{K})/q(\mathsf{L}))| \;=\; [q(\mathsf{L}){:}q(\mathsf{F})] \;=\; [\mathsf{L}{:}\mathsf{F}]. \tag{5.10}$$

Hence, every fundamental neighborhood of id_{K} in G maps to a fundamental neighborhood of $id_{q(\mathsf{K})}$ in $\mathcal{G}$. On the other hand, let E be a field with $q(\mathsf{F}) \subseteq E \subseteq q(\mathsf{K})$. Then, $q(\mathsf{K})$ is Galois over E. So $E = q(\mathsf{K})^{\mathcal{G}(q(\mathsf{K})/E)}$, and hence

$$E \cap \mathsf{K} \;=\; q(\mathsf{K})^{\mathcal{G}(q(\mathsf{K})/E)} \cap \mathsf{K} \;=\; \mathsf{K}^{\psi^{-1}(\mathcal{G}(q(\mathsf{K})/E))}. \tag{5.11}$$

Since $E \cap \mathsf{K}$ is thus the fixed ring of K under a group of graded automorphisms, the ring $E \cap \mathsf{K}$ is a graded field. Every element of $q(\mathsf{K})$ has the form ab^{-1} with $a \in \mathsf{K}$ and $b \in \mathsf{F} \setminus \{0\}$. Such an ab^{-1} lies in E if and only if $a \in E \cap \mathsf{K}$. Hence, $q(E \cap \mathsf{K}) = E$. Therefore, for $E \cap \mathsf{K}$ just as for L above, $\psi\big(\mathcal{G}(\mathsf{K}/(E \cap \mathsf{K}))\big) = \mathcal{G}(q(\mathsf{K})/E)$ and $[\mathsf{K}{:}(E \cap \mathsf{K})] = [q(\mathsf{K}){:}E]$. Therefore, ψ^{-1} takes every fundamental neighborhood of $id_{q(\mathsf{K})}$ in $\mathcal{G}$ to a fundamental neighborhood of id_{K} in G. Thus, the group isomorphism ψ is also a homeomorphism. This completes the proof of Prop. 5.30. It follows also that $\mathcal{G}(\mathsf{K}/\mathsf{L})$ is closed in G since the corresponding group $\mathcal{G}(q(\mathsf{K})/q(\mathsf{L}))$ is closed in $\mathcal{G}$. The last assertion of (i) follows from (5.10). All of (i) is now proved. The argument for E shows bijectivity of the injective map $\mathsf{L} \mapsto q(\mathsf{L})$ from the set of graded fields L with $\mathsf{F} \subseteq \mathsf{L} \subseteq \mathsf{K}$ to the set of fields E with $q(\mathsf{F}) \subseteq E \subseteq q(\mathsf{K})$. With this and the topological group isomorphism $G \to \mathcal{G}$ and Prop. 5.30, assertions (ii)–(iv) follow at once from Galois theory for the Galois extension of fields $q(\mathsf{K})/q(\mathsf{F})$. (iv) also follows from (ii) by taking $\mathsf{L} = \mathsf{K}$. □

Corollary 5.33. *Let* K/F *be an algebraic graded field extension. Then,* K *is Galois over* F *if and only if* K *is both tame and normal over* F.

Proof. Since K is tame (resp. normal, resp. Galois) over F if and only if $q(\mathsf{K})$ is separable (resp. normal, resp. Galois) over $q(\mathsf{F})$ by Prop. 5.19 (resp. Prop. 5.28, resp. Prop. 5.30), this follows from the corresponding ungraded result. □

Example 5.34. For any graded field F, the inertial closure F_{in} and the tame closure F_{tr} are Galois graded field extensions of F. This is particularly obvious for F_{in}, since we observed in Def. 5.23 that $q(\mathsf{F}_{in}) = (\mathsf{F}_0)_{sep} \otimes_{\mathsf{F}_0} q(\mathsf{F})$, so there is a canonical identification

$$\mathcal{G}(\mathsf{F}_{in}/\mathsf{F}) \;=\; \mathcal{G}\big((\mathsf{F}_0)_{sep}/\mathsf{F}_0\big).$$

The fact that $\mathsf{F}_{tr}/\mathsf{F}$ is a Galois extension of graded fields or, equivalently, that $q(\mathsf{F}_{tr})/q(\mathsf{F})$ is a Galois extension of fields, follows from the fact that $q(\mathsf{F}_{tr})$ is the separable closure of $q(\mathsf{F})$ in $q(\mathsf{F}_{alg})$, and from Prop. 5.10(v), which shows that $q(\mathsf{F}_{alg})$ is a normal extension of $q(\mathsf{F})$.

In view of Prop. 5.30, we may identify the Galois group of any Galois extension of graded fields K/F with the Galois group of the extension $q(\mathsf{K})/q(\mathsf{F})$. We use the Galois correspondence to show that the correspondence (5.8) between graded subextensions of a tame graded field extension and subextensions of the quotient fields is bijective.

Proposition 5.35. *Let M/F be a tame graded field extension. The mapping $\mathsf{L} \mapsto q(\mathsf{L})$ from graded fields L with $\mathsf{F} \subseteq \mathsf{L} \subseteq \mathsf{M}$ to fields E with $q(\mathsf{F}) \subseteq E \subseteq q(\mathsf{M})$ is bijective. The inverse map is given by $E \mapsto E \cap \mathsf{K}$.*

Proof. We already know from Cor. 5.14 that the map $\mathsf{L} \mapsto q(\mathsf{L})$ is injective. For surjectivity, choose a tame closure F_{tr} of F containing M. Take any field E with $q(\mathsf{F}) \subseteq E \subseteq q(\mathsf{M})$. Since $\mathsf{F}_{tr}/\mathsf{F}$ is a Galois graded field extension (see Ex. 5.34), the proof of Prop. 5.32 (see (5.11)) showed that $E \cap \mathsf{F}_{tr}$ is a graded field with $q(E \cap \mathsf{F}_{tr}) = E$. Also, as $q(\mathsf{F}) \subseteq E \subseteq q(\mathsf{M})$, we have

$$\mathsf{F} = q(\mathsf{F}) \cap \mathsf{F}_{tr} \subseteq E \cap \mathsf{F}_{tr} \subseteq q(\mathsf{M}) \cap \mathsf{F}_{tr} = \mathsf{M},$$

with the equalities given by Prop. 5.3. This proves surjectivity of the map $\mathsf{L} \mapsto q(\mathsf{L})$. Since $E \cap \mathsf{F}_{tr} \subseteq \mathsf{M}$, we have $E \cap \mathsf{F}_{tr} = E \cap \mathsf{M}$. So, the inverse of the surjective map is given by $E \mapsto E \cap \mathsf{M}$. □

Remark 5.36. There is a graded version of Kummer theory: Fix a graded field F and a positive integer n such that F contains a primitive n-th root of unity (which necessarily lies in F_0). Let μ_n denote the group of n-th roots of unity in F. A finite-degree graded field extension L of F is Galois with $\mathcal{G}(\mathsf{L}/\mathsf{F})$ abelian of exponent dividing n if and only if $\mathsf{L} = \mathsf{F}(\sqrt[n]{c_1}, \ldots, \sqrt[n]{c_k})$ for some $c_1 \ldots, c_k \in \mathsf{F}^\times$. Such an L is called a *graded n-Kummer extension* of F. (Toward proving the "only if" assertion, note that the hypotheses on roots of unity and $\mathcal{G}(\mathsf{K}/\mathsf{F})$ assure that $q(\mathsf{L})$ is an n-Kummer extension of $q(\mathsf{F})$. If $b \in q(\mathsf{L})^\times$ and $b^n \in q(\mathsf{F})$, then for each $\sigma \in \mathcal{G}(q(\mathsf{L})/q(\mathsf{F})) = \mathcal{G}(\mathsf{L}/\mathsf{F})$ we have $\sigma(b)b^{-1} = \omega_\sigma$ for some $\omega_\sigma \in \mu_n$. Write $b = ad^{-1}$ for some $a \in \mathsf{L} \setminus \{0\}$, $d \in \mathsf{F} \setminus \{0\}$. Then, $\sigma(a)a^{-1} = \omega_\sigma$. Since σ is a graded automorphism of L and $\omega_\sigma \in \mathsf{F}_0^\times$, it follows that $\sigma(a_\gamma)a_\gamma^{-1} = \omega_\sigma$ for each nonzero homogeneous component a_γ of a. Hence, $a_\gamma^n \in \mathsf{F}$ for all such a_γ. Taking enough such b to generate $q(\mathsf{L})$ over $q(\mathsf{F})$ will yield enough such a_γ to generate L over F.) When L is a graded n-Kummer extension of F, let $\mathcal{K}(\mathsf{L}/\mathsf{F}) = \{a \in \mathsf{L}^\times \mid a^n \in \mathsf{F}\}/\mathsf{F}^\times$. Then, the n-th power map yields an isomorphism

$$\mathcal{K}(\mathsf{L}/\mathsf{F}) \xrightarrow{\sim} \{c \in \mathsf{F}^\times \mid \sqrt[n]{c} \in \mathsf{L}^\times\}/\mathsf{F}^{\times n}$$

and there is a perfect pairing

$$\mathcal{K}(\mathsf{L}/\mathsf{F}) \times \mathcal{G}(\mathsf{L}/\mathsf{F}) \longrightarrow \mu_n \qquad \text{given by} \qquad (a\mathsf{F}^\times, \sigma) \mapsto \sigma(a)a^{-1}.$$

Thus, $\mathcal{K}(\mathsf{L}/\mathsf{F}) \cong \mathit{Hom}(\mathcal{G}(\mathsf{L}/\mathsf{F}), \mu_n)$ canonically and $\mathcal{K}(\mathsf{L}/\mathsf{F}) \cong \mathcal{G}(\mathsf{L}/\mathsf{F})$ non-canonically. In particular, $|\mathcal{K}(\mathsf{L}/\mathsf{F})| = |\,\mathcal{G}(\mathsf{L}/\mathsf{F})| = [\mathsf{L}{:}\mathsf{F}]$. The group $\mathcal{K}(\mathsf{L}/\mathsf{F})$ is a homogeneous armature of the graded F-algebra L in the terminology of §7.2.2 below.

A major feature of the Galois theory of graded fields is that the Galois group of any Galois extension contains a distinguished normal closed subgroup, which we introduce next.

Definition 5.37. Let K/F be an arbitrary Galois extension of graded fields, and let $\mathsf{L} = \mathsf{K}_0 \otimes_{\mathsf{F}_0} \mathsf{F}$ be the unramified closure[2] of F in K. Every $\sigma \in \mathcal{G}(\mathsf{K}/\mathsf{F})$ preserves the grading of K, hence carries L to itself. Therefore, the graded field extension L/F and the field extension $\mathsf{K}_0/\mathsf{F}_0$ are Galois. Restricting automorphisms of L to K_0 defines a canonical isomorphism

$$\mathcal{G}(\mathsf{L}/\mathsf{F}) \xrightarrow{\sim} \mathcal{G}(\mathsf{K}_0/\mathsf{F}_0),$$

which we use to identify $\mathcal{G}(\mathsf{K}_0/\mathsf{F}_0)$ with $\mathcal{G}(\mathsf{L}/\mathsf{F})$. The *inertia subgroup* $\mathcal{G}^{in}(\mathsf{K}/\mathsf{F})$ of $\mathcal{G}(\mathsf{K}/\mathsf{F})$ is defined as the kernel of the map that carries every automorphism of K/F to its restriction to K_0 or, equivalently, to L:

$$\mathcal{G}^{in}(\mathsf{K}/\mathsf{F}) = \mathit{ker}\big(\mathcal{G}(\mathsf{K}/\mathsf{F}) \to \mathcal{G}(\mathsf{K}_0/\mathsf{F}_0)\big) = \mathcal{G}(\mathsf{K}/\mathsf{L}).$$

Thus, we have a canonical exact sequence:

$$1 \longrightarrow \mathcal{G}^{in}(\mathsf{K}/\mathsf{F}) \longrightarrow \mathcal{G}(\mathsf{K}/\mathsf{F}) \longrightarrow \mathcal{G}(\mathsf{K}_0/\mathsf{F}_0) \longrightarrow 1. \tag{5.12}$$

For the following statement, note that if K/F is a Galois extension of graded fields, then for any $\sigma \in \mathcal{G}(\mathsf{K}/\mathsf{F})$ and any $u \in \mathsf{K}^\times$ we have $\mathit{deg}\,\sigma(u) = \mathit{deg}\,u$, hence $\sigma(u)u^{-1} \in \mathsf{K}_0^\times$.

Lemma 5.38. *Let K/F be an arbitrary Galois extension of graded fields. For $\sigma \in \mathcal{G}^{in}(\mathsf{K}/\mathsf{F})$ and $u \in \mathsf{K}^\times$, the element $\sigma(u)u^{-1}$ depends only on σ and on the image of $\mathit{deg}\,u$ in $\Gamma_\mathsf{K}/\Gamma_\mathsf{F}$. The map $\mathit{deg}(u) + \Gamma_\mathsf{F} \mapsto \sigma(u)u^{-1}$ is a group homomorphism*

$$f_\sigma \colon \Gamma_\mathsf{K}/\Gamma_\mathsf{F} \longrightarrow \mathsf{K}_0^\times,$$

and the map $\sigma \mapsto f_\sigma$ is an injective group homomorphism

$$\mathcal{G}^{in}(\mathsf{K}/\mathsf{F}) \hookrightarrow \mathit{Hom}(\Gamma_\mathsf{K}/\Gamma_\mathsf{F}, \mathsf{K}_0^\times).$$

Proof. As in Def. 5.37, we let $\mathsf{L} = \mathsf{K}_0 \otimes_{\mathsf{F}_0} \mathsf{F}$ be the unramified (or inertial) closure of F in K. Let u, $u' \in \mathsf{K}^\times$ be such that $\mathit{deg}\,u \equiv \mathit{deg}\,u' \bmod \Gamma_\mathsf{F}$. Since $\mathsf{K}_0 = \mathsf{L}_0$, we have $u'u^{-1} \in \mathsf{L}^\times$, hence $\sigma(u'u^{-1}) = u'u^{-1}$ for all $\sigma \in \mathcal{G}^{in}(\mathsf{K}/\mathsf{F})$. It follows that $\sigma(u')u'^{-1} = \sigma(u)u^{-1}$ for all $\sigma \in \mathcal{G}^{in}(\mathsf{K}/\mathsf{F})$, hence we may define

$$f_\sigma(\mathit{deg}(u) + \Gamma_\mathsf{F}) = \sigma(u)u^{-1} \in \mathsf{K}_0^\times.$$

---

[2] The graded field extension K/F is tame since the field extension $q(\mathsf{K})/q(\mathsf{F})$ is Galois, hence separable. Therefore, L is also the inertial closure of F in K.

The fact that f_σ is a group homomorphism is a consequence of the following computation: for u, $v \in \mathsf{K}^\times$ we have

$$\begin{aligned} f_\sigma(deg(u) + deg(v) + \Gamma_\mathsf{F}) \;&=\; \sigma(uv)\cdot(uv)^{-1} \;=\; \sigma(u)u^{-1}\cdot\sigma(v)v^{-1} \\ &=\; f_\sigma(deg(u)+\Gamma_\mathsf{F})\cdot f_\sigma(deg(v)+\Gamma_\mathsf{F}). \end{aligned}$$

Now, for σ, $\tau \in \mathcal{G}^{in}(\mathsf{K}/\mathsf{F})$ and $u \in \mathsf{K}^\times$ we have $\sigma(\tau(u)u^{-1}) = \tau(u)u^{-1}$ since $\tau(u)u^{-1} \in \mathsf{K}_0^\times$. Moreover, $deg\,\tau(u) = deg\,u$, hence $f_\sigma(deg(u)+\Gamma_\mathsf{F}) = \sigma\tau(u)\tau(u)^{-1}$. Therefore,

$$\begin{aligned} f_{\sigma\tau}(deg(u)+\Gamma_\mathsf{F}) \;&=\; \sigma\tau(u)u^{-1} \;=\; \sigma\tau(u)\tau(u)^{-1}\cdot\tau(u)u^{-1} \\ &=\; f_\sigma(deg(u)+\Gamma_\mathsf{F})\cdot f_\tau(deg(u)+\Gamma_\mathsf{F}). \end{aligned}$$

It follows that $f_{\sigma\tau} = f_\sigma\cdot f_\tau$. Moreover, if $f_\sigma = 1$, then $\sigma(u)u^{-1} = 1$ for all $u \in \mathsf{K}^\times$, hence $\sigma = id$. Thus, the proof is complete. □

Lemma 5.38 readily shows that the inertia subgroup $\mathcal{G}^{in}(\mathsf{K}/\mathsf{F})$ is abelian. Therefore, we may use the exact sequence (5.12) to define a canonical action of $\mathcal{G}(\mathsf{K}_0/\mathsf{F}_0)$ on $\mathcal{G}^{in}(\mathsf{K}/\mathsf{F})$: for $\overline{\sigma} \in \mathcal{G}(\mathsf{K}_0/\mathsf{F}_0)$ and $\tau \in \mathcal{G}^{in}(\mathsf{K}/\mathsf{F})$, we may consider an arbitrary lift $\sigma \in \mathcal{G}(\mathsf{K}/\mathsf{F})$ of $\overline{\sigma}$ and set

$$\overline{\sigma} * \tau \;=\; \sigma\tau\sigma^{-1}.$$

The term on the right does not depend on the choice of the lift σ, because if σ' is another lift of $\overline{\sigma}$, then $\sigma^{-1}\sigma' \in \mathcal{G}^{in}(\mathsf{K}/\mathsf{F})$, hence $\sigma^{-1}\sigma'$ commutes with τ. Thus, $\mathcal{G}^{in}(\mathsf{K}/\mathsf{F})$ is a $\mathbb{Z}\big[\mathcal{G}(\mathsf{K}_0/\mathsf{F}_0)\big]$-module.

Using the notation of Lemma 5.38, note that the image of each f_σ lies in the torsion subgroup of $\mathsf{K}_0^\times$, since $\Gamma_\mathsf{K}/\Gamma_\mathsf{F}$ is a torsion group. We let $\mu(\mathsf{K}_0)$ denote the torsion subgroup of $\mathsf{K}_0^\times$, which is the group of roots of unity in K_0, and let

$$\langle\sigma,\overline{\gamma}\rangle \;=\; f_\sigma(\overline{\gamma}) \in \mu(\mathsf{K}_0) \qquad \text{for } \sigma \in \mathcal{G}^{in}(\mathsf{K}/\mathsf{F}) \text{ and } \overline{\gamma} \in \Gamma_\mathsf{K}/\Gamma_\mathsf{F}.$$

Thus,

$$\langle\sigma, deg(u)+\Gamma_\mathsf{F}\rangle \;=\; \sigma(u)u^{-1} \qquad \text{for } \sigma \in \mathcal{G}^{in}(\mathsf{K}/\mathsf{F}) \text{ and } u \in \mathsf{K}^\times.$$

Theorem 5.39. *Let K/F be an arbitrary Galois extension of graded fields. The map*

$$\langle\cdot,\cdot\rangle\colon\ \mathcal{G}^{in}(\mathsf{K}/\mathsf{F})\times(\Gamma_\mathsf{K}/\Gamma_\mathsf{F}) \longrightarrow \mu(\mathsf{K}_0)$$

is a $\mathbb{Z}$-bilinear pairing, which is $\mathcal{G}(\mathsf{K}_0/\mathsf{F}_0)$-equivariant for the canonical action on $\mathcal{G}^{in}(\mathsf{K}/\mathsf{F})$ and $\mu(\mathsf{K}_0)$, and the trivial action on $\Gamma_\mathsf{K}/\Gamma_\mathsf{F}$; i.e.,

$$\langle\overline{\sigma} * \tau,\overline{\gamma}\rangle \;=\; \overline{\sigma}\big(\langle\tau,\overline{\gamma}\rangle\big) \tag{5.13}$$

for all $\overline{\sigma} \in \mathcal{G}(\mathsf{K}_0/\mathsf{F}_0)$, $\tau \in \mathcal{G}^{in}(\mathsf{K}/\mathsf{F})$, and $\overline{\gamma} \in \Gamma_\mathsf{K}/\Gamma_\mathsf{F}$. The pairing is nondegenerate on each side, in the following sense:

(i) *if $\sigma \in \mathcal{G}^{in}(\mathsf{K}/\mathsf{F})$ satisfies $\langle\sigma,\overline{\gamma}\rangle = 1$ for all $\overline{\gamma} \in \Gamma_\mathsf{K}/\Gamma_\mathsf{F}$, then $\sigma = id$;*
(ii) *if $\overline{\gamma} \in \Gamma_\mathsf{K}/\Gamma_\mathsf{F}$ satisfies $\langle\sigma,\overline{\gamma}\rangle = 1$ for all $\sigma \in \mathcal{G}^{in}(\mathsf{K}/\mathsf{F})$, then $\overline{\gamma} = 0$.*

Moreover, the pairing is a continuous map for the product of the Krull topology on $\mathcal{G}^{in}(\mathsf{K}/\mathsf{F})$ *and the discrete topology on* $\Gamma_{\mathsf{K}}/\Gamma_{\mathsf{F}}$, *and for the discrete topology on* $\mu(\mathsf{K}_0)$.

We call $\langle \cdot, \cdot \rangle$ the *ramification pairing* of the Galois extension of graded fields K/F.

Proof. As in Def. 5.37, let $\mathsf{L} = \mathsf{K}_0 \otimes_{\mathsf{F}_0} \mathsf{F}$ be the unramified (or inertial) closure of F in K. Bilinearity of the pairing $\langle \cdot, \cdot \rangle$ and (i) readily follow from Lemma 5.38. Suppose $\gamma \in \Gamma_{\mathsf{K}}$ satisfies $\langle \sigma, \gamma + \Gamma_{\mathsf{F}} \rangle = 1$ for all $\sigma \in \mathcal{G}^{in}(\mathsf{K}/\mathsf{F})$. For any $u \in \mathsf{K}^{\times}$ with $deg\, u = \gamma$ we then have $\sigma(u)u^{-1} = 1$ for all $\sigma \in \mathcal{G}^{in}(\mathsf{K}/\mathsf{F})$. Since $\mathcal{G}^{in}(\mathsf{K}/\mathsf{F}) = \mathcal{G}(\mathsf{K}/\mathsf{L})$, it follows that $u \in \mathsf{L}^{\times}$, hence $\gamma \in \Gamma_{\mathsf{L}} = \Gamma_{\mathsf{F}}$. This proves (ii).

We next show that the pairing is $\mathcal{G}(\mathsf{K}_0/\mathsf{F}_0)$-equivariant: let $\overline{\sigma} \in \mathcal{G}(\mathsf{K}_0/\mathsf{F}_0)$ and let $\sigma \in \mathcal{G}(\mathsf{K}/\mathsf{F})$ be a lift of $\overline{\sigma}$. For $\tau \in \mathcal{G}^{in}(\mathsf{K}/\mathsf{F})$ and $u \in \mathsf{K}^{\times}$ we have

$$\begin{aligned}\langle \overline{\sigma} * \tau, deg(u) + \Gamma_{\mathsf{F}} \rangle &= \sigma\tau\sigma^{-1}(u)u^{-1} = \sigma(\tau\sigma^{-1}(u)\sigma^{-1}(u)^{-1}) \\ &= \sigma\big(\langle \tau, deg\, \sigma^{-1}(u) + \Gamma_{\mathsf{F}} \rangle\big).\end{aligned}$$

Since $deg\, \sigma^{-1}(u) = deg\, u$, we thus have (5.13).

To prove that the pairing is continuous, it suffices to show that for any $\overline{\gamma} \in \Gamma_{\mathsf{K}}/\Gamma_{\mathsf{F}}$ the set

$$U(\overline{\gamma}) = \{\tau \in \mathcal{G}^{in}(\mathsf{K}/\mathsf{F}) \mid \langle \tau, \overline{\gamma} \rangle = 1\}$$

is an open subgroup of $\mathcal{G}^{in}(\mathsf{K}/\mathsf{F})$. Let $u \in \mathsf{K}^{\times}$ be such that $\overline{\gamma} = deg(u) + \Gamma_{\mathsf{F}}$. Then $U(\overline{\gamma}) = \mathcal{G}^{in}(\mathsf{K}/\mathsf{F}) \cap \mathcal{G}\big(\mathsf{K}/\mathsf{F}(u)\big)$. Since $\mathsf{F}(u)/\mathsf{F}$ is a finite-degree extension, $\mathcal{G}\big(\mathsf{K}/\mathsf{F}(u)\big)$ is an open subgroup in $\mathcal{G}(\mathsf{K}/\mathsf{F})$, hence $U(\overline{\gamma})$ is open. □

In the situation above, each $\tau \in \mathcal{G}^{in}(\mathsf{K}/\mathsf{F})$ yields a homomorphism

$$\langle \tau, \cdot \rangle : \Gamma_{\mathsf{K}}/\Gamma_{\mathsf{F}} \longrightarrow \mu(\mathsf{K}_0), \qquad \overline{\gamma} \mapsto \langle \tau, \overline{\gamma} \rangle.$$

We thus have a group homomorphism

$$\mathcal{G}^{in}(\mathsf{K}/\mathsf{F}) \longrightarrow Hom\big(\Gamma_{\mathsf{K}}/\Gamma_{\mathsf{F}}, \mu(\mathsf{K}_0)\big) \quad \text{given by} \quad \tau \mapsto \langle \tau, \cdot \rangle. \tag{5.14}$$

Corollary 5.40. *For an arbitrary Galois extension of graded fields* K/F, *the map* (5.14) *is an isomorphism of topological* $\mathcal{G}(\mathsf{K}_0/\mathsf{F}_0)$*-modules, for the action of* $\mathcal{G}(\mathsf{K}_0/\mathsf{F}_0)$ *on* $Hom\big(\Gamma_{\mathsf{K}}/\Gamma_{\mathsf{F}}, \mu(\mathsf{K}_0)\big)$ *given by*

$$\overline{\sigma} * \chi \colon \ \overline{\gamma} \mapsto \overline{\sigma}\big(\chi(\overline{\gamma})\big)$$

for $\overline{\sigma} \in \mathcal{G}(\mathsf{K}_0/\mathsf{F}_0)$, $\chi \in Hom\big(\Gamma_{\mathsf{K}}/\Gamma_{\mathsf{F}}, \mu(\mathsf{K}_0)\big)$, *and* $\overline{\gamma} \in \Gamma_{\mathsf{K}}/\Gamma_{\mathsf{F}}$, *and for the topology of pointwise convergence on* $Hom\big(\Gamma_{\mathsf{K}}/\Gamma_{\mathsf{F}}, \mu(\mathsf{K}_0)\big)$ (*with the discrete topology on* $\mu(\mathsf{K}_0)$).

Recall (e.g. from Kelley [114, p. 92]) that if $\mathcal{S}$ is a set of functions from a set X to a topological space Y, then the topology of pointwise convergence on $\mathcal{S}$ is obtained by viewing it as a subspace of Y^X, where $Y^X = \prod_{x \in X} Y$

is given the product topology using the topology on Y. Here, $\mu(\mathsf{K}_0)$ is given the discrete topology, and in the topology of pointwise convergence on $Hom(\Gamma_{\mathsf{K}}/\Gamma_{\mathsf{F}}, \mu(\mathsf{K}_0))$ a subgroup is open if and only if it contains all the homomorphisms that map a given finite set of elements in $\Gamma_{\mathsf{K}}/\Gamma_{\mathsf{F}}$ to 1.

Proof. Since the ramification pairing is $\mathcal{G}(\mathsf{K}_0/\mathsf{F}_0)$-equivariant and continuous, it readily follows that the map (5.14) is a continuous homomorphism of $\mathcal{G}(\mathsf{K}_0/\mathsf{F}_0)$-modules. It is also injective since the ramification pairing is nondegenerate, and it maps closed sets to closed sets since $\mathcal{G}^{in}(\mathsf{K}/\mathsf{F})$ is compact and $Hom(\Gamma_{\mathsf{K}}/\Gamma_{\mathsf{F}}, \mu(\mathsf{K}_0))$ is Hausdorff. To complete the proof, it thus suffices to prove that this map is onto.

For any group homomorphism $\varphi\colon \Gamma_{\mathsf{K}}/\Gamma_{\mathsf{F}} \to \mu(\mathsf{K}_0)$, we define a homomorphism $\sigma_\varphi\colon \mathsf{K}^\times \to \mathsf{K}^\times$ by

$$\sigma_\varphi(u) = \varphi(deg(u) + \Gamma_{\mathsf{F}}) \cdot u \qquad \text{for } u \in \mathsf{K}^\times,$$

and we extend σ_φ by additivity to a map $\sigma_\varphi\colon \mathsf{K} \to \mathsf{K}$. We then have $\sigma_\varphi \in \mathcal{G}(\mathsf{K}/\mathsf{L}) = \mathcal{G}^{in}(\mathsf{K}/\mathsf{F})$ and $\varphi(\overline{\gamma}) = \langle \sigma_\varphi, \overline{\gamma}\rangle$ for all $\overline{\gamma} \in \Gamma_{\mathsf{K}}/\Gamma_{\mathsf{F}}$, hence the map (5.14) is onto. □

Of course, we may also use the ramification pairing to attach to each $\overline{\gamma} \in \Gamma_{\mathsf{K}}/\Gamma_{\mathsf{F}}$ a group homomorphism

$$\langle \cdot, \overline{\gamma}\rangle\colon \mathcal{G}^{in}(\mathsf{K}/\mathsf{F}) \longrightarrow \mu(\mathsf{K}_0), \qquad \tau \mapsto \langle \tau, \overline{\gamma}\rangle.$$

Since the pairing is continuous, this homomorphism is continuous. It is also a homomorphism of $\mathcal{G}(\mathsf{K}_0/\mathsf{F}_0)$-modules since the pairing is $\mathcal{G}(\mathsf{K}_0/\mathsf{F}_0)$-equivariant. We thus have a map from $\Gamma_{\mathsf{K}}/\Gamma_{\mathsf{F}}$ to the group of continuous $\mathcal{G}(\mathsf{K}_0/\mathsf{F}_0)$-homomorphisms $\mathcal{G}^{in}(\mathsf{K}/\mathsf{F}) \to \mu(\mathsf{K}_0)$:

$$\Gamma_{\mathsf{K}}/\Gamma_{\mathsf{F}} \longrightarrow Hom^c_{\mathcal{G}(\mathsf{K}_0/\mathsf{F}_0)}\big(\mathcal{G}^{in}(\mathsf{K}/\mathsf{F}), \mu(\mathsf{K}_0)\big), \quad \text{given by} \quad \overline{\gamma} \mapsto \langle \cdot, \overline{\gamma}\rangle. \tag{5.15}$$

Corollary 5.41. *For an arbitrary Galois extension of graded fields* K/F*, the map* (5.15) *is a group isomorphism.*

Proof. Since the ramification pairing is nondegenerate, we already know that the map is an injective homomorphism of groups; it suffices to show that it is onto. Consider a continuous $\mathcal{G}(\mathsf{K}_0/\mathsf{F}_0)$-homomorphism $\varphi\colon \mathcal{G}^{in}(\mathsf{K}/\mathsf{F}) \to \mu(\mathsf{K}_0)$ and let $\mathsf{L} = \mathsf{K}_0 \otimes_{\mathsf{F}_0} \mathsf{F}$ be the unramified (or inertial) closure of F in K. The kernel of φ is an open subgroup of $\mathcal{G}^{in}(\mathsf{K}/\mathsf{F}) = \mathcal{G}(\mathsf{K}/\mathsf{L})$, so we may find a finite-degree graded field extension M of L in K such that $ker\,\varphi = \mathcal{G}(\mathsf{K}/\mathsf{M})$. The map φ induces an injective $\mathcal{G}(\mathsf{K}_0/\mathsf{F}_0)$-homomorphism

$$\overline{\varphi}\colon \mathcal{G}(\mathsf{M}/\mathsf{L}) = \mathcal{G}(\mathsf{K}/\mathsf{L})/\mathcal{G}(\mathsf{K}/\mathsf{M}) \hookrightarrow \mu(\mathsf{K}_0).$$

Since φ is a $\mathcal{G}(\mathsf{K}_0/\mathsf{F}_0)$-homomorphism, $\mathcal{G}(\mathsf{K}/\mathsf{M})$ is preserved under the action of $\mathcal{G}(\mathsf{K}_0/\mathsf{F}_0)$, hence it is a normal subgroup of $\mathcal{G}(\mathsf{K}/\mathsf{F})$ and therefore M/F is a Galois extension of graded fields. We have $\mathsf{L} \subseteq \mathsf{M} \subseteq \mathsf{K}$ and $\mathsf{L}_0 = \mathsf{K}_0$, hence $\mathsf{M}_0 = \mathsf{K}_0$ and L is the unramified closure of F in M. Therefore,

$\mathcal{G}^{in}(\mathsf{M}/\mathsf{F}) = \mathcal{G}(\mathsf{M}/\mathsf{L})$. Consider the map (5.15) for the extension M/F instead of K/F:

$$\Gamma_{\mathsf{M}}/\Gamma_{\mathsf{F}} \hookrightarrow Hom_{\mathcal{G}(\mathsf{M}_0/\mathsf{F}_0)}(\mathcal{G}^{in}(\mathsf{M}/\mathsf{F}), \mu(\mathsf{M}_0)). \tag{5.16}$$

The group $\mathcal{G}^{in}(\mathsf{M}/\mathsf{F})$ is finite since $[\mathsf{M}{:}\mathsf{L}]$ is finite. Let e be the exponent of this finite abelian group. The group $\mu_e(\mathsf{M}_0)$ of e-th roots of unity in M_0 is a cyclic subgroup of $\mathsf{M}_0^\times$, whose order d divides e. Choosing a generator of this group, we have an isomorphism $\mu_e(\mathsf{M}_0) \cong \mathbb{Z}/d\mathbb{Z} \hookrightarrow \mathbb{Q}/\mathbb{Z}$, hence

$$\begin{aligned} Hom_{\mathcal{G}(\mathsf{M}_0/\mathsf{F}_0)}(\mathcal{G}^{in}(\mathsf{M}/\mathsf{F}), \mu(\mathsf{M}_0)) \subseteq\ & Hom(\mathcal{G}^{in}(\mathsf{M}/\mathsf{F}), \mu_e(\mathsf{M}_0)) \\ & \hookrightarrow Hom(\mathcal{G}^{in}(\mathsf{M}/\mathsf{F}), \mathbb{Q}/\mathbb{Z}). \end{aligned} \tag{5.17}$$

The rightmost group is the dual of the finite abelian group $\mathcal{G}^{in}(\mathsf{M}/\mathsf{F})$; its order is $|\mathcal{G}^{in}(\mathsf{M}/\mathsf{F})|$. Therefore,

$$\big|\, Hom_{\mathcal{G}(\mathsf{M}_0/\mathsf{F}_0)}(\mathcal{G}^{in}(\mathsf{M}/\mathsf{F}), \mu(\mathsf{M}_0))\big| \;\le\; |\mathcal{G}^{in}(\mathsf{M}/\mathsf{F})| \;=\; [\mathsf{M}{:}\mathsf{L}] \;=\; |\Gamma_{\mathsf{M}}/\Gamma_{\mathsf{F}}|,$$

hence the map (5.16) is surjective. In particular, we may find $\gamma \in \Gamma_{\mathsf{M}}$ such that

$$\overline{\varphi}(\tau) \;=\; \langle \tau, \gamma + \Gamma_{\mathsf{F}} \rangle \qquad \text{for all } \tau \in \mathcal{G}^{in}(\mathsf{M}/\mathsf{F}).$$

For $\sigma \in \mathcal{G}^{in}(\mathsf{K}/\mathsf{F})$ we have

$$\varphi(\sigma) \;=\; \overline{\varphi}(\sigma|_{\mathsf{M}}) \;=\; \langle \sigma|_{\mathsf{M}}, \gamma + \Gamma_{\mathsf{F}} \rangle \;=\; \langle \sigma, \gamma + \Gamma_{\mathsf{F}} \rangle,$$

hence $\varphi = \langle \cdot, \gamma + \Gamma_{\mathsf{F}} \rangle$. This shows that φ lies in the image of the map (5.15) and completes the proof. □

Remark 5.42. Since the map (5.16) is a bijection, the inclusions in (5.17) are equalities. Therefore, M_0 must contain a primitive e-th root of unity, and every homomorphism $\mathcal{G}^{in}(\mathsf{M}/\mathsf{F}) \to \mu(\mathsf{M}_0)$ is $\mathcal{G}(\mathsf{M}_0/\mathsf{F}_0)$-equivariant. In particular, if $\mathcal{G}^{in}(\mathsf{M}/\mathsf{F})$ is central in $\mathcal{G}(\mathsf{M}/\mathsf{F})$, the $\mathcal{G}(\mathsf{M}_0/\mathsf{F}_0)$-action on $\mathcal{G}^{in}(\mathsf{M}/\mathsf{F})$ is trivial and therefore the image of each $\mathcal{G}(\mathsf{M}_0/\mathsf{F}_0)$-equivariant homomorphism $\mathcal{G}^{in}(\mathsf{M}/\mathsf{F}) \to \mu(\mathsf{M}_0)$ lies in $\mu(\mathsf{F}_0)$.

The proof of Cor. 5.41 highlights the importance of the roots of unity: there must be sufficiently many roots of unity in K_0 if the inertia subgroup $\mathcal{G}^{in}(\mathsf{K}/\mathsf{F})$ is large. In particular, we have the following consequence for totally ramified extensions:

Corollary 5.43. *Let K/F be a totally ramified graded field extension of finite degree, and let e be the exponent of the abelian group $\Gamma_{\mathsf{K}}/\Gamma_{\mathsf{F}}$. The extension K/F is Galois if and only if K_0 contains a primitive e-th root of unity.*

Proof. If K/F is Galois, Cor. 5.41 yields an isomorphism

$$\Gamma_{\mathsf{K}}/\Gamma_{\mathsf{F}} \xrightarrow{\sim} Hom(\mathcal{G}(\mathsf{K}/\mathsf{F}), \mu(\mathsf{K}_0)).$$

Since $\Gamma_{\mathsf{K}}/\Gamma_{\mathsf{F}}$ has exponent e, it follows that $\mu(\mathsf{K}_0)$ contains an element of order e.

Conversely, if K_0 contains a primitive e-th root of unity we have

$$\left|Hom(\Gamma_{\mathsf{K}}/\Gamma_{\mathsf{F}}, \mu(\mathsf{K}_0))\right| = |\Gamma_{\mathsf{K}}/\Gamma_{\mathsf{F}}| = [\mathsf{K}{:}\mathsf{F}].$$

Each homomorphism $\varphi\colon \Gamma_{\mathsf{K}}/\Gamma_{\mathsf{F}} \to \mu(\mathsf{K}_0)$ yields an automorphism σ_φ of K/F as in the proof of Cor. 5.40, hence K/F has $[\mathsf{K}{:}\mathsf{F}]$ automorphisms, and is therefore Galois. □

5.2 Valued fields from a graded perspective

In this section, we show how the classical ramification theory of valued field extensions relates to the ramification in graded field extensions through the gr construction. We show that the extension of graded fields associated to a tamely ramified Galois extension of valued fields is Galois, and that the Galois group of the valued field extension and of the corresponding graded field extension are canonically isomorphic. The canonical isomorphism preserves the inertia subgroup and the ramification pairing, see Cor. 5.52. As a main application, we obtain for Henselian-valued fields a lifting property for tame graded field extensions, similar to the inertial lifting property of Prop. A.17 in Appendix A.

Throughout this section, let v be a valuation on a field F with values in a given divisible ordered abelian group Γ.

5.2.1 Uniform polynomials

We first discuss a technical notion that plays for valued fields a role comparable to homogenizable polynomials for graded fields. We use it to characterize Henselian valuations in Th. 5.49.

Throughout, we let F_{alg} be an algebraic closure of F.

Proposition 5.44. *Let* $f = \sum_{i=0}^{n} a_i X^i \in F[X]$ *with* $a_0 a_n \neq 0$. *Let* $\lambda = \frac{1}{n}\big(v(a_0) - v(a_n)\big) \in \mathbb{H}(\Gamma_F)$. *The following conditions are equivalent:*

(a) *For some extension of v to F_{alg}, all the roots of f in F_{alg} have the same value.*
(b) *For every extension of v to F_{alg}, all the roots of f in F_{alg} have value λ.*
(c) $v(a_i) \geq (n-i)\lambda + v(a_n)$ *for* $i = 1, 2, \ldots, n$.
(d) *Let w be any extension of v to F_{alg}, take any $c \in F_{\mathrm{alg}}$ with $w(c) = \lambda$, and let $h = \frac{1}{a_n c^n} f(cX)$. Then, h is a monic polynomial in $\mathcal{O}_{F_{\mathrm{alg}},w}[X]$.*

Proof. (b) ⇒ (a) This is clear.

(a) ⇒ (c) Let $f = a_n(X - u_1)\ldots(X - u_n)$ in $F_{\mathrm{alg}}[X]$, and let s_j be the j-th elementary symmetric polynomial in $u_1, \ldots, u_n$ for $j = 1, 2, \ldots, n$. Suppose all the u_i have the same value for some extension w of v to F_{alg}.

Then, $w(u_i) = \lambda$ for all i, as $a_0 = (-1)^n a_n u_1 \dots u_n$. Since each s_j is a sum of products of j of the u_i, we have $w(s_j) \geq j\lambda$. Hence, as $a_i = (-1)^{n-i} s_{n-i} a_n$ for $i < n$,

$$v(a_i) = w(s_{n-i}) + v(a_n) \geq (n-i)\lambda + v(a_n),$$

proving (c).

(c) ⇒ (b) Let w be any extension of v to F_{alg}, and let u be any root of f in F_{alg}. Since $\sum_{i=0}^n a_i u^i = 0$, in the list of values $w(a_0), w(a_1 u), \dots, w(a_i u^i), \dots, w(a_n u^n)$, the least value must occur more than once. If $w(u) > \lambda$, then (c) yields for $i > 0$,

$$w(a_i u^i) > v(a_i) + i\lambda \geq (n-i)\lambda + v(a_n) + i\lambda = n\lambda + v(a_n) = v(a_0).$$

This is not possible, since then the least value on the list would be $w(a_0)$, occurring only once. Similarly, if $w(u) < \lambda$, then for $i < n$,

$$\begin{aligned} w(a_n u^n) &= v(a_n) + i\,w(u) + (n-i)w(u) \\ &< v(a_n) + i\,w(u) + (n-i)\lambda \leq v(a_i) + i\,w(u) = w(a_i u^i). \end{aligned}$$

This is also ruled out, since the least value on the list would then be $w(a_n u^n)$, occurring only once. Therefore, $w(u) = \lambda$ for every root u of f and every extension w of v to F_{alg}.

(c) ⇔ (d) Clearly, h is a monic polynomial of degree n. Write $h = \sum_{i=0}^n b_i X^i$ in $F_{alg}[X]$, where each $b_i = a_i a_n^{-1} c^{i-n}$. Then, $w(b_i) = v(a_i) - v(a_n) + (i-n)\lambda$. Hence, $w(b_i) \geq 0$ if and only if $v(a_i) \geq (n-i)\lambda + v(a_n)$. □

One can check that the conditions on f in Prop. 5.44 are also equivalent to: the Newton polygon for f with respect to v consists of a single line segment.

Definition 5.45. If $f = \sum_{i=0}^n a_i X^i \in F[X]$ satisfies the equivalent conditions of Prop. 5.44, then we call f a *uniform polynomial*, or more precisely a *λ-uniform polynomial*, where $\lambda = \frac{1}{n}\big(v(a_0) - v(a_n)\big)$ is the common value of all the roots of f. We then write

$$\widetilde{f}^{(\lambda)} = \sum_{i=0}^{n} \overline{a}_i^{(\lambda)} X^i \qquad \text{where } \overline{a}_i^{(\lambda)} \text{ is the image of } a_i \text{ in } \mathsf{gr}(F)_{(n-i)\lambda + v(a_n)}.$$

Note that $\overline{a}_i^{(\lambda)} = \widetilde{a}_i$ in $\mathsf{gr}(F)[X]$ if $v(a_i) = (n-i)\lambda + v(a_n)$, while $\overline{a}_i^{(\lambda)} = 0$ if $v(a_0) > (n-i)\lambda + v(a_n)$. Thus, $\widetilde{f}^{(\lambda)}$ is a λ-homogenizable polynomial in $\mathsf{gr}(F)[X]$ with $deg\, \widetilde{f}^{(\lambda)} = v(a_0)$ in the grading on $\mathsf{gr}(F)[X]^{(\lambda)}$, and $deg_X\big(\widetilde{f}^{(\lambda)}\big) = deg_X(f) = n$.

Lemma 5.46. *Let $f = \sum_{i=0}^n a_i X^i$ be a λ-uniform polynomial in $F[X]$ with $a_n \neq 0$. Let K be an algebraic extension field of F such that f splits over K, say*

$$f = a_n(X - u_1) \dots (X - u_n) \quad \text{in } K[X].$$

Let w be any extension of v to K. Then,

$$\widetilde{f}^{(\lambda)} = \widetilde{a}_n\,(X - \widetilde{u}_1) \dots (X - \widetilde{u}_n) \quad \text{in } \mathsf{gr}_w(K)[X].$$

Proof. We have

$$f = \sum_{i=0}^{n} a_i X^i = a_n \prod_{i=1}^{n} (X - u_i) = \sum_{j=0}^{n} (-1)^j a_n s_j X^{n-j},$$

where $s_0 = 1$ and s_j is the j-th elementary symmetric polynomial in $u_1, \ldots, u_n$, for $j = 1, 2, \ldots, n$. Likewise, let

$$g = \widetilde{a}_n \prod_{i=1}^{n} (X - \widetilde{u}_i) = \sum_{j=0}^{n} (\widetilde{-1})^j \, \widetilde{a}_n \, t_j X^{n-j} \in \mathsf{gr}(K)[X],$$

where $t_0 = \widetilde{1}$ and t_j is the j-th elementary symmetric polynomial in $\widetilde{u}_1, \ldots, \widetilde{u}_n$, for $j = 1, 2, \ldots, n$. By Prop. 5.44, $w(u_i) = \lambda$ for each i. By Def. 5.45 we have $\widetilde{f}^{(\lambda)} = \sum_{j=0}^{n} \overline{a}_{n-j}^{(\lambda)} X^{n-j}$, where $\overline{a}_{n-j}^{(\lambda)}$ is the image of a_{n-j} in $\mathsf{gr}(K)_{j\lambda+v(a_n)}$. Since s_j is a sum of monomials of degree j in the u_i, we have $w(s_j) \geq j\lambda$. Note that $(u_{k_1} \ldots u_{k_j})^{\sim} = \widetilde{u}_{k_1} \ldots \widetilde{u}_{k_j}$ for all indices $k_1, \ldots, k_j$. When $w(s_j) = j\lambda$, $\widetilde{s}_j$ is the sum of the images of its monomials in $\mathsf{gr}(K)$, i.e., $\widetilde{s}_j = t_j \neq 0$ in $\mathsf{gr}(K)_{j\lambda}$. When this occurs, $v(a_{n-j}) = v(a_n) + j\lambda$, so

$$\overline{a}_{n-j}^{(\lambda)} = \widetilde{a}_{n-j} = (\widetilde{-1})^j \, \widetilde{a}_n \, \widetilde{s}_j = (\widetilde{-1})^j \, \widetilde{a}_n \, t_j,$$

which is the $(n-j)$-th coefficient of g. On the other hand, when $w(s_j) > j\lambda$, the sum of the images of its monomials in $\mathsf{gr}(K)$ is 0, i.e., $t_j = 0$. When this occurs,

$$v(a_{n-j}) = v(a_n) + v(s_j) > v(a_n) + j\lambda,$$

so $\overline{a}_{n-j}^{(\lambda)} = 0 = t_j$, Thus, $\widetilde{f}^{(\lambda)} = g$. □

Lemma 5.47. *Let* $f = \sum_{i=0}^{n} a_i X^i \in F[X]$ *with* $a_0 a_n \neq 0$. *If* f *is a* λ*-uniform polynomial and* $f = gh$ *in* $F[X]$, *then* g *and* h *are* λ*-uniform polynomials and* $\widetilde{f}^{(\lambda)} = \widetilde{g}^{(\lambda)} \widetilde{h}^{(\lambda)}$ *in* $\mathsf{gr}(F)[X]$. *So, if* $\widetilde{f}^{(\lambda)}$ *is irreducible in* $\mathsf{gr}(F)[X]$, *then* f *is irreducible in* $F[X]$.

Proof. Let w be any extension of v to F_{alg}. If u is any root of g or h in F_{alg}, then u is also a root of the λ-uniform polynomial f. By Prop. 5.44, $w(u) = \lambda$. Hence, g and h are λ-uniform polynomials by Prop. 5.44 again. Let K be any algebraic extension of F over which f splits, so g and h also split. In $K[X]$ write $g = b \prod_{i=1}^{r} (X - u_i)$ and $h = c \prod_{i=r+1}^{n} (X - u_i)$. Then, the leading coefficient of f is bc, and $\widetilde{bc} = \widetilde{b}\, \widetilde{c}$ in $\mathsf{gr}(K)$. By applying Lemma 5.46 to f, g, and h, we obtain $\widetilde{f}^{(\lambda)} = \widetilde{g}^{(\lambda)} \widetilde{h}^{(\lambda)}$. The assertion about irreducibility follows at once. □

Example 5.48. (*Generalized Eisenstein polynomials*) Let

$$f = X^n + a_{n-1} X^{n-1} + \ldots + a_0 \in F[X] \quad \text{with} \quad n > 1.$$

Suppose the image of $v(a_0)$ has order n in $\Gamma_F / n\Gamma_F$ and $v(a_i) > \frac{n-i}{n} v(a_0)$ for $i = 1, 2, \ldots, n-1$. Such an f is called a *generalized Eisenstein polynomial.* We will show that f is irreducible in $F[X]$ and that if u is a root of f in some field containing F, then v has a unique extension to $F(u)$, which is totally ramified over F. For this, observe that the conditions on the $v(a_i)$ imply that f is a λ-uniform polynomial for $\lambda = \frac{1}{n}v(a_0)$, and

$$\widetilde{f}^{(\lambda)} = X^n + \widetilde{a_0}.$$

If u is any root of f, and w is any extension of v to $F(u)$, then Prop. 5.44 shows that $w(u) = \frac{1}{n}v(a_0)$. The hypothesis on $v(a_0)$ shows that $w(u) + \Gamma_F$ has order n in $\mathbb{H}(\Gamma_F)/\Gamma_F$. Hence,

$$n \geq [F(u):F] \geq |\Gamma_{F(u),w}:\Gamma_F| \geq |(\langle w(u)\rangle + \Gamma_F):\Gamma_F| = n.$$

This shows that $[F(u):F] = n$, so f is irreducible in $F[X]$, and also $|\Gamma_{F(u),w}:\Gamma_F| = n$. Hence, by the Fundamental Inequality w is the only extension of v to $F(u)$, and the extension is totally ramified. This is clearly visible in $\mathsf{gr}_w(F(u))$ since $\widetilde{u}$ is a root of $\widetilde{f}^{(\lambda)}$, from which it is apparent that $\mathsf{gr}(F)(\widetilde{u})$ is a totally ramified graded field extension of $\mathsf{gr}(F)$ of degree n, so $\widetilde{f}^{(\lambda)}$ is irreducible in $\mathsf{gr}(F)[X]$.

In the classical case where $\Gamma_F = \mathbb{Z}$ and $v(a_0) = 1$, the hypotheses above reduce to $v(a_i) > 0$ for $i = 1, 2, \ldots, n-1$; so, f is an Eisenstein polynomial.

Theorem 5.49. *Let v be a valuation on field F. Then, the following conditions are equivalent:*

(a) *v is Henselian.*

(b) *If f is irreducible in $F[X]$, then f is a λ-uniform polynomial for some $\lambda \in \mathbb{H}(\Gamma_F)$, unless $f = cX$ for some $c \in K^\times$.*

(c) *If f is an irreducible λ-uniform polynomial in $F[X]$, then $\widetilde{f}^{(\lambda)} = c\,q^s$, where $c \in \mathsf{gr}(F)^\times$, q is an irreducible and monic λ-homogenizable polynomial in $\mathsf{gr}(F)[X]$, and s is a positive integer.*

(d) *If f is any λ-uniform polynomial in $F[X]$ and $\widetilde{f}^{(\lambda)} = \ell m$ in $\mathsf{gr}(F)[X]$ with $\gcd(\ell, m) = 1$, then there exist λ-uniform polynomials g, $h \in F[X]$ such that $f = gh$, $\widetilde{g}^{(\lambda)} = \ell$, and $\widetilde{h}^{(\lambda)} = m$.*

Moreover, these conditions imply (e) *below, and* (e) *implies* (a)–(d) *if $|\overline{F}| > 2$.*

(e) *If f is a λ-uniform polynomial in $F[X]$ and $\widetilde{f}^{(\lambda)}$ has a simple root b in $\mathsf{gr}(F)$, then f has a unique root u in $F^\times$ with $\widetilde{u} = b$, and u is a simple root of f.*

Proof. (a) $\Rightarrow$ (b) and (c): Suppose v is Henselian. Take any irreducible $f \in F[X]$, and let K be a normal extension of F over which f splits, say $f = a(X - u_1)\ldots(X - u_n)$ in $K[X]$. Let w be the unique extension to K of the Henselian valuation v on F. Since f is irreducible, for each i there is an F-automorphism σ_i of K with $u_i = \sigma_i(u_1)$. Because $w \circ \sigma_i$ is a valuation on K extending v, we have $w \circ \sigma_i = w$. Hence, $w(u_i) = w\big(\sigma_i(u_1)\big) = w(u_1)$ for $i = 1, 2, \ldots, n$. This shows that f is a λ-uniform polynomial for $\lambda = w(u_1)$, by Prop. 5.44(a). This proves (b). Moreover, as σ_i preserves w, it induces a $\mathsf{gr}(F)$-

automorphism $\widetilde{\sigma}_i$ of $\mathsf{gr}(K)$ for which $\widetilde{u}_i = \widetilde{\sigma}_i(\widetilde{u}_1)$. Let q be the minimal polynomial of $\widetilde{u}_1$ over $q(\mathsf{gr}(F))$. Since each $\widetilde{\sigma}_i$ extends to a $q(\mathsf{gr}(F))$-automorphism of $q(\mathsf{gr}(K))$, this q is also the minimal polynomial of $\widetilde{u}_i$ over $q(\mathsf{gr}(F))$. But $\widetilde{u}_1, \widetilde{u}_2, \ldots, \widetilde{u}_n$ are all the roots of $\widetilde{f}^{(\lambda)}$ by Lemma 5.46; hence, $\widetilde{f}^{(\lambda)} = \widetilde{a}\, q^s$ for some integer s. By Prop. 5.7, q lies in $\mathsf{gr}(F)[X]$ and is homogenizable, as $\widetilde{u}_1$ is homogeneous in $\mathsf{gr}(F)[X]$. Furthermore, q is irreducible in $\mathsf{gr}(F)[X]$ by Prop. 5.6 since it is homogenizable and irreducible in $q(\mathsf{gr}(F))[X]$. This proves (c).

(b) $\Rightarrow$ (a) We prove the contrapositive. Suppose v is not Henselian. Then, there is a finite-degree field extension L of F such that v has at least two different extensions to L, say w_1 and w_2. Take u in L with $w_2(u) \neq w_1(u)$. By enlarging L if necessary, we may assume that L is normal over F. Since the automorphism group $Aut(L/F)$ acts transitively on the extensions of v to L (see Th. A.6), we have $w_2 = w_1 \circ \sigma$ for some $\sigma \in Aut(L/F)$. Let f be the minimal polynomial of u over F. Then f is irreducible in $F[X]$, and f has roots u and $\sigma(u)$ with $w_1(\sigma(u)) = w_2(u) \neq w_1(u)$. Hence, f is not a λ-uniform polynomial for any λ, by Prop. 5.44(b).

(c) $\Rightarrow$ (d) Let f, ℓ, m be as in (d). Without loss of generality, we may assume that f, ℓ, m are monic. Since $\widetilde{f}^{(\lambda)}$ is λ-homogenizable, ℓ and m are also λ-homogenizable by Prop. 5.6. Write $f = \prod_{i=1}^{k} p_i^{r_i}$, where the p_i are distinct monic irreducibles in $F[X]$. By Lemma 5.47, the p_i are each λ-uniform polynomials, and by (c) each $\widetilde{p_i}^{(\lambda)} = q_i^{s_i}$ for some monic irreducible λ-homogenizable polynomial q_i in $\mathsf{gr}(F)[X]$. By Lemma 5.47,

$$\prod_{i=1}^{k} q_i^{s_i r_i} = \widetilde{f}^{(\lambda)} = \ell m \qquad \text{in } \mathsf{gr}(F)[X].$$

The unique factorization for λ-homogenizable polynomials noted in Prop. 5.5 yields that the product of the $q_i^{s_i r_i}$ for the q_i dividing ℓ is ℓ, and the product of the other $q_i^{s_i r_i}$ is m. Let g be the product of those $p_i^{r_i}$ with $q_i \mid \ell$, and let h be the product of the remaining $p_i^{r_i}$. Then, $f = gh$ with $\widetilde{g}^{(\lambda)} = \ell$ and $\widetilde{h}^{(\lambda)} = m$.

(d) $\Rightarrow$ (e) If b is a simple root in $\mathsf{gr}(F)$ of the λ-homogenizable polynomial $\widetilde{f}^{(\lambda)}$, then we can write $\widetilde{f}^{(\lambda)} = (X-b)m$ in $\mathsf{gr}(F)[X]$, with $(X-b) \nmid m$. By Prop. 5.6, $X-b$ and m are λ-homogenizable in $\mathsf{gr}(F)[X]$; since $X-b$ is irreducible in $\mathsf{gr}(F)[X]$, we must have $gcd(X-b, m) = 1$ in $\mathsf{gr}(F)[X]$. So, by (d) we have $f = gh$ for λ-uniform polynomials $g, h \in F[X]$ with $\widetilde{g}^{(\lambda)} = X - b$ and $\widetilde{h}^{(\lambda)} = m$. Write $g = c_1 X + c_0$; so, $\widetilde{c}_1 = \widetilde{1}$ and $\widetilde{c}_0 = -b$. Then, $u = -c_0 c_1^{-1}$ is a root of g, so of f, and $\widetilde{u} = -\widetilde{c}_0 \widetilde{c}_1^{-1} = b$. If u were a repeated root of f or if f had another root $u' \neq u$ with $\widetilde{u'} = b$, then by Lemma 5.46 b would be a root of m. This cannot occur as m is prime to $X - b$. Hence, u is a simple root of f and the only root of f with $\widetilde{u} = b$.

(e) $\Rightarrow$ (a) if $|\overline{F}| > 2$: We prove the contrapositive. Suppose v is not Henselian. Then, there is a finite-degree separable field extension K of F such that v has more than one extension to K. By enlarging K if necessary, we may assume that K is Galois over F. Let $G = \mathcal{G}(K/F)$, and let $w_1, \ldots, w_k$ be all the valuations on K extending v. Let $H = \{\sigma \in G \mid w_1 \circ \sigma = w_1\}$, the decomposition group, and let $L = K^H$, the decomposition field, for w_1 over v. Recall from Th. A.7 that $w_1|_L$ is an immediate extension of v and w_1 is the unique extension of $w_1|_L$ to K. Since $|\overline{F}| > 2$ there is $s \in \overline{F}^\times$ with $s \neq 1$. Because $w_j|_L \neq w_1|_L$ for $j > 1$, the approximation theorem for valuations ([73, Th. 3.2.7(3), p. 64], or Prop. 4.34) applied over L to the valuations $w_j|_L$ for $j > 1$ (some of which may be the same for different j) and $w_1|_L$ shows that there is $c \in L$ with $w_i(c) = 0$ for all i, $\overline{c}^{w_1} = s$, and $\overline{c}^{w_j} = 1$ for $j > 1$. Let $\{\sigma(c) \mid \sigma \in G\} = \{c_1, \ldots, c_n\}$, with the c_i distinct, and let

$$f = (X - c_1) \ldots (X - c_n) \in F[X],$$

which is the minimal polynomial of c over F. If $\sigma \in H$, then $\sigma(c) = c$, so $w_1(\sigma(c)) = w_1(c) = 0$ and $\overline{\sigma(c)}^{w_1} = \overline{c}^{w_1} = s$. If instead $\sigma \in G \setminus H$, then $w_1 \circ \sigma = w_j$ for some $j > 1$, and σ maps $\mathcal{O}_{K,w_j}$ to $\mathcal{O}_{K,w_1}$ and induces an $\overline{F}$-isomorphism $\overline{\sigma}\colon \overline{K}^{w_j} \xrightarrow{\sim} \overline{K}^{w_1}$. In this case, $w_1(\sigma(c)) = w_j(c) = 0$ and

$$\overline{\sigma(c)}^{w_1} = \overline{\sigma}(\overline{c}^{w_j}) = \overline{\sigma}(1) = 1.$$

Since the roots of f all lie in $\mathcal{O}_{K,w_1}^\times$, its coefficients lie in $\mathcal{O}_{K,w_1} \cap F = \mathcal{O}_{F,v}$ and $v(f(0)) = 0$. Thus, f is a λ-uniform polynomial with $\lambda = 0$, so $\widetilde{f}^{(0)} = \overline{f}$, the image of f in $\overline{F}[X]$. Moreover in $\overline{K}^{w_1}[X]$, so in $\overline{F}[X] \subseteq \mathsf{gr}(F)[X]$ we have $\widetilde{f}^{(0)} = (X - s)(X - 1)^{n-1}$ (see Lemma 5.46). Thus $\widetilde{f}^{(0)}$ has the simple root $s \in \mathsf{gr}(F)$. But f is irreducible in $F[X]$ of degree at least 2, so it has no root in F. Thus, (e) does not hold.

(d) $\Rightarrow$ (a) Since we have already proved (d) $\Rightarrow$ (e) $\Rightarrow$ (a) if $|\overline{F}| > 2$, we may assume that $|\overline{F}| = 2$. The proof is very similar to the one just given for (e) $\Rightarrow$ (a): Assume v is not Henselian. Since $\overline{F}$ is not separably closed we can find a finite-degree Galois extension K of F so that there is more than one valuation of K extending v and these valuations have residue field strictly containing $\overline{F}$. As above, let $G = \mathcal{G}(K/F)$, $w_1, \ldots, w_k$ be the valuations of K extending v, $H = \{\sigma \in G \mid w_1 \circ \sigma = w_1\}$, and $L = K^H$. Choose any $t \in \overline{K}^{w_1} \setminus \overline{F}$. By the approximation theorem for the K-valuations $w_1, \ldots, w_k$, there is a $c \in K$ with $w_i(c) = 0$ for each i and $\overline{c}^{w_1} = t$ while $\overline{c}^{w_j} = 1$ for each $j > 1$. Let $\{\sigma(c) \mid \sigma \in G\} = \{c_1, \ldots, c_n\}$ with the c_i distinct, and let the c_i be numbered so that $\{\sigma(c) \mid \sigma \in H\} = \{c_1, \ldots, c_r\}$ with $r \leq n$. Let

$$f = (X - c_1) \ldots (X - c_n) \in F[X] \quad \text{and} \quad z = (X - c_1) \ldots (X - c_r) \in L[X].$$

So, f is the minimal polynomial of c over F and z is the minimal polynomial of c over L. If $\sigma \in H$, then $w_1(\sigma(c)) = w_1(c) = 0$ and

$$\overline{\sigma(c)}^{w_1} = \overline{\sigma}(\overline{c}^{w_1}) = \overline{\sigma}(t) \in \overline{K}^{w_1} \setminus \overline{F}.$$

If instead $\sigma \in G \setminus H$, then $w_1 \circ \sigma = w_j$ for some $j > 1$, so $w_1(\sigma(c)) = 0$ and

$$\overline{\sigma(c)}^{w_1} = \overline{\sigma}(\overline{c}^{w_j}) = \overline{\sigma}(1) = 1.$$

Hence, $\sigma(c) \notin \{c_1, \ldots, c_r\}$ showing that $n > r$. The roots of f all lie in $\mathcal{O}_{K,w_1}^\times$, so its coefficients all lie in $\mathcal{O}_{F,v}$ and $f(0) \in \mathcal{O}_{F,v}^\times$. Thus f is a 0-uniform polynomial in $F[X]$ for v with $\widetilde{f}^{(0)} = \overline{f}$, the image of f in $\overline{F}[X]$. Likewise, z is a 0-uniform polynomial for $w_1|_L$ in $L[X]$ with $\widetilde{z}^{(0)} = \overline{z}$ in $\overline{L}^{w_1}[X]$. But because L is the decomposition field for w_1, we have $\overline{L}^{w_1} = \overline{F}$ by Th. A.7. In $\overline{K}^{w_1}[X]$ we have (see Lemma 5.46)

$$\overline{f} = (X - \overline{c_1}^{w_1}) \ldots (X - \overline{c_n}^{w_1}) = \overline{z} \cdot (X-1)^{n-r}.$$

This gives a nontrivial factorization of $\widetilde{f}^{(0)}$ in $\overline{F}[X] \subseteq \mathsf{gr}(F)[X]$ with $gcd(\overline{z}, (X-1)^{n-r}) = 1$ since the roots of $\overline{z}$ all lie in $\overline{K}^{w_1} \setminus \overline{F}$. But f is irreducible in $F[X]$. Thus, (d) does not hold, completing the proof. □

5.2.2 Tamely ramified extensions of valued fields

Let (F, v) be an arbitrary valued field. For any valued field extension (K, v_K) of (F, v), we may consider the graded field $\mathsf{gr}_{v_K}(K)$ as an extension of $\mathsf{gr}_v(F)$. Our purpose in this subsection is to compare the properties of this graded field extension to those of the valued field extension, particularly in the case where K is Galois over F. The main result is Prop. 5.55, which shows that if v is Henselian and K is a tamely ramified extension of F, then K is determined up to isomorphism by $\mathsf{gr}_{v_K}(K)$.

We write simply $\mathsf{gr}(K)$ for $\mathsf{gr}_{v_K}(K)$ and $\mathsf{gr}(F)$ for $\mathsf{gr}_v(F)$. Clearly, $\mathsf{gr}(K)$ is a finite-degree extension of $\mathsf{gr}(F)$ if $[K{:}F] < \infty$, and in fact by Cor. 3.6 we have

$$[\mathsf{gr}(K){:}\,\mathsf{gr}(F)] \leq [K{:}F].$$

Equality holds if and only if v_K is a v-norm, which means that v_K is the unique extension of v to K and v is defectless in K, see (3.3). Throughout the subsection, we will always assume this, so that $\mathsf{gr}(K)$ holds as much information as possible about K.

If K is algebraic over F (possibly of infinite degree), then for every $u \in K$ the extension $F(u)/F$ is finite-dimensional, hence $[\mathsf{gr}(F)(\widetilde{u}){:}\,\mathsf{gr}(F)] < \infty$. Therefore, every homogeneous element in $\mathsf{gr}(K)$ is algebraic over $\mathsf{gr}(F)$, hence $\mathsf{gr}(K)$ is an algebraic graded field extension of $\mathsf{gr}(F)$.

For the rest of this section, assume that K is algebraic over F, v is defectless in K, and v_K is the unique extension of v to K. This means (see Def. A.3) that for every field L with $F \subseteq L \subseteq K$ and $[L{:}F] < \infty$ we have $[L{:}F] = [\overline{L}{:}\overline{F}] \cdot |\Gamma_L{:}\Gamma_F|$. Let $\mathcal{L}$ be the set of fields L with $F \subseteq L \subseteq K$ and $[L{:}F] < \infty$. Recall (see Def. A.4) that the defectless extension K/F is said to be

- *inertial* if for every field $L \in \mathcal{L}$ we have $[L{:}F] = [\overline{L}{:}\overline{F}]$ and $\overline{L}$ is separable over $\overline{F}$;
- *totally ramified* if for every field $L \in \mathcal{L}$ we have $[L{:}F] = |\Gamma_L{:}\Gamma_F|$;

– *tamely ramified* if for every field $L \in \mathcal{L}$ we have $\overline{L}$ is separable over $\overline{F}$ and $char\,\overline{F} \nmid |\Gamma_L{:}\Gamma_F|$.

Since v is assumed to be defectless in K and to extend uniquely to K, we have for every field $L \in \mathcal{L}$,

$$[L{:}F] = [\overline{L}{:}\overline{F}] \quad \Longleftrightarrow \quad \Gamma_L = \Gamma_F$$

and

$$[L{:}F] = |\Gamma_L{:}\Gamma_F| \quad \Longleftrightarrow \quad \overline{L} = \overline{F}.$$

Since $\mathsf{gr}(L)_0 = \overline{L}$ and $\Gamma_{\mathsf{gr}(L)} = \Gamma_L$, the following proposition readily follows:

Proposition 5.50. *Suppose (K, v_K) is an algebraic extension of (F, v) such that v is defectless in K and v_K is the unique extension of v to K. Then*

(i) *K is inertial over F if and only if $\mathsf{gr}(K)$ is inertial over $\mathsf{gr}(F)$;*
(ii) *K is totally ramified over F if and only if $\mathsf{gr}(K)$ is totally ramified over $\mathsf{gr}(F)$;*
(iii) *K is tamely ramified over F if and only if $\mathsf{gr}(K)$ is tame over $\mathsf{gr}(F)$.*

Now, we make the further assumption that K is a Galois extension of F. Since v_K is the unique valuation on K extending v, each $\sigma \in \mathcal{G}(K/F)$ satisfies $v_K \circ \sigma = v_K$, hence σ preserves the valuation filtration on K and induces an automorphism $\widetilde{\sigma}$ of $\mathsf{gr}(K)$ such that

$$\widetilde{\sigma}(\widetilde{x}) = \widetilde{\sigma(x)} \qquad \text{for all } x \in K.$$

Recall from Appendix A (see §A.1) that the ramification subgroup of $\mathcal{G}(K/F)$ is defined by

$$\mathcal{G}^{\mathrm{ram}}(K/F) = \{\sigma \in \mathcal{G}(K/F) \mid v_K(\sigma(x) - x) > v_K(x) \text{ for all } x \in K\}.$$

Proposition 5.51. *Let K be a field extension of F, and assume v is defectless in K and extends uniquely to K. If K is Galois over F, then $\mathsf{gr}(K)$ is a normal graded field extension of $\mathsf{gr}(F)$, and the canonical map*

$$\rho\colon \mathcal{G}(K/F) \longrightarrow Aut\big(\mathsf{gr}(K)/\mathsf{gr}(F)\big) \quad \textit{given by} \quad \sigma \mapsto \widetilde{\sigma}$$

is a continuous group homomorphism. Moreover, there is a short exact sequence:

$$1 \longrightarrow \mathcal{G}^{\mathrm{ram}}(K/F) \longrightarrow \mathcal{G}(K/F) \stackrel{\rho}{\longrightarrow} Aut(\mathsf{gr}(K)/\mathsf{gr}(F)) \longrightarrow 1. \tag{5.18}$$

Proof. Let $G = \mathcal{G}(K/F)$ and $\mathcal{G} = Aut(\mathsf{gr}(K)/\mathsf{gr}(F))$. Take any nonzero homogeneous b in $\mathsf{gr}(K)$ and any a in $K^\times$ with $\widetilde{a} = b$, and let f be the minimal polynomial of a over F. Because K is Galois over F, we have $f = (X - a_1)\dots(X - a_n)$ in $K[X]$ with $a = a_1$ and each a_i has the form $a_i = \tau_i(a)$ for some $\tau_i \in G$. Since v_K is the unique extension of v to K, we have

$$v_K(a_i) = v_K(a) \qquad \text{for } i = 1, \dots, n,$$

hence f is a λ-uniform polynomial with $\lambda = v_K(a)$. By Lemma 5.46, it follows that

$$\widetilde{f}^{(\lambda)} = (X - \widetilde{a}_1)\dots(X - \widetilde{a}_n) \qquad \text{in } \mathsf{gr}(K)[X].$$

Since b is a root of $\widetilde{f}^{(\lambda)} \in \mathsf{gr}(F)[X]$, the minimal polynomial of b over $q(\mathsf{gr}(F))$ divides $\widetilde{f}^{(\lambda)}$, so it splits over $q(\mathsf{gr}(K))$. This is true for every nonzero homogeneous b in K; hence, $\mathsf{gr}(K)$ is normal over $\mathsf{gr}(F)$, by Prop. 5.28.

Suppose further that our homogeneous b in $\mathsf{gr}(K)$ is chosen so that $\widetilde{\sigma}(b) = b$ for every $\sigma \in G$. Then,

$$\widetilde{a}_i = \widetilde{\tau}_i(\widetilde{a}) = \widetilde{\tau}_i(b) = b.$$

Thus, $\widetilde{f}^{(\lambda)} = (X - \widetilde{a}_1)\dots(X - \widetilde{a}_n) = (X - b)^n$. Since $\widetilde{f}^{(\lambda)} \in \mathsf{gr}(F)[X]$, any $\eta \in \mathcal{G}$ permutes the roots of $\widetilde{f}^{(\lambda)}$, so $\eta(b) = b$. Thus, any homogeneous $b \in \mathsf{gr}(K)^{im\,\rho}$ lies in the fixed graded field $\mathsf{gr}(K)^{\mathcal{G}}$. Since $\mathsf{gr}(K)^{im\,\rho}$ is a graded field, it follows that

$$\mathsf{gr}(K)^{im\,\rho} = \mathsf{gr}(K)^{\mathcal{G}}. \tag{5.19}$$

Once we show below that $im\,\rho$ is a closed subgroup of $\mathcal{G}$, this will yield that $im\,\rho = \mathcal{G}$.

The map $\rho\colon G \to \mathcal{G}$, $\sigma \mapsto \widetilde{\sigma}$, is clearly a group homomorphism, and we have

$$\begin{aligned} ker\,\rho &= \{\sigma \in G \mid \widetilde{\sigma}(\widetilde{a}) = \widetilde{a} \ \text{ for all } a \in K^{\times}\} \\ &= \{\sigma \in G \mid v(\sigma(a) - a) > v(a) \ \text{ for all } a \in K^{\times}\} = \mathcal{G}^{\text{ram}}(K/F). \end{aligned}$$

To see that the homomorphism ρ is continuous for the Krull topologies on G and $\mathcal{G}$, it suffices to show that $\rho^{-1}(\mathcal{H})$ is open in G for each open subgroup $\mathcal{H}$ of $\mathcal{G}$. Take any such $\mathcal{H}$. Let $\mathsf{M} = \mathsf{gr}(K)^{\mathcal{G}}$. Then, $\mathsf{gr}(K)$ is Galois over M, by definition, and $\mathcal{G} = \mathcal{G}(\mathsf{gr}(K)/\mathsf{M})$. Let $\mathsf{L} = \mathsf{gr}(K)^{\mathcal{H}}$, which is a graded subfield of $\mathsf{gr}(K)$ with $\mathsf{M} \subseteq \mathsf{L}$. Since $\mathcal{H}$ is open in $\mathcal{G}$, it is also closed in $\mathcal{G}$ with $|\mathcal{G}{:}\mathcal{H}| < \infty$. Hence, $\mathcal{H} = \mathcal{G}(\mathsf{gr}(K)/\mathsf{L})$ by the Galois correspondence, and $[\mathsf{L}{:}\mathsf{M}] = |\mathcal{G}{:}\mathcal{H}| < \infty$. Because this degree is finite, there is a finite-degree Galois extension S of F in K with $\mathsf{L} \subseteq \mathsf{M} \cdot \mathsf{gr}(S)$. Let $N = \mathcal{G}(K/S)$, which is an open normal subgroup of G with $|G{:}N| = [S{:}F] < \infty$. For any $\tau \in N$, we have $\widetilde{\tau}|_{\mathsf{gr}(S)} = id$ since $\tau|_S = id$, and $\widetilde{\tau}|_{\mathsf{M}} = id$ since every element of $\mathcal{G}$ fixes M; hence, $\widetilde{\tau}|_{\mathsf{L}} = id$. Therefore,

$$\rho(N) \subseteq \mathcal{G}(\mathsf{gr}(K)/\mathsf{L}) = \mathcal{H}.$$

Thus, $N \subseteq \rho^{-1}(\mathcal{H})$. Because N is open in G, so $|G{:}N| < \infty$, the group $\rho^{-1}(\mathcal{H})$ is thus open in G, since it is a finite union of cosets of N. Hence, ρ is continuous.

Since G is compact in the Krull topology, $\mathcal{G} = \mathcal{G}(\mathsf{gr}(K)/\mathsf{M})$ is Hausdorff, and ρ is continuous, $im\,\rho$ is closed in $\mathcal{G}$. Therefore, by the Galois correspondence and (5.19) above,

$$im\,\rho = \mathcal{G}\left(\mathsf{gr}(K)/\,\mathsf{gr}(K)^{im\,\rho}\right) = \mathcal{G}\left(\mathsf{gr}(K)/\,\mathsf{gr}(K)^{\mathcal{G}}\right) = \mathcal{G}.$$

Thus, the sequence of the proposition is exact. □

Note that in the proposition $\mathsf{gr}(K)$ is normal but not necessarily Galois over $\mathsf{gr}(F)$. For example, an extension of the form $K = F(\sqrt[p]{a})$ is Galois if F contains a primitive p-th root of unity, but the corresponding graded field $\mathsf{gr}(K)$ is a purely inseparable extension of $\mathsf{gr}(F)$ if $\widetilde{a} \notin \mathsf{gr}(F)^{\times p}$ and $char\,\overline{F} = p$. In order that $\mathsf{gr}(K)$ be Galois over $\mathsf{gr}(F)$, we need the further condition that $\mathsf{gr}(K)$ be a tame extension of $\mathsf{gr}(F)$. For the next statement, recall from Appendix A (§A.1) that the inertia subgroup of $\mathcal{G}(K/F)$ is

$$\mathcal{G}^{\text{in}}(K/F) = \{\sigma \in \mathcal{G}(K/F) \mid v_K(\sigma(x) - x) > 0 \text{ for all } x \in \mathcal{O}_K\}.$$

Corollary 5.52. *Suppose K is a Galois extension of F, and v_K is the unique extension of v to K. If K is tamely ramified over F, then $\mathsf{gr}(K)$ is Galois over $\mathsf{gr}(F)$, and the homomorphism ρ of Prop. 5.51 is an isomorphism of topological groups. This isomorphism maps $\mathcal{G}^{\text{in}}(K/F)$ onto $\mathcal{G}^{\text{in}}\big(\mathsf{gr}(K)/\mathsf{gr}(F)\big)$, and fits into the following commutative diagram, in which the upper row is the exact sequence of Th.* A.7(iv) *and the lower row is the exact sequence* (5.12) (*with* $\mathsf{gr}(K)$ *and* $\mathsf{gr}(F)$ *for* K *and* F):

$$\begin{array}{ccccccccc}
1 & \longrightarrow & \mathcal{G}^{\text{in}}(K/F) & \longrightarrow & \mathcal{G}(K/F) & \xrightarrow{\ \psi\ } & \mathcal{G}(\overline{K}/\overline{F}) & \longrightarrow & 1 \\
 & & \rho\big\downarrow\cong & & \rho\big\downarrow\cong & & \big\| & & \\
1 & \longrightarrow & \mathcal{G}^{\text{in}}\big(\mathsf{gr}(K)/\mathsf{gr}(F)\big) & \longrightarrow & \mathcal{G}\big(\mathsf{gr}(K)/\mathsf{gr}(F)\big) & \longrightarrow & \mathcal{G}(\overline{K}/\overline{F}) & \longrightarrow & 1
\end{array} \tag{5.20}$$

(*The map ψ carries $\sigma \in \mathcal{G}(K/F)$ to the automorphism $\overline{\sigma}$ of $\overline{K}$ such that $\overline{\sigma}(\overline{x}) = \overline{\sigma(x)}$ for all $x \in \mathcal{O}_K$.*)

Likewise, the following diagram commutes, where the upper map is the bilinear pairing of Th. A.8(vi) *and the lower map is the bilinear pairing of Th. 5.39:*

$$\begin{array}{ccc}
\mathcal{G}^{\text{in}}(K/F) \times (\Gamma_K/\Gamma_F) & \longrightarrow & \mu(\overline{K}) \\
\widetilde{\rho}\times id\big\downarrow & & \big\| \\
\mathcal{G}^{\text{in}}\big(\mathsf{gr}(K)/\mathsf{gr}(F)\big) \times (\Gamma_K/\Gamma_F) & \longrightarrow & \mu(\overline{K})
\end{array} \tag{5.21}$$

Proof. Note that v is defectless in K by the definition of tame ramification. By Prop. 5.51, $\mathsf{gr}(K)$ is normal over $\mathsf{gr}(F)$. Also, $\mathsf{gr}(K)$ is tame over $\mathsf{gr}(F)$ since K is tamely ramified over F. Hence, $\mathsf{gr}(K)$ is Galois over $\mathsf{gr}(F)$ by Cor. 5.33. Furthermore, the tame ramification implies that $\mathcal{G}^{\text{ram}}(K/F)$ is trivial (see Th. A.9(xi)). Therefore, Prop. 5.51 shows that $\rho\colon \mathcal{G}(K/F) \to \mathcal{G}\big(\mathsf{gr}(K)/\mathsf{gr}(F)\big)$ is a continuous group isomorphism. Because $\mathcal{G}(K/F)$ is compact and $\mathcal{G}\big(\mathsf{gr}(K)/\mathsf{gr}(F)\big)$ is Hausdorff, ρ also maps closed sets to closed sets, so it is a homeomorphism.

For $\sigma \in \mathcal{G}(K/F)$ and $x \in \mathcal{O}_K$ we have $\widetilde{x} = \overline{x} \in \overline{K}$ and

$$\widetilde{\sigma}(\overline{x}) = \widetilde{\sigma(x)} = \overline{\sigma(x)},$$

so the right square of diagram (5.20) commutes. Since the rows are exact sequences, it follows that ρ restricts to an isomorphism $\mathcal{G}^{in}(K/F) \xrightarrow{\sim} \mathcal{G}^{in}(\mathsf{gr}(K)/\mathsf{gr}(F))$, which shows that diagram (5.20) commutes. If $x \in K^\times$ satisfies $v_K(x) = \gamma$, then the image of $(\sigma, \gamma + \Gamma_F)$ under the upper horizontal map of (5.21) is $\overline{\sigma(x)/x}$, whereas the image of $(\widetilde{\sigma}, \gamma + \Gamma_F)$ under the lower horizontal map is $\widetilde{\sigma}(\widetilde{x})\widetilde{x}^{-1}$. Since $\overline{\sigma(x)/x} = \widetilde{\sigma}(\widetilde{x})\widetilde{x}^{-1}$, the proof is complete. □

We consider in particular the case where K is an inertial extension of F:

Corollary 5.53. *Suppose K is a Galois extension of F, and v_K is the unique extension of v to K. If K is inertial over F, then $\mathsf{gr}(K)$ is an inertial Galois graded field extension of $\mathsf{gr}(F)$, $\overline{K}$ is Galois over $\overline{F}$, and we have canonical isomorphisms*

$$\mathcal{G}(K/F) \cong \mathcal{G}\big(\mathsf{gr}(K)/\mathsf{gr}(F)\big) \cong \mathcal{G}(\overline{K}/\overline{F})$$

mapping $\sigma \in \mathcal{G}(K/F)$ to $\widetilde{\sigma} \in \mathcal{G}\big(\mathsf{gr}(K)/\mathsf{gr}(F)\big)$ and to $\overline{\sigma} \in \mathcal{G}(\overline{K}/\overline{F})$.

Proof. This readily follows from Cor. 5.52 since $\mathcal{G}^{in}(K/F)$ is trivial when K is an inertial extension of F. □

Here is another corollary to Prop. 5.51, which will be used in the proof of Th. 9.27 below.

Corollary 5.54. *Let (F, v) be an arbitrary valued field of characteristic $p \neq 0$, and let K be a Galois field extension of F of degree p^n for some n. Assume that v is defectless in K and extends uniquely to K. If $\overline{K}/\overline{F}$ is a separable extension, then $\mathsf{gr}(K)$ contains a purely inseparable totally ramified graded field extension of $\mathsf{gr}(F)$ of degree $|\Gamma_K{:}\Gamma_F|$.*

Proof. Recall the canonical exact sequence (5.18). The ramification subgroup $\mathcal{G}^{ram}(K/F)$ of $\mathcal{G}(K/F)$ appearing there is the unique p-Sylow subgroup of the inertia subgroup $\mathcal{G}^{in}(K/F)$, see Th. A.9(ii). Since $\mathcal{G}(K/F)$ is a p-group, we have $\mathcal{G}^{ram}(K/F) = \mathcal{G}^{in}(K/F)$. On the other hand, the extension $\overline{K}/\overline{F}$ is Galois since $\overline{K}/\overline{F}$ is separable (see Th. A.8), and we have an exact sequence

$$1 \longrightarrow \mathcal{G}^{in}(K/F) \longrightarrow \mathcal{G}(K/F) \longrightarrow \mathcal{G}(\overline{K}/\overline{F}) \longrightarrow 1. \qquad (5.22)$$

By comparing the exact sequences (5.18) and (5.22), we see that the restriction of automorphisms of $\mathsf{gr}(K)$ to the homogeneous component $\mathsf{gr}(K)_0 = \overline{K}$ is an isomorphism

$$Aut(\mathsf{gr}(K)/\mathsf{gr}(F)) \xrightarrow{\sim} \mathcal{G}(\overline{K}/\overline{F}).$$

Let $\mathsf{T} = \mathsf{gr}(K)^{Aut(\mathsf{gr}(K)/\mathsf{gr}(F))}$ be the fixed graded subfield. We have

$$[\mathsf{T}{:}\,\mathsf{gr}(F)] = \frac{[\mathsf{gr}(K){:}\,\mathsf{gr}(F)]}{|Aut(\mathsf{gr}(K)/\mathsf{gr}(F))|} = \frac{[K{:}F]}{[\overline{K}{:}\overline{F}]} = |\Gamma_K{:}\Gamma_F|.$$

Moreover,

$$\mathsf{T}_0 = \mathsf{gr}(K)_0^{Aut(\mathsf{gr}(K)/\mathsf{gr}(F))} = \overline{K}^{\mathcal{G}(\overline{K}/\overline{F})} = \overline{F}.$$

Therefore, T is a totally ramified graded field extension of $\mathsf{gr}(F)$. It is also purely inseparable over $\mathsf{gr}(F)$, since $|\Gamma_{\mathsf{T}}:\Gamma_F|$ is a power of p, see Def. 5.25. □

Note that the graded field extension K/F of the corollary is not necessarily Galois, but it is always normal by Prop. 5.51.

For the rest of this section, we assume that the valuation v on F is Henselian. In a given algebraic closure F_{alg} of F, there is then a unique maximal tamely ramified extension of F, the *tame closure of* F, which we denote F_{tr}. See Th. A.24 and Cor. A.25 for the properties of F_{tr} that we recall here. This F_{tr} is the ramification field $R_{F_{sep}/F}$ for the Galois extension F_{sep} over F. For any field L with $F \subseteq L \subseteq F_{alg}$, L is tamely ramified over F if and only if $L \subseteq F_{tr}$. Also,

$$\overline{F_{tr}} = \overline{F}_{sep}, \quad \text{the separable closure of } \overline{F},$$

and

$$\Gamma_{F_{tr}} \text{ is the "prime-to-}char\,\overline{F}\text{ closure" of } \Gamma_{\mathsf{F}} \text{ in its divisible hull } \mathbb{H}(\Gamma_F);$$

that is, if $char\,\overline{F} = 0$, then $\Gamma_{F_{tr}} = \mathbb{H}(\Gamma_F)$, while if $char\,\overline{F} = \overline{p} \neq 0$, then $\Gamma_{F_{tr}}/\Gamma_F$ has no $\overline{p}$-torsion, while $\mathbb{H}(\Gamma_F)/\Gamma_{F_{tr}}$ is a $\overline{p}$-primary torsion group. It follows that when we identify $\mathsf{gr}(F_{alg})$ with the graded algebraic closure $\mathsf{gr}(F)_{alg}$ of $\mathsf{gr}(F)$, we have

$$\mathsf{gr}(F_{tr}) = \mathsf{gr}(F)_{tr}, \quad \text{the tame closure of } \mathsf{gr}(F).$$

Indeed, the descriptions of $\overline{F_{tr}}$ and $\Gamma_{F_{tr}}$ show that $\mathsf{gr}(F_{tr})$ is tame over $\mathsf{gr}(F)$, and hence $\mathsf{gr}(F_{tr}) \subseteq \mathsf{gr}(F)_{tr}$ by Prop. 5.22; the inclusion is an equality because these graded fields have the same degree-0 component and the same grade group.

By applying Cor. 5.52 with $K = F_{tr}$, we obtain:

Proposition 5.55. *For any Henselian field F, mapping $\sigma \in \mathcal{G}(F_{tr}/F)$ to $\widetilde{\sigma}$ defines a canonical isomorphism between the Galois group of F_{tr}/F and the absolute Galois group of* $\mathsf{gr}(F)$,

$$\rho\colon \mathcal{G}(F_{tr}/F) \xrightarrow{\sim} \mathcal{G}(\mathsf{gr}(F)).$$

Moreover, the map $L \mapsto \mathsf{gr}(L)$ gives a one-to-one inclusion-preserving and degree-preserving correspondence between tamely ramified field extensions of F in F_{alg} and tame graded field extensions of $\mathsf{gr}(F)$ in $\mathsf{gr}(F)_{alg}$. Furthermore, L is Galois over F if and only if $\mathsf{gr}(L)$ is Galois over $\mathsf{gr}(F)$.

Proof. Since F_{tr} is Galois over F and $\mathsf{gr}(F_{tr}) = \mathsf{gr}(F)_{tr}$, Cor. 5.52 shows that ρ defines an isomorphism

$$\mathcal{G}(F_{tr}/F) \xrightarrow{\sim} \mathcal{G}(\mathsf{gr}(F_{tr})/\mathsf{gr}(F)) = \mathcal{G}(\mathsf{gr}(F)_{tr}/\mathsf{gr}(F)) = \mathcal{G}(\mathsf{gr}(F)).$$

For a graded field L with $F \subseteq L \subseteq F_{tr}$, the corresponding closed subgroup of $\mathcal{G}(F_{\rm tr}/F)$ is $H = \mathcal{G}(F_{\rm tr}/L)$. The group isomorphism $\mathcal{G}(F_{\rm tr}/F) \cong \mathcal{G}(\mathsf{gr}(F_{\rm tr})/\mathsf{gr}(F))$ maps H to $\mathcal{H} = \{\widetilde{\sigma} \mid \sigma \in H\} = \mathcal{G}(\mathsf{gr}(F_{\rm tr})/\mathsf{gr}(L))$, which is the closed subgroup of $\mathcal{G}(\mathsf{gr}(F_{\rm tr})/\mathsf{gr}(F))$ corresponding to $\mathsf{gr}(L)$ under the graded Galois correspondence (see Prop. 5.32). From the Galois correspondences (ungraded and graded) between fields and closed subgroups and the isomorphism of Galois groups, the map $L \mapsto \mathsf{gr}(L)$ is a one-to-one correspondence between fields L with $F \subseteq L \subseteq F_{tr}$ and graded fields E with $\mathsf{gr}(F) \subseteq \mathsf{E} \subseteq \mathsf{gr}(F_{tr}) = \mathsf{gr}(F)_{tr}$. The fields L are all the tamely ramified extensions of F in $F_{\rm alg}$, and the fields E are all the tame extensions of $\mathsf{gr}(F)$ in $\mathsf{gr}(F)_{\rm alg}$, by Prop. 5.22. The one-to-one correspondence is clearly inclusion-preserving, and is degree-preserving as F_{tr} is defectless over F. The usual Galois theory and the graded version in Prop. 5.32 show that L is Galois over F if and only if H is a normal subgroup, if and only if $\mathcal{H}$ is a normal subgroup, if and only if $\mathsf{gr}(L)$ is Galois over $\mathsf{gr}(F)$. □

The core of the preceding proposition can be reformulated as the following lifting property for tame graded field extensions of $\mathsf{gr}(F)$. We will see an analogous lifting property for graded division algebras tame over $\mathsf{gr}(F)$ in Th. 8.13 below.

Corollary 5.56. *Let* M *be any tame graded field extension of* $\mathsf{gr}(F)$ *for the Henselian field* F*. Then, there is a tamely ramified field extension* L *of* F *with* $\mathsf{gr}(L) \cong_g \mathsf{M}$*. Moreover,* L *is unique up to* F*-isomorphism, and* L *is Galois over* F *if and only if* M *is Galois over* $\mathsf{gr}(F)$*.*

Such a field L is called a *tame lift* of M over F.

Proof. This is clear from Prop. 5.55. □

As a particular case of tame lifts, we recover classical results on inertial lifts.

Corollary 5.57. *Let* S *be any separable algebraic field extension of* $\overline{F}$ *for the Henselian field* F*. Then, there is an inertial field extension* K *of* F *with* $\overline{K} \cong S$*. Moreover,* K *is unique up to isomorphism, and* K *is Galois over* F *if and only if* S *is Galois over* $\overline{F}$*.*

Such a field K is called an *inertial lift* of S over F.

Proof. Let $\mathsf{M} = S \otimes_{\overline{F}} \mathsf{gr}(F)$, which is an inertial, hence tame, graded field extension of $\mathsf{gr}(F)$, and let K be a tame lift of M over F, as in Cor. 5.56. Then, K is inertial over F, as $\mathsf{gr}(K) = \mathsf{M}$ is inertial over $\mathsf{gr}(F)$, and $\overline{K} \cong \mathsf{M}_0 = S$. Moreover, K is unique up to isomorphism by Cor. 5.56, as M is determined uniquely up to isomorphism by S. Again by Cor. 5.56, K is Galois over F if and only if M is Galois over $\mathsf{gr}(F)$. If M is Galois over $\mathsf{gr}(F)$, then $S = \mathsf{M}_0$ is Galois over $\overline{F} = \mathsf{gr}(F)_0$. Conversely, if S is Galois over $\overline{F}$, then the field $q(\mathsf{M}) = S \otimes_{\overline{F}} q(\mathsf{gr}(F))$ is Galois over $q(\mathsf{gr}(F))$, so M is Galois over $\mathsf{gr}(F)$ by Prop. 5.30; hence, K is Galois over F. □

Exercises

Exercise 5.1. (A graded field not of group-ring type) Let k be an arbitrary field and let F be a graded field with $\mathsf{F}_0 = k$ and $\Gamma_{\mathsf{F}} = \mathbb{Z}[\frac{1}{2}]$. For $i \in \mathbb{N}$, let $y_i \in \mathsf{F}$ be a nonzero homogeneous element of degree 2^{-i}, and let $u_i = y_{i+1}^2 y_i^{-1} \in k^\times$. Consider the sequence

$$(u_0, u_0u_1^2, u_0u_1^2u_2^4, \ldots, \prod_{i=0}^{n} u_i^{2^i}, \ldots) \in \prod_{i=0}^{\infty} \big(k^\times/(k^\times)^{2^{i+1}}\big).$$

This sequence defines an element u_∞ in the projective limit $\varprojlim\big(k^\times/(k^\times)^{2^{i+1}}\big)$ for the canonical maps $k^\times/(k^\times)^{2^i} \leftarrow k^\times/(k^\times)^{2^j}$ that arise from the inclusions $(k^\times)^{2^i} \supseteq (k^\times)^{2^j}$ for $i \le j$. Let $k^\times/k^{\times 2^\infty} = \varprojlim\big(k^\times/(k^\times)^{2^{i+1}}\big)$ and let $\varphi\colon k^\times \to k^\times/k^{\times 2^\infty}$ be the canonical map.

(a) Show that modulo the image of φ the element u_∞ does not depend on the choice of the homogeneous elements y_i, and that F is of group-ring type if and only if u_∞ lies in the image of φ.

(b) (Hwang–Wadsworth) Let $k = \mathbb{Q}$, fix a prime number p, and let $u_i = p^{2^i}$ for $i \in \mathbb{N}$. Show that the corresponding element $u_\infty \in \mathbb{Q}^\times/\mathbb{Q}^{\times 2^\infty}$ satisfies $u_\infty^3 = \varphi(p)^{-1}$. Use u_∞ to construct a graded field F with $\mathsf{F}_0 = \mathbb{Q}$ and $\Gamma_{\mathsf{F}} = \mathbb{Z}[\frac{1}{2}]$ that is not of group-ring type.

Exercise 5.2. (An extension of the quotient field of a graded field that is not the quotient field of a graded field) (Van Geel–Van Oystaeyen) Let F_0 be any field of characteristic different from 2, and let F be the Laurent polynomial ring $\mathsf{F} = \mathsf{F}_0[t, t^{-1}]$, with the $\mathbb{Z}$-grading by the degree of t, so $\mathsf{F}_\gamma = \mathsf{F}_0 t^\gamma$ for each $\gamma \in \mathbb{Z}$, and $q(\mathsf{F}) = \mathsf{F}_0(t)$. Let $E = q(\mathsf{F})(\sqrt{t+1})$. Show that there is no graded field extension K of F such that $q(\mathsf{K}) \cong E$. As a consequence, show that there is no algebraic graded field extension K of F such that $q(\mathsf{K})$ contains an isomorphic copy of E. [Hint: if $E \hookrightarrow q(\mathsf{K})$, show that one can assume K/F is tame, and use Prop. 5.35.]

Exercise 5.3. Let K be an algebraic graded field extension of a graded field F.

(i) Show that there is a one-to-one correspondence between the unramified extensions L of F in K and the field extensions E of F_0 in K_0, which maps L to L_0 and E to $E \otimes_{\mathsf{F}_0} \mathsf{F}$.

(ii) Show that if K/F is totally ramified, there is a one-to-one correspondence between the graded field extensions L of F in K and the subgroups Δ such that $\Gamma_{\mathsf{F}} \subseteq \Delta \subseteq \Gamma_{\mathsf{K}}$, which maps L to Γ_{L} and Δ to $\bigoplus_{\delta\in\Delta} \mathsf{K}_\delta$.

(iii) Give an example of an extension K/F containing totally ramified extensions L_1, L_2 of F with $\mathsf{L}_1 \neq \mathsf{L}_2$ and $\Gamma_{\mathsf{L}_1} = \Gamma_{\mathsf{L}_2}$. (Thus, the correspondence in (ii) does not generally hold if K/F is not totally ramified.)

Exercise 5.4. There are characterizations of normal graded field extensions analogous to characterizations for ungraded fields. Let F and K be graded fields with $\mathsf{F} \subseteq \mathsf{K} \subseteq \mathsf{F}_{alg}$. Prove that the following conditions are equivalent:

(a) K is normal over F;
(b) for any homogeneous $a \in \mathsf{K}$, the minimal polynomial of a over $q(\mathsf{F})$ splits over K;
(c) K is a graded splitting field over F, i.e., there is a family of homogenizable polynomials $\{f_i\}_{i \in I}$ in $\mathsf{F}[X]$ such that each f_i splits over K and K is generated over F by the roots of the f_i;
(d) for every F-homomorphism $\eta \colon \mathsf{K} \to \mathsf{F}_{alg}$, we have $\eta(\mathsf{K}) \subseteq \mathsf{K}$.

Exercise 5.5. Let K be a normal graded field extension of a graded field F. Let T be the tame closure of F in K, and let I be the purely inseparable closure of F in K, i.e., the graded subfield of K generated over F by all the homogeneous elements of K which are purely inseparable over F. Prove that T is Galois over F with $\mathcal{G}(\mathsf{T}/\mathsf{F}) \cong Aut(\mathsf{K}/\mathsf{F})$; I is the fixed graded field $\mathsf{K}^{Aut(\mathsf{K}/\mathsf{F})}$; and $\mathsf{K} = \mathsf{T} \otimes_{\mathsf{F}} \mathsf{I}$.

Exercise 5.6. Let (K, w) be an extension of the valued field (F, v) with K normal over F. Prove that $\mathsf{gr}_w(K)$ is normal over $\mathsf{gr}_v(F)$ and $\overline{K}$ is normal over $\overline{F}$.

Exercise 5.7. Let (F, v) be a valued field and $f \in F[X]$ be a monic λ-uniform polynomial. Assume $\widetilde{f}^{(\lambda)}$ is irreducible, hence f is irreducible by Lemma 5.47, and let $L = F[X]/(f)$. Let w be any valuation on L extending v. Show that

$$\mathsf{gr}_w(L) \cong_g \mathsf{gr}(F)[X]^{(\lambda)}/(\widetilde{f}^{(\lambda)}).$$

Deduce that w is the unique extension of v to L and that v is defectless in L.

Exercise 5.8. Let F be a graded field. A finite-dimensional commutative graded algebra S over F is said to be *separable* if S is a direct product of graded fields, each tame over F. Prove:

(i) A finite-dimensional commutative graded algebra S is separable over F if and only if $q(\mathsf{S})$ is separable over $q(\mathsf{F})$ (i.e., a direct product of fields each separable over $q(F)$).
(ii) Let S and S' be finite-dimensional graded field extensions of F. Then, $\mathsf{S} \otimes_{\mathsf{F}} \mathsf{S}'$ is separable over F if and only if S and S' are each separable over F.
(iii) Suppose S is a finite-dimensional commutative semisimple graded F-algebra and K is a graded field extension of F. Then $\mathsf{S} \otimes_{\mathsf{F}} \mathsf{K}$ is separable over K if and only if S is separable over F.

Exercise 5.9. Let $F \subseteq K$ be fields with $[K{:}F] < \infty$ and K Galois over F. Let $v_1, \ldots, v_r$ be the valuations on K extending a valuation v on F. Let $\alpha = min(v_1, \ldots, v_r)$, which is a surmultiplicative v-value function on K, with

$\mathsf{gr}_\alpha(K) \cong_g \mathsf{gr}_{v_1}(K) \times \ldots \times \mathsf{gr}_{v_r}(K)$ by Th. 4.36. Let $G = \mathcal{G}(K/F)$. Since G acts transitively on the v_i, all the graded fields $\mathsf{gr}_{v_i}(K)$ are isomorphic. Moreover, for any $\sigma \in \mathcal{G}(K/F)$, since σ permutes the v_i, we have $\alpha \circ \sigma = \alpha$. Hence, σ induces a graded $\mathsf{gr}(F)$-automorphism $\widetilde{\sigma}$ of $\mathsf{gr}_\alpha(K)$ defined on homogeneous elements by $\widetilde{\sigma}(\widetilde{c}) = \widetilde{\sigma(c)}$ for all $c \in K$. Thus, there is a group homomorphism $\Xi\colon \mathcal{G}(K/F) \to Aut_{\mathsf{gr}(F)}(\mathsf{gr}_\alpha(K))$ given by $\sigma \mapsto \widetilde{\sigma}$.

(i) Let $G = \mathcal{G}(K/F)$, and let

$$G^{\mathrm{ram}} = \{\sigma \in G \mid v_1(\sigma(c) - c)) > v_1(c) \text{ for all } c \in K^\times\},$$

which is the ramification group for the extension v_1 of v. Prove that

$$ker(\Xi) = \bigcap_{\sigma \in G} \sigma\, G^{\mathrm{ram}}\, \sigma^{-1}.$$

(ii) Prove that the fixed ring $\mathsf{gr}(K)^{\Xi(G)}$ is a graded field isomorphic to the purely inseparable closure of $\mathsf{gr}(F)$ in $\mathsf{gr}_{v_i}(K)$.
(iii) Prove that α is a tame v-gauge (see Prop. 6.41 in the next chapter) if and only if Ξ is injective and $\mathsf{gr}_\alpha(K)$ is $\Xi(G)$-Galois over $\mathsf{gr}(F)$.

Exercise 5.10. Let (K, v_K) be a valued field extension of a valued field (F, v). Assume K is a Galois extension of F and v_K is the unique extension of v to K. Show that the following conditions are equivalent:

(a) K is tamely ramified over F;
(b) $\mathcal{G}^{\mathrm{ram}}(K/F) = \{id\}$;
(c) K is defectless over F and $\mathsf{gr}(K)$ is Galois over $\mathsf{gr}(F)$;
(d) The ρ of Prop. 5.51 defines an isomorphism $\mathcal{G}(K/F) \xrightarrow{\sim} \mathcal{G}(\mathsf{gr}(K)/\mathsf{gr}(F))$.

Exercise 5.11. Let F be a field with valuation v and let K be a cyclic Galois field extension of F such that v has a unique and tamely ramified extension to K. Let $e = |\Gamma_K{:}\Gamma_F|$. Prove that Γ_K/Γ_F is a cyclic group and that $\overline{F}$ contains a primitive e-th root of unity.

Notes

§5.1: This section is mostly an expanded version of Hwang–Wadsworth [102] and Mounirh–Wadsworth [167, §1]. There was earlier work by Van Geel–Van Oystaeyen [250], who considered algebraic extensions of graded fields, but only with $\mathbb{Z}$-gradings. Another precursor is the paper by Boulagouaz [24]. He considered graded field extensions, and showed that the minimal polynomial of a homogeneous element in a graded field extension is homogenizable. For a graded field F, he also pointed out the valuation on $q(\mathsf{F})$ determined by a total ordering on Γ_{F}. He proved Prop. 5.19 and Prop. 5.50(iii) in [24, Th. 4, Th. 5]. Subsequently, Boulagouaz [28] investigated the Galois theory of graded field extensions and proved portions of Prop. 5.32 for finite-degree Galois extensions.

§5.2: Much of subsection 5.2.1 comes from Mounirh–Wadsworth [167, §1]. Preceding [167], Boulagouaz [27] had considered uniform polynomials over Henselian fields and introduced $\widetilde{f}^{(\lambda)}$ for λ-uniform polynomials; he proved a version of Lemma 5.46 and (a) $\Rightarrow$ (b) and (a) $\Rightarrow$ (e) of Th. 5.49. Cor. 5.56 was proved by Hwang–Wadsworth [102, Th. 5.2], with a better proof in [167, Cor. 1.13].

Exercise 5.1 is based on Hwang–Wadsworth [102, Ex. 1.2]. Exercise 5.2 is adapted from Van Geel–Van Oystaeyen [250, Ex. 3.10.2]. Exercise 5.6 comes from Mounirh–Wadsworth [167, Th. 1.5].

Chapter 6
Brauer Groups

This chapter has two parts. In the first section, we organize the central simple graded algebras over a given graded field into a Brauer group and obtain analogues of the general results on the Brauer group of fields. Graded fields have a special structure however, since the degree 0 component is a canonically-defined subfield. This structure is reflected in a canonical filtration of the Brauer group of graded fields. The second section aims to reproduce this canonical filtration for the Brauer group of valued fields. For this, special types of gauges are defined on central simple algebras over valued fields. They lead to the definition of the tame and the inertial part of the Brauer group. These developments culminate with Th. 6.64, which establishes a canonical index-preserving isomorphism between the tame part of the Brauer group of a Henselian-valued field and the Brauer group of its graded field. This result is fundamental for the use of graded algebras to study division algebras over Henselian fields. An application is given to obtain a split exact sequence for the inertially split part of the Brauer group of a Henselian field. Another application of the theorem yields Witt's description of the Brauer group of a field with a complete discrete rank 1 valuation.

6.1 The Brauer group of a graded field

The classical approach to the Brauer group of fields can be used as a model for the definition of the Brauer group of graded fields, as we show in §6.1.1. The central quotient construction allows us to identify this Brauer group with a subgroup of the Brauer group of the quotient field, from which analogues of the classical structural results are easily derived. In §6.1.2, we discuss the homological description of the Brauer group of a graded field in terms of 2-cocycles and crossed products. Canonical subgroups in the Brauer group of a graded field F, the inertial part and the inertially split part, are defined in

J.-P. Tignol, A.R. Wadsworth, *Value Functions on Simple Algebras, and Associated Graded Rings*, Springer Monographs in Mathematics,
DOI 10.1007/978-3-319-16360-4_6

the final §6.1.3. The main result in this section is Th. 6.24, which establishes a functorial isomorphism between the inertially split part modulo the inertial part and a group of continuous characters of the absolute Galois group of F_0. These characters take their values in $\mathbb{T}(\Gamma_{\mathsf{F}}) = \Gamma_{\mathsf{F}} \otimes_{\mathbb{Z}} (\mathbb{Q}/\mathbb{Z})$.

6.1.1 *Definition of* Br(F)

Mimicking the classical definition of the Brauer group of a field, we define a group structure on equivalence classes of central simple graded algebras over a given graded field F.

Definition 6.1. Let A and B be central simple graded algebras over F. The graded algebras A and B are said to be *Brauer-equivalent* if there exist finite-dimensional graded F-vector spaces V and W such that

$$\mathsf{A} \otimes_{\mathsf{F}} \mathrm{End}_{\mathsf{F}} \mathsf{V} \cong_g \mathsf{B} \otimes_{\mathsf{F}} \mathrm{End}_{\mathsf{F}} \mathsf{W}.$$

To see that this relation is indeed an equivalence relation, observe that if U and V are finite-dimensional graded F-vector spaces, then the canonical map

$$(\mathrm{End}_{\mathsf{F}} \mathsf{V}) \otimes_{\mathsf{F}} (\mathrm{End}_{\mathsf{F}} \mathsf{U}) \longrightarrow \mathrm{End}_{\mathsf{F}}(\mathsf{V} \otimes_{\mathsf{F}} \mathsf{W})$$

is an isomorphism of graded F-algebras. Therefore, if A, B, C are central simple graded F-algebras and U, V, W, P are finite-dimensional graded F-vector spaces such that

$$\mathsf{A} \otimes_{\mathsf{F}} \mathrm{End}_{\mathsf{F}} \mathsf{V} \cong_g \mathsf{B} \otimes_{\mathsf{F}} \mathrm{End}_{\mathsf{F}} \mathsf{W} \quad \text{and} \quad \mathsf{B} \otimes_{\mathsf{F}} \mathrm{End}_{\mathsf{F}} \mathsf{U} \cong_g \mathsf{C} \otimes_{\mathsf{F}} \mathrm{End}_{\mathsf{F}} \mathsf{P},$$

then we also have

$$\mathsf{A} \otimes_{\mathsf{F}} \mathrm{End}_{\mathsf{F}}(\mathsf{V} \otimes_{\mathsf{F}} \mathsf{U}) \cong_g \mathsf{B} \otimes_{\mathsf{F}} \mathrm{End}_{\mathsf{F}}(\mathsf{W} \otimes_{\mathsf{F}} \mathsf{U}) \cong_g \mathsf{C} \otimes_{\mathsf{F}} \mathrm{End}_{\mathsf{F}}(\mathsf{W} \otimes_{\mathsf{F}} \mathsf{P}).$$

We let $\sim_g$ denote the Brauer-equivalence relation on central simple graded F-algebras. For each such algebra A, we use the notation [A] for the Brauer-equivalence class of A. It is clear that Brauer-equivalence classes of central simple graded F-algebras form a set. This set has a group structure, as we proceed to show; it is called the *Brauer group* of F and is denoted by[1] Br(F).

Note that, in contrast to the ungraded theory, we may have central simple graded F-algebras A, B such that $\mathsf{A} \sim_g \mathsf{B}$ and $[\mathsf{A}{:}\mathsf{F}] = [\mathsf{B}{:}\mathsf{F}]$, and yet $\mathsf{A} \not\cong_g \mathsf{B}$. For example, this happens with $\mathsf{A} = \mathsf{M}_n(\mathsf{F})(\gamma_1, \ldots, \gamma_n)$ and $\mathsf{B} = \mathsf{M}_n(\mathsf{F})(\delta_1, \ldots, \delta_n)$ for $\gamma_1, \ldots, \delta_n \in \Gamma$ such that there is no ε in Γ and per-

[1] Note that the grading is a crucial part of the structure, so there is an important distinction to make between Br(F), the Brauer group of central simple *graded* F-algebras, and $Br(\mathsf{F}^\natural)$, the Brauer group of Azumaya algebras over the (ungraded) ring $\mathsf{F}^\natural$. See however Remark 6.5.

mutation σ of $\{1,2,\ldots,n\}$ satisfying $\delta_i - \gamma_{\sigma(i)} \in \varepsilon + \Gamma_{\mathsf{F}}$ for all $i = 1,2,\ldots,n$: see Prop. 2.27.

Proposition 6.2. *Let* F *be a graded field.*

(i) *Every Brauer-equivalence class of central simple graded* F*-algebras contains a central graded division algebra, which is uniquely determined up to graded isomorphism; it is the graded division algebra associated by Wedderburn's Theorem to any central simple graded* F*-algebra in the Brauer-equivalence class (see Th. 2.26).*

(ii) *The tensor product induces a well-defined operation making* $\mathsf{Br}(\mathsf{F})$ *an abelian group with* 0 *element* $[\mathsf{F}]$.

Proof. (i) Let A be a central simple graded F-algebra. If D is the central graded division algebra over F associated to A, we have $\mathsf{A} \cong_g \mathsf{End}_{\mathsf{D}}\,\mathsf{J}$ for some finite-dimensional graded D-vector space J. The F-span $\mathsf{V} \subseteq \mathsf{J}$ of a homogeneous D-base of J is a graded F-vector space such that

$$\mathsf{J} = \mathsf{V} \otimes_{\mathsf{F}} \mathsf{D} \qquad \text{and} \qquad \mathsf{End}_{\mathsf{D}}\,\mathsf{J} \cong_g \mathsf{D} \otimes_{\mathsf{F}} \mathsf{End}_{\mathsf{F}}\,\mathsf{V}.$$

Therefore, $\mathsf{A} \sim_g \mathsf{D}$. Conversely, if $\mathsf{A} \sim_g \mathsf{E}$ for some central graded division algebra E over F, then there are finite-dimensional graded F-vector spaces U, W such that

$$\mathsf{A} \otimes_{\mathsf{F}} \mathsf{End}_{\mathsf{F}}\,\mathsf{U} \cong_g \mathsf{E} \otimes_{\mathsf{F}} \mathsf{End}_{\mathsf{F}}\,\mathsf{W}.$$

Since $\mathsf{A} \cong_g \mathsf{End}_{\mathsf{D}}(\mathsf{V} \otimes_{\mathsf{F}} \mathsf{D})$, it follows that

$$\mathsf{End}_{\mathsf{D}}(\mathsf{U} \otimes_{\mathsf{F}} \mathsf{V} \otimes_{\mathsf{F}} \mathsf{D}) \cong_g \mathsf{End}_{\mathsf{E}}(\mathsf{W} \otimes_{\mathsf{F}} \mathsf{E}).$$

Therefore, $\mathsf{E} \cong_g \mathsf{D}$ by the uniqueness part of Wedderburn's Theorem; see Prop. 2.27.

(ii) Corollary 2.31 and Prop. 2.32 show that the tensor product of central simple graded F-algebras is again a central simple graded F-algebra. Moreover, if A, A', B, B' are central simple graded F-algebras such that $\mathsf{A} \sim_g \mathsf{A}'$ and $\mathsf{B} \sim_g \mathsf{B}'$, then $\mathsf{A} \otimes_{\mathsf{F}} \mathsf{B} \sim_g \mathsf{A}' \otimes_{\mathsf{F}} \mathsf{B}'$: indeed, if V, V', W, W' are finite-dimensional F-vector spaces such that

$$\mathsf{A} \otimes_{\mathsf{F}} \mathsf{End}_{\mathsf{F}}\,\mathsf{V} \cong_g \mathsf{A}' \otimes_{\mathsf{F}} \mathsf{End}_{\mathsf{F}}\,\mathsf{V}' \quad \text{and} \quad \mathsf{B} \otimes_{\mathsf{F}} \mathsf{End}_{\mathsf{F}}\,\mathsf{W} \cong_g \mathsf{B}' \otimes_{\mathsf{F}} \mathsf{End}_{\mathsf{F}}\,\mathsf{W}',$$

then

$$\mathsf{A} \otimes_{\mathsf{F}} \mathsf{B} \otimes_{\mathsf{F}} \mathsf{End}_{\mathsf{F}}(\mathsf{V} \otimes_{\mathsf{F}} \mathsf{W}) \cong_g \mathsf{A}' \otimes_{\mathsf{F}} \mathsf{B}' \otimes_{\mathsf{F}} \mathsf{End}(\mathsf{V}' \otimes_{\mathsf{F}} \mathsf{W}').$$

Therefore, the tensor product induces a well-defined operation on $\mathsf{Br}(\mathsf{F})$: for any central simple graded F-algebras A, B, we set

$$[\mathsf{A}] + [\mathsf{B}] = [\mathsf{A} \otimes_{\mathsf{F}} \mathsf{B}].$$

Clearly, this operation is commutative and $[\mathsf{F}]$ is the 0 element. To complete the proof that $\mathsf{Br}(\mathsf{F})$ is a group, define for each graded F-algebra A the *opposite* graded algebra A^{op} by

$$\mathsf{A}^{\mathrm{op}} = \{a^{\mathrm{op}} \mid a \in \mathsf{A}\}$$

with the operations

$$a^{\text{op}} + b^{\text{op}} = (a+b)^{\text{op}}, \qquad a^{\text{op}} \cdot b^{\text{op}} = (ba)^{\text{op}}, \quad \text{and} \quad c \cdot a^{\text{op}} = (ca)^{\text{op}}$$

for a, $b \in \mathsf{A}$ and $c \in \mathsf{F}$, and the grading for which $(\mathsf{A}^{\text{op}})_\gamma = \{a^{\text{op}} \mid a \in \mathsf{A}_\gamma\}$, for all $\gamma \in \Gamma$. Clearly, A^{op} is graded central simple if A is graded central simple. The next result shows that $[\mathsf{A}^{\text{op}}] = -[\mathsf{A}]$ in $\mathsf{Br}(\mathsf{F})$, which completes the proof of the proposition.

Lemma 6.3. *For any central simple graded algebra* A, *there is a canonical isomorphism of graded algebras*

$$\mathsf{A} \otimes_{\mathsf{F}} \mathsf{A}^{\text{op}} \xrightarrow{\sim} \mathsf{End}_{\mathsf{F}}\, \mathsf{A}$$

which carries $a \otimes b^{\text{op}}$ *to the linear transformation* $x \mapsto axb$ *for* a, b, $x \in \mathsf{A}$.

Proof. The map is clearly a homomorphism of graded F-algebras. It is injective because $\mathsf{A} \otimes_{\mathsf{F}} \mathsf{A}^{\text{op}}$ is graded simple by Prop. 2.32, hence it is surjective by dimension count. □

In view of Prop. 6.2, the index (i.e., the degree of the associated graded division algebra, see §2.2.3) of all the central simple graded algebras in a given Brauer class is the same; we call it the *index* of the Brauer class.

The next proposition shows that $\mathsf{Br}(\mathsf{F})$ may be viewed as a subgroup of the Brauer group of the quotient field $q(\mathsf{F})$:

Proposition 6.4. *Extension of scalars defines an index-preserving embedding*

$$q_{\mathsf{F}} \colon \mathsf{Br}(\mathsf{F}) \hookrightarrow Br\big(q(\mathsf{F})\big), \qquad [\mathsf{A}] \mapsto [q(\mathsf{A})] = [\mathsf{A}^\natural \otimes_{\mathsf{F}^\natural} q(\mathsf{F})].$$

Proof. This follows from Lemma 2.14, Lemma 2.15, and Prop. 2.28. □

Remark 6.5. Let A be a graded F-algebra. Then, A is graded central simple if and only if the ungraded algebra $\mathsf{A}^\natural$ is Azumaya over $\mathsf{F}^\natural$. Indeed, if A is graded central simple, then $\mathsf{A}^\natural$ is a free module of finite rank over $\mathsf{F}^\natural$; hence, Lemma 6.3 and DeMeyer–Ingraham [59, Th. 3.4, p. 52] show that $\mathsf{A}^\natural$ is Azumaya over $\mathsf{F}^\natural$. Conversely, suppose $\mathsf{A}^\natural$ is Azumaya over $\mathsf{F}^\natural$. We then have $Z(\mathsf{A}^\natural) = \mathsf{F}^\natural$ and, by [59, Prop. 21, p. 47 and Cor. 3.7, p. 54], the $\mathsf{F}^\natural$-module $\mathsf{A}^\natural$ is finitely generated, so $[\mathsf{A}{:}\mathsf{F}] < \infty$, and every ideal $I \subseteq \mathsf{A}^\natural$ is generated by $I \cap \mathsf{F}^\natural$. If I is homogeneous, then $I \cap \mathsf{F}^\natural$ is a homogeneous ideal of F, hence it is $\{0\}$ or F as F is a graded field. Therefore, A is a central simple graded F-algebra. We thus obtain a well-defined canonical homomorphism

$$\natural \colon \mathsf{Br}(\mathsf{F}) \longrightarrow Br(\mathsf{F}^\natural), \qquad [\mathsf{A}] \mapsto [\mathsf{A}^\natural],$$

where $Br(\mathsf{F}^\natural)$ denotes the Brauer group of (equivalence classes) of Azumaya algebras over $\mathsf{F}^\natural$, as in, e.g., Demeyer–Ingraham [59, p. 60]. The map $\natural$ is injective, since it fits in the following commutative diagram:

$$\begin{array}{ccc} \mathsf{Br}(\mathsf{F}) & \xhookrightarrow{\ q_{\mathsf{F}}\ } & Br(q(\mathsf{F})) \\ & \searrow^{\natural} \qquad \nearrow_{q_{\mathsf{F}}} & \\ & Br(\mathsf{F}^{\natural}). & \end{array}$$

However, $\natural$ is not in general surjective, as illustrated in Exercise 8.11 below. A description of the image of $\natural$ is given in Exercise 6.1 and in the notes at the end of this chapter.

For any central simple graded F-algebra A, we define the *exponent* of A (or of its Brauer class $[\mathsf{A}]$) to be

$$exp\ \mathsf{A} \ = \ \text{the order of } [\mathsf{A}] \text{ in the group } \mathsf{Br}(\mathsf{F}).$$

This is analogous to the definition of the exponent of a central simple algebra over an ungraded field.

Corollary 6.6. $\mathsf{Br}(\mathsf{F})$ *is a torsion group, and for every central simple graded* F-*algebra* A, *we have*

$$exp\ \mathsf{A} = exp\ q(\mathsf{A}) \qquad \textit{and} \qquad exp\ \mathsf{A} \mid ind\,\mathsf{A}.$$

Moreover, exp A *and ind* A *have the same prime factors.*

Proof. The group $\mathsf{Br}(\mathsf{F})$ is torsion, since it embeds in $Br(q(\mathsf{F}))$, which is a torsion group by Pierce [178, §14.4, Prop. a, p. 260]. The embedding shows that $exp\ \mathsf{A} = exp\ q(\mathsf{A})$. Moreover, since $q_{\mathsf{F}} \colon \mathsf{Br}(\mathsf{F}) \to Br(q(\mathsf{F}))$ is index-preserving and $exp\ q(\mathsf{A}) \mid ind\,q(\mathsf{A})$ by [178, §14.4, Lemma a, p. 259], we have

$$exp\ \mathsf{A} \ = \ exp\ q(\mathsf{A}) \mid ind\,q(\mathsf{A}) \ = \ ind\,\mathsf{A}.$$

Since $exp\ q(\mathsf{A})$ and $ind\,q(\mathsf{A})$ have the same prime factors by [178, §14.4, Prop. b(ii), p. 261], this must likewise be true for $exp\ \mathsf{A}$ and $ind\,\mathsf{A}$. □

Remark 6.7. *Primary Decomposition.* The injectivity of the map q_{F} of Prop. 6.4 also allows one to see that every graded division algebra D with center F has a primary decomposition. Specifically, let $deg\,\mathsf{D} = p_1^{r_1} \dots p_k^{r_k}$ and $exp\ \mathsf{D} = p_1^{s_1} \dots p_k^{s_k}$, where $p_1, \dots, p_k$ are distinct primes and $1 \leq s_i \leq r_i$ for all i. Then, there exist graded division algebras $\mathsf{D}_{(p_1)}, \dots, \mathsf{D}_{(p_k)}$ each with center F such that each $\mathsf{D}_{(p_i)}$ has exponent a power of p_i and

$$\mathsf{D} \cong_g \mathsf{D}_{(p_1)} \otimes_{\mathsf{F}} \dots \otimes_{\mathsf{F}} \mathsf{D}_{(p_k)}.$$

Moreover, the $\mathsf{D}_{(p_i)}$ with these properties are unique up to graded isomorphism. In addition, $deg(\mathsf{D}_{(p_i)}) = p_i^{r_i}$, $exp(\mathsf{D}_{(p_i)}) = p_i^{s_i}$, and $q(\mathsf{D}_{(p_i)}) \cong q(\mathsf{D})_{(p_i)}$, the p_i-primary component of $q(\mathsf{D})$. To obtain the p_i-component, take any positive integer t_i with $t_i \equiv 1 \ (mod\ p_i^{s_i})$ and $p_j^{s_j} \mid t_i$ for all $j \neq i$, and let $\mathsf{D}_{(p_i)}$ be the graded division algebra associated to the t_i-th tensor power $\mathsf{D}^{\otimes t_i}$. Then, as q_{F} is an index-preserving homomorphism, $q(\mathsf{D}_{(p_i)})$ is the division algebra associated to $q(\mathsf{D})^{\otimes t_i}$, which is $q(\mathsf{D})_{(p_i)}$. The properties asserted for the $\mathsf{D}_{(p_i)}$ all follow from the corresponding properties for the $q(\mathsf{D})_{(p_i)}$.

We may also extend scalars in a graded field extension:

Proposition 6.8. *Let K/F be an extension of graded fields. For any graded central simple F-algebra A, the K-algebra $\mathsf{A} \otimes_\mathsf{F} \mathsf{K}$ is central simple. The map $[\mathsf{A}] \mapsto [\mathsf{A} \otimes_\mathsf{F} \mathsf{K}]$ defines a homomorphism*

$$ext_{\mathsf{K}/\mathsf{F}} \colon \mathsf{Br}(\mathsf{F}) \longrightarrow \mathsf{Br}(\mathsf{K}).$$

This homomorphism fits in the following commutative diagram:

$$\begin{array}{ccc} \mathsf{Br}(\mathsf{F}) & \xrightarrow{\;ext_{\mathsf{K}/\mathsf{F}}\;} & \mathsf{Br}(\mathsf{K}) \\ {\scriptstyle q_\mathsf{F}}\big\downarrow & & \big\downarrow{\scriptstyle q_\mathsf{K}} \\ Br\big(q(\mathsf{F})\big) & \xrightarrow{\;ext_{q(\mathsf{K})/q(\mathsf{F})}\;} & Br\big(q(\mathsf{K})\big). \end{array}$$

Proof. For any graded F-algebra A, we have

$$q_\mathsf{K}(\mathsf{A} \otimes_\mathsf{F} \mathsf{K}) \;=\; (\mathsf{A} \otimes_\mathsf{F} \mathsf{K})^\natural \otimes_{\mathsf{K}^\natural} q(\mathsf{K}) \;=\; \mathsf{A}^\natural \otimes_{\mathsf{F}^\natural} q(\mathsf{K})$$

and also

$$q_\mathsf{F}(\mathsf{A}) \otimes_{q(\mathsf{F})} q(\mathsf{K}) \;=\; \big(\mathsf{A}^\natural \otimes_{\mathsf{F}^\natural} q(\mathsf{F})\big) \otimes_{q(\mathsf{F})} q(\mathsf{K}) \;=\; \mathsf{A}^\natural \otimes_{\mathsf{F}^\natural} q(\mathsf{K}),$$

hence

$$q_\mathsf{F}(\mathsf{A}) \otimes_{q(\mathsf{F})} q(\mathsf{K}) \;=\; q_\mathsf{K}(\mathsf{A} \otimes_\mathsf{F} \mathsf{K}). \tag{6.1}$$

If A is graded central simple over F, then Prop. 2.28 shows that $q_\mathsf{F}(\mathsf{A})$ is central simple over $q(\mathsf{F})$, hence $q_\mathsf{K}(\mathsf{A} \otimes_\mathsf{F} \mathsf{K})$ is central simple over $q(\mathsf{K})$. It follows that $\mathsf{A} \otimes_\mathsf{F} \mathsf{K}$ is graded central simple over K by Prop. 2.28 (or Lemma 2.14). It is clear that the map $ext_{\mathsf{K}/\mathsf{F}}$ is a well-defined homomorphism, and commutativity of the diagram follows from (6.1). □

In the situation of Prop. 6.8, we let $\mathsf{Br}(\mathsf{K}/\mathsf{F})$ denote the kernel of $ext_{\mathsf{K}/\mathsf{F}}$,

$$\mathsf{Br}(\mathsf{K}/\mathsf{F}) \;=\; ker\big(ext_{\mathsf{K}/\mathsf{F}} \colon \mathsf{Br}(\mathsf{F}) \to \mathsf{Br}(\mathsf{K})\big) \;\subseteq\; \mathsf{Br}(\mathsf{F}).$$

We say that a central simple graded F-algebra A is *split by* K if $[\mathsf{A}] \in \mathsf{Br}(\mathsf{K}/\mathsf{F})$, which means that $\mathsf{A} \otimes_\mathsf{F} \mathsf{K} \cong_g \mathsf{End}_\mathsf{K} \mathsf{V}$ for some finite-dimensional graded K-vector space V.

Remark 6.9. Since $q_\mathsf{F} \colon \mathsf{Br}(\mathsf{F}) \to Br(q(\mathsf{F}))$ is injective, a central simple graded F-algebra A is split by K if and only if the central simple $q(\mathsf{F})$-algebra $q(\mathsf{A})$ is split by $q(\mathsf{K})$. This happens in particular if K is a graded subfield of A with $[\mathsf{K}{:}\mathsf{F}] = deg\,\mathsf{A}$, because then $q(\mathsf{K}) = q_\mathsf{F}(\mathsf{K})$ is a subfield of $q(\mathsf{A})$ with $[q(\mathsf{K}){:}q(\mathsf{F})] = deg\,q(\mathsf{A})$, since $deg\,q(\mathsf{A}) = deg\,\mathsf{A}$ and $[q(\mathsf{K}){:}q(\mathsf{F})] = [\mathsf{K}{:}\mathsf{F}]$. Notably, if A is a graded division algebra and K is a maximal graded subfield of A, then $[\mathsf{K}{:}\mathsf{F}] = deg\,\mathsf{A}$ by Cor. 2.36, so K splits A.

6.1.2 Homological interpretation of Br(F)

The Brauer group Br(F) of a graded field has a homological interpretation analogous to the homological interpretation of the Brauer group of an ungraded field. In this subsection we develop the homological formulation of Br(F).

Just as in the classical theory, the bridge between central simple algebras and Brauer classes is provided by crossed product algebras. We now describe crossed products in the graded setting, referring to Pierce's monograph [178] for the classical theory of crossed products and the definition of cohomology groups.

Let $\mathsf{F} \subseteq \mathsf{K}$ be a finite-degree Galois extension of graded fields. Let $G = \mathcal{G}(\mathsf{K}/\mathsf{F})$, and let $n = |G| = [\mathsf{K}{:}\mathsf{F}] < \infty$. Recall from Prop. 5.30 that $q(\mathsf{K})$ is then Galois over $q(\mathsf{F})$, and G is canonically identified with $\mathcal{G}(q(\mathsf{K})/q(\mathsf{F}))$. Fix any 2-cocycle $f \in Z^2(G, \mathsf{K}^\times)$. Since f lies in $Z^2(G, q(\mathsf{K})^\times)$, it determines the following crossed product algebra over $q(\mathsf{F})$:

$$B = \big(q(\mathsf{K})/q(\mathsf{F}), G, f\big) = \bigoplus_{\sigma\in G} q(\mathsf{K})x_\sigma,$$

whose multiplication is given by

$$(cx_\sigma)(dx_\tau) = c\sigma(d)f(\sigma,\tau)x_{\sigma\tau} \qquad \text{for all } c,d \in q(\mathsf{K}), \text{ and } \sigma,\tau \in G. \tag{6.2}$$

This B is a central simple $q(\mathsf{F})$-algebra with multiplicative identity $1_B = f(id, id)^{-1}x_{id}$. Within B we have

$$\mathsf{A} = \bigoplus_{\sigma\in G} \mathsf{K}x_\sigma,$$

which is clearly a subring and $\mathsf{F}^\natural$-subalgebra of B as $f(\sigma,\tau) \in \mathsf{K}$ for all $\sigma, \tau \in G$. We define a grading on A by first setting

$$deg\, x_\sigma = \tfrac{1}{n} \sum_{\rho\in G} deg\, f(\sigma,\rho) \qquad \text{for all } \sigma \in G. \tag{6.3}$$

Then define

$$\mathsf{A}_\gamma = \bigoplus_{\sigma\in G} \mathsf{K}_{\gamma - deg\, x_\sigma} x_\sigma \qquad \text{for all } \gamma \in \Gamma. \tag{6.4}$$

Hence,

$$deg(cx_\sigma) = deg\, c + deg\, x_\sigma \qquad \text{for all } c \in \mathsf{K}^\times,\ \sigma \in G. \tag{6.5}$$

It is clear that A thus becomes a graded F-vector space. To see that it is also a graded ring, take the basic cocycle equation $f(\sigma,\tau)f(\sigma\tau,\rho) = \sigma(f(\tau,\rho))f(\sigma,\tau\rho)$, compute the degree on each side (recalling that σ is degree-preserving), and sum over all ρ in G. Thus, after dividing by n in the torsion-free group Γ,

$$deg\, f(\sigma,\tau) + deg\, x_{\sigma\tau} = deg\, x_\tau + deg\, x_\sigma. \tag{6.6}$$

This yields, for all $c,\, d \in \mathsf{K}^\times$ and $\sigma,\, \tau \in G$,

$$\begin{aligned} deg[(cx_\sigma)(dx_\tau)] &= deg[c\sigma(d)f(\sigma,\tau)x_{\sigma\tau}] \\ &= deg\, c + deg\, d + deg\, f(\sigma,\tau) + deg\, x_{\sigma\tau} \\ &= deg\, c + deg\, d + deg\, x_\sigma + deg\, x_\tau \\ &= deg(cx_\sigma) + deg(dx_\tau). \end{aligned}$$

Hence, $\mathsf{A}_\gamma \cdot \mathsf{A}_\delta \subseteq \mathsf{A}_{\gamma+\delta}$ for all $\gamma,\, \delta \in \Gamma$, showing that A is a graded ring. Clearly, A is a graded F-algebra, with multiplicative identity $1_\mathsf{A} = 1_B = f(id, id)^{-1} x_{id}$. Moreover,

$$q(\mathsf{A}) \;=\; \mathsf{A}^\natural \otimes_{\mathsf{F}^\natural} q(\mathsf{F}) \;=\; \bigoplus_{\sigma\in G} q(\mathsf{K})x_\sigma \;=\; B. \tag{6.7}$$

Since B is a central simple $q(\mathsf{F})$-algebra, by Prop. 2.28 and Lemma 2.14 A is a central simple graded F-algebra with $[\mathsf{A}{:}\mathsf{F}] = n^2$. This A is called the *graded crossed product algebra* determined by f, and denoted $(\mathsf{K}/\mathsf{F}, G, f)$. Thus,

$$(\mathsf{K}/\mathsf{F}, G, f) \;=\; \bigoplus_{\sigma\in G} \mathsf{K}x_\sigma, \tag{6.8}$$

with the multiplication given by (6.2) and the grading given by (6.3)–(6.5). From the definition of the grading on A it is clear that the group $\{cx_\sigma \mid c \in \mathsf{K}^\times,\, \sigma \in G\}$ consists of homogeneous units of A, and the degrees of these elements fill out Γ_A. Thus, Γ_A is a group and

$$\Gamma_\mathsf{A} \;=\; \Gamma_\mathsf{A}^\times \;=\; \bigcup_{\sigma\in G} (deg\, x_\sigma + \Gamma_\mathsf{K}). \tag{6.9}$$

Note that the map $c \mapsto c\cdot 1_\mathsf{A} = cf(id, id)^{-1}x_{id}$ gives a graded field isomorphism $\mathsf{K} \xrightarrow{\sim} \mathsf{K}x_{id} \subseteq \mathsf{A}$; moreover, $C_\mathsf{A}(\mathsf{K}x_{id}) = \mathsf{K}x_{id}$ by dimension count using the graded Double Centralizer Theorem 2.35(ii).

Lemma 6.10. *If f and f' are cohomologous in $Z^2(G, \mathsf{K}^\times)$, then*

$$(\mathsf{K}/\mathsf{F}, G, f) \;\cong_g\; (\mathsf{K}/\mathsf{F}, G, f').$$

Proof. Let

$$\mathsf{A} = (\mathsf{K}/\mathsf{F}, G, f) \;=\; \bigoplus_{\sigma\in G} \mathsf{K}x_\sigma \quad \text{and} \quad \mathsf{A}' = (\mathsf{K}/\mathsf{F}, G, f') \;=\; \bigoplus_{\sigma\in G} \mathsf{K}x'_\sigma.$$

Since f and f' are cohomologous, there are $b_\sigma \in \mathsf{K}^\times$ for all $\sigma \in G$ such that

$$f(\sigma,\tau) \;=\; b_\sigma \sigma(b_\tau) b_{\sigma\tau}^{-1}\, f'(\sigma,\tau) \qquad \text{for all } \sigma, \tau \in G. \tag{6.10}$$

Hence, the definition of $deg\, x_\sigma$ in (6.3) yields

$$\begin{aligned} deg\, x_\sigma &= \tfrac{1}{n}\big[\textstyle\sum_{\rho\in G} deg\, b_\sigma + \sum_{\rho\in G} deg\, \sigma(b_\rho) - \sum_{\rho\in G} deg\, b_{\sigma\rho} + \sum_{\rho\in G} deg\, f'(\sigma,\rho)\big] \\ &= deg\, b_\sigma + deg\, x'_\sigma. \end{aligned} \tag{6.11}$$

The bijective function $\varphi\colon \mathsf{A} \to \mathsf{A}'$ given by $\sum_{\sigma\in G} c_\sigma x_\sigma \mapsto \sum_{\sigma\in G} c_\sigma b_\sigma x'_\sigma$ is a ring homomorphism in view of (6.10), and is a left $\mathsf{K}^\natural$-module homomorphism.

Furthermore, since $deg\,\varphi(x_\sigma) = deg\,x_\sigma$ by (6.11), this φ is a graded F-algebra isomorphism. □

Since every 2-cocycle is cohomologous to a *normalized* 2-cocycle, i.e., a cocycle f such that $f(\sigma, id) = f(id, \sigma) = 1$ for all $\sigma \in G$, Lemma 6.10 shows that every graded crossed product algebra $\mathsf{A} = (\mathsf{K}/\mathsf{F}, \sigma, f)$ can be defined by a normalized 2-cocycle; thus, $1_\mathsf{A} = x_{id}$. From now on we work exclusively with normalized cocycles.

A notable special case of graded crossed products is given by graded cyclic algebras. Suppose $\mathcal{G}(\mathsf{K}/\mathsf{F})$ is a cyclic group of order n with generator σ. Take any $b \in \mathsf{F}^\times$. Then we have the *graded cyclic algebra*

$$\mathsf{C} = \big(\mathsf{K}/\mathsf{F}, \sigma, b\big) = \bigoplus_{i=0}^{n-1} \mathsf{K}y^i \quad \text{where} \quad ycy^{-1} = \sigma(c) \text{ for all } c \in \mathsf{K} \text{ and } y^n = b. \tag{6.12}$$

This C is the graded crossed product algebra $\mathsf{C} = \big(\mathsf{K}/\mathsf{F}, G, f\big)$, where $G = \mathcal{G}(\mathsf{K}/\mathsf{F}) = \langle\sigma\rangle$ and $f \in Z^2(G, \mathsf{K}^\times)$ is given by

$$f(\sigma^i, \sigma^j) = \begin{cases} 1 & \text{if } i+j \leq n-1, \\ b & \text{if } i+j \geq n, \end{cases} \qquad \text{for } i, j = 0, 1, \ldots, n-1.$$

So, C is a central simple graded F-algebra of degree n, with grading given as in (6.3) and (6.5) by

$$deg\,y = \tfrac{1}{n}[(n-1)\,deg\,1 + deg\,b] = \tfrac{1}{n}\,deg\,b \tag{6.13}$$

and

$$deg(cy^i) = deg\,c + \tfrac{i}{n}\,deg\,b \quad \text{for all } c \in \mathsf{K}^\times,\ i \in \mathbb{Z}. \tag{6.14}$$

Note also that

$$q(\mathsf{C}) = \big(q(\mathsf{K})/q(\mathsf{F}), \sigma, b\big)$$

which is an (ungraded) cyclic algebra over $q(\mathsf{F})$.

Theorem 6.11. *Let $\mathsf{F} \subseteq \mathsf{K}$ be a finite-degree Galois extension of graded fields, and let $G = \mathcal{G}(\mathsf{K}/\mathsf{F})$. There is a group isomorphism*

$$\Delta_{\mathsf{K}/\mathsf{F}} \colon H^2(G, \mathsf{K}^\times) \xrightarrow{\sim} \mathsf{Br}(\mathsf{K}/\mathsf{F}) \qquad \textit{given by} \qquad [f] \mapsto \big[(\mathsf{K}/\mathsf{F}, G, f)\big].$$

This map fits into a commutative diagram with injective vertical maps:

$$\begin{array}{ccc} H^2(G, \mathsf{K}^\times) & \xrightarrow{\Delta_{\mathsf{K}/\mathsf{F}}} & \mathsf{Br}(\mathsf{K}/\mathsf{F}) \\ \big\downarrow & & \big\downarrow{\scriptstyle q_\mathsf{F}} \\ H^2(G, q(\mathsf{K})^\times) & \xrightarrow{\cong} & Br(q(\mathsf{K})/q(\mathsf{F})) \end{array} \tag{6.15}$$

where the bottom horizontal arrow is given by the (ungraded) crossed product construction.

The proof of the theorem follows the next two lemmas.

Lemma 6.12. *Let* D *be a graded division algebra finite-dimensional over its center, and let* $\iota\colon \mathsf{D}^\times \to q(\mathsf{D})^\times$ *be the inclusion map. For any automorphism* ξ *of* D*, let* ξ *also denote the extension of* ξ *to* $q(\mathsf{D})$*. There is a group homomorphism* $\lambda\colon q(\mathsf{D})^\times \to \mathsf{D}^\times$ *such that* $\lambda\circ\iota = \mathrm{id}_{\mathsf{D}^\times}$ *and for every (degree-preserving) graded ring automorphism* ξ *of* D*, we have* $\lambda\circ\xi = \xi\circ\lambda$.

Proof. We argue as in §3.2.5, with a little more care required as D can be noncommutative. Since Γ_{D} is a torsion-free abelian group, there is a total ordering $\leq$ on Γ_{D} so that $(\Gamma_{\mathsf{D}}, \leq)$ is a totally ordered abelian group. Choose and fix any such ordering. Then, define $\lambda\colon \mathsf{D}\setminus\{0\} \to \mathsf{D}^\times$ as follows: for any $c = \sum_{\gamma\in\Gamma_{\mathsf{D}}} c_\gamma$ with each $c_\gamma \in D_\gamma$,

$$\lambda(c) \;=\; c_\delta \quad \text{where } \delta \text{ is minimal in } \{\gamma \mid c_\gamma \neq 0\}.$$

Clearly,

$$\lambda(cd) \;=\; \lambda(c)\cdot\lambda(d) \qquad \text{for any } c,d\in\mathsf{D}\setminus\{0\}. \tag{6.16}$$

Note that $\lambda|_{\mathsf{D}^\times} = id$ and $\lambda\big(Z(\mathsf{D})\setminus\{0\}\big) \subseteq Z(\mathsf{D})^\times$. Extend λ to $q(\mathsf{D})^\times$ as follows: Since $[\mathsf{D}{:}Z(\mathsf{D})] < \infty$ any element of $q(\mathsf{D})^\times$ is expressible as cz^{-1} with $c\in\mathsf{D}\setminus\{0\}$ and $z\in Z(\mathsf{D})\setminus\{0\}$; define

$$\lambda(cz^{-1}) \;=\; \lambda(c)\lambda(z)^{-1}.$$

If $cz^{-1} = c'z'^{-1}$ with $c'\in\mathsf{D}\setminus\{0\}$ and $z'\in Z(\mathsf{D})\setminus\{0\}$, then it follows from (6.16) that $\lambda(c)\lambda(z)^{-1} = \lambda(c')\lambda(z')^{-1}$. Hence, λ is well-defined on $q(\mathsf{D})^\times$, and clearly formula (6.16) holds for any $c,d\in q(\mathsf{D})^\times$. Thus, $\lambda\colon q(\mathsf{D})^\times \to \mathsf{D}^\times$ is a group homomorphism with $\lambda\circ\iota = id_{\mathsf{D}^\times}$.

Let ξ be any graded automorphism of D. Of course, ξ restricts to a graded automorphism of $Z(\mathsf{D})$. For $c = \sum_{\gamma\in\Gamma_{\mathsf{D}}} c_\gamma \in \mathsf{D}\setminus\{0\}$ with each $c_\gamma\in\mathsf{D}_\gamma$ we have $\xi(c) = \sum_{\gamma\in\Gamma_{\mathsf{D}}}\xi(c_\gamma)$ and each $\xi(c_\gamma)\in\mathsf{D}_\gamma$. Hence, $\lambda(\xi(c)) = \xi(\lambda(c))$ for every $c\in\mathsf{D}\setminus\{0\}$. Therefore, for all $c\in\mathsf{D}\setminus\{0\}$, $z\in Z(\mathsf{D})\setminus\{0\}$,

$$\lambda\xi(cz^{-1}) = \lambda\big(\xi(c)\xi(z)^{-1}\big) = \lambda\xi(c)\,\lambda\xi(z)^{-1} = \xi\lambda(c)\xi\lambda(z)^{-1} = \xi\lambda(cz^{-1}),$$

showing that $\lambda\circ\xi = \xi\circ\lambda$. □

The next lemma is the graded analogue of a standard result for ungraded division algebras. We need only part (i) here, but will use part (ii) in §8.1.

Lemma 6.13. *Let* D *be a graded division algebra finite-dimensional over its center* F*. Let* L *be a finite-degree graded field extension of* F*. Then,*

(i) L *splits* D *if and only if there is a central simple graded* F*-algebra* A *Brauer-equivalent to* D *such that* L *embeds in* A *and* $[\mathsf{L}{:}\mathsf{F}] = \deg\mathsf{A}$.
(ii) *Let* $t = \operatorname{ind}\mathsf{D}/\operatorname{ind}(\mathsf{D}\otimes_{\mathsf{F}}\mathsf{L})$*. Then,* $t\in\mathbb{Z}$*,* $t\,\big|\,[\mathsf{L}{:}\mathsf{F}]$*, and* L *embeds* F*-isomorphically into* D *if and only if* $t = [\mathsf{L}{:}\mathsf{F}]$.

Proof. Recall from Prop. 2.32 and Cor. 2.31 that $\mathsf{D}\otimes_{\mathsf{F}}\mathsf{L}$ is a central simple graded L-algebra. Let S be a simple right graded $\mathsf{D}\otimes_{\mathsf{F}}\mathsf{L}$-module, and let $\mathsf{B} = \mathsf{End}_{\mathsf{D}\otimes_{\mathsf{F}}\mathsf{L}}(\mathsf{S})$, which is a graded division algebra by Lemma 2.25 with $\mathsf{End}_{\mathsf{B}}(\mathsf{S}) \cong_g \mathsf{D} \otimes_{\mathsf{F}} \mathsf{L}$ by Prop. 2.34. So, B is the graded division algebra Brauer-equivalent to $\mathsf{D} \otimes_{\mathsf{F}} \mathsf{L}$. Set

$$\ell = [\mathsf{L}:\mathsf{F}], \quad d = deg\,\mathsf{D} = ind\,\mathsf{D}, \quad \text{and} \quad b = deg\,\mathsf{B} = ind(\mathsf{D} \otimes_{\mathsf{F}} \mathsf{L}),$$

so $t = d/b$. Let $s = dim_{\mathsf{D}}\,\mathsf{S}$. Then,

$$dim_{\mathsf{F}}\,\mathsf{S} = (dim_{\mathsf{D}}\,\mathsf{S})\cdot[\mathsf{D}:\mathsf{F}] = sd^2 \text{ and } dim_{\mathsf{F}}\,\mathsf{S} = (dim_{\mathsf{B}}\,\mathsf{S})\cdot[\mathsf{B}:\mathsf{F}] = b^2\ell\, dim_{\mathsf{B}}\,\mathsf{S}.$$

Moreover,

$$d^2\ell = [\mathsf{D} \otimes_{\mathsf{F}} \mathsf{L} : \mathsf{F}] = [\mathsf{End}_{\mathsf{B}}(\mathsf{S}) : \mathsf{F}] = b^2\ell(dim_{\mathsf{B}}\,\mathsf{S})^2.$$

This shows that $d = b\,dim_{\mathsf{B}}\,\mathsf{S}$, which with the preceding equalities yields

$$b^2\ell\, dim_{\mathsf{B}}\,\mathsf{S} = sd^2 = sdb\, dim_{\mathsf{B}}\,\mathsf{S};$$

hence,

$$b\ell = sd. \tag{6.17}$$

(i) Suppose L splits D, i.e., $b = 1$. Let $\mathsf{A} = \mathsf{End}_{\mathsf{D}}(\mathsf{S})$, which is Brauer-equivalent to D in $\mathsf{Br}(\mathsf{F})$. Since S is a $\mathsf{D} \otimes_{\mathsf{F}} \mathsf{L}$-module, L embeds in A. From (6.17) and $b = 1$, we have $deg\,A = s\,deg\,\mathsf{D} = \ell$, so A satisfies the conditions in (i). Conversely, suppose there is a central simple graded F-algebra A' Brauer-equivalent to D such that $deg\,\mathsf{A}' = \ell$ and L embeds in A'. Write $\mathsf{A}' = \mathsf{End}_{\mathsf{D}}(\mathsf{S}')$ for some right graded D-vector space S', and let $s' = dim_{\mathsf{D}}\,\mathsf{S}'$. Since L embeds in $\mathsf{End}_{\mathsf{D}}(\mathsf{S}')$, S' is a right $\mathsf{D} \otimes_{\mathsf{F}} \mathsf{L}$-module. Then, by Prop. 2.24 and Th. 2.26 S' is a direct sum of shifts of the simple $\mathsf{D} \otimes_{\mathsf{F}} \mathsf{L}$-module S; so $s' \geq s$. Hence, (6.17) yields

$$b\ell = sd \leq s'd = deg\,\mathsf{A}' = \ell.$$

Thus, $b = 1$, i.e., L splits D.

(ii) Since we saw above that $d = b\,dim_{\mathsf{B}}\,\mathsf{S}$, we have $t = d/b = dim_{\mathsf{B}}\,\mathsf{S} \in \mathbb{Z}$. Also, (6.17) says $\ell = st$, showing that $t \mid \ell$. Suppose $t = \ell$. Then, $s = 1$, so, the 1-dimensional D-vector space S must be a shift of D, i.e., $\mathsf{S} \cong_g \mathsf{D}(\gamma)$ for some $\gamma \in \Gamma$. Hence,

$$\mathsf{End}_{\mathsf{D}}(\mathsf{S}) \cong_g \mathsf{End}_{\mathsf{D}}(\mathsf{D}(\gamma)) \cong_g \mathsf{End}_{\mathsf{D}}(\mathsf{D}) \cong_g \mathsf{D}.$$

Now, as S is a $\mathsf{D} \otimes_{\mathsf{F}} \mathsf{L}$-module, L embeds in $\mathsf{End}_{\mathsf{D}}(\mathsf{S})$; the preceding isomorphisms then show L embeds in D. Conversely, suppose L embeds in D. Then, the left multiplication action of L on D and the right multiplication action of D on D combine to make D into a right $\mathsf{D} \otimes_{\mathsf{F}} \mathsf{L}^{\mathrm{op}} = \mathsf{D} \otimes_{\mathsf{F}} \mathsf{L}$-module. It is a simple module, since D is already a simple right D-module. Thus, we may take $\mathsf{S} = \mathsf{D}$. Then, $s = 1$, so $t = \ell$ by (6.17). □

Proof of Th. 6.11. Since Lemma 6.10 shows that $(\mathsf{K}/\mathsf{F}, G, f)$ depends up to graded isomorphism only on the cohomology class of f, our $\Delta_{\mathsf{K}/\mathsf{F}}$ is well-defined as a map into $\mathsf{Br}(\mathsf{F})$. Take any graded crossed product algebra $(\mathsf{K}/\mathsf{F}, G, f)$. Because it contains a copy of K with $[\mathsf{K}:\mathsf{F}] = deg\,(\mathsf{K}/\mathsf{F}, G, f)$, it is split by K (see Remark 6.9). Hence, $im\,\Delta_{\mathsf{K}/\mathsf{F}} \subseteq \mathsf{Br}(\mathsf{K}/\mathsf{F})$. The isomorphism noted in (6.7), $(\mathsf{K}/\mathsf{F}, G, f) \otimes_{\mathsf{F}} q(\mathsf{F}) \cong (q(\mathsf{K})/q(\mathsf{F}), G, f)$, shows that diagram (6.15) is commutative. To see that the left map in (6.15) is injective take a homomorphism $\lambda\colon q(\mathsf{K}^\times) \to \mathsf{K}^\times$ as in Lemma 6.12. Since each $\sigma \in G$ is a graded automorphism of K, the lemma shows that $\lambda \circ \sigma = \sigma \circ \lambda$; so, λ is a G-equivariant splitting map for the inclusion $\iota\colon \mathsf{K}^\times \hookrightarrow q(\mathsf{K})^\times$. Thus, $\mathsf{K}^\times$ is a direct summand of $q(\mathsf{K})^\times$ as G-modules, and hence the maps $H^i(G, \mathsf{K}^\times) \to H^i(G, q(\mathsf{K})^\times)$ are injective for all i. In particular, the left map in diagram (6.15) is injective. The bottom map in the diagram is also injective (in fact, an isomorphism). Hence, $\Delta_{\mathsf{K}/\mathsf{F}}$ is injective. Since the right map $q(\mathsf{F})$ in diagram (6.15) is also injective and the composition $q(\mathsf{F}) \circ \Delta_{\mathsf{K}/\mathsf{F}}$ is a group homomorphism, $\Delta_{\mathsf{K}/\mathsf{F}}$ must also be a group homomorphism.

It remains to verify that $\Delta_{\mathsf{K}/\mathsf{F}}$ is surjective. For this, take any graded division algebra D with $[\mathsf{D}] \in \mathsf{Br}(\mathsf{K}/\mathsf{F})$. By Lemma 6.13(i) there is a central simple graded F-algebra A Brauer-equivalent to D and containing K with $deg\,\mathsf{A} = [\mathsf{K}:\mathsf{F}]$. Then by dimension count, the graded Double Centralizer Theorem, Th. 2.35(ii), shows that $C_{\mathsf{A}}(\mathsf{K}) = \mathsf{K}$. Take any $\sigma \in G$, and view σ as a graded F-algebra homomorphism $\mathsf{K} \to \mathsf{A}$. Because $C_{\mathsf{A}}(\mathsf{K}) = \mathsf{K}$, this σ is of course also a graded homomorphism $C_{\mathsf{A}}(\mathsf{K}) \to \mathsf{A}$ and $\sigma(C_{\mathsf{A}}(\mathsf{K}))$ centralizes $\sigma(\mathsf{K})$. Therefore, we can apply the graded Skolem–Noether Theorem, Th. 2.37(iii). It yields a homogeneous element $x_\sigma \in \mathsf{A}^\times$ such that $x_\sigma c x_\sigma^{-1} = \sigma(c)$ for all $c \in \mathsf{K}$. Let $B = \mathsf{A} \otimes_{\mathsf{F}} q(\mathsf{F})$, which is a central simple $q(\mathsf{F})$-algebra. It contains the field $\mathsf{K} \otimes_{\mathsf{F}} q(\mathsf{F}) = q(\mathsf{K})$ which is Galois over $q(\mathsf{F})$ with group G. Furthermore, $[q(\mathsf{K}){:}q(\mathsf{F})] = deg\,B$. For each $\sigma \in G$ we have $x_\sigma a x_\sigma^{-1} = \sigma(a)$ for all $a \in q(\mathsf{K})$. Hence, from ungraded Brauer group theory, see e.g. Pierce [178, §14.1, Lemma, p. 251], the sum $\sum_{\sigma \in G} q(\mathsf{K}) x_\sigma$ is direct and is all of B; so, $B = (q(\mathsf{K})/q(\mathsf{F}), G, f)$, where $f \in Z^2(G, q(\mathsf{K})^\times)$ is given by $f(\sigma,\tau) = x_\sigma x_\tau x_{\sigma\tau}^{-1}$ for all $\sigma, \tau \in G$. Since each $f(\sigma,\tau)$ lies in $\mathsf{A}^\times$ and centralizes K, we have $f(\sigma,\tau) \in C_{\mathsf{gr}(A)}(\mathsf{K})^\times = \mathsf{K}^\times$; hence, $f \in Z^2(G, \mathsf{K}^\times)$. As the sum $\sum_{\sigma \in G} \mathsf{K} x_\sigma$ is direct, it is all of A by dimension count. Hence, $\mathsf{A} = (\mathsf{K}/\mathsf{F}, G, f)$, and $[\mathsf{D}] = [\mathsf{A}] = \Delta_{\mathsf{K}/\mathsf{F}}[f]$. Thus, $im\,\Delta_{\mathsf{K}/\mathsf{F}} = \mathsf{Br}(\mathsf{K}/\mathsf{F})$, completing the proof. □

Proposition 6.14. *Let $\mathsf{F} \subseteq \mathsf{L} \subseteq \mathsf{K}$ be graded fields with $[\mathsf{K}:\mathsf{F}] < \infty$ and K and L each Galois over F. Let $G = \mathcal{G}(\mathsf{K}/\mathsf{F})$ and $N = \mathcal{G}(\mathsf{K}/\mathsf{L}) \subseteq G$. Then there is a commutative diagram:*

$$\begin{array}{ccc} H^2(G/N, \mathsf{L}^\times) & \xrightarrow{\Delta_{\mathsf{L}/\mathsf{F}}} & \mathsf{Br}(\mathsf{L}/\mathsf{F}) \\ {\scriptstyle inf}\downarrow & & \downarrow \\ H^2(G, \mathsf{K}^\times) & \xrightarrow{\Delta_{\mathsf{K}/\mathsf{F}}} & \mathsf{Br}(\mathsf{K}/\mathsf{F}). \end{array} \tag{6.18}$$

In this diagram inf is the inflation map and the right map is inclusion.

Proof. We have a cube of maps

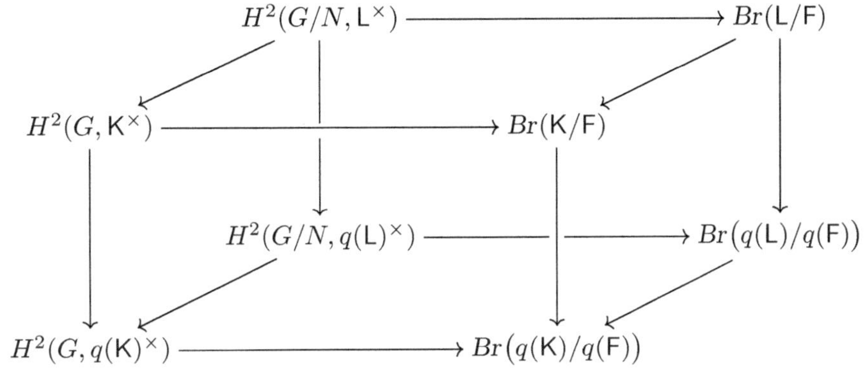

with obvious downward maps on the vertical edges. In this cube, the left and right faces are clearly commutative, the front and back faces are commutative by Th. 6.11, and the bottom face is commutative from the homological theory of the ungraded Brauer group (see Pierce [178, Prop. p. 263]). Since the right front vertical map $q_{\mathsf{F}} \colon \mathsf{Br}(\mathsf{K}/\mathsf{F}) \to Br(q(\mathsf{K})/q(\mathsf{F}))$ is injective (see Prop. 6.4), it follows that the top face of the cube is also commutative, as desired. □

Proposition 6.14 shows that the crossed product map $\Delta_{\mathsf{K}/\mathsf{F}}$ is compatible with the enlargement of K. This compatibility allows us to define a map $\Delta_{\mathsf{K}/\mathsf{F}}$ for K an infinite Galois extension, by a limit process. If G is the Galois group of some infinite-dimensional Galois extension of graded fields K/F, the cocycles in $Z^2(G, \mathsf{K}^\times)$ are assumed to be continuous for the Krull topology on G and the discrete topology on $\mathsf{K}^\times$, which means they are inflations of cocycles from quotients of G by open subgroups. Likewise, $H^2(G, \mathsf{K}^\times)$ denotes the group of cohomology classes of continuous cocycles. Thus, we have

$$H^2(G, \mathsf{K}^\times) \;=\; \varinjlim_{L \in \mathcal{L}} H^2(\mathcal{G}(\mathsf{L}/\mathsf{F}), \mathsf{L}^\times)$$

where $\mathcal{L}$ is the set of finite-degree Galois graded field extensions L of F lying in K. This $\mathcal{L}$ is partially ordered by inclusion; for $\mathsf{L}, \mathsf{M} \in \mathcal{L}$, with $\mathsf{L} \subseteq \mathsf{M}$ we have a canonical isomorphism $\mathcal{G}(\mathsf{L}/\mathsf{F}) \cong \mathcal{G}(\mathsf{M}/\mathsf{F})/\,\mathcal{G}(\mathsf{M}/\mathsf{L})$, and the map $H^2(\mathcal{G}(\mathsf{L}/\mathsf{F}), \mathsf{L}^\times) \to H^2(\mathcal{G}(\mathsf{M}/\mathsf{F}), \mathsf{M}^\times)$ is inflation.

Likewise, $\mathsf{Br}(\mathsf{K}/\mathsf{F}) = \varinjlim_{L \in \mathcal{L}} \mathsf{Br}(\mathsf{L}/\mathsf{F})$, where for $\mathsf{L}, \mathsf{M} \in \mathcal{L}$ with $\mathsf{L} \subseteq \mathsf{M}$, the map $\mathsf{Br}(\mathsf{L}/\mathsf{F}) \to \mathsf{Br}(\mathsf{M}/\mathsf{F})$ is inclusion. In view of Prop. 6.14, we may define a group homomorphism

$$\Delta_{\mathsf{K}/\mathsf{F}} \colon H^2(G, \mathsf{K}^\times) \longrightarrow \mathsf{Br}(\mathsf{K}/\mathsf{F})$$

as the direct limit of the isomorphisms $\Delta_{\mathsf{L}/\mathsf{F}}$ for $\mathsf{L} \in \mathcal{L}$.

Corollary 6.15. *Let $\mathsf{F} \subseteq \mathsf{K}$ be a Galois extension of graded fields, possibly of infinite degree, and let $G = \mathcal{G}(\mathsf{K}/\mathsf{F})$. The map $\Delta_{\mathsf{K}/\mathsf{F}}$ is an isomorphism, and diagram* (6.15) *is still commutative when* $[\mathsf{K}{:}\mathsf{F}] = \infty$.

Proof. Since each $\Delta_{\mathsf{L}/\mathsf{F}}$ for $\mathsf{L} \in \mathcal{L}$ is an isomorphism, $\Delta_{\mathsf{K}/\mathsf{F}}$ is also an isomorphism. The diagram (6.15) for K/F with $[\mathsf{K}{:}\mathsf{F}] = \infty$ is commutative because the corresponding diagram for L/F is commutative for each $\mathsf{L} \in \mathcal{L}$. □

6.1.3 The canonical Brauer group filtration (graded case)

For an arbitrary graded field F, scalar extension from F_0 to F defines an index-preserving embedding of $Br(\mathsf{F}_0)$ into $\mathsf{Br}(\mathsf{F})$,

$$Br(\mathsf{F}_0) \hookrightarrow \mathsf{Br}(\mathsf{F}), \qquad [A] \mapsto [A \otimes_{\mathsf{F}_0} \mathsf{F}].$$

(See Prop. 2.16.) We use the notation $\mathsf{Br}_{in}(\mathsf{F})$ for the image of $Br(\mathsf{F}_0)$ in $\mathsf{Br}(\mathsf{F})$ and call $\mathsf{Br}_{in}(\mathsf{F})$ the *inertial part* of the Brauer group $\mathsf{Br}(\mathsf{F})$. The central graded division algebras D over F such that $[\mathsf{D}] \in \mathsf{Br}_{in}(\mathsf{F})$ satisfy $\mathsf{D} = \mathsf{D}_0 \otimes_{\mathsf{F}_0} \mathsf{F}$, by Prop. 2.52; they are the central graded division[2] algebras that are inertial over F, as per Def. 2.50.

Let F_{in} be the inertial closure of the graded field F (see Def. 5.23). We set $\mathsf{Br}_{is}(\mathsf{F}) = \mathsf{Br}(\mathsf{F}_{in}/\mathsf{F})$ and call this subgroup the *inertially split part* of $\mathsf{Br}(\mathsf{F})$. The central simple graded F-algebras A such that $[\mathsf{A}] \in \mathsf{Br}_{is}(\mathsf{F})$ are said to be *inertially split*; indeed, they are split by an inertial extension of their center. In particular, the inertial graded division algebras are clearly inertially split, so we get the following canonical filtration of $\mathsf{Br}(\mathsf{F})$:

$$Br(\mathsf{F}_0) \cong \mathsf{Br}_{in}(\mathsf{F}) \subseteq \mathsf{Br}_{is}(\mathsf{F}) \subseteq \mathsf{Br}(\mathsf{F}). \tag{6.19}$$

The first quotient in this filtration will be determined below (see Th. 6.24) and the second quotient will be determined in Th. 7.69. By using the results on inertial scalar extension from §2.3.2, we can already single out a significant case where the second quotient is trivial:

Proposition 6.16. *If* $dim_{\mathbb{Q}}(\Gamma_{\mathsf{F}} \otimes_{\mathbb{Z}} \mathbb{Q}) = 1$ *(for example, if* $\Gamma_{\mathsf{F}} \cong \mathbb{Z}$*), then* $\mathsf{Br}(\mathsf{F}) = \mathsf{Br}_{is}(\mathsf{F})$.

The proof follows the next lemma, which gives a sufficient (but not necessary) condition for a graded division algebra to be inertially split:

Lemma 6.17. *Let* D *be a central graded division algebra over a graded field* F. *If* $\Gamma_{\mathsf{D}}/\Gamma_{\mathsf{F}}$ *is cyclic, then* D *is inertially split.*

Proof. Let E be the central graded division algebra over F_{in} associated to $\mathsf{D} \otimes_{\mathsf{F}} \mathsf{F}_{in}$. Since $Z(\mathsf{D}_0)$ is Galois over F_0 by Prop. 2.40, it can be iden-

---

[2] Note however that there are central simple graded algebras A such that $[\mathsf{A}] \in \mathsf{Br}_{in}(\mathsf{F})$ but A is *not* inertial: for example, $\mathsf{A} = M_n(\mathsf{F})(\gamma_1, \ldots, \gamma_n)$ where $\gamma_1, \ldots, \gamma_n \in \Gamma$ are not all in the same coset of Γ_{F}.

tified with a subfield of $(\mathsf{F}_{in})_0$, which is a separable closure of F_0. Therefore, we have

$$Z(\mathsf{D}_0) \otimes_{\mathsf{F}_0} (\mathsf{F}_{in})_0 \cong (\mathsf{F}_{in})_0 \times \ldots \times (\mathsf{F}_{in})_0 = (\mathsf{F}_0)_{sep} \times \ldots \times (\mathsf{F}_0)_{sep}.$$

From Prop. 2.58, it follows that $\mathsf{E}_0 = (\mathsf{F}_{in})_0$ and $\Gamma_{\mathsf{E}} \subseteq \Gamma_{\mathsf{D}}$, so $\Gamma_{\mathsf{E}}/\Gamma_{\mathsf{F}_{in}}$ is cyclic. Therefore, E is generated over its center F_{in} by a single element. It must be commutative, so $\mathsf{E} = \mathsf{F}_{in}$ and D is split by F_{in}. □

Proof of Prop. 6.16. Recall that $dim_{\mathbb{Q}}(\Gamma_{\mathsf{F}} \otimes_{\mathbb{Z}} \mathbb{Q}) = rr(\Gamma_{\mathsf{F}})$, the rational rank of Γ_{F} (see (A.18) in Appendix A). Take any central graded division algebra D over F. Since $\Gamma_{\mathsf{D}}/\Gamma_{\mathsf{F}}$ is finite, Prop. A.36 shows that

$$trk(\Gamma_{\mathsf{D}}/\Gamma_{\mathsf{F}}) \leq rr(\Gamma_{\mathsf{F}}) = 1,$$

where $trk(\Gamma_{\mathsf{D}}/\Gamma_{\mathsf{F}})$ is the minimal number of generators of $\Gamma_{\mathsf{D}}/\Gamma_{\mathsf{F}}$, see (A.23). Hence, $\Gamma_{\mathsf{D}}/\Gamma_{\mathsf{F}}$ is a cyclic group. By Lemma 6.17, D is inertially split. □

The following proposition shows that insertion of $\mathsf{Br}(\mathsf{F}_{tr})$ does not yield an additional step in the filtration (6.19):

Proposition 6.18. *Every central graded division algebra* D *over a graded field* F *contains a maximal graded subfield* K *such that the extension* K/F *is tame. Every central simple graded* F*-algebra is split by the tame closure* F_{tr} *of* F.

The key fact needed for proving Prop. 6.18 is given in the following lemma. We say that a central graded division algebra D over a graded field F is *totally ramified* if $\mathsf{D}_0 = \mathsf{F}_0$.

Lemma 6.19. *Let* F *be a graded field with* $\operatorname{char} \mathsf{F} = p \neq 0$*, and let* D *be a graded division algebra with center* F*. If* $\deg \mathsf{D}$ *is a power of* p *and* D *is totally ramified over* F*, then* $\mathsf{D} = \mathsf{F}$.

Proof. Let $p^r = [\mathsf{D}{:}\mathsf{F}] = |\Gamma_{\mathsf{D}}{:}\Gamma_{\mathsf{F}}|$, and suppose $r \geq 1$. Let $(a_1, \ldots, a_{p^r})$ be any homogeneous base of D as a graded F-vector space. Then, for each i we have $deg(a_i^{p^r}) = p^r \deg a_i \in \Gamma_{\mathsf{F}}$; so, $a_i^{p^r} \in \mathsf{F}$, as D is totally ramified over F. Since a_i is thus purely inseparable over $q(\mathsf{F})$, its trace $tr_{q(\mathsf{F})(a_i)/q(\mathsf{F})}(a_i) = 0$ if $a_i \notin q(\mathsf{F})$. Let $Trd_{q(\mathsf{D})} \colon q(\mathsf{D}) \to q(\mathsf{F})$ be the reduced trace of $q(\mathsf{D})$ as a central simple $q(\mathsf{F})$-algebra. Then, $Trd_{q(\mathsf{D})}(a_i) = 0$ if $a_i \notin q(\mathsf{F})$, while $Trd_{q(\mathsf{D})}(a_i) = p^r a_i = 0$ if $a_i \in q(\mathsf{F})$. Hence, as the a_i span the $q(\mathsf{F})$-vector space $q(\mathsf{D})$ and $Trd_{q(\mathsf{D})}$ is $q(\mathsf{F})$-linear, $Trd_{q(\mathsf{D})}$ must vanish identically. However, the reduced trace is always nontrivial on any central simple algebra over a field, see Reiner [199, Th. 9.9, p. 116]. This contradiction can only be avoided if $r = 0$, i.e., $\mathsf{D} = \mathsf{F}$. □

We will see another way to prove this lemma in Cor. 7.53 below by using the nondegenerate canonical pairing associated to a totally ramified graded division algebra.

Proof of Prop. 6.18. It suffices to prove the first statement, which readily implies the second. For if K is a maximal graded subfield of D, then K splits D (see Remark 6.9).

If *char* $\mathsf{F} = 0$, then every algebraic graded field extension of F is tame; so, we may take for K any maximal graded subfield of D. Now assume that *char* $\mathsf{F} = p \neq 0$. The division algebra D_0 contains a maximal subfield M that is a separable extension of $Z(\mathsf{D}_0)$, by Pierce [178, §13.5, Prop., p. 245]. Let $\mathsf{L} = M \otimes_{\mathsf{F}_0} \mathsf{F}$, which is the compositum of M and F in D, and is a graded subfield of D with $\mathsf{L}_0 = M$ and $\Gamma_{\mathsf{L}} = \Gamma_{\mathsf{F}}$. Let $\mathsf{N} = C_{\mathsf{D}}(\mathsf{L})$, which is a graded division algebra with $\mathsf{L} \subseteq \mathsf{N} \subseteq \mathsf{D}$; also, $\mathsf{L} = Z(\mathsf{N})$ by the graded Double Centralizer Theorem 2.35(i). Then, $\mathsf{N}_0 = M = \mathsf{L}_0$, as $C_{\mathsf{D}_0}(M) = M$; hence N is totally ramified over L. The p-primary component $\mathsf{N}_{(p)}$ of N (cf. Remark 6.7) is then also totally ramified over L. By Lemma 6.19, $\mathsf{N}_{(p)} = \mathsf{L}$. Hence, $p \nmid [\mathsf{N}{:}\mathsf{L}]$. Let K be any maximal graded subfield of N. By Cor. 2.36, $[\mathsf{K}{:}\mathsf{L}] = deg\,\mathsf{N}$, so $[\mathsf{K}{:}\mathsf{F}] = deg\,\mathsf{N} \cdot [\mathsf{L}{:}\mathsf{F}] = deg\,\mathsf{D}$. Hence, K is a maximal graded subfield of D by Cor. 2.36. Since $M = \mathsf{L}_0 \subseteq \mathsf{K}_0 \subseteq \mathsf{N}_0 = M$, we have $\mathsf{K}_0 = M$, which is separable over $Z(\mathsf{D}_0)$, which in turn is separable over F_0. Hence, K_0 is separable over F_0. Also, $|\Gamma_{\mathsf{K}}{:}\Gamma_{\mathsf{F}}| = |\Gamma_{\mathsf{K}}{:}\Gamma_{\mathsf{L}}| \mid [\mathsf{N}{:}\mathsf{L}]$, and $[\mathsf{N}{:}\mathsf{L}]$ is prime to p. Thus, K is tame over F. □

Corollary 6.20. *Suppose char* $\mathsf{F} = p \neq 0$. *Then, the p-primary component of* $\mathsf{Br}(\mathsf{F})$ *lies in* $\mathsf{Br}_{is}(\mathsf{F})$.

Proof. Let D be a central simple graded division algebra over F with $exp(\mathsf{D})$ a power of p. Then, $deg\,\mathsf{D} = p^r$ for some integer r by Cor. 6.6. By Prop. 6.18 there is a graded subfield K of D with K tame over F and $[\mathsf{K}{:}\mathsf{F}] = deg\,\mathsf{D}$. Since $|\Gamma_{\mathsf{K}}{:}\Gamma_{\mathsf{F}}|$ divides $[\mathsf{K}{:}\mathsf{F}]$, it is a power of p. Hence, $\Gamma_{\mathsf{K}} = \Gamma_{\mathsf{F}}$, as K is tame over F; so, K is inertial over F. Hence, D is inertially split, since it is split by K, cf. Remark 6.9. □

In the rest of this subsection, we calculate the inertially split part $\mathsf{Br}_{is}(\mathsf{F})$ of the Brauer group of an arbitrary graded field F. The main result (Th. 6.24) is an exact sequence relating $\mathsf{Br}_{is}(\mathsf{F})$ to the inertial part $\mathsf{Br}_{in}(\mathsf{F})$ and to the absolute Galois group of F_0.

We first introduce some notation: for any torsion-free abelian group Γ, we define the *torus of* Γ to be

$$\mathbb{T}(\Gamma) = \Gamma \otimes_{\mathbb{Z}} (\mathbb{Q}/\mathbb{Z}). \tag{6.20}$$

(Note that $\mathbb{T}(\mathbb{Z}^n) = (\mathbb{Q}/\mathbb{Z})^n$; this is why we use the term torus.) Recall that the divisible hull of Γ, which is $\Gamma \otimes_{\mathbb{Z}} \mathbb{Q} = \varinjlim_n \frac{1}{n}\Gamma$, is denoted $\mathbb{H}(\Gamma)$. Since Γ is torsion-free, the following sequence is exact:

$$0 \longrightarrow \Gamma \longrightarrow \mathbb{H}(\Gamma) \longrightarrow \mathbb{T}(\Gamma) \longrightarrow 0.$$

Therefore, we identify $\mathbb{T}(\Gamma)$ with the quotient $\mathbb{H}(\Gamma)/\Gamma$ and write its elements in the form $\frac{\gamma}{n} + \Gamma$, where $\gamma \in \Gamma$ and $n \in \mathbb{Z}$ ($n \neq 0$).

Let $\mathcal{G}(\mathsf{F}_0)$ denote the absolute Galois group of F_0,

$$\mathcal{G}(\mathsf{F}_0) = \mathcal{G}\big((\mathsf{F}_0)_{sep}/\mathsf{F}_0\big).$$

Any Galois extension K of F_0 can be identified with a subfield of the separable closure $(\mathsf{F}_0)_{sep}$, hence we have a restriction map

$$res_K\colon \mathcal{G}(\mathsf{F}_0) \longrightarrow \mathcal{G}(K/\mathsf{F}_0).$$

We will only use this notation in the case where $\mathcal{G}(K/\mathsf{F}_0)$ is abelian; then res_K does not depend on the choice of the embedding $K \hookrightarrow (\mathsf{F}_0)_{sep}$.

For any central graded division algebra D over F, the grade group Γ_{D} lies in the divisible hull $\mathbb{H}(\Gamma_{\mathsf{F}})$ of Γ_{F}, hence we have an induced embedding

$$\Gamma_{\mathsf{D}}/\Gamma_{\mathsf{F}} \hookrightarrow \mathbb{T}(\Gamma_{\mathsf{F}}).$$

Recall from Prop. 2.40 that $Z(\mathsf{D}_0)$ is a Galois field extension of F_0 with abelian Galois group, and we have a canonical surjective homomorphism

$$\theta_{\mathsf{D}}\colon \Gamma_{\mathsf{D}} \longrightarrow \mathcal{G}(Z(\mathsf{D}_0)/\mathsf{F}_0).$$

Assuming D is inertially split, we define a map $\chi_{\mathsf{D}}\colon \mathcal{G}(\mathsf{F}_0) \to \mathbb{T}(\Gamma_{\mathsf{F}})$ as follows: for $\rho \in \mathcal{G}(\mathsf{F}_0)$, we pick $\gamma \in \Gamma_{\mathsf{D}}$ such that $res_{Z(\mathsf{D}_0)}(\rho) = \theta_{\mathsf{D}}(\gamma)$, and set

$$\chi_{\mathsf{D}}(\rho) = \gamma + \Gamma_{\mathsf{F}} \in \Gamma_{\mathsf{D}}/\Gamma_{\mathsf{F}} \subseteq \mathbb{T}(\Gamma_{\mathsf{F}}). \tag{6.21}$$

We call the map χ_{D} the *ramification character* of D. At this point, it is not clear that χ_{D} is well-defined, since $ker\,\theta_{\mathsf{D}}$ might be larger than Γ_{F}. We will prove in Prop. 6.22 that indeed $ker\,\theta_{\mathsf{D}} = \Gamma_{\mathsf{F}}$ when D is inertially split, and moreover that χ_{D} is a continuous homomorphism from $\mathcal{G}(\mathsf{F}_0)$ to $\mathbb{T}(\Gamma_{\mathsf{F}})$ for the Krull topology on $\mathcal{G}(\mathsf{F}_0)$ and the discrete topology on $\mathbb{T}(\Gamma_{\mathsf{F}})$. For this we will use a graded crossed product algebra Brauer-equivalent to D.

Fix a central graded division algebra D over F, which we assume to be inertially split. We may then find a finite-degree inertial Galois graded field extension K of F that splits D. Let $G = \mathcal{G}(\mathsf{K}/\mathsf{F})$. By Th. 6.11, D is Brauer-equivalent to a graded crossed product algebra $\mathsf{A} = \big(\mathsf{K}/\mathsf{F}, G, f\big)$ for some $f \in Z^2(G, \mathsf{K}^\times)$. Write $\mathsf{A} = \bigoplus_{\sigma\in G} \mathsf{K}x_\sigma$ with multiplication as in (6.2) and grading as in (6.3) and (6.4), and consider the map

$$\eta_f\colon G \longrightarrow \mathbb{T}(\Gamma_{\mathsf{F}}), \qquad \sigma \mapsto deg(x_\sigma) + \Gamma_{\mathsf{F}}.$$

Since $\Gamma_{\mathsf{K}} = \Gamma_{\mathsf{F}}$, we have $deg\, f(\sigma,\tau) \in \Gamma_{\mathsf{F}}$ for all σ, $\tau \in G$; hence, (6.6) shows that η_f is a group homomorphism. We let $H = ker\,\eta_f$ and let $\overline{\eta}_f\colon G/H \hookrightarrow \mathbb{T}(\Gamma_{\mathsf{F}})$ be the injective homomorphism induced by η_f. Note that since K is inertial over F, by Cor. 5.53 the field extension $\mathsf{K}_0/\mathsf{F}_0$ is Galois with group G. Recall from (2.29) the canonical homomorphism

$$\theta_{\mathsf{A}}\colon \Gamma_{\mathsf{A}}^\times \longrightarrow Aut\big(Z(\mathsf{A}_0)/\mathsf{F}_0\big).$$

Lemma 6.21. *The algebra A_0 is simple and $Z(\mathsf{A}_0) = \mathsf{K}_0^H$, the subfield of K_0 fixed under H, so*

$$Aut\big(Z(\mathsf{A}_0)/\mathsf{F}_0\big) = \mathcal{G}\big(Z(\mathsf{A}_0)/\mathsf{F}_0\big) = G/H.$$

Moreover, for $\gamma \in \Gamma_{\mathsf{A}}^{\times}$,

$$\overline{\eta}_f(\theta_{\mathsf{A}}(\gamma)) = \gamma + \Gamma_{\mathsf{F}}. \tag{6.22}$$

Proof. For $\sigma \in H$, we may find a homogeneous element $a_\sigma \in \mathsf{F}^\times$ such that

$$deg\, x_\sigma = deg\, a_\sigma.$$

Let $y_\sigma = a_\sigma^{-1} x_\sigma$, so $deg\, y_\sigma = 0$ and $\mathsf{K}x_\sigma = \mathsf{K}y_\sigma$. For $\sigma \notin H$, the graded vector space $\mathsf{K}x_\sigma$ does not contain any homogeneous element of degree 0. Therefore, it follows from the definition of the grading on A in (6.4) that

$$\mathsf{A}_0 = \bigoplus_{\sigma \in H} \mathsf{K}_0 y_\sigma.$$

Since for $\sigma \in H$ conjugation by y_σ induces σ on K_0, and different choices of σ yield different automorphisms of K_0, the algebra A_0 is a crossed product; hence it is a simple algebra, and its center is K_0^H. By (6.9) we have

$$\Gamma_{\mathsf{A}} = \Gamma_{\mathsf{A}}^{\times} = \bigcup_{\sigma \in G} (deg(x_\sigma) + \Gamma_{\mathsf{F}}).$$

By definition of θ_{A} and η_f, we have for $\gamma \in deg(x_\sigma) + \Gamma_{\mathsf{F}}$

$$\theta_{\mathsf{A}}(\gamma) = int(x_\sigma)|_{Z(\mathsf{A}_0)} = \sigma|_{Z(\mathsf{A}_0)} \quad \text{and} \quad \eta_f(\sigma) = \gamma + \Gamma_{\mathsf{F}}.$$

The lemma follows. □

Proposition 6.22. *Let* D *be an inertially split central graded division algebra over* F. *Then,* $ker\,\theta_{\mathsf{D}} = \Gamma_{\mathsf{F}}$ *and the ramification character* χ_{D} *of* (6.21) *is a well-defined continuous homomorphism with*

$$im\,\chi_{\mathsf{D}} = \Gamma_{\mathsf{D}}/\Gamma_{\mathsf{F}} \qquad \textit{and} \qquad ker\,\chi_{\mathsf{D}} = \mathcal{G}(Z(\mathsf{D}_0)).$$

Moreover, $\chi_{\mathsf{D}} = 0$ *if and only if* D *is inertial over* F.

Proof. Use the notation in Lemma 6.21. Since A_0 is simple, Prop. 2.47 shows that $\Gamma_{\mathsf{A}} = \Gamma_{\mathsf{D}}$, $Z(\mathsf{A}_0) \cong Z(\mathsf{D}_0)$, and $\theta_{\mathsf{A}} = \theta_{\mathsf{D}}$. If $\gamma \in ker\,\theta_{\mathsf{D}}$, then (6.22) shows that $\gamma \in \Gamma_{\mathsf{F}}$. Thus, $ker\,\theta_{\mathsf{D}} = \Gamma_{\mathsf{F}}$, hence χ_{D} is well-defined. Moreover, θ_{D} induces a group isomorphism

$$\overline{\theta}_{\mathsf{D}} \colon \Gamma_{\mathsf{D}}/\Gamma_{\mathsf{F}} \xrightarrow{\sim} \mathcal{G}(Z(\mathsf{D}_0)/\mathsf{F}_0),$$

and χ_{D} may be described as the composition

$$\mathcal{G}(\mathsf{F}_0) \xrightarrow{res_{Z(\mathsf{D}_0)}} \mathcal{G}(Z(\mathsf{D}_0)/\mathsf{F}_0) \xrightarrow{\overline{\theta}_{\mathsf{D}}^{-1}} \Gamma_{\mathsf{D}}/\Gamma_{\mathsf{F}} \hookrightarrow \mathbb{T}(\Gamma_{\mathsf{F}}). \tag{6.23}$$

Therefore, χ_{D} is a group homomorphism. It is continuous since it factors through $\mathcal{G}(Z(\mathsf{D}_0)/\mathsf{F}_0)$. We have $\chi_{\mathsf{D}} = 0$ if and only if $\Gamma_{\mathsf{D}}/\Gamma_{\mathsf{F}} = \{0\}$, which occurs if and only if D is inertial over F. The description of $im\,\chi_{\mathsf{D}}$ and $ker\,\chi_{\mathsf{D}}$ is clear from (6.23), since $res_{Z(\mathsf{D}_0)}$ is surjective with kernel $\mathcal{G}(Z(\mathsf{D}_0))$ and $\overline{\theta}_{\mathsf{D}}^{-1}$ is an isomorphism. □

Another proof that $\ker\theta_{\mathsf{D}} = \Gamma_{\mathsf{F}}$ when D is inertially split will be given in Prop. 8.49 below.

The ramification character χ_{D} depends only on the isomorphism class of D, hence on the Brauer class $[\mathsf{D}] \in Br_{is}(\mathsf{F})$. Letting $Hom^c\big(\mathcal{G}(\mathsf{F}_0), \mathbb{T}(\Gamma_{\mathsf{F}})\big)$ denote the group of continuous homomorphisms from $\mathcal{G}(\mathsf{F}_0)$ to $\mathbb{T}(\Gamma_{\mathsf{F}})$ (also known as *characters of* $\mathcal{G}(\mathsf{F}_0)$), we may therefore define the *ramification map*

$$\chi\colon Br_{is}(\mathsf{F}) \longrightarrow Hom^c\big(\mathcal{G}(\mathsf{F}_0), \mathbb{T}(\Gamma_{\mathsf{F}})\big), \qquad [\mathsf{D}] \mapsto \chi_{\mathsf{D}}. \tag{6.24}$$

Our next goal is to show that the ramification map is a group homomorphism. We prove this by describing χ cohomologically.

Let F_{in} be the inertial closure of F, so

$$\mathcal{G}(\mathsf{F}_{in}/\mathsf{F}) = \mathcal{G}(\mathsf{F}_0).$$

Note that $\Gamma_{\mathsf{F}_{in}} = \Gamma_{\mathsf{F}}$, and since the action of $\mathcal{G}(\mathsf{F}_0)$ on $\mathsf{F}_{in}^{\times}$ is by degree-preserving automorphisms, the induced action of $\mathcal{G}(\mathsf{F}_0)$ on Γ_{F} is trivial. The degree map yields a homomorphism

$$deg\colon H^2(\mathcal{G}(\mathsf{F}_0), \mathsf{F}_{in}^{\times}) \longrightarrow H^2(\mathcal{G}(\mathsf{F}_0), \Gamma_{\mathsf{F}}).$$

On the other hand, the exact sequence of trivial $\mathcal{G}(\mathsf{F}_0)$-modules

$$0 \longrightarrow \Gamma_{\mathsf{F}} \longrightarrow \mathbb{H}(\Gamma_{\mathsf{F}}) \longrightarrow \mathbb{T}(\Gamma_{\mathsf{F}}) \longrightarrow 0$$

yields an exact sequence

$$H^1\big(\mathcal{G}(\mathsf{F}_0), \mathbb{H}(\Gamma_{\mathsf{F}})\big) \longrightarrow H^1\big(\mathcal{G}(\mathsf{F}_0), \mathbb{T}(\Gamma_{\mathsf{F}})\big) \xrightarrow{\ \partial\ } H^2(\mathcal{G}(\mathsf{F}_0), \Gamma_{\mathsf{F}}) \longrightarrow H^2\big(\mathcal{G}(\mathsf{F}_0), \mathbb{H}(\Gamma_{\mathsf{F}})\big).$$

Since $\mathbb{H}(\Gamma_{\mathsf{F}})$ is divisible and torsion-free, we have

$$H^1\big(\mathcal{G}(\mathsf{F}_0), \mathbb{H}(\Gamma_{\mathsf{F}})\big) = H^2\big(\mathcal{G}(\mathsf{F}_0), \mathbb{H}(\Gamma_{\mathsf{F}})\big) = 0;$$

hence, ∂ is an isomorphism. Moreover, since $\mathcal{G}(\mathsf{F}_0)$ acts trivially on $\mathbb{T}(\Gamma_{\mathsf{F}})$,

$$H^1\big(\mathcal{G}(\mathsf{F}_0), \mathbb{T}(\Gamma_{\mathsf{F}})\big) = Hom^c\big(\mathcal{G}(\mathsf{F}_0), \mathbb{T}(\Gamma_{\mathsf{F}})\big).$$

Thus, we have a homomorphism

$$\partial^{-1} \circ deg\colon H^2(\mathcal{G}(\mathsf{F}_0), \mathsf{F}_{in}^{\times}) \longrightarrow Hom^c\big(\mathcal{G}(\mathsf{F}_0), \mathbb{T}(\Gamma_{\mathsf{F}})\big).$$

Proposition 6.23. *The ramification map* χ *is a group homomorphism and the following diagram is commutative:*

$$\begin{array}{ccc} H^2(\mathcal{G}(\mathsf{F}_0), \mathsf{F}_{in}^{\times}) & \xrightarrow{\ \Delta_{\mathsf{F}_{in}/\mathsf{F}}\ } & \mathsf{Br}_{is}(\mathsf{F}) \\ & {\scriptstyle \partial^{-1}\circ deg} \searrow \qquad \swarrow {\scriptstyle \chi} & \\ & Hom^c\big(\mathcal{G}(\mathsf{F}_0), \mathbb{T}(\Gamma_{\mathsf{F}})\big) & \end{array}$$

Proof. Since $\Delta_{\mathsf{F}_{in}/\mathsf{F}}$ is a group isomorphism by Cor. 6.15, and since $\partial^{-1}\circ deg$ is a group homomorphism, it suffices to prove the commutativity of the diagram. Let D be an inertially split central graded division algebra over F, and consider a Brauer-equivalent graded crossed product A as in Lemma 6.21. We may assume $\mathsf{A} = (\mathsf{K}/\mathsf{F}, G, f)$ with $\mathsf{K} \subseteq \mathsf{F}_{in}$. We then have the restriction map

$$\pi\colon\ \mathcal{G}(\mathsf{F}_0) \;=\; \mathcal{G}(\mathsf{F}_{in}/\mathsf{F}) \;\longrightarrow\; G \;=\; \mathcal{G}(\mathsf{K}/\mathsf{F}),$$

and we can define a continuous cocycle $\widehat{f} \in Z^2(\mathcal{G}(\mathsf{F}_0), \mathsf{F}_{in}^{\times})$ such that $\Delta_{\mathsf{F}_{in}/\mathsf{F}}[\widehat{f}] = [\mathsf{D}]$ by inflation:

$$\widehat{f}(\rho,\tau) \;=\; f\big(\pi(\rho),\pi(\tau)\big) \qquad \text{for } \rho,\tau \in \mathcal{G}(\mathsf{F}_0).$$

To show that $\partial^{-1}\big(deg[\widehat{f}]\big) = \chi_{\mathsf{D}}$, we compute $\partial(\chi_{\mathsf{D}})$. We saw in the proof of Prop. 6.22 that $\theta_{\mathsf{D}} = \theta_{\mathsf{A}}$, hence for $\rho \in \mathcal{G}(\mathsf{F}_0)$,

$$\overline{\theta}_{\mathsf{D}}(deg(x_{\pi(\rho)}) + \Gamma_{\mathsf{F}}) \;=\; \theta_{\mathsf{A}}(deg\, x_{\pi(\rho)}) \;=\; \pi(\rho)|_{Z(\mathsf{A}_0)} \;=\; res_{Z(\mathsf{D}_0)}(\rho).$$

From (6.23) it also follows that

$$\chi_{\mathsf{D}}(\rho) \;=\; \overline{\theta}_{\mathsf{D}}^{\,-1}\big(res_{Z(\mathsf{D}_0)}(\rho)\big) \qquad \text{for } \rho \in \mathcal{G}(\mathsf{F}_0),$$

hence

$$\chi_{\mathsf{D}}(\rho) \;=\; deg(x_{\pi(\rho)}) + \Gamma_{\mathsf{F}}.$$

Thus, $deg\, x_{\pi(\rho)}$ is a representative in $\mathbb{H}(\Gamma_{\mathsf{F}})$ of $\chi_{\mathsf{D}}(\rho) \in \mathbb{T}(\Gamma_{\mathsf{F}})$. Therefore, the cohomology class $\partial(\chi_{\mathsf{D}})$ in $H^2(\mathcal{G}(\mathsf{F}_0), \Gamma_{\mathsf{F}})$ is represented by the cocycle g defined by

$$g(\rho,\tau) \;=\; deg\, x_{\pi(\rho)} - deg\, x_{\pi(\rho\tau)} + deg\, x_{\pi(\tau)} \qquad \text{for } \rho,\ \tau \in \mathcal{G}(\mathsf{F}_0).$$

By (6.6), the right displayed expression equals $deg\, f\big(\pi(\rho),\pi(\tau)\big)$, hence $g = deg\,\widehat{f}$. Thus, $\partial(\chi_{\mathsf{D}}) = [g] = [deg\,\widehat{f}]$. □

Theorem 6.24. *For an arbitrary graded field* F*, the following sequence is exact:*

$$0 \longrightarrow \mathsf{Br}_{in}(\mathsf{F}) \longrightarrow \mathsf{Br}_{is}(\mathsf{F}) \xrightarrow{\ \chi\ } Hom^c\big(\mathcal{G}(\mathsf{F}_0), \mathbb{T}(\Gamma_{\mathsf{F}})\big) \longrightarrow 0.$$

This exact sequence is actually split exact, as we will see in Prop. 6.27 below.

Proof. It was shown in Prop. 6.22 that $\chi_{\mathsf{D}} = 0$ for an inertially split central graded division algebra D if and only if D is inertial. Thus, $ker\,\chi = \mathsf{Br}_{in}(\mathsf{F})$, and it only remains to prove that χ is onto.

For any $\varphi \in Hom^c\big(\mathcal{G}(\mathsf{F}_0), \mathbb{T}(\Gamma_{\mathsf{F}})\big)$, the kernel $ker\,\varphi$ is an open subgroup, hence the fixed field $\big((\mathsf{F}_0)_{sep}\big)^{ker\,\varphi}$, which we designate simply by F_0^{φ}, is a finite Galois extension of F_0 with Galois group isomorphic to $im\,\varphi$. We call φ a *cyclic character* if $im\,\varphi$ is a cyclic subgroup of $\mathbb{T}(\Gamma_{\mathsf{F}})$. In general, not every $\varphi \in Hom^c\big(\mathcal{G}(\mathsf{F}_0), \mathbb{T}(\Gamma_{\mathsf{F}})\big)$ is a cyclic character, but it is a sum of cyclic characters. Indeed, the finite abelian group $im\,\varphi$ is a direct sum, $im\,\varphi = C_1 \oplus \ldots \oplus C_k$, where each C_i is a finite cyclic subgroup of $\mathbb{T}(\Gamma_{\mathsf{F}})$. Let π_i be the projection onto the i-th component in this direct sum decomposition. Then $\varphi = \sum_{i=1}^{k} \pi_i \circ \varphi$ and each $\pi_i \circ \varphi$ is a cyclic character.

To prove that χ is onto, it now suffices to show that every cyclic character φ lies in $im\,\chi$. Let

$$n = [\mathsf{F}_0^{\varphi} : \mathsf{F}_0] = |\mathcal{G}(\mathsf{F}_0^{\varphi}/\mathsf{F}_0)| = |im\,\varphi|.$$

Choose a generator σ of $\mathcal{G}(\mathsf{F}_0^{\varphi}/\mathsf{F}_0)$. All the $\rho \in \mathcal{G}(\mathsf{F}_0)$ such that $\rho|_{\mathsf{F}_0^{\varphi}} = \sigma$ have the same image under φ, and this image generates $im\,\varphi$. Let $\gamma \in \Gamma_{\mathsf{F}}$ satisfy

$$\varphi(\rho) = \tfrac{1}{n}\gamma + \Gamma_{\mathsf{F}} \qquad \text{for all } \rho \in \mathcal{G}(\mathsf{F}_0) \text{ such that } \rho|_{\mathsf{F}_0^{\varphi}} = \sigma.$$

Now, choose any nonzero $b \in \mathsf{F}_{\gamma}$. Let $\mathsf{K} = \mathsf{F}_0^{\varphi} \otimes_{\mathsf{F}_0} \mathsf{F}$, which is an inertial cyclic Galois graded field extension of F with $\mathsf{K}_0 = \mathsf{F}_0^{\varphi}$ and $\mathcal{G}(\mathsf{K}/\mathsf{F}) = \mathcal{G}(\mathsf{F}_0^{\varphi}/\mathsf{F}_0)$ generated by σ. Let D be the graded cyclic F-algebra of degree n

$$\mathsf{D} = (\mathsf{K}/\mathsf{F}, \sigma, b) = \bigoplus_{i=0}^{n-1} \mathsf{K}y^i \quad \text{where} \quad ycy^{-1} = \sigma(c) \quad \text{for } c \in \mathsf{K} \text{ and } y^n = b,$$

with $deg\,y = \frac{1}{n}\gamma$, as in (6.12) and (6.13) above. Since $\Gamma_{\mathsf{K}} = \Gamma_{\mathsf{F}}$, the homogeneous elements of $\mathsf{K}y^i$ have degrees in $\Gamma_{\mathsf{F}} + \frac{i}{n}\gamma$. But $\frac{1}{n}\gamma + \Gamma_F$ has order n in $\mathbb{T}(\Gamma_{\mathsf{F}})$ as $|\,im\,\varphi| = n$; hence, $\mathsf{D}_0 = \mathsf{K}y^0 = \mathsf{F}_0^{\varphi}$. Therefore, the finite-dimensional simple graded F-algebra D is a division algebra by Cor. 2.43 since D_0 is a division ring. Also, D is inertially split, since K is a maximal graded subfield of D that is inertial over F. We have $\theta_{\mathsf{D}}(\frac{1}{n}\gamma) = int(y)|_{\mathsf{D}_0} = \sigma$. Therefore, if $\rho \in \mathcal{G}(\mathsf{F}_0)$ satisfies $res_{Z(\mathsf{D}_0)}(\rho) = \sigma^i$ for some $i \in \{0, \ldots, n-1\}$, then

$$\chi_{\mathsf{D}}(\rho) = \tfrac{i}{n}\gamma + \Gamma_{\mathsf{F}} = \varphi(\rho).$$

This proves that $\varphi = \chi_D$. Thus, every cyclic character lies in $im\,\chi$, proving that χ is surjective. □

We next show that the exact sequence in Th. 6.24 is functorial. If F is a graded subfield of some graded field K, then $\mathsf{F}_0 \subseteq \mathsf{K}_0$ and we may take for $(\mathsf{F}_0)_{sep}$ the separable closure of F_0 in $(\mathsf{K}_0)_{sep}$. Thus, $(\mathsf{F}_0)_{sep} \subseteq (\mathsf{K}_0)_{sep}$ and we may consider the restriction map

$$\pi_{\mathsf{K}_0/\mathsf{F}_0} \colon \mathcal{G}(\mathsf{K}_0) \longrightarrow \mathcal{G}(\mathsf{F}_0), \qquad \rho \mapsto \rho|_{(\mathsf{F}_0)_{sep}}.$$

This map induces a homomorphism

$$\begin{aligned} \pi^*_{\mathsf{K}_0/\mathsf{F}_0} \colon Hom^c(\mathcal{G}(\mathsf{F}_0), \mathbb{T}(\Gamma_{\mathsf{F}})) &\longrightarrow Hom^c(\mathcal{G}(\mathsf{K}_0), \mathbb{T}(\Gamma_{\mathsf{K}})), \\ \chi &\mapsto \tau_{\mathsf{K}/\mathsf{F}} \circ \chi \circ \pi_{\mathsf{K}_0/\mathsf{F}_0}, \end{aligned} \tag{6.25}$$

where $\tau_{\mathsf{K}/\mathsf{F}} \colon \mathbb{T}(\Gamma_{\mathsf{F}}) \to \mathbb{T}(\Gamma_{\mathsf{K}})$ is the canonical map induced by the inclusion $\Gamma_{\mathsf{F}} \hookrightarrow \Gamma_{\mathsf{K}}$. On the other hand, we may also consider the extension of scalars map

$$ext_{\mathsf{K}/\mathsf{F}} \colon \mathsf{Br}(\mathsf{F}) \longrightarrow \mathsf{Br}(\mathsf{K}).$$

The inclusions $\mathsf{F} \subseteq \mathsf{K}$ and $(\mathsf{F}_0)_{sep} \subseteq (\mathsf{K}_0)_{sep}$ yield a graded ring homomorphism

$$\mathsf{F}_{in} = (\mathsf{F}_0)_{sep} \otimes_{\mathsf{F}_0} \mathsf{F} \longrightarrow (\mathsf{K}_0)_{sep} \otimes_{\mathsf{F}_0} \mathsf{K} = \mathsf{K}_{in}.$$

This map is injective as F_{in} is a graded field. Therefore, we have $\mathsf{F}_{in} \subseteq \mathsf{K}_{in}$, and the following diagram, where all the maps are scalar extensions, is commutative:

$$\begin{array}{ccccc}
Br(\mathsf{F}_0) & \longrightarrow & \mathsf{Br}(\mathsf{F}) & \longrightarrow & \mathsf{Br}(\mathsf{F}_{in}) \\
\downarrow & & \downarrow & & \downarrow \\
Br(\mathsf{K}_0) & \longrightarrow & \mathsf{Br}(\mathsf{K}) & \longrightarrow & \mathsf{Br}(\mathsf{K}_{in})
\end{array}$$

Therefore, $ext_{\mathsf{K}/\mathsf{F}}$ maps $\mathsf{Br}_{in}(\mathsf{F})$ to $\mathsf{Br}_{in}(\mathsf{K})$ and $\mathsf{Br}_{is}(\mathsf{F})$ to $\mathsf{Br}_{is}(\mathsf{K})$.

Proposition 6.25. *Let $\mathsf{F} \subseteq \mathsf{K}$ be graded fields. There is a commutative diagram with exact rows:*

$$\begin{array}{ccccccccc}
0 & \longrightarrow & \mathsf{Br}_{in}(\mathsf{F}) & \longrightarrow & \mathsf{Br}_{is}(\mathsf{F}) & \xrightarrow{\ \chi\ } & Hom^c(\mathcal{G}(\mathsf{F}_0), \mathbb{T}(\Gamma_\mathsf{F})) & \longrightarrow & 0 \\
 & & \big\downarrow {\scriptstyle ext_{\mathsf{K}/\mathsf{F}}} & & \big\downarrow {\scriptstyle ext_{\mathsf{K}/\mathsf{F}}} & & \big\downarrow {\scriptstyle \pi^*_{\mathsf{K}_0/\mathsf{F}_0}} & & \\
0 & \longrightarrow & \mathsf{Br}_{in}(\mathsf{K}) & \longrightarrow & \mathsf{Br}_{is}(\mathsf{K}) & \xrightarrow{\ \chi\ } & Hom^c(\mathcal{G}(\mathsf{K}_0), \mathbb{T}(\Gamma_\mathsf{K})) & \longrightarrow & 0
\end{array} \tag{6.26}$$

Proof. Commutativity of the left square is clear. To see that the right square commutes, we use the cohomological description of χ in Prop. 6.23. We have the compatible exact sequences of discrete modules for the profinite groups $\mathcal{G}(\mathsf{F}_0)$ and $\mathcal{G}(\mathsf{K}_0)$:

$$\begin{array}{ccccccccc}
0 & \longrightarrow & \mathsf{F}^\times_{in,0} & \longrightarrow & \mathsf{F}^\times_{in} & \longrightarrow & \Gamma_\mathsf{F} & \longrightarrow & 0 \\
 & & \downarrow & & \downarrow & & \downarrow & & \\
0 & \longrightarrow & \mathsf{K}^\times_{in,0} & \longrightarrow & \mathsf{K}_{in} & \longrightarrow & \Gamma_\mathsf{K} & \longrightarrow & 0
\end{array}$$

The functorial properties of Galois cohomology thus yield the commutative diagram:

$$\begin{array}{ccccccc}
\mathsf{Br}_{is}(\mathsf{F}) & \xleftarrow{\Delta_{\mathsf{F}_{in}/\mathsf{F}}} & H^2(\mathcal{G}(\mathsf{F}_0), \mathsf{F}^\times_{in}) & \xrightarrow{deg} & H^2(\mathcal{G}(\mathsf{F}_0), \Gamma_\mathsf{F}) & \xleftarrow{\ \partial\ } & Hom^c(\mathcal{G}(\mathsf{F}_0), \mathbb{T}(\Gamma_\mathsf{F})) \\
\big\downarrow {\scriptstyle ext_{\mathsf{K}/\mathsf{F}}} & & \big\downarrow {\scriptstyle res} & & \big\downarrow {\scriptstyle res} & & \big\downarrow {\scriptstyle \pi^*_{\mathsf{K}_0/\mathsf{F}_0}} \\
\mathsf{Br}_{is}(\mathsf{K}) & \xleftarrow{\Delta_{\mathsf{K}_{in}/\mathsf{K}}} & H^2(\mathcal{G}(\mathsf{K}_0), \mathsf{K}^\times_{in}) & \xrightarrow{deg} & H^2(\mathcal{G}(\mathsf{K}_0), \Gamma_\mathsf{K}) & \xleftarrow{\ \partial\ } & Hom^c(\mathcal{G}(\mathsf{K}_0), \mathbb{T}(\Gamma_\mathsf{K}))
\end{array}$$

By Prop. 6.23, the composition $\partial^{-1} \circ deg \circ \Delta^{-1}_{\mathsf{F}_{in}/\mathsf{F}}$ is the ramification map χ defined on $\mathsf{Br}_{is}(\mathsf{F})$, and likewise $\partial^{-1} \circ deg \circ \Delta^{-1}_{\mathsf{K}_{in}/\mathsf{K}}$ is the ramification map on $\mathsf{Br}_{is}(\mathsf{K})$. Therefore, the outer commutative diagram is the right square of (6.26). □

Corollary 6.26. *Let* T *be a totally ramified algebraic graded field extension of* F*. Then, the ramification map* χ *of* (6.24) *restricts to an isomorphism*

$$\mathsf{Br}_{is}(\mathsf{F}) \cap \mathsf{Br}(\mathsf{T}/\mathsf{F}) \;\cong\; \mathit{Hom}^c(\mathcal{G}(\mathsf{F}_0), \Gamma_{\mathsf{T}}/\Gamma_{\mathsf{F}}).$$

Proof. Since T is totally ramified over F, we have $\mathsf{T}_0 = \mathsf{F}_0$; hence

$$\mathsf{Br}_{in}(\mathsf{T}) \;\cong\; Br(\mathsf{T}_0) \;=\; Br(\mathsf{F}_0) \;\cong\; \mathsf{Br}_{in}(\mathsf{F}).$$

Therefore, in commutative diagram (6.26) with T replacing K in the second row, since the left vertical map is an isomorphism the Snake Lemma yields a short exact sequence of kernels of the vertical maps,

$$0 \longrightarrow 0 \longrightarrow \mathsf{Br}_{is}(\mathsf{F}) \cap \mathsf{Br}(\mathsf{T}/\mathsf{F}) \longrightarrow \mathit{Hom}^c(\mathcal{G}(\mathsf{F}_0), \Gamma_{\mathsf{T}}/\Gamma_{\mathsf{F}}) \longrightarrow 0.$$

This yields the asserted isomorphism. □

Proposition 6.27. *For any graded field* F*, the exact sequence of Th. 6.24 is split exact. Thus, there is a non-canonical direct product decomposition*

$$\mathsf{Br}_{is}(\mathsf{F}) \;\cong\; \mathsf{Br}_{in}(\mathsf{F}) \times \mathit{Hom}^c\big(\mathcal{G}(\mathsf{F}_0), \mathbb{T}(\Gamma_{\mathsf{F}})\big).$$

Proof. Let T be a maximal totally ramified algebraic graded field extension of F in some graded algebraic closure of F. (Such a T exists by Zorn's Lemma.) We claim that $\Gamma_{\mathsf{T}} = \mathbb{H}(\Gamma_{\mathsf{F}})$. If not, there is a prime number p and a $\gamma \in \Gamma_{\mathsf{T}}$ with $\frac{1}{p}\gamma \notin \Gamma_{\mathsf{T}}$. Take any $b \in \mathsf{T}_\gamma \setminus \{0\}$, and let $\mathsf{L} = \mathsf{T}(\sqrt[p]{b})$, which is a graded field extension of F in F_{alg}. Since $\frac{1}{p}\gamma \in \Gamma_{\mathsf{L}}$, we have $|\Gamma_{\mathsf{L}}{:}\Gamma_{\mathsf{F}}| = [\mathsf{L}{:}\mathsf{F}] = p$, so L is totally ramified over T. Then, $\mathsf{L}_0 = \mathsf{T}_0 = \mathsf{F}_0$, showing that L is totally ramified over F. This contradicts the maximality of T, proving the claim. Hence, $\Gamma_{\mathsf{T}}/\Gamma_{\mathsf{F}} = \mathbb{T}(\Gamma_{\mathsf{F}})$. Therefore, the isomorphism

$$\mathsf{Br}_{is}(\mathsf{F}) \cap \mathsf{Br}(\mathsf{T}/\mathsf{F}) \xrightarrow{\;\sim\;} \mathit{Hom}^c(\mathcal{G}(\mathsf{F}_0), \mathbb{T}(\Gamma_{\mathsf{F}}))$$

given by Cor. 6.26 has an inverse map sending $\mathit{Hom}^c(\mathcal{G}(\mathsf{F}_0), \mathbb{T}(\Gamma_{\mathsf{F}}))$ into $\mathsf{Br}_{is}(\mathsf{F})$ which provides a splitting map for χ. □

6.2 The Brauer group of a valued field

The canonical filtration of the Brauer group of a graded field discussed in §6.1.3 above is a major feature that we aim to reproduce for valued fields. The correspondence between central simple algebras over a valued field and central simple graded algebras over its graded field is obtained by means of gauges. Special types of gauges are defined for this purpose; their definition is given in the more general context of semisimple algebras over valued fields. The inertial gauges discussed in §6.2.1 are those that do not allow any ramification, while the tame gauges of §6.2.2 exclude only ramification that may be problematic in relation to the residue characteristic. The collections of central simple algebras carrying these special types of gauges define subgroups of the Brauer group, which are related to corresponding subgroups of the Brauer group of the associated graded field in the final §6.2.3.

Throughout this section, F is a valued field with valuation v. We write simply $\mathsf{gr}(F)$ for $\mathsf{gr}_v(F)$.

6.2.1 Inertial gauges

Inertial gauges defined in this subsection give a common rubric for inertial extensions of valued fields and inertial valued division algebras. After the general definition, which applies to any semisimple (finite-dimensional) algebra, we consider in turn various special cases: simple algebras, field extensions, and central simple algebras. For central division algebras that carry a valuation, inertial gauges lead to the definition of inertial valuations.

For any v-gauge α on a semisimple F-algebra A, the $\mathsf{gr}(F)$-algebra $\mathsf{gr}(A)$ is graded semisimple by definition of a gauge, but the number of its graded simple components may be larger than the number of simple components of A; see Prop. 3.36.

Definition 6.28. A v-gauge α on a semisimple F-algebra A is said to be *inertial* (or, more precisely, *inertial over* v) if the semisimple graded algebra $\mathsf{gr}_\alpha(A)$ is inertial over $\mathsf{gr}(F)$ (see Def. 2.50) and the number of graded simple components of $\mathsf{gr}_\alpha(A)$ equals the number of simple components of A.

If $A = A_1 \times \ldots \times A_n$ is the decomposition of A into simple components, then by Prop. 3.36 every v-gauge α on A has the form $\alpha = \alpha_1 \times \ldots \times \alpha_n$ for some v-gauges $\alpha_1, \ldots, \alpha_n$ on $A_1, \ldots, A_n$ respectively, and

$$\mathsf{gr}_\alpha(A) = \mathsf{gr}_{\alpha_1}(A_1) \times \ldots \times \mathsf{gr}_{\alpha_n}(A_n).$$

It follows that α is inertial if and only if each α_i is inertial. To make the definition of inertial v-gauges more explicit, it therefore suffices to consider simple algebras.

Proposition 6.29. *A v-gauge α on a simple F-algebra A is inertial if and only if it is residually simple (i.e., A_0 is simple), $Z(A_0)$ is a separable field extension of $\overline{F}$, and any of the following equivalent conditions holds:*

(a) $\mathsf{gr}_\alpha(A) = A_0 \otimes_{\overline{F}} \mathsf{gr}(F)$;
(b) $[A_0{:}\overline{F}] = [A{:}F]$;
(c) $\Gamma_A = \Gamma_F$.

Proof. The equivalence of (a), (b), (c) follows from Lemma 2.49 applied to $\mathsf{gr}_\alpha(A)$, since the hypothesis that α is a norm implies $[A{:}F] = [\mathsf{gr}_\alpha(A){:}\mathsf{gr}(F)]$. Any of the conditions (a), (b), (c) together with the separability of $Z(A_0)$ over $\overline{F}$ is equivalent by definition to $\mathsf{gr}_\alpha(A)$ being inertial. Moreover, when

(a) holds, A_0 is simple if and only if $\mathsf{gr}_\alpha(A)$ is simple: see Prop. 2.16 and Prop. 2.47. The proposition follows. □

We next consider the case of field extensions. Recall that if K is a field containing F with $[K{:}F] < \infty$ then a valuation v_K on K extending v is said to be *inertial over* v if $[\overline{K}{:}\overline{F}] = [K{:}F]$ and $\overline{K}$ is separable over $\overline{F}$ (see Def. A.4).

Proposition 6.30. *Let K be a finite-degree field extension of F. Then, K has an inertial v-gauge α if and only if there is a valuation v_K of K that is inertial over v. When this occurs, v_K is the only valuation on K extending v, $\alpha = v_K$, and K is separable over F.*

Proof. Suppose α is an inertial v-gauge on K. As K_0^α is simple, Cor. 4.37 and Th. 4.36 show that α is a valuation on K, and is the unique extension of v to K. Hence $\overline{K}^\alpha = K_0^\alpha$, which is separable over $F_0 = \overline{F}$, and, as α is a v-norm,

$$[\overline{K}{:}\overline{F}] = [K_0{:}F_0] = [\mathsf{gr}_\alpha(K){:}\,\mathsf{gr}(F)] = [K{:}F].$$

Thus, the valuation α is inertial over v.

Conversely, if v_K is inertial over v, then v is defectless in K and v_K is the unique extension of v to K. Therefore, Cor. 4.37 shows that v_K is the unique v-gauge on K. Moreover, $\overline{K}$ is a separable field extension of $\overline{F}$ with $[\overline{K}{:}\overline{F}] = [K{:}F]$, hence condition (b) of Prop. 6.29 holds for $A = K$ and v_K is therefore an inertial v-gauge.

It remains to see that K is a separable extension of F when K carries an inertial valuation v_K extending v. Let S be the separable closure of F in K, and let v_S be the restriction of v_K to S. Then, $\overline{K}$ is purely inseparable over $\overline{S}$ as K is purely inseparable over S; but $\overline{K}$ is also separable over $\overline{S}$ since it is separable over $\overline{F}$, hence $\overline{S} = \overline{K}$. So,

$$[S{:}F] \geq [\overline{S}{:}\overline{F}] = [\overline{K}{:}\overline{F}] = [K{:}F],$$

implying that $K = S$, which is separable over F. □

The following result reduces the discussion of inertial gauges on simple algebras to the case of central simple algebras.

Proposition 6.31. *Let A be a simple F-algebra and let $K = Z(A)$.*

(i) *If a v-gauge α on A is inertial, then its restriction to K is an inertial valuation v_K extending v, and α is an inertial v_K-gauge.*
(ii) *Let v_K be a valuation on K extending v. If v_K is inertial over v, then every inertial v_K-gauge on A is an inertial v-gauge.*

Proof. (i) Since α is a v-gauge its restriction v_K is a surmultiplicative v-norm on K. Moreover, as α is inertial, we have $\Gamma_A = \Gamma_F$, hence also $\Gamma_K = \Gamma_F$. Now, $\mathsf{gr}_{v_K}(K) \subseteq Z(\mathsf{gr}_\alpha(A))$, hence $K_0 \subseteq Z(A_0)$, which is a field separable over $\overline{F}$; so K_0 is also a field separable over $\overline{F}$. Therefore, v_K is an inertial v-gauge on K. By Prop. 6.30, v_K is a valuation on K that is inertial over v, and v_K is the unique extension of v to K. Therefore, by Prop. 4.41 the v-gauge α is also a v_K-gauge. Moreover, the simple graded algebra $\mathsf{gr}_\alpha(A)$ is clearly inertial over $\mathsf{gr}(K)$ since it is inertial over $\mathsf{gr}(F)$. Hence, α is an inertial v_K-gauge.

(ii) Suppose v_K is inertial over v and α is an inertial v_K-gauge on A. It is therefore also a v-gauge, and A_0 is simple. Since $\mathsf{gr}_\alpha(A) = A_0 \otimes_{\overline{K}} \mathsf{gr}(K)$ and $\mathsf{gr}(K) = \overline{K} \otimes_{\overline{F}} \mathsf{gr}(F)$, we have

$$\mathsf{gr}_\alpha(A) = A_0 \otimes_{\overline{F}} \mathsf{gr}(F).$$

Moreover, $Z(A_0)$ is separable over $\overline{K}$ and $\overline{K}$ is separable over $\overline{F}$, hence $Z(A_0)$ is separable over $\overline{F}$. Therefore, α is an inertial v-gauge. □

For the rest of this subsection, we focus on *central* simple algebras, for which we have the following refinement of Prop. 6.29:

Proposition 6.32. *A v-gauge on a central simple F-algebra A is inertial over v if and only if A_0 is central simple over $\overline{F}$ and any of the following equivalent conditions holds:*

(a) $\mathsf{gr}_\alpha(A) = A_0 \otimes_{\overline{F}} \mathsf{gr}(F)$;
(b) $[A_0{:}\overline{F}] = [A{:}F]$;
(c) $\Gamma_A = \Gamma_F$.

Proof. In view of Prop. 6.29, it suffices to prove that if α is an inertial v-gauge, then $Z(A_0) = \overline{F}$. Let $\mathsf{Z} = Z(\mathsf{gr}_\alpha(A))$. Since $\mathsf{gr}(A) = A_0 \otimes_{\overline{F}} \mathsf{gr}(F)$, we have $\mathsf{Z} = Z(A_0) \otimes_{\overline{F}} \mathsf{gr}(F)$ by Cor. 2.31; hence, $\mathsf{Z}_0 = Z(A_0)$. By Cor. 4.32, Z is a graded field purely inseparable over $\mathsf{gr}(F)$. So, Z_0 is purely inseparable over $F_0 = \overline{F}$. But Z_0 is also separable over $\overline{F}$, since α is an inertial v-gauge. Hence, $\mathsf{Z}_0 = \overline{F}$. Thus, A_0 is central over $\overline{F}$. □

We single out the case of valuations on central division algebras: if D is a central division algebra over F and v_D is a valuation on D extending v, we say v_D is *inertial* (*over* v) if v_D is an inertial v-gauge on D.

Corollary 6.33. *Let v_D be a valuation on a central division algebra D over F with $v_D|_F = v$. The following conditions are equivalent:*

(a) *v_D is inertial;*
(b) *$[\overline{D}{:}\overline{F}] = [D{:}F]$ and $Z(\overline{D})$ is separable over $\overline{F}$;*
(c) *$[\overline{D}{:}\overline{F}] = [D{:}F]$ and $Z(\overline{D}) = \overline{F}$.*

In particular, if $\operatorname{char} \overline{F} \nmid \deg D$, then the valuation v_D is inertial if and only if $[\overline{D}{:}\overline{F}] = [D{:}F]$. (See Ex. 6.36(ii) *for an example where this equality holds for a valuation v_D that is* not *inertial.)*

Proof. The equivalence (a) $\Leftrightarrow$ (c) readily follows from Prop. 6.29, and (c) $\Rightarrow$ (b) is clear. Therefore, it suffices to show (b)$\Rightarrow$(c). When $[\overline{D}{:}\overline{F}] = [D{:}F]$, the Fundamental Inequality (Prop. 1.3) shows that $\Gamma_D = \Gamma_F$, hence it follows from Prop. 1.5 that $Z(\overline{D})$ is the purely inseparable closure of $\overline{F}$ in $Z(\overline{D})$. We thus have $Z(\overline{D}) = \overline{F}$ if $Z(\overline{D})$ is separable over $\overline{F}$. □

A valuation on a division algebra Brauer-equivalent to an algebra with an inertial gauge is inertial:

Proposition 6.34. *Let v_D be a valuation on a central division algebra D over F with $v_D|_F = v$. If some $M_n(D)$ has an inertial v-gauge, then the valuation v_D is inertial over v.*

Proof. Let (F_h, v_h) be the Henselization of (F, v), and let $D_h = D \otimes_F F_h$. By Th. 4.1, D_h is a division algebra and for the valuation $v_{D,h}$ on D_h extending v_h we have $\mathsf{gr}_{v_{D,h}}(D_h) \cong_g \mathsf{gr}_{v_D}(D)$. Let α be an inertial v-gauge on $A = M_n(D)$, and let $\alpha_h = \alpha \otimes v_h$ be the induced inertial v_h-gauge on $A_h = A \otimes_F F_h$. We have

$$\mathsf{gr}_{\alpha_h}(A_h) \cong_g \mathsf{gr}_\alpha(A) \otimes_{\mathsf{gr}_v(F)} \mathsf{gr}_{v_h}(F_h) \cong_g \mathsf{gr}_\alpha(A).$$

Write $A_h = End_{D_h}(M)$ for some D_h-vector space M. By Th. 4.26, $v_{D,h}$ is defectless over v_h and there is a $v_{D,h}$-norm η on M such that $\alpha = End(\eta)$ and $\mathsf{gr}_{\alpha_h}(A_h) \cong_g \mathsf{End}_{\mathsf{gr}_{v_{D,h}}(D_h)}(\mathsf{gr}_\eta(M))$. The valuation v_D is defectless over v since $v_{D,h}$ is defectless over v_h; hence, v_D is a v-gauge. Since $\mathsf{gr}_{\alpha_h}(A_h)$ is an inertial simple graded $\mathsf{gr}_{v_h}(F_h)$-algebra, by Prop. 2.52 its associated graded division ring $\mathsf{gr}_{v_{D,h}}(D_h)$ is also an inertial $\mathsf{gr}_{v_h}(F_h)$-algebra. Hence, $\mathsf{gr}_{v_D}(D) = \mathsf{gr}_{v_{D,h}}(D_h)$ is an inertial $\mathsf{gr}_v(F) = \mathsf{gr}_{v_h}(F_h)$-algebra. Since $\mathsf{gr}_v(D)_0$ is the division ring $\overline{D}$, the v-gauge v_D is therefore inertial. Hence, the valuation v_D is inertial over v. □

Remark 6.35. There is a stronger result than the preceding proposition, provable using the theory of Dubrovin valuation rings: Let F be a field with valuation v, and let D be a central division algebra over F. If $A = M_n(D)$ has an inertial v-gauge α, then D also has an inertial v-gauge β. For since the inertial v-gauge α is residually simple, its gauge ring $\mathcal{O}_{A,\alpha}$ is a Dubrovin valuation ring of $M_n(D)$ by Tignol–Wadsworth [246, Prop. 2.5], and $\mathcal{O}_{A,\alpha}$ is integral over $\mathcal{O}_{F,v}$ by Lemma 4.24. By Marubayashi et al. [132, Th. 9.4, Prop. 5.14(1), Th. 9.8] there is a Dubrovin valuation ring R of D, and $M_n(R) \cong \mathcal{O}_{A,\alpha}$; this R is integral over $\mathcal{O}_{F,v}$ since $M_n(R)$ is. Therefore, by [132, Th. 12.3, Th. 23.2] there is a Morandi v-value function β_R on D determined by R whose associated ring $\mathcal{O}_{\beta_R}$ is R. Since $\mathcal{O}_{A,\alpha} \cong M_n(\mathcal{O}_{\beta_R})$ and α is a v-gauge it follows easily that β_R is defectless over v and hence a v-gauge; from the ring isomorphism and the fact that α is an inertial v-gauge it follows that β_R is an inertial v-gauge on D, as needed. This yields Prop. 6.34, since Prop. 4.33 shows that if D has a valuation v_D extending v, then $v_D = \beta_R$. Also, given the inertial v-gauge β_R on D it is easy to build an inertial v-gauge on any matrix algebra $M_m(D)$.

We now give some examples of inertial gauges on central simple algebras. We consider first symbol algebras, then crossed products.

Example 6.36. (*Symbol algebras*) Suppose F contains a primitive n-th root of unity ω, for some integer $n > 1$. Let a, $b \in F$ with $v(a) = v(b) = 0$, and consider the symbol algebra

$$S = \big(a, b/F\big)_n$$

with generators i, j satisfying

$$i^n = a, \qquad j^n = b, \qquad ij = \omega ji.$$

Define a v-norm α on S by

$$\alpha\Big(\sum_{k=0}^{n-1}\sum_{\ell=0}^{n-1} c_{k\ell}\, i^k j^\ell\Big) = \min_{k,\,\ell}\big(v(c_{k\ell})\big) \qquad \text{for all } c_{k\ell} \in F.$$

Since $v(a) = v(b) = 0$, this norm is the one defined in §3.2.3. As pointed out there, the graded algebra $\mathsf{gr}_\alpha(S)$ has the homogeneous $\mathsf{gr}(F)$-base $(\widetilde{i}^k\widetilde{j}^\ell)_{k,\ell=0}^{n-1}$, and $\widetilde{i}$, $\widetilde{j}$ satisfy

$$\widetilde{i}^n = \overline{a}, \qquad \widetilde{j}^n = \overline{b}, \quad \text{and} \quad \widetilde{i}\,\widetilde{j} = \overline{\omega}\,\widetilde{j}\,\widetilde{i}.$$

We divide the rest of the discussion into two cases, depending on the residue characteristic.

(i) If $\mathit{char}\,\overline{F} \nmid n$, then the polynomial $X^n - 1$ is separable in $\overline{F}[X]$, hence the powers $\overline{\omega}^k$ for $k = 0, \ldots, n-1$ are all distinct. Therefore, $\overline{\omega}$ is a primitive n-th root of unity in $\overline{F}$, and $\mathsf{gr}_\alpha(S)$ is the graded symbol algebra

$$\mathsf{gr}_\alpha(S) = \big(\overline{a}, \overline{b}/\,\mathsf{gr}(F)\big)_n.$$

This algebra is graded simple and inertial over $\mathsf{gr}(F)$ (see Ex. 2.53), hence α is an inertial v-gauge on S.

(ii) By contrast, if $\mathit{char}\,F = 0$ and $\mathit{char}\,\overline{F} = \overline{p} \neq 0$, and n is a power of $\overline{p}$, then $\overline{\omega} = 1$ and $\mathsf{gr}_\alpha(S)$ is commutative. Indeed,

$$\mathsf{gr}_\alpha(S) = S_0 \otimes_{\overline{F}} \mathsf{gr}(F) \quad \text{with} \quad S_0 \cong \overline{F}[X,Y]/(X^n - \overline{a}, Y^n - \overline{b}).$$

If $[\overline{F}(\sqrt[n]{\overline{a}}, \sqrt[n]{\overline{b}}):\overline{F}] = n^2$, then S_0 is the field $\overline{F}(\sqrt[n]{\overline{a}}, \sqrt[n]{\overline{b}})$ which is purely inseparable over $\overline{F}$. The map α is a v-gauge on S, and also a valuation, see Prop. 3.39, but it is not inertial since $Z(S_0)$ is not separable over $\overline{F}$. If $[\overline{F}(\sqrt[n]{\overline{a}}, \sqrt[n]{\overline{b}}):\overline{F}] < n^2$ (e.g., if $\overline{a}$ or $\overline{b}$ lies in $\overline{F}^p$), then the local ring S_0 has a nontrivial nilradical, so $\mathsf{gr}_\alpha(S)$ is not graded semisimple, and α is not a gauge.

Example 6.37. (*Inertial crossed product algebras*) We show how to build crossed product algebras with inertial gauges by using inertial field extensions. Let (L, v_L) be a finite-degree Galois inertial valued field extension of (F, v), and let $G = \mathcal{G}(L/F)$. By Cor. 5.53, $\overline{L}$ is Galois over $\overline{F}$, and $\mathsf{gr}(L)$ is Galois over $\mathsf{gr}(F)$. We identify G with the canonically isomorphic Galois groups $\mathcal{G}(\overline{L}/\overline{F})$ and $\mathcal{G}\big(\mathsf{gr}(L)/\,\mathsf{gr}(F)\big)$. Because v_L is the only extension of v to L, every $\sigma \in G$ maps $\mathcal{O}_L^\times$ to itself. Take any normalized 2-cocycle $f \in Z^2(G, \mathcal{O}_L^\times)$, and let B be the central simple crossed product F-algebra

$$B = (L/F, G, f) = \bigoplus_{\sigma \in G} L x_\sigma,$$

where

$$(cx_\sigma)(dx_\tau) = c\sigma(d)f(\sigma,\tau)\,x_{\sigma\tau} \quad \text{for all} \quad c,d \in L,\ \sigma,\tau \in G.$$

Define a v_L-norm β on B by

$$\beta\Big(\sum_{\sigma\in G} c_\sigma x_\sigma\Big) = \min_{\sigma\in G}\big(v_L(c_\sigma)\big).$$

Then, as v_L is a v-norm on L, β is also a v-norm on B. We will show that β is an inertial v-gauge on B and that $\mathsf{gr}_\beta(B)$ is a graded crossed product algebra.

The definition of β shows that $(x_\sigma)_{\sigma\in G}$ is a splitting base for the v_L-norm β on B as a left L-vector space. So,

$$\mathsf{gr}_\beta(B) = \bigoplus_{\sigma\in G} \mathsf{gr}(L)\,\widetilde{x_\sigma} \qquad \text{with each } \widetilde{x_\sigma} \in B_0,$$

and for any $c \in L$, $\sigma \in G$, $\widetilde{c}\,\widetilde{x_\sigma} = \widetilde{cx_\sigma}$ in $\mathsf{gr}_\beta(B)$. Thus, for all c, $d \in L$ and σ, $\tau \in G$, as $v_L \circ \sigma = v_L$ we have

$$\begin{aligned}\beta\big((cx_\sigma)(dx_\tau)\big) &= \beta\big(c\sigma(d)f(\sigma,\tau)x_{\sigma\tau}\big) = v_L(c) + v_L(d)\\ &= \beta(cx_\sigma) + \beta(dx_\tau).\end{aligned} \tag{6.27}$$

Hence,

$$\begin{aligned}(\widetilde{c}\,\widetilde{x_\sigma})(\widetilde{d}\,\widetilde{x_\tau}) &= \widetilde{cx_\sigma}\;\widetilde{dx_\tau} = (cx_\sigma dx_\tau)^{\sim} = \big(c\sigma(d)f(\sigma,\tau)x_{\sigma\tau}\big)^{\sim}\\ &= \big(c\sigma(d)f(\sigma,\tau)\big)^{\sim}\,\widetilde{x_{\sigma\tau}} = \widetilde{c}\,\widetilde{\sigma(d)}\,\widetilde{f(\sigma,\tau)}\,\widetilde{x_{\sigma\tau}}.\end{aligned} \tag{6.28}$$

Let $(a_i)_{i=1}^n$ be any splitting base of L for the v-norm v_L. Then, from the definition of β it is clear that $\big(a_i x_\sigma\big)_{i=1,\,\sigma\in G}^{n}$ is a splitting base of B for β as a v-norm. Since $\beta\big((a_ix_\sigma)(a_jx_\tau)\big) = \beta(a_ix_\sigma) + \beta(a_jx_\tau)$ by (6.27), Lemma 3.23 shows that β is surmultiplicative. Let $\widetilde{f} \in Z^2(G,\overline{L}^\times)$ be defined by $\widetilde{f}(\sigma,\tau) = \overline{f(\sigma,\tau)}$. Formula (6.28) is additive in $\widetilde{c}$ and in $\widetilde{d}$, so it still holds when $\widetilde{c},\widetilde{d}$ are replaced by arbitrary elements of $\mathsf{gr}(L)$. This shows that $\mathsf{gr}_\beta(B)$ is the graded crossed product algebra $\big(\mathsf{gr}(L)/\mathsf{gr}(F), G, \widetilde{f}\,\big)$, which is a central simple graded $\mathsf{gr}(F)$-algebra, cf. §6.1.2. Hence, β is a v-gauge. Moreover, since each $\widetilde{x_\sigma} \in B_0$ we have $B_0 = \bigoplus_{\sigma\in G} \overline{L}\,\widetilde{x_\sigma}$, and formula (6.28) shows that B_0 is the central simple crossed product $\overline{F}$-algebra $(\overline{L}/\overline{F}, G, \widetilde{f})$. Since $[B_0{:}\overline{F}] = [\overline{L}{:}\overline{F}]^2 = [L{:}F]^2 = [\mathsf{gr}_\beta(B){:}\,\mathsf{gr}(F)]$ and $Z(B_0) = \overline{F}$, $\mathsf{gr}_\beta(B)$ is an inertial graded $\mathsf{gr}(F)$-algebra. Hence, β is an inertial v-norm on B.

Further examples of inertial gauges on central simple algebras can be obtained by tensor products and scalar extensions, as we now show.

Proposition 6.38. *Let A be a central simple F-algebra with an inertial v-gauge α, and let B be a simple F-algebra with a v-gauge β. Let $C = A \otimes_F B$, which is a simple F-algebra. Then,*

(i) *$\alpha \otimes \beta$ is a v-gauge on C with $C_0 = A_0 \otimes_{\overline{F}} B_0$, $\Gamma_C = \Gamma_B$, and*

$$\mathsf{gr}_{\alpha\otimes\beta}(C) \cong_g A_0 \otimes_{\overline{F}} \mathsf{gr}_\beta(B).$$

(ii) *$\alpha \otimes \beta$ is an inertial v-gauge if and only if β is inertial.*

Proof. Since α is inertial, Prop. 6.32 shows that $\mathsf{gr}_\alpha(A) = A_0 \otimes_{\overline{F}} \mathsf{gr}(F)$ and A_0 is central simple over $\overline{F}$. Therefore, $\mathsf{gr}_\alpha(A)$ is a central simple graded $\mathsf{gr}(F)$-algebra by Prop. 2.16. It follows from Prop. 3.41 that $\alpha \otimes \beta$ is a v-gauge on C, and from Prop. 3.27 that

$$\mathsf{gr}_{\alpha\otimes\beta}(C) = \mathsf{gr}_\alpha(A) \otimes_{\mathsf{gr}(F)} \mathsf{gr}_\beta(B) = A_0 \otimes_{\overline{F}} \mathsf{gr}_\beta(B).$$

Thus, $C_0 = A_0 \otimes_{\overline{F}} B_0$ and $\Gamma_C = \Gamma_B$, proving (i).

Since $Z(A_0) = \overline{F}$, we also see that C_0 is simple with $Z(C_0) = Z(B_0)$; moreover, $\Gamma_C = \Gamma_F$ if and only if $\Gamma_B = \Gamma_F$, hence Prop. 6.29 shows that $\alpha \otimes \beta$ is inertial if and only if β is inertial. □

Proposition 6.39. *Let A be a central simple F-algebra with an inertial v-gauge α, and let (L, v_L) be any valued field extension of (F, v). Then $\alpha\otimes v_L$ is an inertial v_L-gauge on $A \otimes_F L$ with $(A \otimes_F L)_0 = A_0 \otimes_{\overline{F}} \overline{L}$.*

Proof. Let $B = A \otimes_F L$ and $\beta = \alpha \otimes v_L$. As in the first lines of the proof of Prop. 6.38, we have $\mathsf{gr}_\alpha(A)$ is a central simple graded $\mathsf{gr}(F)$-algebra, hence β is a v_L-gauge on B by Prop. 3.42, and

$$\mathsf{gr}_\beta(B) = \mathsf{gr}_\alpha(A) \otimes_{\mathsf{gr}(F)} \mathsf{gr}(L) = A_0 \otimes_{\overline{F}} \mathsf{gr}(L).$$

It follows that $\Gamma_B = \Gamma_L$ and $B_0 = A_0 \otimes_{\overline{F}} \overline{L}$, which is a central simple $\overline{L}$-algebra since A_0 is a central simple $\overline{F}$-algebra. Therefore, β is an inertial v_L-gauge by Prop. 6.32. □

6.2.2 Tame gauges

Throughout this subsection, we develop the idea of a tame gauge, which generalizes both tamely ramified field extensions and tame valued division algebras. We will see that if *char* $\overline{F} = 0$ then all v-gauges are tame, see Cor. 6.54. When *char* $\overline{F} \neq 0$, tame gauges form a large class which avoid the pathologies that sometimes occur with prime characteristic. Just as for inertial gauges, we give a general definition that applies to any semisimple (finite-dimensional) algebra, and then specialize to simple algebras, field extensions, and central simple algebras.

For any v-gauge α on a semisimple F-algebra A, recall from Remark 3.33 that $\alpha|_{Z(A)}$ is a v-gauge on $Z(A)$.

Definition 6.40. A v-gauge α on a semisimple (finite-dimensional) F-algebra A is said to be *tame* (or, more precisely, *tame over* v) if the following conditions hold:

(i) $Z(\mathsf{gr}_\alpha(A)) = \mathsf{gr}_{\alpha|_{Z(A)}}(Z(A))$; and

(ii) $\mathsf{gr}_{\alpha|_{Z(A)}}(Z(A))$ is a separable graded algebra over $\mathsf{gr}(F)$, i.e., a direct product of tame graded field extensions of $\mathsf{gr}(F)$.

Since condition (ii) only involves $Z(A)$, and since condition (i) always holds if A is commutative, condition (ii) is equivalent to: $\alpha|_{Z(A)}$ is a tame v-gauge on $Z(A)$.

Note that Ex. 6.36(ii) and Ex. 8.37 below yield examples of non-tame gauges in every characteristic.

If $A = A_1 \times \ldots \times A_n$ is the decomposition of A into simple components, then every v-gauge α on A has the form $\alpha = min(\alpha_1, \ldots, \alpha_n)$ for some v-gauges $\alpha_1, \ldots, \alpha_n$ on $A_1, \ldots, A_n$ respectively (see Prop. 3.36), and we have

$$Z(\mathsf{gr}_\alpha(A)) = Z(\mathsf{gr}_{\alpha_1}(A_1)) \times \ldots \times Z(\mathsf{gr}_{\alpha_n}(A_n))$$

and

$$\mathsf{gr}_{\alpha|_{Z(A)}}(Z(A)) = \mathsf{gr}_{\alpha_1|_{Z(A_1)}}(Z(A_1)) \times \ldots \times \mathsf{gr}_{\alpha_n|_{Z(A_n)}}(Z(A_n)).$$

Therefore, α is tame over v if and only if each $\alpha_1, \ldots, \alpha_n$ is tame. The discussion of tame gauges is thus reduced to the case of simple algebras.

We next characterize tame gauges on (finite-degree) field extensions of F.

Proposition 6.41. *Let K be a finite-degree field extension of F, and let $v_1, \ldots, v_n$ be the various extensions of v to K. Then, K has a tame v-gauge α if and only if v is defectless in K and the following conditions hold for $i = 1, \ldots, n$:*

(i) *$\overline{K}^{v_i}$ is separable over $\overline{F}$;*

(ii) *$char\,\overline{F} \nmid |\Gamma_{K,v_i} : \Gamma_F|$.*

When this occurs, $\alpha = min(v_1, \ldots, v_n)$ and K is separable over F. In particular, if v extends to a unique valuation v_K on K, then the valuation v_K is a tame v-gauge on K if and only if K is tamely ramified over F as in Def. A.4.

Proof. Corollary 4.37 shows that K has a v-gauge if and only if v is defectless in K, and that this v-gauge is then $\alpha = min(v_1, \ldots, v_n)$. For this α, we have

$$\mathsf{gr}_\alpha(K) \cong_g \mathsf{gr}_{v_1}(K) \times \ldots \times \mathsf{gr}_{v_n}(K).$$

Therefore, the v-gauge α is tame if and only if each $\mathsf{gr}_{v_i}(K)$ is a tame graded field extension of $\mathsf{gr}(F)$, which occurs if and only if (i) and (ii) hold.

It remains to show that K is separable over F when α is a tame v-gauge. Suppose $char\,\overline{F} = \overline{p} \neq 0$, and let S be the separable closure of F in K. Since K is purely inseparable over S, every valuation on S extends uniquely to K. Therefore, the restrictions of $v_1, \ldots, v_n$ to S are all distinct; they are the various valuations on S extending v. Since K/S is purely inseparable,

$\overline{K}^{v_i}/\overline{S}^{v_i}$ is purely inseparable and $|\Gamma_{K,v_i}:\Gamma_{S,v_i}|$ is a power of $\overline{p}$ for $i = 1, \ldots, n$. Therefore, (i) and (ii) imply

$$\overline{K}^{v_i} = \overline{S}^{v_i} \qquad \text{and} \qquad \Gamma_{K,v_i} = \Gamma_{S,v_i} \qquad \text{for } i = 1, \ldots, n.$$

Since v is defectless in K and the Fundamental Inequality (A.1) holds for S/F, we have

$$[K:F] = \sum_{i=1}^{n} [\overline{K}^{v_i}:\overline{F}] \cdot |\Gamma_{K,v_i}:\Gamma_F| = \sum_{i=1}^{n} [\overline{S}^{v_i}:\overline{F}] \cdot |\Gamma_{S,v_i}:\Gamma_F| \leq [S:F].$$

Since $S \subseteq K$, it follows that $S = K$, hence K is separable over F. □

We next consider simple algebras. Let A be a simple F-algebra, and let $K = Z(A)$. Suppose α is a v-gauge on A. Then $\alpha|_K$ is a v-gauge on K, see Remark 3.33, hence v is defectless in K and $\alpha|_K = min(v_1, \ldots, v_n)$ with $v_1, \ldots, v_n$ the various valuations on K that extend v, see Cor. 4.37. For $i = 1, \ldots, n$ let α_i be the v_i-component of α as in Th. 4.42.

Proposition 6.42. *With the notation above, the v-gauge α on A is tame if and only if the following conditions hold:*

(i) *α_i is a tame v_i-gauge on A for each $i = 1, \ldots, n$;*
(ii) *$\alpha|_K$ is a tame v-gauge on K.*

Proof. Theorem 4.42 and Cor. 2.31 yield

$$Z(\mathsf{gr}_\alpha(A)) \cong_g Z(\mathsf{gr}_{\alpha_1}(A)) \times \ldots \times Z(\mathsf{gr}_{\alpha_n}(A)).$$

On the other hand, by Cor. 4.37 we have

$$\mathsf{gr}_{\alpha|_K}(K) \cong_g \mathsf{gr}_{v_1}(K) \times \ldots \times \mathsf{gr}_{v_n}(K).$$

Therefore, we have $Z(\mathsf{gr}_\alpha(A)) = \mathsf{gr}_{\alpha|_K}(K)$ if and only if (i) holds. Since condition (ii) in the definition of tame gauges (Def. 6.40) is equivalent to condition (ii) of the proposition, the proof is complete. □

Proposition 6.42 reduces the study of tame gauges to the case of central simple algebras, which will be the focus of our discussion for the rest of this subsection (except for the tameness criterion in Th. 6.52).

Proposition 6.43. *A v-gauge α on a central simple F-algebra A is tame if and only if $\mathsf{gr}_\alpha(A)$ is a central simple graded $\mathsf{gr}(F)$-algebra.*

Proof. Condition (ii) in the definition of a tame gauge (Def. 6.40) holds trivially when $Z(A) = F$, hence a v-gauge α on A is tame if and only if the graded $\mathsf{gr}(F)$-algebra $\mathsf{gr}_\alpha(A)$ is central. Since $\mathsf{gr}_\alpha(A)$ is graded semisimple for any v-gauge α, the latter condition holds if and only if $\mathsf{gr}_\alpha(A)$ is a central simple graded $\mathsf{gr}(F)$-algebra. □

Consider in particular the case of a central division algebra D over F. We say that a valuation v_D on D extending v is *tame* (*over* v) if v_D is a tame v-gauge on D. After the next lemma we give a characterization of tame valuations.

Lemma 6.44. *Let D be a central division algebra over F with a valuation v_D extending v on F. Let $\overline{p} = \operatorname{char} \overline{F}$ and let $\mathsf{Z} = Z(\mathsf{gr}_{v_D}(D))$. If $\overline{p} = 0$, then $\mathsf{Z} = \mathsf{gr}(F)$. If $\overline{p} \neq 0$, then Z_0 is the purely inseparable closure of $\overline{F}$ in $Z(\overline{D})$ and $\Gamma_{\mathsf{Z}}/\Gamma_F$ is the $\overline{p}$-primary component of $\ker(\theta_D)/\Gamma_F$.*

Proof. We saw in Lemma 4.31 that Z is purely inseparable over $\mathsf{gr}(F)$. If $\overline{p} = 0$, this shows that $\mathsf{Z} = \mathsf{gr}(F)$. If $\overline{p} \neq 0$, this yields that Z_0 is purely inseparable over $\overline{F}$ and $\Gamma_{\mathsf{Z}}/\Gamma_{\mathsf{gr}(F)}$ is a $\overline{p}$-torsion group. On the other hand, we have seen in Prop. 2.40 that $Z(\overline{D})$ is Galois over Z_0. Hence, Z_0 is the purely inseparable closure of $\overline{F}$ in $Z(\overline{D})$.

To complete the proof, it remains to show that $ker(\theta_D)/\Gamma_{\mathsf{Z}}$ has no $\overline{p}$-torsion. By definition of θ_D and $\theta_{\mathsf{gr}(D)}$, the following diagram commutes:

$$\begin{array}{ccc} \Gamma_{\mathsf{gr}(D)} & \xrightarrow{\theta_{\mathsf{gr}(D)}} & \mathcal{G}(Z(\overline{D})/\mathsf{Z}_0) \\ \| & & \| \\ \Gamma_D & \xrightarrow{\theta_D} & Aut(Z(\overline{D})/\overline{F}) \end{array}$$

Therefore, $ker(\theta_D) = ker(\theta_{\mathsf{gr}(D)})$. We work with $\theta_{\mathsf{gr}(D)}$. Let $\mathsf{E} = C_{\mathsf{gr}(D)}(Z(\overline{D}))$ and $\mathsf{C} = C_{\mathsf{gr}(D)}(\overline{D})$ be the centralizers of $Z(\overline{D})$ and of $\overline{D}$ respectively (or of $Z(\overline{D}) \otimes_{\mathsf{Z}_0} \mathsf{Z}$ and $\overline{D} \otimes_{\mathsf{Z}_0} \mathsf{Z}$ respectively) in $\mathsf{gr}(D)$. By the Double Centralizer Theorem (Th. 2.35) we have $Z(\mathsf{C}) = Z(\mathsf{E}) = Z(\overline{D}) \otimes_{\mathsf{Z}_0} \mathsf{Z}$ and

$$\mathsf{E} \cong_g (\overline{D} \otimes_{\mathsf{Z}_0} \mathsf{Z}) \otimes_{\mathsf{Z}} \mathsf{C} = \overline{D} \otimes_{\mathsf{Z}_0} \mathsf{C}.$$

Therefore, $\Gamma_{Z(\mathsf{C})} = \Gamma_{\mathsf{Z}}$ and $\Gamma_{\mathsf{E}} = \Gamma_{\mathsf{C}}$. By definition of $\theta_{\mathsf{gr}(D)}$, we have $ker(\theta_{\mathsf{gr}(D)}) = \Gamma_{\mathsf{E}}$. Thus, we have to show that $\Gamma_{\mathsf{C}}/\Gamma_{Z(\mathsf{C})}$ has no $\overline{p}$-torsion. From the definition of C, it is clear that $\mathsf{C}_0 = Z(\overline{D})$, hence C is totally ramified over its center. By Lemma 6.19, it follows that its $\overline{p}$-primary component $\mathsf{C}_{(\overline{p})}$ is trivial, hence $\overline{p} \nmid [\mathsf{C}{:}Z(\mathsf{C})]$, and therefore $\Gamma_{\mathsf{C}}/\Gamma_{Z(\mathsf{C})}$ has no $\overline{p}$-torsion. □

Proposition 6.45. *Let D be a central division algebra over F with a valuation v_D extending v on F. Let $\overline{p} = \operatorname{char} \overline{F}$. If $\overline{p} = 0$, then v_D is tame. If $\overline{p} \neq 0$, then v_D is tame if and only if v_D is defectless over F, $Z(\overline{D})$ is separable over $\overline{F}$, and $\ker(\theta_D)/\Gamma_F$ has no $\overline{p}$-torsion. In particular, v_D is tame whenever $\overline{p} \nmid \deg D$.*

Proof. Let $\mathsf{Z} = Z(\mathsf{gr}(D))$. Note that v_D is a v-gauge if and only if v_D is defectless over F. Since $F = Z(D)$, v_D is a tame v-gauge if and only if it is a v-gauge and $\mathsf{Z} = \mathsf{gr}(F)$. When $\overline{p} = 0$, v_D is defectless over F by Th. 4.3 and $\mathsf{Z} = \mathsf{gr}(F)$ by Lemma 6.44. Hence, v_D is tame in this case.

For the rest of the proof, assume $\overline{p} \neq 0$. We have $\mathsf{Z} = \mathsf{gr}(F)$ if and only if $\mathsf{Z}_0 = \overline{F}$ and $\Gamma_{\mathsf{Z}} = \Gamma_F$. Recall from Prop. 1.5(iii) that $Z(\overline{D})$ is normal over $\overline{F}$. Hence, $\overline{F}$ is purely inseparably closed in $Z(\overline{D})$ if and only if $Z(\overline{D})$ is separable over $\overline{F}$. Therefore, by Lemma 6.44, the conditions for Z to equal $\mathsf{gr}(F)$ are equivalent to: $Z(\overline{D})$ is separable over $\overline{F}$ and $ker(\theta_D)/\Gamma_F$ has no $\overline{p}$-torsion. This yields the stated conditions for v_D to be tame.

Now, suppose $\overline{p} \nmid deg\, D$. Recall that $(deg\, D)^2 = \partial_{D/F}[\overline{D}{:}\overline{F}]|\Gamma_D{:}\Gamma_F|$ and that $\partial_{D/F}$ is a power of $\overline{p}$ by Th. 4.3. Hence, $\partial_{D/F} = 1$, so v_D is defectless over F; also, $\overline{p} \nmid [Z(\overline{D}){:}\overline{F}]$, so $Z(\overline{D})$ is separable over $\overline{F}$; finally, $\overline{p} \nmid |\Gamma_D{:}\Gamma_F|$, so $ker(\theta_D)/\Gamma_F$ has no $\overline{p}$-torsion. Therefore, by the preceding paragraph v_D is tame. □

We next show how to obtain tame gauges on crossed product algebras built from tamely ramified Galois extensions of valued fields. This will yield a large family of examples of tame gauges on central simple algebras.

Let K be a finite-degree Galois extension field of F; let $G = \mathcal{G}(K/F)$, and let $n = |G|$. Suppose that v_K is the unique valuation on K extending v on F, and that v_K is a tame v-gauge, i.e., K is tamely ramified over F. Let $f \in Z^2(G, K^\times)$ be any normalized 2-cocycle, and let A be the associated crossed product algebra over F:

$$A = (K/F, G, f) = \bigoplus_{\sigma\in G} Kx_\sigma,$$

where

$$(cx_\sigma)(dx_\tau) = c\sigma(d)f(x_\sigma, x_\tau)x_{\sigma\tau} \quad \text{for all } c,\, d \in K,\ \sigma,\ \tau \in G.$$

Define a v-value function α on A by setting

$$\alpha(x_\sigma) = \tfrac{1}{n}\sum_{\rho\in G} v_K(f(\sigma,\rho)) \quad \text{for} \quad \sigma \in G, \tag{6.29}$$

then defining for any $\sum_{\sigma\in G} c_\sigma x_\sigma \in A$ with all $c_\sigma \in K$,

$$\alpha\big(\sum_{\sigma\in G} c_\sigma x_\sigma\big) = \min_{\sigma\in G}\big(v_K(c_\sigma) + \alpha(x_\sigma)\big). \tag{6.30}$$

Proposition 6.46. *α is a tame v-gauge on A with*

$$\mathsf{gr}_\alpha(A) = \big(\mathsf{gr}_{v_K}(K)/\,\mathsf{gr}(F), G, \widetilde{f}\big),$$

where $\widetilde{f} \in Z^2(G, \mathsf{gr}(K)^\times)$ is given by $\widetilde{f}(\sigma,\tau) = \widetilde{f(\sigma,\tau)}$ for all $\sigma,\tau \in G$.

Proof. Let $(a_i)_{i=1}^n$ be any splitting base of v_K on K as a v-norm. It follows from the definition of α in (6.30) that $\big(a_i x_\sigma\big)_{1\le i\le n,\ \sigma\in G}$ is a v-splitting base for α on A; so, α is a v-norm on A. Since v_K is the unique extension of v to K, we have $v_K \circ \sigma = v_K$ for all $\sigma \in G$. Thus, by applying v_K to the basic cocycle equation $f(\sigma,\tau)\, f(\sigma\tau,\rho) = \sigma(f(\tau,\rho))\, f(\sigma,\tau\rho)$ and summing over all $\rho \in G$, we obtain in Γ,

$$|G|\, v_K(f(\sigma,\tau)) + |G|\,\alpha(x_{\sigma\tau}) = |G|\,\alpha(x_\tau) + |G|\,\alpha(x_\sigma) \quad \text{for all } \sigma,\tau \in G.$$

Since Γ is torsion-free, this yields

$$\alpha(x_\sigma) + \alpha(x_\tau) = v_K(f(\sigma,\tau)) + \alpha(x_{\sigma\tau}) \qquad \text{for all } \sigma,\tau \in G.$$

Therefore, for all c, $d \in K$, $\sigma, \tau \in G$,

$$\begin{aligned} \alpha\big((cx_\sigma)(dx_\tau)\big) &= \alpha\big(c\sigma(d)f(\sigma,\tau)x_{\sigma\tau}\big) \\ &= v_K(c) + v_K(d) + v_K(f(\sigma,\tau)) + \alpha(x_{\sigma\tau}) \\ &= v_K(c) + v_K(d) + \alpha(x_\sigma) + \alpha(x_\tau) \\ &= \alpha(cx_\sigma) + \alpha(dx_\tau). \end{aligned} \tag{6.31}$$

It follows by Lemma 3.23 applied to the splitting base $(a_i x_\sigma)_{1\le i\le n,\ \sigma\in G}$ of A that α is surmultiplicative.

Now consider $\mathsf{gr}_\alpha(A)$. The definition of α shows that it is a v_K-norm on A with $(x_\sigma)_{\sigma\in G}$ a splitting base. Hence, $\widetilde{c}\,\widetilde{x_\sigma} = \widetilde{cx_\sigma}$ for all $c \in K$, $\sigma \in G$, and

$$\mathsf{gr}_\alpha(A) \;=\; \bigoplus_{\sigma\in G} \mathsf{gr}(K)\,\widetilde{x_\sigma}.$$

Furthermore, equation (6.31) shows that $\widetilde{cx_\sigma}\,\widetilde{dx_\tau} = (cx_\sigma dx_\tau)^\sim$ for all $c, d \in K$, $\sigma, \tau \in G$. Thus, letting $\widetilde{\sigma}$ denote the graded automorphism of $\mathsf{gr}(K)$ induced by $\sigma \in G$, we have

$$\begin{aligned} (\widetilde{c}\,\widetilde{x_\sigma})\,(\widetilde{d}\,\widetilde{x_\tau}) &= \widetilde{cx_\sigma}\,\widetilde{dx_\tau} = (cx_\sigma dx_\tau)^\sim = \big(c\sigma(d)f(\sigma,\tau)x_{\sigma\tau}\big)^\sim \\ &= \big(c\sigma(d)f(\sigma,\tau)\big)^\sim\,\widetilde{x_{\sigma\tau}} = \widetilde{c}\,\widetilde{\sigma}(\widetilde{d})\,\widetilde{f(\sigma,\tau)}\,\widetilde{x_{\sigma\tau}}. \end{aligned}$$

This equation is additive in $\widetilde{c}$ and in $\widetilde{d}$, so it continues to hold if $\widetilde{c}$, $\widetilde{d}$ are replaced by arbitrary elements of $\mathsf{gr}(K)$. Therefore, as $\mathsf{gr}(K)$ is G-Galois over $\mathsf{gr}(F)$ by Cor. 5.52, we have

$$\mathsf{gr}_\alpha(A) \;\cong_g\; (\mathsf{gr}(K)/\mathsf{gr}(F), G, \widetilde{f}\,) \tag{6.32}$$

i.e., $\mathsf{gr}_\alpha(A)$ is a graded crossed product algebra as in (6.8), so it is graded central simple over $\mathsf{gr}(F)$. Hence, α is a tame v-gauge on A by Prop. 6.43. □

Remark 6.47. Suppose v on F is Henselian and K is any finite-degree tamely ramified Galois field extension of F; let

$$G = \mathcal{G}(K/F) \cong \mathcal{G}\big(\mathsf{gr}(K)/\mathsf{gr}(F)\big).$$

The proof of Prop. 6.46 shows that we have a commutative diagram

$$\begin{array}{ccc} H^2(G, K^\times) & \xrightarrow{\ \eta\ } & H^2(G, \mathsf{gr}(K)^\times) \\ \cong\big\downarrow & & \cong\big\downarrow \Delta_{\mathsf{gr}(K)/\mathsf{gr}(F)} \\ Br(K/F) & \xrightarrow{\ \mathsf{gr}\ } & \mathsf{Br}\big(\mathsf{gr}(K)/\mathsf{gr}(F)\big) \end{array} \tag{6.33}$$

where the map η is induced by the map $K^\times \to \mathsf{gr}(K)^\times$ carrying $x \in K^\times$ to $\widetilde{x} \in \mathsf{gr}(K)^\times$, and gr maps $[(K/F, G, f)]$ to $[(\mathsf{gr}(K)/\mathsf{gr}(F), G, \widetilde{f})]$, for $f \in Z^2(G, K^\times)$. The right map is an isomorphism by Th. 6.11, and the left map is the corresponding isomorphism from the classical theory of Brauer groups of (ungraded) fields. The map gr is well-defined because the map η is well-defined and the vertical maps are isomorphisms. The map gr is an isomorphism via Th. 6.64 below. (It is the isomorphism of kernels of the scalar extension maps in the right diagram of Prop. 6.59 (with K replacing L), whose horizontal maps are isomorphisms by Th. 6.64. Recall that F, hence also K, is Henselian here.) Thus, the map η in diagram (6.33) must also be an isomorphism. This isomorphism of cohomology groups is not readily apparent by homological arguments if $char\,\overline{F}$ divides $[K{:}F]$. Note also that by direct limit arguments (cf. Cor. 6.15) the diagram is still commutative if we assume that K is algebraic over F instead of finite-dimensional.

Remark 6.48. In the setting of Prop. 6.46, suppose further that $\mathcal{G}(K/F)$ is cyclic, say with generator σ, and A is the cyclic algebra $(K/F,\sigma,a)$ for some $a\in F^\times$. Then, $A=(K/F,G,f)$, where $f\in Z^2(G,K^\times)$ is given by

$$f(\sigma^i,\sigma^j) \;=\; \begin{cases}1 & \text{if } i+j\le n-1,\\ a & \text{if } i+j\ge n,\end{cases} \qquad \text{for all } i,j\in\{0,1,\ldots,n-1\}.$$

Then, $\widetilde{f}$ is given by

$$\widetilde{f}(\widetilde{\sigma}^i,\widetilde{\sigma}^j) \;=\; \begin{cases}1 & \text{if } i+j\le n-1,\\ \widetilde{a} & \text{if } i+j\ge n,\end{cases} \qquad \text{for all } i,j\in\{0,1,\ldots,n-1\}.$$

Thus, for the tame gauge α on A of Prop. 6.46, $\mathsf{gr}_\alpha(A)$ is the graded cyclic algebra $\big(\mathsf{gr}(K)/\mathsf{gr}(F),\widetilde{\sigma},\widetilde{a}\big)$.

Further examples of tame gauges can be obtained by tensor products, as we now show.

Proposition 6.49. *Let A be a central simple F-algebra with a tame v-gauge α and let B be a simple F-algebra with a v-gauge β. Let $C=A\otimes_F B$, which is a simple F-algebra. Then:*

(i) *$\alpha\otimes\beta$ is a v-gauge on C;*
(ii) *$\alpha\otimes\beta$ is a tame v-gauge if and only if β is a tame v-gauge.*

Proof. Since $\mathsf{gr}_\alpha(A)$ is a central simple graded $\mathsf{gr}(F)$-algebra, Prop. 3.41 yields (i), and Prop. 3.27 shows that

$$\mathsf{gr}_{\alpha\otimes\beta}(C) \;=\; \mathsf{gr}_\alpha(A)\otimes_{\mathsf{gr}(F)}\mathsf{gr}_\beta(B).$$

It follows that $Z\big(\mathsf{gr}_{\alpha\otimes\beta}(C)\big)=Z\big(\mathsf{gr}_\beta(B)\big)$. Since $Z(C)=Z(B)$, assertion (ii) follows. □

In considering possible tensor decompositions of valued division algebras in Ch. 10 the following corollary will be useful:

Corollary 6.50. *Let D be a division algebra with center F, and suppose D has a valuation v. Suppose also that $D=A\otimes_F B$ with $v|_A$ and $v|_B$ tame* *over $v|_F$. Then, v on D is tame over $v|_F$, and*

$$\mathsf{gr}(D) = \mathsf{gr}(A) \otimes_{\mathsf{gr}(F)} \mathsf{gr}(B).$$

Hence, $\Gamma_D = \Gamma_A + \Gamma_B$ and $\overline{A} \otimes_{\overline{F}} \overline{B}$ embeds in $\overline{D}$.

Proof. Since $v|_A$ and $v|_B$ are tame valuations, they are by definition tame $v|_F$-gauges. Hence, by Prop. 6.49(ii) $v|_A \otimes v|_B$ is a tame $v|_F$-gauge on D, which by Prop. 4.33(i) must coincide with v. Therefore, by Prop. 3.27

$$\mathsf{gr}_v(D) = \mathsf{gr}_{v|_A \otimes v|_B}(A \otimes_F B) = \mathsf{gr}(A) \otimes_{\mathsf{gr}(F)} \mathsf{gr}(B).$$

Thus,

$$\Gamma_D = \Gamma_{\mathsf{gr}(D)} = \Gamma_{\mathsf{gr}(A)} + \Gamma_{\mathsf{gr}(B)} = \Gamma_A + \Gamma_B.$$

Also, $\overline{A} \otimes_{\overline{F}} \overline{B} = \mathsf{gr}(A)_0 \otimes_{\mathsf{gr}(F)_0} \mathsf{gr}(B)_0$, and this ring clearly embeds in $\big(\mathsf{gr}(A) \otimes_{\mathsf{gr}(F)} \mathsf{gr}(B)\big)_0$, which equals $\mathsf{gr}(D)_0 = \overline{D}$. □

Tame gauges also behave well with respect to scalar extensions:

Proposition 6.51. *Let A be a central simple F-algebra with a v-gauge α, and let (L, v_L) be any valued field extension of (F, v). Then $\alpha \otimes v_L$ is a tame v_L-gauge on $A \otimes_F L$ if and only if α is a tame v-gauge on A.*

Proof. Since α is a v-gauge on A, we have seen in Cor. 3.28 that $\alpha \otimes v_L$ is a surmultiplicative v_L-norm on $A \otimes_F L$ and that there is a $\mathsf{gr}_{v_L}(L)$-algebra isomorphism

$$\mathsf{gr}_{\alpha\otimes v_L}(A \otimes_F L) \cong_g \mathsf{gr}_\alpha(A) \otimes_{\mathsf{gr}(F)} \mathsf{gr}_{v_L}(L).$$

If $\mathsf{gr}_\alpha(A)$ is a central simple graded $\mathsf{gr}(F)$-algebra, it follows from this isomorphism that $\mathsf{gr}_{\alpha\otimes v_L}(A \otimes_F L)$ is a central simple graded $\mathsf{gr}_{v_L}(L)$-algebra; hence, the surmultiplicative v_L-norm $\alpha \otimes v_L$ is a tame v_L-gauge on $A \otimes_F L$. Conversely, if $\alpha \otimes v_L$ is a tame v_L-gauge, then the isomorphism above shows that $\mathsf{gr}_\alpha(A)$ is a central $\mathsf{gr}(F)$-algebra, hence the v-gauge α is tame. □

To complete this subsection, we give a tameness criterion in terms of Henselizations, from which it follows that every gauge is tame if the residue characteristic is 0, and that whenever a semisimple F-algebra A carries a tame v-gauge, then every other v-gauge on A is tame.

Let A be a semisimple F-algebra, and let (F_h, v_h) be a Henselization of (F, v). Let $A_h = A \otimes_F F_h$, and let $A_1, \ldots, A_n$ be the simple components of A_h:

$$A_h = A \otimes_F F_h = A_1 \times \ldots \times A_n.$$

For each i, let D_i be the division algebra associated to A_i and let $Z_i = Z(A_i) = Z(D_i)$. Thus, $Z_1, \ldots, Z_n$ are the Henselizations of $Z(A)$ for the various valuations extending v, see Th. A.32.

Theorem 6.52. *In the setting just described, the following conditions are equivalent:*

(a) *A carries a tame v-gauge;*
(b) *for $i = 1, 2, \ldots, k$, the unique valuation w_i on D_i extending v_h is tame, and Z_i is tamely ramified over F_h.*

If A has a tame v-gauge, then every v-gauge on A is tame.

Note that in view of Prop. 6.42, condition (b) can be rephrased as: for each i, the valuation w_i is a tame v_h-gauge.

Proof. (a)⇒(b) Suppose α is a tame v-gauge on A, and let $\alpha_h = \alpha \otimes v_h$. By Prop. 3.42, α_h is a v_h-gauge on A_h, and since (F_h, v_h) is an immediate extension of (F, v) we have $\mathsf{gr}_{v_h}(F_h) = \mathsf{gr}(F)$ hence also

$$\mathsf{gr}_{\alpha_h}(A_h) = \mathsf{gr}_{\alpha}(A), \qquad \mathsf{gr}_{\alpha_h|_{Z(A_h)}}(Z(A_h)) = \mathsf{gr}_{\alpha|_{Z(A)}}(Z(A)).$$

Therefore, tameness of α implies that the v_h-gauge α_h is tame. Each factor $\alpha_i = \alpha_h|_{A_i}$ is then a tame v_i-gauge on A_i, hence $\alpha_i|_{Z_i}$ is a tame v_h-gauge on Z_i. By Prop. 6.41, we have $\alpha_i|_{Z_i} = w_i|_{Z_i}$, and Z_i is tamely ramified over F_h. Moreover, since α_i is an *End*-gauge by Th. 4.26, Prop. 2.17 shows that $Z(\mathsf{gr}_{w_i}(D_i)) = Z(\mathsf{gr}_{\alpha_i}(A_i))$. The tameness of α_i also yields

$$Z(\mathsf{gr}_{\alpha_i}(A_i)) = \mathsf{gr}_{\alpha_i}(Z_i).$$

But $\mathsf{gr}_{\alpha_i}(Z_i) = \mathsf{gr}_{w_i}(Z_i)$ since $\alpha_i|_{Z_i} = w_i|_{Z_i}$, hence $Z(\mathsf{gr}_{w_i}(D_i)) = \mathsf{gr}_{w_i}(Z_i)$, proving that w_i is tame.

(b)⇒(a) Suppose each w_i is a tame v_h-gauge. Then, as w_i is a v_h-norm on D_i, this w_i is defectless over v_h. Hence, v is defectless in A, by definition (Def. 4.19(ii) and (iii)); so Th. 4.50 shows that A has a v-gauge, call it β. Since F_h/F is an immediate extension, $\beta \otimes v_h$ is a v_h-gauge on A_h by Prop. 3.42, and this gauge is tame if and only if β is tame by Prop. 6.51. To complete the proof, it thus suffices to show that every v_h-gauge on A_h is tame when (b) holds.

Let α' be a v_h-gauge on A_h, and let $\alpha'_i = \alpha'|_{A_i}$, which is a v_h-gauge on A_i for $i = 1, \ldots, n$. Then, $\alpha'_i|_{Z_i}$ is a v_h-gauge on Z_i by Remark 3.33. But Z_i has the valuation $w_i|_{Z_i}$ extending v_h, which by Prop. 4.33(i) is its only v_h-gauge; so, $\alpha'_i|_{Z_i} = w_i|_{Z_i}$. Hence, $\mathsf{gr}_{\alpha'_i}(Z_i) = \mathsf{gr}_{w_i}(Z_i)$, which is a tame graded field extension of $\mathsf{gr}_{v_h}(F_h)$, as Z_i is tamely ramified over F_h. We can view $A_i = End_{D_i}(M_i)$ for some finite-dimensional right D_i-vector space M_i. By Th. 4.26, there is a w_i-norm η_i on M_i such that $\alpha'_i = End(\eta_i)$ so that $\mathsf{gr}_{\alpha'_i}(A_i) \cong_g \mathsf{End}_{\mathsf{gr}(D_i)}(\mathsf{gr}_{\eta_i}(M_i))$; hence, $Z(\mathsf{gr}_{\alpha'_i}(A_i)) \cong_g Z(\mathsf{gr}_{w_i}(D_i))$, by Prop. 2.17. Now, as w_i is a tame valuation on D_i,

$$Z(\mathsf{gr}_{w_i}(D_i)) = \mathsf{gr}_{w_i}(Z_i) = \mathsf{gr}_{\alpha'_i}(Z_i).$$

Thus, $Z(\mathsf{gr}_{\alpha'_i}(A_i)) = \mathsf{gr}_{\alpha'_i}(Z_i)$, showing that α'_i is a tame v_h-gauge on A_i, for each i; hence α' is a tame v_h-gauge on A_h.

The argument also shows that if A carries a tame v-gauge, then every v-gauge on A is tame. This is also immediate from (a)⇔(b), since (b) is independent of the choice of gauge on A. □

We single out the case of central simple algebras:

Corollary 6.53. *Let A be a central simple F-algebra and let D be the central division algebra over F_h associated to A_h. Let w be the valuation on D extending v_h. The algebra A carries a tame v-gauge if and only if the valuation w is tame. For any tame v-gauge α on A, the central simple graded $\mathsf{gr}(F)$-algebra $\mathsf{gr}_\alpha(A)$ is Brauer-equivalent to the graded division algebra $\mathsf{gr}_w(D)$ when we identify $\mathsf{gr}(F)$ with $\mathsf{gr}(F_h)$.*

Proof. The first assertion is just the case of Th. 6.52 for A central simple. We saw in the proof of that theorem that for any tame v-gauge α on A we have $\mathsf{gr}_{\alpha_h}(A_h) \cong_g \mathsf{End}_{\mathsf{gr}(D)}(\mathsf{gr}_\eta(M))$ for some finite-dimensional right D-vector space M and some w-norm η on M, hence $\mathsf{gr}_{\alpha_h}(A_h)$ is Brauer-equivalent to $\mathsf{gr}_w(D)$. Since (F_h, v_h) is an immediate extension of (F, v), we have $\mathsf{gr}_{\alpha_h}(A_h) = \mathsf{gr}_\alpha(A)$. The corollary follows. □

Corollary 6.54. *Suppose char $\overline{F} = 0$. Then, every v-gauge on a semisimple F-algebra is tame.*

Proof. Let A be a semisimple v-algebra with a v-gauge α. We use the notation introduced just before Th. 6.52. Because *char* $\overline{F} = 0$, the field extension Z_i/F_h is tamely ramified, and the valuation w_i on D_i extending v_h is tame by Prop. 6.45. Theorem 6.52 shows that α is a tame v_h-gauge. □

6.2.3 The canonical Brauer group filtration (valued case)

The special types of gauges defined in §6.2.1 and §6.2.2 are now used to define subgroups of the Brauer group of a valued field, which are compared to subgroups of the Brauer group of the corresponding graded field. In the case where the valuation is Henselian, the correspondence is bijective, so we derive from the exact sequence describing the inertially split part of the Brauer group of a graded field an analogous exact sequence for the inertially split part of the Brauer group of the valued field, see Th. 6.66.

Definition 6.55. Let F be a field with an arbitrary valuation. We define the *inertial part* of the Brauer group $Br(F)$ as the set of Brauer-equivalence classes of central simple F-algebras A on which an inertial v-gauge can be defined, and the *tame* (or *tamely ramified*) *part of* $Br(F)$ likewise with tame gauges instead of inertial gauges:

$$\begin{aligned} Br_{in}(F) &= \{[A] \in Br(F) \mid A \text{ has an inertial } v\text{-gauge}\};\\ Br_{tr}(F) &= \{[A] \in Br(F) \mid A \text{ has a tame } v\text{-gauge}\}. \end{aligned}$$

When it is necessary to specify the valuation, we write $Br_{in}(F, v)$ and $Br_{tr}(F, v)$.

Note that for any central simple F-algebra A, if $[A] \in Br_{in}(F)$ then A carries an inertial v-gauge. See Remark 6.35. Similarly, Cor. 6.53 shows that if $[A] \in Br_{tr}(F)$, then A carries a tame gauge. Corollary 6.53 also shows that if A has a tame v-gauge, then every v-gauge on A is tame. By contrast, a central simple F-algebra that carries an inertial v-gauge may also carry v-gauges that are not inertial. For an easy example, recall that every v-norm α on an F-vector space V yields a v-gauge $End(\alpha)$ on $End_F V$; that gauge is inertial if and only if Γ_V consists of only one coset modulo Γ_F. Thus, every split central simple F-algebra of degree at least 2 carries non-inertial v-gauges as well as inertial v-gauges.

Proposition 6.38(ii) shows that $Br_{in}(F)$ is closed under the operation in $Br(F)$; hence, $Br_{in}(F)$ is a subgroup of the torsion group $Br(F)$. Similarly, $Br_{tr}(F)$ is a subgroup of $Br(F)$ by Prop. 6.49(ii). If α is an inertial v-gauge on a central simple F-algebra A, then A_0 is a central simple $\overline{F}$-algebra and $\mathsf{gr}_\alpha(A) = A_0 \otimes_{\overline{F}} \mathsf{gr}(F)$, see Prop. 6.32; hence $\mathsf{gr}_\alpha(A)$ is a central simple graded $\mathsf{gr}(F)$-algebra by Prop. 2.16. Therefore, every inertial v-gauge is tame, and we have a sequence of subgroups:

$$Br_{in}(F) \subseteq Br_{tr}(F) \subseteq Br(F).$$

The tame part fills $Br(F)$ completely when the residue characteristic is 0:

Proposition 6.56. *If $char\,\overline{F} = 0$, then $Br_{tr}(F) = Br(F)$. If $char\,\overline{F} = \overline{p} \neq 0$, then $Br_{tr}(F)$ contains the prime-to-$\overline{p}$ part of $Br(F)$.*

Proof. Let A be a central simple F-algebra. If $char\,\overline{F} = 0$, then by Th. 4.50 A has a v-gauge, and every v-gauge is tame by Cor. 6.54. So, $[A] \in Br_{tr}(F)$.

Now, suppose $char\,\overline{F} = \overline{p} \neq 0$. Let (F_h, v_h) be the Henselization of (F, v), and let D be the central division algebra over F_h associated to $A \otimes_F F_h$. If $exp\,A$ is prime to $\overline{p}$, then $deg\,D$ is not divisible by $\overline{p}$, hence the valuation on D extending v_h is tame by Prop. 6.45. It then follows from Cor. 6.53 that A carries a tame v-gauge, so $[A] \in Br_{tr}(F)$. □

We now relate $Br_{in}(F)$ and $Br_{tr}(F)$ to the Brauer group of $\mathsf{gr}(F)$.

Proposition 6.57. *There is a well-defined group homomorphism*

$$\mathsf{gr}\colon Br_{tr}(F) \longrightarrow \mathsf{Br}(\mathsf{gr}(F)) \qquad \textit{given by} \quad [A] \mapsto [\mathsf{gr}_\alpha(A)]$$

where α is an arbitrary tame v-gauge on A. This homomorphism maps $Br_{in}(F)$ into $\mathsf{Br}_{in}(\mathsf{gr}(F))$, so we have a commutative diagram

$$\begin{array}{ccc} Br_{in}(F) & \xrightarrow{\mathsf{gr}} & \mathsf{Br}_{in}(\mathsf{gr}(F)) \\ \big\downarrow & & \big\downarrow \\ Br_{tr}(F) & \xrightarrow{\mathsf{gr}} & \mathsf{Br}(\mathsf{gr}(F)) \end{array}$$

Proof. Let (F_h, v_h) be the Henselization of (F, v). For any tame v-gauge on a central simple F-algebra A, Cor. 6.53 shows that $\mathsf{gr}_\alpha(A)$ is Brauer-equivalent to the graded division algebra $\mathsf{gr}_w(D)$, where D is the division algebra associated to $A \otimes_F F_h$ and w is the valuation extending v_h. Therefore, the map gr is well-defined. It is a group homomorphism since for tame v-gauges α, β on central simple F-algebras A, B the product $\alpha \otimes \beta$ is a tame v-gauge on $A \otimes_F B$ and

$$\mathsf{gr}_{\alpha\otimes\beta}(A \otimes_F B) = \mathsf{gr}_\alpha(A) \otimes_{\mathsf{gr}(F)} \mathsf{gr}_\beta(B),$$

see Prop. 6.49.

If α is inertial, then $\mathsf{gr}_\alpha(A) = A_0 \otimes_{\overline{F}} \mathsf{gr}(F)$, hence $[\mathsf{gr}_\alpha(A)] \in \mathsf{Br}_{in}(\mathsf{gr}(F))$. $\square$

Recall from (6.19) the canonical isomorphism $\iota\colon Br(\mathsf{gr}(F)_0) \xrightarrow{\sim} \mathsf{Br}_{in}(\mathsf{gr}(F))$, which is scalar extension from $\mathsf{gr}(F)_0 = \overline{F}$ to $\mathsf{gr}(F)$. We use ι to define a map on $Br_{in}(F)$:

Corollary 6.58. *There is a well-defined group homomorphism*

$$\mathsf{gr}_0\colon Br_{in}(F) \longrightarrow Br(\overline{F}) \qquad \textit{given by} \quad [A] \mapsto [A_0^\alpha]$$

where α is an arbitrary inertial v-gauge on A and A_0^α is the homogeneous component of degree 0 in $\mathsf{gr}_\alpha(A)$. This homomorphism fits in the following commutative diagram:

$$\begin{array}{ccc} Br_{in}(F) & \xrightarrow{\ \mathsf{gr}\ } & \mathsf{Br}_{in}(\mathsf{gr}(F)) \\ {\scriptstyle \mathsf{gr}_0}\big\downarrow & & \big\uparrow{\scriptstyle \iota} \\ Br(\overline{F}) & =\!=\!=\!= & Br(\mathsf{gr}(F)_0) \end{array}$$

Proof. For any inertial v-gauge α on a central simple F-algebra A we have

$$\mathsf{gr}_\alpha(A) = A_0^\alpha \otimes_{\overline{F}} \mathsf{gr}(F),$$

see Prop. 6.32. Therefore, the map gr_0 is well-defined as the composition of maps

$$Br_{in}(F) \xrightarrow{\mathsf{gr}} \mathsf{Br}_{in}(\mathsf{gr}(F)) \xrightarrow{\iota^{-1}} Br(\overline{F}),$$

and it is a group homomorphism that makes the diagram above commute. $\square$

The maps gr and gr_0 are functorial with respect to valued field extensions:

Proposition 6.59. *Let (L, v_L) be any extension of the valued field (F, v). The scalar extension map $\mathrm{ext}_{L/F}\colon \mathrm{Br}(F) \to Br(L)$ maps $Br_{in}(F)$ into $Br_{in}(L)$, $Br_{tr}(F)$ into $Br_{tr}(L)$, and there are commutative diagrams*

$$\begin{array}{ccc} Br_{in}(F) & \xrightarrow{\mathsf{gr}_0} & Br(\overline{F}) \\ {\scriptstyle \mathrm{ext}_{L/F}}\downarrow & & \downarrow{\scriptstyle \mathrm{ext}_{\overline{L}/\overline{F}}} \\ Br_{in}(L) & \xrightarrow{\mathsf{gr}_0} & Br(\overline{L}) \end{array} \quad \textit{and} \quad \begin{array}{ccc} Br_{tr}(F) & \xrightarrow{\mathsf{gr}} & \mathsf{Br}(\mathsf{gr}(F)) \\ {\scriptstyle \mathrm{ext}_{L/F}}\downarrow & & \downarrow{\scriptstyle \mathrm{ext}_{\mathsf{gr}(L)/\mathsf{gr}(F)}} \\ Br_{tr}(L) & \xrightarrow{\mathsf{gr}} & \mathsf{Br}(\mathsf{gr}(L)) \end{array}$$

Proof. Let A be a central simple F-algebra. If α is an inertial v-gauge on A, Prop. 6.39 shows that $\alpha \otimes v_L$ is an inertial v_L-gauge on $A \otimes_F L$, for which

$$(A \otimes_F L)_0 \;=\; A_0 \otimes_{\overline{F}} \overline{L}.$$

Similarly, if α is a tame v-gauge on A, Prop. 6.51 yields that $\alpha \otimes v_L$ is a tame v_L-gauge on $A \otimes_F L$, for which, by Cor. 3.28,

$$\mathsf{gr}_{\alpha\otimes v_L}(A \otimes_F L) \;=\; \mathsf{gr}_\alpha(A) \otimes_{\mathsf{gr}(F)} \mathsf{gr}_{v_L}(L). \qquad \square$$

For the rest of this subsection (except Cor. 6.61 and Remark 6.62), we focus on the case where the valuation v on F is Henselian. Note that there is then another description of $Br_{in}(F)$: Whenever v is Henselian,

$$Br_{in}(F) \;=\; \left\{ [D] \;\middle|\; \begin{array}{l} D \text{ is a division algebra with center } F \\ \text{and its valuation } v_D \text{ is inertial over } v \end{array} \right\}. \tag{6.34}$$

This is immediate from Prop. 6.34.

Our goal is to show that for Henselian fields the homomorphism gr of Prop. 6.57 is an isomorphism. Injectivity is readily proved:

Lemma 6.60. *If (F, v) is Henselian, the homomorphism*

$$\mathsf{gr}\colon\, Br_{tr}(F) \longrightarrow \mathsf{Br}\big(\mathsf{gr}(F)\big)$$

is injective and index-preserving.

Proof. Let D be a central division algebra over F. If $[D] \in Br_{tr}(F)$, then by Cor. 6.53 the valuation w on D extending v is tame. We have $\mathsf{gr}[D] = [\mathsf{gr}_w(D)]$ by definition. Since $\mathsf{gr}_w(D)$ is a central graded division algebra over $\mathsf{gr}(F)$ with the same degree as D, the map gr is index-preserving and therefore injective. $\square$

Corollary 6.61. *Let (L,v_L) be an extension of an arbitrary valued field (F, v). Suppose v_L is Henselian and let $c \in Br_{tr}(F)$. Then, L splits c if and only if $\mathsf{gr}(L)$ splits $\mathsf{gr}(c)$. In particular, for (F_h, v_h) the Henselization of (F, v) we have*

$$\ker\big(\mathsf{gr}\colon Br_{tr}(F) \to \mathsf{Br}(\mathsf{gr}(F))\big) \;=\; Br(F_h/F) \cap Br_{tr}(F).$$

Proof. In view of Lemma 6.60, the map $\mathsf{gr}\colon\, Br_{tr}(L) \to \mathsf{Br}\big(\mathsf{gr}(L)\big)$ is injective. The corollary then readily follows from the functoriality of the map gr (see Prop. 6.59). For $L = F_h$ we have $\mathsf{gr}(L) = \mathsf{gr}(F)$, hence c is split by F_h if and only if $\mathsf{gr}(c)$ is split. $\square$

Remark 6.62. If v has rank 1, then in fact $Br(F_h/F) \subseteq Br_{in}(F)$, and

$$ker(\mathsf{gr}) = ker(\mathsf{gr}) \cap Br_{in}(F) = Br(F_h/F).$$

Indeed, if a central simple F-algebra A is split by F_h, then there is clearly an inertial v_h-gauge α on the split algebra $A \otimes_F F_h$. By Prop. 4.57, $\alpha|_A$ is an inertial v-gauge on A. So, $[A] \in Br_{in}(F)$, showing that $Br(F_h/F) \subseteq Br_{in}(F)$. Moreover, $\mathsf{gr}_{\alpha|_A}(A) = \mathsf{gr}_\alpha(A \otimes_F F_h)$, so $[A] \in ker(\mathsf{gr})$.

However, the inclusion $Br(F_h/F) \subseteq Br_{in}(F)$ does not hold in general. A counterexample with $rk(v) = 2$ is given in Exercise 6.12 below.

Given a field F with a Henselian valuation v, fix a separable closure F_{sep} of F and consider in F_{sep} the maximal inertial extension F_{in} of F and the maximal tamely ramified extension F_{tr} of F, see §A.2 in Appendix A:

$$F \subseteq F_{in} \subseteq F_{tr} \subseteq F_{sep}.$$

We have $\overline{F_{in}} = \overline{F}_{sep}$ and $\Gamma_{F_{in}} = \Gamma_F$ (see Th. A.23), hence

$$\mathsf{gr}(F_{in}) = \overline{F}_{sep} \otimes_{\overline{F}} \mathsf{gr}(F).$$

The graded field $\mathsf{gr}(F_{in})$ is therefore the maximal inertial graded field extension of $\mathsf{gr}(F)$,

$$\mathsf{gr}(F_{in}) = \mathsf{gr}(F)_{in}. \tag{6.35}$$

As in the graded case (see §6.1.3), we define the *inertially split part of* $Br(F)$ by

$$Br_{is}(F) = Br(F_{in}/F).$$

Accordingly, central simple F-algebras split by F_{in} are said to be *inertially split.*

The tame part $Br_{tr}(F)$ has a similar description:

Proposition 6.63. *For (F, v) Henselian, we have*

$$Br_{tr}(F) = Br(F_{tr}/F).$$

Moreover, if $char\,\overline{F} = \overline{p} \neq 0$, *the $\overline{p}$-primary component of $Br_{tr}(F)$, for which we use the notation $Br_{tr}(F)_{(\overline{p})}$, lies in $Br_{is}(F)$:*

$$Br_{tr}(F)_{(\overline{p})} \subseteq Br_{is}(F).$$

Proof. If $char\,\overline{F} = 0$, we have $Br_{tr}(F) = Br(F)$ by Prop. 6.56, and $F_{tr} = F_{sep}$ (see Th. A.24), so the proposition is clear. For the rest of the proof, we assume $char\,\overline{F} = \overline{p} \neq 0$.

We first establish the second assertion. Let D be a central division algebra over F with $[D] \in Br_{tr}(F)_{(\overline{p})}$. Corollary 6.53 shows that the valuation w on D extending v is tame, so by Prop. 6.43 the graded division algebra $\mathsf{gr}_w(D)$ has center $\mathsf{gr}(F)$. Since $deg\,\mathsf{gr}_w(D) = deg\,D$, we have $[\mathsf{gr}_w(D)] \in \mathsf{Br}\big(\mathsf{gr}(F)\big)_{(\overline{p})}$, hence $[\mathsf{gr}_w(D)] \in \mathsf{Br}_{in}\big(\mathsf{gr}(F)\big)$ by Cor. 6.20. Now, since $\mathsf{gr}(F_{in}) = \mathsf{gr}(F)_{in}$, functoriality of gr yields the following commutative diagram:

$$\begin{array}{ccc} Br_{tr}(F) & \xrightarrow{\mathsf{gr}} & \mathsf{Br}(\mathsf{gr}(F)) \\ {\scriptstyle \mathrm{ext}_{F_{in}/F}} \downarrow & & \downarrow {\scriptstyle \mathrm{ext}_{\mathsf{gr}(F)_{in}/\mathsf{gr}(F)}} \\ Br_{tr}(F_{in}) & \xrightarrow{\mathsf{gr}} & \mathsf{Br}(\mathsf{gr}(F)_{in}) \end{array} \tag{6.36}$$

The horizontal arrows are injective maps by Lemma 6.60; hence, $[D] \in Br(F_{in}/F) = Br_{is}(F)$. This proves the second assertion.

Since $F_{in} \subseteq F_{tr}$, we have $Br_{is}(F) \subseteq Br(F_{tr}/F)$. Hence to prove the inclusion $Br_{tr}(F) \subseteq Br(F_{tr}/F)$ it now suffices to consider the prime-to-$\overline{p}$ components of $Br_{tr}(F)$. Let D be a central division algebra over F of degree prime to $\overline{p}$, and let M be a maximal subfield of D. Then $[M:F]$ is prime to $\overline{p}$, hence M is tamely ramified over F. Therefore, we may identify M with a subfield of F_{tr} (see Th. A.24). Since M splits D, we have $[D] \in Br(F_{tr}/F)$.

We have thus shown $Br_{tr}(F) \subseteq Br(F_{tr}/F)$. To prove the reverse inclusion, consider a Brauer class $c \in Br(F_{tr}/F)$. We may find a tamely ramified Galois extension K of F that splits c. Then c is the Brauer class of a crossed product $(K/F, \mathcal{G}(K/F), f)$ for some normalized cocycle f. Proposition 6.46 shows that this crossed product carries a tame v-gauge, hence $c \in Br_{tr}(F)$. □

As a result of this proposition and the inclusion $F_{in} \subseteq F_{tr}$, we have

$$Br_{is}(F) \subseteq Br_{tr}(F).$$

From the commutative diagram (6.36) and the injectivity of gr (Lemma 6.60) it follows that gr maps $Br_{is}(F)$ to $\mathsf{Br}_{is}(\mathsf{gr}(F))$.

We can now give the main theorem of this section, which is a key ingredient in our program of studying valued division algebras via their associated graded division algebras. This theorem and its consequences will be invoked repeatedly in Ch. 8 in analyzing the structure of division algebras over Henselian fields. In particular, the fact that the isomorphism $\mathsf{gr}\colon Br_{tr}(F) \to \mathsf{Br}(\mathsf{gr}(F))$ is index-preserving and functorial with respect to field extensions (see Prop. 6.59) will be used in §8.1 below to show a very strong correspondence between the tame subalgebras of a division algebra D over a Henselian field F and the graded subalgebras of $\mathsf{gr}(D)$.

Theorem 6.64. *For (F, v) Henselian, the maps*

$$\mathsf{gr}_0\colon Br_{in}(F) \longrightarrow Br(\overline{F}) \qquad \textit{and} \qquad \mathsf{gr}\colon Br_{tr}(F) \longrightarrow \mathsf{Br}(\mathsf{gr}(F))$$

are index-preserving isomorphisms. The map gr *sends $Br_{is}(F)$ onto $\mathsf{Br}_{is}(\mathsf{gr}(F))$ and $Br_{in}(F)$ onto $\mathsf{Br}_{in}(\mathsf{gr}(F))$, so we have the following commutative diagram, where the horizontal maps are inclusions:*

$$\begin{array}{ccccccc} Br_{in}(F) & \hookrightarrow & Br_{is}(F) & \hookrightarrow & Br_{tr}(F) & \hookrightarrow & Br(F) \\ {\scriptstyle \mathsf{gr}} \downarrow {\scriptstyle \cong} & & {\scriptstyle \mathsf{gr}} \downarrow {\scriptstyle \cong} & & {\scriptstyle \mathsf{gr}} \downarrow {\scriptstyle \cong} & & \\ \mathsf{Br}_{in}(\mathsf{gr}(F)) & \hookrightarrow & \mathsf{Br}_{is}(\mathsf{gr}(F)) & \hookrightarrow & \mathsf{Br}(\mathsf{gr}(F)) & & \end{array} \tag{6.37}$$

Proof. We already know from Lemma 6.60 that gr is an index-preserving injection. Since the scalar extension map

$$\iota\colon\ Br(\overline{F}) \ = \ Br(\mathsf{gr}(F)_0) \hookrightarrow \mathsf{Br}\big(\mathsf{gr}(F)\big)$$

is likewise index-preserving, it follows from Cor. 6.58 that gr_0 is an index-preserving injection. Thus, to prove the theorem it suffices to show that gr_0 is surjective, and that gr maps $Br_{in}(F)$ onto $\mathsf{Br}_{in}\big(\mathsf{gr}(F)\big)$, $Br_{is}(F)$ onto $\mathsf{Br}_{is}\big(\mathsf{gr}(F)\big)$, and $Br_{tr}(F)$ onto $\mathsf{Br}\big(\mathsf{gr}(F)\big)$.

Step 1: gr_0 *is surjective.* We must see how to "lift" generators of $Br(\overline{F})$ to realize them as the degree 0 parts of central simple F-algebras with inertial gauges. For this, first take any positive integer n prime to $char\,\overline{F}$ (with n arbitrary if $char\,\overline{F} = 0$). Consider the separable closure $\overline{F}_{sep}$ of $\overline{F}$, and let $G_{\overline{F}} = \mathcal{G}(\overline{F}_{sep}/\overline{F})$. Let ${}_nBr(\overline{F})$ denote the n-torsion subgroup of $Br(\overline{F})$. By the Kummer sequence in Galois cohomology, see Gille–Szamuely [84, Cor. 4.4.9, p. 99], we have

$${}_nBr(\overline{F}) \ \cong \ H^2(G_{\overline{F}}, \mu_n),$$

where μ_n is the group of all n-th roots of unity in $\overline{F}_{sep}$. This implies that every Brauer class in ${}_nBr(\overline{F})$ is represented by a crossed product algebra $B = (S/\overline{F}, G, f)$, where S is a finite-degree Galois field extension of $\overline{F}$ with $\mu_n(S) = \mu_n$ (i.e., S contains all the n-th roots of unity in $\overline{F}_{sep}$), $G = \mathcal{G}(S/\overline{F})$, and $f \in Z^2(G, \mu_n(S))$. By Cor. 5.57, as v is Henselian, there is an inertial lift of S over F, i.e., a field $K \supseteq F$ such that the extension v_K of v to K is inertial over v with $\overline{K} \cong S$, so $[K{:}F] = [S{:}\overline{F}] < \infty$. Since S is Galois over $\overline{F}$, by Cor. 5.57 and Cor. 5.53 K is Galois over F and the canonical map $\mathcal{G}(K/F) \to \mathcal{G}(S/\overline{F}) = G$ is an isomorphism; we identify $\mathcal{G}(K/F)$ with G. Because the polynomial $X^n - 1$ splits over S with no repeated roots and v_K is Henselian, $X^n - 1$ splits over K with no repeated roots, and distinct roots in K map to distinct roots in S. That is, the group $\mu_n(K)$ of n-th roots of unity in K has order n, and the G-module homomorphism $\mu_n(K) \to \mu_n(S)$ given by $\omega \mapsto \overline{\omega}$ is injective, hence an isomorphism. Let $g \in Z^2(G, \mu_n(K))$ be the inverse image of $f \in Z^2(G, \mu_n(S))$ under the induced isomorphism $Z^2(G, \mu_n(K)) \cong Z^2(G, \mu_n(S))$; then let $A = (K/F, G, g)$, which is a central simple crossed product F-algebra. Since $v_K(\omega) = 0$ for each $\omega \in \mu_n(K)$, we can define an inertial v-gauge α on A as in Ex. 6.37 above. As pointed out in that example,

$$A_0^{\alpha} \cong (\overline{K}/\overline{F}, G, \widetilde{g}) \ = \ (S/\overline{F}, G, f) \ = \ B.$$

So, $[B] = \mathsf{gr}_0[A]$. Thus, the entire prime-to-$char\,\overline{F}$ part of $Br(\overline{F})$ lies in the image of gr_0.

If $char\,\overline{F} = 0$, the proof of surjectivity of gr_0 is complete. Consider now the remaining case that $char\,\overline{F} = \overline{p} \neq 0$. By Albert's Theorem (see Jacobson [108, Th. 4.5.7]), any Brauer class in $Br(\overline{F})$ of exponent a power of $\overline{p}$ is represented by a cyclic algebra $C = (M/\overline{F}, \sigma, c)$, for some cyclic Galois extension M of $\overline{F}$

with $\langle\sigma\rangle = \mathcal{G}(M/\overline{F})$ and some $c \in \overline{F}^\times$. By Cor. 5.57 there is an inertial lift N of M over F. Then, as above, since $\overline{N} = M$ which is Galois over $\overline{F}$, N is Galois over F and we may identify $\mathcal{G}(N/F)$ with $\mathcal{G}(M/\overline{F}) = \langle\sigma\rangle$. Let a be any inverse image of c in $\mathcal{O}_F^\times$, and let $T = (N/F, \sigma, a)$ which is a central simple cyclic F-algebra. The usual $f \in Z^2(\mathcal{G}(N/F), N^\times)$ describing T as a crossed product algebra has $im(f) = \{1, a\} \subseteq \mathcal{O}_F^\times$. Therefore, by Ex. 6.37, T has an inertial gauge η with $T_0^\eta \cong (\overline{N}/\overline{F}, \sigma, \overline{a}) = C$; so $\mathsf{gr}_0([T]) = [C]$. We have now shown that every element of $Br(\overline{F})$ of $\overline{p}$-power exponent or prime-to-$\overline{p}$ exponent lies in $im\,\mathsf{gr}_0$. Since such elements generate $Br(\overline{F})$, the homomorphism gr_0 is surjective.

Step 2: gr *maps* $Br_{in}(F)$ *onto* $\mathsf{Br}_{in}(\mathsf{gr}(F))$. This readily follows from Step 1 and Cor. 6.58, since the scalar extension map $\iota\colon Br(\overline{F}) \to \mathsf{Br}_{in}(\mathsf{gr}(F))$ is an isomorphism.

Step 3: gr *maps* $Br_{is}(F)$ *onto* $\mathsf{Br}_{is}(\mathsf{gr}(F))$. We already know from Step 2 that $\mathsf{Br}_{in}(\mathsf{gr}(F))$ lies in the image of $Br_{is}(F)$ under gr. The proof of Th. 6.24 (surjectivity of χ) shows that $\mathsf{Br}_{is}(\mathsf{gr}(F))/\mathsf{Br}_{in}(\mathsf{gr}(F))$ is generated by the images of graded cyclic algebras $\mathsf{C} = (\mathsf{L}/\mathsf{gr}(F), \sigma, c)$ with L inertial over $\mathsf{gr}(F)$. By Cor. 5.57, for any such L we may find an inertial lift K of L_0 over F. Thus, K is an inertial extension of F with $\overline{K} = \mathsf{L}_0$, hence $\mathsf{gr}(K) = \mathsf{L}$, and K is a Galois extension of F with $\mathcal{G}(K/F) = \mathcal{G}(\mathsf{L}/\mathsf{gr}(F)) = \langle\sigma\rangle$ by Cor. 5.53. Take any $a \in K^\times$ with $\widetilde{a} = c$, and let $A = (K/F, \sigma, a)$. As noted in Remark 6.48, there is a tame gauge α on the cyclic algebra A such that $\mathsf{gr}_\alpha(A) \cong_g \mathsf{C}$. Hence, $[\mathsf{C}] = \mathsf{gr}[A]$. The algebra A is split by the inertial extension K of F, so $[A] \in Br_{is}(F)$. This completes Step 3.

Step 4: gr *maps* $Br_{tr}(F)$ *onto* $\mathsf{Br}(\mathsf{gr}(F))$. If $char\,\overline{F} = \overline{p} \neq 0$, let $Br'_{tr}(F)$ denote the prime-to-$\overline{p}$ part of $Br_{tr}(F)$, which is the direct sum of the q-primary components of $Br_{tr}(F)$ for all primes $q \neq \overline{p}$. If $char\,\overline{F} = 0$, let $Br'_{tr}(F) = Br_{tr}(F)$. Define $\mathsf{Br}'(\mathsf{gr}(F))$ analogously.

We first prove homologically that $\mathsf{Br}'(\mathsf{gr}(F)) \subseteq im\,\mathsf{gr}$. Let $\mathfrak{m}_{F_{tr}}$ be the maximal ideal of the valuation ring $\mathcal{O}_{F_{tr}}$ of F_{tr}. Let

$$G = \mathcal{G}(F_{tr}/F) = \mathcal{G}\left(\mathsf{gr}(F)_{tr}/\mathsf{gr}(F)\right)$$

(cf. Prop. 5.55). Then, the short exact sequence of discrete G-modules

$$1 \longrightarrow 1 + \mathfrak{m}_{F_{tr}} \longrightarrow F_{tr}^\times \longrightarrow \mathsf{gr}(F)_{tr}^\times \longrightarrow 1$$

yields the exact sequence:

$$H^2(G, 1 + \mathfrak{m}_{F_{tr}}) \longrightarrow H^2(G, F_{tr}^\times) \xrightarrow{\eta} H^2(G, \mathsf{gr}(F)_{tr}^\times) \longrightarrow H^3(G, 1 + \mathfrak{m}_{F_{tr}}). \tag{6.38}$$

For any prime number $q \neq char\,\overline{F}$ and any $m \in \mathfrak{m}_{F_{tr}}$, the polynomial $f = X^q - (1 + m)$ in $\mathcal{O}_{F_{tr}}[X]$ has image $\overline{f} = X^q - 1$ in $\overline{F_{tr}}[X]$. Since the valuation on F_{tr} is Henselian and 1 is a simple root of $\overline{f}$ as q is prime to $char\,\overline{F}$, this f has a unique root $u \in \mathcal{O}_{F_{tr}}$ with $\overline{u} = 1$. This shows that

$1 + \mathfrak{m}_{F_{tr}}$ is uniquely q-divisible. Hence, $H^j(G, 1+\mathfrak{m}_{F_{tr}})$ has trivial q-primary component for all $j \geq 1$. This shows that the image of the map η in the sequence (6.38) contains the entire q-primary component of $H^2(G, \mathsf{gr}(F)^{\times}_{tr})$ for every $q \neq char\,\overline{F}$. When we identify cohomology groups with Brauer groups, the map η corresponds to gr, as noted in Remark 6.47. Hence, $\mathsf{Br}'(\mathsf{gr}(F)) \subseteq im\,\mathsf{gr}$.

If $char(\overline{F}) = 0$, then the proof is complete. If $char(\overline{F}) = \overline{p} \neq 0$, then $\mathsf{Br}(\mathsf{gr}(F)) = \mathsf{Br}'(\mathsf{gr}(F)) \oplus \mathsf{Br}(\mathsf{gr}(F))_{(\overline{p})}$ where the second summand is the $\overline{p}$-primary component of $\mathsf{Br}(\mathsf{gr}(F))$. Since $\mathsf{Br}'(\mathsf{gr}(F)) \subseteq im\,\mathsf{gr}$, and by Prop. 6.63 and Step 3

$$\mathsf{Br}(\mathsf{gr}(F))_{(\overline{p})} \subseteq \mathsf{Br}_{is}\big(\mathsf{gr}(F)\big) \subseteq im\,\mathsf{gr},$$

we have $\mathsf{Br}\big(\mathsf{gr}(F)\big) \subseteq im\,\mathsf{gr}$. Thus, in all cases gr is surjective, completing Step 4 and the proof of Th. 6.64. □

As an initial application of Th. 6.64, we transfer to $Br_{is}(F)$ the description of $\mathsf{Br}_{is}\big(\mathsf{gr}(F)\big)$ given in Th. 6.24 and Prop. 6.27. The first step for this is to define the analogue of the ramification map χ.

Lemma 6.65. *Assume (F, v) is Henselian. For every inertially split central division algebra D over F, the field $Z(\overline{D})$ is a Galois extension of $\overline{F}$ and $ker\,\theta_D = \Gamma_F$; hence, θ_D induces an isomorphism*

$$\overline{\theta}_D \colon \Gamma_D/\Gamma_F \xrightarrow{\sim} \mathcal{G}(Z(\overline{D})/\overline{F}).$$

Proof. Let w be the valuation on D extending v. Since D is inertially split and $Br_{is}(F) \subseteq Br_{tr}(F)$, the valuation w is tame. Therefore, by Prop. 1.5 and Prop. 6.45, $Z(\overline{D})$ is a Galois extension of $\overline{F}$ and θ_D maps Γ_D onto $\mathcal{G}(Z(\overline{D})/\overline{F})$. Moreover, $\mathsf{gr}(D)$ is a central graded division algebra over $\mathsf{gr}(F)$, and it is inertially split since gr maps $Br_{is}(F)$ to $\mathsf{Br}_{is}\big(\mathsf{gr}(F)\big)$. From the definitions of θ_D and $\theta_{\mathsf{gr}(D)}$, it is clear that the following diagram commutes:

$$\begin{array}{ccc} \Gamma_{\mathsf{gr}(D)} & \xrightarrow{\theta_{\mathsf{gr}(D)}} & \mathcal{G}(Z(\mathsf{gr}(D)_0)/\,\mathsf{gr}(F)_0) \\ \| & & \| \\ \Gamma_D & \xrightarrow{\quad\theta_D\quad} & \mathcal{G}(Z(\overline{D})/\overline{F}) \end{array}$$

Therefore, $ker\,\theta_D = ker\,\theta_{\mathsf{gr}(D)}$. Since $\Gamma_{\mathsf{gr}(F)} = \Gamma_F$, we have $ker\,\theta_D = \Gamma_F$ by Prop. 6.22, and the lemma follows. □

Now, let $\mathcal{G}(\overline{F}) = \mathcal{G}(\overline{F}_{sep}/\overline{F})$ be the absolute Galois group of $\overline{F}$. For any inertially split central division algebra D over F, define the *ramification character of D*

$$\chi_D \colon \mathcal{G}(\overline{F}) \longrightarrow \mathbb{T}(\Gamma_F) \tag{6.39}$$

as the composition of the following homomorphisms:

$$\mathcal{G}(\overline{F}) \xrightarrow{\mathrm{res}_{Z(\overline{D})}} \mathcal{G}(Z(\overline{D})/\overline{F}) \xrightarrow{\overline{\theta}_D^{\,-1}} \Gamma_D/\Gamma_F \hookrightarrow \mathbb{T}(\Gamma_F).$$

Since $res_{Z(\overline{D})}$ is surjective and $\overline{\theta}_D^{\,-1}$ is an isomorphism, we have

$$im\,\chi_D = \Gamma_D/\Gamma_F \qquad \text{and} \qquad ker\,\chi_D = \mathcal{G}(Z(\overline{D})). \tag{6.40}$$

The ramification character χ_D is a continuous homomorphism, since it factors through $\mathcal{G}(Z(\overline{D})/\overline{F})$. We may therefore define the *ramification map*

$$\chi\colon\ Br_{is}(F) \longrightarrow Hom^c(\mathcal{G}(\overline{F}), \mathbb{T}(\Gamma_F)) \qquad \text{by} \qquad [D] \mapsto \chi_D. \tag{6.41}$$

As observed in the proof of Lemma 6.65, we have $\theta_D = \theta_{\mathsf{gr}(D)}$; hence, $\chi_D = \chi_{\mathsf{gr}(D)}$ and the following diagram commutes:

$$\begin{array}{ccccccccc} 0 & \longrightarrow & Br_{in}(F) & \hookrightarrow & Br_{is}(F) & \xrightarrow{\chi} & Hom^c(\mathcal{G}(\overline{F}), \mathbb{T}(\Gamma_F)) & \longrightarrow & 0 \\ & & \downarrow{\scriptstyle \mathsf{gr}}\,\cong & & \downarrow{\scriptstyle \mathsf{gr}}\,\cong & & \| & & \\ 0 & \longrightarrow & \mathsf{Br}_{in}(\mathsf{gr}(F)) & \hookrightarrow & \mathsf{Br}_{is}(\mathsf{gr}(F)) & \xrightarrow{\chi} & Hom^c(\mathcal{G}(\overline{F}), \mathbb{T}(\Gamma_F)) & \longrightarrow & 0 \end{array} \tag{6.42}$$

It follows that the upper map χ is a group homomorphism, and since the lower sequence is split exact by Th. 6.24 and Prop. 6.27, so is the upper sequence. By combining this exact sequence with the inverse of the isomorphism $\mathsf{gr}_0\colon Br_{in}(F) \xrightarrow{\sim} Br(\overline{F})$ of Th. 6.64, we readily obtain:

Theorem 6.66. *For (F, v) Henselian, the following sequence is split exact:*

$$0 \longrightarrow Br(\overline{F}) \xrightarrow{\mathsf{gr}_0^{-1}} Br_{is}(F) \xrightarrow{\chi} Hom^c(\mathcal{G}(\overline{F}), \mathbb{T}(\Gamma_F)) \longrightarrow 0.$$

Hence, there is a non-canonical direct product decomposition

$$Br_{is}(F) \cong Br(\overline{F}) \times Hom^c(\mathcal{G}(\overline{F}), \mathbb{T}(\Gamma_F)).$$

The exact sequence in Th. 6.66 is functorial with respect to scalar extensions of Henselian-valued fields. To make this statement precise, we need the following observation:

Lemma 6.67. *Let (K, w) be a Henselian-valued extension of a Henselian-valued field (F, v), and let K_{sep} be a separable closure of K. The inertial closure K_{in} of K in K_{sep} contains the inertial closure of F:*

$$F_{in} \subseteq K_{in}.$$

Therefore, the scalar extension map $\mathrm{ext}_{K/F}\colon Br(F) \to Br(K)$ maps $Br_{is}(F)$ to $Br_{is}(K)$.

Proof. It suffices to prove the inclusion $F_{in} \subseteq K_{in}$, since the rest follows easily. By Zorn's Lemma, there is a maximal inertial extension M of F in K_{in}. If $\overline{M} \neq \overline{F}_{sep}$, then there is a proper finite-degree separable extension S of $\overline{M}$ in the separably closed field $\overline{K_{in}}$. By Prop. A.17, there is a field L that is the inertial lift of S over M in K_{in}; so, $[L{:}M] = [S{:}\overline{M}] > 1$. Since L is inertial over M, which is inertial over F, it follows by Prop. A.5(ii) that

L is inertial over F, contradicting the maximality of M. Hence, $\overline{M} = \overline{F}_{sep}$, which implies that $M = F_{in}$, since inertial field extensions are determined by their associated graded fields (see Cor. 5.56), which are determined by their degree 0 components. Thus, $F_{in} \subseteq K_{in}$. □

Proposition 6.68. *For any Henselian-valued extension (K, w) of a Henselian-valued field (F, v), the following diagram commutes, where the map $\pi^*_{\overline{K}/\overline{F}}$ is defined as in (6.25):*

$$\begin{array}{ccccccccc} 1 & \longrightarrow & Br(\overline{F}) & \xrightarrow{\mathsf{gr}_0^{-1}} & Br_{is}(F) & \xrightarrow{\chi} & Hom^c(\mathcal{G}(\overline{F}), \mathbb{T}(\Gamma_F)) & \longrightarrow & 1 \\ & & \downarrow{\scriptstyle ext_{\overline{K}/\overline{F}}} & & \downarrow{\scriptstyle ext_{K/F}} & & \downarrow{\scriptstyle \pi^*_{\overline{K}/\overline{F}}} & & \\ 1 & \longrightarrow & Br(\overline{K}) & \xrightarrow{\mathsf{gr}_0^{-1}} & Br_{is}(K) & \xrightarrow{\chi} & Hom^c(\mathcal{G}(\overline{K}), \mathbb{T}(\Gamma_K)) & \longrightarrow & 1 \end{array} \tag{6.43}$$

Proof. In the following diagram, commutativity of the right square is clear, and commutativity of the left square follows from Prop. 6.59:

$$\begin{array}{ccccc} Br(\overline{F}) & \xleftarrow[\cong]{\mathsf{gr}_0} & Br_{in}(F) & \hookrightarrow & Br_{is}(F) \\ \downarrow{\scriptstyle ext_{\overline{K}/\overline{F}}} & & \downarrow{\scriptstyle ext_{K/F}} & & \downarrow{\scriptstyle ext_{K/F}} \\ Br(\overline{K}) & \xleftarrow[\cong]{\mathsf{gr}_0} & Br_{in}(K) & \hookrightarrow & Br_{is}(K) \end{array}$$

Commutativity of the left square of (6.43) follows. To see that the right square also commutes, we compare it to the corresponding diagram in the graded case by means of the following diagram:

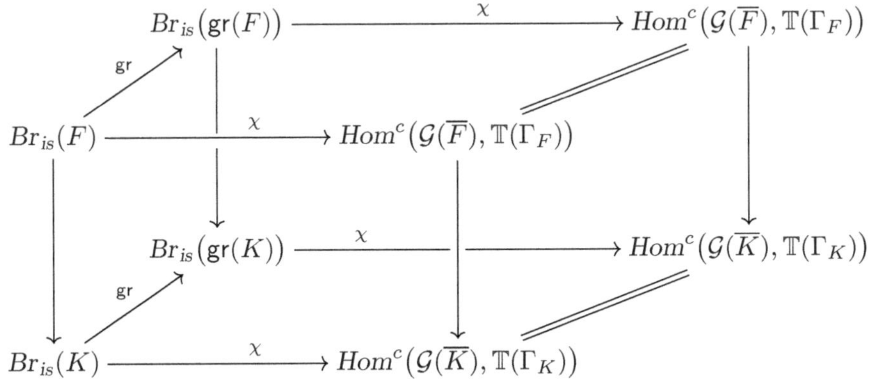

Commutativity of the top and bottom faces was observed in (6.42). The left face commutes by Prop. 6.59, and the right face obviously commutes. Since the maps gr are isomorphisms by Th. 6.64, commutativity of the front face follows from commutativity of the back face, which was proved in Prop. 6.25. □

Remark 6.69. It follows from Prop. 6.68 that for any totally ramified algebraic extension T of a Henselian field F

$$Br_{is}(F) \cap Br(T/F) \cong Hom^c(\mathcal{G}(\overline{F}), \Gamma_T/\Gamma_F) \tag{6.44}$$

(cf. the proof of Cor. 6.26). Now take for T any maximal totally ramified extension of F in F_{alg}. One can show that $\Gamma_T = \mathbb{H}(\Gamma_F)$ (see Exercise 6.18) and then deduce, as in the proof of Prop. 6.27, that

$$Br_{is}(F) \cong Br(\overline{F}) \times \big(Br_{is}(F) \cap Br(T/F)\big). \tag{6.45}$$

To close this section, we briefly discuss special cases where the rational rank[3] is 1, i.e., $dim_{\mathbb{Q}}\, \mathbb{H}(\Gamma_F) = 1$.

Proposition 6.70. *Let (F, v) be a Henselian-valued field such that $\mathrm{rr}(\Gamma_F) = 1$ (this holds for example when $\Gamma_F \cong \mathbb{Z}$), and let $\overline{p} = \mathrm{char}\,\overline{F}$. We have $Br_{is}(F) = Br_{tr}(F)$, and for every prime $q \neq \overline{p}$, the q-primary component of $Br(F)$ satisfies*

$$Br(F)_{(q)} \subseteq Br_{is}(F).$$

Moreover, we have a split exact sequence

$$0 \longrightarrow Br(\overline{F})_{(q)} \xrightarrow{\mathsf{gr}_0^{-1}} Br(F)_{(q)} \xrightarrow{\chi} Hom^c\big(\mathcal{G}(\overline{F}), \mathbb{T}(\Gamma_F)\big)_{(q)} \longrightarrow 0.$$

Proof. Since $rr(\Gamma_F) = 1$, Prop. 6.16 yields $\mathsf{Br}\big(\mathsf{gr}(F)\big) = \mathsf{Br}_{is}\big(\mathsf{gr}(F)\big)$. The following diagram, extracted from (6.37), then shows that $Br_{is}(F) = Br_{tr}(F)$:

$$\begin{array}{ccc} Br_{is}(F) & \hookrightarrow & Br_{tr}(F) \\ \mathsf{gr}\downarrow\cong & & \mathsf{gr}\downarrow\cong \\ \mathsf{Br}_{is}\big(\mathsf{gr}(F)\big) & = & \mathsf{Br}\big(\mathsf{gr}(F)\big) \end{array}$$

If $q \neq \overline{p}$, we have $Br(F)_{(q)} \subseteq Br_{tr}(F)$ by Prop. 6.56, hence $Br(F)_{(q)} \subseteq Br_{is}(F)$. We may then take the q-primary component of each term in the exact sequence of Th. 6.66 to complete the proof. □

Our final result is a classical theorem of Witt:

Theorem 6.71 (Witt). *Let (F, v) be a Henselian-valued field such that $\Gamma_F \cong \mathbb{Z}$ and $\overline{F}$ is perfect. Then $Br(F) = Br_{is}(F)$ and the following sequence is split exact:*

$$0 \longrightarrow Br(\overline{F}) \xrightarrow{\mathsf{gr}_0^{-1}} Br(F) \xrightarrow{\chi} Hom^c\big(\mathcal{G}(\overline{F}), \mathbb{Q}/\mathbb{Z}\big) \longrightarrow 0.$$

Proof. In view of Th. 6.66, it suffices to prove the first assertion, which follows immediately once we show that $Br(F_{in}) = 0$. For this, suppose D is a central division algebra over F_{in}. Since F_{in} is Henselian, its valuation extends to a

[3] See §A.4 in Appendix A.

valuation v_D on D, and we have $\Gamma_D \cong \mathbb{Z}$ since $\Gamma_{F_{in}} = \Gamma_F \cong \mathbb{Z}$. Therefore, by Prop. 4.16, v_D is defectless, and we thus have

$$[D{:}F_{in}] = [\overline{D}{:}\overline{F_{in}}] \cdot |\Gamma_D{:}\Gamma_F|.$$

Now, $\overline{F_{in}} = \overline{F}_{sep}$. This field is algebraically closed since $\overline{F}$ is perfect. Therefore, $\overline{D} = \overline{F_{in}}$ and the equation above reduces to $[D{:}F] = |\Gamma_D{:}\Gamma_F|$. Since $\Gamma_F \cong \mathbb{Z}$, the quotient Γ_D/Γ_F is cyclic. Choose $x \in D$ so that $v_D(x)$ maps to a generator of Γ_D/Γ_F. Then,

$$[F_{in}(x){:}F_{in}] \geq |\Gamma_D{:}\Gamma_F| = [D{:}F_{in}].$$

On the other hand, as $F_{in}(x)$ is a subfield of D we have $[F_{in}(x){:}F_{in}] \leq deg\, D$, so

$$deg\, D \geq [D{:}F_{in}] = (deg\, D)^2.$$

It follows that $deg\, D = 1$, hence D is split. □

Note that the hypothesis that $\overline{F}$ be perfect cannot be dispensed with: Ex. 6.36(ii) yields examples of division algebras that are not inertially split. The center can be chosen to be Henselian for a discrete valuation.

Exercises

Exercise 6.1. Let F be a graded field, and let $\mathsf{F}^\natural$ be F with the grading ignored. This exercise explores the relation between Brauer groups $\mathsf{Br}(\mathsf{F})$ and $Br(\mathsf{F}^\natural)$. Recall first (see Matsumura [137, pp. 166–167]) that if R is an integrally closed Noetherian integral domain, then the canonical map from its Picard group $Pic(R)$ of finitely generated rank 1 projective modules to its divisor class group $\mathcal{C}\ell(R)$ is injective. If R is also a unique factorization domain, it follows that $Pic(R) = \{0\}$, since $\mathcal{C}\ell(R) = \{0\}$. Because $\mathsf{F}^\natural$ is a direct limit of such rings (see Lemma 5.1), $Pic(\mathsf{F}^\natural) = \{0\}$. Recall also the theorem of Auslander–Goldman [18, Th. 7.2] that if R is a regular ring then the scalar extension map $q_R\colon Br(R) \to Br(q(R))$ is injective. (For another proof of this, see Orzech–Small [174, Th. 6.19, p. 67].) Since $\mathsf{F}^\natural$ is a direct limit of regular rings, it follows that $Br(\mathsf{F}^\natural)$ embeds in $Br(q(\mathsf{F}^\natural))$.

(i) Let K be a finite-degree Galois graded field extension of F. Prove that

$$Br(\mathsf{K}^\natural/\mathsf{F}^\natural) \cong H^2(\mathcal{G}(\mathsf{K}/\mathsf{F}), \mathsf{K}^\times) \cong \mathsf{Br}(\mathsf{K}/\mathsf{F}).$$

(Use the 6-term exact sequence for the Brauer group of a Galois extension of rings, see Demeyer–Ingraham [59, Th. 1.1, p. 116].)

(ii) Prove that $\mathsf{Br}(\mathsf{F}) \cong Br((\mathsf{F}_{tr})^\natural/\mathsf{F}^\natural)$.

(iii) Let k be a perfect field, and let $\mathsf{F} = k[X, X^{-1}]$, with its usual $\mathbb{Z}$-grading. Prove that

$$Br(\mathsf{F}^\natural) = Br(\mathsf{F}^\natural_{in}/\mathsf{F}^\natural) \cong \mathsf{Br}(\mathsf{F}).$$

(Whenever k is not perfect, so has prime characteristic, $\mathsf{Br}(\mathsf{F})$ is a proper subgroup of $Br(\mathsf{F}^\natural)$—see Exercise 8.11 of Chapter 8.)

Exercise 6.2. Let D be a graded division algebra finite-dimensional over its center F and let L be a graded field with $\mathsf{F} \subseteq \mathsf{L} \subseteq \mathsf{D}$. Let C be the centralizer $C_{\mathsf{D}}(\mathsf{L})$. Make D into a (simple) graded right $\mathsf{D} \otimes_{\mathsf{F}} \mathsf{L}$-module by $d' \cdot \left(\sum d_i \otimes \ell_i\right) = \sum \ell_i d' d_i$. Prove that $\mathsf{End}_{\mathsf{D}\otimes_{\mathsf{F}}\mathsf{L}}(\mathsf{D}) \cong_g \mathsf{C}$. Deduce that $[\mathsf{D} \otimes_{\mathsf{F}} \mathsf{L}] = [\mathsf{C}]$ in $\mathsf{Br}(\mathsf{L})$. (The proof for this is analogous to the ungraded case.)

Exercise 6.3. This is a generalization of Lemma 6.21. Let $\mathsf{F} \subseteq \mathsf{K}$ be a finite-degree Galois extension of graded fields, and let $G = \mathcal{G}(\mathsf{K}/\mathsf{F})$. Let $f \in Z^2(G, \mathsf{K}^\times)$ be a normalized 2-cocycle, and let

$$\mathsf{A} = (\mathsf{K}/\mathsf{F}, G, f) = \bigoplus_{\sigma \in G} \mathsf{K} x_\sigma$$

be the associated graded crossed product algebra, as in (6.2)–(6.4). Let $\eta_f \colon G \to \mathbb{T}(\Gamma_{\mathsf{K}})$ be given by $\sigma \mapsto deg\, x_\sigma + \Gamma_{\mathsf{K}}$, which is a well-defined group homomorphism by (6.6), and let $H = ker\, \eta_f$. Prove that if each $\sigma \in H$ restricts to a different automorphism of K_0, then $Z(\mathsf{A}_0) = \mathsf{K}_0^H$ and $\mathsf{A}_0 = (\mathsf{K}_0/\mathsf{K}_0^H, H, g_0)$, a (simple) crossed product algebra, for some $g_0 \in Z^2(H, \mathsf{K}_0^\times)$.

Exercise 6.4. Let D, D' be inertially split central graded division algebras over a graded field F and let $\mathsf{A} = \mathsf{D} \otimes_{\mathsf{F}} \mathsf{D}'$. Show that $\Gamma_{\mathsf{A}}^\times = \Gamma_{\mathsf{A}}$ and that $Z(\mathsf{A}_0)$ is the F_0-subalgebra of $Z(\mathsf{D}_0) \otimes_{\mathsf{F}_0} Z(\mathsf{D}'_0)$ fixed under the automorphisms of the form $\theta_{\mathsf{D}}(\gamma) \otimes \theta_{\mathsf{D}'}(\gamma)^{-1}$, for $\gamma \in \Gamma_{\mathsf{D}} \cap \Gamma_{\mathsf{D}'}$. Use this information to relate θ_{A} to $\theta_{\mathsf{D}} \times \theta_{\mathsf{D}'} \colon \Gamma_{\mathsf{D}} \times \Gamma_{\mathsf{D}'} \to Aut_{\mathsf{F}_0}\big(Z(\mathsf{D}_0) \otimes_{\mathsf{F}_0} Z(\mathsf{D}'_0)\big)$ in order to prove by nonhomological arguments that the ramification map $\chi \colon \mathsf{Br}_{is}(\mathsf{F}) \to Hom^c\left(\mathcal{G}(\mathsf{F}_0), \mathbb{T}(\Gamma_F)\right)$ is a group homomorphism. Give a nonhomological proof of Prop. 6.25 in the same style.

Exercise 6.5. For a graded division algebra D with center F and any integer $j > 0$, let $\mathsf{D}^{\otimes j} = \mathsf{D} \otimes_{\mathsf{F}} \ldots \otimes_{\mathsf{F}} \mathsf{D}$ (j times), and let D^j denote the associated graded division algebra of $\mathsf{D}^{\otimes j}$. Assuming D is inertially split, show that for any integer $j > 0$,

$$\Gamma_{\mathsf{D}^j} = j\Gamma_{\mathsf{D}} + \Gamma_{\mathsf{F}},$$

and $Z((\mathsf{D}^j)_0)$ is the subfield in $Z(\mathsf{D}_0)$ fixed under the j-torsion subgroup of $\mathcal{G}(Z(\mathsf{D}_0)/\mathsf{F}_0)$. (See also Prop. 8.47 below.)

Exercise 6.6. Let $\mathsf{F} \subseteq \mathsf{K}$ be graded fields, let D be an inertially split graded division algebra with center F, and let E be the associated graded division algebra of $\mathsf{D} \otimes_{\mathsf{F}} \mathsf{K}$. Let Q be the fixed field $Z(\mathsf{D}_0)^{\theta_{\mathsf{D}}(\Gamma_{\mathsf{D}} \cap \Gamma_{\mathsf{K}})}$; so, Q is the field with $\mathsf{F}_0 \subseteq Q \subseteq Z(\mathsf{D}_0)$ such that $\theta_{\mathsf{D}}(\Gamma_{\mathsf{D}} \cap \Gamma_{\mathsf{K}}) = \mathcal{G}(Z(\mathsf{D}_0)/Q)$. Prove that E is inertially split, $Z(\mathsf{E}_0) = Q \cdot \mathsf{K}_0$,

$$\Gamma_{\mathsf{E}} = \theta_{\mathsf{D}}^{-1}\big(\mathcal{G}(Z(\mathsf{D}_0)/[\mathsf{K}_0 \cap Z(\mathsf{D}_0)])\big) + \Gamma_{\mathsf{K}} \subseteq \Gamma_{\mathsf{D}} + \Gamma_{\mathsf{K}},$$

and θ_{E} is the map such that $\Gamma_{\mathsf{K}} \subseteq \mathit{ker}\,\theta_{\mathsf{E}}$ and the following diagram commutes:

$$\begin{array}{ccc}
\theta_{\mathsf{D}}^{-1}\big(\mathcal{G}(Z(\mathsf{D}_0)/[\mathsf{K}_0 \cap Z(\mathsf{D}_0)])\big) & \longrightarrow & \theta_{\mathsf{D}}^{-1}\big(\mathcal{G}(Z(\mathsf{D}_0)/[\mathsf{K}_0 \cap Z(\mathsf{D}_0)])\big) + \Gamma_{\mathsf{K}} \\
\downarrow{\scriptstyle \theta_{\mathsf{D}}} & & \downarrow{\scriptstyle \theta_{\mathsf{E}}} \\
\mathcal{G}(Z(\mathsf{D}_0)/[\mathsf{K}_0 \cap Z(\mathsf{D}_0)]) & & \mathcal{G}((Q \cdot \mathsf{K}_0)/\mathsf{K}_0) \\
\searrow & & \nearrow{\scriptstyle \cong} \\
& \mathcal{G}(Q/(\mathsf{K}_0 \cap Q)) &
\end{array}$$

Exercise 6.7. Let (L, v_L) be an inertial valued field extension of (F, v). Let A be a central simple F-algebra with a residually simple v-gauge α. Let B be a central simple L-algebra Brauer-equivalent to $L \otimes_F A$, and suppose B has a residually simple v_L-gauge β. Show that

(i) B_0 is Brauer-equivalent to $\overline{L}\cdot Z(A_0) \otimes_{Z(A_0)} A_0$.
(ii) $\Gamma_B = \Gamma_B^\times = \theta_A^{-1}\big(\mathcal{G}(Z(A_0)/\,[\overline{L} \cap Z(A_0)])\big) \subseteq \Gamma_A$.
(iii) There is a commutative diagram:

$$\begin{array}{ccc}
\Gamma_B^\times & \hookrightarrow & \Gamma_A^\times \\
\downarrow{\scriptstyle \theta_B} & & \downarrow{\scriptstyle \theta_A} \\
\mathcal{G}\big(\overline{L}\cdot Z(A_0)/\overline{L}\big) & \xrightarrow{\text{res}} & \mathcal{G}(Z(A_0)/\overline{F}\,)
\end{array}$$

[Note that since $Z(A_0)$ is normal over $\overline{F}$ by Cor. 4.32, the fields $\overline{L}\cdot Z(A_0)$ and $\overline{L} \cap Z(A_0)$ are independent of the choice of embedding of $\overline{L}$ and $Z(A_0)$ in a separable closure of $\overline{F}$. Moreover, $\overline{L} \cap Z(A_0)$ is uniquely determined as a subfield of $Z(A_0)$, since $\mathcal{G}(Z(A_0)/\overline{F})$ is abelian. Use Prop. 2.58.]

Exercise 6.8. Let (F, v) be a valued field, let A be a central simple F-algebra with an inertial v-gauge α, and let B be a simple F-algebra with a v-gauge β. Let $C = A \otimes_F B$, with its v-gauge $\alpha \otimes \beta$ (see Prop. 6.38).

(i) Prove that $\Gamma_C^\times = \Gamma_B^\times$ and that $\theta_C = \theta_B$.
(ii) Prove that there is a one-to-one-correspondence between the simple components of C_0 and those of B_0, and that the action of $\Gamma_C^\times$ on the components of C_0 coincides with the action of $\Gamma_B^\times$ on the components of B_0.

Exercise 6.9. Let v be a valuation on a field F with valuation ring $\mathcal{O}_F$, and let A be a central simple F-algebra. Let R be an $\mathcal{O}_F$-subalgebra of A with $F \cdot R = A$. Prove that R is the gauge ring $\mathcal{O}_{A,\alpha}$ for some inertial v-gauge α on A if and only if R is an Azumaya algebra over $\mathcal{O}_F$.

Exercise 6.10. Let (F, v) be a valued field, and let α be a v-gauge on a simple F-algebra A. Let β be any coarsening of α as in §3.2.7, let $w = \beta|_F$, and let $u = v/w$, the valuation on $\overline{F}^w$ induced by v. Prove that α is an inertial v-gauge if and only if β is a w-gauge and α_0 is an inertial u-gauge on A_0^β (the degree 0 component of $\mathsf{gr}_\beta(A)$). (Use Exercise 6.9.)

Exercise 6.11. Let (F, v) be a valued field, and let A be a central simple F-algebra. Prove that A has an inertial gauge if and only if $A \otimes_F F_{h,w}$ has an inertial gauge for each coarsening w of v. Here, $F_{h,w}$ denotes the Henselization of F with respect to w. (Note: It is possible for $A \otimes_F F_{h,v}$ to have an inertial gauge while A does not. See the following exercise.)

Exercise 6.12. Let F be the rational function field $k(x, y)$ for any field k with $char(k) \neq 2$, and let v be the restriction to F of the standard rank 2 Henselian valuation on $k((x))((y))$. Let A be the quaternion division algebra $\big(1 + x,\, y/F\big)$, as in Ex. 4.58 above. The Henselization F_h of F re v splits A since $1 + x \in F_h^2$. Show that $[A] \notin Br_{in}(F, v)$. (Hint: Let w be the rank 1 coarsening of v on F. Prove that $[A] \notin Br_{in}(F, w)$ and invoke Exercise 6.10.)

Exercise 6.13. Let (F, v) be a valued field and let A be a semisimple F-algebra. Prove that every inertial v-gauge on A is a tame gauge.

Exercise 6.14. Let (F, v) be an arbitrary valued field and A, B be semisimple F-algebras with respective v-gauges α and β. Prove:

(i) If α is a tame v-gauge, then $\alpha \otimes \beta$ is a v-gauge on $A \otimes_F B$.
(ii) $\alpha \otimes \beta$ is a tame v-gauge if and only if α and β are each tame v-gauges.

Exercise 6.15. Let F be a Henselian field, let A and B be tame division algebras with center F, and let D be the associated division algebra of $A \otimes_F B$. Prove that $\Gamma_D \subseteq \Gamma_A + \Gamma_B$. (The tameness hypothesis cannot be omitted here: see Exercise 8.8 below.)

Exercise 6.16. Let (F, v) be a Henselian-valued field and let D be a central division algebra over F. Also let $D_{(p)}$ be the p-primary component of D, for each prime factor p of $deg\, D$; i.e., $D \cong \bigotimes_p D_{(p)}$ and $deg\, D_{(p)}$ is a power of p. Show that the following conditions are equivalent:

(a) the valuation v_D on D extending v is inertial;
(b) $[D] \in Br_{in}(F)$;
(c) for each prime factor p of $deg\, D$, the valuation $v_{D_{(p)}}$ on $D_{(p)}$ extending v is inertial.

Exercise 6.17. Let (F, v) be a Henselian-valued field and let D be a central division algebra over F. With the notation as in the preceding problem, show that the following conditions are equivalent:

(a) the valuation v_D on D extending v is tame;
(b) $[D] \in Br_{tr}(F)$;
(c) for each prime factor p of $deg\, D$, the valuation $v_{D_{(p)}}$ on $D_{(p)}$ extending v is tame;

(d) either $char\,\overline{F} = 0$ or $char\,\overline{F} = \overline{p} \neq 0$ and $[D_{(\overline{p})}] \in Br_{is}(F)$;
(e) D has a maximal subfield tamely ramified over F.

Exercise 6.18. Let F be a field with a valuation, and let T be a maximal totally ramified extension of F in an algebraic closure of F. (Such a T exists by Zorn's Lemma.) Prove that $\Gamma_T = \mathbb{H}(\Gamma_F)$. (Hint: Use Prop. A.5.)

Exercise 6.19. Let F be a field with Henselian valuation v, let K be a cyclic Galois field extension of F of degree n and let $\mathcal{G}(K/F) = \langle\sigma\rangle$. Assume that $char\,\overline{F} \nmid n$, so K is tamely ramified over F. Let $e = |\Gamma_K{:}\Gamma_F|$ and $f = [\overline{K}{:}\overline{F}]$. Take some $u \in \mathcal{O}_F^\times$ such that $[\overline{F}(\sqrt[n]{\overline{u}}\,){:}\overline{F}] = n$, and let A be the cyclic F-algebra

$$A = (K/F, \sigma, u) = \bigoplus_{i=0}^{n-1} Kx^i \quad \text{where } xcx^{-1} = \sigma(c) \text{ for } c \in K \text{ and } x^n = u.$$

(i) Show that A is inertially split.
(ii) Define a v-value function α on A by

$$\alpha\Big(\sum_{i=0}^{n-1} c_i x^i\Big) = min\,\big(v(c_0), \ldots, v(c_{n-1})\big) \qquad \text{for } c_i \in K.$$

Prove that α is a tame v-gauge on A. Note that $\Gamma_A = \Gamma_A^\times = \Gamma_K$.
(iii) Prove that

$$A_0 = \bigoplus_{i=0}^{n-1} \overline{K}\,\widetilde{x}^i \qquad \text{and} \qquad Z(A_0) = \bigoplus_{j=0}^{e-1} \overline{F}\,\widetilde{x}^{fj} \cong \overline{F}(\sqrt[e]{\overline{u}}\,).$$

(iv) If $\overline{K} \otimes_{\overline{F}} \overline{F}(\sqrt[e]{\overline{u}}\,)$ is a field, prove that A_0 is the cyclic algebra of degree f

$$A_0 = \big((\overline{K} \otimes_{\overline{F}} \overline{F}(\sqrt[e]{\overline{u}}))/\overline{F}(\sqrt[e]{\overline{u}}\,), \sigma', \sqrt[e]{\overline{u}}\,\big),$$

where $\sigma' = \overline{\sigma} \otimes id$, with $\overline{\sigma} \in \mathcal{G}(\overline{K}/\overline{F})$ the automorphism induced by σ.
(v) Let L be the inertial lift of $\overline{K}$ over F in K. So, $\overline{L} = \overline{K}$ and $\Gamma_L = \Gamma_F$. Moreover, K is totally and tamely ramified and Galois over L with $[K{:}L] = e$ and $\mathcal{G}(K/L) = \langle\sigma^f\rangle$. Recall (see Prop. A.22) that K is an e-Kummer extension of L, and there is a $t \in K^\times$ such that $K = L(t)$, $t^e \in L$, and $\Gamma_K/\Gamma_L = \langle v(t) + \Gamma_L\rangle$. So, $\sigma^f(t) = \omega t$ for some primitive e-th root of unity ω. Prove that $\theta_A(v(t))$ is the $\overline{F}$-automorphism τ of $\overline{F}(\sqrt[e]{\overline{u}}\,)$ such that $\tau(\sqrt[e]{\overline{u}}\,) = \overline{\omega}^{-1}\sqrt[e]{\overline{u}}$.

Notes

§6.1.1: The Brauer group $\mathsf{Br}(\mathsf{F})$ of central simple graded algebras over a graded field F was defined by Boulagouaz [25] who also proved that $\mathsf{Br}(\mathsf{F})$ embeds in $Br(\mathsf{F}^\natural)$ and $Br(q(\mathsf{F}))$. Earlier, Caenepeel and van Oystaeyen [47] had treated the Brauer group of graded Azumaya algebras A over a graded ring R with $\Gamma_{\mathsf{R}} = \mathbb{Z}$, but limited consideration to the case where $\Gamma_{\mathsf{A}} = \mathbb{Z}$.

For any graded field F, we have seen in Exercise 6.1 that we may identify $\mathsf{Br}(\mathsf{F})$ with $Br(\mathsf{F}^{\natural}_{tr}/\mathsf{F}^{\natural})$ in $Br(\mathsf{F}^{\natural})$. While $\mathsf{Br}(\mathsf{F})$ can be a proper subgroup of $Br(\mathsf{F}^{\natural})$, as Exercise 8.11 below demonstrates, these Brauer groups can differ only in their *char* F_0-primary components: Hoobler [101, §2] has analyzed the n-torsion group ${}_nBr(R[X,X^{-1}])$ in terms of ${}_nBr(R)$ for an integrally closed Noetherian commutative ring R with no nontrivial idempotents and n prime to all the residue characteristics of R. (See Ford [80] for a different proof with fewer restrictions on R.) Hoobler then gives a formula for ${}_nBr(R[X_1,X_1^{-1},\dots,X_m,X_m^{-1}])$ for all m in terms of ${}_nBr(R)$. Since a graded field F is a direct limit of Laurent polynomial rings over F_0, it follows from Hoobler's result and the corresponding information about $\mathsf{Br}(\mathsf{F})$ that

$$ {}_n\mathsf{Br}(\mathsf{F}) \;=\; {}_nBr(\mathsf{F}^{\natural}) \qquad \text{for all } n \text{ prime to } char\,\mathsf{F}_0. $$

§6.1.2: The homological interpretation of $\mathsf{Br}(\mathsf{F})$ for a graded field F was given in Hwang–Wadsworth [103]. Earlier, Caenepeel–van Oystaeyen [47] treated homologically the Brauer group of $\mathbb{Z}$-graded algebras.

§6.1.3 For any graded field F, homological arguments using the exact sequence

$$ 1 \longrightarrow (\mathsf{F}_0)^{\times}_{sep} \longrightarrow (\mathsf{F}_{in})^{\times} \xrightarrow{deg} \Gamma_{\mathsf{F}} \longrightarrow 0, $$

and the isomorphism $H^2(\mathcal{G}(\mathsf{F}_0),\Gamma_{\mathsf{F}}) \cong H^1(\mathcal{G}(\mathsf{F}_0),\mathbb{T}(\Gamma_{\mathsf{F}}))$ as $\mathbb{H}(\Gamma_{\mathsf{F}})$ is uniquely divisible, yield the exact sequence

$$ 0 \longrightarrow \mathsf{Br}_{in}(\mathsf{F}) \longrightarrow \mathsf{Br}_{is}(\mathsf{F}) \xrightarrow{\partial^{-1}\circ deg} Hom^c\big(\mathcal{G}(\mathsf{F}_0),\mathbb{T}(\Gamma_{\mathsf{F}})\big). $$

If Γ_{F} is finitely generated, then it is a free $\mathbb{Z}$-module, hence there are homomorphisms $\Gamma_{\mathsf{F}} \to (\mathsf{F}_{in})^{\times}$ that split the degree map. It is then clear that the map $\partial^{-1}\circ deg$ is a split epimorphism, providing a short proof of Th. 6.24 and Prop. 6.27 in that case.

The exact sequence for $\mathsf{Br}_{is}(\mathsf{F})$ in Th. 6.24 was motivated by the corresponding result for $Br_{is}(F)$ for a Henselian field F given in Scharlau [225] and Jacob–Wadsworth [106]. Likewise, the description in Prop. 6.22 and Prop. 6.23 of the ramification character χ_{D} of an inertially split graded division algebra D was suggested by the corresponding results in [106] for inertially split division algebras over a Henselian field.

§6.2.2: Tame gauges were introduced in Tignol–Wadsworth [246].

Lemma 6.44 on $Z(\mathsf{gr}(D))$ for a valued division algebra D and Prop. 6.45 on the graded characterization of a tame valuation on D were given by Boulagouaz [25, Th. 4.3, Cor. 4.4].

The construction of tame gauges on crossed products in Prop. 6.46 can be generalized to eliminate the hypothesis that v has a unique extension to K: suppose K is a finite-degree Galois extension field of F and that the valuation v on F is defectless in K. Let $\eta = min(v_1,\dots,v_r)$ be the unique

v-gauge on K, where $v_1, \dots, v_r$ are all the valuations on K extending v. Suppose η is a tame v-gauge on K. We have $\mathsf{gr}_\eta(K) \cong_g \prod_{i=1}^r \mathsf{gr}_{v_i}(K)$, which is a separable $\mathsf{gr}(F)$-algebra, though not a graded field if $r > 1$. Moreover, $\mathsf{gr}_\eta(K)$ is G-Galois over $\mathsf{gr}(F)$, where $G = \mathcal{G}(K/F)$. This can be shown, e.g., by scalar extension to the Henselization F_h. Furthermore, if $f \in Z^2(G, K^\times)$ is any 2-cocycle, then the crossed product algebra $A = (K/F, G, f)$ has a tame v-gauge α with $\mathsf{gr}_\alpha(A) \cong_g (\mathsf{gr}_\eta(K)/\mathsf{gr}(F), G, \widetilde{f})$.

§6.2.3: Let F be a Henselian field. The key isomorphism

$$\mathsf{gr}\colon\ Br(F_{tr}/F) \xrightarrow{\sim} \mathsf{Br}(\mathsf{gr}(F))$$

given in Th. 6.64 was originally proved by Hwang–Wadsworth [103, Th. 5.3], with a much better proof using gauges given in Tignol–Wadsworth [246, Th. 3.8]. This isomorphism is compatible not only with scalar extensions (see Prop. 6.59), but also with corestrictions. That is, if L is any finite-degree tame field extension of F, then there is a commutative diagram

$$\begin{array}{ccc} Br_{tr}(L) & \xrightarrow[\cong]{\mathsf{gr}} & \mathsf{Br}(\mathsf{gr}(L)) \\ \Big\downarrow{\scriptstyle cor_{L/F}} & & \Big\downarrow{\scriptstyle cor_{\mathsf{gr}(L)/\mathsf{gr}(F)}} \\ Br_{tr}(F) & \xrightarrow[\cong]{\mathsf{gr}} & \mathsf{Br}(\mathsf{gr}(f)) \end{array}$$

This was proved in [103, Th. 6.1].

Diagram (6.37) has a cohomological version, which is obtained as follows: let $G = \mathcal{G}(F_{in}/F) \cong \mathcal{G}(\mathsf{gr}(F)_{in}/\mathsf{gr}(F)) \cong \mathcal{G}(\overline{F}_{sep})$, and let $\mathcal{O}_{F_{in}}^\times$ denote the group of units of the valuation ring $\mathcal{O}_{F_{in}}$. Also, let $\mathcal{G}(F) = \mathcal{G}(F_{sep}/F)$ and let $\mathcal{G}(\mathsf{gr}(F)) = \mathcal{G}(\mathsf{gr}(F)_{tr}/\mathsf{gr}(F))$. Then, there is a commutative diagram of cohomology groups, with the obvious maps,

$$\begin{array}{ccccccc} H^2(G, \mathcal{O}_{F_{in}}^\times) & \longrightarrow & H^2(G, F_{in}^\times) & \longrightarrow & H^2(\mathcal{G}(F_{tr}/F), F_{tr}^\times) & \longrightarrow & H^2(\mathcal{G}(F), F_{sep}^\times) \\ \Big\downarrow & & \Big\downarrow & & \Big\downarrow & & \\ H^2(G, \overline{F}_{sep}^\times) & \to & H^2(G, \mathsf{gr}(F_{in})^\times) & \to & H^2(\mathcal{G}(\mathsf{gr}(F)), \mathsf{gr}(F)_{tr}^\times) & & \end{array} \tag{6.46}$$

This diagram is actually isomorphic to the Brauer group diagram (6.37). For each cohomology group in (6.46) maps isomorphically to the Brauer group in the corresponding position in (6.37). To see this for the left maps, note that $H^2(G, \mathcal{O}_{F_{in}}^\times)$ maps into $Br_{in}(F)$ since Ex. 6.37 shows that crossed product algebras for inertial Galois field extensions determined by cocycles with values of degree 0 are inertial. This map is surjective since its image contains a generating set of $Br_{in}(F)$ (see the proof of surjectivity of gr in the proof of Th. 6.64). It is injective because the map $H^2(G, F_{in}^\times) \xrightarrow{\sim} Br_{is}(F)$ is injective and the map $H^2(G, \mathcal{O}_{F_{in}}^\times) \to H^2(G, F_{in}^\times)$ is injective, as $H^1(G, \Gamma_F) = Hom^c(G, \Gamma_F) = 0$ since G acts trivially on Γ_F, which is torsion-free. The map $H^2(G, \overline{F}_{sep}^{\times}) \xrightarrow{\sim} \mathsf{Br}_{in}(\mathsf{gr}(F))$ is the composition of isomorphisms $H^2(G, \overline{F}_{sep}^{\times}) \xrightarrow{\sim} Br(\overline{F}) \xrightarrow{\sim} \mathsf{Br}_{in}(\mathsf{gr}(F))$. See Remark 6.47 for commutativity of the diagram mapping (6.37) to (6.46). Because the vertical maps in (6.37) are isomorphisms, it follows that the vertical maps in (6.46) are isomorphisms. This is not evident by cohomological arguments.

The ramification map χ of (6.41) is definable homologically working from the exact sequence:

$$0 \longrightarrow \mathcal{O}_{F_{in}}^{\times} \longrightarrow F_{in} \longrightarrow \Gamma_F \longrightarrow 0.$$

When v is not Henselian, it is not known in general whether the map gr_0 of Cor. 6.58 is surjective. On this question, note that if n is a positive integer prime to *char* $\overline{F}$ and F contains a primitive n-th root of unity, then $\overline{F}$ contains a primitive n-th root of unity, and degree n symbol algebras over $\overline{F}$ can be lifted to inertial degree n symbol algebras over F—see Ex. 6.36(i) above. Then, ${}_nBr(\overline{F}) \subseteq im(\mathsf{gr}_0)$, since ${}_nBr(\overline{F})$ is generated by degree n symbol algebras by the Merkurjev–Suslin Theorem. It follows further from work of Merkurjev in [142] that (i) if n is any odd integer, then ${}_nBr(\overline{F}) \subseteq im(\mathsf{gr}_0)$; (ii) if $q = 2^k$ for some $k \in \mathbb{N}$, then ${}_qBr(\overline{F}) \subseteq im(\mathsf{gr}_0)$ if *char* $\overline{F} = 2$ or if *char* $\overline{F} \neq 2$ and $\overline{F}(\omega_q)$ is cyclic Galois over $\overline{F}$, where ω_q is a primitive q-th root of unity.

For any valuation v on F, let $Br(\mathcal{O}_F)$ be the Brauer group of equivalence classes of Azumaya algebras over $\mathcal{O}_F$. Exercise 6.9 shows that $Br_{in}(F)$ is the image of the scalar extension map $ext_{F/\mathcal{O}_F} \colon Br(\mathcal{O}_F) \to Br(F)$, $[R] \mapsto [R \otimes_{\mathcal{O}_F} F]$. There is another scalar extension map,

$$ext_{\overline{F}/\mathcal{O}_F} \colon Br(\mathcal{O}_F) \longrightarrow Br(\overline{F}), \quad [R] \mapsto [R \otimes_{\mathcal{O}_F} (\mathcal{O}_F/\mathfrak{m}_F)].$$

It is easy to see that $ext_{\overline{F}/\mathcal{O}_F} = \mathsf{gr}_0 \circ ext_{F/\mathcal{O}_F}$. Since $ext_{F/\mathcal{O}_F}$ is known to be injective for any valuation ring $\mathcal{O}_F$, see, e.g., Saltman [222, Lemma 1.2] or Jacob–Wadsworth [106, Prop. 2.5], these observations yield a different proof of the well-definition of gr_0 than the one given in Prop. 6.57.

Witt proved Th. 6.71 in [260] for a field F with a complete discrete rank 1 valuation with perfect residue field. More precisely, he proved the unique decomposition $D \sim I \otimes_F N_\psi$ of Prop. 8.77 below for any division algebra D with center F.

Exercises 6.6 and 6.7 are based on Jacob–Wadsworth [106, Th. 5.10, Th. 3.1].

Chapter 7
Total Ramification

Alternating pairings on finite abelian groups arise in two different ways in connection with central simple algebras: If the central simple F-algebra A has a decomposition into a tensor product of symbol algebras, then the factor group $A^\times/F^\times$ contains a finite abelian subgroup generated by the images of symbol algebra generators, and this subgroup, which we call an *armature*, carries an alternating pairing with values in the group of roots of unity in F, the *commutator pairing*. On the other hand, when a valued division algebra D is totally ramified over its center F, the factor group Γ_D/Γ_F carries a canonical alternating pairing with values in the group of roots of unity in the residue field $\overline{F}$. When F is strictly Henselian, this canonical pairing will allow us to set up a bijection between tame central division algebras over F and nondegenerate pairings on subgroups of $\mathbb{T}(\Gamma_F)$.

To prepare for their subsequent use, we first study alternating pairings on finite abelian groups in a general framework, without reference to central simple algebras. In §7.1, we analyze the collection $\mathit{Symp}(\Omega)$ of nondegenerate alternating pairings on finite subgroups of an arbitrary torsion abelian group Ω, and show that $\mathit{Symp}(\Omega)$ has a canonical torsion abelian group structure. If Γ is a torsion-free abelian group, then for the torsion group $\mathbb{T}(\Gamma) = \Gamma \otimes_{\mathbb{Z}} (\mathbb{Q}/\mathbb{Z})$ we prove in Th. 7.22 that $\mathit{Symp}(\mathbb{T}(\Gamma)) \cong \mathbb{T}(\wedge^2\Gamma)$. We next consider in §7.2 armatures in algebras and their homogeneous counterparts in graded algebras. Of particular interest is the construction of gauges on central simple algebras over valued fields from armatures of the algebras, see §7.2.3.

In the last two sections, §7.3 and §7.4, the focus is on totally ramified graded or valued division algebras. The case of graded division algebras is considered first. For a totally ramified graded division algebra D over a graded field F, the factor group $\mathsf{D}^\times/\mathsf{F}^\times$ is a homogeneous armature isomorphic to $\Gamma_{\mathsf{D}}/\Gamma_{\mathsf{F}}$ via the degree map. This map carries the commutator pairing on the armature $\mathsf{D}^\times/\mathsf{F}^\times$ to the canonical pairing on $\Gamma_{\mathsf{D}}/\Gamma_{\mathsf{F}}$. The results of §7.2 therefore provide a decomposition of the graded division algebra into

J.-P. Tignol, A.R. Wadsworth, *Value Functions on Simple Algebras, and Associated Graded Rings*, Springer Monographs in Mathematics,
DOI 10.1007/978-3-319-16360-4_7

a tensor product of symbol algebras. Moreover, when the graded field F is inertially closed (i.e., F_0 is separably closed), we obtain an isomorphism between $\mathsf{Br}(\mathsf{F})$ and the part of $\mathit{Symp}(\mathbb{T}(\Gamma_{\mathsf{F}}))$ of torsion prime to $\mathit{char}\,\mathsf{F}_0$, by mapping the Brauer class of each graded division algebra D to the canonical pairing on $\Gamma_{\mathsf{D}}/\Gamma_{\mathsf{F}}$. For an arbitrary graded field F, this isomorphism, applied to its inertial closure F_{in}, yields a description of the top layer $\mathsf{Br}(\mathsf{F})/\mathsf{Br}_{is}(\mathsf{F})$ in the canonical filtration of the Brauer group of F. Analogues of these results for division algebras over valued fields are obtained in §7.4. When the center is Henselian, we obtain a fairly accurate reproduction of the graded case situation, provided we restrict to tame totally ramified division algebras. When the valuation on the center is not Henselian, however, the results are significantly weaker; for example, we show in Ex. 7.77 that a tame totally ramified division algebra need not be a tensor product of symbol algebras.

7.1 Symplectic modules

This section consists of three parts. In §7.1.1 we establish a structure theorem for finite abelian groups (i.e., $\mathbb{Z}$-modules of finite length) with an alternating pairing with values in $\mathbb{Q}/\mathbb{Z}$. When the pairing is nondegenerate the module has a symplectic base and therefore decomposes into an orthogonal direct sum of submodules $\mathbb{S}_n$ of a standard type (see (7.1)). The set of alternating pairings on finite subgroups of a given torsion $\mathbb{Z}$-module Ω is studied in §7.1.2: We show that this set carries a functorial group structure, which turns it into a torsion abelian group $\mathit{Symp}(\Omega)$. The special case where $\Omega = \mathbb{T}(\Gamma)$ $(= \Gamma \otimes_{\mathbb{Z}} (\mathbb{Q}/\mathbb{Z}))$ for some torsion-free abelian group Γ is considered in §7.1.3: We establish in Th. 7.22 a canonical isomorphism between $\mathit{Symp}(\mathbb{T}(\Gamma))$ and $\mathbb{T}(\wedge^2\Gamma)$.

7.1.1 Alternating pairings on torsion abelian groups

Throughout this section, we use the notation $\mathbb{T} = \mathbb{Q}/\mathbb{Z}$. Let M be a torsion abelian group. A $\mathbb{Z}$-bilinear map

$$a\colon M \times M \longrightarrow \mathbb{T}$$

is said to be an *alternating pairing* on M if $a(x,x) = 0$ for all $x \in M$. This condition implies $a(y,x) = -a(x,y)$ for all x, $y \in M$, as is easily seen by expanding the left side of the equation $a(x+y, x+y) = 0$. A morphism of alternating pairings, $\psi\colon (M,a) \to (N,b)$, is a group homomorphism $\psi\colon M \to N$ such that $b(\psi(x),\psi(y)) = a(x,y)$ for all $x, y \in M$. A bijective morphism of alternating pairings is called an *isometry*.

We use for alternating pairings on torsion abelian groups the usual terminology associated with inner product spaces, defining the *orthogonal* of a subgroup $N \subseteq M$ to be the subgroup

$$\begin{aligned} N^{\perp} &= \{x \in M \mid a(x,y) = 0 \text{ for all } y \in N\} \\ &= \{x \in M \mid a(y,x) = 0 \text{ for all } y \in N\} \end{aligned}$$

and the *radical* of a to be

$$rad(a) = M^{\perp}.$$

The alternating pairing is said to be *nondegenerate* if its radical is $\{0\}$. Every alternating pairing a on a torsion abelian group M induces on the factor group $M/rad(a)$ a nondegenerate alternating pairing $\overline{a}$ defined by

$$\overline{a}\big(x + rad(a), y + rad(a)\big) = a(x,y) \qquad \text{for all } x,\, y \in M.$$

The following easy observation reduces the study of alternating pairings on torsion abelian groups to the case of abelian groups with primary torsion:

Proposition 7.1. *Let a be an alternating pairing on a torsion abelian group M. The primary decomposition $M = \bigoplus_p M_{(p)}$ is an orthogonal sum:*

$$M = \perp_p M_{(p)}.$$

Proof. It suffices to show that for any two distinct prime numbers p, q, the primary components $M_{(p)}$ and $M_{(q)}$ are orthogonal. Let $x \in M_{(p)}$ and $y \in M_{(q)}$, and suppose $p^n x = q^m y = 0$ for some integers n, $m \geq 0$. Then,

$$p^n\, a(x,y) = a(p^n x, y) = 0 \qquad \text{and} \qquad q^m\, a(x,y) = a(x, q^m y) = 0.$$

Since p^n and q^m are relatively prime, it follows that $a(x,y) = 0$. □

For nondegenerate alternating pairings on finite abelian groups, we have the following precise structure theorem:

Theorem 7.2. *Let a be a nondegenerate alternating pairing on a finite abelian group M of exponent $e > 1$. There are positive integers $n_1, \ldots, n_r$ and elements $u_1, v_1, \ldots, u_r, v_r \in M$ with the following properties:*

- $n_1 = e$, $n_r > 1$, *and* $n_j \mid n_i$ *for* $i \leq j$;
- $M = \mathbb{Z}u_1 \oplus \mathbb{Z}v_1 \oplus \ldots \oplus \mathbb{Z}u_r \oplus \mathbb{Z}v_r$;
- u_i *and* v_i *have order* n_i *for* $i = 1, \ldots, r$;
- $a(u_i, v_i) = \frac{1}{n_i} + \mathbb{Z}$ *for* $i = 1, \ldots, r$;
- $a(u_i, u_j) = a(v_i, v_j) = a(u_i, v_j) = 0$ *for* $i, j = 1, \ldots, r$ *with* $i \neq j$.

In particular,

$$M \cong (\mathbb{Z}/n_1\mathbb{Z})^2 \times \ldots \times (\mathbb{Z}/n_r\mathbb{Z})^2,$$

so the invariant factors of M are $n_1, n_1, \ldots, n_r, n_r$, and $|M| = (n_1 \ldots n_r)^2$.

Proof. Let $u_1 \in M$ be an element of order e. Let $H = \{a(u_1, y) \mid y \in M\}$. Since H is a finite subgroup of $\mathbb{T}$, we have $H = \frac{1}{d}\mathbb{Z}/\mathbb{Z}$ where $d = |H|$. Since

$a(du_1, y) = d\,a(u_1, y) = 0$ for all $y \in M$ and a is nondegenerate, $du_1 = 0$; hence, $e \mid d$. Therefore, there is a $v_1 \in M$ with $a(u_1, v_1) = \frac{1}{e} + \mathbb{Z}$. Let q be the order of v_1. Then $q \mid exp(M) = e$. But also $0 = a(u_1, qv_1) = \frac{q}{e} + \mathbb{Z}$; so $e \mid q$. Hence, v_1 has order e.

Now, consider the subgroup $N \subseteq M$ orthogonal to u_1 and v_1:

$$N = \{x \in M \mid a(u_1, x) = a(v_1, x) = 0\}.$$

For ℓ, $m \in \mathbb{Z}$ we have

$$a(u_1, \ell u_1 + mv_1) = \tfrac{m}{e} + \mathbb{Z} \qquad \text{and} \qquad a(v_1, \ell u_1 + mv_1) = \tfrac{-\ell}{e} + \mathbb{Z}.$$

Therefore, if $\ell u_1 + mv_1 \in N$, then ℓ, $m \in e\mathbb{Z}$, hence $\ell u_1 = mv_1 = 0$. This applies in particular when $\ell u_1 + mv_1 = 0$, and shows that the sum $\mathbb{Z}u_1 + \mathbb{Z}v_1$ is direct. We thus have

$$(\mathbb{Z}u_1 \oplus \mathbb{Z}v_1) \cap N = \{0\}.$$

We claim that

$$M = (\mathbb{Z}u_1 \oplus \mathbb{Z}v_1) \oplus N,$$

and that the restriction of a to N is a nondegenerate pairing. The proposition then follows by induction on $|M|$.

To prove the claim, observe that since $eM = \{0\}$ we have $a(M, M) \subseteq \frac{1}{e}\mathbb{Z}/\mathbb{Z}$. Therefore, for any $x \in M$ we may find ℓ, $m \in \mathbb{Z}$ such that

$$a(u_1, x) = \tfrac{m}{e} + \mathbb{Z} \qquad \text{and} \qquad a(v_1, x) = \tfrac{\ell}{e} + \mathbb{Z},$$

hence $x + \ell u_1 - mv_1 \in N$. We have thus shown that $M = \mathbb{Z}u_1 \oplus \mathbb{Z}v_1 \oplus N$. It follows that if $x \in N$ satisfies $a(x, y) = 0$ for all $y \in N$, then $a(x, z) = 0$ for all $z \in M$, hence $x = 0$. We have thus proved the claim. □

Definition 7.3. A finite abelian group M with a nondegenerate alternating pairing a is called a *symplectic module.* A *symplectic base* of a symplectic module (M, a) is a collection of elements u_1, v_1, ..., u_r, $v_r \in M$ satisfying the following conditions:

- $M = \mathbb{Z}u_1 \oplus \mathbb{Z}v_1 \oplus \ldots \oplus \mathbb{Z}u_r \oplus \mathbb{Z}v_r$;
- for $i = 1, \ldots, r$, the elements u_i and v_i have the same order;
- for $i = 1, \ldots, r$, letting n_i denote the order of u_i and v_i, we have

$$a(u_i, v_i) = \tfrac{1}{n_i} + \mathbb{Z};$$

- for i, $j = 1, \ldots, r$ with $i \neq j$, we have

$$a(u_i, u_j) = a(v_i, v_j) = a(u_i, v_j) = 0.$$

Thus, Th. 7.2 states that every symplectic module has a symplectic base, for which we additionally have $n_j \mid n_i$ for $i \leq j$. Other symplectic bases, which do not necessarily satisfy this extra divisibility condition, can be obtained by combining symplectic bases of the various primary components, in view of Prop. 7.1.

For any integer $n > 1$, we introduce the following symplectic module, which we call the *standard symplectic module of exponent* n:

$$\mathbb{S}_n = \left((\tfrac{1}{n}\mathbb{Z}/\mathbb{Z})^2, s_n\right) \tag{7.1}$$

where for all $x_1, x_2, y_1, y_2 \in \mathbb{Z}$,

$$s_n\left((\tfrac{x_1}{n}+\mathbb{Z}, \tfrac{x_2}{n}+\mathbb{Z}), (\tfrac{y_1}{n}+\mathbb{Z}, \tfrac{y_2}{n}+\mathbb{Z})\right) = \tfrac{1}{n}(x_1y_2 - x_2y_1) + \mathbb{Z}. \tag{7.2}$$

Thus, the standard generators $(\frac{1}{n}+\mathbb{Z}, 0)$ and $(0, \frac{1}{n}+\mathbb{Z})$ of $(\frac{1}{n}\mathbb{Z}/\mathbb{Z})^2$ form a symplectic base of $\mathbb{S}_n$. Using this notation, we may rephrase Th. 7.2 as follows:

Corollary 7.4. *Every nontrivial symplectic module* (M, a) *is isometric to an orthogonal direct sum of standard symplectic modules*

$$(M, a) \cong \mathbb{S}_{n_1} \perp \ldots \perp \mathbb{S}_{n_r} \qquad \textit{for some } n_1, \ldots, n_r > 1.$$

The integers $n_1, \ldots, n_r$ *may be chosen so that* $n_r \mid n_{r-1} \mid \ldots \mid n_1$.

As a consequence, a symplectic module is uniquely determined up to isometry by its underlying finite abelian group: symplectic modules (M, a) and (N, b) are isometric if and only if the abelian groups M and N are isomorphic. But not every group isomorphism is a symplectic module isometry, and a finite abelian group M may carry several (isometric but not identical) nondegenerate alternating pairings.

We conclude this subsection with a few observations on totally isotropic submodules of symplectic modules, for use in §9.1 and §9.1.2.

For any torsion abelian group M, we let $M^\vee = \mathit{Hom}(M, \mathbb{T})$ denote the (*toric*) *dual*, or *character group*, of M. For any subgroup $L \subseteq M$, the restriction of homomorphisms yields a map $M^\vee \to L^\vee$; this map is surjective since $\mathbb{T}$ is an injective $\mathbb{Z}$-module. If M is finite, it is easily seen from a direct sum decomposition of M into cyclic groups that $M^\vee \cong M$ (noncanonically). In particular, for any symplectic module (M, a), the injective map

$$\widehat{a}\colon M \longrightarrow M^\vee, \qquad x \mapsto a(x, \cdot)$$

is an isomorphism. Given a subgroup $L \subseteq M$, the composition of $\widehat{a}$ with the canonical map $M^\vee \to L^\vee$ is a surjective map $M \to L^\vee$ which fits into the following exact sequence:

$$0 \longrightarrow L^\perp \longrightarrow M \longrightarrow L^\vee \longrightarrow 0. \tag{7.3}$$

It follows that $|M| = |L| \cdot |L^\perp|$. The subgroup L is said to be *totally isotropic* if $a(x, y) = 0$ for all $x, y \in L$. This condition is equivalent to $L \subseteq L^\perp$. Therefore, L is a maximal totally isotropic subgroup if and only if $L = L^\perp$. The maximal totally isotropic subgroups of a symplectic module are called *Lagrangians*.

For any finite abelian group M, we let $\mathit{trk}(M)$ denote the minimal number of generators of M, i.e., the number of invariant factors of M, see §A.4 in

Appendix A. For any subgroup $N \subseteq M$ we have $trk(N) \leq trk(M)$, see (A.24).

Proposition 7.5. *For any Lagrangian L in a symplectic module (M, a), we have*

$$|M| = |L|^2, \quad trk(L) \leq trk(M) \leq 2\, trk(L), \quad \text{and} \quad exp(M) \mid exp(L)^2.$$

Proof. The equality $|M| = |L|^2$ follows readily from the exact sequence (7.3) since $L^\perp = L$ and $|L^\perp| = |L|$. Since $L \subseteq M$, we have $trk(L) \leq trk(M)$. To see that $trk(M) \leq 2\, trk(L)$, we consider (7.3): since a system of generators of M can be obtained by combining a system of generators of $L = L^\perp$ with a preimage of a system of generators of $L^\vee$, we have

$$trk(M) \leq trk(L) + trk(L^\vee) = 2\, trk(L).$$

To prove the last relation, let $e = exp(L)$. For each $x \in M$ we have $e \cdot x \in L^\perp = L$ in view of the exactness of (7.3). Therefore, $e^2 \cdot x = 0$. □

In the standard symplectic module $\mathbb{S}_n$, the submodule $L_1 = (\frac{1}{n}\mathbb{Z}/\mathbb{Z}) \times \{0\}$ is a Lagrangian with $trk(L_1) = 1 = \frac{1}{2}\, trk(\mathbb{S}_n)$ and $exp(L_1) = n = exp(\mathbb{S}_n)$. If $n = rs$ with $r, s > 1$, the submodule L_2 spanned by $(\frac{r}{n}+\mathbb{Z}, 0)$ and $(0, \frac{s}{n}+\mathbb{Z})$ is a Lagrangian with $trk(L_2) = 2 = trk(\mathbb{S}_n)$ and $exp(L_2) = lcm(r, s)$.

The following restrictions on the type of Lagrangians in a symplectic module follow readily from Prop. 7.5:

Corollary 7.6. *Let (M, a) be a symplectic module.*

(i) *If $trk(M) > 2$, then M contains no cyclic Lagrangian.*
(ii) *If $exp(M) = p^n$ for some prime p and some $n \geq 3$, then M contains no elementary abelian Lagrangian.*

7.1.2 The group $Symp(\Omega)$

Symplectic modules naturally arise from value groups of totally ramified division algebras, as we will see in §7.3.1 and §7.4.1. On the other hand, to describe Brauer groups of strictly Henselian fields or inertially closed graded fields, it is useful to consider the set of symplectic modules whose underlying finite abelian group is contained in a given (usually infinite) torsion abelian group (see §7.3.2 and §7.4.2). Therefore, we introduce the following definition:

Definition 7.7. For an arbitrary torsion abelian group Ω, let $Symp(\Omega)$ denote the set of symplectic modules (M, a) such that $M \subseteq \Omega$.

We will show in this subsection that $Symp(\Omega)$ depends functorially on the group Ω, and will then use this to define an abelian group structure on $Symp(\Omega)$.

Let $f \colon \Omega_1 \to \Omega_2$ be a homomorphism of torsion abelian groups, and let $(M, a) \in Symp(\Omega_1)$. We wish to use f to build from (M, a) a corresponding symplectic module in $Symp(\Omega_2)$. Let $f_M\colon M \to \Omega_2$ be the restriction of f to M. We have $im\, f_M \cong M/\ker f_M$, but a on M does not in general induce a well-defined pairing on $M/\ker f_M$. So, there is no natural way to define an alternating pairing (let alone a nondegenerate one) on $im\, f_M$ using f_M and a. To get around these difficulties, consider $(\ker f_M)^\perp \subseteq M$, the orthogonal of $\ker f_M$ for the alternating pairing a, and its image under f_M,

$$f_M\big((\ker f_M)^\perp\big) \subseteq \Omega_2.$$

Then, f_M induces an isomorphism

$$\begin{aligned} f_M\big((\ker f_M)^\perp\big) &\cong (\ker f_M)^\perp/\big((\ker f_M)^\perp \cap \ker f_M\big) \\ &= (\ker f_M)^\perp/rad\big((\ker f_M)^\perp\big). \end{aligned}$$

The alternating pairing a on M induces a nondegenerate alternating pairing $\overline{a}$ on $(\ker f_M)^\perp/rad\big((\ker f_M)^\perp\big)$ which we can transfer to $f_M\big((\ker f_M)^\perp\big)$. Specifically, note that for $x, y \in (\ker f_M)^\perp$, the element $a(x,y) \in \mathbb{T}$ does not change if we add to x or y an element in $\ker f_M$; therefore, this element depends only on $f_M(x)$ and $f_M(y)$, and we define an alternating pairing f_*a on $f_M\big((\ker f_M)^\perp\big)$ by

$$f_*a\big(f_M(x), f_M(y)\big) = a(x,y) \qquad \text{for all } x,\, y \in (\ker f_M)^\perp. \tag{7.4}$$

If $f_M(x) \in rad(f_*a)$, then $x \in \big((\ker f_M)^\perp\big)^\perp = \ker f_M$, hence $f_M(x) = 0$. The pairing f_*a is therefore nondegenerate, and we define a map

$$f_\diamond\colon Symp(\Omega_1) \to Symp(\Omega_2)$$

by

$$f_\diamond(M,a) = \big(f_M\big((\ker f_M)^\perp\big), f_*a\big) \in Symp(\Omega_2). \tag{7.5}$$

Thus f induces an isometry of symplectic modules

$$\big((\ker f_M)^\perp/((\ker f_M)^\perp \cap \ker f_M), \overline{a}\big) \cong f_\diamond(M,a). \tag{7.6}$$

In particular if f_M is injective, then it yields an isometry of symplectic modules $(M,a) \xrightarrow{\sim} f_\diamond(M,a)$.

Remark 7.8. Let $f\colon \Omega_1 \to \Omega_3$ and $g\colon \Omega_2 \to \Omega_3$ be group homomorphisms of torsion abelian groups. Suppose there is an isometry $\psi\colon (M,a) \xrightarrow{\sim} (N,b)$ between $(M,a) \in Symp(\Omega_1)$ and $(N,b) \in Symp(\Omega_2)$ which is compatible with f and g in that $f(x) = g(\psi(x))$ for all $x \in M$. Then,

$$f_\diamond(M,a) = g_\diamond(N,b) \qquad \text{in } Symp(\Omega_3).$$

This is an actual equality, not merely an isometry of symplectic modules. It follows easily from the definitions, since $\psi\big((\ker f_M)^\perp\big) = (\ker g_N)^\perp$.

The construction of $f_\diamond$ is compatible with direct products: If $\Omega_1, \ldots, \Omega_k$ are torsion abelian groups and $(M_i, a_i) \in Symp(\Omega_i)$, then we write

$$(M_1, a_1) \times \ldots \times (M_k, a_k) = \big(M_1 \times \ldots \times M_k,\; a_1 \times \ldots \times a_k\big)$$

where $a_1 \times \ldots \times a_k$ is the alternating pairing given by

$$(a_1 \times \ldots \times a_k)\big((x_1, \ldots, x_k), (y_1, \ldots, y_k)\big) = a_1(x_1, y_1) + \ldots + a_k(x_k, y_k).$$

Since this pairing is clearly nondegenerate, we have $(M_1, a_1) \times \ldots \times (M_k, a_k) \in \mathit{Symp}(\Omega_1 \times \ldots \times \Omega_k)$. Indeed, this symplectic module is the orthogonal sum of its canonical submodules isometric to the (M_i, a_i).

Lemma 7.9. *Let $f_i \colon \Omega_i \to \Omega_i'$ be group homomorphisms of torsion abelian groups for $i = 1, 2, \ldots, k$, and let $f_1 \times \ldots \times f_k \colon \Omega_1 \times \ldots \times \Omega_k \to \Omega_1' \times \ldots \times \Omega_k'$ be the induced group homomorphism. Let $(M_i, a_i) \in \mathit{Symp}(\Omega_i)$ for each i. Then*

$$(f_1 \times \ldots \times f_k)_\diamond\big((M_1, a_1) \times \ldots \times (M_k, a_k)\big) = f_{1\diamond}(M_1, a_1) \times \ldots \times f_{k\diamond}(M_k, a_k).$$

Proof. To simplify the notation, we consider just the typical case $k = 2$. Let f_{iM_i} be the restriction of f_i to M_i, so that $f_{1M_1} \times f_{2M_2}$ is the restriction of $f_1 \times f_2$ to $M_1 \times M_2$. We have $\mathit{ker}(f_{1M_1} \times f_{2M_2}) = \mathit{ker}\, f_{1M_1} \times \mathit{ker}\, f_{2M_2}$, so the definition of $a_1 \times a_2$ yields

$$\big(\mathit{ker}(f_{1M_1} \times f_{2M_2})\big)^\perp = (\mathit{ker}\, f_{1M_1})^\perp \times (\mathit{ker}\, f_{2M_2})^\perp.$$

Also, for any (u_1, u_2), $(u_1', u_2') \in (f_{1M_1} \times f_{2M_2})\big(\mathit{ker}(f_{1M_1} \times f_{2M_2})^\perp\big)$ take any s_i, $s_i' \in (\mathit{ker}\, f_{iM_i})^\perp$ with $f_{iM_i}(s_i) = u_i$ and $f_{iM_i}(s_i') = u_i'$. Then,

$$\begin{aligned}
(f_1 \times f_2)_*(a_1 \times a_2)&\big((u_1, u_2), (u_1', u_2')\big) \\
&= (a_1 \times a_2)\big((s_1, s_2), (s_1', s_2')\big) \\
&= a_1(s_1, s_1') + a_2(s_2, s_2') = f_{1*}a_1(u_1, u_1') + f_{2*}a_2(u_2, u_2') \\
&= (f_{1*}a_1 \times f_{2*}a_2)\big((u_1, u_2), (u_1', u_2')\big).
\end{aligned}$$

Hence, $(f_1 \times f_2)_*(a_1 \times a_2) = f_{1*}a_1 \times f_{2*}a_2$, which proves the equality of symplectic modules asserted in the lemma. □

We next prove the key property that passage from f to $f_\diamond$ is compatible with composition of group homomorphisms. This functorial property will allow us to define a group structure on $\mathit{Symp}(\Omega)$. The fact that $\mathit{Symp}(\Omega)$ has a natural group structure is not immediately evident. However this follows at once when one observes that there is a canonical bijection $\mathit{Symp}(\Omega) \leftrightarrow \mathit{Alt}^c(\Omega^\vee)$, where $\Omega^\vee = \mathit{Hom}(\Omega, \mathbb{T})$, which is a profinite abelian group, and $\mathit{Alt}^c(\Omega^\vee)$ is the group of continuous alternating pairings $\Omega^\vee \times \Omega^\vee \to \mathbb{T}$. Since $\mathit{Alt}^c(\Omega^\vee)$ is functorial with respect to Ω and is a group, these properties transfer over to $\mathit{Symp}(\Omega)$. This will be described further in the notes at the end of the chapter. We give here a direct approach to the properties of $\mathit{Symp}(\Omega)$, since it is somewhat shorter and less notationally cumbersome than working via $\mathit{Alt}^c(\Omega^\vee)$.

Proposition 7.10. *Let $f\colon \Omega_1 \to \Omega_2$ and $g\colon \Omega_2 \to \Omega_3$ be homomorphisms of torsion abelian groups, and let $(M,a) \in Symp(\Omega_1)$. Then,*

$$(g \circ f)_\diamond(M,a) \;=\; g_\diamond \circ f_\diamond(M,a). \tag{7.7}$$

Proof. Let f_M (resp. gf_M) be the restriction of f (resp. gf) to M, and let $K = \ker f_M$ and $L = \ker gf_M$. So, $K \subseteq L \subseteq M$, hence $L^\perp \subseteq K^\perp$. We have $f_\diamond(M,a) = (f_M(K^\perp), f_*a)$. Let

$$P \;=\; \big(f_M(K^\perp) \cap \ker g\big)^\perp \;\subseteq\; f_M(K^\perp).$$

Then,

$$g_\diamond \circ f_\diamond(M,a) \;=\; \big(g(P), g_*f_*a\big).$$

To determine $g(P)$, note that as $P \subseteq f_M(K^\perp)$, we have

$$g(P) = g\big(f_M\big(f_M^{-1}(P) \cap K^\perp\big)\big),$$

with

$$\begin{aligned} &f_M^{-1}(P) \cap K^\perp \\ &\quad= \left\{ s \in K^\perp \;\middle|\; \begin{array}{l} 0 = f_*a\big(f_M(s), f_M(r)\big) = a(s,r) \\ \text{for all } r \in K^\perp \text{ with } gf_M(r) = 0 \end{array} \right\} \\ &\quad= K^\perp \cap (K^\perp \cap L)^\perp \;=\; K^\perp \cap (K + L^\perp) \;=\; (K^\perp \cap K) + L^\perp; \end{aligned}$$

the last equality holds because $L^\perp \subseteq K^\perp$. Since $f_M(K^\perp \cap K) = \{0\}$, this shows that

$$g(P) \;=\; gf_M\big((K^\perp \cap K) + L^\perp\big) \;=\; gf_M(L^\perp);$$

hence $g_\diamond \circ f_\diamond(M,a)$ has the same underlying abelian group as $(gf)_\diamond(M,a)$. It remains to check that they have the same alternating pairing.

Take any $t_1, t_2 \in gf_M(L^\perp) \;=\; g(P)$, and choose any $p_1, p_2 \in P$ with $g(p_i) = t_i$. Then $p_i \;=\; f_M(s_i)$ with $s_i \in (K^\perp \cap K) + L^\perp \subseteq K^\perp$. Write $s_i = d_i + u_i$ with $d_i \in K^\perp \cap K$ and $u_i \in L^\perp$. So, $p_i = f_M(u_i)$ as $d_i \in K = \ker f_M$. Hence,

$$\begin{aligned} g_*f_*a(t_1,t_2) \;&=\; g_*f_*a\big(g(p_1), g(p_2)\big) \;=\; f_*a(p_1,p_2) \\ &=\; f_*a\big(f_M(u_1), f_M(u_2)\big) \;=\; a(u_1,u_2). \end{aligned}$$

But since each $u_i \in L^\perp$ with $gf_M(u_i) = t_i$, we have

$$(gf)_*a(t_1,t_2) \;=\; (gf)_*a\big(gf_M(u_1), gf_M(u_2)\big) \;=\; a(u_1,u_2).$$

Thus, the pairings g_*f_*a and $(gf)_*a$ agree. Hence, (7.7) holds. □

We now define an operation $+$ on $Symp(\Omega)$, which we will show makes $Symp(\Omega)$ into an abelian group. We use the addition function

$$\Sigma\colon \Omega \times \Omega \longrightarrow \Omega \qquad \text{given by} \qquad (x,y) \mapsto x+y.$$

For (M,a), $(N,b) \in \mathit{Symp}(\Omega)$ set

$$(M,a) + (N,b) \;=\; \Sigma_\diamond\big((M,a) \times (N,b)\big) \in \mathit{Symp}(\Omega), \tag{7.8}$$

where $\Sigma_\diamond$ is the map functorially associated to Σ as in (7.5). Explicitly, we consider the map

$$\Sigma_{M\times N}\colon M \times N \longrightarrow \Omega, \qquad (m,n) \mapsto m+n$$

and the following subgroup of $M \times N$:

$$\begin{aligned}&(\mathit{ker}\,\Sigma_{M\times N})^{\perp}\\ &\;= \{(x,y) \in M \times N \mid a(x,m) + b(y,n) = 0 \;\text{ for all } (m,n) \in \mathit{ker}\,\Sigma_{M\times N}\}.\end{aligned}$$

Then, by definition,

$$(M,a) + (N,b) \;=\; (P,c) \tag{7.9}$$

where $P = \Sigma_{M\times N}\big((\mathit{ker}\,\Sigma_{M\times N})^{\perp}\big)$, i.e.,

$$P \;=\; \{x + y \mid (x,y) \in (\mathit{ker}\,\Sigma_{M\times N})^{\perp}\}, \tag{7.10}$$

and for $p = x + y$ and $p' = x' + y'$ with (x,y), $(x',y') \in (\mathit{ker}\,\Sigma_{M\times N})^{\perp}$,

$$c(p,p') \;=\; a(x,x') + b(y,y'). \tag{7.11}$$

Example 7.11. (1) We use the notation 0 for the trivial symplectic module $\big(\{0\},0\big)$. Plainly, this module is a zero for the $+$ operation on $\mathit{Symp}(\Omega)$.

(2) For every symplectic module $(M,a) \in \mathit{Symp}(\Omega)$,

$$(M,a) + (M,-a) \;=\; 0.$$

Indeed, we have

$$\mathit{ker}\,\Sigma_{M\times M} \;=\; \{(m,-m) \mid m \in M\},$$

hence $(x,y) \in (\mathit{ker}\,\Sigma_{M\times M})^{\perp}$ if and only if

$$a(x,m) - a(y,-m) \;=\; 0 \qquad \text{for all } m \in M.$$

Since a is nondegenerate, this equation holds if and only if $x + y = 0$, hence $(\mathit{ker}\,\Sigma_{M\times M})^{\perp} = \mathit{ker}\,\Sigma_{M\times M}$ and therefore $\Sigma_{M\times M}\big((\mathit{ker}\,\Sigma_{M\times M})^{\perp}\big) = \{0\}$.

(3) If (M_1,a_1), $(M_2,a_2) \in \mathit{Symp}(\Omega)$ are such that $M_1 \cap M_2 = \{0\}$, then

$$(M_1,a_1) + (M_2,a_2) \;=\; (M_1 \oplus M_2, a_1 \perp a_2).$$

To see this, observe that $\Sigma_{M_1\times M_2}$ is injective since $M_1 \cap M_2 = \{0\}$. Therefore, we have $(\mathit{ker}\,\Sigma_{M_1\times M_2})^{\perp} = M_1 \times M_2$ and

$$\Sigma_{M_1\times M_2}\big((\mathit{ker}\,\Sigma_{M_1\times M_2})^{\perp}\big) \;=\; M_1 \oplus M_2.$$

It is clear from the definition that the alternating pairing on $M_1 \oplus M_2$ defined by the sum of (M_1,a_1) and (M_2,a_2) is just the orthogonal sum $a_1 \perp a_2$.

Lemma 7.12. *Let $f\colon \Omega_1 \to \Omega'$ and $g\colon \Omega_2 \to \Omega'$ be homomorphisms of torsion abelian groups, and let $(M,a) \in Symp(\Omega_1)$ and $(N,b) \in Symp(\Omega_2)$. Then,*

$$f_\diamond(M,a) + g_\diamond(N,b) = h_\diamond\big(M \times N, a \times b\big)$$

where $h\colon \Omega_1 \times \Omega_2 \to \Omega'$ is given by $(x,y) \mapsto f(x) + g(y)$.

Proof. Let $\Sigma'\colon \Omega' \times \Omega' \to \Omega'$ be the sum map $(x,y) \mapsto x+y$. Since $h = \Sigma' \circ (f,g)$ as maps $\Omega_1 \times \Omega_2 \to \Omega'$, we have by Prop. 7.10 and Lemma 7.9,

$$\begin{aligned} h_\diamond(M \times N, a \times b) &= \Sigma'_\diamond \circ (f,g)_\diamond(M \times N, a \times b) \\ &= \Sigma'_\diamond\big(f_\diamond(M,a) \times g_\diamond(N,b)\big) = f_\diamond(M,a) + g_\diamond(N,b). \quad \square \end{aligned}$$

We next prove that the operation on $Symp(\Omega)$ is associative. For any torsion abelian group Ω and any positive integer k, let

$$\Sigma^{(k)}\colon \Omega^k \longrightarrow \Omega \qquad \text{be given by} \qquad (x_1, \ldots, x_k) \mapsto x_1 + \ldots + x_k.$$

Lemma 7.13. *Let $(M_i, a_i) \in Symp(\Omega)$ for $i = 1,2,3$. Then,*

$$\begin{aligned} \big[(M_1,a_1) + (M_2,a_2)\big] + (M_3,a_3) &= \Sigma^{(3)}_\diamond\big(M_1 \times M_2 \times M_3, a_1 \times a_2 \times a_3\big) \\ &= (M_1,a_1) + \big[(M_2,a_2) + (M_3,a_3)\big]. \end{aligned}$$

Proof. We prove the first equality. Since

$$\Sigma^{(3)} = \Sigma^{(2)} \circ \big(\Sigma^{(2)} \times id_\Omega\big)$$

as maps $(\Omega \times \Omega) \times \Omega \to \Omega$, we have by Prop. 7.10 and Lemma 7.9,

$$\begin{aligned} &\Sigma^{(3)}_\diamond(M_1 \times M_2 \times M_3, a_1 \times a_2 \times a_3) \\ &\quad = \Sigma^2_\diamond \circ (\Sigma^{(2)} \times id_\Omega)_\diamond\big((M_1 \times M_2, a_1 \times a_2) \times (M_3, a_3)\big) \\ &\quad = \Sigma^{(2)}_\diamond\big([\Sigma^{(2)}_\diamond(M_1 \times M_2, a_1 \times a_2)] \times id_{\Omega\diamond}(M_3,a_3)\big) \\ &\quad = \Sigma^{(2)}_\diamond\big([(M_1,a_1) + (M_2,a_2)] \times (M_3,a_3)\big) \\ &\quad = \big[(M_1,a_1) + (M_2,a_2)\big] + (M_3,a_3). \end{aligned}$$

The second equality of the proposition is proved analogously using the equality of group homomorphisms $\Sigma^{(3)} = \Sigma^{(2)} \circ \big(\, id_\Omega \times \Sigma^{(2)}\big)$. $\square$

The formula in Lemma 7.13 for sums of three symplectic modules generalizes to arbitrary finite sums:

Lemma 7.14. *Let $(M_1,a_1), \ldots, (M_k,a_k)$ be symplectic modules in $Symp(\Omega)$. Then,*

$$(M_1,a_1) + \ldots + (M_k,a_k) = \Sigma^{(k)}_\diamond(M_1 \times \ldots \times M_k, a_1 \times \ldots \times a_k).$$

Proof. This is proved in the same way as Lemma 7.13 by induction on k, using the identity $\Sigma^{(k)} = \Sigma^{(2)} \circ (\Sigma^{(k-1)} \times id_\Omega)$. $\square$

This yields an explicit description of the k-fold sum of a symplectic module with itself.

Proposition 7.15. *Let Ω be an arbitrary torsion abelian group and let $(M,a) \in Symp(\Omega)$ be a nontrivial symplectic module. For every positive integer k, we have*

$$k \cdot (M,a) = (kM, a^{(k)}) \qquad \textit{in } Symp(\Omega),$$

where

$$a^{(k)}(kx,ky) = k \cdot a(x,y) \qquad \textit{for all } x,\ y \in M.$$

Proof. Consider the symplectic module $(M^k, a \times \ldots \times a)$, and the group homomorphism $\Sigma_{M^k}\colon M^k \to \Omega$ given by $(m_1,\ldots,m_k) \mapsto m_1 + \ldots + m_k$. In M^k we have

$$\big(\ker \Sigma_{M^k}\big)^{\perp} = \Big\{(x_1,\ldots,x_k) \mid \sum_{i=1}^{k} a(x_i,m_i) = 0 \text{ whenever } m_1+\ldots+m_k = 0\Big\}.$$

Thus, $(x_1,\ldots,x_k) \in \big(\ker \Sigma_{M^k}\big)^{\perp}$ if and only if, for any $m_1,\ldots,m_{k-1} \in M$,

$$\begin{aligned} 0 &= a(x_1,m_1) + \ldots + a(x_{k-1},m_{k-1}) + a(x_k, -m_1 - \ldots - m_{k-1}) \\ &= a(x_1 - x_k, m_1) + \ldots + a(x_{k-1} - x_k, m_{k-1}). \end{aligned}$$

For any $j \in \{1,\ldots,k-1\}$, by choosing $m_i = 0$ for $i \neq j$, this yields $a(x_j - x_k, m_j) = 0$ for all $m_j \in M$, which, as a is nondegenerate, implies $x_j = x_k$. Thus,

$$\big(\ker \Sigma_{M^k}\big)^{\perp} = \{(x,\ldots,x) \mid x \in M\} \subseteq M^k \text{ and } \Sigma_{M^k}\big(\big(\ker \Sigma_{M^k}\big)^{\perp}\big) = kM.$$

By Lemma 7.14, $\sum_{i=1}^{k}(M,a) = (kM, a^{(k)})$, where for $(x,\ldots,x),(y,\ldots,y)$ in $\big(\ker \Sigma_{M^k}\big)^{\perp}$,

$$a^{(k)}\Big(\sum_{i=1}^{k} x, \sum_{i=1}^{k} y\Big) = \sum_{i=1}^{k} a(x,y).$$

That is, $a^{(k)}(kx,ky) = k \cdot a(x,y)$ for all $x,\ y \in M$. □

Example 7.16. We show that $k \cdot \mathbb{S}_{kn} = \mathbb{S}_n$ for all positive integers k,n. Recall that $\mathbb{S}_{kn} = \big((\frac{1}{kn}\mathbb{Z}/\mathbb{Z})^2, s_{kn}\big)$, with symplectic base (u,v), where $u = (\frac{1}{kn} + \mathbb{Z}, 0)$ and $v = (0, \frac{1}{kn} + \mathbb{Z})$. The nondegenerate alternating pairing s_{kn} is determined by $s_{kn}(u,v) = \frac{1}{kn} + \mathbb{Z}$. By Prop. 7.15,

$$k \cdot \mathbb{S}_{kn} = \big(k(\tfrac{1}{kn}\mathbb{Z}/\mathbb{Z})^2), s_{kn}^{(k)}\big) = \big((\tfrac{1}{n}\mathbb{Z}/\mathbb{Z})^2, s_{kn}^{(k)}\big),$$

where the nondegenerate alternating pairing $s_{kn}^{(k)}$ is determined by

$$s_{kn}^{(k)}(ku,kv) = k \cdot s_{kn}(u,v) = \tfrac{1}{n} + \mathbb{Z}.$$

Since (ku, kv) is the standard symplectic base of $\mathbb{S}_n$, this shows that $s_{kn}^{(k)} = s_n$, and hence, $k \cdot \mathbb{S}_{kn} = \mathbb{S}_n$ as symplectic modules in $\mathit{Symp}(\mathbb{T})$.

Theorem 7.17. *For any torsion abelian group Ω, the set $\mathit{Symp}(\Omega)$ with the operation $+$ is a torsion abelian group. If $f\colon \Omega \to \Omega'$ is a homomorphism of torsion abelian groups, then $f_\diamond\colon \mathit{Symp}(\Omega) \to \mathit{Symp}(\Omega')$ is a group homomorphism. Moreover, the assignment $\Omega \mapsto \mathit{Symp}(\Omega)$ and $f \mapsto f_\diamond$ defines a functor from the category $\mathfrak{Tors}$ of torsion abelian groups to itself.*

Proof. Lemma 7.13 shows that the operation on $\mathit{Symp}(\Omega)$ is associative, and Ex. 7.11(1) and (2) yield the identity element and the existence of inverses. Hence, $\mathit{Symp}(\Omega)$ is a group. To see that it is abelian we use the switch homomorphism $sw\colon \Omega \times \Omega \to \Omega \times \Omega$ given by $(x, y) \mapsto (y, x)$. For the sum map $\Sigma\colon \Omega \times \Omega \to \Omega$, $(x, y) \mapsto x + y$, we have $\Sigma \circ sw = \Sigma$ since Ω is abelian. For any $(M, a), (N, b) \in \mathit{Symp}(\Omega)$, the map sw restricts to an isometry

$$(M \times N, a \times b) \xrightarrow{\sim} (N \times M, b \times a).$$

Since this isometry is compatible with Σ, by Remark 7.8

$$\Sigma_\diamond(M \times N, a \times b) = \Sigma_\diamond(N \times M, b \times a).$$

That is, $(M, a)+(N, b)=(N, b)+(M, a)$ in $\mathit{Symp}(\Omega)$. It follows from Prop. 7.15 that the abelian group $\mathit{Symp}(\Omega)$ is torsion. Indeed, Prop. 7.15 shows that for any $(M, a) \in \mathit{Symp}(\Omega)$, the exponent of (M, a) in $\mathit{Symp}(\Omega)$ equals the exponent of M as an abelian group.

Now, take any group homomorphism $f\colon \Omega \to \Omega'$, and let $\Sigma'\colon \Omega' \times \Omega' \to \Omega'$ be the sum map for Ω'. Since f is a homomorphism, we have $f \circ \Sigma = \Sigma' \circ (f \times f)$ as functions $\Omega \times \Omega \to \Omega'$. Hence, by Prop. 7.10 and Lemma 7.9, for any $(M, a), (N, b) \in \mathit{Symp}(\Omega)$,

$$\begin{aligned} f_\diamond\big((M, a) + (N, b)\big) &= f_\diamond \circ \Sigma_\diamond(M \times N, a \times b) = (f \circ \Sigma)_\diamond(M \times N, a \times b) \\ &= (\Sigma' \circ (f \times f))_\diamond(M \times N, a \times b) \\ &= \Sigma'_\diamond \circ (f \times f)_\diamond(M \times N, a \times b) \\ &= \Sigma'_\diamond\big(f_\diamond(M, a) \times f_\diamond(N, b)\big) \\ &= f_\diamond(M, a) + f_\diamond(N, b). \end{aligned}$$

Thus, $f_\diamond$ is a group homomorphism.

It follows from Prop. 7.10 that $\Omega \to \mathit{Symp}(\Omega)$, $f \to f_\diamond$ is a functor on the category $\mathfrak{Tors}$. □

7.1.3 Symplectic modules and second exterior powers

In applications in later sections, groups $\mathit{Symp}(\Omega)$ are considered mostly when Ω is obtained from a *torsion-free* abelian group Γ by the $\mathbb{T}$-operation defined in (6.20):

$$\mathbb{T}(\Gamma) = \varinjlim\big(\tfrac{1}{n}\Gamma\big)/\Gamma = \Gamma \otimes_{\mathbb{Z}} \mathbb{T}. \tag{7.12}$$

Throughout this subsection, we fix a torsion-free abelian group Γ. Our goal is to give a description of the group $\mathit{Symp}(\mathbb{T}(\Gamma))$ in terms of the second exterior power $\wedge^2\Gamma = \Gamma \wedge \Gamma$. Note that $\wedge^2\Gamma$ is torsion-free since Γ is torsion-free. (For Γ is the direct limit of its finitely-generated, hence free abelian subgroups A_j. Then, $\wedge^2\Gamma$ is the direct limit of the free abelian groups $\wedge^2 A_j$.) We will prove that there is a canonical group isomorphism

$$\Upsilon_\Gamma \colon \mathbb{T}(\wedge^2\Gamma) \xrightarrow{\ \sim\ } \mathit{Symp}(\mathbb{T}(\Gamma)).$$

We build this map from compatible maps ${}_n\Upsilon_\Gamma \colon (\frac{1}{n}\wedge^2\Gamma)/\wedge^2\Gamma \to \mathit{Symp}(\mathbb{T}(\Gamma))$ for all n. Fix a positive integer n. To obtain ${}_n\Upsilon_\Gamma$ we first define an alternating function ${}_n\Xi_\Gamma$ on $\Gamma \times \Gamma$. Recall from (7.1) and (7.2) the standard symplectic module $\mathbb{S}_n = ((\frac{1}{n}\mathbb{Z}/\mathbb{Z})^2, s_n)$ with its symplectic base $((\frac{1}{n}+\mathbb{Z}, 0), (0, \frac{1}{n}+\mathbb{Z}))$. We view $\mathbb{S}_n$ as an element of $\mathit{Symp}(\mathbb{T}\times\mathbb{T})$. For any γ, $\delta \in \Gamma$, we have a group homomorphism

$$f(\gamma,\delta) \colon \mathbb{T}\times\mathbb{T} \longrightarrow \mathbb{T}(\Gamma) \quad \text{given by} \quad (\tfrac{i}{k}+\mathbb{Z}, \tfrac{j}{\ell}+\mathbb{Z}) \mapsto \tfrac{i}{k}\gamma + \tfrac{j}{\ell}\delta + \Gamma. \tag{7.13}$$

Then define

$${}_n\Xi_\Gamma \colon \Gamma\times\Gamma \longrightarrow \mathit{Symp}(\mathbb{T}(\Gamma)) \qquad \text{by} \qquad (\gamma,\delta) \mapsto f(\gamma,\delta)_\diamond(\mathbb{S}_n). \tag{7.14}$$

Thus, ${}_n\Xi_\Gamma(\gamma,\delta)$ is a symplectic module whose underlying group is contained in the subgroup of $\mathbb{T}(\Gamma)$ generated by $\frac{1}{n}\gamma+\Gamma$ and $\frac{1}{n}\delta+\Gamma$; see (7.19) for a significant particular case.

Proposition 7.18. *For all $\gamma, \delta, \beta, \varepsilon \in \Gamma$,*

$$\begin{aligned} {}_n\Xi_\Gamma(\gamma,\delta) + {}_n\Xi_\Gamma(\beta,\varepsilon) &= {}_n\Xi_\Gamma(\gamma,\delta+\varepsilon) + {}_n\Xi_\Gamma(\beta-\gamma,\varepsilon) \\ &= {}_n\Xi_\Gamma(\gamma+\beta,\delta) + {}_n\Xi_\Gamma(\beta,\varepsilon-\delta). \end{aligned}$$

Proof. Let (u, v, u', v') be the symplectic base of $\mathbb{S}_n \times \mathbb{S}_n$ obtained from the standard symplectic base of $\mathbb{S}_n$, i.e., $u = (\frac{1}{n}+\mathbb{Z}, 0, 0, 0), \ldots, v' = (0, 0, 0, \frac{1}{n}+\mathbb{Z})$. By Lemma 7.12, we have

$${}_n\Xi_\Gamma(\gamma,\delta) + {}_n\Xi_\Gamma(\beta,\varepsilon) = f(\gamma,\delta)_\diamond(\mathbb{S}_n) + f(\beta,\varepsilon)_\diamond(\mathbb{S}_n) = g_\diamond(\mathbb{S}_n\times\mathbb{S}_n), \tag{7.15}$$

where $g\colon (\frac{1}{n}\mathbb{Z}/\mathbb{Z})^4 \to \mathbb{T}(\Gamma)$ is defined as follows: for all i, j, k, $\ell \in \mathbb{Z}$,

$$\begin{aligned} g(iu + jv + ku' + \ell v') &= f(\gamma,\delta)(iu, jv) + f(\beta,\varepsilon)(ku', \ell v') \\ &= \tfrac{1}{n}\big(i\gamma + j\delta + k\beta + \ell\varepsilon\big) + \Gamma. \end{aligned}$$

Likewise,

$${}_n\Xi_\Gamma(\gamma,\delta+\varepsilon) + {}_n\Xi_\Gamma(\beta-\gamma,\varepsilon) = h_\diamond(\mathbb{S}_n\times\mathbb{S}_n), \tag{7.16}$$

where

$$h(iu + jv + ku' + \ell v') = \tfrac{1}{n}\big(i\gamma + j(\delta+\varepsilon) + k(\beta-\gamma) + \ell\varepsilon\big) + \Gamma.$$

Since (u, v, u', v') and $(u, v+v', u'-u, v')$ are each symplectic bases of $\mathbb{S}_n\times\mathbb{S}_n$, there is an isometry ψ of $\mathbb{S}_n\times\mathbb{S}_n$ mapping the first symplectic base to the second. For all $t = iu + jv + ku' + \ell v' \in (\frac{1}{n}\mathbb{Z}/\mathbb{Z})^4$, we have

$$\begin{aligned} h(t) &= \tfrac{1}{n}\big(i\gamma + j(\delta+\varepsilon) + k(\beta-\gamma) + \ell\varepsilon\big) + \Gamma \\ &= \tfrac{1}{n}\big((i-k)\gamma + j\delta + k\beta + (\ell+j)\varepsilon\big) + \Gamma \\ &= g\big((i-k)u + jv + ku' + (\ell+j)v'\big) \\ &= g\big(iu + j(v+v') + k(u'-u) + \ell v'\big) \\ &= g(\psi(t)). \end{aligned}$$

It follows by Remark 7.8 that $g_\diamond(\mathbb{S}_n \times \mathbb{S}_n) = h_\diamond(\mathbb{S}_n \times \mathbb{S}_n)$. This equality along with (7.15) and (7.16) yields the first equality of the proposition.

For the equality of the first and last terms in the proposition, one can argue as above, using the symplectic bases (u, v, u', v') and $(u+u', v, u', v'-v)$. Alternatively, this equality is deducible from the one already proved, as follows:

$$\begin{aligned} {}_n\Xi_\Gamma(\beta, \varepsilon-\delta) + {}_n\Xi_\Gamma(\gamma+\beta, \delta) &= {}_n\Xi_\Gamma(\beta, (\varepsilon-\delta)+\delta) + {}_n\Xi_\Gamma((\gamma+\beta)-\beta, \delta) \\ &= {}_n\Xi_\Gamma(\beta,\varepsilon) + {}_n\Xi_\Gamma(\gamma,\delta). \end{aligned}$$ □

Corollary 7.19. *The map ${}_n\Xi_\Gamma$ is $\mathbb{Z}$-bilinear and alternating, and it induces a homomorphism*

$${}_n\Upsilon_\Gamma \colon \big(\tfrac{1}{n} \wedge^2 \Gamma\big)/\wedge^2\Gamma \longrightarrow \mathit{Symp}(\mathbb{T}(\Gamma))$$

determined by

$$\tfrac{1}{n}(\gamma\wedge\delta) + \wedge^2\Gamma \mapsto f(\gamma,\delta)_\diamond(\mathbb{S}_n) \qquad \textit{for all } \ \gamma,\delta\in\Gamma.$$

Proof. We first verify a few easy identities for ${}_n\Xi_\Gamma$:

$${}_n\Xi_\Gamma(\gamma,\gamma) = 0 \qquad \text{for all } \gamma\in\Gamma. \tag{7.17}$$

To see this, note that $f(\gamma,\gamma)$ maps $\frac{1}{n}\mathbb{Z}/\mathbb{Z} \times \frac{1}{n}\mathbb{Z}/\mathbb{Z}$ to the cyclic subgroup $\langle \frac{1}{n}\gamma + \Gamma\rangle$ of $\mathbb{T}(\Gamma)$. The underlying group of $f(\gamma,\gamma)_\diamond(\mathbb{S}_n)$ is a subgroup of this group, so it is also cyclic. Since the only alternating pairing on a cyclic group is the trivial one but the pairing on $f(\gamma,\gamma)_\diamond(\mathbb{S}_n)$ is nondegenerate, this symplectic module must be trivial. That is, ${}_n\Xi_\Gamma(\gamma,\gamma) = f(\gamma,\gamma)_\diamond(\mathbb{S}_n) = 0$ in $\mathit{Symp}(\mathbb{T}(\Gamma))$.

Similarly, we have

$${}_n\Xi_\Gamma(n\gamma,\delta) = 0 = {}_n\Xi_\Gamma(\gamma, n\delta) \qquad \text{for all } \gamma,\delta\in\Gamma. \tag{7.18}$$

These equalities hold for the same reason as (7.17) since $f(n\gamma,\delta)$ and $f(\gamma,n\delta)$ map $\frac{1}{n}\mathbb{Z}/\mathbb{Z}\times\frac{1}{n}\mathbb{Z}/\mathbb{Z}$ respectively to the cyclic groups $\langle\frac{1}{n}\delta+\Gamma\rangle$ and $\langle\frac{1}{n}\gamma+\Gamma\rangle$. In particular, (7.18) shows that ${}_n\Xi_\Gamma(0,\delta) = {}_n\Xi_\Gamma(\gamma,0) = 0$ for all $\gamma,\delta\in\Gamma$.

These observations allow us to deduce properties of ${}_n\Xi_\Gamma$ from Prop. 7.18. For all $\gamma,\delta,\varepsilon\in\Gamma$, we have

$${}_n\Xi_\Gamma(\gamma,\delta) + {}_n\Xi_\Gamma(\gamma,\varepsilon) = {}_n\Xi_\Gamma(\gamma,\delta+\varepsilon) + {}_n\Xi_\Gamma(\gamma-\gamma,\varepsilon) = {}_n\Xi_\Gamma(\gamma,\delta+\varepsilon) + 0.$$

Thus, ${}_n\Xi_\Gamma$ is $\mathbb{Z}$-linear in the second variable, and another application of Prop. 7.18 shows that ${}_n\Xi_\Gamma$ is also $\mathbb{Z}$-linear in the first variable. Equation (7.17) shows that ${}_n\Xi_\Gamma$ is alternating. Therefore, ${}_n\Xi_\Gamma$ induces a map $\wedge^2\Gamma \to \mathit{Symp}(\mathbb{T}(\Gamma))$ which when composed with the multiplication-by-n isomorphism $\frac{1}{n}\wedge^2\Gamma \xrightarrow{\sim} \wedge^2\Gamma$ yields a homomorphism ${}_n\mathcal{Z}_\Gamma \colon \frac{1}{n}\wedge^2\Gamma \to \mathit{Symp}(\mathbb{T}(\Gamma))$ given by $\frac{1}{n}(\gamma \wedge \delta) \mapsto f(\gamma,\delta)_\diamond(\mathbb{S}_n)$. Since $\gamma \wedge \delta = \frac{1}{n}(n\gamma \wedge \delta))$, it follows from (7.18) that $\wedge^2\Gamma \subseteq \ker {}_n\mathcal{Z}_\Gamma$. Hence, there is a map

$$(\tfrac{1}{n} \wedge^2 \Gamma)/ \wedge^2 \Gamma \longrightarrow \mathit{Symp}(\mathbb{T}(\Gamma))$$

induced by ${}_n\mathcal{Z}_\Gamma$; this map is the ${}_n\Upsilon_\Gamma$ of the proposition. □

Proposition 7.20. *For any positive integers n and k, the maps ${}_{kn}\Upsilon_\Gamma$ and ${}_n\Upsilon_\Gamma$ agree on $(\frac{1}{n} \wedge^2 \Gamma)/ \wedge^2 \Gamma$. Therefore, the maps ${}_n\Upsilon_\Gamma$ combine for all $n \in \mathbb{N}$ to yield a well-defined homomorphism*

$$\Upsilon_\Gamma \colon \mathbb{T}(\wedge^2\Gamma) \longrightarrow \mathit{Symp}(\mathbb{T}(\Gamma)) \quad \textit{given by} \quad \tfrac{1}{n}(\gamma\wedge\delta)+\wedge^2\Gamma \mapsto f(\gamma,\delta)_\diamond(\mathbb{S}_n).$$

Proof. Take any $\gamma, \delta \in \Gamma$. Recall from Ex. 7.16 that $k \cdot \mathbb{S}_{kn} = \mathbb{S}_n$. Since $f(\gamma,\delta)_\diamond$ is a group homomorphism (see Th. 7.17), we have

$$\begin{aligned}
{}_{kn}\Upsilon_\Gamma\big(\tfrac{1}{n}(\gamma \wedge \delta) + \wedge^2\Gamma\big) &= {}_{kn}\Upsilon_\Gamma\big(\tfrac{1}{kn}(k\gamma \wedge \delta) + \wedge^2\Gamma\big) \\
&= k \cdot {}_{kn}\Upsilon_\Gamma\big(\tfrac{1}{kn}(\gamma \wedge \delta) + \wedge^2\Gamma\big) \\
&= k \cdot f(\gamma,\delta)_\diamond(\mathbb{S}_{kn}) = f(\gamma,\delta)_\diamond(k \cdot \mathbb{S}_{kn}) \\
&= f(\gamma,\delta)_\diamond(\mathbb{S}_n) = {}_n\Upsilon_\Gamma\big(\tfrac{1}{n}(\gamma \wedge \delta) + \wedge^2\Gamma\big).
\end{aligned}$$

Since, as γ, δ range over Γ, the elements $\frac{1}{n}(\gamma \wedge \delta) + \wedge^2\Gamma$ generate the group $(\frac{1}{n} \wedge^2 \Gamma)/ \wedge^2 \Gamma$, the maps ${}_{kn}\Upsilon_\Gamma$ and ${}_n\Upsilon_\Gamma$ agree on the entire domain $(\frac{1}{n} \wedge^2 \Gamma)/ \wedge^2 \Gamma$ of ${}_n\Upsilon_\Gamma$. Hence, the map Υ_Γ built from the ${}_n\Upsilon_\Gamma$ is well-defined on all of $\mathbb{T}(\wedge^2\Gamma)$. □

Remark 7.21. The symplectic module $f(\gamma,\delta)_\diamond(\mathbb{S}_n)$ has as its underlying group some subgroup of $f(\gamma,\delta)\big((\frac{1}{n}\mathbb{Z}/\mathbb{Z})^2\big) = \big\langle \frac{1}{n}\gamma + \Gamma, \frac{1}{n}\delta + \Gamma \big\rangle$ in $\mathbb{T}(\Gamma)$, but the particular subgroup and its pairing depend on the structure of the group $\big\langle \frac{1}{n}\gamma + \Gamma, \frac{1}{n}\delta + \Gamma \big\rangle$. We note a significant case where the symplectic module can be described very explicitly: Suppose γ and δ are *n-independent* in Γ, i.e., for all $k, \ell \in \mathbb{Z}$, if $k\gamma + \ell\delta \in n\Gamma$ then $n \mid k$ and $n \mid \ell$; equivalently, in $\Gamma/n\Gamma$ we have $\langle \gamma + n\Gamma, \delta + n\Gamma \rangle \cong (\mathbb{Z}/n\mathbb{Z})^2$. We set

$$S_n(\gamma,\delta) = f(\gamma,\delta)_\diamond(\mathbb{S}_n) \qquad \text{when } \gamma \text{ and } \delta \text{ are } n\text{-independent.} \tag{7.19}$$

The n-independence implies that the map $(\frac{1}{n}\mathbb{Z}/\mathbb{Z})^2 \to \mathbb{T}(\Gamma)$ given by $f(\gamma,\delta)$ is injective. Therefore, the underlying group of $S_n(\gamma,\delta)$ is $\big\langle \frac{1}{n}\gamma + \Gamma, \frac{1}{n}\delta + \Gamma \big\rangle$, of order n^2, and $(\frac{1}{n}\gamma + \Gamma, \frac{1}{n}\delta + \Gamma)$ is a symplectic base of the pairing.

Theorem 7.22. *For every torsion-free abelian group Γ, the map*

$$\Upsilon_\Gamma \colon \mathbb{T}(\wedge^2\Gamma) \longrightarrow \mathit{Symp}(\mathbb{T}(\Gamma))$$

is an isomorphism. Moreover, it is functorial in that for any homomorphism of torsion-free abelian groups $g \colon \Gamma_1 \to \Gamma_2$ there is a commutative diagram:

$$\begin{array}{ccc} \mathbb{T}(\wedge^2\Gamma_1) & \xrightarrow{\mathbb{T}(\wedge^2 g)} & \mathbb{T}(\wedge^2\Gamma_2) \\ \Upsilon_{\Gamma_1}\big\downarrow & & \big\downarrow\Upsilon_{\Gamma_2} \\ \mathit{Symp}(\mathbb{T}(\Gamma_1)) & \xrightarrow{\mathbb{T}(g)_\diamond} & \mathit{Symp}(\mathbb{T}(\Gamma_2)). \end{array}$$

Proof. We first establish the functorial property of Υ. Let $g\colon \Gamma_1 \to \Gamma_2$ be any homomorphism of torsion-free abelian groups. Then g induces the map $\mathbb{T}(g)\colon \mathbb{T}(\Gamma_1) \to \mathbb{T}(\Gamma_2)$ given by $\frac{1}{n}\gamma \mapsto \frac{1}{n}g(\gamma)$, and analogous maps $\wedge^2 g\colon \wedge^2\Gamma_1 \to \wedge^2\Gamma_2$ and $\mathbb{T}(\wedge^2 g)\colon \mathbb{T}(\wedge^2\Gamma_1) \to \mathbb{T}(\wedge^2\Gamma_2)$. Note that for $\gamma, \delta \in \Gamma_1$, if $f_1(\gamma,\delta)\colon \mathbb{T}^2 \to \mathbb{T}(\Gamma_1)$ is the map of (7.13) used in defining Υ_{Γ_1} and $f_2(\gamma',\delta')$ correspondingly for Υ_{Γ_2}, then $\mathbb{T}(g)\circ f_1(\gamma,\delta) = f_2\big(g(\gamma), g(\delta)\big)$. Hence, for any γ, $\delta \in \Gamma_1$ and $n \in \mathbb{N}$, we have, using the functorial property of *Symp* (see Prop. 7.10),

$$\begin{aligned} \mathbb{T}(g)_\diamond \circ \Upsilon_{\Gamma_1}\big(\tfrac{1}{n}(\gamma\wedge\delta) + \wedge^2\Gamma_1\big) \;&=\; \mathbb{T}(g)_\diamond \circ f_1(\gamma,\delta)_\diamond(\mathbb{S}_n) \\ &=\; (\mathbb{T}(g)\circ f_1(\gamma,\delta))_\diamond(\mathbb{S}_n) \;=\; f_2\big(g(\gamma), g(\delta)\big)_\diamond(\mathbb{S}_n) \\ &=\; \Upsilon_{\Gamma_2}\big(\tfrac{1}{n}(g(\gamma)\wedge g(\delta)) + \wedge^2\Gamma_2\big) \\ &=\; \Upsilon_{\Gamma_2} \circ \mathbb{T}(\wedge^2 g)\big(\tfrac{1}{n}.(\gamma\wedge\delta) + \wedge^2\Gamma_1\big). \end{aligned}$$

Thus, the diagram of the theorem is commutative.

To see that Υ_Γ is surjective, take any symplectic module (M,a) in $\mathit{Symp}(\mathbb{T}(\Gamma))$ with (M,a) isometric to $\mathbb{S}_n$, and let (u,v) be any symplectic base of (M,a). Since $M \subseteq {}_n\mathbb{T}(\Gamma) = \frac{1}{n}\Gamma/\Gamma$ we can write $u = \frac{1}{n}\gamma + \Gamma$ and $v = \frac{1}{n}\delta + \Gamma$ for some $\gamma, \delta \in \Gamma$. Moreover, as $\langle u, v\rangle = M \cong \big(\frac{1}{n}\mathbb{Z}/\mathbb{Z}\big)^2$, the elements γ and δ are n-independent in Γ. Therefore, in the notation of (7.19),

$$(M,a) \cong S_n(\gamma,\delta) \;=\; \Upsilon_\Gamma\big(\tfrac{1}{n}(\gamma\wedge\delta) + \Gamma\big).$$

Since the group $\mathit{Symp}(\mathbb{T}(\Gamma))$ is generated by symplectic modules such as (M,a) by Cor. 7.4 and Ex. 7.11(3), the map Υ_Γ is surjective.

To prove injectivity of Υ_Γ it suffices to verify injectivity of its restriction ${}_p\Upsilon_\Gamma$ to the p-torsion subgroup ${}_p\mathbb{T}(\wedge^2\Gamma)$ for each prime p, because if $\xi \in \mathbb{T}(\wedge^2\Gamma)$ is a nonzero element of $\ker \Upsilon_\Gamma$, then some nonzero multiple of ξ lies in ${}_p\mathbb{T}(\wedge^2\Gamma) \cap \ker \Upsilon_\Gamma = \ker {}_p\Upsilon_\Gamma$ for some prime p.

For the rest of the proof, we fix some prime p. Since Γ and $\wedge^2\Gamma$ are torsion-free, we have canonical isomorphisms

$${}_p\mathbb{T}(\wedge^2\Gamma) \;=\; \big(\tfrac{1}{p}\wedge^2\Gamma\big)/\wedge^2\Gamma \;\cong\; \wedge^2\Gamma/p\wedge^2\Gamma \;\cong\; \wedge^2(\Gamma/p\Gamma) \;\cong\; \wedge^2(\tfrac{1}{p}\Gamma/\Gamma).$$

(The second isomorphism is the composition

$$\wedge^2\Gamma/p\wedge^2\Gamma \;\cong\; (\wedge^2\Gamma)\otimes_\mathbb{Z}(\mathbb{Z}/p\mathbb{Z}) \;\cong\; \wedge^2(\Gamma\otimes_\mathbb{Z}(\mathbb{Z}/p\mathbb{Z})) \;\cong\; \wedge^2(\Gamma/p\Gamma).)$$

The composition of the isomorphisms above is an isomorphism

$$\psi\colon {}_p\mathbb{T}(\wedge^2\Gamma) \xrightarrow{\sim} \wedge^2(\tfrac{1}{p}\Gamma/\Gamma)$$

mapping

$$\tfrac{1}{p}(\gamma\wedge\delta)\mapsto\big(\tfrac{1}{p}\gamma+\Gamma\big)\wedge\big(\tfrac{1}{p}\delta+\Gamma\big)\qquad\text{for all }\ \gamma,\delta\in\Gamma.$$

To simplify notation, let $X=\frac{1}{p}\Gamma/\Gamma={}_p\mathbb{T}(\Gamma)$. The image of ${}_p\Upsilon_\Gamma$ lies in ${}_p\mathit{Symp}(\mathbb{T}(\Gamma))$. But the exponent of a symplectic module (M,a) is the exponent of the group M by Prop. 7.15, so

$${}_p\mathit{Symp}(\mathbb{T}(\Gamma))=\mathit{Symp}({}_p\mathbb{T}(\Gamma))=\mathit{Symp}(X).$$

Therefore, we may view ${}_p\Upsilon_\Gamma$ as a map into $\mathit{Symp}(X)$. By composing this map with ψ^{-1}, we obtain a homomorphism

$$\Upsilon_X={}_p\Upsilon_\Gamma\circ\psi^{-1}\colon\ \wedge^2X\longrightarrow\mathit{Symp}(X)$$

given by

$$\big(\tfrac{1}{p}\gamma+\Gamma\big)\wedge\big(\tfrac{1}{p}\delta+\Gamma\big)\mapsto f(\gamma,\delta)_\diamond(\mathbb{S}_p)\qquad\text{for all }\ \gamma,\delta\in\Gamma.$$

Since ψ is an isomorphism, for injectivity of ${}_p\Upsilon_\Gamma$ it suffices to prove that Υ_X is injective. For this, pick a base $(x_i)_{i\in I}$ of X as an $\mathbb{F}_p$-vector space. For each finite subset $J\subseteq I$, let X_J denote the span of $(x_j)_{j\in J}$ in X. Note that for each generator $x\wedge y$ of $\wedge^2(X_J)$, Remark 7.21 shows that $\Upsilon_X(x\wedge y)=(M,a)$ for some subgroup M of $\langle x,y\rangle$, so $\Upsilon_X(x\wedge y)\in\mathit{Symp}(X_J)$; hence, as Υ_X is a group homomorphism, $\Upsilon_X(\wedge^2X_J)\subseteq\mathit{Symp}(X_J)$.

Since $\wedge^2X$ is the union of the subspaces $\wedge^2(X_J)$ as J runs over the finite subsets of I, to verify that Υ_X is injective it suffices to show that each restriction $\Upsilon_X|_{\wedge^2X_J}$ is injective. We do this by induction on $|J|$. If $|J|=1$, then $\wedge^2(X_J)=\{0\}$, so the result is clear. Suppose $|J|\geq 2$; pick $\ell\in J$ and let $J'=J\setminus\{\ell\}$. Thus, $J=J'\cup\{\ell\}$ and $X_J=X_{J'}\oplus x_\ell\mathbb{F}_p$. Take $\xi\in\wedge^2(X_J)\cap\ker\Upsilon_X$. We write $\xi=\xi'+(x_\ell\wedge y)$ for some $\xi'\in\wedge^2(X_{J'})$ and some $y\in X_{J'}$. Then,

$$\Upsilon_X(x_\ell\wedge y)=-\Upsilon_X(\xi')\qquad\text{in }\ \mathit{Symp}(X_J).\tag{7.20}$$

Let $\gamma,\delta\in\Gamma$ satisfy $x_\ell=\frac{1}{p}\gamma+\Gamma$ and $y=\frac{1}{p}\delta+\Gamma$. If γ and δ were p-independent in Γ, then Remark 7.21 shows that $\Upsilon_X(x_\ell\wedge y)=f(\gamma,\delta)_\diamond(\mathbb{S}_p)$ has underlying group $\langle x_\ell,y\rangle$; this is not a subgroup of $X_{J'}$; so $\Upsilon_X(x_\ell\wedge y)\notin\mathit{Symp}(X_{J'})$. However, since $\xi'\in X_{J'}$, we saw above that $\Upsilon_X(\xi')\in\mathit{Symp}(X_{J'})$. This contradicts the equality (7.20). Hence, γ and δ must be p-dependent in Γ, so $(\gamma+p\Gamma)\wedge(\delta+p\Gamma)=0$ in $\wedge^2(\Gamma/p\Gamma)$, yielding $x_\ell\wedge y=0$ in $\wedge^2(X_J)$. Therefore, $\xi=\xi'\in X_{J'}$. Hence, $\xi=0$, as $\Upsilon_X|_{\wedge^2X_{J'}}$ is injective by induction. Thus, Υ_X is injective, as is ${}_p\Upsilon_\Gamma$. □

As an illustration of Th. 7.22, we consider the particular case where $\Gamma=\mathbb{Z}^n$ for some $n\geq 1$. We may then use coordinates and relate our construction to alternating matrices. For any abelian group M and any integer $n\geq 1$, we let $\mathit{Alt}_n(M)$ denote the group of alternating matrices of order n with entries in M, i.e., skew-symmetric matrices with zero diagonal entries. Note that the

$\mathbb{Z}$-module structure on M allows us to multiply matrices with entries in M by matrices of suitable size with entries in $\mathbb{Z}$.

Until the end of this section, we let $\Gamma = \mathbb{Z}^n$ and let $(\gamma_1, \ldots, \gamma_n)$ be the standard base of $\mathbb{Z}^n$. Slightly abusing notation, for $u = q+\mathbb{Z} \in \mathbb{T}$ (with $q \in \mathbb{Q}$) and $\gamma \in \Gamma$ we write $u\gamma$ for $q\gamma + \Gamma \in \mathbb{T}(\Gamma)$. Every element in $\mathbb{T}(\Gamma)$ thus has a unique expression of the form $\sum_{i=1}^n u_i\gamma_i$ with $u_i \in \mathbb{T}$; hence, $\mathbb{T}(\Gamma) = \mathbb{T}^n$. Similarly, every element $\xi \in \mathbb{T}(\wedge^2\Gamma)$ has a unique expression of the form $\xi = \sum_{1\leq i<j\leq n} u_{ij}\gamma_i \wedge \gamma_j$ with $u_{ij} \in \mathbb{T}$. Letting $u_{ij} = -u_{ji}$ for $i > j$ and $u_{ii} = 0$, we associate to ξ the alternating matrix $(u_{ij})_{i,j=1}^n \in Alt_n(\mathbb{T})$. Thus, we have an identification

$$\mathbb{T}(\wedge^2\Gamma) = Alt_n(\mathbb{T}), \qquad \sum_{1\leq i<j\leq n} u_{ij}\gamma_i \wedge \gamma_j \longleftrightarrow (u_{ij})_{i,j=1}^n. \tag{7.21}$$

If the elements in $\mathbb{Z}^n$ are written as column vectors, the alternating matrix associated to any $u\gamma \wedge \delta$ with $u \in \mathbb{T}$ and γ, $\delta \in \Gamma$ is $u(\gamma\delta^t - \delta\gamma^t) \in Alt_n(\mathbb{T})$. Under the identification (7.21), the isomorphism Υ_Γ takes the form of a map

$$\Upsilon_n\colon\ Alt_n(\mathbb{T}) \xrightarrow{\sim} Symp(\mathbb{T}^n).$$

By definition, this map carries the alternating matrix $(\frac{a_{ij}}{d} + \mathbb{Z})_{i,j=1}^n$, for $a_{ij} \in \mathbb{Z}$ and $d \in \mathbb{N}$, to the symplectic module $\sum_{1\leq i<j\leq n} a_{ij}S_d(\gamma_i, \gamma_j)$, where $S_d(\gamma_i, \gamma_j)$ is the symplectic module with symplectic base $(\frac{1}{d}\gamma_i + \Gamma,\ \frac{1}{d}\gamma_j + \Gamma)$, which is isomorphic to $\mathbb{S}_d$; see (7.19). Since the definition of the addition in $Symp(\mathbb{T}^n)$ is somewhat involved, it is desirable to have a more explicit description of the image of a given alternating matrix. We will obtain such a description by reducing the matrix to a normal form as follows:

Proposition 7.23. *For every $A \in Alt_n(\mathbb{T})$, there exists a matrix $P \in GL_n(\mathbb{Z})$ and nonzero elements $u_1, \ldots, u_r \in \mathbb{T}$ with $u_1\mathbb{Z} \supseteq u_2\mathbb{Z} \supseteq \ldots \supseteq u_r\mathbb{Z}$ such that P^tAP is a block-diagonal matrix of the form*

$$P^tAP = diag\Big(\left(\begin{smallmatrix} 0 & u_1 \\ -u_1 & 0 \end{smallmatrix}\right), \ldots, \left(\begin{smallmatrix} 0 & u_r \\ -u_r & 0 \end{smallmatrix}\right), 0, \ldots, 0\Big).$$

Proof. Choose $a_{ij} \in \mathbb{Z}$, $d \in \mathbb{N}$ such that $A = \left(\frac{a_{ij}}{d} + \mathbb{Z}\right)_{i,j=1}^n$ with the matrix $(a_{ij})_{i,j=1}^n \in Alt_n(\mathbb{Z})$, and let $A' = (a_{ij})_{i,j=1}^n$. We will find a matrix $P \in GL_n(\mathbb{Z})$ and nonzero integers $m_1, \ldots, m_r$ such that $m_1 \mid m_2 \mid \ldots \mid m_r$ and

$$P^tA'P = diag\Big(\left(\begin{smallmatrix} 0 & m_1 \\ -m_1 & 0 \end{smallmatrix}\right), \ldots, \left(\begin{smallmatrix} 0 & m_r \\ -m_r & 0 \end{smallmatrix}\right), 0, \ldots, 0\Big).$$

Then

$$P^tAP = diag\Big(\left(\begin{smallmatrix} 0 & \frac{m_1}{d}+\mathbb{Z} \\ -\frac{m_1}{d}+\mathbb{Z} & 0 \end{smallmatrix}\right), \ldots, \left(\begin{smallmatrix} 0 & \frac{m_r}{d}+\mathbb{Z} \\ -\frac{m_r}{d}+\mathbb{Z} & 0 \end{smallmatrix}\right), 0, \ldots, 0\Big),$$

and with $u_i = \frac{m_i}{d} + \mathbb{Z}$ we have $u_1\mathbb{Z} \supseteq \ldots \supseteq u_r\mathbb{Z}$ because $m_1 \mid \ldots \mid m_r$.

Finding P amounts to finding a symplectic base of $\mathbb{Z}^n$ for the alternating pairing defined by A': Letting $(\gamma|\delta) = \gamma^tA'\delta \in \mathbb{Z}$ for γ, $\delta \in \mathbb{Z}^n$ (viewed as column vectors), we look for a base $(\varepsilon_1, \ldots, \varepsilon_n)$ of $\mathbb{Z}^n$ and integers $m_1, \ldots, m_r$

with $m_1 \mid m_2 \mid \ldots \mid m_r$ such that

$$(\varepsilon_{2i-1}|\varepsilon_{2i}) = m_i \text{ for } i = 1, \ldots, r, \quad (\varepsilon_{2i}|\varepsilon_{2i+1}) = 0 \text{ for } i = 1, \ldots, r-1,$$

and $(\varepsilon_i|\varepsilon_j) = 0$ whenever $|i - j| \geq 2$ or $j \geq 2r$. Then $\varepsilon_1, \ldots, \varepsilon_n$ are the columns of a matrix P satisfying the requirements.

The existence of a symplectic base $\varepsilon_1, \ldots, \varepsilon_n$ is proved below by a slight variation on the arguments that establish the existence of symplectic bases in vector spaces with alternating pairings (or in symplectic modules; see Th. 7.2):

For each $\varepsilon \in \mathbb{Z}^n$, the set $(\varepsilon|\mathbb{Z}^n) = \{(\varepsilon|\delta) \mid \delta \in \mathbb{Z}^n\}$ is an ideal in $\mathbb{Z}$. Consider the set of these ideals: $\mathcal{I} = \{(\varepsilon|\mathbb{Z}^n) \mid \varepsilon \in \mathbb{Z}^n\}$. Pick $\varepsilon_1 \in \mathbb{Z}^n$ such that $(\varepsilon_1|\mathbb{Z}^n)$ is maximal in $\mathcal{I}$. Let $m_1 \in \mathbb{Z}$ be such that $(\varepsilon_1|\mathbb{Z}^n) = m_1\mathbb{Z}$ and pick $\varepsilon_2 \in \mathbb{Z}^n$ such that $(\varepsilon_1|\varepsilon_2) = m_1$. We may assume $A' \neq 0$, hence $m_1 \neq 0$. Then $\varepsilon_1\mathbb{Z} \cap \varepsilon_2\mathbb{Z} = \{0\}$, hence the sum $\varepsilon_1\mathbb{Z} + \varepsilon_2\mathbb{Z}$ is a direct sum. Because $m_1\mathbb{Z}$ is maximal in $\mathcal{I}$ and $(\varepsilon_2|\varepsilon_1) = -m_1$, we have $(\varepsilon_2|\mathbb{Z}^n) = m_1\mathbb{Z}$. We may therefore consider the map

$$p\colon \mathbb{Z}^n \longrightarrow \varepsilon_1\mathbb{Z} \oplus \varepsilon_2\mathbb{Z} \quad \text{given by} \quad \gamma \mapsto \varepsilon_1\frac{(\varepsilon_2|\gamma)}{-m_1} + \varepsilon_2\frac{(\varepsilon_1|\gamma)}{m_1}.$$

We have $p^2 = p$, hence $\mathbb{Z}^n = (\varepsilon_1\mathbb{Z} \oplus \varepsilon_2\mathbb{Z}) \oplus \mathit{ker}\, p$. Now use induction to find a symplectic base in $\mathit{ker}\, p$. To see that the integer m_2 found at the next step is a multiple of m_1, let $m_2 = (\varepsilon_3|\varepsilon_4)$ with $\varepsilon_3, \varepsilon_4 \in \mathit{ker}\, p$, and let $\ell = \mathit{gcd}(m_1, m_2)$. We have $\ell = m_1\ell_1 + m_2\ell_2$ for some $\ell_1, \ell_2 \in \mathbb{Z}$. Because $\varepsilon_3, \varepsilon_4 \in \mathit{ker}\, p$ we also have $(\varepsilon_i|\varepsilon_j) = 0$ for $i = 1, 2$ and $j = 3, 4$. Therefore, from $\ell = m_1\ell_1 + m_2\ell_2$ we obtain $\ell = (\varepsilon_1 + \varepsilon_3|\varepsilon_2\ell_1 + \varepsilon_4\ell_2) \in (\varepsilon_1 + \varepsilon_3|\mathbb{Z}^n)$. Since $\ell \mid m_1$, we thus have

$$m_1\mathbb{Z} \subseteq \ell\mathbb{Z} \subseteq (\varepsilon_1 + \varepsilon_3|\mathbb{Z}^n) \in \mathcal{I}.$$

But $m_1\mathbb{Z}$ is maximal in $\mathcal{I}$, so $m_1\mathbb{Z} = \ell\mathbb{Z}$ and therefore $m_1 \mid m_2$. □

As a consequence of Prop. 7.23, every $A \in \mathit{Alt}_n(\mathbb{T})$ can be decomposed into a product

$$A = Q^t D Q \quad \text{with} \quad D = \mathit{diag}\Big(\big(\begin{smallmatrix} 0 & u_1 \\ -u_1 & 0 \end{smallmatrix}\big), \ldots, \big(\begin{smallmatrix} 0 & u_r \\ -u_r & 0 \end{smallmatrix}\big), 0, \ldots, 0\Big) \tag{7.22}$$

for some $Q \in GL_n(\mathbb{Z})$ and some $u_1, \ldots, u_r \in \mathbb{T}$ such that

$$u_1\mathbb{Z} \supseteq u_2\mathbb{Z} \supseteq \ldots \supseteq u_r\mathbb{Z} \neq \{0\}.$$

The image of A under Υ_n is explicitly described as follows in terms of this decomposition:

Proposition 7.24. *Let $A \in \mathit{Alt}_n(\mathbb{T})$ be decomposed as in (7.22), and let $\delta_1, \ldots, \delta_n \in \mathbb{Z}^n$ be the columns of Q^t. For $i = 1, \ldots, r$, let $n_i \in \mathbb{N}$ be the order of u_i in $\mathbb{T}$, and let $u_i' = \frac{1}{n_i} + \mathbb{Z} \in \mathbb{T}$. We have $n_i \mid n_j$ for $i \leq j$, and the symplectic module $\Upsilon_n(A)$ has symplectic base $(u_1\delta_1, u_1'\delta_2, u_2\delta_3, u_2'\delta_4, \ldots, u_r\delta_{2r-1}, u_r'\delta_{2r})$; it is isomorphic to $\mathbb{S}_{n_1} \perp \ldots \perp \mathbb{S}_{n_r}$.*

Proof. The relation $n_i \mid n_j$ for $i \leq j$ follows from $u_i\mathbb{Z} \supseteq u_j\mathbb{Z}$. Decomposing Q^tD into columns, we have

$$Q^tD = (-\delta_2 u_1, \delta_1 u_1, -\delta_4 u_2, \delta_3 u_2, \dots, -\delta_{2r}u_r, \delta_{2r-1}u_r, 0, \dots, 0).$$

From $A = Q^tDQ$, it follows that

$$A = u_1(\delta_1\delta_2^t - \delta_2\delta_1^t) + \dots + u_r(\delta_{2r-1}\delta_{2r}^t - \delta_{2r}\delta_{2r-1}^t).$$

Under the identification $\mathbb{T}(\wedge^2\Gamma) = Alt_n(\mathbb{T})$, the matrix A therefore corresponds to $\sum_{i=1}^r u_i\delta_{2i-1} \wedge \delta_{2i}$. Write $u_i = \frac{a_i}{n_i} + \mathbb{Z}$ with $a_i \in \mathbb{Z}$ prime to n_i. By definition, Υ_n maps $\sum_{i=1}^r u_i\delta_{2i-1} \wedge \delta_{2i}$ to $\sum_{i=1}^r S_{n_i}(a_i\delta_{2i-1}, \delta_{2i})$. Each $S_{n_i}(a_i\delta_{2i-1}, \delta_{2i})$ has symplectic base $(\frac{a_i}{n_i}\delta_{2i-1}+\Gamma, \frac{1}{n_i}\delta_{2i}+\Gamma) = (u_i\delta_{2i-1}, u_i'\delta_{2i})$, and is isomorphic to $\mathbb{S}_{n_i}$, see Remark 7.21. Note that $(\delta_1, \dots, \delta_n)$ is a base of $\mathbb{Z}^n$ since Q is invertible; therefore, the underlying modules of the various $S_{n_i}(a_i\delta_{2i-1}, \delta_{2i})$ are disjoint in $\mathbb{T}^n$, and it follows that in $Symp(\mathbb{T}^n)$

$$\sum_{i=1}^{n} S_{n_i}(a_i\delta_{2i-1}, \delta_{2i}) = S_{n_1}(a_1\delta_1, \delta_2) \perp \dots \perp S_{n_r}(a_r\delta_{2r-1}, \delta_{2r}),$$

see Ex. 7.11(3). The proposition follows. □

Example 7.25. Let

$$A = \begin{pmatrix} 0 & \frac{1}{5}+\mathbb{Z} & \frac{3}{5}+\mathbb{Z} \\ -\frac{1}{5}+\mathbb{Z} & 0 & \frac{1}{5}+\mathbb{Z} \\ -\frac{3}{5}+\mathbb{Z} & -\frac{1}{5}+\mathbb{Z} & 0 \end{pmatrix} \in Alt_3(\mathbb{T}).$$

We have the following factorization:

$$A = \begin{pmatrix} 3 & -1 & -1 \\ 1 & 4 & 1 \\ 0 & 3 & 1 \end{pmatrix} \cdot \begin{pmatrix} 0 & \frac{2}{5}+\mathbb{Z} & 0 \\ -\frac{2}{5}+\mathbb{Z} & 0 & 0 \\ 0 & 0 & 0 \end{pmatrix} \cdot \begin{pmatrix} 3 & 1 & 0 \\ -1 & 4 & 3 \\ -1 & 1 & 1 \end{pmatrix}.$$

Therefore, the symplectic module $\Upsilon_3(A) \in Symp(\mathbb{T}^3)$ has symplectic base

$$\left(\frac{2}{5}\begin{pmatrix} 3 \\ 1 \\ 0 \end{pmatrix} + \mathbb{Z}^3, \frac{1}{5}\begin{pmatrix} -1 \\ 4 \\ 3 \end{pmatrix} + \mathbb{Z}^3\right) = \left(\begin{pmatrix} \frac{1}{5}+\mathbb{Z} \\ \frac{2}{5}+\mathbb{Z} \\ 0 \end{pmatrix}, \begin{pmatrix} -\frac{1}{5}+\mathbb{Z} \\ -\frac{1}{5}+\mathbb{Z} \\ -\frac{2}{5}+\mathbb{Z} \end{pmatrix}\right).$$

It is isomorphic to $\mathbb{S}_5$.

7.2 Tensor products of symbol algebras

The generators i, j of a symbol algebra S of degree n over a field F satisfy $i^n, j^n \in F^\times$ and $ij \equiv ji \ (mod\ F^\times)$. Therefore, their images in the factor group $S^\times/F^\times$ generate an abelian subgroup isomorphic to $(\mathbb{Z}/n\mathbb{Z})^2$. More generally, F-algebras A that are tensor products of symbol F-algebras can be characterized by the presence in $A^\times/F^\times$ of certain finite abelian subgroups, which we call *armatures*. These finite abelian groups carry an alternating pairing induced by the commutator map. They are introduced in §7.2.1 for algebras over arbitrary fields (without valuation or grading). The main result there describes the structure of algebras that are spanned by an armature; see Th. 7.36. The case of graded algebras is considered in §7.2.2. The degree map restricts to homogeneous armatures and thus provides additional structure. In the final §7.2.3, we show how armatures in algebras over valued fields may be used to construct (under suitable conditions) a gauge in the algebra. We thus generalize the definition of symbol algebra gauges given in §3.2.3.

7.2.1 Armatures

Let A be an arbitrary algebra over a field F. We consider the factor group of multiplicative groups $A^\times/F^\times$ and use the following notation: for $a \in A^\times$, we let

$$[a] = a \cdot F^\times \in A^\times/F^\times.$$

For any subgroup $\mathcal{U} \subseteq A^\times/F^\times$, we let $F[\mathcal{U}] \subseteq A$ denote the additive subgroup generated by the elements $u \in A^\times$ such that $[u] \in \mathcal{U}$:

$$F[\mathcal{U}] = \Big\{ \sum_{\text{finite}} u_i \;\Big|\; [u_i] \in \mathcal{U} \text{ for all } i \Big\} \subseteq A.$$

Clearly, $F[\mathcal{U}]$ is an F-vector space; it is even an F-subalgebra of A since $\mathcal{U}$ is a group. For u, $v \in \mathcal{U}$, we choose x, $y \in A^\times$ such that $[x] = u$, $[y] = v$, and let

$$\langle u, v\rangle = xyx^{-1}y^{-1} \in A^\times.$$

This element does not depend on the choice of the representatives x, y of u and v. We thus obtain a map $\langle\cdot,\cdot\rangle\colon \mathcal{U}\times\mathcal{U} \to A^\times$, which we call the *commutator map*, and sometimes designate by $b_{\mathcal{U}}$.

Proposition 7.26. *For any abelian subgroup $\mathcal{U} \subseteq A^\times/F^\times$, the commutator map is an alternating bilinear pairing*

$$\langle\cdot,\cdot\rangle\colon \mathcal{U}\times\mathcal{U} \longrightarrow F^\times.$$

If it is nondegenerate[1]*, then any system $(x_u)_{u\in\mathcal{U}}$ of representatives in $A^\times$ of the elements of $\mathcal{U}$ is linearly independent over F.*

Proof. For x, $y \in A^\times$ such that $[x] = u$ and $[y] = v$, we have

$$[xyx^{-1}y^{-1}] = [x]\cdot[y]\cdot[x]^{-1}\cdot[y]^{-1} = uvu^{-1}v^{-1},$$

[1] The pairing is nondegenerate if for each $u \neq 1$ there exists v such that $\langle u, v\rangle \neq 1$.

and the term on the right equals 1 since $\mathcal{U}$ is abelian. Therefore, the values of the commutator map lie in $F^\times$. From the equation

$$x(yz)x^{-1}(yz)^{-1} = (xyx^{-1}y^{-1}) \cdot y(xzx^{-1}z^{-1})y^{-1},$$

which holds for any x, y, $z \in A^\times$, we deduce that

$$\langle u, vw\rangle = \langle u, v\rangle \cdot \langle u, w\rangle \qquad \text{for any } u,\ v,\ w \in \mathcal{U}.$$

Since moreover $\langle u, u\rangle = 1$ for all $u \in \mathcal{U}$, the commutator map is an alternating bilinear pairing on $\mathcal{U}$.

Now, suppose the commutator pairing is nondegenerate and let $(x_u)_{u\in\mathcal{U}}$ be any system of representatives in $A^\times$ of the elements of $\mathcal{U}$. Suppose the x_u's are not linearly independent over F. We may then pick a vanishing linear combination with a minimal number of nonzero coefficients $\sum_{u\in\mathcal{U}} \lambda_u x_u = 0$. This linear combination has at least two nonzero coefficients, so we may find v, $w \in \mathcal{U}$ such that $\lambda_v \neq 0$, $\lambda_w \neq 0$, and $v \neq w$. Since the commutator pairing is assumed to be nondegenerate, we may also find $t \in \mathcal{U}$ such that

$$\langle t, v\rangle \neq \langle t, w\rangle.$$

We have

$$x_t\big(\sum_{u\in\mathcal{U}} \lambda_u x_u\big)x_t^{-1} - \langle t, w\rangle\big(\sum_{u\in\mathcal{U}} \lambda_u x_u\big) = \sum_{u\in\mathcal{U}} \lambda_u\big(\langle t, u\rangle - \langle t, w\rangle\big)x_u.$$

The left expression vanishes since $\sum_{u\in\mathcal{U}} \lambda_u x_u = 0$, and the right expression has strictly fewer nonzero coefficients than $\sum_{u\in\mathcal{U}} \lambda_u x_u$. This is a contradiction, which shows that the x_u's are linearly independent over F. □

Definition 7.27. An *armature* in an F-algebra A is a finite abelian subgroup $\mathcal{A} \subseteq A^\times/F^\times$ whose elements can be represented by linearly independent elements in $A^\times$. Any system of representatives $(x_a)_{a\in\mathcal{A}}$ is then an F-base of $F[\mathcal{A}]$, so $|\mathcal{A}| = dim_F\, F[\mathcal{A}]$. The F-algebra $F[\mathcal{A}]$ is said to be *spanned* by the armature $\mathcal{A}$.

As Prop. 7.26 shows, the linear independence condition necessarily holds when the commutator pairing on $\mathcal{A}$ is nondegenerate, but we make no assumption on the pairing, which may even be trivial. Observe however that since $\mathcal{A}$ is finite, the values of the commutator pairing have finite order in $F^\times$, which means that they lie in the group $\mu(F)$ of roots of unity of F. A coherent choice of primitive roots of unity allows us to embed (noncanonically) $\mu(F)$ into $\mathbb{T}$. Therefore, the results of §7.1.1 apply to commutator pairings on armatures (with a shift from the additive notation to multiplicative notation). In particular, we may consider the radical $rad\mathcal{A}$ of an armature $\mathcal{A}$. The commutator pairing induces a nondegenerate alternating pairing on $\mathcal{A}/rad\mathcal{A}$, which turns this group into a symplectic module. As a result, the field F necessarily contains a primitive e-th root of unity, for $e = exp(\mathcal{A}/rad\mathcal{A})$.

Example 7.28. (i) Let A be the factor ring $F[X]/(X^n - \alpha)$ for some integer $n \geq 1$ (which may be divisible by *char* F) and some $\alpha \in F^\times$, and let $x \in A$ be the image of X. The subgroup of $A^\times/F^\times$ generated by $[x]$ is an armature $\mathcal{A}$ of order n in A, with trivial commutator pairing. We have $F[\mathcal{A}] = A$.

(ii) *Kummer extensions.* Fix an arbitrary integer $s \geq 2$, and let F be a field containing a primitive s-th root of unity. Let $\mu_s \subseteq F^\times$ be the group of s-th roots of unity in F. A finite-degree field extension L of F is said to be *Kummer* if it is Galois with abelian Galois group of exponent dividing s. For such an extension, we let $\mathcal{K}(L/F)$ denote the group of s-torsion elements in $L^\times/F^\times$:

$$\mathcal{K}(L/F) = {}_s(L^\times/F^\times) = \{x \in L^\times \mid x^s \in F^\times\}/F^\times.$$

Kummer theory shows that this group does not depend on the choice of s (so long as F contains a primitive s-th root of unity and the exponent of $\mathcal{G}(L/F)$ divides s). It is in canonical duality with the Galois group by the pairing

$$\mathcal{G}(L/F) \times \mathcal{K}(L/F) \longrightarrow \mu_s, \qquad (\sigma, xF^\times) \mapsto \sigma(x)x^{-1},$$

see for example Morandi [160, §11]. In particular, $\mathcal{G}(L/F)$ and $\mathcal{K}(L/F)$ are (noncanonically) isomorphic. Thus, $|\mathcal{K}(L/F)| = |\mathcal{G}(L/F)| = [L{:}F]$, and the subring $F[\mathcal{K}(L/F)]$ of L must be all of L, by the nondegeneracy of the duality pairing. Hence, representatives of the elements of $\mathcal{K}(L/F)$ must be F-linearly independent. Therefore, $\mathcal{K}(L/F)$ is an armature of L viewed as an F-algebra, and this armature spans L. It is also the unique armature of exponent dividing s spanning the F-algebra L. For if $\mathcal{A}$ were another such armature, then $\mathcal{A} \subseteq \mathcal{K}(L/F)$ by definition and $|\mathcal{A}| = [F[\mathcal{A}]{:}F] = [L{:}F] = |\mathcal{K}(L/F)|$. The commutator pairing on $\mathcal{K}(L/F)$ is clearly trivial, since L is commutative. Let $\mathcal{K}^s(L/F)$ be the image of the canonical embedding

$$\mathcal{K}(L/F) \hookrightarrow F^\times/F^{\times s} \quad \text{given by} \quad xF^\times \mapsto x^s F^{\times s}.$$

The group $\mathcal{K}^s(L/F) \subseteq F^\times/F^{\times s}$ determines the extension L of F uniquely up to isomorphism. Indeed, for every finite subgroup $\mathcal{L} \subseteq F^\times/F^{\times s}$ there exists a Kummer extension L of F with $\mathcal{K}^s(L/F) = \mathcal{L}$, and this extension is uniquely determined up to isomorphism: if $\mathcal{L}$ is generated by $\lambda_1 F^{\times s}, \ldots, \lambda_n F^{\times s}$ for some $\lambda_1, \ldots, \lambda_n \in F^\times$, then

$$L \cong F(\sqrt[s]{\lambda_1}, \ldots, \sqrt[s]{\lambda_n}).$$

We use the notation $F(\sqrt[s]{\mathcal{L}})$ for this field extension.

(iii) *Symbol algebras.* Suppose F contains a primitive n-th root of unity ω for some integer $n \geq 2$ (necessarily prime to *char* F) and let A be an n-symbol algebra over F, say $A = \big(c, d/F\big)_{\omega,n}$ with $c, d \in F^\times$. Then, there are elements i, $j \in A$ such that

$$A = \bigoplus_{k,\ell=0}^{n-1} F i^k j^\ell, \quad i^n = c, \quad j^n = d, \quad \text{and} \quad ij = \omega ji.$$

Then $[i]$ and $[j] \in A^\times/F^\times$ generate an armature $\mathcal{A}$ of order n^2 in A. We have $F[\mathcal{A}] = A$ and

$$\langle [i], [j] \rangle = \omega,$$

so the commutator pairing is nondegenerate, and in fact $\mathcal{A}$ is isomorphic as a symplectic module to the standard symplectic module $\mathbb{S}_n$. The isomorphism depends on the identification of the group $\mu_n(F)$ of n-th roots of unity with $(\frac{1}{n}\mathbb{Z})/\mathbb{Z}$: if ω is identified with $\frac{k}{n} + \mathbb{Z}$ for some integer k prime to n, then we may pick an integer ℓ such that $k\ell \equiv 1 \ (mod\ n)$ and map $[i]^\ell$, $[j]$ to the standard base of $(\frac{1}{n}\mathbb{Z}/\mathbb{Z})^2$.

Clearly, every subgroup of an armature is an armature. More examples are obtainable from the ones above by tensor products: note that for any F-algebras A, B we have a well-defined map

$$\otimes\colon (A^\times/F^\times)\times(B^\times/F^\times) \longrightarrow (A\otimes_F B)^\times/F^\times \text{ given by } \big([x],[y]\big) \mapsto [x\otimes y].$$

Proposition 7.29. *Let A and B be F-algebras with armatures $\mathcal{A}$, $\mathcal{B}$ respectively. Then $\mathcal{A}\otimes\mathcal{B}$ is an armature of $A\otimes_F B$ with $|\mathcal{A}\otimes\mathcal{B}| = |\mathcal{A}|\cdot|\mathcal{B}|$. For a_1, $a_2 \in \mathcal{A}$ and b_1, $b_2 \in \mathcal{B}$ we have*

$$\langle a_1\otimes b_1, a_2\otimes b_2\rangle = \langle a_1, a_2\rangle\cdot\langle b_1, b_2\rangle.$$

Proof. The set $\mathcal{A}\otimes\mathcal{B}$ is clearly a finite abelian subgroup of $(A\otimes_F B)^\times/F^\times$, with $|\mathcal{A}\otimes\mathcal{B}| = |\mathcal{A}|\cdot|\mathcal{B}|$.

For each $a\in\mathcal{A}$, choose $x_a \in A^\times$ such that $[x_a] = a$. Likewise, let $(y_b)_{b\in\mathcal{B}}$ be a collection of representatives in $B^\times$ of the elements in $\mathcal{B}$. From the hypothesis that $\mathcal{A}$ is an armature, we know that the x_a's are F-linearly independent. Similarly, the y_b's are F-linearly independent; hence, the products $x_a\otimes y_b$ are linearly independent for $a\in\mathcal{A}$ and $b\in\mathcal{B}$. Therefore, $\mathcal{A}\otimes\mathcal{B}$ is an armature of $A\otimes_F B$. For a_1, $a_2\in\mathcal{A}$ and b_1, $b_2\in\mathcal{B}$ we have

$$\begin{aligned}\langle a_1\otimes b_1, a_2\otimes b_2\rangle &= x_{a_1}\otimes y_{b_1}\cdot x_{a_2}\otimes y_{b_2}\cdot(x_{a_1}\otimes y_{b_1})^{-1}\cdot(x_{a_2}\otimes y_{b_2})^{-1}\\ &= (x_{a_1}x_{a_2}x_{a_1}^{-1}x_{a_2}^{-1})\otimes(y_{b_1}y_{b_2}y_{b_1}^{-1}y_{b_2}^{-1})\\ &= \langle a_1, a_2\rangle\cdot\langle b_1, b_2\rangle.\end{aligned}$$ □

After taking into account the shift from the additive notation in $\mathbb{T}$ to the multiplicative notation in $\mu(F)$, we may rephrase the conclusion of Prop. 7.29 as the statement that, as abelian groups with alternating forms,

$$\mathcal{A}\otimes\mathcal{B} = \mathcal{A}\perp\mathcal{B}.$$

Combining Prop. 7.29 with Ex. 7.28(iii), we see that every tensor product of symbol F-algebras is spanned by an armature with nondegenerate commutator pairing:

Corollary 7.30. *Suppose A is a central simple F-algebra, which is isomorphic to a tensor product of symbol F-algebras of degrees $n_1, \ldots, n_r$:*

$$A \cong S_1 \otimes_F \ldots \otimes S_r.$$

Then $A = F[\mathcal{A}]$ for some armature $\mathcal{A}$ isomorphic to the orthogonal sum of standard symplectic modules of exponents $n_1, \ldots, n_r$:

$$\mathcal{A} \cong \mathbb{S}_{n_1} \perp \ldots \perp \mathbb{S}_{n_r}.$$

Proof. This follows readily from Prop. 7.29, since we saw in Ex. 7.28(iii) that each symbol algebra of degree n has an armature isomorphic to $\mathbb{S}_n$. □

The existence of an armature with nondegenerate commutator pairing actually characterizes tensor products of symbols, as we will see in Cor. 7.34.

Proposition 7.31. *Let $\mathcal{A}$ be an armature in an algebra A. For any subgroup $\mathcal{U} \subseteq \mathcal{A}$, the centralizer of $F[\mathcal{U}]$ in $F[\mathcal{A}]$ is $F[\mathcal{U}^\perp]$:*

$$C_{F[\mathcal{A}]}\big(F[\mathcal{U}]\big) = F[\mathcal{U}^\perp].$$

In particular, $Z\big(F[\mathcal{A}]\big) = F[\operatorname{rad}\mathcal{A}]$, so $F[\mathcal{A}]$ is commutative if and only if the commutator pairing on $\mathcal{A}$ is trivial, and $F[\mathcal{A}]$ is central if and only if the commutator pairing on $\mathcal{A}$ is nondegenerate.

Moreover, if $\mathcal{A} = \mathcal{B} \perp \mathcal{C}$ for some subgroups $\mathcal{B}$, $\mathcal{C}$, then

$$F[\mathcal{A}] = F[\mathcal{B}] \otimes_F F[\mathcal{C}].$$

Proof. For each $a \in \mathcal{A}$, choose $x_a \in A^\times$ such that $[x_a] = a$. Then $(x_a)_{a\in\mathcal{A}}$ is an F-base of $F[\mathcal{A}]$. An arbitrary element $c = \sum_{a\in\mathcal{A}} \lambda_a x_a \in F[\mathcal{A}]$ (with $\lambda_a \in F$ for $a \in \mathcal{A}$) lies in the centralizer of $F[\mathcal{U}]$ if and only if it commutes with each x_u, for $u \in \mathcal{U}$. By definition of the commutator pairing, we have

$$x_u c x_u^{-1} = \sum_{a\in\mathcal{A}} \lambda_a \langle u, a\rangle x_a.$$

Therefore, c centralizes $F[\mathcal{U}]$ if and only if $\langle u, a\rangle = 1$ for all $u \in \mathcal{U}$ and for all $a \in \mathcal{A}$ such that $\lambda_a \neq 0$. This condition means that c is a linear combination of $(x_a)_{a\in\mathcal{U}^\perp}$.

The second assertion follows from the observations that $Z\big(F[\mathcal{A}]\big) = C_{F[\mathcal{A}]}\big(F[\mathcal{A}]\big)$ and $\mathcal{A}^\perp = \operatorname{rad}\mathcal{A}$.

If $\mathcal{A} = \mathcal{B} \perp \mathcal{C}$, then it follows from the first part of the proof that $F[\mathcal{B}]$ and $F[\mathcal{C}]$ centralize each other in $F[\mathcal{A}]$. Therefore, we have a canonical map $F[\mathcal{B}] \otimes_F F[\mathcal{C}] \to F[\mathcal{A}]$. This map is surjective since every element in $\mathcal{A}$ is the product of an element in $\mathcal{B}$ with an element in $\mathcal{C}$; it is therefore injective by dimension count. □

Our goal for the rest of this subsection is a structure theorem for algebras spanned by an armature. (See Th. 7.36.)

Lemma 7.32. *Let $\mathcal{A}$ be an armature in an F-algebra A.*

(i) *Every nonzero two-sided ideal of $F[\mathcal{A}]$ contains a nonzero element in $F[\mathrm{rad}\mathcal{A}]$.*
(ii) *If char $F \nmid |\mathcal{A}|$ and the commutator pairing on $\mathcal{A}$ is trivial, then $F[\mathcal{A}]$ is an étale F-algebra.*

Proof. (i) Let $I \subseteq F[\mathcal{A}]$ be a nonzero two-sided ideal. For each $a \in \mathcal{A}$, choose $x_a \in A^\times$ with $[x_a] = a$; so, $(x_a)_{a\in\mathcal{A}}$ is an F-base of $F[\mathcal{A}]$. Let $y = \sum_{a\in\mathcal{A}} \lambda_a x_a$ (with $\lambda_a \in F$) be a nonzero element in I with a minimal number of nonzero terms, i.e., we assume the cardinality of the set $\{a \in \mathcal{A} \mid \lambda_a \neq 0\}$ to be minimal. By multiplying y by some $x_{a^{-1}}$, we may assume $\lambda_1 \neq 0$. If there is a $c \in \mathcal{A}$ such that $\lambda_c \neq 0$ and $c \notin rad\mathcal{A}$, pick $b \in \mathcal{A}$ such that $\langle b, c\rangle \neq 1$, and consider

$$y - x_b y x_b^{-1} = \sum_{a\in\mathcal{A}} \lambda_a(1 - \langle b, a\rangle)x_a.$$

We have $y - x_b y x_b^{-1} \in I$, and the number of nonzero terms in the expression on the right is strictly less than in y since $\lambda_1(1 - \langle b, 1\rangle) = 0$. On the other hand, we have $\lambda_c(1 - \langle b, c\rangle) \neq 0$, so $y - x_b y x_b^{-1} \neq 0$. This contradicts the conditions on y. Therefore, we have $\lambda_a = 0$ if $a \notin rad\mathcal{A}$, and it follows that y lies in $F[rad\mathcal{A}]$.

(ii) Consider a direct sum decomposition of $\mathcal{A}$ into cyclic groups:

$$\mathcal{A} = \mathcal{A}_1 \oplus \ldots \oplus \mathcal{A}_r \qquad \text{with each } \mathcal{A}_k \text{ cyclic.}$$

Since the commutator pairing is trivial on $\mathcal{A}$, this direct sum is an orthogonal sum, hence Prop. 7.31 yields

$$F[\mathcal{A}] = F[\mathcal{A}_1] \otimes_F \ldots \otimes_F F[\mathcal{A}_r].$$

To show that $F[\mathcal{A}]$ is an étale F-algebra, it now suffices to show that each $F[\mathcal{A}_k]$ is an étale F-algebra. Pick $x \in F[\mathcal{A}_k]$ so that $[x]$ generates $\mathcal{A}_k$, and let $n = |\mathcal{A}_k|$. Then $[x]^n = 1$, so $x^n \in F^\times$. Let $\alpha = x^n$. We have $\mathcal{A}_k = \{[x]^i \mid i = 0, \ldots, n-1\}$, hence

$$F[\mathcal{A}_k] = \bigoplus_{i=0}^{n-1} Fx^i \cong F[X]/(X^n - \alpha).$$

Since *char* F does not divide $|\mathcal{A}|$, it does not divide n; hence, the polynomial $X^n - \alpha$ is separable. Therefore, the F-algebra $F[\mathcal{A}_k]$ is étale. □

By contrast, if *char* F divides $|\mathcal{A}|$, then $F[\mathcal{A}]$ may contain nonzero nilpotent elements: see for example Ex. 7.28(i) with $n =$ *char* $F \neq 0$ and $\alpha = 1$.

We now describe the structure of algebras spanned by an armature, at least under some additional hypotheses.

Proposition 7.33. *Let $\mathcal{A}$ be an armature in an F-algebra A. Suppose $A = F[\mathcal{A}]$.*

(i) *If* char $F \nmid |\mathrm{rad}\mathcal{A}|$, *then* A *is a semisimple* F*-algebra and* $Z(A)$ *is an étale* F*-algebra.*

(ii) *Suppose* $\mathrm{rad}\mathcal{A} = \{0\}$. *Then* A *is a central simple* F*-algebra. If* $\mathcal{S}_1, \ldots, \mathcal{S}_r$ *are symplectic submodules of* $\mathcal{A}$ *such that*

$$\mathcal{A} = \mathcal{S}_1 \perp \ldots \perp \mathcal{S}_r \qquad \text{and} \qquad \mathcal{S}_i \cong \mathbb{S}_{n_i} \quad \text{for } i = 1, \ldots, r,$$

then

$$A = F[\mathcal{S}_1] \otimes_F \ldots \otimes_F F[\mathcal{S}_r],$$

where each $F[\mathcal{S}_i]$ *is a symbol* F*-algebra of degree* n_i.

Proof. (i) By Prop. 7.31, we have $Z(A) = F[\mathrm{rad}\mathcal{A}]$. Therefore, it follows readily from Lemma 7.32(ii) that $Z(A)$ is an étale F-algebra. Suppose $I \subseteq A$ is a nonzero nilpotent two-sided ideal. Lemma 7.32(i) shows that I contains a nonzero element in $Z(A)$. This is a contradiction since étale algebras have no nonzero nilpotent elements. Therefore, A is semisimple.

(ii) Since $rad\mathcal{A} = \{0\}$, Prop. 7.31 shows that $Z(A) = F$. By Lemma 7.32(i), every nonzero two-sided ideal $I \subseteq A$ contains a nonzero element in F. This element is invertible since F is a field, hence $I = A$. Thus, A is simple. By Prop. 7.31, the orthogonal sum decomposition of $\mathcal{A}$ yields a tensor product decomposition

$$A = F[\mathcal{S}_1] \otimes_F \ldots \otimes_F F[\mathcal{S}_r].$$

It remains to show that $F[\mathcal{S}_k]$ is a symbol F-algebra of degree n_k. Suppose $[x]$, $[y] \in \mathcal{S}_k$ are the generators corresponding to the standard base of $\mathbb{S}_{n_k}$. Then $[x]^{n_k} = [y]^{n_k} = 1$ and $\langle [x], [y] \rangle$ is a primitive n_k-th root of unity $\omega \in F$. It follows that $x^{n_k} \in F^\times$, $y^{n_k} \in F^\times$, and $xy = \omega yx$, hence the subalgebra $F[\mathcal{S}_k] \subseteq A$ generated by x and y is a symbol algebra of degree n_k over F. □

We thus obtain a converse to Cor. 7.30:

Corollary 7.34. *An* F*-algebra is a tensor product of* F*-symbol algebras if and only if it is spanned by an armature with nondegenerate commutator pairing.*

Proof. The "if" part follows from Prop. 7.33(ii), and the "only if" part from Cor. 7.30. □

For the rest of this subsection, we analyze the general case where the commutator pairing may be degenerate. We make a simplifying hypothesis however: we assume that the base field contains enough roots of unity; this allows us to define a Galois structure on the center of the spanned algebra.

Fix an armature $\mathcal{A}$ in an F-algebra A. Since our results concern only the subalgebra spanned by $\mathcal{A}$, we assume that $A = F[\mathcal{A}]$. It follows that $Z(A) = F[rad\mathcal{A}]$ by Prop. 7.31. To simplify the notation,

$$\text{let } \mathcal{Z} = rad\mathcal{A}, \text{ and let } s \text{ be the exponent of } \mathcal{Z} \text{ as an abelian group.} \tag{7.23}$$

Then $\mathcal{Z}$ is an armature of $Z(A)$. Assume that F contains a primitive s-th root of unity; so $char\, F \nmid s$ and, by Lemma 7.32, $Z(A)$ is an étale F-algebra. Let $\mu_s \subseteq F^\times$ be the group of s-th roots of unity. Since $\mu_s \cong \mathbb{Z}/s\mathbb{Z}$, the group homomorphisms $\mathcal{Z} \to \mu_s$ may be viewed as characters of $\mathcal{Z}$. Let G be the group of these characters:

$$G = Hom(\mathcal{Z}, \mu_s).$$

Lemma 7.35. *The F-algebra $Z(A)$ is G-Galois for the action of G determined by the following condition: for $g \in G$, $z \in \mathcal{Z}$ and $x \in Z(A)^\times$ such that $[x] = z$,*

$$g * x = g(z)\, x.$$

Proof. Since $Z(A)$ is spanned by $\mathcal{Z}$, the condition is sufficient to define the G-action on $Z(A)$: if $(x_z)_{z\in\mathcal{Z}}$ is a collection of representatives of the elements of $\mathcal{Z}$, hence a base of $Z(A)$, we must have

$$g * \big(\sum_{z\in\mathcal{Z}} \lambda_z x_z\big) = \sum_{z\in\mathcal{Z}} \lambda_z g(z) x_z \qquad \text{for all } g \in G \text{ and } \lambda_z \in F. \tag{7.24}$$

This formula gives a well-defined group action of G on $Z(A)$ by F-vector space automorphisms, which is independent of the choice of the representatives $(x_z)_{z\in\mathcal{Z}}$. It is actually an action by F-algebra automorphisms, since for w, $z \in \mathcal{Z}$ there exists a $\rho \in F^\times$ such that $x_w x_z = \rho x_{wz}$; hence, for $g \in G$,

$$\begin{aligned}(g * x_w)\cdot(g * x_z) &= g(w)g(z)x_w x_z \\ &= g(wz)\rho x_{wz} = \rho(g * x_{wz}) = g * (x_w x_z).\end{aligned} \tag{7.25}$$

Suppose $\sum_{z\in\mathcal{Z}} \lambda_z x_z \in Z(A)$ (with $\lambda_z \in F$ for $z \in \mathcal{Z}$) is fixed under the action of G. Then,

$$\lambda_z g(z) = \lambda_z \qquad \text{for all } g \in G \text{ and } z \in \mathcal{Z}.$$

This equation implies $\lambda_z = 0$ if $z \neq 1$, since for each $z \neq 1$ in $\mathcal{Z}$ there is a $g \in G$ such that $g(z) \neq 1$, as G is the group of characters of $\mathcal{Z}$. Therefore, the subalgebra of $Z(A)$ fixed under the G-action is F. Since in addition $[Z(A){:}F] = |\mathcal{Z}| = |G|$, the proposition follows (see Knus et al. [115, §18.B]). □

By the Galois theory of étale algebras (see, e.g., Knus et al. [115, §18.B]), it follows from Lemma 7.35 that $Z(A) \cong L \times \ldots \times L$ for some Galois field extension L of F. More specifically, let $e_1, \ldots, e_r$ be the primitive idempotents of $Z(A)$. Let $Z_i = e_i Z(A)$ and $A_i = e_i A$ for $i = 1, \ldots, r$. Thus, each Z_i is a field and $Z_i = Z(A_i)$, and we have

$$A = A_1 \times \ldots \times A_r \qquad \text{and} \qquad Z(A) = Z_1 \times \ldots \times Z_r.$$

By Prop. 7.33 the algebra A is semisimple, hence $A_1, \ldots, A_r$ are semisimple. They are the simple components of A since each has center a field.

Note that the G-action on $Z(A)$ permutes the primitive idempotents $e_1, \ldots, e_r$, and G acts transitively on the e_i. For if b is the sum of those e_i

in the G-orbit of e_1, then $b \in Z(A)^G = F$, so the nonzero idempotent b of the field F must equal $1 = e_1 + \ldots + e_r$. Hence, all the e_i lie in the G-orbit of e_1. Because the action of G on the e_i is transitive, the stabilizers of the e_i are conjugate. Since G is abelian, these stabilizers actually coincide. Let

$$H = \{g \in G \mid g * e_1 = e_1\} = \ldots = \{g \in G \mid g * e_r = e_r\}. \tag{7.26}$$

Then $Z_1, \ldots, Z_r$ are isomorphic Galois field extensions of F with Galois group H. Since $\exp H \mid s$ and F contains a primitive s-th root of unity, the extensions Z_i/F are Kummer. To determine them more precisely, consider the group

$$X = \{x \in Z(A)^\times \mid [x] \in \mathcal{Z}\}.$$

This group is abelian and contains $F^\times$. It maps onto $\mathcal{Z}$ by $x \mapsto [x]$. Since $s = \exp \mathcal{Z}$, we have $x^s \in F^\times$ for all $x \in X$. Therefore, there is a commutative diagram with exact rows, where the central vertical map is $x \mapsto x^s$:

$$\begin{array}{ccccccccc} 1 & \longrightarrow & F^\times & \longrightarrow & X & \longrightarrow & \mathcal{Z} & \longrightarrow & 1 \\ & & \big\downarrow{\scriptstyle s} & & \big\downarrow{\scriptstyle s} & & \big\downarrow{\scriptstyle \sigma} & & \\ 1 & \longrightarrow & F^{\times s} & \longrightarrow & F^\times & \longrightarrow & F^\times/F^{\times s} & \longrightarrow & 1 \end{array} \tag{7.27}$$

Let

$$\mathcal{K} = \ker \sigma \subseteq \mathcal{Z}.$$

Theorem 7.36. *For the algebra $A = F[\mathcal{A}]$ with notation as above starting from* (7.23)*, we have the following:*

(i) *The primitive idempotents $e_1, \ldots, e_r$ of $Z(A)$ span the F-algebra $F[\mathcal{K}]$, which is a split étale F-algebra of dimension r. In particular, $|\mathcal{K}| = r$. Moreover,*

$$\mathcal{K} = \{z \in \mathcal{Z} \mid h(z) = 1 \text{ for all } h \in H\}.$$

(ii) *$Z_1, \ldots, Z_r$ are each isomorphic to the Kummer extension $F\big(\sqrt[s]{\sigma(\mathcal{Z})}\big)$ (see Ex. 7.28*(ii) *for the notation).*

(iii) *For each $j \in \{1 \ldots, r\}$ the projection $A \to A_j$ induces a map $\mathcal{A} \to A_j^\times/Z_j^\times$, whose image is an armature $\mathcal{A}_j$ of A_j viewed as a Z_j-algebra. The armature $\mathcal{A}_j$ satisfies $A_j = Z_j[\mathcal{A}_j]$ and $\mathcal{A}_j \cong \mathcal{A}/\mathcal{Z}$ as symplectic modules.*

(iv) *Let $\mathcal{C}$ be any subgroup of $\mathcal{A}$. For any $j \in \{1, \ldots, r\}$ let $\mathcal{C}_j$ be the image of $\mathcal{C}$ in $A_j^\times/e_jF^\times$. Then, $\mathcal{C}_j$ is an armature for $e_jF[\mathcal{C}] = e_jF[\mathcal{C}_j]$ and $\mathcal{C}_j \cong \mathcal{C}\cdot\mathcal{K}/\mathcal{K}$ with isometric alternating pairings.*

(v) *Let $t = \exp \mathcal{A}$. If F contains a primitive t-th root of unity, then*

$$A_1 \cong A_2 \cong \ldots \cong A_r.$$

Proof. (i) and (ii) We first show that $F[\mathcal{K}]$ is a split étale F-algebra. For this, consider ${}_sX = \{x \in X \mid x^s = 1\}$. By applying the Snake Lemma to the diagram (7.27) we obtain the exact sequence

$$1 \longrightarrow \mu_s \longrightarrow {}_sX \longrightarrow \mathcal{K} \longrightarrow 1.$$

Since the abelian group ${}_sX$ has exponent s, this exact sequence is split. We may therefore find representatives $x_k \in Z(A)^\times$ for each $k \in \mathcal{K}$ satisfying

$$x_k x_\ell = x_{k\ell} \qquad \text{for } k,\, \ell \in \mathcal{K}.$$

With this choice of representatives, the algebra $F[\mathcal{K}]$ is isomorphic to the group algebra of $\mathcal{K}$. Since F contains a primitive s-th root of unity and $\mathcal{K}$ is an abelian group with $exp\, \mathcal{K} \mid s$, this group algebra is a split étale F-algebra. Thus, $F[\mathcal{K}]$ is spanned as an F-vector space by its primitive idempotents. It is not clear at this point that these idempotents are primitive in $Z(A)$, but they are disjoint sums of $e_1, \ldots, e_r$, so $|\mathcal{K}| \leq r$. We next show that this inequality is an equality, from which it follows that $e_1, \ldots, e_r$ are the primitive idempotents of $F[\mathcal{K}]$.

For any $j \in \{1, 2, \ldots, r\}$, we have

$$Z_j = e_j F[\mathcal{Z}] = e_j F[X] = (e_j F)[e_j X]$$

with $(e_j X)^s = e_j X^s \subseteq e_j F$. When we identify $e_j F$ with F, the group $(e_j X)^s$ has image $\sigma(\mathcal{Z})$ in $F^\times / F^{\times s}$. Hence, by Kummer theory,

$$Z_j \cong F\big(\sqrt[s]{\sigma(\mathcal{Z})}\big) \qquad \text{and} \qquad [Z_j{:}F] = |\sigma(\mathcal{Z})|.$$

Since $Z_1 \times \ldots \times Z_r = Z(A) = F[\mathcal{Z}]$, it follows that

$$r \cdot |\sigma(\mathcal{Z})| = [Z(A){:}F] = |\mathcal{Z}| = |\mathcal{K}| \cdot |\sigma(\mathcal{Z})|,$$

proving that $|\mathcal{K}| = r$, as claimed.

We now prove $\mathcal{K} = H^\perp$, where by definition

$$H^\perp = \{z \in \mathcal{Z} \mid h(z) = 1 \text{ for all } h \in H\}.$$

By the duality theory of finite abelian groups, we have $|H^\perp| = |G{:}H|$. Since $Z(A)$ is G-Galois over F and each Z_j is H-Galois over F, we have

$$|G| = [Z(A){:}F] = \sum_{j=1}^{r} [Z_j{:}F] = r|H|.$$

Hence, $|G{:}H| = r$, and therefore $|H^\perp| = |\mathcal{K}|$. Thus, to prove $\mathcal{K} = H^\perp$, it suffices to show $F[\mathcal{K}] \subseteq F[H^\perp]$. Since $F[\mathcal{K}]$ is spanned by $e_1, \ldots, e_r$, we need only prove that $e_j \in F[H^\perp]$ for $j = 1, \ldots, r$. Let $(y_z)_{z \in \mathcal{Z}}$ be a system of representatives in $Z(A)^\times$ of the elements of $\mathcal{Z}$, and let

$$e_j = \sum_{z \in \mathcal{Z}} \lambda_z y_z \qquad \text{with } \lambda_z \in F \text{ for } z \in \mathcal{Z}.$$

For $h \in H$ we have $h * e_j = e_j$, hence

$$\sum_{z \in \mathcal{Z}} \lambda_z h(z) y_z = \sum_{z \in \mathcal{Z}} \lambda_z y_z.$$

Therefore, $h(z) = 1$ if $\lambda_z \neq 0$; hence, e_j is an F-linear combination of the y_z for $z \in H^\perp$. This completes the proof of (i) and (ii).

For (iii) we now fix some index $j \in \{1, \dots, r\}$ and consider the projection $A \to A_j$ mapping $x \in A$ to $e_j x$, which is a ring homomorphism. It induces a map

$$\varepsilon_j \colon A^\times/F^\times \longrightarrow A_j^\times/Z_j^\times .$$

Let $\mathcal{A}_j = \varepsilon_j(\mathcal{A})$. The kernel of ε_j contains $Z(A)^\times$, hence ε_j induces a surjective map $\overline{\varepsilon}_j \colon \mathcal{A}/\mathcal{Z} \to \mathcal{A}_j$. For x, $y \in A^\times$ such that $[x]$, $[y] \in \mathcal{A}$, we have $xyx^{-1}y^{-1} \in F^\times$, hence under the identification $F = e_jF \subseteq Z_j$ we have

$$(e_jx)(e_jy)(e_jx^{-1})(e_jy^{-1}) \,=\, xyx^{-1}y^{-1}.$$

Therefore, ε_j preserves the commutator pairings:

$$\langle \varepsilon_j(a), \varepsilon_j(b)\rangle \,=\, \langle a, b\rangle \qquad \text{for } a,\, b \in \mathcal{A};$$

hence also,

$$\langle \overline{\varepsilon}_j(a\mathcal{Z}), \overline{\varepsilon}_j(b\mathcal{Z})\rangle \,=\, \langle a\mathcal{Z}, b\mathcal{Z}\rangle \qquad \text{for } a\mathcal{Z},\, b\mathcal{Z} \in \mathcal{A}/\mathcal{Z}.$$

Since the pairing on $\mathcal{A}/\mathcal{Z}$ is nondegenerate, it follows that $\overline{\varepsilon}_j$ is injective. It is therefore an isometry of symplectic modules $\mathcal{A}/\mathcal{Z} \xrightarrow{\sim} \mathcal{A}_j$. To see that $\mathcal{A}_j$ is an armature in the Z_j-algebra A_j, it only remains to prove that any system of representatives in $A_j^\times$ of the elements in $\mathcal{A}_j$ is linearly independent over Z_j. This follows from Prop. 7.26 since the commutator pairing on $\mathcal{A}_j$ is nondegenerate. That $\mathcal{A}_j$ spans A_j is clear since $\mathcal{A}$ spans A.

For (iv) let $\mathcal{C}$ be any subgroup of $\mathcal{A}$. For the moment, assume that $\mathcal{C} \supseteq \mathcal{K}$. Fix $j \in \{1, 2, \dots, r\}$. The projection $A \to A_j$ given by $a \mapsto e_ja$ induces a group homomorphism $\eta_j \colon A^\times/F^\times \to A_j^\times/e_jF^\times$. Note that $\mathcal{K} \subseteq \mathit{ker}\,\eta_j$ since $e_jF[\mathcal{K}] = e_jF$. By definition, $\mathcal{C}_j = \eta_j(\mathcal{C})$. So, η_j induces a surjection $\overline{\eta}_j \colon \mathcal{C} \to \mathcal{C}_j$ with $\mathcal{K} \subseteq \mathit{ker}\,\overline{\eta}_j$. Hence,

$$|\mathcal{C}_j| \,\le\, |\mathcal{C}|/|\mathcal{K}| \,=\, |\mathcal{C}|/r.$$

Take a set of representatives y_c in $A^\times$ for each $c \in \mathcal{C}$. Since

$$\mathcal{C}_j \,=\, \{(e_jy_c)e_jF^\times \mid c \in \mathcal{C}\},$$

we have

$$e_jF[\mathcal{C}] \,=\, \sum_{c\in\mathcal{C}} e_jFy_c \,=\, \sum_{c\in\mathcal{C}} e_jF(e_jy_c) \,=\, (e_jF)[\mathcal{C}_j].$$

To see that $\mathcal{C}_j$ is actually an armature of $e_jF[\mathcal{C}]$ we count dimensions. We have

$$[e_jF[\mathcal{C}]\!:\!e_jF] \,=\, [e_jF[\mathcal{C}_j]\!:\!e_jF] \,\le\, |\mathcal{C}_j| \,\le\, |\mathcal{C}|/r. \tag{7.28}$$

But as $F[\mathcal{C}] = e_1F[\mathcal{C}] \times \ldots \times e_rF[\mathcal{C}]$, we also have

$$|\mathcal{C}| \,=\, [F[\mathcal{C}]\!:\!F] \,=\, \sum_{i=1}^{r} [e_iF[\mathcal{C}]\!:\!e_iF] \le r|\mathcal{C}|/r \,=\, |\mathcal{C}|. \tag{7.29}$$

Therefore, equality must hold throughout (7.29) and (7.28). Thus, for each j, $[e_jF[\mathcal{C}]\!:\!e_jF] = |\mathcal{C}_j| = |\mathcal{C}|/r$. This shows that $\mathcal{C}_j$ is an armature of $e_jF[\mathcal{C}]$ and also that $\mathcal{K}$ must be the full kernel of $\overline{\eta}_j$. So $\mathcal{C}_j \cong \mathcal{C}/\mathcal{K} = \mathcal{C}\cdot\mathcal{K}/\mathcal{K}$, and this isomorphism is an isometry of the alternating pairings on these groups by an argument analogous to the one for (iii).

We have so far assumed that $\mathcal{K} \subseteq C$. But if $\mathcal{K} \not\subseteq \mathcal{C}$ observe that $\mathcal{C}_j = \eta_j(\mathcal{C} \cdot \mathcal{K})$ as $\mathcal{K} \subseteq \ker \eta_j$, and $e_j F[\mathcal{C}] = e_j F[\mathcal{C} \cdot \mathcal{K}]$. Hence, the preceding argument applied to $\mathcal{C} \cdot \mathcal{K}$ shows that $\mathcal{C}_j$ is an armature for $e_j F[\mathcal{C}]$ and $\mathcal{C}_j$ is isometric to $\mathcal{C} \cdot \mathcal{K} / \mathcal{K}$.

(v) Let μ_t denote the group of t-th roots of unity in F, where $t = \exp \mathcal{A}$. Suppose $|\mu_t| = t$. Let $\mathcal{G} = \mathit{Hom}(\mathcal{A}, \mu_t)$. Observe that there is a group action of $\mathcal{G}$ on A analogous to the action of G on $Z(A)$: Take any set of representatives $(x_a)_{a \in \mathcal{A}}$ of the elements of $\mathcal{A}$ and define:

$$g * \Big(\sum_{a \in \mathcal{A}} \lambda_a x_a \Big) = \sum_{a \in \mathcal{A}} \lambda_a g(a) x_a \qquad \text{for all } g \in \mathcal{G} \text{ and } \lambda_a \in F$$

(cf. (7.24)). A calculation like (7.25) shows that $\mathcal{G}$ acts on A by F-algebra automorphisms. The duality for finite abelian groups shows that the restriction map $\rho \colon \mathit{Hom}(\mathcal{A}, \mu_t) \to \mathit{Hom}(\mathcal{Z}, \mu_t)$ is surjective. Since $\exp \mathcal{Z} = s$, we have $s \mid t$ and $\mathit{Hom}(\mathcal{Z}, \mu_t) = \mathit{Hom}(\mathcal{Z}, \mu_s) = G$. Thus, ρ maps $\mathcal{G}$ onto G, and the definition of the group actions shows that any $g \in \mathcal{G}$ has the same action on $Z(A)$ as $\rho(g)$. We noted earlier (preceding (7.26)) that G acts transitively on $\{e_1, \ldots, e_r\}$; so $\mathcal{G}$ also acts transitively on this set. If $g * e_1 = e_j$ for $g \in \mathcal{G}$, then $g * A_1 = g * (e_1 A) = e_j A = A_j$, so $A_1 \cong A_j$. The transitivity of the action assures that the A_i are all isomorphic F-algebras. □

As a consequence of Th. 7.36 and Prop. 7.33, all the simple components $A_1, \ldots, A_r$ of A have tensor product decompositions into symbol algebras mirroring the decomposition of $\mathcal{A}/\mathcal{Z}$ into an orthogonal sum of standard symplectic modules. In particular, they have the same degree:

$$[A_i{:}Z_i] = \tfrac{1}{r} |\mathcal{A}{:}\mathcal{Z}| = [A{:}F] / (r[Z(A){:}F]) \qquad \text{for } i = 1, \ldots, r.$$

However, the algebras need not be isomorphic when F does not contain a primitive t-th root of unity, as the following example illustrates.

Example 7.37. Take any field F with $\mathit{char}\, F \neq 2$, let $a, b \in F^\times$, and consider the quaternion algebras

$$A_1 = \big(a, b/F\big)_2, \qquad A_2 = \big(a, -b/F\big)_2$$

with respective quaternion bases $1, i_1, j_1, k_1$, and $1, i_2, j_2, k_2$. In the F-algebra $A = A_1 \times A_2$, the following is an armature that spans A:

$$\mathcal{A} = \big\{(1,1)F^\times, (1,-1)F^\times, (i_1, i_2)F^\times, (i_1, -i_2)F^\times, \\ (j_1, j_2)F^\times, (j_1, -j_2)F^\times, (k_1, k_2)F^\times, (k_1, -k_2)F^\times\big\}.$$

In the notation of Th. 7.36, we have

$$\mathcal{Z} = \mathcal{K} = \big\{(1,1)F^\times, (1,-1)F^\times\big\}.$$

The theorem applies since $s = \exp \mathcal{Z} = 2$ and F contains the primitive square root of unity -1. Note that $A_1 \not\cong A_2$ if and only if the quaternion algebra

$(a, -1/F)_2$ is not split. This occurs whenever a is not a sum of two squares of elements of F. But observe that $t = exp\,\mathcal{A} = 4$. If F contains a primitive fourth root of unity, then $A_1 \cong A_2$.

7.2.2 Armatures in graded algebras

We use for graded structures the same notation as in the previous chapters: Γ is a torsion-free divisible abelian group, and all the graded algebras we consider are graded by subsets of Γ. Throughout this subsection, we fix a graded field F.

Definition 7.38. Let A be a graded algebra over a graded field F. A *homogeneous armature* in A is a finite abelian subgroup $\mathcal{A} \subseteq \mathsf{A}^\times/\mathsf{F}^\times$ whose elements can be represented by homogeneous elements in $\mathsf{A}^\times$ that are linearly independent over F. Clearly, the commutator pairing can be defined for homogeneous armatures just as for the armatures in Def. 7.27. Note that the values of this pairing all have degree 0 in F. If $\mathcal{A}$ is a homogeneous armature in a graded F-algebra A, we let $\mathsf{F}[\mathcal{A}]$ denote the F-span of homogeneous elements in $\mathsf{A}^\times$ representing the elements of $\mathcal{A}$. It is a graded F-subalgebra of A with $dim_{\mathsf{F}}\, \mathsf{F}[\mathcal{A}] = |\mathcal{A}|$.

The proofs of Prop. 7.29, Prop. 7.31, Lemma 7.32, and Prop. 7.33 can easily be adapted to yield corresponding results for homogeneous armatures in graded algebras. We state only the analogue of Prop. 7.33:

Proposition 7.39. *Let $\mathcal{A}$ be a homogeneous armature in a graded F-algebra A. Suppose $\mathsf{A} = \mathsf{F}[\mathcal{A}]$.*

(i) *If char $\mathsf{F}_0 \nmid |\mathrm{rad}\mathcal{A}|$, then A is a semisimple graded F-algebra and $Z(\mathsf{A}) = \mathsf{F}[\mathrm{rad}\mathcal{A}]$ is an étale graded F-algebra, i.e., a direct product of tame graded field extensions of F.*
(ii) *If $\mathrm{rad}\mathcal{A} = \{0\}$, then A is a central simple graded F-algebra, which has a tensor product decomposition into graded symbol F-algebras reflecting the decomposition of $\mathcal{A}$ into an orthogonal sum of standard symplectic modules.*

Remark 7.40. Let $\mathsf{A}^\times_{\mathrm{hom}}$ denote the multiplicative group of homogeneous invertible elements in a graded F-algebra A. Let $\mathcal{A}$ be a homogeneous armature in A, and let $q(\mathcal{A})$ denote the image of $\mathcal{A}$ under the natural monomorphism $\mathsf{A}^\times_{\mathrm{hom}}/\mathsf{F}^\times \to q(\mathsf{A})^\times/q(\mathsf{F})^\times$. Then $q(\mathcal{A})$ is clearly an armature of the $q(\mathsf{F})$-algebra $q(\mathsf{A})$. Since a set of representatives in $\mathsf{A}^\times_{\mathrm{hom}}$ for the elements of $\mathcal{A}$ is also a set of representatives for the elements of $q(\mathcal{A})$, the symplectic modules $\mathcal{A}$ and $q(\mathcal{A})$ are isomorphic. If $\mathsf{A} = \mathsf{F}[\mathcal{A}]$, then $q(\mathsf{A}) = q(\mathsf{F})[q(\mathcal{A})]$.

Remark 7.41. Let A be an algebra over a graded field F with a homogeneous armature $\mathcal{A}$ such that $\mathsf{A} = \mathsf{F}[\mathcal{A}]$. If F contains a primitive s-th root of unity where $s = exp\,\mathcal{A}$, then there is a graded analogue to each part of Th. 7.36. In this context, $\mathcal{Z} = rad\mathcal{A}$ is a homogeneous armature of $Z(\mathsf{F}) = \mathsf{F}[\mathcal{Z}]$; the primitive idempotents $e_1, \ldots, e_r$ of $Z(\mathsf{A})$ are homogeneous (necessarily of degree 0); each $\mathsf{Z}_j = e_j Z(\mathsf{A})$ is a graded field and a graded Kummer extension of $e_j\mathsf{F}$ (cf. Remark 5.36); each $\mathsf{A}_j = e_j\mathsf{A}$ is a graded algebra over its center Z_j with homogeneous armature $\mathcal{A}_j \cong \mathcal{A}/\mathcal{Z}$ such that $\mathsf{A}_j = \mathsf{Z}_j[\mathcal{A}_j]$; and each $\mathcal{C}_j$ is a homogeneous armature of the graded algebra $e_j\mathsf{F}[\mathcal{C}] = e_j\mathsf{F}[\mathcal{C}_j]$. The proofs are analogous to those for Th. 7.36. We clarify one part of the argument: To obtain the idempotents $e_1, \ldots, e_r$, note that the group action of $G = Hom(\mathcal{Z}, \mu_s)$ on $Z(\mathsf{A})$ described in Lemma 7.35 is by graded automorphisms and extends to the group action of G on $q(Z(\mathsf{A})) = Z(q(\mathsf{A}))$ associated to the armature $q(\mathcal{Z})$ of the $q(\mathsf{F})$-algebra $Z(q(\mathsf{A}))$ (cf. the preceding remark). Since $q(Z(\mathsf{A}))$ is G-Galois, hence étale, over $q(\mathsf{F})$ by Lemma 7.35, $Z(\mathsf{A})$ is a semisimple commutative graded F-algebra by Prop. 2.28(iii). Hence $Z(\mathsf{A}) = \mathsf{L}_1 \times \ldots \times \mathsf{L}_r$, a direct product of graded fields, and $q(Z(\mathsf{A})) = q(\mathsf{L}_1) \times \ldots \times q(\mathsf{L}_r)$ with each $q(\mathsf{L}_i)$ a field. The homogeneous primitive idempotents $e_1, \ldots, e_r$ of $Z(\mathsf{A})$ thus coincide with the primitive idempotents of $Z(q(\mathsf{A}))$; so G acts transitively on $e_1, \ldots, e_r$. The map $\mathsf{F}^\times/\mathsf{F}^{\times s} \to q(\mathsf{F})^\times/q(\mathsf{F})^{\times s}$ is injective as $\mathsf{F}^\natural$ is integrally closed and divisors of homogeneous elements are homogeneous (cf. Prop. 5.2 and Prop. 2.3(ii)). Hence, $q(\mathcal{K})$ is the kernel of the map $q(\mathcal{Z}) \to q(\mathsf{F})^\times/q(\mathsf{F})^{\times s}$; so $|\mathcal{K}| = |q(\mathcal{K})| = r$. The $e_1, \ldots, e_r$ are the primitive idempotents of the étale $q(\mathsf{F})$-algebra $q(\mathsf{F})[q(\mathcal{K})] = q(\mathsf{F}[\mathcal{K}])$ by Th. 7.36 for $q(\mathsf{A})$, hence they are also the primitive homogeneous idempotents of $\mathsf{F}[\mathcal{K}]$.

The grading on a homogeneous armature yields some additional structure, which we discuss next.

Let $\mathcal{A}$ be a homogeneous armature in a graded F-algebra A. Since the armature yields information only on the subalgebra that it spans, we focus on the algebra $\mathsf{F}[\mathcal{A}]$ and assume that $\mathsf{A} = \mathsf{F}[\mathcal{A}]$. Note that there is a natural group action of $\mathcal{A}$ on A by conjugation. Specifically, for $a \in \mathcal{A}$ and $y \in \mathsf{A}$, choose any homogeneous $x_a \in \mathsf{A}^\times$ with $[x_a] = a$, and define

$$a \cdot y = x_a y x_a^{-1}. \tag{7.30}$$

If $y = x_b$ for some $b \in \mathcal{A}$, then $a \cdot y = \langle a, b\rangle y \in \mathsf{F}y$. Hence, for any subgroup $\mathcal{B}$ of $\mathcal{A}$, the group $\mathcal{A}$ acts by graded F-algebra automorphisms on $\mathsf{F}[\mathcal{B}]$, so it acts by F_0-automorphisms on $\mathsf{F}[\mathcal{B}]_0$. We will see below that this reduces to some familiar actions when $\mathsf{F}[\mathcal{B}]_0 = Z(\mathsf{A}_0)$.

By putting together the exact sequence

$$1 \longrightarrow \mathsf{A}_0^\times \longrightarrow \mathsf{A}_{\mathrm{hom}}^\times \xrightarrow{deg} \Gamma_{\mathsf{A}}^\times \longrightarrow 0$$

with the corresponding exact sequence for F, we obtain an exact sequence

$$1 \longrightarrow \mathsf{A}_0^\times/\mathsf{F}_0^\times \longrightarrow \mathsf{A}_{\mathrm{hom}}^\times/\mathsf{F}^\times \xrightarrow{\overline{deg}} \Gamma_{\mathsf{A}}^\times/\Gamma_{\mathsf{F}} \longrightarrow 0. \tag{7.31}$$

Since $\mathcal{A} \subseteq \mathsf{A}_{\text{hom}}^\times/\mathsf{F}^\times$, we may consider

$$\mathcal{A}_0 = \mathcal{A} \cap (\mathit{ker}\,\overline{\mathit{deg}}) \subseteq \mathsf{A}_{\text{hom}}^\times/\mathsf{F}^\times \quad \text{and its inverse image} \quad \mathcal{A}_0' \subseteq \mathsf{A}_0^\times/\mathsf{F}_0^\times.$$

Since elements of $\mathcal{A}_0$ have representatives in $\mathsf{A}_0^\times$ which also represent the corresponding elements of $\mathcal{A}_0'$, the obvious group isomorphism $\mathcal{A}_0' \cong \mathcal{A}_0$ is also an isomorphism of symplectic modules.

Proposition 7.42. *The degree map* $\overline{\mathit{deg}}\colon \mathcal{A} \to \Gamma_{\mathsf{A}}^\times/\Gamma_{\mathsf{F}}$ *is surjective and* $\Gamma_{\mathsf{A}}^\times = \Gamma_{\mathsf{A}}$*; hence, the following sequence is exact:*

$$1 \longrightarrow \mathcal{A}_0' \longrightarrow \mathcal{A} \xrightarrow{\overline{\mathit{deg}}} \Gamma_{\mathsf{A}}/\Gamma_{\mathsf{F}} \longrightarrow 0.$$

Moreover, $\mathcal{A}_0'$ *is an armature of the* F_0*-algebra* A_0 *with* $\mathsf{F}_0[\mathcal{A}_0'] = \mathsf{A}_0$*, and* $\mathsf{F}[\mathcal{A}_0] = \mathsf{F}_0[\mathcal{A}_0'] \otimes_{\mathsf{F}_0} \mathsf{F}$.

Proof. Since A is spanned by $\mathcal{A}$, every homogeneous component of A contains a representative of some element of $\mathcal{A}$. It follows that $\Gamma_{\mathsf{A}}^\times = \Gamma_{\mathsf{A}}$ and $\overline{\mathit{deg}}$ is surjective. For each $a \in \mathcal{A}_0$ choose a representative x_a lying in $\mathsf{A}_0^\times$. Each x_a is then also a representative of the corresponding element of $\mathcal{A}_0'$. The x_a are F_0-linearly independent, since they are F-linearly independent. Hence, $\mathcal{A}_0'$ is an armature for the F_0-algebra $\mathsf{F}_0[\mathcal{A}_0']$. Moreover,

$$\mathsf{F}_0[\mathcal{A}_0'] \otimes_{\mathsf{F}_0} \mathsf{F} = \Big(\bigoplus_{a\in\mathcal{A}_0} \mathsf{F}_0 x_a \Big) \otimes_{\mathsf{F}_0} \mathsf{F} = \bigoplus_{a\in\mathcal{A}_0} \mathsf{F} x_a = \mathsf{F}[\mathcal{A}_0].$$

Since A is spanned by $\mathcal{A}$, the homogeneous elements of degree 0 in A lie in the F_0-span of $\mathcal{A}_0'$; hence, $\mathsf{A}_0 = \mathsf{F}_0[\mathcal{A}_0']$. □

For the rest of this subsection, we also assume[2] that $\mathit{rad}\mathcal{A} = \{0\}$; hence, A is a central simple graded F-algebra and F_0 contains a primitive root of unity of order $\mathit{exp}\,\mathcal{A}$. Clearly, the commutator pairing on $\mathcal{A}_0$ is the restriction of the commutator pairing on $\mathcal{A}$. In particular, $\mathit{rad}\mathcal{A}_0 = \mathcal{A}_0 \cap \mathcal{A}_0^\perp$. We write $\mathcal{Z}_0$ for this group and $\mathcal{Z}_0'$ for its isomorphic (and isometric) copy in $\mathcal{A}_0'$; so,

$$\mathcal{Z}_0 = \mathit{rad}\mathcal{A}_0 = \mathcal{A}_0 \cap \mathcal{A}_0^\perp, \quad \mathcal{Z}_0' = \mathit{rad}\mathcal{A}_0', \quad \text{and} \quad \mathcal{Z}_0^\perp = \mathcal{A}_0 \cdot \mathcal{A}_0^\perp.$$

Since A_0 is spanned by $\mathcal{A}_0'$, we have $Z(\mathsf{A}_0) = \mathsf{F}_0[\mathcal{Z}_0']$ by Prop. 7.31.

Let s be the exponent of $\mathcal{Z}_0$. Since F_0 contains a primitive root of unity of order $\mathit{exp}\,\mathcal{A}$, it also contains a primitive s-th root of unity; hence, we may apply Th. 7.36 to the armature $\mathcal{A}_0$ to obtain information on A_0. In contrast to the case discussed in Th. 7.36, the group of characters on $\mathcal{Z}_0$ can here be described in terms of the armature $\mathcal{A}$: the commutator pairing on $\mathcal{A}$ defines a homomorphism

$$\mathcal{A} \longrightarrow \mathit{Hom}(\mathcal{Z}_0, \mu_s), \qquad a \mapsto \langle a, \cdot \rangle|_{\mathcal{Z}_0}.$$

[2] Note that when $\mathsf{F}[\mathit{rad}\mathcal{A}]$ is a graded field, we may always reduce to this case by substituting $\mathcal{A}/\mathit{rad}\mathcal{A}$ for $\mathcal{A}$ and considering A as an algebra over $\mathsf{F}[\mathit{rad}\mathcal{A}]$, see Exercise 7.6.

This map is surjective because the pairing on $\mathcal{A}$ is nondegenerate. Its kernel is $\mathcal{Z}_0^\perp$ (by definition). Since $Hom(\mathcal{Z}_0, \mu_s) \cong Hom(\mathcal{Z}_0', \mu_s)$, we obtain a canonical isomorphism

$$\mathcal{A}/\mathcal{Z}_0^\perp \xrightarrow{\sim} Hom(\mathcal{Z}_0', \mu_s).$$

The Galois action of $Hom(\mathcal{Z}_0', \mu_s)$ on $Z(\mathsf{A}_0)$ defined in Lemma 7.35 then yields a map

$$\psi\colon \mathcal{A}/\mathcal{Z}_0^\perp \longrightarrow Aut_{\mathsf{F}_0}\big(Z(\mathsf{A}_0)\big).$$

On the other hand, since $\Gamma_{\mathsf{A}}^\times = \Gamma_{\mathsf{A}}$ we know that Γ_{A} also acts on $Z(\mathsf{A}_0)$ through the canonical map θ_{A} of (2.29). We have $\Gamma_{\mathsf{F}} \subseteq ker\,\theta_{\mathsf{A}}$, hence θ_{A} induces a map $\overline{\theta}_{\mathsf{A}}\colon \Gamma_{\mathsf{A}}/\Gamma_{\mathsf{F}} \to Aut_{\mathsf{F}_0}\big(Z(\mathsf{A}_0)\big)$. The maps ψ and $\overline{\theta}_{\mathsf{A}}$ are related as follows:

Proposition 7.43. *The following diagram is commutative, where the left vertical map is the canonical projection:*

$$\begin{array}{ccc}
\mathcal{A} & \xrightarrow{\;\overline{deg}\;} & \Gamma_{\mathsf{A}}/\Gamma_{\mathsf{F}} \\
\big\downarrow & & \big\downarrow{\scriptstyle \overline{\theta}_{\mathsf{A}}} \\
\mathcal{A}/\mathcal{Z}_0^\perp & \xrightarrow{\;\psi\;} & Aut_{\mathsf{F}_0}\big(Z(\mathsf{A}_0)\big)
\end{array}$$

Proof. Let $a \in \mathcal{A}$ and $x_a \in \mathsf{A}_{\mathrm{hom}}^\times$ be such that $[x_a] = a$. For $z \in Z(\mathsf{A}_0)$ such that $[z] \in \mathcal{Z}_0$ we have

$$\psi(a\mathcal{Z}_0^\perp)\colon z \mapsto \langle a, [z]\rangle z.$$

On the other hand, we have

$$\overline{\theta}_{\mathsf{A}}(a\mathcal{Z}_0^\perp) = \theta_{\mathsf{A}}(deg\, x_a)\colon z \mapsto x_a z x_a^{-1}.$$

Since $\langle a, [z]\rangle = x_a z x_a^{-1}$, and since $Z(\mathsf{A}_0)$ is spanned by the elements z such that $[z] \in \mathcal{Z}_0$, the diagram commutes. □

Recall that $s = exp(\mathcal{Z}_0)$. Let

$$\sigma\colon \mathcal{Z}_0' \longrightarrow \mathsf{F}_0^\times/\mathsf{F}_0^{\times s}$$

be the map induced by $x \mapsto x^s$ on $\{x \in Z(\mathsf{A}_0)^\times \mid [x] \in \mathcal{Z}_0'\}$. Let $\mathcal{K}_0' = ker\,\sigma \subseteq \mathcal{Z}_0'$; let $\mathcal{K}_0$ be the isomorphic image of $\mathcal{K}_0'$ in $\mathcal{A}$ and let $\mathcal{K}_0^\perp$ be the orthogonal of $\mathcal{K}_0$ in $\mathcal{A}$. Let $r = |\mathcal{K}_0'| = |\mathcal{K}_0|$. By Th. 7.36 applied to $\mathcal{A}_0'$, the commutative F_0-algebra $Z(\mathsf{A}_0) = \mathsf{F}_0[\mathcal{Z}_0']$ has the same r primitive idempotents $e_1, \ldots, e_r$ as the split étale F_0-algebra $\mathsf{F}_0[\mathcal{K}_0']$, which are also the primitive homogeneous idempotents of $\mathsf{F}[\mathcal{K}_0] = \mathsf{F}_0[\mathcal{K}_0'] \otimes_{\mathsf{F}_0} \mathsf{F}$.

Theorem 7.44. *With the hypotheses and notation as above, beginning after Prop. 7.42, we have:*

(i) $\mathsf{F}[\mathcal{K}_0^\perp] = e_1\mathsf{A}e_1 \times \ldots \times e_r\mathsf{A}e_r$.

(ii) *Let* $\mathsf{B}_j = e_j\mathsf{A}e_j$ *for* $j = 1, \ldots, r$. *Then, each* B_j *is a central simple graded* $e_j\mathsf{F}$*-algebra, and the* B_j *are all graded isomorphic. Moreover, for each* j, $\Gamma_{\mathsf{B}_j}/\Gamma_{e_j\mathsf{F}} = \overline{deg}(\mathcal{K}_0^{\perp})$.

(iii) *Let* $\mathcal{B}_j$ *be the image of* $\mathcal{K}_0^{\perp}$ *under the projection*

$$\mathsf{F}[\mathcal{K}_0^{\perp}]_{\text{hom}}^{\times}/\mathsf{F} \longrightarrow \mathsf{B}_{j,\text{hom}}^{\times}/e_j\mathsf{F}^{\times}.$$

Then, $\mathcal{B}_j$ *is a homogeneous armature for* B_j *as a graded* $e_j\mathsf{F}$*-algebra, it spans* B_j*, and* $\mathcal{B}_j \cong \mathcal{K}_0^{\perp}/\mathcal{K}_0$ *as symplectic modules.*

(iv) *Let* $\mathsf{B}_{j,0}$ *be the degree* 0 *component of* B_j. *Then,* $\mathsf{B}_{1,0}, \ldots, \mathsf{B}_{r,0}$ *are isomorphic simple algebras each with center isomorphic to* $\mathsf{F}_0\big(\sqrt[s]{\sigma(\mathcal{Z}_0')}\big)$, *and*

$$\mathsf{A}_0 = \mathsf{B}_{1,0} \times \ldots \times \mathsf{B}_{r,0}.$$

Proof. (i) Since $\mathcal{K}_0 \subseteq \mathcal{Z}_0 = rad\mathcal{A}_0$, the commutator pairing is trivial on $\mathcal{K}_0$; hence, $\mathcal{K}_0 \subseteq \mathcal{K}_0^{\perp}$. The group action of $\mathcal{A}$ on A given in (7.30) restricts to an action by F_0-automorphisms on the split étale algebra $\mathsf{F}_0[\mathcal{K}_0'] = \mathsf{F}[\mathcal{K}_0]_0$. Hence, this action restricts to an action on the set $\{e_1, \ldots, e_r\}$ of primitive idempotents of $\mathsf{F}_0[\mathcal{K}_0']$. Indeed, $\mathcal{A}$ acts transitively on these idempotents, since $Hom(\mathcal{Z}_0', \mu_s)$ acts transitively as we saw preceding (7.26), and the surjection $\mathcal{A} \to Hom(\mathcal{Z}_0', \mu_s)$ is compatible with the actions of these groups on $Z(\mathsf{A}_0)$ by Prop. 7.43. Since $\mathsf{F}[\mathcal{K}_0^{\perp}]$ centralizes $\mathsf{F}[\mathcal{K}_0]$, the group $\mathcal{K}_0^{\perp}$ acts trivially on the e_i. But $|\mathcal{A}{:}\mathcal{K}_0^{\perp}| = |\mathcal{K}_0| = r$ as $rad\mathcal{A} = \{0\}$. Since each e_i has orbit size r, the full stabilizer of e_i under the $\mathcal{A}$ action is $\mathcal{K}_0^{\perp}$, for each i. Take a set of representatives x_a in $\mathsf{A}_{\text{hom}}^{\times}$ for the a in $\mathcal{A}$. Then, for any $i \in \{1, 2, \ldots, r\}$ and $a \in \mathcal{A} \setminus \mathcal{K}_0^{\perp}$, we have $a \cdot e_i = e_j$ for some $j \neq i$, so,

$$e_i\mathsf{F}x_a e_i = \mathsf{F}e_i(a \cdot e_i)x_a = \mathsf{F}e_i e_j x_a = \{0\}.$$

Hence,

$$e_i\mathsf{A}e_i = \sum_{a\in\mathcal{A}} e_i\mathsf{F}x_a e_i = \sum_{a\in\mathcal{K}_0^{\perp}} e_i\mathsf{F}x_a e_i + \{0\} \subseteq \mathsf{F}[\mathcal{K}_0^{\perp}].$$

On the other hand, since the orthogonal idempotents $e_1, \ldots, e_r$ are central in $\mathsf{F}[\mathcal{K}_0^{\perp}]$ and $e_1 + \ldots + e_r = 1$, we have $e_i\mathsf{F}[\mathcal{K}_0^{\perp}]e_j = \{0\}$ whenever $j \neq i$, and

$$\mathsf{F}[\mathcal{K}_0^{\perp}] = \sum_{i=1}^{r}\sum_{j=1}^{r} e_i\mathsf{F}[\mathcal{K}_0^{\perp}]e_j = \sum_{i=1}^{r} e_i\mathsf{F}[\mathcal{K}_0^{\perp}]e_i \subseteq \sum_{i=1}^{r} e_i\mathsf{A}e_i = e_1\mathsf{A}e_1 \times \ldots \times e_r\mathsf{A}e_r.$$

Since each $e_i\mathsf{A}e_i$ lies in $\mathsf{F}[\mathcal{K}_0^{\perp}]$, the displayed inclusion is an equality.

(ii) Let $\mathsf{B}_j = e_j\mathsf{A}e_j = e_j\mathsf{F}[\mathcal{K}_0^{\perp}]$, which is a graded algebra since $e_j \in \mathsf{A}_0$. By the graded Wedderburn Theorem 2.26 the central simple graded F-algebra A is isomorphic to $\mathsf{End}_{\mathsf{D}}(\mathsf{J})$ for some graded division algebra D with center F and finite-dimensional graded right D-vector space J. Since e_j is homogeneous, $e_j\mathsf{J}$ is a graded D-subspace of J; so, as in the ungraded case, $\mathsf{B}_j \cong_g \mathsf{End}_{\mathsf{D}}(e_j\mathsf{J})$, which is a central simple graded F-algebra. The isomorphism shows that $Z(\mathsf{B}_j) = e_j\mathsf{F}$. Since the action of $\mathcal{A}$ on A is by graded algebra automorphisms and $\mathcal{A}$ acts transitively on the e_j, it also acts transitively on the B_j.

So, the B_j are all isomorphic graded algebras. Hence, $\Gamma_{\mathsf{B}_1} = \ldots = \Gamma_{\mathsf{B}_r}$. Since $\Gamma_{\mathsf{F}[\mathcal{K}_0^\perp]} = \sum_{j=1}^{r} \Gamma_{\mathsf{B}_j}$, it follows that $\Gamma_{\mathsf{B}_j}/\Gamma_{e_j\mathsf{F}} = \Gamma_{\mathsf{F}[\mathcal{K}_0^\perp]}/\Gamma_{\mathsf{F}} = \overline{deg}(\mathcal{K}_0^\perp)$ for all j.

(iii) Consider $\mathsf{F}[\mathcal{K}_0^\perp]$ with its homogeneous armature $\mathcal{K}_0^\perp$. We have

$$rad(\mathcal{K}_0^\perp) = \mathcal{K}_0^\perp \cap \mathcal{K}_0 = \mathcal{K}_0.$$

Let $s' = exp(\mathcal{K}_0)$, which divides $exp\,\mathcal{A}$; so, F_0 contains a primitive s'-th root of unity. Therefore, the graded version of Th. 7.36 (see Remark 7.41) applies to $\mathcal{K}_0^\perp$, and the $\mathcal{Z}$ for that theorem is $rad(\mathcal{K}_0^\perp) = \mathcal{K}_0$. Also, the e_j of that theorem are the primitive homogeneous idempotents of $\mathsf{F}[\mathcal{K}_0]$, which are the same as our $e_1, \ldots, e_r$ here. Thus,

$$Z(\mathsf{F}[\mathcal{K}_0^\perp]) = \mathsf{F}[\mathcal{K}_0] = e_1\mathsf{F}[\mathcal{K}_0] \times \ldots \times e_r\mathsf{F}[\mathcal{K}_0] = e_1\mathsf{F} \times \ldots \times e_r\mathsf{F},$$

where the last equality holds because $\mathsf{F}[\mathcal{K}_0] = \mathsf{F}_0[\mathcal{K}_0'] \otimes_{\mathsf{F}_0} \mathsf{F}$ and each $e_j\mathsf{F}[\mathcal{K}_0'] = \mathsf{F}$. Thus, the Z_j in the graded Th. 7.36 is $e_j\mathsf{F}$ and the A_j is $e_j\mathsf{F}[\mathcal{K}_0^\perp] = \mathsf{B}_j$, and the $\mathcal{A}_j$ is $\mathcal{B}_j$. Hence, part (iii) here is given by part (iii) of the graded Th. 7.36.

(iv) By Prop. 7.42, $\mathcal{A}_0'$ is an armature for the F_0-algebra $\mathsf{A}_0 = \mathsf{F}_0[\mathcal{A}_0']$. Theorem 7.36 applied to $\mathcal{A}_0'$ shows that $\mathsf{A}_0 = e_1\mathsf{A}_0 \times \ldots \times e_r\mathsf{A}_0$ and each $e_j\mathsf{A}_0$ is a simple algebra with center isomorphic to $\mathsf{F}_0\big(\sqrt[s]{\sigma(\mathcal{Z}_0')}\big)$. Since $\mathcal{K}_0 \subseteq rad\mathcal{A}_0 \subseteq \mathcal{A}_0^\perp$, we have $\mathcal{A}_0 \subseteq \mathcal{K}_0^\perp$. Hence,

$$\mathsf{A}_0 = \mathsf{F}_0[\mathcal{A}_0'] \subseteq \mathsf{F}[\mathcal{A}_0]_0 \subseteq \mathsf{F}[\mathcal{K}_0^\perp]_0 \subseteq \mathsf{A}_0;$$

so,

$$\mathsf{A}_0 = \mathsf{F}[\mathcal{K}_0^\perp]_0 = \mathsf{B}_{1,0} \times \ldots \times \mathsf{B}_{r,0},$$

with $\mathsf{B}_{j,0} = e_j\mathsf{A}_0$ for all j. □

When $r = 1$ in the preceding theorem, there are canonical subalgebras of A which have natural descriptions in terms of $\mathcal{A}$:

Proposition 7.45. *Suppose $r = 1$ in Th. 7.44. This occurs if and only if A_0 is simple, if and only if $\mathcal{K}_0 = \{0\}$.*

(i) *Let $\mathsf{U} = \mathsf{A}_0 \otimes_{\mathsf{F}_0} \mathsf{F}$. Then, $\mathsf{U} = \mathsf{F}[\mathcal{A}_0]$ and U is a simple graded algebra inertial over F.*

(ii) *Let $\mathsf{Z} = Z(\mathsf{U})$. Then, Z is a graded field with $\mathsf{Z} = \mathsf{F}[\mathcal{Z}_0] = \mathsf{F}\big(\sqrt[s]{\sigma(\mathcal{Z}_0')}\big)$, and Z is inertial and abelian Galois over F.*

(iii) *Let $\mathsf{C} = C_{\mathsf{A}}(\mathsf{U})$. Then, C is a graded division algebra totally ramified over its center Z with $\Gamma_{\mathsf{C}}/\Gamma_{\mathsf{Z}} = \overline{deg}(\mathcal{A}_0^\perp)$ and $\Gamma_{\mathsf{C}}/\Gamma_{\mathsf{Z}} \cong \mathcal{A}_0^\perp/\mathcal{A}_0$.*

(iv) *Suppose $Z(\mathsf{A}_0) = \mathsf{F}_0$. Then, $\mathsf{A} = \mathsf{U} \otimes_{\mathsf{F}} \mathsf{C}$ and $\Gamma_{\mathsf{A}}/\Gamma_{\mathsf{F}} = \overline{deg}(\mathcal{A}_0^\perp)$.*

Proof. The equivalence of $r = 1$ with the other conditions is clear from Th. 7.44.

(i) We have seen in Prop. 7.42 that $\mathsf{A}_0 = \mathsf{F}_0[\mathcal{A}_0']$. Hence,

$$\mathsf{F}[\mathcal{A}_0] = \mathsf{F}_0[\mathcal{A}_0'] \otimes_{\mathsf{F}_0} \mathsf{F} = \mathsf{A}_0 \otimes_{\mathsf{F}_0} \mathsf{F} = \mathsf{U}.$$

Since $r = 1$, the F_0-algebra A_0 is simple with center $\mathsf{F}_0\big(\sqrt[s]{\sigma(\mathcal{Z}_0')}\big)$ by Th. 7.44(iv), and this field is separable over F_0. Therefore, U is graded simple and inertial over F.

(ii) Since $Z(\mathsf{A}_0) = \mathsf{F}_0[\mathcal{Z}_0'] = \mathsf{F}_0\big(\sqrt[s]{\sigma(\mathcal{Z}_0')}\big)$, which is a Kummer field extension of F_0, we have

$$Z(\mathsf{U}) = Z(\mathsf{A}_0) \otimes_{\mathsf{F}_0} \mathsf{F} = \mathsf{F}_0[\mathcal{Z}_0'] \otimes_{\mathsf{F}_0} \mathsf{F} = \mathsf{F}\big(\sqrt[s]{\sigma(\mathcal{Z}_0')}\big),$$

which is an inertial Kummer graded field extension of F.

(iii) Since $\mathsf{U} = \mathsf{F}[\mathcal{A}_0]$, by the graded version of Prop. 7.31 we have $\mathsf{C} = \mathsf{F}[\mathcal{A}_0^\perp]$ and

$$Z(\mathsf{C}) = \mathsf{F}[rad(\mathcal{A}_0^\perp)] = \mathsf{F}[\mathcal{Z}_0] = \mathsf{Z}.$$

Moreover, C is graded simple by the graded Double Centralizer Theorem 2.35(i), as U is graded simple. The map $\overline{deg}\colon \mathcal{A} \to \Gamma_\mathsf{A}/\Gamma_\mathsf{F}$ has kernel $\mathcal{A}_0$. Hence, for the restriction of $\overline{deg}$ to $\mathcal{A}_0^\perp$, the kernel is $\mathcal{A}_0^\perp \cap \mathcal{A}_0 = \mathcal{Z}_0$ while the image is $\Gamma_\mathsf{C}/\Gamma_\mathsf{F}$ by Prop. 7.42. Thus,

$$\Gamma_\mathsf{C}/\Gamma_\mathsf{F} = \overline{deg}(\mathcal{A}_0^\perp) \cong \mathcal{A}_0^\perp/(\mathcal{A}_0^\perp \cap \mathcal{A}_0).$$

Proposition 7.42 also shows that $\mathsf{C}_0 = \mathsf{F}_0[\mathcal{Z}_0'] = \mathsf{Z}_0$. Since C is graded simple with C_0 a division algebra, by Cor. 2.43 C is a graded division algebra. Moreover, C is totally ramified over $Z(\mathsf{C}) = \mathsf{Z}$ since $\mathsf{C}_0 = \mathsf{Z}_0$.

(iv) Suppose $Z(\mathsf{A}_0) = \mathsf{F}_0$. Since $Z(\mathsf{A}_0) = \mathsf{F}_0[\mathcal{Z}_0']$, this implies that $\mathcal{Z}_0' = \{0\}$, so $\{0\} = \mathcal{Z}_0 = \mathcal{A}_0 \cap \mathcal{A}_0^\perp$. Hence, $\mathcal{A} = \mathcal{A}_0 \perp \mathcal{A}_0^\perp$ as $\mathcal{A}$ is nondegenerate. It follows by the graded version of Prop. 7.31 that

$$\mathsf{A} = \mathsf{F}[\mathcal{A}_0] \otimes_\mathsf{F} \mathsf{F}[\mathcal{A}_0^\perp] = \mathsf{U} \otimes_\mathsf{F} \mathsf{C}.$$

Also, as $\mathcal{A} = \mathcal{A}_0 \cdot \mathcal{A}_0^\perp$ with $\mathcal{A}_0 = ker\, \overline{deg}$, we have $\Gamma_\mathsf{A}/\Gamma_\mathsf{F} = \overline{deg}(\mathcal{A}) = \overline{deg}(\mathcal{A}_0^\perp)$. □

7.2.3 Armature gauges

In this subsection, F is a field with a valuation v. As usual, we assume v takes its values in a divisible totally ordered abelian group Γ. Let A be an arbitrary F-algebra and $\mathcal{A} \subseteq A^\times/F^\times$ be an armature. Let $e = \exp \mathcal{A}$, so $x^e \in F^\times$ for all $x \in A^\times$ such that $[x] \in \mathcal{A}$. Assuming $A = F[\mathcal{A}]$, we define a map

$$\alpha_\mathcal{A}\colon A \longrightarrow \Gamma \cup \{\infty\}$$

as follows: we choose an F-base $(x_a)_{a\in\mathcal{A}}$ of A with $[x_a] = a$ for $a \in \mathcal{A}$, and set

$$\alpha_\mathcal{A}\Big(\sum_{a\in\mathcal{A}} \lambda_a x_a\Big) = \min_{a\in\mathcal{A}} \big(v(\lambda_a) + \tfrac{1}{e} v(x_a^e)\big) \in \Gamma \cup \{\infty\} \quad \text{for } \lambda_a \in F. \tag{7.32}$$

We also define a map

$$\overline{\alpha}_\mathcal{A}\colon \mathcal{A} \longrightarrow \Gamma/\Gamma_F \quad \text{by } \overline{\alpha}_\mathcal{A}(a) = \alpha_\mathcal{A}(x_a) + \Gamma_F. \tag{7.33}$$

Since x_a is uniquely determined by a up to a factor in $F^\times$, this map $\overline{\alpha}_{\mathcal{A}}$ is well-defined.

Lemma 7.46. *The map $\overline{\alpha}_{\mathcal{A}}$ is a group homomorphism, and the map $\alpha_{\mathcal{A}}$ is a surmultiplicative v-norm on A, which depends only on $\mathcal{A}$ and not on the choice of representatives x_a.*

The v-norm $\alpha_{\mathcal{A}}$ is called the *armature v-norm* associated to $\mathcal{A}$.

Proof. For any $a,\ b \in \mathcal{A}$ there exists a $\rho \in F^\times$ such that $x_a x_b = \rho x_{ab}$, hence

$$\alpha_{\mathcal{A}}(x_a x_b) \;=\; v(\rho) + \tfrac{1}{e} v(x_{ab}^e).$$

Now, $x_{ab}^e = \rho^{-e}(x_a x_b)^e$, and since $x_a x_b x_a^{-1} x_b^{-1}$ is a root of unity in F we have $(x_a x_b)^e \equiv x_a^e x_b^e \pmod{\mu(F)}$, so

$$v(x_{ab}^e) \;=\; v\big((x_a x_b)^e\big) - e\, v(\rho) \;=\; v(x_a^e) + v(x_b^e) - e\, v(\rho).$$

Therefore,

$$\alpha_{\mathcal{A}}(x_a x_b) \;=\; \tfrac{1}{e}\big(v(x_a^e) + v(x_b^e)\big) \;=\; \alpha_{\mathcal{A}}(x_a) + \alpha_{\mathcal{A}}(x_b) \tag{7.34}$$

and

$$\alpha_{\mathcal{A}}(x_{ab}) \;=\; \tfrac{1}{e} v(x_{ab}^e) \;\equiv\; \alpha_{\mathcal{A}}(x_a) + \alpha_{\mathcal{A}}(x_b) \pmod{\Gamma_F}.$$

Thus, $\overline{\alpha}_{\mathcal{A}}$ is a group homomorphism.

It is clear from the definition in (7.32) that the map $\alpha_{\mathcal{A}}$ is a v-norm on A with splitting base $(x_a)_{a\in\mathcal{A}}$. A different choice of representatives for the elements in $\mathcal{A}$ yields the same v-norm; for if $[y_a] = [x_a]$ then $y_a = \lambda x_a$ for some $\lambda \in F^\times$, and

$$\tfrac{1}{e} v(y_a^e) \;=\; v(\lambda) + \tfrac{1}{e} v(x_a^e).$$

Thus, the norm $\alpha_{\mathcal{A}}$ depends only on the armature $\mathcal{A}$, and not on the choice of the representatives x_a. This norm is surmultiplicative by Lemma 3.23 in view of (7.34). □

Continuing with the same notation, observe that the homogeneous elements $\widetilde{x_a} \in \mathsf{gr}_{\alpha_{\mathcal{A}}}(A)$ for $a \in \mathcal{A}$ are invertible since in view of the equation $e\,\alpha_{\mathcal{A}}(x_a) = v(x_a^e)$ we have

$$\widetilde{x_a}^{\,e} \;=\; \widetilde{x_a^e} \;\in\; \mathsf{gr}(F)^\times.$$

For $a \in \mathcal{A}$, we may thus define

$$\widetilde{a} \;=\; [\widetilde{x_a}] \;\in\; \mathsf{gr}_{\alpha_{\mathcal{A}}}(A)^\times_{\mathrm{hom}} / \mathsf{gr}(F)^\times$$

and let $\widetilde{\mathcal{A}} = \{\widetilde{a} \mid a \in \mathcal{A}\}$. Recall from (7.31) the map

$$\overline{deg}\colon\ \mathsf{gr}_{\alpha_{\mathcal{A}}}(A)^\times_{\mathrm{hom}} \longrightarrow \Gamma_{\mathsf{gr}(A)}/\Gamma_{\mathsf{gr}(F)}$$

induced by the degree map.

Proposition 7.47. *The set $\widetilde{\mathcal{A}}$ is a homogeneous armature of the graded $\mathsf{gr}(F)$-algebra $\mathsf{gr}_{\alpha_{\mathcal{A}}}(A)$, and*

$$\mathsf{gr}_{\alpha_{\mathcal{A}}}(A) = \mathsf{gr}(F)[\widetilde{\mathcal{A}}].$$

The map $a \mapsto \widetilde{a}$ is an isomorphism of abelian groups $\mathcal{A} \xrightarrow{\sim} \widetilde{\mathcal{A}}$, which makes the following diagram commute:

$$\begin{array}{ccc} \mathcal{A} & \xrightarrow{\overline{\alpha}_{\mathcal{A}}} & \Gamma_A/\Gamma_F \\ {\scriptstyle\cong}\big\downarrow & & \big\| \\ \widetilde{\mathcal{A}} & \xrightarrow{\overline{deg}} & \Gamma_{\mathsf{gr}(A)}/\Gamma_{\mathsf{gr}(F)} \end{array}$$

Moreover, the commutator pairings on $\mathcal{A}$ and $\widetilde{\mathcal{A}}$ are related as follows:

$$\langle \widetilde{a}, \widetilde{b} \rangle = \widetilde{\langle a, b\rangle} \qquad \textit{for } a,\, b \in \mathcal{A}. \tag{7.35}$$

Proof. For a, $b \in \mathcal{A}$, let $\omega \in F^\times$ satisfy $x_a x_b = \omega x_{ab}$. In view of (7.34) we have

$$\widetilde{x_a}\,\widetilde{x_b} = \widetilde{x_a x_b} = \widetilde{\omega}\widetilde{x_{ab}},$$

hence $[\widetilde{x_a}] \cdot [\widetilde{x_b}] = [\widetilde{x_{ab}}]$. Therefore, $\widetilde{\mathcal{A}}$ is a group isomorphic to $\mathcal{A}$ under the map $[x_a] \mapsto [\widetilde{x_a}]$. Moreover, the elements $\widetilde{x_a}$ form a homogeneous base of $\mathsf{gr}_{\alpha_{\mathcal{A}}}(A)$ over $\mathsf{gr}(F)$ since $(x_a)_{a\in\mathcal{A}}$ is a splitting base of A. Therefore, $\widetilde{\mathcal{A}}$ is a homogeneous armature in $\mathsf{gr}_{\alpha_{\mathcal{A}}}(A)$, and

$$\mathsf{gr}_{\alpha_{\mathcal{A}}}(A) = \mathsf{gr}(F)[\widetilde{\mathcal{A}}].$$

The diagram commutes because $deg\,\widetilde{x_a} = \alpha_{\mathcal{A}}(x_a)$ for $a \in \mathcal{A}$. Finally, (7.35) follows readily from the equation

$$\widetilde{x_a}\,\widetilde{x_b}\,\widetilde{x_a}^{-1}\,\widetilde{x_b}^{-1} = \widetilde{x_a x_b x_a^{-1} x_b^{-1}}. \qquad \square$$

As in Prop. 7.42, we let $\widetilde{\mathcal{A}}_0$ denote the kernel of $\overline{deg}\colon \widetilde{\mathcal{A}} \to \Gamma_A/\Gamma_F$; let $\widetilde{\mathcal{A}}_0'$ be the isomorphic inverse image of $\widetilde{\mathcal{A}}_0$ in $A_0^{\alpha_{\mathcal{A}}\times}/\overline{F}^\times$, where $A_0^{\alpha_{\mathcal{A}}}$ is the degree 0 component of $\mathsf{gr}_{\alpha_{\mathcal{A}}}(A)$. By Prop. 7.42, $\widetilde{\mathcal{A}}_0'$ is an armature of $A_0^{\alpha_{\mathcal{A}}}$ (as an $\overline{F}$-algebra), and

$$A_0^{\alpha_{\mathcal{A}}} = \overline{F}[\widetilde{\mathcal{A}}_0']. \tag{7.36}$$

Corollary 7.48. *The map $\overline{\alpha}_{\mathcal{A}}\colon \mathcal{A} \to \Gamma_A/\Gamma_F$ is surjective, and $\Gamma_A^\times = \Gamma_A$. The kernel $\ker\overline{\alpha}_{\mathcal{A}}$ is isomorphic under the canonical map $\widetilde{\ }$ to $\widetilde{\mathcal{A}}_0$.*

Proof. Proposition 7.42 shows that $\overline{deg}\colon \widetilde{\mathcal{A}} \to \Gamma_{\mathsf{gr}(A)}^\times/\Gamma_{\mathsf{gr}(F)}$ is surjective and $\Gamma_{\mathsf{gr}(A)}^\times = \Gamma_{\mathsf{gr}(A)}$. Therefore, $\Gamma_A^\times = \Gamma_A$, and surjectivity of $\overline{\alpha}_{\mathcal{A}}$ follows from the commutativity of the diagram in Prop. 7.47. It also follows from this commutative diagram that the canonical isomorphism $\mathcal{A} \to \widetilde{\mathcal{A}}$ maps $\ker\overline{\alpha}_{\mathcal{A}}$ to $\widetilde{\mathcal{A}}_0$. $\square$

Note that the isomorphism $\mathcal{A} \xrightarrow{\sim} \widetilde{\mathcal{A}}$ does not necessarily preserve the commutator pairings: if $char\,\overline{F} = \overline{p} > 0$, formula (7.35) shows that the commutator pairing on the $\overline{p}$-primary component of $\widetilde{\mathcal{A}}$ is trivial. If we avoid this case, the norm $\alpha_{\mathcal{A}}$ turns out to be a tame v-gauge, as the following theorem shows:

Theorem 7.49. *Let $e = \exp \mathcal{A}$. Suppose $\mathrm{char}\,\overline{F} \nmid e$. Then $\alpha_{\mathcal{A}}$ is a tame v-gauge on A, and the canonical isomorphism $\mathcal{A} \to \widetilde{\mathcal{A}}$ preserves the commutator pairings (under the identification of $\mu_e(F)$ with $\mu_e(\overline{F})$ through the residue map).*

The v-gauge $\alpha_{\mathcal{A}}$ is called the *armature v-gauge* associated to $\mathcal{A}$.

Proof. The last assertion follows readily from (7.35). Since $\widetilde{\mathcal{A}}$ is a homogeneous armature of exponent e such that $\mathsf{gr}_{\alpha_{\mathcal{A}}}(A) = \mathsf{gr}(F)[\widetilde{\mathcal{A}}]$, Prop. 7.39(i) shows that $\mathsf{gr}_{\alpha_{\mathcal{A}}}(A)$ is a semisimple graded $\mathsf{gr}(F)$-algebra, hence the surmultiplicative v-norm $\alpha_{\mathcal{A}}$ is a v-gauge. The same proposition further shows that

$$Z\big(\mathsf{gr}_{\alpha_{\mathcal{A}}}(A)\big) = \mathsf{gr}(F)[rad\widetilde{\mathcal{A}}],$$

and that this graded $\mathsf{gr}(F)$-algebra is separable. On the other hand, we have $Z(A) = F[rad\mathcal{A}]$ by Prop. 7.31; hence, by applying Prop. 7.47 to $Z(A)$ we obtain

$$\mathsf{gr}_{\alpha_{\mathcal{A}}|Z(A)}\big(Z(A)\big) = \mathsf{gr}(F)[\widetilde{rad\mathcal{A}}].$$

Since the canonical isomorphism $\mathcal{A} \xrightarrow{\sim} \widetilde{\mathcal{A}}$ preserves the commutator pairings, we have $\widetilde{rad\mathcal{A}} = rad\widetilde{\mathcal{A}}$, hence $Z\big(\mathsf{gr}_{\alpha_{\mathcal{A}}}(A)\big) = \mathsf{gr}_{\alpha_{\mathcal{A}}|Z(A)}\big(Z(A)\big)$. Therefore, $\alpha_{\mathcal{A}}$ is a tame v-gauge. □

Example 7.50. Suppose A is a symbol algebra of degree n with standard generators i, j, and let $\mathcal{A} = \{[i^k j^\ell] \mid k, l = 0, \ldots, n-1\}$. Then $\mathcal{A}$ is an armature in A with $A = F[\mathcal{A}]$ (see Ex. 7.28(iii)) and the map $\alpha_{\mathcal{A}}$ is the v-norm of §3.2.3. It was already observed in §3.2.3 that $\alpha_{\mathcal{A}}$ is a v-gauge if $char\,\overline{F} \nmid n$. Examples where $\alpha_{\mathcal{A}}$ fails to be a tame v-gauge—or even a v-gauge at all—were given in §3.2.3.

We have seen that an armature $\mathcal{A}$ on an F-algebra A induces a homogeneous armature $\widetilde{\mathcal{A}}$ on $\mathsf{gr}_{\alpha_{\mathcal{A}}}(A)$. When v is Henselian and A is tame and central simple over F, we can go in the other direction, obtaining an armature on A when $\mathsf{gr}(A)$ has a homogeneous armature:

Proposition 7.51. *Let A be a central simple algebra over a field with Henselian valuation v. Suppose A has a tame v-gauge β and that $\mathsf{gr}_\beta(A)$ is spanned by a homogeneous $\mathsf{gr}(F)$-armature $\mathcal{B}$. Then, A is spanned by an armature $\mathcal{A}$ such that there is an isomorphism $\mathsf{gr}_{\alpha_{\mathcal{A}}}(A) \cong_g \mathsf{gr}_\beta(A)$ inducing a symplectic module isomorphism $\widetilde{\mathcal{A}} \cong \mathcal{B}$. Hence, $\mathcal{A} \cong \mathcal{B}$ as symplectic modules.*

Proof. Let $\mathsf{B} = \mathsf{gr}_\beta(A)$. Since the gauge β is tame, the center of B is $\mathsf{gr}(F)$, hence the commutator pairing on $\mathcal{B}$ is nondegenerate. Let $\mathcal{B} \cong \mathbb{S}_{n_1} \perp \ldots \perp \mathbb{S}_{n_r}$ be a decomposition of $\mathcal{B}$ into an orthogonal sum of standard symplectic modules, and let $\mathsf{B} = \mathsf{T}_1 \otimes_{\mathsf{gr}(F)} \cdots \otimes_{\mathsf{gr}(F)} \mathsf{T}_r$ be the corresponding symbol algebra decomposition, where, say $\mathsf{T}_i = \big(s_i, t_i/\mathsf{gr}(F)\big)_{n_i}$. Thus, $\mathcal{B} = \mathcal{T}_1 \otimes \ldots \otimes \mathcal{T}_r$ where each $\mathcal{T}_i$ is the homogeneous symbol algebra armature for T_i. For each i, choose $a_i,\ b_i \in F^\times$ with $\widetilde{a}_i = s_i$ and $\widetilde{b}_i = t_i$ in $\mathsf{gr}(F)$. Since $\mathsf{gr}(F)$ contains a primitive n_i-th root of unity, we have $char\,\overline{F} \nmid n_i$; so, F contains a primitive n_i-th root of unity, as v is Henselian. Thus, there is a symbol algebra $S_i = \big(a_i, b_i/F\big)_{n_i}$ so that with its usual nondegenerate armature $\mathcal{S}_i$, we have $\mathsf{gr}_{\alpha_{\mathcal{S}_i}}(S_i) \cong_g \mathsf{T}_i$. This isomorphism induces a symplectic module isomorphism $\widetilde{\mathcal{S}_i} \cong \mathcal{T}_i$. Let $C = S_1 \otimes_F \ldots \otimes_F S_r$, which is spanned by the armature $\mathcal{C} = \mathcal{S}_1 \otimes \ldots \otimes \mathcal{S}_r$ that satisfies $\widetilde{\mathcal{C}} = \widetilde{\mathcal{S}_1} \otimes \ldots \otimes \widetilde{\mathcal{S}_r}$. The armature gauge $\alpha_{\mathcal{C}}$ on C is tame since $char\,\overline{F} \nmid n_1 \ldots n_r = deg\,C$. Since $\alpha_{\mathcal{C}} = \alpha_{\mathcal{S}_1} \otimes \ldots \otimes \alpha_{\mathcal{S}_r}$ as tame gauges, we have

$$\mathsf{gr}_{\alpha_{\mathcal{C}}}(C) \cong_g \mathsf{gr}_{\alpha_{\mathcal{S}_1}}(S_1) \otimes_{\mathsf{gr}(F)} \cdots \otimes_{\mathsf{gr}(F)} \mathsf{gr}_{\alpha_{\mathcal{S}_r}}(S_r) \cong_g \mathsf{B}.$$

These graded algebra isomorphisms induce symplectic module isomorphisms

$$\widetilde{\mathcal{C}} = \widetilde{\mathcal{S}_1} \otimes \ldots \otimes \widetilde{\mathcal{S}_r} \cong \mathcal{T}_1 \otimes \ldots \otimes \mathcal{T}_r = \mathcal{B}.$$

Since v is Henselian, the Brauer group homomorphism $\mathsf{gr}\colon Br_{tr}(F) \to \mathsf{Br}(\mathsf{gr}(F))$ is injective by Lemma 6.60. It follows that $C \sim A$ in $Br_{tr}(F)$ and hence $C \cong A$ by dimension count. Any F-isomorphism $C \xrightarrow{\sim} A$ sends $\mathcal{C}$ to an armature $\mathcal{A}$ of A with $\mathsf{gr}_{\alpha_{\mathcal{A}}}(A) \cong_g \mathsf{gr}_{\alpha_{\mathcal{C}}}(C) \cong_g \mathsf{B}$ inducing symplectic module isomorphisms $\widetilde{\mathcal{A}} \cong \widetilde{\mathcal{C}} \cong \mathcal{B}$. Since $\mathcal{A} \cong \widetilde{\mathcal{A}}$ by Prop. 7.47, it follows that $\mathcal{A} \cong \mathcal{B}$. □

7.3 Total ramification in graded algebras

We study total ramification in graded algebras from two perspectives: in §7.3.1 we consider the structure of individual graded division algebras, while in §7.3.2 and §7.3.3 we discuss Brauer group aspects.

The structure of totally ramified graded division algebras is particularly transparent: for each such graded algebra D over a graded field F, the factor group $\mathsf{D}^\times/\mathsf{F}^\times$ is a homogeneous armature isomorphic to $\Gamma_{\mathsf{D}}/\Gamma_{\mathsf{F}}$; by the results of §7.2, it is then easy to see that D is a tensor product of graded symbol algebras. In addition, we set up in Prop. 7.54 a one-to-one correspondence between subalgebras of D and subgroups of Γ_{D} containing Γ_{F}, and we define on $\Gamma_{\mathsf{D}}/\Gamma_{\mathsf{F}}$ a canonical alternating pairing induced by the commutator pairing on D.

When the graded field F is inertially closed (i.e., F_0 is separably closed), all the central graded division algebras over F are totally ramified. We use their commutator pairings to define a group homomorphism

$$\Pi_{\mathsf{F}} : \mathsf{Br}(\mathsf{F}) \longrightarrow \mathit{Symp}(\mathbb{T}(\Gamma_{\mathsf{F}})).$$

On the other hand, we may formally associate to each element in $\mathbb{T}(\wedge^2\Gamma_{\mathsf{F}})$ of order prime to *char* F_0 a sum of symbol algebra classes in $\mathsf{Br}(\mathsf{F})$. We thus obtain a double description of $\mathsf{Br}(\mathsf{F})$, in the form of two canonical isomorphisms (see Th. 7.64):

$$\mathbb{T}(\wedge^2\Gamma_{\mathsf{F}})' \xrightarrow{\Sigma_{\mathsf{F}}} \mathsf{Br}(\mathsf{F}) \xrightarrow{\Pi_{\mathsf{F}}} \mathit{Symp}(\mathbb{T}(\Gamma_{\mathsf{F}}))'.$$

The composition of these two maps turns out to be the (restriction of the) canonical isomorphism $\Upsilon_{\Gamma_{\mathsf{F}}}$ of Th. 7.22. We use these results in §7.3.3 to determine the top layer in the canonical filtration of the Brauer group of an arbitrary graded field F: We show in Th. 7.69 that the quotient $\mathsf{Br}(\mathsf{F})/\mathsf{Br}_{is}(\mathsf{F})$ is canonically isomorphic to the part of $\mathbb{T}(\wedge^2\Gamma_{\mathsf{F}})$ "of $\mu(\mathsf{F}_0)$-torsion," i.e., to the subgroup consisting of all the elements with an order n such that F_0 contains a primitive n-th root of unity; see (7.46).

7.3.1 Totally ramified graded division algebras

A (finite-dimensional) graded division algebra D over a graded field F is said to be *totally ramified* if $|\Gamma_{\mathsf{D}}{:}\Gamma_{\mathsf{F}}| = [\mathsf{D}{:}\mathsf{F}]$ (or, equivalently, $\mathsf{D}_0 = \mathsf{F}_0$).

Theorem 7.52. *Let* D *be a graded division algebra over a graded field* F *and let* $\mathsf{Z} = Z(\mathsf{D})$. *If* D *is totally ramified over* F, *then* $\mathsf{D}^\times/\mathsf{F}^\times$ *is the unique homogeneous armature in* D *that spans* D. *The radical of the commutator pairing is* $\mathsf{Z}^\times/\mathsf{F}^\times$. *Moreover, the degree map defines a group isomorphism* $\mathsf{D}^\times/\mathsf{F}^\times \xrightarrow{\sim} \Gamma_{\mathsf{D}}/\Gamma_{\mathsf{F}}$.

Proof. We have $\mathsf{D}^\times_{\text{hom}} = \mathsf{D}^\times$ by Prop. 2.3. Since $\mathsf{D}_0 = \mathsf{F}_0$, the exact sequence (7.31) (which holds whether or not D has an armature) boils down to an isomorphism

$$\overline{deg}\colon \mathsf{D}^\times/\mathsf{F}^\times \xrightarrow{\sim} \Gamma_{\mathsf{D}}/\Gamma_{\mathsf{F}}.$$

Thus, $\mathsf{D}^\times/\mathsf{F}^\times$ is a finite abelian group. Representatives of its elements have degrees in different cosets modulo Γ_{F}; hence, they are linearly independent over F. Therefore, $\mathsf{D}^\times/\mathsf{F}^\times$ is a homogeneous armature in D, and it is clear that this armature spans D. Every other homogeneous armature $\mathcal{A}$ of D is a proper subgroup of $\mathsf{D}^\times/\mathsf{F}^\times$, and $\mathcal{A}$ cannot span D since $|\mathcal{A}| < |\mathsf{D}^\times/\mathsf{F}^\times| = [\mathsf{D}{:}\mathsf{F}]$. Obviously, $\mathsf{Z}^\times/\mathsf{F}^\times$ lies in the radical of the commutator pairing. To see the reverse inclusion, pick $x \in \mathsf{D}^\times$. If $x \notin \mathsf{Z}^\times$, then there exists a $y \in \mathsf{D}^\times$ such that $xy \neq yx$, hence $\langle [x], [y] \rangle \neq 1$. □

For any graded division algebra D that is totally ramified over F, we use the isomorphism $\overline{deg}$ to transfer to $\Gamma_{\mathsf{D}}/\Gamma_{\mathsf{F}}$ the commutator pairing $b_{\mathcal{D}}$ of the homogeneous armature $\mathcal{D} = \mathsf{D}^\times/\mathsf{F}^\times$. We define the *canonical pairing*

$$c_\mathsf{D}\colon \Gamma_\mathsf{D} \times \Gamma_\mathsf{D} \longrightarrow \mu(\mathsf{F}_0) \tag{7.37}$$

by

$$c_\mathsf{D}(deg\, x, deg\, y) = xyx^{-1}y^{-1} \qquad \text{for } x,\, y \in \mathsf{D}^\times. \tag{7.38}$$

Theorem 7.52 shows that the radical of this pairing is $\Gamma_{Z(\mathsf{D})}$, which contains Γ_F. We let $\overline{c}_\mathsf{D}$ denote the induced pairing

$$\overline{c}_\mathsf{D}\colon (\Gamma_\mathsf{D}/\Gamma_\mathsf{F}) \times (\Gamma_\mathsf{D}/\Gamma_\mathsf{F}) \longrightarrow \mu(\mathsf{F}_0). \tag{7.39}$$

Suppose further that $Z(\mathsf{D}) = \mathsf{F}$. Then the commutator pairing $b_\mathcal{D}$ on $\mathcal{D}$ is nondegenerate, and since $\overline{deg}\colon \mathcal{D} \to \Gamma_\mathsf{D}/\Gamma_\mathsf{F}$ is an isomorphism with

$$b_\mathcal{D} = \overline{c}_\mathsf{D} \circ (\overline{deg} \times \overline{deg})\colon \mathcal{D} \times \mathcal{D} \longrightarrow \mu(\mathsf{F}_0),$$

the pairing $\overline{c}_\mathsf{D}$ on $\Gamma_\mathsf{D}/\Gamma_\mathsf{F}$ is nondegenerate. Furthermore, as symplectic modules,

$$\big(\Gamma_\mathsf{D}/\Gamma_\mathsf{F}, \overline{c}_\mathsf{D}\big) = \overline{deg}_\diamond(\mathcal{D}, b_\mathcal{D}) \cong (\mathcal{D}, b_\mathcal{D}). \tag{7.40}$$

Corollary 7.53. *Let* D *be a central graded division algebra over a graded field* F. *If* D *is totally ramified over* F, *then the invariant factors of* $\Gamma_\mathsf{D}/\Gamma_\mathsf{F}$ *occur in pairs. If* $n_1, \ldots, n_r$ *are integers such that*

$$\Gamma_\mathsf{D}/\Gamma_\mathsf{F} \cong (\mathbb{Z}/n_1\mathbb{Z})^2 \oplus \ldots \oplus (\mathbb{Z}/n_r\mathbb{Z})^2,$$

and if $m = lcm\,(n_1, \ldots, n_r)$ *(i.e.,* $m = exp(\Gamma_\mathsf{D}/\Gamma_\mathsf{F})$*), then* F_0 *contains a primitive* m*-th root of unity (in particular* char $\mathsf{F}_0 \nmid m$*), and there are graded symbol algebras* $\mathsf{S}_1, \ldots, \mathsf{S}_r$ *such that*

$$\mathsf{D} \cong_g \mathsf{S}_1 \otimes_\mathsf{F} \ldots \otimes_\mathsf{F} \mathsf{S}_r \quad \text{and} \quad deg\, \mathsf{S}_i = n_i \text{ for } i = 1, \ldots, r.$$

Hence, if $\mathsf{D} \neq \mathsf{F}$, *then* $rr(\Gamma_\mathsf{F}) \geq 2$ *(so* $\Gamma_\mathsf{F} \not\cong \mathbb{Z}$*).*

Proof. Since $(\Gamma_\mathsf{D}/\Gamma_\mathsf{F}, \overline{c}_\mathsf{D})$ is a symplectic module, it follows from Th. 7.2 that the invariant factors of $\Gamma_\mathsf{D}/\Gamma_\mathsf{F}$ occur in pairs. If $n_1, \ldots, n_r$ are such that $\Gamma_\mathsf{D}/\Gamma_\mathsf{F} \cong (\mathbb{Z}/n_1\mathbb{Z})^2 \oplus \ldots \oplus (\mathbb{Z}/n_r\mathbb{Z})^2$, then there is an isomorphism of symplectic modules

$$\mathsf{D}^\times/\mathsf{F}^\times \cong \mathbb{S}_{n_1} \perp \ldots \perp \mathbb{S}_{n_r},$$

since both sides have the same invariant factors (see Cor. 7.4). Therefore, Prop. 7.39 yields a tensor decomposition of D into graded symbol algebras of degrees $n_1, \ldots, n_r$. In particular, F_0 contains a primitive n_i-th root of unity for each $i = 1, \ldots, r$, hence also a primitive m-th root of unity for $m = lcm\,(n_1, \ldots, n_r)$. If $rr(\Gamma_\mathsf{F}) \leq 1$, then by Prop. A.36 $\Gamma_\mathsf{D}/\Gamma_\mathsf{F}$ is a cyclic group. Since $\Gamma_\mathsf{D}/\Gamma_\mathsf{F}$ has an even number of invariant factors, we must then have $\Gamma_\mathsf{D} = \Gamma_\mathsf{F}$. Then $[\mathsf{D}{:}\mathsf{F}] = |\Gamma_\mathsf{D}{:}\Gamma_\mathsf{F}| = 1$, so $\mathsf{D} = \mathsf{F}$. □

Since D is totally ramified over F, every subalgebra is determined by its grade group, and the canonical pairing holds all the information about centralizers. We make this observation precise in the following proposition:

Proposition 7.54. *Let* D *be a totally ramified central graded division algebra over a graded field* F.

(i) *There is a one-to-one correspondence between graded subalgebras* $\mathsf{A} \subseteq \mathsf{D}$ *and subgroups* $\Delta \subseteq \Gamma_{\mathsf{D}}$ *containing* Γ_{F}, *which maps every subalgebra* A *to its grade group* Γ_{A}, *and every subgroup* Δ *to the graded algebra* D_{Δ} *defined by*

$$\mathsf{D}_{\Delta} = \bigoplus_{\delta \in \Delta} \mathsf{D}_{\delta}.$$

If the subalgebra A *corresponds to the subgroup* Δ, *then*

$$[\mathsf{A}:\mathsf{F}] = |\Delta:\Gamma_{\mathsf{F}}|.$$

(ii) *For any subgroup* Δ *with* $\Gamma_{\mathsf{F}} \subseteq \Delta \subseteq \Gamma_{\mathsf{D}}$, *let*

$$\Delta^{\perp} = \{\gamma \in \Gamma_{\mathsf{D}} \mid c_{\mathsf{D}}(\gamma, \delta) = 1 \text{ for all } \delta \in \Delta\}.$$

The centralizer of D_{Δ} *in* D *is*

$$C_{\mathsf{D}}(\mathsf{D}_{\Delta}) = \mathsf{D}_{\Delta^{\perp}},$$

and the restriction of c_{D} *to* Δ *is the canonical pairing* $c_{D_{\Delta}}$. *In particular, the graded division algebra* D_{Δ} *is central over* F *if and only if* $\Delta \cap \Delta^{\perp} = \Gamma_{\mathsf{F}}$.

(iii) D_{Δ} *is a graded field if and only if* $\Delta \subseteq \Delta^{\perp}$. *When this occurs,* D_{Δ} *is a graded e-Kummer extension of* F, *where* $e = exp(\Gamma_{\mathsf{D}}/\Gamma_{\mathsf{F}})$.

Proof. (i) For any subgroup Δ in Γ_{D} containing Γ_{F}, it is clear that D_{Δ} is a graded subalgebra of D with $\Gamma_{\mathsf{D}_{\Delta}} = \Delta$. Moreover, for any graded subalgebra $\mathsf{A} \subseteq \mathsf{D}$ and any $\gamma \in \Gamma_{\mathsf{A}}$, the homogeneous component A_{γ} is an F_0-vector subspace of D_{γ}. But D_{γ} is 1-dimensional over F_0 since D is totally ramified over F, hence $\mathsf{A}_{\gamma} = \mathsf{D}_{\gamma}$, and therefore $\mathsf{A} = \bigoplus_{\gamma \in \Gamma_{\mathsf{A}}} \mathsf{D}_{\gamma} = \mathsf{D}_{\Gamma_{\mathsf{A}}}$. This shows that the correspondence between subalgebras of D and subgroups of Γ_{D} is bijective.

(ii) The equality $C_{\mathsf{D}}(\mathsf{D}_{\Delta}) = \mathsf{D}_{\Delta^{\perp}}$ follows readily from the definition of the canonical pairing, and the rest of (ii) is clear.

(iii) D_{Δ} is a graded field if and only if $\mathsf{D}_{\Delta} \subseteq C_{\mathsf{D}}(\mathsf{D}_{\Delta}) = \mathsf{D}_{\Delta^{\perp}}$, which holds if and only if $\Delta \subseteq \Delta^{\perp}$. When this occurs, every (homogeneous) element $a \in D_{\Delta}^{\times}$ satisfies $deg(a^e) \in \Gamma_{\mathsf{F}}$, hence $a^e \in \mathsf{F}$. Since in addition F contains a primitive e-th root of unity (see Cor. 7.53), it follows that D_{Δ} is a graded e-Kummer extension of F, as described in Remark 5.36. □

Remark 7.55. The classification in Prop. 7.54 of the F-subalgebras of the totally ramified graded division algebra D by subgroups of $\Gamma_{\mathsf{D}}/\Gamma_{\mathsf{F}}$ can also be seen in terms of armatures: By Th. 7.52, the unique homogeneous armature spanning D is $\mathsf{D}^{\times}/\mathsf{F}^{\times}$, and $\mathsf{D}^{\times}/\mathsf{F}^{\times} \cong \Gamma_{\mathsf{D}}/\Gamma_{\mathsf{F}}$ via $\overline{deg}$. Every subgroup $\mathcal{A}$ of $\mathsf{D}^{\times}/\mathsf{F}^{\times}$ is a homogeneous armature spanning the graded F-subalgebra $\mathsf{F}[\mathcal{A}]$ of D, and $\overline{deg}(\mathcal{A}) = \Gamma_{\mathsf{F}[\mathcal{A}]}/\Gamma_{\mathsf{F}}$. For any graded F-subalgebra A of D, since A is totally ramified over F, by Th. 7.52 the subgroup $\mathsf{A}^{\times}/\mathsf{F}^{\times}$ of $\mathsf{D}^{\times}/\mathsf{F}^{\times}$ is the unique homogeneous armature spanning A.

Corollary 7.56. *With the same notation as in Prop. 7.54, let* $\Delta_1, \ldots, \Delta_r$ *be subgroups of* Γ_{D} *containing* Γ_{F} *that yield an orthogonal decomposition of the symplectic module* $(\Gamma_{\mathsf{D}}/\Gamma_{\mathsf{F}}, \overline{c}_{\mathsf{D}})$*:*

$$\Gamma_{\mathsf{D}}/\Gamma_{\mathsf{F}} = (\Delta_1/\Gamma_{\mathsf{F}}) \perp \ldots \perp (\Delta_r/\Gamma_{\mathsf{F}}). \tag{7.41}$$

Then $\mathsf{D} = \mathsf{D}_{\Delta_1} \otimes_{\mathsf{F}} \ldots \otimes_{\mathsf{F}} \mathsf{D}_{\Delta_r}$*. In particular, if* $[\mathsf{D}{:}\mathsf{F}] = p_1^{m_1} \ldots p_r^{m_r}$ *for some distinct primes* $p_1, \ldots, p_r$ *and some integers* $m_1, \ldots, m_r$ *(which are necessarily even), then there is a canonical primary decomposition*

$$\mathsf{D} = \mathsf{D}_1 \otimes_{\mathsf{F}} \ldots \otimes_{\mathsf{F}} \mathsf{D}_r$$

where $[\mathsf{D}_i{:}\mathsf{F}] = p_i^{m_i}$ *and* $\Gamma_{\mathsf{D}_i}/\Gamma_{\mathsf{F}}$ *is the* p_i*-primary component of* $\Gamma_{\mathsf{D}}/\Gamma_{\mathsf{F}}$ *for* $i = 1, \ldots, r$*.*

Proof. If $\delta \in \Delta_1$ satisfies $c_{\mathsf{D}}(\delta, \varepsilon) = 1$ for every $\varepsilon \in \Delta_1$, then $c_{\mathsf{D}}(\delta, \gamma) = 1$ for every $\gamma \in \Gamma_{\mathsf{D}}$ since γ is a sum of elements in $\Delta_1, \ldots, \Delta_r$, and δ is orthogonal to $\Delta_2, \ldots, \Delta_r$. It follows that $\delta \in \Gamma_{\mathsf{F}}$ since $rad c_{\mathsf{D}} = \Gamma_{\mathsf{F}}$. Therefore, Prop. 7.54 shows that D_{Δ_1} is central over F. By Th. 2.35, we have

$$\mathsf{D} = \mathsf{D}_{\Delta_1} \otimes_{\mathsf{F}} C_{\mathsf{D}}(\mathsf{D}_{\Delta_1}).$$

Proposition 7.54 also shows that $C_{\mathsf{D}}(\mathsf{D}_{\Delta_1}) = \mathsf{D}_{\Delta_1^{\perp}}$. From the orthogonal decomposition (7.41), it is clear that $\Delta_1^{\perp} = \Delta_2 + \ldots + \Delta_r$, so

$$\Delta_1^{\perp}/\Gamma_{\mathsf{F}} = (\Delta_2/\Gamma_{\mathsf{F}}) \perp \ldots \perp (\Delta_r/\Gamma_{\mathsf{F}}),$$

and induction on r yields the decomposition of D. In particular, we may take for $\Delta_1, \ldots, \Delta_r$ the inverse images in Γ_{D} of the primary components of $\Gamma_{\mathsf{D}}/\Gamma_{\mathsf{F}}$. By Prop. 7.1 we have an orthogonal decomposition

$$\Gamma_{\mathsf{D}}/\Gamma_{\mathsf{F}} = (\Delta_1/\Gamma_{\mathsf{F}}) \perp \ldots \perp (\Delta_r/\Gamma_{\mathsf{F}}),$$

hence, by the first part of the proof, a tensor decomposition

$$\mathsf{D} = \mathsf{D}_{\Delta_1} \otimes_{\mathsf{F}} \ldots \otimes_{\mathsf{F}} \mathsf{D}_{\Delta_r}.$$

Since D_{Δ_i} corresponds to Δ_i under the bijection of Prop. 7.54, we have

$$[\mathsf{D}_{\Delta_i}{:}\mathsf{F}] = |\Delta_i/\Gamma_{\mathsf{F}}| = p_i^{m_i}. \qquad \square$$

Here is an explicit description of which symbol algebras and which tensor products are totally ramified over a graded field:

Proposition 7.57. *Let* F *be a graded field.*

(i) *For* $a, b \in \mathsf{F}^{\times}$, $n \in \mathbb{N}$, *and* ω *a primitive* n*-th root of unity in* $\mathsf{F}_0^{\times}$*, the graded symbol algebra* $\mathsf{S} = \big(a, b/\mathsf{F}\big)_{\omega,n}$ *is a graded division algebra totally ramified over* F *if and only if the images of* $\deg a, \deg b$ *generate a subgroup of order* n^2 *in* $\Gamma_{\mathsf{F}}/n\Gamma_{\mathsf{F}}$*. When this occurs,*

$$\Gamma_{\mathsf{S}} = \big\langle \tfrac{1}{n}\deg a, \tfrac{1}{n}\deg b \big\rangle + \Gamma_{\mathsf{F}},$$

and the canonical pairing c_{S} *on* Γ_{S} *is determined by*

$$c_{\mathsf{S}}(\tfrac{1}{n}\, deg\, a, \tfrac{1}{n}\, deg\, b) \;=\; \omega.$$

(ii) *Let* $\mathsf{A}_1, \ldots, \mathsf{A}_r$ *be graded algebras over* F. *Then,* $\mathsf{A}_1 \otimes_{\mathsf{F}} \ldots \otimes_{\mathsf{F}} \mathsf{A}_r$ *is a graded division algebra totally ramified over* F *if and only if each* A_i *is a graded division algebra totally ramified over* F *and the sum* $\sum_{i=1}^{r} \Gamma_{\mathsf{A}_i}/\Gamma_{\mathsf{F}}$ *in* $\mathbb{T}(\Gamma_{\mathsf{F}})$ *is a direct sum.*

Proof. (i) Let i, j be the standard homogeneous generators of S satisfying $i^n = a$, $j^n = b$, and $ij = \omega ji$. The grading on S is determined by $deg\, i = \frac{1}{n}\, deg\, a$ and $deg\, j = \frac{1}{n}\, deg\, b$, and we have $\Gamma_{\mathsf{S}} = \langle \frac{1}{n}\, deg\, a, \frac{1}{n}\, deg\, b\rangle + \Gamma_{\mathsf{F}}$ (see (2.9)). So, $n\Gamma_{\mathsf{S}} = \langle deg\, a, deg\, b\rangle + n\Gamma_{\mathsf{F}}$. If S is a graded division algebra totally ramified over F, then

$$n^2 \;=\; [\mathsf{S}{:}\mathsf{F}] \;=\; |\Gamma_{\mathsf{S}}{:}\Gamma_{\mathsf{F}}| \;=\; |n\Gamma_{\mathsf{S}}{:}n\Gamma_{\mathsf{F}}|,$$

so the images of $deg\, a$, $deg\, b$ generate a subgroup of order n^2 in $\Gamma_{\mathsf{F}}/n\Gamma_{\mathsf{F}}$. Also, we have

$$c_{\mathsf{S}}(\tfrac{1}{n}\, deg\, a, \tfrac{1}{n}\, deg\, b) \;=\; iji^{-1}j^{-1} = \omega,$$

and this completely determines the canonical symplectic pairing c_{S} since Γ_{S} is generated by $\frac{1}{n}\, deg\, a$, $\frac{1}{n}\, deg\, b$, and Γ_{F}.

Conversely, suppose $\big|\langle deg\, a, deg\, b\rangle + n\Gamma_{\mathsf{F}} : n\Gamma_{\mathsf{F}}\big| = n^2$. Then

$$|\Gamma_{\mathsf{S}}{:}\Gamma_{\mathsf{F}}| \;=\; |n\Gamma_{\mathsf{S}}{:}n\Gamma_{\mathsf{F}}| \;=\; n^2 \;=\; [\mathsf{S}{:}\mathsf{F}],$$

so $\mathsf{S}_0 = \mathsf{F}_0$ by (2.3). Since S_0 is a division ring, by Cor. 2.43 the simple graded algebra S is a graded division algebra, which is totally ramified over F as $\mathsf{S}_0 = \mathsf{F}_0$.

(ii) Let $\mathsf{B} = \mathsf{A}_1 \otimes_{\mathsf{F}} \ldots \otimes_{\mathsf{F}} \mathsf{A}_r$. Then, $\Gamma_{\mathsf{B}} = \Gamma_{\mathsf{A}_1} + \ldots + \Gamma_{\mathsf{A}_r}$ by the definition of the grading on tensor products. Suppose B is a graded division algebra totally ramified over F. Then each graded subalgebra A_i is also a graded division algebra totally ramified over F. Moreover,

$$\Big|\sum_{i=1}^{r} \Gamma_{\mathsf{A}_i}/\Gamma_{\mathsf{F}}\Big| \;=\; |\Gamma_{\mathsf{B}}{:}\Gamma_{\mathsf{F}}| \;=\; [\mathsf{B}{:}\mathsf{F}] \;=\; \sum_{i=1}^{r} [\mathsf{A}_i{:}\mathsf{F}] \;=\; \sum_{i=1}^{r} |\Gamma_{\mathsf{A}_i}/\Gamma_{\mathsf{F}}|.$$

Hence, the sum $\sum_{i=1}^{r} \Gamma_{\mathsf{A}_i}/\Gamma_{\mathsf{F}}$ in $\mathbb{T}(\Gamma_{\mathsf{F}})$ must be direct.

Conversely, suppose each A_i is a graded division algebra totally ramified over F and the sum $\sum_{i=1}^{r} \Gamma_{\mathsf{A}_i}/\Gamma_{\mathsf{F}}$ is direct. Then,

$$|\Gamma_{\mathsf{B}}{:}\Gamma_{\mathsf{F}}| \;=\; \Big|\sum_{i=1}^{r} \Gamma_{\mathsf{A}_i}/\Gamma_{\mathsf{F}}\Big| \;=\; \sum_{i=1}^{r} |\Gamma_{\mathsf{A}_i}{:}\Gamma_{\mathsf{F}}| \;=\; \sum_{i=1}^{r} [\mathsf{A}_i{:}\mathsf{F}] \;=\; [\mathsf{B}{:}\mathsf{F}].$$

Consequently, every nonzero homogeneous component B_γ of B must be 1-dimensional over F_0 by (2.3); in particular $\mathsf{B}_0 = \mathsf{F}_0$. Also, because $\Gamma_{\mathsf{A}_i}^{\times} = \Gamma_{\mathsf{A}_i}$ for each graded division algebra A_i, we have $\Gamma_{\mathsf{B}}^{\times} = \Gamma_{\mathsf{B}}$. Hence, by Prop. 2.4 B is a graded division algebra; it is totally ramified over F as $\mathsf{B}_0 = \mathsf{F}_0$. □

When a central simple graded F-algebra with an armature has associated graded division algebra D totally ramified over F, we show that the armature can be used to determine Γ_{D} and its canonical pairing. This will be applied to tensor products of totally ramified division algebras.

Proposition 7.58. *Let A be a central simple graded F-algebra spanned by a homogeneous armature $\mathcal{A}$, and let D be the associated graded division algebra of A. Let $\mathcal{A}_0$ be the kernel of $\overline{deg}\colon \mathcal{A} \to \Gamma_{\mathsf{A}}/\Gamma_{\mathsf{F}}$. If D is totally ramified over F, then*

$$\overline{deg}_{\diamond}(\mathcal{A}, b_{\mathcal{A}}) = \big(\Gamma_{\mathsf{D}}/\Gamma_{\mathsf{F}}, \overline{c}_{\mathsf{D}}\big).$$

Hence, $\Gamma_{\mathsf{D}}/\Gamma_{\mathsf{F}} = \overline{deg}(\mathcal{A}_0^{\perp})$ and $\big(\Gamma_{\mathsf{D}}/\Gamma_{\mathsf{F}}, \overline{c}_{\mathsf{D}}\big) \cong \big(\mathcal{A}_0^{\perp}/(\mathcal{A}_0^{\perp} \cap \mathcal{A}_0), \overline{b}_{\mathcal{A}}\big)$ as symplectic modules, where $\overline{b}_{\mathcal{A}}$ is the pairing induced by the armature pairing $b_{\mathcal{A}}$ of $\mathcal{A}$.

Note that the hypothesis that D be totally ramified over F can be replaced with the following equivalent hypothesis expressed directly in terms of A:

$$\mathsf{A}_0 \cong M_{n_1}(\mathsf{F}_0) \times \ldots \times M_{n_r}(\mathsf{F}_0) \qquad \text{for some integers } n_1, \ldots, n_r.$$

(Recall that D is totally ramified over F if and only if $\mathsf{D}_0 = \mathsf{F}_0$, and that by Prop. 2.41 every simple component of A_0 is Brauer-equivalent to D_0.)

Proof. It suffices to prove the displayed equality of symplectic modules, since the remaining assertions follow from the definition of $\overline{deg}_{\diamond}$ (see (7.6)).

As in Prop. 7.42–Th. 7.44, let $\mathcal{Z}_0 = rad\mathcal{A}_0 = \mathcal{A}_0 \cap \mathcal{A}_0^{\perp}$, let $s = exp\,\mathcal{Z}_0$, and let $\mathcal{K}_0 = \{[x] \in \mathcal{Z}_0 \mid x^s \in \mathsf{F}^s\}$; so $\mathsf{F}[\mathcal{K}_0]$ is a direct product of copies of F by the graded version of Th. 7.36 (see Remark 7.41). Let e be any of the primitive homogeneous idempotents of $\mathsf{F}[\mathcal{K}_0]$, and let $\mathsf{B} = e\mathsf{A}e$. Note that $\mathsf{B} \sim \mathsf{A}$ in $\mathsf{Br}(\mathsf{F})$ when we identify $\mathsf{F}e$ with F. For if $\mathsf{A} = \mathsf{End}_{\mathsf{D}}(\mathsf{J})$ for some right D-vector space J, then $\mathsf{B} \cong_g \mathsf{End}_{\mathsf{D}}(e\mathsf{J})$. Hence, D is the associated graded division algebra of B. The projection $\rho\colon \mathsf{F}[\mathcal{K}_0^{\perp}] \to \mathsf{B} = \mathsf{F}[\mathcal{K}_0^{\perp}]e = e\mathsf{F}[\mathcal{K}_0^{\perp}]$ given by $s \mapsto se$ is a graded ring homomorphism inducing a group homomorphism $\rho^{\times}\colon \mathsf{F}[\mathcal{K}_0^{\perp}]^{\times}_{\mathrm{hom}}/\mathsf{F}^{\times} \to \mathsf{B}^{\times}_{\mathrm{hom}}/\mathsf{F}e^{\times}$. Let $\mathcal{B} = \rho^{\times}(\mathcal{K}_0^{\perp})$. By Th. 7.44, $\mathcal{B}$ is a homogeneous armature spanning B. The degree map $deg_{\mathsf{B}}\colon \mathsf{B}_{\mathrm{hom}} \setminus \{0\} \to \Gamma_{\mathsf{B}}$ induces $\overline{deg_{\mathsf{B}}}\colon \mathcal{B} \to \Gamma_{\mathsf{B}}/\Gamma_{\mathsf{F}}$. For clarity, for the rest of the proof we write $\overline{deg}_{\mathsf{A}}$ instead of $\overline{deg}$ for the map $\mathcal{A} \to \Gamma_{\mathsf{A}}/\Gamma_{\mathsf{F}}$.

We claim that

$$\overline{deg}_{\mathsf{A}\,\diamond}(\mathcal{A}, b_{\mathcal{A}}) = \overline{deg}_{\mathsf{B}\,\diamond}(\mathcal{B}, b_{\mathcal{B}}).$$

This will allow us to work in B rather than A. To prove the claim, observe first that since $\mathcal{K}_0 \subseteq \mathcal{A}_0 = ker\,\overline{deg}_{\mathsf{A}}$, there is an induced map $\overline{deg}'$ on $\mathcal{A}/\mathcal{K}_0$ such that $\overline{deg}_{\mathsf{A}} = \overline{deg}' \circ \pi$ where $\pi\colon \mathcal{A} \to \mathcal{A}/\mathcal{K}_0$ is the projection. Since $\mathcal{K}_0^{\perp} \supseteq \mathcal{K}_0$ as $\mathcal{K}_0 \subseteq rad\mathcal{A}_0$, we have $\pi_{\diamond}(\mathcal{A}, b_{\mathcal{A}}) = (\mathcal{K}_0^{\perp}/\mathcal{K}_0, \overline{b}_{\mathcal{A}})$. Hence, by the functorial property of $Symp$ given in Prop. 7.10,

$$\overline{deg}_{\mathsf{A}\,\diamond}(\mathcal{A}, b_{\mathcal{A}}) = \overline{deg}'_{\diamond} \circ \pi_{\diamond}(\mathcal{A}, b_{\mathcal{A}}) = \overline{deg}'_{\diamond}(\mathcal{K}_0^{\perp}/\mathcal{K}_0, \overline{b}_{\mathcal{A}}).$$

Since $\mathcal{K}_0 = ker(\rho^\times) \cap \mathcal{K}_0^\perp$ (see Th. 7.44(iii)), the map $\rho^\times$ induces an isomorphism $\overline{\rho}\colon \mathcal{K}_0^\perp/\mathcal{K}_0 \xrightarrow{\sim} \mathcal{B}$. Because the ring homomorphism ρ maps (multiplicative) commutators to commutators, the map $\overline{\rho}$ is an isometry of symplectic modules: $(\mathcal{K}_0^\perp/\mathcal{K}_0, \overline{b}_\mathcal{A}) \cong (\mathcal{B}, b_\mathcal{B})$. Moreover, as ρ is a degree-preserving homomorphism, we have $\overline{deg}'|_{\mathcal{K}_0^\perp/\mathcal{K}_0} = \overline{deg}_\mathsf{B} \circ \overline{\rho}$. Hence, (see Remark 7.8),

$$\overline{deg}'_\diamond(\mathcal{K}_0^\perp/\mathcal{K}_0, \overline{b}_\mathcal{A}) \;=\; \overline{deg}_\mathsf{B}(\mathcal{B}, b_\mathcal{B}).$$

The displayed equalities prove the claim.

Now consider B. Note that B_0 is a simple algebra by Th. 7.44(iv). Since $\mathsf{B} \sim \mathsf{D}$ in $\mathsf{Br}(\mathsf{F})$, by Prop. 2.41 $Z(\mathsf{B}_0) \cong Z(\mathsf{D}_0) = \mathsf{F}_0$. Therefore, by Prop. 7.45(iv), for $\mathcal{B}_0 = ker\, \overline{deg}_\mathsf{B}$,

$$\mathsf{B} \;=\; \mathsf{F}[\mathcal{B}_0] \otimes_\mathsf{F} \mathsf{C} \qquad \text{where} \qquad \mathsf{C} \;=\; \mathsf{F}[\mathcal{B}_0^\perp],$$

with $\mathsf{F}[\mathcal{B}_0]$ a central simple inertial graded F-algebra and C a graded division algebra totally ramified over its center F. Let E be the associated graded division algebra of $\mathsf{F}[\mathcal{B}_0]$. Since E is inertial over F by Prop. 2.56, we have

$$(\mathsf{E} \otimes_\mathsf{F} \mathsf{C})_0 \;=\; \mathsf{E}_0 \otimes_{\mathsf{F}_0} \mathsf{C}_0 \;=\; \mathsf{E}_0 \otimes_{\mathsf{F}_0} \mathsf{F}_0 \;=\; \mathsf{E}_0.$$

This is a division algebra over F_0; hence, $\mathsf{E} \otimes_\mathsf{F} \mathsf{C}$ is a graded division algebra, by Cor. 2.43. Since

$$\mathsf{E} \otimes_\mathsf{F} \mathsf{C} \sim \mathsf{F}[\mathcal{B}_0] \otimes_\mathsf{F} \mathsf{C} = \mathsf{B} \sim \mathsf{D} \quad \text{in } \mathsf{Br}(\mathsf{F}),$$

the graded division algebras $\mathsf{E} \otimes_\mathsf{F} \mathsf{C}$ and D are isomorphic. Thus,

$$\mathsf{E}_0 = (\mathsf{E} \otimes_\mathsf{F} \mathsf{C})_0 = \mathsf{D}_0 = \mathsf{F}_0.$$

Hence, as E is inertial over F, we have $\mathsf{E} \cong_g \mathsf{E}_0 \otimes_{\mathsf{F}_0} \mathsf{F} \cong_g \mathsf{F}$; so

$$\mathsf{D} \cong_g \mathsf{C} \;=\; \mathsf{F}[\mathcal{B}_0^\perp].$$

Since any graded isomorphism $\mathsf{D} \xrightarrow{\sim} \mathsf{C}$ induces an isomorphism of their homogeneous armatures $\mathsf{D}^\times/\mathsf{F}^\times \xrightarrow{\sim} \mathsf{C}^\times/\mathsf{F}^\times$ that is compatible with the armature pairings and the degree map, we have an equality of symplectic modules

$$(\Gamma_\mathsf{D}/\Gamma_\mathsf{F}, \overline{c}_\mathsf{D}) \;=\; (\Gamma_\mathsf{C}/\Gamma_\mathsf{F}, \overline{c}_\mathsf{C}).$$

Now, since $\mathcal{B}_0 = ker\, \overline{deg}_\mathsf{B}$, by definition (see (7.5))

$$\overline{deg}_{\mathsf{B}\diamond}(\mathcal{B}, b_\mathcal{B}) \;=\; \big(\overline{deg}_\mathsf{B}(\mathcal{B}_0^\perp), \overline{deg}_{\mathsf{B}*} b_\mathsf{B}\big).$$

Here, $\overline{deg}_\mathsf{B}(\mathcal{B}_0^\perp) = \Gamma_{\mathsf{F}[\mathcal{B}_0^\perp]}/\Gamma_\mathsf{F} = \Gamma_\mathsf{C}/\Gamma_\mathsf{F}$ by Prop. 7.45(iii). Also, $\mathcal{B}_0^\perp = \mathsf{C}^\times/\mathsf{F}^\times$ since the totally ramified graded F-algebra C has a unique homogeneous armature. For any $s, t \in \mathsf{C}^\times$, we have $\overline{deg}_\mathsf{B}[s] = deg\, s + \Gamma_\mathsf{F}$, so

$$\begin{aligned}\overline{deg}_{\mathsf{B}*} b_\mathcal{B}(deg\, s + \Gamma_\mathsf{F}, deg\, t + \Gamma_\mathsf{F}) \;&=\; \overline{deg}_{\mathsf{B}*} b_\mathcal{B}\big(\overline{deg}_\mathsf{B}[s], \overline{deg}_\mathsf{B}[t]\big)\\ &=\; b_\mathcal{B}([s],[t]) \;=\; sts^{-1}t^{-1} \;=\; \overline{c}_\mathsf{C}(deg\, s + \Gamma_\mathsf{F}, deg\, t + \Gamma_\mathsf{F}).\end{aligned}$$

Hence, $\overline{deg}_{\mathsf{B}*} b_{\mathcal{B}} = \overline{c}_{\mathsf{C}}$, and we have

$$\left(\overline{deg}_{\mathsf{B}}(\mathcal{B}_0^{\perp}), \overline{deg}_{\mathsf{B}*} b_{\mathsf{B}}\right) \;=\; (\Gamma_{\mathsf{C}}/\Gamma_{\mathsf{F}}, \overline{c}_{\mathsf{C}}).$$

The claim and the last three displayed equalities of symplectic modules show that $\overline{deg}_{\mathsf{A}\diamond}(\mathcal{A}, b_{\mathcal{A}}) = (\Gamma_{\mathsf{D}}/\Gamma_{\mathsf{F}}, \overline{c}_{\mathsf{D}})$, as desired. □

Proposition 7.59. *Let* S *and* T *be graded division algebras with center* F *and each totally ramified over* F. *Let* D *be the associated graded division algebra of* $\mathsf{S} \otimes_{\mathsf{F}} \mathsf{T}$. *If* D *is totally ramified over* F, *then, as symplectic modules,*

$$(\Gamma_{\mathsf{D}}/\Gamma_{\mathsf{F}}, \overline{c}_{\mathsf{D}}) \;=\; (\Gamma_{\mathsf{S}}/\Gamma_{\mathsf{F}}, \overline{c}_{\mathsf{S}}) + (\Gamma_{\mathsf{T}}/\Gamma_{\mathsf{F}}, \overline{c}_{\mathsf{T}}).$$

Proof. Let $\mathcal{S} = \mathsf{S}^{\times}/\mathsf{F}^{\times}$ and $\mathcal{T} = \mathsf{T}^{\times}/\mathsf{F}^{\times}$, which are the homogeneous armatures spanning S and T. Let $\mathsf{A} = \mathsf{S} \otimes_{\mathsf{F}} \mathsf{T}$ and let $\mathcal{A} = \mathcal{S} \otimes \mathcal{T} \subseteq \mathsf{A}^{\times}_{\text{hom}}/\mathsf{F}^{\times}$ as in Prop. 7.29. So, $\mathcal{A}$ is a homogeneous armature spanning A. There is an isomorphism $\psi \colon \mathcal{S} \times \mathcal{T} \to \mathcal{A}$ given by $(s,t) \mapsto s \otimes t$, where for $s = [x] \in \mathcal{A}$ and $t = [y] \in \mathcal{T}$ we have $s \otimes t = [x \otimes y]$ in $\mathcal{A}$. Let $b_{\mathcal{S}}, b_{\mathcal{T}}$, and $b_{\mathcal{A}}$ be the commutator pairings on $\mathcal{S}, \mathcal{T}$, and $\mathcal{A}$. By Prop. 7.29 the map ψ is an isometry of symplectic modules

$$(\mathcal{S} \times \mathcal{T}, b_{\mathcal{S}} \times b_{\mathcal{T}}) \;\cong\; (\mathcal{A}, b_{\mathcal{A}}).$$

For homogeneous elements $x \in \mathsf{S}^{\times}, y \in \mathsf{T}^{\times}$, the product $x \otimes y$ is homogeneous in A with $deg_{\mathsf{A}}(x \otimes y) = deg_{\mathsf{S}}\, x + deg_{\mathsf{T}}\, y$. Hence, for the induced maps on armatures, $\overline{deg}_{\mathsf{S}} \colon \mathcal{S} \to \mathbb{T}(\Gamma)$, $\overline{deg}_{\mathsf{T}} \colon \mathcal{T} \to \mathbb{T}(\Gamma)$, and $\overline{deg}_{\mathsf{A}} \colon \mathcal{A} \to \mathbb{T}(\Gamma)$, we have

$$\overline{deg}_{\mathsf{A}}(s \otimes t) \;=\; \overline{deg}_{\mathsf{S}}(s) + \overline{deg}_{\mathsf{T}}(t) \qquad \text{for all } s \in \mathcal{S},\ t \in \mathcal{T}.$$

That is, the following diagram is commutative

$$\begin{array}{ccc}
\mathcal{S} \times \mathcal{T} & \xrightarrow[\cong]{\ \psi\ } & \mathcal{A} \\
{\scriptstyle \overline{deg}_{\mathsf{S}} \times \overline{deg}_{\mathsf{T}}} \downarrow & & \downarrow {\scriptstyle \overline{deg}_{\mathsf{A}}} \\
\mathbb{T}(\Gamma_{\mathsf{F}}) \times \mathbb{T}(\Gamma_{\mathsf{F}}) & \xrightarrow{\ \Sigma\ } & \mathbb{T}(\Gamma_{\mathsf{F}})
\end{array}$$

where the bottom map Σ is the sum, $(\gamma, \delta) \mapsto \gamma + \delta$. Since ψ is an isometry, by Remark 7.8, Lemma 7.12, and (7.40) we have

$$\begin{aligned}
\overline{deg}_{\mathsf{A}\diamond}(\mathcal{A}, b_{\mathcal{A}}) &= (\overline{deg}_{\mathsf{A}} \circ \psi)_{\diamond}(\mathcal{S} \times \mathcal{T}, b_{\mathcal{S}} \times b_{\mathcal{T}}) \\
&= \left(\Sigma \circ (\overline{deg}_{\mathsf{S}} \times \overline{deg}_{\mathsf{T}})\right)_{\diamond}(\mathcal{S} \times \mathcal{T}, b_{\mathcal{S}} \times b_{\mathcal{T}}) \\
&= \overline{deg}_{\mathcal{S}\diamond}(\mathcal{S}, b_{\mathcal{S}}) + \overline{deg}_{\mathcal{T}\diamond}(\mathcal{T}, b_{\mathcal{T}}) \\
&= (\Gamma_{\mathsf{S}}/\Gamma_{\mathsf{F}}, \overline{c}_{\mathsf{S}}) + (\Gamma_{\mathsf{T}}/\Gamma_{\mathsf{F}}, \overline{c}_{\mathsf{T}}).
\end{aligned}$$

Since $\overline{deg}_{\mathsf{A}\diamond}(\mathcal{A}, b_{\mathcal{A}}) = (\Gamma_{\mathsf{D}}/\Gamma_{\mathsf{F}}, \overline{c}_{\mathsf{D}})$ by Prop. 7.58, the equality of the proposition follows. □

For a graded division algebra D with center F and an integer $j \in \mathbb{N}$, we write D^j for the associated graded division algebra of the j-th tensor power $\mathsf{D}^{\otimes j}$ of D over F. When D is totally ramified over its center, the symbol algebra decomposition of D allows us to see that each D^j is also totally ramified over F; the preceding proposition then determines its canonical pairing:

Proposition 7.60. *Let D be a central graded division algebra over a graded field F. If D is totally ramified over F, then for every $j \in \mathbb{N}$ the graded division algebra D^j is totally ramified over F and $\Gamma_{\mathsf{D}^j} = j\Gamma_{\mathsf{D}} + \Gamma_{\mathsf{F}}$. Moreover, as symplectic modules,*

$$(\Gamma_{\mathsf{D}^j}/\Gamma_{\mathsf{F}}, \overline{c}_{\mathsf{D}^j}) = \big(j(\Gamma_{\mathsf{D}}/\Gamma_{\mathsf{F}}), \overline{c}_{\mathsf{D}}^{(j)}\big)$$

in the notation of Prop. 7.15. Hence, $\exp \mathsf{D} = \exp(\Gamma_{\mathsf{D}}/\Gamma_{\mathsf{F}})$.

Proof. For the total ramification of D^j, it suffices to treat the case that j prime. The result for arbitrary $j \in \mathbb{N}$ then follows by induction. So, assume j is prime.

Assume first that D is a graded symbol algebra, say $\mathsf{D} = \big(a, b/\mathsf{F}\big)_n$ for some $a, b \in \mathsf{F}^\times$. By Prop. 7.57(i), $\Gamma_{\mathsf{D}} = \big\langle \frac{1}{n}\, deg\, a, \frac{1}{n}\, deg\, b\big\rangle + \Gamma_{\mathsf{F}}$ and

$$\big|\langle deg\, a, deg\, b\rangle + n\Gamma_{\mathsf{F}} : n\Gamma_{\mathsf{F}}\big| = n^2.$$

Suppose $j \mid n$. Then $\mathsf{D}^j \sim (a, b/\mathsf{F})_{n/j}$ in $\mathsf{Br}(\mathsf{F})$ by the identities for symbol algebras (see Draxl [63, Lemma 6, p. 80]). Since

$$\big(\langle deg\, a, deg\, b\rangle + n\Gamma_{\mathsf{F}}\big)/n\Gamma_{\mathsf{F}} \cong \mathbb{Z}/n\mathbb{Z} \times \mathbb{Z}/n\mathbb{Z},$$

we have

$$\big(\langle deg\, a, deg\, b\rangle + \tfrac{n}{j}\Gamma_{\mathsf{F}}\big)/\tfrac{n}{j}\Gamma_{\mathsf{F}} \cong \mathbb{Z}/\tfrac{n}{j}\mathbb{Z} \times \mathbb{Z}/\tfrac{n}{j}\mathbb{Z}.$$

By Prop. 7.57(i), $(a, b/\mathsf{F})_{n/j}$ is therefore a graded division algebra totally ramified over F. Hence, $\mathsf{D}^j \cong_g (a, b/\mathsf{F})_{n/j}$ and

$$\Gamma_{\mathsf{D}^j} = \big\langle \tfrac{j}{n}\, deg\, a, \tfrac{j}{n}\, deg\, b\big\rangle + \Gamma_{\mathsf{F}} = j\Gamma_{\mathsf{D}} + \Gamma_{\mathsf{F}}.$$

Now suppose instead that $j \nmid n$. Then, $gcd(j, n) = 1$, as j is prime. We have $\mathsf{D}^j = \big(a, b/\mathsf{F}\big)_n^j \sim \big(a, b^j/\mathsf{F}\big)_n$ in $\mathsf{Br}(\mathsf{F})$. Since j is prime to n, the images of $deg\, b$ and $deg(b^j)$ generate the same subgroup of $\Gamma_{\mathsf{F}}/n\Gamma_{\mathsf{F}}$. Hence,

$$\big|\langle deg\, a, deg(b^j)\rangle + n\Gamma_{\mathsf{F}} : n\Gamma_{\mathsf{F}}\big| = \big|\langle deg\, a, deg\, b\rangle + n\Gamma_{\mathsf{F}} : n\Gamma_{\mathsf{F}}\big| = n^2.$$

By Prop. 7.57(i) it follows that $\big(a, b^j/\mathsf{F}\big)_n$ is a graded division algebra totally ramified over F. Hence, $\mathsf{D}^j \cong_g \big(a, b^j/\mathsf{F}\big)_n$ and Prop. 7.57(i) says that

$$\Gamma_{\mathsf{D}^j} = \big\langle \tfrac{1}{n}\, deg\, a, \tfrac{j}{n}\, deg\, b\big\rangle + \Gamma_{\mathsf{F}} = \big\langle \tfrac{j}{n}\, deg\, a, \tfrac{j}{n}\, deg\, b\big\rangle + \Gamma_{\mathsf{F}} = j\Gamma_{\mathsf{D}} + \Gamma_{\mathsf{F}}.$$

When D is not a graded symbol algebra, we have by Cor. 7.53 and Cor. 7.56 $\mathsf{D} = \mathsf{S}_1 \otimes_{\mathsf{F}} \ldots \otimes_{\mathsf{F}} \mathsf{S}_r$ where each S_i is a graded symbol algebra totally ramified over F and

$$\Gamma_{\mathsf{D}}/\Gamma_{\mathsf{F}} = \Gamma_{\mathsf{S}_1}/\Gamma_{\mathsf{F}} \perp \ldots \perp \Gamma_{\mathsf{S}_r}/\Gamma_{\mathsf{F}}.$$

Then $\mathsf{D}^j \sim \mathsf{S}_1^j \otimes_{\mathsf{F}} \ldots \otimes_{\mathsf{F}} \mathsf{S}_r^j$ in $\mathsf{Br}(\mathsf{F})$, and we have just seen that each S_i^j is totally ramified over F with $\Gamma_{\mathsf{S}_i^j} = j\Gamma_{\mathsf{S}_i} + \Gamma_{\mathsf{F}} \subseteq \Gamma_{\mathsf{S}_i}$. Because the sum $\sum_{i=1}^r \Gamma_{\mathsf{S}_i}/\Gamma_{\mathsf{F}}$ is direct in $\mathbb{T}(\Gamma_{\mathsf{F}})$, the sum of subgroups $\sum_{i=1}^r \Gamma_{\mathsf{S}_i^j}/\Gamma_{\mathsf{F}}$ is also direct. By Prop. 7.57(ii), $\mathsf{S}_1^j \otimes_{\mathsf{F}} \ldots \otimes_{\mathsf{F}} \mathsf{S}_r^j$ is a graded division algebra totally ramified over F. Hence, $\mathsf{D}^j \cong_g \mathsf{S}_1^j \otimes_{\mathsf{F}} \ldots \otimes_{\mathsf{F}} \mathsf{S}_r^j$, and

$$\Gamma_{\mathsf{D}^j} = \sum_{i=1}^r \Gamma_{\mathsf{S}_i^j} = \sum_{i=1}^r j\Gamma_{\mathsf{S}_i} + \Gamma_{\mathsf{F}} = j\Gamma_{\mathsf{D}} + \Gamma_{\mathsf{F}}.$$

Since each D^j is totally ramified over F, it follows by induction on j using Prop. 7.59 and Prop. 7.15 that

$$(\Gamma_{\mathsf{D}^j}/\Gamma_{\mathsf{F}}, \overline{c}_{\mathsf{D}^j}) = \sum_{i=1}^{j} (\Gamma_{\mathsf{D}}/\Gamma_{\mathsf{F}}, \overline{c}_{\mathsf{D}}) = \big(j(\Gamma_{\mathsf{D}}/\Gamma_{\mathsf{F}}), \overline{c}_{\mathsf{D}}^{(j)}\big).$$

Because D^j is totally ramified over F, it follows that D^j is split if and only if $\Gamma_{\mathsf{D}^j} = \Gamma_{\mathsf{F}}$, which holds if and only if $j\Gamma_{\mathsf{D}} \subseteq \Gamma_{\mathsf{F}}$. Hence, $exp\, \mathsf{D} = exp(\Gamma_{\mathsf{D}}/\Gamma_{\mathsf{F}})$. □

7.3.2 The Brauer group of an inertially closed graded field

The graded field F is said to be *inertially closed* if it has no proper inertial extension, which amounts to saying that the field F_0 is separably closed. We assume throughout this subsection that F is inertially closed. Earlier results in this chapter will be applied to give a rather concrete description of $\mathsf{Br}(\mathsf{F})$, showing that it is completely determined by Γ_{F} and $char\, \mathsf{F}_0$. The preceding analysis of total ramification applies to *all* the graded division algebras over F:

Lemma 7.61. *If* F *is inertially closed and* D *is a graded division algebra with center* F*, then* D *is totally ramified over* F *and char* $\mathsf{F}_0 \nmid exp\, \mathsf{D}$.

Proof. Recall from Prop. 2.3(v) and Prop. 2.40 that D_0 is a division algebra with $Z(\mathsf{D}_0)$ Galois over F_0. Since F_0 is separably closed, it follows that $Z(\mathsf{D}_0) = \mathsf{F}_0$ and D_0 is split. Hence, $\mathsf{D}_0 = \mathsf{F}_0$, so D is totally ramified over F. Let $e = exp(\Gamma_{\mathsf{D}}/\Gamma_{\mathsf{F}}) = exp\, \mathsf{D}$ (see Prop. 7.60). The nondegeneracy of the canonical pairing $\overline{c}_{\mathsf{D}}$ implies that F_0 contains a primitive e-th root of unity (see Cor. 7.53); hence, $char\, \mathsf{F}_0 \nmid e$. □

Since F is inertially closed, the separably closed field F_0 then contains a primitive n-th root of unity for each integer n prime to $char\, \mathsf{F}_0$. By a suitable choice of primitive roots, we identify the group of roots of unity $\mu(\mathsf{F}_0)$ with the subgroup $\mathbb{T}' \subseteq \mathbb{T} = \mathbb{Q}/\mathbb{Z}$ of elements whose torsion is prime to $char\, \mathsf{F}_0$. Thus, we fix an isomorphism

$$\omega \colon \mathbb{T}' = \bigoplus_{p \neq char\, \mathsf{F}_0} \mathbb{T}_{(p)} \xrightarrow{\sim} \mu(\mathsf{F}_0),$$

where $\mathbb{T}_{(p)} = \mathbb{Z}[\frac{1}{p}]/\mathbb{Z} \cong \mathbb{Q}_p/\mathbb{Z}_p$. Fix an integer $n > 1$ prime to *char* F_0. The degree map induces a surjective group homomorphism

$$deg_n \colon \mathsf{F}^\times/\mathsf{F}^{\times n} \longrightarrow \Gamma_\mathsf{F}/n\Gamma_\mathsf{F}.$$

Lemma 7.62. *If* F *is inertially closed, then* deg_n *is an isomorphism.*

Proof. Suppose $x \in \mathsf{F}^\times$ represents an element in the kernel of deg_n, so $deg\, x = deg(y^n)$ for some $y \in \mathsf{F}^\times$. Then $xy^{-n} \in \mathsf{F}_0$. Since F_0 is separably closed, the polynomial $X^n - xy^{-n} \in \mathsf{F}_0[X]$ has a root $z \in \mathsf{F}_0$. It follows that $x = (yz)^n \in \mathsf{F}^{\times n}$. Therefore, deg_n is injective. Since it is clearly surjective, the lemma follows. □

As in §7.1.3, we consider the torsion group

$$\mathbb{T}(\wedge^2\Gamma_\mathsf{F}) = (\wedge^2\Gamma_\mathsf{F}) \otimes_\mathbb{Z} \mathbb{T} = \varinjlim\big(\tfrac{1}{n} \wedge^2 \Gamma_\mathsf{F}\big)/(\wedge^2\Gamma_\mathsf{F}).$$

Let $\mathbb{T}(\wedge^2\Gamma_\mathsf{F})'$ denote the subgroup of elements in $\mathbb{T}(\wedge^2\Gamma_\mathsf{F})$ whose torsion is prime to *char* F_0. Using Lemma 7.62, we define a map

$$\Sigma_\mathsf{F} \colon \mathbb{T}(\wedge^2\Gamma_\mathsf{F})' \longrightarrow \mathsf{Br}(\mathsf{F})$$

as follows: For a given integer $n > 1$ prime to *char* F_0 and for $\gamma,\ \delta \in \Gamma_\mathsf{F}$, we pick elements $c, d \in \mathsf{F}^\times$ such that $deg\, c = \gamma$, $deg\, d = \delta$, and consider the symbol algebra

$$(c, d/\mathsf{F})_{\omega(1/n),n}.$$

The Brauer class of this algebra depends linearly on c and d, and vanishes when c or d lies in $F^{\times n}$. Therefore, by Lemma 7.62, the Brauer class depends only on γ and δ, not on the choice of c and d, and it depends linearly on γ and δ modulo $n\Gamma_\mathsf{F}$. Moreover, we have for all $c \in \mathsf{F}^\times$

$$(c, c/\mathsf{F})_{\omega(1/n),n} = (c, -1/\mathsf{F})_{\omega(1/n),n}$$

(see Draxl [63, Lemma 11, p. 82]). We have $-1 \in \mathsf{F}_0^{\times n}$ since F_0 is separably closed,[3] hence the right term is a split central simple graded F-algebra. Thus, we have a well-defined group homomorphism $(\Gamma_\mathsf{F} \wedge \Gamma_\mathsf{F})/n(\Gamma_\mathsf{F} \wedge \Gamma_\mathsf{F}) \to \mathsf{Br}(\mathsf{F})$. Using the multiplication-by-n isomorphism

$${}_n\mathbb{T}(\wedge^2\Gamma_\mathsf{F}) = \big(\tfrac{1}{n} \wedge^2 \Gamma_\mathsf{F}\big)/ \wedge^2 \Gamma_\mathsf{F} \xrightarrow{\sim} (\Gamma_\mathsf{F} \wedge \Gamma_\mathsf{F})/n(\Gamma_\mathsf{F} \wedge \Gamma_\mathsf{F}),$$

we obtain a group homomorphism

$$\Sigma_n \colon {}_n\mathbb{T}(\wedge^2\Gamma_\mathsf{F}) \longrightarrow \mathsf{Br}(\mathsf{F}) \ \text{ given by } \ \tfrac{1}{n}(\gamma \wedge \delta) + \wedge^2\Gamma_\mathsf{F} \mapsto \big[(c, d/\mathsf{F})_{\omega(1/n),n}\big], \tag{7.42}$$

for any $c, d \in \mathsf{F}^\times$ with $deg\, c = \gamma$ and $deg\, d = \delta$. For any integer $m > 1$ prime to *char* F_0, we have $\omega(\frac{1}{mn})^m = \omega(\frac{1}{n})$, hence by a known property of symbol algebras (see Draxl [63, Lemma 6, p. 80])

$$\big[(c, d/\mathsf{F})_{\omega(1/n),n}\big] = \big[(c^m, d/\mathsf{F})_{\omega(1/(mn)),mn}\big] = m\,\Sigma_{mn}\big(\tfrac{1}{mn}(\gamma \wedge \delta) + \wedge^2\Gamma_\mathsf{F}\big).$$

---

[3] Of course, if n is odd it suffices to observe that $-1 = (-1)^n$.

Hence, the following diagram is commutative:

$$\begin{array}{ccc} {}_n\mathbb{T}(\wedge^2\Gamma_{\mathsf{F}}) & \hookrightarrow & {}_{mn}\mathbb{T}(\wedge^2\Gamma_{\mathsf{F}}) \\ & {\scriptstyle\Sigma_n}\searrow \quad \swarrow{\scriptstyle\Sigma_{mn}} & \\ & \mathsf{Br}(\mathsf{F}) & \end{array}$$

Thus, the homomorphisms Σ_n are compatible with the inductive system $\big({}_n\mathbb{T}(\wedge^2\Gamma_{\mathsf{F}})\big)_{\mathit{char}\,\mathsf{F}_0\nmid n}$; they define a group homomorphism $\Sigma_{\mathsf{F}}\colon \mathbb{T}(\wedge^2\Gamma_{\mathsf{F}})' \to \mathsf{Br}(\mathsf{F})$ defined on generators as in (7.42).

Now consider the group $\mathit{Symp}\big(\mathbb{T}(\Gamma_{\mathsf{F}})\big)$ defined as in §7.1.3; we define a map

$$\Pi_{\mathsf{F}}\colon \mathsf{Br}(\mathsf{F}) \longrightarrow \mathit{Symp}\big(\mathbb{T}(\Gamma_{\mathsf{F}})\big)$$

as follows: For any central graded division algebra D over F, we have $\Gamma_{\mathsf{D}} \subseteq \mathbb{H}(\Gamma_{\mathsf{F}}) = \Gamma_{\mathsf{F}} \otimes_{\mathbb{Z}} \mathbb{Q}$, hence $\Gamma_{\mathsf{D}}/\Gamma_{\mathsf{F}} \subseteq \mathbb{T}(\Gamma_{\mathsf{F}})$. Because D is totally ramified over F by Lemma 7.61, the canonical pairing $\overline{c}_{\mathsf{D}}$ on $\Gamma_{\mathsf{D}}/\Gamma_{\mathsf{F}}$ turns $\Gamma_{\mathsf{D}}/\Gamma_{\mathsf{F}}$ into a symplectic module[4], which is thus an element in $\mathit{Symp}\big(\mathbb{T}(\Gamma_{\mathsf{F}})\big)$. We define

$$\Pi_{\mathsf{F}}\colon [\mathsf{D}] \mapsto (\Gamma_{\mathsf{D}}/\Gamma_{\mathsf{F}}, \overline{c}_{\mathsf{D}}) \in \mathit{Symp}\big(\mathbb{T}(\Gamma_{\mathsf{F}})\big).$$

It follows from Lemma 7.61 and Prop. 7.59 that Π_{F} is a group homomorphism. When $\mathsf{D} \neq \mathsf{F}$, the group $\Gamma_{\mathsf{D}}/\Gamma_{\mathsf{F}}$ is nontrivial. Hence, $\mathit{ker}\,\Pi_{\mathsf{F}} = \{[\mathsf{F}]\}$, which shows that Π_{F} is injective. Since $\mathsf{Br}(\mathsf{F})$ has no $\mathit{char}\,\mathsf{F}_0$-primary torsion by Lemma 7.61 or Lemma 6.19, the image of Π_{F} lies in the subgroup $\mathit{Symp}\big(\mathbb{T}(\Gamma_{\mathsf{F}})\big)'$ of elements with torsion prime to $\mathit{char}\,\mathsf{F}_0$. The effect of Π_{F} on any central simple graded F-algebra spanned by a homogeneous F-armature $\mathcal{A}$ is given as follows: The degree function $\mathit{deg}\colon \mathsf{A}^{\times}_{\mathrm{hom}} \to \mathbb{H}(\Gamma_{\mathsf{F}})$ induces a map $\overline{\mathit{deg}}\colon \mathcal{A} \to \mathbb{T}(\Gamma_{\mathsf{F}})$. By Prop. 7.58,

$$\Pi_{\mathsf{F}}[\mathsf{A}] = \overline{\mathit{deg}}_{\diamond}(\mathcal{A}). \tag{7.43}$$

Remark 7.63. The fact that Π_{F} is a monomorphism implies that a graded division algebra D with center an inertially closed graded field F is determined up to graded isomorphism by $\Gamma_{\mathsf{D}}/\Gamma_{\mathsf{F}}$ and the canonical pairing $\overline{c}_{\mathsf{D}}$. This can also be proved directly without invoking Π_{F} by using the decomposition of D into a tensor product of graded symbol algebras.

Recall from Th. 7.22 the group isomorphism

$$\Upsilon_{\Gamma_{\mathsf{F}}}\colon \mathbb{T}(\wedge^2\Gamma_{\mathsf{F}}) \xrightarrow{\sim} \mathit{Symp}\big(\mathbb{T}(\Gamma_{\mathsf{F}})\big).$$

We use the same notation to designate the restriction of this isomorphism to the subgroups of elements with torsion prime to $\mathit{char}\,\mathsf{F}_0$.

[4] We use the isomorphism ω to view the values of the canonical pairing as elements in $\mathbb{T}$.

Theorem 7.64. *For every inertially closed graded field* F, *the maps* Σ_{F} *and* Π_{F} *are group isomorphisms, and the following diagram is commutative:*

$$\begin{array}{ccc} \mathbb{T}(\wedge^2\Gamma_{\mathsf{F}})' & \xrightarrow{\quad\Sigma_{\mathsf{F}}\quad} & \mathsf{Br}(\mathsf{F}) \\ & \searrow^{\Upsilon_{\Gamma_{\mathsf{F}}}} \quad {}^{\Pi_{\mathsf{F}}}\swarrow & \\ & \mathit{Symp}\big(\mathbb{T}(\Gamma_{\mathsf{F}})\big)' & \end{array}$$

Proof. We first show that the diagram commutes. Take a generator $\frac{1}{n}(\gamma \wedge \delta) + \wedge^2\Gamma_{\mathsf{F}}$ of $\mathbb{T}(\wedge^2\Gamma_{\mathsf{F}})'$, where $\gamma, \delta \in \Gamma_{\mathsf{F}}$ and n is prime to *char* F_0. By Prop. 7.20,

$$\Upsilon_{\Gamma_{\mathsf{F}}}\big(\tfrac{1}{n}(\gamma \wedge \delta) + \wedge^2\Gamma_{\mathsf{F}}\big) \;=\; f(\gamma,\delta)_\diamond(\mathbb{S}_n), \tag{7.44}$$

where $\mathbb{S}_n = \big((\frac{1}{n}\mathbb{Z}/\mathbb{Z})^2, s_n\big)$ is the standard symplectic module of exponent n and $f(\gamma,\delta)$ is defined in (7.13).

Working around the diagram along the other path, pick $c,\ d \in \mathsf{F}^\times$ such that $\mathit{deg}\, c = \gamma$ and $\mathit{deg}\, d = \delta$, and consider the graded symbol algebra $\mathsf{S} = \big(c, d/\mathsf{F}\big)_{\omega(1/n),n}$. We have (see (7.42))

$$\Sigma_{\mathsf{F}}\big(\tfrac{1}{n}(\gamma \wedge \delta) + \wedge^2\Gamma_{\mathsf{F}}\big) \;=\; [\mathsf{S}] \,\in\, \mathsf{Br}(\mathsf{F}). \tag{7.45}$$

Let i, j be the standard homogeneous generators of S with $i^n = a$, $j^n = b$, $ij = \omega(\frac{1}{n})ji$, $\mathit{deg}\, i = \frac{1}{n}\,\mathit{deg}\, c$, and $\mathit{deg}\, j = \frac{1}{n}\,\mathit{deg}\, c$. Let $\mathcal{S} = \langle [i], [j]\rangle \subseteq \mathsf{S}_{\mathrm{hom}}^\times/\mathsf{F}^\times$ be the usual homogeneous F-armature spanning S, and let $b_{\mathcal{S}}$ be the commutator pairing on $\mathcal{S}$. There is an isometry of symplectic modules $\mathbb{S}_n \cong (\mathcal{S}, b_{\mathcal{S}})$ given explicitly by the group homomorphism

$$\psi\colon (\tfrac{1}{n}\mathbb{Z}/\mathbb{Z})^2 \longrightarrow \mathcal{S}, \qquad (\tfrac{1}{n} + \mathbb{Z}, 0) \mapsto [i],\ (0, \tfrac{1}{n} + \mathbb{Z}) \mapsto [j].$$

The map ψ is an isometry from s_n to $b_{\mathcal{S}}$ (given the identification of $\omega(\frac{1}{n})$ with $\frac{1}{n} + \mathbb{Z}$) since $s_n\big((\frac{1}{n} + \mathbb{Z}, 0), (0, \frac{1}{n} + \mathbb{Z})\big) = \frac{1}{n} + \mathbb{Z}$ and $b_{\mathcal{S}}([i],[j]) = \omega(\frac{1}{n})$. Thus, $\psi_\diamond(\mathbb{S}_n) = (\mathcal{S}, b_{\mathcal{S}})$. The degree map $\mathit{deg}\colon \mathsf{S}_{\mathrm{hom}}^\times \to \mathbb{H}(\Gamma_{\mathsf{F}})$ induces $\overline{\mathit{deg}}\colon \mathcal{S} \to \mathbb{T}(\Gamma_{\mathsf{F}})$ determined by $\overline{\mathit{deg}}[i] = \frac{1}{n}\gamma + \Gamma_{\mathsf{F}}$ and $\overline{\mathit{deg}}[j] = \frac{1}{n}\delta + \Gamma_{\mathsf{F}}$. The formula for the map $f(\gamma,\delta)$ in (7.13) yields

$$\overline{\mathit{deg}} \circ \psi \;=\; f(\gamma,\delta)|_{(\frac{1}{n}\mathbb{Z}/\mathbb{Z})^2}.$$

From the functorial property of *Symp* (see Prop. 7.10) together with (7.43), it follows that

$$f(\gamma,\delta)_\diamond(\mathbb{S}_n) \;=\; (\overline{\mathit{deg}} \circ \psi)_\diamond(\mathbb{S}_n) \;=\; \overline{\mathit{deg}}_\diamond \circ \psi_\diamond(\mathbb{S}_n) \;=\; \overline{\mathit{deg}}_\diamond(\mathcal{S}, b_{\mathcal{S}}) \;=\; \Pi_{\mathsf{F}}[\mathsf{S}].$$

This combined with (7.44) and (7.45) shows that the diagram of the theorem is commutative.

Since the diagram is commutative and $\Upsilon_{\Gamma_{\mathsf{F}}}$ is an isomorphism by Th. 7.22, the map Π_{F} must be surjective. We noted with the definition that Π_{F} is a monomorphism; hence, it is an isomorphism. Therefore, $\Sigma_{\mathsf{F}} = \Pi_{\mathsf{F}}^{-1} \circ \Upsilon_{\Gamma_{\mathsf{F}}}$, and this composition of isomorphisms is also an isomorphism. $\square$

To conclude this section, we show that the maps Σ_{F} and Π_{F} are functorial, and use this property to obtain information on the behavior of central graded division algebras over inertially closed fields under scalar extension.

Proposition 7.65. *Let K/F be an extension of inertially closed graded fields. The following diagram, where the vertical maps are the scalar extension maps, is commutative:*

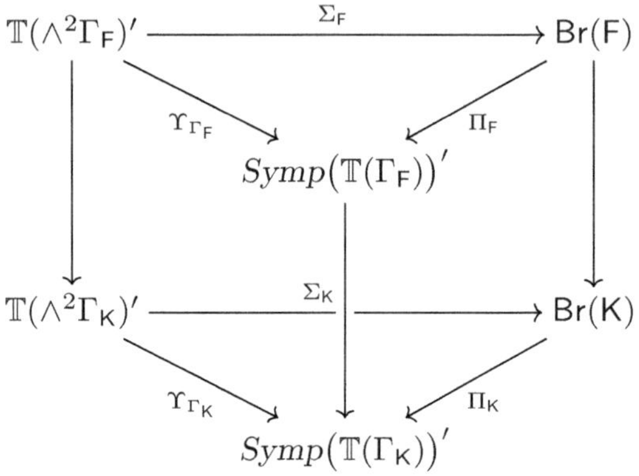

The maps Σ_{F} and Σ_{K} are understood to be defined with the same isomorphism $\omega\colon \mathbb{T}' \xrightarrow{\sim} \mu(\mathsf{F}_0) = \mu(\mathsf{K}_0)$.

Proof. Functoriality of the map Υ was proved in Th. 7.22. To see that Σ is functorial, choose γ, $\delta \in \Gamma_{\mathsf{F}}$ and an integer $n > 1$ prime to $\mathit{char}\,\mathsf{F}_0$, and let $c, d \in \mathsf{F}^\times$ satisfy $\mathit{deg}\, c = \gamma$ and $\mathit{deg}\, d = \delta$. We have

$$\Sigma_{\mathsf{F}}\big(\tfrac{1}{n}(\gamma \wedge \delta) + \wedge^2\Gamma_{\mathsf{F}}\big) \;=\; [(c, d/\mathsf{F})_{\omega(1/n),n}]$$

and

$$[(c, d/\mathsf{F})_{\omega(1/n),n} \otimes_{\mathsf{F}} \mathsf{K}] \;=\; \Sigma_{\mathsf{K}}\big(\tfrac{1}{n}(\gamma \wedge \delta) + \wedge^2\Gamma_{\mathsf{K}}\big).$$

This equation shows that Σ is functorial. Since Th. 7.64 shows that $\Pi_{\mathsf{F}} = \Upsilon_{\Gamma_{\mathsf{F}}} \circ \Sigma_{\mathsf{F}}^{-1}$ (and similarly over K), functoriality of Π follows from that of Υ and Σ. $\square$

Corollary 7.66. *Let K/F be an extension of inertially closed graded fields, let D be a central graded division algebra over F, and let E be the central graded division algebra over K associated to $\mathsf{D} \otimes_{\mathsf{F}} \mathsf{K}$. Consider $\Gamma_{\mathsf{D}} \cap \Gamma_{\mathsf{K}} \subseteq \Gamma_{\mathsf{D}}$ and its orthogonal $(\Gamma_{\mathsf{D}} \cap \Gamma_{\mathsf{K}})^\perp \subseteq \Gamma_{\mathsf{D}}$ for the canonical pairing c_{D}. We have*

$$\Gamma_{\mathsf{E}} \;=\; (\Gamma_{\mathsf{D}} \cap \Gamma_{\mathsf{K}})^\perp + \Gamma_{\mathsf{K}}$$

and

$$c_{\mathsf{E}}(\gamma, \delta) \;=\; c_{\mathsf{D}}(\gamma, \delta) \qquad \textit{for } \gamma,\ \delta \in (\Gamma_{\mathsf{D}} \cap \Gamma_{\mathsf{K}})^\perp.$$

In particular,

(i) K *splits* D *if and only if* $(\Gamma_{\mathsf{D}} \cap \Gamma_{\mathsf{K}})^{\perp} \subseteq \Gamma_{\mathsf{K}}$*;*
(ii) $\mathsf{D} \otimes_{\mathsf{F}} \mathsf{K}$ *is a division algebra if and only if* $\Gamma_{\mathsf{D}} \cap \Gamma_{\mathsf{K}} = \Gamma_{\mathsf{F}}$.

Proof. Let $\varepsilon\colon \mathbb{T}(\Gamma_{\mathsf{F}}) \to \mathbb{T}(\Gamma_{\mathsf{K}})$ be the scalar extension map, which is induced by the inclusion $\Gamma_{\mathsf{F}} \subseteq \Gamma_{\mathsf{K}}$. By the functoriality of Π, we have

$$(\Gamma_{\mathsf{E}}/\Gamma_{\mathsf{K}}, \overline{c}_{\mathsf{E}}) \;=\; \varepsilon_{\diamond}(\Gamma_{\mathsf{D}}/\Gamma_{\mathsf{F}}, \overline{c}_{\mathsf{D}}),$$

where $\varepsilon_{\diamond}$ is the canonical map induced by ε, see §7.1.2. The definition of $\varepsilon_{\diamond}$ yields the description above for Γ_{E} and c_{E}, since $(\mathit{ker}\,\varepsilon) \cap (\Gamma_{\mathsf{D}}/\Gamma_{\mathsf{F}}) = (\Gamma_{\mathsf{D}} \cap \Gamma_{\mathsf{K}})/\Gamma_{\mathsf{F}}$.

Now, K splits D if and only if $\Gamma_{\mathsf{E}} = \Gamma_{\mathsf{K}}$, and (i) follows. For (ii), observe that $\mathsf{D} \otimes_{\mathsf{F}} \mathsf{K}$ is a division algebra if and only if $|\Gamma_{\mathsf{E}}{:}\Gamma_{\mathsf{K}}| = |\Gamma_{\mathsf{D}}{:}\Gamma_{\mathsf{F}}|$. From the description of Γ_{E}, we have

$$|\Gamma_{\mathsf{E}}{:}\Gamma_{\mathsf{K}}| \;=\; \big|(\Gamma_{\mathsf{D}} \cap \Gamma_{\mathsf{K}})^{\perp} : (\Gamma_{\mathsf{D}} \cap \Gamma_{\mathsf{K}}) \cap (\Gamma_{\mathsf{D}} \cap \Gamma_{\mathsf{K}})^{\perp}\big|.$$

Therefore, $|\Gamma_{\mathsf{E}}{:}\Gamma_{\mathsf{K}}| = |\Gamma_{\mathsf{D}}{:}\Gamma_{\mathsf{F}}|$ if and only if $\Gamma_{\mathsf{D}} \cap \Gamma_{\mathsf{K}} = \Gamma_{\mathsf{F}}$. □

7.3.3 The top level of the Brauer group filtration

Let F be any graded field, and let F_{in} be its inertial closure. Thus, $\mathsf{F}_{in} = (\mathsf{F}_0)_{sep} \otimes_{\mathsf{F}_0} \mathsf{F}$ and $\Gamma_{\mathsf{F}_{in}} = \Gamma_{\mathsf{F}}$. In Th. 6.24 and Prop. 6.27 we have described the inertially split part $\mathsf{Br}_{is}(\mathsf{F}) = \mathsf{Br}(\mathsf{F}_{in}/\mathsf{F})$ of $\mathsf{Br}(\mathsf{F})$. Since F_{in} is inertially closed, the structure of $\mathsf{Br}(\mathsf{F}_{in})$ was described in the preceding subsection. We now put this information together to get a picture of the whole of $\mathsf{Br}(\mathsf{F})$.

We will be considering subgroups of abelian groups with torsion determined by roots of unity in a field. For this we will use the following special notation: Let A be an abelian group and let K be a field. We write $\mu_n \subseteq K$ if K contains a primitive n-th root of unity. Let $\mu(K)$ denote the group of all roots of unity in K. Then let ${}_{\mu(K)}A$ denote the "$\mu(K)$-torsion subgroup of A," i.e.,

$${}_{\mu(K)}A \;=\; \sum_{\mu_n \subseteq K} {}_nA. \tag{7.46}$$

Thus, the sum is taken over all $n \in \mathbb{N}$ such that K contains a primitive n-th root of unity. Note that for any prime p, the p-primary component of ${}_{\mu(K)}A$ is

$${}_{\mu(K)}A_{(p)} \;=\; \begin{cases} {}_{p^m}A, & \text{if } m \text{ is maximal such that } \mu_{p^m} \subseteq K, \\ A_{(p)}, & \text{if } \mu_{p^n} \subseteq K \text{ for every } n \in \mathbb{N}. \end{cases}$$

Proposition 7.67. *Let* F *be a graded field. The image of the scalar extension map* $\mathrm{ext}_{\mathsf{F}_{in}/\mathsf{F}}\colon \mathsf{Br}(\mathsf{F}) \to \mathsf{Br}(\mathsf{F}_{in})$ *is the* $\mu(\mathsf{F}_0)$*-torsion subgroup* ${}_{\mu(\mathsf{F}_0)}\mathsf{Br}(\mathsf{F}_{in})$*. Moreover, for any graded division algebra* E *with center* F_{in}*, if* $[\mathsf{E}]$ *lies in* *im*$(ext_{\mathsf{F}_{in}/\mathsf{F}})$, *then there is a graded division algebra* B *totally ramified over its center* F *such that* $\mathsf{B} \otimes_{\mathsf{F}} \mathsf{F}_{in} \cong_g \mathsf{E}$.

The proof will use the following lemma:

Lemma 7.68. *Let* T *be a graded division algebra with center* F. *Then,* T *is totally ramified over* F *if and only if* $\mathsf{T} \otimes_{\mathsf{F}} \mathsf{F}_{in}$ *is a graded division algebra. When this occurs,*

$$(\Gamma_{\mathsf{T}}/\Gamma_{\mathsf{F}}, \overline{c}_{\mathsf{T}}) = (\Gamma_{\mathsf{T}\otimes_{\mathsf{F}}\mathsf{F}_{in}}/\Gamma_{\mathsf{F}}, \overline{c}_{\mathsf{T}\otimes_{\mathsf{F}}\mathsf{F}_{in}}) \qquad \textit{as symplectic modules.}$$

Proof. Let $\mathsf{T}' = \mathsf{T} \otimes_{\mathsf{F}} \mathsf{F}_{in}$. Since F_{in} is inertial over F, by Prop. 2.57 $\mathsf{T}'_0 \cong \mathsf{T}_0 \otimes_{\mathsf{F}_0} \mathsf{F}_{in,0}$. Thus, T is totally ramified over F if and only if $\mathsf{T}_0 = \mathsf{F}_0$, if and only if $\mathsf{T}'_0 = \mathsf{F}_{in,0}$. If this occurs, then T' is a graded division algebra by Cor. 2.43. Conversely, if T' is a graded division algebra, then it is totally ramified over the inertially closed graded field F_{in} by Lemma 7.61; so, $\mathsf{T}'_0 = \mathsf{F}_{in,0}$ and hence T is totally ramified over F. Assume this holds. Then $\Gamma_{\mathsf{T}'} = \Gamma_{\mathsf{T}}$ by Prop. 2.57. Moreover, since $\overline{c}_{\mathsf{T}'}$ can be computed using any representatives in $\mathsf{T}'^{\times}$, those representatives can be chosen from $\mathsf{T}^{\times}$. Hence, the canonical pairing $\overline{c}_{\mathsf{T}'}$ coincides with $\overline{c}_{\mathsf{T}}$. □

Proof of Prop. 7.67. We first show that *im*$(ext_{\mathsf{F}_{in}/\mathsf{F}}) \subseteq {}_{\mu(\mathsf{F}_0)}\mathsf{Br}(\mathsf{F}_{in})$ by using the Galois action on $\mathsf{Br}(\mathsf{F}_{in})$. Let $G = \mathcal{G}(\mathsf{F}_{in}/\mathsf{F})$. For any $\sigma \in G$ and any central simple graded F_{in}-algebra A define the twisted graded algebra ${}^{\sigma}\mathsf{A}$ as follows:

$${}^{\sigma}\mathsf{A} = \{{}^{\sigma}a \mid a \in \mathsf{A}\}, \qquad ({}^{\sigma}\mathsf{A})_{\gamma} = \{{}^{\sigma}a \mid a \in \mathsf{A}_{\gamma}\} \quad \text{for } \gamma \in \Gamma,$$

with the operations

$${}^{\sigma}a + {}^{\sigma}b = {}^{\sigma}(a+b) \quad \text{and} \quad {}^{\sigma}a \cdot {}^{\sigma}b = {}^{\sigma}(ab) \quad \text{for } a,\, b \in \mathsf{A},$$

and

$$\lambda \cdot {}^{\sigma}a = {}^{\sigma}(\sigma^{-1}(\lambda)a) \qquad \text{for } a \in \mathsf{A} \text{ and } \lambda \in \mathsf{F}_{in}.$$

Since σ is a graded automorphism of F_{in}, the ring ${}^{\sigma}\mathsf{A}$ is a central simple graded F-algebra. There is a well-defined group action of G on $\mathsf{Br}(\mathsf{F}_{in})$ given by $\sigma[\mathsf{A}] = [{}^{\sigma}\mathsf{A}]$. Just as in the ungraded case (cf. Draxl [63, (7), p. 51]), it is easy to see that if $[\mathsf{A}] \in$ *im*$(ext_{\mathsf{F}_{in}/\mathsf{F}})$ then $\sigma[\mathsf{A}] = [\mathsf{A}]$.

Now suppose D is a graded division algebra with center F_{in} satisfying $[\mathsf{D}] \in$ *im*$(ext_{\mathsf{F}_{in}/\mathsf{F}})$. Since F_{in} is inertially closed, D must be totally ramified over F_{in}, by Lemma 7.61. So, by Cor. 7.53, $\mathsf{D} = \mathsf{S}_1 \otimes_{\mathsf{F}_{in}} \ldots \otimes_{\mathsf{F}_{in}} \mathsf{S}_k$ where each S_i is a graded symbol algebra totally ramified over F_{in}, say $\mathsf{S}_i = \big(a_i, b_i/\mathsf{F}_{in}\big)_{\omega_i, r_i}$ with $a_i, b_i \in \mathsf{F}_{in}^{\times}$. There is then an orthogonal sum decomposition,

$$(\Gamma_{\mathsf{D}}/\Gamma_{\mathsf{F}}, \overline{c}_{\mathsf{D}}) = (\Gamma_{\mathsf{S}_1}/\Gamma_{\mathsf{F}}, \overline{c}_{\mathsf{S}_1}) \perp \ldots \perp (\Gamma_{\mathsf{S}_k}/\Gamma_{\mathsf{F}}, \overline{c}_{\mathsf{S}_k}).$$

For any $\sigma \in G$, since $\sigma[\mathsf{D}] = [\mathsf{D}]$, we have ${}^{\sigma}\mathsf{D} \cong_g \mathsf{D}$, hence $\Gamma_{{}^{\sigma}\mathsf{D}} = \Gamma_{\mathsf{D}}$ and $\overline{c}_{{}^{\sigma}\mathsf{D}} = \overline{c}_{\mathsf{D}}$. For each i, we have

$${}^{\sigma}\mathsf{S}_i \cong_g \big(\sigma(a_i), \sigma(b_i)/\mathsf{F}_{in}\big)_{\sigma(\omega_i), r_i}.$$

Note that $deg\,\sigma(a_i) = deg\,a_i$ and $deg\,\sigma(b_i) = deg\,b_i$, as σ is a graded automorphism of F_{in}; hence, $\Gamma_{^\sigma\mathsf{S}_i} = \Gamma_{\mathsf{S}_i}$. Since ${}^\sigma\mathsf{D} = {}^\sigma\mathsf{S}_1 \otimes_{\mathsf{F}_{in}} \dots \otimes_{\mathsf{F}_{in}} {}^\sigma\mathsf{S}_k$, we have

$$\overline{c}_{^\sigma\mathsf{S}_i} = \overline{c}_{^\sigma\mathsf{D}}\big|_{(\Gamma_{^\sigma\mathsf{S}_i}/\Gamma_\mathsf{F})^2} = \overline{c}_\mathsf{D}\big|_{(\Gamma_{\mathsf{S}_i}/\Gamma_\mathsf{F})^2} = \overline{c}_{\mathsf{S}_i}.$$

Hence,

$$\begin{aligned}\sigma(\omega_i) &= \overline{c}_{^\sigma\mathsf{S}_i}\big(\tfrac{1}{r_i}\, deg\,\sigma(a_i) + \Gamma_\mathsf{F}, \tfrac{1}{r_i}\, deg\,\sigma(b_i) + \Gamma_\mathsf{F}\big)\\ &= \overline{c}_{\mathsf{S}_i}\big(\tfrac{1}{r_i}\, deg\,a_i + \Gamma_\mathsf{F}, \tfrac{1}{r_i}\, deg\,b_i + \Gamma_\mathsf{F}\big) = \omega_i.\end{aligned}$$

Thus, $\sigma(\omega_i) = \omega_i$ for all $\sigma \in G$; so $\omega_i \in \mathsf{F}_0$ for $i = 1, 2, \dots, k$. Since ω_i is a primitive r_i-th root of unity, we have ${}_{r_i}\mathsf{Br}(\mathsf{F}_{in}) \subseteq {}_{\mu(\mathsf{F}_0)}\mathsf{Br}(\mathsf{F}_{in})$. As $[\mathsf{S}_i] \in {}_{r_i}\mathsf{Br}(\mathsf{F}_{in})$ for each i, it follows that $[\mathsf{D}] = [\mathsf{S}_1] \cdot \ldots \cdot [\mathsf{S}_k] \in {}_{\mu(\mathsf{F}_0)}\mathsf{Br}(\mathsf{F}_{in})$. Thus, $im(ext_{\mathsf{F}_{in}/\mathsf{F}}) \subseteq {}_{\mu(\mathsf{F}_0)}\mathsf{Br}(\mathsf{F}_{in})$. (Another proof of this inclusion will be available after Prop. 8.16(v) below is proved, by showing that for a graded division algebra B with center F and canonical subalgebras Z and C as in that proposition, we have $[\mathsf{B} \otimes_\mathsf{F} \mathsf{F}_{in}] = [\mathsf{C} \otimes_\mathsf{Z} \mathsf{F}_{in}]$ in $\mathsf{Br}(\mathsf{F}_{in})$.)

For the reverse inclusion, take any graded division algebra E with center F_{in} such that $[\mathsf{E}] \in {}_{\mu(\mathsf{F}_0)}\mathsf{Br}(\mathsf{F}_{in})$; so F_0 contains a primitive $exp\,\mathsf{E}$-th root of unity. Since F_{in} is inertially closed, E must be totally ramified over F_{in} by Lemma 7.61. So, we can express $\mathsf{E} = \mathsf{T}_1 \otimes_{\mathsf{F}_{in}} \dots \otimes_{\mathsf{F}_{in}} \mathsf{T}_m$ where each T_i is a totally ramified graded symbol algebra, say $\mathsf{T}_i = \big(c_i, d_i/\mathsf{F}_{in}\big)_{\zeta_i, s_i}$. Then, by Prop. 7.57(i) and Prop. 7.60,

$$s_i = exp(\Gamma_{\mathsf{T}_i}/\Gamma_\mathsf{F}) \mid exp(\Gamma_\mathsf{E}/\Gamma_\mathsf{F}) = exp\,\mathsf{E} \qquad \text{for } i = 1, 2, \dots, m.$$

Hence, ζ_i, which is a primitive s_i-th root of unity, lies in F_0. Because $\Gamma_{\mathsf{F}_{in}} = \Gamma_\mathsf{F}$, for each i there are c_i', $d_i' \in \mathsf{F}^\times$ and y_i, $z_i \in \mathsf{F}_{in,0}^\times$ with $c_i = c_i' y_i$ and $d_i = d_i' z_i$. Let $\mathsf{T}_i' = \big(c_i', d_i'/\mathsf{F}_{in}\big)_{\zeta_i, s_i}$. Since $deg\,y_i = deg\,z_i = 0$ and F_{in} is inertially closed, we have y_i, $z_i \in \mathsf{F}_{in}^{\times s_i}$, hence $c_i \equiv c_i'$ $(mod\ \mathsf{F}_{in}^{\times s_i})$ and $d_i \equiv d_i'$ $(mod\ \mathsf{F}_{in}^{\times s_i})$, and therefore $\mathsf{T}_i \cong_g \mathsf{T}_i'$. Let

$$\mathsf{B} = \big(c_1', d_1'/\mathsf{F}\big)_{\zeta_1, s_1} \otimes_\mathsf{F} \dots \otimes_\mathsf{F} \big(c_m', d_m'/\mathsf{F}\big)_{\zeta_m, s_m},$$

which is a central simple graded F-algebra. Then,

$$\mathsf{B} \otimes_\mathsf{F} \mathsf{F}_{in} \cong_g \mathsf{T}_1' \otimes_{\mathsf{F}_{in}} \dots \otimes_{\mathsf{F}_{in}} \mathsf{T}_m' \cong_g \mathsf{T}_1 \otimes_{\mathsf{F}_{in}} \dots \otimes_{\mathsf{F}_{in}} \mathsf{T}_m \cong_g \mathsf{E}.$$

Hence, $[\mathsf{E}] \in im(ext_{\mathsf{F}_{in}/\mathsf{F}})$. Furthermore, B is a graded division algebra as it embeds in the graded division algebra E, and Lemma 7.68 shows that B is totally ramified over F. □

Theorem 7.69. *For any graded field* F, *there is a split short exact sequence*

$$0 \longrightarrow \mathsf{Br}_{is}(\mathsf{F}) \longrightarrow \mathsf{Br}(\mathsf{F}) \xrightarrow{\ \Phi_\mathsf{F}\ } {}_{\mu(\mathsf{F}_0)}\mathbb{T}(\wedge^2\Gamma_\mathsf{F}) \longrightarrow 0,$$

where Φ_F *is the composition of the scalar extension map* $\mathrm{ext}_{\mathsf{F}_{in}/\mathsf{F}}$ *with the restriction of* $\Sigma_{\mathsf{F}_{in}}^{-1}$ *to* ${}_{\mu(\mathsf{F}_0)}\mathsf{Br}(\mathsf{F}_{in})$.

See Cor. 8.44 for an explicit description of $\Phi_{\mathsf{F}}([\mathsf{D}])$, for D a central graded division algebra over F. The exact sequence is canonical but not canonically split. The construction of the splitting map will use the following lemma:

Lemma 7.70. *Let Γ be a torsion-free abelian group and p a prime number. Let $(\varepsilon_i)_{i\in I}$ be a family of elements in Γ such that $(\varepsilon_i+p\Gamma)_{i\in I}$ is a base of $\Gamma/p\Gamma$ as a vector space over the field $\mathbb{F}_p$ with p elements. For every integer $n \geq 1$, $\frac{1}{p^n}\Gamma/\Gamma$ is a free $\mathbb{Z}/p^n\mathbb{Z}$-module with base $\left(\frac{1}{p^n}\varepsilon_i + \Gamma\right)_{i\in I}$.*

Proof. We use induction on n. If $n = 1$, the result is clear since multiplication by p yields an isomorphism

$$\tfrac{1}{p}\Gamma/\Gamma \cong \Gamma/p\Gamma.$$

Suppose the lemma holds for $n-1$. For each $\gamma \in \Gamma$, the element $\frac{1}{p^{n-1}}\gamma + \Gamma$ has a representation of the form

$$\tfrac{1}{p^{n-1}}\gamma + \Gamma = \sum_{i\in I}(\ell_i + p^{n-1}\mathbb{Z}) \cdot \left(\tfrac{1}{p^{n-1}}\varepsilon_i + \Gamma\right)$$

for some $\ell_i \in \mathbb{Z}$ such that $\{i \in I \mid \ell_i \neq 0\}$ is finite. We may then define $\delta \in \Gamma$ by the equation

$$\tfrac{1}{p^{n-1}}\left(\gamma - \sum_{i\in I}\ell_i\varepsilon_i\right) = \delta.$$

The hypothesis yields a family of integers $(m_i)_{i\in I}$ such that

$$\delta + p\Gamma = \sum_{i\in I}(m_i + p\mathbb{Z}) \cdot (\varepsilon_i + p\Gamma) \quad \text{and} \quad \{i \in I \mid m_i \neq 0\} \text{ is finite.}$$

Then,

$$\tfrac{1}{p^n}\gamma + \Gamma = \sum_{i\in I}(\ell_i + p^{n-1}m_i + p^n\mathbb{Z}) \cdot \left(\tfrac{1}{p^n}\varepsilon_i + \Gamma\right);$$

moreover, the set $\{i \in I \mid \ell_i + p^{n-1}m_i \neq 0\}$ is finite, since it lies in $\{i \in I \mid \ell_i \neq 0\} \cup \{i \in I \mid m_i \neq 0\}$.

Now, suppose $(\ell_i)_{i\in I}$ is a family of integers such that $\{i \in I \mid \ell_i \neq 0\}$ is finite and

$$\sum_{i\in I}(\ell_i + p^n\mathbb{Z}) \cdot \left(\tfrac{1}{p^n}\varepsilon_i + \Gamma\right) = 0 \quad \text{in } \tfrac{1}{p^n}\Gamma/\Gamma. \tag{7.47}$$

Then

$$\sum_{i\in I}(\ell_i + p^{n-1}\mathbb{Z}) \cdot \left(\tfrac{1}{p^{n-1}}\varepsilon_i + \Gamma\right) = 0 \quad \text{in } \tfrac{1}{p^{n-1}}\Gamma/\Gamma,$$

hence the induction hypothesis yields for each $i \in I$ an integer m_i such that $\ell_i = p^{n-1}m_i$. From (7.47) it follows that

$$\tfrac{1}{p} \cdot \sum_{i\in I} m_i\varepsilon_i \in \Gamma,$$

hence $\sum_{i\in I} m_i\varepsilon_i \in p\Gamma$. Since $(\varepsilon_i + p\Gamma)_{i\in I}$ is a base of $\Gamma/p\Gamma$ over $\mathbb{F}_p$, we must have $m_i \in p\mathbb{Z}$ for all $i \in I$, hence $\ell_i \in p^n\mathbb{Z}$ for all $i \in I$. □

Proof of Th. 7.69. By definition, $\mathsf{Br}_{is}(\mathsf{F}) = \mathit{ker}(\mathit{ext}_{\mathsf{F}_{in}/\mathsf{F}})$. Since $\Sigma_{\mathsf{F}_{in}}$ is an isomorphism, we also have $\mathsf{Br}_{is}(\mathsf{F}) = \mathit{ker}\,\Phi_{\mathsf{F}}$. From Prop. 7.67, it follows that

$im(ext_{\mathsf{F}_{in}/\mathsf{F}}) = {}_{\mu(\mathsf{F}_0)}\mathsf{Br}(\mathsf{F}_{in})$, hence the diagram of the theorem is a short exact sequence. We define a splitting map $\mathcal{Y}\colon {}_{\mu(\mathsf{F}_0)}\mathbb{T}(\wedge^2\Gamma_\mathsf{F}) \to \mathsf{Br}(\mathsf{F})$ separately on each primary component of the domain.

Fix a prime p. Choose a collection of elements $(\varepsilon_i)_{i\in I}$ in Γ_F such that $(\varepsilon_i + p\Gamma_\mathsf{F})_{i\in I}$ is an $\mathbb{F}_p$-vector space base of $\Gamma_\mathsf{F}/p\Gamma_\mathsf{F}$. Fix some total ordering on the index set I. Then, $\big((\varepsilon_i\wedge\varepsilon_j) + p\wedge^2\Gamma_\mathsf{F}\big)_{i,j\in I,\, j<i}$ is an $\mathbb{F}_p$-vector space base of $(\wedge^2\Gamma_\mathsf{F})/p\wedge^2\Gamma_\mathsf{F} = \wedge^2(\Gamma_\mathsf{F}/p\Gamma_\mathsf{F})$. Take some p-power $q = p^n$ such that $\mu_q \subseteq \mathsf{F}_0$. By Lemma 7.70, $\big(\tfrac{1}{q}(\varepsilon_i\wedge\varepsilon_j) + \wedge^2\Gamma_\mathsf{F}\big)_{i,j\in I,\, j<i}$ is a base for the free $\mathbb{Z}/q\mathbb{Z}$-module ${}_q\mathbb{T}(\wedge^2\Gamma_\mathsf{F}) = \big(\tfrac{1}{q}\wedge^2\Gamma_\mathsf{F}\big)/\wedge^2\Gamma_\mathsf{F}$. Therefore, to define a group homomorphism ${}_q\mathcal{Y}\colon {}_q\mathbb{T}(\wedge^2\Gamma_\mathsf{F}) \to \mathsf{Br}(\mathsf{F})$, it suffices to specify ${}_q\mathcal{Y}$ for each element of the base. For each $i\in I$, choose and fix some $c_i \in \mathsf{F}^\times$ such that $deg\, c_i = \varepsilon_i$. Then define ${}_q\mathcal{Y}$ by

$$\tfrac{1}{q}(\varepsilon_i\wedge\varepsilon_j) + \wedge^2\Gamma_\mathsf{F} \mapsto \big[(c_i, c_j/\mathsf{F})_{\omega(1/q),\, q}\big] \qquad \text{for all } i,\, j\in I,\ j<i.$$

By the definition of $\Sigma_{\mathsf{F}_{in}}$ (see (7.42)),

$$\begin{aligned}\Sigma_{\mathsf{F}_{in}}\big(\tfrac{1}{q}(\varepsilon_i\wedge\varepsilon_j) + \wedge^2\Gamma_\mathsf{F}\big) &= \big[(c_i, c_j/\mathsf{F}_{in})_{\omega(1/q),\, q}\big]\\ &= ext_{\mathsf{F}_{in}/\mathsf{F}}\big({}_q\mathcal{Y}\big(\tfrac{1}{q}(\varepsilon_i\wedge\varepsilon_j) + \wedge^2\Gamma_\mathsf{F}\big)\big).\end{aligned}$$

Hence, $ext_{\mathsf{F}_{in}/\mathsf{F}}\circ({}_q\mathcal{Y}) = \Sigma_{\mathsf{F}_{in}}|_{{}_q\mathbb{T}(\wedge^2\Gamma_\mathsf{F})}$.

We check also that the maps ${}_q\mathcal{Y}$ are compatible as the exponent n in $q = p^n$ increases: If $\mu_{p^{n+1}} \subseteq \mathsf{F}_0$, then

$$\begin{aligned}{}_{p^{n+1}}\mathcal{Y}\big(\tfrac{1}{p^n}(\varepsilon_i\wedge\varepsilon_j) + \wedge^2\Gamma_\mathsf{F}\big) &= {}_{p^{n+1}}\mathcal{Y}\big(p\big(\tfrac{1}{p^{n+1}}(\varepsilon_i\wedge\varepsilon_j) + \wedge^2\Gamma_\mathsf{F}\big)\big)\\ = \big[(c_i, c_j/\mathsf{F})_{\omega(1/p^{n+1}),\, p^{n+1}}\big]^p &= \big[(c_i, c_j/\mathsf{F})_{\omega(1/p^n),\, p^n}\big]\\ &= {}_{p^n}\mathcal{Y}\big(\tfrac{1}{p^n}(\varepsilon_i\wedge\varepsilon_j) + \wedge^2\Gamma_\mathsf{F}\big).\end{aligned}$$

Hence ${}_{p^{n+1}}\mathcal{Y}$ and ${}_{p^n}\mathcal{Y}$ agree on the domain of ${}_{p^n}\mathcal{Y}$.

If there is a largest $m\in\mathbb{N}$ such that $\mu(p^m)\subseteq\mathsf{F}_0$, then the p-primary component ${}_{\mu(\mathsf{F}_0)}\mathbb{T}(\wedge^2\Gamma_\mathsf{F})_{(p)}$ is ${}_{p^m}\mathbb{T}(\wedge^2\Gamma_\mathsf{F})$; then define $\mathcal{Y}_{(p)} = {}_{p^m}\mathcal{Y}$. If there is no such m, then ${}_{\mu(\mathsf{F}_0)}\mathbb{T}(\wedge^2\Gamma_\mathsf{F})_{(p)} = \mathbb{T}(\wedge^2\Gamma_\mathsf{F})_{(p)}$; then define $\mathcal{Y}_{(p)}$ as the union of the compatible maps ${}_{p^n}\mathcal{Y}$ for all $n\in\mathbb{N}$. In either case, $\mathcal{Y}_{(p)}$ maps ${}_{\mu(\mathsf{F}_0)}\mathbb{T}(\wedge^2\Gamma_\mathsf{F})_{(p)}$ to $\mathsf{Br}(\mathsf{F})$ and satisfies $ext_{\mathsf{F}_{in}/\mathsf{F}}\circ\mathcal{Y}_{(p)} = \Sigma|_{{}_{\mu(\mathsf{F}_0)}\mathbb{T}(\wedge^2\Gamma_\mathsf{F})_{(p)}}$.

We now combine these maps for all primes p. The map

$$\mathcal{Y} = \oplus\mathcal{Y}_{(p)}\colon {}_{\mu(\mathsf{F}_0)}\mathbb{T}(\wedge^2\Gamma_\mathsf{F}) \longrightarrow \mathsf{Br}(\mathsf{F})$$

satisfies $ext_{\mathsf{F}_{in}/\mathsf{F}}\circ\mathcal{Y} = \Sigma|_{{}_{\mu(\mathsf{F}_0)}\mathbb{T}(\wedge^2\Gamma_\mathsf{F})}$, which is what is needed to have a splitting map in the diagram of the theorem. □

By combining the preceding theorem with (6.19), Th. 6.24, and Prop. 6.27, we can see explicitly how $\mathsf{Br}(\mathsf{F})$ is completely determined by F_0 and Γ_F:

Corollary 7.71. *For any graded field* F, *we have a noncanonical group isomorphism*

$$\mathsf{Br}(\mathsf{F}) \cong Br(\mathsf{F}_0)\times Hom^c\big(\mathcal{G}(\mathsf{F}_0), \mathbb{T}(\Gamma_\mathsf{F})\big)\times {}_{\mu(\mathsf{F}_0)}\mathbb{T}(\wedge^2\Gamma_\mathsf{F}).$$

7.4 Total ramification in algebras over valued fields

We now consider total ramification of valued division algebras, first from the viewpoint of individual algebras (in §7.4.1), then from the Brauer group perspective (in §7.4.2). When the center is Henselian and the division algebras are tame, the results closely duplicate those of §7.3 for graded division algebras: The totally ramified division algebras decompose in tensor products of symbol algebras (see Prop. 7.74), and we obtain in §7.4.2 a double description of the tame part of the Brauer group of a strictly Henselian field F (see Th. 7.80):

$$\mathbb{T}(\wedge^2\Gamma_F)' \cong Br_{tr}(F) \cong Symp(\mathbb{T}(\Gamma_F))'.$$

This result enables us to determine the top layer in the canonical filtration of the tame part of the Brauer group of any Henselian field F, in the form of a canonical isomorphism $Br_{tr}(F)/\,Br_{is}(F) \xrightarrow{\sim} {}_{\mu(\overline{F})}\mathbb{T}(\wedge^2\Gamma_F)$, where the target group is the subgroup of all elements in $\mathbb{T}(\wedge^2\Gamma_F)$ with an order n such that $\overline{F}$ contains a primitive n-th root of unity; see Th. 7.84.

When the center is not Henselian, the structure results are far less complete: In §7.4.1 we still obtain from the commutator map of a totally ramified division algebra D a canonical alternating pairing on Γ_D, which imposes restrictions on the possible decompositions of D as a tensor product of symbol algebras (see Prop. 7.75), but there are examples of tame totally ramified division algebras that have no such decompositions; see Ex. 7.77.

Throughout this section, F is a field with a valuation v. We use the same notation as in earlier chapters.

7.4.1 Totally ramified division algebras

Let D be a division algebra (finite-dimensional) over F, and assume D carries a valuation v_D extending v. We say v_D (or D) is *totally ramified* over v (or over F) if $|\Gamma_D{:}\Gamma_F| = [D{:}F]$ (or, equivalently, $\overline{D} = \overline{F}$ and D is defectless over F). When this occurs, the graded division algebra $\mathsf{gr}_{v_D}(D)$ is totally ramified as a $\mathsf{gr}(F)$-algebra, hence we have a canonical alternating pairing $c_{\mathsf{gr}(D)}$ on $\Gamma_{\mathsf{gr}(D)} = \Gamma_D$, see (7.37). This pairing can be defined directly in terms of D as follows: we define the *canonical pairing*

$$c_D \colon \Gamma_D \times \Gamma_D \longrightarrow \mu(\overline{F})$$

by

$$c_D\big(v_D(x), v_D(y)\big) = \overline{xyx^{-1}y^{-1}} \qquad \text{for } x,\, y \in D^\times.$$

Thus, for x, $y \in D^\times$ we have

$$c_D\big(v_D(x), v_D(y)\big) = c_{\mathsf{gr}(D)}(\deg \widetilde{x}, \deg \widetilde{y}),$$

hence c_D is a well-defined alternating pairing on Γ_D. Its radical clearly contains Γ_F, hence we have an induced pairing

$$\overline{c}_D\colon (\Gamma_D/\Gamma_F)\times(\Gamma_D/\Gamma_F)\longrightarrow \mu(\overline{F}), \tag{7.48}$$

and

$$\overline{c}_D \text{ on } \Gamma_D/\Gamma_F \quad \text{coincides with} \quad \overline{c}_{\mathsf{gr}(D)} \text{ on } \Gamma_{\mathsf{gr}(D)}/\Gamma_{\mathsf{gr}(F)}. \tag{7.49}$$

Proposition 7.72. *If D is central over F and v_D is tame and totally ramified over v, then $\operatorname{rad} c_D=\Gamma_F$, hence the pairing $\overline{c}_D$ is nondegenerate. Therefore, $\overline{F}$ contains a primitive root of unity of order $\exp(\Gamma_D/\Gamma_F)$, and the invariant factors of Γ_D/Γ_F occur in pairs. Hence, if $D\neq F$, then $\dim_{\mathbb{Q}}\mathbb{H}(\Gamma_F)\geq 2$ (so $\Gamma_{\mathsf{F}}\not\cong\mathbb{Z}$).*

Proof. It is clear from the definition that $rad c_D = rad c_{\mathsf{gr}(D)}$, hence, by Th. 7.52, $rad c_D=\Gamma_{Z(\mathsf{gr}(D))}$. Since v_D is tame, we have $Z\big(\mathsf{gr}_{v_D}(D)\big)=\mathsf{gr}(F)$, yielding the first sentence of the proposition. The second sentence follows from the fact that $(\Gamma_D/\Gamma_F,\overline{c}_D)$ is a symplectic module, see Th. 7.2. The final sentence follows from Cor. 7.53 since $\mathsf{gr}(D)$ is totally ramified over its center $\mathsf{gr}(F)$ and $\Gamma_F=\Gamma_{\mathsf{gr}(F)}$. □

Tame totally ramified central division algebras D over F need not be tensor products of symbol algebras, as we will see in Ex. 7.77 below. But when D is such a tensor product, we now show that the tensor powers of D behave analogously to what was seen for totally ramified graded division algebras in Prop. 7.60.

Lemma 7.73. *Let A be a finite-dimensional F-algebra spanned by an F-armature $\mathcal{A}$. Suppose $\operatorname{char}(\overline{F})\nmid\exp\mathcal{A}$, so the armature gauge $\alpha_{\mathcal{A}}$ is tame. Then, the following conditions are equivalent:*

(a) *The map $\overline{\alpha}_{\mathcal{A}}\colon\mathcal{A}\to\Gamma_A/\Gamma_F$ is injective (hence an isomorphism).*
(b) *A is a division algebra over F and $\alpha_{\mathcal{A}}$ is a valuation on A with A tame and totally ramified over F.*
(c) *$\mathsf{gr}_{\alpha_{\mathcal{A}}}(A)$ is a graded division algebra totally ramified over $\mathsf{gr}(F)$.*

Proof. (a) ⇒ (c) Suppose $\overline{\alpha}_{\mathcal{A}}$ is injective. Then $\overline{\alpha}_{\mathcal{A}}$ is an isomorphism because it is surjective by Cor. 7.48. For the induced armature $\widetilde{\mathcal{A}}$ of $\mathsf{gr}_{\alpha_{\mathcal{A}}}A$ the commutative diagram of Prop. 7.47 yields that $\widetilde{\mathcal{A}}_0 = ker\,\overline{deg} = \{[1]\}$. Hence (see (7.36)) $A_0^{\alpha_{\mathcal{A}}}=\overline{F}[\widetilde{\mathcal{A}}_0']=\overline{F}$. Also, $\Gamma^{\times}_{\mathsf{gr}(A)}=\Gamma^{\times}_{A}=\Gamma_A=\Gamma_{\mathsf{gr}(A)}$ by Cor. 7.48. Therefore, by Prop. 2.4 $\mathsf{gr}(A)$ is a graded division algebra. Since $\alpha_{\mathcal{A}}$ is a tame gauge, A is defectless over F. Thus,

$$[\mathsf{gr}(A)\colon\mathsf{gr}(F)] = [A{:}F] = |\mathcal{A}| = |\Gamma_A{:}\Gamma_F| = |\Gamma_{\mathsf{gr}(A)}{:}\Gamma_{\mathsf{gr}(F)}|,$$

which shows that $\mathsf{gr}(A)$ is totally ramified over $\mathsf{gr}(F)$.

(c) ⇒ (b) Suppose $\mathsf{gr}(A)$ is a graded division algebra totally ramified over $\mathsf{gr}(F)$. By Cor. 3.30, A is a division algebra and $\alpha_{\mathcal{A}}$ is a valuation on A

extending v on F. Since $\mathsf{gr}(A)$ is totally ramified over $\mathsf{gr}(F)$ and A is defectless over F as $\alpha_{\mathcal{A}}$ is a tame gauge, it follows that A is tame and totally ramified over F.

(b) ⇒ (a) Suppose (b) holds. The total ramification of A over F implies that $|\Gamma_A{:}\Gamma_F| = [A{:}F] = |\mathcal{A}|$. Hence the surjective map $\overline{\alpha}_{\mathcal{A}} \colon \mathcal{A} \to \Gamma_A/\Gamma_F$ (see Cor. 7.48) must also be injective. □

The conditions of the lemma hold whenever the valuation on F is Henselian:

Proposition 7.74. *Let D be a central division algebra over a Henselian field F. If D is tame and totally ramified over F, then D is spanned by an armature. If $\mathcal{A}$ is any armature spanning D, then the armature gauge $\alpha_{\mathcal{A}}$ coincides with the valuation on D, and the map $\overline{\alpha}_{\mathcal{A}}$ of (7.33) yields an isomorphism of symplectic modules*

$$(\mathcal{A}, b_{\mathcal{A}}) \cong (\Gamma_D/\Gamma_F, \overline{c}_D)$$

when the $\exp(\Gamma_D/\Gamma_F)$-roots of unity of F are identified with their images in $\overline{F}$.

If $\Gamma_D/\Gamma_F \cong (\mathbb{Z}/n_1\mathbb{Z})^2 \times \ldots \times (\mathbb{Z}/n_r\mathbb{Z})^2$ for some integers $n_1, \ldots, n_r \geq 2$, and $\gamma_1, \delta_1, \ldots, \gamma_r, \delta_r \in \Gamma_D$ map to a symplectic base $(\overline{\gamma}_1, \overline{\delta}_1, \ldots, \overline{\gamma}_r, \overline{\delta}_r)$ of Γ_D/Γ_F for the pairing $\overline{c}_D$, with $\overline{\gamma}_i = \gamma_i + \Gamma_F$ and $\overline{\delta}_i = \delta_i + \Gamma_F$ of order n_i, and $\overline{c}_D(\overline{\gamma}_i, \overline{\delta}_i) = \overline{\omega}_i$ for some primitive n_i-th root of unity $\omega_i \in F$, then there are $x_1, y_1, \ldots, x_r, y_r \in F^{\times}$ with $v(x_i) = n_i\gamma_i$ and $v(y_i) = n_i\delta_i$ such that

$$D \cong (x_1, y_1/F)_{\omega_1, n_1} \otimes_F \ldots \otimes_F (x_r, y_r/F)_{\omega_r, n_r}.$$

Proof. Since D is tame and totally ramified over F, the graded division algebra $\mathsf{gr}(D)$ is totally ramified over its center $\mathsf{gr}(F)$. Hence, by Th. 7.52 the unique homogeneous armature spanning $\mathsf{gr}(D)$ is $\mathcal{B} = \mathsf{gr}(D)^{\times}/\mathsf{gr}(F)^{\times}$. By Prop. 7.51, D itself is therefore spanned by an armature. Let $\mathcal{A}$ be any armature spanning D. By Prop. 4.33(i), the armature gauge $\alpha_{\mathcal{A}}$ coincides with the valuation on D extending the Henselian valuation on F. The homogeneous armature $\widetilde{\mathcal{A}}$ induced by $\mathcal{A}$ that spans $\mathsf{gr}(D)$ (see Prop. 7.47) must coincide with $\mathcal{B}$. Since $\mathcal{A} \cong \widetilde{\mathcal{A}} = \mathcal{B}$ and the pairing $\overline{c}_D$ on $\mathcal{B}$ is nondegenerate, we have *char*$(\overline{F}) \nmid$ *exp* $\mathcal{B} =$ *exp* $\mathcal{A}$. Thus, condition (b) of the preceding lemma holds for D. The lemma shows that the map $\overline{\alpha}_{\mathcal{A}} \colon \mathcal{A} \to \Gamma_D/\Gamma_F$ is a group isomorphism. It is clear from the definitions that this map is compatible with the armature pairing $b_{\mathcal{A}}$ on $\mathcal{A}$ and the pairing $\overline{c}_D$ on Γ_D/Γ_F when we identify the *exp* $\mathcal{A}$-roots of unity in F with their images in $\overline{F}$. Hence, $\overline{\alpha}_{\mathcal{A}}$ is an isometry of symplectic modules.

Now, the symplectic base $(\overline{\gamma}_1, \overline{\delta}_1, \ldots, \overline{\gamma}_r, \overline{\delta}_r)$ of Γ_D/Γ_F pulls back to a symplectic base of $\mathcal{A}$, which we use to decompose $\mathcal{A} = \mathcal{A}_1 \perp \ldots \perp \mathcal{A}_m$ where each $\mathcal{A}_1$ maps to $\langle\overline{\gamma_i}, \overline{\delta_i}\rangle$. Set $S_i = F[\mathcal{A}_i]$ for $i = 1, 2, \ldots, r$. Then, for each i,

$$S_i = (x_i, y_i/F)_{\omega_i, n_i}, \quad \text{with} \quad x_i,\ y_i \in F^{\times},\ v(x_i) = n_i\gamma_i, v(y_i) = n_i\delta_i,$$

and

$$D \cong S_1 \otimes_F \ldots \otimes_F S_r.$$

□

The next proposition shows that for a tame totally ramified central division algebra D over F, the structure of Γ_D/Γ_F imposes restrictions on the possible decompositions of D into a tensor product of symbols. Note that such a decomposition does not always exist, see Ex. 7.77 below, though we have just seen that there is a decomposition whenever the valuation on F is Henselian. Recall that if D is a division algebra with center F and j is a positive integer, we write D^j for the associated division algebra of the central simple F-algebra $D^{\otimes j} = D \otimes_F \ldots \otimes_F D$ (j factors).

Proposition 7.75. *Let D be a central division algebra over F. Suppose v extends to a valuation v_D on D that is tame and totally ramified over v. If $S_1, \ldots, S_r$ are symbol algebras such that $D \cong S_1 \otimes_F \ldots \otimes_F S_r$, then*

$$\Gamma_D/\Gamma_F \cong (\mathbb{Z}/n_1\mathbb{Z})^2 \times \ldots \times (\mathbb{Z}/n_r\mathbb{Z})^2 \quad \textit{where } n_i = \deg S_i \textit{ for } i = 1, \ldots, r.$$

Moreover, for each $j \in \mathbb{N}$, the valuation v extends to a valuation on D^j with D^j tame and totally ramified over F, $\Gamma_{D^j} = j\Gamma_D + \Gamma_F$, and as symplectic modules,

$$\big(\Gamma_{D^j}/\Gamma_F, \overline{c}_{D^j}\big) = \big(j(\Gamma_D/\Gamma_F), \overline{c}_D^{(j)}\big). \tag{7.50}$$

Hence, $\exp D = \exp(\Gamma_D/\Gamma_F)$.

Proof. Suppose D has a tensor product decomposition as above, and let $S_i = \big(a_i, b_i/F\big)_{n_i}$. Then v_D restricts to a valuation v_{S_i} on S_i that is tame and totally ramified over v; so $char\,\overline{F} \nmid n_i$. Each S_i has an armature $\mathcal{A}_i$ isomorphic to $(\mathbb{Z}/n_i\mathbb{Z})^2$, by Ex. 7.28(iii). Since a valuation extending v on a division algebra is its only v-gauge by Prop. 4.33(i), v_{S_i} coincides with the armature gauge $\alpha_{\mathcal{A}_i}$ on S_i. Hence, $\mathsf{gr}_{v_{S_i}}(S_i) \cong_g \big(\widetilde{a}_i, \widetilde{b}_i/\mathsf{gr}(F)\big)_{n_i}$ (cf. (3.27)). Also, the tame gauge $v_{S_1} \otimes \ldots \otimes v_{S_r}$ on D must coincide with its valuation v_D and with the armature gauge $\alpha_{\mathcal{A}_1 \perp \ldots \perp \mathcal{A}_r}$. Therefore, by Prop. 3.27

$$\mathsf{gr}(S_1) \otimes_{\mathsf{gr}(F)} \cdots \otimes_{\mathsf{gr}(F)} \mathsf{gr}(S_r) \cong_g \mathsf{gr}_{v_{S_1} \otimes \ldots \otimes v_{S_r}}(D) = \mathsf{gr}_{v_D}(D),$$

which is a graded division algebra totally ramified over its center $\mathsf{gr}(F)$. It follows from Lemma 7.73 that v_D induces an isomorphism

$$\overline{v}_D \colon \mathcal{A}_1 \perp \ldots \perp \mathcal{A}_r \xrightarrow{\sim} \Gamma_D/\Gamma_F,$$

which proves the first statement.

The rest of the proof builds on the corresponding graded result given in Prop. 7.60. We assume that j is prime. Once this case is completed, the proposition follows in general by induction on j.

By the identities for symbol algebras, $S_i^j \sim T_i$ in $Br(F)$, where

$$T_i = \big(a_i, b_i/F\big)_{n_i/j} \text{ if } j \mid n_i \qquad \text{and} \qquad T_i = \big(a_i, b_i^j/F\big)_{n_i} \text{ if } j \nmid n_i.$$

Let β_i be the v-gauge on T_i from its realization as a symbol algebra. Then,

$$\mathsf{gr}_{\beta_i}(T_i) = \big(\widetilde{a}_i, \widetilde{b}_i/F\big)_{n_i/j} \text{ if } j \mid n_i \quad \text{and} \quad \mathsf{gr}_{\beta_i}(T_i) = \big(\widetilde{a}_i, \widetilde{b}_i^j/F\big)_{n_i} \text{ if } j \nmid n_i.$$

In either case, since $\mathsf{gr}(D)$ is totally ramified over $\mathsf{gr}(F)$ we saw in the proof of Prop. 7.60 that

$$\mathsf{gr}(D)^j \cong_g \mathsf{gr}(T_1) \otimes_{\mathsf{gr}(F)} \dots \otimes_{\mathsf{gr}(F)} \mathsf{gr}(T_r),$$

and this graded division algebra is totally ramified over $\mathsf{gr}(F)$.

Let $E = T_1 \otimes_F \dots \otimes_F T_r$ with its tame armature gauge $\varepsilon = \beta_1 \otimes \dots \otimes \beta_r$. Since each $T_i \sim S_i^j$, we have $E \sim D^j$ in $Br(F)$. Also, by Prop. 3.27

$$\mathsf{gr}_\varepsilon(E) \cong_g \mathsf{gr}(T_1) \otimes_{\mathsf{gr}(F)} \dots \otimes_{\mathsf{gr}(F)} \mathsf{gr}(T_r) \cong_g \mathsf{gr}(D)^j.$$

Since $\mathsf{gr}(E)$ is thus a graded division algebra, by Lemma 7.73 ε is a valuation on E extending v, and E is totally and tamely ramified over F. Hence, $D^j \cong E$, so D^j has a valuation totally and tamely ramified over F, and we have equalities of symplectic modules

$$\begin{aligned}(\Gamma_{D^j}/\Gamma_F, \overline{c}_{D^j}) &= (\Gamma_E/\Gamma_F, \overline{c}_E) \\ &= (\Gamma_{\mathsf{gr}(E)}/\Gamma_{\mathsf{gr}(F)}, \overline{c}_{\mathsf{gr}(E)}) = (\Gamma_{\mathsf{gr}(D)^j}/\Gamma_{\mathsf{gr}(F)}, \overline{c}_{\mathsf{gr}(D)^j}).\end{aligned}$$

By Prop. 7.60 applied to $\mathsf{gr}(D)$, we also have

$$(\Gamma_{\mathsf{gr}(D)^j}/\Gamma_{\mathsf{gr}(F)}, \overline{c}_{\mathsf{gr}(D)^j}) = (j(\Gamma_{\mathsf{gr}(D)}/\Gamma_{\mathsf{gr}(F)}), \overline{c}^{(j)}_{\mathsf{gr}(D)}) = (j(\Gamma_D/\Gamma_F), \overline{c}_D^{(j)}),$$

proving (7.50). In particular, we have $\Gamma_{D^j} = j\Gamma_D + \Gamma_F$. Since D^j is totally ramified over F, it is split if and only if $\Gamma_{D^j} = \Gamma_F$. Therefore, $\exp D = \exp(\Gamma_D/\Gamma_F)$. □

We will see in Ex. 7.77 below how the result in Prop. 7.75 may fail when D is not a tensor product of symbol algebras. This situation does not arise when v is Henselian:

Corollary 7.76. *Let D be a central division algebra over an arbitrary valued field (F, v). Suppose v extends to a valuation v_D on D, and that v_D is tame and totally ramified over v. Then $\exp(\Gamma_D/\Gamma_F) \mid \exp D$. If (F, v) is Henselian, then $\exp(\Gamma_D/\Gamma_F) = \exp D$.*

Proof. We first consider the Henselian case. In this case, Prop. 7.74 shows that D is a tensor product of symbol algebras. The equation $\exp(\Gamma_D/\Gamma_F) = \exp D$ then follows from Prop. 7.75.

In the general case, choose a Henselization (F_h, v_h) of (F, v) and let $D_h = D \otimes_F F_h$. By Th. 4.1, D_h is a division algebra and $\mathsf{gr}(D_h) = \mathsf{gr}(D)$. Therefore, the valuation on D_h extending v_h is tame and totally ramified. Thus, by the Henselian case,

$$\exp(\Gamma_D/\Gamma_F) = \exp(\Gamma_{D_h}/\Gamma_{F_h}) = \exp D_h.$$

Since $\exp D_h \mid \exp D$, the proof is complete. □

If D is totally ramified but not tame over F, then $\exp(\Gamma_D/\Gamma_F)$ need not divide $\exp D$, even if F is Henselian. See Exercise 8.8 for an example.

Example 7.77. *A totally ramified division algebra that is not a symbol algebra nor a tensor product of symbol algebras.* Let F be a field with $\operatorname{char} F \neq 2$ and let M be a biquadratic field extension of F, i.e., a Galois extension of F with Galois group $\mathcal{G}(M/F)$ isomorphic to $(\mathbb{Z}/2\mathbb{Z})^2$. Let σ_1, σ_2 be generators of $\mathcal{G}(M/F)$ and let $M = F(m_1, m_2)$ where

$$\sigma_1(m_1) = -m_1, \ \ \sigma_1(m_2) = m_2, \ \ \sigma_2(m_1) = m_1, \ \ \sigma_2(m_2) = -m_2. \tag{7.51}$$

Choose $u \in M^\times$ such that $N_{M/F}(u) = 1$, and define a collection $\boldsymbol{u} = (u_{i,j})_{i,j=1}^2$ of elements of $M^\times$ by

$$u_{1,1} = u_{2,2} = 1, \qquad u_{1,2} = u^{-1}, \qquad u_{2,1} = u.$$

Also, let $\boldsymbol{\sigma} = (\sigma_1, \sigma_2)$. As in §1.1.3, we consider the division ring

$$D = \mathcal{L}((M; \boldsymbol{\sigma}, \boldsymbol{u})).$$

Its elements are iterated Laurent series in two indeterminates x_1, x_2 with coefficients in M, with multiplication subject to the following relations: for $m \in M^\times$,

$$x_1 m = \sigma_1(m) x_1, \quad x_2 m = \sigma_2(m) x_2, \quad \text{and} \quad x_1 x_2 = u x_2 x_1. \tag{7.52}$$

As shown in §1.1.3, this division ring carries the (x_1,x_2)-adic valuation $v_{(x_1,x_2)}$, which we designate by w. We have $\Gamma_{D,w} = \mathbb{Z}^2$ with $w(x_1) = (1,0)$ and $w(x_2) = (0,1)$. Moreover, $\overline{D}^w = M$ and $\overline{Z(D)}^w = F$, and Prop. 1.13 shows that $[D{:}Z(D)] = 16$ and $|\Gamma_{D,w}{:}\Gamma_{Z(D),w}| = 4$.

Now assume that M carries a valuation v with $\Gamma_{M,v} = \mathbb{Z}^2$ spanned by $v(m_1)$ and $v(m_2)$, such that $\Gamma_{F,v} = (2\mathbb{Z})^2$ and $\operatorname{char} \overline{F} \neq 2$. Thus, $|\Gamma_{M,v}{:}\Gamma_{F,v}| = 4$, hence v is the unique extension of its restriction to F. By Prop. 1.8, there is a composite valuation $\eta = v * w$ on D, and by Remark 1.9 we have

$$\overline{D}^\eta = \overline{M}^v = \overline{F}^v = \overline{Z(D)}^\eta.$$

Thus, D is totally ramified over $Z(D)$ for the valuation η. This valuation is tame since the residue characteristic does not divide $[D{:}Z(D)]$, hence $|\Gamma_{D,\eta}{:}\Gamma_{Z(D),\eta}| = 16$. The value group $\Gamma_{D,\eta}$ fits into the exact sequence

$$0 \longrightarrow \Gamma_{M,v} \longrightarrow \Gamma_{D,\eta} \longrightarrow \Gamma_{D,w} \longrightarrow 0,$$

see (1.11). Therefore, $\Gamma_{D,\eta}$ is spanned by $\eta(x_1)$, $\eta(x_2)$, $\eta(m_1)$, and $\eta(m_2)$. In view of (7.51) and (7.52), the canonical pairing satisfies

$$c_D\big(\eta(x_1), \eta(x_2)\big) = \overline{u}^v, \quad c_D\big(\eta(x_1), \eta(m_1)\big) = c_D\big(\eta(x_2), \eta(m_2)\big) = -1,$$
$$c_D\big(\eta(x_1), \eta(m_2)\big) = c_D\big(\eta(x_2), \eta(m_1)\big) = c_D\big(\eta(m_1), \eta(m_2)\big) = 1.$$

Therefore, if $\overline{u}^v = \pm 1$ we have $2\gamma \in \operatorname{rad} c_D$ for all $\gamma \in \Gamma_{D,\eta}$. Since $\operatorname{rad} c_D = \Gamma_{Z(D),\eta}$ by Prop. 7.72, it follows that the factor group $\Gamma_{D,\eta}/\Gamma_{Z(D),\eta}$ has exponent 2; hence,

$$\Gamma_{D,\eta}/\Gamma_{Z(D),\eta} \cong (\mathbb{Z}/2\mathbb{Z})^4.$$

On the other hand, if $\overline{u}^{\,v}$ is a primitive 4-th root of unity, then $\eta(x_1)$ and $\eta(x_2)$ have order 4 modulo $rad c_D = \Gamma_{Z(D),\eta}$. Hence, the exponent of $\Gamma_{D,\eta}/\Gamma_{Z(D),\eta}$ is 4, and therefore, since its invariant factors occur in pairs (see Prop. 7.72),

$$\Gamma_{D,\eta}/\Gamma_{Z(D),\eta} \cong (\mathbb{Z}/4\mathbb{Z})^2.$$

We now construct a single extension M/F for which each case occurs; i.e., there are two valuations on M satisfying the conditions above, one for which $\overline{u} = \pm 1$, and another for which $\overline{u}$ is a primitive 4-th root of unity. Then D is totally ramified over $Z(D)$ for two valuations, one for which $\Gamma_D/\Gamma_{Z(D)} \cong (\mathbb{Z}/2\mathbb{Z})^4$ and another for which $\Gamma_D/\Gamma_{Z(D)} \cong (\mathbb{Z}/4\mathbb{Z})^2$. In view of Prop. 7.75, this implies that D is not a symbol algebra, nor a tensor product of symbol algebras.

We start with an arbitrary field k containing a primitive 4-th root of unity i (hence $char\, k \neq 2$) and adjoin to k two indeterminates m_1, m_2: let

$$M_0 = k(m_1, m_2) \qquad \text{and} \qquad F_0 = k(m_1^2, m_2^2).$$

We let

$$y = N_{M_0/F_0}(1 - m_1 - m_2) = 1 - 2m_1^2 - 2m_2^2 + m_1^4 - 2m_1^2 m_2^2 + m_2^4.$$

Note that y is not a square in M_0 since it is a product of four nonassociate irreducibles in $k[m_1, m_2]$. Define M and F by adjoining to M_0 and F_0 a 4-th root of y, which we designate by z:

$$M = M_0(z) \quad \text{and} \quad F = F_0(z) \quad \text{where } z^4 = y.$$

So, $[M{:}M_0] = 4$. Let v_0 be the (m_1, m_2)-adic valuation on M_0, which is the restriction of the standard valuation on $k((m_1))((m_2))$. For this valuation we have $\overline{y} = 1$, hence $\overline{z}^{\,4} = 1$ for any extension v of v_0 to M. Let $\tau \in \mathcal{G}(M/M_0)$ be the automorphism such that $\tau(z) = iz$. We have

$$\overline{z}^{\,v\circ\tau} = i\,\overline{z}^{\,v};$$

hence, v_0 has four extensions $v_1, \ldots, v_4$ to M, which can be indexed in such a way that

$$\overline{z}^{\,v_\ell} = i^\ell \qquad \text{for } \ell = 1, 2, 3, 4.$$

Set $u = (1 - m_1 - m_2) z^{-1} \in M^\times$. We have

$$N_{M/F}(u) = N_{M_0/F_0}(1 - m_1 - m_2) \cdot z^{-4} = 1,$$

and

$$\overline{u}^{\,v_1} = i^{-1}, \qquad \overline{u}^{\,v_2} = -1, \qquad \overline{u}^{\,v_3} = i, \qquad \overline{u}^{\,v_4} = 1.$$

It follows from Cor. 7.76 that the division algebra D of degree 4 constructed in Ex. 7.77 has exponent 4, since it is tame and totally ramified over its center for the valuation $\eta_1 = v_1 * w$, for which $exp(\Gamma_{D,\eta_1}/\Gamma_{Z(D),\eta_1}) = 4$. Note however that this division algebra is also tame and totally ramified for the valuation $\eta_2 = v_2 * w$, for which $exp(\Gamma_{D,\eta_2}/\Gamma_{Z(D),\eta_2}) = 2$. Thus, the equality

$exp(\Gamma_D/\Gamma_{Z(D)}) = exp\, D$ does not always hold when $Z(D)$ is not Henselian. Also, if $Z(D)_{h,2}$ denotes a Henselization of $Z(D)$ for the restriction of η_2, we have by Cor. 7.76

$$exp(D \otimes_{Z(D)} Z(D)_{h,2}) \;=\; exp(\Gamma_{D,\eta_2}/\Gamma_{Z(D),\eta_2}) \;=\; 2.$$

Therefore, $Z(D)_{h,2}$ splits D^2, which means by Morandi's theorem 4.1 that $\eta_2|_{Z(D)}$ does not extend to a valuation on D^2.

7.4.2 The tame Brauer group of a Henselian field

In this subsection we describe the tame part of the Brauer group of a field with Henselian or strictly Henselian valuation. We build on the analogous results for a graded field given in §§7.3.2–7.3.3. We first consider strictly Henselian valuations.

A valued field (F, v) is said to be *strictly Henselian* if it is Henselian and inertially closed, which means that the residue field $\overline{F}$ is separably closed. Its associated graded field $\mathsf{gr}(F)$ is then inertially closed; the Brauer group of $\mathsf{gr}(F)$ is described in Th. 7.64. By invoking the isomorphism $\mathsf{gr}\colon Br_{tr}(F) \xrightarrow{\sim} \mathsf{Br}\big(\mathsf{gr}(F)\big)$ of Th. 6.64 we will obtain an analogue to Th. 7.64 for the tame part of $Br(F)$. See Th. 7.80 below.

For any integer $n > 1$, the valuation induces a map

$$v_n\colon F^\times/F^{\times n} \longrightarrow \Gamma_F/n\Gamma_F.$$

We compare it to the map $deg_n\colon \mathsf{gr}(F)^\times/\mathsf{gr}(F)^{\times n} \longrightarrow \Gamma_{\mathsf{gr}(F)}/n\Gamma_{\mathsf{gr}(F)}$ of Lemma 7.62, using the canonical map $F^\times \to \mathsf{gr}(F)^\times$ that carries $x \in F^\times$ to $\widetilde{x} \in \mathsf{gr}(F)^\times$.

Lemma 7.78. *If F is strictly Henselian and n is prime to char $\overline{F}$, then v_n is an isomorphism and the following diagram commutes, where the vertical maps are canonical:*

$$\begin{array}{ccc} F^\times/F^{\times n} & \xrightarrow{\;v_n\;} & \Gamma_F/n\Gamma_F \\ \big\downarrow & & \big\| \\ \mathsf{gr}(F)^\times/\mathsf{gr}(F)^{\times n} & \xrightarrow{\;deg_n\;} & \Gamma_{\mathsf{gr}(F)}/n\Gamma_{\mathsf{gr}(F)} \end{array}$$

Proof. Since $v(x) = deg\,\widetilde{x}$ for $x \in F^\times$, the diagram commutes. Since Lemma 7.62 shows that deg_n is injective, $ker\, v_n$ is the kernel of the canonical map $F^\times/F^{\times n} \to \mathsf{gr}(F)^\times/\mathsf{gr}(F)^{\times n}$. If $xF^{\times n}$ lies in this kernel, we may find $y \in F^\times$ such that $\widetilde{x} = \widetilde{y}^n$, hence $v(xy^{-n}) = 0$ and $\overline{xy^{-n}} = 1$. Then $X - xy^{-n} \in \mathcal{O}_v[X]$ has residue $X^n - 1 \in \overline{F}[X]$. Since n is prime to $char\,\overline{F}$, the residue polynomial has 1 as a simple root. Hensel's Lemma allows us to lift this root to a root $z \in \mathcal{O}_v^\times$ of $X^n - xy^{-n}$. Then $x = (yz)^n \in F^{\times n}$, hence v_n is injective. Since it is clearly surjective, the proof is complete. □

As in §7.3.2, we let $\mathbb{T}'$ denote the subgroup of $\mathbb{T} = \mathbb{Q}/\mathbb{Z}$ consisting of the elements with torsion prime to $char\,\overline{F}$, and we fix an isomorphism

$$\omega\colon \mathbb{T}' \xrightarrow{\sim} \mu(F)',$$

where $\mu(F)'$ is the group of roots of unity in F of order prime to $char\,\overline{F}$. We also use the notation $\mathbb{T}(\wedge^2\Gamma_F)'$ (resp. $Symp\big(\mathbb{T}(\Gamma_F)\big)'$) for the subgroup of $\mathbb{T}(\wedge^2\Gamma_F)$ (resp. $Symp\big(\mathbb{T}(\Gamma_F)\big)$) whose torsion is prime to $char\,\overline{F}$.

Assume F is strictly Henselian. We next define homomorphisms

$$\Pi_F\colon Br_{tr}(F) \longrightarrow Symp\big(\mathbb{T}(\Gamma_F)\big)' \quad \text{and} \quad \Sigma_F\colon \mathbb{T}(\wedge^2\Gamma_F)' \longrightarrow Br_{tr}(F).$$

By Cor. 6.53, every element in $Br_{tr}(F)$ is the Brauer class of a central division algebra over F on which v extends to a tame valuation. This tame valuation is necessarily totally ramified since F is inertially closed. Therefore, for any such division algebra D, there is a canonical pairing $\overline{c}_D$, which provides Γ_D/Γ_F with a symplectic module structure; also, $\Gamma_D/\Gamma_F \subseteq \mathbb{T}(\Gamma_F)'$ since $char\,\overline{F} \nmid [D{:}F]$ by the nondegeneracy of $\overline{c}_D$ (see Prop. 7.72). We may thus set

$$\Pi_F\big([D]\big) = (\Gamma_D/\Gamma_F, \overline{c}_D).$$

Proposition 7.79. *If F is strictly Henselian, there is a homomorphism*

$$\Sigma_F\colon \mathbb{T}(\wedge^2\Gamma_F)' \longrightarrow Br_{tr}(F)$$

carrying $\frac{1}{n}(v(c)\wedge v(d)) + \wedge^2\Gamma_F$ to $\big[(c,d/F)_{\omega(1/n),n}\big]$ for all c, $d \in F^\times$ and $n > 1$ prime to $char\,\overline{F}$. This homomorphism fits into the following commutative diagram:

$$\begin{array}{ccccc}
\mathbb{T}(\wedge^2\Gamma_F)' & \xrightarrow{\ \Sigma_F\ } & Br_{tr}(F) & \xrightarrow{\ \Pi_F\ } & Symp\big(\mathbb{T}(\Gamma_F)\big)' \\
\Big\| & & \Big\downarrow{\scriptstyle \mathsf{gr}} & & \Big\| \\
\mathbb{T}(\wedge^2\Gamma_{\mathsf{gr}(F)})' & \xrightarrow{\ \Sigma_{\mathsf{gr}(F)}\ } & \mathsf{Br}(\mathsf{gr}(F)) & \xrightarrow{\ \Pi_{\mathsf{gr}(F)}\ } & Symp\big(\mathbb{T}(\Gamma_{\mathsf{gr}(F)})\big)'
\end{array}$$

Note that the definition of $\Sigma_{\mathsf{gr}(F)}$ uses an isomorphism $\mathbb{T}' \xrightarrow{\sim} \mu(\overline{F})$; we choose for this isomorphism the composition of ω with the residue map $\mu(F)' \xrightarrow{\sim} \mu(\overline{F})$.

Proof. The existence of a homomorphism Σ_F making the left square commute is clear, since gr is an isomorphism. For c, $d \in F^\times$ and $n > 1$ prime to $char\,\overline{F}$, the symbol algebra $\big(c,d/F\big)_{\omega(1/n),n}$ satisfies

$$\mathsf{gr}\big[\big(c,d/F\big)_{\omega(1/n),n}\big] = \big[\big(\widetilde{c},\widetilde{d}/\,\mathsf{gr}(F)\big)_{\overline{\omega(1/n)},n}\big],$$

see (3.27). Therefore, $\big[(c,d/F)_{\omega(1/n),n}\big] \in Br_{tr}(F)$. By definition of $\Sigma_{\mathsf{gr}(F)}$, we have

$$\Sigma_{\mathsf{gr}(F)}\big(\tfrac{1}{n}(v(c)\wedge v(d)) + \wedge^2\Gamma_{\mathsf{gr}(F)}\big) = \big[\big(\widetilde{c},\widetilde{d}/\,\mathsf{gr}(F)\big)_{\overline{\omega(1/n)},n}\big].$$

Therefore, Σ_F carries $\frac{1}{n}(v(c)\wedge v(d)) + \wedge^2\Gamma_F$ to $\big[\big(c,d/F\big)_{\omega(1/n),n}\big]$.

Commutativity of the right square follows from the fact that for any central division algebra D over F on which v extends to a tame valuation v_D (necessarily totally ramified), we have

$$\mathsf{gr}[D] = [\mathsf{gr}_{v_D}(D)] \quad \text{and} \quad (\Gamma_D/\Gamma_F, \overline{c}_D) = (\Gamma_{\mathsf{gr}(D)}/\Gamma_{\mathsf{gr}(F)}, \overline{c}_{\mathsf{gr}(D)}). \qquad \square$$

In view of the preceding proposition, the next result follows easily from Th. 7.64:

Theorem 7.80. *For every strictly Henselian field F, the maps Σ_F and Π_F are group isomorphisms, and the following diagram is commutative:*

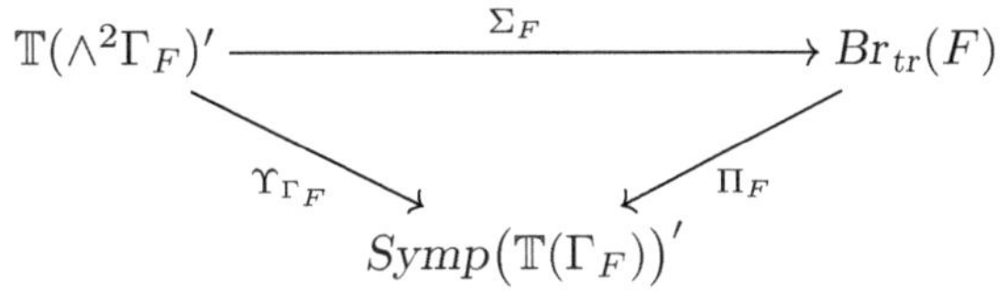

As in the case of inertially closed graded fields, the maps Σ and Π are functorial. The following proposition and its corollary are proved by the same arguments as Prop. 7.65 and Cor. 7.66. We omit the proofs.

Proposition 7.81. *Let K/F be an extension of strictly Henselian fields. The following diagram, where the vertical maps are the scalar extension maps, is commutative:*

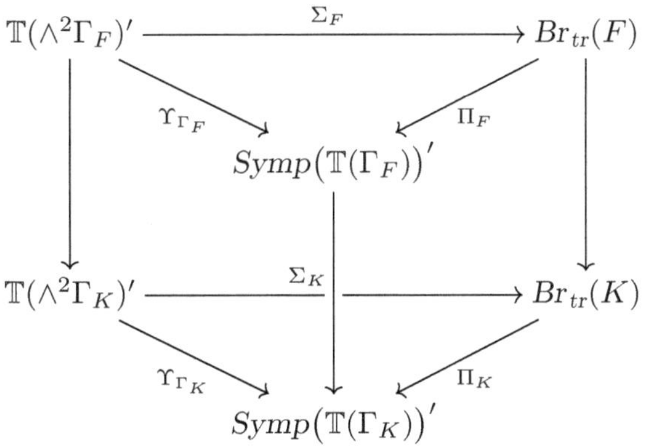

The maps Σ_F and Σ_K are understood to be defined with the same isomorphism $\omega\colon \mathbb{T}' \xrightarrow{\sim} \mu(F)' = \mu(K)'$.

Corollary 7.82. *Let K/F be an extension of strictly Henselian fields, let D be a central division algebra over F, and let E be the central division algebra over K associated to $D \otimes_F K$. Assume the extension of the valuation of F to D is tame. Consider $\Gamma_D \cap \Gamma_K \subseteq \Gamma_D$ and its orthogonal $(\Gamma_D \cap \Gamma_K)^\perp \subseteq \Gamma_D$ for the canonical pairing c_D. We have*

$$\Gamma_E = (\Gamma_D \cap \Gamma_K)^\perp + \Gamma_K$$

and

$$c_E(\gamma,\delta) = c_D(\gamma,\delta) \qquad \textit{for } \gamma,\ \delta \in (\Gamma_D \cap \Gamma_K)^{\perp}.$$

In particular,

(i) *K splits D if and only if $(\Gamma_D \cap \Gamma_K)^{\perp} \subseteq \Gamma_K$;*
(ii) *$D \otimes_F K$ is a division algebra if and only if $\Gamma_D \cap \Gamma_K = \Gamma_F$.*

In the particular case where $\Gamma_F = \mathbb{Z}^n$, we may use the identification $\mathbb{T}(\wedge^2\Gamma_F) = Alt_n(\mathbb{T})$ of (7.21) to substitute the map $\Upsilon_n\colon Alt_n(\mathbb{T}) \xrightarrow{\sim} Symp(\mathbb{T}^n)$ for Υ_{Γ_F}. The reduction of alternating matrices to block-diagonal form (see Prop. 7.23) translates via the isomorphism Σ_F into a decomposition of Brauer classes into symbol algebra classes, as the following example shows:

Example 7.83. Let $F = \mathbb{C}((x_1))((x_2))((x_3))$ with the (x_1,x_2,x_3)-adic valuation. We identify Γ_F with $\mathbb{Z}^3$ (with the right-to-left lexicographic ordering) and $\mu(F)$ with $\mathbb{T}$ by mapping $e^{2i\pi a/b}$ to $\frac{a}{b} + \mathbb{Z}$, for $a,\ b \in \mathbb{Z}$ $(b \neq 0)$. Let $\omega = e^{2i\pi/5}$. Consider the algebra

$$S = (x_1, x_2/F)_{\omega,5} \otimes_F (x_1, x_3/F)_{\omega,5}^{\otimes 3} \otimes_F (x_2, x_3/F)_{\omega,5}.$$

Its Brauer class is the image under Σ_F of the alternating matrix

$$A = \begin{pmatrix} 0 & \frac{1}{5}+\mathbb{Z} & \frac{3}{5}+\mathbb{Z} \\ -\frac{1}{5}+\mathbb{Z} & 0 & \frac{1}{5}+\mathbb{Z} \\ -\frac{3}{5}+\mathbb{Z} & -\frac{1}{5}+\mathbb{Z} & 0 \end{pmatrix}.$$

We saw in Ex. 7.25 that the symplectic module $\Upsilon_3(A) \in Symp(\mathbb{T}^3)$ is isomorphic to $\mathbb{S}_5$, with symplectic base $(\frac{1}{5}+\mathbb{Z}, \frac{2}{5}+\mathbb{Z}, 0)$, $(-\frac{1}{5}+\mathbb{Z}, -\frac{1}{5}+\mathbb{Z}, -\frac{2}{5}+\mathbb{Z})$; therefore, the division algebra D Brauer-equivalent to S is

$$D = (x_1x_2^2, x_1^{-1}x_2^{-1}x_3^{-2}/F)_{\omega,5}.$$

This follows because $\Pi_F([D]) = \Upsilon_3(A)$.

The preceding results for strictly Henselian fields lead to a better picture of the tame part of the Brauer group of any Henselian field. We now prove analogues for Henselian fields to the properties of Brauer groups of graded fields given in §7.3.3. For a field F with Henselian valuation v, the inertial closure F_{in} of F is strictly Henselian. We use the notation ${}_{\mu(\overline{F})}Br_{tr}(F_{in})$ for the "$\mu(\overline{F})$-torsion" of $Br_{tr}(F_{in})$, as defined in (7.46).

Theorem 7.84. *Let F be a Henselian field. Then, the image of the scalar extension map $\mathrm{ext}_{F_{in}/F}\colon Br_{tr}(F) \to Br_{tr}(F_{in})$ is ${}_{\mu(\overline{F})}Br_{tr}(F_{in})$, and there is a split exact sequence:*

$$0 \longrightarrow Br_{is}(F) \longrightarrow Br_{tr}(F) \xrightarrow{\Psi_F} {}_{\mu(\overline{F})}\mathbb{T}(\wedge^2\Gamma_F) \longrightarrow 0$$

where Ψ_F is the composition of $\mathrm{ext}_{F_{in}/F}$ with the restriction of $\Sigma_{F_{in}}^{-1}$ to ${}_{\mu(\overline{F})}Br_{tr}(F_{in})$.

See Cor. 8.61 below for an explicit description of $\Psi_F([D])$, for D a tame central division algebra over F. The exact sequence is canonical, but not canonically split.

Proof. Recall from (6.35) that $\mathsf{gr}(F_{in}) = \mathsf{gr}(F)_{in}$, the inertial closure of the graded field $\mathsf{gr}(F)$. Theorem 6.64 yields isomorphisms $\mathsf{gr}\colon Br_{tr}(F) \xrightarrow{\sim} \mathsf{Br}(\mathsf{gr}(F))$ and also $\mathsf{gr}\colon Br_{tr}(F_{in}) \xrightarrow{\sim} \mathsf{Br}(\mathsf{gr}(F)_{in})$ that are compatible with the scalar extension maps (see commutative diagram (6.36)). Since $ext_{\mathsf{gr}(F)_{in}/\mathsf{gr}(F)}$ maps $\mathsf{Br}(\mathsf{gr}(F))$ onto ${}_{\mu(F_0)}\mathsf{Br}(\mathsf{gr}(F)_{in})$ by Prop. 7.67 and $\overline{F} = F_0$ (the degree 0 component of $\mathsf{gr}(F)$), the isomorphisms show that $ext_{F_{in}/F}$ must map $Br_{tr}(F)$ onto ${}_{\mu(\overline{F})}Br_{tr}(F_{in})$. Hence, Ψ_F carries $Br_{tr}(F)$ onto ${}_{\mu(\overline{F})}\mathbb{T}(\wedge^2\Gamma_F)$. Consider the following diagram:

$$\begin{array}{ccccccccc} 0 & \longrightarrow & Br_{is}(F) & \longrightarrow & Br_{tr}(F) & \xrightarrow{\Psi_F} & {}_{\mu(\overline{F})}\mathbb{T}(\wedge^2\Gamma_F) & \longrightarrow & 0 \\ && \Big\downarrow{\scriptstyle \mathsf{gr}\,\cong} && \Big\downarrow{\scriptstyle \mathsf{gr}\,\cong} && \Big\| && \\ 0 & \longrightarrow & \mathsf{Br}_{is}(\mathsf{gr}(F)) & \longrightarrow & \mathsf{Br}\big(\mathsf{gr}(F)\big) & \xrightarrow{\Phi_{\mathsf{gr}(F)}} & {}_{\mu(F_0)}\mathbb{T}(\wedge^2\Gamma_{\mathsf{gr}(F)}) & \longrightarrow & 0 \end{array}$$

The left square is commutative by Th. 6.64, as is the right square by commutative diagram (6.36) and Prop. 7.79. The left and center vertical maps are isomorphisms by Th. 6.64. Since the bottom row is split exact by Th. 7.69, the top row must also be split exact. □

The following corollary is immediate from the isomorphism of Th. 6.64

$$\mathsf{gr}\colon\ Br_{tr}(F) \xrightarrow{\sim} \mathsf{Br}(\mathsf{gr}(F))$$

for any Henselian field F combined with the corresponding result for $\mathsf{Br}(\mathsf{gr}(F))$ given in Cor. 7.71.

Corollary 7.85. *For any Henselian field F, we have a noncanonical group isomorphism*

$$Br_{tr}(F) \cong Br(\overline{F}) \times Hom^c\left(\mathcal{G}(\overline{F}), \mathbb{T}(\Gamma_F)\right) \times {}_{\mu(\overline{F})}\mathbb{T}(\wedge^2\Gamma_F).$$

Exercises

Exercise 7.1. For an arbitrary integer $n > 1$, show that

$$Symp\big((\mathbb{Z}/n\mathbb{Z})^2\big) \cong \mathbb{Z}/n\mathbb{Z}.$$

Exercise 7.2. Let Γ be a torsion-free abelian group. Given γ, $\delta \in \Gamma$ and an integer $n \geq 1$, prove that there exist an integer $m \geq 1$ and elements ξ, $\eta \in \Gamma$ such that

$$\tfrac{1}{n}(\gamma \wedge \delta) = \tfrac{1}{m}(\xi \wedge \eta)$$

and ξ, η are m-independent (i.e., for a, $b \in \mathbb{Z}$, we have $a\xi + b\eta \in m\Gamma$ if and only if $a \equiv b \equiv 0 \ (mod\ m)$).

Exercise 7.3. Let Ω and Ω' be torsion abelian groups, and let $f : \Omega \to \Omega'$ be a group homomorphism. Let $(M, a) \in Symp(\Omega)$ with $M \cong (\mathbb{Z}/m\mathbb{Z})^2$. Suppose that $\ker f \cap M \cong \mathbb{Z}/r_1\mathbb{Z} \times \mathbb{Z}/r_2\mathbb{Z}$. Prove that the underlying abelian group of $f_\diamond(M, a)$ has invariant factors $m/(r_1 r_2), m/(r_1 r_2)$.

Exercise 7.4. Let Γ be an arbitrary torsion-free abelian group and let p be a prime number. Show that if $(\varepsilon_i)_{i \in I_p}$ is a family of elements in Γ such that $(\varepsilon_i + p\Gamma)_{i \in I_p}$ is an $\mathbb{F}_p$-vector space base of $\Gamma/p\Gamma$, then the p-primary component $\mathbb{T}(\Gamma)_{(p)}$ is a direct sum of copies of $\mathbb{Q}_p/\mathbb{Z}_p$ indexed by I_p. If the cardinalities $|I_p|$ are the same for all p, show that $\mathbb{T}(\Gamma)$ is a direct sum of $|I_p|$ copies of $\mathbb{T}$.

Exercise 7.5. Let D be a division algebra over a field F, and assume F contains a primitive s-th root of unity for some integer s. Show that every finite abelian subgroup of exponent dividing s in $D^\times/F^\times$ is an armature.

Exercise 7.6. Let $\mathcal{A}$ be an armature in an F-algebra A. Show that if $F[rad\mathcal{A}]$ is a field, then $\mathcal{A}/rad\mathcal{A}$ can be canonically identified with an armature in the $F[rad\mathcal{A}]$-algebra $F[\mathcal{A}]$.

Exercise 7.7. Let F be an inertially closed graded field, and let F_{alg} be a graded algebraic closure of F. Define a map η from the set of finite-degree graded field extensions of F to $\mathbb{H}(\Gamma_\mathsf{F})$ by $\mathsf{K} \mapsto \Gamma_\mathsf{K}$. Show that η gives a one-to-one correspondence between (i) finite-degree tame graded field extensions of F in F_{alg}; and (ii) subgroups $\Delta \subseteq \mathbb{H}(\Gamma_\mathsf{F})$ containing Γ_F as a subgroup of finite index prime to $char\,\mathsf{F}_0$. [Hint: for injectivity, use induction on the degree, Prop. 5.18, and Lemma 7.62.]

Exercise 7.8. Let (F, v) be a valued field, and let A be a central simple F-algebra with $char\,\overline{F} \nmid deg\,A$. Let D be the associated division algebra of A, and suppose v extends to a valuation v_D on D that is totally ramified over v. Suppose A is spanned by an armature $\mathcal{A}$, and consider the map $\overline{\alpha}_\mathcal{A} : \mathcal{A} \to \Gamma^\times_A/\Gamma_F$ induced by the armature gauge $\alpha_\mathcal{A}$ (see §7.2.3). Let $\mathcal{K} = \ker \overline{\alpha}_\mathcal{A}$ and let $\mathcal{K}^\perp$ be the orthogonal of $\mathcal{K}$ in $\mathcal{A}$ with respect to the commutator pairing $b_\mathcal{A}$. Prove that as symplectic modules

$$(\Gamma_D/\Gamma_F, \overline{c}_D) = \overline{\alpha}_{\mathcal{A}\diamond}(\mathcal{A}, b_\mathcal{A}).$$

Deduce that

$$[D{:}F] = |\mathcal{K}^\perp{:}(\mathcal{K} \cap \mathcal{K}^\perp)| \qquad \text{and} \qquad \Gamma_D/\Gamma_F = \overline{deg}(\mathcal{K}^\perp).$$

Exercise 7.9. Let $F = \mathbb{R}((x))((y))$ be the field of iterated Laurent series in two indeterminates over the field $\mathbb{R}$ of real numbers, with the (x, y)-adic valuation. Show that $Br(F) \cong (\mathbb{Z}/2\mathbb{Z})^4$ and $Br(F_{in}) \cong \mathbb{T}$.

Notes

§7.1.1: Symplectic modules seem to have been first investigated in relation to classification problems in topology. The earliest references where their structure is established are de Rham [201, §19] and C.T.C. Wall [259]. The structure of their Lagrangians is discussed by Tignol–Amitsur [243] for the purpose of obtaining lower bounds for the degree of Galois splitting fields of universal division algebras. Another approach to these lower bounds, also using Lagrangians in symplectic modules, is due to Reichstein–Youssin [197].

§7.1.2: The definition of the group structure on $Symp(\Omega)$ and the map $f_\diamond$ determined by a homomorphism $f\colon \Omega_1 \to \Omega_2$ can be explained by relating $Symp(\Omega)$ to alternating forms on the dual of Ω. Here is a sketch of the connection: Let G be an abelian topological group of one of the following types: Type I: a torsion group with the discrete topology; Type II: a profinite group. For G of either type, the *dual* of G is

$$G^\vee = Hom^c(G, \mathbb{T}),$$

the group of continuous homomorphisms from G to $\mathbb{T} = \mathbb{Q}/\mathbb{Z}$, where $\mathbb{T}$ has the discrete topology; $G^\vee$ is given the compact open topology. It is known that if G is of Type I or Type II, then $G^\vee$ is of the other type. Moreover, the canonical monomorphism $G \to G^{\vee\vee}$ is an isomorphism of topological groups. This is a special case of the Pontrjagin duality for locally compact abelian topological groups, and it is described thoroughly in Ribes–Zaleskiĭ [205]. Further, if G_1 and G_2 are groups of the same type then for any continuous group homomorphism $f\colon G_1 \to G_2$, the induced map $f^\vee\colon G_2^\vee \to G_1^\vee$ is a continuous group homomorphism. The assignment $f \mapsto f^\vee$ is a contravariant functor preserving exact sequences, and $f^{\vee\vee} = f$ under the canonical identification $G_i^{\vee\vee} = G_i$. For any closed subgroup A of $G^\vee$, the annihilator of A in G is

$$A^0 = \{s \in G \mid \psi(s) = 0 \ \text{ for all } \psi \in A\},$$

which is a closed subgroup of G. The canonical map $A^0 \to (G^\vee/A)^\vee$ is a topological group isomorphism.

Now, let Ω be any torsion abelian group. Let $Alt^c(\Omega^\vee)$ denote the group of continuous alternating bilinear functions $\Omega^\vee \times \Omega^\vee \to \mathbb{T}$. Define a map from $Alt^c(\Omega^\vee)$ to $Symp(\Omega)$ as follows: Given $\alpha \in Alt^c(\Omega^\vee)$, the continuity of α and the compactness of $\Omega^\vee$ imply that $rad\,\alpha$ is a closed subgroup of finite index in $\Omega^\vee$. The pairing α induces a nondegenerate alternating pairing $\overline{\alpha}$ on the finite abelian group $\Omega^\vee/rad\,\alpha$. We use $\overline{\alpha}$ to define an isomorphism $\overline{\alpha}_*\colon \Omega^\vee/rad\,\alpha \to \big(\Omega^\vee/rad\,\alpha\big)^\vee$ by $\overline{\alpha}_*(\eta)(\rho) = \overline{\alpha}(\eta,\rho)$ for all $\eta, \rho \in \Omega^\vee/rad\,\alpha$. The pairing $\overline{\alpha}$ transfers via this isomorphism to a pairing on $(\Omega^\vee/rad\,\alpha)^\vee$, which in turn transfers to a pairing α^0 on $rad\,\alpha^0$ using the canonical isomorphism $(\Omega^\vee/rad\,\alpha)^\vee \cong rad\,\alpha^0$; the alternating pairing α^0 is nondegenerate since $\overline{\alpha}$ is nondegenerate. We thus have a map

$$\mathcal{V}_\Omega\colon Alt^c(\Omega^\vee) \longrightarrow Symp(\Omega) \qquad \text{given by} \qquad \alpha \mapsto (rad\,\alpha^0, \alpha^0).$$

One can show that $\mathcal{V}_\Omega$ is a bijection. (For the inverse map, take any $(M,a) \in \mathit{Symp}(\Omega)$. Then a induces an isomorphism $a_*\colon M \to M^\vee$ and thereby a corresponding nondegenerate pairing $a^\vee \in \mathit{Alt}(M^\vee)$ given by $a^\vee(a_*(m), a_*(n)) = a(m,n)$ for all $m,n \in M$. The inclusion $M \hookrightarrow \Omega$ yields an epimorphism $\Omega^\vee \to M^\vee$, and $a^\vee$ on $M^\vee$ lifts to a continuous alternating pairing α_a on $\Omega^\vee$. One can check that $\mathcal{V}_\Omega(a_\alpha) = (M,a)$.) Fundamentally, this shows that there is a bijection between nondegenerate alternating pairings on finite factor groups of $\Omega^\vee$ (modulo closed subgroups) and nondegenerate alternating pairings on finite subgroups of Ω.

Suppose Ω_1 and Ω_2 are torsion abelian groups, and $f\colon \Omega_1 \to \Omega_2$ is a group homomorphism. The map f induces a continuous homomorphism $f^\vee\colon \Omega_2^\vee \to \Omega_1^\vee$, and thereby a map $\mathit{Alt}(f)\colon \mathit{Alt}^c(\Omega_1^\vee) \to \mathit{Alt}^c(\Omega_2^\vee)$. One can then show that there is a commutative diagram

$$\begin{array}{ccc} \mathit{Alt}^c(\Omega_1^\vee) & \xrightarrow{\mathit{Alt}(f)} & \mathit{Alt}^c(\Omega_2^\vee) \\ {\scriptstyle \mathcal{V}_{\Omega_1}}\big\downarrow & & \big\downarrow{\scriptstyle \mathcal{V}_{\Omega_2}} \\ \mathit{Symp}(\Omega_1) & \xrightarrow{f_\diamond} & \mathit{Symp}(\Omega_2). \end{array}$$

Since the assignment $f \mapsto \mathit{Alt}(f)$ is clearly compatible with composition of group homomorphisms, the same must be true for $f \mapsto f_\diamond$. Also, the group operation on $\mathit{Alt}^c(\Omega)$ transfers via $\mathcal{V}_\Omega$ to a group operation on $\mathit{Symp}(\Omega)$, and this coincides with the group operation defined in §7.1.2.

The group $\mathit{Alt}^c(G)$ for G an arbitrary profinite abelian group has been studied by Brussel, who established in [44, Th. 2.4] canonical isomorphisms

$$H^2(G,\mathbb{T}) \cong \mathit{Alt}^c(G) \cong \wedge^2(G^\vee).$$

Brussel shows in [44, Th. 3.6] how for any given $\alpha \in \mathit{Alt}^c(G)$ the order of the group $G/\mathit{rad}\alpha$ can be computed from the pfaffian of a matrix representing α with respect to a base of $G/\mathit{rad}\alpha$. See also [42, §2], where Brussel proves a similar result relating pfaffians of submatrices of an alternating matrix $A \in \mathit{Alt}_n(\mathbb{T})$ to the order of the corresponding symplectic module $\Upsilon_n(A) \in \mathit{Symp}(\mathbb{T}^n)$. Note however that Brussel's result is not expressed in terms of symplectic modules, but in terms of indices of central simple algebras over a strictly Henselian field with residue field of characteristic 0 and value group $\mathbb{Z}^n$. These two viewpoints are equivalent by Th. 7.80.

§7.2: Armatures were first defined by Tignol in [236] for division algebras. When the base field contains enough roots of unity, the linear independence of representatives of the armature elements is automatic in this case, see [236, Lemme 1.5] and Exercise 7.5. The definition for central simple algebras was given by Tignol–Wadsworth in [245]. The notion of armature was inspired by the *q-generating subsets* of Amitsur et al. [11, §1] and the *p-central sets* of Rowen [213, §2].

If $\mathcal{A}$ is a spanning armature of a central F-algebra A, then $\mathcal{A}$ is a maximal abelian subgroup of $A^\times/F^\times$. To see this, consider the centralizer $C(\mathcal{A})$ of $\mathcal{A}$ in $A^\times/F^\times$. The commutator pairing defines a group homomorphism $C(\mathcal{A}) \to \mathcal{A}^\vee$. This homomorphism is injective because the center of $F[\mathcal{A}]$ is F, hence $|C(\mathcal{A})| \leq |\mathcal{A}^\vee|$. But $|\mathcal{A}^\vee| = |\mathcal{A}|$ and $\mathcal{A} \subseteq C(\mathcal{A})$, so we must have $C(\mathcal{A}) = \mathcal{A}$, and therefore $\mathcal{A}$ is a maximal abelian subgroup. When F is algebraically closed of characteristic 0 and $A = M_n(F)$, we have $A^\times/F^\times = \mathrm{PGL}_n(F)$. Reichstein–Youssin [197, Lemma 7.5] show that a finite abelian subgroup $\mathcal{A} \subseteq \mathrm{PGL}_n(F)$ is toral (i.e., embeddable in a torus) if and only if the commutator pairing is trivial on $\mathcal{A}$. They also prove [197, Lemma 7.8] that every symplectic module of order n^2 embeds as an armature in $\mathrm{PGL}_n(F)$.

§7.4: The first result on the structure of tame totally ramified central division algebras over Henselian fields is due to Draxl [64, Th. 1], who showed that these algebras are tensor products of symbol algebras. Tame totally ramified division algebras (over fields that are not necessarily Henselian) were further studied by Tignol–Wadsworth in [245], where the canonical pairing on the value group is defined. By contrast, it had already been observed by Scharlau that for a Henselian field F, the part of $Br(F_{in})$ of torsion prime to $char\,\overline{F}$ is generated by symbol algebras [225, Kor. 3.4]. In the same paper, Scharlau purports to describe the quotients of the filtration of $Br(F)$ as in Th. 6.66 and Th. 7.84. However, his Satz 4.1 is flawed: the image of $Br_{tr}(F)$ in $Br_{tr}(F_{in})$ is only the $\mu(\overline{F})$-torsion part of $Br_{tr}(F_{in})$, not the full $Br_{tr}(F_{in})$; see Th. 7.84 and Exercise 7.9.

That symbol algebras generate the Brauer group of a strictly Henselian field of residue characteristic 0 can also be derived by an argument of Saltman [220, Th. 2.1] from the fact that the absolute Galois group of these fields is abelian. For F a strictly Henselian field with $char\,\overline{F} = 0$ and $\Gamma_F \cong \mathbb{Z}^n$, Brussel sets up in [42, §1] an isomorphism $Br(F) \cong Alt_n(\mathbb{T})$ depending on a base of Γ_F; his map is the isomorphism Σ_F of Prop. 7.79 given the identification $\mathbb{T}(\wedge^2\Gamma_F) \cong Alt_n(\mathbb{T})$ of (7.21). In [44], Brussel uses the same ideas more generally to describe the relative Brauer group of a Galois extension with abelian Galois group. The isomorphism $Br(F) \cong Alt_n(\mathbb{T})$ for F strictly Henselian with residue field k of characteristic zero and $\Gamma_F \cong \mathbb{Z}^n$ can also be obtained as follows: Since every central division algebra over F is tame, Th. 6.64 yields an isomorphism $\mathsf{gr}\colon Br(F) \xrightarrow{\sim} \mathsf{Br}(\mathsf{gr}(F))$. As observed in the notes to Ch. 6, we have $\mathsf{Br}(\mathsf{gr}(F)) = Br(\mathsf{gr}(F)^\natural)$ because $char\,\mathsf{gr}(F) = 0$. But $\mathsf{gr}(F)^\natural \cong k[t_1, t_1^{-1}, \dots, t_n, t_n^{-1}]$ is the coordinate ring of an n-dimensional torus over k, so $Br(\mathsf{gr}(F)^\natural) \cong (\mathbb{Q}/\mathbb{Z})^{n(n-1)/2}$ by Magid's computation of the Brauer group of a torus [129, Cor. 7]. (See also Gille–Pianzola [83, Prop. 3.1(2), (4.2)] and Gille–Semenov [85, Th. 2.8] for other approaches to Magid's computation.)

Example 7.77 was given in a more general form in Tignol–Wadsworth [245]. For any integers r, $s > 1$, and $t \in \{1, \dots, rs\}$, Prop. 5.8 of that paper yields a division algebra of degree and exponent rs that is tame and totally ramified over its center, which contains a primitive rs-th root of unity, with relative value group isomorphic to $(\mathbb{Z}/\ell\mathbb{Z})^2 \times (\mathbb{Z}/d\mathbb{Z})^2$ where $\ell = lcm(rs, rt, st)/t$ and $d = rs/\ell$. In particular, if r and s are not relatively prime, then the division algebra does not contain any armature. It is therefore not a symbol algebra, nor a tensor product of symbol algebras.

Chapter 8
Division Algebras over Henselian Fields

The description of the tame part of the Brauer group of a Henselian field is a common theme for Ch. 6 and 7, culminating in Cor. 7.85. In the present chapter, we turn our attention to the division algebras that represent the elements in this tame Brauer group.

Recall from Cor. 1.7 that every (finite-dimensional) division algebra D over a Henselian-valued field (F, v) carries a unique valuation v_D extending v. Valuation properties of D as an F-algebra are always understood to be in reference to v_D over v. Thus, we say that D is *defectless over* F if v_D is a v-norm, i.e., $[D{:}F] = [\overline{D}{:}\overline{F}] \cdot |\Gamma_D{:}\Gamma_F|$; we say D is *inertial over* F if v_D is an inertial v-gauge, i.e., $[D{:}F] = [\overline{D}{:}\overline{F}]$ and $Z(\overline{D})$ is separable over $\overline{F}$; we say D is *tame over* F if v_D is a tame v-gauge, see Def. 8.4 below.

We saw in Cor. 7.85 that the tame Brauer group $Br_{tr}(F)$ decomposes non-canonically into a direct sum of three subgroups. One consists of the Brauer classes of the inertial division algebras, and another of Brauer classes of certain totally ramified division algebras. The third one is represented by a type of division algebras defined in §8.3 below, the *decomposably semiramified division algebras*. The various types of division algebras can be characterized in terms of subalgebras canonically defined by their valuation properties in §8.2.2. Major results in this chapter are Th. 8.60 and Th. 8.65, which yield information on a division algebra over a Henselian field from its decomposition up to Brauer-equivalence in a tensor product of inertial, semiramified, and totally ramified division algebras. Theorem 8.65 shows in particular that if a division algebra D is split by an inertial extension of its center F, then the Brauer class of the residue algebra $\overline{D}$ is in the image of the scalar extension map $Br(\overline{F}) \to Br(Z(\overline{D}))$. Therefore, it has a preimage in the quotient $Br(\overline{F})/\,Br(Z(\overline{D})/\overline{F})$. We show that in fact D canonically determines an element $sp(D)$ in the quotient of $Br(\overline{F})$ by a smaller subgroup $Dec(Z(\overline{D})/\overline{F})$. This element $sp(D)$ gives significant information on D, and plays an important role in the construction of various examples in Ch. 9 and Ch. 12.

J.-P. Tignol, A.R. Wadsworth, *Value Functions on Simple Algebras, and Associated Graded Rings*, Springer Monographs in Mathematics,
DOI 10.1007/978-3-319-16360-4_8

Just as we have seen for totally ramified division algebras, most of the notions discussed in this chapter gain their full relevance when the rank of the valuation on the base field is at least 2. The theory is less rich but still substantial when the valuation on the base field is discrete rank 1; a summary of the main results of this chapter in this special case is given in §8.4.6.

Our approach is consistent with the methods developed throughout this monograph: we first discuss in §8.2.1, §8.3.3, §8.4.1, and §8.4.2 graded analogues of the notions we want to investigate. The results in the Henselian case are then obtained in §8.2.2, §8.3.4, §8.4.3, and §8.4.4 by lifting techniques, which are discussed in §8.1.

8.1 Tame and inertial lifts in valued division algebras

Classically, the idea of lifting refers to the following issue: If (F, v) is a valued field and K is a field extension of $\overline{F}$, find a valued field extension (L, v_L) of (F, v) such that $\overline{L} = K$ and $[L{:}F] = [K{:}\overline{F}]$. The problem is easily solved for any finite-degree extension $K/\overline{F}$, see Prop. 8.2. It is more delicate (and sometimes insoluble) when the field L has to be found within a given extension (M, v_M) such that $\overline{M} \supseteq K$, see Prop. 8.3 and Exercises 8.1 and 8.2. We broaden the classical framework by substituting the graded field $\mathsf{gr}(F)$ for the residue field $\overline{F}$, and by considering division algebras over F instead of field extensions.

The main result in this section is Th. 8.13 which shows that if A is a defectless division algebra over a Henselian field F and E is a subalgebra of $\mathsf{gr}(A)$ which is tame[1] over $\mathsf{gr}(F)$, then there is a lift D of E over F, this lift is tame and unique up to isomorphism, and A contains an isomorphic copy of D. Thus, the isomorphism classes of tame-over-F subalgebras of A are in one-to-one correspondence with the isomorphism classes of tame-over-$\mathsf{gr}(F)$ subalgebras of $\mathsf{gr}(A)$. This is quite striking since $\mathsf{gr}(A)$ has a much simpler structure than A itself, and it indicates the utility of studying $\mathsf{gr}(A)$ to understand properties of A. Notable special cases of tame lifts are inertial lifts: we will show in Cor. 8.15 that inertial F-subalgebras of A are classified by separable $\overline{F}$-subalgebras of $\overline{A}$.

Definition 8.1. Let (F, v) be a valued field and let E be a (finite-dimensional) graded division algebra over $\mathsf{gr}(F)$. Any division algebra D over F with a valuation v_D extending v such that $[\mathsf{gr}(D){:}\mathsf{gr}(F)] = [D{:}F]$ and $\mathsf{gr}(D) \cong_g \mathsf{E}$ as graded $\mathsf{gr}(F)$-algebras is called a *lift* of E over F. These conditions imply that v_D is the unique extension of v to D, and that v_D is defectless over F, hence v_D is a v-gauge.

Lifts of tame graded field extensions over a Henselian field have been considered in §5.2: See Cor. 5.56. Note that when E is commutative (i.e., a graded

[1] See Def. 8.4.

field) the division algebra D need not be commutative: see Ex. 6.36 or Exercise 8.6 for a symbol algebra D such that $\mathsf{gr}(D)$ is a graded field. However, Lemma 4.31 shows that for any valued division algebra D the center $Z(\mathsf{gr}(D))$ is a purely inseparable graded field extension of $\mathsf{gr}(Z(D))$. Therefore, if $Z(\mathsf{E})$ is a tame graded field extension of $\mathsf{gr}(F)$, then $\mathsf{gr}(Z(D)) = Z(\mathsf{E})$ for any lift D of E over F, so $Z(D)$ is a field that is a lift of $Z(\mathsf{E})$. In particular, every lift of a tame graded field extension is a field.

We will assume for most of this section that the valuation v is Henselian. However, this hypothesis is not needed for our first result, which shows that any finite-degree graded field extension can be lifted to a field:

Proposition 8.2. *Let (F, v) be an arbitrary valued field and let K be a finite-degree graded field extension of $\mathsf{gr}(F)$. There exists a field L that is a lift of K over F.*

Proof. Arguing by induction on $[\mathsf{K}\colon \mathsf{gr}(F)]$, we need only consider the case where $\mathsf{K} = \mathsf{gr}(F)[a]$ for some homogeneous element a. Let $n = [\mathsf{K}\colon \mathsf{gr}(F)]$, let $\delta = deg\, a$, and let $g \in q(\mathsf{gr}(F))[X]$ be the minimal polynomial of a over $q(\mathsf{gr}(F))$. By Prop. 5.7 we have $g \in \mathsf{gr}(F)[X]$ and g is homogeneous in the graded polynomial ring $\mathsf{gr}(F)[X]^{(\delta)}$. Moreover, the map $\mathsf{gr}(F)[X] \to \mathsf{K}$ carrying X to a induces an isomorphism

$$\mathsf{gr}(F)[X]^{(\delta)}/(g) \cong_g \mathsf{K}.$$

Choose $b_0, \ldots, b_{n-1} \in F$ such that

$$g = X^n + \widetilde{b_{n-1}}X^{n-1} + \ldots + \widetilde{b_0};$$

then, $deg\, \widetilde{b_i} = \delta(n-i)$ if $b_i \neq 0$. Let

$$f = X^n + b_{n-1}X^{n-1} + \ldots + b_0 \in F[X] \quad \text{and} \quad L = F[X]/(f).$$

The polynomial f is δ-uniform and satisfies $\widetilde{f}^{(\delta)} = g$. Since g is irreducible, it follows from Lemma 5.47 that f is irreducible, hence L is a field. Let w be any valuation on L extending v, and let $u \in L$ be the image of X, which is a root of f. We have $w(u) = \delta$ by Prop. 5.44, and $\widetilde{f}^{(\delta)}(\widetilde{u}) = 0$ by Lemma 5.46. Therefore, we may embed K as a graded subalgebra in $\mathsf{gr}_w(L)$ by mapping a to $\widetilde{u}$. Since $deg_X\, f = n$ we have $[L{:}F] = n$, hence $[\mathsf{gr}_w(L)\colon \mathsf{gr}(F)] \leq n$. But since K embeds in $\mathsf{gr}_w(L)$ we also have $[\mathsf{gr}_w(L)\colon \mathsf{gr}(F)] \geq n$, hence $[\mathsf{gr}_w(L)\colon \mathsf{gr}(F)] = [L{:}F]$ and $\mathsf{gr}_w(L) \cong_g \mathsf{K}$. □

In the notation of Prop. 8.2, we have L is inertial (resp. totally ramified) over F if and only if K is inertial (resp. totally ramified) over $\mathsf{gr}(F)$ and, similarly, L is tamely ramified over F if and only if K is tame over $\mathsf{gr}(F)$; see Prop. 5.50.

From here on until the end of this section, we assume the valuation v on F is Henselian. The lift of a tame graded field extension $\mathsf{K}/\,\mathsf{gr}(F)$ is then unique up to isomorphism by Cor. 5.56. (That corollary yields an alternative proof for the existence of a lift of a tame (possibly infinite-dimensional) graded field extension.) We now show a stronger lifting and uniqueness property, finding a lift of a tame graded field extension in a given Henselian extension of the ground field.

Proposition 8.3. *Suppose (F, v) is a Henselian-valued field, and let K be a finite-degree graded field extension of $\mathsf{gr}(F)$. Let M be a field containing F with a Henselian valuation v_M extending v, and suppose $\mathsf{K} \subseteq \mathsf{gr}(M)$. If K is tame over F, then there is a unique lift L of K over F such that $L \subseteq M$ and $\mathsf{gr}(L) = \mathsf{K}$.*

Proof. As in the proof of Prop. 8.2, we argue by induction on $[\mathsf{K}{:}\,\mathsf{gr}(F)]$ and thus reduce to the case where $\mathsf{K} = \mathsf{gr}(F)[a]$ for some homogeneous element a. We use the same notation as in the proof of Prop. 8.2, so $g \in \mathsf{gr}(F)[X]$ is the minimal polynomial of a over $q\big(\mathsf{gr}(F)\big)$ and $f \in F[X]$ is a δ-uniform monic polynomial such that $\widetilde{f}^{(\delta)} = g$. Since K is tame over $\mathsf{gr}(F)$, the field extension $q(\mathsf{K})/q\big(\mathsf{gr}(F)\big)$ is separable (see Prop. 5.19). Hence, a is a simple root of g. Since v_M is Henselian and $\widetilde{f}^{(\delta)}$ has a simple root in $\mathsf{gr}(M)$, by Th. 5.49 f has a unique root u in M with $\widetilde{u} = a$ in $\mathsf{gr}(M)$. Let $L = F(u) \subseteq M$. The proof of Prop. 8.2 shows that L is a lift of K over F. We have $\mathsf{gr}(L) = \mathsf{K}$ since $\widetilde{u} = a$. Since u is the unique root of f in M with $\widetilde{u} = a$, it follows that L is the unique lift of K over F in M with $\mathsf{gr}(L) = \mathsf{K}$. □

We call the lift L as in Prop. 8.3 the *lift of* K *over* F *in* M, or the *tame lift of* K *over* F *in* M, for emphasis. Note that when K is not tame over $\mathsf{gr}(F)$, Prop. 8.3 can fail in two different ways: there can be no lifts of K over F in $\mathsf{gr}(M)$, or there can be nonisomorphic lifts of K over F; see Exercises 8.1 and 8.2 below.

We next prove analogues of Prop. 8.2 and Prop. 8.3 for lifting graded division algebras. We still assume that the valuation v on F is Henselian, so every finite-dimensional division algebra D over F carries a unique valuation v_D extending v, see Cor. 1.7. The graded division algebra $\mathsf{gr}(D)$ is always defined with respect to v_D. Tameness conditions again play a special role. The most useful are the following:

Definition 8.4. A (finite-dimensional) graded division algebra E over a graded field F is said to be *tame* (over F) if $Z(\mathsf{E})$ is a tame graded field extension of F. A (finite-dimensional) division algebra D over a Henselian field F is said to be *tame* (over F) if v_D is a tame v-gauge, which means that D is defectless over F, $Z\big(\mathsf{gr}(D)\big) = \mathsf{gr}\big(Z(D)\big)$, and $\mathsf{gr}\big(Z(D)\big)$ is a tame graded field extension of $\mathsf{gr}(F)$, see Def. 6.40. Using the results in §6.2.2 (particularly Prop. 6.42 and Prop. 6.45), we may describe this condition more explicitly: let $\overline{p} = char\,\overline{F}$; if $\overline{p} \nmid [D{:}F]$ (in particular if $\overline{p} = 0$), then D is tame. In general, D is tame if and only if D is defectless over F, the extension $Z(\overline{D})/\overline{F}$ is separable and $\overline{p} \nmid |ker(\theta_D){:}\Gamma_F|$. Proposition 6.63 shows that D is tame if and only if it is split by the maximal tamely ramified extension F_{tr} of F (see §A.2 in Appendix A).

Note that by Lemma 4.31 the extension $Z(\mathsf{gr}(D))/\mathsf{gr}(Z(D))$ is purely inseparable for any division algebra D over F. Therefore, we have $Z(\mathsf{gr}(D)) = \mathsf{gr}(Z(D))$ whenever $Z(\mathsf{gr}(D))$ is a tame graded field extension of $\mathsf{gr}(F)$. Thus, a division algebra D is tame over F if and only if the graded division algebra $\mathsf{gr}(D)$ is tame over $\mathsf{gr}(F)$, which means that every lift of a tame graded division algebra is tame. A lift of a tame graded division algebra E over F is also for emphasis called a *tame lift* of E over F.

We next show that graded division algebras lift.

Proposition 8.5. *Let (F, v) be a Henselian-valued field and let E be a finite-dimensional graded division algebra over $\mathsf{gr}(F)$. There exists a division algebra D that is a lift of E over F. If E is tame over $\mathsf{gr}(F)$, the division algebra D is unique up to F-isomorphism.*

Proof. We first prove the existence of D. Let $\mathsf{K} = Z(\mathsf{E})$, and let L be a lift of K over F, as in Prop. 8.2. Identify K with $\mathsf{gr}(L)$. Let D be the L-central division algebra with $\mathsf{gr}_L[D] = [\mathsf{E}]$ where gr_L is the index-preserving isomorphism $Br_{tr}(L) \xrightarrow{\sim} \mathsf{Br}(\mathsf{gr}(L))$ of Th. 6.64 above. Then, $\mathsf{gr}(D)$ and E are Brauer-equivalent central graded division algebras over K; hence, $\mathsf{gr}(D) \cong_g \mathsf{E}$. Also, D is defectless over F since D is defectless over L and L is defectless over F, hence $[\mathsf{gr}(D){:}\, \mathsf{gr}(F)] = [D{:}F]$ and D is a lift of E over F.

Now assume K is tame over $\mathsf{gr}(F)$, and assume D' is a lift of E over F; so, $v_{D'}$ is a v-gauge on D'. Let $\mathsf{Z} = \mathsf{gr}(Z(D'))$; so

$$\mathsf{gr}(F) \subseteq \mathsf{Z} \subseteq Z(\mathsf{gr}(D')) \subseteq \mathsf{gr}(D').$$

Since K is tame over $\mathsf{gr}(F)$ and $Z(\mathsf{gr}(D')) \cong \mathsf{K}$, we have $Z(\mathsf{gr}(D'))$ tame over Z. But recall from Lemma 4.31 that $Z(\mathsf{gr}(D'))$ is purely inseparable over Z. Hence, $Z(\mathsf{gr}(D')) = \mathsf{Z}$. Then, by definition D' is tame. Moreover, $Z(D')$ is a lift of K. As observed in Cor. 5.56, the lift of a tame graded field extension is unique up to F-isomorphism, so we may assume $Z(D') = L$. We have $[D'] \in Br_{tr}(L)$ and $\mathsf{gr}_L[D'] = [\mathsf{E}] = \mathsf{gr}_L[D]$, hence $D \cong D'$ since the map gr_L is injective. □

Corollary 8.6. *Let D and D' be division algebras over a Henselian field F. Assuming that D and D' are tame over F, we have*

$$D \cong D' \quad \textit{if and only if} \quad \mathsf{gr}(D) \cong_g \mathsf{gr}(D').$$

Proof. If $\mathsf{gr}(D) \cong_g \mathsf{gr}(D')$, then D and D' are each tame lifts of $\mathsf{gr}(D)$ over F. Hence, $D \cong D'$ by Prop. 8.5. Conversely, if $\varphi\colon D \to D'$ is any F-isomorphism, then $v_D = v_{D'} \circ \varphi$, since there is a unique extension of the Henselian valuation v to D. Therefore, φ induces a graded $\mathsf{gr}(F)$-isomorphism $\widetilde{\varphi}\colon \mathsf{gr}(D) \to \mathsf{gr}(D')$. □

If E is a central graded division algebra over $\mathsf{gr}(F)$, then E is tame and D is a lift of E over F if and only if the Brauer class $[D]$ is a preimage of $[\mathsf{E}]$ under the isomorphism $\mathsf{gr}\colon Br_{tr}(F) \xrightarrow{\sim} \mathsf{Br}(\mathsf{gr}(F))$. Since gr is a group homomorphism, we have the following noteworthy property:

Proposition 8.7. *Let E_1, E_2, and E_3 be graded division algebras each with center $\mathsf{gr}(F)$, with respective tame lifts D_1, D_2, and D_3 over F. Then, $D_3 \sim D_1 \otimes_F D_2$ in $Br(F)$ if and only if $\mathsf{E}_3 \sim \mathsf{E}_1 \otimes_{\mathsf{gr}(F)} \mathsf{E}_2$ in $\mathsf{Br}(\mathsf{gr}(F))$.*

The following lemma is a graded analogue of a standard result on central simple algebras. It will be used in proving an embedding property for tame lifts of graded division algebras analogous to what was proved for graded fields in Prop. 8.3. Recall that all the simple graded algebras are tacitly assumed to be finite-dimensional.

Lemma 8.8. *Let A be a central graded division algebra over a graded field Z. Let E be a simple graded Z-algebra, and let m be the matrix size $ms(\mathsf{E}^{\mathrm{op}} \otimes_{\mathsf{Z}} \mathsf{A})$. Then, $m \mid [\mathsf{E}{:}\mathsf{Z}]$, and equality holds if and only if E is Z-isomorphic to a graded Z-subalgebra of A.*

Proof. Let S be a simple right graded $\mathsf{E}^{\mathrm{op}} \otimes_{\mathsf{Z}} \mathsf{A}$-module and let $\mathsf{B} = \mathsf{End}_{\mathsf{E}^{\mathrm{op}} \otimes_{\mathsf{Z}} \mathsf{A}}(\mathsf{S})$, which is a graded division algebra. The graded Wedderburn Theorem 2.26 says that $\mathsf{E}^{\mathrm{op}} \otimes_{\mathsf{Z}} \mathsf{A} \cong_g \mathsf{End}_{\mathsf{B}}(\mathsf{S})$, hence $m = dim_{\mathsf{B}}\, \mathsf{S}$ and

$$[\mathsf{E}{:}\mathsf{Z}] \cdot [\mathsf{A}{:}\mathsf{Z}] \;=\; m^2 [\mathsf{B}{:}\mathsf{Z}]. \tag{8.1}$$

Viewing S as a right graded A-vector space, let $s = dim_{\mathsf{A}}\, \mathsf{S}$, so

$$dim_{\mathsf{Z}}\, \mathsf{S} \;=\; m[\mathsf{B}{:}\mathsf{Z}] \;=\; s[\mathsf{A}{:}\mathsf{Z}]. \tag{8.2}$$

From (8.1) and (8.2) we obtain $[\mathsf{E}{:}\mathsf{Z}] \cdot [\mathsf{A}{:}\mathsf{Z}] = ms[\mathsf{A}{:}\mathsf{Z}]$. Thus,

$$ms \;=\; [\mathsf{E}{:}\mathsf{Z}]. \tag{8.3}$$

Hence, $m \mid [\mathsf{E}{:}\mathsf{Z}]$. Since S is a right graded $\mathsf{E}^{\mathrm{op}} \otimes_{\mathsf{Z}} \mathsf{A}$-module, or, equivalently, a graded E–A bimodule, there is a graded Z-algebra homomorphism $\varphi\colon \mathsf{E} \to \mathsf{End}_{\mathsf{A}}(\mathsf{S})$; φ must be injective as E is graded simple. If $m = [\mathsf{E}{:}\mathsf{Z}]$, then $s = 1$, so

$$\mathsf{E} \cong_g \varphi(\mathsf{E}) \subseteq \mathsf{End}_{\mathsf{A}}(\mathsf{S}) \cong_g \mathsf{End}_{\mathsf{A}}(\mathsf{A}) \cong_g \mathsf{A},$$

showing that E embeds in A. Conversely, if E embeds in A, then A is an E–A bimodule, or equivalently a right graded $\mathsf{E}^{\mathrm{op}} \otimes_{\mathsf{Z}} \mathsf{A}$-module. Moreover, A is a simple graded module, since it is already simple as a right graded A-module. Thus, we may take $\mathsf{S} = \mathsf{A}$. Then $s = 1$, so $m = [\mathsf{E}{:}\mathsf{Z}]$ by (8.3). □

Theorem 8.9. *Let A be a central division algebra over a Henselian field F, and let E be a graded division algebra over $\mathsf{gr}(F)$ with $\mathsf{E} \subseteq \mathsf{gr}(A)$. Assume that A is defectless over F and at least one of the following additional conditions holds: A is tame over F or E is tame over $\mathsf{gr}(F)$. Then every lift D of E over F embeds into A, and D may be chosen so that the image of $\mathsf{gr}(D)$ in $\mathsf{gr}(A)$ satisfies $\mathsf{gr}(D) \subseteq \mathsf{E} \cdot Z(\mathsf{gr}(A))$.*

If A is tame over F, then $Z(\mathsf{gr}(A)) = \mathsf{gr}(F)$, so the last inclusion is equivalent to $\mathsf{gr}(D) = \mathsf{E}$ since $\mathsf{gr}(D) \subseteq \mathsf{E} \cdot Z(\mathsf{gr}(A)) = \mathsf{E}$ and $\mathsf{gr}(D) \cong_g \mathsf{E}$.

Proof. Let D be an arbitrary lift of E over F. To prove that D embeds in A, we invoke the ungraded version of Lemma 8.8 (which is provable in the same way as the lemma).

We first give the argument under the hypothesis that A is tame over F. Then $\mathsf{gr}(A)$ is a central graded division algebra over $\mathsf{gr}(F)$ and the inclusion $\mathsf{E} \subseteq \mathsf{gr}(A)$ yields by Lemma 8.8

$$ms\big(\mathsf{E}^{\mathrm{op}} \otimes_{\mathsf{gr}(F)} \mathsf{gr}(A)\big) \;=\; [\mathsf{E} \colon \mathsf{gr}(F)].$$

The valuations $v_{D^{\mathrm{op}}}$ and v_A are v-gauges as D and A are defectless over F, and $\mathsf{gr}(D^{\mathrm{op}})$ and $\mathsf{gr}(A)$ are graded division algebras. Hence, by Prop. 3.27, $v_{D^{\mathrm{op}}} \otimes v_A$ is a surmultiplicative v-norm on $D^{\mathrm{op}} \otimes_F A$, and

$$\mathsf{gr}_{v_{D^{\mathrm{op}}} \otimes v_A}(D^{\mathrm{op}} \otimes_F A) \;\cong_g\; \mathsf{gr}(D^{\mathrm{op}}) \otimes_{\mathsf{gr}(F)} \mathsf{gr}(A) \;\cong_g\; \mathsf{E}^{\mathrm{op}} \otimes_{\mathsf{gr}(F)} \mathsf{gr}(A). \tag{8.4}$$

The term on the right is a simple graded algebra since $\mathsf{gr}(A)$ is a central simple graded algebra over $\mathsf{gr}(F)$. Therefore, $v_{D^{\mathrm{op}}} \otimes v_A$ is a v-gauge on $D^{\mathrm{op}} \otimes_F A$. Since F is Henselian, Prop. 4.41 shows that $v_{D^{\mathrm{op}}} \otimes v_A$ is also a $v_{Z(D^{\mathrm{op}} \otimes A)}$-gauge. Theorem 4.30 then yields

$$ms\big(\mathsf{gr}_{v_{D^{\mathrm{op}}} \otimes v_A}(D^{\mathrm{op}} \otimes_F A)\big) \;=\; ms(D^{\mathrm{op}} \otimes_F A).$$

By collecting the information from the displayed equations above, we obtain

$$ms(D^{\mathrm{op}} \otimes_F A) \;=\; ms\big(\mathsf{E}^{\mathrm{op}} \otimes_{\mathsf{gr}(F)} \mathsf{gr}(A)\big) \;=\; [\mathsf{E} \colon \mathsf{gr}(F)].$$

Now, $[\mathsf{E} \colon \mathsf{gr}(F)] = [D{:}F]$ since D is a lift of E; hence, $ms(D^{\mathrm{op}} \otimes_F A) = [D{:}F]$. Therefore, D embeds into A by the ungraded version of Lemma 8.8.

When E is tame and A is assumed only to be defectless, the argument is similar but slightly more delicate. Let $\mathsf{Z} = Z(\mathsf{gr}(A))$, so Z is a purely inseparable graded field extension of $\mathsf{gr}(F)$ by Lemma 4.31. It is therefore linearly disjoint of $Z(\mathsf{E})$, which is a tame graded field extension of $\mathsf{gr}(F)$, so $Z(\mathsf{E}) \otimes_{\mathsf{gr}(F)} \mathsf{Z}$ is a graded field. The equation

$$\mathsf{E} \otimes_{\mathsf{gr}(F)} \mathsf{Z} \;=\; \mathsf{E} \otimes_{Z(\mathsf{E})} \big(Z(\mathsf{E}) \otimes_{\mathsf{gr}(F)} \mathsf{Z}\big)$$

shows that $\mathsf{E} \otimes_{\mathsf{gr}(F)} \mathsf{Z}$ is a simple graded algebra by Prop. 6.8. The multiplication map $\mathsf{E} \otimes_{\mathsf{gr}(F)} \mathsf{Z} \to \mathsf{gr}(A)$ is therefore injective; hence,

$$\mathsf{E} \otimes_{\mathsf{gr}(F)} \mathsf{Z} \;\cong_g\; \mathsf{E}{\cdot}\mathsf{Z} \subseteq \mathsf{gr}(A) \quad \text{and} \quad [\mathsf{E}{\cdot}\mathsf{Z} : \mathsf{Z}] \;=\; [\mathsf{E} \colon \mathsf{gr}(F)]. \tag{8.5}$$

Since $\mathsf{E}{\cdot}\mathsf{Z}$ is a graded Z-subalgebra of $\mathsf{gr}(A)$, Lemma 8.8 yields

$$ms\big((\mathsf{E}{\cdot}\mathsf{Z})^{\mathrm{op}} \otimes_{\mathsf{Z}} \mathsf{gr}(A)\big) \;=\; [\mathsf{E}{\cdot}\mathsf{Z} : \mathsf{Z}]. \tag{8.6}$$

To compute $ms(D^{\mathrm{op}} \otimes_F A)$, we again use the surmultiplicative v-norm $v_{D^{\mathrm{op}}} \otimes v_A$, which satisfies (8.4). Since $Z(\mathsf{E}) \otimes_{\mathsf{gr}(F)} \mathsf{Z}$ is a graded field, the algebra $\mathsf{E}^{\mathrm{op}} \otimes_{\mathsf{gr}(F)} \mathsf{gr}(A)$ is graded simple, and we have

$$\mathsf{E}^{\mathrm{op}} \otimes_{\mathsf{gr}(F)} \mathsf{gr}(A) \;=\; \mathsf{E}^{\mathrm{op}} \otimes_{\mathsf{gr}(F)} \mathsf{Z} \otimes_{\mathsf{Z}} \mathsf{gr}(A) \;=\; (\mathsf{E}{\cdot}\mathsf{Z})^{\mathrm{op}} \otimes_{\mathsf{Z}} \mathsf{gr}(A).$$

Thus, $v_{D^{\mathrm{op}}} \otimes v_A$ is again a $v_{Z(D^{\mathrm{op}} \otimes A)}$-gauge, so

$$\begin{aligned} ms(D^{\mathrm{op}} \otimes_F A) &= ms\big(\mathsf{gr}(D^{\mathrm{op}} \otimes_F A)\big) \\ &= ms\big(\mathsf{E}^{\mathrm{op}} \otimes_{\mathsf{gr}(F)} \mathsf{gr}(A)\big) = ms\big((\mathsf{E}{\cdot}\mathsf{Z})^{\mathrm{op}} \otimes_{\mathsf{Z}} \mathsf{gr}(A)\big). \end{aligned}$$

In view of (8.5) and (8.6), and since $[\mathsf{E}\colon \mathsf{gr}(F)] = [D{:}F]$ as D is a lift of E over F, we obtain $ms(D^{\mathrm{op}} \otimes_F A) = [D{:}F]$, so D embeds into A.

For the rest of the proof, we assume $D \subseteq A$, and modify D by an inner automorphism to push $\mathsf{gr}(D)$ into $\mathsf{E} \cdot Z\big(\mathsf{gr}(A)\big)$. The following arguments apply without change whether A or E is tame. We still write Z for $Z\big(\mathsf{gr}(A)\big)$, so $\mathsf{Z} = \mathsf{gr}(F)$ when A is tame. We have $\mathsf{E} \otimes_{\mathsf{gr}(F)} \mathsf{Z} \cong_g \mathsf{E}{\cdot}\mathsf{Z}$ by (8.5) when E is tame, and trivially if A is tame. Therefore, the isomorphism of graded $\mathsf{gr}(F)$-algebras $\mathsf{gr}(D) \cong_g \mathsf{E}$, which comes from the fact that D is a lift of E over F, extends to an isomorphism of graded Z-algebras

$$\mathsf{gr}(D) \cdot \mathsf{Z} \cong_g \mathsf{E}{\cdot}\mathsf{Z}.$$

By the graded Skolem–Noether Theorem 2.37(ii), which applies as $\mathsf{gr}(A)$ is a graded division ring, there is a $c \in \mathsf{gr}(A)^{\times}$ with $c(\mathsf{gr}(D) \cdot \mathsf{Z})c^{-1} = \mathsf{E}{\cdot}\mathsf{Z}$. Choose any $a \in A^{\times}$ such that $\widetilde{a} = c$, and let $D' = aDa^{-1} \subseteq A$. Then, $D' \cong D$. Also,

$$\mathsf{gr}(D') = c\,\mathsf{gr}(D)c^{-1} \cong_g \mathsf{gr}(D) \cong_g \mathsf{E}$$

and

$$\mathsf{gr}(D') \cdot \mathsf{Z} = c\,\mathsf{gr}(D)c^{-1} \cdot \mathsf{Z} = c(\mathsf{gr}(D) \cdot \mathsf{Z})c^{-1} = \mathsf{E}{\cdot}\mathsf{Z}.$$

So, $\mathsf{gr}(D') \subseteq \mathsf{E}{\cdot}\mathsf{Z}$. □

A major consequence of Th. 8.9 is that for a tame central division algebra A, the graded division algebra $\mathsf{gr}(A)$ carries complete information about the subalgebras of A.

Corollary 8.10. *Let A be a central division algebra over a Henselian field F. Assume that A is tame over F, and let D be a (finite-dimensional) division algebra over F. Then,*

(i) *if D is not defectless over F, then D is not F-isomorphic to a subalgebra of A;*
(ii) *if D is defectless over F, then D is F-isomorphic to a subalgebra of A if and only if $\mathsf{gr}(D)$ is $\mathsf{gr}(F)$-isomorphic to a graded subalgebra of $\mathsf{gr}(A)$.*

Proof. (i) Since v_A is tame, it is a v-norm. Hence, for any subalgebra D of A, the unique extension v_D of v to D is also a v-norm, as $v_D = v_A|_D$. That is, D is defectless over F.

(ii) Now, suppose D is defectless over F. Clearly, if D is F-isomorphic to a subalgebra D' of A, then $\mathsf{gr}(D) \cong_g \mathsf{gr}(D')$, which is a graded subalgebra of $\mathsf{gr}(A)$. Conversely, suppose $\mathsf{gr}(D)$ embeds in $\mathsf{gr}(A)$. Then, as D is a lift of $\mathsf{gr}(D)$, Th. 8.9 shows that D embeds in A. □

When A is totally ramified, the graded subalgebras of $\mathsf{gr}(A)$ are classified by their grade group, see Prop. 7.54. Theorem 8.9 then yields a striking classification of subalgebras of A by their value groups:

Corollary 8.11. *Let A be a central division algebra over a Henselian field F. Assume that A is tame and totally ramified; then, mapping every subalgebra $D \subseteq A$ to its value group Γ_D yields a one-to-one correspondence between isomorphism classes of F-subalgebras of A and subgroups of Γ_A containing Γ_F. In this correspondence, the centralizer of an algebra D corresponds to the orthogonal $\Gamma_D^\perp$ relative to the canonical pairing c_A. Hence, D is a field if and only if Γ_D is totally isotropic.*

Proof. Since A is totally ramified over F, every subalgebra of A is totally ramified over F. Since moreover A is tame, we have $char\,\overline{F} \nmid |\Gamma_A{:}\Gamma_F|$, hence every subalgebra of A is tame over F. If D, D' are subalgebras of A such that $\Gamma_D = \Gamma_{D'}$, then in the notation of Prop. 7.54 we have

$$\mathsf{gr}(D) \;=\; \mathsf{gr}(A)_{\Gamma_D} \;=\; \mathsf{gr}(D').$$

Therefore, $D \cong D'$ by Cor. 8.6. Moreover, for every subgroup $\Delta \subseteq \Gamma_A$ with $\Gamma_F \subseteq \Delta$, the graded algebra $\mathsf{gr}(A)_\Delta$ lifts to a subalgebra $D \subseteq A$ by Th. 8.9, and we have $\Gamma_D = \Delta$. Thus, the correspondence between isomorphism classes of subalgebras and subgroups is bijective.

For the centralizer $C_A(D)$ of a subalgebra $D \subseteq A$ we have $\Gamma_{C_A(D)} \subseteq \Gamma_D^\perp$ by definition of the canonical pairing, and

$$|\Gamma_{C_A(D)}{:}\Gamma_F| \;=\; [C_A(D){:}F] \;=\; [A{:}F]/[D{:}F] \;=\; |\Gamma_A{:}\Gamma_F|/|\Gamma_D{:}\Gamma_F|.$$

Therefore, $|\Gamma_{C_A(D)}{:}\Gamma_F| \;=\; |\Gamma_D^\perp{:}\Gamma_F|$, so the inclusion $\Gamma_{C_A(D)} \subseteq \Gamma_D^\perp$ is an equality. □

One notable aspect of Cor. 8.10 is that—in contrast to the particular case of totally ramified algebras considered in Cor. 8.11—a tame central division algebra A may contain subalgebras D that are *not* tame. Then $\mathsf{gr}(D)$ may have nonisomorphic lifts, and each of these lifts embeds in A. This is illustrated in the following example:

Example 8.12. Let k be a field of characteristic 0 that is Henselian for a dyadic valuation u and has an inertial biquadratic extension ℓ. Choose $a_1, a_2 \in \mathcal{O}_k^\times$ so that $\overline{\ell} \;=\; \overline{k}(\wp^{-1}(\overline{a}_1), \wp^{-1}(\overline{a}_2))$; so, $\ell \;=\; k(y_1, y_2)$ where $y_\alpha^2 - y_\alpha = a_\alpha$ for $\alpha = 1,\,2$. We also have $\ell = k(x_1, x_2)$ where $x_\alpha = 1 - 2y_\alpha$ satisfies $x_\alpha^2 = 1 + 4a_\alpha$ for $\alpha = 1,\,2$.

By adjoining two Laurent series indeterminates to k, we define a field F and a biquaternion F-algebra A by

$$F = k((t_1))((t_2)) \quad \text{and} \quad A = (1 + 4a_1, t_1/F) \otimes_F (1 + 4a_2, t_2/F).$$

Let v be the valuation on F obtained by composing the (t_1, t_2)-adic valuation with u. It is easily seen that A is a division algebra over F (this follows, e.g., from Prop. 8.41 below, as A is an example of a decomposably semiramified central simple algebra), and that $\mathsf{gr}(A)$ is a biquaternion algebra

$$\mathsf{gr}(A) = [\widetilde{a}_1, \widetilde{t}_1) \otimes_{\mathsf{gr}(F)} [\widetilde{a}_2, \widetilde{t}_2) \quad \text{with} \quad \mathsf{gr}(F) = \mathsf{gr}(k)[\widetilde{t}_1, \widetilde{t}_1^{-1}, \widetilde{t}_2, \widetilde{t}_2^{-1}].$$

Here, $[\widetilde{a}_\alpha, \widetilde{t}_\alpha)$ denotes the graded cyclic "semisymbol algebra":

$$[\widetilde{a}_\alpha, \widetilde{t}_\alpha) = \big(\mathsf{gr}(F)[b_\alpha]/\mathsf{gr}(F), \sigma_\alpha, \widetilde{t}_\alpha\big),$$

where b_α is a root of $X^2 - X = \widetilde{a}_\alpha \in \mathsf{gr}(F)[X]$ and $\sigma_\alpha(b_\alpha) = b_\alpha + 1$. Thus, the valuation v_A on A extending v is a tame v-gauge, which means that A is tame over F.

For $\alpha = 1, 2$, let i_α and $j_\alpha \in A$ be the standard generators of the quaternion subalgebra $(1+4a_\alpha, t_\alpha/F)$. The graded subfield $\mathsf{E} = \mathsf{gr}(F)[\widetilde{j}_1, \widetilde{j}_2] \subseteq \mathsf{gr}(A)$ is not tame over $\mathsf{gr}(F)$ since $\mathsf{E} \cong \mathsf{gr}(F)[\sqrt{\widetilde{t}_1}, \sqrt{\widetilde{t}_2}]$, which is purely inseparable over $\mathsf{gr}(F)$. This graded subfield obviously lifts to the field $F(j_1, j_2) \subseteq A$, but it also lifts to the quaternion algebra $(t_1, t_2/F)$, which must embed in A by Cor. 8.10.

To construct an explicit embedding of $(t_1, t_2/F)$ into A, observe that the subfield $k(x_1) \subseteq \ell$ contains an element ξ such that $N_{k(x_1)/k}(\xi) = 1 + 4a_2$ and $\overline{\xi} = 1$: This follows by Hensel's Lemma since for $\xi = 1 + y_1\xi_1$ with $\xi_1 \in k$ the equation $N_{k(x_1)/k}(\xi) = 1 + 4a_2$ takes the form

$$1 + \xi_1 - b_1\xi_1^2 = 1 + 4a_2,$$

and 0 is a simple root of the residue equation. By identifying $k(x_1)$ with $k(i_1) \subseteq A$, we may view the element $\xi i_2^{-1} j_1$ as lying in A. We have

$$(\xi i_2^{-1} j_1)^2 = t_1 \qquad \text{and} \qquad (\xi i_2^{-1} j_1) j_2 = -j_2(\xi i_2^{-1} j_1).$$

Therefore, $\xi i_2^{-1} j_1$ and j_2 generate a quaternion subalgebra $D \subseteq A$ isomorphic to $(t_1, t_2/F)$. Note that $\widetilde{\xi} = \widetilde{i}_2 = 1$ in $\mathsf{gr}(A)$, so $\mathsf{gr}(D) = \mathsf{E}$.

We now develop the other side of Th. 8.9, concerning the lifting of tame graded division algebras in division algebras that may not be tame. The next result extends Th. 8.9 by allowing lifts in noncentral division algebras.

Theorem 8.13. *Let A be a defectless division algebra over a Henselian field F and let E be a graded division algebra over $\mathsf{gr}(F)$ with $\mathsf{E} \subseteq \mathsf{gr}(A)$. If E is tame over $\mathsf{gr}(F)$, then there is a lift D of E over F such that $D \subseteq A$ and $\mathsf{gr}(D) \subseteq \mathsf{E} \cdot Z\big(\mathsf{gr}(A)\big)$.*

Proof. Let $\mathsf{K} = Z(\mathsf{E})$ and $\mathsf{L} = \mathsf{gr}\big(Z(A)\big)$. Since K is tame over $\mathsf{gr}(F)$, the graded field compositum $\mathsf{K}{\cdot}\mathsf{L}$ in $\mathsf{gr}(A)$ is tame over L. We have $\mathsf{K}{\cdot}\mathsf{L} = Z(\mathsf{E}{\cdot}\mathsf{L})$; hence $\mathsf{E}{\cdot}\mathsf{L}$ is a tame graded division algebra over L. Theorem 8.9 yields a lift D' of $\mathsf{E}{\cdot}\mathsf{L}$ over $Z(A)$ with $D' \subseteq A$ and

$$\mathsf{gr}(D') \subseteq \mathsf{E} \cdot \mathsf{L} \cdot Z\big(\mathsf{gr}(A)\big) = \mathsf{E} \cdot Z\big(\mathsf{gr}(A)\big).$$

We will find the desired D inside D'.

Note that D' is tame over $Z(A)$, hence $Z(D')$ is a lift of $Z(\mathsf{E}{\cdot}\mathsf{L}) = \mathsf{K}{\cdot}\mathsf{L}$ over $Z(A)$. Consider an arbitrary lift B of E over F. Since E is tame over $\mathsf{gr}(F)$, the center $Z(B)$ of B is a lift of K over F. But K also lifts to a subfield of $Z(D')$ by Prop. 8.3; since the lift of a tame graded field extension is unique up to isomorphism by Cor. 5.56, we may assume $Z(B) \subseteq Z(D')$.

We claim that $B \otimes_{Z(B)} Z(D')$ is a lift of $\mathsf{E}{\cdot}\mathsf{L}$ over $Z(A)$. To see this, observe that $v_B \otimes v_{Z(D')}$ is a surmultiplicative $v_{Z(D')}$-norm and there is a canonical graded $\mathsf{gr}\big(Z(D')\big)$-algebra isomorphism

$$\mathsf{gr}_{v_B \otimes v_{Z(D')}}\big(B \otimes_{Z(B)} Z(D')\big) \cong_g \mathsf{gr}(B) \otimes_{\mathsf{gr}(Z(B))} \mathsf{gr}\big(Z(D')\big) = \mathsf{E} \otimes_{\mathsf{K}} \mathsf{K}{\cdot}\mathsf{L}, \tag{8.7}$$

see Cor. 3.28. The right term can be identified with the subalgebra $\mathsf{E}{\cdot}\mathsf{L} \subseteq \mathsf{gr}(A)$, hence it is a graded division algebra. It follows by Cor. 3.30 that $B \otimes_{Z(B)} Z(D')$ is a division algebra with $v_B \otimes v_{Z(D')}$ a valuation; it is necessarily the unique valuation extending $v_{Z(D')}$. The isomorphism (8.7) shows that $B \otimes_{Z(B)} Z(D')$ is a lift of $\mathsf{E}{\cdot}\mathsf{L}$ over $Z(D')$. It is therefore also a lift of $\mathsf{E}{\cdot}\mathsf{L}$ over $Z(A)$, proving the claim.

To complete the proof, we use the uniqueness property of the lift of a tame graded division algebra: Proposition 8.5 yields a $Z(A)$-algebra isomorphism

$$\varphi \colon B \otimes_{Z(B)} Z(D') \xrightarrow{\sim} D'.$$

The image of B in D' via φ is the division algebra D we seek. □

Corollary 8.14. *Let A be a division algebra finite-dimensional and defectless over a Henselian field F. The map $D \mapsto \mathsf{gr}(D)$ gives a one-to-one correspondence between isomorphism classes of F-subalgebras of A tame over F and isomorphism classes of graded $\mathsf{gr}(F)$-subalgebras E of $\mathsf{gr}(D)$ that are tame over $\mathsf{gr}(F)$.*

Proof. This is immediate from Prop. 8.5 and Th. 8.13. □

A significant case of lifting occurs with inertial gauges: Let S be a division algebra finite-dimensional over $\overline{F}$ with $Z(S)$ separable over $\overline{F}$. Let $\mathsf{E} = S \otimes_{\overline{F}} \mathsf{gr}(F)$, which is a graded division algebra inertial over $\mathsf{gr}(F)$. Its center $Z(\mathsf{E}) \cong_g Z(S) \otimes_{\overline{F}} \mathsf{gr}(F)$ is a graded field tame over $\mathsf{gr}(F)$. By Prop. 8.5 there is a division algebra D tame over F with $\mathsf{gr}(D) \cong_g \mathsf{E}$. Then, $\overline{D} = D_0 \cong \mathsf{E}_0 \cong S$ and

$$\big[\overline{D}{:}\overline{F}\big] = [D_0{:}F_0] = [\mathsf{E}{:}\mathsf{gr}(F)] = [D{:}F].$$

Also, $Z(\overline{D}) \cong Z(S)$, which is separable over $\overline{F}$. Thus, D is an inertial division algebra over F with $\overline{D} \cong S$. Proposition 8.5 shows that D is uniquely determined up to F-isomorphism by S. We call Dan *inertial lift* of S over F.

Corollary 8.15. *Let (F, v) be a Henselian-valued field, let S be a division algebra finite-dimensional over $\overline{F}$ with $Z(S)$ separable over $\overline{F}$, and let D be an inertial lift of S over F. Then,*

(i) *If D' is a division algebra inertial over F with $\overline{D'} \cong S$, then $D' \cong D$.*
(ii) *Let A be any division algebra finite-dimensional and defectless over F. If S embeds $\overline{F}$-isomorphically in $\overline{A}$, then there is an F-monomorphism $D \hookrightarrow A$.*

Proof. (i) $\mathsf{gr}(D')$ is a graded division algebra which is inertial over $\mathsf{gr}(F)$, so

$$\mathsf{gr}(D') \cong_g D'_0 \otimes_{F_0} \mathsf{gr}(F) \cong_g S \otimes_{\overline{F}} \mathsf{gr}(F).$$

Thus, D' is a tame lift of $S \otimes_{\overline{F}} \mathsf{gr}(F)$ over F. Hence, $D' \cong D$ by Prop. 8.5.

(ii) Since S embeds in $\overline{A} = A_0$, $S \otimes_{\overline{F}} \mathsf{gr}(F)$ is a graded $\mathsf{gr}(F)$-subalgebra of $\mathsf{gr}(A)$. Since D is a tame lift of $S \otimes_{\overline{F}} \mathsf{gr}(F)$, D embeds in A by Th. 8.13. □

8.2 Canonical subalgebras of graded and valued division algebras

If D is a graded division algebra, there are four graded subalgebras of D determined canonically by the grading, which highlight different aspects of the graded structure. We describe these subalgebras in §8.2.1. Then in §8.2.2 we describe the analogous subalgebras for a division algebra that is tame with respect to a Henselian valuation on its center. The particular case where the valuation is discrete of rank 1 is simpler; it is discussed in §8.4.6.

8.2.1 Canonical subalgebras of a graded division algebra

Throughout this subsection, fix an arbitrary graded field F and let D denote a (finite-dimensional) central graded division algebra over F. Its degree 0 component D_0 is a division algebra over F_0, but it is not necessarily central. Recall from Prop. 2.40 that the center $Z(\mathsf{D}_0)$ is a Galois extension of F_0 with an abelian Galois group, which is a homomorphic image of Γ_{D} under the canonical homomorphism θ_{D} of (2.23). The following fundamental equality was obtained in Cor. 2.7:

$$[\mathsf{D}{:}\mathsf{F}] = [\mathsf{D}_0{:}\mathsf{F}_0] \cdot |\Gamma_{\mathsf{D}}{:}\Gamma_{\mathsf{F}}|.$$

The central graded division algebra D over F is said to be

- *unramified* or *inertial* if $\Gamma_{\mathsf{D}} = \Gamma_{\mathsf{F}}$ or, equivalently, $[\mathsf{D}{:}\mathsf{F}] = [\mathsf{D}_0{:}\mathsf{F}_0]$, see Def. 2.50;
- *semiramified* if $Z(\mathsf{D}_0) = \mathsf{D}_0$ (i.e., D_0 is a field) and $[\mathsf{D}_0{:}\mathsf{F}_0] = |\Gamma_{\mathsf{D}}{:}\Gamma_{\mathsf{F}}|$; this type of algebra will be further discussed in §8.4.2, see Def. 8.26;
- *totally ramified* if $\mathsf{D}_0 = \mathsf{F}_0$ or, equivalently, $[\mathsf{D}{:}\mathsf{F}] = |\Gamma_{\mathsf{D}}{:}\Gamma_{\mathsf{F}}|$, see §7.3.1.

If D is unramified, then $\mathsf{D} = \mathsf{D}_0 \otimes_{\mathsf{F}_0} \mathsf{F}$ since any F_0-base of D_0 is an F-base of D; see the proof of Prop. 2.5. By contrast, in the totally ramified case the structure of D is determined by the extension of grade groups $\Gamma_\mathsf{F} \subseteq \Gamma_\mathsf{D}$; it is quite simple, as we have seen in §7.3.1. In the general case, D contains four canonically-defined subalgebras, which exhibit various ramification types. We now describe those subalgebras.

Let

- $\mathsf{U} = \mathsf{D}_0 \otimes_{\mathsf{F}_0} \mathsf{F}$, the maximal unramified subalgebra of D;
- $\mathsf{C} = C_\mathsf{D}(\mathsf{U})$, the centralizer of U;
- $\mathsf{Z} = Z(\mathsf{D}_0) \otimes_{\mathsf{F}_0} \mathsf{F} = \mathsf{U} \cap \mathsf{C}$, the center of U;
- $\mathsf{E} = C_\mathsf{D}(\mathsf{Z})$, the centralizer of Z.

Since $\mathsf{Z} \subseteq \mathsf{U}$, we have $\mathsf{C} \subseteq \mathsf{E}$, hence C is also the centralizer of U in E. By the graded Double Centralizer Theorem 2.35, Z is also the center of C and E, and we have

$$\mathsf{E} \cong_g \mathsf{U} \otimes_\mathsf{Z} \mathsf{C}. \tag{8.8}$$

Let $\mu(\mathsf{F}_0)$ denote the group of roots of unity in F_0, and μ_e the group of e-th roots of unity in $(\mathsf{F}_0)_{sep}$.

Proposition 8.16. *With the notation above,*

(i) *the extension Z/F is an inertial Galois extension of graded fields with an abelian Galois group, and the canonical homomorphism θ_D of* (2.23) *induces a surjective homomorphism*

$$\theta_\mathsf{D} \colon \Gamma_\mathsf{D} \longrightarrow \mathcal{G}(\mathsf{Z}_0/\mathsf{F}_0) = \mathcal{G}(\mathsf{Z}/\mathsf{F});$$

(ii) *the central division algebra U over Z is unramified, the central division algebra C over Z is totally ramified, and*

$$\mathsf{E}_0 = \mathsf{U}_0 = \mathsf{D}_0, \qquad \Gamma_\mathsf{E} = \Gamma_\mathsf{C} = \mathit{ker}\, \theta_\mathsf{D}.$$

(iii) $[\mathsf{D}{:}\mathsf{E}] = |\Gamma_\mathsf{D}{:}\Gamma_\mathsf{E}| = [\mathsf{Z}{:}\mathsf{F}] = [\mathsf{Z}_0{:}\mathsf{F}_0]$.

(iv) *The canonical pairing $\overline{c}_\mathsf{C}$ of the totally ramified Z-algebra C is a nondegenerate symplectic pairing*

$$a_\mathsf{D} \colon (\mathit{ker}\, \theta_\mathsf{D})/\Gamma_\mathsf{F} \times (\mathit{ker}\, \theta_\mathsf{D})/\Gamma_\mathsf{F} \longrightarrow \mu(\mathsf{F}_0).$$

Hence, char $\mathsf{F}_0 \nmid |\mathit{ker}\, \theta_\mathsf{D}{:}\Gamma_\mathsf{F}| = [\mathsf{C}{:}\mathsf{Z}]$.

(v) *Let e be the exponent of the finite abelian group $(\mathit{ker}\, \theta_\mathsf{D})/\Gamma_\mathsf{F}$. Then, $\mu_e \subseteq \mathsf{F}_0^\times$.*

Proof. (i) By definition of Z, we have $\mathsf{Z}_0 = Z(\mathsf{D}_0)$. Proposition 2.40 shows that $\mathsf{Z}_0/\mathsf{F}_0$ is a Galois field extension, and that its Galois group is a homomorphic image of Γ_D under θ_D. So, the unramified extension Z of F is actually an inertial extension. Moreover, every automorphism of $\mathsf{Z}_0/\mathsf{F}_0$ extends uniquely to an automorphism of Z/F, and we have $\mathcal{G}(\mathsf{Z}/\mathsf{F}) = \mathcal{G}(\mathsf{Z}_0/\mathsf{F}_0)$. Thus, (i) is proved.

(ii) By definition of U and Z, we have $\Gamma_{\mathsf{U}} = \Gamma_{\mathsf{Z}}$, hence U is unramified. The equality $\mathsf{U}_0 = \mathsf{D}_0$ also follows from the definition of U. Since $\mathsf{U} \subseteq \mathsf{E} \subseteq \mathsf{D}$, we have $\mathsf{U}_0 \subseteq \mathsf{E}_0 \subseteq \mathsf{D}_0$; these inclusions are equalities since $\mathsf{U}_0 = \mathsf{D}_0$.

Since C centralizes U by definition, the algebra C_0 centralizes $\mathsf{U}_0 = \mathsf{D}_0$, hence $\mathsf{C}_0 \subseteq \mathsf{Z}_0$. The reverse inclusion is clear since $\mathsf{Z} \subseteq \mathsf{C}$, hence $\mathsf{C}_0 = \mathsf{Z}_0$ and C is therefore a totally ramified Z-algebra. The definition of θ_{D} shows that $\mathit{ker}\,\theta_{\mathsf{D}}$ consists of the degrees of homogeneous elements centralizing $Z(\mathsf{D}_0)$. Since centralizing $Z(\mathsf{D}_0)$ amounts to centralizing Z, it follows that $\mathit{ker}\,\theta_{\mathsf{D}} = \Gamma_{\mathsf{E}}$. Clearly, we have $\Gamma_{\mathsf{C}} \subseteq \Gamma_{\mathsf{E}}$ since $\mathsf{C} \subseteq \mathsf{E}$. To prove that $\Gamma_{\mathsf{C}} = \Gamma_{\mathsf{E}}$, we count dimensions: by (8.8) we have

$$[\mathsf{E}{:}\mathsf{Z}] = [\mathsf{U}{:}\mathsf{Z}] \cdot [\mathsf{C}{:}\mathsf{Z}].$$

Since U is unramified, $\mathsf{U}_0 = \mathsf{E}_0$, and C is totally ramified, it follows that

$$[\mathsf{E}{:}\mathsf{Z}] = [\mathsf{E}_0{:}\mathsf{Z}_0] \cdot |\Gamma_{\mathsf{C}}{:}\Gamma_{\mathsf{Z}}|.$$

On the other hand, we have the fundamental equality of Cor. 2.7:

$$[\mathsf{E}{:}\mathsf{Z}] = [\mathsf{E}_0{:}\mathsf{Z}_0] \cdot |\Gamma_{\mathsf{E}}{:}\Gamma_{\mathsf{Z}}|.$$

By comparing the last two equations, we obtain that $|\Gamma_{\mathsf{C}}{:}\Gamma_{\mathsf{Z}}| = |\Gamma_{\mathsf{E}}{:}\Gamma_{\mathsf{Z}}|$, hence $\Gamma_{\mathsf{C}} = \Gamma_{\mathsf{E}}$.

(iii) Since we have seen in (ii) that $\mathsf{D}_0 = \mathsf{E}_0$, we have $[\mathsf{D}{:}\mathsf{E}] = |\Gamma_{\mathsf{D}}{:}\Gamma_{\mathsf{E}}|$ by the fundamental equality of Cor. 2.7. From (i) and (ii) it follows that θ_{D} induces an isomorphism

$$\Gamma_{\mathsf{D}}/\Gamma_{\mathsf{E}} \xrightarrow{\sim} \mathcal{G}(\mathsf{Z}_0/\mathsf{F}_0) = \mathcal{G}(\mathsf{Z}/\mathsf{F});$$

hence, $|\Gamma_{\mathsf{D}}{:}\Gamma_{\mathsf{E}}| = [\mathsf{Z}{:}\mathsf{F}] = [\mathsf{Z}_0{:}\mathsf{F}_0]$.

(iv) and (v) Since C is totally ramified over $Z(\mathsf{C}) = \mathsf{Z}$, there is a canonical nondegenerate alternating pairing $\overline{c}_{\mathsf{C}}\colon \Gamma_{\mathsf{C}}/\Gamma_{\mathsf{Z}} \times \Gamma_{\mathsf{C}}/\Gamma_{\mathsf{Z}} \to \mu(\mathsf{Z}_0)$ given by

$$\overline{c}_{\mathsf{C}}(\gamma + \Gamma_{\mathsf{Z}}, \delta + \Gamma_{\mathsf{Z}}) = s_\gamma s_\delta s_\gamma^{-1} s_\delta^{-1} \quad \text{for any nonzero} \quad s_\gamma \in \mathsf{C}_\gamma,\ s_\delta \in \mathsf{C}_\delta$$

(see (7.38) and (7.39)). This is the pairing a_{D} of (iv). It has the domain stated in (iv) as $\Gamma_{\mathsf{C}} = \mathit{ker}\,\theta_{\mathsf{D}}$ and $\Gamma_{\mathsf{Z}} = \Gamma_{\mathsf{F}}$. We claim also that the image of the pairing lies in $\mathsf{F}_0^\times$. To see this, take any $\gamma, \delta \in \Gamma_{\mathsf{C}}$, and any nonzero $s_\gamma \in \mathsf{C}_\gamma$, $s_\delta \in \mathsf{C}_\delta$, and let

$$\omega = s_\gamma s_\delta s_\gamma^{-1} s_\delta^{-1} = \overline{c}_{\mathsf{C}}(\gamma + \Gamma_{\mathsf{Z}}, \delta + \Gamma_{\mathsf{Z}}) \in \mathsf{Z}_0^\times.$$

Also take any $\varepsilon \in \Gamma_{\mathsf{D}}$ and nonzero $d_\varepsilon \in \mathsf{D}_\varepsilon$, and let $\psi = \mathit{int}(d_\varepsilon)$, which is a graded automorphism of D. Then $\psi(\mathsf{D}_0) = \mathsf{D}_0$, so $\psi(\mathsf{U}) = \mathsf{U}$, hence $\psi(\mathsf{C}) = \mathsf{C}$, so $\psi(s_\gamma) \in \mathsf{C}_\gamma$ and $\psi(s_\delta) \in \mathsf{C}_\delta$. Thus,

$$\theta_{\mathsf{D}}(\varepsilon)(\omega) = \psi(\omega) = \psi(s_\gamma)\psi(s_\delta)\psi(s_\gamma)^{-1}\psi(s_\delta)^{-1} = \overline{c}_{\mathsf{C}}(\gamma{+}\Gamma_{\mathsf{Z}}, \delta{+}\Gamma_{\mathsf{Z}}) = \omega.$$

Since θ_{D} maps onto $\mathcal{G}(\mathsf{Z}_0/\mathsf{F}_0)$ and Z_0 is Galois over F_0, it follows that $\omega \in \mathsf{F}_0$. This establishes the claim. Because the pairing $\overline{c}_{\mathsf{C}}$ is nondegenerate its image is the cyclic group of order $e = \mathit{exp}((\mathit{ker}\,\theta_{\mathsf{D}})/\Gamma_{\mathsf{F}})$ in $\mu(\mathsf{Z}_0) \cap \mathsf{F}_0^\times = \mu(\mathsf{F}_0)$, which is μ_e. Thus, $\mu_e \subseteq \mathsf{F}_0^\times$, proving (v). Therefore $char(\mathsf{F}_0) \nmid e$; since e and $|ker\,\theta_\mathsf{D}{:}\Gamma_\mathsf{F}|$ have the same prime factors, it follows that $char(\mathsf{F}_0) \nmid |ker\,\theta_\mathsf{D}{:}\Gamma_\mathsf{F}|$. Also,

$$|ker\,\theta_\mathsf{D}{:}\Gamma_\mathsf{F}| \;=\; |\Gamma_\mathsf{C}{:}\Gamma_\mathsf{Z}| \;=\; [\mathsf{C}{:}\mathsf{Z}],$$

as C is totally ramified over Z, completing the proof of (iv). □

The inclusion relations among the algebras D, E, U, C, Z, F, and among their degree zero components and their grade groups are shown in the following diagram:

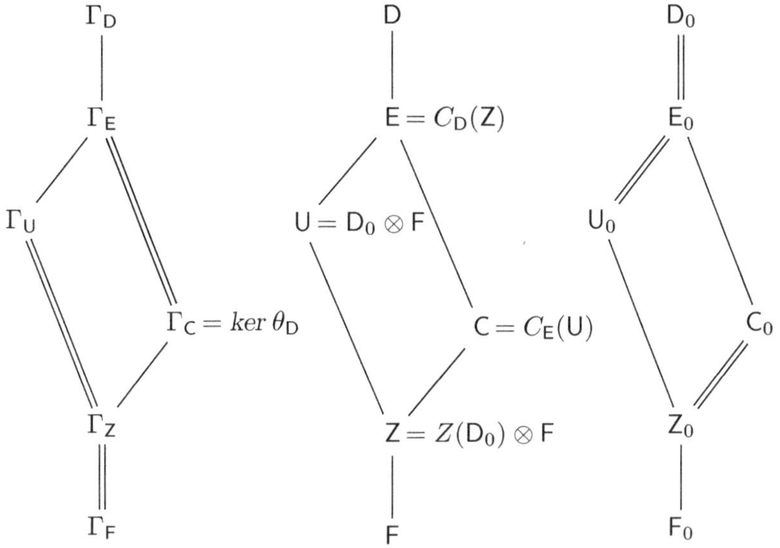

(8.9)

Observe that the diagram simplifies when D satisfies further hypotheses. Notably, $\mathsf{C} = \mathsf{Z}$ (so $\mathsf{E} = \mathsf{U}$) if and only if D is split by an inertial extension of F (see Prop. 8.49 below); this occurs whenever $rk\,\Gamma_\mathsf{F} = 1$. Of course, D is inertial if and only if $\mathsf{U} = \mathsf{E} = \mathsf{D}$ (so $\mathsf{C} = \mathsf{Z} = \mathsf{F}$); D is totally ramified if and only if $\mathsf{U} = \mathsf{Z} = \mathsf{F}$ (so $\mathsf{C} = \mathsf{E} = \mathsf{D}$); and D is semiramified if and only if the central diamond collapses, i.e., $\mathsf{U} = \mathsf{C} = \mathsf{Z} = \mathsf{E}$.

8.2.2 Canonical subalgebras of a tame division algebra

In §8.2.1 we described a fundamental family of canonical graded subalgebras of a graded division algebra. We now obtain the corresponding subalgebras for a tame division algebra over a Henselian field F. In the graded setting the subalgebras were entirely unique. In the valued setting, the subalgebras are unique only up to conjugacy.

Let D be a central division algebra over F, and assume that D is tame over F with respect to the Henselian valuation v on F. The tameness of D

implies that $Z(\overline{D})$ is separable over $\overline{F}$. Consider the following F-subalgebras of D:

- U is an inertial lift of $\overline{D}$ over F;
- $C = C_D(U)$, the centralizer of U in D;
- $Z = Z(U) = U \cap C$, the center of U;
- $E = C_D(Z)$, the centralizer of Z in D.

As we have seen by Cor. 8.15 and the remarks preceding the corollary, since $Z(\overline{D})$ is separable over $\overline{F}$, there is an inertial lift of $\overline{D}$ over F, and there is a copy of this lift within D. This proves the existence of U in D. Corollary 8.15 also shows that U is unique up to F-isomorphism; so, by Skolem–Noether U is unique up to conjugacy in D. Hence, C, Z, and E are also unique up to conjugacy in D. (They are absolutely unique once a choice of U is made.) Also, by the Double Centralizer Theorem, $Z = Z(U) = Z(C)$ and

$$E = U \otimes_Z C.$$

Proposition 8.17. *With the notation above, U, C, Z, and E are all tame over F, and their associated graded algebras are the canonical subalgebras of $\mathsf{gr}(D)$ described in Prop. 8.16. Moreover,*

(i) *the field Z is an inertial lift of $Z(\overline{D})$ over F, and Z is Galois over F with abelian Galois group; the canonical homomorphism θ_D of (1.1) induces a surjective homomorphism*

$$\theta_D \colon \Gamma_D \longrightarrow \mathcal{G}(Z(\overline{D})/\overline{F}) = \mathcal{G}(Z/F);$$

(ii) *the central division algebra U over Z is unramified, the central division algebra C over Z is tame and totally ramified, and*

$$\overline{E} = \overline{U} = \overline{D}, \qquad \Gamma_E = \Gamma_C = \ker \theta_D.$$

(iii) $[D{:}E] = |\Gamma_D{:}\Gamma_E| = |Z{:}F| = [\overline{Z}{:}\overline{F}]$.

(iv) *The canonical pairing $\overline{c}_C$ of the totally ramified graded Z-algebra C is a nondegenerate symplectic pairing*

$$a_D \colon (\ker \theta_D)/\Gamma_F \times (\ker \theta_D)/\Gamma_F \longrightarrow \mu(\overline{F}).$$

Hence, $\operatorname{char} \overline{F} \nmid |\ker \theta_D{:}\Gamma_F| = [C{:}Z]$. *The pairing a_D coincides with the pairing $a_{\mathsf{gr}(D)}$ on $(\ker \theta_{\mathsf{gr}(D)})/\Gamma_{\mathsf{gr}(F)}$, and is therefore independent of the choice of C.*

(v) *Let e be the exponent of $(\ker \theta_D)/\Gamma_F$. Then, $\mu_e \subseteq \overline{F}^{\times}$.*

The proof of Prop. 8.17 will use the following lemma:

Lemma 8.18. *Let D be a tame central division algebra over F, and let B be any F-subalgebra of D. Then $\mathsf{gr}(C_D(B)) = C_{\mathsf{gr}(D)}(\mathsf{gr}(B))$.*

Proof. Note that $Z(\mathsf{gr}(D)) = \mathsf{gr}(Z(D)) = \mathsf{gr}(F)$, as D is tame over F. Let $C = C_D(B)$. Clearly, $\mathsf{gr}(C) \subseteq C_{\mathsf{gr}(D)}(\mathsf{gr}(B))$. We prove equality by counting dimensions over $\mathsf{gr}(F)$. For this, observe that

$$[D{:}F] \;=\; [B{:}F] \cdot [C{:}F]$$

by the Double Centralizer Theorem, and likewise

$$[\mathsf{gr}(D){:}\,\mathsf{gr}(F)] \;=\; [\mathsf{gr}(B){:}\,\mathsf{gr}(F)] \cdot [C_{\mathsf{gr}(D)}(\mathsf{gr}(B)){:}\,\mathsf{gr}(F)]$$

by the graded Double Centralizer Theorem 2.35(ii). Since D is defectless over F, the valuation on D is a norm, hence its restrictions to B and to C are also norms, by Prop. 3.14. Thus, B and C are defectless over F, and we have

$$[D{:}F] \;=\; [\mathsf{gr}(D){:}\,\mathsf{gr}(F)],\ [B{:}F] \;=\; [\mathsf{gr}(B){:}\,\mathsf{gr}(F)],\ [C{:}F] \;=\; [\mathsf{gr}(C){:}\,\mathsf{gr}(F)].$$

By comparing the displayed equations above, we see that

$$[\mathsf{gr}(C){:}\,\mathsf{gr}(F)] \;=\; [C_{\mathsf{gr}(D)}(\mathsf{gr}(B)){:}\,\mathsf{gr}(F)].$$

Therefore, $\mathsf{gr}(C) = C_{\mathsf{gr}(D)}(\mathsf{gr}(B))$. □

Proof of Prop. 8.17. D is defectless over F since it is tame over F; hence, its subalgebras Z, C, and E are also defectless over F. Also, v_D is a tame v-gauge, so $Z(\mathsf{gr}(D)) = \mathsf{gr}(Z(D)) = \mathsf{gr}(F)$. Hence, $Z(\mathsf{gr}(D)_0) = Z(\overline{D})$ is Galois over $\mathsf{gr}(F)_0 = \overline{F}$. Because U is an inertial lift of $\overline{D}$ over F, it is a tame lift of $\overline{D} \otimes_{\overline{F}} \mathsf{gr}(F)$. So, the graded subalgebras $\mathsf{gr}(U)$ and $\overline{D} \otimes_{\overline{F}} \mathsf{gr}(F)$ are isomorphic graded subalgebras of $\mathsf{gr}(D)$. Hence, $\mathsf{gr}(U) = \overline{D} \otimes_{\overline{F}} \mathsf{gr}(F)$, since this is the unique maximal unramified graded subalgebra of $\mathsf{gr}(D)$. Because U is tame over F, for $Z = Z(U)$ we have

$$\mathsf{gr}(Z) \;=\; Z(\mathsf{gr}(U)) \;=\; Z(\overline{D}) \otimes_{\overline{F}} \mathsf{gr}(F).$$

Thus, $\mathsf{gr}(U)$ and $\mathsf{gr}(Z)$ are the canonical subalgebras U and Z of $\mathsf{gr}(D)$ as in Prop. 8.16, and U and Z are their tame lifts over F, and Z is also an inertial lift of $Z(\overline{D})$ over F. Since $\mathsf{gr}(Z)$ is Galois over $\mathsf{gr}(F)$ by Prop. 8.16(i), its tame lift Z is Galois over F by Cor. 5.56, and

$$\mathcal{G}(Z/F) \cong \mathcal{G}(\mathsf{gr}(Z)/\,\mathsf{gr}(F)) \cong \mathcal{G}(Z(\overline{D})/\overline{F}) \;=\; im(\theta_D)$$

by Cor. 5.52, Prop. 8.16(i), and Prop. 1.5(iv). These groups are abelian, since the domain of θ_D is abelian. Thus, (i) is proved.

By the preceding lemma, $\mathsf{gr}(C) = C_{\mathsf{gr}(D)}(\mathsf{gr}(U))$. Hence, C is tame over F, as $Z(\mathsf{gr}(C)) = Z(\mathsf{gr}(U)) = \mathsf{gr}(Z(C))$, which is tame over $\mathsf{gr}(F)$. The lemma also shows that $\mathsf{gr}(E) = \mathsf{gr}(C_D(Z)) = C_{\mathsf{gr}(D)}(\mathsf{gr}(Z))$; hence, E is tame over F, as $Z(\mathsf{gr}(E)) = Z(\mathsf{gr}(U)) = \mathsf{gr}(Z)$, which is tame over $\mathsf{gr}(F)$. Thus, U, Z, C, and E are tame lifts over F of $\mathsf{gr}(U)$, $Z(\mathsf{gr}(U))$, $C_{\mathsf{gr}(D)}(\mathsf{gr}(U))$, and $C_{\mathsf{gr}(D)}(Z(\mathsf{gr}(U)))$, which are the canonical subalgebras of $\mathsf{gr}(D)$ in Prop. 8.16. Parts (ii), (iii), (iv), and (v) now follow from the corresponding assertions in Prop. 8.16, as $\theta_D = \theta_{\mathsf{gr}(D)}$. It is clear from the definitions and (7.49) that the pairing $a_D = \overline{c}_Z$ on $(\mathit{ker}\,\theta_D)/\Gamma_F$ is the same as the nondegenerate symplectic pairing $a_{\mathsf{gr}(D)} = \overline{c}_{\mathsf{gr}(C)}$ on $(\mathit{ker}\,\theta_{\mathsf{gr}(D)})/\Gamma_{\mathsf{gr}(F)}$. □

The inclusion relations among the algebras D, U, Z, C, E, and F and among their residue algebras and value groups are shown in the following diagrams, which closely mirror the graded case shown in (8.9).

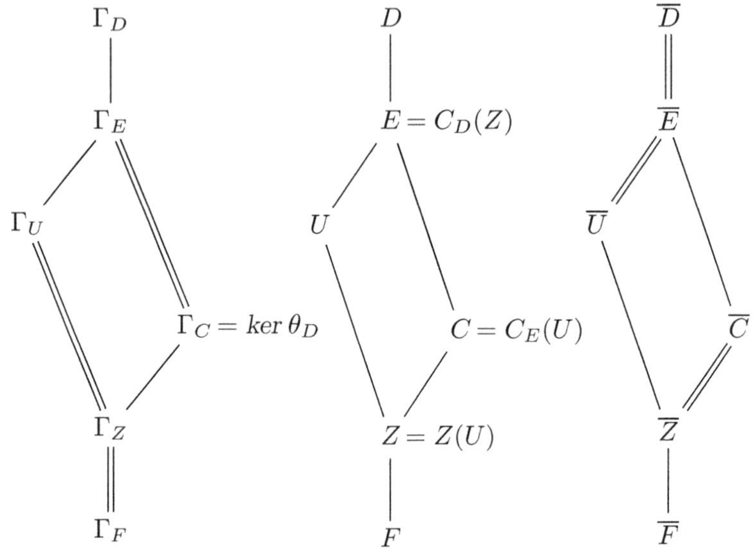

(8.10)

The diagram simplifies when D satisfies further hypotheses in addition to being tame over F. Notably, $C = Z$ (so $E = U$) if and only if D is inertially split (see Prop. 8.64 below); this holds whenever $rr(\Gamma_F) = 1$, by Prop. 6.70 (see §8.4.6 for the case where $\Gamma_F \cong \mathbb{Z}$). Also, D is inertial if and only if $U = E = D$ (so $C = Z = F$); it is totally ramified if and only if $U = Z = F$ (so $D = E = C$); and it is semiramified (see Def. 8.36 below) if and only if the central diamond collapses, i.e., $U = Z = C = E$.

8.3 Semiramified algebras

The main purpose of this section is to introduce the notion of decomposably semiramified central division algebras over a Henselian field. Division algebras of this type have a special kind of decomposition into tensor products of cyclic algebras; they are also characterized by their splitting properties: we show in Prop. 8.40 that a central division algebra over a Henselian field F is decomposably semiramified exactly when it is split by some inertial extension of F and also by some totally ramified extension of F. As a result, we will see in §8.4 that these algebras represent the images of the maps that were shown in §6.2.3 to split the ramification map $\chi\colon\ \mathit{Br}_{\mathit{is}}(F) \to \mathit{Hom}^c(\mathcal{G}(\overline{F}), \mathbb{T}(\Gamma_F))$

(see Remark 6.69). Therefore, they are, together with the inertial algebras and the totally ramified algebras, one of the three special types of division algebras into which every tame division algebra over a Henselian field decomposes up to Brauer equivalence; see §8.4.

In preparation for the discussion of decomposably semiramified division algebras, we define for any abelian Galois field extension K of an arbitrary field F a subgroup $Dec(K/F)$ of the relative Brauer group $Br(K/F)$: it is the subgroup generated by the Brauer classes that are split by cyclic extensions of F contained in K; thus, $Dec(K/F) = Br(K/F)$ when K is a cyclic extension of F; but in general the quotient $Br(K/F)/Dec(K/F)$ captures fairly subtle properties of the extension K/F. The group $Dec(K/F)$ will play a crucial role for the construction of indecomposable division algebras in Ch. 10 and of division algebras with large essential dimension in Ch. 12, and it will be indispensable for the definition of the specialization coset of an inertially split division algebra over a Henselian field in §8.4.4. We will see in §8.3.2 that it has a description as an $\widehat{H}^{-1}$ cohomology group when $\mathcal{G}(K/F)$ is bicyclic, i.e., generated by two elements.

Following our usual approach, we first develop the theory of decomposably semiramified division algebras in the context of graded algebras (§8.3.3) before considering the parallel theory for algebras over Henselian fields in §8.3.4.

8.3.1 The Dec subgroup of the Brauer group

In this subsection, we define (without reference to a valuation) in the Brauer group of an arbitrary field F a subgroup $Dec(K/F)$ attached to any finite-degree abelian Galois extension K of F.

Let K be a finite-degree abelian Galois field extension of an arbitrary (ungraded) field F. Let $G = \mathcal{G}(K/F)$, and let $G^\vee$ be the character group of G,

$$G^\vee = Hom(G, \mathbb{T}).$$

For any $\chi \in G^\vee$, let L_χ denote the fixed field $L_\chi = K^{\ker \chi}$, which is a cyclic Galois extension of F. Further, let $r_\chi = |\langle \chi \rangle| = [L_\chi{:}F]$, let $\overline{\chi}\colon \mathcal{G}(L_\chi/F) \xrightarrow{\sim} \frac{1}{r_\chi}\mathbb{Z}/\mathbb{Z}$ be the isomorphism induced by χ, and let σ_χ be the generator of $\mathcal{G}(L_\chi/F)$ such that $\overline{\chi}(\sigma_\chi) = \frac{1}{r_\chi} + \mathbb{Z}$. Recall (e.g. from Serre, [229, Ch. 14, §1]) that there is a $\mathbb{Z}$-bilinear pairing

$$\xi\colon G^\vee \times F^\times \to Br(K/F) \qquad \text{given by} \quad (\chi, a) \mapsto \big[(L_\chi/F, \sigma_\chi, a)\big].$$

Definition 8.19. $Dec(K/F)$ is the subgroup of $Br(K/F)$ generated by the $\xi(\chi, a)$ for $\chi \in G^\vee$ and $a \in F^\times$. Thus,

$$Dec(K/F) = \sum_{\substack{F \subseteq L \subseteq K \\ L/F \text{ cyclic}}} Br(L/F) \subseteq Br(K/F),$$

where the sum is taken over the fields L with $F \subseteq L \subseteq K$ and L cyclic Galois over F. Take any cyclic decomposition $G = \langle\sigma_1\rangle \times \ldots \times \langle\sigma_n\rangle$, and let L_i be the fixed field of $\langle\sigma_1\rangle \times \ldots \times \langle\sigma_{i-1}\rangle \times \langle\sigma_{i+1}\rangle \times \ldots \times \langle\sigma_n\rangle$, for each i. Then each L_i is cyclic Galois over F with $\mathcal{G}(L_i/F) = \langle\sigma_i|_{L_i}\rangle$, and $K = L_1 \otimes_F \ldots \otimes_F L_n$. Let $r_i = [L_i{:}F] = |\langle\sigma_i\rangle|$, and let $\chi_i \in G^\vee$ be the character defined by $\chi_i(\sigma_i) = \frac{1}{r_i} + \mathbb{Z}$ and $\chi_i(\sigma_j) = 0$ for $j \neq i$. Thus, $L_{\chi_i} = L_i$, $r_{\chi_i} = r_i$, and $\sigma_{\chi_i} = \sigma_i|_{L_i}$, for every i. Clearly, $G^\vee = \langle\chi_1\rangle \times \ldots \times \langle\chi_n\rangle$. Since every $\chi \in G^\vee$ is a $\mathbb{Z}$-linear combination of $\chi_1, \ldots, \chi_n$, and the pairing ξ is $\mathbb{Z}$-bilinear, it follows that

$$\begin{aligned} Dec(K/F) &= \sum_{i=1}^{n} Br(L_i/F) \\ &= \big\{\big[(L_1,\sigma_1,a_1) \otimes_F \ldots \otimes_F (L_n,\sigma_n,a_n)\big] \bigm| a_1,\ldots,a_n \in F^\times\big\}. \end{aligned} \tag{8.11}$$

Every central simple algebra in $Br(K/F)$ is Brauer-equivalent to a central simple algebra A containing K as a maximal subfield with

$$deg(A) = [K{:}F] = r_1 \ldots r_n.$$

From (8.11), $[A] \in Dec(K/F)$ if and only if A is Brauer-equivalent to an algebra $E = (L_1,\sigma_1,a_1) \otimes_F \ldots \otimes_F (L_n,\sigma_n,a_n)$ for some $a_i \in F^\times$. Since $[A{:}F] = [E{:}F]$, we then have $A \cong E$. Because of this explicit description of A as a tensor product of cyclic subalgebras determined by cyclic subextensions of F in K, the algebra A is then said to "decompose according to K." Thus,

$$Dec(K/F) = \big\{[A] \in Br(F) \mid A \text{ decomposes according to } K\big\}.$$

This explains the name $Dec(K/F)$.

The Dec construction is functorial in the following sense: if M is an arbitrary extension of F in some field containing K, then the field compositum $K{\cdot}M$ is a finite-degree abelian Galois field extension of M. For every cyclic extension L/F with $L \subseteq K$, the field $L{\cdot}M$ is a cyclic extension of M in $K{\cdot}M$, and the scalar extension map $ext_{M/F}$ carries $Br(L/F)$ to $Br(L{\cdot}M\,/M)$. Therefore, $ext_{M/F}$ also carries $Dec(K/F)$ to $Dec(K{\cdot}M\,/M)$, and there are induced maps

$$\begin{aligned} ext_{M/F}&\colon Dec(K/F) \longrightarrow Dec(K{\cdot}M\,/M), \\ ext_{M/F}&\colon Br(K/F)/Dec(K/F) \longrightarrow Br(K{\cdot}M\,/M)/Dec(K{\cdot}M\,/M). \end{aligned}$$

Note that for every cyclic Galois extension L of F in K the exponent of $Br(L/F)$ divides $[L{:}F]$, hence also $exp(\mathcal{G}(K/F))$. Therefore, from the definition of $Dec(K/F)$ it is clear that

$$exp\,\big(Dec(K/F)\big) \bigm| exp\,\big(\mathcal{G}(K/F)\big). \tag{8.12}$$

Thus, letting $e = exp(\mathcal{G}(K/F))$, we have

$$Dec(K/F) \subseteq {}_e Br(K/F), \tag{8.13}$$

where the term on the right is the e-torsion subgroup of $Br(K/F)$. Of course, if K is a cyclic extension of F, then

$$Dec(K/F) = {}_e\,Br(K/F) = Br(K/F).$$

The case where K is a bicyclic extension, i.e., $\mathcal{G}(K/F)$ is generated by two elements, will be discussed in the next subsection.

Example 8.20. Suppose F is a local field (i.e., F is complete for a discrete rank 1 valuation with finite residue field). For any abelian Galois field extension K of F we have

$$Dec(K/F) = {}_e\,Br(K/F) = {}_e\,Br(F) \qquad \text{where } e = exp\big(\mathcal{G}(K/F)\big).$$

This follows because K contains a cyclic extension L of F with $[L{:}F] = e$, and every central division algebra over F with exponent a factor of e is split by L (see Reiner [199, Th. (31.4), Cor. (31.10)]); hence,

$${}_e\,Br(F) \subseteq Br(L/F) \subseteq Dec(K/F).$$

For many field extensions however, including many extensions of global fields, the inclusion in (8.13) is strict. The quotient $Br(K/F)/Dec(K/F)$ for F an algebraic number field and K a Galois extension with group $\mathcal{G}(K/F) \cong (\mathbb{Z}/n\mathbb{Z})^2$ is described in Exercise 8.5. Examples of elementary abelian Galois extensions K/F of exponent p with $Dec(K/F) \subsetneqq {}_p\,Br(K/F)$ will be given in Ch. 10 and Ch. 12.

8.3.2 Bicyclic algebras

When the Galois group of the abelian Galois field extension K/F is generated by two elements (i.e., $n = 2$ in (8.11)), the quotient $Br(K/F)/Dec(K/F)$ has a cohomological description, which will be given in Prop. 8.23 below. In preparation for this, we discuss a simplified description of crossed products with bicyclic groups, which will be used on several occasions in the sequel: See Prop. 9.14 and Th. 11.15.

Let K be a finite-degree Galois extension of an arbitrary field F, and let $G = \mathcal{G}(K/F)$. Suppose G is bicyclic (so abelian), i.e., $G = \langle\sigma_1\rangle \times \langle\sigma_2\rangle$ for some $\sigma_1, \sigma_2 \in G$. Let r_i be the order of σ_i and $\boldsymbol{\sigma} = (\sigma_1, \sigma_2)$. Also, let $L_1 = K^{\sigma_2}$ be the subfield of K fixed under σ_2 and let $L_2 = K^{\sigma_1}$, so

$$K = L_1 \otimes_F L_2 \qquad \text{with} \qquad [L_1{:}F] = r_1, \quad [L_2{:}F] = r_2.$$

Let A be a central simple F-algebra containing K with $deg\,A = r_1 r_2 = [K{:}F]$. Then $C_A(K) = K$ by the Double Centralizer Theorem, and A is a crossed product algebra with respect to K and G. We call such an A a *bicyclic algebra*.

Our presentation of G as $\langle\sigma_1\rangle \times \langle\sigma_2\rangle$ leads to a corresponding presentation of A: By the Skolem–Noether Theorem, there are elements z_1, $z_2 \in A^\times$ such that

$$z_1 a = \sigma_1(a) z_1 \quad\text{and}\quad z_2 a = \sigma_2(a) z_2 \quad \text{for all } a \in K.$$

We then have $z_1^{i_1} z_2^{i_2} a = \sigma_1^{i_1}\sigma_2^{i_2}(a) z_1^{i_1} z_2^{i_2}$ for all $a \in K$, so the crossed product algebra A can be written

$$A = \bigoplus_{i_1=0}^{r_1-1} \bigoplus_{i_2=0}^{r_2-1} K z_1^{i_1} z_2^{i_2}. \tag{8.14}$$

Let

$$b_1 = z_1^{r_1}, \qquad u = z_1 z_2 z_1^{-1} z_2^{-1}, \qquad \text{and} \qquad b_2 = z_2^{r_2}.$$

Because $\sigma_1^{r_1} = \sigma_2^{r_2} = id_K$ and $\sigma_1\sigma_2 = \sigma_2\sigma_1$, we have $b_1, u, b_2 \in C_A(K)^\times = K^\times$. Moreover, as z_i centralizes b_i, we have $b_i \in K^{\sigma_i}$, so $b_1 \in L_2^\times$ and $b_2 \in L_1^\times$. Since any product $(c z_1^{i_1} z_2^{i_2}) \cdot (d z_1^{j_1} z_2^{j_2})$ with c, $d \in K$ is expressible as $e z_1^{k_1} z_2^{k_2}$ for $k_1 \in \{0, \dots, r_1 - 1\}$ and $k_2 \in \{0, \dots, r_2 - 1\}$, where e in K is a product of c, $\sigma_1^{i_1}\sigma_2^{i_2}(d)$ and a computable product of b_1, u, b_2 and their G-conjugates, the multiplication in A is determined completely by K with its G-action and b_1, u, and b_2; therefore, we write

$$A = \mathcal{A}(K/F, \boldsymbol{\sigma}, b_1, u, b_2).$$

We next determine the conditions on the elements b_1, u, b_2 for the existence of a crossed product algebra of the form $\mathcal{A}(K/F, \boldsymbol{\sigma}, b_1, u, b_2)$.

From the identity $z_1 z_2 = u z_2 z_1$ it follows by induction on s that

$$z_1^s z_2 = \prod_{i=0}^{s-1} \sigma_1^i(u) z_2 z_1^s \quad\text{and}\quad z_1 z_2^s = \prod_{i=0}^{s-1} \sigma_2^i(u) z_2^s z_1.$$

Hence, $b_1 z_2 = z_1^{r_1} z_2 = N_{K/L_2}(u) z_2 b_1$ and $z_1 b_2 = N_{K/L_1}(u) b_2 z_1$, which shows that

$$b_1 \sigma_2(b_1)^{-1} = N_{K/L_2}(u) \qquad\text{and}\qquad \sigma_1(b_2) b_2^{-1} = N_{K/L_1}(u). \tag{8.15}$$

We will see in Prop. 8.22 below that there are no other constraints on b_1, b_2, and u.

Definition 8.21. With the notation above, define $Z_{\boldsymbol{\sigma}}^2(G, K^\times) \subseteq L_2^\times \times K^\times \times L_1^\times$ to be the subgroup of triples (b_1, u, b_2) satisfying (8.15), and let

$$B_{\boldsymbol{\sigma}}^2(G,K^\times) = \big\{\big(N_{K/L_2}(a_1),\, a_1\sigma_2(a_1)^{-1}\sigma_1(a_2)a_2^{-1},\, N_{K/L_1}(a_2)\big) \mid a_1, a_2 \in K^\times\big\}.$$

It is readily checked that $B_{\boldsymbol{\sigma}}^2(G, K^\times) \subseteq Z_{\boldsymbol{\sigma}}^2(G, K^\times)$. The notation is motivated by the observation that the elements in $Z_{\boldsymbol{\sigma}}^2(G, K^\times)$ (resp. $B_{\boldsymbol{\sigma}}^2(G, K^\times)$) are the cocycles (resp. coboundaries) in an alternative description of the cohomology group $H^2(G, K^\times)$: See the Notes at the end of this chapter.

Proposition 8.22. *Use the same notation as above.*

(i) *For any* $\zeta = (b_1, u, b_2) \in Z^2_{\boldsymbol{\sigma}}(G, K^\times)$, *the relations* $z_i a = \sigma_i(a) z_i$ *for* $a \in K$ $(i = 1, 2)$ *and*

$$z_1^{r_1} = b_1, \qquad z_1 z_2 = u z_2 z_1, \qquad z_2^{r_2} = b_2$$

define on the K*-vector space* $\bigoplus_{i=0}^{r_1-1} \bigoplus_{j=0}^{r_2-1} K z_1^i z_2^j$ *an* F*-algebra multiplication extending the multiplication in* K. *The algebra thus defined is central simple of degree* $r_1 r_2$.

Moreover, using for the algebra defined in (i) *the notation* $\mathcal{A}(K/F, \boldsymbol{\sigma}, b_1, u, b_2)$ *or* $\mathcal{A}(K/F, \boldsymbol{\sigma}, \zeta)$, *the following hold:*

(ii) *for* ζ, $\zeta' \in Z^2_{\boldsymbol{\sigma}}(G, K^\times)$, *we have* $\mathcal{A}(K/F, \boldsymbol{\sigma}, \zeta) \cong \mathcal{A}(K/F, \boldsymbol{\sigma}, \zeta')$ *if and only if* $\zeta \equiv \zeta' \pmod{B^2_{\boldsymbol{\sigma}}(G, K^\times)}$;

(iii) *for* ζ, $\eta \in Z^2_{\boldsymbol{\sigma}}(G, K^\times)$, *we have in* $\mathrm{Br}(F)$,

$$\mathcal{A}(K/F, \boldsymbol{\sigma}, \zeta) \otimes_F \mathcal{A}(K/F, \boldsymbol{\sigma}, \eta) \sim \mathcal{A}(K/F, \boldsymbol{\sigma}, \zeta\eta).$$

Proof. (i) We prove the existence of a central simple F-algebra A of degree $r_1 r_2$ containing K and two elements z_1, z_2 subject to the stated relations.

Consider in $K^\times$ the collection $\boldsymbol{u} = (u_{i,j})_{i,j=1}^2$ defined by

$$u_{1,1} = u_{2,2} = 1, \quad u_{1,2} = u, \quad u_{2,1} = u^{-1}.$$

These elements satisfy the conditions (1.3); hence, we may consider the division ring of iterated Laurent series $D = \mathcal{L}((K; \boldsymbol{\sigma}, \boldsymbol{u}))$ of (1.4) in §1.1.3. By definition, D consists of Laurent series in two indeterminates x_1, x_2 over K, where the multiplication satisfies

$$x_1 x_2 = u x_2 x_1 \quad \text{and} \quad x_i a = \sigma_i(a) x_i \quad \text{for } i = 1,\, 2 \text{ and } a \in K.$$

Proposition 1.13 shows that D has dimension $(r_1 r_2)^2$ over its center. More precisely, the proof of that proposition (see (1.17)) shows that the center of D is

$$Z(D) = F((b_1^{-1} x_1^{r_1}))((b_2^{-1} x_2^{r_2})).$$

Consider the graded algebra $\mathsf{gr}(D)$ associated to the (x_1, x_2)-adic valuation on D. It consists of Laurent polynomials in two noncommuting indeterminates $\widetilde{x}_1$, $\widetilde{x}_2$ over K, and it is easily seen that its center is

$$Z\big(\mathsf{gr}(D)\big) = \mathsf{gr}\big(Z(D)\big) = F\big[b_1^{-1}\widetilde{x}_1^{r_1}, (b_1^{-1}\widetilde{x}_1^{r_1})^{-1}, b_2^{-1}\widetilde{x}_2^{r_2}, (b_2^{-1}\widetilde{x}_2^{r_2})^{-1}\big].$$

(The first equation also follows from the observation that the (x_1, x_2)-adic valuation on D is tame, since D contains $K \otimes_F Z(D)$ as a maximal subfield inertial over the center.) Since $\mathsf{gr}(D)$ is a central graded division algebra of degree $r_1 r_2$ over $\mathsf{gr}\big(Z(D)\big)$, the ungraded ring $\mathsf{gr}(D)^\natural$ (i.e., the ring $\mathsf{gr}(D)$ with the grading ignored) is an Azumaya algebra of degree $r_1 r_2$ over $\mathsf{gr}\big(Z(D)\big)^\natural$, see Remark 6.5. Consider the maximal ideal $M \subseteq \mathsf{gr}\big(Z(D)\big)^\natural$ generated by

$b_1^{-1}\widetilde{x}_1^{r_1} - 1$ and $b_2^{-1}\widetilde{x}_2^{r_2} - 1$. We have $\mathsf{gr}(Z(D))^{\natural}/M = F$, and by a property of Azumaya algebras (see for example Knus–Ojanguren [116, Th. 5.1]) $\mathsf{gr}(D)^{\natural}/M\,\mathsf{gr}(D)^{\natural}$ is a central simple F-algebra A of degree r_1r_2. This algebra contains K, and the images z_1, z_2 in A of $\widetilde{x}_1$ and $\widetilde{x}_2$ satisfy the stated relations. Thus, $A \cong \mathcal{A}(K/F, \boldsymbol{\sigma}, b_1, u, b_2)$.

Alternatively, the existence of A can be proved by a generalized cyclic algebra construction as follows: Consider the cyclic algebra

$$C = (K/L_2, \sigma_1, b_1) = \bigoplus_{i=0}^{r_1-1} K z_1^i.$$

Using (8.15), one can show as in the construction of $\mathcal{L}((K;\boldsymbol{\sigma},\boldsymbol{u}))$ in §1.1.3 that σ_2 extends to an automorphism $\widehat{\sigma}_2$ of C such that $\widehat{\sigma}_2(z_1) = u^{-1}z_1$. (This reduces to checking that $(u^{-1}z_1)^{r_1} = \sigma_2(b_1)$ and $u^{-1}z_1\sigma_2(c) = \sigma_2\sigma_1(c)u^{-1}z_1$ for all $c \in K$.) Since $\widehat{\sigma}_2^{r_2} = \mathit{int}(b_2)$ on C, we may then define an algebra A by

$$A = C \oplus Cz_2 \oplus \ldots \oplus Cz_2^{r_2-1}$$

where the multiplication is given by

$$z_2^{r_2} = b_2 \qquad \text{and} \qquad z_2c = \widehat{\sigma}_2(c)z_2 \quad \text{for } c \in C.$$

Standard arguments show that A is a central simple F-algebra of degree r_1r_2, see Albert [4, Th. 11.11]. (This is a special case of the generalized crossed product construction described in Tignol [239] and Kursov and Yanchevskiĭ [120].)

(ii) Let $\zeta = (b_1, u, b_2)$ and $\zeta' = (b_1', u', b_2')$. Assume first that $\zeta \equiv \zeta' \ (\mathit{mod}\ B^2_{\boldsymbol{\sigma}}(G, K^\times))$; we may then find a_1, $a_2 \in K^\times$ such that

$$b_1' = N_{K/L_2}(a_1)b_1, \quad u' = a_1\sigma_2(a_1)^{-1}\sigma_1(a_2)a_2^{-1}u, \quad b_2' = N_{K/L_1}(a_2)b_2.$$

In $A = \mathcal{A}(K/F, \boldsymbol{\sigma}, \zeta)$ with z_1, z_2 as above, let $y_1 = a_1z_1$ and $y_2 = a_2z_2$. Then, $\mathit{int}(y_i)|_K = \mathit{int}(z_i)|_K = \sigma_i$ for $i = 1$, 2, and

$$\begin{aligned} y_1^{r_1} &= (a_1z_1)^{r_1} = N_{K/L_2}(a_1)z_1^{r_1} = b_1', \\ y_2^{r_2} &= (a_2z_2)^{r_2} = N_{K/L_1}(a_2)z_2^{r_2} = b_2', \\ y_1y_2y_1^{-1}y_2^{-1} &= (a_1z_1)(a_2z_2)(a_1z_1)^{-1}(a_2z_2)^{-1} \\ &= a_1\sigma_1(a_2)\sigma_2(a_1)^{-1}a_2^{-1}z_1z_2z_1^{-1}z_2^{-1} = u'. \end{aligned}$$

Therefore, the presentation of A using y_1, y_2 instead of z_1,z_2 shows that

$$\mathcal{A}(K/F, \boldsymbol{\sigma}, \zeta) \cong \mathcal{A}(K/F, \boldsymbol{\sigma}, \zeta').$$

Conversely let $A' = \mathcal{A}(K/F, \boldsymbol{\sigma}, \zeta')$ with corresponding generators z_1', z_2', and suppose there is an F-isomorphism $\psi\colon A' \to A$. There is then an F-isomorphism $\psi^{-1}\colon \psi(K) \to K$ of simple F-subalgebras of A; by the Skolem–Noether Theorem there is an (inner) automorphism φ of A with $\varphi|_{\psi(K)} = \psi^{-1}|_{\psi(K)}$. By replacing ψ by $\varphi \circ \psi$, we may assume that $\psi|_K = \mathit{id}_K$.

Then $int(\psi(z_i'))|_K = \sigma_i = int(z_i)|_K$ for $i = 1, 2$, hence $\psi(z_1')z_1^{-1}$ and $\psi(z_2')z_2^{-1}$ centralize K, and therefore lie in K. Let $a_1, a_2 \in K^\times$ be defined by

$$\psi(z_1') = a_1 z_1 \qquad \text{and} \qquad \psi(z_2') = a_2 z_2.$$

From $\psi(z_1')^{r_1} = b_1'$, $\psi(z_1')\psi(z_2')\psi(z_1')^{-1}\psi(z_2')^{-1} = u'$, $\psi(z_2')^{r_2} = b_2'$, and the calculations given above for the y_i, we have

$$b_1' = N_{K/L_2}(a_1)b_1, \ u' = a_1\sigma_1(a_2)\sigma_2(a_1)^{-1}a_2^{-1}u, \text{ and } b_2' = N_{K/L_1}(a_2)b_2.$$

Therefore, $\zeta' \equiv \zeta \ (mod\ B^2_{\boldsymbol{\sigma}}(G, K^\times))$.

(iii) Let $\zeta = (b_1, u, b_2)$ and $\eta = (c_1, v, c_2)$. Also, let $A = \mathcal{A}(K/F, \boldsymbol{\sigma}, \zeta)$ and let $f_A \colon G \times G \to K^\times$ be the 2-cocycle for A viewed as a crossed product using generators $z_1^{i_1}z_2^{i_2}$ for $i_1 \in \{0, \ldots, r_1 - 1\}$ and $i_2 \in \{0, \ldots, r_2 - 1\}$, as in (8.14). Thus,

$$z_1^{i_1}z_2^{i_2} \cdot z_1^{j_1}z_2^{j_2} = f_A(\sigma_1^{i_1}\sigma_2^{i_2}, \sigma_1^{j_1}\sigma_2^{j_2})z_1^{k_1}z_2^{k_2},$$

where $k_\ell \in \{0, \ldots, r_\ell - 1\}$ and $k_\ell \equiv i_\ell + j_\ell \ (mod\ r_\ell)$ for $\ell = 1, 2$. Then, for $\ell = 1, 2$, we have

$$f_A(\sigma_\ell, \sigma_\ell^i) = 1 \quad \text{for } i = 1, 2, \ldots, r_{\ell-2}, \quad \text{while} \quad f_A(\sigma_\ell, \sigma_\ell^{r_\ell - 1}) = b_\ell. \tag{8.16}$$

Also,

$$f_A(\sigma_1, \sigma_2) = 1 \qquad \text{and} \qquad f_A(\sigma_2, \sigma_1) = u^{-1}. \tag{8.17}$$

Likewise for $B = \mathcal{A}(K/F, \boldsymbol{\sigma}, \eta)$, let f_B be the corresponding 2-cocycle satisfying the analogues of (8.16) and (8.17). Let $g = f_A f_B \in Z^2(G, K^\times)$. So, for $\ell = 1, 2$,

$$g(\sigma_\ell, \sigma_\ell^i) = 1 \quad \text{for } i = 1, 2, \ldots, r_{\ell-2}, \quad \text{while} \quad g(\sigma_\ell, \sigma_\ell^{r_\ell - 1}) = b_\ell c_\ell,$$

and

$$g(\sigma_1, \sigma_2) = 1 \qquad \text{and} \qquad g(\sigma_2, \sigma_1) = u^{-1}v^{-1}.$$

Let $C = (K/F, G, g)$. From the isomorphism $Br(K/F) \cong H^2(G, K^\times)$, we have $C \sim A \otimes_F B$ in $Br(F)$. When we write $C = \bigoplus_{\sigma \in G} K x_\sigma$ with $x_\sigma x_\tau = g(\sigma, \tau)\, x_{\sigma\tau}$ and $int(x_\sigma)|_K = \sigma$, the equations above for g show that $x_{\sigma_\ell^i} = (x_{\sigma_\ell})^i$ for $i = 2, \ldots, r_\ell - 1$ and $(x_{\sigma_\ell})^{r_\ell} = b_\ell c_\ell$; also, $x_{\sigma_1}x_{\sigma_2} = x_{\sigma_1\sigma_2}$ and $x_{\sigma_2}x_{\sigma_1} = u^{-1}v^{-1}\, x_{\sigma_1\sigma_2} = u^{-1}v^{-1}\, x_{\sigma_1}x_{\sigma_2}$. These equations show that $C = \mathcal{A}(K/F, \boldsymbol{\sigma}, b_1c_1, uv, b_2c_2)$ (when we take x_{σ_1}, x_{σ_2} for z_1, z_2). This proves (iii). □

Continuing with the same notation, we now turn to the description of the quotient $Br(K/F)/Dec(K/F)$.

Recall (e.g. from Serre [229, Ch. 8, §1]) that the (Tate) cohomology group $\widehat{H}^{-1}(G, K^\times)$ is defined as the quotient

$$\widehat{H}^{-1}(G, K^\times) = ker\, N_{K/F}/I_G K^\times, \tag{8.18}$$

where $N_{K/F}\colon K^\times \to F^\times$ is the norm map and $I_GK^\times$ is the subgroup of $K^\times$ generated by elements of the form $\rho(x)x^{-1}$, for $\rho \in G$ and $x \in K^\times$,

$$I_GK^\times = \langle \rho(x)x^{-1} \mid \rho \in G,\, x \in K^\times \rangle.$$

Since G is generated by σ_1 and σ_2, repeated use of the identity

$$\tau\rho(a)a^{-1} = \tau\big(\rho(a)\big)\rho(a)^{-1} \cdot \rho(a)a^{-1} \qquad \text{for } \tau,\, \rho \in G \text{ and } a \in K^\times$$

shows that

$$I_GK^\times = \{a_1\sigma_2(a_1)^{-1}\sigma_1(a_2)a_2^{-1} \mid a_1, a_2 \in K^\times\}.$$

Proposition 8.23. *Let K be a bicyclic Galois extension of a field F and let $\boldsymbol{\sigma} = (\sigma_1, \sigma_2)$ be a base of $\mathcal{G}(K/F)$. There is a well-defined homomorphism*

$$\upsilon_{\boldsymbol{\sigma}}\colon Br(K/F) \longrightarrow \widehat{H}^{-1}(G, K^\times)$$

such that

$$\upsilon_{\boldsymbol{\sigma}}\big(\big[\mathcal{A}(K/F, \boldsymbol{\sigma}, b_1, u, b_2)\big]\big) = u \cdot I_GK^\times$$

for all $(b_1, u, b_2) \in Z^2_{\boldsymbol{\sigma}}(G, K^\times)$. This homomorphism fits into the following exact sequence, where the map i is the inclusion:

$$0 \longrightarrow Dec(K/F) \xrightarrow{\ i\ } Br(K/F) \xrightarrow{\ \upsilon_{\boldsymbol{\sigma}}\ } \widehat{H}^{-1}(G, K^\times) \longrightarrow 1.$$

Proof. For all $(b_1, u, b_2) \in Z^2_{\boldsymbol{\sigma}}(G, K^\times)$, (8.15) yields

$$N_{K/F}(u) = N_{L_2/F}(b_1\sigma_2(b_1)^{-1}) = 1,$$

so $u \cdot I_GK^\times \in \widehat{H}^{-1}(G, K^\times)$. If $(b_1, u, b_2) \equiv (b_1', u', b_2') \pmod{B^2_{\boldsymbol{\sigma}}(G, K^\times)}$, then there are a_1, $a_2 \in K^\times$ such that

$$u' = a_1\sigma_2(a_1)^{-1}\sigma_1(a_2)a_2^{-1}\, u,$$

hence $u' \cdot I_GK^\times = u \cdot I_GK^\times$. Therefore, it follows by Prop. 8.22(ii) that the map $\upsilon_{\boldsymbol{\sigma}}$ is well-defined. It is completely defined on $Br(K/F)$ because every central simple F-algebra split by K is Brauer-equivalent to a crossed product of K/F, and every such crossed product has the form $\mathcal{A}(K/F, \boldsymbol{\sigma}, \zeta)$ for some $\zeta \in Z^2_{\boldsymbol{\sigma}}(G, K^\times)$. Moreover, Prop. 8.22(iii) shows that $\upsilon_{\boldsymbol{\sigma}}$ is a group homomorphism. Thus, it only remains to prove the exactness of the sequence in the statement.

Surjectivity of $\upsilon_{\boldsymbol{\sigma}}$ is easy to see: Given $u \in K^\times$ such that $N_{K/F}(u) = 1$, we have

$$N_{L_1/F}\big(N_{K/L_1}(u)\big) = 1 \quad \text{and} \quad N_{L_2/F}\big(N_{K/L_2}(u)\big) = 1,$$

hence Hilbert's Theorem 90 (see Draxl [63, p. 35] or Pierce [178, p. 312]) yields $b_2 \in L_1^\times$ and $b_1 \in L_2^\times$ such that

$$N_{K/L_2}(u) = b_1\sigma_2(b_1)^{-1} \quad \text{and} \quad N_{K/L_1}(u) = \sigma_1(b_2)b_2^{-1}.$$

It follows that $(b_1, u, b_2) \in Z^2_{\boldsymbol{\sigma}}(G, K^\times)$, so we may consider the crossed product algebra $A = \mathcal{A}(K/F, \boldsymbol{\sigma}, b_1, u, b_2)$. By definition, $\upsilon_{\boldsymbol{\sigma}}([A]) = u \cdot I_G K^\times$, hence $\upsilon_{\boldsymbol{\sigma}}$ is surjective.

By (8.11), every Brauer class in $Dec(K/F)$ is represented by an algebra A of the form

$$A = (L_1, \sigma_1, b_1) \otimes_F (L_2, \sigma_2, b_2)$$

for some b_1, $b_2 \in F^\times$. We may choose z_1, $z_2 \in A^\times$ satisfying $int(z_i)|_K = \sigma_i$ $(i = 1, 2)$ with $z_1 \in (L_1, \sigma_1, b_1)$ and $z_2 \in (L_2, \sigma_2, b_2)$ subject to $z_1^{r_1} = b_1$ and $z_2^{r_2} = b_2$; then z_1 and z_2 commute, so we obtain for A a crossed product representation of the form $A \cong \mathcal{A}(K/F, \boldsymbol{\sigma}, b_1, 1, b_2)$. It follows that $\upsilon_{\boldsymbol{\sigma}}([A]) = 1$, hence $Dec(K/F) \subseteq ker\, \upsilon_{\boldsymbol{\sigma}}$, and it only remains to prove the reverse inclusion.

Let $A = \mathcal{A}(K/F, \boldsymbol{\sigma}, b_1, u, b_2)$ and suppose $\upsilon_{\boldsymbol{\sigma}}([A]) = 1$. Then $u \in I_G K^\times$, so $u = a_1 \sigma_2(a_1)^{-1} \sigma_1(a_2) a_2^{-1}$ for some a_1, $a_2 \in K^\times$. Letting $b_1' = N_{K/L_2}(a_1)^{-1} b_1$ and $b_2' = N_{K/L_1}(a_2)^{-1} b_2$, we then have $(b_1, u, b_2) \equiv (b_1', 1, b_2') \, (mod\, B^2_{\boldsymbol{\sigma}}(G, K^\times))$, so by Prop. 8.22(ii) we have

$$A \cong \mathcal{A}(K/F, \boldsymbol{\sigma}, b_1', 1, b_2'). \tag{8.19}$$

Note that the cocycle conditions (8.15) imply that b_1', $b_2' \in F^\times$. Therefore, for the distinguished generators z_1', z_2' of A in the crossed product representation (8.19), the subalgebra generated by L_i and z_i' is the cyclic algebra $(L_i/F, \sigma_i, b_i')$, for $i = 1, 2$. Moreover, these two subalgebras centralize each other, as z_2' centralizes $L_1 = K^{\sigma_2}$, z_1' centralizes $L_2 = K^{\sigma_1}$, and $z_1' z_2' = z_2' z_1'$. The tensor product of these subalgebras has the same dimension as A; hence,

$$A \cong (L_1/F, \sigma_1, b_1') \otimes_F (L_2/F, \sigma_2, b_2').$$

Therefore, $[A]$ lies in $Dec(K/F)$; we have thus shown that $ker\, \upsilon_{\boldsymbol{\sigma}} \subseteq Dec(K/F)$. □

Example 8.24. *Symbol algebras viewed as bicyclic algebras.* Let r_1, $r_2 \in \mathbb{N}$ be arbitrary nonzero integers and let $n = r_1 r_2$. Let F be a field containing a primitive n-th root of unity ω (so $char\, F \nmid n$) and let a_1, $a_2 \in F^\times$. Assume that $K = F(\sqrt[r_1]{a_1}, \sqrt[r_2]{a_2})$ is a field with $[K{:}F] = n$. Then K is an abelian Galois extension of F with Galois group $\mathcal{G}(K/F) \cong \mathbb{Z}/r_1\mathbb{Z} \times \mathbb{Z}/r_2\mathbb{Z}$. Consider the symbol algebra

$$A = \big(a_1, a_2/F\big)_{\omega, n}.$$

The standard generators i, j of A satisfy $i^n = a_1$, $j^n = a_2$, and $ij = \omega ji$. Therefore, $(i^{r_2})^{r_1} = a_1$, $(j^{r_1})^{r_2} = a_2$, and $i^{r_2} j^{r_1} = \omega^{r_1 r_2} j^{r_1} i^{r_2}$, so i^{r_2} and j^{r_1} commute. We may therefore identify K with the subalgebra of A generated by i^{r_2} and j^{r_1}. Let $\boldsymbol{\sigma} = (\sigma_1, \sigma_2)$ where σ_1, σ_2 are the generators of $\mathcal{G}(K/F)$ defined by

$$\sigma_1(i^{r_2}) = \omega^{r_2} i^{r_2}, \ \sigma_1(j^{r_1}) = j^{r_1}, \quad \text{and} \quad \sigma_2(i^{r_2}) = i^{r_2}, \ \sigma_2(j^{r_1}) = \omega^{r_1} j^{r_1}.$$

From $ij = \omega ji$ we have $j^{-1} ij = \omega i$ and $iji^{-1} = \omega j$, so

$$int(j^{-1})|_K = \sigma_1 \qquad \text{and} \qquad int(i)|_K = \sigma_2.$$

We may therefore represent A as a crossed product of K with distinguished generators $z_1 = j^{-1}$ and $z_2 = i$. Writing $\sqrt[r_1]{a_1}$ for i^{r_2} and $\sqrt[r_2]{a_2}$ for j^{r_1}, and noting that $j^{-1}i = \omega i j^{-1}$, we thus have

$$A = \mathcal{A}(K/F, \boldsymbol{\sigma}, \sqrt[r_2]{a_2^{-1}}, \omega, \sqrt[r_1]{a_1}).$$

By the definition of $\upsilon_{\boldsymbol{\sigma}}$ in Prop. 8.23, we thus have

$$\upsilon_{\boldsymbol{\sigma}}\big(\big[(a_1, a_2/F)_{\omega,n}\big]\big) = \omega \cdot I_G K^\times \in \widehat{H}^{-1}(G, K^\times).$$

It follows from Prop. 8.23 that the image of ω in $\widehat{H}^{-1}(G, K^\times)$ is trivial if and only if $\big[(a_1, a_2/F)_{\omega,n}\big] \in Dec(K/F)$. If this condition holds, then since $exp\big(\mathcal{G}(K/F)\big) = lcm(r_1, r_2)$ it follows that $exp\,(a_1, a_2/F)_{\omega,n} \mid lcm(r_1, r_2)$, see (8.12).

The simplified description of bicyclic crossed products in Prop. 8.22 has a graded version, which we now outline in the case where the bicyclic graded field extension is inertial. (This is the only case that will be used in the sequel.)

Let F be a graded field, let K be an inertial abelian Galois graded field extension of F. Let $G = \mathcal{G}(\mathsf{K}/\mathsf{F}) = \mathcal{G}(\mathsf{K}_0/\mathsf{F}_0)$. Suppose that G is bicyclic, i.e., $G = \langle\sigma_1\rangle \times \langle\sigma_2\rangle$ for some $\sigma_i \in G$, so $\boldsymbol{\sigma} = (\sigma_1, \sigma_2)$ is a base of G. Let $\mathsf{L}_1 = \mathsf{K}^{\sigma_2}$ and $\mathsf{L}_2 = \mathsf{K}^{\sigma_1}$. The L_i are graded subfields of K, each of which is Galois over F with $\mathcal{G}(\mathsf{L}_i/\mathsf{F}) = \langle\sigma_i|_{\mathsf{L}_i}\rangle$; also $\mathsf{K} = \mathsf{L}_1 \otimes_{\mathsf{F}} \mathsf{L}_2$. Let $r_i = [\mathsf{L}_i{:}\mathsf{F}] = |\langle\sigma_i\rangle|$. Let A be a central simple graded F-algebra with $\mathsf{K} \subseteq \mathsf{A}$ and $deg\,\mathsf{A} = [\mathsf{K}{:}\mathsf{F}]$. Then $C_{\mathsf{A}}(\mathsf{K}) = \mathsf{K}$ by the graded Double Centralizer Theorem 2.35, and by the graded Skolem–Noether Theorem 2.37(iii) (which applies since $C_{\mathsf{A}}(\mathsf{K}) = \mathsf{K}$) there exist homogeneous $z_1, z_2 \in \mathsf{A}^\times$ with $z_i c z_i^{-1} = \sigma_i(c)$ for all $c \in \mathsf{K}$, $i = 1, 2$. Let $b_i = z_i^{r_i}$ and $u = z_1 z_2 z_1^{-1} z_2^{-1}$. Just as in the ungraded case (see (8.15)), we have $b_1 \in \mathsf{L}_2^\times$, $b_2 \in \mathsf{L}_1^\times$, $u \in \mathsf{K}^\times$, and

$$N_{\mathsf{K}/\mathsf{L}_2}(u) = b_1\sigma_2(b_1)^{-1} \qquad \text{and} \qquad N_{\mathsf{K}/\mathsf{L}_1}(u) = \sigma_1(b_2)b_2^{-1},$$

so $(b_1, u, b_2) \in Z^2_{\boldsymbol{\sigma}}(G, \mathsf{K}^\times)$. We may then represent A as a graded crossed product algebra:

$$\mathsf{A} = \bigoplus_{i_1=0}^{r_1-1} \bigoplus_{i_2=0}^{r_2-1} \mathsf{K} z_1^{i_1} z_2^{i_2}.$$

The grading on A is determined by $deg\, z_i = \frac{1}{r_i}\, deg(z_i^{r_i}) = \frac{1}{r_i}\, deg\, b_i$, and we have

$$\Gamma_{\mathsf{A}} = \big\langle \tfrac{1}{r_1}\, deg\, b_1, \tfrac{1}{r_2}\, deg\, b_2 \big\rangle + \Gamma_{\mathsf{F}}.$$

Clearly $deg\, u = 0$. Since the multiplication and grading on A are completely determined by K and $\zeta = (b_1, u, b_2)$, we write

$$\mathsf{A} = \mathsf{A}(\mathsf{K}/\mathsf{F}, \boldsymbol{\sigma}, \zeta) = \mathsf{A}(\mathsf{K}/\mathsf{F}, \boldsymbol{\sigma}, b_1, u, b_2). \tag{8.20}$$

Note that

$$q(\mathsf{A}(\mathsf{K}/\mathsf{F},\boldsymbol{\sigma},b_1,u,b_2)) \;=\; \mathcal{A}(q(\mathsf{K})/q(\mathsf{F}),\boldsymbol{\sigma},b_1,u,b_2).$$

Also, if $deg\, b_1 = deg\, b_2 = 0$, then

$$\mathsf{A}(\mathsf{K}/\mathsf{F},\boldsymbol{\sigma},b_1,u,b_2) \;\cong\; \mathcal{A}(\mathsf{K}_0/\mathsf{F}_0,\boldsymbol{\sigma},b_1,u,b_2)\otimes_{\mathsf{F}_0}\mathsf{F}.$$

Proposition 8.25. *Use the same notation as above.*

(i) *For any* $\zeta = (b_1,u,b_2) \in Z^2_{\boldsymbol{\sigma}}(G,\mathsf{K}^\times)$, *there is a central simple graded* F*-algebra* $\mathsf{A}(\mathsf{K}/\mathsf{F},\boldsymbol{\sigma},\zeta)$.

(ii) *For* ζ, $\zeta' \in Z^2_{\boldsymbol{\sigma}}(G,\mathsf{K}^\times)$ *we have* $\mathsf{A}(\mathsf{K}/\mathsf{F},\boldsymbol{\sigma},\zeta) \cong_g \mathsf{A}(\mathsf{K}/\mathsf{F},\boldsymbol{\sigma},\zeta')$ *if and only if* $\zeta \equiv \zeta'$ (*mod* $B^2_{\boldsymbol{\sigma}}(G,\mathsf{K}^\times)$).

(iii) *For* ζ, $\eta \in Z^2_{\boldsymbol{\sigma}}(G,\mathsf{K}^\times)$ *we have in* $\mathsf{Br}(\mathsf{F})$

$$\mathsf{A}(\mathsf{K}/\mathsf{F},\boldsymbol{\sigma},\zeta)\otimes_{\mathsf{F}}\mathsf{A}(\mathsf{K}/\mathsf{F},\boldsymbol{\sigma},\eta) \;\sim\; \mathsf{A}(\mathsf{K}/\mathsf{F},\boldsymbol{\sigma},\zeta\eta).$$

Proof. (i) Given $\zeta = (b_1,u,b_2) \in Z^2_{\boldsymbol{\sigma}}(G,\mathsf{K}^\times) \subseteq Z^2_{\boldsymbol{\sigma}}(G,q(\mathsf{K})^\times)$, Prop. 8.22(i) yields the abelian crossed product

$$B = \mathcal{A}(q(\mathsf{K})/q(\mathsf{F}),\boldsymbol{\sigma},\zeta) = \bigoplus_{i_1=0}^{r_1-1}\bigoplus_{i_2=0}^{r_2-1} q(\mathsf{K})\, z_1^{i_1} z_2^{i_2}.$$

Within B, take the free left $\mathsf{K}^\natural$-module $\mathsf{A} = \bigoplus_{i_1=0}^{r_1-1}\bigoplus_{i_2=0}^{r_2-1}\mathsf{K}\, z_1^{i_1} z_2^{i_2}$, which is made into a graded K-vector space by setting

$$deg\, z_1^{i_1} z_2^{i_2} \;=\; \tfrac{i_1}{r_1}\, deg\, b_1 + \tfrac{i_2}{r_2}\, deg\, b_2.$$

It is easy to check that A is a ring and a graded F-algebra. (Observe that when we write $z_1^{i_1} z_2^{i_2}\cdot z_1^{j_1} z_2^{j_2} = c z_1^{k_1} z_2^{k_2}$ with $i_\ell, j_\ell, k_\ell \in \{0,\ldots,r_\ell - 1\}$ $(\ell = 1,2)$ and $c \in q(\mathsf{K}^\times)$, then as $deg\, u = 0$, we have $c \in \mathsf{K}^\times$ with

$$deg\, c \;=\; deg\, z_1^{i_1} z_2^{i_2} + deg\, z_1^{j_1} z_2^{j_2} - deg\, z_1^{k_1} z_2^{k_2}.)$$

Furthermore, A is a simple graded algebra with center F by Lemma 2.14 since $q(\mathsf{A}) = B$, which is a central simple $q(\mathsf{F})$-algebra. Thus, $\mathsf{A} = \mathsf{A}(\mathsf{K}/\mathsf{F},\boldsymbol{\sigma},\zeta)$.

(ii) The same proof as in Prop. 8.22(ii) applies, using the graded Skolem–Noether Theorem 2.37(iii).

(iii) This follows from the ungraded version (Prop. 8.22(iii)) applied over $q(\mathsf{F})$, by injectivity of the map $q_{\mathsf{F}}\colon \mathsf{Br}(\mathsf{F}) \to Br(q(\mathsf{F}))$ (see Prop. 6.4). □

8.3.3 Semiramified graded division algebras

We begin our discussion of decomposably semiramified division algebras with the graded case. After defining decomposably semiramified graded division algebras, we give in Prop. 8.29 a general construction of this type of algebra.

The main result of this subsection is Prop. 8.33, where decomposably semiramified graded division algebras are characterized by their splitting properties.

Throughout this subsection, F is an arbitrary graded field.

Definition 8.26. A central graded division algebra D over F is said to be *semiramified* if D_0 is a field and

$$[\mathsf{D}_0{:}\mathsf{F}_0] \;=\; |\Gamma_{\mathsf{D}}{:}\Gamma_{\mathsf{F}}| \;(=\; deg\,\mathsf{D}\,).$$

When this occurs D_0 is abelian Galois over F_0 and $\Gamma_{\mathsf{D}}/\Gamma_{\mathsf{F}} \cong \mathcal{G}(\mathsf{D}_0/\mathsf{F}_0)$ via θ_{D}, and $\mathsf{D}_0 \otimes_{\mathsf{F}_0} \mathsf{F}$ is a maximal graded subfield of D which is inertial and abelian Galois over F. So, D is inertially split; indeed we have

$$[\mathsf{D}] \;\in\; \mathrm{Br}(\mathsf{D}_0 \otimes_{\mathsf{F}_0} \mathsf{F}\,/\mathsf{F}).$$

We say that D is *decomposably semiramified* if D is semiramified and $[\mathsf{D}]$ lies in the subgroup of $\mathrm{Br}(\mathsf{D}_0 \otimes_{\mathsf{F}_0} \mathsf{F}\,/\mathsf{F})$ generated by all the $\mathrm{Br}(\ell \otimes_{\mathsf{F}_0} \mathsf{F}\,/\mathsf{F})$, where ℓ is a cyclic extension of F_0 in D_0:

$$[\mathsf{D}] \;\in \sum_{\substack{\mathsf{F}_0 \subseteq \ell \subseteq \mathsf{D}_0 \\ \ell/\mathsf{F}_0 \text{ cyclic}}} \mathrm{Br}(\ell \otimes_{\mathsf{F}_0} \mathsf{F}\,/\mathsf{F}) \;\subseteq\; \mathrm{Br}(\mathsf{D}_0 \otimes_{\mathsf{F}_0} \mathsf{F}\,/\mathsf{F});$$

here the sum is taken over the fields ℓ with $\mathsf{F}_0 \subseteq \ell \subseteq \mathsf{D}_0$ and ℓ cyclic Galois over F_0. (Compare this condition with the definition of the *Dec* group in Def. 8.19.) The condition on $[\mathsf{D}]$ necessarily holds when D_0 is a cyclic extension of F_0 (equivalently, when $\Gamma_{\mathsf{D}}/\Gamma_{\mathsf{F}}$ is cyclic). We then say that the semiramified graded division algebra D is *cyclically semiramified.* Such a D is clearly a cyclic graded algebra.

Remark 8.27. If D is an inertially split central graded division algebra over F, then $|\Gamma_{\mathsf{D}}{:}\Gamma_{\mathsf{F}}| \;=\; [Z(\mathsf{D}_0){:}\mathsf{F}_0]$ because θ_{D} induces an isomorphism $\Gamma_{\mathsf{D}}/\Gamma_{\mathsf{F}} \xrightarrow{\sim} \mathcal{G}(Z(\mathsf{D}_0)/\mathsf{F}_0)$. Therefore, the equation $[\mathsf{D}{:}\mathsf{F}] \;=\; [\mathsf{D}_0{:}\mathsf{F}_0] \cdot |\Gamma_{\mathsf{D}}{:}\Gamma_{\mathsf{F}}|$ yields

$$deg\,\mathsf{D} \;=\; deg\,\mathsf{D}_0 \cdot |\Gamma_{\mathsf{D}}{:}\Gamma_{\mathsf{F}}|.$$

From this it is easy to see that the following conditions are equivalent for D inertially split:

(a) D is semiramified;
(b) $|\Gamma_{\mathsf{D}}{:}\Gamma_{\mathsf{F}}| = deg\,\mathsf{D}$;
(c) $[Z(\mathsf{D}_0){:}\mathsf{F}_0] = deg\,\mathsf{D}$;
(d) D_0 is a field.

Decomposably semiramified graded division algebras are so named because they have a very concrete description as tensor products of cyclically semiramified graded algebras:

Proposition 8.28. *Let* N *be a semiramified central graded division algebra over* F*, and let*

$$\mathsf{N}_0 \;=\; \ell_1 \otimes_{\mathsf{F}_0} \ldots \otimes_{\mathsf{F}_0} \ell_n$$

be a decomposition of N_0 *into a tensor product of cyclic field extensions of* F_0. *For* $i = 1, \ldots, n$, *let* σ_i *be a generator of* $\mathcal{G}(\ell_i/\mathsf{F}_0)$, *and let* $r_i = [\ell_i{:}\mathsf{F}_0]$, *so*

$$\mathit{deg}\,\mathsf{N} = [\mathsf{N}_0{:}\mathsf{F}_0] = |\Gamma_\mathsf{N}{:}\Gamma_\mathsf{F}| = r_1 \ldots r_n.$$

For $i = 1, \ldots, n$, *let* $\mathsf{L}_i = \ell_i \otimes_{\mathsf{F}_0} \mathsf{F}$, *and write* σ_i *again for the* F*-automorphism of* L_i *extending* σ_i. *Suppose* N *is decomposably semiramified. Then,*

(i) *there are* $b_1, \ldots, b_n \in \mathsf{F}^\times$ *such that* N *decomposes into a tensor product of graded cyclic algebras*

$$\mathsf{N} \cong_g \big(\mathsf{L}_1/\mathsf{F}, \sigma_1, b_1\big) \otimes_\mathsf{F} \ldots \otimes_\mathsf{F} \big(\mathsf{L}_n/\mathsf{F}, \sigma_n, b_n\big);$$

(ii) *for any such* $b_1, \ldots, b_n$, *each* $\frac{1}{r_i}\,\mathit{deg}\, b_i + \Gamma_\mathsf{F}$ *has order* r_i *in* $\mathbb{T}(\Gamma_\mathsf{F})$ *and*

$$\Gamma_\mathsf{N}/\Gamma_\mathsf{F} = \big\langle \tfrac{1}{r_1}\,\mathit{deg}\, b_1 + \Gamma_\mathsf{F}\big\rangle \oplus \ldots \oplus \big\langle \tfrac{1}{r_n}\,\mathit{deg}\, b_n + \Gamma_\mathsf{F}\big\rangle; \qquad (8.21)$$

(iii) *each* $\big(\mathsf{L}_i/\mathsf{F}, \sigma_i, b_i\big)$ *is a cyclically semiramified graded division algebra with grade group* $\langle \frac{1}{r_i}\,\mathit{deg}\, b_i\rangle + \Gamma_\mathsf{F}$ *and* 0*-component* ℓ_i;

(iv) N *contains a maximal graded subfield* T *that is totally ramified over* F *with* $\Gamma_\mathsf{T} = \Gamma_\mathsf{F}$ *and* $\mathsf{T} \cong_g \mathsf{F}(\sqrt[r_1]{b_1}, \ldots, \sqrt[r_n]{b_n})$.

Proof. The same arguments as for the *Dec* subgroup (see (8.11)) show that if N is decomposably semiramified, then $[\mathsf{N}] \in \sum_{i=1}^n \mathsf{Br}(\mathsf{L}_i/\mathsf{F})$. Since $\mathsf{Br}(\mathsf{L}_i/\mathsf{F})$ consists of Brauer classes of graded cyclic algebras (see Th. 6.11), there are $b_1, \ldots, b_n \in \mathsf{F}^\times$ such that

$$[\mathsf{N}] = \big[\big(\mathsf{L}_1/\mathsf{F}, \sigma_1, b_1\big) \otimes_\mathsf{F} \ldots \otimes_\mathsf{F} \big(\mathsf{L}_n/\mathsf{F}, \sigma_n, b_n\big)\big] \qquad \text{in } \mathsf{Br}(\mathsf{F}).$$

Let $\mathsf{C}_i = \big(\mathsf{L}_i/\mathsf{F}, \sigma_i, b_i\big)$ for $i = 1, \ldots, n$. The displayed equation shows that N is the associated graded division algebra of $\mathsf{C}_1 \otimes_\mathsf{F} \ldots \otimes_\mathsf{F} \mathsf{C}_n$. But the degree of this tensor product is $r_1 \ldots r_n = \mathit{deg}\,\mathsf{N}$, so $\mathsf{C}_1 \otimes_\mathsf{F} \ldots \otimes_\mathsf{F} \mathsf{C}_n$ is a graded division algebra. It follows that

$$\mathsf{N} \cong_g \mathsf{C}_1 \otimes_\mathsf{F} \ldots \otimes_\mathsf{F} \mathsf{C}_n \qquad (8.22)$$

because the graded division algebra in any given Brauer class is uniquely determined up to isomorphism (see Prop. 6.2). For the rest of the proof, we consider the isomorphism (8.22) as an identification. Let $\mathsf{K} = \mathsf{L}_1 \otimes_\mathsf{F} \ldots \otimes_\mathsf{F} \mathsf{L}_n \subseteq \mathsf{N}$ and write $\widehat{\sigma}_i$ for the F-automorphism of K that restricts to σ_i on L_i and to the identity on L_j for $j \neq i$. Write $\widehat{\sigma}_i$ also for the restriction of $\widehat{\sigma}_i$ to $\mathsf{K}_0 = \ell_1 \otimes_{\mathsf{F}_0} \ldots \otimes_{\mathsf{F}_0} \ell_n = \mathsf{N}_0$. From the definition of graded cyclic algebras, there are $z_i \in \mathsf{C}_i^\times \subseteq \mathsf{N}^\times$ such that

$$z_i x = \widehat{\sigma}_i(x) z_i \ \text{ for all } x \in \mathsf{K} \qquad \text{and} \qquad z_i^{r_i} = b_i.$$

Then $\theta_\mathsf{N}(\mathit{deg}\, z_i) = \widehat{\sigma}_i$. Now, θ_N induces an isomorphism $\overline{\theta}_\mathsf{N}\colon \Gamma_\mathsf{N}/\Gamma_\mathsf{F} \xrightarrow{\sim} \mathcal{G}(\mathsf{N}_0/\mathsf{F}_0)$ and $\mathcal{G}(\mathsf{N}_0/\mathsf{F}_0) = \langle\widehat{\sigma}_1\rangle \times \ldots \times \langle\widehat{\sigma}_n\rangle$; it follows that each $\mathit{deg}\, z_i + \Gamma_\mathsf{N}$ has order r_i in $\mathbb{T}(\Gamma_\mathsf{N})$ and

$$\Gamma_\mathsf{N}/\Gamma_\mathsf{F} = \langle \mathit{deg}\, z_1 + \Gamma_\mathsf{F}\rangle \oplus \ldots \oplus \langle \mathit{deg}\, z_n + \Gamma_\mathsf{F}\rangle.$$

This proves (8.21) because $deg\, z_i = \frac{1}{r_i} deg\, b_i$, as $z_i^{r_i} = b_i$. We see also that the grade group of each C_i contains $\frac{1}{r_i} deg\, b_i$, which has order r_i modulo Γ_{F}. Therefore, $|\Gamma_{\mathsf{C}_i}{:}\Gamma_{\mathsf{F}}| \geq r_i$. On the other hand, the degree zero component of C_i contains ℓ_i, which is a cyclic extension of F_0 of degree r_i; hence, $[(\mathsf{C}_i)_0{:}\mathsf{F}_0] \geq r_i$. But

$$r_i^2 = [\mathsf{C}_i{:}\mathsf{F}] = [(\mathsf{C}_i)_0{:}\mathsf{F}_0] \cdot |\Gamma_{\mathsf{C}_i}{:}\Gamma_{\mathsf{F}}|,$$

so we must have $|\Gamma_{\mathsf{C}_i}{:}\Gamma_{\mathsf{F}}| = [(\mathsf{C}_i)_0{:}\mathsf{F}_0] = r_i$, hence $\Gamma_{\mathsf{C}_i} = \langle \frac{1}{r_i} deg\, b_i \rangle + \Gamma_{\mathsf{F}}$ and $(\mathsf{C}_i)_0 = \ell_i$. This shows that C_i is a cyclically semiramified graded division algebra.

To complete the proof, let $\mathsf{T} = \mathsf{F}(z_1, \ldots, z_n) \subseteq \mathsf{N}$. Since the z_i commute pairwise, T is a graded subfield of N. Also, we have

$$\Gamma_{\mathsf{T}}/\Gamma_{\mathsf{F}} \supseteq \langle deg\, z_1 + \Gamma_{\mathsf{F}} \rangle + \ldots + \langle deg\, z_n + \Gamma_{\mathsf{F}} \rangle = \Gamma_{\mathsf{N}}/\Gamma_{\mathsf{F}} \supseteq \Gamma_{\mathsf{T}}/\Gamma_{\mathsf{F}},$$

so $\Gamma_{\mathsf{T}} = \Gamma_{\mathsf{N}}$ and $|\Gamma_{\mathsf{T}}{:}\Gamma_{\mathsf{F}}| = r_1 \ldots r_n = deg\, \mathsf{N}$. But by Cor. 2.36(i) the dimension of a graded subfield of N cannot exceed $deg\, \mathsf{N}$, so $[\mathsf{T}{:}\mathsf{F}] = |\Gamma_{\mathsf{T}}{:}\Gamma_{\mathsf{F}}|$ and T is totally ramified over F. We have $\mathsf{T} \cong \mathsf{F}(\sqrt[r_1]{b_1}, \ldots, \sqrt[r_n]{b_n})$ since $z_i^{r_i} = b_i$ for $i = 1, \ldots, n$. □

As a converse to Prop. 8.28, the following proposition shows that every central simple graded algebra with a tensor decomposition as in Prop. 8.28 is a decomposably semiramified graded division algebra:

Proposition 8.29. *Let $\ell_1, \ldots, \ell_n$ be cyclic field extensions of F_0, which are linearly disjoint, so $\ell = \ell_1 \otimes_{\mathsf{F}_0} \ldots \otimes_{\mathsf{F}_0} \ell_n$ is a field. For $i = 1, \ldots, n$, let σ_i be a generator of $\mathcal{G}(\ell_i/\mathsf{F}_0)$, let $r_i = [\ell_i{:}\mathsf{F}_0]$, and let $\mathsf{L}_i = \ell_i \otimes_{\mathsf{F}_0} \mathsf{F}$. Suppose that $b_1, \ldots, b_n \in \mathsf{F}^\times$ satisfy the following condition: $\frac{1}{r_1} deg\, b_1 + \Gamma_{\mathsf{F}}, \ldots, \frac{1}{r_n} deg\, b_n + \Gamma_{\mathsf{F}}$ have order $r_1, \ldots, r_n$ respectively and generate independent subgroups of $\mathbb{T}(\Gamma_{\mathsf{F}})$, in the sense that the sum*

$$\langle \tfrac{1}{r_1} deg\, b_1 + \Gamma_{\mathsf{F}} \rangle + \ldots + \langle \tfrac{1}{r_n} deg\, b_n + \Gamma_{\mathsf{F}} \rangle \subseteq \mathbb{T}(\Gamma_{\mathsf{F}})$$

is a direct sum. Then the graded algebra

$$\mathsf{N} = \big(\mathsf{L}_1/\mathsf{F}, \sigma_1, b_1\big) \otimes_{\mathsf{F}} \ldots \otimes_{\mathsf{F}} \big(\mathsf{L}_n/\mathsf{F}, \sigma_n, b_n\big)$$

is a decomposably semiramified graded division algebra with $\mathsf{N}_0 = \ell$ and

$$\Gamma_{\mathsf{N}}/\Gamma_{\mathsf{F}} = \langle \tfrac{1}{r_1} deg\, b_1 + \Gamma_{\mathsf{F}} \rangle \oplus \ldots \oplus \langle \tfrac{1}{r_n} deg\, b_n + \Gamma_{\mathsf{F}} \rangle.$$

Proof. Let $\mathsf{L} = \mathsf{L}_1 \otimes_{\mathsf{F}} \ldots \otimes_{\mathsf{F}} \mathsf{L}_n = \ell \otimes_{\mathsf{F}_0} \mathsf{F}$, so L is a maximal graded subfield of N and an inertial extension of F. For $i = 1, \ldots, n$, let $z_i \in \big(\mathsf{L}_i/\mathsf{F}, \sigma_i, b_i\big)^\times$ satisfy

$$z_i x = \sigma_i(x) z_i \text{ for all } x \in \mathsf{L}_i \quad \text{and} \quad z_i^{r_i} = b_i.$$

Then z_i is homogeneous as b_i is homogeneous, and $(z_i^j)_{j=0}^{r_i-1}$ is a base of $\big(\mathsf{L}_i/\mathsf{F}, \sigma_i, b_i\big)$ as a left graded vector space over L_i. To obtain an L-base of N, we form tensor products of the $z_1^{j_1}, \ldots, z_n^{j_n}$: Let

$$\mathcal{J} = \{(j_1, \ldots, j_n) \in \mathbb{Z}^n \mid 0 \leq j_i \leq r_i - 1 \text{ for } i = 1, \ldots, n\}$$

and, for $\boldsymbol{j} = (j_1, \ldots, j_n) \in \mathcal{J}$, let $z^{\boldsymbol{j}} = z_1^{j_1} \otimes \ldots \otimes z_n^{j_n}$. Then,

$$\mathsf{N} = \bigoplus_{\boldsymbol{j} \in \mathcal{J}} \mathsf{L}\, z^{\boldsymbol{j}}.$$

We have $deg\, z_i = \frac{1}{r_i} deg\, b_i$ because $z_i^{r_i} = b_i$; hence, the hypothesis on the degrees of $b_1, \ldots, b_n$ implies that the degrees of the $z^{\boldsymbol{j}}$ are pairwise distinct modulo Γ_{F}. It follows that $\mathsf{N}_0 = \mathsf{L}_0 = \ell$ and

$$\Gamma_{\mathsf{N}}/\Gamma_{\mathsf{F}} = \langle \tfrac{1}{r_1} deg\, b_1 + \Gamma_{\mathsf{F}} \rangle \oplus \ldots \oplus \langle \tfrac{1}{r_n} deg\, b_n + \Gamma_{\mathsf{F}} \rangle.$$

Therefore,

$$[\mathsf{N}_0 : \mathsf{F}_0] = |\Gamma_{\mathsf{N}} : \Gamma_{\mathsf{F}}| = r_1 \ldots r_n = deg\, \mathsf{N}.$$

Moreover, the simple graded algebra N is a graded division algebra since N_0 is a field (see Cor. 2.43), hence N is a semiramified graded division algebra. For $i = 1, \ldots, n$ we have

$$[(\mathsf{L}_i/\mathsf{F}, \sigma_i, b_i)] \in \mathsf{Br}(\ell_i \otimes_{\mathsf{F}_0} \mathsf{F}/\mathsf{F});$$

hence, $[\mathsf{N}] \in \sum_{i=1}^n \mathsf{Br}(\ell_i \otimes_{\mathsf{F}_0} \mathsf{F}/\mathsf{F})$, which shows that N is decomposably semiramified. □

Examples of decomposably semiramified graded division algebras are thus easy to build. We refer to Ex. 8.56 for an example of a semiramified graded division algebra that is *not* decomposably semiramified.

The main result of this subsection is Prop. 8.33, which yields equivalent conditions for a graded division algebra to be decomposably semiramified. We start with a few observations about characters of the absolute Galois group $\mathcal{G}(\mathsf{F}_0)$ with values in $\mathbb{T}(\Gamma_{\mathsf{F}})$, which will allow us to characterize decomposably semiramified graded division algebras; see Prop. 8.32.

Definition 8.30. Let $\varphi \in Hom^c(\mathcal{G}(\mathsf{F}_0), \mathbb{T}(\Gamma_{\mathsf{F}}))$, and let K_φ be the fixed field of $ker\, \varphi$. Then K_φ is a finite-degree abelian Galois field extension of F_0, and φ induces a group isomorphism $\overline{\varphi}\colon \mathcal{G}(K_\varphi/\mathsf{F}_0) \xrightarrow{\sim} im\, \varphi$. We say that φ is a *cyclic character* if $im\, \varphi$ is a (finite) cyclic group, or, equivalently, K_φ is a cyclic Galois extension of F_0. Let $\varphi_1, \ldots, \varphi_n \in Hom^c(\mathcal{G}(\mathsf{F}_0), \mathbb{T}(\Gamma_{\mathsf{F}}))$. We say that $\varphi_1, \ldots, \varphi_n$ are *independent characters* if

$$im(\varphi_1 + \ldots + \varphi_n) = im\, \varphi_1 \oplus \ldots \oplus im\, \varphi_n.$$

This condition is equivalent to the following: $im\, \varphi_1, \ldots, im\, \varphi_n$ are independent subgroups of $\mathbb{T}(\Gamma_{\mathsf{F}})$, in the sense that their sum is direct, and $K_{\varphi_1}, \ldots, K_{\varphi_n}$ are linearly disjoint over F_0 (i.e., $K_{\varphi_1} \otimes_{\mathsf{F}_0} \ldots \otimes_{\mathsf{F}_0} K_{\varphi_n}$ is a field).

Lemma 8.31. *Every character* φ *in* $Hom^c(\mathcal{G}(\mathsf{F}_0), \mathbb{T}(\Gamma_\mathsf{F}))$ *is expressible as a sum of independent cyclic characters.*

Proof. The finite abelian group $im\,\varphi$ is a direct sum, $im\,\varphi = C_1 \oplus \ldots \oplus C_k$, where each C_i is a finite cyclic subgroup of $\mathbb{T}(\Gamma_\mathsf{F})$. Let π_i be the projection onto the i-th component in this direct sum decomposition. Then $\varphi = \sum_{i=1}^k \pi_i \circ \varphi$, and each $\pi_i \circ \varphi$ is a cyclic character as $im(\pi_i \circ \varphi) = C_i$. Since

$$im\,\varphi \;=\; C_1 + \ldots + C_k \;=\; im(\pi_1 \circ \varphi) + \ldots + im(\pi_k \circ \varphi),$$

it follows that the $\pi_i \circ \varphi$ are independent cyclic characters. □

It is easy to see that all decompositions of φ into a sum of independent cyclic characters are obtainable in the manner described in the preceding proof.

We now apply the observations above to the characters obtained from the ramification map χ of (6.24). In the following proposition, we write χ_A for the image under χ of the Brauer class of an inertially split central simple graded F-algebra A.

Proposition 8.32. *A central simple graded algebra* A *over* F *is a decomposably semiramified graded division algebra if and only if* $\mathsf{A} \cong_g \mathsf{C}_1 \otimes_\mathsf{F} \ldots \otimes_\mathsf{F} \mathsf{C}_n$ *where each* C_i *is a cyclically semiramified graded division algebra with center* F *and* $\chi_{\mathsf{C}_1}, \ldots, \chi_{\mathsf{C}_n}$ *are independent cyclic characters.*

Proof. Let A be a decomposably semiramified graded division algebra. Proposition 8.28 yields a decomposition of A into a tensor product of cyclically semiramified graded division algebras $\mathsf{A} \cong_g \mathsf{C}_1 \otimes_\mathsf{F} \ldots \otimes_\mathsf{F} \mathsf{C}_n$ such that

$$\Gamma_\mathsf{A}/\Gamma_\mathsf{F} \;=\; (\Gamma_{\mathsf{C}_1}/\Gamma_\mathsf{F}) \oplus \ldots \oplus (\Gamma_{\mathsf{C}_n}/\Gamma_\mathsf{F}). \tag{8.23}$$

Since $im\,\chi_{\mathsf{C}_i} = \Gamma_{\mathsf{C}_i}/\Gamma_\mathsf{F}$ and $im\,\chi_\mathsf{A} = \Gamma_\mathsf{A}/\Gamma_\mathsf{F}$ by Prop. 6.22, and since $\chi_\mathsf{A} = \chi_{\mathsf{C}_1} + \ldots + \chi_{\mathsf{C}_n}$ because χ is a group homomorphism, (8.23) can be rewritten as

$$im(\chi_{\mathsf{C}_1} + \ldots + \chi_{\mathsf{C}_n}) \;=\; im\,\chi_{\mathsf{C}_1} \oplus \ldots \oplus im\,\chi_{\mathsf{C}_n}.$$

Therefore, the characters $\chi_{\mathsf{C}_1}, \ldots, \chi_{\mathsf{C}_n}$ are independent. They are also cyclic because each $\Gamma_{\mathsf{C}_i}/\Gamma_\mathsf{F}$ is a cyclic group.

For the converse, suppose $\mathsf{A} = \mathsf{C}_1 \otimes_\mathsf{F} \ldots \otimes_\mathsf{F} \mathsf{C}_n$ where each C_i is cyclically semiramified and $\chi_{\mathsf{C}_1}, \ldots, \chi_{\mathsf{C}_n}$ are independent characters. Let $r_i = deg\,\mathsf{C}_i$ and let ℓ_i be the fixed field of $ker\,\chi_{\mathsf{C}_i}$. So, ℓ_i is a cyclic Galois extension of F_0 of degree r_i, and ℓ_i can be identified with the zero component of C_i. By applying Prop. 8.28 to each C_i, we see that

$$\mathsf{C}_i \;=\; \big((\ell_i \otimes_{\mathsf{F}_0} \mathsf{F})/\mathsf{F}, \sigma_i, b_i\big)$$

for some $b_i \in \mathsf{F}^\times$ with $\frac{1}{r_i}\, deg\, b_i + \Gamma_\mathsf{F}$ of order r_i in $\mathbb{T}(\Gamma_\mathsf{F})$, and we have

$$im\,\chi_{\mathsf{C}_i} \;=\; \Gamma_{\mathsf{C}_i}/\Gamma_\mathsf{F} \;=\; \big\langle \tfrac{1}{r_i}\, deg\, b_i + \Gamma_\mathsf{F} \big\rangle.$$

The condition that $\chi_{\mathsf{C}_1}, \ldots, \chi_{\mathsf{C}_n}$ are independent characters implies that the sum

$$\big\langle \tfrac{1}{r_1}\, deg\, b_1 + \Gamma_{\mathsf{F}} \big\rangle + \ldots + \big\langle \tfrac{1}{r_n}\, deg\, b_n + \Gamma_{\mathsf{F}} \big\rangle$$

is a direct sum, and that $\ell_1, \ldots, \ell_n$ are linearly disjoint extensions of F_0. We may then apply Prop. 8.29 to see that A is a decomposably semiramified graded division algebra. □

Proposition 8.33. *Let* N *be a graded division algebra with center* F. *The following conditions are equivalent:*

(a) N *is decomposably semiramified;*
(b) N *is semiramified and* N *is split by some totally ramified graded field extension of* F*;*
(c) N *contains a maximal graded subfield that is inertial over* F *and another maximal graded subfield that is totally ramified over* F*;*
(d) N *is inertially split and* N *is also split by some totally ramified graded field extension of* F.

Proof. (a) ⇒ (b) and (c): It is clear from the definition that a decomposably semiramified graded division algebra is semiramified, and contains a maximal graded subfield inertial over the center. The existence of a maximal graded subfield that is totally ramified over center was proved in Prop. 8.28. (For (b), recall that maximal graded subfields of a graded division algebra split the algebra, see Remark 6.9.)

(b) ⇒ (d) and (c) ⇒ (d): These are clear since semiramified graded division algebras are inertially split and maximal graded subfields of a graded division algebra split the algebra.

(d) ⇒ (a): Suppose N is inertially split and that N is also split by a graded field S totally ramified over F. Then, by Cor. 6.26 χ_{N} lies in $Hom^c(\mathcal{G}(\mathsf{F}_0), \Gamma_{\mathsf{S}}/\Gamma_{\mathsf{F}})$. Thus, $im\, \chi_{\mathsf{N}} \subseteq \Gamma_{\mathsf{S}}/\Gamma_{\mathsf{F}}$.

By Lemma 8.31 there is a decomposition $\chi_{\mathsf{N}} = \varphi_1 + \ldots + \varphi_n$, where $\varphi_1, \ldots, \varphi_n$ are independent cyclic characters. Because of the independence, we have $im\, \varphi_i \subseteq im\, \chi_{\mathsf{N}} \subseteq \Gamma_{\mathsf{S}}/\Gamma_{\mathsf{F}}$, for each i. We construct cyclically semiramified graded division algebras C_i such that $\chi_{\mathsf{C}_i} = \varphi_i$ and S splits C_i. For this, let ℓ_i be the fixed field of $ker\, \varphi_i$; so, ℓ_i is a cyclic Galois extension of F_0, and φ_i induces an isomorphism $\overline{\varphi}_i \colon \mathcal{G}(\ell_i/\mathsf{F}_0) \xrightarrow{\sim} im\, \varphi_i$. Let σ_i be a generator of $\mathcal{G}(\ell_i/\mathsf{F}_0)$, and let $r_i = [\ell_i{:}\mathsf{F}_0] = |\, im\, \varphi_i|$. Since $im\, \varphi_i \subseteq \Gamma_{\mathsf{S}}/\Gamma_{\mathsf{F}}$, there is a $\gamma_i \in \Gamma_{\mathsf{S}}$ with $\overline{\varphi}_i(\sigma_i) = \gamma_i + \Gamma_{\mathsf{F}}$. Then, $im\, \varphi_i = \langle \gamma_i + \Gamma_{\mathsf{F}} \rangle$, so $\gamma_i + \Gamma_{\mathsf{F}}$ has order r_i in $\Gamma_{\mathsf{S}}/\Gamma_{\mathsf{F}}$; hence, $r_i\gamma_i \in \Gamma_{\mathsf{F}}$. Therefore, $\mathsf{S}_{r_i\gamma_i} = \mathsf{F}_{r_i\gamma_i}$, as S is totally ramified over F. Choose any $z_i \in \mathsf{S}_{\gamma_i} \setminus \{0\}$, let $b_i = z_i^{r_i} \in \mathsf{S}_{r_i\gamma_i} \subseteq \mathsf{F}$, and, writing σ_i again for the F-automorphism of $\ell_i \otimes_{\mathsf{F}_0} \mathsf{F}$ extending σ_i, let

$$\mathsf{C}_i \;=\; \big((\ell_i \otimes_{\mathsf{F}_0} \mathsf{F})/\mathsf{F}, \sigma_i, b_i \big).$$

By Prop. 8.29, C_i is a semiramified graded division algebra. There is a $z'_i \in \mathsf{C}_i^{\times}$ with $z'_i x = \sigma_i(x) z'_i$ for all $x \in \ell_i$ and $z'^{r_i}_i = b_i = z_i^{r_i}$. Then

$$deg\, z'_i \;=\; \tfrac{1}{r_i}\, deg\, b_i \;=\; deg\, z_i \;=\; \gamma_i.$$

The map θ_{C_i} carries $\gamma_i = deg\, z_i'$ to σ_i, hence $\chi_{\mathsf{C}_i} = \varphi_i$. Moreover, since the image of γ_i in $\mathbb{T}(\Gamma_{\mathsf{F}})$ has order r_i, the graded fields $\mathsf{F}(z_i')$ and $\mathsf{F}(z_i)$ are totally ramified of degree r_i over F, and

$$\mathsf{F}(z_i') \cong_g \mathsf{F}[X]/(X^{r_i} - b_i) \cong_g \mathsf{F}(z_i) \subseteq \mathsf{S}.$$

Because $[\mathsf{F}(z_i'){:}\mathsf{F}] = r_i = deg\, \mathsf{C}_i$, the graded field $\mathsf{F}(z_i')$ splits C_i; hence, $\mathsf{F}(z_i)$ splits C_i, and therefore S also splits C_i.

Let $\mathsf{N}' = \mathsf{C}_1 \otimes_{\mathsf{F}} \ldots \otimes_{\mathsf{F}} \mathsf{C}_n$, which is a decomposably semiramified graded division algebra by Prop. 8.32 since the $\chi_{\mathsf{C}_i} = \varphi_i$ are independent characters and the C_i are cyclically semiramified graded division algebras. Note that

$$\chi_{\mathsf{N}'} = \chi_{\mathsf{C}_1} + \ldots + \chi_{\mathsf{C}_n} = \varphi_1 + \ldots + \varphi_n = \chi_{\mathsf{N}}$$

and S splits N' since it splits each C_i. We saw in the proof of Cor. 6.26 that the map $\mathsf{Br}_{is}(\mathsf{F}) \cap \mathsf{Br}(\mathsf{S}/\mathsf{F}) \to Hom^c(\mathcal{G}(\mathsf{F}), \mathbb{T}(\Gamma_{\mathsf{F}}))$ is injective. Therefore, $[\mathsf{N}] = [\mathsf{N}']$ in $\mathsf{Br}(\mathsf{F})$, and since N and N' are each graded division algebras, we have $\mathsf{N} \cong_g \mathsf{N}'$. Hence, N is decomposably semiramified. □

The last part of the proof of Prop. 8.33 shows that a decomposably semiramified graded division algebra N is determined up to isomorphism by χ_{N} and any totally ramified graded field extension S of F that splits N. For such an S, the unique graded field T with $\mathsf{F} \subseteq \mathsf{T} \subseteq \mathsf{S}$ and $\Gamma_{\mathsf{T}} = \Gamma_{\mathsf{N}}$ is a maximal graded subfield of N.

Remark 8.34. The proof of Prop. 8.33 also shows that if N is a decomposably semiramified graded F-algebra and we take *any* decomposition $\chi_{\mathsf{N}} = \varphi_1 + \ldots + \varphi_n$ where the φ_i are independent cyclic characters, then N can be decomposed as $\mathsf{N} = \mathsf{C}_1 \otimes_{\mathsf{F}} \ldots \otimes_{\mathsf{F}} \mathsf{C}_n$ where each C_i is cyclically semiramified and $\chi_{\mathsf{C}_i} = \varphi_i$. Furthermore, since any direct sum decomposition $\Gamma_{\mathsf{N}}/\Gamma_{\mathsf{F}} = \Gamma_1/\Gamma_{\mathsf{F}} \oplus \Gamma_2\Gamma_{\mathsf{F}}$ yields a (unique) corresponding decomposition $\chi_{\mathsf{N}} = \chi_1 + \chi_2$ with independent characters χ_1 and χ_2 satisfying $im\, \chi_i = \Gamma_i/\Gamma_{\mathsf{F}}$, there is a corresponding tensor decomposition $\mathsf{N} = \mathsf{N}_1 \otimes_{\mathsf{F}} \mathsf{N}_2$ with $\chi_{\mathsf{N}_i} = \chi_i$, so $\Gamma_{\mathsf{N}_i} = \Gamma_i$.

If two decomposably semiramified graded division algebras N and N' with center F are split by the same totally ramified graded field extension S of F, then $\mathsf{N} \otimes_{\mathsf{F}} \mathsf{N}'$ is inertially split and is also split by S. The preceding proposition therefore shows that the associated graded division algebra of $\mathsf{N} \otimes_{\mathsf{F}} \mathsf{N}'$ is again decomposably semiramified. However, the following example shows that this does not hold in general without the assumption of a common totally ramified graded splitting field.

Example 8.35. *A graded division algebra that is a tensor product of cyclically semiramified graded division algebras but is not decomposably semiramified.* Let k be a field, and let $\mathsf{F} = k[x, x^{-1}]$, the Laurent polynomial ring over k with its usual grading with $\Gamma_{\mathsf{F}} = \mathbb{Z}$ and $\mathsf{F}_j = k\,x^j$ for all $j \in \mathbb{Z}$. For some integer $n \geq 2$, assume that k contains a primitive n-th root of unity ω.

Take $a, b, c \in k^\times$ such that $[k(\sqrt[n]{a}, \sqrt[n]{b}):k] = n^2$ and the symbol algebra $\big(b, c/k(\sqrt[n]{ab})\big)_n$ is a division algebra. Let

$$\mathsf{D} = \mathsf{C} \otimes_{\mathsf{F}} \mathsf{C}' \qquad \text{where} \qquad \mathsf{C} = \big(a, x/\mathsf{F}\big)_n \quad \text{and} \quad \mathsf{C}' = \big(b, cx/\mathsf{F}\big)_n.$$

Thus, C and C' are graded symbol algebras over F, and D is a simple graded algebra with center F. Note that $\Gamma_{\mathsf{C}} = \Gamma_{\mathsf{C}'} = \frac{1}{n}\mathbb{Z}$. Also,

$$\mathsf{C}_0 \cong \mathsf{F}_0[X]/(X^n - a) \cong F_0(\sqrt[n]{a})$$

since $[F_0(\sqrt[n]{a}):\mathsf{F}_0] = n$ by hypothesis; likewise, $\mathsf{C}'_0 \cong \mathsf{F}_0(\sqrt[n]{b})$. By Prop. 8.29, C and C' are cyclically semiramified graded division algebras over F. Let σ (resp. σ') be the generator of $\mathcal{G}(\mathsf{F}_0(\sqrt[n]{a})/\mathsf{F}_0)$ (resp. $\mathcal{G}(\mathsf{F}_0(\sqrt[n]{b})/\mathsf{F}_0)$) with $\sigma(\sqrt[n]{a}) = \omega\sqrt[n]{a}$ (resp. $\sigma'(\sqrt[n]{b}) = \omega\sqrt[n]{b}$). Then, $\overline{\chi}_{\mathsf{C}}(\sigma) = \overline{\chi}_{\mathsf{C}'}(\sigma') = \frac{-1}{n} + \mathbb{Z}$. The cyclic characters χ_{C} and $\chi_{\mathsf{C}'}$ are clearly not independent, as $im\,\chi_{\mathsf{C}} = \frac{1}{n}\mathbb{Z} = im\,\chi_{\mathsf{C}'}$. The standard identities for symbol algebras hold also for graded symbol algebras, since the map $\mathsf{Br}(\mathsf{F}) \to Br(q(\mathsf{F}))$ is injective by Prop. 6.4 and it sends graded symbol algebras to symbol algebras. Thus, we have in $\mathsf{Br}(\mathsf{F})$,

$$\begin{aligned} \mathsf{D} = \mathsf{C} \otimes_{\mathsf{F}} \mathsf{C}' &= \big(a, x/\mathsf{F}\big)_n \otimes_{\mathsf{F}} \big(b, cx/\mathsf{F}\big)_n \\ &\sim \big(a, x/\mathsf{F}\big)_n \otimes_{\mathsf{F}} \big(b, c/\mathsf{F}\big)_n \otimes_{\mathsf{F}} \big(b, x/\mathsf{F}\big)_n \sim \big(ab, x/\mathsf{F}\big)_n \otimes_{\mathsf{F}} \big(b, c/\mathsf{F}\big)_n. \end{aligned}$$

The symbol algebra $\big(ab, x/\mathsf{F}\big)_n$ is clearly a cyclically semiramified graded division algebra with degree 0 component $\mathsf{F}_0(\sqrt[n]{ab})$ and grade group $\frac{1}{n}\mathbb{Z}$. The symbol algebra $\big(b, c/\mathsf{F}\big)_n$ is inertial over F with degree 0 component $\big(b, c/\mathsf{F}_0\big)_n$. By Prop. 2.56 the degree 0 component of $\big(ab, x/\mathsf{F}\big)_n \otimes_{\mathsf{F}} \big(b, c/\mathsf{F}\big)_n$ is $\big(b, c/\mathsf{F}_0(\sqrt[n]{ab})\big)_n$, which is a division algebra, by hypothesis. Hence, the tensor product $\big(ab, x/\mathsf{F}\big)_n \otimes_{\mathsf{F}} \big(b, c/\mathsf{F}\big)_n$ is a graded division algebra, by Cor. 2.43. Since this tensor product is a graded division algebra Brauer-equivalent to D and of the same dimension over F as D, it must be graded isomorphic to D. Hence, D is a graded division algebra that is inertially split but not semiramified, with $\Gamma_{\mathsf{D}} = \frac{1}{n}\mathbb{Z}$.

8.3.4 Semiramified division algebras over Henselian fields

In this subsection, we turn to the valuation-theoretic analogues of the notions introduced in the context of graded fields in the preceding subsection. Let (F, v) be a valued field. Through most of this subsection, the valuation v will be assumed to be Henselian. This hypothesis is not needed for the following definition, however:

Definition 8.36. A central division algebra D over F with a valuation extending v is said to be *semiramified* if $\overline{D}$ is a field and

$$[\overline{D}{:}\overline{F}] = |\Gamma_D{:}\Gamma_F| = \mathit{deg}\, D.$$

So, D is defectless over F. If D is defectless over F and $\overline{D}$ is a field, then any of the equations $[\overline{D}{:}\overline{F}] = |\Gamma_D{:}\Gamma_F|$, $|\Gamma_D{:}\Gamma_F| = \mathit{deg}\, D$, $\mathit{deg}\, D = [\overline{D}{:}\overline{F}]$ characterizes semiramified division algebras.

Unlike what we have seen in the graded context, semiramified division algebras over a Henselian field need not be inertially split, nor even tame, as the following example illustrates:

Example 8.37. *A semiramified but not tame division algebra.* Let (F, v) be a Henselian valued field with $\mathit{char}\, F = p \neq 0$ such that $\Gamma_F = \mathbb{Z}$ and $\overline{F}$ is not perfect. (For example, we could take for k any field with $\mathit{char}\, k = p$, let x be transcendental over k and let $F = k(x)((y))$, the Laurent power series field in y over $k(x)$, and let v be the complete discrete rank 1 y-adic valuation on F. Then $\overline{F} = k(x)$, which is not perfect, and $\Gamma_F = \mathbb{Z}$.) Take any $t, u \in F^\times$ with $v(t) = -1, v(u) = 0$, and $\overline{u} \notin \overline{F}^p$. Let

$$f = X^p - X - t \in F[X],$$

let s be a root of f in an algebraic closure of F, and let $K = F(s)$. Note that f is a λ-uniform polynomial for $\lambda = \frac{-1}{p}$ with $\widetilde{f}^{(\lambda)} = X^p - \widetilde{t}$. Indeed, f is a generalized Eisenstein polynomial as in Ex. 5.48. It was pointed out in that example that f is irreducible in $F[X]$ and with respect to the extension v_K of v, K is totally ramified over F with $\Gamma_K = \frac{1}{p}\mathbb{Z}$. Moreover, $\mathsf{gr}(K) = \mathsf{gr}(F)(\widetilde{s}) = \mathsf{gr}(F)(\sqrt[p]{\widetilde{t}}\,)$, which is totally but not tamely ramified over $\mathsf{gr}(F)$. Since f is an Artin–Schreier polynomial, K is Galois over F with $\mathcal{G}(K/F) = \langle\sigma\rangle$, where $\sigma(s) = s + 1$. Let D be the cyclic algebra

$$D = (K/F, \sigma, u) = \bigoplus_{j=0}^{p-1} Kz^j,$$

where $zcz^{-1} = \sigma(c)$ for all $c \in K$ and $z^p = u$. Define a v_K-norm α on the left K-vector space D by

$$\alpha\Big(\sum_{j=0}^{p-1} c_j z^j\Big) = \mathit{min}\big(v_K(c_0), \ldots, v_K(c_{p-1})\big), \qquad \text{where all } c_j \in K.$$

Thus, $\big(z^j\big)_{j=0}^{p-1}$ is a splitting base for the v_K-norm α with $\alpha(z^j) = 0$ for $j = 0, 1, \ldots, p-1$. Since $\alpha(z^p) = v_K(u) = 0$, it is easy to see that

$$\alpha(cz^j) = v_K(c) \qquad \text{for all } c \in K \text{ and } j \in \mathbb{Z}.$$

Because $\big(\widetilde{s^i}\big)_{i=0}^{p-1}$ is a $\mathsf{gr}(F)$-base of $\mathsf{gr}(K)$, we have $\big(s^i\big)_{i=0}^{p-1}$ is a splitting base of v_K as a v-norm on K. Hence, $\big(s^i z^j\big)_{i,j=0}^{p-1}$ is splitting base of D for the v-norm α. We have

$$\alpha(s^i z^j) \;=\; v_K(s^i) \;=\; \tfrac{-i}{p} \qquad \text{for all } i,\, j \in \mathbb{Z}.$$

Note that for all $i, j, i', j' \in \mathbb{Z}$, since $v_K(s+j) = v_K(s) = \frac{-1}{p}$, we have

$$\begin{aligned}\alpha\big((s^i z^j)(s^{i'} z^{j'})\big) \;&=\; \alpha\big(s^i \sigma^j(s^{i'}) z^j z^{j'}\big)\\ &= v_K\big(s^i (s+j)^{i'}\big) \;=\; \tfrac{-(i+i')}{p} \;=\; \alpha(s^i z^j) + \alpha(s^{i'} z^{j'}).\end{aligned}$$

Hence, by Lemma 3.23, α is a surmultiplicative v-norm on D. Also,

$$\alpha(zs) \;=\; \alpha(sz) \;=\; \alpha(z) + \alpha(s) = \tfrac{-1}{p} \;<\; 0 \;=\alpha(z).$$

Since $zs = sz + z$ with $\alpha(sz) < \alpha(z)$ we have $\widetilde{zs} \;=\; (sz+z)^{\sim} = \widetilde{sz}$. Therefore, $\widetilde{z}\,\widetilde{s} = \widetilde{zs} \;=\; \widetilde{sz} \;= \widetilde{s}\,\widetilde{z}$. Hence, $\mathsf{gr}_\alpha(D)$ is the commutative graded ring $\mathsf{gr}(F)[\widetilde{s}, \widetilde{z}]$, and there is a graded homomorphism

$$\psi\colon\ \mathsf{gr}(F)[X,Y]/(X^p - \widetilde{t},\ Y^p - \widetilde{u}) \;\longrightarrow\; \mathsf{gr}_\alpha(D).$$

Since the totally ramified extension $\mathsf{gr}(F)(\sqrt[p]{\widetilde{t}}\,)$ and the unramified extension $\mathsf{gr}(F)(\sqrt[p]{\widetilde{u}})$ are linearly disjoint over $\mathsf{gr}(F)$, the domain of ψ is the graded field $\mathsf{L} \;=\; \mathsf{gr}(F)\big(\sqrt[p]{\widetilde{t}},\, \sqrt[p]{\widetilde{u}}\big)$. Hence, ψ must be a graded isomorphism, and $\mathsf{gr}_\alpha(D) \cong_g \mathsf{L}$. Since $\mathsf{gr}_\alpha(D)$ has no zero divisors, by Cor. 3.30 D is a division ring and α is a valuation on D, necessarily the extension of v to D. We have $\overline{D} \;=\; \mathsf{gr}(D)_0 \;\cong\; \mathsf{L}_0 \;\cong\; \overline{F}(\sqrt[p]{\overline{u}})$, which is purely inseparable over $\overline{F}$ of degree p. Also, $\Gamma_D \;=\; \Gamma_{\mathsf{gr}(D)} \;=\; \Gamma_{\mathsf{L}} \;=\; \frac{1}{p}\mathbb{Z}$, showing that $|\Gamma_D{:}\Gamma_F| \;=\; p$. Thus, D is semiramified over F, but it is not tame over F, as $\overline{D}$ is not separable over $\overline{F}$ (and $p \,\big|\, |\Gamma_D{:}\Gamma_F|$).

For the rest of this section we assume the valuation v on F is Henselian. When we add a tameness assumption, we will see that the theory for semiramified division algebras mirrors what has appeared in the graded setting in §8.3.3.

Proposition 8.38. *For a central division algebra D over a Henselian field F, the following conditions are equivalent.*

(a) *D is semiramified and tame over F;*
(b) *D is semiramified and inertially split;*
(c) *D is semiramified and the field $\overline{D}$ is separable over $\overline{F}$;*
(d) *D is tame over F and $\mathsf{gr}(D)$ is semiramified over $\mathsf{gr}(F)$;*
(e) *D is inertially split and $\overline{D}$ is a field.*

When these conditions hold, the field $\overline{D}$ is an abelian Galois extension of $\overline{F}$, and the map θ_D induces an isomorphism $\overline{\theta}_D\colon \Gamma_D/\Gamma_F \xrightarrow{\sim} \mathcal{G}(\overline{D}/\overline{F})$.

Proof. (a) $\Rightarrow$ (b) and (c): If D is tame over F, then $Z(\overline{D})$ is separable over $\overline{F}$, and D contains an inertial lift Z of $Z(\overline{D})$ over F. If D is also semiramified, then $\overline{D} = Z(\overline{D})$, and $[Z{:}F] = [Z(\overline{D}){:}\overline{F}] = [\overline{D}{:}\overline{F}] = deg\, D$. So Z splits D, showing that D is inertially split.

(b) $\Rightarrow$ (a) is clear since inertially split division algebras are tame.

(c) $\Rightarrow$ (b): Suppose D is semiramified and $\overline{D}$ is separable over $\overline{F}$. Then by Prop. 1.5, $\overline{D}$ is Galois over $\overline{F}$ and θ_D maps onto $\mathcal{G}(Z(\overline{D})/\overline{F})$ with kernel containing Γ_F. So,

$$|\Gamma_D{:}\Gamma_F| \geq |\Gamma_D{:}\, ker\,\theta_D| = \big|im\,\theta_D\big| = \big|\mathcal{G}(Z(\overline{D})/\overline{F})\big| = [\overline{D}{:}\overline{F}] = |\Gamma_D{:}\Gamma_F|.$$

Therefore, $ker(\theta_D) = \Gamma_F$, and hence D is inertially split by Prop. 8.49.

(a) $\Leftrightarrow$ (d): Suppose D is tame over F. Then $Z(\mathsf{gr}(D)) = \mathsf{gr}(F)$, so $deg\,\mathsf{gr}(D) = deg\, D$. Since $D_0 = \overline{D}$ and $\Gamma_{\mathsf{gr}(D)} = \Gamma_D$, $\mathsf{gr}(D)$ is semiramified if and only if D is semiramified.

(b) $\Rightarrow$ (e) is clear from the definition of semiramified.

(e) $\Rightarrow$ (a) If D is inertially split, then D is tame over F, hence it is defectless. Moreover, $\overline{\theta}_D\colon \Gamma_D/\Gamma_F \to \mathcal{G}(Z(\overline{D})/\overline{F})$ is an isomorphism by Lemma 6.65, hence

$$[Z(\overline{D}){:}\overline{F}] = \big|\mathcal{G}(Z(\overline{D})/\overline{F})\big| = |\Gamma_D{:}\Gamma_F|.$$

When in addition $\overline{D}$ is a field, the left displayed term is $[\overline{D}{:}\overline{F}]$. We then have $[\overline{D}{:}\overline{F}] = |\Gamma_D{:}\Gamma_F| = deg\, D$ since D is defectless. Hence, D is semiramified. □

For a tame semiramified central division algebra over a Henselian field, the diagram (8.10) of canonical subalgebras simplifies considerably: The inertial lift U of $\overline{D}$ over F is a field, hence it coincides with its center Z, and it is a maximal subfield because its degree is equal to the degree of D. Thus,

$$Z = U = C = E,$$

and we have $[D] \in Br(Z/F)$, as observed in the first part of the proof of Prop. 8.38.

Definition 8.39. A central division algebra D over a Henselian field F is said to be *decomposably semiramified* if it is tame and semiramified, and $[D] \in Dec(Z/F)$ for Z the inertial lift of $\overline{D}$ over F. This last condition necessarily holds when $\overline{D}$ is a cyclic extension of $\overline{F}$; the tame division algebra D is then said to be *cyclically semiramified.*

Proposition 8.40. *Let N be a central division algebra over a Henselian field F. The following conditions are equivalent:*

(a) *N is decomposably semiramified;*
(b) *N is semiramified and tame over F and N is split by some totally ramified field extension of F;*
(c) *N contains a maximal subfield inertial over F and another maximal subfield totally ramified over F;*

(d) *N is inertially split and N is split by some totally ramified field extension of F;*
(e) *N is tame over F and* $\mathsf{gr}(N)$ *is decomposably semiramified.*

Proof. (a) $\Leftrightarrow$ (e): As N is tame, we know from Prop. 8.38 that N is semiramified if and only if $\mathsf{gr}(N)$ is semiramified. Therefore, assuming that N is tame and semiramified, and letting Z denote the inertial lift of $\overline{N}$ over F, it suffices to prove that

$$[N] \in \mathit{Dec}(Z/F) \quad \text{if and only if} \quad [\mathsf{gr}(N)] \in \sum_{\substack{\overline{F}\subseteq \ell\subseteq \overline{D} \\ \ell/\overline{F}\text{ cyclic}}} \mathsf{Br}(\ell \otimes_{\overline{F}} \mathsf{F}/\mathsf{F}).$$

Assume first that $[N] = \sum_i [C_i]$ where each C_i is a division algebra split by some cyclic extension L_i of F in Z. Since Z is inertial over F, the extension L_i/F is also inertial, hence C_i is inertially split and therefore tame. By taking the image of $[C_i]$ under the homomorphism $\mathsf{gr}\colon \mathit{Br}_{tr}(F) \to \mathsf{Br}\big(\mathsf{gr}(F)\big)$ of Prop. 6.57, we find that

$$[\mathsf{gr}(N)] = \sum_i [\mathsf{gr}(C_i)] \qquad \text{in } \mathsf{Br}\big(\mathsf{gr}(F)\big).$$

In view the functoriality of gr (see Prop. 6.59), the graded division algebra $\mathsf{gr}(C_i)$ is split by $\mathsf{gr}(L_i) = \overline{L_i} \otimes_{\overline{F}} \mathsf{gr}(F)$. The extension $\overline{L_i}/\overline{F}$ is cyclic because L_i/F is cyclic; hence, we see from the last equation that $[\mathsf{gr}(N)]$ lies in $\sum_{\substack{\overline{F}\subseteq \ell\subseteq \overline{D} \\ \ell/\overline{F}\text{ cyclic}}} \mathsf{Br}(\ell \otimes_{\overline{F}} \mathsf{F}/\mathsf{F})$.

For the converse, suppose $[\mathsf{gr}(N)] = \sum_i [\mathsf{C}_i]$ where each C_i is a central graded division algebra over $\mathsf{gr}(F)$ split by some $\ell_i \otimes_{\overline{F}} \mathsf{gr}(F)$, with ℓ_i a cyclic extension of $\overline{F}$ in $\overline{D}$. Since the map gr is an isomorphism (see Th. 6.64), we may find tame central simple F-algebras A_i such that $\mathsf{gr}[A_i] = [\mathsf{C}_i]$, hence

$$[N] = \sum_i [A_i] \qquad \text{in } \mathit{Br}(F). \tag{8.24}$$

The field ℓ_i lifts to some subfield $L_i \subseteq Z$, which is a cyclic extension of F because ℓ_i is cyclic over $\overline{F}$. The field L_i splits A_i because $\ell_i \otimes_{\overline{F}} \mathsf{gr}(F)$ splits C_i, hence (8.24) shows that $[N] \in \mathit{Dec}(Z/F)$.

Now that we have proved the equivalence of (a) and (e), the equivalence of these conditions to (b)–(d) follows from the corresponding graded result, Prop. 8.33. For this note that if T is a field totally ramified over F, then $\mathsf{gr}(T)$ is a graded field totally ramified over $\mathsf{gr}(F)$. Also, if K is a graded field totally ramified and of finite degree over $\mathsf{gr}(F)$, then there is a lift L of K over F by Prop. 8.2, and L is totally ramified over F. If T is a maximal subfield of N, (resp. T splits N), then $\mathsf{gr}(T)$ is a maximal graded subfield of $\mathsf{gr}(N)$ (resp. $\mathsf{gr}(T)$ splits N). Moreover, if K is a maximal subfield of $\mathsf{gr}(N)$ (resp. K splits $\mathsf{gr}(N)$) then its lift L is a maximal subfield of N by Cor. 8.10 (resp. L splits N by Cor. 6.61). Likewise, if T is inertial over F, then $\mathsf{gr}(T)$ is inertial over $\mathsf{gr}(F)$ and if K is inertial over $\mathsf{gr}(F)$, then its tame lift L is inertial over F; the same observations about maximal subfields and splitting fields apply as in the totally ramified case. □

Many results about division algebras over Henselian fields were first proved without using graded algebras. However, the fact that conditions (b), (c), or (d) imply (a) in Prop. 8.40 is not apparent without recourse to the graded machinery.

In view of the equivalence of (a) and (e) in Prop. 8.40, it is clear that decomposably semiramified division algebras over Henselian fields have properties that faithfully reflect those of decomposably semiramified graded division algebras; in particular, analogues of Prop. 8.28 and Prop. 8.29 hold in the context of central simple algebras over Henselian fields. For later use, we prove the analogue of Prop. 8.29:

Proposition 8.41. *Let (F, v) be a Henselian-valued field and let $\ell_1, \ldots, \ell_n$ be cyclic Galois field extensions of $\overline{F}$, which are assumed to be linearly disjoint, so $\ell = \ell_1 \otimes_{\overline{F}} \ldots \otimes_{\overline{F}} \ell_n$ is a field. For $i = 1, \ldots, n$, let L_i be the inertial lift of ℓ_i over F, so L_i is a cyclic extension of F. Write σ_i for a generator of $\mathcal{G}(L_i/F)$ and let $r_i = [L_i{:}F] = [\ell_i{:}\overline{F}]$. Suppose $b_1, \ldots, b_n \in F^\times$ satisfy the following condition: $\frac{1}{r_1}v(b_1) + \Gamma_F, \ldots, \frac{1}{r_n}v(b_n) + \Gamma_F$ have order respectively $r_1, \ldots, r_n$ and generate independent subgroups of $\mathbb{T}(\Gamma_F)$. Then the central simple F-algebra*

$$N = \big(L_1/F, \sigma_1, b_1\big) \otimes_F \ldots \otimes_F \big(L_n/F, \sigma_n, b_n\big)$$

is a decomposably semiramified division algebra, with $\overline{N} = \ell$ and

$$\Gamma_N/\Gamma_F = \big\langle \tfrac{1}{r_1}v(b_1) + \Gamma_F\big\rangle \oplus \ldots \oplus \big\langle \tfrac{1}{r_n}v(b_n) + \Gamma_F\big\rangle.$$

Proof. The proposition can be established by mimicking the arguments in the proof of Prop. 8.29. Alternatively, since each L_i is inertial over F we may use Prop. 6.46 to find on each $\big(L_i/F, \sigma_i, b_i\big)$ a tame gauge α_i such that

$$\mathsf{gr}_{\alpha_i}\big(L_i, \sigma_i, b_i\big) = \big(\mathsf{gr}(L_i), \widetilde{\sigma}_i, \widetilde{b}_i\big) = \big(\ell_i \otimes_{\overline{F}} \mathsf{gr}(F), \widetilde{\sigma}_i, \widetilde{b}_i\big),$$

where $\widetilde{\sigma}_i$ is the automorphism of $\mathsf{gr}(L_i)$ induced by σ_i. By Prop. 6.49, the tensor product $\alpha = \alpha_1 \otimes \ldots \otimes \alpha_n$ is a tame gauge on N, and we have by Prop. 3.27

$$\mathsf{gr}_\alpha(N) = \big(\ell_1 \otimes_{\overline{F}} \mathsf{gr}(F), \widetilde{\sigma}_1, \widetilde{b}_1\big) \otimes_{\mathsf{gr}(F)} \cdots \otimes_{\mathsf{gr}(F)} \big(\ell_n \otimes_{\overline{F}} \mathsf{gr}(F), \widetilde{\sigma}_n, \widetilde{b}_n\big).$$

Since $deg\, \widetilde{b}_i = v(b_i)$ we may now apply Prop. 8.29 to see that $\mathsf{gr}_\alpha(N)$ is a graded division algebra with center $\mathsf{gr}(F)$, which is decomposably semiramified with $\mathsf{gr}_\alpha(N)_0 = \ell$ and

$$\Gamma_{\mathsf{gr}_\alpha(N)}/\Gamma_{\mathsf{gr}(F)} = \big\langle \tfrac{1}{r_1}v(b_1) + \Gamma_F\big\rangle \oplus \ldots \oplus \big\langle \tfrac{1}{r_n}v(b_n) + \Gamma_F\big\rangle.$$

Therefore, N is a division algebra and α is the valuation extending v, see Cor. 3.30. This division algebra is tame because α is a tame gauge, hence it is decomposably semiramified because $\mathsf{gr}(N)$ is decomposably semiramified (see Prop. 8.40). Finally, we have $\overline{N} = \mathsf{gr}(N)_0$ and $\Gamma_N = \Gamma_{\mathsf{gr}(N)}$, so $\overline{N}$ and Γ_N are as stated. □

8.4 Brauer class factorizations

In this section we obtain noncanonical factorizations of the Brauer classes of tame central division algebras D over a Henselian field F in the form

$$[D] = [I] + [N] + [T]$$

where I is an inertial division algebra, N is decomposably semiramified, and T is totally ramified. (Equivalently, to use simpler notation, $D \sim I \otimes_F N \otimes_F T$, where $\sim$ denotes Brauer-equivalence.)

These factorizations are obtained in two steps. First, we produce a decomposition $D \sim S \otimes_F T$ where S is a central division algebra over F that is inertially split and T is a totally ramified central division algebra over F, see Prop. 8.59. This decomposition is not canonical; we will see in Th. 8.60 to what extent the canonical subalgebras of D are related to those of S and T. Then, we take a closer look at inertially split division algebras. We have seen in Th. 6.66 that there is a canonical exact sequence:

$$0 \longrightarrow Br_{in}(F) \longrightarrow Br_{is}(F) \xrightarrow{\chi} Hom^c(\mathcal{G}(\overline{F}), \mathbb{T}(\Gamma_F)) \longrightarrow 0.$$

This exact sequence is noncanonically split. We will see that the division algebras in the images of $Hom^c(\mathcal{G}(\overline{F}), \mathbb{T}(\Gamma_F))$ under the splitting maps given in §6.2.3 are the decomposably semiramified algebras defined in Def. 8.39. This then leads to a decomposition of every inertially split F-central division algebra S as $S \sim I \otimes_F N$ in $Br(F)$ where I is inertial and N is decomposably semiramified. We show how the valuation invariants of S are related to those of I and N, and analyze the different $I \otimes_F N$ decompositions possible for a given S. As a result, we will attach to every inertially split division algebra S a coset $sp(S)$ in $Br(F)/Dec(Z(\overline{S})/\overline{F})$. In subsequent chapters, this coset will prove to hold important information on the structure of S.

We will conclude this section with two short subsections: §8.4.5 yields a typical construction of inertially split division algebras over iterated Laurent series fields, which serves as an illustration of the results of §8.4.4 and also will provide in subsequent chapters examples of division algebras exhibiting various special properties; the final §8.4.6 gives a summary of the structure of central division algebras over Henselian fields in the case of discrete (rank 1) valuations.

Before considering division algebras over Henselian fields, we look at graded division algebras, where the theory is parallel but more transparent.

8.4.1 The $\mathsf{S} \otimes \mathsf{T}$ decomposition (graded case)

Let F be an arbitrary graded field. We show first that every central graded division algebra over F decomposes up to Brauer-equivalence into the tensor product of an inertially split and a totally ramified graded division algebra.

Proposition 8.42. *Let D be a graded division algebra with center F. There are graded division algebras S and T with center F such that S is inertially split, T is totally ramified over F, and*

$$\mathsf{D} \sim \mathsf{S} \otimes_{\mathsf{F}} \mathsf{T} \qquad \textit{in } \mathsf{Br}(\mathsf{F}).$$

Proof. By Prop. 7.67, there is a graded division algebra T totally ramified over its center F satisfying $\mathsf{T} \otimes_{\mathsf{F}} \mathsf{F}_{in} \sim \mathsf{D} \otimes_{\mathsf{F}} \mathsf{F}_{in}$ in $\mathsf{Br}(\mathsf{F}_{in})$. Let S be the associated graded division algebra of $\mathsf{D} \otimes_{\mathsf{F}} \mathsf{T}^{\mathrm{op}}$. Then, F_{in} splits S and $\mathsf{D} \sim \mathsf{S} \otimes_{\mathsf{F}} \mathsf{T}$ in $\mathsf{Br}(\mathsf{F})$. □

Theorem 8.43. *Let S and T be graded division algebras with center F such that S is inertially split and T is totally ramified over F. Let D be the associated graded division algebra of $\mathsf{S} \otimes_{\mathsf{F}} \mathsf{T}$. Then,*

$$Z(\mathsf{D}_0) \;=\; Z(\mathsf{S}_0)^{\theta_{\mathsf{S}}(\Gamma_{\mathsf{S}} \cap \Gamma_{\mathsf{T}})} \quad \textit{and} \quad \mathsf{D}_0 \otimes_{Z(\mathsf{D}_0)} Z(\mathsf{S}_0) \;\sim\; \mathsf{S}_0 \textit{ in } Br(Z(\mathsf{S}_0)).$$

Moreover, $\Gamma_{\mathsf{D}} = \Gamma_{\mathsf{S}} + \Gamma_{\mathsf{T}}$, and θ_{D} is determined by $\Gamma_{\mathsf{T}} = \ker \theta_{\mathsf{D}}$ and commutativity of the diagram:

$$\begin{array}{ccc}
\Gamma_{\mathsf{S}} & \hookrightarrow & \Gamma_{\mathsf{D}} \\
\big\downarrow{\scriptstyle \theta_{\mathsf{S}}} & & \big\downarrow{\scriptstyle \theta_{\mathsf{D}}} \\
\mathcal{G}(Z(\mathsf{S}_0)/\mathsf{F}_0) & \xrightarrow{\ res\ } & \mathcal{G}(Z(\mathsf{D}_0)/\mathsf{F}_0)
\end{array}$$

Also,

$$\big((\ker \theta_{\mathsf{D}})/\Gamma_{\mathsf{F}}, a_{\mathsf{D}}\big) \;=\; \big(\Gamma_{\mathsf{T}}/\Gamma_{\mathsf{F}}, \overline{c}_{\mathsf{T}}\big) \qquad \textit{as symplectic modules.}$$

Proof. Let $\mathsf{A} = \mathsf{S} \otimes_{\mathsf{F}} \mathsf{T}$. We will show that

$$Z(\mathsf{A}_0) \;=\; Z(\mathsf{S}_0)^{\theta_{\mathsf{S}}(\Gamma_{\mathsf{S}} \cap \Gamma_{\mathsf{T}})}, \tag{8.25}$$

after which all the assertions of the proposition will follow quickly.

Let $\gamma_1, \ldots, \gamma_m$ in Γ_{T} be coset representatives for the distinct cosets in $\Gamma_{\mathsf{T}}/\Gamma_{\mathsf{F}}$, with the indexing chosen so that $\gamma_1, \ldots, \gamma_n$ represent the cosets of $(\Gamma_{\mathsf{S}} \cap \Gamma_{\mathsf{T}})/\Gamma_{\mathsf{F}}$. Choose any nonzero $t_i \in \mathsf{T}_{\gamma_i}$ for $i = 1, 2, \ldots, m$. Since $\mathsf{T} = \bigoplus_{i=0}^{m} \mathsf{F}\, t_i$, we have $\mathsf{A} = \bigoplus_{i=0}^{m} \mathsf{S} \otimes t_i$ with each $\mathsf{S} \otimes t_i$ isomorphic to the shifted graded F-vector space $\mathsf{S}(\gamma_i)$. Hence,

$$\mathsf{A}_0 \;=\; \bigoplus_{i=1}^{m} (\mathsf{S} \otimes t_i)_0 \;=\; \bigoplus_{i=1}^{m} \mathsf{S}_{-\gamma_i} \otimes t_i.$$

For $i = 1, 2, \ldots, n$, we have $\gamma_i \in \Gamma_{\mathsf{S}}$. Choose any nonzero $s_i \in S_{-\gamma_i}$ and let $y_i = s_i \otimes t_i \in \mathsf{A}_0^{\times}$ and $\sigma_i = \theta_{\mathsf{S}}(-\gamma_i) \in \mathcal{G}(Z(\mathsf{S}_0)/\mathsf{F}_0)$. For convenience, assume

$\gamma_1 = 0$, $t_1 = 1$, and $s_1 = 1$, and identify S_0 with $\mathsf{S}_0 \otimes t_1 \subseteq \mathsf{A}_0$. Note that for $i = n+1, \ldots, m$, we have $\gamma_i \notin \Gamma_{\mathsf{S}}$, so $\mathsf{S}(\gamma_i)_0 = \{0\}$. Hence,

$$\mathsf{A}_0 \;=\; \bigoplus_{i=1}^{n} \big((\mathsf{S}_0 s_i) \otimes t_i\big) \;\oplus\; \bigoplus_{i=n+1}^{m} \{0\} \;=\; \bigoplus_{i=1}^{n} \mathsf{S}_0 y_i. \tag{8.26}$$

Take any $a \in Z(\mathsf{A}_0)$, and write $a = \sum_{i=1}^{n} c_i y_i$ with each $c_i \in \mathsf{S}_0$. For any $z \in Z(\mathsf{S}_0)$, we have

$$y_i\, z \;=\; (s_i \otimes t_i)(z \otimes 1) \;=\; \sigma_i(z) s_i \otimes t_i \;=\; \sigma_i(z)\, y_i.$$

Hence,

$$0 \;=\; z\big(\sum_{i=1}^{n} c_i y_i\big) \;-\; \big(\sum_{i=1}^{n} c_i y_i\big) z \;=\; \sum_{i=1}^{n} (z - \sigma_i(z))\, c_i y_i.$$

Since the sum in (8.26) is direct, it follows that $(z - \sigma_i(z))c_i = 0$ for $i = 1, 2, \ldots, n$ and all $z \in Z(\mathsf{S}_0)$. Note that $\mathit{ker}\,\theta_{\mathsf{S}} = \Gamma_{\mathsf{F}}$ by Prop. 8.49 as S is inertially split; hence, for $i > 1$, since $\gamma_i \notin \Gamma_{\mathsf{F}}$ we have $\sigma_i \neq id_{Z(\mathsf{S}_0)}$. This forces $c_i = 0$ for $i > 1$, i.e., $a = c_1 y_1 = c_1 \in \mathsf{S}_0$. Indeed, $a \in Z(\mathsf{S}_0)$, as a centralizes $\mathsf{S}_0 \subseteq \mathsf{A}_0$. Furthermore, $0 = y_i a - a y_i = (\sigma(a) - a) y_i$ for $i = 1, 2, \ldots, n$. Thus, $\sigma_i(a) = a$ for all i. Since $\{\sigma_1, \ldots, \sigma_n\} = \theta_{\mathsf{S}}(\Gamma_{\mathsf{S}} \cap \Gamma_{\mathsf{T}})$, this shows that $Z(\mathsf{A}_0) \subseteq Z(\mathsf{S}_0)^{\theta_{\mathsf{S}}(\Gamma_{\mathsf{S}} \cap \Gamma_{\mathsf{T}})}$. The reverse inclusion is clear, since any element of $Z(\mathsf{S}_0)^{\theta_{\mathsf{S}}(\Gamma_{\mathsf{S}} \cap \Gamma_{\mathsf{T}})}$ centralizes S_0 and also $y_1, \ldots, y_n$. This proves (8.25).

Since $Z(\mathsf{A}_0)$ is a field, the semisimple ring A_0 must actually be simple. Moreover, $Z(\mathsf{A}_0) \subseteq Z(\mathsf{S}_0) \subseteq \mathsf{A}_0$, and the calculation of the preceding paragraph shows that $C_{\mathsf{A}_0}(Z(\mathsf{S}_0)) = \mathsf{S}_0$. Hence, $\mathsf{A}_0 \otimes_{Z(\mathsf{A}_0)} Z(\mathsf{S}_0) \sim \mathsf{S}_0$ in $\mathit{Br}(Z(\mathsf{S}_0))$. Also, $\Gamma_{\mathsf{A}} = \Gamma_{\mathsf{S}} + \Gamma_{\mathsf{T}}$. For any $\delta \in \Gamma_{\mathsf{S}}$ and $\varepsilon \in \Gamma_{\mathsf{T}}$, choose nonzero $s_\delta \in \mathsf{S}_\delta$ and $t_\varepsilon \in \mathsf{T}_\varepsilon$; then, $s_\gamma \otimes t_\delta \in \mathsf{A}_{\gamma+\delta} \cap \mathsf{A}^\times$. Hence, $\Gamma_{\mathsf{A}} = \Gamma_{\mathsf{A}}^\times$. Moreover, for any $a \in Z(\mathsf{A}_0)$ we have

$$\theta_{\mathsf{A}}(\delta + \varepsilon)(a) \;=\; (s_\delta \otimes t_\varepsilon)(a \otimes 1)(s_\delta \otimes t_\varepsilon)^{-1} \;=\; \theta_{\mathsf{S}}(\delta)(a) \otimes 1. \tag{8.27}$$

By taking $\varepsilon = 0$ this yields a commutative diagram:

$$\begin{array}{ccc} \Gamma_{\mathsf{S}} & \hookrightarrow & \Gamma_{\mathsf{A}}^\times \\ {\scriptstyle \theta_{\mathsf{S}}}\downarrow & & \downarrow{\scriptstyle \theta_{\mathsf{A}}} \\ \mathcal{G}(Z(\mathsf{S}_0)/\mathsf{F}_0) & \xrightarrow{\ res\ } & \mathcal{G}(Z(\mathsf{A}_0)/\mathsf{F}_0) \end{array}$$

Furthermore, (8.27) shows that $\delta + \varepsilon \in \mathit{ker}\,\theta_{\mathsf{A}}$ if and only if $\theta_{\mathsf{S}}(\delta)|_{Z(\mathsf{A}_0)} = id$, which in view of (8.25) holds if and only if $\theta_{\mathsf{S}}(\delta) \in \theta_{\mathsf{S}}(\Gamma_{\mathsf{S}} \cap \Gamma_{\mathsf{T}})$. Since $\mathit{ker}\,\theta_{\mathsf{S}} = \Gamma_{\mathsf{F}}$, this condition is equivalent to $\delta \in \Gamma_{\mathsf{S}} \cap \Gamma_{\mathsf{T}}$. Thus, $\mathit{ker}\,\theta_{\mathsf{A}} = \Gamma_{\mathsf{T}}$. This fact and the commutative diagram completely determine θ_{A}.

Because $Z(\mathsf{A}_0)$ is a field, Prop. 2.47 allows us to work back from A to its associated graded division algebra D. That proposition shows that D_0 is Brauer-equivalent to A_0, $\Gamma_{\mathsf{D}} = \Gamma_{\mathsf{A}}$, and $\theta_{\mathsf{D}} = \theta_{\mathsf{A}}$. Thus, the properties we have proved about A yield all but the last of the assertions of the theorem about D.

We now prove the symplectic module equality. Let Z, C, U, and E be the canonical subalgebras of D as in Prop. 8.16. Since Z is inertial over F, we have $\mathsf{F} \subseteq \mathsf{Z} \subseteq \mathsf{F}_{in}$. By Exercise 6.2,

$$\mathsf{D} \otimes_{\mathsf{F}} \mathsf{Z} \sim C_{\mathsf{D}}(\mathsf{Z}) = \mathsf{U} \otimes_{\mathsf{Z}} \mathsf{C} \qquad \text{in } \mathsf{Br}(\mathsf{Z}).$$

The graded algebra U is inertial over Z, hence inertially split. Thus, in $\mathsf{Br}(\mathsf{F}_{in})$,

$$\mathsf{T} \otimes_{\mathsf{F}} \mathsf{F}_{in} \sim \mathsf{D} \otimes_{\mathsf{F}} \mathsf{F}_{in} \sim (\mathsf{D} \otimes_{\mathsf{F}} \mathsf{Z}) \otimes_{\mathsf{Z}} \mathsf{F}_{in} \sim (\mathsf{U} \otimes_{\mathsf{Z}} \mathsf{C}) \otimes_{\mathsf{Z}} \mathsf{F}_{in} \sim \mathsf{C} \otimes_{\mathsf{Z}} \mathsf{F}_{in}.$$

Let $\mathsf{T}' = \mathsf{T} \otimes_{\mathsf{F}} \mathsf{F}_{in}$ and $\mathsf{C}' = \mathsf{C} \otimes_{\mathsf{Z}} \mathsf{F}_{in}$. Since T and C are totally ramified over F, Lemma 7.68 shows that T' and C' are graded division algebras; so, the Brauer-equivalence implies that $\mathsf{T}' \cong_g \mathsf{C}'$. Therefore, $\big(\Gamma_{\mathsf{T}'}/\Gamma_{\mathsf{F}_{in}}, \overline{c}_{\mathsf{T}'}\big) = \big(\Gamma_{\mathsf{C}'}/\Gamma_{\mathsf{F}_{in}}, \overline{c}_{\mathsf{C}'}\big)$ as symplectic modules. Thus, by Lemma 7.68 applied to T and C and Prop. 8.16(iv), we have equalities of symplectic modules

$$\begin{aligned}\big(\Gamma_{\mathsf{T}}/\Gamma_{\mathsf{F}}, \overline{c}_{\mathsf{T}}\big) &= \big(\Gamma_{\mathsf{T}'}/\Gamma_{\mathsf{F}_{in}}, \overline{c}_{\mathsf{T}'}\big) = \big(\Gamma_{\mathsf{C}'}/\Gamma_{\mathsf{F}_{in}}, \overline{c}_{\mathsf{C}'}\big)\\ &= \big(\Gamma_{\mathsf{C}}/\Gamma_{\mathsf{F}}, \overline{c}_{\mathsf{C}}\big) = \big((\ker\theta_{\mathsf{D}})/\Gamma_{\mathsf{F}}, a_{\mathsf{D}}\big). \qquad \square\end{aligned}$$

Recall from Th. 7.69 the map $\Phi_{\mathsf{F}}\colon \mathsf{Br}(\mathsf{F}) \to {}_{\mu(\mathsf{F}_0)}\mathbb{T}(\wedge^2\Gamma_{\mathsf{F}})$. By using Th. 8.43, we may describe the image of the Brauer class of any central graded division algebra D, as we show next. Note that the symplectic module $\big((\ker\theta_{\mathsf{D}})/\Gamma_{\mathsf{F}}, a_{\mathsf{D}}\big)$ may be viewed as an element in ${}_{\mu(\mathsf{F}_0)}\mathit{Symp}\big(\mathbb{T}(\Gamma_{\mathsf{F}})\big)$ (using an identification of $\mu(\mathsf{F}_0)$ with ${}_{\mu(\mathsf{F}_0)}\mathbb{T}$). Therefore, we may consider its image under the inverse of the isomorphism of Th. 7.22,

$$\Upsilon_{\Gamma_{\mathsf{F}}}^{-1}\colon \mathit{Symp}\big(\mathbb{T}(\Gamma_{\mathsf{F}})\big) \xrightarrow{\sim} \mathbb{T}(\wedge^2\Gamma_{\mathsf{F}}).$$

Corollary 8.44. *For any central graded division algebra* D *over* F*, we have*

$$\Phi_{\mathsf{F}}([\mathsf{D}]) = \Upsilon_{\Gamma_{\mathsf{F}}}^{-1}\big((\ker\theta_{\mathsf{D}})/\Gamma_{\mathsf{D}}, a_{\mathsf{D}}\big).$$

Proof. Let F_{in} be the inertial closure of F, so by definition

$$\Phi_{\mathsf{F}}([\mathsf{D}]) = \Sigma_{\mathsf{F}_{in}}\big(ext_{\mathsf{F}_{in}/\mathsf{F}}([\mathsf{D}])\big).$$

Since $\Upsilon_{\mathsf{F}_{in}} = \Pi_{\mathsf{F}_{in}} \circ \Sigma_{\mathsf{F}_{in}}$ by Th. 7.64, and since $\Gamma_{\mathsf{F}} = \Gamma_{\mathsf{F}_{in}}$, we have

$$\Upsilon_{\Gamma_{\mathsf{F}}}\big(\Phi_{\mathsf{F}}([\mathsf{D}])\big) = \Pi_{\mathsf{F}_{in}}\big(ext_{\mathsf{F}_{in}/\mathsf{F}}([\mathsf{D}])\big).$$

Therefore, to prove the corollary we have to show that the right displayed term equals $\big((\ker\theta_{\mathsf{D}})/\Gamma_{\mathsf{F}}, a_{\mathsf{D}}\big)$.

To see this, consider a decomposition $\mathsf{D} \sim \mathsf{S} \otimes_{\mathsf{F}} \mathsf{T}$ as in Prop. 8.42. Since S is inertially split, we have $ext_{\mathsf{F}_{in}/\mathsf{F}}([\mathsf{D}]) = ext_{\mathsf{F}_{in}/\mathsf{F}}([\mathsf{T}])$. Now, T is totally ramified, hence it follows from Prop. 2.57 and Cor. 2.43 that $\mathsf{T} \otimes_{\mathsf{F}} \mathsf{F}_{in}$ is a totally ramified central graded division algebra over F_{in}, with $\Gamma_{\mathsf{T}\otimes\mathsf{F}_{in}} = \Gamma_{\mathsf{T}}$. The canonical pairings $c_{\mathsf{T}\otimes\mathsf{F}_{in}}$ and c_{T} coincide; hence,

$$\Pi_{\mathsf{F}_{in}}\big(ext_{\mathsf{F}_{in}/\mathsf{F}}([\mathsf{T}])\big) = \big(\Gamma_{\mathsf{T}\otimes\mathsf{F}_{in}}/\Gamma_{\mathsf{F}_{in}}, \overline{c}_{\mathsf{T}\otimes\mathsf{F}_{in}}\big) = \big(\Gamma_{\mathsf{T}}/\Gamma_{\mathsf{F}}, \overline{c}_{\mathsf{T}}\big).$$

By Th. 8.43, the rightmost term here equals $\big((\ker\theta_{\mathsf{D}})/\Gamma_{\mathsf{F}}, a_{\mathsf{D}}\big)$. The corollary follows. $\square$

Remark 8.45. When $\mathsf{D} \sim \mathsf{S} \otimes_{\mathsf{F}} \mathsf{T}$ as above, the canonical subalgebra $\mathsf{Z} = Z(\mathsf{D}_0) \otimes_{\mathsf{F}_0} \mathsf{F}$ of D as in Prop. 8.16 is described by the preceding theorem. But the other canonical subalgebras of D are not so easy to determine from S and T. While C and $\mathsf{T} \otimes_{\mathsf{F}} \mathsf{Z}$ are each totally ramified over Z and they have the same canonical pairing, they need not be isomorphic. Also, while $\mathsf{D}_0 \otimes_{Z(\mathsf{D}_0)} Z(\mathsf{S}_0) \sim \mathsf{S}_0$, the proof indicates that A_0 (hence D_0) is built from other parts of S as well as S_0; also, A_0 depends on T but not merely on $\Gamma_{\mathsf{T}}/\Gamma_{\mathsf{F}}$ and its canonical pairing. This is illustrated in the following example.

Example 8.46. Let p be a prime number and let k be a field containing a primitive p^2-root of unity. Let x and y be commuting indeterminates over k, and let $\mathsf{F} = k[x, x^{-1}, y, y^{-1}]$ with its usual grading with $\Gamma_{\mathsf{F}} = \mathbb{Z} \times \mathbb{Z}$ and $\mathsf{F}_{(i,j)} = k\, x^i y^j$ for all $(i,j) \in \mathbb{Z} \times \mathbb{Z}$. Take any $a \in k^\times$ such that $[k(\sqrt[p^2]{a}):k] = p^2$ and any $b \in k^\times$. Let

$$\mathsf{S} = \big(a, y/\mathsf{F}\big)_{p^2} \qquad \text{and} \qquad \mathsf{T} = \big(x, by/\mathsf{F}\big)_p,$$

and let D be the graded division algebra associated to $\mathsf{S} \otimes_{\mathsf{F}} \mathsf{T}$. We determine the Brauer class of D_0.

Observe that S is a cyclically semiramified (hence inertially split) graded division algebra with $\mathsf{S}_0 = Z(\mathsf{S}_0) = k(\sqrt[p^2]{a})$ and $\Gamma_{\mathsf{S}} = \mathbb{Z} \times \frac{1}{p^2}\mathbb{Z}$. Also, T is totally ramified with $\Gamma_{\mathsf{T}} = \frac{1}{p}\mathbb{Z} \times \frac{1}{p}\mathbb{Z}$. Since $\Gamma_{\mathsf{S}} \cap \Gamma_{\mathsf{T}} = \mathbb{Z} \times \frac{1}{p}\mathbb{Z}$, Th. 8.43 yields that $Z(\mathsf{D}_0) = k(\sqrt[p]{a})$ and D_0 is split by $k(\sqrt[p^2]{a})$, and

$$\Gamma_{\mathsf{D}} = \Gamma_{\mathsf{S}} + \Gamma_{\mathsf{T}} = \tfrac{1}{p}\mathbb{Z} \times \tfrac{1}{p^2}\mathbb{Z}.$$

Knowing $Z(\mathsf{D}_0)$, we can now determine D_0. Let

$$\mathsf{Z} = Z(\mathsf{D}_0) \otimes_k \mathsf{F} = k(\sqrt[p]{a})[x, x^{-1}, y, y^{-1}] \subseteq \mathsf{D},$$

and let $\mathsf{E} = C_{\mathsf{D}}(\mathsf{Z})$. So, $\mathsf{D}_0 = \mathsf{E}_0$ (see Prop. 8.16(ii)), and in $\mathsf{Br}(\mathsf{Z})$,

$$\mathsf{S} \otimes_{\mathsf{F}} \mathsf{Z} = \big((\sqrt[p]{a})^p, y/\mathsf{Z}\big)_{p^2} \sim \big(\sqrt[p]{a}, y/\mathsf{Z}\big)_p.$$

Hence, in $\mathsf{Br}(\mathsf{Z})$,

$$\begin{aligned}\mathsf{E} \sim \mathsf{D} \otimes_{\mathsf{F}} \mathsf{Z} &\sim \big(\sqrt[p]{a}, y/\mathsf{Z}\big)_p \otimes_{\mathsf{Z}} \big(x, by/\mathsf{Z}\big)_p \\ &\sim \big(\sqrt[p]{a}, b^{-1}/\mathsf{Z}\big)_p \otimes_{\mathsf{Z}} \big(\sqrt[p]{a}\,x, by/\mathsf{Z}\big)_p = \mathsf{I} \otimes_{\mathsf{Z}} \mathsf{T}',\end{aligned}$$

where $\mathsf{I} = \big(\sqrt[p]{a}, b^{-1}/\mathsf{Z}\big)_p$ and $\mathsf{T}' = \big(\sqrt[p]{a}\,x, by/\mathsf{Z}\big)_p$. Note that I is inertial over Z with $\mathsf{I}_0 = \big(\sqrt[p]{a}, b^{-1}/k(\sqrt[p]{a})\big)_p$ and T' is totally ramified over Z with $\Gamma_{\mathsf{T}'} = \frac{1}{p}\mathbb{Z} \times \frac{1}{p}\mathbb{Z}$. Since I is inertial, by Prop. 2.56 $\big(\mathsf{I} \otimes \mathsf{T}'\big)_0 = \mathsf{I}_0 \otimes_{\mathsf{Z}_0} \mathsf{T}'_0 = \mathsf{I}_0$. Thus, $\big(\mathsf{I} \otimes \mathsf{T}'\big)_0$ is simple, which by Prop. 2.47 implies that

$$\mathsf{D}_0\ (= \mathsf{E}_0) \text{ is the associated division algebra of } \big(\sqrt[p]{a}, b^{-1}/k(\sqrt[p]{a})\big)_p.$$

Also,

$$deg\, \mathsf{D} = [\mathsf{D}_0{:}\mathsf{F}_0]|\Gamma_{\mathsf{D}}{:}\Gamma_{\mathsf{F}}| = (ind\, \mathsf{D}_0)^2\, [Z(\mathsf{D}_0){:}\mathsf{F}_0]\, p^3 = (ind\, \mathsf{D}_0)^2\, p^4,$$

so $ind\,\mathsf{D} = p^2\, ind\big(\sqrt[p]{a}\,, b^{-1}/k(\sqrt[p]{a})\big)_p$. One can check also that (whether or not I_0 is a division ring) the canonical subalgebra $\mathsf{C} = C_{\mathsf{D}}(\mathsf{D}_0)$ of D is isomorphic to T' but not isomorphic to $\mathsf{T} \otimes_{\mathsf{F}} \mathsf{Z}$.

For any graded division algebra D with center F and any $j \in \mathbb{N}$, recall that D^j denotes the associated division algebra of the j-th tensor power $\mathsf{D}^{\otimes j} = \mathsf{D} \otimes_{\mathsf{F}} \ldots \otimes_{\mathsf{F}} \mathsf{D}$ (j factors) of D. We illustrate the use of the $\mathsf{S} \otimes \mathsf{T}$ decomposition by determining the grade group of D^j. The formula for Γ_{D^j} is easy to prove if D is inertially split (see Exercise 6.5) or totally ramified over F, but would be quite difficult to establish for arbitrary graded division algebras without using Th. 8.43.

Proposition 8.47. *Let* D *be a graded division algebra with center* F. *For any* $j \in \mathbb{N}$, *we have* $\Gamma_{\mathsf{D}^j} = j\Gamma_{\mathsf{D}} + \Gamma_{\mathsf{F}}$.

Proof. Write $\mathsf{D} \sim \mathsf{S} \otimes_{\mathsf{F}} \mathsf{T}$ in $\mathsf{Br}(\mathsf{F})$, where S is an inertially split graded division algebra and T is a totally ramified graded division algebra over F. Then, $\mathsf{D}^j \sim \mathsf{S}^j \otimes_{\mathsf{F}} \mathsf{T}^j$ in $\mathsf{Br}(\mathsf{F})$. By Prop. 7.75, T^j is tame and totally ramified over F with $\Gamma_{\mathsf{T}^j} = j\Gamma_{\mathsf{T}} + \Gamma_{\mathsf{F}}$. Also, S^j is inertially split since S is inertially split, and by Prop. 6.22

$$\Gamma_{\mathsf{S}^j}/\Gamma_{\mathsf{F}} \;=\; im\,\chi_{\mathsf{S}^j} \;=\; im(j\cdot\chi_{\mathsf{S}}) \;=\; j(\Gamma_{\mathsf{S}}/\Gamma_{\mathsf{F}}).$$

Hence, $\Gamma_{\mathsf{S}^j} = j\Gamma_{\mathsf{S}} + \Gamma_{\mathsf{F}}$. Theorem 8.43, applied to D^j and to D, yields

$$\Gamma_{\mathsf{D}^j} \;=\; \Gamma_{\mathsf{S}^j} + \Gamma_{\mathsf{T}^j} \;=\; (j\Gamma_{\mathsf{S}} + \Gamma_{\mathsf{F}}) + (j\Gamma_{\mathsf{T}} + \Gamma_{\mathsf{F}}) \;=\; j\Gamma_{\mathsf{D}} + \Gamma_{\mathsf{F}}. \qquad \square$$

Corollary 8.48. *For any graded division algebra* D *with center* F, *we have* $exp(\Gamma_{\mathsf{D}}/\Gamma_{\mathsf{F}}) \mid exp\,\mathsf{D}$. *If* $\mathsf{D} \sim \mathsf{S} \otimes_{\mathsf{F}} \mathsf{T}$ *with* S *inertially split and* T *totally ramified, then*

$$exp\,\mathsf{D} \;=\; lcm\big(exp\,\mathsf{S}, exp(\Gamma_{\mathsf{D}}/\Gamma_{\mathsf{F}})\big).$$

Proof. Let $e = exp\,\mathsf{D}$ and $\ell = exp(\Gamma_{\mathsf{D}}/\Gamma_{\mathsf{F}})$. Since $\mathsf{D}^e = \mathsf{F}$, Prop. 8.47 yields

$$e\Gamma_{\mathsf{D}} + \Gamma_{\mathsf{F}} \;=\; \Gamma_{\mathsf{D}^e} \;=\; \Gamma_{\mathsf{F}}.$$

Hence, $e\cdot(\Gamma_{\mathsf{D}}/\Gamma_{\mathsf{F}}) = \{0\}$, showing that $\ell \mid e$. This proves the first statement, and implies that

$$e \;=\; \ell\cdot exp\,\mathsf{D}^{\ell}.$$

For any decomposition $\mathsf{D} \sim \mathsf{S} \otimes_{\mathsf{F}} \mathsf{T}$ as above, we have $\Gamma_{\mathsf{T}} \subseteq \Gamma_{\mathsf{D}}$ by Th. 8.43 and $exp\,\mathsf{T} = exp(\Gamma_{\mathsf{T}}/\Gamma_{\mathsf{F}})$ by Prop. 7.60, hence $\mathsf{T}^{\ell} = \mathsf{F}$. Therefore, $\mathsf{D}^{\ell} \cong \mathsf{S}^{\ell}$, so $exp\,\mathsf{D}^{\ell} = exp\,\mathsf{S}^{\ell}$. Now, $exp\,\mathsf{S}^{\ell} = exp\,\mathsf{S}/gcd(\ell, exp\,\mathsf{S})$, so

$$e \;=\; \ell\cdot exp\,\mathsf{S}/gcd(\ell, exp\,\mathsf{S}) \;=\; lcm(\ell, exp\,\mathsf{S}). \qquad \square$$

The exponent of an inertially split graded division algebra will be described in Th. 8.50(iv) below.

8.4.2 Inertially split graded algebras

In this subsection, F is again an arbitrary graded field. We obtain a decomposition up to Brauer-equivalence of any inertially split graded division algebra D into a tensor product of an inertial graded division algebra and a decomposably semiramified graded division algebra. This decomposition is not unique, but it leads to an invariant of D that we call its *specialization coset.*

Our first result describes when a central graded division algebra is inertially split.

Proposition 8.49. *Let D be a central graded division algebra over a graded field f F, and let E, U, C, Z be the canonically-defined graded subalgebras of D as in §8.2.1. The following conditions are equivalent:*

(a) *D is inertially split;*
(b) *D contains a maximal graded subfield that is inertial over F;*
(c) *$\mathsf{C} = \mathsf{Z}$;*
(d) *$\ker\theta_{\mathsf{D}} = \Gamma_{\mathsf{F}}$;*
(e) *$ind\,\mathsf{D} = ind\,\mathsf{D}_0 \cdot [Z(\mathsf{D}_0){:}\mathsf{F}_0]$.*

These conditions hold whenever $dim_{\mathbb{Q}}\,\mathbb{H}(\Gamma_{\mathsf{F}}) = 1$ (in particular, when $\Gamma_{\mathsf{F}} \cong \mathbb{Z}$).

Proof. (a) $\Rightarrow$ (c): Since $\mathsf{C} = C_{\mathsf{D}}(\mathsf{Z})$, Exercise 6.2 shows that $\mathsf{D} \otimes_{\mathsf{F}} \mathsf{Z} \sim \mathsf{C}$ in $\mathsf{Br}(\mathsf{Z})$. Recall that the graded field extension Z/F is inertial, since $\mathsf{Z}_0/\mathsf{F}_0$ is separable by Prop. 2.40. It follows that $\mathsf{D} \otimes_{\mathsf{F}} \mathsf{F}_{in} \sim \mathsf{C} \otimes_{\mathsf{Z}} \mathsf{F}_{in}$ in $\mathsf{Br}(\mathsf{F}_{in})$. Since C is totally ramified over Z, each homogeneous component of $\mathsf{C} \otimes_{\mathsf{Z}} \mathsf{F}_{in}$ has the form $c \otimes (\mathsf{F}_{in})_0$ for some $c \in \mathsf{C}^{\times}$. Therefore, the nonzero homogeneous elements of $\mathsf{C} \otimes_{\mathsf{Z}} \mathsf{F}_{in}$ are invertible; hence, this algebra is a graded division algebra. But if D is inertially split, then $\mathsf{C} \otimes_{\mathsf{Z}} \mathsf{F}_{in}$ is also split, and hence $\mathsf{C} = \mathsf{Z}$.

The implication (b) $\Rightarrow$ (a) is clear since any maximal graded subfield of D splits D, by Remark 6.9. The equivalence (c) $\Leftrightarrow$ (d) follows from Prop. 8.16(ii), as $\Gamma_{\mathsf{Z}} = \Gamma_{\mathsf{F}}$. For the equivalence of (d) and (e), note that

$$[\mathsf{D}{:}\mathsf{F}] \;=\; [\mathsf{D}_0{:}\mathsf{F}_0]\,|\Gamma_{\mathsf{D}}{:}\Gamma_{\mathsf{F}}| \;=\; [\mathsf{D}_0{:}Z(\mathsf{D}_0)]\,[Z(\mathsf{D}_0){:}\mathsf{F}_0]\,|\Gamma_{\mathsf{D}}{:}\,\ker\theta_{\mathsf{D}}|\,|\ker\theta_{\mathsf{D}}{:}\Gamma_{\mathsf{F}}|. \tag{8.28}$$

Since $Z(\mathsf{D}_0)$ is Galois over F_0 and θ_{D} is surjective, we have

$$|\Gamma_{\mathsf{D}}{:}\,\ker\theta_{\mathsf{D}}| \;=\; |\,\mathcal{G}(Z(\mathsf{D}_0)/\mathsf{F}_0)| \;=\; [Z(\mathsf{D}_0){:}\mathsf{F}_0].$$

It follows by taking square roots in (8.28) that

$$ind\,\mathsf{D} \;=\; ind\,\mathsf{D}_0\,[Z(\mathsf{D}_0){:}\mathsf{F}_0]\,\sqrt{|\ker\theta_{\mathsf{D}}{:}\Gamma_{\mathsf{F}}|}.$$

Thus, (d) $\Leftrightarrow$ (e).

To show the equivalence of (a)–(e) it remains only to prove that (d) $\Rightarrow$ (b). As in the proof of Prop. 6.18, let M be a maximal subfield of D_0 which is separable over $\mathsf{Z}(\mathsf{D}_0)$; then M is also separable over F_0 since $\mathsf{Z}(\mathsf{D}_0)/\mathsf{F}_0$ is separable by Prop. 2.40. If (d) holds, then

$$[Z(\mathsf{D}_0):\mathsf{F}_0] = |\Gamma_\mathsf{D}: \mathit{ker}\,\theta_\mathsf{D}| = |\Gamma_\mathsf{D}:\Gamma_\mathsf{F}|.$$

Then $\mathsf{K} = M \otimes_{\mathsf{F}_0} \mathsf{F}$ is a maximal graded subfield of D by Cor. 2.36, as

$$\begin{aligned}[\mathsf{K}:\mathsf{F}]^2 = [M:\mathsf{F}_0]^2 &= [M:Z(\mathsf{D}_0)]^2\,[Z(\mathsf{D}_0):\mathsf{F}_0]^2 \\ &= [\mathsf{D}_0:Z(\mathsf{D}_0)]\,[Z(\mathsf{D}_0):\mathsf{F}_0]\,|\Gamma_\mathsf{D}:\Gamma_\mathsf{F}| = [\mathsf{D}:\mathsf{F}].\end{aligned}$$

Since K is also inertial over F, it satisfies the conditions in (b).

If $\mathit{dim}_\mathbb{Q}\,\mathbb{H}(\Gamma_\mathsf{F}) = 1$, then Z is not the center of a nontrivial totally ramified graded division algebra, by Cor. 7.53, hence (c) holds. □

Note that (a) ⇒ (d) of Prop. 8.49 was proved in a different way in Prop. 6.22 above.

Proposition 6.27 above shows that the canonical short exact sequence for $\mathsf{Br}_{is}(\mathsf{F})$,

$$0 \longrightarrow \mathsf{Br}_{in}(\mathsf{F}) \longrightarrow \mathsf{Br}_{is}(\mathsf{F}) \xrightarrow{\ \chi\ } \mathit{Hom}^c\big(\mathcal{G}(\mathsf{F}_0), \mathbb{T}(\Gamma_\mathsf{F})\big) \longrightarrow 0,$$

is split exact. The proof of this proposition shows more specifically that for any maximal totally ramified algebraic graded field extension T of F, there is an isomorphism $\mathit{Hom}^c(\mathcal{G}(\mathsf{F}_0), \mathbb{T}(\Gamma_\mathsf{F})) \xrightarrow{\sim} \mathsf{Br}_{is}(\mathsf{F}) \cap \mathsf{Br}(\mathsf{T}/\mathsf{F})$ that splits the short exact sequence. By Prop. 8.33, the graded division algebras in the image of this isomorphism are decomposably semiramified; moreover, every decomposably semiramified division algebra N lies in the image of such an isomorphism for T chosen to split N. The splitting map yields a direct product decomposition

$$\mathsf{Br}_{is}(\mathsf{F}) = \mathsf{Br}_{in}(\mathsf{F}) \times \big(\mathsf{Br}_{is}(\mathsf{F}) \cap \mathsf{Br}(\mathsf{T}/\mathsf{F})\big). \tag{8.29}$$

Thus, any inertially split graded division algebra D with center F has a decomposition up to Brauer-equivalence

$$\mathsf{D} \sim \mathsf{I} \otimes_\mathsf{F} \mathsf{N} \tag{8.30}$$

where I is an inertial graded division algebra and N is decomposably semiramified. This decomposition is not unique (see Prop. 8.51 below), but is still useful in that it expresses D in terms of very tractable types of graded division algebras. The next theorem shows that many properties of D are directly expressible in terms of those of I and N.

Theorem 8.50. *Let D be an inertially split graded division algebra with center F, and suppose $\mathsf{D} \sim \mathsf{I} \otimes_\mathsf{F} \mathsf{N}$ in $\mathsf{Br}(\mathsf{F})$, where I is an inertial graded division algebra and N is decomposably semiramified. Then,*

(i) $\chi_\mathsf{D} = \chi_\mathsf{N}$; *hence,* $\Gamma_\mathsf{D} = \Gamma_\mathsf{N}$, $Z(\mathsf{D}_0) = Z(\mathsf{N}_0) = \mathsf{N}_0$, *and* $\theta_\mathsf{D} = \theta_\mathsf{N}$.
(ii) D_0 *is the associated division algebra of the central simple* N_0*-algebra* $\mathsf{I}_0 \otimes_{\mathsf{F}_0} \mathsf{N}_0$.

(iii) $ind\,\mathsf{D} = ind(\mathsf{I}_0 \otimes_{\mathsf{F}_0} \mathsf{N}_0) \cdot |\Gamma_{\mathsf{N}}{:}\Gamma_{\mathsf{F}}|$.
(iv) $exp\ \mathsf{D} = lcm(exp\ \mathsf{I}, exp\ \mathsf{N}) = lcm\big(exp(\mathsf{I}_0), exp(\Gamma_{\mathsf{N}}/\Gamma_{\mathsf{F}})\big)$.
(v) D *is semiramified if and only if* N_0 *splits* I_0.

Proof. (i) $\chi_{\mathsf{I}} = 0$ by Th. 6.24 as I is inertially split. Thus, by Prop. 6.23 $\chi_{\mathsf{D}} = \chi_{\mathsf{I}} + \chi_{\mathsf{N}} = \chi_{\mathsf{N}}$. The remaining assertions in (i) follow from this equality, since χ_{D} determines Γ_{D}, $Z(\mathsf{D}_0)$, and θ_{D}. Alternatively, these assertions follow by Prop. 2.56, as I is inertial over F.

(ii) Let $\mathsf{A} = \mathsf{I} \otimes_{\mathsf{F}} \mathsf{N}$. Then, $\mathsf{A}_0 \cong \mathsf{I}_0 \otimes_{\mathsf{F}_0} \mathsf{N}_0$ by Prop. 2.56 as I is inertial over F; so, A_0 is simple, as $Z(\mathsf{I}_0) = \mathsf{F}_0$ and N_0 is a field. Since D and A are Brauer-equivalent, the division algebra D_0 is Brauer-equivalent to A_0.

(iii) In view of (i) and (ii) and the fact that N is semiramified, we have

$$\begin{aligned}(ind\,\mathsf{D})^2 &= [\mathsf{D}{:}\mathsf{F}] = [\mathsf{D}_0{:}\mathsf{F}_0] \cdot |\Gamma_{\mathsf{D}}{:}\Gamma_{\mathsf{F}}| = [\mathsf{D}_0{:}Z(\mathsf{D}_0)] \cdot [Z(\mathsf{D}_0){:}\mathsf{F}_0] \cdot |\Gamma_{\mathsf{D}}{:}\Gamma_{\mathsf{F}}| \\ &= (ind\,\mathsf{D}_0)^2 \cdot [\mathsf{N}_0{:}\mathsf{F}_0] \cdot |\Gamma_{\mathsf{N}}{:}\Gamma_{\mathsf{F}}| = ind(\mathsf{I}_0 \otimes_{\mathsf{F}_0} \mathsf{N}_0)^2 \cdot |\Gamma_{\mathsf{N}}{:}\Gamma_{\mathsf{F}}|^2.\end{aligned}$$

(iv) By Prop. 8.33 there is a totally ramified graded field extension K of F that splits N. Let T be a maximal totally ramified algebraic graded field extension of K. Then, T is also a maximal totally ramified extension of F, and T splits N. Hence, $[\mathsf{I}]$ and $[\mathsf{N}]$ are the components of $[\mathsf{D}]$ in the direct product decomposition (8.29) of $\mathsf{Br}_{is}(\mathsf{F})$ determined by T. Thus, $exp(\mathsf{D}) = lcm\big(exp\ \mathsf{I}, exp\ \mathsf{N}\big)$. For the second equality in (iv), note that $exp\ \mathsf{I} = exp(\mathsf{I}_0)$ from the isomorphism $\mathsf{Br}_{in}(\mathsf{F}) \cong Br(\mathsf{F}_0)$. It remains only to check that $exp\ \mathsf{N} = exp(\Gamma_{\mathsf{N}}/\Gamma_{\mathsf{F}})$. For this, let N^j be the associated graded division algebra of the j-th tensor power $\mathsf{N}^{\otimes j}$ of N, for any integer $j > 0$. Since K and F_{in} split N, they each split N^j for every j. Hence, each power N^j is decomposably semiramified, by Prop. 8.33. Therefore, N^j is split if and only if χ_{N^j} is trivial. This occurs if and only if we have triviality of $im\,\chi_{\mathsf{N}^j} = im(j\chi_{\mathsf{N}}) = j\ im\,\chi_{\mathsf{N}} = j(\Gamma_{\mathsf{N}}/\Gamma_{\mathsf{F}})$. Hence, $exp\ \mathsf{N} = exp(\Gamma_{\mathsf{N}}/\Gamma_{\mathsf{F}})$.

(v) This follows from (ii), since the inertially split graded division algebra D is semiramified if and only if D_0 is a field. □

The $\mathsf{I} \otimes_{\mathsf{F}} \mathsf{N}$ decomposition of an inertially split graded division algebra is not unique; the variation in the possible decompositions is measured by a certain *Dec* group:

Proposition 8.51. *Let* D *be an inertially split graded division algebra with center* F. *Suppose* $\mathsf{D} \sim \mathsf{I} \otimes_{\mathsf{F}} \mathsf{N}$ *in* $\mathsf{Br}(\mathsf{F})$, *where* I *is an inertial graded division algebra and* N *is a decomposably semiramified graded division algebra. Let* I' *be another inertial graded division algebra with center* F. *Then, there is a decomposably semiramified graded division algebra* N' *such that* $\mathsf{D} \sim \mathsf{I}' \otimes_{\mathsf{F}} \mathsf{N}'$ *if and only if* $[\mathsf{I}_0] \equiv [\mathsf{I}'_0]$ $(mod\ Dec(Z(\mathsf{D}_0)/\mathsf{F}_0))$.

Proof. Suppose $\mathsf{D} \sim \mathsf{I}' \otimes_{\mathsf{F}} \mathsf{N}'$ as above. Then, $\chi_{\mathsf{N}'} = \chi_{\mathsf{D}} = \chi_{\mathsf{N}}$ by Th. 8.50(i). Let $\chi_{\mathsf{N}} = \chi_1 + \ldots + \chi_n$, where the χ_i are independent cyclic characters. Then (see Remark 8.34) there are cyclically semiramified graded division algebras $\mathsf{C}_1, \ldots, \mathsf{C}_n$ and $\mathsf{C}'_1, \ldots, \mathsf{C}'_n$ such that $\mathsf{N} = \mathsf{C}_1 \otimes_{\mathsf{F}} \ldots \otimes_{\mathsf{F}} \mathsf{C}_n$, $\mathsf{N}' = \mathsf{C}'_1 \otimes_{\mathsf{F}} \ldots \otimes \mathsf{C}'_n$, and $\chi_{\mathsf{C}_i} = \chi_i = \chi_{\mathsf{C}'_i}$ for all i. Let ℓ_i be the fixed field of $\ker \chi_i$. Then, ℓ_i is a cyclic Galois extension of F_0; also, $\ell_i \subseteq Z(\mathsf{D}_0)$, as $\ker \chi_i \supseteq \ker \chi_{\mathsf{D}}$ since the χ_i are independent. Let $\mathsf{J}_i = \mathsf{C}_i \otimes_{\mathsf{F}} \mathsf{C}_i'^{\mathrm{op}}$. Then, $\mathsf{J}_i \in \mathsf{Br}_{in}(\mathsf{F})$ since $\chi_{\mathsf{J}_i} = \chi_{\mathsf{C}_i} - \chi_{\mathsf{C}_i'} = 0$. Furthermore, the inertial extension $\ell_i \otimes_{\mathsf{F}_0} \mathsf{F}$ of F is a graded subfield of maximal degree in C_i and C_i' (see the proof of Prop. 8.33), hence it splits J_i. Since the isomorphism $\Xi\colon \mathsf{Br}_{in}\mathsf{F} \xrightarrow{\sim} Br(\mathsf{F}_0)$ is functorial with respect to graded field extensions by Prop. 2.55, we have $[\mathsf{J}_{i,0}] \in Br(\ell_i/\mathsf{F}_0) \subseteq Dec(Z(\mathsf{D}_0)/\mathsf{F}_0)$ for each i. In $\mathsf{Br}_{in}(\mathsf{F})$ we have

$$[\mathsf{I}'] - [\mathsf{I}] = [\mathsf{N}] - [\mathsf{N}'] = \sum_{i=1}^{n} [\mathsf{C}_i] - \sum_{i=1}^{n} [\mathsf{C}_i'] = \sum_{i=1}^{n} [\mathsf{J}_i].$$

Hence, in $Br(\mathsf{F}_0)$

$$[\mathsf{I}_0'] - [\mathsf{I}_0] = \Xi([\mathsf{I}'] - [\mathsf{I}]) = \sum_{i=1}^{n} \Xi[\mathsf{J}_i] = \sum_{i=1}^{n} [\mathsf{J}_{i,0}] \in Dec(Z(\mathsf{D}_0)/\mathsf{F}_0).$$

Conversely, suppose $\mathsf{I}_0 \equiv \mathsf{I}_0'$ $(mod\ Dec(Z(\mathsf{D}_0)/\mathsf{F}_0))$. Let $\mathsf{A} = \mathsf{I} \otimes_{\mathsf{F}} \mathsf{I}'^{\mathrm{op}}$. Then, A is inertial over F and $\mathsf{A}_0 \cong \mathsf{I}_0 \otimes_{\mathsf{F}_0} \mathsf{I}_0'^{\mathrm{op}}$ (see Prop. 2.54). So, in $Br(\mathsf{F}_0)$, $[\mathsf{A}_0] = [\mathsf{I}_0] - [\mathsf{I}_0'] \in Dec(Z(\mathsf{D}_0)/\mathsf{F}_0)$. Let N' be the associated graded division algebra of $\mathsf{A} \otimes_{\mathsf{F}} \mathsf{N}$. We claim that N' is decomposably semiramified. Given this, we have in $\mathsf{Br}(\mathsf{F})$

$$[\mathsf{D}] = [\mathsf{I}] + [\mathsf{N}] = [\mathsf{I}'] + [\mathsf{A}] + [\mathsf{N}] = [\mathsf{I}'] + [\mathsf{N}'],$$

showing that the decomposition $\mathsf{D} \sim \mathsf{I}' \otimes_{\mathsf{F}} \mathsf{N}'$ has the desired form.

To prove the claim, let $\chi_1, \ldots, \chi_n$ and $\ell_1, \ldots, \ell_n$ be as above. Then

$$Z(\mathsf{D}_0) = \ell_1 \otimes_{\mathsf{F}_0} \ldots \otimes_{\mathsf{F}_0} \ell_n$$

by the independence of the χ_i. Let $G = \mathcal{G}(Z(\mathsf{D}_0)/\mathsf{F}_0) = \mathcal{G}\big((Z(\mathsf{D}_0) \otimes_{\mathsf{F}_0} \mathsf{F})/\mathsf{F}\big)$, and let $\sigma_1, \ldots, \sigma_n \in G$ be automorphisms such that $\sigma_i|_{\ell_j} = id$ whenever $j \neq i$ and $\sigma_i|_{\ell_i}$ is a generator of $\mathcal{G}(\ell_i/\mathsf{F}_0)$. So, $G = \langle \sigma_1 \rangle \times \ldots \times \langle \sigma_n \rangle$. Let $r_i = [\ell_i{:}\mathsf{F}_0] = \big|\langle \sigma_i \rangle\big|$.

Since $\mathsf{A}_0 \in Dec(Z(\mathsf{D}_0)/\mathsf{F}_0)$ we have in $Br(\mathsf{F}_0)$

$$\mathsf{A}_0 \sim \big(\ell_1/\mathsf{F}_0, \sigma_1, a_1\big) \otimes_{\mathsf{F}_0} \ldots \otimes_{\mathsf{F}_0} \big(\ell_n/\mathsf{F}_0, \sigma_n, a_n\big) \quad \text{for some } a_1, \ldots, a_n \in \mathsf{F}_0^\times$$

(see (8.11)). The cyclic algebra $\big(\ell_i/\mathsf{F}_0, \sigma_i, a_i\big)$ is the degree 0 component of an inertial cyclic graded cyclic algebra $\big((\ell_i \otimes_{\mathsf{F}_0} \mathsf{F})/\mathsf{F}, \sigma_i, a_i\big)$ (cf. Ex. 2.53). Hence, $\mathsf{A} \sim \bigotimes_{i=1}^{n}{}_{\mathsf{F}} \big((\ell_i \otimes_{\mathsf{F}_0} \mathsf{F})/\mathsf{F}, \sigma_i, a_i\big)$ in $\mathsf{Br}_{in}(\mathsf{F})$. Because N is decomposably semiramified and $\mathsf{N}_0 = Z(\mathsf{D}_0) = \ell_1 \otimes_{\mathsf{F}_0} \ldots \otimes_{\mathsf{F}_0} \ell_n$, we have

$$\mathsf{N} \cong_g \bigotimes_{i=1}^{n}{}_{\mathsf{F}} \big((\ell_i \otimes_{\mathsf{F}_0} \mathsf{F})/\mathsf{F}, \sigma_i, b_i\big)$$

by Prop. 8.28, for some $b_i \in \mathsf{F}^\times$ with $\frac{1}{r_i} \deg b_i + \Gamma_{\mathsf{F}} = \overline{\chi}_i(\sigma_i)$. Thus, in $\mathsf{Br}(\mathsf{F})$,

$$\mathsf{A} \otimes_{\mathsf{F}} \mathsf{N} \sim \bigotimes_{i=1}^{n}{}_{\mathsf{F}} \big((\ell_i \otimes_{\mathsf{F}_0} \mathsf{F})/\mathsf{F}, \sigma_i, a_i b_i\big).$$

Since $deg(a_i b_i) = deg\, b_i$, the tensor product $\bigotimes_{i=1}^{n} \big((\ell_i \otimes_{\mathsf{F}_0} \mathsf{F})/\mathsf{F}, \sigma_i, a_i b_i\big)$ is a decomposably semiramified graded division algebra by Prop. 8.29. This tensor product is Brauer-equivalent and hence isomorphic to the graded division algebra N'. Thus, N' is decomposably semiramified, as claimed. □

Definition 8.52. Let D be an inertially split graded division algebra with center F. From Prop. 8.51 it follows that the set of Brauer classes $[\mathsf{I}_0] \in Br(\mathsf{F}_0)$, for I an inertial graded division algebra in a decomposition $\mathsf{D} \sim \mathsf{I} \otimes_{\mathsf{F}} \mathsf{N}$ with N decomposably semiramified, is a coset modulo $Dec(Z(\mathsf{D}_0)/\mathsf{F}_0)$. We call this coset the *specialization coset* of D. (This terminology is motivated by Remark 8.74 below.) Thus, the specialization coset of D is

$$sp(\mathsf{D}) \;=\; \left\{ [\mathsf{I}_0] \;\middle|\; \begin{array}{l} \mathsf{I} \text{ is inertial and } \mathsf{D} \otimes_{\mathsf{F}} \mathsf{I}^{\mathrm{op}} \text{ is associated to a decompos-} \\ \text{ably semiramified graded division algebra} \end{array} \right\}$$
$$\in Br(\mathsf{F}_0)/Dec(Z(\mathsf{D}_0)/\mathsf{F}_0).$$

Note that if D is semiramified then Th. 8.50(i–ii) shows that in any $\mathsf{I} \otimes \mathsf{N}$ decomposition the algebra I_0 is split by $Z(\mathsf{D}_0) = \mathsf{D}_0$, hence

$$sp(\mathsf{D}) \in Br(\mathsf{D}_0/\mathsf{F}_0)/Dec(\mathsf{D}_0/\mathsf{F}_0).$$

Corollary 8.53. *An inertially split graded division algebra* D *with center* F *is decomposably semiramified if and only if* $sp(\mathsf{D}) = Dec(Z(\mathsf{D}_0)/\mathsf{F}_0)$.

Proof. The algebra D is decomposably semiramified if and only if we can take $\mathsf{I} = \mathsf{F}$ in the definition of $sp(\mathsf{D})$. □

Example 8.54. *The case where* $\Gamma_{\mathsf{D}}/\Gamma_{\mathsf{F}}$ *is cyclic.* Suppose D is an inertially split central graded division algebra over F and $\Gamma_{\mathsf{D}}/\Gamma_{\mathsf{F}}$ is cyclic. The extension $Z(\mathsf{D}_0)/\mathsf{F}_0$ is then cyclic since θ_{D} induces an isomorphism $\Gamma_{\mathsf{D}}/\Gamma_{\mathsf{F}} \cong \mathcal{G}(Z(\mathsf{D}_0)/\mathsf{F}_0)$, see Prop. 6.22. By the definition of the *Dec* group, it follows that $Dec(Z(\mathsf{D}_0)/\mathsf{F}_0) = Br(Z(\mathsf{D}_0)/\mathsf{F}_0)$. Then there is a central simple F_0-algebra A such that $A \otimes_{\mathsf{F}_0} Z(\mathsf{D}_0) \sim \mathsf{D}_0$. For any such A we have

$$sp(D) \;=\; [A] + Dec(Z(\mathsf{D}_0)/\mathsf{F}_0) \;=\; [A] + Br(Z(\mathsf{D}_0)/\mathsf{F}_0). \tag{8.31}$$

This follows because Th. 8.50(ii) shows that in any decomposition $\mathsf{D} \sim \mathsf{I} \otimes \mathsf{N}$ with I inertial and N decomposably semiramified we have $\mathsf{I}_0 \otimes_{\mathsf{F}_0} Z(\mathsf{D}_0) \sim \mathsf{D}_0$ and, by definition, $sp(\mathsf{D}) = [\mathsf{I}_0] + Dec(Z(\mathsf{D}_0)/\mathsf{F}_0)$. If A satisfies $A \otimes_{\mathsf{F}_0} Z(\mathsf{D}_0) \sim \mathsf{D}_0$, then $[A] \equiv [\mathsf{I}_0]$ $(mod\ Br(Z(\mathsf{D}_0)/\mathsf{F}_0))$, yielding (8.31).

As a further example, consider a semiramified graded division algebra D with $\Gamma_{\mathsf{D}}/\Gamma_{\mathsf{F}}$ bicyclic. Let $G = \mathcal{G}(Z(\mathsf{D}_0)/\mathsf{F}_0)$. Since $\Gamma_{\mathsf{D}}/\Gamma_{\mathsf{F}} \cong G$ under the isomorphism induced by θ_{D}, we also have G bicyclic. As D is assumed to be semiramified, we have $\mathsf{D}_0 = Z(\mathsf{D}_0)$, and $\mathsf{D}_0 \otimes_{\mathsf{F}_0} \mathsf{F}$ is a maximal subfield of D. Therefore, fixing a base $\boldsymbol{\sigma}$ of G, we have a bicyclic crossed product representation as in (8.20):

$$\mathsf{D} \cong_g \mathsf{A}(\mathsf{D}_0 \otimes_{\mathsf{F}_0} \mathsf{F}/\mathsf{F}, \boldsymbol{\sigma}, b_1, u, b_2) \tag{8.32}$$

for some $(b_1, u, b_2) \in Z^2_{\boldsymbol{\sigma}}(G, (\mathsf{D}_0 \otimes_{\mathsf{F}_0} \mathsf{F})^\times)$. Recall from Prop. 8.23 the map $\upsilon_{\boldsymbol{\sigma}}\colon Br(\mathsf{D}_0/\mathsf{F}_0) \to \widehat{H}^{-1}(G, \mathsf{D}_0^\times)$, which induces an isomorphism

$$\overline{\upsilon}_{\boldsymbol{\sigma}}\colon Br(\mathsf{D}_0/\mathsf{F}_0)/Dec(\mathsf{D}_0/\mathsf{F}_0) \xrightarrow{\sim} \widehat{H}^{-1}(G, \mathsf{D}_0^\times).$$

Proposition 8.55. *In the setting just described, with* $\Gamma_{\mathsf{D}}/\Gamma_{\mathsf{F}}$ *bicyclic, let* $z_1, z_2 \in \mathsf{D}^\times$ *satisfy* $\theta_{\mathsf{D}}(deg\, z_i) = \sigma_i$ *for* $i = 1, 2$. *Then* $N_{\mathsf{D}_0/\mathsf{F}_0}(z_1 z_2 z_1^{-1} z_2^{-1}) = 1$ *and*

$$\overline{\upsilon}_{\boldsymbol{\sigma}}(sp(D)) \;=\; z_1 z_2 z_1^{-1} z_2^{-1} \cdot I_G \mathsf{D}_0^\times \;\in\; \widehat{H}^{-1}(G, \mathsf{D}_0^\times).$$

Proof. To simplify notation, let $\mathsf{K} = \mathsf{D}_0 \otimes_{\mathsf{F}_0} \mathsf{F}$. Let $\mathsf{L}_1 = \mathsf{K}^{\sigma_2}$, $\mathsf{L}_2 = \mathsf{K}^{\sigma_1}$, and $r_i = [\mathsf{L}_i{:}\mathsf{F}] = |\langle \sigma_i \rangle|$ $(i = 1, 2)$, as in §8.3.2. Since $\theta_{\mathsf{D}}(deg\, z_i) = \sigma_i$, we have $int(z_i)|_{\mathsf{K}} = \sigma_i$; hence, we may take z_1 and z_2 as distinguished generators in the crossed product representation (8.32). Then,

$$b_1 \;=\; z_1^{r_1}, \qquad u \;=\; z_1 z_2 z_1^{-1} z_2^{-1}, \qquad b_2 \;=\; z_2^{r_2}.$$

As noted in (8.15), we have $N_{q(\mathsf{K})/q(\mathsf{L}_2)}(u) = b_1 \sigma_2(b_1)^{-1}$, so $N_{q(\mathsf{K})/q(\mathsf{F})}(u) = 1$. Since $deg\, u = 0$ and $q(\mathsf{K}) = \mathsf{D}_0 \otimes_{\mathsf{F}_0} q(\mathsf{F})$, it follows that $N_{\mathsf{D}_0/\mathsf{F}_0}(u) = 1$, proving the first assertion.

To prove the second assertion, we construct an $\mathsf{I} \otimes \mathsf{N}$ decomposition for D. Observe that $r_i\, deg\, z_i \in ker\, \theta_{\mathsf{D}} = \Gamma_{\mathsf{F}}$ because $\theta_{\mathsf{D}}(deg\, z_i) = \sigma_i$ has order r_i. We may therefore pick $c_i \in \mathsf{F}^\times$ with $deg\, c_i = r_i\, deg\, z_i = deg\, b_i$. Then $(c_1, 1, c_2) \in Z^2_{\boldsymbol{\sigma}}(G, \mathsf{K}^\times)$, so we may consider the graded abelian crossed product

$$\mathsf{N} \;=\; \mathsf{A}(\mathsf{K}/\mathsf{F}, \boldsymbol{\sigma}, c_1, 1, c_2)$$

(see Prop. 8.25(i)). Because the distinguished generators of this crossed product commute, we have, as in the proof of Prop. 8.23,

$$\mathsf{N} \;=\; \big(\mathsf{L}_1/\mathsf{F}, \sigma_1, c_1\big) \otimes_{\mathsf{F}} \big(\mathsf{L}_2/\mathsf{F}, \sigma_2, c_2\big).$$

Since $\boldsymbol{\sigma}$ is a base of $G \cong \Gamma_{\mathsf{D}}/\Gamma_{\mathsf{F}}$, the elements $deg\, z_1 = \frac{1}{r_1} deg\, c_1$ and $deg\, z_2 = \frac{1}{r_2} deg\, c_2$ generate independent subgroups of $\mathbb{T}(\Gamma_{\mathsf{F}})$; hence, N is a decomposably semiramified graded division algebra by Prop. 8.29. The product formula for crossed products (Prop. 8.25(iii)) yields in $\mathsf{Br}(\mathsf{F})$

$$\mathsf{D} \sim \mathsf{A}(\mathsf{K}/\mathsf{F}, \boldsymbol{\sigma}, b_1 c_1^{-1}, u, b_2 c_2^{-1}) \otimes_{\mathsf{F}} \mathsf{N}.$$

Since $deg\, b_i = deg\, c_i$ $(i = 1, 2)$, the first factor on the right here is an inertial central simple graded algebra, and we have

$$\mathsf{A}(\mathsf{K}/\mathsf{F}, \boldsymbol{\sigma}, b_1 c_1^{-1}, u, b_2 c_2^{-1}) \;=\; \mathcal{A}(\mathsf{D}_0/\mathsf{F}_0, \boldsymbol{\sigma}, b_1 c_1^{-1}, u, b_2 c_2^{-1}) \otimes_{\mathsf{F}_0} \mathsf{F}.$$

Let I be the associated graded division algebra of this algebra. By Prop. 2.52, I is inertial over F and

$$\mathsf{I}_0 \sim \mathcal{A}(\mathsf{D}_0/\mathsf{F}_0, \boldsymbol{\sigma}, b_1 c_1^{-1}, u, b_2 c_2^{-1}) \quad \text{in } Br(\mathsf{F}_0).$$

Therefore, by definition of the specialization coset,

$$sp(\mathsf{D}) \;=\; [\mathcal{A}(\mathsf{D}_0/\mathsf{F}_0, \boldsymbol{\sigma}, b_1 c_1^{-1}, u, b_2 c_2^{-1})] + Dec(\mathsf{D}_0/\mathsf{F}_0).$$

By definition of $\upsilon_{\boldsymbol{\sigma}}$, it follows that

$$\overline{\upsilon}_{\boldsymbol{\sigma}}\big(sp(\mathsf{D})\big) \;=\; u \cdot I_G \mathsf{D}_0^{\times}.$$

This completes the proof since $u = z_1 z_2 z_1^{-1} z_2^{-1}$. □

Example 8.56. *A semiramified but not decomposably semiramified graded division algebra.* Let k be a field containing a primitive n^2-root of unity ω for some integer $n > 1$, and let F be the Laurent polynomial ring in two variables $\mathsf{F} = k[X, X^{-1}, Y, Y^{-1}]$, graded as usual with $\Gamma_{\mathsf{F}} = \mathbb{Z} \times \mathbb{Z}$ and $\mathsf{F}_{(i,j)} = k\, X^i Y^j$ for every $(i,j) \in \mathbb{Z} \times \mathbb{Z}$; so, F is a graded field. Take any a, $b \in k^{\times}$ such that the field $K = k(\sqrt[n]{a}, \sqrt[n]{b}\,)$ has degree n^2 over k. Let D be the following graded symbol algebra of degree n^2 over F:

$$\mathsf{D} \;=\; \big(aX^n, bY^n/\mathsf{F}\big)_{\omega,n^2}.$$

We will show that D is a semiramified graded division algebra, and that D is decomposably semiramified if and only if $\big[(a, b/k)_{\omega,n^2}\big] \in \mathit{Dec}(K/k)$.

Let i, j be the standard generators of D, such that $i^{n^2} = aX^n$, $j^{n^2} = bY^n$, and $ij = \omega ji$. Then $i^n X^{-1}$ and $j^n Y^{-1}$ commute and generate in D_0 a subfield that we identify with K, hence $[\mathsf{D}_0{:}k] \geq n^2$. On the other hand, the grade set of a graded symbol algebra is determined in (2.9): We have

$$\Gamma_{\mathsf{D}} \;=\; \Gamma_{\mathsf{D}}^{\times} \;=\; \big\langle \tfrac{1}{n^2}\, \mathit{deg}(aX^n), \tfrac{1}{n^2}\, \mathit{deg}(bY^n) \big\rangle + \Gamma_{\mathsf{F}} \;=\; \tfrac{1}{n}\mathbb{Z} \times \tfrac{1}{n}\mathbb{Z},$$

and the first equation implies $[\mathsf{D}{:}\mathsf{F}] = [\mathsf{D}_0{:}k] \cdot |\Gamma_{\mathsf{D}}{:}\Gamma_{\mathsf{F}}|$ by Prop. 2.48. Therefore, the inequality $[\mathsf{D}_0{:}k] \geq n^2$ cannot be strict, hence $\mathsf{D}_0 = K$. As D_0 is a field and $\Gamma_{\mathsf{D}} = \Gamma_{\mathsf{D}}^{\times}$, Prop. 2.4 shows that D is a graded division algebra. Moreover, $[\mathsf{D}_0{:}k] = \mathit{deg}\, \mathsf{D}$; hence, D is semiramified.

Now, observe that $\mathcal{G}(\mathsf{D}_0/k)$ is bicyclic, so we may apply Prop. 8.55: A base of $G = \mathcal{G}(K/k)$ is given by $\boldsymbol{\sigma} = (\sigma_1, \sigma_2)$ where

$$\sigma_1 \;=\; \mathit{int}(j^{-1})|_K \;=\; \theta_{\mathsf{D}}\big((0, -\tfrac{1}{n})\big), \qquad \sigma_2 \;=\; \mathit{int}(i)|_K \;=\; \theta_{\mathsf{D}}\big((\tfrac{1}{n}, 0)\big).$$

(Compare Ex. 8.24.) By Prop. 8.55, we have

$$\overline{\upsilon}_{\boldsymbol{\sigma}}\big(sp(D)\big) \;=\; j^{-1} i j i^{-1} \cdot I_G K^{\times} \;=\; \omega \cdot I_G K^{\times} \;\in\; \widehat{H}^{-1}(G, K^{\times}).$$

On the other hand, Ex. 8.24 yields $\upsilon_{\boldsymbol{\sigma}}\big((a, b/k)_{\omega,n^2}\big) = \omega \cdot I_G K^{\times}$. As $\overline{\upsilon}_{\boldsymbol{\sigma}}$ is bijective, it follows that

$$sp(D) \;=\; [\,(a, b/k\,)_{\omega,n^2}\,] + \mathit{Dec}(K/k) \;\in\; \mathit{Br}(K/k)/\mathit{Dec}(K/k). \tag{8.33}$$

Corollary 8.53 then shows that D is decomposably semiramified if and only if $[\,(a, b/k\,)_{\omega,n^2}\,] \in \mathit{Dec}(K/k)$.

Alternatively, (8.33) can be obtained directly from the following decomposition of D, which is derived from the identities for graded symbol algebras:

$$\mathsf{D} \sim \big(a, b/\,\mathsf{F}\,\big)_{\omega,n^2} \otimes_{\mathsf{F}} \mathsf{N}$$

with $\mathsf{N} = \big(a, Y/\mathsf{F}\big)_{\omega^n,n} \otimes_{\mathsf{F}} \big(X, b/\mathsf{F}\big)_{\omega^n,n}$ decomposably semiramified. Details are left to the reader.

For a specific example where D is not decomposably semiramified, one could take $k = k'(a,b)$, where k' is any field containing a primitive n^2-root of unity and a and b are algebraically independent over k'. Then, $A = \big(a, b\,/k\big)_{\omega,n^2}$ is a division algebra of exponent n^2. One can see this by verifying that b^i is not a norm from $k(\sqrt[n^2]{a}\,)$ to k for $i = 1, 2, \ldots, n^2-1$. Alternatively, note that $A' = A \otimes_k k'((a))((b))$ is a division algebra totally ramified over $k'((a))((b))$ and

$$n^2 = deg\,A \geq exp\,A \geq exp\,A' = exp\,\big(\Gamma_{A'}/\Gamma_{k'((a))((b))}\big) = n^2,$$

cf. Prop. 7.75. However, since $Dec(K/k) = Br(k(\sqrt[n]{a}\,)/k) \cdot Br(k(\sqrt[n]{b}\,)/k)$, every algebra class in $Dec(K/k)$ has exponent dividing n (see (8.12)). Hence, $[A] \notin Dec(K/k)$ and therefore D is not decomposably semiramified.

We conclude this subsection with some remarks on the functoriality of the $\mathsf{I} \otimes \mathsf{N}$ decomposition. We observed in Prop. 6.25 that if E is an arbitrary graded field extension of F and D is an inertially split graded division algebra with center F, then the graded division algebra D_{E} associated to $\mathsf{D} \otimes_{\mathsf{F}} \mathsf{E}$ is inertially split. The same proposition also shows that D_{E} is inertial if D is inertial. However, if D is decomposably semiramified, it does *not* follow that D_{E} is decomposably semiramified; it may not even be semiramified, as the following example shows: Use the notation of Ex. 8.56 and let $\mathsf{D} = \big(X, b/\mathsf{F}\big)_n$ and $\mathsf{E} = \mathsf{F}(\sqrt[n]{a^{-1}X}\,)$. Then $\mathsf{D}_{\mathsf{E}} = \big(a, b/\mathsf{E}\big)_n$, and this algebra is inertial over E. The $\mathsf{I}\otimes\mathsf{N}$ decomposition as in (8.30) is therefore not functorial under arbitrary graded field extension. However, it is functorial under inertial extensions, by virtue of the following proposition:

Proposition 8.57. *Let N be a decomposably semiramified central graded division algebra over a graded field F, and let E be a graded field extension of F with $\Gamma_{\mathsf{N}} \cap \Gamma_{\mathsf{E}} = \Gamma_{\mathsf{F}}$. Then, the graded division algebra N_{E} over E associated to $\mathsf{N} \otimes_{\mathsf{F}} \mathsf{E}$ is decomposably semiramified.*

Proof. N_{E} is inertially split since N is inertially split and the scalar extension map $\mathsf{Br}(\mathsf{F}) \to \mathsf{Br}(\mathsf{E})$ maps $\mathsf{Br}_{is}(\mathsf{F})$ into $\mathsf{Br}_{is}(\mathsf{E})$ as noted preceding Prop. 6.25. Since N is decomposably semiramified, by Prop. 8.33(c) it has a maximal graded subfield T that is totally ramified over F. So, $\Gamma_{\mathsf{T}} \subseteq \Gamma_{\mathsf{N}}$; hence, $\Gamma_{\mathsf{T}} \cap \Gamma_{\mathsf{E}} = \Gamma_{\mathsf{F}}$. Let $\mathsf{T}\cdot\mathsf{E}$ be a compositum of T and E in a graded algebraic closure of E. Then, $\Gamma_{\mathsf{T}\cdot\mathsf{E}} \supseteq \Gamma_{\mathsf{T}} + \Gamma_{\mathsf{E}}$ and

$$\begin{aligned}[\mathsf{T}{:}\mathsf{F}] \geq [\mathsf{T}\cdot\mathsf{E} : \mathsf{E}] \geq |\Gamma_{\mathsf{T}\cdot\mathsf{E}}{:}\Gamma_{\mathsf{E}}| &\geq |\Gamma_{\mathsf{T}} + \Gamma_{\mathsf{E}} : \Gamma_{\mathsf{E}}| \\ &= |\Gamma_{\mathsf{T}} : \Gamma_{\mathsf{T}} \cap \Gamma_{\mathsf{E}}| = |\Gamma_{\mathsf{T}}{:}\Gamma_{\mathsf{F}}| = [\mathsf{T}{:}\mathsf{F}].\end{aligned}$$

Hence, equality holds throughout, showing that the graded field $\mathsf{T}\cdot\mathsf{E}$ is totally ramified over E. Thus, N_{E} is inertially split and also split by the totally ramified extension $\mathsf{T}\cdot\mathsf{E}$ of E; hence, N_{E} is decomposably semiramified, by Prop. 8.33(d). □

From Prop. 8.57, we can readily see that the specialization coset construction is functorial for inertial extensions.

Corollary 8.58. *Let* D *be an inertially split graded division algebra with center* F, *and let* E *be an inertial graded field extension of* F. *Let* D_E *be the graded division algebra over* E *associated to* $\mathsf{D} \otimes_\mathsf{F} \mathsf{E}$. *Then,*

$$sp(\mathsf{D}_\mathsf{E}) = ext_{\mathsf{E}_0/\mathsf{F}_0}(sp(\mathsf{D})) \in Br(\mathsf{E}_0)/Dec(\mathsf{E}_0 \cdot Z(\mathsf{D}_0)\,/\mathsf{E}_0).$$

Proof. Recall from Prop. 2.58 that $Z(\mathsf{D}_\mathsf{E}) = \mathsf{E}_0 \cdot Z(\mathsf{D}_0)$. Let I and N be central graded division algebras over F with I inertial and N decomposably semiramified such that $\mathsf{D} \sim \mathsf{I} \otimes_\mathsf{F} \mathsf{N}$, and let I_E, N_E be the graded division algebras over E associated respectively to $\mathsf{I} \otimes_\mathsf{F} \mathsf{E}$ and to $\mathsf{N} \otimes_\mathsf{F} \mathsf{E}$. We have $\mathsf{D}_\mathsf{E} \sim \mathsf{I}_\mathsf{E} \otimes_\mathsf{E} \mathsf{N}_\mathsf{E}$ with I_E inertial by Prop. 6.25, and N_E decomposably semiramified by Prop. 8.57, so

$$sp(\mathsf{D}_\mathsf{E}) = \big[(\mathsf{I}_\mathsf{E})_0\big] + Dec(\mathsf{E}_0 \cdot Z(\mathsf{D}_0)\,/\mathsf{E}_0).$$

By Prop. 2.58 we have $(\mathsf{I}_\mathsf{E})_0 \sim \mathsf{I}_0 \otimes_{\mathsf{F}_0} \mathsf{E}_0$. Since $sp(\mathsf{D})$ is the coset represented by $[\mathsf{I}_0]$, the corollary follows. □

8.4.3 The $S \otimes T$ decomposition (valued case)

We now return to the case of a Henselian field F, and describe analogues of the results on graded division algebras obtained in the previous two subsections. We first establish a tensor product decomposition of the Brauer class of any tame division algebra into an inertially split and a totally ramified part.

Proposition 8.59. *Let D be a division algebra tame over its center F. There are division algebras S and T with center F such that S is inertially split, T is tame and totally ramified over F, and*

$$D \sim S \otimes_F T \qquad \textit{in} \quad Br(F).$$

Proof. Because D is tame, the graded division algebra $\mathsf{gr}(D)$ has center $\mathsf{gr}(F)$. By Prop. 8.42 there are graded division algebras A and B with center $\mathsf{gr}(F)$ such that A is inertially split, B is totally ramified over $\mathsf{gr}(F)$, and $\mathsf{gr}(D) \sim \mathsf{A} \otimes_{\mathsf{gr}(F)} \mathsf{B}$ in $\mathsf{Br}(\mathsf{gr}(F))$. Let S and T be respective tame lifts of A and B over F as in §8.1. Then, S is inertially split and T is tame and totally ramified over F since T is defectless over F and $\overline{T} = \mathsf{B}_0 = \mathsf{gr}(F)_0 = \overline{F}$. Also, $D \sim S \otimes_F T$, by Prop. 8.7. □

Theorem 8.60. *Let S and T be central division algebras over F with S inertially split and T tame and totally ramified over F. Let D be the division algebra associated to $S \otimes_F T$. Then,*

$$Z(\overline{D}) = Z(\overline{S})^{\theta_S(\Gamma_S \cap \Gamma_T)} \quad \textit{and} \quad \overline{D} \otimes_{Z(\overline{D})} Z(\overline{S}) \sim \overline{S} \textit{ in } Br(Z(\overline{S})).$$

Moreover, $\Gamma_D = \Gamma_S + \Gamma_T$*, and* θ_D *is determined by* $\Gamma_T = \ker(\theta_D)$ *and commutativity of the diagram:*

$$\begin{array}{ccc} \Gamma_S & \hookrightarrow & \Gamma_D \\ {\scriptstyle \theta_S}\downarrow & & \downarrow{\scriptstyle \theta_D} \\ \mathcal{G}(Z(\overline{S})/\overline{F}) & \xrightarrow{\text{res}} & \mathcal{G}(Z(\overline{D})/\overline{F}) \end{array}$$

Also,

$$\big((\ker \theta_D)/\Gamma_F, a_D\big) \;=\; \big(\Gamma_T/\Gamma_F, \overline{c}_T\big) \qquad \text{as symplectic modules.}$$

Proof. This follows at once from Th. 8.43 since D is tame over F and $\mathsf{gr}(D) \sim \mathsf{gr}(S) \otimes_{\mathsf{gr}(F)} \mathsf{gr}(T)$ in $\mathsf{Br}(\mathsf{gr}(F))$ with $\mathsf{gr}(S)$ inertially split, $\Gamma_{\mathsf{gr}(S)} = \Gamma_S$, $S_0 = \overline{S}$, $\theta_{\mathsf{gr}(S)} = \theta_S$, and $\mathsf{gr}(T)$ totally ramified over its center $\mathsf{gr}(F)$, $\Gamma_{\mathsf{gr}(T)} = \Gamma_T$, and $D_0 = \overline{D}$, $\Gamma_{\mathsf{gr}(D)} = \Gamma_D$, $\theta_{\mathsf{gr}(D)} = \theta_D$. Moreover,

$$\big(\Gamma_{\mathsf{gr}(T)}/\Gamma_{\mathsf{gr}(F)}, \overline{c}_{\mathsf{gr}(T)}\big) \;=\; \big(\Gamma_T/\Gamma_F, \overline{c}_T\big)$$

by (7.49), and

$$\big((\ker \theta_{\mathsf{gr}(D)})/\Gamma_{\mathsf{gr}(F)}, a_{\mathsf{gr}(D)}\big) \;=\; \big((\ker \theta_D)/\Gamma_F, a_D\big)$$

by Prop. 8.17(iv). □

Just as in the graded case, we may use Th. 8.60 to obtain an explicit description of the map $\Psi_F\colon Br_{tr}(F) \to {}_{\mu(\overline{F})}\mathbb{T}(\wedge^2\Gamma_F)$ of Th. 7.84. The following corollary can be proved by mimicking the arguments in the proof of Cor. 8.44 or, alternatively, by using Cor. 8.44 and the commutative diagram of Prop. 7.79:

Corollary 8.61. *For any tame central division algebra D over F, we have*

$$\Psi_F([D]) \;=\; \Upsilon_{\Gamma_F}^{-1}\big((\ker \theta_D)/\Gamma_F, a_D\big).$$

We next determine the value group of the division algebra D^j associated to the j-th tensor power $D^{\otimes j}$, by working from the corresponding result in the graded case. We are still assuming that the valuation on F is Henselian.

Proposition 8.62. *Let D be a tame division algebra with center F. For any $j \in \mathbb{N}$, we have $\Gamma_{D^j} = j\Gamma_D + \Gamma_F$.*

Proof. Since the map $\mathsf{gr}\colon Br_{tr}(F) \to \mathsf{Br}\big(\mathsf{gr}(F)\big)$ is a homomorphism (see Prop. 6.57), we have $\mathsf{gr}(D^j) \cong \mathsf{gr}(D)^j$, so $\Gamma_{D^j} = \Gamma_{\mathsf{gr}(D)^j}$. Likewise, $\Gamma_D = \Gamma_{\mathsf{gr}(D)}$ and $\Gamma_F = \Gamma_{\mathsf{gr}(F)}$, so the proposition readily follows from Prop. 8.47. □

There are also explicit formulas for $Z(\overline{D^j})$ and θ_{D^j}, given in Exercise 8.19 below.

Corollary 8.63. *Let D be a tame division algebra with center F, and let S, T be division algebras with center F such that $D \sim S \otimes_F T$ in $Br(F)$ with S inertially split and T tame and totally ramified. Then,*

$$exp\ D = lcm\big(exp\ S, exp(\Gamma_D/\Gamma_F)\big).$$

Proof. In view of the isomorphism $\mathsf{gr}\colon Br_{tr}(F) \xrightarrow{\sim} \mathsf{Br}\big(\mathsf{gr}(F)\big)$ of Th. 6.64, the corollary follows from Cor. 8.48. Alternatively, it can be derived from Prop. 8.62 by the same arguments that were used to derive Cor. 8.48 from Prop. 8.47. □

The exponent of an inertially split division algebra over a Henselian field is determined in Th. 8.65(iv) below; see also Exercise 8.17.

8.4.4 Inertially split algebras over a Henselian field

Throughout this subsection, F is a field with a Henselian valuation v. We complete the program sketched in the introduction to this section by describing the inertially split division algebras over F and the decompositions of such algebras up to Brauer-equivalence as a product of inertial and decomposably semiramified division algebras. In view of the index-preserving isomorphism $Br_{is}(F) \cong \mathsf{Br}_{is}(\mathsf{gr}(F))$ of Th. 6.64, it is unsurprising that the theory closely parallels the results for inertially split graded division algebras given in §8.4.2. Indeed, much of what is given here will be deduced from the corresponding graded results.

We first give characterizations of inertially split division algebras over F (cf. Prop. 8.49).

Proposition 8.64. *Let D be a tame central division algebra over a Henselian field F, and let U, C, Z, E be the canonical (up to conjugacy) F-subalgebras of D as in §8.2.2. The following conditions are equivalent:*

(a) *D is inertially split;*
(b) *D contains a maximal subfield that is inertial over F;*
(c) *$C \cong Z$;*
(d) *$ker\ \theta_D = \Gamma_F$;*
(e) *$ind\ D = ind\ \overline{D} \cdot [Z(\overline{D}){:}\overline{F}]$;*
(f) *$\mathsf{gr}(D)$ is an inertially split $\mathsf{gr}(F)$-algebra.*

These conditions hold whenever $dim_{\mathbb{Q}}\, \mathbb{H}(\Gamma_F) = 1$ (in particular, when $\Gamma_F \cong \mathbb{Z}$).

Proof. (a) $\Leftrightarrow$ (f) follows from the bijection $Br_{is}(F) \leftrightarrow \mathsf{Br}_{is}(\mathsf{gr}(F))$ of Th. 6.64, in which $[D] \leftrightarrow [\mathsf{gr}(D)]$. The remaining equivalences and the last sentence here follow from the corresponding results in the graded setting given in Prop. 8.49. For (b) note that if K is a subfield of D which is inertial over F, then $\mathsf{gr}(K)$ is a graded subfield of $\mathsf{gr}(D)$ which is inertial over $\mathsf{gr}(F)$, and $[\mathsf{gr}(K) : \mathsf{gr}(F)] = [K:F]$. Moreover, K is a maximal subfield of D if and only if $\mathsf{gr}(K)$ is a maximal graded subfield of $\mathsf{gr}(D)$, by Cor. 2.36 and the corresponding ungraded result. Conversely, if L is a graded subfield of $\mathsf{gr}(D)$ with L inertial over $\mathsf{gr}(F)$ and K is an inertial lift of L over F (which exists by Cor. 5.57) then $\mathsf{gr}(K) \cong_g \mathsf{L}$ and K is isomorphic to a subfield of D by Cor. 8.10. For (c) recall that by Prop. 8.17 $\mathsf{gr}(C)$ and $\mathsf{gr}(Z)$ are the canonical subalgebras C and Z of $\mathsf{gr}(D)$. For (d) note that $\theta_D = \theta_{\mathsf{gr}(D)}$. □

From Prop. 8.40 and Remark 6.69 above, it follows that the division algebras in the images of the splitting maps $Hom^c(\mathcal{G}(\overline{F}), \mathbb{T}(\Gamma_F)) \to Br_{is}(F)$ are all decomposably semiramified. Therefore, by Th. 6.66 any inertially split division algebra D with center F has a decomposition

$$D \sim I \otimes_F N, \tag{8.34}$$

where I and N are division algebras with $[I] \in Br_{in}(F)$ (so I is inertial over F by (6.34)) and N is decomposably semiramified. This can also be seen by taking tame lifts in the decomposition $\mathsf{gr}(D) \sim \mathsf{I} \otimes_{\mathsf{gr}(F)} \mathsf{N}$ given in (8.30) where I and N are graded division algebras with I inertial over $\mathsf{gr}(F)$ and N decomposably semiramified. On the other hand, every division algebra D associated to a tensor product $I \otimes_F N$ as above is inertially split, since I and N are inertially split. Parallel to the graded case, significant properties of D can be read off from those of I and N:

Theorem 8.65. *Let D be any inertially split division algebra with center F. Let I be any inertial division algebra with center F and N any decomposably semiramified division algebra with center F such that D is the associated division algebra of $I \otimes_F N$,*

$$D \sim I \otimes_F N \quad \textit{in } Br(F).$$

Then,

(i) $\chi_D = \chi_N$; *hence* $\Gamma_D = \Gamma_N$, $Z(\overline{D}) = Z(\overline{N}) = \overline{N}$ *and* $\theta_D = \theta_N$.
(ii) $\overline{D}$ *is the associated division algebra of the central simple* $\overline{N}$*-algebra* $\overline{I} \otimes_{\overline{F}} \overline{N}$.
(iii) $ind\, D = ind(\overline{I} \otimes_{\overline{F}} \overline{N}) \cdot |\Gamma_N{:}\Gamma_F|$.
(iv) $exp\ D = lcm\big(exp\ I, exp\ N\big) = lcm\big(exp\ \overline{I}, exp(\Gamma_N/\Gamma_F)\big)$.
(v) *D is semiramified if and only if $\overline{N}$ splits $\overline{I}$.*

Proof. Since $D \sim I \otimes_F N$ with D, I, and N each tame over F, we have

$$\mathsf{gr}(D) \sim \mathsf{gr}(I) \otimes_{\mathsf{gr}(F)} \mathsf{gr}(N) \qquad \text{in } \mathsf{Br}(\mathsf{gr}(F)),$$

with $\mathsf{gr}(I)$ inertial over $\mathsf{gr}(F)$ and $\mathsf{gr}(N)$ decomposably semiramified by Prop. 8.40. (i)–(iv) follow from Th. 8.50 since $\chi_D = \chi_{\mathsf{gr}(D)}$ (see (6.42)), $\Gamma_D = \Gamma_{\mathsf{gr}(D)}$, $\overline{D} = \mathsf{gr}(D)_0$, $ind\, D = ind(\mathsf{gr}(D))$, and $exp\ D = exp(\mathsf{gr}(D))$ by Th. 6.64, and likewise for I, N, and F. Similarly, (v) follows from Th. 8.50(v), since, as D is tame over F, by Prop. 8.38 D is semiramified if and only if $\mathsf{gr}(D)$ is semiramified. □

Just as in the graded case, the variation in the possible $I \otimes N$ decompositions of an inertially split division algebra is measured by a *Dec* group.

Proposition 8.66. *Let D be an inertially split division algebra with center F. Suppose $D \sim I \otimes_F N$ in $\mathrm{Br}(F)$, where I is an inertial division algebra and N is a decomposably semiramified division algebra. Let I' be another inertial division algebra with center F. Then, there is a decomposably semiramified division algebra N' such that $D \sim I' \otimes_F N'$ if and only if $[\overline{I}] \equiv [\overline{I'}]$ (mod $\mathrm{Dec}(Z(\overline{D})/\overline{F})$).*

Proof. Since $D \sim I \otimes_F N$, we have $\mathsf{gr}(D) \sim \mathsf{gr}(I) \otimes_{\mathsf{gr}(F)} \mathsf{gr}(N)$ with $\mathsf{gr}(I)$ inertial and $\mathsf{gr}(N)$ decomposably semiramified, see Prop. 8.40. If $D \sim I' \otimes_F N'$ for some inertial division algebra I' and some decomposably semiramified division algebra N', then

$$\mathsf{gr}(D) \sim \mathsf{gr}(I) \otimes_{\mathsf{gr}(F)} \mathsf{gr}(N) \sim \mathsf{gr}(I') \otimes_{\mathsf{gr}(F)} \mathsf{gr}(N'),$$

so $[\mathsf{gr}(I)_0] \equiv [\mathsf{gr}(I')_0]$ (*mod* $Dec(Z(\mathsf{gr}(D)_0)/\mathsf{gr}(F)_0)$) by Prop. 8.51. This proves the "only if" part, since $\mathsf{gr}(I)_0 = \overline{I}$, $\mathsf{gr}(I')_0 = \overline{I'}$, $\mathsf{gr}(D)_0 = \overline{D}$, and $\mathsf{gr}(F)_0 = \overline{F}$.

Conversely, if I' is an inertial division algebra such that $[\overline{I}] \equiv [\overline{I'}]$ (*mod* $Dec(Z(\overline{D})/\overline{F})$), then Prop. 8.51 shows that $\mathsf{gr}(D) \sim \mathsf{gr}(I') \otimes_{\mathsf{gr}(F)} \mathsf{M}$ for some decomposably semiramified graded division algebra M. Let N' be the tame lift of M over F. Then N' is decomposably semiramified by Prop. 8.40 and $D \sim I' \otimes_F N'$ by Prop. 8.7. □

Definition 8.67. Just as in the graded case (see Def. 8.52), we define the *specialization coset* of an inertially split division algebra D with center F: we set

$$sp(D) = \left\{ [\overline{I}] \;\middle|\; \begin{array}{l} I \text{ is inertial and } D \otimes_F I^{\mathrm{op}} \text{ is associated to a decom-} \\ \text{posably semiramified graded division algebra} \end{array} \right\}$$
$$\in Br(\overline{F})/Dec(Z(\overline{D})/\overline{F}).$$

From the proof of Prop. 8.66, it is clear that

$$sp(D) = sp\big(\mathsf{gr}(D)\big). \tag{8.35}$$

When D is tame semiramified, Th. 8.65(i)–(ii) shows that in any $I \otimes N$ decomposition the algebra $\overline{I}$ is split by $Z(\overline{D}) = \overline{D}$; hence,

$$sp(D) \in Br(\overline{D}/\overline{F})/Dec(\overline{D}/\overline{F}).$$

As in the graded case (see Cor. 8.53), we have:

Corollary 8.68. *An inertially split central division algebra D over F is decomposably semiramified if and only if $sp(D) = \mathrm{Dec}(Z(\overline{D})/\overline{F})$.*

Proof. The algebra D is decomposably semiramified if and only if we can take $I = F$ in the definition of $sp(D)$. □

The following analogues of Ex. 8.54 and Prop. 8.55 are easy consequences of those results, via (8.35).

Proposition 8.69. *Let D be an inertially split central division algebra over F.*

(i) *If Γ_D/Γ_F is cyclic, then*

$$sp(D) = [A] + Dec(Z(\overline{D})/\overline{F})$$

for any central simple $\overline{F}$-algebra A such that $A \otimes_{\overline{F}} Z(\overline{D}) \sim \overline{D}$.

(ii) *Suppose D is semiramified and Γ_D/Γ_F is bicyclic. Take any $z_1, z_2 \in D^\times$ such that $\theta_D(v(z_1)), \theta_D(v(z_2))$ is a base $\boldsymbol{\sigma}$ of $\mathcal{G}(\overline{D}/\overline{F})$, then*

$$\overline{v}_{\boldsymbol{\sigma}}(sp(D)) = \overline{z_1 z_2 z_1^{-1} z_2^{-1}} \cdot I_{\mathcal{G}(\overline{D}/\overline{F})}\overline{D}^\times \in \widehat{H}^{-1}(\mathcal{G}(\overline{D}/\overline{F}), \overline{D}^\times).$$

Another analogy with the graded case is that the specialization coset construction is functorial with respect to inertial field extensions:

Proposition 8.70. *Let N be any decomposably semiramified division algebra with center F, let (K, w) be any Henselian-valued field extending (F, v), and let N_K be the associated division algebra of $N \otimes_F K$. If $\Gamma_K \cap \Gamma_N = \Gamma_F$, then N_K is decomposably semiramified.*

Proof. Proposition 6.59 shows that N_K is tame over K, as N is tame over F; it also shows that $\mathsf{gr}(N_K)$ is the graded division algebra associated to $\mathsf{gr}(N) \otimes_{\mathsf{gr}(F)} \mathsf{gr}(K)$. Since N is decomposably semiramified, by Prop. 8.40 $\mathsf{gr}(N)$ is a decomposably semiramified graded division algebra with center $\mathsf{gr}(F)$. Moreover,

$$\Gamma_{\mathsf{gr}(K)} \cap \Gamma_{\mathsf{gr}(N)} = \Gamma_K \cap \Gamma_N = \Gamma_F = \Gamma_{\mathsf{gr}(F)}.$$

Therefore, the graded version of our proposition, Prop. 8.57, shows that $\mathsf{gr}(N_K)$ is a semiramified graded division algebra. Since N_K is tame, another application of Prop. 8.40 shows that N_K is decomposably semiramified. □

Proposition 8.71. *Let D be an inertially split central division algebra over F, and let E/F be an inertial field extension. Let D_E be the central division algebra over E associated to $D \otimes_F E$. Then,*

$$sp(D_E) = ext_{\overline{E}/\overline{F}}(sp(D)) \in Br(\overline{E})/Dec(\overline{E} \cdot Z(\overline{D})/\overline{E}).$$

Proof. We have $sp(D) = sp(\mathsf{gr}(D))$ and $sp(D_E) = sp(\mathsf{gr}(D_E))$ by (8.35). But $\mathsf{gr}(D_E) \sim \mathsf{gr}(D) \otimes_{\mathsf{gr}(F)} \mathsf{gr}(E)$ by Prop. 6.59, so Cor. 8.58 yields

$$sp(\mathsf{gr}(D_E)) = ext_{\overline{E}/\overline{F}}(sp(\mathsf{gr}(D))).$$

The proposition follows. □

The connections between D, I, and N, shown in Th. 8.65, provide a toolkit for building inertially split division algebras D with desired properties from selected inertially split and decomposably semiramified division algebras. We know the subalgebras of D inertial over F: they are precisely the inertial lifts of the F-subalgebras of $\overline{D} \sim \overline{I} \otimes_{\overline{F}} \overline{N}$ that are separable over $\overline{F}$. But the influence of N is often to "rigidify" D and to limit the possible subalgebras of D that have some ramification over F. See, e.g., Prop. 9.12, Th. 9.24, and Th. 10.26.

When $D \sim I \otimes_F N$ as above and D itself is semiramified, Th. 8.65(v) shows that $\overline{N}$ splits $\overline{I}$, so $\overline{D} = \overline{N}$. Thus, I is not directly visible within D. Nonetheless, I can have significant influence on the structure of D, as will be illustrated in Th. 9.27 and Prop. 9.37 below.

8.4.5 Example: the $\mathcal{S}$ construction

We construct in this subsection a typical example of an inertially split division algebra, which will be used in several examples in Ch. 9, Ch. 10, and Ch. 12.

Let A be a central simple algebra over an arbitrary field k, and let $\ell_1, \ldots, \ell_n$ be linearly disjoint cyclic extensions of k. Thus, the tensor product

$$\ell = \ell_1 \otimes_k \ldots \otimes_k \ell_n$$

is a field. For $i = 1, \ldots, n$, let $r_i = [\ell_i{:}k]$ and let σ_i be a generator of $\mathcal{G}(\ell_i/k)$. Let also $t_1, \ldots, t_n$ be independent indeterminates over k. Consider the fields

$$F' = k(t_1, \ldots, t_n) \qquad \text{and} \qquad F = k((t_1)) \ldots ((t_n)).$$

We form the following tensor products of cyclic algebras over F' and F respectively:

$$\begin{aligned} N' &= \big(\ell_1 \otimes_k F', \sigma_1 \otimes id, t_1\big) \otimes_{F'} \ldots \otimes_{F'} \big(\ell_n \otimes_k F', \sigma_n \otimes id, t_n\big), \\ N &= \big(\ell_1 \otimes_k F, \sigma_1 \otimes id, t_1\big) \otimes_F \ldots \otimes_F \big(\ell_n \otimes_k F, \sigma_n \otimes id, t_n\big). \end{aligned}$$

Define

$$\mathcal{S}(A; \ell_1, \ldots, \ell_n) = \text{the associated division algebra of } A \otimes_k N, \qquad (8.36)$$

and

$$\mathcal{S}'(A; \ell_1, \ldots, \ell_n) = \text{the associated division algebra of } A \otimes_k N'. \qquad (8.37)$$

(Note that these division algebras also depend on the choice of the automorphisms $\sigma_1, \ldots, \sigma_n$.)

The field F is Henselian for the $(t_1, \ldots, t_n)$-adic valuation, with $\overline{F} = k$ and $\Gamma_F = \mathbb{Z}^d$ ordered lexicographically from right to left. It is the center of $\mathcal{S}(A; \ell_1, \ldots, \ell_n)$. Since the field extensions $\ell_1, \ldots, \ell_n$ are linearly disjoint over k and the subgroups of $\mathbb{T}(\Gamma_F)$ generated by $\frac{1}{r_1}v(t_1), \ldots, \frac{1}{r_n}v(t_n)$ are independent of order $r_1, \ldots, r_n$ respectively, it follows from Prop. 8.41 that N is a decomposably semiramified division algebra over F, with

$$\overline{N} = \ell \quad \text{and} \quad \Gamma_N = \left(\tfrac{1}{r_1}\mathbb{Z}\right) \times \ldots \times \left(\tfrac{1}{r_n}\mathbb{Z}\right).$$

If D is the central division algebra over k associated to A, the division algebra $D \otimes_k F$ is inertial, and we have in $Br(F)$

$$\mathcal{S}(A; \ell_1, \ldots, \ell_n) \sim (D \otimes_k F) \otimes_F N.$$

This algebra is inertially split since $D \otimes_k F$ and N are each inertially split. We may thus apply Th. 8.65 with $I = D \otimes_k F$, and we obtain:

Proposition 8.72. *Let* $\mathcal{S} = \mathcal{S}(A; \ell_1, \ldots, \ell_n)$. *We have*

$$\Gamma_{\mathcal{S}} = \Gamma_N = \left(\tfrac{1}{r_1}\mathbb{Z}\right) \times \ldots \times \left(\tfrac{1}{r_n}\mathbb{Z}\right) \quad \text{and} \quad \overline{\mathcal{S}} \sim D \otimes_k \overline{N} \sim A \otimes_k \ell.$$

Thus, $\mathcal{S}$ *is semiramified if and only if* ℓ *splits* A. *Moreover,*

$$\deg \mathcal{S} = r_1 \ldots r_n \cdot \operatorname{ind}(A \otimes_k \ell) \quad \text{and} \quad \exp \mathcal{S} = \operatorname{lcm}(r_1, \ldots, r_n, \exp A).$$

The specialization coset of $\mathcal{S}$ *in* $Br(k)/Dec(\ell/k)$ *is represented by* $[A]$.

The exponent and index of $\mathcal{S}'(A; \ell_1, \ldots, \ell_n)$ are easily derived:

Corollary 8.73. *Let* $\mathcal{S}' = \mathcal{S}'(A; \ell_1, \ldots, \ell_n)$ *and* $\mathcal{S} = \mathcal{S}(A; \ell_1, \ldots, \ell_n)$. *Then,* $\deg \mathcal{S}' = \deg \mathcal{S}$ *and* $\exp \mathcal{S}' = \exp \mathcal{S}$. *Hence,* $\mathcal{S} \cong \mathcal{S}' \otimes_{F'} F$.

Proof. We have $\mathcal{S} \sim \mathcal{S}' \otimes_{F'} F$ in $Br(F)$; hence,

$$\deg \mathcal{S} \mid \deg \mathcal{S}' \quad \text{and} \quad \exp \mathcal{S} \mid \exp \mathcal{S}'.$$

Therefore, by Prop. 8.72 it suffices to show that

$$\deg \mathcal{S}' \leq r_1 \ldots r_n \operatorname{ind}(A \otimes_k \ell) \quad \text{and} \quad \exp \mathcal{S}' \mid \operatorname{lcm}(r_1, \ldots, r_n, \exp A).$$

The latter is clear because each cyclic algebra $\left(\ell_i \otimes_k F', \sigma_i \otimes \mathrm{id}, t_i\right)$ has exponent dividing r_i, and because in $Br(F')$

$$\mathcal{S}' \sim A \otimes_k \left(\ell_1 \otimes_k F', \sigma_1 \otimes \mathrm{id}, t_1\right) \otimes_{F'} \ldots \otimes_{F'} \left(\ell_n \otimes_k F', \sigma_n \otimes \mathrm{id}, t_n\right).$$

To bound $\deg \mathcal{S}'$, observe that

$$\mathcal{S}' \otimes_k \ell \sim A \otimes_k \ell \otimes_k F';$$

hence,

$$\operatorname{ind}(\mathcal{S}' \otimes_k \ell) = \operatorname{ind}(A \otimes_k \ell).$$

On the other hand, by definition of the matrix size we have

$$\deg(\mathcal{S}' \otimes_k \ell) = \operatorname{ind}(\mathcal{S}' \otimes_k \ell) \cdot \operatorname{ms}(\mathcal{S}' \otimes_k \ell).$$

Since the left term equals $\deg \mathcal{S}'$, it follows that

$$\deg \mathcal{S}' = \operatorname{ind}(A \otimes_k \ell) \cdot \operatorname{ms}(\mathcal{S}' \otimes_k \ell).$$

Now, the ungraded version of Lemma 8.8 yields

$$ms(\mathcal{S}' \otimes_k \ell) \leq [\ell : k] = r_1 \dots r_n,$$

so the proof of the degree formula is complete. The isomorphism $\mathcal{S} \cong \mathcal{S}' \otimes_{F'} F$ follows from the equality of degrees, since $\mathcal{S} \sim \mathcal{S}' \otimes_{F'} F$ in $Br(F)$. □

Alternatively, one can show that

$$\mathcal{S}' \cong q(\mathsf{gr}(\mathcal{S}))$$

(see Exercise 8.21). This provides another way to prove the degree and exponent formulas of the corollary.

Remark 8.74. By specializing the indeterminates $t_1, \dots, t_n$ to arbitrary elements in $k^\times$, we may specialize N' into any Brauer class in $Dec(\ell/k)$, hence the Brauer class of $\mathcal{S}'(A; \ell_1, \dots, \ell_n)$ is specialized to any Brauer class in the coset represented by $[A]$ in $Br(k)/Dec(\ell/k)$. This explains the term "specialization coset" for $sp\big(\mathcal{S}(A; \ell_1, \dots, \ell_n)\big)$.

8.4.6 The case of discrete rank 1 valuations

In this subsection, we still assume (F, v) is a Henselian-valued field, but we now assume in addition that v is discrete of rank 1, i.e., $\Gamma_F = \mathbb{Z}$. This hypothesis entails significant simplifications, which make it worthwhile to restate the main results obtained in this chapter for this special case.

Proposition 8.75. *Suppose F is Henselian and $\Gamma_F = \mathbb{Z}$, and let D be a central division algebra over F. Then D is defectless, and if $\overline{D} = \overline{F}$, then $D = F$. Moreover, the following conditions are equivalent:*

(a) *the field extension $Z(\overline{D})/\overline{F}$ is separable;*
(b) *D is tame;*
(c) *D is inertially split;*
(d) *D contains a maximal subfield that is an inertial extension of F.*

These conditions hold whenever $\operatorname{char} \overline{F} \nmid \deg D$; in particular, they hold when $\operatorname{char} \overline{F} = 0$.

Proof. That D is defectless was already observed in Prop. 4.16.

Suppose we have $\overline{D} = \overline{F}$. Then, $|\Gamma_D{:}\Gamma_F| = [D{:}F]$ since D is defectless. But $\Gamma_D \cong \mathbb{Z}$ since $\Gamma_F = \mathbb{Z}$, hence if $x \in D$ is chosen so that $v(x)$ generates Γ_D, we have

$$[D{:}F] \geq [F(x){:}F] \geq |\Gamma_{F(x)}{:}\Gamma_F| = |\Gamma_D{:}\Gamma_F| = [D{:}F].$$

Therefore, $D = F(x)$, so D is commutative. Then $D = F$ since D is central over F.

Since Γ_D/Γ_F is a cyclic group, its subgroup $ker(\theta_D)/\Gamma_F$ cannot carry a nondegenerate alternating pairing if it is not trivial, hence $ker(\theta_D) = \Gamma_F$ by Prop. 8.17(iv). The tameness condition in Def. 8.4 then reduces to the separability of $Z(\overline{D})/\overline{F}$, so we have (a) $\Leftrightarrow$ (b). The implications (b) $\Rightarrow$ (c) $\Leftrightarrow$ (d) are proved in Prop. 8.64, and (c) $\Rightarrow$ (b) is clear since a division algebra is tame when it is split by a tame field extension, see Prop. 6.63. (Alternatively, (b) $\Leftrightarrow$ (c) is clear from Prop. 6.70.) □

Under the hypotheses of Prop. 8.75, when D is tame its canonical subalgebras U, C, Z, E of §8.2.2 satisfy $E = U$ and $C = Z$ since D is inertially split, so the inclusion diagrams (8.10) reduce to

$$\begin{array}{ccccc}
\Gamma_D & \qquad & D & \qquad & \overline{D} \\
| & & | & & \| \\
\Gamma_E & & E & & \overline{E} \\
\| & & | & & | \\
\Gamma_Z & & Z & & \overline{Z} \\
\| & & | & & | \\
\Gamma_F & & F & & \overline{F}
\end{array} \tag{8.38}$$

Here, E is an inertial lift of $\overline{D}$ over F, and Z is its center, which is an inertial lift of $Z(\overline{D})$. The canonical map θ_D induces an isomorphism

$$\overline{\theta}_D \colon \Gamma_D/\Gamma_F \xrightarrow{\sim} \mathcal{G}(Z(\overline{D})/\overline{F})$$

(see Prop. 8.17), so $|\Gamma_D{:}\Gamma_E| = [\overline{Z}{:}\overline{F}]$.

Corollary 8.76. *Suppose F is Henselian and $\Gamma_F = \mathbb{Z}$. For a tame central division algebra D over F the following conditions are equivalent:*

(a) $E = Z$ *in* (8.38);
(b) $[\overline{D}{:}\overline{F}] = |\Gamma_D{:}\Gamma_F|$;
(c) D *is semiramified;*
(d) D *is cyclically semiramified;*
(e) D *is decomposably semiramified.*

Proof. The equivalence (a) $\Leftrightarrow$ (b) is clear because

$$[E{:}Z] = [\overline{E}{:}\overline{Z}], \quad |\Gamma_D{:}\Gamma_F| = [\overline{Z}{:}\overline{F}], \quad \text{and} \quad [\overline{D}{:}\overline{F}] = [\overline{E}{:}\overline{Z}] \cdot [\overline{Z}{:}\overline{F}].$$

We have (d) $\Rightarrow$ (e) $\Rightarrow$ (c) $\Rightarrow$ (b) from the definitions, so it only remains to prove (a) $\Rightarrow$ (d).

Since D is tame, we already know from Prop. 8.75 that it is inertially split. Assuming (a), we have $[\overline{D}{:}\overline{F}] = |\Gamma_D{:}\Gamma_F| = deg\, D$ and $\overline{D} = \overline{Z}$, so $\overline{D}$ is a field. Since moreover Γ_D/Γ_F is a cyclic group, (d) follows. □

If D is a central division algebra over F and if D is tame, hence inertially split, by Prop. 8.75, the extension $Z(\overline{D})/\overline{F}$ is cyclic, hence

$$Dec(Z(\overline{D})/\overline{F}) = Br(Z(\overline{D})/\overline{F}) \ \big(= ker\big(ext_{Z(\overline{D})/\overline{F}} \colon Br(\overline{F}) \to Br(Z(\overline{D}))\big)\big).$$

Therefore, the specialization coset $sp(D)$ is just the preimage of $[\overline{D}]$ in $Br(\overline{F})$ under $ext_{Z(\overline{D})/\overline{F}}$.

In light of Prop. 8.75, the Witt sequence of Th. 6.71 can be restated, for any Henselian field F with $\Gamma_F = \mathbb{Z}$, as a split exact sequence:

$$0 \longrightarrow Br(\overline{F}) \longrightarrow Br_{tr}(F) \xrightarrow{\ \chi\ } Hom^c\big(\mathcal{G}(\overline{F}), \mathbb{Q}/\mathbb{Z}\big) \longrightarrow 0. \tag{8.39}$$

To describe a splitting of this exact sequence, choose and fix a uniformizer π for $\mathcal{O}_F$, i.e., a nonzero element of $F^\times$ with $v(\pi) = 1$. For any $\psi \in Hom^c\big(\mathcal{G}(\overline{F}), \mathbb{Q}/\mathbb{Z}\big)$, let L be the fixed field $\overline{F}^{ker\,\psi}$ and let

$$n = [L{:}\overline{F}] = |\,im\,\psi| < \infty.$$

Take any $\tau \in \mathcal{G}(\overline{F}_{sep})$ satisfying $\psi(\tau) = \frac{1}{n} + \mathbb{Z} \in \mathbb{Q}/\mathbb{Z}$. Let K be the inertial lift of L over F. By Cor. 5.57 and Cor. 5.53, K is unique up to isomorphism and, as L is Galois over $\overline{F}$, K is Galois over F with $\mathcal{G}(K/F) \cong \mathcal{G}(L/\overline{F}) \cong \langle \tau|_L \rangle$. Choose a generator σ of $\mathcal{G}(K/F)$ mapping to $\tau|_L$ in $\mathcal{G}(L/\overline{F})$, and let N_ψ be the cyclic F-algebra

$$N_\psi = \big(K/F, \sigma, \pi\big). \tag{8.40}$$

Then, N_ψ is a cyclically semiramified division algebra with $\overline{N_\psi} = \overline{K} = L$, $\Gamma_{N_\psi} = \frac{1}{n}\mathbb{Z}$, and ramification character $\chi_{N_\psi} = \psi$. The map

$$Hom^c\big(\mathcal{G}(\overline{F}), \mathbb{Q}/\mathbb{Z}\big) \to Br_{tr}(F) \quad \text{given by} \quad \psi \mapsto [N_\psi]$$

is a group homomorphism splitting the short exact sequence (8.39).

Still assuming F is Henselian with $\Gamma_F = \mathbb{Z}$, now take any division algebra D with center F and D tame over F. Let $n = |\Gamma_D{:}\Gamma_F|$. Pick any $d \in D^\times$ with $v(d) = \frac{1}{n}$, and let $\sigma = \theta_D(\frac{1}{n})$, which is the automorphism of $Z(\overline{D})$ induced by conjugation by d on $\mathcal{O}_D$. Then, θ_D induces the isomorphism $\overline{\theta}_D$ from $\Gamma_D/\Gamma_F = \frac{1}{n}\mathbb{Z}/\mathbb{Z}$ to $\mathcal{G}(\overline{D}/\overline{F})$ determined by $\frac{1}{n} + \mathbb{Z} \mapsto \sigma$. The ramification character χ_D is the composition of maps

$$\mathcal{G}(\overline{F}_{sep}) \longrightarrow \mathcal{G}(Z(\overline{D})/\overline{F}) \xrightarrow{\ \overline{\theta}_D^{\,-1}\ } \tfrac{1}{n}\mathbb{Z}/\mathbb{Z} \longrightarrow \mathbb{Q}/\mathbb{Z}.$$

Proposition 8.77. *Let $\psi = \chi_D$ and N_ψ as in (8.40). Then there is a unique inertial division algebra I with center F such that in $Br(F)$*

$$D \sim I \otimes_F N_\psi.$$

Moreover,

(i) $Z(\overline{D}) \cong \overline{N_\psi}$ *and* $\overline{D} \sim \overline{I} \otimes_{\overline{F}} \overline{N_\psi}$;
(ii) $\Gamma_D = \Gamma_{N_\psi} = \frac{1}{n}\mathbb{Z}$ *and* $\theta_D = \theta_{N_\psi}$;
(iii) $exp\,D = lcm\big(exp\,\overline{I}, |\Gamma_D/\Gamma_F|\big)$;

(iv) $ind\, D = ind(\overline{I} \otimes_{\overline{F}} \overline{N_\psi}) \cdot |\Gamma_D/\Gamma_F|$;
(v) $sp(D) = [\overline{I}\,] + Br(Z(\overline{D})/\overline{F}) \in Br(\overline{F})/\,Br(Z(\overline{D})/\overline{F})$.

Proof. Since $\chi_D = \psi = \chi_{N_\psi}$, the existence and uniqueness of I follow from the split exact sequence (8.39). The assertions (i)–(iv) follow from Th. 8.65, and (v) from Def. 8.67, taking into account the equality $Dec(Z(\overline{D})/\overline{F}) = Br(Z(\overline{D})/\overline{F})$. □

Exercises

Exercise 8.1. Let (k, u) be a valued field of characteristic 0 with a residue field of characteristic $p > 0$, and let $F = k(x, y)$, where x and y are independent indeterminates over k. The field F carries a valuation v such that $v(\sum a_{ij}x^iy^j) = min(u(a_{ij}))$ for all $a_{ij} \in k$. It residue field is $\overline{F} = \overline{k}(\overline{x}, \overline{y})$ with $\overline{x}$ and $\overline{y}$ algebraically independent over $\overline{k}$. Let (F_h, v_h) be a Henselization of (F, v) and let $M = F_h(\sqrt[p]{x}, \sqrt[p]{y})$. The field M is defectless and unramified over F_h, but it is not inertial since $\overline{M}$ is a purely inseparable extension of $\overline{F_h} = \overline{k}(\overline{x}, \overline{y})$. We have $\mathsf{gr}(M) = \mathsf{gr}(F)\big[\sqrt[p]{\widetilde{x}}, \sqrt[p]{\widetilde{y}}\,\big]$ with $deg\, \sqrt[p]{\widetilde{x}} = deg\, \sqrt[p]{\widetilde{y}} = 0$, so for $\mathsf{K} = \mathsf{gr}(F)\big[\sqrt[p]{\widetilde{x}} + \sqrt[p]{\widetilde{y}}\,\big]$ we have $\mathsf{K} \subseteq \mathsf{gr}(M)$ and $[\mathsf{K}\colon \mathsf{gr}(F)] = p$. Show that there is no lift of K over F_h in M.

Exercise 8.2. This exercise illustrates the non-uniqueness of lifts in the non-tame setting. Let $\mathbb{Q}_2$ be the 2-adic completion of $\mathbb{Q}$, and let u be the complete discrete 2-adic valuation on $\mathbb{Q}_2$, with $\Gamma_{\mathbb{Q}_2} = \mathbb{Z}$. Let $L = \mathbb{Q}_2(\sqrt{2}, \sqrt{3}\,)$, with the unique extension of u to L again denoted by u.

(i) Show that $u(1 + \sqrt{2} + \sqrt{3}\,) = \frac{3}{4}$. (Hint: Let
$$f \;=\; X^2 - 2(1 + \sqrt{2}\,)X + 2\sqrt{2} \in \mathbb{Q}(\sqrt{2}\,)[X].$$
Show that f is a λ-uniform polynomial for $\lambda = \frac{3}{4}$, and that $1+\sqrt{2}+\sqrt{3}$ is a root of f.)

(ii) Deduce that $[L{:}\mathbb{Q}_2] = 4$ and that L is totally (but not tamely) ramified over $\mathbb{Q}_2$. Deduce that $\mathsf{gr}(L) \cong_g \mathsf{gr}(\mathbb{Q}_2(\sqrt[4]{2}\,))$, even though $L \not\cong \mathbb{Q}_2(\sqrt[4]{2})$ (e.g., L is Galois over $\mathbb{Q}_2$, but $\mathbb{Q}_2(\sqrt[4]{2}\,)$ is not).

(iii) For the three nonisomorphic quadratic extensions of $\mathbb{Q}_2$ within L, prove that $\mathsf{gr}(\mathbb{Q}_2(\sqrt{2}\,)) = \mathsf{gr}(\mathbb{Q}_2(\sqrt{3}\,)) = \mathsf{gr}(\mathbb{Q}_2(\sqrt{6}\,))$ within $\mathsf{gr}(L)$.

Exercise 8.3. Here is a generalization of Exercise 7.8. Let F be a Henselian field, let A be a central simple F-algebra with $char\,\overline{F} \nmid deg\, A$, and let D be the division algebra associated to A. Suppose that A is spanned by an armature $\mathcal{A}$. Let $\overline{\alpha}_{\mathcal{A}}\colon \mathcal{A} \to \Gamma_A^\times/\Gamma_F$ be the map induced by the armature gauge $\alpha_{\mathcal{A}}$, and let a_D be the nondegenerate pairing on $ker(\theta_D)/\Gamma_F$ of Prop. 8.17(iv). Prove that as symplectic modules
$$\overline{\alpha}_{\mathcal{A}\diamond}(\mathcal{A}, b_{\mathcal{A}}) \;=\; (ker(\theta_D)/\Gamma_F, a_D).$$

[Hint: Reduce to the graded case by Prop. 7.47, then extend scalars to the inertial closure of $\mathsf{gr}(F)$ and use Prop. 7.58.]

Exercise 8.4. Let K/F be an elementary abelian Galois field extension of degree 4. Show that $Dec(K/F) = {}_2 Br(K/F)$.

Exercise 8.5. Let F be an algebraic number field, let $n \in \mathbb{Z}$ with $n \geq 2$, and let L_1, L_2 be cyclic Galois field extensions of F that are linearly disjoint over F with $[L_1{:}F] = [L_2{:}F] = n$. Let K be the field $L_1 \otimes_F L_2$, so $\mathcal{G}(K/F) \cong \mathbb{Z}/n\mathbb{Z} \times \mathbb{Z}/n\mathbb{Z}$. For each discrete valuation ring $\mathcal{O}_\mathfrak{p}$ of F, let $v_\mathfrak{p}$ be the associated discrete valuation, and let $v_\mathfrak{P}$ be any extension of $v_\mathfrak{p}$ to K. Let $k_\mathfrak{p}$ be the completion of F with respect to $v_\mathfrak{p}$, and likewise $K_\mathfrak{P}$ and $L_{i,\mathfrak{P}}$ the completions of K and L_i with respect to $v_\mathfrak{P}$. Let $d_\mathfrak{p} = [K_\mathfrak{P}{:}F_\mathfrak{p}]$, and let $s_\mathfrak{p} = lcm\big([L_{1,\mathfrak{P}}{:}F_\mathfrak{p}], [L_{2,\mathfrak{P}}{:}F_\mathfrak{p}]\big)$.

(i) Let $\mathfrak{p}_1, \ldots, \mathfrak{p}_m$ be those $\mathfrak{p}$ such that $v_\mathfrak{p}$ ramifies in K or $char\,\overline{F}^{v_\mathfrak{p}} \,\big|\, n$. Prove that there is an exact sequence

$$1 \longrightarrow Br(K/F)/Dec(K/F) \longrightarrow \prod_{i=1}^{m} \tfrac{1}{d_{\mathfrak{p}_i}}\mathbb{Z}/\tfrac{1}{s_{\mathfrak{p}_i}}\mathbb{Z} \longrightarrow \tfrac{1}{n^2}\mathbb{Z}/\tfrac{1}{n}\mathbb{Z}. \tag{8.41}$$

Thus, $Br(K/F)/Dec(K/F)$ is a finite group.

(ii) Prove that if there are at least two $\mathfrak{p}_i$ with $d_{\mathfrak{p}_i} = n^2$ then the right map in (8.41) is surjective.

(iii) Suppose there is a $\mathfrak{p}$ with $d_\mathfrak{p} = n^2$. Let L be a finite-degree Galois extension of F with $L \subseteq F_\mathfrak{p}$. Prove that $[K{\cdot}L : L] = n^2$ and

$$\big|\, Br(K{\cdot}L\,/L)/Dec(K{\cdot}L\,/L)\big| \geq n^{[L:F]-1}.$$

Exercise 8.6. Here is another example, besides Ex. 8.37, of a semiramified but not tame division algebra. Let (F, v) be a Henselian valued field satisfying $\Gamma_F = \mathbb{Z}$, $char\, F = 0$, $char\,\overline{F} = \overline{p} \neq 0$, F contains a primitive $\overline{p}$-th root of unity, and $\overline{F}$ is not perfect. Take any $u \in \mathcal{O}_F$ with $\overline{u} \notin \overline{F}^{\overline{p}}$, and any $\pi \in \mathcal{O}_F$ with $v(\pi) = 1$, and let S be the symbol algebra of degree $\overline{p}$

$$S = \big(u, \pi\,/F\big)_{\overline{p}}.$$

With the surmultiplicative v-norm α on S built using the standard F-base of S as in §3.2.3, show that (cf. (3.28)) $\mathsf{gr}_\alpha(S)$ is a graded field with

$$\mathsf{gr}_\alpha(S) \cong_g \mathsf{gr}(F)[X, Y]/(X^{\overline{p}} - \overline{u},\, Y^{\overline{p}} - \widetilde{\pi}),$$

where $\mathsf{gr}(F) = \overline{F}[\widetilde{\pi}, \widetilde{\pi}^{-1}]$ with $\widetilde{\pi}$ transcendental over $\overline{F}$. Deduce that $S_0 = \overline{F}(\sqrt[\overline{p}]{\overline{u}})$ and $\Gamma_{\mathsf{gr}(S)} = \frac{1}{\overline{p}}\mathbb{Z}$. Deduce that S is a division algebra over F, α is the extension of v to S, and S is semiramified but not tame.

Exercise 8.7. Let F be a Henselian field. Let A and B be tame division algebras with center F, and let D be the associated division algebra of $A \otimes_F B$. Prove that $\Gamma_D \subseteq \Gamma_A + \Gamma_B$. (Hint: Consider the associated graded algebras.)

The tameness assumption here is not superfluous, as the next exercise illustrates.

Exercise 8.8. Let $\mathbb{Q}_2$ be the 2-adic completion of $\mathbb{Q}$, and let $F = \mathbb{Q}_2((x))((y))$. Let w be the usual Henselian valuation on F with $\Gamma_{F,w} = \mathbb{Z} \times \mathbb{Z}$ with right-to-left lexicographic ordering and $\overline{F}^{w} = \mathbb{Q}_2$. Let u be the complete discrete 2-adic valuation on $\mathbb{Q}_2$, and let $v = u * w$ be the composite valuation on F as in §1.2.3. So, $\overline{F}^{v} = \mathbb{F}_2$ and $\Gamma_{F,v} = \Gamma_{\mathbb{Q}_2,u} \times \Gamma_{F,w} = \mathbb{Z} \times \mathbb{Z} \times \mathbb{Z}$ with right-to-left lexicographic ordering. Then v is Henselian since w and u are Henselian, see Prop. A.15 in Appendix A. Let A and B be the quaternion algebras

$$A = \big(2, x/F\big) \qquad \text{and} \qquad B = \big(3, y/F\big),$$

and let $D = A \otimes_F B$.

(i) Prove that, with respect to w, A and B are cyclically semiramified division algebras and D is a decomposably semiramified division algebra.
(ii) Prove that the extensions v_A and v_B of v to A and B are non-tame gauges. Prove that $v_A \otimes v_B$ is a surmultiplicative v-value function but not a gauge on D. So, $v_A \otimes v_B \neq v_D$ as v-value functions on D.
(iii) Prove that $\Gamma_{D,v_D} = \frac{1}{4}\mathbb{Z} \times \frac{1}{2}\mathbb{Z} \times \frac{1}{2}\mathbb{Z}$. (Hint: Use Exercise 8.2 above.) Deduce that D is totally ramified over F with respect to v and that $\mathit{exp}(\Gamma_{D,v_D}/\Gamma_{F,v}) > \mathit{exp}\, D$.
(iv) Prove that $\Gamma_{D,v_D} \not\subseteq \Gamma_{A,v_A} + \Gamma_{B,v_B}$. Prove also that $\mathsf{gr}_{v_D}(D)$, $\mathsf{gr}_{v_A}(A)$, and $\mathsf{gr}_{v_B}(B)$ are graded fields, but $\mathsf{gr}_{v_D}(D) \not\cong_g \mathsf{gr}_{v_A}(A) \otimes_{\mathsf{gr}_v(F)} \mathsf{gr}_{v_B}(B)$.

Exercise 8.9. Let (F, v) be a Henselian valued field, and let T be a finite-degree field extension of F. If T is totally ramified over F, then T is said to be *of radical type* over F if the surjective homomorphism $T^\times/F^\times \to \Gamma_T/\Gamma_F$ induced by v is split.

(i) Prove that if T is totally and tamely ramified over F, then it is of radical type over F.
(ii) Let $L = \mathbb{Q}_2(\sqrt{2}, \sqrt{3}\,)$, as in Exercise 8.2 above. Prove that L is totally ramified, but not of radical type over $\mathbb{Q}_2$.
(iii) Let T be any finite-degree totally ramified extension of F. Prove that there is a totally ramified extension T' of radical type over F with $\mathsf{gr}(T') \cong_g \mathsf{gr}(T)$.

Exercise 8.10. Let D be the semiramified but not decomposably semiramified graded division algebra of Ex. 8.56. Prove that when n is a prime number D contains no proper totally ramified graded field extensions of F and D is indecomposable, i.e., it has no proper graded subalgebras D' and D'' such that $\mathsf{D} = \mathsf{D}' \otimes_{\mathsf{F}} \mathsf{D}''$.

Exercise 8.11. The example considered here is much like Ex. 8.37, but we start with a graded field rather than a Henselian valued field. Let k be an imperfect field with $\mathit{char}\, k = p \neq 0$, let t be an indeterminate over k, and let $\mathsf{F} = k[t, t^{-1}]$ with its usual grading, where $\Gamma_{\mathsf{F}} = \mathbb{Z}$ and $\mathsf{F}_j = k\,t^j$ for all $j \in \mathbb{Z}$; so, F is a graded field. Let $f = X^p - X - t \in \mathsf{F}[X]$, let s be a root of f in $q(\mathsf{F})_{alg}$, and let $K = \mathsf{F}^\natural[s]$.

(i) Prove that f is irreducible in $q(\mathsf{F})[X]$, and that K is a free $\mathsf{F}^\natural$- module of rank p, and K is a Galois ring extension of $\mathsf{F}^\natural$ with $\mathcal{G}(K/\mathsf{F}^\natural) = \langle\sigma\rangle$, where $\sigma(s) = s+1$. Prove also that the grading on F cannot be extended to K.

(ii) Take any $u \in k \setminus k^p$ and let $A = \big(K/\mathsf{F}^\natural, \sigma, u\big)$, which is a cyclic algebra over the ring $\mathsf{F}^\natural$. Prove that A is an Azumaya algebra over $\mathsf{F}^\natural$, and that $A \otimes_{\mathsf{F}^\natural} q(\mathsf{F})$ is a division ring. Prove also that $[A \otimes_{\mathsf{F}^\natural} q(\mathsf{F})]$ does not lie in the image of the scalar extension map $\mathsf{Br}(\mathsf{F}) \to Br(q(\mathsf{F}))$. Deduce that $[A] \in Br(\mathsf{F}^\natural) \setminus \mathsf{Br}(\mathsf{F})$.

Exercise 8.12. Let T and T' be tame and totally ramified division algebras with center a Henselian field F such that $\Gamma_T = \Gamma_{T'}$ and T and T' have the same canonical pairing. Prove that if $T \equiv T' \ (mod\ \ Br_{in}(F))$ then $T \cong T'$.

Exercise 8.13. Let A be a central simple algebra over a Henselian field F. Assume A is inertially split and let $\chi_A\colon \mathcal{G}(\overline{F}) \to \mathbb{T}(\Gamma_F)$ be the ramification character of (the Brauer class of) A, i.e., the image of $[A]$ under the ramification map χ of (6.41). Show that $deg\, A \geq |im\, \chi_A|$, and that equality holds if and only if A is a tame semiramified division algebra. Prove an analogue of this statement for central simple graded algebras over graded fields. [Hint: Consider the associated division algebra and use (6.40).]

Exercise 8.14. Let D be a division algebra with center a Henselian field F. Prove that $D \sim I \otimes_F B$ in $Br(F)$ where I is inertial over F and B is a division algebra with $\overline{B}$ a field if and only if $\big[\,\overline{D}\,\big] \in im\big(Br(\overline{F}) \to Br(Z(\overline{D}))\big)$.

Exercise 8.15. This exercise gives an example of a tame division algebra D over a Henselian field F such that $\big[\,\overline{D}\,\big] \notin im\big(Br(\overline{F}) \to Br(Z(\overline{D}))\big)$. Let p be a prime number, and let k be a field containing a primitive p^2-root of unity ω. Let K be the iterated Laurent series field $k((a))((b))$; let F be the iterated Laurent series field $K((x))((y))$, and let v be the Henselian (x, y)-adic valuation on F; so, $\overline{F} = K$. Let

$$D = \big(a, x/F\big)_{\omega, p^2} \otimes_F \big(bx^p, y/F\big)_{\omega, p^2}.$$

(i) Prove that D is a division algebra, so v extends to a valuation v_D on D. Prove also that

$$Z(\overline{D}) \cong \overline{F}(\sqrt[p]{a}, \sqrt[p]{b}), \qquad \overline{D} \cong \big(\sqrt[p]{b}, \sqrt[p]{a}\,/\,\overline{F}(\sqrt[p]{a}, \sqrt[p]{b})\big)_{\omega^p, p},$$

$\Gamma_D = \frac{1}{p^2}\mathbb{Z} \times \frac{1}{p^2}\mathbb{Z}$, and $ker\, \theta_D = \frac{1}{p}\mathbb{Z} \times \frac{1}{p}\mathbb{Z}$; hence, D is not inertially split. (Hint: Let $i_1, j_1; i_2, j_2$ be standard generators for the two symbol algebras, and let α be the associated armature v-gauge on D. Show that D_0 is generated over $F_0 = K$ by $\widetilde{i_1}$ and $\widetilde{j_1}^{\,-p}\widetilde{i_2}$, and that

$\Gamma^{\times}_{\mathsf{gr}(D)} = \Gamma_{\mathsf{gr}(D)} = \frac{1}{p^2}\mathbb{Z} \times \frac{1}{p^2}\mathbb{Z}$. Deduce that $\mathsf{gr}(D)$ is a graded division algebra, hence D is a division algebra and $v_D = \alpha$.)

(ii) Let ζ be a p-th root of ω in k_{sep}, and let $K' = K(\zeta) = k(\zeta)((a))((b))$. Let

$$A = \big(b, a \,/\, K'\big)_{\zeta, p^3}.$$

A standard symbol algebra identity shows that in $Br(K'(\sqrt[p]{a}, \sqrt[p]{b}))$,

$$A \otimes_{K'} K'(\sqrt[p]{a}, \sqrt[p]{b}) \sim \overline{D} \otimes_{K(\sqrt[p]{a}, \sqrt[p]{b})} K'(\sqrt[p]{a}, \sqrt[p]{b}).$$

Hence, if $\zeta \in k$, then $K' = K = \overline{F}$, so

$$\big[\overline{D}\big] = \big[A \otimes_K Z(\overline{D})\big] \in im\big(Br(\overline{F}) \to Br(Z(\overline{D}))\big).$$

However, if $\zeta \notin k$, show that $\big[\overline{D}\big] \notin im\big(Br(\overline{F}) \to Br(Z(\overline{D}))\big)$. (Hint: If $\big[\overline{D}\big] \in im\big(Br(\overline{F}) \to Br(Z(\overline{D}))\big)$, then there exist $[B] \in Br(K)$ and $[C] \in Br(K'(\sqrt[p]{a}, \sqrt[p]{b})/K')$ with $[A] = [B \otimes_K K'] + [C]$ in $Br(K')$. Let K'_{in} (resp. K_{in}) be the inertial closure of K' (resp. K) with respect to the Henselian (a, b)-adic valuation. Prove that $exp(A \otimes_{K'} K'_{in}) = p^3$, while $exp(B \otimes_K K'_{in}) \mid exp(B \otimes_K K_{in}) \mid p^2$ as $\zeta \notin K$ (see Th. 7.84), and $exp(C \otimes_{K'} K'_{in}) \mid ind\, C \mid p^2$.)

Exercise 8.16. Let D be a tame semiramified division algebra with center a Henselian field F. Suppose E and E' are inertially split F-subalgebras of D with $D = E \otimes_F E'$. Prove that E and E' are semiramified with $\overline{D} = \overline{E} \otimes_{\overline{F}} \overline{E'}$ and $\Gamma_D/\Gamma_F = (\Gamma_E/\Gamma_F) \times (\Gamma_{E'}/\Gamma_F)$.

Exercise 8.17. Let D be a tame division algebra with center a Henselian field F, and let I, N, T be division algebras with center F such that $D \sim I \otimes_F N \otimes_F T$ in $Br(F)$ with I inertial, N decomposably semiramified, and T tame and totally ramified. Show that $exp\, D = lcm\big(exp\,\overline{I}, exp(\Gamma_D/\Gamma_F)\big)$.

Exercise 8.18. Let S be an inertially split division algebra with center a Henselian field F, and take any $j \in \mathbb{N}$. By Prop. 8.62, $\Gamma_{S^j} = j\Gamma_S + \Gamma_F$. Let $\mathcal{J}$ be the j-torsion subgroup of $\mathcal{G}(Z(\overline{S})/\overline{F})$. Prove that $Z(\overline{S^j}) = Z(\overline{S})^{\mathcal{J}}$ and that there is a commutative diagram

$$\begin{array}{ccc}
\Gamma_S & \xrightarrow{\;j\cdot\;} & \Gamma_{S^j} \\
\Big\downarrow{\scriptstyle \theta_S} & & \Big\downarrow{\scriptstyle \theta_{S^j}} \\
\mathcal{G}(Z(\overline{S})/\overline{F}) & \xrightarrow{\;res\;} & \mathcal{G}(Z(\overline{S^j})/\overline{F})
\end{array}$$

where the top map is multiplication by j.

Exercise 8.19. Let D be a tame division algebra with center a Henselian field F, and take any $j \in \mathbb{N}$. Let $\overline{\Gamma_D} = \Gamma_D/\Gamma_F$, and let $\overline{\theta}_D \colon \overline{\Gamma_D} \to \mathcal{G}(Z(\overline{D})/\overline{F})$ be the epimorphism induced by θ_D. By Prop. 8.62, $\overline{\Gamma_{D^j}} = j\overline{\Gamma_D}$. Let J be the

j-torsion subgroup of $\overline{\Gamma_D}$. Prove that $Z(\overline{D^j}) = Z(\overline{D})^{\overline{\theta}_D(J)}$ and there is a commutative diagram

$$\begin{array}{ccc} \overline{\Gamma_D} & \xrightarrow{\ j\cdot\ } & \overline{\Gamma_{D^j}} \\ \Big\downarrow{\scriptstyle \overline{\theta}_D} & & \Big\downarrow{\scriptstyle \overline{\theta}_{D^j}} \\ \mathcal{G}(Z(\overline{D})/\overline{F}) & \xrightarrow{\ res\ } & \mathcal{G}(Z(\overline{D^j})/\overline{F}) \end{array}$$

where the top map is multiplication by j. (Note: $\overline{\theta}_D(J)$ may be a proper subgroup of the j-torsion subgroup of $\mathcal{G}(Z(\overline{D})/\overline{F})$. See Ex. 8.46 with $j = p$ for an example of this in the graded setting.)

Exercise 8.20. Let K/F be a bicyclic field extension, and let $\boldsymbol{\sigma} = (\sigma_1, \sigma_2)$ be a base of its Galois group. As in §8.3.2, let $L_1 = K^{\sigma_2}$ and $L_2 = K^{\sigma_1}$, and let $(b_1, u, b_2) \in Z^2_{\boldsymbol{\sigma}}(G, K^\times)$. Also, let $\boldsymbol{u} = (u_{i,j})_{i,j=1}^2$ be defined by $u_{1,1} = u_{2,2} = 1$, $u_{1,2} = u$, $u_{2,1} = u^{-1}$. Using the notation of §1.1.3 (see (1.4)), show that

$$\mathcal{S}(\mathcal{A}(K/F, \boldsymbol{\sigma}, b_1, u, b_2); L_1, L_2) \cong \mathcal{L}((K; \boldsymbol{\sigma}, \boldsymbol{u})).$$

(Compare the proof of Prop. 8.22.)

Exercise 8.21. Let $\mathcal{S} = \mathcal{S}(A; \ell_1, \ldots, \ell_n)$ be the semiramified division algebra of (8.36), and let $\mathcal{S}' = \mathcal{S}'(A; \ell_1, \ldots, \ell_n)$ be the algebra of (8.37).

(i) Prove that $\mathcal{S}' \cong q(\mathsf{gr}(\mathcal{S}))$.

(ii) Since $\mathcal{S} \cong \mathcal{S}' \otimes_{F'} F$ (see Cor. 8.73) we may view $\mathcal{S}'$ as a subalgebra of $\mathcal{S}$. Prove that the valuation on $\mathcal{S}$ is an immediate extension of its restriction to $\mathcal{S}'$.

Notes

§8.1: While tame lifts of graded division algebras had not been considered before in their full generality, the special case of inertial lifts of division algebras over the residue field (Cor. 8.15) has been discussed by Platonov–Yanchevskiĭ [192] and by Jacob–Wadsworth [106, Th. 2.8(a), Th. 2.9].

The correspondence between subalgebras of totally ramified central division algebras over a Henselian field and subgroups of the value group (Cor. 8.11) was established by Tignol–Wadsworth [245, Th. 3.8]. It generalizes the classification of subfields of totally ramified Mal'cev–Neumann division algebras given by Tignol–Amitsur in [242, Th. 4.2].

§8.2: The canonical subalgebras of a graded division algebra appear in Hwang–Wadsworth [103, (2.3)]. Proposition 8.16 gathers results proved by Boulagouaz [25, Prop. 2.4, Prop. 2.6] and also by Hwang–Wadsworth [103,

Prop. 2.1, Prop. 2.3]. The canonical[2] subalgebras of a division algebra over a Henselian field are defined in Platonov–Yanchevskiĭ [192, Th. 3].

§8.3: The *Dec* subgroup of the Brauer group was defined by Tignol [235]. For a different approach to this group, recall that the cohomology of a finite cyclic group $C = \langle\sigma\rangle$ can be computed by using the following free resolution of $\mathbb{Z}$ as a trivial module over the group ring $\mathbb{Z}[C]$:

$$W_*^\sigma(C)\colon \qquad \ldots \longrightarrow W_2^\sigma(C) \xrightarrow{\partial_2} W_1^\sigma(C) \xrightarrow{\partial_1} W_0^\sigma(C) \xrightarrow{\varepsilon} \mathbb{Z} \longrightarrow 0$$

where $W_i^\sigma(C) = \mathbb{Z}[C]$ for all i, the map ε is the augmentation, and ∂_i is multiplication by $\sigma - 1$ if i is odd, and by $N = \sum_{\rho\in C}\rho$ if i is even: see, e.g., Gille–Szamuely [84, Ex. 3.2.9]. For an arbitrary C-module M, the cohomology groups $H^i(C,M)$ are then described as follows:

$$H^i(C,M) = \begin{cases} \{m\in M \mid Nm = 0\}/\{\sigma(m) - m \mid m \in M\} & \text{for } i \text{ odd, } i \geq 1,\\ \{m\in M \mid \sigma(m) - m = 0\}/\{Nm \mid m\in M\} & \text{for } i \text{ even, } i\geq 2.\end{cases}$$

For an arbitrary finite abelian group G, given a base $\boldsymbol{\sigma} = (\sigma_1,\ldots,\sigma_n)$, i.e., a decomposition $G = \langle\sigma_1\rangle \times \ldots \times \langle\sigma_n\rangle$ into a direct product of finite cyclic subgroups $C_i = \langle\sigma_i\rangle$, we have $\mathbb{Z}[G] = \mathbb{Z}[C_1]\otimes_{\mathbb{Z}}\ldots\otimes_{\mathbb{Z}}\mathbb{Z}[C_n]$. Tensoring the resolutions $W_*^{\sigma_i}(C_i)$ yields a free resolution $W_*^{\boldsymbol{\sigma}}(G)$ of $\mathbb{Z}$ as a trivial G-module, depending on the base $\boldsymbol{\sigma}$ (see Gille–Szamuely [84, Constr. 3.4.1, Lemma 3.4.4] for the tensor product of resolutions). Using the resolution $W_*^{\boldsymbol{\sigma}}(G)$, we obtain for an arbitrary G-module M a simplified description of the cohomology groups $H^i(G,M)$: For all $m \geq 1$, let $C^m_{\boldsymbol{\sigma}}(G,M)$ denote the set of sequences $(a_{i_1,\ldots,i_m})_{1\leq i_1\leq\ldots\leq i_m\leq n}$ in M, and let $C^0_{\boldsymbol{\sigma}}(G,M) = M$. For $i = 1, \ldots, n$, let $N_i \in \mathbb{Z}[C_i]$ be defined by

$$N_i = \sum_{\rho\in C_i} \rho.$$

For $m = 0, 1, 2$, let $H^m_{\boldsymbol{\sigma}}(G,M)$ be the homology at $C^m_{\boldsymbol{\sigma}}(G,M)$ of the following 0-sequence:

$$0 \xrightarrow{d_0} C^0_{\boldsymbol{\sigma}}(G,M) \xrightarrow{d_1} C^1_{\boldsymbol{\sigma}}(G,M) \xrightarrow{d_2} C^2_{\boldsymbol{\sigma}}(G,M) \xrightarrow{d_3} C^3_{\boldsymbol{\sigma}}(G,M)$$

where the maps are defined as follows:

$$d_1(a) = \big((\sigma_i - 1)a\big)_{1\leq i\leq n} \qquad \text{for } a \in C^0_{\boldsymbol{\sigma}}(G,M);$$

$$d_2\big((a_i)_{1\leq i\leq n}\big) = (b_{i,j})_{1\leq i\leq j\leq n} \qquad \text{for } (a_i)_{1\leq i\leq n} \in C^1_{\boldsymbol{\sigma}}(G,M),$$

where $b_{i,i} = N_i a_i$ and

$$b_{i,j} = (\sigma_i - 1)a_j - (\sigma_j - 1)a_i \quad \text{for } i < j;$$

$$d_3\big((a_{i,j})_{1\leq i\leq j\leq n}\big) = (b_{i,j,k})_{1\leq i\leq j\leq k\leq n} \qquad \text{for } (a_{i,j})_{1\leq i\leq j\leq n} \in C^2_{\boldsymbol{\sigma}}(G,M),$$

[2] up to conjugacy

where $b_{i,i,i} = (\sigma_i - 1)a_{i,i}$,

$$b_{i,i,j} = (\sigma_j - 1)a_{i,i} + N_i a_{i,j}, \quad b_{i,j,j} = (\sigma_i - 1)a_{j,j} - N_j a_{i,j}, \quad \text{for } i < j,$$

and

$$b_{i,j,k} = (\sigma_i - 1)a_{j,k} - (\sigma_j - 1)a_{i,k} + (\sigma_k - 1)a_{i,j} \qquad \text{for } i < j < k.$$

For $m = 0, 1, 2$, let $Z^m_{\boldsymbol{\sigma}}(G,M) = \mathit{ker}\, d_{m+1}$ and $B^m_{\boldsymbol{\sigma}}(G,M) = \mathit{im}\, d_m$, so $H^m_{\boldsymbol{\sigma}}(G,M) = Z^m_{\boldsymbol{\sigma}}(G,M)/B^m_{\boldsymbol{\sigma}}(G,M)$. By a standard result of homological algebra (see, e.g., Gille–Szamuely [84, Prop. 3.1.9]), the cohomology groups $H^n_{\boldsymbol{\sigma}}(G,M)$ defined via the resolution $W^{\boldsymbol{\sigma}}_*(G,M)$ are canonically isomorphic to the cohomology groups $H^n(G,M)$ defined via the standard resolution: For $m = 1$, the isomorphism $\phi^1 \colon H^1(G,M) \xrightarrow{\sim} H^1_{\boldsymbol{\sigma}}(G,M)$ is induced by the map carrying each 1-cochain $f \colon G \to M$ to the sequence $\big(f(\sigma_i)\big)_{1 \le i \le n} \in C^1_{\boldsymbol{\sigma}}(G,M)$; for $m = 2$, the canonical isomorphism $\phi^2 \colon H^2(G,M) \xrightarrow{\sim} H^2_{\boldsymbol{\sigma}}(G,M)$ maps the cohomology class of each cocycle $f \colon G \times G \to M$ to the coset modulo $B^2_{\boldsymbol{\sigma}}(G,M)$ of the sequence $(a_{i,j})_{1 \le i \le j \le n}$ defined by

$$a_{i,i} = \sum_{\rho \in C_i} f(\rho, \sigma_i), \qquad a_{i,j} = f(\sigma_i, \sigma_j) - f(\sigma_j, \sigma_i) \quad \text{for } i < j.$$

In the particular case where G is the Galois group of a Galois field extension K/F and M is the multiplicative group $K^\times$, Hilbert's Theorem 90 takes the form $Z^1_{\boldsymbol{\sigma}}(G,K^\times) = B^1_{\boldsymbol{\sigma}}(G,K^\times)$. This equation is exactly Lemma 1.14. Also, we have an isomorphism $\Delta^{\boldsymbol{\sigma}}_{K/F} \colon H^2_{\boldsymbol{\sigma}}(G,K^\times) \xrightarrow{\sim} \mathit{Br}(K/F)$ that fits in the following commutative diagram, together with the crossed product isomorphism $\Delta_{K/F} \colon H^2(G,K^\times) \xrightarrow{\sim} \mathit{Br}(K/F)$ mapping $[f]$ to $[(K/F,G,f)]$ (see Th. 6.11 for the graded analogue of $\Delta_{K/F}$):

$$\begin{array}{ccc} H^2(G,K) & \xrightarrow{\ \phi^2\ } & H^2_{\boldsymbol{\sigma}}(G,K^\times) \\ & {\scriptstyle \Delta_{K/F}} \searrow \quad \swarrow {\scriptstyle \Delta^{\boldsymbol{\sigma}}_{K/F}} & \\ & \mathit{Br}(K/F) & \end{array}$$

It is convenient to associate to each $(a_{i,j})_{1 \le i \le j \le n} \in Z^2_{\boldsymbol{\sigma}}(G,K^\times)$ the matrix $\boldsymbol{u} = (u_{i,j})_{i,j=1}^n$ defined by

$$u_{i,i} = 1, \quad u_{i,j} = a_{i,j} \text{ for } i < j \quad \text{and} \quad u_{i,j} = a_{j,i}^{-1} \text{ for } i > j.$$

The cocycle condition on $(a_{i,j})_{1 \le i \le j \le n}$ then implies that $\boldsymbol{u}$ satisfies the conditions (1.3) and also $N_i N_j u_{i,j} = 1$ (cf. Prop. 1.13). The cohomology classes of cocycles with associated matrix $u_{i,j} = 1$ form a subgroup in $H^2_{\boldsymbol{\sigma}}(G,K^\times)$. This subgroup is in bijection with $\mathit{Dec}(K/F)$ through $\Delta^{\boldsymbol{\sigma}}_{K/F}$; see Tignol [235, Prop. 1.5]. Therefore, the matrix $\boldsymbol{u}$ associated to a cocycle $\alpha \in Z^2_{\boldsymbol{\sigma}}(G,K^\times)$ captures the same information as the image of the Brauer class of $\Delta^{\boldsymbol{\sigma}}_{K/F}([\alpha])$ in the quotient $\mathit{Br}(K/F)/\mathit{Dec}(K/F)$. This result generalizes Prop. 8.23, because for $n = 2$ the definition of $Z^2_{\boldsymbol{\sigma}}(G, K^\times)$ and $B^2_{\boldsymbol{\sigma}}(G, K^\times)$ given above coincides with the definition given in Def. 8.21, when sequences $(a_{i,j})_{1\le i\le j\le 2}$ are written as triples $(a_{1,1}, a_{1,2}, a_{2,2})$, and the map $\Delta^{\boldsymbol{\sigma}}_{K/F}$ carries the cohomology class of $(a_{1,1}, a_{1,2}, a_{2,2}) \in Z^2_{\boldsymbol{\sigma}}(G, K^\times)$ to the Brauer class of $\mathcal{A}(K/F, \boldsymbol{\sigma}, a_{1,1}, a_{1,2}, a_{2,2})$.

The simplified description of crossed products for an arbitrary abelian Galois field extension K/F afforded by the isomorphism $\Delta^{\boldsymbol{\sigma}}_{K/F}$ was first noticed by Cecioni [49, §2] and rediscovered by Dickson [60, §4], Cauchon [48, Ch. IX], and Amitsur–Saltman [12]. Amitsur and Saltman used it to define division algebras that they called "generic abelian crossed products." In our approach, the Amitsur–Saltman construction can be viewed as a special case of the $\mathcal{S}$ construction of §8.4.5: Let $\alpha \in Z^2_{\boldsymbol{\sigma}}(G, K^\times)$ and let $\boldsymbol{u}$ be the matrix associated to α as above. Let also A be the crossed product of K/F with cocycle α, and let $K = \ell_1 \otimes_F \ldots \otimes_F \ell_n$ be the decomposition of K into a tensor product of cyclic extensions of F corresponding to the base $\boldsymbol{\sigma}$ (as in Def. 8.19). Then, referring to the iterated Laurent series construction of §1.1.3 (see (1.4)), we have

$$\mathcal{S}(A; \ell_1, \ldots, \ell_n) = \mathcal{L}((K; \boldsymbol{\sigma}, \boldsymbol{u}))$$

and

$$\mathcal{S}'(A; \ell_1, \ldots, \ell_n) = q\big(\mathsf{gr}\big(\mathcal{L}((K; \boldsymbol{\sigma}, \boldsymbol{u}))\big)\big),$$

see Exercises 8.20 and 8.21. The algebras $\mathcal{S}'(A; \ell_1, \ldots, \ell_n)$ are the Amitsur–Saltman generic abelian crossed products.

Decomposably semiramified division algebras over Henselian fields were introduced in Jacob–Wadsworth [106, §4], where they are dubbed "nicely semiramified" or "NSR." Their role in the structure theory of division algebras over Henselian fields was already acknowledged in Platonov–Yanchevskiĭ [191], where they appear under the notation $A\mu$. Their characterization in Prop. 8.40 is partly in [106, Th. 4.4] (which uses the notion of "totally ramified field extension of radical type" as in Exercise 8.9), but (c) or (d) $\Rightarrow$ (a) of Prop. 8.40 is due to Mounirh [164, Th. 2.4].

§8.4: The results concerning valued division algebras in this section are mostly due to Jacob–Wadsworth [106]; some of them were previously obtained by Platonov–Yanchevskiĭ [191]: The decomposition $D \sim S \otimes T$ of tame division algebras over Henselian fields (Prop. 8.59) and the description of $\overline{D}$, Γ_D, and θ_D in terms of the inertially split algebra S and the totally ramified algebra T (Th. 8.60) are given in [106, Lemma 6.2, Th. 6.3]. Also, Prop. 8.62 is [106, Prop. 6.9]. Likewise, the decomposition of inertially split division algebras D as $D \sim I \otimes N$ with I inertial and N decomposably semiramified appears in [191, Prop. 4] and [106, Lemma 5.14], and the calculation of $\mathit{exp}\, D$, $\mathit{ind}\, D$, and the relations between $\overline{D}$, Γ_D and $\overline{I}$, $\overline{N}$, Γ_N (Th. 8.65) are established in [106, Th. 5.15] and partly in [191, Prop. 5]. However, the techniques of proof in [106] are substantially different from those of §8.4, as they do not involve the associated graded rings. The analogues of the $S \otimes T$ and $I \otimes N$ decompositions for graded division algebras were observed by Boulagouaz [26, Prop. 6.12, Th. 6.14].

Corollary 8.63 strengthens a result of Platonov and Yanchevskiĭ [190, (3.19)] that if D is a central division algebra tame over a Henselian field F, then $exp(\Gamma_D/\Gamma_F) \mid exp\ D$. This property does not hold for non-tame division algebras: Jacob–Wadsworth [106, Ex. 7.5] give an example of a central division algebra D of exponent 2 and degree 4 over a Henselian field F such that $\Gamma_D/\Gamma_F \cong \mathbb{Z}/4\mathbb{Z}$. (See also Exercise 8.7.)

Exercise 8.4 is a result of Albert, see Knus et al. [115, Th. (16.1)].

Exercise 8.5(i) is a variation of a result of Nganou. He proves in [173, Th. 1.13] a formula for ${}_e Br(K/F)/Dec(K/F)$, where F is a global field, K is a finite-degree abelian Galois extension of F, $e = exp\,\mathcal{G}(K/F)$, and ${}_e Br(K/F)$ is the e-torsion subgroup of $Br(K/F)$.

It follows from Exercise 8.9 and Cor. 8.10 that if D is any tame division algebra with center F and D has a subfield T totally ramified over F, then D has another subfield T' totally ramified of radical type over F with $\mathsf{gr}(T') \cong_g \mathsf{gr}(T)$. This property is not apparent using just valuation theory without the graded theory.

Exercise 8.11 was used by Caenepeel and van Oystaeyen [47, Th. IV.1.11, Prop. IV.1.12, pp. 139–140] to prove that if F is a graded field with $\Gamma_{\mathsf{F}} = \mathbb{Z}$, then $Br(\mathsf{F}) = Br(\mathsf{F}^{\natural})$ if and only if the field F_0 is perfect. Earlier, Auslander and Goldman in [18, Th. 7.5] used the polynomial version of this example, i.e., $\big(k[t,s]/k[t],\sigma,u\big)$ with s and u as in Exercise 8.11, to prove that $Br(k[t]) = Br(k)$ if and only if k is perfect.

Exercise 8.14 was pointed out in Jacob–Wadsworth [106, pp. 172–173]. The example in Exercise 8.15 of a tame central division algebra D over a Henselian field F with $[\overline{D}] \notin im\big(Br(\overline{F}) \to Br(Z(\overline{D}))\big)$ comes from [106, Ex. 7.1]. The absence of roots of unity turns out to be the only obstruction to $[\overline{D}]$ lying in the image of $Br(\overline{F})$: For any Henselian field F and any central division algebra D over F, we have $[\overline{D}]$ in the image of $Br(\overline{F})$ when $\overline{F}$ contains a primitive $\big(exp(\Gamma_D/\Gamma_F)\big)^2$-root of unity, as was shown by Wadsworth [254].

Exercise 8.16 is taken from Mounirh [165, Prop. 1.3(3)].

In conclusion, we make some remarks on *wild* (i.e., non-tame) division algebras, which have been mostly left out of our exposition.

As Def. 8.4 shows, wild division algebras arise only when the residue characteristic is nonzero, and their degrees are divisible by the residue characteristic. Moreover, if the center F has value group $\Gamma_F \cong \mathbb{Z}$, then the residue field $\overline{F}$ must be nonperfect, see Prop. 8.75. An example of wild division algebra of prime degree is given in Ex. 8.37, and another one in Exercise 8.6. Wild defectless division algebras of prime degree were extensively analyzed by Saltman [221] in the case where the center F is complete for a discrete rank 1 valuation, and by Tignol [240] (using the associated graded ring) for the case where F is Henselian of arbitrary rank. In the course of this analysis, they identified a key element of structure, called the *level* of the division algebra by Saltman, and the *height* by Tignol, defined by

$$h(D) = \min\{v_D(xy - yx) - v_D(x) - v_D(y) \mid x,\ y \in D^\times\} \in \Gamma_D.$$

Viewing the reduced trace $\mathit{Trd}_D\colon D \to F$ as an element in $\mathit{End}_F\, D$, we may consider its value under the norm $\mathit{End}\, v_D$. Tignol shows in [240, Th. 3.12] that $(\mathit{End}\, v_D)(\mathit{Trd}_D) = (p-1)h(D)$, from which it follows that

$$0 \le h(D) \begin{cases} \le v(p)/(p-1), & \text{if } \mathit{char}\, F = 0, \\ < \infty, & \text{if } \mathit{char}\, F = p. \end{cases}$$

Moreover, $h(D) = 0$ if and only if D is tame, see [240, Th. 3.5], and D is cyclic when the inequalities are strict, see [240, §4]. A detailed classification is given in Tignol [241].

Central simple algebras that are tensor products of cyclic algebras of degree p over a maximally complete valued field F of characteristic p with $\overline{F}$ perfect were considered by Aravire–Jacob [14], who showed that every such algebra is Brauer-equivalent to a tensor product of cyclic p-algebras in a somewhat complicated standard form. They used this result to show that for k a finite field, every central division algebra over $k((x))((y))$ is cyclic with index equal to the exponent [14, Cor. 3.5].

Chapter 9
Subfields and Splitting Fields of Division Algebras

When a central division algebra contains a maximal subfield that is a Galois extension of the center, the algebra has a very useful presentation as a crossed product. Therefore, the existence of Galois maximal subfields in division algebras has been intensively investigated since very early in the development of the theory.

The scope of this chapter extends beyond Galois maximal subfields. Under the hypothesis that the base field carries a Henselian valuation, we obtain in §9.1 information on the splitting fields of tame division algebras. This section concludes with a construction of tame totally ramified division algebras for which precise information on the Galois groups of Galois splitting fields is available.

In §9.2 we discuss a particular type of subfield. Following Albert [4], we say that a field extension K/F is *pure* when K is generated by a single element with minimal polynomial of the form $X^n - a$. By definition, every cyclic division algebra contains a pure maximal subfield. Albert [4, Th. 11.4] proved that this property actually characterizes cyclic division algebras when the degree is prime. In §9.2.1, we show after Matzri et al. [138] that Albert's theorem does not extend to the case where the degree is the square of a prime. The examples we give are semiramified division algebras. Their pure totally ramified subfields are related to properties of their specialization coset in §9.2.2.

In §9.3 we consider the special case of central simple algebras whose degree is a power of their (nonzero) characteristic. These algebras are called *p-algebras*. We show after Mounirh [165] that tame semiramified division p-algebras over Henselian fields are cyclic if and only if they are decomposably semiramified. This leads to constructions of noncyclic division p-algebras that are remarkably easy (using the valuation theoretic tools now at our disposal) given that the existence of such algebras was an open question for several decades. Then, following McKinnie [140] we show that the Galois subfields of a tame semiramified division p-algebra D are all inertial, under a "nonde-

J.-P. Tignol, A.R. Wadsworth, *Value Functions on Simple Algebras, and Associated Graded Rings*, Springer Monographs in Mathematics,
DOI 10.1007/978-3-319-16360-4_9

generacy" hypothesis on the specialization coset $sp(D)$. These p-algebras are therefore crossed products for only one type of Galois group.

Finally, in §9.4 we discuss noncrossed product division algebras, i.e., division algebras that are not crossed products. We give particularly easy constructions of such algebras in any degree divisible by 8 or by the square of an odd prime.

9.1 Splitting fields of tame division algebras

Throughout this section, F is a field with a Henselian valuation v. We use the functorial property of the exact sequences describing $Br_{is}(F)$ and $Br_{tr}(F)$ to obtain information on the splitting fields of tame central division algebras over F. We consider only splitting fields that are algebraic over F; their valuation properties are understood to be in reference to their unique valuation extending v.

9.1.1 Inertial and inertially split algebras

We first discuss in particular inertial and inertially split division algebras. The case of inertial division algebras is easy:

Proposition 9.1. *Let D be a central division algebra over F, and L an algebraic field extension of F. If D is inertial over F, then L splits D if and only if $\overline{L}$ splits $\overline{D}$.*

Proof. Since F is Henselian, the map $\mathsf{gr}_0\colon Br_{in}(F) \to Br(\overline{F})$, which carries $[D]$ to $[\overline{D}]$, is an isomorphism by Th. 6.64. Since L is algebraic over F, it is also Henselian, so we also have an isomorphism $\mathsf{gr}_0\colon Br_{in}(L) \xrightarrow{\sim} Br(\overline{L})$. The proposition follows from the commutativity of the diagram of Prop. 6.59:

$$\begin{array}{ccc} Br_{in}(F) & \xrightarrow[\cong]{\mathsf{gr}_0} & Br(\overline{F}) \\ \Big\downarrow{\scriptstyle \mathrm{ext}_{L/F}} & & \Big\downarrow{\scriptstyle \mathrm{ext}_{\overline{L}/\overline{F}}} \\ Br_{in}(L) & \xrightarrow[\cong]{\mathsf{gr}_0} & Br(\overline{L}) \end{array}$$ □

We next consider inertially split division algebras. Recall from Lemma 6.65 that if D is an inertially split central division algebra over F, then $Z(\overline{D})$ is a Galois extension of $\overline{F}$ with abelian Galois group isomorphic to Γ_D/Γ_F under the map induced by θ_D. If $\varepsilon\colon Z(\overline{D}) \hookrightarrow \overline{F}_{alg}$ is an embedding of $Z(\overline{D})$ into an algebraic closure of $\overline{F}$, then every other embedding of $Z(\overline{D})$ into $\overline{F}_{alg}$ has the form $\varepsilon \circ \sigma$ for some $\sigma \in \mathcal{G}\left(Z(\overline{D})/\overline{F}\right)$; hence, all the embeddings of $Z(\overline{D})$ into $\overline{F}_{alg}$ have the same image. Therefore, for every extension K of $\overline{F}$ in $\overline{F}_{alg}$, the field compositum $Z(\overline{D}) \cdot K \subseteq \overline{F}_{alg}$ is unambiguously defined, independent of the embedding ε. Moreover, the intersection $Z(\overline{D}) \cap K = \varepsilon^{-1}(K)$ is also independent of the choice of embedding ε. For, given any σ in the abelian group $\mathcal{G}(Z(\overline{D})/\overline{F})$, conjugation by σ^{-1} fixes $\mathcal{G}\big(Z(\overline{D})/\varepsilon^{-1}(K)\big)$; hence,

$$(\varepsilon \circ \sigma)^{-1}(K) \;=\; \sigma^{-1}\big(\varepsilon^{-1}(K)\big) \;=\; \varepsilon^{-1}(K).$$

Proposition 9.2. *Let D be a central division algebra over F, and let L be an algebraic field extension of F. Assume that D is inertially split. Then,*

(i) $[D \otimes_F L] \subseteq Br_{in}(L)$ *if and only if*

$$\mathcal{G}\left(Z(\overline{D})/\, Z(\overline{D}) \cap \overline{L}\right) \subseteq \theta_D(\Gamma_D \cap \Gamma_L); \tag{9.1}$$

(ii) *if L splits D, then $\overline{L} \cdot Z(\overline{D})$ splits $\overline{D}$ and* (9.1) *holds;*

(iii) *if $Br(\overline{L}) = \{0\}$, then L splits D if and only if* (9.1) *holds.*

Proof. (i) We first prove the criterion for $D \otimes_F L$ to be inertial. Recall the functorial ramification map $\chi\colon Br_{is}(F) \to Hom^c\big(\mathcal{G}(\overline{F}), \mathbb{T}(\Gamma_F)\big)$ of (6.41) (see Prop. 6.68). We defined in (6.25) a map $\pi^*_{\overline{L}/\overline{F}}$ that fits in the following commutative diagram:

$$\begin{array}{ccc}
Br_{is}(F) & \xrightarrow{\;\chi\;} & Hom^c\big(\mathcal{G}(\overline{F}), \mathbb{T}(\Gamma_F)\big) \\
\Big\downarrow{\scriptstyle \mathrm{ext}_{L/F}} & & \Big\downarrow{\scriptstyle \pi^*_{\overline{L}/\overline{F}}} \\
Br_{is}(L) & \xrightarrow{\;\chi\;} & Hom^c\big(\mathcal{G}(\overline{L}), \mathbb{T}(\Gamma_L)\big)
\end{array}$$

We show that the inclusion (9.1) is equivalent to $\pi^*_{\overline{L}/\overline{F}}(\chi_D) = 0$. This in turn is equivalent to $[D \otimes_F L] \in \ker \chi = Br_{in}(L)$ by exactness of (6.42).

By definition, $\pi^*_{\overline{L}/\overline{F}}$ carries $\chi_D\colon \mathcal{G}(\overline{F}) \to \mathbb{T}(\Gamma_F)$ to the composition

$$\pi^*_{\overline{L}/\overline{F}}(\chi_D) \;=\; \tau_{L/F} \circ \chi_D \circ \pi_{\overline{L}/\overline{F}}\colon \mathcal{G}(\overline{L}) \;\longrightarrow\; \mathbb{T}(\Gamma_L),$$

where $\pi_{\overline{L}/\overline{F}}\colon \mathcal{G}(\overline{L}) \to \mathcal{G}(\overline{F})$ is the restriction map and $\tau_{L/F}\colon \mathbb{T}(\Gamma_F) \to \mathbb{T}(\Gamma_L)$ is the canonical map induced by the inclusion $\Gamma_F \hookrightarrow \Gamma_L$. We have $\ker \tau_{L/F} = \Gamma_L/\Gamma_F$, so $\pi^*_{\overline{L}/\overline{F}}(\chi_D) = 0$ if and only if $\chi_D \circ \pi_{\overline{L}/\overline{F}}\big(\mathcal{G}(\overline{L})\big) \subseteq \Gamma_L/\Gamma_F$. By definition, χ_D is the composite map

$$\chi_D\colon \mathcal{G}(\overline{F}) \xrightarrow{\;res_{Z(\overline{D})}\;} \mathcal{G}(Z(\overline{D})/\overline{F}) \xrightarrow[\cong]{\;\overline{\theta}_D^{\,-1}\;} \Gamma_D/\Gamma_F \hookrightarrow \mathbb{T}(\Gamma_F).$$

We have $res_{Z(\overline{D})} \circ \pi_{\overline{L}/\overline{F}}\big(\mathcal{G}(\overline{L})\big) = \mathcal{G}\left(Z(\overline{D})/\, Z(\overline{D}) \cap \overline{L}\right)$, hence

$$\chi_D \circ \pi_{\overline{L}/\overline{F}}\big(\mathcal{G}(\overline{L})\big) \;=\; \overline{\theta}_D^{\,-1}\big(\mathcal{G}(Z(\overline{D})/\, Z(\overline{D}) \cap \overline{L})\big).$$

Therefore, $\pi^*_{\overline{L}/\overline{F}}(\chi_D) = 0$ if and only if the right displayed term lies in Γ_L/Γ_F. This inclusion is equivalent to (9.1), since $\overline{\theta}_D^{\,-1}$ is a bijection $\mathcal{G}(Z(\overline{D})/\overline{F}) \xrightarrow{\sim} \Gamma_D/\Gamma_F$. Thus, (i) is proved.

(ii) Now consider splitting of D. If L splits D, then the preceding argument shows that (9.1) holds. If $Br(\overline{L}) = \{0\}$, then Th. 6.66 shows that $\chi\colon Br_{is}(L) \to Hom^c(\mathcal{G}(\overline{L}), \mathbb{T}(\Gamma_L))$ is an isomorphism. It follows that $ext_{L/F}([D]) = 0$ if and only if $\pi^*_{\overline{L}/\overline{F}}(\chi_D) = 0$.

(iii) For the residue condition when L splits D, consider a decomposition $D \sim I \otimes_F N$ with I an inertial central division algebra over F and N a decomposably semiramified central division algebra over F, as in (8.34). By Prop. 8.40(c), N contains a maximal subfield Z inertial over F. We have

$$[Z{:}F]^2 = [N{:}F] = [\overline{N}{:}\overline{F}]^2$$

because Z is a maximal subfield and N is semiramified. Since Z is inertial over F, it follows that $\overline{Z} = \overline{N}$. But $\overline{N} = Z(\overline{D})$ by Th. 8.65(i), hence Z is an inertial lift of $Z(\overline{D})$ over F. Therefore, Z is an abelian Galois extension of F by Th. A.23(vi), so the field compositum $L \cdot Z$ is unambiguously defined. Since Z splits N and L splits D, it follows that $L \cdot Z$ splits I; hence, $\overline{L \cdot Z}$ splits $\overline{I}$ by Prop. 9.1. According to Th. 8.65(ii), the division algebra $\overline{D}$ is associated to $\overline{I} \otimes_{\overline{F}} \overline{Z}$; hence, $\overline{L \cdot Z}$ also splits $\overline{D}$. But, as Z is inertial over F, Prop. A.18 shows that $\overline{L \cdot Z} = \overline{L} \cdot \overline{Z} = \overline{L} \cdot Z(\overline{D})$. □

Corollary 9.3. *If a totally ramified extension L of F splits an inertially split central division algebra D over F, then $\Gamma_D \subseteq \Gamma_L$.*

Proof. Since $\overline{L} = \overline{F}$ and θ_D is surjective, (9.1) yields

$$\theta_D(\Gamma_D) \subseteq \theta_D(\Gamma_D \cap \Gamma_L).$$

Since moreover $ker\,\theta_D = \Gamma_F \subseteq \Gamma_L$, it follows that $\Gamma_D \subseteq \Gamma_L$. □

9.1.2 Tame algebras

We now consider the general case of a tame division algebra D. Recall from Prop. 8.17(iv) the symplectic module $\big((ker\,\theta_D)/\Gamma_F, a_D\big)$.

Proposition 9.4. *For any tame central division algebra D over F, and any algebraic field extension L of F, define*

$$\Delta(D, L) = \big((ker\,\theta_D) \cap \Gamma_L\big)/\Gamma_F \subseteq (ker\,\theta_D)/\Gamma_F, \qquad (9.2)$$

and let $\Delta(D, L)^\perp \subseteq (ker\,\theta_D)/\Gamma_F$ be the orthogonal of $\Delta(D, L)$ for a_D. Then,

(i) *$[D \otimes_F L] \in Br_{is}(L)$ if and only if*

$$\Delta(D, L)^\perp \subseteq \Delta(D, L); \qquad (9.3)$$

in particular, if L splits D, then (9.3) holds;

(ii) *if L is strictly Henselian, then L splits D if and only if (9.3) holds.*

Proof. Consider the canonical map $\Psi_F\colon Br_{tr}(F) \to {}_{\mu(\overline{F})}\mathbb{T}(\wedge^2\Gamma_F)$ of Th. 7.84 and the canonical isomorphism $\Upsilon_{\Gamma_F}\colon \mathbb{T}(\wedge^2\Gamma_F) \to \mathit{Symp}(\mathbb{T}(\Gamma_F))$ of Th. 7.22. Let $\tau\colon \mathbb{T}(\Gamma_F) \to \mathbb{T}(\Gamma_L)$ be the map induced by the inclusion $\Gamma_F \subseteq \Gamma_L$. By functoriality (see Prop. 7.81 and Lemma 6.67), we have a commutative diagram

$$\begin{array}{ccc} Br_{tr}(F) & \xrightarrow{\Upsilon_{\Gamma_F}\circ\Psi_F} & {}_{\mu(\overline{F})}\mathit{Symp}(\mathbb{T}(\Gamma_F)) \\ {\scriptstyle ext_{L/F}}\big\downarrow & & \big\downarrow{\scriptstyle \tau_\diamond} \\ Br_{tr}(L) & \xrightarrow{\Upsilon_{\Gamma_L}\circ\Psi_L} & {}_{\mu(\overline{L})}\mathit{Symp}(\mathbb{T}(\Gamma_L)) \end{array}$$

Corollary 8.61 yields $\Upsilon_{\Gamma_F} \circ \Psi_F([D]) = \big((\ker\theta_D)/\Gamma_F, a_D\big)$. Since

$$(\ker\tau) \cap \big((\ker\theta_D)/\Gamma_F\big) = \Delta(D,L),$$

the description of $\tau_\diamond$ (see (7.5)) shows that

$$\tau_\diamond\big((\ker\theta_D)/\Gamma_F, a_D\big) = 0 \quad \text{if and only if} \quad \Delta(D,L)^\perp \subseteq \Delta(D,L).$$

(Cf. Cor. 7.82.) Commutativity of the diagram shows that this condition holds if and only if $ext_{L/F}([D])$ lies in $\ker(\Upsilon_{\Gamma_L} \circ \Psi_L) = \ker\Psi_L$. By Th. 7.84 this kernel is $Br_{is}(L)$. This yields (i).

(ii) If L is strictly Henselian, then $Br_{is}(L) = \{0\}$. Hence, (ii) follows from (i). □

We single out the particular case where F itself is strictly Henselian:

Corollary 9.5. *Let D be a tame central division algebra over a strictly Henselian field F, and let L be an algebraic field extension of F. The field L splits D if and only if $(\Gamma_D \cap \Gamma_L)^\perp \subseteq \Gamma_D \cap \Gamma_L$, where the orthogonality is relative to the canonical pairing c_D on Γ_D. When this condition holds, Γ_L contains a maximal totally isotropic subgroup of Γ_D, and L contains an isomorphic copy of a maximal subfield of D.*

Proof. Since F is strictly Henselian, D is totally ramified, hence $\ker\theta_D = \Gamma_D$ and $a_D = \overline{c}_D$. Moreover, L also is strictly Henselian, so the first assertion follows from Prop. 9.4. If $(\Gamma_D \cap \Gamma_L)^\perp \subseteq \Gamma_D \cap \Gamma_L$, the subgroup $(\Gamma_D \cap \Gamma_L)^\perp$ is totally isotropic in Γ_D with respect to c_D. Let $\Lambda \subseteq \Gamma_D$ be a maximal totally isotropic subgroup containing $(\Gamma_D \cap \Gamma_L)^\perp$; then,

$$(\Gamma_D \cap \Gamma_L)^\perp \subseteq \Lambda = \Lambda^\perp \subseteq (\Gamma_D \cap \Gamma_L)^{\perp\perp} = \Gamma_D \cap \Gamma_L.$$

(The equality on the right holds because c_D has radical Γ_F, which lies in $\Gamma_D \cap \Gamma_L$.) Thus, Γ_L contains the maximal totally isotropic subgroup Λ of Γ_D.

It remains only to verify that L contains a subfield isomorphic to a maximal subfield of D. By Cor. 8.11 there is a subfield $K \subseteq D$ with $\Gamma_K = \Lambda$, and K is a maximal subfield of D, as Λ is a maximal totally isotropic subgroup of Γ_D.

We show that L contains a copy of K. Consider the following graded subfield of $\mathsf{gr}(L)$:

$$\mathsf{E} = \bigoplus_{\gamma\in\Lambda} \mathsf{gr}(L)_\gamma.$$

We have $\Lambda = \Gamma_{\mathsf{E}} = \Gamma_{\mathsf{gr}(K)}$, and $|\Lambda{:}\Gamma_F|$ is prime to $char\,\overline{F}$ because D is tame. Since $\mathsf{gr}(F)$ is inertially closed, it follows that $\mathsf{gr}(K) \cong_g \mathsf{E}$ (see Ex. 7.7). Now E is a tame graded field extension of $\mathsf{gr}(F)$; hence, it lifts to a subfield $K' \subseteq L$ by Prop. 8.3. By the uniqueness of tame lifts (Cor. 5.56), we have $K \cong K'$. □

As an application of Prop. 9.4, we obtain conditions on the value group of a splitting field L of a tame division algebra D, and on the Galois group when L is Galois over $Z(D)$. We will see in particular that a tame central division algebra D over F cannot be split by a cyclic extension of F unless the abelian group $(\mathit{ker}\,\theta_D)/\Gamma_F$ is generated by at most two elements.

Recall from §7.1.1 that for any finite abelian group G we let $\mathit{trk}(G)$ denote the minimal number of generators of G, i.e., the number of invariant factors of G (see §A.4 in the Appendix), and that maximal isotropic subgroups of a symplectic module are called Lagrangians.

Proposition 9.6. *If a finite-degree field extension L of a Henselian field F splits the tame central division algebra D over F, then for $\Delta(D,L)$ as in (9.2),*

$$\mathit{trk}(\Gamma_L/\Gamma_F) \geq \mathit{trk}\big(\Delta(D,L)\big) \geq \tfrac{1}{2}\,\mathit{trk}\big((\mathit{ker}\,\theta_D)/\Gamma_F\big). \tag{9.4}$$

Moreover, if L is Galois over F, then $\mathcal{G}(L/F)$ contains an isomorphic copy of a Lagrangian of $\big((\ker\theta_D)/\Gamma_F, a_D\big)$.

Proof. The first inequality in (9.4) follows from the inclusion of $\Delta(D,L)$ in Γ_L/Γ_F. To obtain the second inequality, recall that $\Delta(D,L)^\perp \subseteq \Delta(D,L)$ by Prop. 9.4. So, $\Delta(D,L)^\perp$ is totally isotropic, and hence it lies in some Lagrangian Λ of $\big((\mathit{ker}\,\theta_D)/\Gamma_F, a_D\big)$. Therefore, as in the proof of Cor. 9.5, we have

$$\Delta(D,L)^\perp \subseteq \Lambda = \Lambda^\perp \subseteq \Delta(D,L)^{\perp\perp} = \Delta(D,L),$$

hence $\mathit{trk}\big(\Delta(D,L)\big) \geq \mathit{trk}(\Lambda)$. The second inequality in (9.4) follows from the inequality $2\,\mathit{trk}(\Lambda) \geq \mathit{trk}\big((\mathit{ker}\,\theta_D)/\Gamma_F, a_D\big)$ established in Prop. 7.5.

Now assume L is Galois over F, and let F_{in} be the inertial closure of F in a given separable closure containing L. The field compositum $L\cdot F_{in}$ is unambiguously defined, and it is Galois over F_{in}. Let $G = \mathcal{G}(L\cdot F_{in}/F_{in})$. Since G embeds in $\mathcal{G}(L/F)$, it suffices to show that G contains an isomorphic copy of a Lagrangian of $\big((\mathit{ker}\,\theta_D)/\Gamma_F, a_D\big)$.

Let $M \subseteq L\cdot F_{in}$ be the ramification field of $L\cdot F_{in}$ over F_{in}, and let $P = \mathcal{G}(L\cdot F_{in}/M)$, which is the ramification subgroup of G. So, P is trivial if $char\,\overline{F} = 0$, while P is the unique $\overline{p}$-Sylow subgroup of G if $char\,\overline{F} = \overline{p} \neq 0$, see Th. A.9(i) and (ii). In either case, P is normal in G and $\mathit{gcd}\big(|P|,|G/P|\big) = 1$.

By the Schur–Zassenhaus Lemma (see for instance Rotman [211, p. 151]), it follows that $G/P = \mathcal{G}(M/F_{in})$ is isomorphic to a subgroup of G. Note that since F_{in} has no proper inertial extensions in $L\cdot F_{in}$, it is the inertia field for $L\cdot F_{in}/F_{in}$. Therefore, M is totally and tamely ramified over F_{in}, and (with the notation of Th. A.9)

$$G/P = G^{ine}/G^{ram} \cong \Gamma_M/\Gamma_{F_{in}},$$

see Th. A.9(viii) and (x). Therefore, it suffices to prove that $\Gamma_M/\Gamma_{F_{in}}$ contains a subgroup isomorphic to a Lagrangian of $\big((ker\,\theta_D)/\Gamma_F, a_D\big)$.

Consider a decomposition $D \sim S \otimes_F T$, where S is inertially split and T is tame and totally ramified, as in Prop. 8.59. Since D is split by L and S by F_{in}, it follows that T is split by $L\cdot F_{in}$. Since T is tame, its degree is prime to $[L\cdot F_{in} : M]$, so T is already split by M. On the other hand, Th. 3.43 shows that $T \otimes_F F_{in}$ is a division algebra with $\Gamma_{T\otimes_F F_{in}} = \Gamma_T$. Thus, $T \otimes_F F_{in}$ is tame with center the strictly Henselian field F_{in}. Corollary 9.5 shows that Γ_M contains a maximal totally isotropic subgroup of $\Gamma_{T\otimes_F F_{in}} = \Gamma_T$ for the canonical pairing $c_{T\otimes_F F_{in}} = c_T$; hence, Γ_M/Γ_F contains a Lagrangian of $(\Gamma_T/\Gamma_F, \overline{c}_T)$. But Th. 8.60 shows that $(\Gamma_T/\Gamma_F, \overline{c}_T) = \big((ker\,\theta_D)/\Gamma_F, a_D\big)$, so the proof is complete. □

Corollary 9.7. *Let D be a tame central division algebra over F. If D is split by a cyclic Galois extension of F, then either D is inertially split or $trk\big((ker\,\theta_D)/\Gamma_F\big) = 2$.*

Proof. When the condition in the statement holds, Prop. 9.6 shows that the symplectic module $\big((ker\,\theta_D)/\Gamma_F, a_D\big)$ contains a cyclic Lagrangian. By Cor. 7.6(i) it follows that $trk\big((ker\,\theta_D)/\Gamma_F\big) \le 2$. If $trk\big((ker\,\theta_D)/\Gamma_F\big) = 0$, then $ker\,\theta_D = \Gamma_F$, so D is inertially split in view of Prop. 8.64. To complete the proof, observe that $trk\big((ker\,\theta_D)/\Gamma_F\big)$ is even, since $(ker\,\theta_D)/\Gamma_F$ carries the nondegenerate alternating pairing a_D, see Th. 7.2. □

9.1.3 Example: the $\mathfrak{T}$ construction

We now give specific examples of division algebras for which the results in §9.1.2 yield precise information on the maximal subfields and on the algebraic splitting fields. Let k be an arbitrary field and let $n_1, \ldots, n_r$ be integers with $n_i \ge 2$ for all i. Let

$$n = n_1 \ldots n_r \quad \text{and} \quad m = lcm(n_1, \ldots, n_r).$$

Assume that k contains a primitive m-th root of unity ω, and let $\omega_i = \omega^{m/n_i}$ for $i = 1, \ldots, r$; so ω_i is a primitive n_i-th root of unity. Let $x_1, y_1, \ldots, x_r, y_r$ be $2r$ independent indeterminates over k. Define

$$F = k((x_1))((y_1))\ldots((x_r))((y_r))$$

and set

$$\mathcal{T}(k;n_1,\ldots,n_r) = (x_1,y_1/F)_{\omega_1,n_1} \otimes_F \ldots \otimes_F (x_r,y_r/F)_{\omega_r,n_r}. \tag{9.5}$$

The field F is Henselian (indeed maximally complete) for the $(x_1,y_1,\ldots,x_r,y_r)$-adic valuation v, with residue field k and value group $\Gamma_F = \mathbb{Z}^{2r}$ lexicographically ordered from right to left.

Proposition 9.8. *The F-algebra $\mathcal{T}(k;n_1,\ldots,n_r)$ is a central division algebra of degree n and exponent m. It is tame and totally ramified with value group $\left(\frac{1}{n_1}\mathbb{Z}\right)^2\times\ldots\times\left(\frac{1}{n_r}\mathbb{Z}\right)^2$, and it is spanned by an armature $\mathcal{A}$ that is isomorphic to $\mathbb{S}_{n_1}\perp\ldots\perp\mathbb{S}_{n_r}$.*

Proof. Write $\mathcal{T}$ for $\mathcal{T}(k;n_1,\ldots,n_r)$. Since it is a tensor product of symbol algebras of degrees n_1, …, n_r, the algebra $\mathcal{T}$ carries an armature $\mathcal{A}$ isomorphic to $\mathbb{S}_{n_1}\perp\ldots\perp\mathbb{S}_{n_r}$, by Cor. 7.30. This armature has exponent m, and *char* $k\nmid m$ since k contains a primitive m-th root of unity. Therefore, $\mathcal{T}$ also carries a tame armature v-gauge $\alpha_{\mathcal{A}}$, by Th. 7.49. We have

$$\Gamma_{\mathcal{T}} = \left(\tfrac{1}{n_1}\mathbb{Z}\right)^2\times\ldots\times\left(\tfrac{1}{n_r}\mathbb{Z}\right)^2,$$

so the map $\overline{\alpha}_{\mathcal{A}}$ induced by $\alpha_{\mathcal{A}}$ is an isomorphism $\mathcal{A}\xrightarrow{\sim}\Gamma_{\mathcal{T}}/\Gamma_F$. It then follows from Lemma 7.73 that $\mathcal{T}$ is a division algebra, which is tame and totally ramified over F, and that $\alpha_{\mathcal{A}}$ is the valuation of $\mathcal{T}$ extending v. We have $\exp\mathcal{T} = \exp(\Gamma_{\mathcal{T}}/\Gamma_F)$ by Prop. 7.75, so $\exp\mathcal{T} = m$. It is clear from the definition that $\deg\mathcal{T} = n$, so the proof is complete. □

The algebra $\mathcal{T}(k;n_1,\ldots,n_r)$ can also be obtained by an iterated twisted Laurent series construction: in the notation of §1.2.6, we have

$$\mathcal{T}(k;n_1,\ldots,n_r) \cong \mathcal{L}((k;\boldsymbol{I},\boldsymbol{u}))$$

where $\boldsymbol{u} = (u_{i,j})_{i,j=1}^{2r}$ is defined as follows:

$$u_{i,j} = \begin{cases} \omega_\ell & \text{if } i = 2\ell \text{ and } j = 2\ell-1 \text{ for some } \ell\in\{1,\ldots,r\};\\ \omega_\ell^{-1} & \text{if } i = 2\ell-1 \text{ and } j = 2\ell \text{ for some } \ell\in\{1,\ldots,r\};\\ 1 & \text{in every other case.}\end{cases}$$

Proposition 9.9. *If L is a finite-degree Galois field extension of F that splits $\mathcal{T}(k;n_1,\ldots,n_r)$, then $\mathcal{G}(L/F)$ contains an isomorphic copy of a Lagrangian of $\mathbb{S}_{n_1}\perp\ldots\perp\mathbb{S}_{n_r}$. Moreover, every Lagrangian of $\mathbb{S}_{n_1}\perp\ldots\perp\mathbb{S}_{n_r}$ is isomorphic to the Galois group over F of some maximal subfield of $\mathcal{T}(k;n_1,\ldots,n_r)$.*

Proof. Write $\mathcal{T}$ for $\mathcal{T}(k;n_1,\ldots,n_r)$. Since $\mathcal{T}$ is tame and totally ramified, we have $\ker\theta_{\mathcal{T}} = \Gamma_{\mathcal{T}}$, and we saw in the proof of Prop. 9.8 that

$$(\Gamma_{\mathcal{T}}/\Gamma_F, a_{\mathcal{T}}) \cong \mathbb{S}_{n_1}\perp\ldots\perp\mathbb{S}_{n_r}.$$

Therefore, the first statement follows from Prop. 9.6.

For the second statement, choose an arbitrary Lagrangian Λ of $\Gamma_{\mathcal{T}}/\Gamma_F$. Under the inverse of the isomorphism $\overline{\alpha}_{\mathcal{A}}$ associated to the armature $\mathcal{A}$, this Λ is isomorphic to a Lagrangian $\mathcal{L} \subseteq \mathcal{A}$. We may thus consider the algebra $L = F[\mathcal{L}]$ spanned by $\mathcal{L}$, which is a subalgebra of $\mathcal{T} = F[\mathcal{A}]$. By Prop. 7.31, the centralizer $C_{\mathcal{T}}(F[\mathcal{L}])$ satisfies

$$C_{\mathcal{T}}(F[\mathcal{L}]) = F[\mathcal{L}^{\perp}] = F[\mathcal{L}].$$

Therefore, L is a maximal subfield of $\mathcal{T}$. Let $s = exp(\mathcal{L})$. Since F contains a primitive s-th root of unity, we may apply Th. 7.36(ii) to see that L is a Kummer extension of F. More precisely, we have $L \cong F[\sqrt[s]{\sigma(\mathcal{L})}]$ where $\sigma\colon \mathcal{L} \to F^{\times}/F^{\times s}$ maps $xF^{\times}$ to $x^s F^{\times s}$, see (7.27). By Kummer theory, L is a Galois extension of F with $\mathcal{G}(L/F) \cong \mathcal{L}$, see Ex. 7.28(ii). Thus, $\mathcal{G}(L/F) \cong \Lambda$. This completes the proof. □

Corollary 9.10.

(i) *If $n_r \mid n_{r-1} \mid \ldots \mid n_1$ and $r \geq 2$, then the division algebra $\mathcal{T}(k; n_1, \ldots, n_r)$ is not split by any cyclic Galois extension of F.*

(ii) *If $n_1, \ldots, n_r$ are powers of some prime p and $n_i \geq p^3$ for some i, then the division algebra $\mathcal{T}(k; n_1, \ldots, n_r)$ is not split by any elementary abelian Galois extension of F.*

Proof. In case (i), $trk(\mathbb{S}_{n_1} \perp \ldots \perp \mathbb{S}_{n_r}) = 2r \geq 4$; hence, $\mathbb{S}_{n_1} \perp \ldots \perp \mathbb{S}_{n_r}$ does not contain any cyclic Lagrangian, by Cor. 7.6(i). From Prop. 9.9, it follows that $\mathcal{T}(k; n_1, \ldots, n_r)$ is not split by any cyclic Galois extension of F.

In case (ii), Cor. 7.6(ii) shows that $\mathbb{S}_{n_1} \perp \ldots \perp \mathbb{S}_{n_r}$ does not contain any elementary abelian Lagrangian, hence by Prop. 9.9 $\mathcal{T}(k; n_1, \ldots, n_r)$ is not split by any elementary abelian Galois extension of F. □

9.2 Pure subfields in semiramified division algebras

Recall from the introduction to this chapter that a field extension K/F is said to be *pure* when K is generated by a single element with minimal polynomial of the form $X^n - a$. We discuss pure subfields of division algebras in two different contexts: in §9.2.1 we will construct, for any prime p, division algebras of degree p^2 that contain a pure maximal subfield, even though they are not cyclic. Our discussion is based on Matzri et al. [138]; it uses tame semiramified division algebras obtained by the $\mathcal{S}$-construction of §8.4.5. Note that the noncyclic division algebra $\mathcal{T}(k; p, p)$ of (9.5) is not a candidate here, since it does not contain any pure maximal subfield: such a subfield would become cyclic after adjoining to k a primitive p^2-root of unity ζ, whereas Cor. 9.10 shows that $\mathcal{T}(k(\zeta); p, p)$ is not cyclic.

Then, in §9.2.2, we will relate the existence of pure totally ramified extensions in a semiramified graded or valued division algebra to properties of its specialization coset, for use in §9.3.

9.2.1 Noncyclic algebras with pure maximal subfields

Let p be a prime number, and let k be a field containing a primitive p-th root of unity ω (so $char\,k \neq p$). To discuss cyclic extensions of k, we use the following extension criterion:

Lemma 9.11 (Albert). *Let L be a field extension of k with $[L{:}k] = p^n > 1$. Assume L is Galois over k with cyclic Galois group. There exists an extension M of L with $[M{:}L] = p$ and M Galois over k with cyclic Galois group if and only if ω is a norm from L to k.*

Proof. Suppose M exists, and let σ be a generator of $\mathcal{G}(M/k)$. Then σ^{p^n} generates $\mathcal{G}(M/L)$, so we may write $M = L(\delta)$ for some δ such that $\sigma^{p^n}(\delta) = \omega\delta$. Let $\lambda = \sigma(\delta)\delta^{-1}$. Then

$$\sigma^{p^n}(\lambda) = \sigma^{p^n+1}(\delta)\sigma^{p^n}(\delta^{-1}) = \sigma(\omega\delta)\,(\omega\delta)^{-1} = \lambda,$$

so $\lambda \in L^\times$. Moreover,

$$N_{L/k}(\lambda) = \prod_{i=0}^{p^n-1} \sigma^i(\lambda) = \prod_{i=0}^{p^n-1} \left(\sigma^{i+1}(\delta)\sigma^i(\delta)^{-1}\right) = \sigma^{p^n}(\delta)\delta^{-1} = \omega.$$

Therefore, ω is a norm from L to k.

Conversely, given $\lambda \in L^\times$ such that $N_{L/k}(\lambda) = \omega$, Hilbert's Theorem 90 yields an element $d \in L^\times$ such that $\lambda^p = \sigma(d)d^{-1}$. If $d = d_0^p$ for some $d_0 \in L^\times$, then $\lambda = \omega^r\sigma(d_0)d_0^{-1}$ for some power ω^r of ω, hence $N_{L/k}(\lambda) = \omega^{p^n r} = 1$, a contradiction. We may then set $M = L(\delta)$ with $\delta^p = d$ and extend a generator σ of $\mathcal{G}(L/k)$ to M by setting $\sigma(\delta) = \lambda\delta$. Details are left to the reader. □

Now, let A be a central simple k-algebra of degree p^2. Assume that A contains a maximal subfield K of the form $K = L_1 \otimes_k L_2$ where L_1 and L_2 are cyclic Galois extensions of k of degree p. We let $\mathcal{S}$ denote the semiramified division algebra $\mathcal{S}(A; L_1, L_2)$ of (8.36), with center $F = k((t_1))((t_2))$. By Prop. 8.72, this division algebra has degree p^2, as K splits A, and exponent $lcm(p, ind\,A)$.

Proposition 9.12. *If $\mathcal{S}$ is a cyclic algebra, then there is a subfield $K' \subseteq K$ such that $[K'{:}k] = p$ and ω is a norm from K' to k.*

Proof. We use the (t_1, t_2)-adic valuation v on F. Let $M \subseteq \mathcal{S}$ be a maximal subfield which is a cyclic Galois extension of F. Let $L \subseteq M$ be the inertial closure of F in M; so, $\mathcal{G}(M/L)$ is the inertia subgroup of $\mathcal{G}(M/F)$ and we have an exact sequence (see Th. A.8)

$$1 \longrightarrow \mathcal{G}(M/L) \longrightarrow \mathcal{G}(M/F) \longrightarrow \mathcal{G}(\overline{M}/\overline{F}) \longrightarrow 1.$$

If $M = L$, then $\overline{M}$ would be a cyclic extension of $\overline{F} = k$ of degree p^2. This is impossible since we have $\overline{\mathcal{S}} = K$, and $\mathcal{G}(K/k)$ is elementary abelian. If $L = F$, then M would be totally and tamely ramified over F, hence Γ_M/Γ_F would be a cyclic group of degree p^2 since it is in duality with $\mathcal{G}(M/F)$, see Th. A.9. This is impossible since $\Gamma_{\mathcal{S}}/\Gamma_F$ is elementary abelian. Therefore, we must have $[M{:}L] = [L{:}F] = p$. By Lemma 9.11, it follows that ω is a norm from L to F. Let σ be a generator of $\mathcal{G}(L/F)$, and choose $x \in L$ satisfying

$$\omega = N_{L/F}(x) = x\sigma(x)\ldots\sigma^{p-1}(x).$$

Since $v(\omega) = 0$, we have $v(x) = 0$, hence we may reduce the equation to the residue field:

$$\omega = \overline{x\sigma(x)}\ldots\overline{\sigma^{p-1}(x)} = \overline{x}\,\overline{\sigma}(\overline{x})\ldots\overline{\sigma}^{p-1}(\overline{x}).$$

This shows that ω is a norm from $\overline{L}$ to $\overline{F} = k$. The proposition follows by taking $K' = \overline{L}$. □

Remark 9.13. A necessary and sufficient condition for the cyclicity of $\mathcal{S}$ is given in Exercise 9.3.

We next show that if ω is a norm from K to L_1, then for a suitable choice of A the algebra $\mathcal{S} = \mathcal{S}(A; L_1, L_2)$ contains a pure maximal subfield.

We fix the notation: for $i = 1$, 2, we have $L_i = k(\sqrt[p]{a_i})$ for some $a_i \in k^\times$. Let $\alpha_i \in L_i$ satisfy $\alpha_i^p = a_i$. The Galois group $\mathcal{G}(K/k)$ is generated by automorphisms σ_1, σ_2 such that

$$\sigma_1(\alpha_1) = \omega\alpha_1,\ \sigma_1(\alpha_2) = \alpha_2 \qquad \text{and} \qquad \sigma_2(\alpha_1) = \alpha_1,\ \sigma_2(\alpha_2) = \omega\alpha_2.$$

We also write σ_1 and σ_2 for the restrictions of σ_1 and σ_2 to L_1 and L_2 respectively, so σ_i generates $\mathcal{G}(L_i/k)$ for $i = 1$, 2.

Proposition 9.14. *If there exists a $u \in K^\times$ such that $N_{K/L_1}(u) = \omega$, then there is a central simple k-algebra A of degree p^2 such that $K \subseteq A$ and $C_A(L_1) \cong (\alpha_1, a_2/L_1)_{\omega,p}$. For any such algebra A, the semiramified division algebra $\mathcal{S} = \mathcal{S}(A; L_1, L_2)$ with center $F = k((t_1))((t_2))$ contains an element s with $s^{p^2} = a_1 t_2^p$ and $[F(s){:}F] = p^2$. Hence, $F(s)$ is a pure maximal subfield of $\mathcal{S}$.*

Proof. Suppose $u \in K^\times$ is such that $N_{K/L_1}(u) = \omega$. Then

$$N_{K/F}(u) = N_{L_1/F}(\omega) = \omega^p = 1,$$

hence $N_{L_2/F}\big(N_{K/L_1}(u)\big) = 1$. By Hilbert's Theorem 90 we may find $b_1 \in L_2^\times$ such that $b_1\sigma_2(b_1)^{-1} = N_{K/L_2}(u)$. On the other hand, we have

$$\sigma_1(\alpha_1)\alpha_1^{-1} = \omega = N_{K/L_1}(u),$$

so $(b_1, \alpha_1, u) \in Z^2_{\boldsymbol{\sigma}}(G, K^\times)$ in the notation of Def. 8.21. By Prop. 8.22(i), we may consider the crossed product algebra

$$A = \mathcal{A}(K/k, \boldsymbol{\sigma}, b_1, u, \alpha_1),$$

with distinguished generators z_1, z_2 such that $z_1^p = b_1$, $z_2^p = \alpha_1$, and $z_1 z_2 = u z_2 z_1$. This k-algebra is central simple of degree p^2 and contains K. The elements z_2 and α_2 centralize L_1 and generate in A an L_1-algebra isomorphic to $(\alpha_1, a_2/L_1)_{\omega,p}$. Dimension count shows that this subalgebra must be the centralizer of L_1. The algebra A thus meets the requirements.

Now fix any central simple k-algebra A containing L_1 with $C_A(L_1) \cong B$. By the definition of $\mathcal{S}$ in (8.36), $\mathcal{S}$ has center $F = k((t_1))((t_2))$, and we have in $Br(F)$

$$\begin{aligned}\mathcal{S} &\sim A \otimes_k (L_1 \otimes_k F/F, \sigma_1 \otimes id, t_1) \otimes_F \left(L_2 \otimes_k F/F, \sigma_2 \otimes id, t_2\right)\\ &= A \otimes_k \left(a_1, t_1/F\right)_{\omega^{-1},p} \otimes_F \left(a_2, t_2/F\right)_{\omega^{-1},p}.\end{aligned}$$

Therefore, $\mathcal{S}$ is split by $K \otimes_k F = F(\sqrt[p]{a_1}, \sqrt[p]{a_2})$, and we may consider α_1, $\alpha_2 \in \mathcal{S}$. The centralizer of α_1 in $\mathcal{S}$ is the division algebra associated to $\mathcal{S} \otimes_k L_1$ whose center is $L_1 \otimes_k F = F(\alpha_1)$. By the choice of A we have in $Br(F(\alpha_1))$

$$\begin{aligned}\mathcal{S} \otimes_k L_1 &\sim (A \otimes_k L_1) \otimes_{L_1} \left(a_2, t_2/F(\alpha_1)\right)_{\omega^{-1},p}\\ &\sim \left(\alpha_1, a_2/L_1\right)_{\omega,p} \otimes_{L_1} \left(t_2, a_2/F(\alpha_1)\right)_{\omega,p}.\end{aligned}$$

Therefore, $C_{\mathcal{S}}(\alpha_1) \cong \left(\alpha_1 t_2, a_2/F(\alpha_1)\right)_{\omega,p}$. Thus, $C_{\mathcal{S}}(\alpha_1)$—hence also $\mathcal{S}$—contains an element $s \notin F(\alpha_1)$ satisfying $s^p = \alpha_1 t_2$. So, $[F(\alpha_1)(s){:}F(\alpha_1)] = p$. Note that $s^{p^2} = a_1 t_2^p \in F$. Since $F(s^p) = F(\alpha_1 t_2) = F(\alpha_1)$, we have

$$[F(s){:}F] = [F(s^p)(s){:}F(s^p)]\,[F(s^p){:}F] = [F(\alpha_1)(s){:}F(\alpha_1)]\,[F(\alpha_1){:}F] = p^2.$$

Thus, $F(s)$ is a pure maximal subfield of $\mathcal{S}$. □

Example 9.15. In view of Prop. 9.12 and Prop. 9.14, to construct an example of a noncyclic division algebra $\mathcal{S}(A; L_1, L_2)$ of degree p^2 with a pure maximal subfield, it suffices to find a field extension $K = k(\sqrt[p]{a_1}, \sqrt[p]{a_2})$ such that ω is a norm from K to $k(\sqrt[p]{a_1})$, but not a norm to k from any subfield $K' \subseteq K$ of codimension p. To obtain an explicit example, consider an arbitrary field k_0 containing the primitive p-th root of unity ω but no primitive p^2-root of unity, and let $k = k_0(x)$, where x is an indeterminate over k_0. Choose

$$a_1 = \omega(1-x) \quad \text{and} \quad a_2 = x.$$

Then, in $Br(L_1) = Br\left(k(\sqrt[p]{\omega(1-x)}\,)\right)$,

$$\begin{aligned}\left(\omega, a_2/L_1\right)_p &= \left(\omega, x/\,k(\sqrt[p]{\omega(1-x)}\,)\right)_p\\ &\sim \left(\omega(1-x), x/\,k(\sqrt[p]{\omega(1-x)}\,)\right)_p \otimes_{L_1} \left(x, 1-x/\,k(\sqrt[p]{\omega(1-x)}\,)\right)_p\\ &\sim L_1 \otimes_{L_1} L_1 = L_1.\end{aligned}$$

Since this symbol algebra is split, ω is a norm from $K = L_1(\sqrt[p]{a_2}\,)$ to L_1.

On the other hand, consider the symbol algebra $(\omega, a_1/k)_p$. Let v be the discrete $(x-1)$-adic valuation on $k = k_0(x)$, and let (k', v') be the completion (or the Henselization) of (k, v). Because $\overline{k'}^{v'} = \overline{k}^{v} = k_0$ and $[k_0(\sqrt[p]{\omega}):k_0] = p$ by hypothesis, the field $k'(\sqrt[p]{\omega})$ is an inertial extension of k' of degree p. Since in addition $v'(a_1) = 1 \notin p\Gamma_{k'}$, Prop. 8.41 shows that $(\omega, a_1/k')_p$ is a (cyclically semiramified) division algebra. Because $(\omega, a_1/k)_p$ is therefore nonsplit, ω is not a norm from $k(\sqrt[p]{a_1})$ to k. Similarly, for any $i = 0, 1, \ldots, p-1$, we can see that the symbol algebra $(\omega, a_1^i a_2/k)_p$ is a division algebra by using the discrete x-adic valuation v_x on k, since $\overline{k}^{v_x} = k_0$ and $v_x(a_1^i a_2) = 1$. Hence, ω is not a norm from $k(\sqrt[p]{a_1^i a_2})$ to k. Since $k(\sqrt[p]{a_1})$ and the $k(\sqrt[p]{a_1^i a_2})$ are all the degree-p field extensions of k within K, this field K satisfies all the required conditions.

Remark 9.16. Recall, e.g. from Serre [229, Prop. 6, p. 119], that the extension of scalars and corestriction maps

$$ext_{L_1/k}\colon Br(k) \to Br(L_1) \quad \text{and} \quad cor_{L_1/k}\colon Br(L_1) \to Br(k)$$

satisfy $cor_{L_1/k} \circ ext_{L_1/k} = p \cdot id$. When A is chosen as in Prop. 9.14, we have $ext_{L_1/k}([A]) = \big[(\alpha_1, a_2/L_1)_{\omega,p}\big]$, hence

$$\begin{aligned} p[A] &= cor_{L_1/k}\big(\big[(\alpha_1, a_2/L_1)_{\omega,p}\big]\big) \\ &= \big[(N_{L_1/k}(\alpha_1), a_2/k)_{\omega,p}\big] = \begin{cases} [(-a_1, a_2/k)_2] & \text{if } p = 2, \\ [(a_1, a_2/k)_{\omega,p}] & \text{if } p \neq 2. \end{cases} \end{aligned}$$

In the preceding example, where $k = k_0(x)$ and $a_1 = \omega(1-x)$ and $a_2 = x$, we have

$$p[A] = \begin{cases} [(1-x, x/k)_2] = 0 & \text{if } p = 2, \\ [(\omega(1-x), x)_{\omega,p}] = [(\omega, x/k)_{\omega,p}] \neq 0 & \text{if } p \neq 2. \end{cases}$$

For $p \neq 2$, the symbol algebra representing $p[A]$ was shown to be nonsplit in the discussion of the example. By Prop. 8.72 we have

$$exp\, \mathcal{S}(A; L_1, L_2) = lcm(p, exp\, A). \tag{9.6}$$

Therefore, the division algebra $\mathcal{S}(A; L_1, L_2)$ constructed with this example has exponent 2 if $p = 2$, and p^2 if $p \neq 2$. In the next two examples, we show how different choices of A, a_1, a_2 yield noncyclic division algebras $\mathcal{S}(A; L_1, L_2)$ with pure maximal subfields where the algebra has degree and exponent 4 or degree p^2 and exponent p for $p \neq 2$.

Example 9.17. *A noncyclic division algebra of degree and exponent* 4 *with pure maximal subfield.* Let $p = 2$, $k = \mathbb{Q}$, $a_1 = -3$, and $a_2 = -1$; so $K = \mathbb{Q}(\sqrt{-3}, \sqrt{-1})$. The quaternion algebra $(-1, -1/\mathbb{Q})_2$ is split by $\mathbb{Q}(\sqrt{-3})$

since the quaternion $i+j+ij$ satisfies $(i+j+ij)^2 = -3$; hence $\big(-1,-1/\mathbb{Q}(\sqrt{-3})\big)_2$ is split, which means that -1 is a norm from $K=\mathbb{Q}(\sqrt{-3},\sqrt{-1})$ to $\mathbb{Q}(\sqrt{-3})$. On the other hand, -1 is not a norm to $\mathbb{Q}$ from any of the quadratic extensions $\mathbb{Q}(\sqrt{-3})$, $\mathbb{Q}(\sqrt{-1})$, $\mathbb{Q}(\sqrt{3})$ of $\mathbb{Q}$ in K, since -3, -1, and 3 are not sums of two squares in $\mathbb{Q}$. So, the division algebra $\mathcal{S}(A;L_1,L_2)$ as in Prop. 9.14 is not cyclic and contains a pure maximal subfield. But the quaternion algebra $\big(-a_1,a_2/\mathbb{Q}\big)_2=\big(3,-1/\mathbb{Q}\big)_2$ is not split, so the computation in Remark 9.16 yields $2[A]\neq 0$, so $\mathcal{S}(A;L_1,L_2)$ has exponent 4.

Example 9.18. *A noncyclic division algebra of degree p^2 and exponent p with pure maximal subfield, for p an odd prime.* Let k_0 be any field containing the primitive p-th root of unity ω but no primitive p^2-root of unity, and let $k_1=k_0(a_1,a_2,b)$, where a_1, a_2, b are independent indeterminates. Consider the following symbol algebras:

$$S_1=\big(a_1,\omega/k_1\big)_{\omega,p}, \qquad S_2=\big(a_2,\omega/k_1\big)_{\omega,p},$$
$$R=\big(a_1,a_2/k_1\big)_{\omega,p}, \qquad T=\big(a_1,b/k_1\big)_{\omega,p}.$$

Let k be the function field over k_1 of the product of the Severi–Brauer varieties of $S_2\otimes_{k_1}T$ and of R (see Gille–Szamuely [84, Ch. 5]). Finally, let $K=k(\sqrt[p]{a_1},\sqrt[p]{a_2})$, $L_1=k(\sqrt[p]{a_1})$, $L_2=k(\sqrt[p]{a_2})$. We claim that ω is a norm from K to L_1, but not from any of the codimension p subfields of K to k.

For this, let v be the discrete a_1-adic valuation on $k_1=k_0(a_2,b)(a_1)$, and let (k_1',v') be the completion or Henselization of (k_1,v); so $\overline{k_1'}=k_0(a_2,b)$ and $\Gamma_{k_1'}=\mathbb{Z}$. Because k_0 has no p^2-root of unity and a_2 and b are algebraically independent over k_0, we have $\big[\,\overline{k_1'}(\sqrt[p]{\omega},\sqrt[p]{a_2},\sqrt[p]{b}\,):\overline{k_1'}\,\big]=p^3$. Therefore, the ramification characters of the inertially split k_1'-algebras $S_1\otimes_{k_1}k_1'$, $R\otimes_{k_1}k_1'$, and $T\otimes_{k_1}k_1'$ are $\mathbb{Z}/p\mathbb{Z}$-independent in $\mathit{Hom}^c\big(\mathcal{G}(\overline{k_1'}),\mathbb{Q}/\mathbb{Z}\big)$. Hence $[S_1]$, $[R]$, and $[T]$ are $\mathbb{Z}/p\mathbb{Z}$-independent in $\mathit{Br}(k_1)$. All three algebras are split by $k_1(\sqrt[p]{a_1}\,)$. But $S_2\otimes_{k_1}k_1(\sqrt[p]{a_1}\,)$ is nonsplit, as we can see using the discrete a_2-adic valuation v_2 on $k_1(\sqrt[p]{a_1}\,)$. (Note that $v_2(a_2)=1\notin p\mathbb{Z}$ and $\overline{k_1(\sqrt[p]{a_1}\,)}^{v_2}=k_0(\sqrt[p]{a_1},b)$, which contains no p^2-root of unity.) Therefore $[S_1]$, $[R]$, $[T]$, and $[S_2]$ are $\mathbb{Z}/p\mathbb{Z}$-independent in $\mathit{Br}(k_1)$.

Recall from Gille–Szamuely [84, Th. 5.4.1] that the kernel of the scalar extension map $\mathit{ext}_{k/k_1}\colon \mathit{Br}(k_1)\to\mathit{Br}(k)$ is the subgroup generated by $[S_2]+[T]$ and by $[R]$. Since T splits over L_1, it follows that S_2 also splits over L_1, which means that the symbol algebra $(a_2,\omega/L_1)_{\omega,p}$ is split. Therefore, ω is a norm from $K=L_1(\sqrt[p]{a_2}\,)$ to L_1. On the other hand, from the $\mathbb{Z}/p\mathbb{Z}$-independence noted above of the four Brauer classes in $\mathit{Br}(k_1)$, the group $\langle[S_1],[S_2]\rangle$ intersects $\mathit{ker}\,\mathit{ext}_{k/k_1}$ trivially. Therefore, for any integers z_1,z_2 that are not both multiples of p,

$$\big[\big(a_1^{z_1}a_2^{z_2},\omega/k\big)_p\big]=\mathit{ext}_{k/k_1}\big(z_1[S_1]+z_2[S_2]\big)\neq 0\quad\text{in }\mathit{Br}(k).$$

Thus, ω is not a norm from the field $k\big(\sqrt[p]{a_1^{z_1}a_2^{z_2}}\,\big)$ to k whenever z_1 and z_2 are not both multiples of p. Such fields are all the subfields of K of degree p over k. This completes the proof of the claim above.

Finally, for the A of Prop. 9.14 the computation in Remark 9.16 shows that in $Br(k)$ we have

$$p\,[A] \;=\; \big[\big(a_1, a_2/k\big)_{\omega,p}\big] \;=\; [R \otimes_{k_1} k].$$

Since k was chosen to split R, we have $\exp A = p$, hence $\mathcal{S}(A; L_1, L_2)$ has exponent p by (9.6).

9.2.2 Pure totally ramified subfields

Let D be a semiramified graded division algebra with center a graded field F. Recall from Def. 8.26 that D_0 is a field, and that the field extension $\mathsf{D}_0/\mathsf{F}_0$ is Galois with Galois group canonically isomorphic to $\Gamma_{\mathsf{D}}/\Gamma_{\mathsf{F}}$ via θ_{D}.

Proposition 9.19. *Suppose the semiramified graded division algebra* D *contains a nonzero homogeneous element* x *such that* $\deg x \notin \Gamma_{\mathsf{F}}$ *and* $x^p \in \mathsf{F}^\times$ *for some prime* p *(possibly* $p = \operatorname{char}\mathsf{F}$*). Let* $K \subseteq \mathsf{D}_0$ *be the subfield fixed under* $\theta_{\mathsf{D}}(\deg x)$*. Then, every* $\sigma \in \mathcal{G}(\mathsf{D}_0/\mathsf{F}_0)$ *restricts to an automorphism of* K*, and*

$$\operatorname{ext}_{K^\sigma/\mathsf{F}_0}\big(sp(\mathsf{D})\big) \;=\; Dec(\mathsf{D}_0/K^\sigma) \qquad \textit{for every } \sigma \in \mathcal{G}(\mathsf{D}_0/\mathsf{F}_0),$$

where K^σ *is the fixed subfield of* K *under* σ*.*

Proof. That every $\sigma \in \mathcal{G}(\mathsf{D}_0/\mathsf{F}_0)$ restricts to an automorphism of K is clear from Galois theory, since $\mathcal{G}(\mathsf{D}_0/\mathsf{F}_0)$ is abelian, hence K/F_0 is a Galois extension. Let $\mathsf{L} = K^\sigma \otimes_{\mathsf{F}_0} \mathsf{F}$, which is an inertial graded field extension of F lying in D. By functoriality of the specialization coset construction for inertial extensions, (see Cor. 8.58), we have

$$\operatorname{ext}_{K^\sigma/\mathsf{F}_0}\big(sp(\mathsf{D})\big) \;=\; sp(\mathsf{D}_{\mathsf{L}}),$$

where D_{L} is the graded division algebra associated to $\mathsf{D} \otimes_{\mathsf{F}} \mathsf{L}$. Since $\mathsf{L} \subseteq \mathsf{D}$, the algebra D_{L} is the centralizer of L in D,

$$\mathsf{D}_{\mathsf{L}} \;=\; C_{\mathsf{D}}(\mathsf{L}).$$

Therefore, we have to show that $sp\big(C_{\mathsf{D}}(\mathsf{L})\big) = Dec(\mathsf{D}_0/K^\sigma)$. By Cor. 8.53, this is equivalent to showing that $C_{\mathsf{D}}(\mathsf{L})$ is decomposably semiramified.

Let n be the order of σ, and let $\tau = \theta_{\mathsf{D}}(\deg x)$, so $K = \mathsf{D}_0^\tau$. Since $x^p \in \mathsf{F}^\times$, the order of τ is p; hence, either τ lies in the subgroup of $\mathcal{G}(\mathsf{D}_0/\mathsf{F}_0)$ generated by σ, or the sum of the subgroups generated by σ and τ is direct. In the first case, we have $K^\sigma = \mathsf{D}_0^\sigma$, so the field extension D_0/K^σ is cyclic; hence $Dec(\mathsf{D}_0/K^\sigma) = Br(\mathsf{D}_0/K^\sigma)$. Because $C_{\mathsf{D}}(\mathsf{L})_0 = \mathsf{D}_0$, the algebra $C_{\mathsf{D}}(\mathsf{L})$ is semiramified, so $sp(C_{\mathsf{D}}(\mathsf{L})) \subseteq Br(\mathsf{D}_0/K^\sigma)$ by Th. 8.50(v). Hence, $sp(C_{\mathsf{D}}(\mathsf{L})) = Dec(\mathsf{D}_0/K^\sigma)$.

For the rest of the proof, we may assume the second case, that the sum of the subgroups of $\mathcal{G}(\mathsf{D}_0/\mathsf{F}_0)$ generated by σ and by τ is direct. We then have

$$\mathsf{D}_0 = K \otimes_{K^\sigma} \mathsf{D}_0^\sigma.$$

Let $\mathsf{C} = C_{\mathsf{D}}(x)$ be the centralizer of x. Since $[\mathsf{F}(x){:}\mathsf{F}] = p$, the Double Centralizer Theorem 2.35 yields

$$[\mathsf{C}{:}\mathsf{F}] \cdot p = [\mathsf{D}{:}\mathsf{F}].$$

We have $\mathsf{C}_0 = \mathsf{D}_0^\tau = K$ and $[\mathsf{D}_0{:}K] = p$, so it follows from the preceding equation that $\Gamma_{\mathsf{C}} = \Gamma_{\mathsf{D}}$. Therefore, we may find a homogeneous element $y \in \mathsf{C}^\times$ such that $\theta_{\mathsf{D}}(deg\, y) = \sigma$. We then have $deg(y^n) \in ker\, \theta_{\mathsf{D}} = \Gamma_{\mathsf{F}}$, so $y^n c \in \mathsf{D}_0^\times$ for some $c \in \mathsf{F}^\times$. Since $y^n c$ commutes with both x and y, we have $y^n c \in K^\sigma$, hence $y^n \in K^\sigma \otimes_{\mathsf{F}_0} \mathsf{F} = \mathsf{L}$. Let A be the graded F-subalgebra of D generated by K and y, and let B be the graded F-subalgebra generated by D_0^σ and x. Then A and B are each cyclic graded algebras with center L, so they are subalgebras of $C_{\mathsf{D}}(\mathsf{L})$:

$$\mathsf{A} = (K \otimes_{\mathsf{F}_0} \mathsf{F}/\mathsf{L}, \sigma, y^n), \qquad \mathsf{B} = (\mathsf{D}_0^\sigma \otimes_{\mathsf{F}_0} \mathsf{F}/\mathsf{L}, \tau, x^p).$$

Because x centralizes K and y centralizes D_0^σ, the algebras A and B centralize each other in $C_{\mathsf{D}}(\mathsf{L})$. Hence, the graded subalgebra $\mathsf{A} \cdot \mathsf{B}$ of $C_{\mathsf{D}}(\mathsf{L})$ is graded isomorphic to $\mathsf{A} \otimes_{\mathsf{L}} \mathsf{B}$. By dimension count, $C_{\mathsf{D}}(\mathsf{L}) \cong_g \mathsf{A} \otimes_{\mathsf{L}} \mathsf{B}$. To conclude, observe that $deg\, y + \Gamma_{\mathsf{F}}$ and $deg\, x + \Gamma_{\mathsf{F}}$ have order n and p respectively, and generate independent subgroups of $\mathbb{T}(\Gamma_{\mathsf{F}})$ since they correspond under the isomorphism $\overline{\theta}_{\mathsf{D}}$ to τ and σ, which generate independent subgroups of $\mathcal{G}(\mathsf{D}_0/\mathsf{F}_0)$. Therefore, Prop. 8.29 shows that $C_{\mathsf{D}}(\mathsf{L})$ is decomposably semiramified, as desired. □

There is an analogue of Prop. 9.19 for semiramified division algebras over Henselian fields:

Proposition 9.20. *Let D be a tame semiramified division algebra over a Henselian-valued field (F, v), and let p be a prime. Suppose D contains an element x such that $v(x) \notin \Gamma_F$ and $x^p \in F^\times$, and let $K \subseteq \overline{D}$ be the subfield fixed under $\theta_D(v(x))$. Then, for every $\sigma \in \mathcal{G}(\overline{D}/\overline{F})$, we have*

$$\mathrm{ext}_{K^\sigma/\overline{F}}(sp(D)) = Dec(\overline{D}/K^\sigma).$$

Proof. Consider the graded division algebra $\mathsf{gr}(D)$ associated to the valuation on D. This graded division algebra is semiramified by Prop. 8.38. We have $\widetilde{x}^p \in \mathsf{gr}(F)$ and $deg\, \widetilde{x} = v(x) \notin \Gamma_F$, so Prop. 9.19 yields

$$\mathrm{ext}_{K^\sigma/\overline{F}}(sp(\mathsf{gr}(D))) = Dec(\overline{D}/K^\sigma).$$

The proposition then follows since $sp(\mathsf{gr}(D)) = sp(D)$, see Prop. 8.71. □

For convenience of reference, we give a name to the condition that appears in Prop. 9.19 and Prop. 9.20:

Definition 9.21. Let M be an abelian Galois extension of an arbitrary field F. A coset $s \in Br(M/F)/Dec(M/F)$ is said to be *weak* if there exists a subfield $K \subseteq M$ of prime codimension such that for all $\sigma \in \mathcal{G}(M/F)$

$$\mathrm{ext}_{K^\sigma/F}(s) = Dec(M/K^\sigma).$$

Otherwise, the coset s is said to be *strong.*

Note that since $Br(M/F) = Dec(M/F)$ whenever $\mathcal{G}(M/F)$ is cyclic, if s is a strong coset, then necessarily $trk\,\mathcal{G}(M/F) \geq 2$.

Corollary 9.22.

(i) *Let* D *be a semiramified graded division algebra over a graded field* F. *If the coset* $sp(\mathsf{D}) \in Br(\mathsf{D}_0/\mathsf{F}_0)/Dec(\mathsf{D}_0/\mathsf{F}_0)$ *is strong, then* D *contains no nontrivial totally ramified graded field extension of* F.

(ii) *Let* D *be a tame semiramified central division algebra over a Henselian-valued field* F, *and let* $x \in D^\times$ *satisfy* $x^p \in F^\times$ *for some prime* p. *If the coset* $sp(D)$ *is strong in* $Br(\overline{D}/\overline{F})/Dec(\overline{D}/\overline{F})$, *then* $v(x) \in \Gamma_F$.

Proof. Note that every nontrivial totally ramified graded field extension of F contains a homogeneous element x such that $deg\, x \notin \Gamma_{\mathsf{F}}$ and $x^p \in \mathsf{F}^\times$ for some prime p; see Prop. 5.18. Thus, the corollary is a direct consequence of Prop. 9.19 and Prop. 9.20. □

Examples of strong cosets are easy to obtain from the following observation:

Proposition 9.23. *Suppose* M *is an abelian Galois extension of a field* F *with* $[M{:}F] = p^n$ *for some prime* p *and some integer* $n \geq 2$. *If* $trk\,\mathcal{G}(M/F) \geq 2$, *then every Brauer class of exponent* p^n *in* $Br(M/F)$ *represents a strong coset modulo* $Dec(M/F)$.

Proof. Let $K \subseteq M$ be a subfield of prime codimension. Since $\mathcal{G}(M/F)$ is an abelian p-group, its p-torsion subgroup has the same torsion rank as $\mathcal{G}(M/F)$. As this rank is at least 2, we may find an element $\sigma \in \mathcal{G}(M/F)$ of order p generating a subgroup distinct from $\mathcal{G}(M/K)$. Thus, $[M{:}K^\sigma] = p^2$ and $exp\,\mathcal{G}(M/K^\sigma) = p$. Suppose $\alpha \in Br(M/F)$ satisfies $\mathrm{ext}_{K^\sigma/F}(\alpha) \in Dec(M/K^\sigma)$. Then $p\,\mathrm{ext}_{K^\sigma/F}(\alpha) = 0$ since $Dec(M/K^\sigma)$ is killed by p (see (8.12)). By taking the corestriction from K^σ to F, and using

$$cor_{K^\sigma/F} \circ \mathrm{ext}_{K^\sigma/F} = [K^\sigma{:}F] \cdot id = p^{n-2} \cdot id,$$

we obtain $p^{n-1}\alpha = 0$. Therefore, if $exp\,\alpha = p^n$, the coset represented by α is strong. □

For a specific example, one may take $F = k(a, b)$ where k is any field containing a primitive p^2-root of unity, and a and b are algebraically independent over k, and let $M = F(\sqrt[p]{a}, \sqrt[p]{b})$. The Brauer class of the symbol algebra $(a,b/F)_{p^2}$ is split by M and has exponent p^2, see Ex. 8.56. See also Ex. 9.26 for an example in characteristic p. For other examples note that if F is a global field, then every central simple F-algebra has exponent equal to its index (see Pierce [178, p. 359] or Reiner [199, Th. (32.19)]); therefore, every division algebra with maximal subfield M has exponent $[M:F]$.

9.3 Galois subfields in p-algebras

Recall from the introduction to this chapter that a *p-algebra* is a central simple algebra whose degree is a power of its (nonzero) characteristic. In his penetrating work on such algebras in the 1930s, Albert proved that every p-algebra is split by some cyclic Galois field extension of its center. But he left unresolved the question whether every division algebra D which is a p-algebra must be a cyclic algebra, i.e., whether D must have a maximal subfield which is cyclic Galois over $Z(D)$. This question was finally settled for $p \neq 2$ in 1978 when Amitsur and Saltman in [12] gave examples of noncyclic division p-algebras of degree p^n for any prime p and any $n \geq 2$. (An example of a noncyclic division p-algebra of degree 4 had already been given by Gordon [86] in 1940.) The Amitsur–Saltman examples are generic abelian crossed products, which are rings of quotients of semiramified graded division algebras and have the form $\mathcal{S}'(A;\ell_1,\ldots,\ell_n)$ as in (8.37). A little later, Saltman showed in [218] that there are strong constraints on the Galois groups of the subfields of these algebras. This is relevant for the construction of noncrossed product p-algebras, as we will see in §9.4. Later in this section we will give an account of the Henselian field version of these results based on McKinnie [140]. But first we prove a theorem of Mounirh that captures the essence of the Amitsur–Saltman approach by providing a criterion for cyclicity of tame semiramified p-algebras over a Henselian field.

Theorem 9.24 (Mounirh). *Let F be a Henselian field with char $F = p > 0$, and let D be a tame and semiramified division p-algebra with center F. Then, D is a cyclic algebra if and only if D is decomposably semiramified.*

Proof. Let $deg\, D = m = p^d$. Suppose D is a cyclic algebra, say

$$D = \big(K/F,\sigma,a\big) = \bigoplus_{i=0}^{m-1} Kz^i,$$

where K is a cyclic Galois field extension of F with $\mathcal{G}(K/F) = \langle\sigma\rangle$, $zcz^{-1} = \sigma(c)$ for all $c \in K$, and $z^m = a \in F^\times$. The maximal subfield $T = F(z)$ of D is purely inseparable over F, as $z^{p^d} = a$; hence $\overline{T}$ is purely inseparable over $\overline{F}$. But $\overline{T} \subseteq \overline{D} = Z(\overline{D})$ as D is semiramified, and $Z(\overline{D})$ is separable over $\overline{F}$, as D is tame over F; hence, $\overline{T} = \overline{F}$. Because T is defectless over F (since the tame division algebra D is defectless over F), T must therefore be totally ramified over F. Since the maximal subfield T splits D, Prop. 8.40 shows that D is decomposably semiramified.

Conversely, suppose that D is decomposably semiramified. Then, by definition D is a tensor product of cyclic algebras. A result of Albert, see Albert [4, Lemma 13, p. 109] or Jacobson [108, Th. 4.5.1, p. 171], says that any tensor product of cyclic p-algebras is itself cyclic. Hence, D is a cyclic algebra. □

To obtain explicit examples, we use the $\mathcal{S}$ and $\mathcal{S}'$ constructions of §8.4.5:

Corollary 9.25. *Let k be a field of characteristic $p \neq 0$, and let K be an abelian but noncyclic Galois field extension of k with $[K{:}k] = p^s$ for $s \geq 2$. Fix a decomposition of K into a tensor product of cyclic Galois extensions of k,*

$$K = K_1 \otimes_k \ldots \otimes_k K_r.$$

For any central simple algebra A over k with $[A] \in Br(K/k) \setminus Dec(K/k)$, the division algebras $\mathcal{S}(A; K_1, \ldots, K_r)$ and $\mathcal{S}'(A; K_1, \ldots, K_r)$ are noncyclic p-algebras of degree p^s.

Proof. Write $\mathcal{S}$ for $\mathcal{S}(A; K_1, \ldots, K_r)$ and $\mathcal{S}'$ for $\mathcal{S}'(A; K_1, \ldots, K_r)$. Then, $\mathcal{S}$ is semiramified and tame over its Henselian center, and since A is split by K we have $deg\, \mathcal{S} = [K{:}k] = p^s$, see Prop. 8.72. Therefore, $\mathcal{S}$ is a p-algebra. But, in view of Cor. 8.68, $\mathcal{S}$ is not decomposably ramified because $sp(\mathcal{S})$ is represented by $[A]$, which does not lie in $Dec(K/k)$. Therefore, $\mathcal{S}$ is noncyclic by Th. 9.24. Since $\mathcal{S}' \otimes_{Z(\mathcal{S}')} Z(\mathcal{S}) \cong \mathcal{S}$ by Cor. 8.73, $\mathcal{S}'$ must also be noncyclic. □

Here is a specific example:

Example 9.26. *A noncyclic p-algebra.* Let p be a prime number, let q be a power of p, and let $\mathbb{F}_q$ be the finite field with q elements. Let k be the local field $\mathbb{F}_q((t))$, with its usual complete discrete t-adic valuation w, with $\overline{k} = \mathbb{F}_q$ and $\Gamma_k = \mathbb{Z}$. Let $K_1 = \mathbb{F}_{q^p}((t))$ and let $M = \mathbb{F}_{q^{p^2}}((t))$, which is the unramified field extension of k of degree p^2; let $\mathcal{G}(M/k) = \mathcal{G}(\mathbb{F}_{q^{p^2}}/\mathbb{F}_q) = \langle \varphi \rangle$. Let $f = X^p - X - t^{-1} \in k[X]$, let s be a root of f in k_{alg}, and let $K_2 = k(s)$ and $K = K_1(s)$. Note that f is a generalized Eisenstein polynomial as in Ex. 5.48, and f is also an Artin–Schreier polynomial. Hence, f is irreducible in $k[X]$, K_2 is totally ramified over k with $[K_2{:}k] = p$, $w(s) = \frac{1}{p} w(t^{-1}) = \frac{-1}{p}$, and K_2 is Galois over k. Since K_1 is unramified over k, the fields K_1 and K_2 are linearly disjoint over k. So, $[K{:}k] = p^2$ and $K = K_1 \otimes_k K_2$ is Galois over k with $\mathcal{G}(K/k) \cong \mathbb{Z}/p\mathbb{Z} \times \mathbb{Z}/p\mathbb{Z}$.

Let $A = (M/k, \varphi, t)$, which is a cyclically semiramified division algebra over k by Prop. 8.41, with $deg\, A = p^2$ and $\Gamma_A = \langle \frac{1}{p^2} w(t) \rangle + \Gamma_k = \frac{1}{p^2}\mathbb{Z}$. Also, $exp\ A = p^2$, since

$$p^2 \;=\; exp(\Gamma_A/\Gamma_k) \;=\; exp(\chi_A) \mid exp\ A \mid deg\, A \;=\; p^2.$$

Observe that K splits A. Indeed, since k is a local field, K splits every central simple k-algebra of index dividing $[K{:}k] = p^2$, see, e.g., Reiner [199, Cor. (31.10)]. But we can see the splitting of A more directly as follows: Let $J = C_A(K_1) \cong \big(M/K_1, \varphi^p, t\big)$, so $A \otimes_k K_1 \sim J$ in $Br(K_1)$. Then, in $Br(K)$,

$$\begin{aligned} A \otimes_k K \sim J \otimes_{K_1} K &\sim \big(M(s)/K, \varphi^p, t\big) \\ &\sim \big(M(s)/K, \varphi^p, ts^p\big) \otimes_K \big(M(s)/K, \varphi^p, s^{-1}\big)^{\otimes p}. \end{aligned}$$

Since $M(s) = \mathbb{F}_{q^{p^2}} \cdot K$ is inertial over K and $w(ts^p) = 0$, the first term on the second line has an inertial gauge (cf. Ex. 6.37); hence, it is split, as $Br_{in}(K) \cong Br(\overline{K}) = Br(\mathbb{F}_{q^p}) = \{0\}$. The second term on the second displayed line is split since $exp\,\big(M(s)/K, \varphi^p, s\big) \,\big|\, [M(s){:}K] = p$. Thus, K splits A. However, $Dec(K/k)$ is a p-torsion group, since $\mathcal{G}(K/k)$ has exponent p (cf. (8.12)). Because $exp\, A = p^2$, we have $[A] \in Br(K/k) \setminus Dec(K/k)$. Therefore, $\mathcal{S}(A; K_1, K_2)$ and $\mathcal{S}'(A; K_1, K_2)$ are noncyclic p-algebras of degree p^2. Variations of this construction give examples of noncyclic p-algebras of degree p^s and exponent p^r for any $s \geq r \geq 2$.

These examples are remarkably easy, given that the existence of noncyclic p-algebras for odd p was a major open question for some decades.

We now turn to the result of Saltman as adapted for division algebras over Henselian fields by McKinnie. If G is a finite group, we say that a division algebra D is a *crossed product with group* G if D has a maximal subfield L Galois over the center with $\mathcal{G}(L/Z(D)) \cong G$. The Saltman–McKinnie approach yields examples of division p-algebras that are crossed products with group G for only a single group G, which is a noncyclic abelian p-group. Such an algebra is of course noncyclic.

Theorem 9.27 (Saltman, McKinnie). *Let (F, v) be a Henselian-valued field with char $F = p > 0$, and let D be a tame semiramified division p-algebra with center F. If the specialization coset $sp(D) \in Br(\overline{D}/\overline{F})/Dec(\overline{D}/\overline{F})$ is strong, as in Def. 9.21, then every subfield L of D Galois over F is inertial. In particular, $\mathcal{G}(L/F)$ is a homomorphic image of Γ_D/Γ_F, and $\mathcal{G}(L/F) \cong \Gamma_D/\Gamma_F$ if L is a maximal subfield.*

Proof. Consider the graded division algebra $\mathsf{gr}(D)$. Proposition 8.71 shows that $sp\big(\mathsf{gr}(D)\big) = sp(D)$. Hence, the hypothesis that $sp(D)$ is strong yields by Cor. 9.22 that $\mathsf{gr}(D)$ contains no nontrivial totally ramified graded field extension of $\mathsf{gr}(F)$. Since v is defectless in D, it is also defectless in L. Moreover, we have $\overline{L} \subseteq \overline{D}$, so $\overline{L}$ is a separable extension of $\overline{F}$ because D is tame semiramified. We also have $\mathsf{gr}(L) \subseteq \mathsf{gr}(D)$, so $\mathsf{gr}(L)$ contains no nontrivial totally ramified graded field extension of $\mathsf{gr}(F)$. By Cor. 5.54 it follows that $|\Gamma_L{:}\Gamma_F| = 1$, so L is an inertial extension of F. We then have a canonical isomorphism $\mathcal{G}(L/F) \cong \mathcal{G}(\overline{L}/\overline{F})$, see Cor. 5.53, and $\mathcal{G}(\overline{L}/\overline{F})$ is a homomorphic image of $\mathcal{G}(\overline{D}/\overline{F})$ because $\overline{L}/\overline{F}$ is a subextension of $\overline{D}/\overline{F}$. To complete the proof, recall that θ_D defines an isomorphism $\Gamma_D/\Gamma_F \xrightarrow{\sim} \mathcal{G}(\overline{D}/\overline{F})$. □

Proposition 9.28. *Let G be an arbitrary noncyclic abelian p-group, and let k be an arbitrary field of characteristic p. There exists a division p-algebra D over a field extension of k such that D is a crossed product with group G, but not a crossed product with any group G' not isomorphic to G. In particular, D is not a cyclic algebra.*

Proof. Let $|G| = p^n$. The first step is to construct a division algebra A over a field containing k such that $exp\,A = deg\,A = p^n$ and A is a crossed product with group G.

Set $\ell = k(t_1, \ldots, t_{p^n})$, the rational function field in p^n indeterminates, and let σ be the k-automorphism of ℓ that permutes the indeterminates cyclically:

$$\sigma : t_1 \mapsto t_2 \mapsto \ldots \mapsto t_{p^n} \mapsto t_1.$$

Let $k_0 = \ell^\sigma$ be the fixed subfield of ℓ under σ. Extend σ to the Laurent series field $\ell((u))$ by setting $\sigma(u) = u$, and let $k_1 = \ell^\sigma = k_0((u))$. We equip $\ell((u))$ with its usual complete discrete rank 1 (so Henselian) u-adic valuation. Let A be the cyclically semiramified division algebra

$$A = \big(l((u))/k_1, \sigma, u\big),$$

of degree p^n. We have $exp\,A = p^n$ by Th. 8.65(i), (iv) since $exp(\Gamma_A/\Gamma_{k_1}) = p^n$ (or by Prop. 8.72 since we can view $A = \mathcal{S}(k_0; \ell)$). By a theorem of Saltman [217, Th. 1] (or see Jacobson [108, Th. 4.4.10, pp. 168–169]), the cyclic p-algebra A is a crossed product with group H for any group H of order p^n. In particular, A contains a Galois extension k_2 of k_1 with $\mathcal{G}(k_2/k_1) \cong G$. Thus, A has the needed properties.

Note that $[A]$ lies in $Br(k_2/k_1)$, and by Prop. 9.23 $[A]$ represents a strong coset modulo $Dec(k_2/k_1)$ since $exp\,A = deg\,A$ and $trk\,G \geq 2$. Consider a decomposition of k_2 into cyclic extensions of k_1,

$$k_2 = \ell_1 \otimes_{k_1} \ldots \otimes_{k_1} \ell_r,$$

and let $D = \mathcal{S}(A; \ell_1, \ldots, \ell_r)$ as in (8.36). By Prop. 8.72 D is a tame division algebra of degree p^n with $sp(D) = [A] + Dec(k_2/k_1)$. Because D is inertially split and $\overline{D} \sim A \otimes_{k_1} k_2 \sim k_2$ by Prop. 8.72, D is semiramified. Since A is split by k_2, D is split by $k_2 \otimes_{k_1} Z(D)$, which has degree p^n over $Z(D)$; so, $k_2 \otimes_{k_1} Z(D)$ embeds into D as a maximal subfield, which is a Galois extension of $Z(D)$ with group isomorphic to G. Th. 9.27 applies, and shows that every other maximal subfield of D Galois over $Z(D)$ has Galois group isomorphic to G. □

Note that the condition that G be noncyclic is essential for Prop. 9.28: the theorem of Saltman used in the proof says that cyclic division p-algebras of degree p^n for $n \geq 2$ are crossed products with group H for *every* group H of order p^n.

9.4 Noncrossed product algebras

Recall from the introduction to this chapter that a central division algebra over a field F is said to be a *noncrossed product* if it does not contain any maximal subfield that is a Galois extension of F. The construction of noncrossed product division algebras is one of the highlights of the theory of finite-dimensional division algebras in the twentieth century. The first examples were given by Amitsur [6] in 1972. They are universal division rings, i.e., rings of quotients of rings of generic matrices (see Jacobson [107] or Saltman [223]), and the proof that they are not crossed products (in suitable degrees) is based on a comparison technique that we sketch below. Other examples, based on substantially different techniques, were given by Jacob–Wadsworth [105] in 1986, by Brussel [37] in 1995, and by Reichstein–Youssin[1] [198] in 2001. Brussel obtained examples of noncrossed product central division algebras over fields that are as well understood as the rational function field in one variable $\mathbb{Q}(t)$ or the Laurent series field $\mathbb{Q}((t))$. We give such examples §9.4.1 and §9.4.2, by adapting explicit constructions due to Hanke [92] and Coyette [57]. We also abstract the essence of Brussel's approach in Prop. 9.34 by proving that a tame central division algebra D over Henselian field F has a maximal subfield tame and Galois over F if and only if $\overline{D}$ has a maximal subfield Galois over $\overline{F}$ (not just over $Z(\overline{D})$). In §9.4.3 we show how to construct examples of large degree from noncrossed product division algebras of small degree, using an idea of Saltman [219].

Let k be an arbitrary finite field. For any integer $n > 0$, there is a *universal division algebra* $UD(k,n)$ of degree n, which is the ring of central quotients of the algebra generated over k by countably many[2] generic matrices of order n, see Jacobson [107, §II.4] or Saltman [223, Ch. 14]. Its center is a transcendental extension of k.

Recall that a division algebra D is said to be a crossed product with group G if D contains a maximal subfield that is Galois over $Z(D)$ with Galois group isomorphic to G.

Theorem 9.29 (Amitsur). *Let G be a group of order n. If $UD(k,n)$ is a crossed product with group G, then every division algebra of degree n with center containing k is a crossed product with group G.*

For the proof, we refer to Jacobson [107, Th. 4, p. 93] or to Saltman [223, Cor. 14.28].

---

1 The Reichstein–Youssin construction seems to be the only one so far that does not rely on valuation theory in an essential way.

2 Saltman considers a universal division algebra $UD(k,n)(r)$ built from a finite number $r \geq 2$ of generic matrices. The specialization properties of $UD(k,n)$ and $UD(k,n)(r)$ are the same; in particular, Th. 9.29 holds for $UD(k,n)(r)$ as well.

Theorem 9.30. *Let k be an arbitrary infinite field.*

(i) *If n is a positive integer divisible by the cube of a prime and $\operatorname{char} k \nmid n$, then $UD(k,n)$ is not a crossed product.*
(ii) *If $\operatorname{char} k = p \neq 0$, then $UD(k,p^\alpha)$ is not a crossed product for any integer $\alpha \geq 3$.*

Proof. (i) Assume that $\operatorname{char} k \nmid n$. Let $n = p^\alpha m$, where p is prime, m is not divisible by p, and $\alpha \geq 3$. The separable closure k_{sep} contains a primitive n-th root of unity, so we may use the $\mathcal{T}$-construction of §9.1.3 to form the following algebras:

$$\mathcal{T}_1 = \mathcal{T}(k_{sep}; p^\alpha, m) \qquad \text{and} \qquad \mathcal{T}_2 = \mathcal{T}(k_{sep}; \underbrace{p, \ldots, p}_{\alpha}, m).$$

They each have degree n, and their centers contain k. Therefore, if $UD(k,n)$ is a crossed product with group G, then by Th. 9.29 $\mathcal{T}_1$ and $\mathcal{T}_2$ also are crossed products with group G. From Prop. 9.9 it then follows that G is isomorphic to a Lagrangian of $\mathbb{S}_{p^\alpha} \perp \mathbb{S}_m$ and also to a Lagrangian of $\mathbb{S}_p^\alpha \perp \mathbb{S}_m$. In particular, G is abelian, and its p-primary component $G_{(p)}$ has order p^α and is isomorphic to a Lagrangian of $\mathbb{S}_{p^\alpha}$ and also to a Lagrangian of $\mathbb{S}_p^\alpha$. But Lagrangians of $\mathbb{S}_{p^\alpha}$ have torsion rank at most 2 by Prop. 7.5, and Lagrangians of $\mathbb{S}_p^\alpha$ are elementary abelian of order p^α, hence they have torsion rank α. These conditions are incompatible as $\alpha \geq 3$. Therefore, there is no such group G, so $UD(k,n)$ is not a crossed product.

(ii) Now, assume $n = p^\alpha$ with $p = \operatorname{char} k$ and $\alpha \geq 3$. The groups $G_1 = (\mathbb{Z}/p\mathbb{Z})^\alpha$ and $G_2 = (\mathbb{Z}/p^{\alpha-1}\mathbb{Z}) \times (\mathbb{Z}/p\mathbb{Z})$ are nonisomorphic noncyclic abelian groups of order p^α. By Prop. 9.28, we may find division algebras D_1, D_2 over extensions of k with the property that D_i is a crossed product with group G_i only. By Amitsur's Theorem 9.29, it follows that $UD(k,n)$ is not a crossed product. □

Remark 9.31. We will see in Cor. 9.45 below that $UD(\mathbb{Q},n)$ is not a crossed product if n is divisible by the square of an odd prime. For p an odd prime, it is not known whether $UD(k,p^2)$ is a crossed product when the field k contains a primitive p-th root of unity, nor when $p = \operatorname{char} k$.

For the rest of this section, we turn to a different kind of construction, pioneered by Brussel, which yields noncrossed products with a center that is easier to describe. The basic observation is the following:

Proposition 9.32. *Let D be a central division algebra over a field F with a Henselian valuation v. Assume D is inertially split and let $M \subseteq D$ be a maximal subfield. Then $\overline{M} \cdot Z(\overline{D})$ is a maximal subfield of $\overline{D}$. Moreover, if M is a Galois extension of F and $\operatorname{char} \overline{F} \nmid \deg D$, then $\overline{M} \cdot Z(\overline{D})$ is a Galois extension of $\overline{F}$.*

Proof. By Prop. 9.2 we know that $\overline{M}\cdot Z(\overline{D})$ splits $\overline{D}$. But $\overline{M}\cdot Z(\overline{D}) \subseteq \overline{D}$, so $\overline{M}\cdot Z(\overline{D})$ is a maximal subfield of $\overline{D}$. If M is Galois over F and $char\,\overline{F} \nmid deg\,D$, then M is tamely ramified over F since $[M{:}F] = deg\,D$, hence $\overline{M}$ is Galois over $\overline{F}$ (see Th. A.8 and Th. A.9). On the other hand, $Z(\overline{D})$ is Galois over $\overline{F}$ by Prop. 1.5, hence $\overline{M}\cdot Z(\overline{D})$ is Galois over $\overline{F}$. □

Thus, if D is a crossed product and $char\,\overline{F} \nmid deg\,D$, then $\overline{D}$ also is a crossed product, but with a maximal subfield $\overline{M}\cdot Z(\overline{D})$ that is not only Galois over $Z(\overline{D})$, but also over $\overline{F}$. This additional constraint is sometimes sufficient to rule out the existence of Galois maximal subfields in D, as we will see in the next two subsections.

When D is not inertially split, then $\overline{M}\cdot Z(\overline{D})$ need not be a maximal subfield of $\overline{D}$ for M a maximal subfield of D. Nonetheless, we next show, by changing the choice of maximal subfield, that if D is tame, then a tame maximal subfield Galois over $F = Z(D)$ yields a maximal subfield of $\overline{D}$ Galois over $\overline{F}$ (and conversely). We first prove the graded version of this.

Proposition 9.33. *Let* D *be a graded division algebra with center* F. *The following are equivalent:*

(a) D *has a maximal graded subfield Galois over* F.
(b) D_0 *has a maximal subfield Galois over* F_0.

Proof. Let $\mathsf{U}, \mathsf{Z}, \mathsf{C}, \mathsf{E}$ be the canonical subalgebras of D described in §8.2.1. We will use repeatedly the properties of these subalgebras given in Prop. 8.16.

(b) ⇒ (a) Let M be a maximal subfield of D_0, with M Galois over F_0. The maximality implies that $M \supseteq Z(\mathsf{D}_0)$. Let $\mathsf{L} = M\cdot\mathsf{F} = M \otimes_{\mathsf{F}_0} \mathsf{F}$, which is a graded subfield of D. This L is the inertial extension of F with $\mathsf{L}_0 = M$; so

$$[\mathsf{L}{:}\mathsf{F}] \;=\; [M{:}\mathsf{F}_0] \;=\; ind\,\mathsf{D}_0 \cdot [Z(\mathsf{D}_0){:}\mathsf{F}_0].$$

Also, $\mathsf{Z} \subseteq \mathsf{L} \subseteq \mathsf{U}$, as $Z(\mathsf{D}_0) \subseteq M \subseteq \mathsf{D}_0$. Moreover, L is Galois over F, since L_0 is Galois over F_0.

Let T be any maximal graded subfield of C. Since C is totally ramified over its center Z, its graded subfield T is also totally ramified over Z. Hence, T is obtainable from C by restricting the grade group:

$$\mathsf{T} = \bigoplus_{\gamma\in\Gamma_{\mathsf{T}}} \mathsf{C}_\gamma, \tag{9.7}$$

and T is Galois over Z by Prop. 7.54(iii). We claim that T is also Galois *over* F. For this, recall that $\mathsf{Z} = Z(\mathsf{D}_0) \otimes_{\mathsf{F}_0} \mathsf{F}$ is Galois over F with $\mathcal{G}(\mathsf{Z}/\mathsf{F}) \cong \mathcal{G}(Z(\mathsf{D}_0)/\mathsf{F}_0)$. Take any $\sigma \in \mathcal{G}(\mathsf{Z}/\mathsf{F})$. Since θ_{D} is surjective, there is a $d \in \mathsf{D}^\times$ with $int(d)|_{Z(\mathsf{D}_0)} = \sigma|_{Z(\mathsf{D}_0)}$; so, $int(d)|_{\mathsf{Z}} = \sigma$. Because the unit d of D is homogeneous, the map $int(d)$ is a degree-preserving automorphism of D. Moreover, $int(d)$ maps C to itself, since C is canonical in D. So, (9.7) shows that $int(d)$ maps T to T. Since Z is Galois over F and every automorphism in $\mathcal{G}(\mathsf{Z}/\mathsf{F})$ extends to a graded automorphism of T, and T is Galois over Z, it follows that T is Galois over F, as claimed. Let

$\mathsf{K} = \mathsf{L} \otimes_{\mathsf{Z}} \mathsf{T} \subseteq \mathsf{U} \otimes_{\mathsf{Z}} \mathsf{C} = \mathsf{E} \subseteq \mathsf{D}$. Then, K is a graded subfield of D, and K is Galois over F, since L and T are each Galois over F. Moreover,

$$[\mathsf{K}:\mathsf{Z}] \,=\, [\mathsf{L}:\mathsf{Z}] \cdot [\mathsf{T}:\mathsf{Z}] \,=\, deg\,\mathsf{U} \cdot deg\,\mathsf{C} \,=\, deg\,\mathsf{E}.$$

This shows that K is a maximal graded subfield of E, hence also a maximal graded subfield of D.

(a) $\Rightarrow$ (b) Suppose K is a maximal graded subfield of D with K Galois over F. The field K_0 is then Galois over F_0, but it need not be a maximal subfield of D_0. We therefore modify K to enlarge the degree-0 part.

Let T be any maximal graded subfield of C. We proved just above that T is Galois over F. Let $\mathsf{S} = C_{\mathsf{D}}(\mathsf{T})$. Since $\mathsf{Z} = Z(\mathsf{C}) \subseteq \mathsf{T}$, we have $\mathsf{S} \subseteq C_{\mathsf{D}}(\mathsf{Z}) = \mathsf{E} = \mathsf{U} \otimes_{\mathsf{Z}} \mathsf{C}$. Hence, by Prop. 2.30

$$\mathsf{S} \,=\, C_{\mathsf{E}}(\mathsf{T}) \,=\, C_{\mathsf{U}\otimes_{\mathsf{Z}}\mathsf{C}}(\mathsf{Z} \otimes_{\mathsf{Z}} \mathsf{T}) \,=\, C_{\mathsf{U}}(\mathsf{Z}) \otimes_{\mathsf{Z}} C_{\mathsf{C}}(\mathsf{T}) \,=\, \mathsf{U} \otimes_{\mathsf{Z}} \mathsf{T}.$$

Therefore, as $\Gamma_{\mathsf{U}} = \Gamma_{\mathsf{F}}$ and $\mathsf{U}_0 = \mathsf{D}_0$, we have $\Gamma_{\mathsf{S}} = \Gamma_{\mathsf{U}} + \Gamma_{\mathsf{T}} = \Gamma_{\mathsf{T}}$ and $\mathsf{S}_0 = \mathsf{D}_0$. The latter equality shows that D is totally ramified over S; hence, for each $\gamma \in \Gamma_{\mathsf{S}}$ we have $\mathsf{S}_\gamma = \mathsf{D}_\gamma$.

Let $\mathsf{N} = \mathsf{K} \cap \mathsf{S}$, a graded subfield of K. Then $\mathsf{N}_0 = \mathsf{K}_0 \cap \mathsf{S}_0 = \mathsf{K}_0 \cap \mathsf{D}_0 = \mathsf{K}_0$. Hence, K is totally ramified over N, so N is obtainable from K by restricting the graded group:

$$\mathsf{N} \,=\, \bigoplus_{\gamma \in \Gamma_{\mathsf{N}}} \mathsf{K}_\gamma. \tag{9.8}$$

Recall that any $\sigma \in \mathcal{G}(\mathsf{K}/\mathsf{F})$ maps each K_γ to itself, so by (9.8) $\sigma(\mathsf{N}) = \mathsf{N}$. Therefore, $\mathsf{N}^{\mathcal{G}(\mathsf{N}/\mathsf{F})} \subseteq \mathsf{N} \cap \mathsf{K}^{\mathcal{G}(\mathsf{K}/\mathsf{F})} = \mathsf{F}$. Hence, N is Galois over F. Note also that for any $\gamma \in \Gamma_{\mathsf{K}} \cap \Gamma_{\mathsf{S}}$ we have $\mathsf{N}_\gamma = \mathsf{K}_\gamma \cap \mathsf{S}_\gamma = \mathsf{K}_\gamma \cap \mathsf{D}_\gamma = \mathsf{K}_\gamma \neq \{0\}$. Therefore, $\Gamma_{\mathsf{N}} = \Gamma_{\mathsf{K}} \cap \Gamma_{\mathsf{S}}$.

Let

$$\mathsf{K}' \,=\, \mathsf{N}\cdot\mathsf{T} \,=\, \big(\mathsf{K} \cap C_{\mathsf{D}}(\mathsf{T})\big)\cdot\mathsf{T},$$

which is a graded field as N centralizes T. Also, K' is Galois over F, since N and T are each Galois over F. It remains to see that K' is a maximal graded subfield of D and K'_0 is a maximal subfield of D_0. For this note first that as K is totally ramified over N, we have

$$[\mathsf{K}:\mathsf{N}] \,=\, |\Gamma_{\mathsf{K}}:\Gamma_{\mathsf{N}}| \,=\, |\Gamma_{\mathsf{K}}:\Gamma_{\mathsf{K}} \cap \Gamma_{\mathsf{S}}| \,=\, |\Gamma_{\mathsf{K}} + \Gamma_{\mathsf{S}}:\Gamma_{\mathsf{S}}|.$$

Since K and S each lie in $C_{\mathsf{D}}(\mathsf{K} \cap \mathsf{T})$, we have $\Gamma_{\mathsf{K}} + \Gamma_{\mathsf{S}} \subseteq \Gamma_{C_{\mathsf{D}}(\mathsf{K}\cap\mathsf{T})}$. Thus,

$$|\Gamma_{\mathsf{K}} + \Gamma_{\mathsf{S}}:\Gamma_{\mathsf{S}}| \,\leq\, |\Gamma_{C_{\mathsf{D}}(\mathsf{K}\cap\mathsf{T})}:\Gamma_{\mathsf{S}}| \,\leq\, |C_{\mathsf{D}}(\mathsf{K} \cap \mathsf{T}):\mathsf{S}| \,=\, [\mathsf{T}:\mathsf{K} \cap \mathsf{T}],$$

with the last equality following from the graded Double Centralizer Theorem 2.35 applied to the graded division subalgebras $\mathsf{K} \cap \mathsf{T}$ and T of D. The displayed inequalities show that $[\mathsf{K}:\mathsf{N}] \leq [\mathsf{T}:\mathsf{K} \cap \mathsf{T}]$. Note that $\mathsf{N} \cap \mathsf{T} = (\mathsf{K} \cap C_{\mathsf{D}}(\mathsf{T})) \cap \mathsf{T} = \mathsf{K} \cap \mathsf{T}$. Because N is Galois over F, we thus have

$$[\mathsf{K}':\mathsf{N}] \,=\, [\mathsf{N}\cdot\mathsf{T}:\mathsf{N}] \,=\, [\mathsf{T}:\mathsf{N} \cap \mathsf{T}] \,=\, [\mathsf{T}:\mathsf{K} \cap \mathsf{T}] \,\geq\, [\mathsf{K}:\mathsf{N}].$$

Since K is a maximal graded subfield of D, it follows that K' is also a maximal graded subfield of D. Moreover, since $\mathsf{T} \subseteq \mathsf{K}' \subseteq \mathsf{S} = C_{\mathsf{D}}(\mathsf{T})$, we have $\Gamma_{\mathsf{T}} \subseteq \Gamma_{\mathsf{K}'} \subseteq \Gamma_{\mathsf{S}} = \Gamma_{\mathsf{T}}$; so $\Gamma_{\mathsf{K}'} = \Gamma_{\mathsf{T}}$. Because K' contains Z, it is a maximal graded subfield of $C_{\mathsf{D}}(\mathsf{Z}) = \mathsf{E}$. Thus,

$$\begin{aligned}[\mathsf{K}_0'{:}\mathsf{Z}_0]\,|\Gamma_{\mathsf{K}'}{:}\Gamma_{\mathsf{Z}}| \;&=\; [\mathsf{K}'{:}\mathsf{Z}] \;=\; deg\,\mathsf{E} \;=\; deg\,\mathsf{U}\cdot deg\,\mathsf{C}\\ &=\; deg\,\mathsf{D}_0\,[\mathsf{T}{:}\mathsf{Z}] \;=\; deg\,\mathsf{D}_0\,|\Gamma_{\mathsf{T}}{:}\Gamma_{\mathsf{Z}}| \;=\; deg\,\mathsf{D}_0\,|\Gamma_{\mathsf{K}'}{:}\Gamma_{\mathsf{Z}}|.\end{aligned}$$

Therefore $[\mathsf{K}_0'{:}\mathsf{Z}_0] = deg\,\mathsf{D}_0$, showing that K_0' is maximal subfield of D_0. Moreover K_0' is Galois over F_0 as K' is Galois over F. □

Proposition 9.34. *Let F be a Henselian field, and let D be a tame division algebra with center F. The following are equivalent:*

(a) *D has a maximal subfield Galois and tamely ramified over F.*
(b) *$\overline{D}$ has a maximal subfield Galois over $\overline{F}$.*

Proof. Let K be a maximal subfield of D that is Galois and tamely ramified over F. The tameness assures that the graded field $\mathsf{gr}(K)$ is Galois over $\mathsf{gr}(F)$, by Prop. 5.55. Moreover, K is defectless over F since the tame division algebra D is defectless over F. Hence,

$$[\mathsf{gr}(K){:}\,\mathsf{gr}(F)] \;=\; [K{:}F] \;=\; deg\,D \;=\; deg\,\mathsf{gr}(D),$$

showing that $\mathsf{gr}(K)$ is a maximal graded subfield of $\mathsf{gr}(D)$ by Cor. 2.36(ii). By Prop. 9.33, $\overline{D} = \mathsf{gr}(D)_0$ has a maximal subfield Galois over $\overline{F} = \mathsf{gr}(F)_0$. Conversely, suppose $\overline{D}$ has a maximal subfield Galois over $\overline{F}$. By Prop. 9.33, $\mathsf{gr}(D)$ has a maximal graded subfield L that is Galois (so also tame) over F. By Th. 8.9 there is a tame lift M in D of L over F. Then,

$$[M{:}F] \;=\; [\mathsf{L}{:}\,\mathsf{gr}(F)] \;=\; deg\,\mathsf{gr}(D) \;=\; deg\,D,$$

so M is a maximal subfield of D. Moreover, K is Galois over F by Prop. 5.55 since $\mathsf{gr}(K) = \mathsf{L}$ is Galois over $\mathsf{gr}(F)$. □

9.4.1 Noncrossed products of degree p^2

In this subsection, p is an odd prime. Let A be a central division algebra of degree p^2 over a field k and let $K \subseteq A$ be a subfield which is a cyclic Galois extension of k of degree p. By Th. 1.4, any valuation on k that extends to A extends uniquely to K.

Lemma 9.35. *Assume k carries two valuations v_1, v_2 with the following properties:*

(i) *the residue field $\overline{k}^{v_1}$ does not contain a primitive p^2-root of unity and $\operatorname{char} \overline{k}^{v_1} \neq p$;*

(ii) *the residue field* $\overline{k}^{v_2}$ *is finite, does not contain a primitive p-th root of unity, and* $\operatorname{char}\overline{k}^{v_2} \neq p$;
(iii) v_2 *extends to a valuation on* A;
(iv) *the extension* K/k *is totally ramified for the unique extension of* v_1 *to* K, *and inertial for the unique extension of* v_2.

Then, A *does not contain any maximal subfield* L *which is Galois over* k *and contains* K.

Proof. Assume the contrary; then $\mathcal{G}(L/F)$ is either cyclic or elementary abelian. We use v_1 to exclude the first case and v_2 for the second.
Case I: $\mathcal{G}(L/k) \cong \mathbb{Z}/p^2\mathbb{Z}$.
For any extension of v_1 to L, the inertia field of L/k cannot contain K, as K is totally ramified over k. Therefore, the inertia field must be k. Also, the ramification field must be L, as $char\,\overline{k}^{v_1} \neq p$ (see Th. A.9(ii)). Hence, L is totally ramified over k for v_1. But this is impossible: There is no cyclic extension of k of degree p^2 that is tamely and totally ramified for v_1, since $\overline{k}^{v_1}$ does not contain a primitive p^2-root of unity, see Th. A.9(vi).
Case II: $\mathcal{G}(L/k) \cong (\mathbb{Z}/p\mathbb{Z})^2$.
Since v_2 extends to A it extends uniquely to every subfield of A, by Th. 1.4. Thus v_2 extends uniquely to L, and the extension is tamely ramified since $char\,\overline{k}^{v_2} \neq p$. The extension cannot be inertial, since otherwise the residue field extension would be elementary abelian Galois of degree p^2; this is impossible because $\overline{k}^{v_2}$ is finite. Therefore, the inertia field is K, and L is totally ramified over K. By Th. A.9(vi), it follows that $\overline{K}^{v_2}$ contains a primitive p-th root of unity. This is impossible since $\overline{k}^{v_2}$ does not contain a p-th root of unity and $[\overline{K}^{v_2}:\overline{k}^{v_2}] = p$. □

Examples satisfying the conditions in Lemma 9.35 are easy to construct with k a global field. Here is a specific construction with $k = \mathbb{Q}$:

Example 9.36. Take $k = \mathbb{Q}$ and choose two primes q_1, q_2 different from p, such that

$$q_1 \equiv 1 \ (mod\ p), \quad q_1 \not\equiv 1 \ (mod\ p^2), \quad \text{and} \quad q_2 \not\equiv 1 \ (mod\ p).$$

(There exist infinitely many such q_i by Dirichlet's theorem on primes in an arithmetic progression.) Let v_1 (resp. v_2) be the q_1-adic (resp. q_2-adic) valuation on $\mathbb{Q}$. The conditions on q_1 and q_2 imply that v_1 and v_2 satisfy (i) and (ii) in Lemma 9.35. Since $q_1 \equiv 1$ $(mod\ p)$, the residue field $\overline{\mathbb{Q}}^{v_1} = \mathbb{F}_{q_1}$ contains a primitive p-th root of unity, hence $\mathbb{Q}_{q_1}(\sqrt[p]{q_1})$ is a cyclic extension of $\mathbb{Q}_{q_1}$. By the Grunwald–Wang Theorem (see Artin–Tate [16, Ch. 10] or Lorenz–Roquette [128]) there exists a cyclic extension K of $\mathbb{Q}$ of degree p such that $K \otimes_{\mathbb{Q}} \mathbb{Q}_{q_1} \cong \mathbb{Q}_{q_1}(\sqrt[p]{q_1})$ and $K \otimes_{\mathbb{Q}} \mathbb{Q}_{q_2}$ is isomorphic to the (unique) inertial extension of degree p of $\mathbb{Q}_{q_2}$, which is cyclic Galois over $\mathbb{Q}_{q_2}$. The field K thus satisfies (iv). Now, by the Hasse–Brauer–Noether–Albert description of the Brauer group $Br(\mathbb{Q})$ (see Pierce [178, §18.5] or Reiner [199, (32.13)]), there is a central division algebra A over $\mathbb{Q}$ with local invariants $\frac{1}{p^2} + \mathbb{Z}$ at q_1, $\frac{-1}{p^2} + \mathbb{Z}$ at q_2, and 0 at every other prime. Thus, $deg\, A = p^2$ and $A \otimes_{\mathbb{Q}} \mathbb{Q}_{q_1}$, $A \otimes_{\mathbb{Q}} \mathbb{Q}_{q_2}$ are division algebras. Therefore, v_1 and v_2 extend to valuations on A by Prop. 1.16. Moreover, by Pierce [178, Lemma, p. 354] or Reiner [199, Th. (31.9)], the invariants of $A \otimes_{\mathbb{Q}} K$ are $\frac{1}{p} + \mathbb{Z}$ at the unique prime of K extending q_1, $\frac{-1}{p} + \mathbb{Z}$ at the unique prime extending q_2, and 0 at every other prime. Therefore, $ind(A \otimes_{\mathbb{Q}} K) = p$, hence $ms(A \otimes_{\mathbb{Q}} K) = p = [K{:}\mathbb{Q}]$, and it follows from the ungraded version of Lemma 8.8 that K embeds in A. Thus, all the hypotheses of Lemma 9.35 hold.

Proposition 9.37. *Assume the hypotheses of Lemma 9.35 hold for k, K, and A. Then, the division algebra $\mathcal{S}(A;K)$ of* (8.36) *with center $k((t))$ has degree p^2 and is not a crossed product.*

Proof. Write $\mathcal{S}$ for $\mathcal{S}(A;K)$. Since $K \subsetneq A$, we have $ind(A \otimes_k K) = p$, so $deg\, \mathcal{S} = p^2$, see Prop. 8.72. Moreover, $\overline{\mathcal{S}} \sim A \otimes_k K$, so $\overline{\mathcal{S}}$ is the centralizer of K in A,

$$\overline{\mathcal{S}} \cong C_A(K), \quad \text{and} \quad Z(\overline{\mathcal{S}}) \cong K.$$

We have $char\, k \neq p$ in view of condition (i) of Lemma 9.35, so Prop. 9.32 applies and shows that if M is a maximal subfield of $\mathcal{S}$ which is Galois over the center, then $\overline{M}{\cdot}K$ is a maximal subfield of $C_A(K)$ (hence also of A) which is Galois over k. This is impossible since Lemma 9.35 shows that there is no such field. Therefore, $\mathcal{S}$ is not a crossed product. □

Remark 9.38. Under the hypotheses of Prop. 9.37, the division algebra $\mathcal{S}'(A;K)$ defined in (8.37) has center $k(t)$ and degree p^2, and is not a crossed product either. This follows because $\mathcal{S}(A;K) = \mathcal{S}'(A;K) \otimes_{k(t)} k((t))$ by Cor. 8.73. Thus, Ex. 9.36 yields examples of noncrossed products over $\mathbb{Q}((t))$ and $\mathbb{Q}(t)$.

9.4.2 Noncrossed products of degree 8

Let A be a central division algebra of degree 8 over a number field k and let $K \subseteq A$ be a subfield which is a biquadratic extension of k, i.e., K is a Galois extension of k with elementary abelian Galois group of order 4. By Th. 1.4, any valuation on k that extends to A extends uniquely to K.

Lemma 9.39. *Assume k is real and K is not real. Assume further that k carries two valuations v_1, v_2 with the following properties:*

(i) *the finite residue fields $\overline{k}^{v_i}$ satisfy* $\left|\overline{k}^{v_i}\right| \equiv 3 \bmod 4$;
(ii) *Γ_{k,v_1} and Γ_{k,v_2} have rational rank* 1;
(iii) *the valuations v_1, v_2 extend to valuations on A;*
(iv) *the inertia fields of v_1 and v_2 in K are distinct quadratic extensions K_1, K_2 of k.*

Then, A does not contain any maximal subfield L that is Galois over k and contains K.

Proof. Groups of order 8 are either quaternionic, dihedral, or abelian. Assuming that L exists, we rule out one-by-one the various options for $\mathcal{G}(L/k)$. We view L as embedded in $\mathbb{C}$, and note that the valuations v_1, v_2 extend uniquely to L because $L \subseteq A$.

Case I: $\mathcal{G}(L/k)$ is quaternionic.
Since the quaternionic group contains a unique subgroup of order 2, K is the unique subfield in L of codimension 2. But the subfield R of L fixed under the complex conjugation is not K since K is not real, and is an extension of k since k is real. This case is thus impossible.

Case II: $\mathcal{G}(L/k)$ is dihedral.
For $i = 1, 2$, let L_i be the inertia field of L over k for v_i; so, L is Galois over L_i and $L_i \cap K$ is the inertia field K_i of K over k for v_i. (This follows by Galois theory, using the exact sequence of Th. A.8(iv) for the extensions L/k and K/k.) Moreover, the group $\mathcal{G}(L_i/k) \cong \mathcal{G}\left(\overline{L_i}^{v_i}/\overline{k}^{v_i}\right)$ is cyclic as $\overline{k}^{v_i}$ is finite. Also, by Prop. A.36

$$trk(\Gamma_{L,v_i}/\Gamma_{L_i,v_i}) \leq rr(\Gamma_{L_i,v_i}) = rr(\Gamma_{k,v_i}) = 1;$$

thus, $\Gamma_{L,v_i}/\Gamma_{L_i,v_i}$ is a cyclic group. Because L/L_i is totally and tamely ramified for v_i, it follows that $\mathcal{G}(L/L_i)$ is also cyclic, see Th. A.9(vi). Since the dihedral group contains a unique normal cyclic subgroup with the property that the factor group is also cyclic, it follows that $L_1 = L_2$. Hence, $K_1 = L_1 \cap K = L_2 \cap K = K_2$, contrary to condition (iv).

Case III: $\mathcal{G}(L/k)$ is abelian.
Let R, L_1, and L_2 be defined as in the two preceding cases. Since $R \neq K$, the intersection $R \cap K$ is a proper subfield of K. Renumbering v_1 and v_2, we may assume $R \cap K \neq K_1$, hence $R \cap K_1 = k$. But since L_1/k is cyclic, K_1 is the unique minimal proper extension of k contained in L_1, hence $R \cap L_1 = k$. It follows that R is totally ramified over k for v_1. It is also a Galois extension of k since every subgroup of $\mathcal{G}(L/k)$ is normal. Therefore, $\overline{k}^{v_1}$ contains a primitive 4-th root of unity, by Th. A.9(vi). Condition (i) shows that this is impossible. □

Example 9.40. To obtain specific examples satisfying the conditions of Lemma 9.39, take $k = \mathbb{Q}$ and $K = \mathbb{Q}(\sqrt{p}, \sqrt{-q})$ for prime numbers p and q satisfying

$$p \equiv q \equiv 3 \,(mod\, 4), \quad p + q\mathbb{Z} \notin (\mathbb{Z}/q\mathbb{Z})^2, \quad -q + p\mathbb{Z} \notin (\mathbb{Z}/p\mathbb{Z})^2$$

(e.g., $p = 3$ and $q = 7$). Let v_1 and v_2 be the p- and the q-adic valuations, and choose for A the central division algebra over $\mathbb{Q}$ with local invariants $\frac{1}{8} + \mathbb{Z}$ at p and $\frac{-1}{8} + \mathbb{Z}$ at q, and 0 at every other place. The same arguments as in Ex. 9.36 show that v_1 and v_2 extend to valuations on A, and that A contains (an isomorphic copy of) K. The inertia field of v_1 in K is $\mathbb{Q}(\sqrt{-q}\,)$, while the inertia field of v_2 is $\mathbb{Q}(\sqrt{p}\,)$.

Proposition 9.41. *Assume the hypotheses of Lemma 9.39 hold for k, K, and A. Then, the division algebra $\mathcal{S}(A;K_1,K_2)$ of (8.36) with center $k((t_1))((t_2))$ has degree and exponent 8 and is not a crossed product.*

Proof. We use the same arguments as in the proof of Prop. 9.37, slightly modified. Write $\mathcal{S}$ for $\mathcal{S}(A;K_1,K_2)$. Since $K\subseteq A$, we have $ind(A\otimes_k K)=2$, so $deg\,\mathcal{S}=8$. Also, $exp\,\mathcal{S}=8$ by Prop. 8.72 since $exp\,A=deg\,A=8$, as k is an algebraic number field. Moreover, $\overline{\mathcal{S}}\sim A\otimes_k K$, so $\overline{\mathcal{S}}$ is the centralizer of K in the division algebra A,

$$\overline{\mathcal{S}}\cong C_A(K), \quad \text{and} \quad Z(\overline{\mathcal{S}})\cong K.$$

Since $char\,k=0$, Prop. 9.32 applies and shows that if M is a maximal subfield of $\mathcal{S}$ that is Galois over the center, then $\overline{M}\cdot K$ is a maximal subfield of $C_A(K)$ (hence also of A) that is Galois over k. This is impossible since Lemma 9.39 shows that there is no such field. Therefore, $\mathcal{S}$ is not a crossed product. □

Remark 9.42. Under the same hypotheses as Prop. 9.41, the division algebra $\mathcal{S}'(A;K_1,K_2)$ has center $k(t_1,t_2)$, degree 8, exponent 8, and is not a crossed product since $\mathcal{S}(A;K_1,K_2)=\mathcal{S}'(A;K_1,K_2)\otimes_{k(t_1,t_2)}k((t_1))((t_2))$ by Cor. 8.73.

9.4.3 Iterative construction of noncrossed products

Given a noncrossed product division algebra of some degree d, we show in this subsection how noncrossed product division algebras of degree any multiple of d can be constructed. We follow an idea of Saltman [219], but simplify the discussion by considering only degrees that are not divisible by the characteristic. (See the Notes for comments on the general case.)

Throughout this subsection, D is a central division algebra over an arbitrary field k. We assume $char\,k\nmid deg\,D$.

Lemma 9.43. *Let t be an indeterminate over k. If $D\otimes_k k(t)$ is a crossed product, then D is a crossed product.*

Proof. Suppose $L\subseteq D\otimes_k k(t)$ is a maximal subfield that is a Galois extension of $k(t)$. Then $L\otimes_{k(t)}k((t))$ is a maximal subfield of the division algebra $D\otimes_k k((t))$ that is a Galois extension of $k((t))$. We apply Prop. 9.32 to $D\otimes_k k((t))$ with the t-adic valuation on $k((t))$ to see that $\overline{L\otimes_{k(t)}k((t))}$ is a maximal subfield of $\overline{D\otimes_k k((t))}=D$ that is a Galois extension of k. □

Now, let n be an arbitrary integer not divisible by $char\, k$. We take n independent indeterminates $t_1, \ldots, t_n$ over k and set $K = k(t_1, \ldots, t_n)$. Define a k-automorphism σ of K by permuting the indeterminates cyclically:

$$\sigma\colon t_1 \mapsto t_2 \mapsto t_3 \mapsto \ldots \mapsto t_n \mapsto t_1.$$

Let $F = K^\sigma$, and let u be an indeterminate over K. After extending σ to $K((u))$ by fixing u, we can build the cyclically semiramified division algebra $(K((u))/F((u)), \sigma, u)$, which has degree n and center $F((u))$.

Proposition 9.44. *The tensor product $B = D \otimes_k (K((u))/F((u)), \sigma, u)$ is a division algebra with center $F((u))$ and $\deg B = n \deg D$ and $\exp B = \operatorname{lcm}(\exp D, n)$. If B is a crossed product, then D is a crossed product.*

Proof. By definition of the algebras built by the $\mathcal{S}$ construction (see §8.4.5), we have

$$\mathcal{S}(D \otimes_k F; K) \sim B.$$

Since $t_1, \ldots, t_n$ are indeterminates over k, the algebra $D \otimes_k K$ is a division algebra, hence it follows from Prop. 8.72 that

$$deg\, \mathcal{S}(D \otimes_k F; K) = n\ deg\, D = deg\, B.$$

So in fact

$$\mathcal{S}(D \otimes_k F; K) = B,$$

and we have $\overline{\mathcal{S}(D \otimes_k F; K)} = D \otimes_k K$. Moreover, by Prop. 8.72,

$$exp(\mathcal{S}(D \otimes_k F; K)) = lcm\big(exp(D \otimes_k K), n\big) = lcm(\exp D, n).$$

If $L \subseteq \mathcal{S}(D \otimes_k F; K)$ is a maximal subfield that is Galois over the center, then Prop. 9.32 shows that $\overline{L} \cdot K$ is a maximal subfield of $D \otimes_k K$ that is a Galois extension of F, hence also of K. Therefore, $D \otimes_k K$ is a crossed product. But then, by applying Lemma 9.43 inductively, we see that D is a crossed product. □

Corollary 9.45. *For any positive integer d divisible by the square of an odd prime or by 8, the universal division algebra $UD(\mathbb{Q}, d)$ is not a crossed product.*

Proof. Suppose $p^2 \mid d$ for some odd prime number p. Take for D a noncrossed product division algebra $\mathcal{S}(A; K)$ as in Prop. 9.37 with A and K as in Ex. 9.36. By applying Prop. 9.44 with this D and $n = d/p^2$, we obtain a noncrossed product division algebra B of degree d containing $\mathbb{Q}$. By Amitsur's Theorem 9.29, $UD(\mathbb{Q}, d)$ is not a crossed product. The case where $8 \mid d$ was given in Th. 9.30(i), but can also be proved in the same way as the p^2 case by taking for D a division algebra $\mathcal{S}(A; K_1, K_2)$ as in Prop. 9.41. □

Saltman's goal in [219] was to produce noncrossed products with degree exceeding the exponent. This is achieved via Prop. 9.44:

Corollary 9.46. *There exist noncrossed product division algebras of exponent d and degree m for any positive integers d and m with the same prime factors and $d \mid m$ such that $p^2 \mid d$ for some odd prime p or $8 \mid d$.*

Proof. Let $D_1 = UD(\mathbb{Q}, d)$ where $p^2 \mid d$ for some odd prime p or $8 \mid d$. By the preceding corollary, D_1 is a noncrossed product division algebra. It is known (see Jacobson [107, Th. 1, p. 113]) that any universal division algebra $UD(k, r)$ has exponent equal to its degree. Thus, $exp\, D_1 = deg\, D_1 = d$. By applying Prop. 9.44 with $D = D_1$ and $n =$ any prime q dividing d, we obtain a noncrossed product division algebra $D_2 = B$ with $exp\, D_2 = d$ and $deg\, D_2 = dq$. By iterating this process we can obtain a noncrossed product division algebra of exponent d and index any multiple of d with the same prime factors as d. □

Exercises

Exercise 9.1. Prove that the degree-p^2 noncrossed product division algebra $\mathcal{S}(A; K)$ of Prop. 9.37 has exponent p^2.

Exercise 9.2. Let M/F be an elementary abelian Galois extension of degree p^2, for some prime p, and let $K \subseteq M$ be a subfield of codimension p. Suppose F contains a primitive p-th root of unity ω (hence $char\, F \neq p$). We say that a cyclic F-algebra A of degree p^2 is *adapted to M through K* if $A \cong (L/F, \sigma, a)$ for some cyclic Galois extension L/F and some $a \in F^\times$ such that $L^{\sigma^p} \cong K$ and $K(\sqrt[p]{a}) \cong M$. Show that there exists a cyclic F-algebra adapted to M through K if and only if ω is a norm from K to F. When this condition holds, the Brauer classes of cyclic F-algebras adapted to M through K form a coset modulo $Dec(M/F)$.

Exercise 9.3. Let F be a Henselian-valued field containing a primitive p-th root of unity for some prime p, with $p \neq char\, \overline{F}$. Let D be a tame semiramified central division algebra over F. Assume $deg\, D = p^2$ and $exp(\Gamma_D/\Gamma_F) = p$, so $\overline{D}/\overline{F}$ is an elementary abelian Galois extension of degree p^2. Show that D is cyclic if and only if $sp(D)$ is the coset of Brauer classes of cyclic $\overline{F}$-algebras adapted to $\overline{D}$ through some subfield of codimension p (see Exercise 9.2). When this condition holds, show that D is a cyclic algebra adapted to any inertial lift of $\overline{D}$.

Notes

§9.1: Proposition 9.2 gives information on the splitting fields of an inertially split central division algebra D over a Henselian field F. In the more specific case where the valuation on F is also discrete of rank 1 and $ind\, E = exp\, E$ for all central division algebras E over $\overline{F}$, Brussel has given in [40] a much more complete description of the splitting fields of D. This applies, e.g., when $F = k((x))$ for k a global or local field.

Corollary 9.5: It was proved in Tignol–Amitsur [244, Th.] (= [10, pp. 565–571]) that for a tame division algebra D over a strictly Henselian field F every splitting field of D finite-dimensional over F contains a maximal subfield of D. The criterion for splitting in terms of the canonical pairing on Γ_D was given by Tignol–Wadsworth [245, Prop. 4.5].

§9.2.1: The first example of a noncyclic division algebra with pure maximal subfield is a degree 4 algebra constructed by Albert [3]. This algebra has exponent 4. Another example, with degree 4 and exponent 2, was given by Dubisch [66]. Examples of degree p^2 for $p \neq 2$ remained elusive until the work of Matzri et al. [138]. Note that, by contrast, a division p-algebra with a pure maximal subfield is necessarily cyclic, by a theorem of Albert [4, Th. VII.26].

Lemma 9.11 is due to Albert [2, Th. 3] (see also [5, p. 468, Th. 3]). It can also be proved by computing the connecting map $H^1(\mathcal{G}(k), \mathbb{Z}/p^n\mathbb{Z}) \to H^2(\mathcal{G}(k), \mathbb{Z}/p\mathbb{Z})$ in the Galois cohomology exact sequence associated to the short exact sequence of trivial Galois modules

$$0 \longrightarrow \mathbb{Z}/p\mathbb{Z} \longrightarrow \mathbb{Z}/p^{n+1}\mathbb{Z} \longrightarrow \mathbb{Z}/p^n\mathbb{Z} \longrightarrow 0.$$

Proposition 9.14 and Ex. 9.15 were given by Matzri–Rowen–Vishne [138]. They raised as an open question the existence of noncyclic division algebras of odd prime exponent with pure maximal subfields [138, Question 4.4]. Example 9.18 settles this question.

§9.2.2: The notion of weak coset is inspired by the degeneracy condition on the matrix (u_{ij}) associated to a 2-cocycle of an abelian Galois group, as defined by Amitsur–Saltman [12] (= [10, pp. 441–452]) and revisited by McKinnie [139, Def. 1.5]. (See the Notes of Ch. 8 for the definition of the matrix (u_{ij}) and the correspondence with cosets modulo the *Dec* subgroup.) Proposition 9.20 is essentially due to Amitsur–Saltman [12, Lemma 3.1]; it has been rephrased by Boulagouaz–Mounirh [29, §3] and by McKinnie [139, Lemma 1.7].

§9.3: Th. 9.24 is due to Mounirh [165, Prop. 1.2]. Th. 9.27 is proved in McKinnie [140, Th. 1.2.1]. It was first established by Saltman [218, Th. 3.2] in the particular case where D is a "generic abelian crossed product," i.e., obtained by the $\mathcal{S}'$ construction of §8.4.5, under a somewhat stronger hypothesis on the specialization coset $sp(D)$.

A finite group G is said to be "rigid" (for a field F) if there is a crossed product algebra over F with group G and no other group. Saltman showed in [218] that every finite noncyclic abelian p-group is rigid for some field of characteristic p (see Prop. 9.28). It is easy to see from Cor. 9.5 that every elementary abelian p-group is rigid for a valued field (F, v) with Γ_F sufficiently large and $\overline{F}$ containing a primitive p-th root of unity: take a division algebra D that is a tensor product of symbol algebras of degree p and with a valuation extending v such that D is totally ramified over F. This was essentially proved by Amitsur in [6] (= [10, pp. 419–431]). Brussel gave in [37], [39] examples of rigid *nonabelian* p-groups for $k(x)$ and $k((x))$, where k is an algebraic number field containing no primitive p-th root of unity.

§9.4: Amitsur proved Th. 9.29 in his landmark noncrossed product paper [6] (= [10, pp. 419–431]), and used it to prove that $UD(\mathbb{Q}, n)$ is not a crossed product if $p^2 \mid n$ for some odd prime p or if $8 \mid n$. His examples of division algebras that are crossed products only with certain groups included some division algebras D over the strictly Henselian field $F = k((x_1))\ldots((x_m))$, where k is algebraically closed. While he noted that such a D has a valuation extending the usual valuation on F, his information about Galois groups of maximal subfields of D was encoded in the diagonal entries of certain integer-valued matrices. In fact, those diagonal entries represent the invariant factors of Γ_D/Γ_F. While the valuation theory was not explicit in Amitsur's paper, the further work on noncrossed products that it spawned was a major impetus in the development of noncommutative valuation theory. (The other major impetus was the work of Platonov and Yanchevskiĭ on SK_1 of division algebras. See Ch. 11 below.)

After Amitsur's paper, several authors obtained other examples of noncrossed product universal division algebras $UD(F, n)$ by combining Amitsur's Theorem 9.29 with further examples of crossed product division algebras with limited groups. This included Schacher–Small [224] (for *char* $F \neq 0$ and certain n prime to *char* F), Amitsur [7] (= [10, pp. 433–439]), Risman [206], Fein–Schacher [78], and Saltman [218] (for p-algebras).

The first examples of noncrossed product algebras not built from generic division algebras were given by Jacob–Wadsworth [105]. Once again the method was that of obtaining incompatible information about possible Galois groups of maximal subfields. But in this case the center F of the noncrossed product division algebra is an intersection $F = K_1 \cap K_2$, where each K_i is a field with Henselian valuation v_i, such that $v_1|_F$ and $v_2|_F$ are independent. For a prime p with $\overline{K_1}$ and $\overline{K_2}$ each containing a primitive p^n-th root of unity, it was shown that there is a canonical isomorphism

$$_{p^n}Br(F) \cong {}_{p^n}Br(K_1) \times {}_{p^n}Br(K_2)$$

that is well-behaved with respect to index. When $n \geq 3$, this allows one to obtain central division algebras D over F such that each $v_i|_F$ extends to D and the division algebras $D \otimes_F K_i$ are tame and totally ramified over K_i, so crossed products, but only with different groups. Thus, D is not a crossed product.

The first examples of noncrossed products over "nice" and relatively "small" fields were given by Brussel [37]. In his examples, the ground field is the Laurent series field $k((t))$ or rational function field $k(t)$ where k is a global field. The examples given in §9.4.1 and §9.4.2 illustrate his approach. The explicit examples given in these subsections are based on Hanke [92], [93], and Coyette [57].

Building on Brussel's approach over $k((t))$, Hanke and Sonn [97] gave a very detailed analysis of which division algebras are noncrossed products over a field F with complete discrete rank 1 valuation with $\overline{F}$ a global field. For such a field there is the Witt decomposition (see (8.39))

$$Br_{tr}(F) \cong Br(\overline{F}) \times Hom^c(\mathcal{G}(\overline{F}), \mathbb{Q}/\mathbb{Z}).$$

For any $\chi \in Hom^c(\mathcal{G}(\overline{F}), \mathbb{Q}/\mathbb{Z})$, they find a lower bound on the indexes of noncrossed products in $Br(\overline{F}) \times \{\chi\}$, and show that above that bound "nearly all" division algebras are noncrossed products. This work was generalized by Hanke–Neftin–Sonn [95] for inertially split division algebras over a Henselian field F with $\overline{F}$ a global field and Γ_F arbitrary; Hanke–Neftin–Wadsworth [96] generalized this still further for tame division algebras over a Henselian field with global residue field. Proposition 9.33 and Prop. 9.34 come from [96].

In later work on noncrossed products, Brussel [43] gave examples of such algebras over the rational function field $F = k(t)$ where k is a local field such as $\mathbb{Q}_p$. For this he worked with the Gaussian extension to F, call it v, of the complete discrete valuation on k; so $\overline{F} = \overline{k}(t)$ where $\overline{k}$ is finite and $\Gamma_F = \mathbb{Z}$. Let (K, v_h) be the Henselization (or the completion) of (F, v), so $\overline{K} = \overline{F} = \overline{k}(t)$, which is a global field. By lifting suitable families of cyclic field extensions and Brauer classes from $\overline{K}$ to K, he was able to build noncrossed product central division algebras E over K such that E has the form $D \otimes_F K$. Then, D is a noncrossed product division algebra over F. More recently, Brussel–McKinnie–Tengan [45] used the same general approach, combined with very substantial algebraic geometric machinery, to prove the existence of noncrossed product algebras over fields F of the following type: For any prime p, let X be a smooth curve over the p-adic ring $\mathbb{Z}_p$, and let F be the function field of X; so, F is an algebraic function field in one variable over the local field $\mathbb{Q}_p$. Let v be the discrete rank 1 valuation on F associated with the special fiber of X, and let $(\widehat{F}, \widehat{v})$ be the completion of (F, v). Then $v|_{\mathbb{Q}_p}$ is the complete discrete p-adic valuation on $\mathbb{Q}_p$; moreover, $\overline{\widehat{F}} = \overline{F}^{\,v}$, which is an algebraic function field in one variable over $\mathbb{F}_p$, so $\overline{\widehat{F}}$ is a global field. They showed that there is a highly noncanonical index-preserving map $s\colon\ Br'(\widehat{F}) \to Br'(F)$ splitting the map $ext_{\widehat{F}/F}\colon\ Br'(F) \to Br'(\widehat{F})$, where Br' denotes the prime-to-p part of Br. They used s to lift noncrossed products over $\widehat{F}$ to noncrossed products over F. Subsequently, Chen [51] proved analogous results by a somewhat different method in the more general situation that $\mathbb{Z}_p$ is replaced by an arbitrary complete discrete rank 1 valuation.

Amitsur generalized his Theorem 9.29 from maximal subfields of $UD(F, n)$ to splitting fields. This was announced in [8] with a proof given in [9] (= [10, pp. 573–582]). Call a finite group H a *splitting group* of a central simple L-algebra A if there is a Galois field extension K of L such that K splits A and $\mathcal{G}(K/L) \cong H$. Amitsur proved that for any infinite field F, if G is a splitting group of $UD(F,n)$, then for every central simple algebra A of degree n over any field L containing F, there is a subgroup H_A of G that is a splitting group of A. Tignol–Amitsur [242], [243] (= [10, pp. 507–537, 539–563]) obtained lower bounds on $|G|$ by analyzing subfields of Mal'cev–Neumann division algebras that are Kummer extensions of the center; they reduced the question of splitting groups to Galois groups of maximal subfields using Cor. 9.5. Tignol [237] thereafter fully characterized the cyclic and Kummer subfields of Mal'cev–Neumann algebras. His results were generalized by Morandi–Sethuraman [161], [162] for division algebras of the form $S \otimes_F T$ over a Henselian field F, where S is inertially split and T is tame and totally ramified over F; these results were later generalized to arbitrary tame division algebras over a Henselian field by Mounirh [166] using graded methods.

In his work on noncrossed products of degree exceeding the exponent, Saltman proved the following general result [219, Lemma 3]: If a division algebra D with center F has a maximal subfield normal over F, then D has a maximal subfield Galois over F (so D is a crossed product). The nontrivial proof of this depends on a delicate analysis of presentations of cyclic p-algebras that is given in [217, Lemma 6] (or see Jacobson [108, Lemma 4.4.16]). Hanke noted in [94, Prop. 2] that Saltman's result holds in the more general setting replacing $F = Z(D)$ with F a subfield of $Z(D)$ over which $Z(D)$ is separable of finite degree. Using this (and Exercise 5.6 above), one can eliminate the hypothesis on $char\,\overline{F}$ in Prop. 9.32—the conclusion then becomes: $\overline{D}$ has maximal subfield Galois over $\overline{F}$. Proposition 9.32 comes from Jacob–Wadsworth [106, Th. 5.15(b)]. Again using Saltman's result, it was shown by Hanke–Neftin–Wadsworth [96] that Prop. 9.34 holds with the "tamely ramified" condition in (a) deleted. This yields a full criterion for when a tame division algebra over a Henselian field is a crossed product, expressed entirely in terms of residue data. By another application of Saltman's result, one can eliminate the hypothesis that $char\, k \nmid deg\, D$ in Lemma 9.43 and Prop. 9.44.

For further discussion of noncrossed products and related questions, see Auel et al. [17].

Chapter 10
Indecomposable Division Algebras

A central division algebra D over a field F is said to be *decomposable* if $D = D_1 \otimes_F D_2$ for some proper subalgebras D_1, D_2 of D. If D has no such decomposition, we say that D is *indecomposable*. For example, if F has a Henselian valuation, we have seen in Prop. 7.74 that every D tame and totally ramified over F has a decomposition into symbol algebras reflecting the decomposition of Γ_D/Γ_F, and that $exp(\Gamma_D/\Gamma_F) = exp\, D$; hence, D is decomposable whenever its degree exceeds its exponent. Similarly, by the analogue of Prop. 8.28, a decomposably semiramified central division algebra N over F is decomposable whenever its residue field $\overline{N}$ is not a cyclic Galois extension of $\overline{F}$. It is clearly desirable to know when a division algebra D decomposes as $D_1 \otimes_F D_2$, since many properties of D can be determined from the D_i, whose smaller size may make them easier to work with than D itself.

In this chapter we give examples of indecomposable algebras, emphasizing constructions that use valuation theory. Since every division algebra has a primary decomposition, the degree of an indecomposable algebra must be a power of some prime p. Therefore, we restrict attention to division algebras of prime power degree. Indecomposable algebras of exponent p^2 or higher are relatively easy to construct as "p-th roots" of other division algebras; the proof of their indecomposability relies just on a simple observation relating the degree of a division algebra of p-power degree and the degree of its p-th power (see Prop. 10.1). Such constructions are discussed in §10.1, where we also give an example of an indecomposable division algebra D that becomes decomposable after a scalar extension of degree prime to $deg\, D$. Indecomposable algebras of prime exponent are more difficult to achieve; they are considered in §10.2. We first give in §10.2.1 a criterion to test the decomposability of a tame semiramified division algebra of prime exponent over a Henselian field. This criterion is then applied to give examples of indecomposable algebras: The case of exponent 2 and degree 8 is treated in §10.2.2, and the case of exponent $p \neq 2$ and degree p^r with $r \geq 2$ in §10.2.4. The last section, §10.3, deals more specifically with decompositions into tensor prod-

J.-P. Tignol, A.R. Wadsworth, *Value Functions on Simple Algebras, and Associated Graded Rings*, Springer Monographs in Mathematics,
DOI 10.1007/978-3-319-16360-4_10

ucts of symbol algebras, using the notion of armature from §7.2.1. We give a necessary and sufficient condition for the existence of such a decomposition for an inertially split division algebra over a Henselian field, and also for some division algebras that are tensor products of an inertial and a totally ramified division algebra.

10.1 Indecomposable algebras of nonprime exponent

Throughout this section, p is a fixed prime number. We consider only division algebras of index (hence of exponent) a power of p. Following a method of Sethuraman (which does not use valuations), we give examples of indecomposable division algebras of degree p^r and exponent p^s for any $r \geq s \geq 2$. Under the additional hypothesis that $2s > r$, we also show (after Brussel) that some central division algebras of degree p^r and exponent p^s over Laurent series fields in one indeterminate over suitable number fields are indecomposable. Finally, we present an example of Brussel showing that indecomposability of division algebras of degree a power of p is not preserved under scalar extension of degree prime to p.

For a division algebra D with center F and $j \in \mathbb{N}$, recall that D^j denotes the associated division algebra of $D^{\otimes j} = D \otimes_F D \otimes_F \ldots \otimes_F D$ (j times).

There are some evident barriers to decomposability of a division algebra pointed out by Saltman in [220, Lemma 3.2]:

Proposition 10.1. *Let D be a division algebra of degree a power of a prime p.*

(i) *If $\exp D = \deg D$, then D is indecomposable.*
(ii) *If $\deg D^p = \frac{1}{p} \deg D$, then D is indecomposable.*

Proof. Suppose $\deg D = p^r$ and $D \cong D_1 \otimes_F D_2$ with $\deg D_i = p^{r_i}$ where $1 \leq r_i < r$ for $i = 1, 2$. We have $\exp D_i \mid \deg D_i = p^{r_i}$. Hence,

$$\exp D \leq \operatorname{lcm}(\exp D_1, \exp D_2) \leq p^{\max(r_1, r_2)} < p^r = \deg D.$$

This proves (i). By Albert's index result [4, Lemma 7, p. 76], $\deg D^p$ divides $\frac{1}{p} \deg D$. Since $D^p \sim D_1^p \otimes_F D_2^p$ in $Br(F)$, we have

$$\begin{aligned} \deg D^p &= \operatorname{ind}(D_1^p \otimes_F D_2^p) \leq \deg D_1^p \cdot \deg D_2^p \\ &\leq \tfrac{1}{p} \deg D_1 \cdot \tfrac{1}{p} \deg D_2 = \tfrac{1}{p^2} \deg D < \tfrac{1}{p} \deg D. \end{aligned}$$

This proves (ii). □

For example, over a local or global field every division algebra of prime power degree is indecomposable, since its exponent equals its index.

Sethuraman observed in [230, Prop. 1] a nice application of (ii) above to symbol algebras. We formulate it in a slightly more general form:

Corollary 10.2 (Sethuraman). *Let L/F be a cyclic Galois extension of fields of degree p^r, for some $r \geq 1$, and let $K \subseteq L$ be the (unique) subfield such that $[L{:}K] = p$. Let σ be a generator of $\mathcal{G}(L/F)$ and let $b \in F^\times$. Consider the cyclic algebras $A = (L/F, \sigma, b)$ and $B = (K/F, \sigma|_K, b)$, of degrees p^r and p^{r-1} respectively. We have $\exp A = p \exp B$. Moreover, if B is a division algebra, then A is an indecomposable division algebra.*

Proof. The equation $\exp A = p \exp B$ readily follows from $A^{\otimes p} \sim B$ (see for example Reiner [199, Th. 30.10, p. 262]). By Albert's index result, $\mathit{ind}\, A \geq p\, \mathit{ind}\, A^{\otimes p}$. Therefore, if B is a division algebra we have $\mathit{ind}\, A \geq p^r$, hence A is a division algebra because $\mathit{deg}\, A = p^r$. Since $A^p = B$, the division algebra A is indecomposable by Prop. 10.1(ii). □

Corollary 10.2 applies in particular to symbol algebras: If F contains a primitive p^r-th root of unity ω, then we may take $L = F(a^{1/p^r})$ for suitable $a \in F^\times$, and thus $A = (a, b/F)_{\omega, p^r}$ and $B = (a, b/F)_{\omega^p, p^{r-1}}$. Examples where B is a division algebra are provided by the following construction:

Example 10.3. *Cyclic division algebras of prescribed degree and exponent.* Let k be a field, and let m, n be positive integers such that $m \mid n$ and m and n have the same prime factors. Let $x, y_1, \ldots, y_{n-1}$ be independent indeterminates over k and $K = k(x, y_1, y_2, \ldots, y_{n-1})$. Let σ be the k-automorphism of K defined by $\sigma(x) = x$, $\sigma(y_i) = y_{i+1}$ for $1 \leq i \leq n-2$ while

$$\sigma(y_{n-1}) = x^m/(y_1 y_2 \ldots y_{n-1}).$$

Then, $\sigma\big(x^m/(y_1 y_2 \ldots y_{n-1})\big) = y_1$, so σ has order n. Let F be the fixed field K^σ, and consider the degree n cyclic algebra

$$U = (K/F, \sigma, x) = \bigoplus_{i=0}^{n-1} K z^i, \quad \text{where } z^n = x \text{ and } zc = \sigma(c)z \text{ for } c \in K.$$

Ousmane shows in [176] following an approach of Rosset in [209] that U is a division algebra of exponent m (and that it is generic for cyclic algebras of degree n and exponent dividing m). Indeed, consider the multiplicative group $G = \langle x, y_1, \ldots, y_{n-1}, z\rangle \subseteq U^\times$. The k-subalgebra of U generated by G is the group ring $k[G]$ of G over k. Let A be the free abelian subgroup $\langle x, y_1, \ldots, y_{n-1}\rangle$ of G. Then A is normal in G and $G/A = \langle \overline{z}\rangle$, where $\overline{z} = zA$; so G/A is cyclic of order n. A short calculation shows that G is torsion-free. Since G has an abelian normal subgroup of finite index, a theorem of Brown (see Passman [177, Th. 4.1, p. 616]) if $\mathit{char}\, k = 0$ or Linnell [123] if $\mathit{char}\, k \neq 0$ shows that $k[G]$ has no zero divisors. Therefore, the ring of central quotients $U = q(k[G])$ is a division algebra. Let $\xi \in H^2(\langle \overline{z}\rangle, A)$ be the cohomology class corresponding to the group extension $A \to G \to \langle \overline{z}\rangle$. Rosset proves in [209, Th. 3] that $\exp U = \exp \xi$. But, as $\langle \overline{z}\rangle$ is cyclic, letting

$N(a) = a\sigma(a)\dots\sigma^{n-1}(a)$ for $a \in A$, we have

$$H^2(\langle\overline{z}\rangle, A) \cong A^{\langle\overline{z}\rangle}/N(A) = \langle x\rangle/\langle x^m, x^n\rangle \cong \mathbb{Z}/m\mathbb{Z}.$$

Since $z^n = x$, ξ maps to the image of x in $\widehat{H}^0(\langle\overline{z}\rangle, A)$; hence $m = \exp\xi = \exp U$.

Example 10.4. *Indecomposable division algebras of degree p^r and exponent p^s, for any $r \geq s \geq 2$.* Let k be any field containing a primitive p^r-th root of unity ω, and let B be the cyclic algebra U described in Ex. 10.3 with $n = p^{r-1}$ and $m = p^{s-1}$. Let $F = Z(B)$. Since k contains enough roots of unity, B is a symbol algebra, say $B = \big(u, x/F\big)_{\omega^p, p^{r-1}}$. Corollary 10.2 shows that $A = \big(u, x/F\big)_{\omega, p^r}$ is an indecomposable division algebra of degree p^r and exponent p^s.

The same construction can be used in characteristic p: First, recall that any cyclic extension of p-power degree of a field F of characteristic p can be embedded as a subfield of codimension p in a cyclic extension of F; see[1] Albert [1]. Now, let k be any field of characteristic p, and, as above, let $B = (K/F, \sigma, x)$ be the cyclic algebra U described in Ex. 10.3 with $n = p^{r-1}$ and $m = p^{s-1}$. There is a cyclic extension L of F of degree p^r containing K as a subfield of codimension p; extend σ to an automorphism σ' of L, and define $A = (L/F, \sigma', x)$. By Cor. 10.2, the algebra A is an indecomposable division algebra of degree p^r and exponent p^s.

The indecomposability criterion in Prop. 10.1(ii) can also be used to obtain indecomposable division algebras over fields with a simple structure. The following construction is essentially due to Brussel: Fix some prime p and integers r, s with

$$2s > r \geq s \geq 1.$$

We will use the $\mathcal{S}$-construction of §8.4.5 to produce indecomposable division algebras of degree p^r and exponent p^s over the field of Laurent series in one indeterminate over a number field k. The only condition on k is the following: If $p = 2$ and $s \geq 3$, assume that -1 or 2 is a square in k. (This hypothesis is needed to satisfy the Wang condition in the Grunwald–Wang theorem.) We use the following terminology: a finite prime $\mathfrak{p}$ of k means a discrete rank 1 valuation ring of k, with associated valuation $v_\mathfrak{p}$; then $k_\mathfrak{p}$ denotes the completion of k with respect to $v_\mathfrak{p}$. Let $\mathfrak{p}_1$ and $\mathfrak{p}_2$ be two finite primes of k. Let A be the division algebra with center k and local invariants $\frac{1}{p^s} + \mathbb{Z}$ at $\mathfrak{p}_1$, $-\frac{1}{p^s} + \mathbb{Z}$ at $\mathfrak{p}_2$, and 0 otherwise. Since the sum of the local invariants is 0 in $\mathbb{Q}/\mathbb{Z}$, such an A exists by the Albert–Brauer–Hasse–Noether theory of $Br(k)$, see Reiner [199, (32.13), p. 277]; moreover, $\deg A = \exp A$, and $\exp A$ is the least common multiple of the orders of its local invariants, which is p^s. By the Grunwald–Wang Theorem, see Lorenz–Roquette [128] or Neukirch et al. [171, Cor. 9.2.3, p. 461], there exists a cyclic Galois field extension ℓ of k of degree p^s with local degrees $[\ell\cdot k_{\mathfrak{p}_1} : k_{\mathfrak{p}_1}] = [\ell\cdot k_{\mathfrak{p}_2} : k_{\mathfrak{p}_2}] = p^{2s-r}$.

[1] This property also follows by an easy cohomological argument from the triviality of $H^2(\mathcal{G}(F), \mathbb{Z}/p\mathbb{Z})$; see Gille–Szamuely [84, (3), p. 260].

We may then consider the division algebra $\mathcal{S}(A;\ell)$ with center $k((x))$ constructed as in §8.4.5 (see (8.36)). By definition, $\mathcal{S}(A;\ell)$ is the division algebra associated to the following tensor product of A with a cyclic algebra:

$$\mathcal{S}(A;\ell) \sim A \otimes_k (\ell((x))/k((x)), \sigma, x),$$

where σ is a generator of $\mathcal{G}(\ell/k)$ extended to $\ell((x))$ by fixing $k((x))$.

Proposition 10.5. *The division algebra $\mathcal{S}(A;\ell)$ just defined is indecomposable and satisfies*

$$\deg \mathcal{S}(A;\ell) = p^r, \quad \exp \mathcal{S}(A;\ell) = p^s, \quad \deg \mathcal{S}(A;\ell)^p = p^{r-1}. \qquad (10.1)$$

Proof. Indecomposability of $\mathcal{S}(A;\ell)$ follows by Prop. 10.1(ii) once we show $\deg \mathcal{S}(A;\ell) = p \cdot \deg \mathcal{S}(A;\ell)^p$. Therefore, it suffices to prove (10.1).

In view of the definition of $\mathcal{S}(A;\ell)$, it is clear that $\mathcal{S}(A;\ell)^p$ is the division algebra associated to $A^p \otimes_k (\ell((x))/k((x)), \sigma, x)^p$. Letting $m \subseteq \ell$ denote the (unique) subfield of codimension p, we have by Reiner [199, Th. 30.10, p. 262]

$$(\ell((x))/k((x)), \sigma, x)^p = (m((x))/k((x)), \sigma|_m, x).$$

Therefore, $\mathcal{S}(A;\ell)^p = \mathcal{S}(A^p;m)$. Now, the exponent and index of algebras obtained by the $\mathcal{S}$ construction are determined in §8.4.5: By Prop. 8.72 we have

$$\begin{aligned} \deg \mathcal{S}(A;\ell) &= ind(A \otimes_k \ell) \cdot [\ell{:}k] = ind(A \otimes_k \ell) \cdot p^s, \\ \exp \mathcal{S}(A;\ell) &= lcm(\exp A, [\ell{:}k]) = p^s, \end{aligned}$$

and

$$\deg \mathcal{S}(A^p;m) = ind(A^p \otimes_k m) \cdot [m{:}k] = ind(A^p \otimes_k m) \cdot p^{s-1}.$$

Therefore, to prove (10.1) it suffices to show

$$ind(A \otimes_k \ell) = ind(A^p \otimes_k m) = p^{r-s}.$$

The indices are determined by local invariants: At each extension $\mathfrak{P}_{1i}$ of $\mathfrak{p}_1$ to ℓ, the algebra $A \otimes_k \ell$ has local invariant

$$\tfrac{1}{p^s}[\ell_{\mathfrak{P}_{1i}}{:}k_{\mathfrak{p}_1}] + \mathbb{Z} = \tfrac{p^{2s-r}}{p^s} + \mathbb{Z} = \tfrac{1}{p^{r-s}} + \mathbb{Z}$$

(see Reiner [199, Th. 31.9, p. 267]). Likewise, $A \otimes_k \ell$ has local invariant $-\frac{1}{p^{r-s}} + \mathbb{Z}$ at each extension $\mathfrak{P}_{2i}$ of $\mathfrak{p}_2$ to ℓ, and all its other local invariants are trivial. Hence, $ind(A \otimes_k \ell) = p^{r-s}$. Likewise, the local invariants of A^p are $\frac{1}{p^{s-1}} + \mathbb{Z}$ at $\mathfrak{p}_1$ and $-\frac{1}{p^{s-1}} + \mathbb{Z}$ at $\mathfrak{p}_2$, and 0 elsewhere. Because ℓ/k has degree p^s but local degree $p^{2s-r} > 1$ at $\mathfrak{p}_1$, the decomposition field for any extension of $\mathfrak{p}_1$ to ℓ must be contained in m. Therefore, for each prime $\mathfrak{q}_{1i}$ of m extending $\mathfrak{p}_1$ we have $[m_{\mathfrak{q}_{1i}}{:}k_{\mathfrak{p}_1}] = p^{2s-r-1}$. By Reiner [199, Th. 31.9, p. 267] again, the local invariant of $A \otimes_k m$ at $\mathfrak{q}_{1i}$ is

$$\tfrac{1}{p^{s-1}}[m_{\mathfrak{q}_{1i}}{:}k_{\mathfrak{p}_1}] + \mathbb{Z} = \tfrac{p^{2s-r-1}}{p^{s-1}} + \mathbb{Z} = \tfrac{1}{p^{r-s}} + \mathbb{Z}.$$

Likewise, $A \otimes_k m$ has local invariant $-\frac{1}{p^{r-s}} + \mathbb{Z}$ at each extension of $\mathfrak{p}_2$ to m, and 0 everywhere else, hence $ind(A \otimes_k m) = p^{r-s}$. □

In the rest of this section, we fix a prime $p \neq 2$ and we construct, following Brussel [41], an indecomposable division algebra of degree p^5 and exponent p^4 that becomes decomposable when a p-th root of unity is adjoined to the center. Since the index of a division algebra of prime-power degree does not change under a scalar extension of degree prime to p, Prop. 10.1 cannot be used to prove the indecomposability of this example.

Until the end of this section, we fix a number field k not containing a primitive p-th root of unity. Let $\mathfrak{p}$, $\mathfrak{q}$, $\mathfrak{r}$ be three finite primes of k such that $k_\mathfrak{p}$ contains no primitive p-th root of unity, while $k_\mathfrak{q}$ and $k_\mathfrak{r}$ each contain a primitive p^2-root of unity. Such finite primes exist by the Tchebotarev Density Theorem (see Janusz [109, Th. 10.4, p. 182] and Rosen [208, Th. 9.13A, p. 125]) applied to the cyclic Galois extensions $k(\zeta)/k$ and $k(\zeta^p)/k$, where ζ is a primitive p^2-root of unity.) Let ℓ_1 and ℓ_2 be cyclic Galois field extensions of k of degree p^2 with the following local behavior: $\ell_1 \cdot k_\mathfrak{p} = k_\mathfrak{p}$ and $\ell_1 \cdot k_\mathfrak{q}$, $\ell_1 \cdot k_\mathfrak{r}$ are totally ramified of degree p^2 over $k_\mathfrak{q}$ and $k_\mathfrak{r}$ respectively; $\ell_2 \cdot k_\mathfrak{p}$ is unramified over $k_\mathfrak{p}$ of degree p and $\ell_2 \cdot k_\mathfrak{q}$, $\ell_2 \cdot k_\mathfrak{r}$ are unramified of degree p^2 over $k_\mathfrak{q}$ and $k_\mathfrak{r}$ respectively. Such fields ℓ_1 and ℓ_2 exist by the Grunwald–Wang Theorem. (The special case of Grunwald–Wang does not arise since p is odd.) Let A be the division algebra with center k and local invariants $\frac{1}{p^2} + \mathbb{Z}$ at $\mathfrak{p}$, $\frac{1}{p^4} + \mathbb{Z}$ at $\mathfrak{q}$, $-\frac{1}{p^4} - \frac{1}{p^2} + \mathbb{Z}$ at $\mathfrak{r}$, and 0 otherwise. Here is a summary of these local hypotheses:

	at $\mathfrak{p}$	at $\mathfrak{q}$	at $\mathfrak{r}$
ℓ_1/k	degree 1	totally ramified degree p^2	totally ramified degree p^2
ℓ_2/k	unramified degree p	unramified degree p^2	unramified degree p^2
A	$\frac{1}{p^2} + \mathbb{Z}$	$\frac{1}{p^4} + \mathbb{Z}$	$-\frac{1}{p^4} - \frac{1}{p^2} + \mathbb{Z}$

An inspection of the local invariants of A shows that

$$deg\, A \;=\; exp\, A \;=\; p^4.$$

The contrasting behavior of ℓ_1 and ℓ_2 at $\mathfrak{q}$ (or at $\mathfrak{r}$) shows that ℓ_1 and ℓ_2 are linearly disjoint extensions of k. Thus, the tensor product

$$\ell \;=\; \ell_1 \otimes_k \ell_2$$

is a field; it is a Galois extension of k with $\mathcal{G}(\ell/k) \cong (\mathbb{Z}/p^2\mathbb{Z})^2$. We may then consider the following division algebra obtained by the $\mathcal{S}$-construction of §8.4.5:

$$D \;=\; \mathcal{S}(A; \ell_1, \ell_2).$$

Thus, the center of D is a field of iterated Laurent series in two indeterminates. We write F for the center of D:

$$Z(D) \;=\; F \;=\; k((x_1))((x_2)).$$

By definition, the division algebra D is associated to $A \otimes_k N$, where N is defined through the choice of generators σ_1 of $\mathcal{G}(\ell_1/k)$ and σ_2 of $\mathcal{G}(\ell_2/k)$ by

$$N \;=\; (\ell_1 \otimes_k F, \sigma_1 \otimes id_F, x_1) \otimes_F (\ell_2 \otimes_k F, \sigma_2 \otimes id_F, x_2). \tag{10.2}$$

For simplicity, in the sequel we write simply σ_i for $\sigma_i \otimes id_F$.

Theorem 10.6 (Brussel). *The division algebra D is indecomposable with $\deg D = p^5$ and $\exp D = p^4$. It becomes decomposable when a primitive p-th root of unity is adjoined to its center.*

Proof. We divide the proof into three parts.

Claim I: $\deg D = p^5$ and $\exp D = p^4$. The degree and exponent of the $\mathcal{S}$-algebras are determined in §8.4.5: By Prop. 8.72 we have

$$\deg \mathcal{S}(A; \ell_1, \ell_2) \;=\; ind(A \otimes_k \ell) \cdot [\ell{:}k] \;=\; ind(A \otimes_k \ell) \cdot p^4$$

and

$$\exp \mathcal{S}(A; \ell_1, \ell_2) \;=\; lcm(\exp A, [\ell_1{:}k], [\ell_2{:}k]) = p^4.$$

The local hypotheses assure that $[\ell \cdot k_{\mathfrak{p}} : k_{\mathfrak{p}}] = p$ and $[\ell \cdot k_{\mathfrak{q}} : k_{\mathfrak{q}}] = [\ell \cdot k_{\mathfrak{r}} : k_{\mathfrak{r}}] = p^4$; hence $A \otimes_k \ell$ has local invariant $\frac{1}{p} + \mathbb{Z}$ at each extension of $\mathfrak{p}$ to ℓ, and 0 at every other prime of ℓ. Therefore, $ind(A \otimes_k \ell) = p$. The first claim is thus proved.

Claim II: D is indecomposable. Suppose the contrary. We may then write $D = B_1 \otimes_F B_2$ with $\deg B_1 \geq \deg B_2 > 1$. Since $\deg D = p^5$, we have either $\deg B_1 = p^3$ and $\deg B_2 = p^2$, or $\deg B_1 = p^4$ and $\deg B_2 = p$. In the former case, $\exp B_1$ and $\exp B_2$ divide p^3; this is incompatible with $\exp D = p^4$. Therefore, $\deg B_1 = p^4$ and $\deg B_2 = p$. For the (x_1, x_2)-adic valuation on F, the division algebras D, B_1, and B_2 are tame since $char\, \overline{F} = 0$. They are also inertially split because k does not contain a primitive p-th root of unity (see Th. 7.84). We next consider the value groups to show that B_1 is semiramified: We have $\Gamma_D = (\frac{1}{p^2}\mathbb{Z})^2$ by Prop. 8.72 and $exp(\Gamma_{B_2}/\Gamma_F) \mid \exp B_2 = p$ by Th. 8.65(iv), so $\Gamma_{B_2} \subseteq (\frac{1}{p}\mathbb{Z})^2 = p\Gamma_D$. But $\Gamma_D = \Gamma_{B_1} + \Gamma_{B_2}$ by Cor. 6.50, hence $\Gamma_D/\Gamma_F = (\Gamma_{B_1}/\Gamma_F) + p(\Gamma_D/\Gamma_F)$. By Nakayama's Lemma (viewing Γ_D/Γ_F as a module over $\mathbb{Z}/p^2\mathbb{Z}$), it follows that $\Gamma_{B_1}/\Gamma_F = \Gamma_D/\Gamma_F$, hence $\Gamma_{B_1} = \Gamma_D$ and $|\Gamma_{B_1}{:}\Gamma_F| = p^4 = \deg B_1$. Therefore, Th. 8.65 shows that B_1 is semiramified, hence $\overline{B_1}$ is a field. We will obtain a contradiction by examining the local behavior of $\overline{B_1}$ at $\mathfrak{p}$.

First, we show that $\overline{B_1} \cdot k_{\mathfrak{p}}$ is inertial over $k_{\mathfrak{p}}$: Let $L \subseteq \overline{B_1} \cdot k_{\mathfrak{p}}$ be the inertial closure of $k_{\mathfrak{p}}$ in $\overline{B_1} \cdot k_{\mathfrak{p}}$; the field L does not contain a primitive p-th root of unity since $[\overline{B_1} \cdot k_{\mathfrak{p}} : k_{\mathfrak{p}}]$ is a power of p and $k_{\mathfrak{p}}$ does not contain a primitive p-th root of unity. But $\overline{B_1} \cdot k_{\mathfrak{p}}$ is a totally ramified and Galois extension of L of p-power degree, hence by Prop. A.22(ii) in Appendix A, $\overline{B_1} \cdot k_{\mathfrak{p}} = L$.

Next, we show that $\overline{B_1}\cdot k_{\mathfrak{p}}$ is an elementary abelian extension of $k_{\mathfrak{p}}$: For this, consider the ramification character $\chi_D\colon \mathcal{G}(k) \to \mathbb{T}(\Gamma_F) = (\mathbb{Q}/\mathbb{Z})^2$ of the inertially split division algebra D as in (6.39). Since the ramification map χ of (6.41) is a homomorphism (see Prop. 6.68), we have

$$\chi_D = \chi_{B_1} + \chi_{B_2},$$

so $\mathit{ker}\,\chi_{B_1} \supseteq \mathit{ker}\,\chi_D \cap \mathit{ker}\,\chi_{B_2}$. Since these kernels are absolute Galois groups, and since $Z(\overline{D}) = \ell$ by Prop. 8.72, this yields

$$\mathcal{G}(\overline{B_1}) \supseteq \mathcal{G}(\ell) \cap \mathcal{G}(Z(\overline{B_2})) = \mathcal{G}\big(\,\ell\cdot Z(\overline{B_2})\big).$$

Therefore, $\overline{B_1} \subseteq \ell\cdot Z(\overline{B_2})$, and hence $\overline{B_1}\cdot k_{\mathfrak{p}} \subseteq \ell\cdot Z(\overline{B_2})\cdot k_{\mathfrak{p}}$. Now, $\ell\cdot k_{\mathfrak{p}}$ and $Z(\overline{B_2})\cdot k_{\mathfrak{p}}$ are abelian Galois extensions of $k_{\mathfrak{p}}$ with $[\ell\cdot k_{\mathfrak{p}} : k_{\mathfrak{p}}] = p$ by hypothesis, and

$$[Z(\overline{B_2})\cdot k_{\mathfrak{p}} : k_{\mathfrak{p}}] \le [Z(\overline{B_2}) : k] \le \mathit{deg}\, B_2 = p.$$

Hence, $\mathcal{G}\big((\ell\cdot Z(\overline{B_2})\cdot k_{\mathfrak{p}})/k_{\mathfrak{p}}\big)$ is elementary abelian, and it follows that its factor group $\mathcal{G}(\overline{B_1}\cdot k_{\mathfrak{p}}/k_{\mathfrak{p}})$ is elementary abelian.

Thus, the field $\overline{B_1}\cdot k_{\mathfrak{p}}$ is an inertial and elementary abelian extension of $k_{\mathfrak{p}}$. But every inertial extension of a local field is cyclic, hence we must have $[\overline{B_1}\cdot k_{\mathfrak{p}} : k_{\mathfrak{p}}] = 1$ or p. We will obtain a contradiction by showing that $\overline{B_1}\cdot k_{\mathfrak{p}}$ splits a central division algebra of exponent p^2 over $k_{\mathfrak{p}}$.

For this, we write $B_1 \sim I_1 \otimes_F N_1$ in $\mathit{Br}(F)$, for some division algebras I_1 inertial over F and N_1 decomposably semiramified. This N_1 can be chosen to be any decomposably semiramified division algebra with ramification character $\chi_{N_1} = \chi_{B_1}$. We make a specific choice of N_1 as follows: Let $K = F(\sqrt[p^2]{x_1}, \sqrt[p^2]{x_2})$ (for any fixed choice of p^2-roots of x_1 and x_2); so, K is a totally ramified field extension of F with $\Gamma_K = \big(\tfrac{1}{p^2}\mathbb{Z}\big)^2 = \Gamma_D = \Gamma_{B_1}$. Since

$$\mathit{im}\,\chi_{B_1} = \Gamma_{B_1}/\Gamma_F = \Gamma_K/\Gamma_F,$$

by invoking Remark 6.69 we may choose for N_1 the division algebra satisfying $\chi_{N_1} = \chi_{B_1}$ and $[N_1] \in \mathit{Br}_{\mathit{is}}(F) \cap \mathit{Br}(K/F)$. Since K splits N and N_1 we have in $\mathit{Br}(K)$

$$A \otimes_k K \sim D \otimes_F K = (B_1 \otimes_F K) \otimes_K (B_2 \otimes_F K) \sim (I_1 \otimes_F K) \otimes_K (B_2 \otimes_F K). \tag{10.3}$$

Since K is totally ramified over F while I_1 is inertial over F, by Th. 3.43 it follows that $I_1 \otimes_F K$ is a division algebra with $\overline{I_1 \otimes K} = \overline{I_1}$. In particular, $I_1 \otimes_F K$ is inertial over K. Since $A \otimes_k K$ is also inertial over K, (10.3) shows that the division algebra $B_{2\,K}$ associated to $B_2 \otimes_F K$ is inertial. As the map $\mathit{Br}_{\mathit{in}}(K) \to \mathit{Br}(\overline{K})$ is a homomorphism and $\overline{K} = k$, (10.3) yields

$$A \sim \overline{I_1} \otimes_k \overline{B_{2\,K}} \qquad \text{in } \mathit{Br}(k).$$

Since $\mathit{exp}(A \otimes_k k_{\mathfrak{p}}) = p^2$ while $\mathit{exp}(\overline{B_{2\,K}} \otimes_k k_{\mathfrak{p}}) \big|\, \mathit{exp}\, B_2 = p$, we must have $\mathit{exp}(\overline{I_1} \otimes_k k_{\mathfrak{p}}) = p^2$. As B_1 is semiramified, the field $\overline{B_1}$ splits $\overline{I_1}$ (see

Th. 8.65(v)), hence $\overline{B_1}\cdot k_{\mathfrak{p}}$ splits $\overline{I_1}\otimes_k k_{\mathfrak{p}}$. This is a contradiction, since we saw above that $[\overline{B_1}\cdot k_{\mathfrak{p}} : k_{\mathfrak{p}}] = p$. Therefore, D is indecomposable.

Claim III: D becomes decomposable when a primitive p-th root of unity is adjoined to F. Let $k' = k(\omega)$, where ω is a primitive p-th root of unity. We use $'$ to designate the result of scalar extension from k to k'; thus we let

$$\ell_i' = \ell_i \otimes_k k', \quad \ell' = \ell \otimes_k k', \quad A' = A \otimes_k k', \quad F' = F \otimes_k k', \quad D' = D \otimes_k k'.$$

As $p \nmid [k':k]$, the algebra ℓ' is a field and D' is a division algebra with center F' and with $deg\, D' = deg\, D = p^5$ and $exp\, D' = exp\, D = p^4$. We will give a nontrivial tensor decomposition of D'.

Let $\mathfrak{p}_j$ for $j = 1, 2, \ldots$ be the primes of k' extending the prime $\mathfrak{p}$ of k, and likewise let $\mathfrak{q}_j$, $\mathfrak{r}_j$ be the primes of k' extending $\mathfrak{q}$ and $\mathfrak{r}$ respectively. The local data in the table above carry over to corresponding data for ℓ_1'/k', ℓ_2'/k', and A' for each of the primes $\mathfrak{p}_j$, $\mathfrak{q}_j$, $\mathfrak{r}_j$. Since $k'_{\mathfrak{p}_j}$ contains ω, it has cyclic totally ramified extensions of degree p. By the Grunwald–Wang theorem, we may then choose a cyclic Galois field extension ℓ_0 of k' with $\ell_0 \cdot k'_{\mathfrak{p}_j}$ totally ramified of degree p over $k'_{\mathfrak{p}_j}$ for all j, and $\ell_0 \cdot k'_{\mathfrak{q}_j} = k'_{\mathfrak{q}_j}$, $\ell_0 \cdot k'_{\mathfrak{r}_j} = k'_{\mathfrak{r}_j}$ for all j. Let σ_0 be a generator of $\mathcal{G}(\ell_0/k')$, and let

$$C = (\ell_0 \otimes_{k'} F'/F', \sigma_0 \otimes id_{F'}, x_1 x_2),$$

a cyclic F'-algebra of degree p. We will show that $D' \cong C \otimes_{F'} E$ for some division algebra E of degree p^4. Since $deg\, D' = p^5$ and $deg\, C = p$, it suffices to show that $D' \otimes_{F'} C^{\mathrm{op}}$ is split by an extension of F' of degree at most p^4.

Let $\mathcal{G}(k')^\vee = Hom^c(\mathcal{G}(k'), \mathbb{T})$ be the character group of the absolute Galois group of k'. For $i = 1, 2$, let $\chi_i \in \mathcal{G}(k')^\vee$ be the character associated to the cyclic extension ℓ_i'/k' and the generator σ_i of $\mathcal{G}(\ell_i'/k')$; i.e., $\chi_i(\rho) = \frac{j}{p^2} + \mathbb{Z}$ if $\rho \in \mathcal{G}(k')$ satisfies $\rho|_{\ell_i'} = \sigma_i^j$. Define similarly the character χ_0 of exponent p associated to ℓ_0/k' and $\sigma_0 \in \mathcal{G}(\ell_0/k')$. We may then consider the characters $\varphi_1 = \chi_1 - \chi_0$, $\varphi_2 = \chi_2 - \chi_0$, and the corresponding cyclic extensions of k':

$$m_1 = (k'_{sep})^{\ker \varphi_1}, \qquad m_2 = (k'_{sep})^{\ker \varphi_2}.$$

Let also m denote the composite field extension $m = m_1 \cdot m_2$. Since $exp\, \chi_1 = exp\, \chi_2 = p^2$ and $exp\, \chi_0 = p$, we have $exp\, \varphi_1 = exp\, \varphi_2 = p^2$; hence,

$$[m_1 : k'] = [m_2 : k'] = p^2,$$

and therefore $[m : k'] \le p^4$. We show below that $m \otimes_{k'} F'$ splits $D' \otimes_{F'} C^{\mathrm{op}}$. This will complete the proof of the theorem.

In preparation for this, recall (e.g. from Serre [229, Ch. 14, §1]; see also the definition of the *Dec* subgroup in Def. 8.19) that there is a bilinear (cup-product) pairing

$$\xi \colon \mathcal{G}(F')^\vee \times F'^\times \longrightarrow Br(F')$$

defined as follows: for $\chi \in \mathcal{G}(F')^\vee$ of exponent n, let $L_\chi = (F'_{sep})^{\ker \chi}$ and let $\sigma_\chi \in \mathcal{G}(L_\chi/F')$ be the restriction to L_χ of any $\rho \in \mathcal{G}(F')$ such that

$\chi(\rho') = \frac{1}{n} + \mathbb{Z}$; then $\xi(\chi, a) = [(L_\chi/F', \sigma_\chi, a)]$ for any $a \in F'^{\times}$. For simplicity, we use the same notation for characters in $\mathcal{G}(k')^\vee$ and for their restriction in $\mathcal{G}(F')^\vee$. We may then write

$$[(\ell_i' \otimes F', \sigma_i \otimes id_{F'}, x_i)] = \xi(\chi_i, x_i) \qquad \text{for } i = 1,\, 2.$$

To simplify notation, we let $L_0 = \ell_0 \otimes_{k'} F'$, $L_i = \ell_i' \otimes_{k'} F'$ for $i = 1$, 2, and we write again σ_i for $\sigma_i \otimes id_{F'}$. We also write $M_i = m_i \otimes_{k'} F'$ for $i = 1$, 2, and $M = m \otimes_{k'} F' = M_1 \cdot M_2$. Scalar extension at the Brauer group level corresponds under ξ to restriction of characters and inclusion of units, so we have

$$ext_{M_i/F'}[(L_i/F', \sigma_i, x_i)] \;=\; \xi(res_{m_i/k'}(\chi_i), x_i) \quad \text{for } i = 1,\, 2.$$

By definition of m_i, the restriction map $res_{m_i/k'} \colon \mathcal{G}(k')^\vee \to \mathcal{G}(m_i)^\vee$ carries φ_i to 0, hence $res_{m_i/k'}(\chi_i) = res_{m_i/k'}(\chi_0)$ for $i = 1$, 2. Therefore,

$$ext_{M_i/F'}[(L_i/F', \sigma_i, x_i)] = \xi(res_{m_i/k'}(\chi_0), x_i) = ext_{M_i/F'}[(L_0/F', \sigma_0, x_i)].$$

Consequently, from the definition of N in (10.2) we have for $N' = N \otimes_k k'$

$$\begin{aligned} ext_{M/F'}([N']) \;&=\; ext_{M/F'}\big([(L_1/F', \sigma_1, x_1) \otimes_{F'} (L_2/F', \sigma_2, x_2)]\big) \\ &=\; ext_{M/F'}\big([(L_0/F', \sigma_0, x_1)] + [(L_0/F', \sigma_0, x_2)]\big) \\ &=\; ext_{M/F'}\big([C]\big). \end{aligned}$$

Therefore, the extension M/F' splits $N' \otimes_{F'} C^{\mathrm{op}}$. Since $D' \sim A' \otimes_{k'} N'$, it now suffices to show that m splits A' to see that M splits $D' \otimes_{F'} C^{\mathrm{op}}$, and thus complete the proof.

For this, we look at the local invariants at $\mathfrak{p}_j$, $\mathfrak{q}_j$, and $\mathfrak{r}_j$. For $i = 1$, 2, the field $m_i \cdot k'_{\mathfrak{p}_j}$ is the cyclic extension of $k'_{\mathfrak{p}_j}$ associated to the image of the character φ_i under the restriction map $res_{k'_{\mathfrak{p}_j}/k'} \colon \mathcal{G}(k')^\vee \to \mathcal{G}(k'_{\mathfrak{p}_j})^\vee$. By definition of ℓ_1 we have $\ell_1 \cdot k'_{\mathfrak{p}_j} = k'_{\mathfrak{p}_j}$, hence $res_{k'_{\mathfrak{p}_j}/k'}(\chi_1) = 0$. Therefore,

$$res_{k'_{\mathfrak{p}_j}/k'}(\varphi_1) \;=\; res_{k'_{\mathfrak{p}_j}/k'}(-\chi_0),$$

which implies that $m_1 \cdot k'_{\mathfrak{p}_j} = \ell_0 \cdot k'_{\mathfrak{p}_j}$. Thus, $m_1 \cdot k'_{\mathfrak{p}_j}$ is a totally ramified extension of $k'_{\mathfrak{p}_j}$ of degree p for all j. On the other hand, we have

$$res_{k'_{\mathfrak{p}_j}/k'}(\varphi_2) \;=\; res_{k'_{\mathfrak{p}_j}/k'}(\chi_2 - \chi_0),$$

and the expression on the right is nonzero because $\ell_2' \cdot k'_{\mathfrak{p}_j} \neq \ell_0 \cdot k'_{\mathfrak{p}_j}$, as $\ell_2' \cdot k'_{\mathfrak{p}_j}/k'_{\mathfrak{p}_j}$ is unramified of degree p whereas $\ell_0 \cdot k'_{\mathfrak{p}_j}/k'_{\mathfrak{p}_j}$ is totally ramified of degree p. Therefore, $m_2 \cdot k'_{\mathfrak{p}_j}$ is a cyclic extension of $k'_{\mathfrak{p}_j}$ of degree p. If $m_2 \cdot k'_{\mathfrak{p}_j} = \ell_0 \cdot k'_{\mathfrak{p}_j}$, then

$$res_{k'_{\mathfrak{p}_j}/k'}(\chi_2 - \chi_0) \;=\; t \cdot res_{k'_{\mathfrak{p}_j}/k'}(\chi_0) \quad \text{for some integer } t \text{ prime to } p.$$

This equation leads to $res_{k'_{\mathfrak{p}_j}/k'}(\chi_2) = (1 + t) res_{k'_{\mathfrak{p}_j}/k'}(\chi_0)$, which is impossible since $\ell_2' \cdot k'_{\mathfrak{p}_j} \neq k'_{\mathfrak{p}_j}$, $\ell_0 \cdot k'_{\mathfrak{p}_j}$. Therefore, $m_2 \cdot k'_{\mathfrak{p}_j} \neq \ell_0 \cdot k'_{\mathfrak{p}_j} = m_1 \cdot k'_{\mathfrak{p}_j}$, and it follows that $[m\cdot k'_{\mathfrak{p}_j}:k'_{\mathfrak{p}_j}] = p^2$. As a result, $m\cdot k'_{\mathfrak{p}_j}$ splits $A' \otimes k'_{\mathfrak{p}_j}$, because the local invariant of A' at $\mathfrak{p}_j$ has order p^2.

We next consider the primes $\mathfrak{q}_j$ and $\mathfrak{r}_j$. For $i = 1, 2$, the field $m_i\cdot k'_{\mathfrak{q}_j}$ is the cyclic extension of $k'_{\mathfrak{q}_j}$ associated to $res_{k'_{\mathfrak{q}_j}/k'}(\varphi_i)$. Since $\ell_0\cdot k'_{\mathfrak{q}_j} = k'_{\mathfrak{q}_j}$ by definition of ℓ_0, we have $res_{k'_{\mathfrak{q}_j}/k'}(\chi_0) = 0$, hence $res_{k'_{\mathfrak{q}_j}/k'}(\varphi_i) = res_{k'_{\mathfrak{q}_j}/k'}(\chi_i)$. Therefore, $m_i\cdot k'_{\mathfrak{q}_j} = \ell'_i\cdot k'_{\mathfrak{q}_j}$ for $i = 1, 2$, and it follows that $m\cdot k'_{\mathfrak{q}_j} = \ell'\cdot k'_{\mathfrak{q}_j}$. In view of the local behavior of ℓ at $\mathfrak{q}$, we have $[\ell'\cdot k'_{\mathfrak{q}_j}:k'_{\mathfrak{q}_j}] = p^4$. Since the local invariant of A' at $\mathfrak{q}_j$ has order p^4, we see that $m\cdot k'_{\mathfrak{q}_j}$ splits $A' \otimes_{k'} k'_{\mathfrak{q}_j}$ for all j. At the primes $\mathfrak{r}_j$ the local behavior of m is the same as at $\mathfrak{q}_j$, and the order of the local invariant of A' is also p^4; therefore, the same arguments apply to prove that $m\cdot k'_{\mathfrak{r}_j}$ splits $A' \otimes_{k'} k'_{\mathfrak{r}_j}$. We have thus shown that m splits A', and the proof is complete. □

10.2 Indecomposable algebras of prime exponent

We turn in this section to the construction of indecomposable algebras of prime exponent. All our examples will be tame semiramified algebras obtained by the $\mathcal{S}$ construction of §8.4.5, but the techniques of proof are completely different from those of the preceding section. A main ingredient is a criterion due to Jacob, which characterizes the possible decompositions of certain tame semiramified division algebras over a Henselian field in terms of their specialization coset. This criterion leads to a sufficient condition on a central simple algebra A and cyclic extensions $\ell_1, \ldots, \ell_r$ of degree p to guarantee the indecomposability of division algebras of the type $\mathcal{S}(A;\ell_1,\ldots,\ell_r)$. We then proceed to give in §10.2.3 and §10.2.4 Jacob's specific examples of algebras satisfying this condition. He gave examples for any odd prime p and any integer $r \geq 2$, thus yielding indecomposable algebras of exponent p and index p^r for p odd and any $r \geq 2$. He also gave similar examples for $p = 2$ and $r \geq 3$. This case involves significant added technical complications; instead of Jacob's examples for $p = 2$, we given in §10.2.2 a construction of Elman et al., which yields indecomposable algebras of exponent 2 and degree 8.

10.2.1 Jacob's indecomposability criterion

In this subsection we prove a criterion of Jacob [104] for the indecomposability of certain semiramified division algebras of prime exponent p and arbitrarily large degree p^r over a Henselian-valued field.

Proposition 10.7 (Jacob). *Let F be a Henselian field, and let D be a tame division algebra with center F. If Γ_D/Γ_F has prime exponent p, then*

$D \cong S \otimes_F T$ for some inertially split division algebra S and tame totally ramified division algebra T with $\overline{D} = \overline{S}$, $\Gamma_S \cap \Gamma_T = \Gamma_F$, and $\Gamma_D = \Gamma_S + \Gamma_T$.

Proof. As per Prop. 8.59 and (8.34), write

$$D \sim I \otimes_F N \otimes_F R,$$

for division algebras I, N, and R over F with I inertial, N decomposably semiramified, and R tame and totally ramified. By Th. 3.43 the tensor product $I \otimes_F R$ is a division algebra, hence we may take $S = I$ and $T = R$ if N is trivial. Also, if R is trivial then D is inertially split, hence we may take $S = D$ and $T = F$. For the rest of the proof we assume that N and R are nontrivial. The main step in the proof is to reduce to the case where $\Gamma_N \cap \Gamma_R = \Gamma_F$.

For the inertially split division algebra E associated to $I \otimes_F N$ we have $\Gamma_E = \Gamma_N$ by Th. 8.65. Then, as $D \sim E \otimes_F R$, Th. 8.60 yields $\Gamma_D = \Gamma_E + \Gamma_R = \Gamma_N + \Gamma_R$; hence $exp(\Gamma_N/\Gamma_F) = exp(\Gamma_R/\Gamma_F) = p$. Therefore, as R is tame and totally ramified, F must contain a primitive p-th root of unity ω, by Prop. A.22(ii).

We claim: $N \otimes_F R \sim I' \otimes_F N' \otimes_F R'$ in $Br(F)$, for some inertial algebra I', decomposably semiramified algebra N', and totally ramified algebra R' such that $\Gamma_{N'} \cap \Gamma_{R'} = \Gamma_F$. The claim is proved by induction on $deg\, N$. If $\Gamma_N \cap \Gamma_R = \Gamma_F$, then we are done by taking $N' = N$ and $R' = R$. Assume not. Then, there is a $\gamma \in (\Gamma_N \cap \Gamma_R) \setminus \Gamma_F$; take any $x \in F^\times$ with $v(x) = p\gamma$. As Γ_N/Γ_F is an elementary abelian p-group, its cyclic subgroup $\langle \gamma + \Gamma_F \rangle$ has a complementary subgroup, say Δ/Γ_F. Since N decomposes according to any direct product decomposition of Γ_N/Γ_F by the valued version of Remark 8.34, we have $N \cong N_1 \otimes_F N_2$ for some decomposably ramified subalgebras N_1 and N_2 of N with $\Gamma_{N_1} = \Delta$ and $\Gamma_{N_2} = \langle \gamma \rangle + \Gamma_F$. Since N_2 is cyclically semiramified, we can write $N_2 = \big(u, x/F\big)_{\omega,p}$ for some $u \in F^\times$ with $v(u) = 0$ and $v(x) = p\gamma$.

For the totally ramified division algebra R, let $\overline{c}_R$ be the nondegenerate pairing on Γ_R/Γ_F as in Prop. 7.72. Choose $\delta \in \Gamma_R$ with $\overline{c}_R(\overline{\gamma}, \overline{\delta}) = \overline{\omega}$, where $\overline{\gamma} = \gamma + \Gamma_F$ and $\overline{\delta} = \delta + \Gamma_F$. Since $\overline{\gamma}$ and $\overline{\delta}$ have order $p = exp\, \Gamma_R/\Gamma_F$, they are part of a symplectic base $(\overline{\gamma}, \overline{\delta}, \overline{\varepsilon}_1, \overline{\varphi}_1, \ldots, \overline{\varepsilon}_r, \overline{\varphi}_r)$ of the symplectic module $(\Gamma_R/\Gamma_F, \overline{c}_R)$. By Prop. 7.74, the tame and totally ramified division algebra R has a tensor product decomposition

$$R \cong (y, z/F)_{\omega,p} \otimes_F (a_1, b_1/F)_{\omega_1,n_1} \otimes_F \ldots \otimes_F (a_r, b_r/F)_{\omega_r,n_r}$$

for some $y, z \in F^\times$ with $v(y) = p\gamma$ and $v(z) = p\delta$, and some $a_1, b_1, \ldots, a_r, b_r \in F^\times$. Write

$$R_1 = (y, z/F)_{\omega,p} \quad \text{and} \quad R_2 = (a_1, b_1/F)_{\omega_1,n_1} \otimes_F \ldots \otimes_F (a_r, b_r/F)_{\omega_r,n_r}.$$

Then in $Br(F)$, by the bimultiplicativity of symbol algebras, we have

$$\begin{aligned} N \otimes_F R &\sim N_1 \otimes_F \big(u, x/F\big)_{\omega,p} \otimes_F \big(y, z/F\big)_{\omega,p} \otimes_F R_2 \\ &\sim \big(u, xy^{-1}/F\big)_{\omega,p} \otimes_F \big(y, u^{-1}z/F\big)_{\omega,p} \otimes_F N_1 \otimes_F R_2. \end{aligned}$$

Since $v(u) = v(xy^{-1}) = 0$, the associated division algebra of $\big(u, xy^{-1}/F\big)_{\omega,p}$ is inertial over F, while N_1 is decomposably semiramified with $deg\, N_1 < deg\, N$. Let $R_3 = \big(y, u^{-1}z/F\big)_{\omega,p}$. Because $R_1 = \big(y, z/F\big)_{\omega,p}$ is totally ramified over F and $v(u^{-1}z) = v(z)$, the algebra R_3 is also totally ramified over F with $\Gamma_{R_3} = \Gamma_{R_1}$. Hence, $\Gamma_{R_3} \cap \Gamma_{R_2} = \Gamma_{R_1} \cap \Gamma_{R_2} = \Gamma_F$. Therefore, e.g. by Th. 3.43, $R_3 \otimes_F R_2$ is a division algebra totally ramified over F. Because $deg\, N_1 < deg\, N$, it follows by induction that

$$N_1 \otimes_F \big(R_3 \otimes_F R_2\big) \sim I_1 \otimes_F N' \otimes_F R'$$

for some division algebras I_1, N', R' with I_1 inertial over F, N' decomposably semiramified and R' totally ramified over F. Since $\big(u, xy^{-1}/F\big)_{\omega,p}$ and I_1 are inertial over F, the associated division algebra I' of $\big(u, xy^{-1}/F\big)_{\omega,p} \otimes_F I_1$ is inertial over F. As $N \otimes_F R \sim I' \otimes_F N' \otimes_F R'$, with $\Gamma_{N'} \cap \Gamma_{R'} = \Gamma_F$, we have thus established the claim.

Now, let S be the associated division algebra of $I \otimes_F I' \otimes_F N'$ and let $T = R'$. Then, in $Br(F)$,

$$D \sim I \otimes_F N \otimes_F R \sim I \otimes_F I' \otimes_F N' \otimes_F R' \sim S \otimes_F T.$$

Moreover, S is inertially split, as I, I', and R are each inertially split, and $\Gamma_S = \Gamma_{N'}$ by Th. 8.65 as I and I' are inertial over F. Also, T is totally ramified over F, and $\Gamma_S \cap \Gamma_T = \Gamma_{N'} \cap \Gamma_{R'} = \Gamma_F$. Furthermore, $\overline{S} \otimes_{\overline{F}} \overline{T} = \overline{S} \otimes_{\overline{F}} \overline{F} = \overline{S}$. By Th. 3.43, $S \otimes_F T$ is a division algebra with $\overline{S \otimes_F T} \cong \overline{S} \otimes_{\overline{F}} \overline{T} = \overline{S}$ and $\Gamma_{S \otimes_F T} = \Gamma_S + \Gamma_T$. Since D and $S \otimes_F T$ are Brauer-equivalent division algebras, they are isomorphic. □

Remark 10.8. The proposition shows that any indecomposable tame division algebra D over a Henselian field F of prime exponent less than its degree must be inertially split. For otherwise, D would have to be tame and totally ramified over F, but then it would be decomposable by Cor. 7.76.

Lemma 10.9. *Let F be a Henselian field, and let T_1 and T_2 be division algebras tame and totally ramified over their center F. If $T_1 \otimes_F T_2$ is inertially split, then the associated division algebra is decomposably semiramified. Moreover, $\Gamma_{T_1} = \Gamma_{T_2}$.*

Proof. Let F_{in} be the inertial closure of F. Since T_i is tame and totally ramified over F, its scalar extension $T_i \otimes_F F_{in}$ is a tame and totally ramified division algebra over F_{in} with $\Gamma_{T_i \otimes_F F_{in}} = \Gamma_{T_i}$ and the same symplectic pairing on each; see Lemma 7.68 for the graded version of this. Because F_{in} splits $T_1 \otimes_F T_2$, the totally ramified division algebras $T_1 \otimes_F F_{in}$ and $T_2^{\mathrm{op}} \otimes_F F_{in}$ are isomorphic. Hence, there are equalities of symplectic modules,

$$\begin{aligned}\big(\Gamma_{T_1}/\Gamma_F, \overline{c}_{T_1}\big) &= \big(\Gamma_{T_1 \otimes_F F_{in}}/\Gamma_{F_{in}}, \overline{c}_{T_1 \otimes_F F_{in}}\big) \\ &= \big(\Gamma_{T_2^{\mathrm{op}} \otimes_F F_{in}}/\Gamma_{F_{in}}, \overline{c}_{T_2^{\mathrm{op}} \otimes_F F_{in}}\big) = \big(\Gamma_{T_2^{\mathrm{op}}}/\Gamma_F, \overline{c}_{T_2^{\mathrm{op}}}\big).\end{aligned}$$

Thus, in particular, $\Gamma_{T_1} = \Gamma_{T_2^{\mathrm{op}}} = \Gamma_{T_2}$. Take any symplectic base $(\overline{\gamma_1}, \overline{\delta_1}, \ldots, \overline{\gamma_m}, \overline{\delta_m})$ of Γ_{T_1}/Γ_F for $\overline{c}_{T_1}$, where $\overline{\gamma_i} = \gamma_i + \Gamma_F$ and $\overline{\delta_i} = \delta_i + \Gamma_F$. Say $\overline{c}_{T_1}(\overline{\gamma_i}, \overline{\delta_i}) = \overline{\omega_i}$, where ω_i is a primitive r_i-th root of unity in F with r_i prime to $char\,\overline{F}$. By Prop. 7.74, T_1 decomposes into a tensor product of symbol algebras

$$T_1 \cong A_1 \otimes_F \ldots \otimes_F A_m$$

where for each $i = 1, \ldots, m$,

$$A_i = (x_{1i}, y_{1i}/F)_{\omega_i, r_i}, \quad \text{with } x_{1i}, y_{1i} \in F^\times,\ v(x_{1i}) = r_i\gamma_i,\ v(y_{1i}) = r_i\delta_i.$$

Since $(\overline{\gamma_1}, \ldots, \overline{\delta_m})$ is also a symplectic base of $\Gamma_{T_2^{\mathrm{op}}}/\Gamma_F$ for $\overline{c}_{T_2^{\mathrm{op}}}$, we have analogously

$$T_2^{\mathrm{op}} \cong B_1 \otimes_F \ldots \otimes_F B_m, \qquad \text{where each} \qquad B_i = (x_{2i}, y_{2i}/F)_{\omega_i, r_i},$$

with $x_{2i}, y_{2i} \in F^\times$ and $v(x_{2i}) = r_i\gamma_i = v(x_{1i})$ and $v(y_{2i}) = r_i\delta_i = v(y_{1i})$. Write $x_{2i} = a_i x_{1i}$ and $y_{2i} = b_i y_{1i}$, with $v(a_i) = v(b_i) = 0$, for all i. Then, in $Br(F)$,

$$\begin{aligned} A_i \otimes_F B_i^{\mathrm{op}} &\sim (x_{1i}, y_{1i}/F)_{\omega_i, r_i} \otimes_F (b_i y_{1i}, a_i x_{1i}/F)_{\omega_i, r_i} \\ &\sim (b_i, a_i x_{1i}/F)_{\omega_i, r_i} \otimes_F (y_{1i}, a_i/F)_{\omega_i, r_i}. \end{aligned}$$

Hence, $T_1 \otimes_F T_2 \sim \bigotimes_{i=1}^m{}_F \big((b_i, a_i x_{1i}/F)_{\omega_i, r_i} \otimes_F (y_{1i}, a_i/F)_{\omega_i, r_i}\big)$. Therefore, $T_1 \otimes_F T_2$ is split by the field extension

$$L = F\big(\sqrt[r_1]{a_1 x_{11}}, \sqrt[r_1]{y_{11}}, \ldots, \sqrt[r_m]{a_m x_{1m}}, \sqrt[r_m]{y_{1m}}\big).$$

Since $v(\sqrt[r_i]{a_i x_{1i}}) = \frac{1}{r_i} v(a_i x_{1i}) = \gamma_i$ and $v(\sqrt[r_i]{y_{1i}}) = \frac{1}{r_i} v(y_{i1}) = \delta_i$, we have

$$\Gamma_L \supseteq \Gamma_F + \sum_{i=1}^m (\langle\gamma_i\rangle + \langle\delta_i\rangle) = \Gamma_{T_1};$$

so,

$$|\Gamma_L{:}\Gamma_F| \geq |\Gamma_{T_1}{:}\Gamma_F| = [T_1{:}F] = r_1^2 \ldots r_m^2 \geq [L{:}F].$$

Hence, L is totally ramified over F. Since L splits $T_1 \otimes_F T_2$, which is also inertially split, the division algebra associated to $T_1 \otimes_F T_2$ is decomposably semiramified by Prop. 8.40. □

Theorem 10.10 (Jacob). *Let F be a Henselian field, and let D be a semiramified inertially split division algebra with center F with Γ_D/Γ_F of prime exponent p. Suppose D is "tamely decomposable," say $D \cong B_1 \otimes_F B_2$ with the B_i tame proper subalgebras of D. Then, $\overline{B_1}$ and $\overline{B_2}$ are proper subfields of $\overline{D}$ that are linearly disjoint over $\overline{F}$, and*

$$sp(D) \in \big[Br(\overline{B_1}/\overline{F}) + Br(\overline{B_2}/\overline{F}) + Dec(\overline{D}/\overline{F})\big] \,/\, Dec(\overline{D}/\overline{F}).$$

Proof. Since B_1 and B_2 are tame, by Cor. 6.50, the tensor product $\overline{B_1} \otimes_{\overline{F}} \overline{B_2}$ embeds in the field $\overline{D}$. Hence $\overline{B_1}$ and $\overline{B_2}$ are fields, and they are linearly disjoint over $\overline{F}$. If $\overline{B_1} = \overline{D}$, then we must have $|\Gamma_{B_1}{:}\Gamma_F| \geq [\overline{D}{:}\overline{F}]$ because the canonical map $\theta_{B_1}: \Gamma_{B_1} \to \mathcal{G}(\overline{D}/\overline{F})$ is surjective and $ker(\theta_{B_1}) \supseteq \Gamma_F$ (see Prop. 8.17(i)). As D is semiramified we have $[\overline{D}{:}\overline{F}] = deg\, D$, hence

$$[B_1{:}F] = [\overline{B_1}{:}\overline{F}] \cdot |\Gamma_{B_1}{:}\Gamma_F| \geq (deg\, D)^2.$$

This implies $B_1 = D$, a contradiction. Therefore, $\overline{B_1} \subsetneqq \overline{D}$. Likewise, $\overline{B_2}$ is a proper subfield of $\overline{D}$.

Since $exp(\Gamma_{B_i}/\Gamma_F) \mid exp(\Gamma_D/\Gamma_F)$ for $i = 1, 2$, by Prop. 10.7 we can write each $B_i = S_i \otimes_F T_i$, where S_i is inertially split, T_i is tame and totally ramified over F with $\Gamma_{S_i} \cap \Gamma_{T_i} = \Gamma_F$, and $\overline{S_i} = \overline{B_i} \subseteq \overline{D}$. Since $\overline{S_i}$ is therefore a field and S_i is inertially split, it follows from Prop. 8.38 that S_i is semiramified. Let $N_3 = T_1 \otimes_F T_2$. Since

$$D \cong S_1 \otimes_F S_2 \otimes_F N_3,$$

N_3 is a division algebra; also N_3 is inertially split, because D, S_1, and S_2 are all inertially split. By Lemma 10.9, N_3 is decomposably semiramified. For $i = 1, 2$, choose an inertial division algebra I_i and decomposably semiramified division algebra N_i such that $S_i \sim I_i \otimes_F N_i$ in $Br(F)$. By Th. 8.65, $\overline{N_i} = Z(\overline{S_i}) = \overline{S_i} = \overline{B_i}$, $\Gamma_{S_i} = \Gamma_{B_i}$, and, as S_i is semiramified, $\overline{I_i} \in Br(\overline{S_i}/\overline{F}) = Br(\overline{B_i}/\overline{F})$. To complete the proof, we show that $[\overline{I_1}] + [\overline{I_2}]$ represents the specialization coset $sp(D)$. Since in $Br(F)$

$$D \sim (I_1 \otimes_F N_1) \otimes_F (I_2 \otimes_F N_2) \otimes_F N_3 \sim (I_1 \otimes_F I_2) \otimes_F (N_1 \otimes_F N_2 \otimes_F N_3),$$

it suffices to show that the division algebra associated to $N_1 \otimes_F N_2 \otimes_F N_3$ is decomposably semiramified. We already know that N_1, N_2, and N_3 are inertially split, so by Prop. 8.40 it only remains to show that $N_1 \otimes_F N_2 \otimes_F N_3$ is split by a totally ramified extension of F.

By Cor. 6.50 we have $\Gamma_D = \Gamma_{S_1} + \Gamma_{S_2} + \Gamma_{N_3}$, hence

$$\Gamma_D/\Gamma_F = (\Gamma_{S_1}/\Gamma_F) + (\Gamma_{S_2}/\Gamma_F) + (\Gamma_{N_3}/\Gamma_F). \tag{10.4}$$

Since D and N_3 are semiramified we have

$$|\Gamma_D{:}\Gamma_F| = deg\, D \quad \text{and} \quad |\Gamma_{N_3}{:}\Gamma_F| = deg\, N_3.$$

Likewise, S_i is semiramified and $\Gamma_{S_i} = \Gamma_{N_i}$ for $i = 1, 2$, hence

$$\Gamma_{N_i}/\Gamma_F = \Gamma_{S_i}/\Gamma_F \quad \text{and} \quad |\Gamma_{S_i}{:}\Gamma_F| = deg\, S_i \quad \text{for } i = 1, 2.$$

Since $deg\, D = deg\, S_1\, deg\, S_2\, deg\, N_3$ it follows that the sum in (10.4) is direct:

$$\Gamma_D/\Gamma_F = (\Gamma_{N_1}/\Gamma_F) \oplus (\Gamma_{N_2}/\Gamma_F) \oplus (\Gamma_{N_3}/\Gamma_F).$$

For $j = 1, 2, 3$, since N_j is decomposably semiramified, by the valued version of Prop. 8.28 it contains a maximal subfield L_j totally ramified over F with $\Gamma_{L_j} = \Gamma_{N_j}$. Let $L = L_1 \otimes_F L_2 \otimes_F L_3$. Since the sum of the Γ_{L_j}/Γ_F is direct, L is a field, and a totally ramified extension of F. It is also a splitting field of $N_1 \otimes_F N_2 \otimes N_3$, hence the proof is complete. □

In the preceding proof, in fact $N_1 \otimes_F N_2 \otimes_F N_3$ is a division algebra, and $\overline{D} = \overline{N_1} \otimes_{\overline{F}} \overline{N_2} \otimes_{\overline{F}} \overline{N_3} = \overline{B_1} \otimes_{\overline{F}} \overline{B_2} \otimes_{\overline{F}} \overline{N_3}$. Note that B_1 and B_2 need not be inertially split: As in Lemma 10.9, they could be totally ramified. Also, a tame semiramified division algebra D as in the preceding proof may have decompositions $D \cong B_1 \otimes B_2$ with B_1, B_2 non-tame, see Ex. 8.12.

Corollary 10.11. *Let $k \subseteq \ell$ be fields with $[\ell{:}k] = p^r$ for some prime p and ℓ elementary abelian Galois over k. Assume $\operatorname{char} k \neq p$ and let A be a central simple k-algebra of exponent p split by ℓ. Suppose that for every pair of intermediate fields E_1, E_2 with $k \subsetneqq E_i \subsetneqq \ell$ and $E_1 \cap E_2 = k$ we have*

$$[A] \notin Br(E_1/k) + Br(E_2/k) + Dec(\ell/k).$$

Write $\ell = \ell_1 \otimes_k \ldots \otimes_k \ell_r$ where each ℓ_i is a cyclic extension of k of degree p. Then the semiramified division algebras $\mathcal{S} = \mathcal{S}(A; \ell_1, \ldots, \ell_r)$ and $\mathcal{S}' = \mathcal{S}'(A; \ell_1, \ldots, \ell_r)$ of (8.36) and (8.37) are indecomposable of exponent p and degree p^r.

Proof. By Prop. 8.72, $\mathcal{S}$ is semiramified and inertially split with respect to the usual rank r Henselian valuation on $Z(\mathcal{S}) = k((t_1))\ldots((t_r))$, with $deg\,\mathcal{S} = deg\,\mathcal{S}' = p^r$ and $exp\,\mathcal{S} = exp\,\mathcal{S}' = p$. Because $char\,\overline{\mathcal{S}} = char\,k \neq p$, every $Z(\mathcal{S})$-subalgebra of $\mathcal{S}$ is tame. Since $sp(\mathcal{S}) = [A] + Dec(\ell/k)$, Th. 10.10 therefore implies that $\mathcal{S}$ is indecomposable. Then $\mathcal{S}'$ is also indecomposable, as $\mathcal{S}' \otimes_{Z(\mathcal{S}')} Z(\mathcal{S}) \cong \mathcal{S}$ by Cor. 8.73. □

The condition simplifies in the low degree cases:

Corollary 10.12. *Let ℓ/k be an elementary abelian Galois extension of fields of degree a power of a prime p and let A be a central simple k-algebra of exponent p split by ℓ such that $[A] \notin Dec(\ell/k)$. Assume $\operatorname{char} k \neq p$.*

(i) *If $[\ell{:}k] = p^2$ for some prime $p \neq 2$ and $\ell = \ell_1 \otimes_k \ell_2$ where each ℓ_i is a cyclic extension of k of degree p, then the division algebras $\mathcal{S}(A; \ell_1, \ell_2)$ and $\mathcal{S}'(A; \ell_1, \ell_2)$ are indecomposable of exponent p and degree p^2.*
(ii) *If $[\ell{:}k] = 8$ and $\ell = \ell_1 \otimes_k \ell_2 \otimes_k \ell_3$ where each ℓ_i is a quadratic extension of k, then the division algebras $\mathcal{S}(A; \ell_1, \ell_2, \ell_3)$ and $\mathcal{S}'(A; \ell_1, \ell_2, \ell_3)$ are indecomposable of exponent 2 and degree 8.*

Specific examples where Cor. 10.12 applies will be given in the next subsection (for $p = 2$) and in Exercise 10.6 for p odd.

Proof. In each case, we verify that the condition $[A] \notin Dec(\ell/k)$ is sufficient to satisfy the hypothesis of Cor. 10.11. This is clear in case (i), because for any intermediate field E with $k \subsetneqq E \subsetneqq \ell$ the extension E/k is cyclic, hence $Br(E/k) \subseteq Dec(\ell/k)$ by definition of the Dec group.

In case (ii) we have to verify that if E_1, E_2 are extensions of k with $k \subsetneqq E_1, E_2 \subsetneqq \ell$ and $E_1 \cap E_2 = k$, then $[A] \notin Br(E_1/k) + Br(E_2/k) + Dec(\ell/k)$. This is clear if $[E_1{:}k] = [E_2{:}k] = 2$, for then $Br(E_1/k)$, $Br(E_2/k) \subseteq Dec(\ell/k)$, while $[A] \notin Dec(\ell/k)$. Suppose then $[E_1 : k] = 4$ and $[E_2 : k] = 2$, and assume, by way of contradiction, that $[A] = [A_1] + [A_2] + [A_3]$ with $[A_i] \in Br(E_i/k)$ for $i = 1, 2$, and $[A_3] \in Dec(\ell/k)$. Since $\mathcal{G}(\ell/k)$ has exponent 2, we have $2[A_3] = 0$ (see (8.12)). Likewise, $2[A_2] = 0$ because $[E_2 : k] = 2$. Since $2[A] = 0$ by hypothesis, it follows that $2[A_1] = 0$. We observed in Exercise 8.4 that, as a consequence of a theorem of Albert (see Knus et al. [115, Th. (16.1)]), ${}_2 Br(E_1/k) = Dec(E_1/k)$, hence $[A_1] \in Dec(E_1/k) \subseteq Dec(\ell/k)$. Thus, $[A_i] \in Dec(\ell/k)$ for all i; hence, $[A] \in Dec(\ell/k)$, a contradiction. □

10.2.2 Indecomposable algebras of exponent 2 and degree 8

We give in this subsection an example of an elementary abelian Galois extension ℓ/k with $[\ell : k] = 8$ and ${}_2 Br(\ell/k) \neq Dec(\ell/k)$. In view of Cor. 10.12, this example yields indecomposable division algebras of exponent 2 and degree 8.

Let k_0 be an arbitrary field of characteristic 0. We let x, y be independent indeterminates over k_0 and set

$$k = k_0(x, y), \quad a_1 = 1 - x, \quad a_2 = y, \quad a_3 = x^2(y+1)^2 - 4xy, \tag{10.5}$$

and also

$$\ell_i = k(\sqrt{a_i}\,) \text{ for } i = 1, 2, 3, \quad \text{and} \quad \ell = k(\sqrt{a_1}, \sqrt{a_2}, \sqrt{a_3}\,).$$

Each ℓ_i is a quadratic extension of k. We will see in Th. 10.13 below that $[\ell : k] = 8$. Consider the following element in k

$$u = (x-1)y = (xy) \cdot \big((x-1)x^{-1}\big) = (1-x) \cdot (-y).$$

Since $u = -a_1 a_2$, the quaternion algebra $\big(a_1 a_2, u/k\big)_2$ is split; hence,

$$\big(a_1, u/k\big)_2 \cong \big(a_2, u/k\big)_2. \tag{10.6}$$

Since the quaternion algebras $\big(1-x, x/k\big)_2$ and $\big(1-x, x-1/k\big)_2$ are split, so is $\big(a_1, (x-1)x^{-1}/k\big)_2$; hence, from $u = (xy) \cdot \big((x-1)x^{-1}\big)$ we have

$$\big(a_1, u/k\big)_2 \cong \big(a_1, xy/k\big)_2. \tag{10.7}$$

Likewise, since $\big(a_2, -y/k\big)_2$ is split, we obtain from $u = (1-x) \cdot (-y)$

$$\big(a_2, u/k\big)_2 \cong \big(a_2, 1-x/k\big)_2. \tag{10.8}$$

Now, note that $xy = \frac{1}{4}(x^2(y+1)^2 - a_3)$, so for $b_1 = \frac{1}{2}(x(y+1) + \sqrt{a_3}\,) \in \ell_3$ we have

$$xy = N_{\ell_3/k}(b_1). \tag{10.9}$$

Note also that $4xy(1-x) = x^2(y-1)^2 - a_3$; hence, for

$$b_2 = \tfrac{1}{2}(x(y-1) + \sqrt{a_3}\,)b_1^{-1} \in \ell_3$$

we have

$$1 - x = N_{\ell_3/k}(b_2). \tag{10.10}$$

By combining (10.6), (10.7), (10.8), (10.9), and (10.10), we obtain

$$\left(a_1, N_{\ell_3/k}(b_1)/k\right)_2 \cong \left(a_1, u/k\right)_2 \cong \left(a_2, u/k\right)_2 \cong \left(a_2, N_{\ell_3/k}(b_2)/k\right)_2. \tag{10.11}$$

Consider then the ℓ_3-algebra

$$B = \left(a_1, b_1/\ell_3\right)_2 \otimes_{\ell_3} \left(a_2, b_2/\ell_3\right)_2. \tag{10.12}$$

Since $a_1, a_2 \in k^\times$, we may use the projection formula (see e.g. Gille–Szamuely [84, Prop. 3.4.10]) to compute its corestriction: In view of (10.11) we have

$$\mathrm{cor}_{\ell_3/k}[B] = \left[\left(a_1, N_{\ell_3/k}(b_1)/k\right)_2 \otimes_k \left(a_2, N_{\ell_3/k}(b_2)/k\right)_2\right] = 0.$$

An easy cohomological argument due to Arason [13, Cor. 4.6] (see also Knus et al. [115, Cor. (30.12)]) shows that the following sequence is exact:

$$_2\,Br(k) \xrightarrow{\mathrm{ext}_{\ell_3/k}} {_2\,Br(\ell_3)} \xrightarrow{\mathrm{cor}_{\ell_3/k}} {_2\,Br(k)}.$$

Therefore, we may find a Brauer class $\alpha \in {_2\,Br(k)}$ such that

$$\mathrm{ext}_{\ell_3/k}(\alpha) = [B]. \tag{10.13}$$

Theorem 10.13. *With the notation above in* (10.5), (10.12), *and* (10.13), *we have* $[\ell{:}k] = 8$ *and* $\alpha \in {_2\,Br(\ell/k)}$ *but* $\alpha \notin Dec(\ell/k)$.

Proof. It is clear from the definition that B is split by ℓ, hence α is split by ℓ. If $[\ell{:}k] \neq 8$, then $[\ell{:}k] = 2$ or 4, hence ${_2\,Br(\ell/k)} = Dec(\ell/k)$: See Exercise 8.4. Thus, it suffices to prove that $\alpha \notin Dec(\ell/k)$.

We will use the following easy result:

Lemma 10.14 (Common Slot Lemma). *Let Q be a quaternion algebra over an arbitrary field E of characteristic different from 2. If a, b, a', $b' \in E^\times$ are such that*

$$Q \cong \left(a, b/E\right)_2 \cong \left(a', b'/E\right)_2,$$

then there exists a $z \in E^\times$ such that

$$Q \cong \left(a, z/E\right)_2 \cong \left(a', z/E\right)_2.$$

Proof. A proof is given in Lam [122, Th. 3.4.13]. We include a slightly different proof for the reader's convenience. If Q is split we may take $z = 1$. If Q is a division algebra, consider quaternions q, $q' \in Q$ such that $q^2 = a$ and $q'^2 = a'$. If q and q' commute, then $q' \in qF^\times$ hence $a' \equiv a \pmod{F^{\times 2}}$ and we may take $z = b$ or b'. If q and q' do not commute, then the quaternion $q'' = qq' - q'q$ is nonzero and anticommutes with q and q', so we may take q and q'' or, alternatively, q' and q'' as standard generators of Q. Thus, for $z = q''^2$ we obtain the desired relations. □

Proof of Th. 10.13, completed. Assume that $\alpha \in Dec(\ell/k)$. Then, there exist c_1, c_2, $c_3 \in k^\times$ such that

$$\alpha = \big[(a_1, c_1/k)_2 \otimes_k (a_2, c_2/k)_2 \otimes_k (a_3, c_3/k)_2\big].$$

By extending scalars to ℓ_3, we obtain

$$B \cong (a_1, c_1/\ell_3)_2 \otimes_{\ell_3} (a_2, c_2/\ell_3)_2,$$

hence

$$(a_1, b_1c_1/\ell_3)_2 \cong (a_2, b_2c_2/\ell_3)_2.$$

The Common Slot Lemma 10.14 yields an element $z \in \ell_3^\times$ such that

$$(a_1, b_1c_1/\ell_3)_2 \cong (a_1, z/\ell_3)_2 \cong (a_2, z/\ell_3)_2 \cong (a_2, b_2c_2/\ell_3)_2.$$

By taking the images under the corestriction map, and using the projection formula, we obtain

$$(a_1, N_{\ell_3/k}(b_1)/k)_2 \cong (a_1, N_{\ell_3/k}(z)/k)_2 \cong (a_2, N_{\ell_3/k}(z)/k)_2.$$

From (10.11), it follows that

$$\big[(a_1, uN_{\ell_3/k}(z)/k)_2\big] = \big[(a_2, uN_{\ell_3/k}(z)/k)_2\big] = 0.$$

Therefore, there exist $z_1 \in \ell_1^\times$ and $z_2 \in \ell_2^\times$ such that

$$uN_{\ell_3/k}(z) = N_{\ell_1/k}(z_1) = N_{\ell_2/k}(z_2). \tag{10.14}$$

To prove that (10.14) is impossible, we use valuation theory: Since *char* $k_0 = 0$, the field k_0 carries a valuation v_0 extending the 2-adic valuation on $\mathbb{Q}$. For this valuation, the value group of k_0 is some ordered group Γ_{k_0}, and the residue field $\overline{k_0}$ has characteristic 2. Consider the Gaussian extension v_1 of v_0 to $k = k_0(x, y)$, with value group Γ_{k_0} and residue field $\overline{k_0}(\overline{x}, \overline{y})$ (see Exercise 1.3), and let v be the valuation on k obtained by composing the valuation v_1 with the $\overline{x}$-adic valuation on $\overline{k_0}(\overline{x}, \overline{y})$ (see §1.2.3 for the composition of valuations). Thus, writing simply x and y for $\overline{x}$ and $\overline{y}$, the value group of v is $\Gamma_{k,v} = \mathbb{Z} \times \Gamma_{k_0}$ with the right-to-left lexicographic ordering, $v(x) = (1, 0)$, and the residue field is $\overline{k} = \overline{k_0}(y)$. It is clear that v extends uniquely to $\ell_2 = k(\sqrt{y})$, and ℓ_2/k is unramified (with purely inseparable residue field extension) for this extension.

Claim: v extends uniquely to ℓ_1 and to ℓ_3, and the extensions ℓ_1/k and ℓ_3/k are totally ramified.

Assuming the claim, we prove that (10.14) is impossible as follows: Since v extends uniquely to ℓ_2 and ℓ_2/k is unramified, we have

$$v\big(N_{\ell_2/k}(z_2)\big) = 2v(z_2) \in 2\Gamma_{k,v},$$

so there exists a $d \in k^\times$ such that $v\big(N_{\ell_2/k}(z_2)\big) = 2v(d)$. By substituting z_1d^{-1} for z_1, z_2d^{-1} for z_2, and zd^{-1} for z, we may assume henceforth that $v\big(N_{\ell_1/k}(z_1)\big) = 0$. Since $u = (x-1)y$ we have $v(u) = 0$ and $\overline{u} = y$, so from (10.14) we also have $v\big(N_{\ell_3/k}(z)\big) = 0$, and, by taking residues,

$$y \cdot \overline{N_{\ell_3/k}(z)} \;=\; \overline{N_{\ell_1/k}(z_1)}.$$

But ℓ_1/k and ℓ_3/k are totally ramified, so $\overline{N_{\ell_3/k}(z)} = \overline{z}^2 \in \overline{k}^{\times 2}$ and, similarly, $\overline{N_{\ell_1/k}(z_1)} \in \overline{k}^{\times 2}$. Therefore, $y \in \overline{k}^{\times 2}$, a contradiction.

Thus, it only remains to prove the claim. Recall that $\ell_1 = k(\sqrt{1-x}\,)$ and

$$\ell_3 \;=\; k\big(\sqrt{x^2(y+1)^2 - 4xy}\,\big) \;=\; k\big(\sqrt{1 - 4x^{-1}y(y+1)^{-1}}\,\big),$$

so ℓ_1 and ℓ_3 both have the form $k(r)$ with $r^2 = 1 - t$ for some $t \in k$ with $0 < v(t) < v(4)$ and $v(t) \notin 2\Gamma_{k,v}$. (For ℓ_3, note that

$$v(4x^{-1}y(y+1)^{-1}) = (-1, v_0(4)) > 0 \text{ and } v(4) = (0, v_0(4)) > (-1, v_0(4)).)$$

Let w be an extension of v to $k(r)$. We have $\overline{r}^2 = 1$, so $r = 1 + m$ for some $m \in k(r)$ with $w(m) > 0$. Then,

$$1 + 2m + m^2 \;=\; 1 - t.$$

If $w(m) \geq v(2)$, then $w(2m + m^2) \geq v(4)$. This is impossible because $v(t) < v(4)$. Therefore, we have $w(m) < v(2)$, hence $v(t) = w(2m+m^2) = 2w(m)$. Since $v(t) \notin 2\Gamma_{k,v}$, it follows that $w(m) \notin \Gamma_{k,v}$, so w is the unique extension of v to $k(r)$, and w is ramified over v. □

Corollary 10.15. *With the notation above in* (10.5), (10.12), *and* (10.13), *for any central simple k-algebra A with $[A] = \alpha$ in $Br(k)$, the division algebras $\mathcal{S}(A;\ell_1,\ell_2,\ell_3)$ and $\mathcal{S}'(A;\ell_1,\ell_2,\ell_3)$ are indecomposable of exponent* 2 *and degree* 8.

Proof. This follows from Cor. 10.12 and Th. 10.13. □

10.2.3 Characteristic p calculations

In preparation for the construction of indecomposable division algebras of odd prime exponent, we give Jacob's characteristic p calculations that will be applied over the residue field of a Henselian field of characteristic 0 in the next subsection. In this subsection, p may be any prime number.

Let L be a field of characteristic p. Let

$$k_1(L) \;=\; \big\{\{a\}_p \mid a \in L^\times\big\}, \quad \text{where } \{a\}_p = \{b\}_p \text{ just when } ab^{-1} \in L^{\times p},$$

with well-defined abelian group operation given by

$$\{a\}_p + \{b\}_p \;=\; \{ab\}_p.$$

Thus $k_1(L) \cong L^\times / L^{\times p}$, but with the operation in $k_1(L)$ written additively. The map $\{\cdot\}_p \colon L^\times \to k_1(L)$ can be considered as a formal *mod-p* logarithm for $L^\times$.

Let Ω_L be the (absolute) differential module of L, i.e., Ω_L is the L-vector space spanned by elements da for $a \in L$, subject to the relations

$$d(a+b) \;=\; da + db \quad \text{and} \quad d(ab) = a\,db + b\,da \qquad \text{for all } a,\, b \in L.$$

For basic properties of Ω_L, see, e.g., Matsumura [137, §§25–26]. Since $char\, L = p$, recall that $da = 0$ in Ω_L if and only if $a \in L^p$. Let $\Omega_L^2 = \Omega_L \wedge \Omega_L$, the second exterior power of Ω_L as an L-vector space. Note that there is a well-defined injective group homomorphism

$$dlog_L \colon k_1(L) \longrightarrow \Omega_L \qquad \text{given by} \qquad \{a\}_p \mapsto \tfrac{da}{a} \;\; (= a^{-1}da).$$

This map induces a homomorphism

$$k_1(L) \otimes_{\mathbb{Z}} k_1(L) \longrightarrow \Omega_L^2 \qquad \text{given by} \qquad \{a\}_p \otimes \{b\}_p \mapsto \tfrac{da}{a} \wedge \tfrac{db}{b};$$

we call this map $dlog_L$, as well.

Now fix an integer $r \geq 1$. For the rest of this subsection, assume that $[L{:}L^p] = p^{r+1}$. A collection $(z_0, z_1, \ldots, z_r)$ of elements of L is called a *p-base* of L if $L = L^p(z_0, \ldots, z_r)$. We then have that every $a \in L$ is expressible uniquely as

$$a \;=\; \sum_{i_0=0}^{p-1} \sum_{i_1=0}^{p-1} \cdots \sum_{i_r=0}^{r-1} (c_{i_0 i_1 \ldots i_r})^p z_0^{i_0} z_1^{i_1} \ldots z_r^{i_r} \qquad \text{with each } \; c_{i_0 i_1 \ldots i_r} \in L.$$

To simplify the notation for such expressions, we use the index set

$$\mathcal{I} \;=\; \{(i_0, \ldots, i_r) \mid \text{each } i_j \in \{0, 1, \ldots, p-1\}\}.$$

For $\mathbf{i} = (i_0, i_1, \ldots, i_r) \in \mathcal{I}$, set $z^{\mathbf{i}} = z_0^{i_0} z_1^{i_1} \ldots z_r^{i_r}$. Thus, each $a \in L$ is uniquely expressible as $a = \sum_{\mathbf{i} \in \mathcal{I}} c_{\mathbf{i}}^{\,p} z^{\mathbf{i}}$ with each $c_{\mathbf{i}} \in L$.

It is easy to check that if $(z_0, \ldots, z_r)$ is a p-base of L, then $\big(\frac{dz_0}{z_0}, \ldots, \frac{dz_r}{z_r}\big)$ is an L-vector space base of Ω_L (cf. Matsumura [137, Th. 26.5, p. 202]). Hence, the $\frac{dz_i}{z_i} \wedge \frac{dz_j}{z_j}$, for all $i, j \in \{0, 1, \ldots, r\}$ with $i < j$, make up an L-vector space base of Ω_L^2. We call this the base of Ω_L^2 corresponding to $(z_0, \ldots, z_r)$.

Now suppose L has a valuation v. When we view L as an algebra over its subfield L^p, this v is a surmultiplicative $v|_{L^p}$-value function on L; since $\mathsf{gr}(L)$ is a graded field, v is a $v|_{L^p}$-gauge if and only if it is a $v|_{L^p}$-norm, i.e., $[\mathsf{gr}(L)\colon \mathsf{gr}(L^p)] = [L{:}L^p] = p^{r+1}$. When this occurs, we say that a p-base $(z_0, \ldots, z_r)$ of L is a *splitting p-base* of L (for v) if $\{z^{\mathbf{i}} \mid \mathbf{i} \in \mathcal{I}\}$ is a splitting base for the $v|_{L^p}$-gauge v on L, i.e., for all $c_{\mathbf{i}} \in L$

$$v\big(\sum_{\mathbf{i} \in \mathcal{I}} c_{\mathbf{i}}^{\,p} z^{\mathbf{i}}\big) \;=\; \min_{\mathbf{i} \in \mathcal{I}} \big(v(c_{\mathbf{i}}^{\,p} z^{\mathbf{i}})\big).$$

This is equivalent to: $(\widetilde{z}_0, \widetilde{z}_1, \ldots, \widetilde{z}_r)$ is a p-base of $q(\mathsf{gr}(L))$. Not every p-base is a splitting p-base, but note that if for some $k \in \{0, 1, \ldots r\}$ we have $v(z_0) = v(z_1) = \ldots = v(z_k) = 0$ with $(\,\overline{z_0}, \overline{z_1}, \ldots, \overline{z_k}\,)$ a p-base of $\overline{L}$ and $\big(v(z_{k+1}) + p\Gamma_L, \ldots, v(z_r) + p\Gamma_L\big)$ is an $\mathbb{F}_p$-vector space base of $\Gamma_L / p\Gamma_L$, then $(z_0, \ldots, z_r)$ is a splitting p-base of L.

Lemma 10.16. *Let $(z_0, \dots, z_r)$ be a splitting p-base of L for v. Take any $b \in L$ and write $\frac{db}{b} = \sum_{j=0}^{r} u_j \frac{dz_j}{z_j}$. Then,*

(i) $v(u_k) \geq 0$, *for each $k \in \{0, 1, \dots, r\}$.*
(ii) *If $v(z_0) = 0$ and $\big(v(z_1) + p\Gamma_L, \dots, v(z_r) + p\Gamma_L\big)$ is an $\mathbb{F}_p$-vector space base of $\Gamma_L/p\Gamma_L$, then $\overline{u_1}, \dots, \overline{u_r} \in \mathbb{F}_p$.*

Proof. (i) For $\mathbf{i} \in \mathcal{I}$, write $\mathbf{i} = (i_0, \dots, i_r)$. Let $b = \sum_{\mathbf{i} \in \mathcal{I}} c_{\mathbf{i}}^p z^{\mathbf{i}}$. Since

$$d(c_{\mathbf{i}}^p z^{\mathbf{i}}) \;=\; c_{\mathbf{i}}^p z^{\mathbf{i}} \sum_{j=0}^{r} i_j \tfrac{dz_j}{z_j},$$

we have

$$\tfrac{db}{b} \;=\; \tfrac{1}{b} \sum_{\mathbf{i} \in \mathcal{I}} d(c_{\mathbf{i}}^p z^{\mathbf{i}}) \;=\; \tfrac{1}{b} \sum_{\mathbf{i} \in \mathcal{I}} \Big(c_{\mathbf{i}}^p z^{\mathbf{i}} \sum_{j=0}^{r} i_j \tfrac{dz_j}{z_j}\Big).$$

Hence,

$$u_k \;=\; \tfrac{1}{b} \sum_{\mathbf{i} \in \mathcal{I}} i_k c_{\mathbf{i}}^p z^{\mathbf{i}}.$$

Therefore,

$$v(bu_k) \;=\; \min_{\mathbf{i} \in \mathcal{I}} \big(v(i_k\, c_{\mathbf{i}}^p z^{\mathbf{i}})\big) \;\geq\; \min_{\mathbf{i} \in \mathcal{I}} \big(v(c_{\mathbf{i}}^p z^{\mathbf{i}})\big) \;=\; v(b).$$

Hence, $v(u_k) \geq 0$.

(ii) Take any $k \in \{1, 2, \dots, r\}$. We have seen that $v(bu_k) \geq v(b)$. If $v(bu_k) > v(b)$, then $v(u_k) > 0$, so $\overline{u_k} = 0 \in \mathbb{F}_p$. Assume now that $v(bu_k) = v(b)$, and let

$$\mathcal{J} \;=\; \{\mathbf{j} = (j_0, \dots, j_r) \in \mathcal{I} \mid v(j_k c_{\mathbf{j}}^p z^{\mathbf{j}}) = v(b)\} \;\subseteq\; \mathcal{I}.$$

Fix any $\mathbf{j} = (j_0, \dots, j_r) \in \mathcal{J}$. Then $j_k \neq 0$ in L, so $v(j_k) = 0$. Take any $\mathbf{i} \in \mathcal{I}$ with $v(c_{\mathbf{i}}^p z^{\mathbf{i}}) = v(b) = v(c_{\mathbf{j}}^p z^{\mathbf{j}})$. Then, $v(z^{\mathbf{i}}) \equiv v(z^{\mathbf{j}})$ $(mod\ p\Gamma_L)$ so the added hypothesis of (ii) implies that $i_1 = j_1, \dots, i_r = j_r$. Then, $v(i_k) = v(j_k) = 0$, so $v(i_k c_{\mathbf{i}}^p z^{\mathbf{i}}) = v(c_{\mathbf{i}}^p z^{\mathbf{i}}) = v(b)$, showing that $\mathbf{i} \in \mathcal{J}$. Therefore, if $\mathbf{i} \notin \mathcal{J}$, then $v(c_{\mathbf{i}}^p z^{\mathbf{i}}) > v(b)$. But also, if $\mathbf{i} \in \mathcal{J}$, then $v(i_k) = 0$, so $v(c_{\mathbf{i}}^p z^{\mathbf{i}}) = v(b)$, which we just saw implies that $i_k = j_k$. Let

$$s \;=\; \sum_{\mathbf{j} \in \mathcal{J}} c_{\mathbf{j}}^p z^{\mathbf{j}}, \qquad t \;=\; \sum_{\mathbf{i} \in \mathcal{I} \setminus \mathcal{J}} c_{\mathbf{i}}^p z^{\mathbf{i}}, \qquad \text{and} \qquad t' \;=\; \sum_{\mathbf{i} \in \mathcal{I} \setminus \mathcal{J}} i_k c_{\mathbf{i}}^p z^{\mathbf{i}}.$$

Then, $v(s) = v(b) < v(t) \leq v(t')$. Note that $bu_k = j_k s + t'$, while $b = s + t$. Hence, for $u_k = bu_k/b = (j_k s + t')/(s + t)$ we have $\overline{u_k} = \overline{j_k} \in \mathbb{F}_p$. □

We now make further hypotheses on L and v, and choose some particular elements of $L^\times$. Suppose $\mathcal{O}_L$ contains a field L' isomorphic to $\overline{L}$ via the residue map $a \mapsto \overline{a}$. We identify $\overline{L}$ with L', thereby viewing $\overline{L}$ as a subfield of L. Suppose L has a splitting p-base $(z_0, \dots, z_r)$ such that $z_0 \in \overline{L}$, (z_0) is a p-base of $\overline{L}$, and $\big(v(z_1) + p\Gamma_L, \dots, v(z_r) + p\Gamma_L\big)$ is an $\mathbb{F}_p$-base of $\Gamma_L/p\Gamma_L$, with $0 = v(z_0) < v(z_1) < \dots < v(z_r)$. For example, we could take $L = \mathbb{F}_p(z_0, \dots, z_r)$ with the z_i algebraically independent over $\mathbb{F}_p$, and let v be the $(z_1, \ldots, z_r)$-adic valuation on L. Then, $\overline{L} = \mathbb{F}_p(z_0)$ and $\Gamma_L = \mathbb{Z}^r$ with right-to-left lexicographic order. In addition choose $a_1, \ldots, a_r \in L$ so that $\{1+z_0\}_p, \{a_1\}_p, \ldots, \{a_r\}_p$ are $\mathbb{F}_p$-linearly independent in $k_1(L)$.

Lemma 10.17. *With L, v, $z_0, \ldots, z_r$, $a_1, \ldots, a_r$ as above, fix some $k \in \{1, 2, \ldots, r\}$. Take any $s_2, s_3, \ldots, s_r \in L^\times$ such that $\big(v(z_k) + p\Gamma_L$, $v(s_2) + p\Gamma_L$, $v(s_3) + p\Gamma_L, \ldots, v(s_r) + p\Gamma_L\big)$ is an $\mathbb{F}_p$-base of $\Gamma_L/p\Gamma_L$. Let $N = L(\sqrt[p]{s_2}, \sqrt[p]{s_3}, \ldots, \sqrt[p]{s_r})$. Then, there do not exist c, $c' \in N^\times$ such that*

$$dlog_N\big[\{c\}_p \otimes \{1+z_0\}_p + \{c'\}_p \otimes \{1+z_k\}_p\big] = dlog_N\big[\sum_{j=1}^{n} \{a_j\}_p \otimes \{z_j\}_p\big] \ \text{ in } \Omega_N^2.$$

Proof. The hypothesis on the $v(s_j)$ assures that v has a unique and totally ramified extension to N with $\Gamma_N = \Gamma_L + \langle \tfrac{1}{p} v(s_2)\rangle + \ldots + \langle \tfrac{1}{p} v(s_r)\rangle$ and $[N{:}L] = |\Gamma_N{:}\Gamma_L| = p^{r-1}$; so $\overline{N} = \overline{L} \subseteq N$, and $(z_0, z_k, \sqrt[p]{s_2}, \ldots, \sqrt[p]{s_r})$ is a splitting p-base of N. Fix c, $c' \in N^\times$ and let

$$\delta_1 = dlog_N\big[\{c\}_p \otimes \{1+z_0\}_p + \{c'\}_p \otimes \{1+z_k\}_p\big].$$

Let also

$$\delta_2 = dlog_N\big[\sum_{j=1}^{n} \{a_j\}_p \otimes \{z_j\}_p\big].$$

We show that $\delta_1 \neq \delta_2$ in Ω_N^2 by comparing their $\frac{dz_0}{z_0} \wedge \frac{dz_k}{z_k}$-components when using the base of Ω_N^2 corresponding to the p-base $(z_0, z_k, \sqrt[p]{s_2}, \ldots, \sqrt[p]{s_r})$ of L.

Since (z_0) is a p-base of $\overline{L}$, we have for each j

$$\frac{da_j}{a_j} = \lambda_j \frac{dz_0}{z_0} \quad \text{in } \Omega_{\overline{L}} \qquad \text{for some } \lambda_j \in \overline{L}.$$

These equations also hold in Ω_N, as $\overline{L} \subseteq N$. Also, $\frac{d(1+z_0)}{1+z_0} = \frac{z_0}{1+z_0}\frac{dz_0}{z_0}$. Since the map $k_1(\overline{L}) \to \Omega_{\overline{L}}$ is injective, the $\mathbb{F}_p$-linear independence of $\{1+z_0\}_p$, $\{a_1\}_p, \ldots, \{a_r\}_p$ is equivalent to: $\frac{z_0}{1+z_0}, \lambda_1, \ldots, \lambda_r$ are $\mathbb{F}_p$-linearly independent in $\overline{L}$. For $j = 1, 2, \ldots, r$, since $z_j \in L = N^p(z_0, z_k)$, we have

$$\frac{dz_j}{z_j} = p_j \frac{dz_0}{z_0} + q_j \frac{dz_k}{z_k} \quad \text{in } \Omega_N \qquad \text{for some } p_j,\ q_j \in N.$$

Note that we can take $q_k = 1$ and $p_k = 0$. By Lemma 10.16, applied using the splitting p-base $(z_0, z_k, \sqrt[p]{s_2}, \ldots, \sqrt[p]{s_r})$ of N, we have $v(q_j) \geq 0$ and $\overline{q_j} \in \mathbb{F}_p$ in $\overline{N} = \overline{L}$, for each j. Moreover,

$$\delta_2 = \sum_{j=1}^{r} \big[\lambda_j \frac{dz_0}{z_0} \wedge \big(p_j \frac{dz_0}{z_0} + q_j \frac{dz_k}{z_k}\big)\big] = \big(\sum_{j=1}^{r} q_j \lambda_j\big) \frac{dz_0}{z_0} \wedge \frac{dz_k}{z_k}.$$

In Ω_N, write

$$\frac{dc}{c} = u\frac{dz_0}{z_0} + y\frac{dz_k}{z_k} + \sum_{j=2}^{r} w_j \frac{d(\sqrt[p]{s_j})}{\sqrt[p]{s_j}} \ \text{ and } \ \frac{dc'}{c'} = u'\frac{dz_0}{z_0} + y'\frac{dz_k}{z_k} + \sum_{j=2}^{r} w_j' \frac{d(\sqrt[p]{s_j})}{\sqrt[p]{s_j}}$$

with all $u, w, y_j, u', w', y_j' \in N$. Note that Lemma 10.16 applied over N shows that u', $y \in \mathcal{O}_N$ and $\overline{y} \in \mathbb{F}_p$ in $\overline{N}$. Since $\frac{d(1+z_j)}{1+z_j} = \frac{z_j}{1+z_j}\frac{dz_j}{z_j}$ for $j = 1, k$, our δ_1

has $\frac{dz_0}{z_0} \wedge \frac{dz_k}{z_k}$-component $-y\frac{z_0}{1+z_0} + u'\frac{z_k}{1+z_k}$. Hence, if $\delta_1 = \delta_2$, then

$$-y\tfrac{z_0}{1+z_0} + u'\tfrac{z_k}{1+z_k} = \sum_{j=1}^{r} q_j\lambda_j \qquad \text{in } N.$$

Since $v(y) \geq 0$, $v(u') \geq 0$, $v(z_k) > 0$, $v(q_j) \geq 0$, and z_0, $\lambda_j \in \overline{L} = \overline{N}$, by taking residues it then follows that

$$-\overline{y}\,\tfrac{z_0}{1+z_0} = \sum_{j=1}^{r} \overline{q_j}\,\lambda_j \qquad \text{in } \overline{N}.$$

But since $\overline{q_k} = 1 \neq 0$ and $\overline{y}, \overline{q_1}, \ldots, \overline{q_r} \in \mathbb{F}_p$, this contradicts the assumed $\mathbb{F}_p$-linear independence of $\frac{z_0}{1+z_0}$, $\lambda_1, \ldots, \lambda_r$. Hence, we cannot have $\delta_1 = \delta_2$. □

Now, let $t_j = 1 + z_j$ for $j = 0, 1, \ldots, r$. As usual, for $\mathbf{i} = (i_0, \ldots, i_r) \in \mathcal{I}$, we set $t^{\mathbf{i}} = t_0^{i_0} t_1^{i_1} \ldots t_r^{i_r}$. For $\mathbf{i} \neq (0, 0, \ldots, 0)$, let $m_{\mathbf{i}}$ be the least j such that $i_j \neq 0$. Since (z_0) is a p-base of $\overline{L}$ and $0 = v(z_0) < v(z_1) < \ldots < v(z_r)$, we claim that

$$t^{\mathbf{i}} = 1 + s_{\mathbf{i}} \qquad \text{where} \qquad v(s_{\mathbf{i}}) = v(z_{m_{\mathbf{i}}}). \tag{10.15}$$

To see this, expand out

$$t^{\mathbf{i}} = \big[(1+z_0)^{i_0}\big] \cdot \big[(1+z_1)^{i_1} \ldots (1+z_r)^{i_r}\big] = [1+s_0] \cdot [1+s'],$$

where $s_0 = (1+z_0)^{i_0} - 1$ and $v(s') > 0$. If $m_{\mathbf{i}} = 0$, i.e., $i_0 \neq 0$, then $v(s_0) = 0$ and $s_{\mathbf{i}} = s_0 + s' + s_0 s'$, which has value $0 = v(z_{m_{\mathbf{i}}})$. If $m_{\mathbf{i}} > 0$, then $s_0 = 0$ and

$$s_{\mathbf{i}} = s' = i_{m_{\mathbf{i}}} z_{m_{\mathbf{i}}} + i_{m_{\mathbf{i}}+1} z_{m_{\mathbf{i}}+1} + \ldots + i_r z_r + \ldots,$$

where the remaining summands each have greater value than $v(z_{m_{\mathbf{i}}})$. So, as $v(i_{m_{\mathbf{i}}}) = 0$, we again have $v(s_{\mathbf{i}}) = v(z_{m_{\mathbf{i}}})$. This establishes the claim.

Let

$$\mathcal{G} = \langle \{t_0\}_p, \{t_1\}_p, \ldots, \{t_r\}_p \rangle,$$

a subgroup of the elementary abelian group $k_1(L)$. Since $[L{:}L^p] = p^{r+1}$ and $L = L^p(z_0, \ldots, z_r) = L^p(t_0, \ldots, t_r)$, the collection $(t_0, \ldots, t_r)$ is a p-base of L (though not a splitting p-base). Therefore, $|\mathcal{G}| = p^{r+1}$. Let

$$\mathcal{S}_j = \langle \{t_j\}_p, \{t_{j+1}\}_p, \ldots, \{t_r\}_p \rangle \qquad \text{for } j = 0, 1, \ldots, r,$$

and let $\mathcal{S}_{r+1} = \{0\}$. Thus, $\mathcal{G} = \mathcal{S}_0 \supsetneqq \mathcal{S}_1 \supsetneqq \ldots \supsetneqq \mathcal{S}_r \supsetneqq \mathcal{S}_{r+1} = \{0\}$ with $|\mathcal{S}_j| = p^{r+1-j}$ for each j.

For a subgroup $\mathcal{H}$ of $\mathcal{G}$, we write $L(\sqrt[p]{\mathcal{H}})$ for the field $L(\{\sqrt[p]{h} \mid \{h\}_p \in \mathcal{H}\})$.

Proposition 10.18. *Let $\mathcal{H}$ be a subgroup of $\mathcal{S}_1$ with $|\mathcal{H}| = p^{r-1}$, and let $N = L(\sqrt[p]{\mathcal{H}})$. Then, $[N{:}L] = p^{r-1}$, and there is some $k \in \{1, 2, \ldots, r\}$ such that $\mathcal{G} = \langle \{1+z_0\}_p, \{1+z_k\}_p \rangle + \mathcal{H}$ and there do not exist c, $c' \in N^\times$ satisfying*

$$dlog_N\big[\{c\}_p \otimes \{1+z_0\}_p + \{c'\}_p \otimes \{1+z_k\}_p\big] = dlog_N\Big[\sum_{j=1}^{n} \{a_j\}_p \otimes \{z_j\}_p\Big] \text{ in } \Omega_N^2.$$

Proof. Since the elementary abelian group $\mathcal{H}$ can be generated by $r-1$ elements, we have $[L(\sqrt[p]{\mathcal{H}}\,){:}L] \le p^{r-1}$. But, as $\mathcal{G}$ can be generated by $\mathcal{H}$ and two more elements and $\mathcal{G}$ contains the images of a p-base of L,

$$\begin{aligned} p^{r+1} &= [L^{1/p}{:}L] = [L(\sqrt[p]{\mathcal{G}}\,){:}L] \\ &= [L(\sqrt[p]{\mathcal{G}}\,){:}L(\sqrt[p]{\mathcal{H}}\,)]\cdot[L(\sqrt[p]{\mathcal{H}}\,){:}L] \le p^2[L(\sqrt[p]{\mathcal{H}}\,){:}L]. \end{aligned}$$

Hence, $[N{:}L] = [L(\sqrt[p]{\mathcal{H}}\,){:}L] = p^{r-1}$.

Let $\mathcal{H}_j = \mathcal{H}\cap\mathcal{S}_j$ for $j = 1, 2, \ldots, r+1$. Thus,

$$\mathcal{H} = \mathcal{H}_1 \supseteq \mathcal{H}_2 \supseteq \ldots \supseteq \mathcal{H}_{r+1} = \{0\}.$$

Since $|\mathcal{S}_j{:}\mathcal{S}_{j+1}| = p$, we have $|\mathcal{H}_j{:}\mathcal{H}_{j+1}| = p$ or 1 for $j = 1, 2, \ldots, r$. But as $|\mathcal{H}| = p^{r-1}$, this index must equal p exactly $r-1$ times. Thus, there is a unique $k \in \{1, 2, \ldots, r\}$ with $\mathcal{H}_k = \mathcal{H}_{k+1}$. For each $j \ne k$ choose an element h_j in $\langle t_1, \ldots, t_r\rangle$ with $\{h_j\}_p \in \mathcal{H}_j \setminus \mathcal{H}_{j+1}$. Then, after multiplying h_j by a p-th power in $L^\times$ if necessary, we have $h_j = t^{\mathbf{i}}$ for some $\mathbf{i} \in \mathcal{I}$ with $i_0 = 0$. Note that $m_{\mathbf{i}} = j$, as $\{h_j\}_p \in \mathcal{H}_j \setminus \mathcal{H}_{j+1}$. Hence, by (10.15) we have

$$h_j = 1 + s_j \quad \text{with} \quad v(s_j) = v(z_j), \qquad \text{for } j = 1, 2, \ldots, r,\ j \ne k.$$

Observe that $\mathcal{H}_j = \langle h_j\rangle + \mathcal{H}_{j+1}$ for $j \ne k$, while $\mathcal{H}_k = \mathcal{H}_{k+1}$. Hence, $\mathcal{H} = \langle\{h_1\}_p, \ldots \{h_{k-1}\}_p, \{h_{k+1}\}_p, \ldots, \{h_r\}_p\rangle$. For convenience, renumber the h_j as h_2, h_3, $\ldots$, h_r, and likewise renumber the s_j. We then have $N = L(\sqrt[p]{h_2}, \ldots, \sqrt[p]{h_r}\,) = L(\sqrt[p]{s_2}, \ldots, \sqrt[p]{s_r}\,)$. Moreover,

$$\{v(z_k), v(s_2), v(s_3), \ldots, v(s_r)\} = \{v(z_1), v(z_2), \ldots, v(z_r)\},$$

which maps onto an $\mathbb{F}_p$-base $\Gamma_L/p\Gamma_L$. The hypotheses of Lemma 10.17 are thus satisfied, and the lemma yields the proposition. □

10.2.4 Indecomposable algebras of odd prime exponent

We can now complete Jacob's construction of an algebra A to which Cor. 10.11 can be applied to yield indecomposable algebras of odd prime exponent.

Let p be an odd prime number and r a positive integer. Let L be a field of characteristic p as in the previous subsection. Thus, L has a valuation v such that $\overline{L} \subseteq L$ and L has a splitting p-base $(z_0, z_1 \ldots, z_r)$ such that (z_0) is a p-base of $\overline{L}$ and $\big(v(z_1) + p\Gamma_L, v(z_2) + p\Gamma_L, \ldots, v(z_r) + p\Gamma_L\big)$ is an $\mathbb{F}_p$-base of $\Gamma_L/p\Gamma_L$ with $0 = v(z_0) < v(z_1) < v(z_2) < \ldots < v(z_r)$. In addition, choose a_1, a_2, $\ldots$, $a_r \in \overline{L}^\times$ such that $\big\{\frac{z_0}{1+z_0}\big\}_p$, $\{a_1\}_p$, $\{a_2\}_p$, $\ldots$, $\{a_r\}_p$ are $\mathbb{F}_p$-linearly independent in $k_1(L)$. Now, let F be a field of characteristic 0 containing a primitive p-th root of unity ω, and suppose that F has a Henselian valuation w such that $\overline{F} = L$ and the value group Γ_F is p-divisible.

Such an F and L are obtainable as follows: Let $F_1 = \mathbb{Q}_p(t_0, \dots, t_r)$, where $t_0, \dots, t_r$ are algebraically independent over the p-adic field $\mathbb{Q}_p$. Let the valuation w_1 be the Gaussian extension to F_1 of the complete discrete p-adic valuation v_p on $\mathbb{Q}_p$ (see Exercise 1.3). Thus, for

$$f = \sum_{i_0,\dots,i_r} c_{i_0\dots i_r} t_0^{i_0} \dots t_r^{i_r} \in \mathbb{Q}_p[t_0, \dots, t_r] \setminus \{0\},$$

we have $w_1(f) = min_{i_0,\dots,i_r}\big(w_1(c_{i_0\dots i_r})\big)$, and for $g \in \mathbb{Q}_p[t_0, \dots, t_r] \setminus \{0\}$, $w_1(f/g) = w_1(f) - w_1(g)$. Let $z_j = \overline{t_j}$. Then, $\overline{F_1} = \mathbb{F}_p(z_0, \dots, z_r)$ with $z_0, \dots, z_r$ algebraically independent over $\mathbb{F}_p = \overline{\mathbb{Q}_p}$. Also, $\Gamma_{F_1} = \Gamma_{\mathbb{Q}_p} = \mathbb{Z}$. Let $F_2 = F_1(\omega)$, where ω is a primitive p-th root of unity. Then $[F_2{:}F_1] = p - 1$, and w_1 extends uniquely to a valuation w_2 on F_2 that is totally ramified over w_1, since $w_2(\omega - 1) = 1/(p-1)$ (see the proof of Lemma 10.20 below). Let (F_3, w_3) be the Henselization (or the completion) of (F_2, w_2); so $\overline{F_3} = \overline{F_2} = \overline{F_1}$ and $\Gamma_{F_3} = \frac{1}{p-1}\mathbb{Z}$. Let $\xi = \omega - 1$, which is a uniformizing parameter for F_3, and let $F = F_3(\xi^{1/p}, \xi^{1/p^2}, \dots, \xi^{1/p^n}, \dots)$, with the roots chosen so that $(\xi^{1/p^{n+1}})^p = \xi^{1/p^n}$, and let w be the unique extension of the Henselian valuation w_3 to F. Each $F_3(\xi^{1/p^n})$ is totally ramified over F_3 with value group $\frac{1}{(p-1)p^n}\mathbb{Z}$. Hence, F is totally ramified over F_3 with p-divisible value group $\frac{1}{p-1}\mathbb{Z}[\frac{1}{p}]$. Also, $\overline{F} = \overline{F_3} = \mathbb{F}_p(z_0, \dots, z_r)$. As noted in the previous subsection, for the valuation v on $L = \overline{F}$, we may take the $(z_1, \dots, z_r)$-adic valuation, and identify $\overline{L}$ with $\mathbb{F}_p(z_0) \subseteq L$. The monic irreducible polynomials in $\mathbb{F}_p[z_0]$ map to an $\mathbb{F}_p$-vector space base of $k_1(\overline{L})$. Thus, for the a_i we may choose any r different monic irreducibles other than z_0 and $z_0 + 1$.

Now choose any $x_0, \dots, x_r \in \mathcal{O}_F$ such that each $\overline{x_j} = z_j$ (e.g., x_j could be the t_j of the preceding paragraph). Additionally, choose particular $b_1, \dots, b_r \in \mathcal{O}_F$ with $\overline{b_j} = a_j$, as follows: Since each $a_j \in \overline{L}$, which has p-base $(1 + z_0)$, we can write $a_j = \sum_{i=0}^{p-1} q_{ij}^{\,p}(1 + z_0)^i$ for some $q_{ij} \in \overline{L} \subseteq L$. Choose $y_{ij} \in \mathcal{O}_F$ with $\overline{y_{ij}} = q_{ij}$, and set

$$b_j = \sum_{i=0}^{p-1} y_{ij}^{\,p}(1 + x_0)^i \in \mathcal{O}_F; \tag{10.16}$$

then $\overline{b_j} = a_j$ for $j = 0, 1, \dots, r$.

Let

$$A = \big(b_1, x_1/F\big)_{\omega,p} \otimes_F \dots \otimes_F \big(b_r, x_r/F\big)_{\omega,p} \tag{10.17}$$

and

$$M = F\big(\sqrt[p]{1 + x_0}\,, \sqrt[p]{1 + x_1}\,, \dots, \sqrt[p]{1 + x_r}\,\big).$$

Theorem 10.19 (Jacob). *$[A] \in Br(M/F)$ with $[M{:}F] = p^{r+1}$, but for all proper subfields E_1, E_2 of M such that $E_1 \cap E_2 = F$, we have*

$$[A] \notin Br(E_1/F) + Br(E_2/F) + Dec(M/F).$$

The next general lemma will be used in proving that M splits A.

Lemma 10.20. *Let $c \in M$ with $w(c) > \frac{p}{p-1}\, w(p)$. Then $1 + c \in M^{\times p}$.*

Proof. Let $\xi = \omega - 1$; so $w(\xi) > 0$. By expanding out $(1+\xi)^p - 1 = 0$, we obtain

$$\xi^{p-1} + p\xi^{p-2} + \ldots + \tbinom{p}{i}\xi^{p-i-1} + \ldots + \tbinom{p}{p-2}\xi + p \;=\; 0.$$

Since all the middle summands have value exceeding $w(p)$, this yields $w(\xi) = \frac{1}{p-1}w(p)$. For the given $c \in M$, consider the polynomial

$$f \;=\; \xi^{-p}\big[(\xi X + 1)^p - 1 - c\big] \;=\; X^p + \sum_{i=2}^{p-1} \tbinom{p}{i}\xi^{i-p}X^i + p\xi^{1-p}X - c.$$

We have $f \in \mathcal{O}[X]$, as $w\binom{p}{i} \geq w(p) = (p-1)w(\xi) > (p-i)w(\xi)$ for $2 \leq i \leq p-1$. Hence, the image $\overline{f} = X^p + \overline{p\xi^{1-p}}X$ of f in $\overline{M}[X]$ has 0 as a simple root in $\overline{M}$. Since w is Henselian, f has a root b in M. Then $(\xi b+1)^p = 1+c$. □

We can now prove the first assertion of the theorem.

Lemma 10.21. $[A] \in Br(M/F)$.

Proof. For $j = 1, 2, \ldots, r$, since $w(\sqrt[p]{1+x_j}\,) = 0 = w(x_j)$, the binomial expansion yields

$$\big(-1 + \sqrt[p]{1+x_j}\,\big)^p \;=\; x_j + ps_j \qquad \text{with} \quad s_j \in \mathcal{O}_M.$$

Hence,

$$x_j/\big(-1 + \sqrt[p]{1+x_j}\,\big)^p \;=\; x_j/(x_j + ps_j) \;=\; 1 + ph_j,$$

where $h_j = -s_j/(x_j + ps_j) \in \mathcal{O}_M$. Similarly, as $b_j = \sum_{i=0}^{p-1} y_{ij}^{\,p}(1+x_0)^i$ with each $y_{ij} \in \mathcal{O}_F$, (see (10.16)), we have

$$\Big(\sum_{i=0}^{p-1} y_{ij}(1+x_0)^{i/p}\Big)^p \;=\; \Big(\sum_{i=0}^{p-1} y_{ij}^{\,p}(1+x_0)^i\Big) + pq_j \;=\; b_j + pq_j \quad \text{with} \;\; q_j \in \mathcal{O}_M.$$

Hence,

$$b_j/\Big(\sum_{i=0}^{p-1} y_{ij}(1+x_0)^{i/p}\Big)^p \;=\; b_j/(b_j + pq_j) \;=\; 1 + pg_j,$$

where $g_j = -q_j/(b_j + pq_j) \in \mathcal{O}_M$. Therefore, in $Br(M)$,

$$\big(b_j, x_j/M\big)_{\omega,p} \sim \big(1 + pg_j, 1 + ph_j/M\big)_{\omega,p}.$$

Recall the symbol algebra identities

$$\big(c, -c/M\big)_{\omega,p} \cong M_p(M) \quad \text{and} \quad \big(c, d/M\big)_{\omega,p} \cong \big(c + d, -(d/c)\,/M\big)_{\omega,p},$$

for all $c, d \in M^\times$ with $d \neq -c$. These are the algebra analogues of the k_2 identities given in (10.18) below, and the algebra identities follow from the k_2 identities in light of the homomorphism $\psi\colon k_2(M) \to Br(M)$ of (10.19) (or by direct calculation). Thus, in $Br(M)$,

$$\begin{aligned}(b_j, x_j/M)_{\omega,p} &\sim (1+pg_j, 1+ph_j/M)_{\omega,p}\\ &\sim (1+pg_j, -pg_j/M)_{\omega,p} \otimes_M (1+pg_j, 1+ph_j/M)_{\omega,p}\\ &\sim (1+pg_j, -pg_j(1+ph_j)/M)_{\omega,p}\\ &\sim (1-p^2g_jh_j, pg_j(1+ph_j)/(1+pg_j)/M)_{\omega,p}.\end{aligned}$$

But the last algebra is split, as $1 - p^2g_jh_j \in M^{\times p}$ by Lemma 10.20, since for the odd prime p, $w(p^2g_jh_j) \geq w(p^2) > \frac{p}{p-1}w(p)$. Because M splits each $(b_j, x_j/F)_{\omega,p}$, it splits A. □

The proof of the final assertion of the theorem entails passage from the Brauer group of a valued field to differentials over the residue field, via k-groups. Before launching into the proof, we describe this path. For any field E, define $k_2(E)$ by

$$k_2(E) = \big(k_1(E) \otimes_{\mathbb{Z}} k_1(E)\big)/\mathcal{R}(E),$$

where $\mathcal{R}(E)$ is the subgroup of $k_1(E) \otimes_{\mathbb{Z}} k_1(E)$ generated by the elements of the form $\{a\}_p \otimes \{1-a\}_p$ for $a \in E^\times \setminus \{1\}$. Thus, $k_2(E) \cong K_2(E)/pK_2(E)$, where $K_2(E)$ is Milnor's second K-group of E. A typical generator $\{a\}_p \otimes \{b\}_p + \mathcal{R}(E)$ for a, $b \in E^\times$ is called a p-symbol, and denoted $\{a,b\}_p$. Such p-symbols are multiplicative in each argument, and also satisfy the following identities:

$$\{a,-a\}_p = 0 \quad \text{and} \quad \{b,c\}_p = \{b+c, -c/b\}_p \quad \text{for all } a,\ b,\ c \in E^\times,\ c \neq -b. \tag{10.18}$$

The first identity is clear if $a = -1$, as $\{1\}_p = 0$; if $a \neq -1$, it is given by

$$0 = \{1+(1/a), -1/a\}_p = -\{a+1, -a\}_p + \{a,-a\}_p = \{a,-a\}_p.$$

For the second identity, note that for $c \neq -b$,

$$0 = \{1+(c/b), -c/b\}_p = \{b+c, -c/b\}_p - \{b,c\}_p + \{b,-b\}_p,$$

and use the first identity. If E contains a primitive p-th root of unity ω, then as $(a, 1-a/E)_{\omega,p}$ is split for all $a \in E^\times \setminus \{1\}$, there is a well-defined "norm residue homomorphism"

$$\psi\colon k_2(E) \longrightarrow {}_pBr(E) \qquad \text{given by} \quad \{a,b\}_p \mapsto (a, b/E)_{\omega,p}. \tag{10.19}$$

By the Merkurjev–Suslin Theorem (see Gille–Szamuely [84, Ch. 8], or Srinivas [232, Ch. 8] for a more complete proof), ψ is an isomorphism.

Now suppose further that $char\, E = 0$ and E has a valuation w with $char\, \overline{E} = p$ and that Γ_E is p-divisible. There is then a residue homomorphism $\rho\colon k_1(E) \to k_1(\overline{E})$ given as follows: If $a \in E^\times$, there is a $b \in E^\times$ with $w(b^p) = w(a)$. Then $w(ab^{-p}) = 0$ and we set $\rho(\{a\}_p) = \{\overline{ab^{-p}}\}_p$. This ρ is well-defined since b is unique up to $\mathcal{O}_E^\times$. The residue map on $k_1(E)$ induces a well-defined homomorphism $\rho'\colon k_1(E) \otimes_{\mathbb{Z}} k_1(E) \to k_2(\overline{E})$ given on generators by $\{a\}_p \otimes \{b\}_p \mapsto \{\overline{a}, \overline{b}\}_p$ whenever $w(a) = w(b) = 0$. We check that ρ' maps $\mathcal{R}(E)$ to 0: Take $a \in E^\times \setminus \{1\}$. If $w(a) > 0$ then $w(1-a) = 0$ and $\overline{1-a} = 1$ in $\overline{E}$, so for any $a' \in E^\times$ with $w(a') = 0$ and $\{a'\}_p = \{a\}_p$, we have $\rho'(\{a\}_p \otimes \{1-a\}_p) = \{\overline{a'}, 1\}_p = 0$; likewise, if $w(1-a) > 0$ then $\rho'(\{a\}_p \otimes \{1-a\}_p) = 0$; if $w(a) = w(1-a) = 0$, then $\rho'(\{a\}_p \otimes \{1-a\}_p) = \{\overline{a}, 1-\overline{a}\}_p = 0$. If $w(a) < 0$, then write $a = ct^p$ with $w(c) = 0$ using the p-divisibility of Γ_E; then $1 - a = t^p(t^{-p} - c)$ with $w(t^{-p}) = -w(a) > 0$, so

$$\rho'(\{a\}_p \otimes \{1-a\}_p) = \rho'(\{c\}_p \otimes \{t^{-p} - c\}_p) = \{\overline{c}, -\overline{c}\}_p = 0.$$

Thus, $\rho'(\mathcal{R}(E)) = \{0\}$. Therefore, ρ' induces a well-defined residue homomorphism again called $\rho\colon k_2(E) \to k_2(\overline{E})$, satisfying $\rho(\{a,b\}_p) = \{\overline{a}, \overline{b}\}_p$ whenever $w(a) = w(b) = 0$.

Note also that the map $dlog_{\overline{E}}\colon k_1(\overline{E}) \otimes_{\mathbb{Z}} k_1(\overline{E}) \to \Omega^2_{\overline{E}}$ factors through $k_2(\overline{E})$: For any $c \in \overline{E}^\times \setminus \{1\}$, we have

$$dlog_{\overline{E}}(\{c\}_p \otimes \{1-c\}_p) = \tfrac{dc}{c} \wedge \tfrac{d(1-c)}{1-c} = \tfrac{-c}{1-c}\,\tfrac{dc}{c} \wedge \tfrac{dc}{c} = 0.$$

Thus, $dlog_{\overline{E}}(\mathcal{R}(\overline{E})) = 0$. Let $dlog'_{\overline{E}}$ denote the resulting homomorphism $k_2(\overline{E}) \to \Omega^2_{\overline{E}}$, which maps $\{c, e\}_p \mapsto \frac{dc}{c} \wedge \frac{de}{e}$. We thus have a sequence of maps

$$_p Br(E) \xrightarrow{\psi^{-1}} k_2(E) \xrightarrow{\rho} k_2(\overline{E}) \xrightarrow{dlog'_{\overline{E}}} \Omega^2_{\overline{E}}. \tag{10.20}$$

The composition of these maps sends $(a, b/E)_{\omega,p}$ to $dlog_{\overline{E}}(\{\overline{a}\}_p \otimes \{\overline{b}\}_p) = \frac{d\overline{a}}{\overline{a}} \wedge \frac{d\overline{b}}{\overline{b}}$ for any $a, b \in E^\times$ with $w(a) = w(b) = 0$.

Proof of Th. 10.19. We have seen above that $[A] \in Br(M/F)$. Suppose now for purposes of contradiction that there are proper subfields E_1, E_2 of M with $E_1 \cap E_2 = F$ and $[A] \in Br(E_1/F) + Br(E_2/F) + Dec(M/F)$. Since $\omega \in F$, the field M is Galois over F with elementary abelian Galois group. By enlarging one of the fields E_i if necessary, we may assume further that neither $E_i = F$ and that $E_1 \cdot E_2 = M$.

The first step in the proof is to find a suitable subfield K of M with $[M{:}K] = p^2$ and $[A \otimes_F K] \in Dec(M/K)$.

Consider the group $G = \langle \{1 + x_0\}_p, \ldots, \{1 + x_r\}_p \rangle \subseteq k_1(F)$. Since $M = F(\{\sqrt[p]{g} \mid \{g\}_p \in G\})$, we write for short $M = F(\sqrt[p]{G})$. By Kummer theory, there are unique subgroups G_1, G_2 of G with $E_i = F(\sqrt[p]{G_i})$ for $i = 1, 2$. Then, the G_i are nontrivial subgroups of G with $G_1 + G_2 = G$ as $M = E_1 \cdot E_2$, and $G_1 \cap G_2 = \{0\}$ as $E_1 \cap E_2 = F$. Since Γ_F is p-divisible there is a residue homomorphism $\rho\colon k_1(F) \to k_1(\overline{F})$ as noted above. Let $\mathcal{G} = \rho(G) = \langle \{1 + z_0\}_p, \ldots, \{1 + z_r\}_p \rangle \subseteq k_1(\overline{F})$. As we saw in the preceding subsection, $|\mathcal{G}| = p^{r+1}$. Hence, $|G| = p^{r+1}$, so $[M{:}F] = p^{r+1}$ and $G \cong \mathcal{G}$ via ρ. Let $\mathcal{G}_i = \rho(G_i) \subseteq \mathcal{G}$. Then, the $\mathcal{G}_i$ are nontrivial subgroups of $\mathcal{G}$ with $\mathcal{G}_1 + \mathcal{G}_2 = \mathcal{G}$ and $\mathcal{G}_1 \cap \mathcal{G}_2 = \{0\}$. As in the preceding subsection, let $\mathcal{S}_1 = \langle \{1 + z_1\}_p, \ldots, \{1 + z_r\}_p \rangle$, a subgroup of $\mathcal{G}$ of index p. For $i = 1, 2$

define subgroups $\mathcal{H}_i \subset \mathcal{G}_i$ by

$$\mathcal{H}_i = \begin{cases} \mathcal{G}_i \cap \mathcal{S}_1, & \text{if } \mathcal{G}_i \not\subseteq \mathcal{S}_1; \\ \text{any subgroup of } \mathcal{G}_i \text{ of index } p, & \text{if } \mathcal{G}_i \subseteq \mathcal{S}_1. \end{cases}$$

Since $|\mathcal{G}{:}\mathcal{S}_1| = p$, we have $|\mathcal{G}_i{:}\mathcal{H}_i| = p$ in all cases. Let $\mathcal{H} = \mathcal{H}_1 + \mathcal{H}_2 \subseteq \mathcal{G}$. Since $\mathcal{H}_1 \cap \mathcal{H}_2 \subseteq \mathcal{G}_1 \cap \mathcal{G}_2 = \{0\}$, we have $|\mathcal{H}| = |\mathcal{H}_1|\cdot|\mathcal{H}_2| = |\mathcal{G}_1|\cdot|\mathcal{G}_2|/p^2 = p^{r-2}$. Also, since the sum $\mathcal{G}_1 + \mathcal{G}_2$ is direct, we have $\mathcal{G}_i \cap \mathcal{H} = \mathcal{H}_i$ for $i = 1, 2$. Let H (resp. H_1, H_2) be the inverse image of $\mathcal{H}$ (resp. $\mathcal{H}_1$, $\mathcal{H}_2$) in G, and let $K = F(\sqrt[p]{H}) \subseteq M$. Then, $[K{:}F] = |H| = |\mathcal{H}| = p^{r-1}$. Additionally, as $\mathcal{H} \subseteq \mathcal{S}_1$, we have $H \subseteq \langle \{1+x_1\}_p, \ldots, \{1+x_r\}_p \rangle$, so $K \subseteq F\big(\sqrt[p]{1+x_1}, \ldots, \sqrt[p]{1+x_r}\big)$. Moreover, by Kummer theory,

$$[E_i{\cdot}K : K] = |G_i + H : H| = |\mathcal{G}_i + \mathcal{H} : \mathcal{H}| = |\mathcal{G}_i : \mathcal{G}_i \cap \mathcal{H}| = p.$$

Hence, $Br(E_i{\cdot}K/K) = Dec(E_i{\cdot}K/K) \subseteq Dec(M/K)$ for $i = 1, 2$. Since $[A] \in Br(E_1/F) + Br(E_2/F) + Dec(M/F)$, it follows that

$$[A \otimes_F K] \in Br(E_1{\cdot}K/K) + Br(E_2{\cdot}K/K) + Dec(M/K) = Dec(M/K).$$

Now, by Prop. 10.18 we have $[\overline{F}(\sqrt[p]{\mathcal{H}}){:}\overline{F}] = p^r$. Since $\overline{K} \supseteq \overline{F}(\sqrt[p]{\mathcal{H}})$ it follows that

$$p^{r-1} = [\overline{F}(\sqrt[p]{\mathcal{H}}){:}\overline{F}] \le [\overline{K}{:}\overline{F}] \le [K{:}F] = p^{r-1};$$

hence, $\overline{K} = \overline{F}(\sqrt[p]{\mathcal{H}})$. Choose $k \in \{1, 2, \ldots, r\}$ as in Prop. 10.18 above (with $N = \overline{K}$). Since $\mathcal{G} = \langle \{1+z_0\}_p, \{1+z_k\}_p \rangle + \mathcal{H}$, we have

$$G = \langle \{1+x_0\}_p, \{1+x_k\}_p \rangle + H,$$

yielding $M = F(\sqrt[p]{G}) = K(\sqrt[p]{1+x_0})\cdot K(\sqrt[p]{1+x_k})$, a compositum of cyclic field extensions of K. We have just proved that

$$[A \otimes_F K] \in Dec(M/K) = Br(K(\sqrt[p]{1+x_0})/K) + Br(K(\sqrt[p]{1+x_k})/K).$$

Therefore, for some $t, t' \in K^\times$, we have in $Br(K)$

$$\begin{aligned}\big(b_1, x_1/K\big)_{\omega,p} \otimes_K \ldots \otimes_K \big(b_r, x_r/K\big)_{\omega,p} &\sim A \otimes_F K \\ &\sim \big(t, 1+x_0/K\big)_{\omega,p} \otimes_K \big(t', 1+x_k/K\big)_{\omega,p}.\end{aligned}$$

Since Γ_K is p-divisible, we may multiply t and t' by suitable elements of $K^{\times p}$, obtaining new t, t' with $w(t) = w(t') = 0$. By applying the map $dlog'_{\overline{K}} \circ \rho \circ \psi^{-1}$ of (10.20) (with K for E) we obtain

$$dlog_{\overline{K}}\Big[\sum_{j=1}^{n} \{a_j\}_p \otimes \{z_j\}_p\Big] = dlog_{\overline{K}}\big[\{\overline{t}\}_p \otimes \{1+z_0\}_p + \{\overline{t'}\}_p \otimes \{1+z_k\}_p\big] \text{ in } \Omega^2_{\overline{K}}.$$

This contradicts Prop. 10.18. Thus, $[A] \notin Br(E_1/F) + Br(E_2/F) + Dec(M/F)$ for any allowed fields E_1 and E_2, completing the proof of the theorem. □

Corollary 10.22. *With the notation of this subsection and the* A *of* (10.17), *the division algebras* $\mathcal{S}(A; F(\sqrt[p]{1+x_0}), \ldots, F(\sqrt[p]{1+x_r}))$ *and* $\mathcal{S}'(A; F(\sqrt[p]{1+x_0}), \ldots, F(\sqrt[p]{1+x_r}))$ *are indecomposable of exponent* p *and degree* p^{r+1}.

Proof. This follows readily from Cor. 10.11 and Th. 10.19. □

Remark 10.23. Examples of indecomposable division algebras of exponent p ($\neq 2$) and degree p^2 can also be obtained from Cor. 10.12 and Exercise 10.6.

10.3 Decompositions into tensor products of symbol algebras

In this section, we obtain necessary and sufficient conditions for the decomposition of certain tame division algebras over a Henselian field into tensor products of symbol algebras. These tensor products are characterized by the existence of an armature, as we saw in §7.2.1. Therefore, our main object of study in this section will be armatures. The main result is Th. 10.26, which relates armatures in an inertially split division algebra to armatures in special representatives of its specialization coset. At the end of the section, we also briefly consider division algebras that are tensor products of an inertial division algebra and a totally ramified division algebra.

Our first objective is to identify for the purposes of Th. 10.26 the special representatives of specialization cosets. For this, we will use the following general result:

Lemma 10.24. *Let* E *be a finite-dimensional division algebra over a field* F, *and let* $Z = Z(E) \supseteq F$. *Suppose* A *is a central simple* F*-algebra with* $A \otimes_F Z \sim E$ *in* $Br(Z)$. *Then,*

(i) $ind\, A \mid deg\, E \cdot [Z{:}F]$;
(ii) $deg\, A = deg\, E \cdot [Z{:}F]$ *if and only if* E *embeds in* A *and* $C_A(E) = Z$.

Concerning (ii), note that if E embeds in A, then $C_A(Z) \sim E$, so $C_A(E) \sim Z$. If $C_A(E) \cong M_r(Z)$, then the Double Centralizer Theorem yields $[A{:}F] = [E{:}F] \cdot [Z{:}F] \cdot r^2$, hence $deg\, E \cdot [Z{:}F] \mid deg\, A$.

Proof. (i) Let $K \subseteq E$ be a maximal subfield. Since K splits A we have $ind\, A \mid [K{:}F]$. Relation (i) follows because $[K{:}F] = deg\, E \cdot [Z{:}F]$.

(ii) If E embeds in A and $C_A(E) = Z$, then we have $deg\, A = deg\, E \cdot [Z{:}F]$ by the Double Centralizer Theorem. Conversely, assume $deg\, A = deg\, E \cdot [Z{:}F]$. Let S be a simple left module over the central simple Z-algebra $A \otimes_F E^{\mathrm{op}}$. This algebra is split because $A \otimes_F E^{\mathrm{op}} = (A \otimes_F Z) \otimes_Z E^{\mathrm{op}}$ and $A \otimes_F Z \sim E$. Therefore,

$$dim_F\, S = deg(A \otimes_F E^{\mathrm{op}}) \cdot [Z{:}F] = (deg\, A)^2. \qquad (10.21)$$

But since $A \otimes_F End_A(S)^{\mathrm{op}} \cong End_F(S)$, we have $deg\, A \cdot deg\, End_A(S) = dim_F\, S$. It follows from this and (10.21) that $deg\, End_A\, S = deg\, A$; hence, $End_A\, S \cong A$ since $End_A\, S$ is Brauer-equivalent to A. The $A \otimes_F E^{\mathrm{op}}$-module structure on S yields an F-algebra homomorphism $E \to End_A\, S$ mapping each $e \in E$ to right multiplication by $1 \otimes e^{\mathrm{op}}$. This homomorphism is injective because E is a division algebra. By composing it with an isomorphism $End_A\, S \cong A$, we obtain an embedding $E \subseteq A$. As Z is the center of E, we have $Z \subseteq C_A(E)$. The Double Centralizer Theorem and dimension count show that this inclusion is an equality. □

We apply the lemma with E the residue division algebra of an inertially split division algebra:

Proposition 10.25. *Let D be a central division algebra over a Henselian-valued field F. Suppose that D is inertially split. Then, there exist central simple $\overline{F}$-algebras A such that*

$$sp(D) = [A] + Dec(Z(\overline{D})/\overline{F}) \qquad \text{and} \qquad deg\, A = deg\, D. \tag{10.22}$$

Any such $\overline{F}$-algebra A contains an isomorphic image of $\overline{D}$ with $C_A(\overline{D}) = Z(\overline{D})$.

Proof. Let $D \sim I \otimes_F N$ with I an inertial division algebra and N a decomposably semiramified division algebra. By definition of the specialization coset, we have $sp(D) = [\overline{I}\,] + Dec(Z(\overline{D})/\overline{F})$, and by Th. 8.65(i), (ii) we have $\overline{D} \sim \overline{I} \otimes_{\overline{F}} Z(\overline{D})$. Therefore, Lemma 10.24(i) shows that there is an integer $m \geq 1$ such that

$$m\, deg\, \overline{I} = deg\, \overline{D} \cdot [Z(\overline{D}){:}\overline{F}].$$

Note that the right expression equals $deg\, D$ in view of Prop. 8.64; hence, the algebra $A = M_m(\overline{I}\,)$ satisfies the conditions (10.22).

For any algebra A satisfying (10.22), we have $A \otimes_{\overline{F}} Z(\overline{D}) \sim \overline{D}$ by Th. 8.65(i), (ii), and $deg\, A = deg\, \overline{D} \cdot [Z(\overline{D}){:}\overline{F}\,] = deg\, D$; the latter equality holds by Prop. 8.64 as D is inertially split. Therefore, Lemma 10.24(ii) shows that $\overline{D}$ embeds in A with $C_A(\overline{D}) = Z(\overline{D})$. □

In the situation of Prop. 10.25, the field $Z(\overline{D})$ is a Galois extension of $\overline{F}$ with Galois group isomorphic to Γ_D/Γ_F. Let $e = exp(\Gamma_D/\Gamma_F)$. Suppose $\overline{F}$ contains a primitive e-th root of unity. The field $Z(\overline{D})$ is then a Kummer extension of $\overline{F}$, and it is spanned by a unique armature $\mathcal{K}(Z(\overline{D})/\overline{F}) \subseteq Z(\overline{D})^\times/\overline{F}^\times$, which is the e-torsion subgroup of $Z(\overline{D})^\times/\overline{F}^\times$, see Ex. 7.28(ii). Of course, the commutator pairing on $\mathcal{K}(Z(\overline{D})/\overline{F})$ is trivial since $Z(\overline{D})$ is a commutative ring. By contrast, if $\mathcal{A} \subseteq D^\times/F^\times$ is an armature such that $D = F[\mathcal{A}]$, then Prop. 7.31 shows that the commutator pairing on $\mathcal{A}$ (with values in the group $\mu(F)$ of roots of unity in F) is nondegenerate; it thus provides $\mathcal{A}$ with a symplectic module structure. The next theorem, which is the main result of this section, establishes a relation between spanning armatures in D and spanning armatures of the special representatives of its specialization coset.

To compare pairings with values in $\mu(F)$ and in $\mu(\overline{F})$, we identify the roots of unity in F of order prime to *char* $\overline{F}$ with their images in $\overline{F}$.

Theorem 10.26. *Let D be a central division algebra over a Henselian-valued field F. Assume that D is inertially split and* char $\overline{F} \nmid \deg D$. *Let A be a central simple $\overline{F}$-algebra representing the specialization coset* sp(D) *and such that $\deg A = \deg D$, as in* (10.22), *and let $\mathcal{S}$ be a symplectic module with $|\mathcal{S}| = [D{:}F]$. Then, the division algebra D is spanned by an armature isomorphic to $\mathcal{S}$ (as a symplectic module) if and only if A is spanned by an armature isomorphic to $\mathcal{S}$ that contains $\mathcal{K}(Z(\overline{D})/\overline{F})$ as a totally isotropic subgroup.*

Proof. Write Z for $Z(\overline{D})$. In view of Prop. 10.25, we identify $\overline{D}$ with a subalgebra of A such that $C_A(\overline{D}) = Z$. By the Double Centralizer Theorem, we then also have $C_A(Z) = \overline{D}$. As observed in (8.35), we have $sp(D) = sp\big(\mathsf{gr}(D)\big)$, hence A also represents the specialization coset of $\mathsf{gr}(D)$. Let N be the decomposably semiramified central graded division algebra over $\mathsf{gr}(F)$ such that

$$\mathsf{gr}(D) \sim A \otimes_{\overline{F}} \mathsf{N}.$$

We have $\mathsf{N}_0 = Z$ and $\Gamma_{\mathsf{N}} = \Gamma_D$ by Th. 8.50(i). Our first goal is to identify $\mathsf{gr}(D)$ with a graded subalgebra of $A \otimes_{\overline{F}} \mathsf{N}$.

Let $G = \mathcal{G}(Z/\overline{F})$. For each $\sigma \in G$, the Skolem–Noether Theorem yields an invertible element $a_\sigma \in A^\times$ such that

$$a_\sigma z = \sigma(z) a_\sigma \qquad \text{for all } z \in Z.$$

Set $a_{id} = 1$. Because $int(a_\sigma)$ preserves Z, it also preserves the centralizer $\overline{D}$ of Z, and we have

$$a_\sigma \cdot \overline{D} = \overline{D} \cdot a_\sigma \quad \text{and} \quad a_\sigma a_\tau a_{\sigma\tau}^{-1} \in \overline{D}^\times \quad \text{for all } \sigma,\, \tau \in G. \tag{10.23}$$

Since N is decomposably semiramified, we know from Prop. 8.33 that it contains a maximal graded subfield T that is totally ramified over $\mathsf{gr}(F)$. We have $\Gamma_{\mathsf{T}} = \Gamma_{\mathsf{N}} = \Gamma_D$. For each $\gamma \in \Gamma_D$, choose $t_\gamma \in \mathsf{T}_\gamma \setminus \{0\}$. Then,

$$\mathsf{T} = \bigoplus_{\gamma \in \Gamma_D} \overline{F} \cdot t_\gamma \quad \text{and} \quad t_\gamma t_\delta \equiv t_{\gamma+\delta} \ \big(mod\ \overline{F}^\times\big) \quad \text{for } \gamma,\, \delta \in \Gamma_D. \tag{10.24}$$

Define

$$\mathsf{E} = \bigoplus_{\gamma \in \Gamma_D} \big(\overline{D} \cdot a_{\theta_D(\gamma)} \otimes t_\gamma\big) \subseteq A \otimes_{\overline{F}} \mathsf{N}.$$

Note that E depends on the choice of T, but not on the choices of t_γ in T_γ. In view of (10.23) and (10.24), it is clear that E is a graded $\mathsf{gr}(F)$-subalgebra of $A \otimes_{\overline{F}} \mathsf{N}$; we have $\mathsf{E}_0 = \overline{D}$, and $\Gamma_{\mathsf{E}} = \Gamma_{\mathsf{E}}^\times = \Gamma_D$ because each $a_{\theta_D(\gamma)} \otimes t_\gamma$ is invertible. Since E_0 is a division algebra and $\Gamma_{\mathsf{E}} = \Gamma_{\mathsf{E}}^\times$, it follows that E is a graded division algebra by Prop. 2.4.

Claim: $\mathsf{E} \cong_g \mathsf{gr}(D)$.

To see this, it suffices to find a homogeneous idempotent $\varepsilon \in A \otimes_{\overline{F}} \mathsf{N}$ such that $\mathsf{E} \cong_g \varepsilon(A \otimes_{\overline{F}} \mathsf{N})\varepsilon$. For then E is Brauer-equivalent to $A \otimes_{\overline{F}} \mathsf{N}$, hence isomorphic to $\mathsf{gr}(D)$, because each Brauer class contains a unique graded division algebra up to isomorphism.

Recall from Pierce [178, Lemma b, p. 256] that, since Z is Galois over $\overline{F}$, the $\overline{F}$-algebra $Z \otimes_{\overline{F}} Z$ is a direct product of copies of Z indexed by G. Moreover, the primitive idempotents $(\varepsilon_\sigma)_{\sigma \in G}$ of $Z \otimes_{\overline{F}} Z$ satisfy

$$\varepsilon_\sigma \cdot (z \otimes 1) = \varepsilon_\sigma \cdot \big(1 \otimes \sigma(z)\big) \qquad \text{for all } \sigma \in G \text{ and } z \in Z,$$

and they are uniquely determined by this property. Hence, for all $\sigma, \rho, \tau \in G$ and $z \in Z$,

$$\begin{aligned} \big((\sigma \otimes \rho)(\varepsilon_\tau)\big) \cdot (z \otimes 1) &= (\sigma \otimes \rho)\big(\varepsilon_\tau \cdot (\sigma^{-1}(z) \otimes 1)\big) \\ &= (\sigma \otimes \rho)\big(\varepsilon_\tau \cdot (1 \otimes \tau\sigma^{-1}(z)\big) = \big((\sigma \otimes \rho)(\varepsilon_\tau)\big) \cdot (1 \otimes \rho\tau\sigma^{-1}(z)). \end{aligned}$$

Therefore, we have

$$(\sigma \otimes \rho)(\varepsilon_\tau) = \varepsilon_{\rho\tau\sigma^{-1}} \qquad \text{for all } \sigma, \rho, \tau \in G.$$

Now, for $\sigma \in G$ and $\gamma \in \Gamma_D$ we have $int(a_\sigma \otimes t_\gamma)|_{Z \otimes Z} = \sigma \otimes \theta_D(\gamma)$; hence, for σ, $\tau \in G$ and $\gamma \in \Gamma_D$,

$$(a_\sigma \otimes t_\gamma) \cdot \varepsilon_\tau = \varepsilon_{\theta_D(\gamma)\tau\sigma^{-1}} \cdot (a_\sigma \otimes t_\gamma).$$

Writing simply ε for ε_{id}, we thus have for $\sigma \in G$ and $\gamma \in \Gamma_D$

$$\varepsilon \cdot (a_\sigma \otimes t_\gamma) \cdot \varepsilon = \begin{cases} \varepsilon \cdot (a_\sigma \otimes t_\gamma) = (a_\sigma \otimes t_\gamma) \cdot \varepsilon & \text{if } \sigma = \theta_D(\gamma); \\ \varepsilon \cdot \varepsilon_{\theta_D(\gamma)\sigma^{-1}} \cdot (a_\sigma \otimes t_\gamma) = 0 & \text{if } \sigma \neq \theta_D(\gamma). \end{cases} \tag{10.25}$$

In particular, it follows that ε centralizes E, hence $\varepsilon\mathsf{E} = \mathsf{E}\varepsilon = \varepsilon\mathsf{E}\varepsilon$. Since ε is homogeneous of degree 0, the map $x \mapsto \varepsilon x \varepsilon$ therefore defines a graded algebra epimorphism $\mathsf{E} \to \varepsilon\mathsf{E}\varepsilon$, which must be an isomorphism since E is a graded division algebra. Thus, $\mathsf{E} \cong_g \varepsilon\mathsf{E}\varepsilon$. On the other hand, a standard argument as in Pierce [178, p. 251] or Reiner [199, p. 247] shows that the sum $\sum_{\sigma \in G} \overline{D} \cdot a_\sigma \subseteq A$ is direct, hence $A = \bigoplus_{\sigma \in G} \overline{D} \cdot a_\sigma$ by dimension count. We also have a direct sum decomposition of N: Since $\mathsf{N}_0 = Z$ and $\Gamma_{\mathsf{N}} = \Gamma_D$, we have $\mathsf{N} = \bigoplus_{\gamma \in \Gamma_D} Z \cdot t_\gamma$. Therefore,

$$A \otimes_{\overline{F}} \mathsf{N} = \bigoplus_{\sigma \in G} \bigoplus_{\gamma \in \Gamma_D} (\overline{D} \otimes_{\overline{F}} Z) \cdot (a_\sigma \otimes t_\gamma).$$

Since $\varepsilon \cdot (Z \otimes_{\overline{F}} Z) = (Z \otimes 1) \cdot \varepsilon$, we obtain from (10.25):

$$\varepsilon \cdot (A \otimes_{\overline{F}} \mathsf{N}) \cdot \varepsilon = \bigoplus_{\gamma \in \Gamma_D} \varepsilon \cdot (\overline{D} \cdot a_{\theta_D(\gamma)} \otimes t_\gamma) \cdot \varepsilon = \varepsilon\mathsf{E}\varepsilon.$$

It follows that $\varepsilon \cdot (A \otimes_{\overline{F}} \mathsf{N}) \cdot \varepsilon \cong_g \mathsf{E}$. We have thus established the claim that $\mathsf{gr}(D) \cong_g \mathsf{E}$.

Note that

$$\mathsf{E}^{\times} \;=\; \bigcup_{\gamma\in\Gamma_D} (\overline{D}^{\times}\cdot a_{\theta_D(\gamma)} \otimes t_\gamma).$$

Since $t_\gamma t_\delta \equiv t_{\gamma+\delta} \; \big(mod\; \overline{F}^{\times}\big)$ by (10.24), there is a homomorphism

$$\varphi\colon \mathsf{E}^{\times} \longrightarrow A^{\times}/\overline{F}^{\times} \quad \text{given by} \quad d\, a_{\theta_D(\gamma)} \otimes t_\gamma \;\mapsto\; d\, a_{\theta_D(\gamma)} \cdot \overline{F}^{\times}$$

for $d \in \overline{D}^{\times}$ and $\gamma \in \Gamma_D$.

Note that the elements of the form $d\, a_\sigma$ with $d \in \overline{D}^{\times}$ lie in the normalizer $N_{A^\times}(Z^\times)$ of $Z^\times$ in $A^\times$, where $N_{A^\times}(Z^\times) = \{a \in A^\times \mid aZ^\times a^{-1} = Z^\times\}$. Moreover, if $c \in N_{A^\times}(Z^\times)$, then $int(c)|_Z = \sigma$ for some $\sigma \in \mathcal{G}(Z/\overline{F})$; then $ca_\sigma^{-1} \in C_A(Z) = \overline{D}$. Hence,

$$N_{A^\times}(Z^\times) \;=\; \bigcup_{\sigma\in G} (\overline{D}^{\times}\cdot a_\sigma) \;\subseteq\; A^\times,$$

showing that $im(\varphi) = N_{A^\times}(Z^\times)/\overline{F}$. Since the kernel of φ is $\mathsf{gr}(F)^\times$, we have an induced isomorphism

$$\overline{\varphi}\colon \mathsf{E}^\times/\,\mathsf{gr}(F)^\times \xrightarrow{\;\sim\;} N_{A^\times}(Z^\times)/\overline{F}^{\times} \;\subseteq\; A^\times/\overline{F}^{\times}. \tag{10.26}$$

Because the t_γ's commute pairwise (as they lie in the graded field T), the isomorphism $\overline{\varphi}$ preserves the commutator pairings.

We now prove the forward implication of the theorem. Assume that D is spanned by an armature $\mathcal{A} \subseteq D^\times/F^\times$ isomorphic to the given symplectic module $\mathcal{S}$. Thus, the canonical pairing on $\mathcal{A}$ is nondegenerate; hence, F contains a primitive s-th root of unity for $s = exp\,\mathcal{S}$. Since $char\,\overline{F}$ does not divide $[D{:}F] = |\mathcal{S}|$, the residue of this root of unity is a primitive s-th root of unity in $\overline{F}$. Moreover, Th. 7.49 yields a tame armature v-gauge $\alpha_{\mathcal{A}}$ on D. But the valuation v_D on D extending v is the only possible v-gauge by Prop. 4.33; hence, $v_D = \alpha_{\mathcal{A}}$. Theorem 7.49 also yields a homogeneous armature $\widetilde{\mathcal{A}} \subseteq \mathsf{gr}(D)^\times/\,\mathsf{gr}(F)^\times$ with $\widetilde{\mathcal{A}}$ spanning $\mathsf{gr}(D)$ and $\widetilde{\mathcal{A}} \cong \mathcal{A} \cong \mathcal{S}$ as symplectic modules. By identifying $\mathsf{gr}(D)$ with E, we may view $\widetilde{\mathcal{A}}$ as a subgroup of $\mathsf{E}^\times/\,\mathsf{gr}(F)^\times$. Therefore, the commutator pairing on $\overline{\varphi}(\widetilde{\mathcal{A}})$ is nondegenerate. Moreover,

$$|\overline{\varphi}(\widetilde{\mathcal{A}})| \;=\; |\widetilde{\mathcal{A}}| \;=\; |\mathcal{A}| \;=\; [D{:}F] \;=\; [A{:}F],$$

hence $\overline{\varphi}(\widetilde{\mathcal{A}})$ is an armature spanning A by Prop. 7.26. This armature is isomorphic to $\widetilde{\mathcal{A}}$, hence also to $\mathcal{S}$. To see that it contains $\mathcal{K}(Z/\overline{F})$, recall the degree map of (7.31), which restricts to a map

$$\overline{deg}\colon \widetilde{\mathcal{A}} \to \Gamma_D/\Gamma_F.$$

This map is surjective by Prop. 7.42, and its kernel $\widetilde{\mathcal{A}}_0$ therefore has order $[D{:}F]/|\Gamma_D{:}\Gamma_F| = [\,\overline{D}{:}\overline{F}\,]$. If $(d\, a_{\theta_D(\gamma)} \otimes t_\gamma)\cdot \mathsf{gr}(F)^\times$ lies in $\widetilde{\mathcal{A}}_0$, then $\gamma \in \Gamma_F$; so, $a_{\theta_D(\gamma)} = a_{id} = 1$ and hence $d\, a_{\theta_D(\gamma)} \cdot \overline{F}^{\times}$ lies in $\overline{D}^{\times}/\overline{F}^{\times}$. Therefore, $\overline{\varphi}(\widetilde{\mathcal{A}}_0)$ is an armature spanning $\overline{D}$, as $|\overline{\varphi}(\widetilde{\mathcal{A}}_0)| = |\widetilde{\mathcal{A}}_0| = [\overline{D}:\overline{F}]$. The radical of this armature is an armature spanning the center of $\overline{D}$, as Prop. 7.31 shows. The center of $\overline{D}$ is Z, so the radical of $\overline{\varphi}(\widetilde{\mathcal{A}}_0)$ must be $\mathcal{K}(Z/\overline{F})$. For Kummer theory shows that, since $\overline{F}$ contains a primitive s-th root of unity, $\mathcal{K}(Z/\overline{F})$ is the unique spanning armature of Z of exponent dividing s, see Ex. 7.28(ii).

We have thus proved that if D is spanned by an armature isomorphic to $\mathcal{S}$, then A is also spanned by an armature isomorphic to $\mathcal{S}$, and this armature contains $\mathcal{K}(Z/\overline{F})$ as a totally isotropic subgroup.

To prove the converse, assume now that A is spanned by an armature $\mathcal{A}$ isomorphic to $\mathcal{S}$ containing $\mathcal{K}(Z/\overline{F})$. For $a \in \mathcal{A}$ and $k \in \mathcal{K}(Z/\overline{F})$, we have $\langle a, k\rangle \in \overline{F}^\times$, which means that if $x_a \in A^\times$ and $x_k \in Z^\times$ satisfy $a = x_a \cdot \overline{F}^\times$ and $k = x_k \cdot \overline{F}^\times$, then

$$x_a x_k x_a^{-1} = c x_k \qquad \text{for some } c \in \overline{F}^\times.$$

Since $\mathcal{K}(Z/\overline{F})$ spans Z, it follows that x_a normalizes $Z^\times$, hence $\mathcal{A} \subseteq N_{A^\times}(Z^\times)/\overline{F}^\times$. We may then use the isomorphism $\overline{\varphi}$ of (10.26) to obtain an isomorphic homogeneous armature $\mathcal{E}$ spanning E. Since $\mathsf{E} \cong_g \mathsf{gr}(D)$, we then have a homogeneous armature $\mathcal{A}'$ spanning $\mathsf{gr}(D)$ with $\mathcal{A}' \cong \mathcal{E} \cong \mathcal{A} \cong \mathcal{S}$. By Prop. 7.51, there is in D a spanning armature isomorphic to $\mathcal{A}'$, and the proof is complete. □

As a first application, we have the following cyclicity criterion for a particular type of semiramified division algebras, which will be used in the proof of Th. 12.11:

Proposition 10.27. *Let p be a prime, and let F be a Henselian-valued field containing a primitive p^2-root of unity, with $\operatorname{char}\overline{F} \neq p$. Let D be a semiramified (hence inertially split) central division algebra of degree p^2 over F. Assume $\mathcal{G}(\overline{D}/\overline{F}) \cong (\mathbb{Z}/p\mathbb{Z})^2$, and let A be a central simple $\overline{F}$-algebra of degree p^2 such that*

$$sp(D) = [A] + Dec(\overline{D}/\overline{F}).$$

Then, the following conditions are equivalent:

(a) *D is a cyclic algebra;*

(b) *$A \cong \big(a, b/\overline{F}\big)_{p^2}$, for some $a,\, b \in \overline{F}^\times$ such that $\overline{D} \cong \overline{F}(\sqrt[p]{a}, \sqrt[p]{b})$.*

When these equivalent conditions hold, every central simple $\overline{F}$-algebra representing $sp(D)$ modulo $Dec(\overline{D}/\overline{F})$ is split by a cyclic Galois extension of $\overline{F}$ of degree dividing p^2.

Proof. Note that the semiramified division algebra D is necessarily tame since $\operatorname{char}\overline{F} \neq p$; see Def. 8.4. It is therefore inertially split by Prop. 8.38, hence $sp(D)$ is defined.

(a) ⇒ (b) Recall from Prop. 10.25 that we may identify $\overline{D}$ with a subfield in A. If D is cyclic, then it is a symbol algebra because F contains a primitive p^2-root of unity. Therefore, D contains an armature isomorphic to the standard symplectic module $\mathbb{S}_{p^2}$. Theorem 10.26 shows that A has an armature $\mathcal{A}$ isomorphic to $\mathbb{S}_{p^2}$, hence it is a symbol algebra. Moreover, $\mathcal{A}$ contains the Kummer armature $\mathcal{K}(\overline{D}/\overline{F})$, which has exponent p and order p^2. Therefore, $\mathcal{K}(\overline{D}/\overline{F})$ must be the p-torsion subgroup of $\mathcal{A}$:

$$\mathcal{K}(\overline{D}/\overline{F}) = {}_p\mathcal{A}.$$

Let $i, j \in A^\times$ be representatives of a symplectic base of $\mathcal{A}$, and let $a = i^{p^2} \in \overline{F}^\times$, $b = j^{p^2} \in \overline{F}^\times$. We then have $A = \big(a, b/\overline{F}\big)_{p^2}$ (see Prop. 7.33). Moreover, i^p, j^p represent a base of $\mathcal{K}(\overline{D}/\overline{F})$, hence

$$\overline{D} \cong \overline{F}(\sqrt[p]{a}, \sqrt[p]{b}).$$

(b) $\Rightarrow$ (a) Let $i, j \in A^\times$ be standard generators of the symbol algebra $\big(a, b/\overline{F}\big)_{p^2}$. We identify $\overline{D}$ with the subfield of A generated by i^p and j^p. The images of i and j in $A^\times/\overline{F}^\times$ generate an armature isomorphic to $\mathbb{S}_{p^2}$ (see Ex. 7.28(iii)), which contains $\mathcal{K}(\overline{D}/\overline{F})$ as a totally isotropic subgroup. Therefore, Th. 10.26 shows that D contains an armature isomorphic to $\mathbb{S}_{p^2}$; hence, D is a symbol algebra by Prop. 7.33.

To prove the last assertion, note that if $\overline{D} = \overline{F}(\sqrt[p]{a}, \sqrt[p]{b})$, then every Brauer class in $\mathit{Dec}(\overline{D}/\overline{F})$ is represented by a tensor product $\big(a, x/\overline{F}\big)_p \otimes_{\overline{F}} \big(b, y/\overline{F}\big)_p$ for some $x, y \in \overline{F}^\times$; see (8.11). The relations for symbol algebras yield in $\mathit{Br}(\overline{F})$

$$\big(a, b/\overline{F}\big)_{p^2} \otimes_{\overline{F}} \big(a, x/\overline{F}\big)_p \otimes_{\overline{F}} \big(b, y/\overline{F}\big)_p \sim \big(ay^{-p}, bx^p/\overline{F}\big)_{p^2}. \qquad (10.27)$$

Therefore, every Brauer class equivalent to the class of $\big(a, b/\overline{F}\big)_{p^2}$ modulo $\mathit{Dec}(\overline{D}/\overline{F})$ is represented by a symbol of degree p^2. □

Here is a specific example where Prop. 10.27 applies:

Example 10.28. *A division algebra of degree* 4 *that is not a cyclic algebra nor a biquaternion algebra.* Let k be the local field $\mathbb{Q}_2(\sqrt{-1})$, and let $\ell_1 = k(\sqrt{2})$, $\ell_2 = k(\sqrt{5})$, and $\ell = \ell_1 \otimes_k \ell_2$. The fields ℓ_1, ℓ_2 are quadratic extensions of k that are linearly disjoint over k (see for instance Lam [122, Cor. VI.2.24]), so ℓ is a field with $[\ell{:}k] = 4$. Let A be a central division algebra of degree 4 over k. Consider the division algebra $\mathcal{S} = \mathcal{S}(A; \ell_1, \ell_2)$ obtained from the construction of §8.4.5. We claim that $\mathcal{S}$ is a division algebra of degree 4 that is neither cyclic nor a tensor product of two quaternion algebras.

The degree and the exponent of $\mathcal{S}$ are given by Prop. 8.72: Since $\mathit{deg}\, A = 4 = [\ell{:}k]$, the algebra A has exponent 4 and is split by ℓ; see Reiner [199, Th. (31.4), Cor. (31.10)]. Therefore, Prop. 8.72 yields $\mathit{deg}\, \mathcal{S} = \mathit{exp}\, \mathcal{S} = 4$. From this, it already follows that $\mathcal{S}$ is not a tensor product of two quaternion algebras. To prove that it is not cyclic, we apply Prop. 10.27. Note that the center of $\mathcal{S}$ is $k((t_1))((t_2))$; for the extension to $\mathcal{S}$ of the (t_1, t_2)-adic valuation on its center, $\mathcal{S}$ is semiramified with $\overline{\mathcal{S}} = \ell$, and $sp(\mathcal{S})$ is represented by A modulo $Dec(\ell/k)$ (see Prop. 8.72). If $\mathcal{S}$ were cyclic, then Prop. 10.27 would show that $A \cong \big(a, b/k\big)_4$ for some a, $b \in k^\times$ such that $k(\sqrt{a}, \sqrt{b}\,) \cong \ell$. Then $A \otimes_k A \sim \big(a, b/k\big)_2$ and the square classes of a, b lie in $\{2k^{\times 2}, 5k^{\times 2}, 10k^{\times 2}\}$. The quaternion algebra $\big(a, b/k\big)_2$ then has a descent to $\mathbb{Q}_2$, i.e., we have $\big(a, b/k\big)_2 \cong H \otimes_{\mathbb{Q}_2} k$ for some quaternion algebra H over $\mathbb{Q}_2$. But every quaternion algebra over $\mathbb{Q}_2$ is split by k (see Reiner [199, Cor. (31.10)]), so $A \otimes_k A$ is split. This is a contradiction since $exp\, A = 4$. Thus, $\mathcal{S}$ is not a cyclic algebra.

Theorem 10.26 also applies to the decomposition of division algebras obtained by the $\mathcal{S}$ construction of §8.4.5 into tensor products of symbol algebras: see Exercise 10.5. To complete this section, we consider another type of construction, combining inertial and totally ramified division algebras.

Let D_0 be a central division algebra over an arbitrary field k, and let $n_1, \ldots, n_r$ be integers with $n_i \geq 2$ for all i. Assume k contains a primitive n_i-th root of unity for all i. Over the field $F = k((x_1))((y_1)) \ldots ((x_r))((y_r))$ consider the division algebra $\mathcal{T} = \mathcal{T}(k; n_1, \ldots, n_r)$ of (9.5):

$$\mathcal{T} = \big(x_1, y_1/F\big)_{\omega_1, n_1} \otimes_F \ldots \otimes_F \big(x_r, y_r/F\big)_{\omega_r, n_r}. \tag{10.28}$$

Let $D = D_0 \otimes_k \mathcal{T}$, and let v be the usual $(x_1, y_1, \ldots, x_r, y_r)$-adic valuation on F, with $\overline{F} = k$ and $\Gamma_F = \mathbb{Z}^{2r}$, which extends to an inertial valuation on $D_0 \otimes_k F$, and to a tame totally ramified valuation on $\mathcal{T}$. By Th. 3.43 the algebra D is a division algebra, which carries a valuation extending v with $\overline{D} = D_0$ and $\Gamma_D = \Gamma_{\mathcal{T}} = (\frac{1}{n_1}\mathbb{Z})^2 \times \ldots \times (\frac{1}{n_r}\mathbb{Z})^2$. Clearly, every decomposition of D_0 into a tensor product of symbol algebras leads to a decomposition of D. The following result yields a converse to this observation:

Proposition 10.29. *If the F-algebra $D = D_0 \otimes_k \mathcal{T}$ is spanned by an armature $\mathcal{A}$, then the k-algebra D_0 is spanned by an armature $\mathcal{A}_0$ such that*

$$\mathcal{A} \cong \mathcal{A}_0 \perp \mathbb{S}_{n_1} \perp \ldots \perp \mathbb{S}_{n_r}.$$

Proof. Since D is central, the commutator pairing on $\mathcal{A}$ is nondegenerate by Prop. 7.31; hence, $char\, F \nmid exp\, \mathcal{A}$. But $\overline{F} = k \subseteq F$, hence $char\, \overline{F} = char\, F$. We may therefore apply Th. 7.49 to see that the armature norm $\alpha_{\mathcal{A}}$ is a tame gauge. It is therefore the unique valuation on D extending the valuation v, see Prop. 4.33. Theorem 7.49 also shows that $\mathcal{A}$ is isomorphic as a symplectic module to the induced homogeneous armature $\widetilde{\mathcal{A}}$ of $\mathsf{gr}(D)$. By Cor. 7.48 the kernel of the map $\overline{deg}\colon \mathcal{A} \to \Gamma_D/\Gamma_F$ is canonically isomorphic to an armature $\mathcal{A}_0$ of $\overline{D} = D_0$. Proposition 7.42 shows that $\overline{deg}$ is surjective, so

$$|\mathcal{A}_0| = [D{:}F]/|\Gamma_D{:}\Gamma_F| = [D_0{:}k].$$

Therefore, $\mathcal{A}_0$ spans D_0. On the other hand, the armature $\mathcal{A}_1$ of $\mathcal{T}$ associated to its tensor product decomposition (10.28) yields a submodule $\widetilde{\mathcal{A}}_1$ of $\widetilde{\mathcal{A}}$ isomorphic to $\mathbb{S}_{n_1} \perp \ldots \perp \mathbb{S}_{n_r}$, and from the definition of the pairing on $\widetilde{\mathcal{A}}$ it is clear that $\widetilde{\mathcal{A}}_1$ is the orthogonal complement of $ker\,\overline{deg}$. Therefore, $\mathcal{A} \cong \mathcal{A}_0 \perp \mathbb{S}_{n_1} \perp \ldots \perp \mathbb{S}_{n_r}$. □

By taking for D_0 an indecomposable division algebra, we thus obtain from Prop. 10.29 examples of division algebras D that are decomposable, although not as a tensor product of symbol algebras. We may also let $D_0 = k$ and obtain restrictions on the possible decompositions of the algebras $\mathcal{T}(k; n_1, \ldots, n_r)$:

Corollary 10.30. *Let p be a prime and s, t integers with $s \geq t \geq 1$. Assume the field k contains a primitive p^s-th root of unity, and let $\mathcal{T} = \mathcal{T}(k; p^s, p^t)$ and $F = Z(\mathcal{T})$. If $\mathcal{T}_1$, $\mathcal{T}_2$ are proper subalgebras of $\mathcal{T}$ such that $\mathcal{T} = \mathcal{T}_1 \otimes_F \mathcal{T}_2$ and $deg\,\mathcal{T}_1 \geq deg\,\mathcal{T}_2$, then $deg\,\mathcal{T}_1 = p^s$ and $deg\,\mathcal{T}_2 = p^t$.*

Proof. Since $\mathcal{T}$ is tame and totally ramified over F, the subalgebras $\mathcal{T}_1$ and $\mathcal{T}_2$ are also tame and totally ramified, hence by Cor. 7.76 they carry spanning armatures $\mathcal{A}_1$ and $\mathcal{A}_2$. Tensoring these armatures, we obtain a spanning armature of $\mathcal{T}$ isomorphic to $\mathcal{A}_1 \perp \mathcal{A}_2$. By Prop. 10.29, we have

$$\mathbb{S}_{p^s} \perp \mathbb{S}_{p^t} \;\cong\; \mathcal{A}_1 \perp \mathcal{A}_2.$$

Comparing the invariant factors of each side, we see that $\mathbb{S}_{p^s} \cong \mathcal{A}_1$ and $\mathbb{S}_{p^t} \cong \mathcal{A}_2$, because

$$p^{2st} = [\mathcal{T}{:}F] > [\mathcal{T}_1{:}F] \geq [\mathcal{T}_2{:}F] \quad \text{and} \quad [\mathcal{T}_i{:}F] = |\mathcal{A}_i| \quad \text{for } i = 1,\, 2.$$

Therefore, $deg\,\mathcal{T}_1 = p^s$ and $deg\,\mathcal{T}_2 = p^t$. □

Exercises

Exercise 10.1 (*Sethuraman*). Here is an example of an indecomposable symbol algebra over a rational function field. Take integers $n \geq m \geq 1$, p a prime number, and k a field containing a primitive p^{n+1}-st root of unity. Let $\ell = n - m$. Let $F = k(x_0, \ldots, x_\ell, y)$, where $x_0, \ldots, x_\ell, y$ are algebraically independent over k. Set

$$A = \big(x_0, x_0^{p^\ell} - y/F\big)_{p^m} \otimes_F \bigotimes_{j=1}^{\ell}{}_F \big(x_j, x_j^{p^{\ell-j}} - y/F\big)_p.$$

Prove that A is a division algebra of degree p^n and exponent p^m and $A \cong \big(a, y/F\big)_{p^n}$ for some $a \in F^\times$. It follows by Cor. 10.2 that $\big(a, y/F\big)_{p^{n+1}}$ is an indecomposable division algebra of degree p^{n+1} and exponent p^{m+1}.

Exercise 10.2. Let D be a division algebra as in Prop. 10.27. Show that the equivalent conditions of that proposition are also equivalent to: $D \cong \big(ax^p, by^p/F\big)_{p^2}$ for some a, $b \in \mathcal{O}_F^\times$ such that $\overline{D} \cong \overline{F}\big(\sqrt[p]{\overline{a}}, \sqrt[p]{\overline{b}}\,\big)$, and some $x, y \in F^\times$ such that the images of $v(x)$ and $v(y)$ are $\mathbb{Z}/p\mathbb{Z}$-independent in $\Gamma_F/p\Gamma_F$.

Exercise 10.3. When a division algebra D has a nontrivial tensor decomposition $D = B \otimes_F C$, the factors B and C have a lower degree than D, but they could have higher exponent, as this exercise illustrates. Let A be a division algebra of degree and exponent 4 with center a field k, such that A has two maximal subfields $\ell = k(\sqrt{b_1}, \sqrt{b_2})$ and $m = k(\sqrt{c_1}, \sqrt{c_2})$ $(b_1, b_2, c_1, c_2 \in k)$ with ℓ and m linearly disjoint over k. (Examples of such A, ℓ, m are easy to construct, e.g., for k any global field using local-global arguments. Or, one could take k to be any 2-adic local field with $|k^\times : k^{\times 2}| \geq 2^4$.) Let $F = k((x_1))((x_2))((y_1))((y_2))$, and let B and C be the division algebras with

$$B \sim A \otimes_k (b_1, x_1/F) \otimes_F (b_2, x_2/F) \text{ and } C \sim A^{\mathrm{op}} \otimes_k (c_1, y_1/F) \otimes_F (c_2, y_2/F).$$

(i) Prove that B and C each have degree and exponent 4.
(ii) Let $D = B \otimes_F C$. Prove that D is a division algebra of exponent 2.

Exercise 10.4. Let p be a prime and let F be a Henselian-valued field containing a primitive p-th root of unity, with $\mathit{char}\,\overline{F} \neq p$. Let D be an inertially split central division algebra of degree p^r over F, for some $r \geq 1$, such that $Z(\overline{D})$ is an elementary abelian Galois extension of degree p^e of $\overline{F}$, for some $e \geq 1$. Show that D decomposes as a tensor product of r symbol algebras of degree p if and only if $\mathit{sp}(D)$ is represented modulo $\mathit{Dec}(Z(\overline{D})/\overline{F})$ by a tensor product of $r - e$ symbol algebras of degree p. [Hint: Use Th. 10.26.]

Exercise 10.5. Let p be a prime, let k be an arbitrary field containing a primitive p-th root of unity, and let ℓ be an elementary abelian Galois extension of k of degree p^r, for some $r \geq 1$. Fix a decomposition $\ell = \ell_1 \otimes_k \ldots \otimes_k \ell_r$ into a tensor product of cyclic extensions of k. If A is a central simple k-algebra such that $[A] \in {}_p\mathit{Br}(\ell/k) \setminus \mathit{Dec}(\ell/k)$, show that the semiramified division algebra $\mathcal{S}(A; \ell_1, \ldots, \ell_r)$ has degree p^r, exponent p, and does not decompose into a tensor product of symbol algebras of degree p. [Hint: Apply Exercise 10.4.]

Exercise 10.6. (cf. Tignol [238]) This exercise gives an example of an elementary abelian Galois extension ℓ/k of degree p^2, for any odd prime p, such that ${}_p\mathit{Br}(\ell/k) \neq \mathit{Dec}(\ell/k)$. It follows by Corollary 10.12(ii) that for $\ell_1 = k\big(\sqrt[p]{(x_1^p - y)(x_2^p - y)}\big)$ and $\ell_2 = k(\sqrt[p]{y})$ and the A of (10.29) below, the division algebras $\mathcal{S}(A; \ell_1, \ell_2)$ and $\mathcal{S}'(A; \ell_1, \ell_2)$ are indecomposable of exponent p and index p^2.

Let p be an odd prime, and let k_0 be an arbitrary field of characteristic 0 containing a primitive p^2-root of unity ω. Let $k = k_0(x_1, x_2, y)$ and

$$\ell = k\big(\sqrt[p]{(x_1^p - y)(x_2^p - y)}, \sqrt[p]{y}\big),$$

where x_1, x_2, y are independent indeterminates. Let $z \in \ell$ be such that $z^p = y$, and consider the symbol algebra

$$A = \big((x_1^p - y)(x_2^p - y), y/k\big)_{\omega,p^2}. \tag{10.29}$$

Prove:

(i) ℓ/k is an elementary abelian Galois extension of degree p^2;
(ii) $A \cong \big(1 - x_1^p y^{-1}, x_1/k\big)_{\omega^p,p} \otimes_k \big(1 - x_2^p y^{-1}, x_2/k\big)_{\omega^p,p}$;
(iii) $[A] \in Dec(\ell/k)$ if and only if there exist $a \in \ell^\times$ and $b \in k^\times$ such that $z = bN_{\ell/k(z)}(a)$.

Next, consider a valuation on k_0 extending the p-adic valuation on $\mathbb{Q}$, and let v denote the Gaussian extension of this valuation to k relative to the transcendence base (x_1, x_2, y) of k over k_0. Let $\tau = \omega^p - 1$ and $\xi = \omega - 1$. Show that:

(iv) $v(\xi) = \frac{1}{p(p-1)} v(p)$ and $\xi^{p(p-1)} = \varepsilon p$, where $\overline{\varepsilon} = -1$ (use the binomial expansion of $(1+\tau)^p$ and $(1+\xi)^p$);
(v) the valuation v extends uniquely to ℓ, and $\overline{\ell} = \overline{k(z)}(\overline{s})$, where

$$s = \gamma(x_2^p - z^p) + \delta(x_1^p - z^p),$$

with

$$\gamma = [(x_1 - z)^p - x_1^p + z^p]/p \quad \text{and} \quad \delta = [(x_2 - z)^p - x_2^p + z^p]/p;$$

so $\overline{\ell}/\overline{k(z)}$ is a purely inseparable extension of degree p;
(vi) there is a $\overline{k}$-linear derivation ∂ of $\overline{k(z)}$ that maps $\overline{z}$ to 1;
(vii) for $u \in \overline{\ell}$, the equation $\partial^2(u^p) = 0$ implies $u^p \in \overline{k}$.

Using (iv)–(vii), show that the condition in (iii) does not hold. Conclude using (ii) that $[A] \in {}_pBr(\ell/k) \setminus Dec(\ell/k)$.

Notes

§10.1: Generic examples of indecomposable division algebras of nonprime exponent were given by Saltman [220]. His approach is analogous to Amitsur's in showing that some universal division algebras are non-crossed products: He shows in [220, Th. 1.13] that for any field k and positive integers n and m with the same prime factors and with $m \mid n$, there is a universal division algebra $U = UD(k,n,m)$ which is generic for central simple algebras of degree n and exponent dividing m over fields containing k. If U has a decomposition $U = S \otimes_{Z(U)} T$, then every simple algebra A with center any field $K \supseteq k$ such that $deg\, A = n$ and $\exp A \mid m$ has a corresponding decomposition $A = B \otimes_K C$, where $deg\, B = deg\, S$ and

$deg\, C = deg\, T$. Taking $n = p^r$ and $m = p^s$ for any prime p and any integers r, s with $r > s \geq \frac{r}{2} + 1 \geq 3$, and for k a field containing a primitive p^s-th root of unity, the algebra $U = \mathrm{UD}(k, p^r, p^s)$ of degree p^r and exponent p^s is then shown to be indecomposable as follows: If r is even, consider $A_1 = \mathcal{T}(k; p^{r/2}, p^{r/2})$ and $A_2 = \mathcal{T}(k; p^s, p^{r-s})$ as in (9.5). Let $F = Z(A_1) = Z(A_2) = k((x_1))((y_1))((x_2))((y_2))$. By Prop. 9.8, we have

$$deg\, A_1 = deg\, A_2 = p^r \quad \text{and} \quad exp\, A_1 = p^{r/2} \mid p^s, \quad exp\, A_2 = p^s.$$

In any nontrivial decomposition $A_1 = B_1 \otimes_F C_1$ we have $deg\, B_1 = deg\, C_1 = p^{r/2}$ by Cor. 10.30. Likewise, in any nontrivial decomposition $A_2 = B_2 \otimes_F C_2$ we have $\{deg\, B_2, deg\, C_2\} = \{p^s, p^{r-s}\}$. Since the degrees of the subalgebras in any decomposition of A_1 do not overlap with the degrees of the subalgebras for A_2, it follows that $\mathrm{UD}(k, p^r, p^s)$ is indecomposable. The case where r is odd is handled analogously, by replacing A_1 by $A_1' = \mathcal{T}(F; p^{r+1/2}, p^{r-1/2})$. In any nontrivial decomposition of A_1' the larger factor has degree $p^{r+1/2}$, while for A_2 the larger factor has degree $p^s \neq p^{r+1/2}$.

As pointed out by Schofield–Van den Bergh [228, Th. 2.2], the algebra $\mathrm{UD}(k, p^r, p^s)$ can actually be proved indecomposable for every field k and all integers $r \geq s \geq 2$ by the index argument of Prop. 10.1. This follows because $\mathrm{UD}(k, p^r, p^s)$ is obtained by extending scalars of the universal division algebra $\mathrm{UD}(k, p^r)$ to the function field of the Severi–Brauer variety of $\mathrm{UD}(k, p^r)^{p^s}$; the Schofield–Van den Bergh index reduction theorem [228, Th. 1.3] shows that $deg\, \mathrm{UD}(k, p^r, p^s)^p = \frac{1}{p}\, deg\, \mathrm{UD}(k, p^r, p^s)$ for all $s \geq 2$ because $deg\, \mathrm{UD}(k, p^r)^p = p^{r-1}$.

Proposition 10.5 is a slight generalization of a result of Brussel, who constructed in [38, Cor. 3.2] an indecomposable division algebra of degree p^3 and exponent p^2 over the field of Laurent series in one indeterminate over a number field. Brussel's paper [38] also contains an in-depth study of the decomposability of central division algebras over a field with a discrete rank one Henselian valuation with residue field a number field. Remarkably, over such a field the indecomposability criterion of Prop. 10.1 is a necessary and sufficient condition: a division algebra D of degree a power of a prime p is decomposable if and only if $deg\, D^p \neq \frac{1}{p}\, deg\, D$, see [38, Cor. 3.6].

§10.2: All of subsections 10.2.1, 10.2.3, and 10.2.4 are based on Jacob [104].

The first example of an indecomposable algebra of prime exponent is an algebra of degree 8 and exponent 2 constructed by Amitsur et al. [11]. The proof of Th. 10.13 is based on Elman et al. [71, Th. 5.1]. This paper also gives the following description of the quotient ${}_2 Br(\ell/k)/Dec(\ell/k)$ for $\ell = k(\sqrt{a_1}, \sqrt{a_2}, \sqrt{a_3})$ an elementary abelian Galois extension of degree 8: letting $N_i \subseteq k^\times$ denote the norm group $N_i = N_{k(\sqrt{a_i})/k}(k(\sqrt{a_i})^\times)$ for $i = 1$, 2, 3, and $N_4 = N_{k(\sqrt{a_1 a_2})/k}(k(\sqrt{a_1 a_2})^\times)$, it is shown in [71, (4.22)] that

$$\frac{{}_2 Br(\ell/k)}{Dec(\ell/k)} \cong \frac{(N_1 \cdot N_3) \cap (N_2 \cdot N_3) \cap (N_4 \cdot N_3)}{(N_1 \cap N_2) \cdot N_3}.$$

The construction of indecomposable algebras of prime exponent $p \neq 2$ was first undertaken by Rowen [213]. Unfortunately, despite several corrections [212], [215, §7.3], the argument is still incomplete. (The proof of [215, Prop. 7.3.27] is flawed: In the next to last paragraph on p. 255 of [215], it is implicitly assumed that f_1 is relatively prime to $N(h_2)$. This flaw invalidates the proof of [215, Th. 7.3.28].)

All the indecomposable division algebras of prime exponent constructed by Jacob [104], as well as the earlier ones given by Amitsur et al. [11], Elman et al. [71], Tignol [238], and those proposed by Rowen [213], have characteristic zero. By contrast, the characteristic is not an issue in the construction of indecomposable algebras of nonprime exponent; see Ex. 10.4. Examples of prime exponent p and arbitrary characteristic different from p were given by Saltman [223, Cor. 12.15] for $p \neq 2$ and degree p^2, and Sivatski [231] for $p = 2$ and degree 2^r, for any $r \geq 3$. A few constructions specific to p-algebras are also found in the literature: indecomposable division algebras of degree 8 and exponent 2 over a field of characteristic 2 (Rowen [214, §3], McKinnie [140, Ex. 3.3.2]), and indecomposable division algebras of degree p^2 and exponent p over a field of characteristic $p \neq 2$ (Aravire–Jacob [15, Ex. 3.3], McKinnie [141, §2]). The Saltman, Sivatski, Rowen, and McKinnie constructions are "generic abelian crossed products," which means that they are obtained by the $\mathcal{S}$ (or $\mathcal{S}'$) construction of §8.4.5.

An indecomposability criterion for tame semiramified division algebras D of prime power degree over a Henselian field F was given by Mounirh [165, Prop. 2.8, Th. 3.5]: If for every field K with $\overline{F} \subseteq K \subseteq \overline{D}$ either $\overline{D}/K$ is cyclic or $ext_{K/\overline{F}}(sp(D)) \notin Dec(\overline{D}/K)$, then D is indecomposable. The same result was proved by McKinnie [140, Th. 2.3.1] in the particular case where D is a p-algebra. When $\overline{D}$ is an elementary abelian extension of $\overline{F}$ of degree p^2 and $sp(D) \notin Dec(\overline{D}/\overline{F})$, it follows from Mounirh's criterion that D is indecomposable. Therefore, this criterion can be used as a substitute to Th. 10.10 in the proof of Cor. 10.12(i). It cannot be used for the proof of Cor. 10.12(ii) however, because when $\overline{D}$ is an elementary abelian extension of degree 8 of $\overline{F}$ we have ${}_2 Br(\overline{D}/K) = Dec(\overline{D}/K)$ for any quadratic extension K of $\overline{F}$ in $\overline{D}$, see Exercise 8.4.

In [113], Karpenko establishes a geometric criterion for the indecomposability of division algebras of prime exponent which is free of any hypothesis on the characteristic (and does not use valuation theory): Given a division algebra D of prime exponent p, he considers its Severi–Brauer variety X and the Chow group $\mathrm{CH}^2(X)$ of codimension 2 cycles on X modulo rational equivalence, and proves that its torsion subgroup $\mathrm{Tors}(\mathrm{CH}^2(X))$ is either trivial or cyclic of order p [113, Prop. 5.1], and that it vanishes when D is decomposable [113, Prop. 5.3]. He also shows that if D is obtained from a division algebra D_0 with $deg\, D_0 = exp\, D_0 = p^r$ by scalar extension to the function field of the Severi–Brauer variety of D_0^p, then $\mathrm{Tors}(\mathrm{CH}^2(X))$ is not trivial if $p \neq 2$ and $r \geq 2$, and also if $p = 2$ and $r \geq 3$ [113, Prop. 5.1]. Thus, these algebras D are indecomposable of exponent p and degree p^r. In an earlier paper, Karpenko used a slightly different approach to prove the same result for $p \neq 2$, and also for $p = 2$ and r odd [112, Th. 3.1]. By using the canonical functors associated with the change of base, it is easy to see that if $\operatorname{Tors}(\mathrm{CH}^2(X))$ has order p, it cannot vanish after a scalar extension of degree prime to p. Therefore, Karpenko's examples of indecomposable algebras of exponent p remain indecomposable after a prime-to-p extension. McKinnie observes in [139, Th. 2.16] that, likewise, Mounirh's condition for indecomposability is preserved under prime-to-p extensions. The same observation holds for the index criterion for indecomposability of algebras of nonprime exponent in Prop. 10.1, since the index of division algebras of degree a power of p does not change under prime-to-p extensions. Therefore, the indecomposable algebras in Ex. 10.4 remain indecomposable under prime-to-p extensions, in contrast with Brussel's examples in Th. 10.6. Likewise, Sivatski's examples of exponent 2 remain indecomposable after an odd-degree extension of the center, see [231, Cor. 10].

§10.3: The proof of Th. 10.26 is based in part on Barry [22, Th. 3.1]. Also, Exercise 10.4 is similar to [22, Cor. 3.2], and Ex. 10.28 to [22, Ex. 3.3]. Exercise 10.1 is a result of Sethuraman [230, Prop. 2]. Exercise 10.6 is from Tignol [238]. Note that, even though $[A] \notin \mathit{Dec}(\ell/k)$, (ii) shows that $[A] \in \mathit{Dec}(k(\sqrt[p]{x_1}, \sqrt[p]{x_2})/k)$.

Chapter 11
Computation of $SK_1(D)$

Throughout this chapter we will be considering multiplicative commutators and commutator groups. Our notation for this is: if G is a group, then for $a, b \in G$ we set $[a, b] = aba^{-1}b^{-1}$. If H and K are subgroups of G, then set

$$[H, K] \;=\; \langle\, [h, k] \mid h \in H,\ k \in K \rangle.$$

Let A be a central simple algebra over a field F. Let $\mathit{Nrd}_A\colon A \to F$ be the reduced norm of A. When restricted to the multiplicative group $A^\times$ the reduced norm gives a group homomorphism $\mathit{Nrd}_A\colon A^\times \to F^\times$. Set

$$A^{(1)} \;=\; \mathit{ker}(\mathit{Nrd}_A) \;=\; \{a \in A^\times \mid \mathit{Nrd}_A(a) = 1\}.$$

Since $\mathit{im}(\mathit{Nrd}_A)$ is commutative, we have $[A^\times, A^\times] \subseteq A^{(1)}$ where $[A^\times, A^\times]$ is the commutator subgroup of $A^\times$. By definition, $SK_1(A)$ is the factor group

$$SK_1(A) \;=\; A^{(1)}/[A^\times, A^\times].$$

Thus, $SK_1(A)$ is the "special" subgroup of $K_1(A) = A^\times/[A^\times, A^\times]$, and there is an exact sequence induced by Nrd_A:

$$1 \longrightarrow SK_1(A) \longrightarrow K_1(A) \longrightarrow F^\times.$$

This can be interpreted as saying that $SK_1(A)$ is the part of $K_1(A)$ not detected within $F^\times$ (via Nrd_A).

The group $SK_1(A)$ is a subtle invariant of $A^\times$ which has been of longstanding interest to those working in central simple algebras, algebraic groups, and K-theory. It was conjectured for several years that $SK_1(A) = 1$ for every A. This was part of the Kneser–Tits conjecture in algebraic groups. It was thus a great surprise in 1975 when Platonov gave in [179] and [180] (with details in [183]) the first examples of algebras with nontrivial SK_1. Platonov's examples were division algebras D over $k((x))((y))$ where k is a global field. His calculations used valuation theory to relate $SK_1(D)$ to properties of $\overline{D}$. Most concrete calculations since then of nontrivial $SK_1(D)$ when $\mathit{ind}\, D > 4$ have used

J.-P. Tignol, A.R. Wadsworth, *Value Functions on Simple Algebras, and Associated Graded Rings*, Springer Monographs in Mathematics,
DOI 10.1007/978-3-319-16360-4_11

valuation theory. Recently it has been shown by Hazrat and Wadsworth [99] that if D is a tame division algebra over a Henselian field then

$$SK_1(D) \cong SK_1(\mathsf{gr}(D)), \tag{11.1}$$

and that calculations in the graded setting are easier and more transparent than in the Henselian valued setting.

In this chapter we follow the approach in [99]: We first prove properties of $SK_1(\mathsf{D})$ for a graded division algebra D. This includes many explicit formulas for $SK_1(\mathsf{D})$ motivated by mostly known formulas for ungraded division algebras over a Henselian field, see Cor. 11.12 and Th. 11.15. We then prove isomorphism (11.1) in Th. 11.21. By applying this isomorphism and the graded results, we obtain in Th. 11.23 computable formulas for $SK_1(D)$ for many classes of tame division algebras D over Henselian fields.

For an excellent introduction to SK_1 for central simple algebras, see the book by Draxl [63].

11.1 SK_1 of graded division algebras

Before considering the analogue of SK_1 for graded division algebras, we recall a few basic properties of the ungraded SK_1.

Proposition 11.1. *Let A be a central simple algebra and let D be the associated division algebra of A. Then,*

$$SK_1(A) \cong SK_1(D).$$

Proof. See Draxl [63, (3), p. 155]. □

Proposition 11.2. *Let A be a central simple algebra. Let $p_1, \ldots, p_n$ be the distinct prime divisors of $\operatorname{ind} A$ and let $e = \operatorname{ind} A/(p_1 \ldots p_k)$. Then, $SK_1(A)$ is an e-torsion group. In particular, if $\operatorname{ind} A$ is square-free, then $SK_1(A) = 1$.*

Proof. See Draxl [63, Th. 4, p. 164]. □

Proposition 11.3. *Let A and B be central simple F-algebras such that $\gcd(\operatorname{ind} A, \operatorname{ind} B) = 1$. Then,*

$$SK_1(A \otimes_F B) \cong SK_1(A) \times SK_1(B).$$

Proof. See Draxl [63, Lemma 6, p. 160]. □

In order to define SK_1 for graded division algebras, we need to describe the reduced norm in the graded setting. For this we first recall a convenient way of characterizing Nrd for ungraded simple algebras that adapts readily to graded division algebras. Let B be a central simple algebra over a field K with $\deg B = n$, and let $\ell_\bullet \colon B \to End_K B$ be the left regular representation;

thus, for $b \in B$ the map ℓ_b is given by $a \mapsto ba$. Recall from (5.4) that the characteristic polynomial of b for B/K is

$$Pchar_{B/K,b} = det\left(id_B\, X - \ell_b\right) \in K[X],$$

i.e., the characteristic polynomial of ℓ_b, which is a monic polynomial of degree n^2. It is known that $Pchar_{B/K,b}$ is an n-th power in $K[X]$ (cf. Reiner [199, Th. 9.5, p. 115]); define the *reduced characteristic polynomial of b for B* to be its n-th root:

$Prd_{B,b}$ is the monic polynomial in $K[X]$ with $(Prd_{B,b})^n = Pchar_{B/K,b}$.

Write

$$Prd_{B,b} = X^n - s_1X^{n-1} + \ldots + (-1)^n s_n.$$

Then the *reduced trace* and *reduced norm* of b for B are defined by

$$Trd_B(b) = s_1 \qquad \text{and} \qquad Nrd_B(b) = s_n.$$

Thus, $tr_{B/K}(b) = n\, Trd_B(b)$ and $N_{B/K}(b) = Nrd_B(b)^n$, where $tr_{B/K}$ and $N_{B/K}$ are the algebra trace and norm defined in (5.3). If $B \cong M_n(K)$, then $Prd_{B,b}$ is the characteristic polynomial of b. The basic properties of Prd_B, Trd_B, and Nrd_B for an arbitrary central simple B are easily obtainable by making a scalar extension to split B and using properties of the characteristic polynomial, trace, and determinant of matrices.

Lemma 11.4. *Let* A *be a simple graded algebra with center* F*. For any* $a \in \mathsf{A}$*, we have* $Prd_{q(\mathsf{A}),a} \in \mathsf{F}[X]$*. Moreover, if* a *is homogeneous of degree* δ*, then* $Prd_{q(\mathsf{A}),a}$ *is* δ*-homogenizable in* $\mathsf{F}[X]$*.*

Proof. Let $n = deg\, \mathsf{A}$. Let p_a be the minimal polynomial of a over $q(\mathsf{F})$, which is also the minimal polynomial over $q(\mathsf{F})$ of ℓ_a in $End_{q(\mathsf{F})}(q(\mathsf{A}))$. By the theory of canonical forms in linear algebra, we have $Pchar_{q(\mathsf{A})/q(\mathsf{F}),a} \mid (p_a)^{n^2}$ in $q(\mathsf{F})[X]$. Because $\mathsf{A}^\natural$ is integral over $\mathsf{F}^\natural$, every root of p_a in a splitting field is integral over $\mathsf{F}^\natural$. Therefore, the coefficients of p_a must be integral over $\mathsf{F}^\natural$. Since $\mathsf{F}^\natural$ is integrally closed by Prop. 5.2, we must have $p_a \in \mathsf{F}[X]$. Likewise, the coefficients of the monic divisor $Prd_{q(\mathsf{A}),a}$ of $(p_a)^{n^2}$ are integral over F; so $Prd_{q(\mathsf{A}),a} \in \mathsf{F}[X]$. If $a \in \mathsf{A}_\delta$, the argument in the proof of (a) $\Rightarrow$ (b) of Prop. 5.3 shows that a is a root of some nonzero δ-homogenizable polynomial f in $\mathsf{F}[X]$. Since $p_a \mid f$ in $q(\mathsf{F})[X]$, Prop. 5.6 shows that p_a is δ-homogenizable in $\mathsf{F}[X]$. So, $(p_a)^{n^2}$ is also δ-homogenizable and another application of Prop. 5.6 shows that $Prd_{q(\mathsf{A}),a}$ is also δ-homogenizable. □

Definition 11.5. Let A be simple graded algebra with center F. For any $a \in \mathsf{A}$, define $Prd_{\mathsf{A},a}$ to be $Prd_{q(\mathsf{A}),a}$ in $\mathsf{F}[X]$. Likewise, set

$$Trd_{\mathsf{A}}(a) = Trd_{q(A)}(a) \in \mathsf{F} \qquad \text{and} \qquad Nrd_{\mathsf{A}}(a) = Nrd_{q(A)}(a) \in \mathsf{F}.$$

Known properties of *Trd* and *Nrd* carry over immediately to the graded setting:

Proposition 11.6. *Let* A *be a simple graded algebra with center* F. *Then,*

(i) $Trd_{\mathsf{A}}\colon \mathsf{A} \to \mathsf{F}$ *is* F*-linear;*
(ii) $Nrd_{\mathsf{A}}(ab) = Nrd_{\mathsf{A}}(a) \cdot Nrd_{\mathsf{A}}(b)$ *for all* $a, b \in \mathsf{A}$*;*
(iii) *if* $c \in \mathsf{F}$, *then* $Nrd_{\mathsf{A}}(c) = c^{\deg \mathsf{A}}$*;*
(iv) $Nrd_{\mathsf{A}}(a) \neq 0$ *if and only if* $a \in \mathsf{A}^\times$*;*
(v) *if* L *is a graded subfield of* A *containing* F *and* $a \in \mathsf{L}$, *then*

$$Prd_{\mathsf{A},a} = (Pchar_{\mathsf{L}/\mathsf{F},a})^{\deg \mathsf{A}/[\mathsf{L}:\mathsf{F}]};$$

hence,

$$Trd_{\mathsf{A}}(a) \;=\; \tfrac{\deg \mathsf{A}}{[\mathsf{L}:\mathsf{F}]}\, tr_{\mathsf{L}/\mathsf{F}}(a) \qquad \textit{and} \qquad Nrd_{\mathsf{A}}(a) \;=\; N_{\mathsf{L}/\mathsf{F}}(a)^{\deg \mathsf{A}/[\mathsf{L}:\mathsf{F}]};$$

(vi) *if* B *is a simple graded* F*-subalgebra of* A *and* $\mathsf{L} = Z(\mathsf{B}) \supseteq \mathsf{F}$, *then for* $b \in \mathsf{B}$,

$$Prd_{\mathsf{A},b} \;=\; N_{\mathsf{L}[X]/\mathsf{F}[X]}(Prd_{\mathsf{B},b})^{\xi} \qquad \textit{where} \quad \xi = \deg \mathsf{A}/([\mathsf{L}:\mathsf{F}] \deg \mathsf{B});$$

hence,

$$Trd_{\mathsf{A}}(b) \;=\; \xi\, tr_{\mathsf{L}/\mathsf{F}}(Trd_{\mathsf{B}}(b)) \qquad \textit{and} \qquad Nrd_{\mathsf{A}}(b) \;=\; N_{\mathsf{L}/\mathsf{F}}(Nrd_{\mathsf{B}}(b))^{\xi}.$$

Proof. Parts (i)–(iv) follow immediately from the corresponding well-known properties of $Trd_{q(\mathsf{A})}$ and $Nrd_{q(\mathsf{A})}$ (see, e.g., Draxl [63, p. 146] or Reiner [199, (9.8), p. 116]). For (iv) note that $\mathsf{A}^\times = q(\mathsf{A})^\times \cap \mathsf{A}$. For (v) we use that A is a graded left L-vector space of dimension $n^2/[\mathsf{L}:\mathsf{F}]$, where $n = \deg \mathsf{A}$. Hence, for any $a \in \mathsf{L}$ we have

$$Pchar_{\mathsf{A}/\mathsf{F},a} \;=\; (Pchar_{\mathsf{L}/\mathsf{F},a})^{n^2/[\mathsf{L}:\mathsf{F}]};$$

hence,

$$Prd_{\mathsf{A},a} \;=\; (Pchar_{\mathsf{A}/\mathsf{F},a})^{1/n} \;=\; (Pchar_{\mathsf{L}/\mathsf{F},a})^{n/[\mathsf{L}:\mathsf{F}]}.$$

The formulas in (v) for $Trd_{\mathsf{A}}(a)$ and $Nrd_{\mathsf{A}}(a)$ follow immediately from this.

For (vi), note that $q(\mathsf{B})$ is a simple subalgebra of $q(\mathsf{A})$ by Prop. 2.28(ii) with $Z(\mathsf{B}) = q(\mathsf{L})$ and $\deg q(\mathsf{A}) = \deg \mathsf{A}$, $\deg q(\mathsf{B}) = \deg \mathsf{B}$, and $[q(\mathsf{L}){:}q(\mathsf{F})] = [\mathsf{L}{:}\mathsf{F}]$. The formulas for $Trd_{q(\mathsf{A})}(b)$ and $Nrd_{q(\mathsf{A})}(b)$ corresponding to those in (vi) are given in the Reduced Tower Formulae in Draxl [63, p. 150]. Since $tr_{\mathsf{L}/\mathsf{F}}$ and $N_{\mathsf{L}/\mathsf{F}}$ are compatible with extensions of the base ring (see (5.7)), the formulas over $q(\mathsf{F})$ immediately imply the ones given in (vi) for $Trd_{\mathsf{A}}(b)$ and $Nrd_{\mathsf{A}}(b)$. The formula for $Prd_B(b)$ is obtainable as follows: Note that $q(\mathsf{A})$ is a free left $q(\mathsf{B})$-module of rank $m = [q(\mathsf{A}){:}q(\mathsf{F})]/[q(\mathsf{B}){:}q(\mathsf{F})] = [\mathsf{A}{:}\mathsf{F}]/[\mathsf{B}{:}\mathsf{F}]$. It is free by counting ranks, because both $q(\mathsf{A})$ and $q(\mathsf{B})$ are each direct sums of the unique simple left $q(\mathsf{B})$-module. Therefore, for $b \in \mathsf{B}$, $Pchar_{q(\mathsf{A})/q(\mathsf{L}),b} = (Pchar_{q(\mathsf{B})/q(\mathsf{L}),b})^m$. Note that this polynomial lies in $\mathsf{L}[X]$ by Lemma 11.4. By Reiner [199, (9.12), p. 117] or Bourbaki [31, Ch. 3, §9.4, Prop. 6],

$$\begin{aligned} Pchar_{q(\mathsf{A})/q(\mathsf{F}),b} \;&=\; N_{q(\mathsf{L})[X]/q(\mathsf{F})[X]}(Pchar_{q(\mathsf{A})/q(\mathsf{L}),b}) \\ &=\; N_{\mathsf{L}[X]/\mathsf{F}[X]}(Pchar_{q(\mathsf{A})/q(\mathsf{L}),b}), \end{aligned}$$

with the second equality given by (5.7). Thus, with $n = deg\,\mathsf{A}$,

$$\begin{aligned} Prd_{\mathsf{A},b} &= (Pchar_{q(\mathsf{A})/q(\mathsf{F}),b})^{1/n} = N_{\mathsf{L}[X]/\mathsf{F}[X]}(Pchar_{q(\mathsf{A})/q(\mathsf{L}),b})^{1/n} \\ &= N_{\mathsf{L}[X]/\mathsf{F}[X]}(Pchar_{q(\mathsf{B})/q(\mathsf{L}),b})^{m/n} = N_{\mathsf{L}[X]/\mathsf{F}[X]}(Prd_{q(\mathsf{B}),b})^{m\,deg\,\mathsf{B}/n} \end{aligned}$$

which is the desired formula for $Prd_{\mathsf{A},b}$ since $\xi = m\,deg\,\mathsf{B}/n$. The formulas for $Trd_{\mathsf{B}}(b)$ and $Nrd_{\mathsf{B}}(b)$ are also deducible from the formula for $Prd_{\mathsf{A},b}$. □

There is further information for homogeneous elements:

Proposition 11.7. *Let* A *be a simple graded algebra with center* F.

(i) *Suppose* $a \in \mathsf{A}$ *is homogeneous of degree* δ. *Then* $\mathrm{Prd}_{\mathsf{A},a}$ *is* δ*-homogenizable; hence,* $\mathrm{Trd}_{\mathsf{A}}(a) \in \mathsf{F}_\delta$ *and* $\mathrm{Nrd}_{\mathsf{A}}(a) \in \mathsf{F}_{n\delta}$ *where* $n = deg\,\mathsf{A}$.

(ii) *Suppose* A_0 *is simple. Then, for any* $s \in \mathsf{A}_0$,

$$Prd_{\mathsf{A},s} = N_{Z(\mathsf{A}_0)[X]/\mathsf{F}_0[X]}(Prd_{\mathsf{A}_0,s})^{\zeta}$$

where $\zeta = deg\,\mathsf{A}/\big([Z(\mathsf{A}_0){:}\mathsf{F}_0]\,deg\,\mathsf{A}_0\big)$; *hence,*

$$Trd_A(s) = \zeta\, tr_{Z(\mathsf{A}_0)/\mathsf{F}_0}(Trd_{\mathsf{A}_0}(s)) \text{ and } Nrd_A(s) = N_{Z(\mathsf{A}_0)/\mathsf{F}_0}(Nrd_{\mathsf{A}_0}(s))^{\zeta}.$$

Proof. (i) If $a \in \mathsf{A}_\delta$, then $Prd_{\mathsf{A},a} = Prd_{q(\mathsf{A}),a}$ is a monic δ-homogenizable polynomial of degree n by Lemma 11.4. This determines the degrees in Γ_{F} of its coefficients which are all homogeneous in F. Hence, $Trd_{\mathsf{A}}(a) \in \mathsf{F}_\delta$ and $Nrd_{\mathsf{A}}(a) \in \mathsf{F}_{n\delta}$.

(ii) Suppose A_0 is simple. We have the graded F-subalgebras $\mathsf{B} = \mathsf{A}_0 \otimes_{\mathsf{F}_0} \mathsf{F}$ and $\mathsf{L} = Z(\mathsf{A}_0) \otimes_{\mathsf{F}_0} \mathsf{F}$ of A. Then, B is a simple graded algebra as A_0 is simple, see Prop. 2.16, and $\mathsf{L} = Z(\mathsf{B})$. For $a \in \mathsf{A}_0 \subseteq \mathsf{B}$, Prop. 11.6(vi) yields

$$Prd_{\mathsf{A},a} = N_{\mathsf{L}[X]/\mathsf{F}[X]}(Prd_{\mathsf{B},b})^{\xi},$$

where

$$\xi = deg\,\mathsf{A}/([\mathsf{L}{:}\mathsf{F}]\,deg\,\mathsf{B}) = deg\,\mathsf{A}/\big([Z(\mathsf{gr}(F_0)){:}\mathsf{F}_0]\,deg\,\mathsf{A}_0\big).$$

But since $a \in \mathsf{A}_0$ and $\mathsf{B} = \mathsf{A}_0 \otimes_{\mathsf{F}_0} \mathsf{F}$, we have $Pchar_{\mathsf{B}/\mathsf{L},a} = Pchar_{\mathsf{A}_0/Z(\mathsf{A}_0),a}$ (cf. (5.7)), so $Prd_{\mathsf{B},a} = Prd_{\mathsf{A}_0,a}$ in $Z(\mathsf{A}_0)[X]$. Moreover, as

$$\mathsf{L}[X] = Z(\mathsf{A}_0)[X] \otimes_{\mathsf{F}_0[X]} \mathsf{F}[X],$$

we have

$$N_{\mathsf{L}[X]/\mathsf{F}[X]}(Prd_{\mathsf{A}_0,a}) = N_{Z(\mathsf{A}_0)[X]/\mathsf{F}_0[X]}(Prd_{\mathsf{A}_0,a}).$$

Hence, $Prd_{\mathsf{A},a} = N_{Z(\mathsf{A}_0)[X]/\mathsf{F}_0[X]}(Prd_{\mathsf{A}_0,a})^{\zeta}$ as desired, and the formulas for $Trd_{\mathsf{A}}(a)$ and $Nrd_{\mathsf{A}}(a)$ follow by comparing coefficients in these polynomials (cf. Reiner [199, (9.11), p. 117]). □

We can now consider SK_1 for graded division algebras. Let D be a graded division algebra with center F. Proposition 11.6(ii) and (iv) and Prop. 11.7(i) show that Nrd_{D} restricts to a group homomorphism $\mathsf{D}^\times \to \mathsf{F}^\times$. Let

$$\mathsf{D}^{(1)} = ker(Nrd\,|_{\mathsf{D}^\times}) = \{a \in \mathsf{D}^\times \mid Nrd(a) = 1\}.$$

Note that since every element of $\mathsf{D}^\times$ is homogeneous, Prop. 11.7(i) shows that $\mathsf{D}^{(1)} \subseteq \mathsf{D}_0^\times$.

Let $[\mathsf{D}^\times, \mathsf{D}^\times]$ be the commutator subgroup of $\mathsf{D}^\times$. Because Nrd_{D} is multiplicative and its image is commutative, we have $[\mathsf{D}^\times, \mathsf{D}^\times] \subseteq \mathsf{D}^{(1)}$. Define

$$SK_1(\mathsf{D}) \;=\; \mathsf{D}^{(1)}/[\mathsf{D}^\times, \mathsf{D}^\times].$$

The fact that $SK_1(\mathsf{D})$ is a subquotient of D_0 will be very helpful for us.

We consider SK_1 in the graded setting only for graded division algebras. For more general central simple graded algebras A, the elements of $\mathsf{A}^\times$, $\mathsf{A}^{(1)}$, and $[\mathsf{A}^\times, \mathsf{A}^\times]$ need not be homogeneous, which makes them very difficult to work with. While $SK_1(\mathsf{A})$ is definable just as for a graded division algebra, it is not known whether the analogue of Prop. 11.1 holds for graded central simple algebras. But we do have the analogue of Prop. 11.2:

Proposition 11.8. *Let D be a graded division algebra. Then, the canonical homomorphism $SK_1(\mathsf{D}) \to SK_1(q(\mathsf{D}))$ is injective. Hence, $SK_1(\mathsf{D})$ is an e-torsion group, where $e = ind\,\mathsf{D}/(p_1 \ldots p_k)$, with $p_1, \ldots, p_k$ the distinct prime divisors of $ind\,\mathsf{D}$.*

Proof. We saw in §3.2.5 that once a total ordering is chosen on Γ_{D}, there is a corresponding "lowest degree term" homomorphism $\lambda\colon q(\mathsf{D})^\times \to \mathsf{D}^\times$ such that $\lambda|_{\mathsf{D}^\times} = id$. Take any

$$a \;=\; \prod_{i=1}^{k} [r_i, s_i] \;\in\; \mathsf{D}^\times \cap [q(\mathsf{D})^\times, q(\mathsf{D})^\times],$$

where $r_i, s_i \in q(\mathsf{D})^\times$ and $[r_i, s_i] = r_i s_i r_i^{-1} s_i^{-1}$. Then,

$$a \;=\; \lambda(a) \;=\; \prod_{i=1}^{k} [\lambda(r_i), \lambda(s_i)] \;\in\; [\mathsf{D}^\times, \mathsf{D}^\times].$$

Thus, $\mathsf{D}^\times \cap [q(\mathsf{D})^\times, q(\mathsf{D})^\times] = [\mathsf{D}^\times, \mathsf{D}^\times]$, which shows that the map

$$SK_1(\mathsf{D}) \;=\; \mathsf{D}^{(1)}/[\mathsf{D}^\times, \mathsf{D}^\times] \;\longrightarrow\; q(\mathsf{D})^{(1)}/[q(\mathsf{D})^\times, q(\mathsf{D})^\times] \;=\; SK_1(q(\mathsf{D}))$$

is injective. Since $ind\,\mathsf{D} \;=\; ind\,q(\mathsf{D})$, it follows from this injectivity and Prop. 11.2 that $SK_1(\mathsf{D})$ is e-torsion. □

In fact, $SK_1(\mathsf{D}) \cong SK_1(q(\mathsf{D}))$; see Hazrat–Wadsworth [99, Th. 5.7]. This is a difficult result that we will not prove here.

In working with SK_1, we will need the compatibility of the reduced norm with automorphisms:

Lemma 11.9. *Let B be a simple algebra with center K, and let η be any automorphism of B. Then, $\mathrm{Pchar}_{B/K,\eta(b)} = \eta(\mathrm{Pchar}_{B/K,b})$ for any $b \in B$; hence, $\mathrm{Prd}_{B,\eta(b)} = \eta(\mathrm{Prd}_{B,b})$, so*

$$Trd_B(\eta(b)) \;=\; \eta(Trd_B(b)) \qquad \textit{and} \qquad Nrd_B(\eta(b)) \;=\; \eta(Nrd_B(b)).$$

Proof. Let $m = [B:K]$. Take any $b \in B$, and let $(c_{ij}) \in M_m(K)$ be the matrix representing the left multiplication map $\ell_b \in End_K B$ relative to a K-vector space base $(y_1, \dots, y_m)$ of B. Then $(\eta(c_{ij}))$ is the matrix of $\ell_{\eta(b)}$ relative to the base $(\eta(y_1), \dots, \eta(y_m))$. Hence, $Pchar_{B/K,\eta(b)} = \eta(Pchar_{B/K,b})$, and the other equalities in the lemma follow immediately from this. □

We recall the definition of the cohomology group appearing in the next theorem. Let G be a finite abelian group and let M be a G-module, with operation written multiplicatively. Let $N_G \colon M \to M$ be the G-norm map given by $c \mapsto \prod_{\sigma \in G} \sigma(c)$. Then, the -1-Tate cohomology group of G with respect to M is

$$\widehat{H}^{-1}(G, M) = \ker N_G / I_G(M), \text{ where } I_G(M) = \langle c\sigma(c)^{-1} \mid c \in M,\ \sigma \in G \rangle.$$

We now give a theorem analyzing SK_1 of a graded division algebra D in terms of pieces of the graded structure on D. This is a graded version of a theorem of Ershov [75] for tame division algebras over a Henselian field. For any integer $n > 1$ and any field K, we write $\mu_n(K)$ for the group of all n-th roots of unity in K.

Theorem 11.10. *Let D be a graded division algebra with center F, and let $G = \mathcal{G}(Z(\mathsf{D}_0)/\mathsf{F}_0)$. Set $\zeta = ind(\mathsf{D})/\big(ind(\mathsf{D}_0)\,[Z(\mathsf{D}_0):\mathsf{F}_0]\big)$. Let*

$$\widetilde{N} = N_{Z(\mathsf{D}_0)/\mathsf{F}_0} \circ Nrd_{\mathsf{D}_0} \colon \mathsf{D}_0^{\times} \longrightarrow \mathsf{F}_0^{\times},$$

and let $\mathcal{K} = \big(\mathsf{D}_0^{(1)} \cap [\mathsf{D}_0^{\times}, \mathsf{D}^{\times}]\big)/[\mathsf{D}_0^{\times}, \mathsf{D}_0^{\times}]$. Then, the following diagram has exact rows and column:

$$\begin{array}{ccccccc}
 & & & & 1 & & \\
 & & & & \downarrow & & \\
\mathcal{K} & \longrightarrow & SK_1(\mathsf{D}_0) & \longrightarrow & \ker \widetilde{N}/[\mathsf{D}_0^{\times}, \mathsf{D}^{\times}] & \xrightarrow{Nrd_{\mathsf{D}_0}} & \widehat{H}^{-1}(G, Nrd_{\mathsf{D}_0}(\mathsf{D}_0^{\times})) \longrightarrow 1 \\
 & & & & \downarrow & & \\
 & & \Gamma_{\mathsf{D}}/\Gamma_{\mathsf{F}} \wedge \Gamma_{\mathsf{D}}/\Gamma_{\mathsf{F}} & \xrightarrow{\kappa} & \mathsf{D}^{(1)}/[\mathsf{D}_0^{\times}, \mathsf{D}^{\times}] & \longrightarrow & SK_1(\mathsf{D}) \longrightarrow 1 \\
 & & & & \downarrow{\scriptstyle \widetilde{N}} & & \\
 & & & & \mu_{\zeta}(\mathsf{F}_0) \cap \widetilde{N}(\mathsf{D}_0^{\times}) & & \\
 & & & & \downarrow & & \\
 & & & & 1 & &
\end{array} \tag{11.2}$$

The map κ in the diagram is given as follows: For any $\gamma, \delta \in \Gamma_{\mathsf{D}}$ choose any nonzero $d_\gamma \in \mathsf{D}_\gamma$ and $d_\delta \in \mathsf{D}_\delta$. Then, $\kappa\big((\gamma + \Gamma_{\mathsf{F}}) \wedge (\delta + \Gamma_{\mathsf{F}})\big) = [d_\gamma, d_\delta]\,[\mathsf{D}_0^{\times}, \mathsf{D}^{\times}]$.

Proof. Consider first the top row of the diagram. Let $M = Nrd_{\mathsf{D}_0}(\mathsf{D}_0^\times) \subseteq Z(\mathsf{D}_0)^\times$. Recall that the map $\theta_{\mathsf{D}}\colon \Gamma_{\mathsf{D}} \to G$ is surjective. For any $\sigma \in G$, choose any $\gamma \in \Gamma_{\mathsf{D}}$ with $\theta_{\mathsf{D}}(\gamma) = \sigma$ and any nonzero $b \in \mathsf{D}_\gamma$. Then, $int(b)|_{\mathsf{D}_0}$ is an automorphism of D_0, so Lemma (11.9) yields that for any $a \in \mathsf{D}_0^\times$,

$$\sigma(Nrd_{\mathsf{D}_0}(a)) = b\,Nrd_{\mathsf{D}_0}(a)b^{-1} = Nrd_{\mathsf{D}_0}(bab^{-1}).$$

Hence, M is closed under the action of G. Note that the norm map N_G for M as a G-module is the field norm $N_{Z(\mathsf{D}_0)/\mathsf{F}_0}$ on M. So, $\widetilde{N} = N_G \circ Nrd_{\mathsf{D}_0}$ and Nrd_{D_0} maps $ker\,\widetilde{N}$ onto $ker\,N_G$. Moreover, we claim that Nrd_{D_0} maps $[\mathsf{D}_0^\times, \mathsf{D}^\times]$ onto $I_G(M) \subseteq ker\,N_G$. For if $a \in \mathsf{D}_0^\times$ and $d \in \mathsf{D}^\times$, then, invoking Lemma 11.9 again,

$$\begin{aligned} Nrd_{\mathsf{D}_0}[a,d] &= Nrd_{\mathsf{D}_0}\big(a \cdot int(d)(a^{-1})\big) = Nrd_{\mathsf{D}_0}(a) \cdot int(d)\big(Nrd_{\mathsf{D}_0}(a)\big)^{-1} \\ &= Nrd_{\mathsf{D}_0}(a) \cdot \theta_{\mathsf{D}}(deg\,d)\big(Nrd_{\mathsf{D}_0}(a)\big)^{-1} \in I_G(M). \end{aligned}$$

Since θ_{D} is surjective, all generators of $I_G(M)$ have the form $Nrd_{\mathsf{D}_0}[a,d]$ for $a \in \mathsf{D}_0^\times$, $d \in \mathsf{D}^\times$. This proves the claimed surjectivity. Therefore, Nrd_{D_0} maps $ker\,\widetilde{N}/[\mathsf{D}_0^\times, \mathsf{D}^\times]$ onto $\widehat{H}^{-1}(G,M)$ with kernel $\mathsf{D}_0^{(1)}[\mathsf{D}_0^\times, \mathsf{D}^\times]/[\mathsf{D}_0^\times, \mathsf{D}^\times]$. Clearly $SK_1(\mathsf{D}_0) = \mathsf{D}_0^{(1)}/[\mathsf{D}_0^\times, \mathsf{D}_0^\times]$ maps onto this, with kernel $\mathcal{K}$. This shows the exactness of the top row of the diagram.

Next, consider the column of the diagram. Recall that $\mathsf{D}^{(1)} \subseteq \mathsf{D}_0^\times$ by Prop. 11.7(i). For $a \in \mathsf{D}_0^\times$, Prop. 11.7(ii) yields

$$Nrd_{\mathsf{D}}(a) = N_{Z(\mathsf{D}_0)/\mathsf{F}_0}(Nrd_{\mathsf{D}_0}(a))^\varsigma = \widetilde{N}(a)^\varsigma.$$

Hence, $ker\,\widetilde{N} \subseteq ker\,Nrd_{\mathsf{D}} = \mathsf{D}^{(1)}$ and $\widetilde{N}(\mathsf{D}^{(1)}) = \mu_\varsigma(\mathsf{F}_0) \cap im\,\widetilde{N}$. Therefore, the column is exact.

For the lower row of the diagram, observe that $[\mathsf{D}^\times, \mathsf{D}^\times] \subseteq \mathsf{D}_0^\times$ and since the degree epimorphism $\mathsf{D}^\times \to \Gamma_{\mathsf{D}}$ has kernel $\mathsf{D}_0^\times$, the induced epimorphism $\mathsf{D}^\times \to \Gamma_{\mathsf{D}}/\Gamma_{\mathsf{F}}$ has kernel $\mathsf{D}_0^\times \cdot \mathsf{F}^\times$; so,

$$\Gamma_{\mathsf{D}}/\Gamma_{\mathsf{F}} \cong \mathsf{D}^\times/(\mathsf{D}_0^\times \cdot \mathsf{F}^\times) = \mathsf{D}^\times/(\mathsf{D}_0^\times \cdot Z(\mathsf{D}^\times)).$$

The following lemma (with $P = \mathsf{D}^\times$ and $S = \mathsf{D}_0^\times$) then yields an epimorphism κ from $\Gamma_{\mathsf{D}}/\Gamma_{\mathsf{F}} \wedge \Gamma_{\mathsf{D}}/\Gamma_{\mathsf{F}}$ onto $[\mathsf{D}^\times, \mathsf{D}^\times]/[\mathsf{D}_0^\times, \mathsf{D}^\times]$, which is the kernel of the epimorphism $\mathsf{D}^{(1)}/[\mathsf{D}_0^\times, \mathsf{D}^\times] \to SK_1(\mathsf{D})$. Therefore, the lower row of the diagram is exact. The definition of the epimorphism in the following lemma yields the description of κ given above. □

Lemma 11.11. *Let P be a group and let S be a subgroup of P with $[P,P] \subseteq S$. Let $Q = P/\big(S \cdot Z(P)\big)$. Then, there is an epimorphism $Q \wedge Q \to [P,P]/[S,P]$ given by $a\,S \cdot Z(P) \wedge b\,S \cdot Z(P) \mapsto aba^{-1}b^{-1}\,[S,P]$ for all $a, b \in P$.*

Proof. Since $[P,P] \subseteq S$, we have $\big[[P,P],[P,P]\big] \subseteq [S,P]$. Hence, $[S,P]$ is a normal subgroup of $[P,P]$ with abelian factor group. Consider the map

$$\eta\colon P \times P \longrightarrow [P,P]/[S,P] \qquad \text{given by} \qquad (a,b) \mapsto aba^{-1}b^{-1}[S,P].$$

For any $a, b, c \in P$, we have the commutator identities

$$[a, bc] = [a, b]\,[b, [a, c]]\,[a, c] \qquad \text{and} \qquad [ab, c] = [a, [b, c]]\,[b, c]\,[a, c].$$

Since $[b, [a, c]]$ and $[a, [b, c]]$ lie in $[P, S] = [S, P]$ and $im\,\eta$ is commutative, it follows that η is bimultiplicative. Hence, η induces a homomorphism $\eta'\colon P \otimes_{\mathbb{Z}} P \to [P, P]/[S, P]$, which is surjective, since its image contains a generating set of $[P, P]/[S, P]$. Because $[S{\cdot}Z(P), P] = [S, P]$ we have $\eta'(a \otimes b) = 1$ if a or b lies in $S \cdot Z(P)$. Hence, η' induces an epimorphism $\eta''\colon Q \otimes_{\mathbb{Z}} Q \to [P, P]/[S, P]$. Since $\eta''(\overline{a} \otimes \overline{a}) = 1$ for all $\overline{a} \in Q$, the map η'' yields an epimorphism $Q \wedge Q \to [P, P]/[S, P]$. □

Corollary 11.12. *Let* D *be a graded division algebra with center* F. *Let* $G = \mathcal{G}(Z(\mathsf{D}_0)/\mathsf{F}_0)$ *and let* $\zeta = ind(\mathsf{D})/\big(ind(\mathsf{D}_0)\,[Z(\mathsf{D}_0){:}\mathsf{F}_0]\big)$.

(i) *If* D *is unramified over* F, *then*

$$SK_1(\mathsf{D}) \cong SK_1(\mathsf{D}_0).$$

(ii) *If* D *is totally ramified over* F, *then*

$$SK_1(\mathsf{D}) \cong \mu_n(\mathsf{F}_0)/\mu_e(\mathsf{F}_0),$$

and $|\mu_e(\mathsf{F}_0)| = e$, *where* $n = ind\,\mathsf{D}$ *and* $e = exp\ \mathsf{D} = exp(\Gamma_{\mathsf{D}}/\Gamma_{\mathsf{F}})$.

(iii) *If* D *is inertially split and* $SK_1(\mathsf{D}_0) = 1$, *then* $\zeta = 1$ *and there is an exact sequence:*

$$\Gamma_{\mathsf{D}}/\Gamma_{\mathsf{F}} \wedge \Gamma_{\mathsf{D}}/\Gamma_{\mathsf{F}} \xrightarrow{\ \psi\ } \widehat{H}^{-1}(G, Nrd_{\mathsf{D}_0}(\mathsf{D}_0^{\times})) \xrightarrow{\ \varphi\ } SK_1(\mathsf{D}) \longrightarrow 1.$$

(iv) *If* D *is semiramified, then* D_0 *is a field,* $G = \mathcal{G}(\mathsf{D}_0/\mathsf{F}_0) \cong \Gamma_{\mathsf{D}}/\Gamma_{\mathsf{F}}$, *and there is an exact sequence:*

$$\Gamma_{\mathsf{D}}/\Gamma_{\mathsf{F}} \wedge \Gamma_{\mathsf{D}}/\Gamma_{\mathsf{F}} \xrightarrow{\ \psi\ } \widehat{H}^{-1}(G, \mathsf{D}_0^{\times}) \xrightarrow{\ \varphi\ } SK_1(\mathsf{D}) \longrightarrow 1.$$

(v) *If* D *is decomposably semiramified, then* D_0 *is a field and the group* $G = \mathcal{G}(\mathsf{D}_0/\mathsf{F}_0) \cong \Gamma_{\mathsf{D}}/\Gamma_{\mathsf{F}}$, *and*

$$SK_1(\mathsf{D}) \cong \widehat{H}^{-1}(G, \mathsf{D}_0^{\times}).$$

In (iii) and (iv), the map ψ is defined by commutators of representatives in $\mathsf{D}^{\times}$, just as for κ in (11.2). The map φ is given as follows: Take any $b \in ker\,N_G$; then, $b = Nrd_{\mathsf{D}_0}(a)$ for some $a \in \mathsf{D}_0^{\times}$, and

$$\varphi\big(b\,I_G(Nrd_{\mathsf{D}_0}(\mathsf{D}_0^{\times}))\big) = a\,[\mathsf{D}^{\times}, \mathsf{D}^{\times}].$$

Proof. Let $\widetilde{N}$ be as in Th. 11.10.

(i) Suppose D is unramified over F. Then, $Z(\mathsf{D}_0) = \mathsf{F}_0$, so $G = \mathcal{G}(Z(\mathsf{D}_0)/\mathsf{F}_0)$ is trivial, as is $\Gamma_{\mathsf{D}}/\Gamma_{\mathsf{F}}$. Also, $\zeta = 1$, as $deg\,\mathsf{D}_0 = deg\,\mathsf{D}$. Since $\widehat{H}^{-1}(G, Nrd_{\mathsf{D}_0}(\mathsf{D}_0^{\times}))$ and $\mu_\zeta(\mathsf{F}_0) \cap \widetilde{N}(\mathsf{D}_0^{\times})$ are trivial, diagram (11.2) yields isomorphisms

$$SK_1(\mathsf{D}_0) \xrightarrow{\ \sim\ } ker\,\widetilde{N}/[\mathsf{D}_0^{\times}, \mathsf{D}^{\times}] \xrightarrow{\ \sim\ } \mathsf{D}^{(1)}/[\mathsf{D}_0^{\times}, \mathsf{D}^{\times}] \xrightarrow{\ \sim\ } SK_1(\mathsf{D}).$$

(ii) Suppose D is totally ramified over F. Then the top row of diagram (11.2) collapses as $\mathsf{D}_0 = \mathsf{F}_0$. Also, $\widetilde{N} = id_{\mathsf{D}_0}$, $\zeta = deg\,\mathsf{D} = n$, and the second row of the diagram becomes the exact sequence:

$$\Gamma_{\mathsf{D}}/\Gamma_{\mathsf{F}} \wedge \Gamma_{\mathsf{D}}/\Gamma_{\mathsf{F}} \xrightarrow{\kappa} \mu_n(\mathsf{F}_0) \longrightarrow SK_1(\mathsf{D}) \longrightarrow 1.$$

The map κ is the nondegenerate canonical pairing $\overline{c}_{\mathsf{D}}$ for D. By Th. 7.2 the image of κ is $\mu_e(\mathsf{F}_0)$ which has order e, where $e = exp(\Gamma_{\mathsf{D}}/\Gamma_{\mathsf{F}})$. Hence, $SK_1(\mathsf{D}) \cong \mu_n(\mathsf{F}_0)/\mu_e(\mathsf{F}_0)$. That $e = exp(\mathsf{D})$ as well was noted in Prop. 7.60.

(iii) Suppose D is inertially split and $SK_1(\mathsf{D}_0) = 1$. Then $\zeta = 1$ by Prop. 8.49 Hence, diagram (11.2) shows that

$$\widehat{H}^{-1}(G, Nrd_{\mathsf{D}_0}(\mathsf{D}_0^\times)) \cong ker\,\widetilde{N}/[\mathsf{D}_0^\times, \mathsf{D}^\times] = \mathsf{D}^{(1)}/[\mathsf{D}_0^\times, \mathsf{D}^\times].$$

By substituting $\widehat{H}^{-1}(G,Nrd_{\mathsf{D}_0}(\mathsf{D}_0^\times))$ into the exact lower row of diagram (11.2), we obtain the exact sequence of (iii). The map ψ is obtained from the κ of diagram (11.2) using the isomorphism above.

(iv) Suppose D is semiramified. Then, D is inertially split as noted in Def. 8.26. Hence, $ker\,\theta_{\mathsf{D}} = \Gamma_{\mathsf{F}}$ by Prop. 8.49, so $G \cong \Gamma_{\mathsf{D}}/\Gamma_{\mathsf{F}}$. Moreover, D_0 is a field, so $SK_1(\mathsf{D}_0) = 1$ and $Nrd_{\mathsf{D}_0}(\mathsf{D}_0^\times) = \mathsf{D}_0^\times$. Thus, the exact sequence of (iii) yields the exact sequence of (iv).

(v) Suppose D is decomposably semiramified. Then, by Prop. 8.33 D has a maximal graded subfield T with T totally ramified over F and $\Gamma_{\mathsf{D}} = \Gamma_{\mathsf{T}}$. Hence, for any $\gamma, \delta \in \Gamma_{\mathsf{D}}$, nonzero representatives $d_\gamma \in \mathsf{D}_\gamma$ and $d_\delta \in \mathsf{D}_\delta$ can be chosen to lie in T. Since they commute, we have

$$\psi\big((\gamma + \Gamma_{\mathsf{F}}) \wedge (\delta + \Gamma_{\mathsf{F}})\big) = [d_\gamma, d_\delta]\, I_G(\mathsf{D}_0^\times) = 1.$$

Thus, ψ is trivial, so the exact sequence of (iv) yields the isomorphism asserted in (v). □

Example 11.13. Let E be a division algebra finite-dimensional over its center k, let X be an indeterminate, and let $\mathsf{D} = E[X, X^{-1}]$ with its usual $\mathbb{Z}$-grading. Then, D is a graded division ring unramified over its center $k[X, X^{-1}]$ with $\mathsf{D}_0 = E$. By Cor. (11.12)(i), $SK_1(\mathsf{D}) \cong SK_1(E)$.

Example 11.14. Let $n_1, \ldots, n_r$ be integers with each $n_i \geq 2$, and let $n = n_1 \ldots n_r$ and $m = lcm(n_1, \ldots, n_r)$. Let k be any field containing a primitive m-th root of unity, and let $\mathcal{T} = \mathcal{T}(k; n_1, \ldots, n_r)$ be the tame totally ramified division algebra of (9.5) with center $F = k((x_1))((y_1)) \ldots ((x_r))((y_r))$. Let $\mathsf{T} = \mathsf{gr}(\mathcal{T})$, which is a graded division algebra totally ramified over its center $\mathsf{F} = \mathsf{gr}(F) = k[x_1, x_1^{-1}, y_1, y_1^{-1}, \ldots, x_r, x_r^{-1}, y_r, y_r^{-1}]$, and is expressible as a tensor product of graded symbol algebras over F analogous to (9.5). Then, $ind\,\mathsf{T} = ind\,\mathcal{T} = n$ and $exp\,\mathsf{T} = exp\,\mathcal{T} = m$ (see Prop. 9.8); so Cor. 11.12(ii) shows that

$$SK_1(\mathsf{T}) \cong \mu_n(k)/\mu_m(k),$$

which is a cyclic group of order $|\mu_n(k)|/m$. Thus, by suitable choice of k and the n_i, any finite cyclic group is realizable as $SK_1(\mathsf{T})$ for some totally ramified graded division algebra T.

For the rest of this section we take a closer look at the semiramified cases (iv) and (v) of Cor. 11.12 when the group $G = \mathcal{G}(\mathsf{D}_0/\mathsf{F}_0) = \Gamma_\mathsf{D}/\Gamma_\mathsf{F}$ has low torsion rank: When G is cyclic, we have $\widehat{H}^{-1}(G, \mathsf{D}_0^\times) \cong H^1(G, \mathsf{D}_0^\times) = 1$ by Hilbert's Theorem 90 (see Draxl [63, p. 35] or Pierce [178, p. 312]). Also, $\Gamma_\mathsf{D}/\Gamma_\mathsf{F} \wedge \Gamma_\mathsf{D}/\Gamma_\mathsf{F} \cong G \wedge G = 1$. Hence, $SK_1(\mathsf{D}) = 1$. This holds in particular whenever $\Gamma_\mathsf{D} = \mathbb{Z}$.

Suppose next that G is bicyclic, i.e., of torsion rank 2. Then Prop. 8.23 yields a Brauer group interpretation of the cohomology group in (iv) and (v) by setting up an isomorphism depending on a base $\boldsymbol{\sigma}$ of G:

$$\overline{\upsilon}_{\boldsymbol{\sigma}} \colon \mathit{Br}(\mathsf{D}_0/\mathsf{F}_0)/\mathit{Dec}(\mathsf{D}_0/\mathsf{F}_0) \cong \widehat{H}^{-1}(G, \mathsf{D}_0^\times).$$

With this isomorphism the cohomology group (hence also $SK_1(\mathsf{D})$) is explicitly computable and visibly nontrivial in many cases, e.g. when F_0 is a global field. One can use this to produce examples of semiramified D where $SK_1(\mathsf{D})$ has arbitrarily large rank as a finite abelian group, or is infinite. By contrast, when D is totally ramified over F, Cor. 11.12(ii) shows that $SK_1(\mathsf{D})$ is always a cyclic group.

Theorem 11.15. *Let D be a semiramified graded division algebra with center F, and suppose that $\mathcal{G}(\mathsf{D}_0/\mathsf{F}_0)$ is bicyclic. Let A be a central simple F_0-algebra representing the specialization coset of D:*

$$\mathit{sp}(\mathsf{D}) = [A] + \mathit{Dec}(\mathsf{D}_0/\mathsf{F}_0) \in \mathit{Br}(\mathsf{D}_0/\mathsf{F}_0)/\mathit{Dec}(\mathsf{D}_0/\mathsf{F}_0).$$

Then,

$$SK_1(\mathsf{D}) \cong \mathit{Br}(\mathsf{D}_0/\mathsf{F}_0)/\big(\langle [A]\rangle + \mathit{Dec}(\mathsf{D}_0/\mathsf{F}_0)\big).$$

In the particular case where D is decomposably semiramified, the algebra A is split in view of Cor. 8.53, hence the result above takes the form

$$SK_1(\mathsf{D}) \cong \mathit{Br}(\mathsf{D}_0/\mathsf{F}_0)/\mathit{Dec}(\mathsf{D}_0/\mathsf{F}_0).$$

Proof. Let $G = \mathcal{G}(\mathsf{D}_0/\mathsf{F}_0)$. Choose $z_1, z_2 \in \mathsf{D}^\times$ so that $\theta_\mathsf{D}(\mathit{deg}\, z_1), \theta_\mathsf{D}(\mathit{deg}\, z_2)$ is a base $\boldsymbol{\sigma}$ of G; equivalently, under the isomorphism induced by θ_D, choose z_1, z_2 so that $\mathit{deg}\, z_1 + \Gamma_\mathsf{F}$, $\mathit{deg}\, z_2 + \Gamma_\mathsf{F}$ is a base of $\Gamma_\mathsf{D}/\Gamma_\mathsf{F}$. Then

$$\Gamma_\mathsf{D}/\Gamma_\mathsf{F} \wedge \Gamma_\mathsf{D}/\Gamma_\mathsf{F} = \big\langle (\mathit{deg}\, z_1 + \Gamma_\mathsf{F}) \wedge (\mathit{deg}\, z_2 + \Gamma_\mathsf{F})\big\rangle,$$

and the map ψ of Cor. 11.12(iv) carries $(\mathit{deg}\, z_1 + \Gamma_\mathsf{F}) \wedge (\mathit{deg}\, z_2 + \Gamma_\mathsf{F})$ to $z_1 z_2 z_1^{-1} z_2^{-1} \cdot I_G \mathsf{D}_0^\times \in \widehat{H}^{-1}(G, \mathsf{D}_0^\times)$. But Prop. 8.55 shows that

$$z_1 z_2 z_1^{-1} z_2^{-1} \cdot I_G \mathsf{D}_0^\times = \overline{\upsilon}_{\boldsymbol{\sigma}}\big(\mathit{sp}(D)\big),$$

so Cor. 11.12(iv) yields

$$SK_1(D) \cong \widehat{H}^{-1}(G, \mathsf{D}_0^\times)/\langle \overline{v}_\sigma(sp(D))\rangle.$$

If $sp(D) = [A] + Dec(\mathsf{D}_0/\mathsf{F}_0)$, the isomorphism $\overline{v}_\sigma : Br(\mathsf{D}_0/\mathsf{F}_0)/Dec(\mathsf{D}_0/\mathsf{F}_0) \xrightarrow{\sim} \widehat{H}^{-1}(G, \mathsf{D}_0^\times)$ induces an isomorphism

$$Br(\mathsf{D}_0/\mathsf{F}_0)/(\langle [A]\rangle + Dec(\mathsf{D}_0/\mathsf{F}_0)) \xrightarrow{\sim} \widehat{H}^{-1}(G, \mathsf{D}_0^\times)/\langle \overline{v}_\sigma(sp(D))\rangle.$$

The theorem follows. □

11.2 SK_1 of division algebras over Henselian fields

Throughout this section, except for Lemma 11.22, F is a field with Henselian valuation v. For every field or division algebra E finite-dimensional over F, the valuation on E is understood to be the unique extension of v to E. We will show that for any division algebra D with center F and tame over F, the canonical map $SK_1(D) \to SK_1(\mathsf{gr}(D))$ is an isomorphism; see Th. 11.21. Thus, all the preceding results for SK_1 of graded division algebras yield corresponding results for $SK_1(D)$.

For any division algebra D finite-dimensional over F, recall that $\mathfrak{m}_D$ denotes the maximal ideal of the valuation ring $\mathcal{O}_D$ of D. There is a short exact sequence of abelian groups

$$1 \longrightarrow 1 + \mathfrak{m}_D \longrightarrow D^\times \xrightarrow{\rho_D} \mathsf{gr}(D)^\times \longrightarrow 1,$$

where $\rho_D(a) = \widetilde{a}$. We will use the right part of the sequence to obtain a map $\widehat{\rho}_D : SK_1(D) \to SK_1(\mathsf{gr}(D))$. But to show that this map is an isomorphism we need to prove more delicate information about the norm behavior of the left term. We prepare for this with two lemmas and corollaries on field norms and reduced norms.

Lemma 11.16. *Let K be a defectless finite-degree field extension of F, and let $a \in K^\times$ with $v(a) = \delta$. Then $\mathrm{Pchar}_{K/F,a}$ is a δ-uniform polynomial in $F[X]$, and $\mathrm{Pchar}_{K/F,a}^{(\delta)} = \mathrm{Pchar}_{\mathsf{gr}(K)/\mathsf{gr}(F),\widetilde{a}}$ in $\mathsf{gr}(F)[X]$. Hence,*

$$\widetilde{N_{K/F}(a)} = N_{\mathsf{gr}(K)/\mathsf{gr}(F)}(\widetilde{a})$$

and

$$tr_{\mathsf{gr}(K)/\mathsf{gr}(F)}(\widetilde{a}) = \begin{cases} \widetilde{tr_{K/F}(a)} & \text{if } v(tr_{K/F}(a)) = \delta, \\ 0 & \text{if } v(tr_{K/F}(a)) > \delta. \end{cases}$$

Proof. Let $n = [K{:}F]$ and let $m = [F(a){:}F]$. Let p_a (resp. $p_{\widetilde{a}}$) be the minimal polynomial of a over F (resp. $\widetilde{a}$ over $q(\mathsf{gr}(F))$). Because v is Henselian, by Th. 5.49 the irreducible polynomial p_a in $F[X]$ is λ-uniform for some $\lambda \in \mathbb{H}(\Gamma_F)$; then $\lambda = v(a) = \delta$, as $p_a(a) = 0$. By Th. 5.49,

${p_a}^{(\delta)} = h^\ell$ for some monic irreducible δ-homogenizable polynomial h in $\mathsf{gr}(F)[X]$; this h is also irreducible in $q(\mathsf{gr}(F))[X]$ by Prop. 5.6. Since $h(\widetilde{a})^\ell = {p_a}^{(\delta)}(\widetilde{a}) = 0$ by Lemma 5.46, we must have $h(\widetilde{a}) = 0$. Hence, $h = p_{\widetilde{a}}$, so $[\mathsf{gr}(F)(\widetilde{a})\colon \mathsf{gr}(F)] = deg_X(h) = m/\ell$. We have $Pchar_{K/F,a} = {p_a}^{n/m}$. Hence, $Pchar_{K/F,a}$ is a δ-uniform polynomial and (cf. Lemma 5.47)

$$Pchar_{K/F,a}{}^{(\delta)} = p_a{}^{(\delta)\,n/m} = p_{\widetilde{a}}^{n\ell/m} = p_{\widetilde{a}}{}^{n/[\mathsf{gr}(F)(\widetilde{a})\colon \mathsf{gr}(F)]} = Pchar_{\mathsf{gr}(K)/\,\mathsf{gr}(F),\widetilde{a}}.$$

If c is the constant term of $Pchar_{K/F,a}$, then $\widetilde{c}$ is the constant term of $Pchar_{K/F,a}{}^{(\delta)}$. Hence,

$$\widetilde{N_{K/F}(a)} = \widetilde{(-1)^n c} = (-1)^n \widetilde{c} = N_{\mathsf{gr}(K)/\,\mathsf{gr}(F)}(\widetilde{a}).$$

The formula for $tr_{\mathsf{gr}(K)/\,\mathsf{gr}(F)}(\widetilde{a})$ follows similarly by considering the X^{n-1}-coefficients of these polynomials. □

Corollary 11.17. *Let D be a division algebra with center F and tame over F. Take any $a \in D^\times$, and let $\delta = v(a)$. Then, $Prd_{D,a}$ is a δ-uniform polynomial in $F[X]$ and $Prd_{D,a}{}^{(\delta)} = Prd_{\mathsf{gr}(D),\widetilde{a}}$ in $\mathsf{gr}(F)[X]$. Hence,*

$$\widetilde{Nrd_D(a)} = Nrd_{\mathsf{gr}(D)}(\widetilde{a}) \ \ \text{and} \ \ Trd_{\mathsf{gr}(D)}(\widetilde{a}) = \begin{cases} \widetilde{Trd_D(a)} & \text{if } v(Trd_D(a)) = \delta, \\ 0 & \text{if } v(Trd_D(a)) > \delta. \end{cases}$$

Proof. Since D is tame over F, we have $Z(\mathsf{gr}(D)) = \mathsf{gr}(F)$ and D is defectless over F. Let $n = deg\, D = deg(\mathsf{gr}(D))$. Since D is defectless over F, its subfield $F(a)$ is also defectless over F. Let $m = [F(a){:}F] = [\mathsf{gr}(F(a))\colon \mathsf{gr}(F)]$. By the ungraded version of Prop. (11.6)(v), we have $Prd_{D,a} = Pchar_{F(a)/F,a}{}^{n/m}$. Since $Pchar_{F(a)/F,a}$ is a δ-uniform polynomial by the preceding lemma, it follows that $Prd_{D,a}$ is also δ-uniform. Moreover, by Lemma 5.47, Lemma 11.16, and Prop. 11.6(v), we have

$$Prd_{D,a}{}^{(\delta)} = Pchar_{F(a)/F}{}^{(\delta)\ n/m} = Pchar_{\mathsf{gr}(F(a))/\,\mathsf{gr}(F),\widetilde{a}}{}^{n/m} = Prd_{\mathsf{gr}(D),\widetilde{a}}.$$

The reduced norm and trace formulas follow by comparing coefficients of these polynomials. □

Lemma 11.18. *Let K be a finite-degree tamely ramified field extension of F. Then, $N_{K/F}(1 + \mathfrak{m}_K) = 1 + \mathfrak{m}_F$.*

Proof. Take any $a \in 1 + \mathfrak{m}_K$. By Lemma 11.16,

$$\widetilde{N_{K/F}(a)} = N_{\mathsf{gr}(F)/\,\mathsf{gr}(K)}(\widetilde{a}) = N_{\mathsf{gr}(K)/\,\mathsf{gr}(F)}(1) = 1.$$

Hence, $N_{K/F}(1 + \mathfrak{m}_K) \subseteq 1 + \mathfrak{m}_F$.

For the reverse inclusion, assume first that there is a nonzero homogeneous $s \in \mathsf{gr}(K)$ with $\mathsf{gr}(K) = \mathsf{gr}(F)[s]$, say $deg\, s = \delta$. Let g be the minimal polynomial of s over $q(\mathsf{gr}(F))$; say $g = X^n + t_{n-1}X^{n-1} + \ldots + t_0$, where $n = [\mathsf{gr}(K)\colon \mathsf{gr}(F)] = [K{:}F]$. Since K is tamely ramified over F, the graded field $\mathsf{gr}(K)$ is tame over $\mathsf{gr}(F)$, so $q(\mathsf{gr}(K))$ is separable over $q(\mathsf{gr}(F))$ by Prop. 5.19. Hence, s is a simple root of g. Also, g is a δ-homogenizable polynomial in $\mathsf{gr}(F)[X]$, by Prop. 5.7. For $i = 0, 1, \ldots, n-1$, choose $c_i \in F^\times$ with $\widetilde{c_i} = t_i$ whenever $t_i \neq 0$, and set $c_i = 0$ whenever $t_i = 0$. Let $f = X^n + c_{n-1}X^{n-1} + \ldots + c_0 \in F[X]$. Then, f is a δ-uniform polynomial with $f^{(\delta)} = g$, and f is irreducible in $F[X]$ by Lemma 5.47 since $f^{(\delta)}$ is irreducible in $\mathsf{gr}(F)[X]$. Since $f^{(\delta)}$ has the simple root s in $\mathsf{gr}(K)$ and K is Henselian, by Th. 5.49 f has a root b in K with $\widetilde{b} = s$. The irreducible polynomial h must be the minimal polynomial of b over F. Now, take any $r \in 1 + \mathfrak{m}_F$, and let $h = X^n + c_{n-1}X^{n-1} + \ldots + c_1X + rc_0 \in F[X]$. This h differs from f only in its constant term, and $\widetilde{rc_0} = \widetilde{c_0}$. Hence, h, like f, is δ-uniform in $F[X]$, and $h^{(\delta)} = f^{(\delta)} = g$. Therefore, just as for f, the polynomial h has a root $d \in K$ with $\widetilde{d} = s = \widetilde{b}$, and h is the minimal polynomial of d over F. Thus, $\widetilde{db^{-1}} = ss^{-1} = 1$, i.e., $db^{-1} \in 1 + \mathfrak{m}_K$. Also,

$$N_{K/F}(db^{-1}) = N_{K/F}(d)N_{K/F}(b)^{-1} = \big((-1)^n rc_0\big)\big((-1)^n c_0\big)^{-1} = r.$$

Thus, $N_{K/F}(1 + \mathfrak{m}_K)$ is all of $1 + \mathfrak{m}_F$ whenever $\mathsf{gr}(K) = \mathsf{gr}(F)[s]$ with s homogeneous.

For the general case, we argue by induction on $[K{:}F]$. If $[K{:}F] = 1$, the result is clear. If $[K{:}F] > 1$, then $\mathsf{gr}(K) \supsetneqq \mathsf{gr}(F)$ since K is defectless over F. So, there is a nonzero homogeneous $s \in \mathsf{gr}(K) \setminus \mathsf{gr}(F)$. Let $\mathsf{M} = \mathsf{gr}(F)[s]$, which is a graded field tame over $\mathsf{gr}(F)$, since $\mathsf{gr}(K)$ is tame over $\mathsf{gr}(F)$. Let L be the tame lift of M over F in K, which exists by Prop. 8.3(iii). Then $\mathsf{gr}(L) = \mathsf{M} \supsetneqq \mathsf{gr}(F)$ and K is tame over L and L is tame over F, since K is tame over F. By the preceding paragraph, we have $N_{L/F}(1+\mathfrak{m}_L) = 1+\mathfrak{m}_F$, and by induction (since $[K{:}L] < [K{:}F]$), $N_{K/L}(1+\mathfrak{m}_K) = 1+\mathfrak{m}_L$. Therefore,

$$N_{K/F}(1 + \mathfrak{m}_K) = N_{L/F}\big(N_{K/L}(1 + \mathfrak{m}_K)\big) = N_{L/F}(1 + \mathfrak{m}_L) = 1 + \mathfrak{m}_F,$$

as desired. □

Corollary 11.19. *Let D be a division algebra tame over its center F. Then $Nrd_D(1 + \mathfrak{m}_D) = 1 + \mathfrak{m}_F$.*

Proof. If $c \in 1 + \mathfrak{m}_D$, then by Cor. 11.17,

$$\widetilde{Nrd_D(c)} = Nrd_{\mathsf{gr}(D)}(\widetilde{c}) = Nrd_{\mathsf{gr}(D)}(1) = 1.$$

Thus, $Nrd_D(1 + \mathfrak{m}_D) \subseteq 1 + \mathfrak{m}_F$.

For the reverse inclusion, note that D has a maximal subfield tamely ramified over F. For $\mathsf{gr}(D)$ contains a maximal graded subfield L tame over $Z(\mathsf{gr}(D)) = \mathsf{gr}(F)$ by Prop. 6.18. Then by Prop. 8.2 and Th. 8.9 D contains a tame lift K of L over F, and K is a maximal subfield of D tamely ramified over F. Using Lemma 11.18, we have

$$Nrd_D(1 + \mathfrak{m}_D) \supseteq Nrd_D(1 + \mathfrak{m}_K) = N_{K/F}(1 + \mathfrak{m}_K) = 1 + \mathfrak{m}_F,$$

proving the corollary. □

Proposition 11.20. *Let D be a division algebra tame over its center F. Then, there is a short exact sequence*

$$1 \to (D^{(1)} \cap (1+\mathfrak{m}_D))/([D^\times, D^\times] \cap (1+\mathfrak{m}_D)) \to SK_1(D) \xrightarrow{\widehat{\rho}_D} SK_1(\mathsf{gr}(D)) \to 1.$$

Proof. The group homomorphism $\rho_D \colon D^\times \to \mathsf{gr}(D^\times)$ given by $a \mapsto \widetilde{a}$ clearly sends $[D^\times, D^\times]$ into $[\mathsf{gr}(D)^\times, \mathsf{gr}(D)^\times]$; by Cor. 11.17 it also sends $D^{(1)}$ into $\mathsf{gr}(D)^{(1)}$. Consider now the diagram with exact rows:

$$\begin{array}{ccccccccc} 1 & \longrightarrow & 1+\mathfrak{m}_D & \longrightarrow & D^\times & \xrightarrow{\rho_D} & \mathsf{gr}(D)^\times & \longrightarrow & 1 \\ & & \big\downarrow{\scriptstyle Nrd_D} & & \big\downarrow{\scriptstyle Nrd_D} & & \big\downarrow{\scriptstyle Nrd_{\mathsf{gr}(D)}} & & \\ 1 & \longrightarrow & 1+\mathfrak{m}_F & \longrightarrow & F^\times & \xrightarrow{\rho_F} & \mathsf{gr}(F)^\times & \longrightarrow & 1 \end{array}$$

The diagram is commutative by Cor. 11.17. Since the left vertical map is surjective by Cor. 11.19, the Snake Lemma shows that $ker(Nrd_D) = D^{(1)}$ maps onto $ker(Nrd_{\mathsf{gr}(D)}) = \mathsf{gr}(D)^{(1)}$. Now consider the commutative diagram:

$$\begin{array}{ccccccccc} 1 & \longrightarrow & [D^\times, D^\times] \cap (1+\mathfrak{m}_D) & \longrightarrow & [D^\times, D^\times] & \xrightarrow{\rho_D} & [\mathsf{gr}(D)^\times, \mathsf{gr}(D)^\times] & \longrightarrow & 1 \\ & & \big\downarrow & & \big\downarrow & & \big\downarrow & & \\ 1 & \longrightarrow & D^{(1)} \cap (1+\mathfrak{m}_D) & \longrightarrow & D^{(1)} & \xrightarrow{\rho_D} & \mathsf{gr}(D)^{(1)} & \longrightarrow & 1 \end{array}$$

The top row is clearly exact, and the bottom row is exact by the surjectivity just proved. Since the right vertical map is injective, by the Snake Lemma there is a short exact sequence of cokernels of the vertical maps, which is the exact sequence of the proposition. □

Theorem 11.21. *Let D be a division algebra tame over its (Henselian) center F. Then,*

(i) (*"Congruence Theorem"*) $D^{(1)} \cap (1+\mathfrak{m}_D) \subseteq [D^\times, D^\times]$;
(ii) *the map $\widehat{\rho}_D \colon SK_1(D) \to SK_1(\mathsf{gr}(D))$ is an isomorphism.*

The proof of the theorem will use the following general lemma:

Lemma 11.22. *Let F be any field, and let D be a division algebra with center F; let L be a finite-degree field extension of F, and $B = D \otimes_F L$. For $a \in D^\times$, if $a \otimes 1 \in [B^\times, B^\times]$ then $a^{[L:F]} \in [D^\times, D^\times]$.*

Proof. (Cf. Draxl [63, Lemma 1, p. 156].) The regular representation yields an embedding of L in $End_F(L) \cong M_\ell(F)$, where $\ell = [L{:}F]$. We have a succession of group homomorphisms:

$$B^\times \hookrightarrow (D \otimes_F M_\ell(F))^\times \xrightarrow{\sim} M_\ell(D)^\times \xrightarrow{det} D^\times/[D^\times, D^\times],$$

where the left map is induced by the inclusion $L \hookrightarrow M_\ell(F)$ and the right map is the Dieudonné determinant (cf. Draxl [63, §20]). Under these maps, $a \otimes 1$ in $B^\times$ maps to $\begin{pmatrix} a & & 0 \\ & \ddots & \\ 0 & & a \end{pmatrix}$ in $M_\ell(D)^\times$ and then to $a^\ell[D^\times, D^\times]$ in $D^\times/[D^\times, D^\times]$. If $a \otimes 1 \in [B^\times, B^\times]$, then its image in the abelian group $D^\times/[D^\times, D^\times]$ is trivial, i.e., $a^\ell \in [D^\times, D^\times]$. □

Proof of Th. 11.21. Observe that Prop. 11.20 shows that assertions (i) and (ii) of the theorem are equivalent. We will prove (i) in two special cases, and use this to prove (ii) in general, thereby also obtaining (i) in general.

Case I. Suppose $char\,\overline{F} \nmid deg\,D$. Let $n = deg\,D$. Take any $a \in D^{(1)} \cap (1 + \mathfrak{m}_D)$, and let $K = F(a) \subseteq D$. Let $f = X^n - a \in \mathcal{O}_K[X]$. Its image in $\overline{K}[X]$ is $\overline{f} = X^n - 1$, which has 1 as a simple root since $char\,\overline{F} \nmid n$. Since the valuation on K is Henselian, f has a root $b \in \mathcal{O}_K$ with $\overline{b} = 1$. That is, $b^n = a$ and $b \in 1 + \mathfrak{m}_K$. Let $e = Nrd_D(b)$. Then, $e = 1 + m$ for some $m \in \mathfrak{m}_F$ by Cor. 11.19. Thus,

$$(1+m)^n = Nrd_D(b^n) = Nrd_D(a) = 1.$$

The binomial expansion gives $(1+m)^n = 1 + nm + dm^2$ for some $d \in \mathcal{O}_F$; hence $nm = -dm^2$. If $m \neq 0$, then as $v(n) = 0$, we have the contradiction $0 < v(m) = v(-bm^2) \geq 2v(m)$. Hence, $m = 0$, i.e., $Nrd_D(b) = 1$.

By the Wedderburn Factorization Theorem (see, e.g., Lam [121, p. 265, Th. 16.9]), the minimal polynomial p_b of b over F has a factorization in $D[X]$,

$$p_b = (X - c_1 b c_1^{-1}) \dots (X - c_\ell b c_\ell^{-1})$$

for some $c_1, \dots, c_\ell \in D^\times$, where $\ell = [F(b){:}F] \mid n$. Thus, modulo $[D^\times, D^\times]$,

$$\begin{aligned} 1 = Nrd_D(b) &= N_{F(b)/F}(b)^{n/\ell} \\ &= (-1)^n p_b(0)^{n/\ell} = \big((c_1 b c_1^{-1}) \dots (c_\ell b c_\ell^{-1})\big)^{n/\ell} \\ &\equiv b^n \big((c_1 b c_1^{-1} b^{-1}) \dots (c_\ell b c_\ell^{-1} b^{-1})\big)^{n/\ell} \equiv b^n = a. \end{aligned}$$

Thus, $a \in [D^\times, D^\times]$, showing that (i) holds in Case I.

Case II. Suppose D is inertially split and $deg\,D = p^m$ for some prime number p. The argument is by induction on $deg\,D$. Since the result is clear when $D = F$, we may assume that $deg\,D > 1$. Take any $a \in D^{(1)} \cap 1 + \mathfrak{m}_D$. Since D is inertially split, by Prop. 8.64 D has a maximal subfield K which is inertial over F. We first reduce to the case that $a \in K$. Because $\overline{K}$ is separable over $\overline{F}$, there is an $s \in \overline{K}^\times$ with $\overline{K} = \overline{F}(s)$. Take any $b \in \mathcal{O}_K^\times$ with $\overline{b} = s$. Then $\overline{ab} = \overline{a}\,\overline{b} = s$. Let f be the minimal polynomial of ab over F. Then,

$$deg\,f = [F(ab){:}F] \leq deg\,D = [K{:}F] = [\overline{K}{:}\overline{F}] = [\overline{F}(s){:}\overline{F}]. \quad (11.3)$$

Also, $f \in \mathcal{O}_F[X]$, as $v(ab) \geq 0$; let $\overline{f}$ be the image of f in $\overline{F}[X]$. Since $f(ab) = 0$, we have $0 = \overline{f}(\overline{ab}) = \overline{f}(s)$ in $\overline{D}$. The inequality in (11.3) shows that $\overline{f}$ must be the minimal polynomial of s over $\overline{F}$. Hence, s is a simple root of $\overline{f}$, as $\overline{K}$ is separable over $\overline{F}$. Because the valuation on K is Henselian, f has a root c in $\mathcal{O}_K$ with $\overline{c} = s$. Since ab and c have the same minimal polynomial f over F there is an F-homomorphism $\eta\colon F(c) \to F(ab)$ with $\eta(c) = ab$. So, by the Skolem–Noether Theorem there is a $d \in D^\times$ with $ab = dcd^{-1}$. Let $a' = cb^{-1} = cdc^{-1}d^{-1}a \in K$. Then, $a' \equiv a \,(mod\ [D^\times, D^\times])$ and $\overline{a'} = \overline{s}\,\overline{s}^{-1} = 1$, so $a' \in 1 + \mathfrak{m}_D$. Therefore, by replacing a by a' we may assume that $a \in K$.

Now assume that K contains a field $M \supsetneqq F$ with M cyclic Galois over F. Let $C = C_D(M)$. Thus, $F \subseteq M \subseteq K \subseteq C \subseteq D$ and $M = Z(C)$ by the Double Centralizer Theorem; also, K is a maximal subfield of C. Let $r = Nrd_C(a) = N_{K/M}(a)$. Since $a \in 1+\mathfrak{m}_C$, Cor. 11.19 shows that $r \in 1+\mathfrak{m}_M$. Also,

$$N_{M/F}(r) \;=\; N_{K/F}(a) \;=\; Nrd_D(a) \;=\; 1.$$

By Hilbert's Theorem 90, $r = t\sigma(t)^{-1}$ for some $t \in M^\times$, where σ is a generator of $\mathcal{G}(M/F)$. Because M is inertial (hence tamely ramified) and Galois over F, by Cor. 5.52 $\mathsf{gr}(M)$ is Galois over $\mathsf{gr}(F)$ with $\mathcal{G}(\mathsf{gr}(M)/\mathsf{gr}(F)) = \langle\widetilde{\sigma}\rangle$ where $\widetilde{\sigma}$ is the graded automorphism of $\mathsf{gr}(M)$ induced by σ. Since $\widetilde{t}\,\widetilde{\sigma}(\widetilde{t}\,)^{-1} = \widetilde{t\sigma(t)^{-1}} = \widetilde{r} = 1$ in $\mathsf{gr}(M)$, we have $\widetilde{\sigma}(\widetilde{t}) = \widetilde{t}$; hence, $\widetilde{t}$ is a homogeneous element in $\mathsf{gr}(M)^{\widetilde{\sigma}} = \mathsf{gr}(F)$. Thus, there is a $y \in F^\times$ with $\widetilde{y} = \widetilde{t}$ in $\mathsf{gr}(M)^\times$. By replacing t by ty^{-1}, we may assume that $t \in 1 + \mathcal{O}_M$, while preserving that $t\sigma(t)^{-1} = r$. Then, by Cor. 11.19 there is an $e \in 1+\mathfrak{m}_C$ with $Nrd_C(e) = t$. By the Skolem–Noether Theorem, there is an $x \in D^\times$ with $int(x)|_M = \sigma$. Then, $int(x)$ maps $C = C_D(M)$ to itself, since it maps M to itself. Let

$$z = xex^{-1}e^{-1} \qquad \text{and} \qquad a'' = az.$$

Clearly, $z \in [D^\times, D^\times]$. Also, $e \in 1 + \mathfrak{m}_C$ implies $e^{-1} \in 1 + \mathfrak{m}_C$ and $xex^{-1} \in 1 + \mathfrak{m}_C$, as $int(x)$ preserves the valuation and sends C to C. Hence, $z \in 1+\mathfrak{m}_C$ and $a'' \in C$. Since $int(x)$ is a (non-inner) automorphism of C, we have by Lemma 11.9,

$$Nrd_C(xex^{-1}) \;=\; x\,Nrd_C(e)x^{-1} \;=\; xtx^{-1} \;=\; \sigma(t).$$

Thus,

$$\begin{aligned} Nrd_C(a'') \;&=\; Nrd_C(a)\,Nrd_C(z) \\ &= r\,Nrd_C(xex^{-1})\,Nrd_C(e^{-1}) \;=\; r\sigma(t)t^{-1} \;=\; 1. \end{aligned}$$

Hence, $a'' \in C^{(1)} \cap (1+\mathfrak{m}_C)$. Since $deg\, C$ is a power of p with $deg\, C < deg\, D$, it follows by induction that $a'' \in [C^\times, C^\times]$. Hence, $a = a''z^{-1} \in [D^\times, D^\times]$, as desired.

We now drop the assumption that K contains a proper cyclic Galois extension of F. Let N be the normal closure of K over F. Then, N lies in the inertial closure F_{in} of F, which is inertial and Galois over F (cf. Th. A.23 in Appendix A). So, N is inertial and Galois over F. Let $L = N^H$, where H is a p-Sylow subgroup of $\mathcal{G}(N/F)$; let $\ell = [L{:}F]$, which is prime to p. Let $D' = D \otimes_F L$ and $K' = K \otimes_F L$. Since ℓ is prime to $deg\, D$, this D' is a division algebra with center L and $deg\, D' = deg\, D$. The field K' is a maximal subfield of D, and we identify K' with its isomorphic copy $K \cdot L$ in N. Thus, K' is inertial over F, so inertial over L, and D' is inertially split. Take any maximal proper subgroup S of the p-group $H = \mathcal{G}(N/L)$ such that $S \supseteq \mathcal{G}(N/K')$. Then S is a normal subgroup of H of index p, so the fixed field $M = N^S$ is (cyclic) Galois over L of degree p with $L \subseteq M \subseteq K'$. By the argument of the preceding paragraph, condition (i) holds for D'. For our $a \in D^{(1)} \cap (1+\mathcal{O}_D)$, its image $a \otimes 1$ in $D \otimes_F L = D'$ satisfies $Nrd_{D'}(a \otimes 1) = Nrd_D(a) = 1$ and $a \otimes 1 \in 1 + \mathfrak{m}_{D'}$ since the valuation on D' extends the valuation on D. By (i) for D' it follows that $a \otimes 1 \in [D'^{\times}, D'^{\times}]$. Hence, by Lemma 11.22 $a^{\ell} \in [D^{\times}, D^{\times}]$. But we also have $a^{\deg D} \in [D^{\times}, D^{\times}]$, since $SK_1(D)$ is $\deg D$-torsion by Prop. 11.2. Since $gcd(\ell, \deg D) = 1$, it follows that $a \in [D^{\times}, D^{\times}]$. Thus, property (i) holds for D in Case II.

Note that by Prop. 11.20, (ii) of the theorem holds for any D covered by Case I or Case II, since (i) holds in those cases.

We now prove (ii) for any D tamely ramified over F. Let $\overline{p} = char\,\overline{D}$. If $\overline{p} = 0$, or if $\overline{p} \neq 0$ and $\overline{p} \nmid \deg D$, then Case I above applies to D; hence, $\widehat{\rho}_D$ is an isomorphism, and we are done. Thus, we may assume that $\overline{p} \neq 0$ and $\overline{p} \mid \deg D$. Using the primary decomposition of D, write $D = P \otimes_F Q$ where $\deg P$ is a power of $\overline{p}$ and $\deg Q$ is prime to $\overline{p}$. Then, P and Q are each tame over F, since $Br_{tr}(F)$ is a group and P and Q are Brauer-equivalent to tensor powers of D. Also, P is inertially split, by Prop. 6.63. Hence, Case I shows that $\widehat{\rho}_Q$ is an isomorphism, and Case II shows that $\widehat{\rho}_P$ is an isomorphism.

We will prove the needed injectivity of $\widehat{\rho}_D$ by passage to $q(\mathsf{gr}(D))$. To simplify notation, write $q\mathsf{g}(D)$ for $q(\mathsf{gr}(D))$. Since $D = P \otimes_F Q$, with P and Q each tame over F, we have $v_P \otimes v_Q$ is a gauge on $P \otimes_F Q$ by Prop. 6.49(i) and $\mathsf{gr}_{v_P \otimes v_Q}(D) = \mathsf{gr}(P) \otimes_{\mathsf{gr}(F)} \mathsf{gr}(Q)$ by Prop. 3.27, and the tensor product gauge $v_P \otimes v_Q$ on D coincides with its valuation v_D by Prop. 4.33(i). Therefore, $q\mathsf{g}(D) = q\mathsf{g}(P) \otimes_{q\mathsf{g}(F)} q\mathsf{g}(Q)$, and $q\mathsf{g}(P)$ is the $\overline{p}$-primary component of $q\mathsf{g}(D)$, as $\deg(q\mathsf{g}(P)) = \deg P$.

Since P is a central simple F-subalgebra of D, for any $a \in P$ we have $Nrd_D(a) = Nrd_P(a)^{\deg D/\deg P}$, by the ungraded version of Prop. 11.6(vi). Therefore, $P^{(1)} \subseteq D^{(1)}$; hence, the inclusion $P \hookrightarrow D$ yields a canonical map $\iota_P \colon SK_1(P) \to SK_1(D)$. Likewise, since $\mathsf{gr}(P)$ is a central simple graded $\mathsf{gr}(F)$-subalgebra of $\mathsf{gr}(D)$ (resp. $q\mathsf{g}(P)$ is a central simple $q\mathsf{g}(F)$-subalgebra of $q\mathsf{g}(D)$) there are maps $\iota_{\mathsf{gr}(P)}$ and $\iota_{q\mathsf{g}(P)}$; analogously, there are maps ι_Q, $\iota_{\mathsf{gr}(Q)}$, and $\iota_{q\mathsf{g}(Q)}$. These fit into a diagram:

$$\begin{array}{ccccc}
SK_1(P)\times SK_1(Q) & \xrightarrow{\cong} & SK_1(\mathsf{gr}(P))\times SK_1(\mathsf{gr}(Q)) & \hookrightarrow & SK_1(q\mathsf{g}(P))\times SK_1(q\mathsf{g}(Q)) \\
\Big\downarrow{\scriptstyle \iota_P\times\iota_Q \;\cong} & & \Big\downarrow{\scriptstyle \iota_{\mathsf{gr}(P)}\times\iota_{\mathsf{gr}(Q)}} & & \Big\downarrow{\scriptstyle \iota_{q\mathsf{g}(P)}\times\iota_{q\mathsf{g}(Q)} \;\cong} \\
SK_1(D) & \xrightarrow{\widehat{\rho}_D} & SK_1(\mathsf{gr}(D)) & \hookrightarrow & SK_1(q\mathsf{g}(D))
\end{array}$$

The left map in the upper row is the isomorphism $\widehat{\rho}_P \times \widehat{\rho}_Q$. It is easy to check from the definitions of the maps that the diagram is commutative. The right horizontal map in each row is injective by Prop. 11.8. The outer vertical maps are isomorphisms by Prop. 11.3. Thus, the composition of maps $SK_1(P) \times SK_1(Q) \to SK_1(q\mathsf{g}(D))$ is injective. Since $\iota_P \times \iota_Q$ is an isomorphism, it follows that the epimorphism $\widehat{\rho}_D$ is an isomorphism. Hence, (ii) holds in general, so (i) also holds in general. □

Theorem 11.23. *Let D be a division algebra tame over its Henselian center F. Let $G = \mathcal{G}(Z(\overline{D})/\overline{F})$.*

(i) *If D is inertial over F, then $SK_1(D) \cong SK_1(\overline{D})$.*

(ii) *If D is totally (and tamely) ramified over F, then*

$$SK_1(D) \;\cong\; \mu_n(\overline{F})/\mu_e(\overline{F}),$$

and $|\mu_e(\overline{F})| = e$, where $n = \deg D$ and $e = \exp D = \exp(\Gamma_D/\Gamma_F)$.

(iii) *Suppose D is inertially split. Then there is an exact sequence:*

$$\Gamma_D/\Gamma_F \wedge \Gamma_D/\Gamma_F \longrightarrow \widehat{H}^{-1}(G, \mathrm{Nrd}_{\overline{D}}(\overline{D}^{\times})) \longrightarrow SK_1(D) \longrightarrow 1.$$

(iv) *Suppose D is semiramified. Then, $G = \mathcal{G}(\overline{D}/\overline{F}) \cong \Gamma_D/\Gamma_F$ and there is an exact sequence:*

$$\Gamma_D/\Gamma_F \wedge \Gamma_D/\Gamma_F \longrightarrow SK_1(D) \longrightarrow \widehat{H}^{-1}(G, \overline{D}^{\times}) \longrightarrow 1.$$

If, further, G is bicyclic, let the specialization coset of D be

$$sp(D) \;=\; [B] + Dec(\overline{D}/\overline{F}) \;\in\; Br(\overline{D}/\overline{F})/Dec(\overline{D}/\overline{F});$$

then,

$$SK_1(D) \;\cong\; Br(\overline{D}/\overline{F})/(\langle [B] \rangle + Dec(\overline{D}/\overline{F})).$$

(v) *Suppose D is decomposably semiramified. Then $G = \mathcal{G}(\overline{D}/\overline{F}) \cong \Gamma_D/\Gamma_F$ and*

$$SK_1(D) \;\cong\; \widehat{H}^{-1}(G, \overline{D}^{\times}).$$

If, further, G is bicyclic, then

$$SK_1(D) \;\cong\; Br(\overline{D}/\overline{F})/Dec(\overline{D}/\overline{F}).$$

Proof. Since $SK_1(D) \cong SK_1(\mathsf{gr}(D))$ by Th. 11.21, everything here follows from Cor. 11.12 and Th. 11.15 applied to $\mathsf{gr}(D)$. Note that if D is inertial over F (resp. totally and tamely ramified over F, inertially split, semiramified, decomposably semiramified), then $\mathsf{gr}(F)$ is inertial over $\mathsf{gr}(F)$ (resp. totally ramified over $\mathsf{gr}(F)$, inertially split, semiramified, decomposably semiramified), with $\mathsf{gr}(D)_0 = \overline{D}$, $\Gamma_{\mathsf{gr}(D)} = \Gamma_D$, and $\theta_{\mathsf{gr}(D)} = \theta_D$. Also, when D is semiramified, $sp(\mathsf{gr}(D)) = sp(D)$ in $Br(\overline{D}/\overline{F})/Dec(\overline{D}/\overline{F})$, by (8.35). □

Example 11.24. Let k be a field, and let $F = k((x))((y))$, with its (x,y)-adic Henselian valuation, with $\overline{F} = k$ and $\Gamma_F = \mathbb{Z} \times \mathbb{Z}$. Let n be any integer with $n \geq 2$.

(i) Let L_1 and L_2 be cyclic Galois field extensions of k of degree n that are linearly disjoint over k. Let $K = L_1 \otimes_k L_2$. So, $L_iF = L_i((x))((y))$, which is cyclic Galois over F. Let $\langle \sigma_i \rangle = \mathcal{G}(L_iF/F) = \mathcal{G}(L_i/k)$ for $i = 1, 2$. Let

$$D = \big(L_1/F, \sigma_1, x\big) \otimes_F \big(L_2/F, \sigma_2, y\big).$$

Then, D is a decomposably semiramified division algebra over F, with $\overline{D} = K$ and $\Gamma_D/\Gamma_F \cong \mathcal{G}(K/k) \cong \mathbb{Z}/n\mathbb{Z} \times \mathbb{Z}/n\mathbb{Z}$. By Th. 11.23(v),

$$SK_1(D) \cong Br(K/k)/Dec(K/k).$$

If k is an algebraic number field, then $SK_1(D)$ is finite and computable via local data. See Exercise 8.5 above. In fact, Draxl showed in [61] that any finite abelian group can be realized as $SK_1(D)$ by suitable choice of cyclic extensions L_i of an algebraic number field k.

(ii) (See Ex. 8.56 above for a graded version of this example.) Suppose k contains a primitive n^2-root of unity. Take any $a, b \in k^\times$ such that $[k(\sqrt[n]{a}, \sqrt[n]{b}):k] = n^2$, and let

$$S = \big(ax^n, by^n/F\big)_{n^2}.$$

Then, in $Br(F)$, we have $S \sim I \otimes_F N$, where

$$I = \big(a, b/F\big)_{n^2} \qquad \text{and} \qquad N = \big(a, y/F\big)_n \otimes_F \big(b^{-1}, x/F\big)_n.$$

Note that $I = \big(a, b/k\big)_{n^2} \otimes_k F$, so I has an inertial gauge with $I_0 = \big(a, b/k\big)_{n^2}$. Let $K = k(\sqrt[n]{a}, \sqrt[n]{b})$. The assumption on $[K:k]$ assures that N is a decomposably semiramified division algebra over F with $\overline{N} = K$. Since K splits I_0, the symbol algebra S is a division algebra semiramified over F with $\overline{S} = \overline{N} = K$ and $\Gamma_S = \Gamma_N = \frac{1}{n}\mathbb{Z} \times \frac{1}{n}\mathbb{Z}$. By Th. 11.23(iv),

$$SK_1(S) \cong Br(K/k)/\big(\langle \big[\big(a, b/k\big)_{n^2}\big] \rangle + Dec(K/k)\big),$$

while by Th. 11.23(v),

$$SK_1(N) \cong Br(K/k)/Dec(K/k).$$

Corollary 8.68 shows that S is decomposably semiramified if and only if $\big[\big(a, b/k\big)_{n^2}\big] \in Dec(K/k)$. This occurs if and only if the obvious epimorphism $SK_1(N) \to SK_1(S)$ is an isomorphism. Note that if $exp\ \big(a, b/k\big)_{n^2} > n$ then $\big[\big(a, b/k\big)_{n^2}\big]$ cannot lie in $Dec(K/k)$; then S is not decomposably semiramified.

Exercise

Exercise 11.1. Let F be a Henselian field with $\overline{F} \cong \mathbb{R}$, and let D be a division algebra with center F.

(i) Prove that $\exp D \le 2$.
(ii) Prove that D is isomorphic to a tensor product of quaternion algebras.
(iii) Prove that $SK_1(D) = 1$.

Notes

Detailed introductions to SK_1 for central simple algebras can be found in Draxl's book [63] and in the Draxl–Kneser lecture notes [65].

For an excellent account of SK_1 for central simple algebras and the connection with the Kneser–Tits conjecture on the triviality of the reduced Whitehead group for isotropic almost simple simply connected algebraic groups see Tits' 1977 Séminaire Bourbaki report [248]. Tits gives a thorough description of what was known on the vanishing of SK_1 prior to Platonov's work. He also describes Platonov's main results and applications of the nontriviality of SK_1 and describes the status at that time of the Kneser–Tits conjecture for the other algebraic groups not associated with SK_1. See Gille's 2008 Séminaire Bourbaki report [82] for an update on the substantial further results on the vanishing of reduced Whitehead groups since the time of Tits' account. Other surveys on SK_1 can be found in Platonov's International Congress address [185], a book chapter by Platonov and Yanchevskiĭ [193, Ch. 4], and a section of the survey article by Wadsworth [256].

The first examples of division algebras with nontrivial SK_1 were given by Platonov in [179] and [180] with proofs in [183]. His examples were bicyclic decomposably semiramified division algebras over twice iterated Laurent series fields over global fields, and his key tool for explicit computation of nontrivial SK_1 was the relative Brauer group formula in Th. (11.23)(v). After producing his examples Platonov gave some striking applications of the nontriviality of SK_1. Notably, he gave in [182] counterexamples to Kneser's question on weak approximation for $SL_1(D)$, where $SL_1(D) = \{a \in D^\times \mid Nrd_D(A) = 1\}$. He reduced this to showing nonsurjectivity of $SK_1(D) \to SK_1(D \otimes_F L)$ where $F = Z(D)$ and L is some finite-degree extension field of F. (Such nonsurjectivity follows easily from Th. 11.23(v) and Exercise (8.5)(iii).) To carry out his argument, he also proved the Stability Theorem, which says that if D is a division algebra finite-dimensional over a field F and x is transcendental over F, then

$$SK_1(D \otimes_F F(x)) \cong SK_1(D). \tag{11.4}$$

Another proof of the Stability Theorem was given by Platonov and Yanchevskiĭ in [188]. Voskresenskiĭ showed in [251] (see also [252, §18.2]) that for any finite-dimensional division algebra D,

$$SK_1(D) \cong SL_1(D)/R, \tag{11.5}$$

where the right expression is the group of R-equivalence classes (in the sense of Manin) on the group $SL_1(D)$, when viewed as the group of rational points of an algebraic geometric variety. This led Platonov to prove in [184] that $SL_1(D)$ is not a rational variety whenever $SK_1(D)$ is nontrivial. His proof again uses the Stability Theorem.

The diagram in Th. 11.10 for graded division algebras comes from Hazrat–Wadsworth [99, Th. 3.4], which was based on an analogous diagram for tame division algebras over Henselian fields given by Ershov in [75]. Ershov's work, sketched earlier in [74], systematizes for general Henselian base fields the general approach of Draxl in [62] for division algebras over twice iterated Laurent series fields.

Theorem 11.21(i), the Congruence Theorem, has been an essential ingredient in all calculations of SK_1 for division algebras D over Henselian fields. It was originally proved by Platonov in [183, Th. 3.12] for D with center a field F with complete discrete (rank 1) valuation such that $\overline{D}$ separable over $\overline{F}$. (Since the valuation is discrete of rank 1, the residue condition is equivalent to: D is inertially split.) A somewhat different proof was given by Draxl in [62, Satz 4]. A number of authors subsequently noted without details that Platonov's proof carries over to more general settings: Yanchevskiĭ in [262, Prop. 1] for inertially split division algebras over fields with discrete rank 1 Henselian valuation; Draxl in [65, (9), p. 98] for inertially split division algebras over a Henselian field; Ershov in [75, Prop. 3] for tame division algebras over a Henselian field. The proof of the Congruence Theorem given here is a somewhat simplified version of the full proof in Hazrat–Wadsworth [99, Th. 4.8].

The formula in Th. 11.23(ii) for $SK_1(D)$ with D tamely and totally ramified over its Henselian center F was given by Lipnitskiĭ in [124] for $F = C((x_1))\dots((x_n))$ with C algebraically closed, and in [127], for F with an "n-fold Henselian discrete valuation," i.e. a Henselian valuation of rank n with $\Gamma_F \cong \mathbb{Z}^n$.

The formula in Th. 11.23(iv) involving the specialization coset comes from Wadsworth [257, Cor. 3.8].

Example 11.24(ii) is the example Platonov gave in [181] of a cyclic algebra with exponent equal to index having nontrivial SK_1. Platonov's calculation of SK_1 in his example was not quite correct, though, since he did not include the term from the specialization coset in his formula.

Exercise 8.5(iii) was used by Platonov in [181] to show that $SK_1(D)$ could be arbitrarily large finite for division algebras D of bounded degree, and can also be infinite (by taking L an infinite degree Galois extension of K). At around the same time Draxl showed in [61] that every finite abelian group appears as $SK_1(D)$ for some division algebra with center $k((x))((y))$, where k is a global field. It was eventually shown (see Gräbe [87]) that any countable abelian group of bounded torsion is isomorphic to some $SK_1(D)$.

Exercise 11.1(iii) was given by Lipnitskiĭ in [124] for $F = \mathbb{R}((x_1))\dots((x_n))$. He later gave in [126] a more difficult result in the same vein: Let D be a division algebra with center $\mathbb{R}(y)((x_1))\dots((x_n))$, where y is transcendental over $\mathbb{R}$. Then $SK_1(D)$ is trivial. (In an earlier paper [125, §4], Lipnitskiĭ published a flawed example of a division algebra D of degree 8 over $\mathbb{R}(y)((x_1))((x_2))$ with nontrivial $SK_1(D)$. The flaw lies in the erroneous claim that $b \notin 1^*$, whereas in fact $b = x_1^{-1}(\sqrt{x_1 b})^2 \in k((u_1))^\times \cdot Nrd(\overline{A}^\times)$.)

It was shown by Hazrat–Wadsworth [99, Th. 5.7] that if D is a finite-dimensional graded division algebra and $q(\mathsf{D})$ its quotient division algebra, then

$$SK_1(q(\mathsf{D})) \cong SK_1(\mathsf{D}). \tag{11.6}$$

The properties of $SK_1(\mathsf{D})$ given in Th. 11.10, Cor. 11.12, and Th. 11.15 above thereby yield quick proofs of the results for SK_1 of iterated twisted rational function field extensions of division algebras given by Platonov and Yanchevskiĭ in [188] and [189]. In particular, the Stability Theorem (11.4) follows at once from (11.6) and Ex. 11.13 above.

In view of Prop. 11.2, the smallest division algebras that might have nontrivial SK_1 are biquaternion algebras, i.e., tensor products of two quaternion algebras. For a biquaternion algebra B with center F, Rost proved (unpublished) that

$$SK_1(B) \cong \mathit{ker}\big(H^4(F,\mu_2) \to H^4(F(X_B),\mu_2)\big), \tag{11.7}$$

where $F(X_B)$ is the function field of the 4-dimensional projective quadric X_B over F determined by the 6-dimensional Albert quadratic form associated to B. A proof of Rost's formula is given in Merkurjev [144, Th. 4] and an equivalent result expressed in terms of quadratic forms is proved in Knus et al. [115, §17A]. Merkurjev gave in [145, Th. 6.6] the following generalization of Rost's result, valid for an arbitrary central simple F-algebra A of degree 4:

$$SK_1(A) \cong \mathit{ker}\big(H^4(F,\mu_2)/(2[A]\cup H^2(F,\mu_2)) \to H^4(F(Y_A),\mu_2)\big), \tag{11.8}$$

where $F(Y_A)$ is the function field of the projective variety Y_A of 8-dimensional left ideals of A. In the same paper Merkurjev showed that Rost's map into $H^4(F,\mu_2)$ is the only degree 4 cohomological invariant for biquaternion algebras. Kahn proved in [111, Th. D] that if A is a central simple F-algebra of index 4 and K is the function field $F(SL_1(A))$, then $SK_1(A\otimes_F K)/\,\mathit{im}(SK_1(A)) \cong \mathbb{Z}/2\mathbb{Z}$.

Yanchevskiĭ showed in [261] that SK_1 is trivial for division algebras over fields of cohomological dimension at most 2. It is an interesting open question[1] whether SK_1 is trivial over fields of cohomological dimension 3. The answer is clearly yes for algebras of degree 4 in view of Merkurjev's formula (11.8). For ground fields of cohomological dimension 4, we have seen above many examples with nontrivial SK_1. Colliot-Thélène raises in [56] the question whether $SK_1(D)$ must always be finite if $Z(D)$ is finitely generated over its

[1] Colliot-Thélène [56, p. 31] attributes this question to Suslin.

prime field. He gives a positive answer for D a biquaternion algebra, for large classes of fields of cohomological dimension 4.

Rost's formula for biquaternion algebras and Merkurjev's generalization for arbitrary algebras of degree 4 involve mapping SK_1 of the algebra into (a factor of) a 4-th cohomology group. Suslin has done substantial work in [233] and [234] on a possible generalization of this to algebras of arbitrary degree. Let D be a division algebra with center F, and let $n = deg\, D$. Assume that $char\, F \nmid n$. Suslin's aim has been to produce a homomorphism

$$f\colon SK_1(D) \longrightarrow H^4(F, \mu_n^{\otimes 3})/\big(H^2(F, \mu_n^{\otimes 2}) \cup [D]\big)$$

such that when F is a twice iterated Laurent series field and D is semiramified the map $SK_1(D)$ to a factor group of ${}_n\, Br(F) = H^2(F, \mu_n)$ given by Th. 11.23(iv) would be obtainable as a composition of f with a succession of residue maps in cohomology associated to discrete valuations. His first map f in [233] built using Chern classes, was not quite right. His second f in [234], built using motivic cohomology, may have the desired property. A further cohomological approach to SK_1, and also SK_2, was given by Kahn in [111].

Suslin conjectured in [233, p. 75] that $SK_1(D)$ is generically nontrivial whenever $ind\, D$ is not square-free, i.e., for any such D, $SK_1(D \otimes_{Z(D)} L) \neq 1$ for some field $L \supseteq Z(D)$. Merkurjev has recently proved Suslin's conjecture, in [152]. Earlier he had used Rost's formula (11.7) to prove Suslin's conjecture when $4 \mid ind\, D$, first in [143] if $char\, D \neq 2$, and later with a different proof in [147] valid in all characteristics. (See Gille [82] for more on Suslin's conjecture.)

There is a unitary SK_1 theory for central simple algebras with unitary involution that parallels the nonunitary theory described in this chapter. An involution τ on a central simple K-algebra A is said to be *unitary* (or of *the second kind*) if $\tau|_K \neq id_K$. If F is the fixed field K^τ, then τ is also called a unitary K/F-involution on A. For such an involution, let

$$\Sigma_\tau(A) = \big\langle \{a \in A^\times \mid \tau(a) = a\} \big\rangle \ \text{ and } \ \Sigma'_\tau(A) = \{a \in A^\times \mid Nrd_A(a) \in F^\times\}.$$

If A is a division algebra then $[A^\times, A^\times] \subseteq \Sigma_\tau(A) \subseteq \Sigma'_\tau(A)$; see Knus et al. [115, Prop. 17.26] for a proof of the first inclusion, which originally appeared in Platonov–Yanchevskiĭ [186]. The *reduced unitary Whitehead group* or unitary SK_1 for τ on A is

$$SK_1(A, \tau) = \Sigma'_\tau(A)/\Sigma_\tau(A).$$

(Other notation in the literature for the unitary SK_1 includes $USK_1(A)$, $USK_1(A, \tau)$, $SUK_1(A, \tau)$, ...) From the perspective of algebraic groups, the similarity between results for $SK_1(A)$ and for $SK_1(A, \tau)$ is unsurprising. For $SL_1(A) = \{a \in A^\times \mid Nrd_A(a) = 1\}$ is a simply connected almost simple algebraic group of inner type $A_{\deg A - 1}$, while the corresponding unitary group

$$SU(A, \tau) = \{a \in A \mid a\tau(a) = 1 \text{ and } Nrd_A(a) = 1\}$$

is of outer type $A_{deg\,A-1}$. The first examples of nontrivial $SK_1(A,\tau)$ were given by Platonov and Yanchevskiĭ in [187]. Most of the early subsequent work on $SK_1(A,\tau)$ was done by Yanchevskiĭ in [263], [264], [265], [266], [267], in which he obtained formulas analogous to the ones obtained mostly by Platonov for the nonunitary SK_1, often via formidable technical calculations. The unitary version of the R-equivalence formula (11.5) says that $SK_1(A,\tau) \cong SU(A,\tau)/R$. This was proved by Monastyrnyĭ and Yanchevskiĭ in [154] and also by Chernousov and Merkurjev in [52]. The unitary analogue of the formula (11.8) for algebras of degree 4 was given by Merkurjev in [146, Th. 4.5]. More recently, Hazrat and Wadsworth showed in [100, Th. 3.5] that if D is a division algebra with center K and unitary K/F-involution τ with F Henselian, D tame over K, and K tame over F, then

$$SK_1(D,\tau) \cong SK_1(\mathsf{gr}(D),\widetilde{\tau}),$$

where $\widetilde{\tau}$ is the unitary graded involution on $\mathsf{gr}(D)$ induced by τ on D. The approach via graded division rings has clarified and simplified the proofs of many of the earlier results on unitary SK_1. Unitary versions of the semiramified results in Th. 11.23(iv) and (v) were given in Wadsworth [257]. The unitary analogue to (11.6) was proved by Wadsworth and Yanchevskiĭ in [258].

If A is a central simple graded algebra over a graded field F, then the underlying ring $\mathsf{A}^\natural$ of A (with grading ignored) is an Azumaya algebra over the integral domain $\mathsf{F}^\natural$. Consequently there is a reduced norm map $Nrd_{\mathsf{A}^\natural}\colon \mathsf{A}^\natural \to \mathsf{F}^\natural$, and we can define $SK_1(\mathsf{A}^\natural) = \{a \in \mathsf{A}^{\natural\times} \mid Nrd_{\mathsf{A}^\natural}(a) = 1\}/[\mathsf{A}^{\natural\times},\mathsf{A}^{\natural\times}]$. If A is a graded division algebra, then $SK_1(\mathsf{A}^\natural) = SK_1(\mathsf{A})$ by definition. But otherwise, A may well have inhomogeneous units; in that case little is known about $SK_1(\mathsf{A}^\natural)$. For example, it is not known whether $SK_1(\mathsf{A}^\natural)$ is a torsion group, nor whether its torsion is bounded. If D is the associated graded division algebra of A, it is unknown what relationship may exist between $SK_1(\mathsf{A}^\natural)$ and $SK_1(\mathsf{D}^\natural)$.

Chapter 12
The Essential Dimension of Central Simple Algebras

Roughly speaking, the essential dimension of an algebraic structure measures the complexity of the structure by giving the number of independent parameters needed to define it. This is expressed as the minimal transcendence degree over a ground field of a field of definition for the structure. For a prime number p, the essential p-dimension is a variation of essential dimension that takes into account simplifications in structure that can occur after field extensions of degree prime to p. (See Def. 12.1 below for precise definitions.) The presence of a valuation on a division algebra D implies some level of complexity to D, and this can be used to give a lower bound on the essential p-dimension of D. This idea has been used by Merkurjev in [150] and Baek and Merkurjev in [21] in computing lower bounds on the essential p-dimension of central simple algebras of a given degree with or without a bound on the exponent. In this chapter we give the valuation-theoretic part of their arguments and sketch their method to obtain lower bounds on the essential p-dimension of central simple algebras. These lower bounds carry structural information on central simple algebras: Since tensor products of n symbol algebras depend on $2n$ parameters, they have a relatively low essential dimension, hence the central simple algebras with high essential dimension constructed by Merkurjev and Baek–Merkurjev are not tensor products of symbol algebras.

The chapter is organized as follows: in §12.1 we define the basic notions of essential dimension and essential p-dimension, focussing almost exclusively on central simple algebras. We also give examples, which emphasize the relation with the decomposability of central simple algebras into tensor products of symbols. The valuation-theoretic ingredients of the Merkurjev and Baek–Merkurjev essential dimension computations form the core of §12.2. The main result we prove in this connection is Th. 12.11, which gives a lower bound for the transcendence degree of a field to which a certain type of semiramified division algebra descends. It is a reformulation by using a valuation of higher rank of results obtained in [150] and [21] by inductive arguments with discrete (rank 1) valuations. We then sketch in §12.3 how Th. 12.11 is used

J.-P. Tignol, A.R. Wadsworth, *Value Functions on Simple Algebras, and Associated Graded Rings*, Springer Monographs in Mathematics,
DOI 10.1007/978-3-319-16360-4_12

by Merkurjev and Baek–Merkurjev (together with some essential dimension computations for torsors under algebraic tori) to obtain lower bounds for the essential dimension of central simple algebras. In the final §12.4, we discuss consequences of these lower bounds for the decomposability of division algebras into tensor products of symbol algebras.

12.1 A brief introduction to essential dimensions

We use the functorial definition of essential dimension suggested by Merkurjev. A very good introduction to this approach can be found in the paper by Berhuy and Favi [23].

Let k be a field, and let $\mathfrak{Fields}_k$ denote the category of fields containing k as a subfield. Morphisms are ring homomorphisms of fields restricting to the identity on k.

Definition 12.1. Let $\mathcal{F}$ be a covariant functor from $\mathfrak{Fields}_k$ to $\mathfrak{Sets}$. For any field $K \supseteq k$ and $a \in \mathcal{F}(K)$, define the *essential dimension of* a to be

$$ed^{\mathcal{F}}(a) \;=\; min\big\{\, trdeg\, K_0/k \mid a \in im\big(\mathcal{F}(K_0) \to \mathcal{F}(K)\big)\big\},$$

where the minimum is taken over all fields K_0 with $k \subseteq K_0 \subseteq K$. Thus, $ed^{\mathcal{F}}(a)$ is the least transcendence degree over k of fields $K_0 \supseteq k$ such that a is defined over K_0. Then define the *essential dimension of* $\mathcal{F}$ to be

$$ed(\mathcal{F}) \;=\; \sup\{ed^{\mathcal{F}}(a) \mid a \in \mathcal{F}(K) \text{ for some field } K \supseteq k\}.$$

Let p be a prime number. For $a \in \mathcal{F}(K)$ the *essential p-dimension* of a is defined to be

$$ed_p^{\mathcal{F}}(a) \;=\; min\left\{ ed^{\mathcal{F}}(b) \;\middle|\; \begin{array}{l} b \,=\, \mathcal{F}_{K\to M}(a) \text{ for some field} \\ M \supseteq K \text{ with } p \nmid [M\!:\!K] < \infty \end{array} \right\}.$$

Then the *essential p-dimension of* $\mathcal{F}$ is defined to be

$$ed_p(\mathcal{F}) \;=\; \sup\{ed_p^{\mathcal{F}}(a) \mid a \in \mathcal{F}(K) \text{ for some field } K \supseteq k\}.$$

Note that for any functor $\mathcal{F}$,

$$ed(\mathcal{F}) \,\geq\, ed_p(\mathcal{F}). \tag{12.1}$$

For a field $L \supseteq k$, write $\mathcal{F}_L$ for the functor $\mathfrak{Fields}_L$ to $\mathfrak{Sets}$ given by $\mathcal{F}_L(K) = \mathcal{F}(K)$ for fields $K \supseteq L$. It is easy to check that

$$ed(\mathcal{F}) \geq ed(\mathcal{F}_L) \qquad \text{and} \qquad ed_p(\mathcal{F}) \geq ed_p(\mathcal{F}_L) \tag{12.2}$$

(cf. Berhuy-Favi [23, Prop. 1.5]).

Essential dimension has been studied for a number of different functors. Our focus here will be on essential dimension and p-dimension of central simple algebras of a given degree and exponent: Let n and m be positive integers with $m \mid n$. Let $\mathcal{A}_k(n)$ be the functor on $\mathfrak{Fields}_k$ given by

$$\mathcal{A}_k(n)(K) = \begin{Bmatrix} \text{isomorphism classes of central} \\ \text{simple } K\text{-algebras of degree } n \end{Bmatrix}, \tag{12.3}$$

for any field $K \supseteq k$. For fields K and L containing k and any k-homomorphism $\eta\colon K \to L$, the associated map $\mathcal{A}_k(n)(\eta)$ sends the K-isomorphism class of A to the L-isomorphism class of $A \otimes_K L$, where L is made a K-algebra via η. Likewise, $\mathcal{A}_k(n,m)$ is the subfunctor of $\mathcal{A}_k(n)$ given by

$$\mathcal{A}_k(n,m)(K) = \begin{Bmatrix} \text{isomorphism classes of central simple } K\text{-algebras} \\ \text{of degree } n \text{ and exponent a factor of } m \end{Bmatrix}. \tag{12.4}$$

Slightly abusing terminology, we will speak of the essential dimension and essential p-dimension of a central simple algebra to mean the essential dimension and essential p-dimension of its isomorphism class. Thus, for a given central simple algebra A of degree n over a field $K \supseteq k$ and any integer d, the inequality

$$ed^{\mathcal{A}_k(n)}(A) \geq d$$

means that A does not have a descent to a subfield of transcendence degree less than d, i.e., if $A \cong B \otimes_{K_0} K$ for some central simple algebra B over a subfield $K_0 \subseteq K$ containing k, then $trdeg(K_0/k) \geq d$. When $\exp A \mid m$, we have $ed^{\mathcal{A}_k(n,m)}(A) \geq d$ if and only if the same condition holds for central simple algebras B of exponent dividing m, hence we always have $ed^{\mathcal{A}_k(n,m)}(A) \geq ed^{\mathcal{A}_k(n)}(A)$. Similarly, for a given prime p and a given central simple algebra A of degree n over a field $K \supseteq k$, and any integer d, the inequality

$$ed_p^{\mathcal{A}_k(n,m)}(A) \geq d$$

means that for any finite-degree extension M/K of degree prime to p and any central simple algebra B over a subfield $M_0 \subseteq M$ containing k such that $B \otimes_{M_0} M \cong A \otimes_K M$, we have $trdeg(M_0/k) \geq d$. If $\exp A \mid m$, restricting to central simple algebras B of exponent dividing m yields the condition for $ed_p^{\mathcal{A}_k(n,m)}(A) \geq d$.

To obtain upper bounds for $ed(\mathcal{A}_k(n))$ or $ed_p(\mathcal{A}_k(n))$ thus requires control on the structure of all central simple algebras of degree n, whereas lower bounds can be obtained by constructing specific examples of central simple algebras for which the essential dimension can be bounded from below. We will see that valuation-theoretic techniques can be very effective for this.

By using the primary decomposition of central simple algebras, it is easy to check that if the integers r and s are maximal such that $p^r \mid n$ and $p^s \mid m$, then

$$ed_p(\mathcal{A}_k(n)) = ed_p(\mathcal{A}_k(p^r))$$

and

$$ed_p(\mathcal{A}_k(n,m)) = ed_p(\mathcal{A}_k(p^r, p^s)).$$

When studying the essential p-dimension of central simple algebras, we may thus restrict to algebras of p-power degree.

From Tsen's Theorem, which implies that the Brauer group of a field of transcendence degree 1 over an algebraically closed field is trivial (see Pierce [178, §19.4]), we can readily derive a lower bound for $ed_p(\mathcal{A}_k(p^r))$ as follows:

Example 12.2. For any prime p, any integer $r \geq 1$, and any field k, we have $ed_p(\mathcal{A}_k(p^r)) \geq 2$. In view of (12.2), it suffices to prove the inequality for k algebraically closed. Assuming this, we consider a central division algebra D of degree p^r over a field K containing k (for example $D = \mathcal{T}(k; p^r)$ as in (9.5) if $char\, k \neq p$). If the essential p-dimension of the isomorphism class of D were at most 1, then we could find a finite-degree field extension M/K of degree prime to p and a central simple algebra E over a subfield $M_0 \subseteq M$ such that $E \otimes_{M_0} M \cong D \otimes_K M$ and $trdeg(M_0/k) \leq 1$. By Tsen's Theorem the algebra E would be split, hence D would be split by M. This is impossible since $deg\, D$ is prime to $[M{:}F]$. Therefore, $ed_p(\mathcal{A}_k(p^r)) \geq 2$.

When $r = 1$ and $char\, k \neq p$, the bound is sharp:

Example 12.3. For any prime p, if $char\, k \neq p$, then $ed_p(\mathcal{A}_k(p)) = 2$. To see this, let D be a nonsplit central simple algebra of degree p over a field $K \supseteq k$. Let L be a maximal subfield of D. Then L is separable over K as $[L{:}K] = p \neq char\, K$. Let ω be a primitive p-th root of unity in a separable closure of L, let N be the Galois closure of $L(\omega)$ over $K(\omega)$, and let M be the fixed field of a p-Sylow subgroup of $\mathcal{G}(N/K(\omega))$. The group $\mathcal{G}(N/K(\omega))$ acts simply transitively on the p roots of the minimal polynomial of any generating element of $L(\omega)$ over $K(\omega)$; hence, the p-Sylow subgroups of $\mathcal{G}(N/K(\omega))$ are cyclic of order p. Thus, $[M{:}K] = [M{:}K(\omega)][K(\omega){:}K]$ is prime to p and $[N{:}M] = p$, and we have $N = L \otimes_K M$. Since N is a maximal subfield of the division algebra $D \otimes_K M$, this algebra is cyclic, hence a symbol algebra, say $D \otimes_K M \cong \big(a, b/M\big)_p$ for some a, $b \in M^\times$; hence,

$$D \otimes_K M \;\cong\; \big(a, b/k(\omega, a, b)\big)_p \otimes_{k(\omega,a,b)} M \qquad \text{with } p \nmid [M{:}F].$$

Thus, we have

$$ed_p^{\mathcal{A}_k(p)}(D) \;\leq\; trdeg(k(\omega, a, b)/k) \;\leq\; 2.$$

On the other hand, the essential p-dimension of the split algebra $M_p(K)$ is easy to determine: for every field $K \supset k$ we have

$$M_p(K) \;=\; M_p(k) \otimes_k K,$$

hence $ed_p^{\mathcal{A}_k(p)}(M_p(K)) = 0$. Because $ed_p^{\mathcal{A}_k(p)}(A) \leq 2$ for every central simple algebra A of degree p over any field $K \supseteq k$, it follows that $ed_p(\mathcal{A}_k(p)) \leq 2$. Example 12.2 shows that this inequality cannot be strict.

Since $ed_p(\mathcal{A}_k(p)) = 2$ it follows that $ed(\mathcal{A}_k(p)) \geq 2$, but the precise value of $ed(\mathcal{A}_k(p))$ is unknown. (If the major open question whether every division algebra of prime degree is a cyclic algebra has an affirmative answer, it would of course follow that $ed(\mathcal{A}_k(p)) = 2$ when k contains a primitive p-th root of unity.)

Merkurjev in [150] (resp. Baek and Merkurjev in [21]) have obtained the best currently known lower bounds on $ed_p(\mathcal{A}_k(p^r))$ (resp. $ed_p(\mathcal{A}_k(p^r, p^s))$) when $char\, k \neq p$. Their arguments used a mixture of valuation theory and calculations of essential dimensions of torsors of tori. Our focus here will be on the valuation-theoretic part of their arguments.

We first consider the easy case of $\mathcal{A}_k(4,2)$ to illustrate how valuation information can be used to obtain lower bounds on essential dimension. Our argument hinges on the next easy lemma:

Lemma 12.4. *Let (F', v') be a valued field extension of a valued field (F, v), and let D be a central division algebra over F. Assume $D \otimes_F F'$ is a division algebra with a valuation w' extending v', and let w be the restriction of w' to D. If w' and w are tame v'- and v-gauges respectively, then $w' = w \otimes v'$ and*

$$\mathsf{gr}_{w'}(D \otimes_F F') \cong_g \mathsf{gr}_w(D) \otimes_{\mathsf{gr}_v(F)} \mathsf{gr}_{v'}(F'). \tag{12.5}$$

Hence, $\Gamma_{D\otimes_F F'} = \Gamma_D + \Gamma_{F'}$, so the canonical map $\Gamma_D/\Gamma_F \to \Gamma_{D\otimes F'}/\Gamma_{F'}$ is surjective with kernel $(\Gamma_D \cap \Gamma_{F'})/\Gamma_F$. Moreover, $\overline{D} \otimes_{\overline{F}} \overline{F'} \subseteq \overline{D \otimes_F F'}$.

Proof. Since w is a tame v-gauge, we have $Z(\mathsf{gr}_w(D)) = \mathsf{gr}_v(F)$ by Prop. 6.43. Therefore, $w \otimes v'$ is a v'-gauge on $D \otimes_F F'$ by Prop. 3.42. But the valuation w' is the unique v'-gauge on $D \otimes_F F'$ by Prop. 4.33(i), hence $w \otimes v' = w'$, and (12.5) follows by Cor. 3.28. From (12.5), it follows readily that $\Gamma_{D\otimes_F F'} = \Gamma_D + \Gamma_{F'}$. Moreover, by comparing the homogeneous components of degree 0 in (12.5), we obtain $\overline{D} \otimes_{\overline{F}} \overline{F'} \subseteq \overline{D \otimes_F F'}$. □

Example 12.5. Suppose $char\, k \neq 2$. We show that

$$ed(\mathcal{A}_k(4,2)) \;=\; ed_2(\mathcal{A}_k(4,2)) \;=\; 4. \tag{12.6}$$

Let A be a central simple algebra of degree 4 and exponent dividing 2 over a field $K \supseteq k$. By Albert's Theorem, see Knus et al. [115, Th. 16.1, p. 233], we have $A \cong \big(a, b/K\big)_2 \otimes_K \big(c, d/K\big)_2$ for some $a, b, c, d \in K^\times$. Let $K_0 = k(a, b, c, d)$, and let $A_0 = \big(a, b/K_0\big)_2 \otimes_{K_0} \big(c, d/K_0\big)_2$. Since $A_0 \otimes_{K_0} K \cong A$, we have

$$ed^{\mathcal{A}_k(4,2)}(A) \;\leq\; trdeg(K_0/k) \;\leq\; 4.$$

This is true for all tensor products A of two quaternion algebras over all fields $K \supseteq k$; hence, using (12.1), we have

$$4 \;\geq\; ed(\mathcal{A}_k(4,2)) \;\geq\; ed_2(\mathcal{A}_k(4,2)). \tag{12.7}$$

To prove equality, consider the totally ramified division algebra $\mathcal{T} = \mathcal{T}(k; 2, 2)$ defined as in (9.5). It has degree 4 and exponent 2 by Prop. 9.8, and its center $\mathcal{Z}$ is a field of iterated Laurent series in four indeterminates. Let v be the usual valuation on $\mathcal{Z}$ with value group $\mathbb{Z}^4$. The algebra $\mathcal{T}$ is tame and totally ramified over $\mathcal{Z}$, with $\Gamma_{\mathcal{T}} = (\frac{1}{2}\mathbb{Z})^2$, see Prop. 9.8. We claim that $ed_2^{\mathcal{A}_k(4,2)}(\mathcal{T}) \geq 4$. To prove this, assume there is an odd-dimensional field extension M of $\mathcal{Z}$ and a central simple algebra B over a subfield M_0 of M containing k such that

$$B \otimes_{M_0} M \cong \mathcal{T} \otimes_{\mathcal{Z}} M.$$

Because $[M{:}\mathcal{Z}]$ is odd, the term on the right is a division algebra, hence B is also a division algebra. The Henselian valuation v extends to $\mathcal{T} \otimes_{\mathcal{Z}} M$, and its extension restricts to B. Since *char* $k \neq 2$, the valuations on $\mathcal{T} \otimes_{\mathcal{Z}} M$ and B are tame gauges, hence Lemma 12.4 yields

$$\Gamma_B + \Gamma_M \;=\; \Gamma_{\mathcal{T}\otimes M} \;=\; \Gamma_{\mathcal{T}} + \Gamma_M.$$

It follows that the following canonical maps induced by inclusions are onto:

$$\Gamma_B/\Gamma_{M_0} \longrightarrow \Gamma_{\mathcal{T}\otimes M}/\Gamma_M \longleftarrow \Gamma_{\mathcal{T}}/\Gamma_{\mathcal{Z}}. \tag{12.8}$$

By Ostrowski's Theorem A.12 (or as $\mathcal{Z}$ is maximally complete, see §3.1.2), $|\Gamma_M{:}\Gamma_{\mathcal{Z}}|$ is odd since $[M{:}\mathcal{Z}]$ is odd. Therefore, $|\Gamma_{\mathcal{T}}{:}\Gamma_{\mathcal{Z}}|$ and $|\Gamma_M{:}\Gamma_{\mathcal{Z}}|$ are relatively prime, so the right map in (12.8) is an isomorphism and it follows that $\Gamma_{\mathcal{T}\otimes M}/\Gamma_M \cong (\mathbb{Z}/2\mathbb{Z})^4$. Since Γ_B/Γ_{M_0} maps surjectively onto $(\mathbb{Z}/2\mathbb{Z})^4$, we have $trk(\Gamma_B/\Gamma_{M_0}) \geq 4$. As $v|_k$ is trivial, Prop. A.36 in Appendix A yields

$$rr(\Gamma_{M_0}/\Gamma_k) \;=\; rr(\Gamma_{M_0}) \;\geq\; trk(\Gamma_B/\Gamma_{M_0}) \;=\; 4.$$

Therefore, by the Transcendence Degree Inequality Th. A.35,

$$trdeg(M_0/k) \;\geq\; rr(\Gamma_{M_0}/\Gamma_k) \;\geq\; 4.$$

This holds for any odd degree extension M of $\mathcal{Z}$ and any subfield M_0 of M containing k with a central simple M_0-algebra B as above; hence $ed_2^{\mathcal{A}_k(4,2)}(\mathcal{T}) \geq 4$, and it follows that

$$ed_2(\mathcal{A}_k(4,2)) \;\geq\; 4.$$

This together with (12.7) proves the equalities in (12.6). Note that since we did not restrict to central simple algebras B of exponent dividing 2, we have in fact proved the stronger inequality $ed_2^{\mathcal{A}_k(4)}(\mathcal{T}) \geq 4$. Therefore, our proof also shows $ed_2(\mathcal{A}_k(4)) \geq 4$. In Th. 12.12 below we will see that $ed_2(\mathcal{A}_k(4)) \geq 5$.

12.2 Descent of semiramified division algebras

The lower bounds for $ed_p(\mathcal{A}(p^r))$ and $ed_p(\mathcal{A}(p^r, p^s))$ established by Merkurjev and Baek–Merkurjev are obtained by bounding the essential p-dimension of a tame semiramified division algebra of the type $\mathcal{S}(A; \ell_1, \ldots, \ell_r)$ as in (8.36), where A is a central simple algebra of degree p^r and $\ell_1, \ldots, \ell_r$ are linearly disjoint cyclic extensions of degree p. In this section, we show that under suitable conditions on the algebra A, a lower bound can be found for the transcendence degree of the subfields to which a prime-to-p extension of $\mathcal{S}(A; \ell_1, \ldots, \ell_r)$ can descend. The general outline of the proof is the same as in Ex. 12.5. The issue is complicated by the fact that the descent of a semiramified division algebra need not be semiramified. We will identify conditions assuring that for certain tame semiramified division algebras, every descent is semiramified (hence inertially split). This will allow us to see that the specialization coset also descends. The full result is given in Th. 12.11.

Recall from §8.4.4 that a division algebra D over a Henselian field F is tame semiramified if and only if $\overline{D}$ is a separable (hence abelian Galois) field extension of $\overline{F}$ and

$$[\overline{D}{:}\overline{F}] = |\Gamma_D{:}\Gamma_F| = \deg D.$$

When this condition holds, the canonical map θ_D induces an isomorphism $\Gamma_D/\Gamma_F \xrightarrow{\sim} \mathcal{G}(\overline{D}/\overline{F})$, and since D is inertially split we have a specialization coset

$$sp(D) \in Br(\overline{D}/\overline{F})/Dec(\overline{D}/\overline{F}),$$

see Def. 8.67.

Our first result yields information on centralizers of subfields:

Proposition 12.6. *Let D be a tame semiramified central division algebra over a Henselian field F, and let $L \subseteq D$ be a subfield containing F. Then its centralizer $C = C_D(L)$ is tame and semiramified with*

$$\Gamma_C = \theta_D^{-1}\big(\mathcal{G}(\overline{D}/\overline{L})\big) \qquad \text{and} \qquad \overline{C} = \overline{D}^{\theta_D(\Gamma_L)}.$$

Moreover, if L is inertial over F we have $\overline{C} = \overline{D}$ and

$$sp(C) = ext_{\overline{L}/\overline{F}}\big(sp(D)\big) \in Br(\overline{L})/Dec(\overline{D}/\overline{L}).$$

Proof. For $c \in C$ we have $\overline{c\ell c^{-1}} = \overline{\ell}$ for all $\ell \in \mathcal{O}_L$, hence $\Gamma_C \subseteq \theta_D^{-1}\big(\mathcal{G}(\overline{D}/\overline{L})\big)$. Since D is tame and semiramified, it is inertially split (see Prop. 8.38). Hence, $ker\,\theta_D = \Gamma_F$, so $|\theta_D^{-1}\big(\mathcal{G}(\overline{D}/\overline{L})\big){:}\Gamma_F| = [\overline{D}{:}\overline{L}]$ and therefore

$$|\Gamma_C{:}\Gamma_F| \le |\theta_D^{-1}\big(\mathcal{G}(\overline{D}/\overline{L})\big){:}\Gamma_F| = [\overline{D}{:}\overline{L}]. \tag{12.9}$$

Let $d = [\overline{D}{:}\overline{F}] = |\Gamma_D{:}\Gamma_F|$. Then, (12.9) yields

$$[\overline{L}{:}\overline{F}] = d/[\overline{D}{:}\overline{L}] \le d/|\Gamma_C{:}\Gamma_F| = |\Gamma_D{:}\Gamma_C|. \tag{12.10}$$

For $\ell \in L$ we have $\overline{\ell c \ell^{-1}} = \overline{c}$ for all $c \in \mathcal{O}_C$, hence $\overline{C} \subseteq \overline{D}^{\theta_D(\Gamma_L)}$, and therefore

$$|\Gamma_L{:}\Gamma_F| = \big|\theta_D(\Gamma_L)\big| = \big[\overline{D}{:}\overline{D}^{\theta_D(\Gamma_L)}\big] \le [\overline{D}{:}\overline{C}]. \tag{12.11}$$

Since the tame division algebra D is defectless over F, we also have L/F and D/C are defectless. Hence, inequalities (12.10) and (12.11) show that

$$[L{:}F] = [\overline{L}{:}\overline{F}] \cdot |\Gamma_L{:}\Gamma_F| \le |\Gamma_D{:}\Gamma_C| \cdot [\overline{D}{:}\overline{C}] = [D{:}C]. \qquad (12.12)$$

But the Double Centralizer Theorem shows that $[D{:}C] = [L{:}F]$. Hence, equality holds throughout (12.12). Therefore, the inequalities in (12.9)–(12.11) cannot be strict. It follows that $\Gamma_C = \theta_D^{-1}(\mathcal{G}(\overline{D}/\overline{L}))$ and $\overline{C} = \overline{D}^{\theta_D(\Gamma_L)}$. Recall that C, the centralizer of L, is the division algebra associated to $D \otimes_F L$. Since D is inertially split, it follows that C is inertially split. Moreover, $\overline{C}$ is a field, as it is a subalgebra of the field $\overline{D}$. Hence, Prop. 8.38 shows that C is semiramified and tame. Now suppose that L is inertial over F. It follows by Prop. 8.71 that $sp(C) = ext_{\overline{L}/\overline{F}}(sp(D))$. Furthermore, since $\Gamma_L = \Gamma_F$, we have $\overline{C} = \overline{D}^{\theta_D(\Gamma_L)} = \overline{D}$. □

In the next two lemmas, we consider the particular case where the Galois extension $\overline{D}/\overline{F}$ is elementary abelian. Our goal is to obtain necessary conditions under which a tame semiramified division algebra has a descent that is not semiramified. (See Ex. 12.9 for a typical example of that situation.)

Lemma 12.7. *Let D be a tame semiramified central division algebra over a Henselian field F. Assume that $\deg D = p^s$ for some prime $p \neq \operatorname{char} \overline{F}$ and some integer $s \geq 2$, and that $\overline{D}$ is an elementary abelian Galois extension of $\overline{F}$. Let L be a subfield of D with L a cyclic Galois extension of F of degree p^2. Then, there is a subfield $P \subseteq D$ with P inertial over F of degree p^{s-2} such that P centralizes L and the centralizer $C_D(P)$ is a cyclic algebra of degree p^2 containing $L{\cdot}P$ as a maximal subfield.*

Proof. Recall that D is inertially split, by Prop. 8.38. Note that since $p \neq char\, \overline{F}$ the field L is necessarily tame over F. If L were totally ramified over F, then by Th. A.9(vi) we would have $\Gamma_L/\Gamma_F \cong \mathbb{Z}/p^2\mathbb{Z}$. This is impossible since, as D is inertially split, $\Gamma_D/\Gamma_F \cong \mathcal{G}(\overline{D}/\overline{F})$ via $\overline{\theta}_D$ by Prop. 8.64; so, Γ_D/Γ_F must be elementary abelian. On the other hand, if L were inertial over F then $\overline{L}$ would be a cyclic Galois extension of $\overline{F}$ of degree p^2, which is impossible since $\overline{D}$ is an elementary abelian Galois extension of $\overline{F}$. Therefore, we must have

$$|\Gamma_L{:}\Gamma_F| = [\overline{L}{:}\overline{F}] = p.$$

Let $C = C_D(L)$. By the Double Centralizer Theorem, we have $deg\, C = p^{s-2}$. Moreover, from Prop. 12.6 we know that C is tame semiramified and $\overline{C} = \overline{D}^{\theta_D(\Gamma_L)}$. Therefore, $\overline{C}$ is an elementary abelian Galois extension of $\overline{F}$ of degree p^{s-1}, and we may find a subfield $E \subseteq \overline{C}$ such that

$$\overline{C} = \overline{L} \otimes_{\overline{F}} E.$$

Thus, $[E{:}\overline{F}] = p^{s-2}$. Take for P an inertial lift of E over F in C (see Cor. 8.15). We have $[P{:}F] = [E{:}\overline{F}] = p^{s-2}$, hence $deg\, C_D(P) = p^2$. Moreover, $L \subseteq C_D(P)$ because $P \subseteq C = C_D(L)$. We have

$$\overline{L\cdot P} \supseteq \overline{L}\cdot\overline{P} = \overline{L}\cdot E = \overline{C} \supsetneqq \overline{P}$$

and $\Gamma_{L\cdot P} \supseteq \Gamma_L + \Gamma_P \supsetneqq \Gamma_F = \Gamma_P$. Therefore,

$$[L\cdot P:P] \geq [\overline{L\cdot P}:\overline{P}]\cdot|\Gamma_{L\cdot P}:\Gamma_P| \geq p^2 = \deg C_D(P),$$

showing that $L\cdot P$ is a maximal subfield of $C_D(P)$. Since L is cyclic Galois over F, the field $L\cdot P$ is cyclic Galois over P. Hence, $C_D(P)$ is a cyclic algebra. □

Lemma 12.8. *Let D be a central division algebra over a field F with a Henselian valuation v, and let (F', v') be a Henselian extension of (F, v). Assume*

(i) *$\deg D = p^r$ for some prime $p \neq \operatorname{char}\overline{F}$ and some integer $r \geq 2$;*
(ii) *D is not semiramified;*
(iii) *$D \otimes_F F'$ is a division algebra, which is semiramified with $\overline{D \otimes_F F'}$ an elementary abelian Galois extension of $\overline{F'}$;*
(iv) *$\Gamma_{F'}$ is a free $\mathbb{Z}$-module of rank r;*
(v) *$\overline{F}$ contains a primitive p^2-root of unity.*

Then $D\otimes_F F'$ contains a subfield P inertial over F' with centralizer $C_{D\otimes F'}(P)$ a cyclic algebra of degree p^2.

Proof. Write D' for $D\otimes_F F'$. Note that D and all its F-subalgebras are tame because $\deg D$ is not divisible by $\operatorname{char}\overline{F}$. The main step of the proof consists in establishing the following claim:

Claim: There exists a subfield $L \subseteq D$ containing F with $\Gamma_L/\Gamma_F \cong \mathbb{Z}/p^2\mathbb{Z}$.

Assuming the claim, we complete the proof as follows: Let M be the inertial closure of F in L. Since L is tame and totally ramified over M and $\overline{F}$ contains a primitive p^2-root of unity (by (v)), by Prop. A.22(ii),(iii) L is Galois over M with $\mathcal{G}(L/M) \cong \Gamma_L/\Gamma_M \cong \mathbb{Z}/p^2\mathbb{Z}$. Because D' is a division algebra, the field L is linearly disjoint from F' over F. We may thus consider the fields $L' = L \otimes_F F' \subseteq D'$ and $M' = M \otimes_F F' \subseteq L'$. The extension L'/M' is cyclic Galois of degree p^2, and M' is inertial over F' as M is inertial over F. Let $C = C_{D'}(M')$ be the centralizer of M' in D'. The division algebra C is tame semiramified over its center M' by Prop. 12.6, and we have $\overline{C} \subseteq \overline{D'}$ while $\overline{M'} \supseteq \overline{F'}$; so, in view of (iii) the field $\overline{C}$ is an elementary abelian Galois extension of $\overline{M'}$. Lemma 12.7 applied to the field L' in the division algebra C yields a subfield $P \subseteq C$ with P inertial over the center M' of C such that $C_C(P)$ is a cyclic algebra of degree p^2. Now, we have $C_C(P) = C_{D'}(P) \cap C$. Since $M' \subseteq P$, we have $C_{D'}(P) \subseteq C_{D'}(M') = C$; hence $C_C(P) = C_{D'}(P)$. Moreover, P is inertial over F' because it is inertial over M' and M' is inertial over F'. The field P therefore satisfies the requirements.

To complete the proof, it thus remains only to establish the claim. First, from (iv) we see that $\Gamma_{D'} \cong \mathbb{Z}^r$, Also, as D' is semiramified,

$$|\Gamma_{D'}/\Gamma_{F'}| = [\overline{D'}:\overline{F'}] = deg\, D' = p^r,$$

and since D' is inertially split (as it is tame and semiramified) $\Gamma_{D'}/\Gamma_{F'} \cong \mathcal{G}(\overline{D'}/\overline{F'})$, which is elementary abelian by (iii). The inclusion $\Gamma_D \subseteq \Gamma_{D'}$ implies that Γ_D is a free $\mathbb{Z}$-module of rank at most r. But since D and D' are tame, Lemma 12.4 shows that $\Gamma_{D'} = \Gamma_D + \Gamma_{F'}$; hence, there is a surjection $\Gamma_D/\Gamma_F \to \Gamma_{D'}/\Gamma_{F'} \cong (\mathbb{Z}/p\mathbb{Z})^r$. This shows that Γ_D/Γ_F requires at least r generators, as does Γ_D. Therefore, $\Gamma_D \cong \mathbb{Z}^r$, so $\Gamma_F \cong \mathbb{Z}^r$, and there are integers $n_1, \ldots, n_r \geq 1$ such that

$$\Gamma_D/\Gamma_F \cong (\mathbb{Z}/p^{n_1}\mathbb{Z}) \times \ldots \times (\mathbb{Z}/p^{n_r}\mathbb{Z}).$$

If $n_1 = \ldots = n_r = 1$, then $|\Gamma_D{:}\Gamma_F| = deg\, D$. Since D is tame, hence defectless, and $\overline{D}$ is commutative as it is a subring of the field $\overline{D'}$, it follows that D is semiramified, contradicting (ii). Therefore, we have $n_i > 1$ for some i, so there exists a $\gamma \in \Gamma_D$ with order p^2 modulo Γ_F.

Consider now the graded division algebra $\mathsf{gr}(D)$, which is central over $\mathsf{gr}(F)$ and has degree p^r since D is tame (see Prop. 6.43), with $\mathsf{gr}(D)_0 = \overline{D}$, a field. Then $\mathsf{gr}(D)$ has the graded subfield

$$\mathsf{R} = \overline{D}^{\theta_D(\gamma)} \otimes_{\overline{F}} \mathsf{gr}(F) \subseteq \mathsf{gr}(D).$$

By definition, R is inertial over $\mathsf{gr}(F)$. Pick $x \in D$ with $v(x) = \gamma$. Since $\theta_{\mathsf{gr}(D)} = \theta_D$, the homogeneous element $\widetilde{x}$ of D_γ centralizes R. Therefore, there is a graded subfield

$$\mathsf{S} = \mathsf{R}(\widetilde{x}) \subseteq \mathsf{gr}(D).$$

Since $\widetilde{x}$ centralizes S_0, we have $\mathsf{S}_0 \subseteq \mathsf{gr}(D)_0^{\theta_{\mathsf{gr}(D)}(\gamma)} = \overline{D}^{\theta_D(\gamma)} = \mathsf{R}_0$; but $\mathsf{R}_0 \subseteq \mathsf{S}_0$, so in fact $\mathsf{S}_0 = \mathsf{R}_0$. It follows that S is a totally ramified graded field extension of R. This extension is generated by $\widetilde{x}$, which has degree γ, and the order of γ modulo $\Gamma_F = \Gamma_{\mathsf{R}}$ is p^2; so, by Prop. 5.18 we have $[\mathsf{S}{:}\mathsf{R}] = p^2$ and $\Gamma_{\mathsf{S}}/\Gamma_{\mathsf{R}}$ is generated by the image of γ. The graded field extension $\mathsf{S}/\mathsf{gr}(F)$ is tame because its degree divides $deg\, \mathsf{gr}(D) = p^r$, and $p \neq char\, \overline{F}$. Therefore, there is a tame lift L of S over F in D (see Th. 8.9). We have

$$\Gamma_L/\Gamma_F = \Gamma_{\mathsf{S}}/\Gamma_{\mathsf{gr}(F)} \cong \mathbb{Z}/p^2\mathbb{Z}.$$

The claim is thus proved, and the proof of Lemma 12.8 is complete. □

The following is a typical example where Lemma 12.8 applies:

Example 12.9. Let k be a field containing a primitive p^2-root of unity ω, and suppose $\ell = k(\sqrt[p]{a}, \sqrt[p]{b})$ is an elementary abelian Galois field extension of k of degree p^2. Let $F = k((x))((y))$ be the field of iterated Laurent series in two indeterminates over k, and $D = \big(x, y/F\big)_{\omega,p^2}$. Consider on F the (x,y)-adic valuation. The symbol algebra D is a tame totally ramified central division algebra over F; it is indeed the algebra $\mathcal{T}(k;p^2)$ in the notation of (9.5). Let $F' = F\big(\sqrt[p]{a^{-1}x},\, \sqrt[p]{b^{-1}y}\big)$ and $D' = D \otimes_F F'$. Taking $x',\, y' \in F'$ with $x'^p = a^{-1}x$ and $y'^p = b^{-1}y$, we have

$$D' \;=\; \big(ax'^p, by'^p/F'\big)_{\omega,p^2}.$$

Hence, as in Ex. 8.56 or Ex. 11.24(ii),

$$D' \;\sim\; \big(a, b/F'\big)_{\omega,p^2} \otimes_{F'} \big(a, y'/F'\big)_{\omega^p,p} \otimes_{F'} \big(x', b/F'\big)_{\omega^p,p}.$$

The division algebra I associated to $\big(a, b/F'\big)_{\omega,p^2}$ is inertial and is split by $\ell \otimes_k F'$; also, as in Ex. 8.56 or Ex. 11.24(ii), $N = \big(a, y'/F'\big)_{\omega^p,p} \otimes_{F'} \big(x', b/F'\big)_{\omega^p,p}$ is a decomposably semiramified division algebra with residue field ℓ. Let E be the associated division algebra of $D' \sim I \otimes_{F'} N$. By Th. 8.65 (v),(ii),(iii) E is a tame semiramified division algebra with residue field ℓ (an elementary abelian Galois extension of $\overline{F'} = k$), and $deg\, E \ge deg\, N = p^2 = deg\, D'$. Hence, $D' \cong E$. Since D' is also a cyclic algebra of degree p^2, Lemma 12.8 holds with $P = F'$.

Observe that every cyclic tame semiramified division algebra of degree p^2 with an elementary abelian residue field extension has the same form as D' above, as Prop. 10.27 shows.

Before embarking on the construction of a semiramified division algebra with high essential dimension, we first make an observation complementing Lemma 12.4 in the case of semiramified division algebras that remain semiramified after scalar extension.

Lemma 12.10. *Let (F', v') be a Henselian-valued extension of a Henselian-valued field (F, v), and let D be a tame semiramified central division algebra over F. If $D \otimes_F F'$ is a semiramified division algebra, then*

$$\overline{D \otimes_F F'} \;=\; \overline{D} \otimes_{\overline{F}} \overline{F'}, \qquad \Gamma_D \cap \Gamma_{F'} \;=\; \Gamma_F,$$

so the canonical map $\Gamma_D/\Gamma_F \to \Gamma_{D \otimes F'}/\Gamma_{F'}$ is an isomorphism, and

$$sp(D \otimes_F F') \;=\; ext_{\overline{F'}/\overline{F}}\big(sp(D)\big) \in Br(\overline{K})/Dec(\overline{D} \otimes_{\overline{F}} \overline{F'}\,/\overline{F'}). \qquad (12.13)$$

The hypothesis that $D \otimes_F F'$ be semiramified holds in particular whenever F' is an algebraic extension of F of degree prime to $deg\, D$.

Proof. Let $D' = D \otimes_F F'$. Note that the division algebra D' is necessarily tame over F', since tameness is preserved under field extension, see Prop. 6.59. Therefore, Lemma 12.4 yields the isomorphism

$$\mathsf{gr}(D') \cong_g \mathsf{gr}(D) \otimes_{\mathsf{gr}(F)} \mathsf{gr}(F'); \qquad (12.14)$$

hence, in particular,

$$\text{the canonical map } \Gamma_D/\Gamma_F \longrightarrow \Gamma_{D'}/\Gamma_{F'} \text{ is surjective.} \qquad (12.15)$$

Assume now that D' is semiramified. Then

$$|\Gamma_{D'}{:}\Gamma_{F'}| = deg\, D' = deg\, D = |\Gamma_D{:}\Gamma_F|,$$

so the canonical surjection of (12.15) is an isomorphism. Therefore, $\Gamma_D \cap \Gamma_{F'} = \Gamma_F$. By comparing the components of degree 0 on each side of the isomorphism (12.14), we see that $\overline{D'} = \overline{D} \otimes_{\overline{F}} \overline{F'}$.

To determine the specialization coset of D', consider a decomposition $D \sim I \otimes_F N$ as in (8.34), with I an inertial division algebra and N a decomposably semiramified division algebra. Letting $I_{F'}$ and $N_{F'}$ denote the division algebras over F' associated to $I \otimes_F F'$ and $N \otimes_F F'$ respectively, we have

$$D' \sim I_{F'} \otimes_{F'} N_{F'}.$$

The division algebra $I_{F'}$ is inertial by Prop. 6.59. On the other hand, since $\Gamma_N = \Gamma_D$ by Th. 8.65(i), we have $\Gamma_N \cap \Gamma_{F'} = \Gamma_F$, hence $N_{F'}$ is decomposably semiramified by Prop. 8.70. Therefore,

$$sp(D') = [\,\overline{I_{F'}}\,] + Dec(\overline{D} \otimes_{\overline{F}} \overline{F'}/\overline{F'}\,).$$

Since $[\,\overline{I_{F'}}\,] = ext_{\overline{F'}/\overline{F}}([\,\overline{I}\,])$ (see Prop. 6.59), equation (12.13) follows.

To finish the proof, assume now that F' is an algebraic extension of F of degree prime to $deg\, D$. Then D' is a division algebra. We must show that it is semiramified. The Ostrowski theorems for division algebras (Th. 4.3) and field extensions (Th. A.12) show that $|\Gamma_D{:}\Gamma_F|$ and $|\Gamma_{F'}{:}\Gamma_F|$ are relatively prime. Therefore, the canonical surjection of (12.15) is also injective; so, $\Gamma_D/\Gamma_F \cong \Gamma_{D'}/\Gamma_{F'}$. Hence,

$$|\Gamma_{D'}{:}\Gamma_{F'}| = deg(D') \quad \text{and} \quad \Gamma_D \cap \Gamma_{F'} = \Gamma_F;$$

it follows from the last equality and (12.14) that $\overline{D'} = \overline{D} \otimes_{\overline{F}} \overline{F'}$. Thus, $\overline{D'}$ is a field, and D' is semiramified. □

We now start with the construction of a semiramified division algebra with high essential dimension. The following notation will remain in force until the end of the section: Let p be a prime number and $r \geq 1$ an integer, and let k be a field of characteristic different from p containing a primitive p^2-root of unity. Let ℓ be an elementary abelian Galois extension of k of degree p^r, which decomposes into a tensor product of cyclic field extensions of k of degree p as follows:

$$\ell = \ell_1 \otimes_k \ldots \otimes_k \ell_r.$$

Let k' be a field extension of k linearly disjoint from ℓ, and write $\ell' = \ell \otimes_k k'$ and $\ell'_i = \ell_i \otimes_k k'$ for $i = 1, \ldots, r$. Let A be a central simple k'-algebra split by ℓ'. We construct as in (8.36) the division algebra

$$\mathcal{S} = \mathcal{S}(A; \ell'_1, \ldots, \ell'_r).$$

Its center $\mathcal{Z} = Z(\mathcal{S})$ is a field of iterated Laurent series in r indeterminates over k'. We let v denote the usual Henselian valuation on $\mathcal{Z}$ with value group $\mathbb{Z}^r$, and use the same notation for its unique extension to any algebraic extension of $\mathcal{Z}$, and also for the induced valuations on their subfields. The discussion in §8.4.5 shows that $\mathcal{S}$ is inertially split. Since ℓ' splits A, by Prop. 8.72 the division algebra $\mathcal{S}$ is semiramified and

$$\Gamma_{\mathcal{S}}/\Gamma_{\mathcal{Z}} \cong (\mathbb{Z}/p\mathbb{Z})^r, \quad \overline{\mathcal{S}} = \ell', \quad deg\, \mathcal{S} = p^r, \quad exp\, \mathcal{S} = lcm(exp\, A, p), \tag{12.16}$$

and

$$sp(\mathcal{S}) = [A] + Dec(\ell'/k').$$

For any algebraic extension M of $\mathcal{Z}$ of degree prime to p, Lemma 12.10 shows that the division algebra $\mathcal{S} \otimes_{\mathcal{Z}} M$ is tame and semiramified with

$$\overline{\mathcal{S} \otimes_{\mathcal{Z}} M} = \ell' \otimes_{k'} \overline{M} = \ell \otimes_k \overline{M}, \quad \Gamma_{\mathcal{S}\otimes_{\mathcal{Z}} M}/\Gamma_M \cong (\mathbb{Z}/p\mathbb{Z})^r, \tag{12.17}$$

and

$$sp(\mathcal{S} \otimes_{\mathcal{Z}} M) = ext_{\overline{M}/k'}([A]) + Dec(\ell \otimes_k \overline{M}\,/\overline{M}). \tag{12.18}$$

Theorem 12.11. *Suppose there is a finite-degree field extension M of $\mathcal{Z}$ of degree prime to p, a subfield $M_0 \subseteq M$ containing k, and a central simple M_0-algebra B satisfying*

$$B \otimes_{M_0} M \cong \mathcal{S} \otimes_{\mathcal{Z}} M.$$

Then B is a division algebra with a valuation extending v on M_0.

If B is semiramified, then

$$trdeg(M_0/k) \geq trdeg(\overline{M_0}/k) + r. \tag{12.19}$$

Moreover, there exists a central simple $\overline{M_0}$-algebra A_0 such that

$$ext_{\overline{M}/k'}([A]) \equiv ext_{\overline{M}/\overline{M_0}}([A_0]) \big(mod\ Dec(\ell \otimes_k \overline{M}\,/\overline{M})\big) \text{ and } exp\, A_0 \mid exp\, B. \tag{12.20}$$

If B is not semiramified, then

(i) *$p^{r-1} \cdot [A] \in im\big(ext_{k'/k} \colon Br(k) \to Br(k')\big)$, and*
(ii) *there exists a field $\ell_* \subseteq \ell$ containing k with $[\ell{:}\ell_*] = p^2$ such that $A \otimes_{k'} (\ell_* {\cdot} \overline{M})$ is split by a cyclic extension of $\ell_* \cdot \overline{M}$ of degree dividing p^2.*

Proof. Because $\mathcal{S} \otimes_{\mathcal{Z}} M$ is a division algebra, it is clear that B is a division algebra, and the valuation extending v on $\mathcal{S} \otimes_{\mathcal{Z}} M$ restricts to a valuation on B. Since v on M is Henselian, it follows from Def. A.26 and Th. A.27 that M contains a Henselization M_{0h} of M_0, and, setting $B_h = B \otimes_{M_0} M_{0h}$, we have

$$B_h \otimes_{M_{0h}} M = B \otimes_{M_0} M \cong \mathcal{S} \otimes_{\mathcal{Z}} M;$$

hence, $deg\, B_h = deg\, \mathcal{S} = p^r$. Since $char\, k \neq p$, it follows that B_h is tame.

Suppose first B is semiramified; then B_h is semiramified, hence we may apply Lemma 12.10 and use (12.17) to obtain

$$\Gamma_{B_h}/\Gamma_{M_{0h}} \cong \Gamma_{\mathcal{S}\otimes_{\mathcal{Z}} M}/\Gamma_M \cong (\mathbb{Z}/p\mathbb{Z})^r.$$

By Prop. A.36, we have

$$rr(\Gamma_{M_{0h}}) \geq trk(\Gamma_{B_h}/\Gamma_{M_{0h}}) = r.$$

Therefore, as Γ_k is trivial, the Transcendence Degree Inequality Th. A.35 yields

$$trdeg(M_{0h}/k) \geq trdeg(\overline{M_{0h}}/k) + rr(\Gamma_{M_{0h}}/\Gamma_k) \geq trdeg(\overline{M_{0h}}/k) + r.$$

We have $\overline{M_{0h}} = \overline{M_0}$ and $trdeg(M_{0h}/k) = trdeg(M_0/k)$ because M_{0h} is an immediate algebraic extension of M_0; hence, we have proved (12.19).

Write $B_h \sim I \otimes_{M_{0h}} N$ for some inertial division algebra I and some decomposably semiramified division algebra N. Set $A_0 = \overline{I}$, which is a division algebra with center $\overline{M_{0h}} = \overline{M_0}$. By definition, $sp(B_h)$ is represented by $[A_0]$, hence Lemma 12.10 shows that $sp(B_h \otimes_{M_{0h}} M)$ is represented by $ext_{\overline{M}/\overline{M_0}}([A_0])$. Since $B_h \otimes_{M_{0h}} M \cong \mathcal{S} \otimes_{\mathcal{Z}} M$, equation (12.18) shows that $sp(B_h \otimes_{M_{0h}} M)$ is also represented by $ext_{\overline{M}/k'}([A])$. Therefore, we have

$$ext_{\overline{M}/k'}([A]) \equiv ext_{\overline{M}/\overline{M_0}}([A_0]) \ \big(mod\ Dec(\ell \otimes_k \overline{M}\,/\overline{M})\big).$$

Moreover, we have $exp\,A_0 = exp\,\overline{I}|exp\,B_h$ by Th. 8.65(iv). Since $exp\,B_h|exp\,B$, we have (12.20) with $A_0 = \overline{I}$.

For the rest of the proof, assume B is *not* semiramified; hence, B_h is also not semiramified. Since M is a finite-degree extension of $\mathcal{Z}$ and $\Gamma_{\mathcal{Z}} = \mathbb{Z}^r$, it follows that Γ_M is a free $\mathbb{Z}$-module of rank r. The conditions (i)–(v) of Lemma 12.8 thus hold for the central division algebra B_h over M_{0h} and the Henselian extension M of M_{0h}, and the lemma shows that $\mathcal{S} \otimes_{\mathcal{Z}} M$ contains a subfield P inertial over M of degree p^{r-2} with centralizer a cyclic algebra of degree p^2. Write C for this centralizer $C_{\mathcal{S}\otimes M}(P)$. From Prop. 12.6 and (12.17), (12.18), we obtain

$$\overline{C} = \overline{\mathcal{S} \otimes_{\mathcal{Z}} M} = \ell \otimes_k \overline{M}$$

and

$$sp(C) = ext_{\overline{P}/\overline{M}}\big(sp(\mathcal{S} \otimes_{\mathcal{Z}} M)\big) = [A \otimes_{k'} \overline{P}] + Dec(\ell \otimes_k \overline{M}\,/\overline{P}). \quad (12.21)$$

In particular, $\overline{C}$ is an elementary abelian Galois extension of $\overline{P}$. Moreover, Prop. 12.6 also shows that C is semiramified. Since C is also cyclic, Prop. 10.27 yields

$$sp(C) = \big[\big(\overline{a}, \overline{b}/\,\overline{P}\big)_{p^2}\big] + Dec(\ell \otimes_k \overline{M}\,/\overline{P}) \quad (12.22)$$

for some $a, b \in \mathcal{O}_P^\times$ such that $\ell \otimes_k \overline{M} = \overline{P}\big(\sqrt[p]{\overline{a}}, \sqrt[p]{\overline{b}}\big)$. Since $A \otimes_{k'} \overline{P}$ represents the specialization coset of C, it also follows from Prop. 10.27 that $A \otimes_{k'} \overline{P}$ is split by a cyclic Galois extension of $\overline{P}$ of degree dividing p^2. The field $\overline{P}$ contains $\overline{M}$ and has codimension p^2 in $\ell \otimes_k \overline{M}$. Since $\mathcal{G}(\ell \otimes_k \overline{M}\,/\overline{M}\,) \cong \mathcal{G}(\ell/k)$, for $\ell_* = \overline{P} \cap \ell$ we have $\overline{P} = \ell_* \otimes_k \overline{M}$ with $[\ell{:}\ell_*] = [\overline{M}{:}P] = p^2$. Therefore, (ii) holds.

To see that (i) also holds, note that Kummer theory yields a', $b' \in k^\times$ such that $\ell = \ell_*(\sqrt[p]{a'}, \sqrt[p]{b'})$. Since $\overline{P} = \ell_* \otimes_k \overline{M}$, we then have

$$\overline{P}(\sqrt[p]{a'}, \sqrt[p]{b'}) = \ell \otimes_k \overline{M} = \overline{P}(\sqrt[p]{\overline{a}}, \sqrt[p]{\overline{b}}).$$

By choosing a' and b' suitably, we may assume that $a' \equiv \overline{a}$ and $b' \equiv \overline{b} \,\big(mod\ \overline{P}^{\times p}\big)$; hence, there exist s, $t \in \overline{P}^\times$ such that $\overline{a} = a's^p$ and $\overline{b} = b't^p$. As in (10.27), we then have

$$\big(\overline{a}, \overline{b}/\overline{P}\big)_{p^2} \sim \big(a', b'/\overline{P}\big)_{p^2} \otimes_{\overline{P}} \big(a', t/\overline{P}\big)_p \otimes_{\overline{P}} \big(s, b'/\overline{P}\big)_p,$$

hence

$$\big[\big(\overline{a}, \overline{b}/\overline{P}\big)_{p^2}\big] \equiv \big[\big(a', b'/\overline{P}\big)_{p^2}\big] \,\big(mod\ Dec(\ell \otimes_k \overline{M} \,/\, \overline{P})\big).$$

Therefore, by (12.21) and (12.22),

$$[A \otimes_{k'} \overline{P}] \equiv \big[\big(a', b'/\overline{P}\big)_{p^2}\big] \,\big(mod\ Dec(\ell \otimes_k \overline{M} \,/\, \overline{P})\big).$$

Note that $Dec(\ell \otimes_k \overline{M} \,/\, \overline{P})$ is p-torsion by (8.12); hence,

$$p \cdot [A \otimes_{k'} \overline{P}] = p \cdot \big[\big(a', b'/\overline{P}\big)_{p^2}\big] = ext_{\overline{P}/k} \big[\big(a', b'/k\big)_p\big].$$

By corestricting from $\overline{P}$ to k' and using

$$cor_{\overline{P}/k'} \circ ext_{\overline{P}/k'} = [\overline{P}{:}k'] \cdot id_{Br(k')} = p^{r-2}[\overline{M}{:}k'] \cdot id_{Br(k')},$$

we obtain

$$p^{r-1} \cdot [\overline{M}{:}k'] \cdot [A] = p^{r-1} \cdot [\overline{M}{:}k'] \cdot ext_{k'/k}[(a', b'/k)_p].$$

Since $[\overline{M}{:}k']$ is prime to $exp\,A$, it follows that $p^{r-1} \cdot [A]$ lies in the image of $Br(k)$ in $Br(k')$, proving (i). □

12.3 Lower bounds

The next two theorems show how the valuation theory in Th. 12.11 was used by Merkurjev (resp. Baek and Merkurjev) to obtain the best currently known lower bounds for $ed_p(\mathcal{A}_k(p^r))$ (resp. $ed_p(\mathcal{A}_k(p^r, p^s))$). The proofs are only sketched, since they require substantial results from the theory of essential dimension of torsors of tori.

Theorem 12.12 (Merkurjev [150, Th. 6.1]). *Let p be a prime number and k a field with $char\, k \neq p$, and let r be any positive integer. Then,*

$$ed_p(\mathcal{A}_k(p^r)) \geq (r-1)p^r + 1.$$

Proof (sketch). Since $ed_p(\mathcal{A}_k(p)) = 2$ (see Ex. 12.3) we may assume that $r \geq 2$. Also, by replacing k by a larger field and invoking (12.2), we may assume that k contains a primitive p^2-root of unity and that there is a field ℓ Galois over k with $\mathcal{G}(\ell/k) \cong (\mathbb{Z}/p\mathbb{Z})^r$. Merkurjev finds a field extension k' of k linearly disjoint from ℓ and a central simple k'-algebra A satisfying the following conditions:

(12.23)

(i) A is split by $\ell \otimes_k k'$;
(ii) $p^{r-1} \cdot [A] \notin im\big(ext_{k'/k} \colon Br(k) \to Br(k')\big)$;
(iii) if there is a finite-degree field extension E of k' of degree prime to p, a subfield $E_0 \subseteq E$ containing k, and a central simple E_0-algebra A_0 such that

$$ext_{E/k'}([A]) \equiv ext_{E/E_0}([A_0]) \ \big(mod\ Dec(\ell \otimes_k E\,/E)\big),$$

then

$$trdeg(E_0/k) \geq (r-1)(p^r - 1).$$

With this choice of A and ℓ, the division algebra $\mathcal{S}$ of Th. 12.11 has degree p^r. We claim that $ed_p^{\mathcal{A}_k(p^r)}(\mathcal{S}) \geq (r-1)p^r + 1$, which suffices to establish the theorem. To prove the claim, assume there is a finite-degree field extension M of $\mathcal{Z}$ of degree prime to p, a subfield $M_0 \subseteq M$ containing k, and a central simple M_0-algebra B such that

$$B \otimes_{M_0} M \cong \mathcal{S} \otimes_{\mathcal{Z}} M.$$

In view of (ii) of (12.23), Th. 12.11 shows that B is a semiramified division algebra; hence, by Th. 12.11

$$trdeg(M_0/k) \geq trdeg(\overline{M_0}/k) + r.$$

Theorem 12.11 also shows that there exists a central simple $\overline{M_0}$-algebra A_0 such that

$$ext_{\overline{M}/k'}([A]) \equiv ext_{\overline{M}/\overline{M_0}}([A_0]) \ \big(mod\ Dec(\ell \otimes_k \overline{M}/\overline{M})\big).$$

Since $[M{:}\mathcal{Z}]$ is prime to p, the residue degree $[\overline{M}{:}k']$ is also prime to p by Ostrowski's Theorem A.12 (or as $\mathcal{Z}$ is maximally complete, see §3.1.2). Therefore, (iii) of (12.23) yields

$$trdeg(\overline{M_0}/k) \geq (r-1)(p^r - 1).$$

Thus, we have $trdeg(M_0/k) \geq (r-1)(p^r - 1) + r = (r-1)p^r + 1$, and the claim is proved.

The algebra A is obtained by a generic construction, which we now sketch. Let $\mathfrak{D}_{\ell/k}$ denote the functor on $\mathfrak{Fields}_k$ given by

$$\mathfrak{D}_{\ell/k}(K) = Br(\ell{\cdot}K\,/K)/Dec(\ell{\cdot}K\,/K) \qquad \text{for any field } K \supseteq k.$$

Merkurjev computes $ed(\mathfrak{D}_{\ell/k})$ by finding a torus S defined over k such that the functor of torsors over S is isomorphic to $\mathfrak{D}_{\ell/k}$. He starts from the simplified description of 2-cocycles of $\mathcal{G}(\ell/k)$ afforded by the choice of a base

$\sigma_1, \ldots, \sigma_r$ of $\mathcal{G}(\ell/k)$ as an $\mathbb{F}_p$-vector space (see Amitsur–Saltman [12, §1], the Notes to Ch. 8, or §8.3.2 for the case $r = 2$). Consider the group ring $R = \mathbb{Z}\mathcal{G}(\ell/k)$. In the free R-module R^r with standard base $e_1, \ldots, e_r$, let M be the submodule generated by the elements $(\sigma_i - 1)e_j - (\sigma_j - 1)e_i$ for $i, j \in \{1, \ldots, r\}$. The k-tori S, P, T with respective character groups $R^r/M, R^r, M$ fit into an exact sequence

$$1 \longrightarrow S \longrightarrow P \longrightarrow T \longrightarrow 1. \tag{12.24}$$

Here, $P = \big(R_{\ell/k}(\mathbb{G}_m)\big)^r$, with $R_{\ell/k}$ denoting the Weil restriction of scalars; thus, P has k-points $P(k) = (\ell^\times)^r$. Let $\mathcal{H}_S$ be the functor on $\mathfrak{Fields}_k$ given by

$$\mathcal{H}_S(K) \;=\; H^1(K, S) \;(= H^1(\mathcal{G}(K), S(K_{sep}))\,) \qquad \text{for any field } K \supseteq k.$$

The cohomology group $\mathcal{H}_S(K)$ classifies S-torsors over K (cf. Knus et al. [115, Prop. (28.14)]). Merkurjev shows in [150, Cor. 3.7] that $\mathcal{H}_S \cong \mathfrak{D}_{\ell/k}$ as functors, and he uses the theory of essential dimension of torsors of tori to prove in [150, (13)] that

$$ed_p(\mathcal{H}_S) \;=\; (r-1)(p^r - 1).$$

Let

$$k' = k(T), \tag{12.25}$$

the function field of T as a variety over k. Then, k' is a regular field extension of k since T is an absolutely irreducible variety; hence $\ell \otimes_k k'$ is a field.

The exact sequence (12.24) shows that P is an S-torsor over T (via the map $P \to T$). Moreover, since $P(k) = \ell^{\times r}$, there is a group representation of P on the k-vector space ℓ^r given by multiplication; this restricts to a representation of S on ℓ^r. Since S acts freely on the Zariski-open subset $\ell^{\times r}$ of ℓ^r, it is known that the S-torsor $P \to T$ has the following generic property: every S-torsor over any infinite field $K \supseteq k$ is obtained by pullback of $P \to T$ with respect to a K-point T, and the K-point can be selected to lie in any chosen nonempty open subset of T. For a proof of this, see Baek–Merkurjev [21, Prop. 2.1] or Garibaldi et al. [81, 5.3, p. 12].

We have a morphism of schemes $Spec(k') \to T$ whose image is the generic point of T. We write k' for $Spec(k')$. The generic fiber of P over T is $P \times_T k'$, which is an S-torsor over k' via the map $P \times_T k' \to T \times_T k' = k'$. Because of the generic property described in the preceding paragraph, this torsor is *versal* in the terminology of [81, §5], or *generic* in the terminology of [150, §3].

Let $\zeta \in H^1(k', S)$ be the cohomology class of the generic S-torsor $P \times_T k'$. Because the torsor is generic, it is known (see Merkurjev [148, Th. 2.9]) that $ed_p(\mathcal{H}_S) = ed_p^{\mathcal{H}_S}(\zeta)$. Under the isomorphism $\mathcal{H}_S(k') \xrightarrow{\sim} \mathfrak{D}_{\ell/k}(k')$ the element ζ maps to $[A] + Dec(\ell{\cdot}k'/k')$ for some central simple k'-algebra A that is split by $\ell{\cdot}k'$. We thus have

$$ed_p^{\mathfrak{D}_{\ell/k}}\big([A] + Dec(\ell{\cdot}k'/k')\big) \;=\; ed_p^{\mathcal{H}_S}(\zeta) \;=\; ed_p(\mathcal{H}_S) \;=\; (r-1)(p^r - 1). \tag{12.26}$$

This implies that A satisfies condition (iii) of (12.23). Merkurjev shows in [150, Cor. 3.9] that A also satisfies condition (ii). □

Remark 12.13. Condition (ii) of (12.23) implies that $exp\, A = p^r$, hence $exp\, \mathcal{S} = p^r$ by Prop. 8.72.

Theorem 12.14 (Baek–Merkurjev [21, Th. 6.1]). *Let p be a prime number and k a field with $\operatorname{char} k \neq p$, and let r and s be integers with $r \geq 2$ and $1 \leq s \leq r$. Then,*

$$ed_p(\mathcal{A}_k(p^r, p^s)) \geq \begin{cases} (r-1)p^r + p^{r-s} & \text{if } p \neq 2 \text{ or } s \geq 2, \\ (r-1)2^{r-1} & \text{if } p = 2 \text{ and } s = 1. \end{cases}$$

The (sketched) proof of the theorem uses the following observation: Let ω be a primitive p^3-root of unity in an algebraic closure of k; consider the following totally ramified division algebra obtained as in (9.5) as a tensor product of r symbol algebras of degree p with independent indeterminates as slots:

$$\mathfrak{T} = \mathfrak{T}(k(\omega); \underbrace{p, \ldots, p}_{r}).$$

We have $deg\, \mathfrak{T} = p^r$ and $exp\, \mathfrak{T} = p$, and the center of $\mathfrak{T}$ is a field of iterated Laurent series:

$$Z(\mathfrak{T}) = k(\omega)((x_1))((y_1))\ldots((x_r))((y_r)),$$

see Prop. 9.8. Let

$$\ell = Z(\mathfrak{T})(\sqrt[p]{x_1}, \ldots, \sqrt[p]{x_r}), \tag{12.27}$$

which we consider as a maximal subfield of $\mathfrak{T}$.

Lemma 12.15. *There are no subfields $\ell_* \subset \ell$ containing $Z(\mathfrak{T})$ with $[\ell{:}\ell_*] = p^2$ and finite-degree field extensions L of ℓ_* with $[L{:}\ell_*]$ prime to p such that $\mathfrak{T} \otimes_{Z(\mathfrak{T})} L$ is split by a cyclic extension of L of degree dividing p^2.*

Proof. Recall from §9.1.3 that for the extension of the $(x_1, \ldots, y_r)$-adic valuation, the division algebra $\mathfrak{T}$ is tame and totally ramified over $Z(\mathfrak{T})$, with $\Gamma_{\mathfrak{T}}/\Gamma_{Z(\mathfrak{T})} \cong (\mathbb{Z}/p\mathbb{Z})^{2r}$. Let $\ell_* \subseteq \ell$ be a subfield containing $Z(\mathfrak{T})$ with $[\ell{:}\ell_*] = p^2$, and let C be the centralizer $C_{\mathfrak{T}}(\ell_*)$; so C is a division algebra of degree p^2 with center ℓ_*, and $C \sim \mathfrak{T} \otimes_{Z(\mathfrak{T})} \ell_*$ in $Br(\ell_*)$. Moreover, C is tame and totally ramified over ℓ_*. Since $p\Gamma_{\mathfrak{T}} \subseteq \Gamma_{Z(\mathfrak{T})}$, we have $p\Gamma_C \subseteq \Gamma_{\ell_*}$, hence $\Gamma_C/\Gamma_{\ell_*} \cong (\mathbb{Z}/p\mathbb{Z})^4$.

Let L be a finite-degree field extension of ℓ_* of degree prime to p. The algebra $C \otimes_{\ell_*} L$ is a division algebra since $[L{:}\ell_*]$ is prime to $deg\, C$; it is the division algebra associated to $\mathfrak{T} \otimes_{Z(\mathfrak{T})} L$. Since $|\Gamma_L{:}\Gamma_{\ell_*}|$ divides $[L{:}\ell_*]$ by Ostrowski's Theorem A.12 (or as ℓ_* is maximally complete, see §3.1.2), we have $|\Gamma_L{:}\Gamma_{\ell_*}|$ is prime to $|\Gamma_C{:}\Gamma_{\ell_*}|$, hence the canonical map $\Gamma_C/\Gamma_{\ell_*} \to \Gamma_{C \otimes L}/\Gamma_L$, which is surjective with kernel $(\Gamma_C \cap \Gamma_L)/\Gamma_{\ell_*}$ by Lemma 12.4, is an isomorphism. Thus, $C \otimes_{\ell_*} L$ is tame and totally ramified over L, with $\Gamma_{C \otimes L}/\Gamma_L \cong (\mathbb{Z}/p\mathbb{Z})^4$. Since

$trk(\Gamma_{C\otimes L}/\Gamma_L) > 2$, Cor. 9.7 shows that $C \otimes_{\ell_*} L$ is not a cyclic algebra; indeed, it is not split by any cyclic Galois field extension of L. □

Proof of Th. 12.14 (sketch). The proof is along the same lines as the proof of Th. 12.12 but with a few differences, which we describe.

Since enlarging k reduces $ed_p(\mathcal{A}_k(p^r, p^s))$ (see (12.2)), we may substitute for k the field $Z(\mathcal{T})$ of Lemma 12.15 above, and thus (still calling the base field k) assume that k contains a primitive p^3-root of unity, has an elementary abelian Galois field extension ℓ of degree p^r, and is the center of the division algebra $\mathcal{T}$ as in Lemma 12.15. As in the proof of Th. 12.12, a crucial step of the proof is to construct an extension k' of k linearly disjoint from ℓ over k and a central simple k'-algebra A of degree p^r subject to three conditions. In this case, we require:

(12.28)

(i) A is split by $\ell \otimes_k k'$ and $exp\, A \mid p^s$;
(ii) there are no subfields $\ell_* \subseteq \ell$ containing k with $[\ell{:}\ell_*] = p^2$ and finite-degree field extensions L of $\ell_* \cdot k'$ with $[L : \ell_* \cdot k']$ prime to p such that $A \otimes_{k'} L$ is split by a cyclic extension of L of degree dividing p^2;
(iii) if there is a finite-degree field extension E of k' of degree prime to p, a subfield $E_0 \subseteq E$ containing k, and a central simple E_0-algebra A_0 such that

$$ext_{E/k'}([A]) \equiv ext_{E/E_0}([A_0]) \ \big(mod\ Dec(\ell \otimes_{k'} E \,/ E)\big) \quad \text{and} \quad exp\, A_0 \,|\, p^s,$$

then

$$trdeg(E_0/k) \geq \begin{cases} (r-1)(p^r - 1) + p^{r-s} - 1 & \text{if } p \neq 2 \text{ or } s \geq 2, \\ (r-1)(2^{r-1} - 1) - 1 & \text{if } p = 2 \text{ and } s = 1. \end{cases}$$

Given such A and k', we can form the division algebra $\mathcal{S}$ with center $\mathcal{Z}$ as in Th. 12.11, and we have $deg\, \mathcal{S} = p^r$ and $exp\, \mathcal{S} \mid p^s$ by (12.16). To establish the theorem, it then suffices to show

$$ed_p^{\mathcal{A}_k(p^r, p^s)}(\mathcal{S}) \geq \begin{cases} (r-1)p^r + p^{r-s} & \text{if } p \neq 2 \text{ or } s \geq 2, \\ (r-1)2^{r-1} & \text{if } p = 2 \text{ and } s = 1. \end{cases} \qquad (12.29)$$

For this, assume there are a finite-degree field extension M of $\mathcal{Z}$ with $[M{:}\mathcal{Z}]$ prime to p, a subfield $M_0 \subseteq M$ containing k, and a central simple M_0-algebra B such that

$$B \otimes_{M_0} M \cong \mathcal{S} \otimes_{\mathcal{Z}} M \qquad \text{and} \qquad exp\, B \mid p^s.$$

Because of condition (ii) of (12.28), Th. 12.11 shows that B is a semiramified division algebra, hence also that

$$trdeg(M_0/k) \geq trdeg(\overline{M_0}/k) + r, \qquad (12.30)$$

and that there exists a central simple $\overline{M_0}$-algebra A_0 such that

$$ext_{\overline{M}/k'}([A]) \equiv ext_{\overline{M}/\overline{M_0}}([A_0]) \ \big(mod\ Dec(\ell \otimes_k \overline{M} \,/\, \overline{M}\,)\big) \quad \text{and} \quad exp\, A_0 \mid p^s.$$

Therefore, by (iii) of (12.28) and (12.30),

$$trdeg(M_0/k) \geq \begin{cases} (r-1)p^r + p^{r-s} & \text{if } p \neq 2 \text{ or } s \geq 2, \\ (r-1)2^{r-1} & \text{if } p = 2 \text{ and } s = 1. \end{cases}$$

This inequality proves (12.29).

To complete this sketch of proof, we give some indications on the generic construction of the algebra A. For any field $K \supseteq k$, let ${}_{p^s}Br(\ell{\cdot}K\,/K)$ denote the p^s-torsion subgroup of $Br(\ell{\cdot}K\,/K)$. Since $\mathcal{G}(\ell{\cdot}K\,/K)$ is a p-torsion group, the group $Dec(\ell{\cdot}K\,/K)$ is p-torsion (see (8.12)), so it lies in ${}_{p^s}Br(\ell{\cdot}K\,/K)$. Let $\mathfrak{D}_{\ell/k,p^s}$ be the functor on $\mathfrak{Fields}_k$ given by

$$\mathfrak{D}_{\ell/k,p^s}(K) = {}_{p^s}Br(\ell{\cdot}K\,/K)/Dec(\ell{\cdot}K\,/K) \qquad \text{for every field } K \supseteq k.$$

Using a description of p^s-torsion 2-cocycles of $\mathcal{G}(\ell/k)$ with the help of an $\mathbb{F}_p$-base of $\mathcal{G}(\ell/k)$, Baek and Merkurjev construct a diagram of free $\mathbb{Z}$-modules with $\mathcal{G}(k)$-action. By duality, they obtain a commutative diagram of tori over k with exact rows (see [21, after (19)]):

$$\begin{array}{ccccccccc} 1 & \longrightarrow & U & \longrightarrow & P & \longrightarrow & V & \longrightarrow & 1 \\ & & \downarrow & & \downarrow & & \| & & \\ 1 & \longrightarrow & S & \longrightarrow & P \times (\mathbb{G}_m)^r & \longrightarrow & V & \longrightarrow & 1 \end{array} \tag{12.31}$$

Here, $P = R_{\ell/k}(\mathbb{G}_m)^{r+1}$. In addition, they show that there is a commutative diagram functorial in K with horizontal maps isomorphisms:

$$\begin{array}{ccc} H^1(K,U) & \xrightarrow{\cong} & {}_{p^s}Br(\ell{\cdot}K\,/K) \\ \downarrow & & \downarrow \\ H^1(K,S) & \xrightarrow{\cong} & {}_{p^s}Br(\ell{\cdot}K\,/K)/Dec(\ell{\cdot}K\,/K) \end{array} \tag{12.32}$$

Hence, if $\mathcal{H}_S$ is the functor on $\mathfrak{Fields}_k$ with $\mathcal{H}_S(K) = H^1(K,S)$ for all fields $K \supseteq k$, then $\mathcal{H}_S \cong \mathfrak{D}_{\ell/k,p^s}$.

Let

$$k' = k(V), \tag{12.33}$$

the function field of the k-variety V. Because $P(k) = \ell^{\times(r+1)}$, which is open in the k-vector space ℓ^{r+1}, the U-torsor $P \to V$ given by the top row of (12.32) has the generic property described in the preceding proof; hence, the induced U-torsor $P \times_V k' \to k'$ on the generic fiber $P \times_V k'$ of P over V is a generic torsor. Let $\zeta \in H^1(k',U)$ be the cohomology class of this generic torsor. Let A be a central simple k'-algebra such that $[A] \in {}_{p^s}Br(\ell{\cdot}k'\,/k')$ is the image of ζ in the top map of (12.32) with $K = k'$; similarly, let $\eta \in H^1(k,U)$ correspond to $[\mathfrak{T}] \in {}_{p^s}Br(\ell/k)$ under the top map of (12.32) with $K = k$. It is shown in [21, Lemma 5.4] that for any subfield $\ell_* \subseteq \ell$ containing k with $[\ell{:}\ell_*] = p^2$, and any finite-degree extension L of $\ell_*{\cdot}k'$ with $[L : \ell_*{\cdot}k']$ prime to p, the algebra $A \otimes_{k'} L$ is not split by any cyclic field extension of L of degree dividing p^2. This is proved using the generic property of the torsor associated to ζ to show that if the asserted property of A fails, then there is a specialization of ζ to η implying that the corresponding property for $\mathcal{T}$ fails, contradicting Lemma 12.15. It follows that A satisfies condition (ii) of (12.28).

Let ζ' be the image of ζ in $H^1(k', S)$. Then ζ' is the cohomology class of the S-torsor $(P \times (\mathbb{G}_m)^r) \times_V k' \to k'$ on the generic fiber determined by the S-torsor $P \times (\mathbb{G}_m)^r \to V$ associated to the lower row of (12.31). Because $(P \times (\mathbb{G}_m)^r)(k) = \ell^{\times(r+1)} \times k^{\times r}$ is open in the k-vector space $\ell^{r+1} \times k^r$, the torsor associated to ζ' is generic; hence, $ed_p(\mathcal{H}_S) = ed_p^{\mathcal{H}_S}(\zeta')$. The commutative diagram (12.32) for $K = k'$ shows that $\zeta' \in H^1(k', S)$ maps to $[A] + Dec(\ell{\cdot}k' /k')$ in $\mathcal{D}_{\ell/k,p^s}(k')$; hence, by the isomorphism of the functors $\mathcal{D}_{\ell/k,p^s}$ and $\mathcal{H}_S$

$$ed_p^{\mathcal{D}_{\ell/k,p^s}}\big([A] + Dec(\ell{\cdot}k' /k')\big) = ed_p^{\mathcal{H}_S}(\zeta') = ed_p(\mathcal{H}_S). \tag{12.34}$$

From the theory of essential dimension of torsors of tori it is shown in [21, Cor. 3.11] that

$$ed_p(\mathcal{H}_S) = \begin{cases} (r-1)(p^r-1) + p^{r-s} - 1 & \text{if } p \neq 2 \text{ or } s \geq 2, \\ (r-1)(2^{r-1}-1) - 1 & \text{if } p = 2 \text{ and } s = 1. \end{cases} \tag{12.35}$$

By combining (12.34) and (12.35), we see that A satisfies condition (iii) of (12.28). □

12.4 Application to decomposability

In this section, we apply the results of this chapter to the construction of central simple algebras that do not decompose into tensor products of symbol algebras. We already observed in §12.1 that tensor product decompositions lead to upper bounds on the essential dimension; see Ex. 12.5. The general result is the following:

Proposition 12.16. *Let p be a prime and let k be a field with $\operatorname{char} k \neq p$. Let A be a central simple algebra of degree p^r and exponent p^s, for some $r \geq s \geq 0$, over a field K containing k.*

(i) *If A decomposes into a tensor product of n symbol algebras, then*

$$ed^{\mathcal{A}_k(p^r)}(A) \leq 2n.$$

(ii) *If A decomposes into a tensor product of n symbol algebras of degree at most p^s, then $ed^{\mathcal{A}_k(p^r,p^s)}(A) \leq 2n$.*

(iii) *If there exists a finite-degree extension M/K of degree prime to p such that $A \otimes_K M$ decomposes into a tensor product of n symbol algebras, then* $\mathrm{ed}_p^{\mathcal{A}_k(p^r)}(A) \le 2n$.

(iv) *If there exists a finite-degree extension M/K of degree prime to p such that $A \otimes_k M$ decomposes into a tensor product of n symbol algebras of degree at most p^s, then* $\mathrm{ed}_p^{\mathcal{A}_k(p^r,p^s)}(A) \le 2n$.

Proof. Each statement readily follows from the definition of the essential (p-) dimension of a central simple algebra, in view of the observation that tensor products of n symbol algebras over a field M containing k are defined over a field M_0 of transcendence degree at most $2n$ over k, because

$$\begin{aligned}(a_1, b_1/M)_{\omega_1,d_1} &\otimes_M \dots \otimes_M (a_n, b_n/M)_{\omega_n,d_n} \\ &\cong (a_1, b_1/M_0)_{\omega_1,d_1} \otimes_{M_0} \dots \otimes_{M_0} (a_n, b_n/M_0)_{\omega_n,d_n} \otimes_{M_0} M\end{aligned}$$

for $M_0 = k(\omega_1, \dots, \omega_n, a_1, b_1, \dots, a_n, b_n)$. □

Consider for example the division algebra $\mathcal{S} = \mathcal{S}(A; \ell_1', \dots, \ell_r')$ of degree p^r constructed in the proof of Th. 12.12 with k' as in (12.25) and A defined preceding (12.26). We saw in that proof that $\mathrm{ed}_p^{\mathcal{A}_k(p^r)}(\mathcal{S}) \ge (r-1)p^r + 1$. If $r \ge 2$, then $(r-1)p^r + 1 \ge 2$, hence $\mathcal{S}$ is not a symbol algebra of degree p^r, and it does not become a symbol algebra over any finite-degree extension of its center of degree prime to p. It is of course not a tensor product of symbol algebras of lower degree, since $\exp \mathcal{S} = p^r$; see Remark 12.13. On the other hand, if $r = 1$ we know that $\mathcal{S}$ becomes a symbol algebra over a finite-degree extension of its center of degree prime to p; see Ex. 12.3.

Corollary 12.17. *Let p be a prime, let k be a field with $\mathrm{char}\, k \ne p$, and let $r, s \in \mathbb{N}$ with $r \ge s \ge 1$ and $r \ge 2$. Consider the division algebra $\mathcal{S} = \mathcal{S}(A; \ell_1', \dots, \ell_r')$ of degree p^r and exponent dividing p^s constructed in the proof of Th. 12.14 with the k' of* (12.33) *and A defined after* (12.33). *If $p = 2$, $r = 2$, and $s = 1$, then $\mathcal{S}$ decomposes into a tensor product of two quaternion algebras. In all the other cases, there is no finite-degree extension $M/Z(\mathcal{S})$ of degree prime to p such that $\mathcal{S} \otimes_{Z(\mathcal{S})} M$ decomposes into a tensor product of symbol algebras of degree at most p^s.*

Proof. According to a theorem of Albert (see Knus et al. [115, Th. 16.1, p. 233]), every division algebra of degree 4 and exponent 2 is a tensor product of two quaternion algebras; the first assertion follows.

If there is a finite-degree extension $M/Z(\mathcal{S})$ of degree prime to p such that $\mathcal{S} \otimes_{Z(\mathcal{S})} M$ decomposes into a tensor product of symbols of degree at most p^s, then the number of factors in the decomposition is at most r since $\deg \mathcal{S} = p^r$, hence by Prop. 12.16(iv) we have $\mathrm{ed}_p^{\mathcal{A}_k(p^r,p^s)}(\mathcal{S}) \le 2r$. Recall from (12.29) that

$$\mathrm{ed}_p^{\mathcal{A}_k(p^r,p^s)}(\mathcal{S}) \ge \begin{cases} (r-1)p^r + p^{r-s} & \text{if } p \ne 2 \text{ or } s \ge 2, \\ (r-1)2^{r-1} & \text{if } p = 2 \text{ and } s = 1. \end{cases}$$

For $r \geq 2$ and $p \neq 2$, we have $(r-1)p^r + p^{r-s} > 3^r$, and the inequality $2r > 3^r$ is impossible. Also, for $r \geq s \geq 2$, the inequality $2r \geq (r-1)2^r + 2^{r-s}$ leads to $2r > 2^r$, which is impossible. Likewise, for $r > 2$ the inequality $2r \geq (r-1)2^{r-1}$ is impossible. Therefore, in these cases there is no finite-degree extension $M/Z(\mathcal{S})$ as above. □

Exercises

Exercise 12.1. Let k be a field containing a primitive p-th root of unity, for some prime p. Show that $ed_p^{\mathcal{A}_k(p^r)}(\mathcal{T}(k;\underbrace{p,\ldots,p}_{r})) = 2r$ for any positive integer r.

Exercise 12.2. Let A be the symbol algebra defined in Exercise 10.6, taking for k_0 the field $\mathbb{C}$ of complex numbers. Show that $ed_p^{\mathcal{A}_{\mathbb{C}}(p^2)}(A) = ed^{\mathcal{A}_{\mathbb{C}}(p^2)}(A) = 2$ while $ed_p^{\mathcal{A}_{\mathbb{C}}(p^2,p)}(A) = ed^{\mathcal{A}_{\mathbb{C}}(p^2,p)}(A) = 3$. [Hint: Use the following result of de Jong [110]: $exp\, B = ind\, B$ for every central simple algebra B over a finitely generated field extension of transcendence degree 2 of $\mathbb{C}$.]

Exercise 12.3. Let D be a central division algebra over a Henselian field F and let F' be a Henselian extension of F as in Lemma 12.8. Show that $p\Gamma_D = \Gamma_D \cap \Gamma_{F'} = (ker\,\theta_D) \cap \Gamma_{F'}$, and that $p\Gamma_D/\Gamma_F$ is a Lagrangian of $(ker\,\theta_D)/\Gamma_F$ for the symplectic pairing a_D of Prop. 8.17(iv). [Hint: Use the description of Γ_D/Γ_F in the proof of Lemma 12.8 and the canonical surjective map $\Gamma_D/\Gamma_F \to \Gamma_{D'}/\Gamma_{F'}$ to show that $\Gamma_D \cap \Gamma_{F'} = p\Gamma_D$. On the other hand, a comparison of θ_D and $\theta_{D'}$ yields $\Gamma_D \cap \Gamma_{F'} \subseteq ker\,\theta_D$. Letting $[\overline{D}{:}\overline{F}] = p^s$, compute $|p\Gamma_D{:}\Gamma_F| = p^{r-s} = \sqrt{|ker\,\theta_D{:}\Gamma_F|}$. Show that $\big((ker\,\theta_D)\cap\Gamma_{F'}\big)/\Gamma_F$ is totally isotropic for a_D by arguing as in the proof of Prop. 9.4.]

Notes

Essential dimension was originally defined by Buhler and Reichstein [46]. The notion was later extended to G-torsors of an algebraic group G by Reichstein [195]. The functorial approach to essential dimension formulated by Merkurjev is well described in Berhuy–Favi [23]. Good surveys on essential dimension are given in Reichstein [196], and Merkurjev [148], [151].

In this chapter we have given lower bounds on essential p-dimension for central simple algebras because those bounds have been obtained using valuation theory. Upper bounds are also known, though in most cases there are large gaps between the lower and the upper bounds. It is known that for any prime p and any field k with $char\, k \neq p$, and any integers r and s with $r \geq 2$ and $1 \leq s \leq r$, except $(p,r,s) = (2,2,1)$,

$$\left.\begin{array}{l}(r-1)p^r+p^{r-s}\\(r-1)2^{r-1}\end{array}\right\}\leq ed_p(\mathcal{A}_k(p^r,p^s))\leq\begin{cases}p^{2r-2}+p^{r-s} & \text{if } p\neq 2 \text{ or } s>1;\\ p^{2r-4}+p^{r-1} & \text{if } p=2 \text{ and } s=1.\end{cases}$$

The lower bounds are given in Th. 12.14. The upper bounds for $r = s$ were obtained by Ruozzi [216], improving previous upper bounds found by Meyer–Reichstein [153]; the upper bounds for $s < r$ in the top line were deduced from Ruozzi's result by Baek–Merkurjev [21, Lemma 7.1]. It follows that for p an odd prime, $p \neq \mathit{char}\, k$,

$$ed_p(\mathcal{A}_k(p^2,p)) = p^2+1,$$

as had been proved earlier by Merkurjev [149]. The upper bound in the lower line was given by Baek [20, Th. 1.1(ii)] In particular, for $\mathit{char}\, k \neq 2$,

$$ed_2(\mathcal{A}_k(8,2)) = 8 \qquad \text{and} \qquad ed_2(\mathcal{A}_k(16,2)) = 24.$$

Some years earlier, Rost [210] had proved that $ed(\mathcal{A}_k(4)) = 5$ for $\mathit{char}\, k \neq 2$. Baek has given bounds on $ed_p((\mathcal{A}_k(p^r,p^s))$ for fields of characteristic p in [19] and [20].

Appendix A
Commutative Valuation Theory

In this appendix we recall a number of known results from the valuation theory of (commutative) fields which are invoked in the main body of the book. We also recall in §A.4 the various kinds of rank associated to valuations. We include only a few proofs, where the proofs are short or not readily available in the standard references such as the books by Engler–Prestel [73], Efrat [70], or Endler [72]. We give references for all omitted proofs.

A.1 Extensions of valuations on fields

Throughout this section, let $F \subseteq K$ be fields, let Γ be a totally ordered divisible abelian group, and let $v\colon F \to \Gamma \cup \{\infty\}$ be a valuation on F. As in Ch. 1, let $\mathcal{O}_{F,v}$ denote the valuation ring of v; $\mathfrak{m}_{F,v}$ the maximal ideal of $\mathcal{O}_{V,v}$; $\overline{F}^{v} = \mathcal{O}_{F,v}/\mathfrak{m}_{F,v}$ the residue field; and $\Gamma_{F,v} = v(F^{\times})$ the value group. When there is no danger of ambiguity, the subscript v may be omitted. An extension of v to K is a valuation $w\colon K \to \Gamma' \cup \{\infty\}$ such that Γ' is a totally ordered divisible abelian group containing Γ as an ordered subgroup and $w|_F = v$. Since we are free to enlarge Γ if necessary, we may always assume that $\Gamma' = \Gamma$.

Theorem A.1 (Chevalley). *For any field K containing F there is at least one extension of any valuation on F to a valuation on K.*

Proof. See, e.g., Engler–Prestel [73, Th. 3.1.1, p. 57]. □

Theorem A.2 (Fundamental Inequality). *Suppose $[K{:}F] < \infty$. Let $v_1, \ldots, v_r$ be distinct extensions of v to K. Then,*

$$\sum_{i=1}^{r} [\overline{K}^{v_i}{:}\overline{F}^{v}] \cdot |\Gamma_{K,v_i}{:}\Gamma_{F,v}| \;\le\; [K{:}F]. \tag{A.1}$$

Proof. See, e.g., Bourbaki [30, Ch. 6, § 8, No. 3, Th. 1] or Engler–Prestel [73, Th. 3.3.4, p. 75]. □

J.-P. Tignol, A.R. Wadsworth, *Value Functions on Simple Algebras, and Associated Graded Rings*, Springer Monographs in Mathematics,
DOI 10.1007/978-3-319-16360-4

Formula (A.1) is called the Fundamental Inequality. It follows immediately from this formula that if $[K{:}F] < \infty$, then v has at most $[K{:}F]$ extensions to K. It follows also that if K is algebraic over F of infinite degree and w is any extension of v to K, then $\overline{K}$ is algebraic over $\overline{F}$, and Γ_K lies in the divisible hull $\mathbb{H}(\Gamma_F) = \mathbb{Q} \otimes_{\mathbb{Z}} \Gamma_F$ of Γ_F.

Definition A.3. Let K be a field algebraic over F, and let v be a valuation on F. Suppose first that $[K{:}F] < \infty$. Then v is said to be *defectless in* K if equality holds in the Fundamental Inequality (A.1) when $v_1, \ldots, v_r$ are all the distinct extensions of v to K. It is easy to check using Th. A.1 and Th. A.2 that if E is a field with $F \subseteq E \subseteq K$, then v is defectless in K if and only if v is defectless in E and each extension of v to E is defectless in K (see Endler [72, (18.1), pp. 136–137]). Now suppose that K is algebraic over F of infinite degree. Then, v is said to be *defectless in* K if v is defectless in L for each field L with $F \subseteq L \subseteq K$ and $[L{:}F] < \infty$. This is equivalent to: there is a family of fields L_j, $j \in J$ such that $F \subseteq L_j \subseteq K$, $[L_j{:}F] < \infty$, and v is defectless in L_j for each $j \in J$, and $K = \bigcup_{j \in J} L_j$.

Recall, e.g. from Bourbaki [30, Ch. VI, §8.5, Cor. 1] or Engler–Prestel [73, Th. 3.3.5, p. 76] that if v on F is discrete of rank 1, i.e., $\Gamma_{F,v} \cong \mathbb{Z}$, then v is defectless in every separable algebraic field extension of F.

Definition A.4. Let K be a field algebraic over F, and let v be a valuation on F with a unique extension to K, call it v_K. Suppose first that $[K{:}F] < \infty$. We say that (with respect to v and v_K)

- K is *inertial*[1] over F if $[\overline{K}{:}\overline{F}] = [K{:}F]$ and $\overline{K}$ is separable over $\overline{F}$;
- K is *totally ramified* over F if $|\Gamma_K{:}\Gamma_F| = [K{:}F]$;
- K is *tamely ramified* over F if $[\overline{K}{:}\overline{F}]|\Gamma_K{:}\Gamma_F| = [K{:}F]$, $\overline{K}$ is separable over $\overline{F}$, and *char* $\overline{F} \nmid |\Gamma_K{:}\Gamma_F|$.

Take any field E with $F \subseteq E \subseteq K$. Then $v_K|_E$ is the unique extension of v to E, and v_K is the unique extension of $v_K|_E$ to K. Note that K is then inertial over F if and only if E is inertial over F with respect to v and $v_K|_E$ and also K is inertial over E with respect to $v_K|_E$ and v_K. Likewise K is totally ramified (resp. tamely ramified) over F if and only if E is totally ramified (resp. tamely ramified) over F and K is totally ramified (resp. tamely ramified) over E.

Now suppose $[K{:}F] = \infty$ (but still assume K algebraic over F with a unique extension of v to v_K on K). We say that K is *inertial* (resp. *totally ramified*, resp. *tamely ramified*) *over* F if L is inertial (resp. totally rami-

[1] We use "inertial" rather than the more widely appearing but somewhat misleading term "unramified," which suggests only that $\Gamma_K = \Gamma_F$. When v has a unique extension to K, the condition that K is inertial over F is equivalent to: $\Gamma_K = \Gamma_F$ *and* v is defectless in K *and* $\overline{K}$ is separable over $\overline{F}$.

fied, resp. tamely ramified) over F for each field L with $F \subseteq L \subseteq K$ and $[L{:}F] < \infty$. Each property holds for K over F if $K = \bigcup_{j \in J} L_j$ for some family of fields L_j, $j \in J$, such that for each j, $F \subseteq L_j \subseteq K$, $[L_j{:}F] < \infty$, and the property holds for L_j over F.

Note that if K is algebraic over F and K has any of these properties over F, then v is defectless in K. Also, if K is inertial or tamely ramified over F, then K is separable over F. For it suffices to check this if $[K{:}F] < \infty$ and $char\, F = p \neq 0$. Then let S be the separable closure of F in K. Since K is purely inseparable over S, we have $\overline{K}$ is purely inseparable over $\overline{S}$ and Γ_K/Γ_S is p-primary torsion, by Prop. A.11 below. Since K is inertial or tamely ramified over S and $v_K|_S$ is defectless in K, we must have $S = K$.

Proposition A.5. *Let $F \subseteq L \subseteq K$ be fields with K algebraic over F. Let v be a valuation on F with a unique extension to K, denoted v_K.*

(i) *If v is defectless in L and $v_K|_L$ is defectless in K, then v is defectless in K.*
(ii) *If L is inertial (resp. totally ramified, resp. tamely ramified) over F and K is inertial (resp. totally ramified, resp. tamely ramified) over L, then K is inertial (resp. totally ramified, resp. tamely ramified) over F.*

Proof. Throughout the proof we write simply v for the restriction of v to any subfield of K.

(i) As noted above, if $[K{:}F] < \infty$, then the desired transitivity follows easily from the Fundamental Inequality. Also, we may assume that $[L{:}F] = \infty$, since otherwise the result is clear from the definitions. Take any field M with $F \subseteq M \subseteq K$ and $[M{:}F] < \infty$. We must show that v on F is defectless in M. For this it suffices to show that there is a field L' with $F \subseteq L' \subseteq L$ and $[F{:}L'] < \infty$ such that v on L' is defectless in $M{\cdot}L'$. For v on F is defectless in L' since it is defectless in L. Then the transitivity of defectlessness for finite-degree extensions shows that v on F is defectless in $M{\cdot}L'$. It then follows by the Fundamental Inequality that v on F is defectless in M.

Note that for any $c_1, \ldots, c_n \in M \cdot L$, there is a finite-degree extension L_1 of F within L such that $c_1, \ldots, c_n \in M{\cdot}L_1$. For each c_i is a finite sum $\sum_j m_{ij}\ell_{ij}$ with the $m_{ij} \in M$ and $\ell_{ij} \in L$; choose L_1 so that all the $\ell_{ij} \in L_1$. Note also that for any field L_2 with $F \subseteq L_2 \subseteq L$ and $[L_2{:}F] < \infty$ we have

$$[M{\cdot}L : L] \leq [M{\cdot}L_2 : L_2] \leq [M{:}F] < \infty,$$

but if L_2 is chosen suitably large, then the first inequality is an equality. To see this, let $\mathcal{K}$ be the kernel of the map $M \otimes_F L \to M{\cdot}L$ given by multiplication, and let $y_1, \ldots, y_m$ be an L-vector space base of $\mathcal{K}$. Choose L_2 finite-dimensional over F but sufficiently large that $y_1, \ldots, y_m$ lie in the kernel $\mathcal{K}_2 = ker\big(M \otimes_F L_2 \to M{\cdot}L_2\big)$. The y_i are L_2-linearly independent since they are L-linearly independent. Then as $dim_{L_2} \mathcal{K}_2 \geq m = dim_L \mathcal{K}$, we have

$$[M\cdot L:L] \le [M\cdot L_2:L_2] = dim_{L_2}(M\otimes_F L_2) - dim_{L_2}\mathcal{K}_2$$
$$= [M{:}F] - dim_{L_2}\mathcal{K}_2 \le dim_L(M\otimes_F L) - dim_L \mathcal{K} = [M\cdot L:L].$$

Hence, $[M\cdot L:L] = [M\cdot L_2:L_2]$.

Now, since $[\overline{M\cdot L}:\overline{L}] < \infty$ and $|\Gamma_{M\cdot L}{:}\Gamma_L| < \infty$, we can choose $a_1,\dots,a_n \in \mathcal{O}_{M\cdot L}$ and $b_1,\dots,b_m \in M^\times$ such that $\overline{M\cdot L} = \overline{L}(\overline{a_1},\dots,\overline{a_n})$ and $\Gamma_{M\cdot L} = \langle v(b_1),\dots,v(b_m)\rangle + \Gamma_L$. Choose a finite-degree extension L' of F in L sufficiently large that $[M\cdot L'{:}L'] = [M\cdot L{:}\,L]$ and $a_1,\dots,a_n,b_1,\dots,b_m \in M\cdot L'$. Since $\overline{M\cdot L'} \supseteq \overline{L'}(\overline{a_1},\dots\overline{a_n})$, we have

$$[\overline{M\cdot L'}:\overline{L'}] \ge [\overline{L'}(\overline{a_1},\dots,\overline{a_n}):\overline{L'}] \ge [\overline{L}(\overline{a_1},\dots,\overline{a_n}):\overline{L}] = [\overline{M\cdot L}:\overline{L}].$$

In addition, let $\Gamma' = \big\langle v(b_1),\dots,v(b_m)\big\rangle \subseteq \Gamma_{M\cdot L'}$. Since $\Gamma_{M\cdot L'} \supseteq \Gamma' + \Gamma_{L'}$ and $\Gamma_{M\cdot L} = \Gamma' + \Gamma_L$, we have

$$|\Gamma_{M\cdot L'}:\Gamma_{L'}| \ge |\Gamma' + \Gamma_{L'}:\Gamma_{L'}| = |\Gamma'{:}\,\Gamma'\cap\Gamma_{L'}|$$
$$\ge |\Gamma'{:}\,\Gamma'\cap\Gamma_L| = |\Gamma'+\Gamma_L:\Gamma_L| = |\Gamma_{M\cdot L}{:}\Gamma_L|.$$

By combining these inequalities with the Fundamental Inequality and recalling that v on L is defectless in $M\cdot L$, we obtain

$$[M\cdot L':L'] \ge [\overline{M\cdot L'}:\overline{L'}]\cdot|\Gamma_{M\cdot L'}{:}\Gamma_{L'}|$$
$$\ge [\overline{M\cdot L}:\overline{L}]\cdot|\Gamma_{M\cdot L}{:}\Gamma_L| = [M\cdot L:L] = [M\cdot L':L'].$$

Thus, equality holds throughout, showing that v on L' is defectless in $M\cdot L'$. As noted at the beginning of the argument, this implies that v on F is defectless in M. Since this is true for every finite-degree extension M of F in K, the valuation v on F is defectless in K.

(ii) All assertions in (ii) follow quickly from (i): Note that K is inertial over F if and only if v on F is defectless in K, $\Gamma_K = \Gamma_F$, and $\overline{K}$ is separable over $\overline{F}$. Thus, if L is inertial over F and K is inertial over L, then v on F is defectless in K by (i), $\Gamma_K = \Gamma_L = \Gamma_F$, and $\overline{K}$ is separable over $\overline{F}$; so K is inertial over F. Similarly, K is totally ramified over F if and only if v on F is defectless in K and $\overline{K} = \overline{F}$. From the transitivity of defectlessness given in (i) the transitivity of total ramification follows immediately. The transitivity of tame ramification follows at once from the definition and (i). □

Before examining the Galois theory of valued field extensions we recall a key property for normal extensions, whether Galois or not.

Theorem A.6. *Suppose K is normal over F. Then, $\mathrm{Aut}(K/F)$ acts transitively on the set of valuations of K extending v on F.*

Proof. See, e.g., Engler–Prestel [73, Th. 3.2.15, p. 69]. □

Now assume that K is Galois over F, possibly of infinite degree, and let $G = \mathcal{G}(K/F)$. Then, G is a profinite topological group with the Krull topology, in which a fundamental system of open neighborhoods of $\{id_K\}$ is the set of subgroups $\mathcal{G}(K/L)$ as L ranges over the fields such that $F \subseteq L \subseteq K$ and $[L{:}F] < \infty$. Fix a valuation w of K extending v on F. There are three distinguished closed subgroups of G determined by the extension w of v:

- $G^{dec} = \{\sigma \in G \mid w \circ \sigma = w\}$, the *decomposition group*;
- $G^{in} = \{\sigma \in G \mid \sigma(c) - c \in \mathfrak{m}_K \text{ for all } c \in \mathcal{O}_K\}$, the *inertia group*;
- $G^{\mathrm{ram}} = \{\sigma \in G \mid w(\sigma(c) - c) > w(c) \text{ for all } c \in K^\times\}$, the *ramification group*.

We also write $\mathcal{G}^{\mathrm{dec}}(K/F)$, $\mathcal{G}^{\mathrm{in}}(K/F)$, $\mathcal{G}^{\mathrm{ram}}(K/F)$ for $\mathcal{G}(K/F)^{\mathrm{dec}}$, $\mathcal{G}(K/F)^{in}$, and $\mathcal{G}(K/F)^{\mathrm{ram}}$ respectively. We have

$$G^{\mathrm{ram}} \subseteq G^{in} \subseteq G^{dec} \subseteq G.$$

The corresponding fixed fields are

$$D_{K/F} = K^{G^{dec}}, \qquad I_{K/F} = K^{G^{in}}, \qquad R_{K/F} = K^{G^{\mathrm{ram}}},$$

which are called the *decomposition field*, *inertia field*, and *ramification field* for the extension w of v. Thus,

$$F \subseteq D_{K/F} \subseteq I_{K/F} \subseteq R_{K/F} \subseteq K.$$

The next three theorems give properties of the fields $D_{K/F}$, $I_{K/F}$, $R_{K/F}$ and the corresponding subgroups of G. For any field L with $F \subseteq L \subseteq K$ the valuation considered on L is understood to be $w|_L$.

Theorem A.7.
(i) *$w|_{D_{K/F}}$ is an immediate extension of v, i.e., $\overline{D_{K/F}} = \overline{F}$ and $\Gamma_{D_{K/F}} = \Gamma_F$.*
(ii) *w is the unique extension of $w|_{D_{K/F}}$ to K.*
(iii) *$D_{K/F} = F$ if and only if w is the unique extension of v to K.*
(iv) *If $[K{:}F] < \infty$, then $[D_{K/F}{:}F]$ equals the number of extensions of v to K.*

Proof. For (i), see Engler–Prestel [73, Cor. 5.3.8(0), pp. 134–135]. Parts (ii)–(iv) follow immediately from Th. A.6, and are proved in Efrat [70, Th. 15.2.2, p. 134]. □

Theorem A.8.
(i) *$\Gamma_{I_{K/F}} = \Gamma_F$.*
(ii) *$\overline{K}$ is normal over $\overline{F}$ and $\overline{I_{K/F}}$ is the separable closure of $\overline{F}$ in $\overline{K}$. So, $\overline{I_{K/F}}$ is Galois over $\overline{F}$.*
(iii) *$I_{K/F}$ is inertial over $D_{K/F}$.*
(iv) *For any $\sigma \in G^{dec}$, since $w \circ \sigma = w$, we have $\sigma(\mathcal{O}_K) = \mathcal{O}_K$ and $\sigma(\mathfrak{m}_K) = \mathfrak{m}_K$; so σ induces $\overline{\sigma} \in Aut(\overline{K}/\overline{F})$. There is a short exact sequence of continuous homomorphisms of profinite groups,*

$$1 \longrightarrow G^{in} \longrightarrow G^{dec} \xrightarrow{\ \psi\ } Aut(\overline{K}/\overline{F}) \longrightarrow 1,$$

where $\psi(\sigma) = \overline{\sigma}$. Hence, G^{in} is a normal subgroup of G^{dec}; so, $I_{K/F}$ is Galois over $D_{K/F}$, and there are topological group isomorphisms

$$\mathcal{G}(I_{K/F}/D_{K/F}) \cong G^{\text{dec}}/G^{\text{in}} \cong \text{Aut}(\overline{K}/\overline{F}) \cong \mathcal{G}(\overline{I_{K/F}}/\overline{F}).$$

(v) *The map $L \mapsto \overline{L}$ gives an inclusion-preserving and degree-preserving one-to-one correspondence between fields L with $D_{K/F} \subseteq L \subseteq I_{K/F}$ and fields M with $\overline{F} \subseteq M \subseteq \overline{I_{K/F}}$. Moreover, L is Galois over $D_{K/F}$ if and only if $\overline{L}$ is Galois over $\overline{F}$.*

Proof. Most of this is given in Efrat [70, pp. 136–142]. Specifically, for (i) combine [70, Prop. 16.1.3(e)] and Th. A.7(i) above; for (ii), $\overline{K}$ is normal over $\overline{F}$ by [70, Cor. 15.2.5], $\overline{I_{K/F}}$ is Galois over $\overline{F}$ by [70, Prop. 16.1.3(b)] and Th. A.7(i) above, and $\overline{K}$ is purely inseparable over $\overline{I_{K/F}}$ by [70, Prop. 16.1.3(a)]; (iv) is given by [70, Th. 16.1.1]. (v) follows from the isomorphism of Galois groups in (iv) (see also Endler [72, Th. (19.13)(b), p. 152]). (iii) follows from the degree-preserving isomorphism in (v) and separability of $\overline{I_{K/F}}$ over $\overline{D_{K/F}}$ given by (ii). □

For a field E, let $\mu(E)$ denote the group of all roots of unity in E.

Theorem A.9. *Let $\overline{p} = \text{char}\,\overline{F}$.*

(i) *If $\overline{p} = 0$, then $G^{\text{ram}} = \{\text{id}_K\}$ and $R_{K/F} = K$.*
(ii) *If $\overline{p} \neq 0$, then G^{ram} is the unique $\overline{p}$-Sylow subgroup of G^{in}. Hence, G^{ram} is a normal subgroup of G^{dec}.*
(iii) *$\overline{K}$ is purely inseparable over $\overline{R_{K/F}}$ and $\overline{R_{K/F}} = \overline{I_{K/F}}$.*
(iv) *$R_{K/F}$ is Galois and tamely ramified over $D_{K/F}$.*
(v) *If $\overline{p} \neq 0$, then $\Gamma_{R_{K/F}}$ is the "prime-to-$\overline{p}$ closure" of Γ_F in Γ_K, i.e., $\Gamma_{R_{K/F}}/\Gamma_F$ has no $\overline{p}$-torsion while $\Gamma_K/\Gamma_{R_{K/F}}$ is a $\overline{p}$-primary group.*
(vi) *There is a well-defined $\mathbb{Z}$-bilinear pairing*

$$\mathcal{G}(R_{K/F}/I_{K/F}) \times \Gamma_{R_{K/F}}/\Gamma_F \longrightarrow \mu(\overline{K})$$

given by $(\sigma, \gamma + \Gamma_F) \mapsto \overline{\sigma(a)/a}$ for any $a \in R_{K/F}^{\times}$ with $w(a) = \gamma$. This pairing has trivial left and right kernels.
(vii) *There is a short exact sequence of continuous homomorphisms of profinite groups:*

$$1 \longrightarrow G^{\text{ram}} \longrightarrow G^{\text{in}} \xrightarrow{\ \varphi\ } \text{Hom}(\Gamma_{R_{K/F}}/\Gamma_F, \mu(\overline{K})) \longrightarrow 1,$$

where φ is induced by the pairing of part (vi).
(viii) *$R_{K/F}$ is totally and tamely ramified and abelian Galois over $I_{K/F}$.*
(ix) *The map $L \mapsto \Gamma_L$ gives an inclusion-preserving one-to-one correspondence between the fields L with $I_{K/F} \subseteq L \subseteq R_{K/F}$ and groups Λ with $\Gamma_F \subseteq \Lambda \subseteq \Gamma_{R_{K/F}}$.*

(x) *If* $[R_{K/F}{:}I_{K/F}] < \infty$, *then*

$$\mathcal{G}(R_{K/F}/I_{K/F}) \cong G^{\text{in}}/G^{\text{ram}} \cong \Gamma_{R_{K/F}}/\Gamma_F.$$

(xi) *If* K *is tamely ramified over* $D_{K/F}$, *then* G^{ram} *is trivial.*

Proof. See Efrat [70, pp. 145–148] for much of this. Specifically, for (i) and (ii) see [70, Th. 16.2.3]; for (iii), for the pure inseparability see [70, Cor. 16.2.4(b)] and for $\overline{R_{K/F}} = \overline{I_{K/F}}$ see [70, Cor. 16.2.7(b)]; for (iv), $R_{K/F}$ is Galois over $D_{K/F}$ by [70, Th. 16.2.3], and $R_{K/F}$ is tamely ramified over $D_{K/F}$ since $I_{K/F}$ is inertial over $D_{K/F}$ by Th. A.8(iii) above and $R_{K/F}$ is totally and tamely ramified over $I_{K/F}$ by (viii) of this theorem; for (v) see [70, Th. 16.2.6]; for (vi), $\Gamma_K/\Gamma_{R_{K/F}}$ is $\overline{p}$-primary by [70, Cor. 16.2.4(a)] and $\Gamma_{R_{K/F}}/\Gamma_F$ has no $\overline{p}$-torsion by (viii) of this theorem and Th. A.7(a); for (vii) see [70, Cor. 16.2.7(c)]; for (viii) see [70, Cor. 16.2.7(a),(d)].

(ix) Let $\mathcal{H}$ denote the contravariant left exact functor of abelian groups $A \mapsto \mathit{Hom}(A, \mu(\overline{K}))$. If A is a finite abelian group of exponent e, let $\mu_e(\overline{K})$ denote the group of e-th roots of unity in $\overline{K}^{\times}$, and let $e' = |\mu_e(\overline{K})|$; then $\mathcal{H}(A) = \mathit{Hom}(A, \mu_e(\overline{K}))$. The duality for finite abelian groups shows that $\mathcal{H}(A) \cong A/e'A$. Moreover, if $|\mathcal{H}(A)| = |A|$, i.e., $e' = e$, then there is a perfect pairing

$$\langle \cdot, \cdot \rangle \colon A \times \mathcal{H}(A) \longrightarrow \mu_e(\overline{K}) \qquad \text{given by} \quad \langle a, f \rangle = f(a).$$

There is then a one-to-one inclusion-reversing correspondence between the subgroups B of A and subgroups of $\mathcal{H}(A)$ given by $B \mapsto \mathcal{H}(A/B)$.

Now, let E be a field with $I_{K/F} \subseteq E \subseteq R_{K/F}$, and let $H = \mathcal{G}(K/E)$, which is a closed subgroup of $G = \mathcal{G}(K/F)$ with $G^{\text{ram}} \subseteq H \subseteq G^{\text{in}}$. For the Galois extension K of E and valuation w as an extension of $w|_E$, there are inertia and ramification groups H^{in} and H^{ram}. It is clear from the definitions that

$$H^{\text{in}} = G^{\text{in}} \cap H = H \qquad \text{and} \qquad H^{\text{ram}} = G^{\text{ram}} \cap H = G^{\text{ram}}.$$

Since $\mathcal{H}$ is left exact, there is an exact sequence

$$0 \longrightarrow \mathcal{H}(\Gamma_{R_{K/F}}/\Gamma_E) \longrightarrow \mathcal{H}(\Gamma_{R_{K/F}}/\Gamma_F) \longrightarrow \mathcal{H}(\Gamma_E/\Gamma_F).$$

We have $G^{\text{in}}/G^{\text{ram}} \cong \mathcal{H}(\Gamma_{R_{K/F}}/\Gamma_F)$ by (vii), and there is a compatible isomorphism of subgroups,

$$H/G^{\text{ram}} = H^{\text{in}}/H^{\text{ram}} \cong \mathcal{H}(\Gamma_{R_{K/F}}/\Gamma_E),$$

by (vii) applied to the extension K/E. Therefore, H is the kernel of the composition of maps

$$G^{\text{in}} \to G^{\text{in}}/G^{\text{ram}} \xrightarrow{\sim} \mathcal{H}(\Gamma_{R_{K/F}}/\Gamma_F) \to \mathcal{H}(\Gamma_E/\Gamma_F).$$

This shows that H (hence also $E = K^H$) is determined by Γ_E/Γ_F. Therefore, the map $L \mapsto \Gamma_L/\Gamma_F$ of (ix) is injective.

For the surjectivity of the map of (ix), it suffices to consider those groups Λ with $\Gamma_F \subset \Lambda \subset \Gamma_{R_{K/F}}$ such that $|\Lambda/\Gamma_F| < \infty$. Indeed, given the result for such Λ, the result for arbitrary Λ then follows by a straightforward direct limit argument. So, assume that $|\Lambda/\Gamma_F| < \infty$, and take a field E with $I_{K/F} \subseteq E \subseteq R_{K/F}$, such that $\Lambda \subseteq \Gamma_E$, and $[E{:}I_{K/F}] < \infty$. Then, E is Galois over $I_{K/F}$ as $R_{K/F}$ is abelian Galois over $I_{K/F}$. The discussion above shows that for $H = \mathcal{G}(K/E)$ there is an injection $G^{in}/H \to \mathcal{H}(\Gamma_E/\Gamma_F)$. But, as E is totally ramified over $I_{K/F}$, we have

$$|\Gamma_E{:}\Gamma_F| = [E{:}F] = |\mathcal{G}(E/F)| = |G^{in}/H| \le |\mathcal{H}(\Gamma_E/\Gamma_F)| \le |\Gamma_E{:}\Gamma_F|. \tag{A.2}$$

Thus, the inequalities here must be equalities; hence, $G^{in}/H \cong \mathcal{H}(\Gamma_E/\Gamma_F)$ and there is a perfect pairing between Γ_E/Γ_F and $\mathcal{H}(\Gamma_E/\Gamma_F)$. Let U be the kernel of the surjective composition $G^{in} \to \mathcal{H}(\Gamma_E/\Gamma_F) \to \mathcal{H}(\Lambda/\Gamma_F)$, and let $L = K^U$. We have $H \subseteq U \subseteq G^{in}$ and U is closed in G as H is closed in G and $|U{:}H| < \infty$. Hence, $I_{K/F} \subseteq L \subseteq E \subseteq R_{K/F}$ and $U = \mathcal{G}(K/L)$. In the isomorphism $G/H \xrightarrow{\sim} \mathcal{H}(\Gamma_E/\Gamma_F)$, U/H maps onto $ker\big(\mathcal{H}(\Gamma_E/\Gamma_F) \to \mathcal{H}(\Gamma_E/\Lambda)\big) = \mathcal{H}(\Gamma_E/\Lambda)$. On the other hand, as $U = \mathcal{G}(K/L)$, we have an isomorphism analogous to $G/H \cong \mathcal{H}(\Gamma_E/\Gamma_F)$ which is $G/U \cong \mathcal{H}(\Gamma_L/\Gamma_F)$. Hence U/H maps onto $ker\big(\mathcal{H}(\Gamma_E/\Gamma_F) \to \mathcal{H}(\Gamma_L/\Gamma_F)\big) = \mathcal{H}(\Gamma_E/\Gamma_L)$. Therefore, $\mathcal{H}(\Gamma_E/\Lambda) = \mathcal{H}(\Gamma_E/\Gamma_L)$ in $\mathcal{H}(\Gamma_E/\Gamma_F)$. From the one-to-one correspondence associated to the perfect pairing, it follows that $\Lambda/\Gamma_F = \Gamma_L/\Gamma_F$, and hence $\Lambda = \Gamma_L$. Thus, the map of (ix) is surjective.

(x) The first isomorphism in (x) is given by Galois theory. For the second, we have $G^{in}/G^{ram} \cong \mathcal{H}(\Gamma_{R_{K/F}}/\Gamma_F)$ by (vii). When $[R_{K/F}{:}I_{K/F}] < \infty$, we can take $E = R_{K/F}$ in the argument for (ix). From this (see after (A.2)), there is a perfect pairing between the finite abelian groups $\Gamma_{R_{K/F}}/\Gamma_F$ and $\mathcal{H}(\Gamma_{R_{K/F}}/\Gamma_F)$, so these groups are noncanonically isomorphic. Thus, $G^{in}/G^{ram} \cong \Gamma_{R_{K/F}}/\Gamma_F$.

(xi) Assume K is tamely ramified over $D_{K/F}$, and suppose G^{ram} is nontrivial. Then there is a finite-degree Galois extension L of $D_{K/F}$ in K with $\mathcal{G}^{ram}(L/D_{K/F})$ nontrivial. Let $M \subseteq L$ be the fixed field under $\mathcal{G}^{ram}(L/D_{K/F})$, which is the ramification field $M = R_{L/D_{K/F}}$ for the Galois extension $L/D_{K/F}$. By (iii) and (v), $\overline{L}$ is purely inseparable over $\overline{M}$ and Γ_L/Γ_M is a $\overline{p}$-primary group. But L is tamely ramified over M, since it is tamely ramified over $D_{K/F}$. Hence, $\overline{L} = \overline{M}$ and $\Gamma_L = \Gamma_M$. Because L is tamely ramified over M, it is defectless over M; hence, $L = M$. Therefore, $\big|\mathcal{G}^{ram}(L/D_{K/F})\big| = [L{:}M] = 1$, a contradiction. Hence, G^{ram} must be trivial. □

The following diagram summarizes the relationships between the residue fields and between the value groups for the intermediate fields determined by the extension w of v when K is Galois over F.

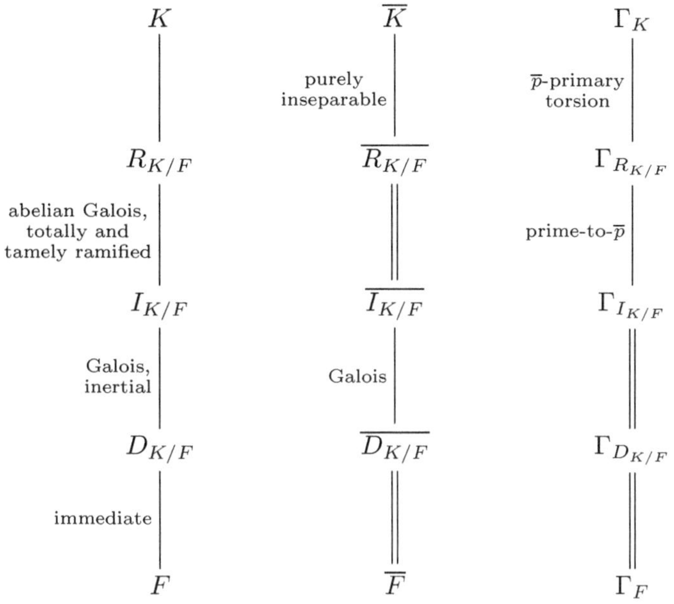

(A.3)

These theorems yield a formula for the defect of a valuation in a Galois extension:

Theorem A.10. *Suppose K is Galois over F and $[K{:}F] < \infty$. Let $\overline{p} = \operatorname{char} \overline{F}$. Let w be any valuation on K extending v on F. Then,*

$$[K{:}F] \;=\; r\,[\overline{K}{:}\overline{F}]\,|\Gamma_K{:}\Gamma_F|\,\delta_{K/F}, \tag{A.4}$$

where r is the number of extensions of v to K and $\delta_{K/F} = 1$ if $\overline{p} = 0$, while $\delta_{K/F} = \overline{p}^d$ for some nonnegative integer d if $\overline{p} \neq 0$. Moreover, v is defectless in K if and only if $\delta_{K/F} = 1$. If $\Gamma_F \cong \mathbb{Z}$, then $\delta_{K/F} = 1$.

Proof. We are writing $\overline{K}$ for $\overline{K}^w$ and Γ_K for $\Gamma_{K,w}$. By Th. A.7(i),(iii), Th. A.8(i),(ii), and Th. A.9(vii), we have

$$\begin{aligned}
[K{:}F] &= [D_{K/F}{:}F]\;[I_{K/F}{:}D_{K/F}]\;\;[R_{K/F}{:}I_{K/F}]\;\;[K{:}R_{K/F}] \\
&= \quad r \qquad [\overline{I_{K/F}} : \overline{D_{K/F}}]\,|\Gamma_{R_{K/F}}{:}\Gamma_{I_{K/F}}|\;[K{:}R_{K/F}] \\
&= \quad r \qquad\quad [\overline{R_{K/F}} : \overline{F}] \qquad |\Gamma_{R_{K/F}}{:}\Gamma_F| \quad [K{:}R_{K/F}].
\end{aligned} \tag{A.5}$$

If $\overline{p} = 0$, then $R_{K/F} = K$ by Th. A.9(i), so (A.5) yields formula (A.4) with $\delta_{K/F} = 1$. Now suppose $\overline{p} \neq 0$. Let

$$\delta \;=\; [K{:}R_{K/F}]\big/\big([\overline{K}{:}\overline{R_{K/F}}]\,|\Gamma_K{:}\Gamma_{R_{K/F}}|\big).$$

Then $\delta \geq 1$ by the Fundamental Inequality (A.1). Also, by Th. A.9(ii),(iii),(v), we have $[K{:}R_{K/F}] = |G^{\mathrm{ram}}| = \overline{p}^a$; $[\overline{K}{:}\overline{R_{K/F}}] = \overline{p}^b$; and $|\Gamma_K{:}\Gamma_{R_{K/F}}| = \overline{p}^c$ for some integers a, b, c. Hence $\delta = \overline{p}^d$, where $d = a - b - c \in \mathbb{Z}$, and $d \geq 0$

as $\delta \geq 1$. Furthermore, from (A.5) we have

$$\begin{aligned}[K{:}F] &= r\,[\overline{R_{K/F}} : \overline{F}]\,|\Gamma_{R_{K/F}}{:}\Gamma_F|\,\big([\overline{K}{:}\overline{R_{K/F}}]\,|\Gamma_K{:}\Gamma_{R_{K/F}}|\,\delta\big)\\ &= r\,[\overline{K} : \overline{F}]\,|\Gamma_K{:}\Gamma_F|\,\delta,\end{aligned}$$

which yields (A.4) with $\delta_{K/F} = \delta = \overline{p}^d$. Thus, (A.4) is proved for all values of $\overline{p}$.

For the last assertion of the theorem, let $v_1, \ldots, v_r$ be all the extensions of v to K, with $w = v_1$. Because G acts transitively on $\{v_1, \ldots, v_r\}$ by Th. A.6, for each i there is a $\sigma_i \in G$ with $w = v_i \circ \sigma_i$. Hence, $\Gamma_{K,v_i} = \Gamma_{K,w}$, for each i. Also, $\sigma_i(\mathcal{O}_{K,w}) = \mathcal{O}_{K,v_i}$ and $\sigma_i(\mathfrak{m}_{K,w}) = \mathfrak{m}_{K,v_i}$. So, σ_i induces an $\overline{F}$-isomorphism $\overline{\sigma}_i \colon \overline{K}^{w} \to \overline{K}^{v_i}$, showing that $[\overline{K}^{v_i}{:}\overline{F}] = [\overline{K}^{w}{:}\overline{F}]$. Formula (A.4) can therefore be rewritten

$$[K{:}F] \;=\; \Big(\sum_{i=1}^{r} [\overline{K}^{v_i}{:}\overline{F}]\,|\Gamma_{K,v_i}{:}\Gamma_F|\Big)\,\delta_{K/F},$$

which shows that v is defectless in K if and only if $\delta_{K/F} = 1$. In particular, v is defectless in K if $\Gamma_F \cong \mathbb{Z}$, as noted in Def. A.3. □

At the other extreme from Galois extensions are purely inseparable extensions. The extension theory for that case is very brief:

Proposition A.11. *Suppose* $\operatorname{char} F = p \neq 0$ *and* E *is a purely inseparable extension of* F*. Then* v *has a unique extension to* E*, and* $\overline{E}$ *is purely inseparable over* $\overline{F}$*, and* Γ_E/Γ_F *is a* p*-primary torsion group.*

Proof. See Engler–Prestel [73, Cor. 3.2.1, p. 66] for the uniqueness of an extension of v to E. The assertions about $\overline{E}$ and Γ_E follow easily from the pure inseparability of E over F. □

The information given above for Galois and purely inseparable extensions yields a proof of the general Ostrowski Theorem on defect for extensions of valuations:

Theorem A.12 ("Ostrowski's Theorem"). *Let* L *be a finite-degree field extension of* F*, and suppose the valuation* v *on* F *has a unique extension to a valuation on* L*. Let* $\overline{p} = \operatorname{char} \overline{F}$*. Then,*

$$[L{:}F] \;=\; [\overline{L}{:}\overline{F}]\,|\Gamma_L{:}\Gamma_F|\,\delta, \tag{A.6}$$

where $\delta = 1$ *if* $\overline{p} = 0$*, and* $\delta = \overline{p}^s$ *for some integer* $s \geq 0$ *if* $\overline{p} \neq 0$*. If* $\Gamma_F \cong \mathbb{Z}$ *and* L *is separable over* F*, then* $\delta = 1$*.*

Proof. Let v_L be the unique extension of v to L. Let S be the separable closure of F in L, and let $v_S = v_L|_S$, which is the unique extension of v to S. Let

$$\partial_{L/F} \;=\; [L{:}F]/\big([\overline{L}{:}\overline{F}]\,|\Gamma_F{:}\Gamma_L|\big),$$

and define $\partial_{S/F}$ and $\partial_{L/S}$ analogously. Thus, $\partial_{L/F}$ is the δ of (A.6), and

$$\partial_{L/F} = \partial_{S/F} \cdot \partial_{L/S}.$$

Let K be the Galois closure of S over F, and let w be any extension of v_S to K. Since v_S is the unique extension of v to S, the extensions of v_S to K are the same as the extensions of v to K; let r be the number of such extensions. By applying Th. A.10 to the Galois extension K of F for w and v and to the Galois extension K of S for w and v_S, we obtain

$$\begin{aligned} [S{:}F] &= [K{:}F] \,/\, [K{:}S] \\ &= \frac{r\,[\overline{K}{:}\overline{F}]\,|\Gamma_K{:}\Gamma_F|\,\delta_{K/F}}{r\,[\overline{K}{:}\overline{S}]\,|\Gamma_K{:}\Gamma_S|\,\delta_{K/S}} \\ &= [\overline{S}{:}\overline{F}]\,|\Gamma_S{:}\Gamma_F|\,\delta_{K/F}/\delta_{K/S}. \end{aligned}$$

Hence,

$$\partial_{S/F} = \delta_{K/F}/\delta_{K/S}.$$

If $\overline{p} = 0$, then $\delta_{K/F} = \delta_{K/L} = 1$ by Th. A.10, and $L = S$ as *char* $F = 0$. Thus, $\delta = \partial_{L/F} = \partial_{S/F} = 1$. If $\overline{p} \neq 0$, then $\delta_{K/F} = \overline{p}^d$ and $\delta_{K/S} = \overline{p}^e$ for some $d, e \in \mathbb{Z}$, by Th. A.10; so, $\partial_{S/F} = \overline{p}^{d-e}$. Also, as L is purely inseparable over S, Prop. A.11 shows that $\partial_{L/S} = \overline{p}^t$ for some integer t. Thus,

$$\delta = \partial_{L/F} = \partial_{S/F} \cdot \partial_{L/S} = \overline{p}^s,$$

where $s = d - e + t \in \mathbb{Z}$. Also, $s \geq 0$ as $\delta \geq 1$ by the Fundamental Inequality (A.1). If $\Gamma_F \cong \mathbb{Z}$ and L is separable over F, then L is defectless over F, as noted in Def. A.3. (The separability condition is genuinely needed here—see Ex. 4.14.) □

A.2 Henselian valuations

In this section we recall some basic properties of Henselian valuations and describe the maximal inertial and maximal tamely ramified extension fields of a Henselian valued field.

Definition A.13. A valuation v on a field F is said to be *Henselian* if v has a unique extension to each field L algebraic over F.

For example, if F is complete with respect to a rank 1 valuation, then the valuation is Henselian by, e.g., Engler–Prestel [73, Prop. 1.2.2, p. 14].

There are several characterizations of the Henselian property for a valuation, some of which are listed in the next theorem. Further characterizations can be found in Ribenboim's paper [204], and others still are given in Th. 5.49 above.

For a valuation v on a field F and $a \in \mathcal{O}_F$ we write $\overline{a}$ for the image of a in $\overline{F}$. For $f = \sum_{i=0}^n a_i X^i \in \mathcal{O}_F[X]$, we write $\overline{f} = \sum_{i=0}^n \overline{a}_i X^i \in \overline{F}[X]$ and f' for the formal derivative of f.

Theorem A.14. *Let v be a valuation on a field F. The following conditions are equivalent:*

(a) *v is Henselian.*
(b) *v has a unique extension to each field L algebraic over F.*
(c) *Take any $f \in \mathcal{O}_F[X] \setminus \mathfrak{m}_F \mathcal{O}_F[X]$ such that $\overline{f} = k\ell$ in $\overline{F}[X]$ with k and ℓ relatively prime. Then, there exist $g, h \in \mathcal{O}_F[X]$ with $f = gh$, $\overline{g} = k$, $\overline{h} = \ell$, and $\deg g = \deg k$.*
(c′) *Take any monic $f \in \mathcal{O}_F[X] \setminus \mathfrak{m}_F \mathcal{O}_F[X]$ such that $\overline{f} = k\ell$ in $\overline{F}[X]$ with k and ℓ relatively prime and monic. Then there exist monic $g, h \in \mathcal{O}_F[X]$ with $f = gh$, $\overline{g} = k$, $\overline{h} = \ell$.*
(d) *For each $f \in \mathcal{O}_F[X] \setminus \mathfrak{m}_F \mathcal{O}_F[X]$, if there is a $c \in \overline{F}$ with $\overline{f}(c) = 0$ but $\overline{f}'(c) \neq 0$, then there is a unique $a \in \mathcal{O}_F$ with $f(a) = 0$ and $\overline{a} = c$.*
(e) *For each monic $f \in \mathcal{O}_F[X]$ with f irreducible in $F[X]$ and $\deg \overline{f} > 0$ there is an irreducible $q \in \overline{F}[X]$ with $\overline{f} = q^s$ for some integer $s \geq 1$.*
(f) *For each $f \in \mathcal{O}_F[X]$ and $b \in \mathcal{O}_F$ such that $v(f(b)) > 2v(f'(b))$ there exists an $a \in \mathcal{O}_F$ such that $f(a) = 0$ and $v(a - b) > v(f'(b))$.*
(g) *Every commutative $\mathcal{O}_F$-algebra which is finitely generated as an $\mathcal{O}_F$-module is a direct product of local rings.*
(h) *Let R be a (possibly noncommutative) $\mathcal{O}_F$-algebra with R integral over $\mathcal{O}_F$, and let I be an ideal of R. If $\widehat{e}$ is an idempotent of R/I, then there is an idempotent e of R such that $e + I = \widehat{e}$.*
(i) *Let R be a (possibly noncommutative) $\mathcal{O}_F$-algebra with R integral over $\mathcal{O}_F$, and let I be an ideal of R. If $\widehat{e}_1, \ldots, \widehat{e}_n$ are idempotents of R/I with $\widehat{e}_i \widehat{e}_j = 0$ whenever $j \neq i$, then there are idempotents $e_1, \ldots, e_n$ of R such that $e_i e_j = 0$ whenever $j \neq i$, and each $e_i + I = \widehat{e}_i$.*
(j) *(Krasner's Lemma) For any $a, b \in F_{\text{alg}}^{\times}$, if*

$$v(b - a) > \max\{v(\sigma(a) - a) \mid \sigma \in \operatorname{Aut}(F_{\text{alg}}/F), \sigma(a) \neq a\},$$

then a is purely inseparable over $F(b)$.

Proof. (a) ⇔ (b) is by the definition of a Henselian valuation. The equivalence of (b), (c), (d), (e), and (f) is given in Engler–Prestel [73, Th. 4.1.3, pp. 87–88]. The uniqueness of a in (d) holds since otherwise c would be a multiple root of $\overline{f}$. (c) ⇔ (c′) is given in Ribenboim [204, Th. 3, (3)⇔(4), p. 13]. (c′) ⇔ (g) is given in Raynaud [194, Prop. 5, pp. 2–3] and also in Nagata [168, Th. (43.15), p. 185]. (c′) ⇔ (i) is given in Marubayashi et al. [132, Th. A.18, p. 180]. (i) ⇒ (h) is clear. (h) ⇒ (c′) follows immediately from Marubayashi et al. [132, Cor. A.14, p. 179]. For (b) ⇔ (j), see Efrat [70, Lemma 18.5.1, p. 171], or see Engler–Prestel [73, Th. 4.1.7, p. 91] for (b) ⇒ (j) and [73, Ex. 4.5.2, pp. 110–111] for (j) ⇒ (b). □

The Henselian property is well-behaved with respect to composites of valuations: See §1.2.3 above for our notation for composite valuations.

Proposition A.15. *Let v be a valuation on F, let w be any coarsening of v, and let u be the valuation v/w on $\overline{F}^{w}$ induced by v; so, $v = u * w$. Then, v is Henselian if any only if w and u are each Henselian.*

Proof. This follows easily from the valuation extension characterization of Henselian valuations in Th. A.14(i), using the fact that for any extension w' of w to the algebraic closure F_{alg} of F, the residue field $\overline{F_{alg}}^{w'}$ is the algebraic closure of $\overline{F}^{w}$. For another approach, see Engler–Prestel [73, Cor. 4.1.4, p. 90]. □

Example A.16. Let k be any field, and let $F = k((x_1))((x_2))\dots((x_n))$, the n-fold iterated Laurent series field over k, with its usual $(x_1, \dots, x_n)$-adic valuation $v_{x_1,\dots,x_n}$, as described in §1.1.3 and Exercise 1.2. Then, $v_{x_1,\dots,x_n}$ is Henselian by Prop. A.15, since it is a composite of n complete discrete rank 1 (so Henselian) valuations.

Now fix a Henselian valuation v on a field F. When we work with algebraic extension fields L of F, it is understood that the valuation considered on L is the unique extension of v to L, which is again denoted v. We next take a closer look at inertial and tamely ramified extensions of F.

A key property of Henselian fields is that simple roots of polynomials over the residue field can be lifted to roots of polynomials over the original field. This leads to a corresponding property of lifting separable field extensions of the residue field:

Proposition A.17. *Let $M \supseteq F$ be a field with a Henselian valuation v_M restricting to the Henselian valuation v on F (e.g., M could be any algebraic extension of F). Let S be a finite-degree separable field extension of $\overline{F}$ with $S \subseteq \overline{M}$. Then, there is a unique field L with $F \subseteq L \subseteq M$, with L inertial over F and $\overline{L} = S$ in $\overline{M}$.*

The field L of this proposition is called the *inertial lift* of S over F in M.

Proof. Since S is separable over $\overline{F}$, we have $S = \overline{F}(s)$ for some s in S. Let $g \in \overline{F}[X]$ be the minimal polynomial of s over $\overline{F}$, and let f be any monic inverse image of g in $\mathcal{O}_F[X]$. Since s is separable over $\overline{F}$, it is a simple root of g. Therefore, as M is Henselian, f has a unique root a in M such that $\overline{a} = s$ (see Th. A.14(d)). Let $L = F(a) \subseteq M$. Then, $S = \overline{F}(s) = \overline{F}(\overline{a}) \subseteq \overline{L}$. Moreover,

$$[L{:}F] \le deg(f) = deg(g) = [\overline{F}(s){:}\overline{F}] = [S{:}\overline{F}] \le [\overline{L}{:}\overline{F}] \le [L{:}F].$$

Thus, equality holds throughout, showing that $\overline{L} = S$ and $[L{:}F] = [\overline{L}{:}\overline{F}]$ and f is the minimal polynomial of a over F. Hence, L is inertial over F. For uniqueness, suppose L' is another subfield of M such that L' is inertial over F with $\overline{L'} = S$ in $\overline{M}$. Since $[L':F] = [S:\overline{F}\,] < \infty$, the valuation on L' is Henselian. Because $\overline{f} = g$ has the simple root s in $S = \overline{L'}$, there is a root a' of f in L'. But then $a' \in M$, and the uniqueness of a implies that $a' = a$. Hence, $L = F(a) \subseteq L'$. Since $[L:F] = [S:\overline{F}\,] = [L':F]$, we have $L' = L$. □

Inertial extensions are also well-behaved with respect to extensions of the ground field:

Proposition A.18. *Let $M \supseteq F$ be a field with Henselian valuation v_M extending the Henselian valuation v on F. Let K be a finite-degree inertial extension field of F, and let $K{\cdot}M$ be a compositum of K and M in M_{alg}. Then, $K{\cdot}M$ is inertial over M and $\overline{K{\cdot}M} = \overline{K}{\cdot}\overline{M}$.*

Proof. Since $\overline{K}$ is separable over $\overline{F}$, there is an $s \in \overline{K}^{\times}$ with $\overline{K} = \overline{F}(s)$. Let g be the minimal polynomial of s over $\overline{F}$, and let f be any monic inverse image of g in $\mathcal{O}_F[X]$. Then, g has no repeated roots as s is separable over $\overline{F}$. As we saw in the proof of Prop. A.17, since K is Henselian there is a unique $a \in \mathcal{O}_K$ such that a is a root of f and $\overline{a} = s$. That proof also shows that f is the minimal polynomial of a over F and $[F(a):F] = [\,\overline{F}(s) : \overline{F}\,] = [\,\overline{K} : \overline{F}\,] = [K:F]$; so, $K = F(a)$. Let h be the (monic) minimal polynomial of a over M; so $f = hk$ for some monic $k \in M[X]$. Since f is monic in $\mathcal{O}_M[X]$, its roots (so the roots of h and k) are all integral over $\mathcal{O}_M$; hence, the coefficients of h and k all lie in the integrally closed valuation ring $\mathcal{O}_M$. By taking images in $\overline{M}[X]$ and $\overline{M}$, we have $g = \overline{f} = \overline{h} \cdot \overline{k}$ and $\overline{h}(s) = \overline{h}(\overline{a}) = \overline{h(a)} = 0$. Since h is irreducible in $\mathcal{O}_M[X]$ and v_M is Henselian, by Th. A.14(e) its image $\overline{h}$ is a power of an irreducible polynomial in $\overline{M}[X]$. But $\overline{h}$ has no repeated roots since $\overline{h} \mid g$. Hence, $\overline{h}$ is irreducible in $\overline{M}[X]$; so $\overline{h}$ is the minimal polynomial of s over $\overline{M}$. Since $K{\cdot}M = F(a){\cdot}M = M(a)$ and $\overline{K{\cdot}M} \supseteq \overline{K}{\cdot}\overline{M} = \overline{F}(s){\cdot}\overline{M} = \overline{M}(s)$, we have

$$\begin{aligned}[K{\cdot}M : M] &= [M(a){:}M] = \deg h = \deg \overline{h} \\ &= [\,\overline{M}(s){:}\overline{M}\,] = [\,\overline{K}{\cdot}\overline{M} : \overline{M}\,] \leq [\,\overline{K{\cdot}M} : \overline{M}\,] \leq [K{\cdot}M : M].\end{aligned}$$

So, equality holds throughout. Hence, $[\,\overline{K{\cdot}M} : \overline{M}\,] = [K{\cdot}M : M]$ and $\overline{K{\cdot}M} = \overline{K}{\cdot}\overline{M} = \overline{M}(s)$, which is separable over $\overline{M}$. Thus, $K \cdot M$ is inertial over F. □

Corollary A.19.

(i) *Let K_1 and K_2 be two finite-degree field extensions in F_{alg} of the Henselian field F. If K_1 and K_2 are each inertial over F, then $K_1{\cdot}K_2$ is inertial over F with $\overline{K_1{\cdot}K_2} = \overline{K_1}{\cdot}\overline{K_2}$.*

(ii) *A field L algebraic over F is inertial over F if and only if L is a compositum of (possibly infinitely many) finite-degree inertial extensions of F.*

Proof. (i) If K_1 is inertial over F, then by Prop. A.18 $K_1{\cdot}K_2$ is inertial over K_2 with $\overline{K_1{\cdot}K_2} = \overline{K_1}{\cdot}\overline{K_2}$. When K_2 is also inertial over F, it follows by transitivity of the inertial extension property that $K_1 \cdot K_2$ is inertial over F.

(ii) If L is inertial over F, then by definition every field N with $F \subseteq N \subseteq L$ and $[N{:}F] < \infty$ is inertial over F. The field L is the compositum of all such fields N. Conversely, suppose L is a compositum of finite-degree inertial extensions of F. If $N_1, \ldots, N_k$ are finite-degree inertial extensions of F in L, then their compositum $N_1 \cdot \ldots \cdot N_k$ is inertial over F by (i). Since L is the union of such composita, it is also inertial over F. □

Definition A.20. Let F be a Henselian field, and M a field containing F with Henselian valuation v_M restricting to the valuation on F. The *inertial closure* of F in M is the compositum of all the finite-degree inertial extension fields of F lying in M. Note that this inertial closure K is inertial over F by Cor. A.19(ii), which also shows that if L is any inertial extension of F lying in M, then $L \subseteq K$. Moreover, $\overline{K}$ coincides with the separable algebraic closure S of $\overline{F}$ in $\overline{M}$. For $\overline{K} \subseteq S$ as K is inertial over F. But for any $s \in S$, by Prop. A.17 there is an inertial lift N of $\overline{F}(s)$ over F in M. Since $N \subseteq K$, we have $s \in \overline{N} \subseteq \overline{K}$; so, $\overline{K} = S$. Note also that if M is Galois over F, then K is the inertia field for the extension v_M of v.

We next describe the tame totally ramified extensions of a Henselian field.

Lemma A.21. *Let K be a totally and tamely ramified extension of F (with F Henselian). Take any $\gamma \in \Gamma_K$, and let m be the order of $\gamma + \Gamma_F$ in Γ_K/Γ_F. Then, there is an $a \in K^\times$ with $v(a) = \gamma$ and $a^m \in F$.*

Proof. Let $\overline{p} = \operatorname{char} \overline{F}$. Take any $b \in K^\times$ with $v(b) = \gamma$. Then, $v(b^m) = m\gamma \in \Gamma_F$. Take any $c \in F^\times$ with $v(c) = m\gamma$. Since $v(b^m/c) = 0$ and $\overline{K} = \overline{F}$ as K is totally ramified over F, there is a $d \in F^\times$ with $v(d) = 0$ and $\overline{d} = \overline{b^m/c}$. Then $\overline{b^m/cd} = \overline{1}$ in $\overline{K}^\times$. Let $f = X^m - b^m/cd \in \mathcal{O}_K[X]$. This f has image $\overline{f} = X^m - \overline{1}$ in $\overline{K}[X]$. Since $\overline{p} \nmid m$ as K is tamely ramified over F, the polynomial $\overline{f}$ has $\overline{1}$ as a simple root in $\overline{K}$. Therefore, as v on K is Henselian, f has a root u in K. Hence, $u^m = b^m/cd$, and $v(u) = \frac{1}{m} v(b^m/cd) = 0$. Let $a = b/u \in K^\times$. Then, $a^m = b^m/u^m = cd \in F$ and $v(a) = v(b) = \gamma$. □

Proposition A.22. *Let K be a finite-degree totally and tamely ramified extension field of F (with F Henselian). Let e be the exponent of Γ_K/Γ_F. Then,*

(i) *K is a radical extension of F, i.e., there are $a_1, \ldots, a_n \in K$ and positive integers $m_1, \ldots, m_n$ such that $K = F(a_1, \ldots, a_n)$ and each $a_i^{m_i} \in F$.*
(ii) *K is Galois over F if and only if $\overline{F}$ contains a primitive e-th root of unity, if and only if F contains a primitive e-th root of unity.*
(iii) *If K is Galois over F, then K is an e-Kummer extension of F with*

$$\Gamma_K/\Gamma_F \cong \{c \in K^\times \mid c^e \in F^\times\}/F^\times \cong \mathcal{G}(K/F).$$

The first isomorphism is canonical; the second is given by Kummer theory and is not canonical.

Proof. (i) Write Γ_K/Γ_F as a direct sum of cyclic subgroups

$$\Gamma_K/\Gamma_F = \langle \gamma_1 + \Gamma_F \rangle \oplus \ldots \oplus \langle \gamma_n + \Gamma_F \rangle, \tag{A.7}$$

and let m_i be the order of $\gamma_i + \Gamma_F$ in Γ_K/Γ_F. So,

$$[K{:}F] = |\Gamma_K/\Gamma_F| = m_1 \ldots m_n.$$

By the preceding lemma, for each i there is an $a_i \in K^\times$ with $a_i^{m_i} \in F$. For $L = F(a_1, \ldots, a_n) \subseteq K$, we have $\gamma_1, \ldots, \gamma_n \in \Gamma_L$. Thus, $\Gamma_L = \Gamma_K$, and hence $L = K$, as K is totally ramified over F, so also totally ramified over L. This shows that K is a radical extension of F.

(ii) Let $\overline{p} = char\,\overline{F}$. For the a_i and γ_i above, we have $\gamma_i \in \Gamma_{F(a_i)}$ and hence

$$[F(a_i){:}F] \geq |\Gamma_{F(a_i)}{:}\Gamma_F| \geq \big|\langle \gamma_i + \Gamma_F \rangle\big| = m_i.$$

But also $[F(a_i){:}F] \leq m_i$ as $a_i^{m_i} \in F$. Therefore, $[F(a_i){:}F] = m_i$, and hence $f_i = X^{m_i} - a_i^{m_i}$ is the minimal polynomial of a_i over F.

Suppose first that K is Galois over F. Then each f_i splits over K with no repeated roots, and hence K contains a primitive m_i-th root of unity. Since $e = exp(\Gamma_K/\Gamma_F) = lcm(m_1, \ldots, m_n)$, K contains a primitive e-th root of unity. Therefore, the polynomial $X^e - 1$ splits over K. Its roots all have value 0. Hence $X^e - \overline{1}$ splits over $\overline{K}$. Since $\overline{K} = \overline{F}$ and $\overline{p} \nmid e$ as K is totally and tamely ramified over F, the polynomial $X^e - \overline{1}$ splits over $\overline{F}$ with no repeated roots. Thus, $\overline{F}$ contains a primitive e-th root of unity.

Suppose next that $\overline{F}$ contains a primitive e-th root of unity. Then, $X^e - \overline{1} \in \overline{F}[X]$ splits, with no repeated root. Because F is Henselian, each different root of $X^e - \overline{1}$ lifts to a root of $X^e - 1 \in \mathcal{O}_F[X]$. These roots are different, since their images in $\overline{F}$ are different. Hence, $X^e - 1$ splits in $F[X]$ with no repeated roots; this shows that F contains a primitive e-th root of unity.

Finally, suppose F contains a primitive e-th root of unity. Then, F contains a primitive m_i-th root of unity for $i = 1, \ldots, n$. Hence, each f_i is separable over F and K is a splitting field for $\{f_1, \ldots, f_n\}$ over F. Hence, K is Galois over F.

Thus, the three conditions in (ii) are equivalent.

(iii) Suppose K is Galois over F. Since K is generated over F by e-th roots of elements of F and F contains a primitive e-th root of unity by (ii), K is an e-Kummer extension of F. For the a_i and m_i of part (i), consider the following subgroups of $K^\times$:

$$A = \langle a_1, \ldots, a_n \rangle, \quad B = \{b \in A \mid v(b) \in \Gamma_F\}, \quad C = \{c \in K^\times \mid c^e \in F^\times\}.$$

Take any $b \in B$, and write $b = \prod_{i=1}^n a_i^{\ell_i}$ with each $\ell_i \in \mathbb{Z}$. Then,

$$\sum_{i=1}^n \ell_i \gamma_i = v(b) \in \Gamma_F.$$

From the direct decomposition of Γ_K/Γ_F in (A.7) it follows that $m_i \,|\, \ell_i$ for each i; hence, $b = \prod_{i=1}^n (a^{m_i})^{\ell_i/m_i} \in F^\times$. Thus, $B = A \cap F^\times$. The composition of epimorphisms $A \xrightarrow{v} \Gamma_K \to \Gamma_K/\Gamma_F$ clearly has kernel B. Hence,

$$\Gamma_K/\Gamma_F \;\cong\; A/B \;=\; A/(A \cap F^\times) \;\cong\; (A{\cdot}F^\times)/F^\times.$$

Clearly $(A{\cdot}F^\times)/F^\times \subset C/F^\times$, and by Kummer theory $|C/F^\times| = [K{:}F]$. So, as K is totally ramified over F,

$$[K{:}F] \;=\; |\Gamma_K/\Gamma_F| \;=\; \big|(A{\cdot}F^\times)/F^\times\big| \;\le\; |C/F^\times| \;=\; [K{:}F].$$

We thus have equality throughout, showing that $C/F^\times = (A{\cdot}F^\times)/F^\times \cong \Gamma_K/\Gamma_F$. The noncanonical isomorphism $\mathcal{G}(K/F) \cong C/F^\times$ follows from the perfect pairing of Kummer theory, $\mathcal{G}(K/F) \times C/F^\times \to \mu(K)$ given by $(\sigma, cF^\times) \mapsto \sigma(c)/c$. □

We now describe the maximal inertial extension and the maximal tamely ramified extension of a Henselian field F. Fix an algebraic closure F_{alg} of F, and let F_{sep} be the separable closure of F in F_{alg}. Then F_{sep} is Galois over F; let

$$\mathcal{G}(F) \;=\; \mathcal{G}(F_{sep}/F), \quad \text{the absolute Galois group of } F.$$

Since F_{sep} is separably closed, if v is nontrivial (i.e., $\mathcal{O}_{F,v} \neq F$), then by Engler–Prestel [73, Th. 3.2.11, p. 66], we have

$$\overline{F_{sep}} \;=\; (\overline{F})_{alg}, \quad \text{the algebraic closure of } \overline{F}, \tag{A.8}$$

and

$$\Gamma_{F_{sep}} \;=\; \mathbb{H}(\Gamma_F), \quad \text{the divisible hull of } \Gamma_F. \tag{A.9}$$

For the Henselian valuation v on F and its unique extension to a valuation on the Galois extension F_{sep} of F we have the distinguished subgroups of $\mathcal{G}(F)$ and corresponding fixed fields described in §A.1 above. The groups are $\mathcal{G}^{dec}(F)$, $\mathcal{G}^{in}(F)$, and $\mathcal{G}^{ram}(F)$. Since v extends uniquely to F_{sep}, for the decomposition group we have $\mathcal{G}^{dec}(F) = \mathcal{G}(F)$, and the decomposition field is F, cf. Th. A.7(iii). Let

$$F_{in} \;=\; (F_{sep})^{\mathcal{G}^{in}(F)},$$

which is called the *inertial closure*[2] *of* F. Thus, F_{in} is the inertia field for the extension from F to F_{sep}. Likewise, let

$$F_{tr} \;=\; (F_{sep})^{\mathcal{G}^{ram}(F)},$$

which is called the *tame closure of* F.

[2] The more common terminology is that F_{in} is the *maximal unramified extension* of F, sometimes abbreviated F_{nr} (for *non ramifiée*). As explained in [1], we avoid using the misleading term unramified in this context.

Thus, F_{tr} is the ramification field for the extension from F to F_{sep}. We have

$$F \subseteq F_{in} \subseteq F_{tr} \subseteq F_{sep}.$$

The names for F_{in} and F_{tr} are justified by Th. A.23(ii) and Th. A.24(ii) below.

Theorem A.23. *Let F be a Henselian field.*

(i) *F_{in} is inertial over F with $\overline{F_{\text{in}}} = (\overline{F})_{\text{sep}}$ (and $\Gamma_{F_{\text{in}}} = \Gamma_F$).*
(ii) *Let E be any field with $F \subseteq E \subseteq F_{\text{alg}}$. Then, E is inertial over F if and only if $E \subseteq F_{\text{in}}$. Thus, F_{in} is the unique maximal inertial extension of F in F_{alg}.*
(iii) *Since v extends uniquely to F_{sep}, every $\sigma \in \mathcal{G}(F)$ induces a map $\overline{\sigma} \in \text{Aut}(\overline{F_{\text{sep}}} \,/\, \overline{F})$. There is a short exact sequence*

$$1 \longrightarrow \mathcal{G}^{\text{in}}(F) \longrightarrow \mathcal{G}(F) \xrightarrow{\ \psi\ } \text{Aut}(\overline{F_{\text{sep}}} \,/\, \overline{F}) \longrightarrow 1,$$

where $\psi(\sigma) = \overline{\sigma}$.
(iv) *F_{in} is Galois over F, and there are topological group isomorphisms*

$$\mathcal{G}(F_{\text{in}}/F) \cong \mathcal{G}(F)/\mathcal{G}^{\text{in}}(F) \cong \text{Aut}(\overline{F_{\text{sep}}} \,/\, \overline{F}) \cong \mathcal{G}((\overline{F})_{\text{sep}}/\overline{F}) = \mathcal{G}(\overline{F}).$$

(v) *The map $L \mapsto \overline{L}$ gives an inclusion-preserving and degree-preserving one-to-one correspondence between the fields L with $F \subseteq L \subseteq F_{\text{in}}$ and the fields N with $\overline{F} \subseteq N \subseteq (\overline{F})_{\text{sep}}$.*
(vi) *For any field L with $F \subseteq L \subseteq F_{\text{in}}$, L is Galois over F if and only if $\overline{L}$ is Galois over $\overline{F}$. When this occurs, the canonical map $\mathcal{G}(L/F) \to \mathcal{G}(\overline{L}/\overline{F})$ is an isomorphism.*

Proof. Since $\overline{F_{sep}} = (\overline{F})_{alg}$ by (A.8) if v is nontrivial and $\overline{F_{sep}} = F_{sep}$ if v is trivial, in all cases the separable closure of $\overline{F}$ in $\overline{F_{sep}}$ is $(\overline{F})_{sep}$. With this, (i), (iii), (iv), and (v) follow from Th. A.8. They are also proved in Engler–Prestel [73, pp. 123–124]. For (ii), see [73, Th. 5.2.9(1), pp. 125–126]. (vi) follows by Galois theory from the isomorphism of Galois groups in (iv). □

Theorem A.24. *Let F be a Henselian field, and let $\overline{p} = \text{char}\,\overline{F}$.*

(i) *F_{tr} is tamely ramified over F with $\overline{F_{tr}} = \overline{F_{in}} = (\overline{F})_{\text{sep}}$.*
(ii) *Let E be any field with $F \subseteq E \subseteq F_{\text{alg}}$. Then E is tamely ramified over F if and only if $E \subseteq F_{\text{tr}}$. Thus, F_{tr} is the unique maximal tamely ramified extension of F in F_{alg}.*
(iii) *If $\overline{p} = 0$, then $F_{\text{tr}} = F_{\text{sep}}$, $\mathcal{G}^{\text{ram}}(F) = \{\text{id}_{F_{\text{sep}}}\}$, and $\Gamma_{F_{\text{tr}}} = \mathbb{H}(\Gamma_F)$.*
(iv) *If $\overline{p} \neq 0$, then $\mathcal{G}^{\text{ram}}(F)$ is the unique $\overline{p}$-Sylow subgroup of $\mathcal{G}^{\text{in}}(F)$, and $\Gamma_{F_{\text{tr}}}$ is the prime-to-$\overline{p}$ closure of Γ_F in $\mathbb{H}(\Gamma_F)$.*
(v) *Let $\mu = \mu((\overline{F})_{\text{sep}})$ be the group of all roots of unity in $(\overline{F})_{\text{sep}}$. For every $\sigma \in \mathcal{G}^{\text{in}}(F)$ there is a well-defined group homomorphism $\widehat{\sigma}\colon \Gamma_{F_{\text{tr}}}/\Gamma_F \to \mu$ given by $\widehat{\sigma}(\gamma + \Gamma_F) = \overline{\sigma(a)/a}$ for any $a \in F_{\text{tr}}$ with $v(a) = \gamma$. There is a*

short exact sequence of continuous homomorphisms of profinite groups

$$1 \longrightarrow \mathcal{G}^{\text{ram}}(F) \longrightarrow \mathcal{G}^{\text{in}}(F) \xrightarrow{\varphi} Hom(\Gamma_{F_{tr}}/\Gamma_F, \mu) \longrightarrow 1,$$

where $\varphi(\sigma) = \widehat{\sigma}$.

(vi) F_{tr} *is totally and tamely ramified and abelian Galois over* F_{in}.

(vii) *The map* $L \mapsto \Gamma_L$ *gives an inclusion-preserving one-to-one correspondence between the fields* L *with* $F_{in} \subseteq L \subseteq F_{tr}$ *and the groups* Λ *with* $\Gamma_F \subseteq \Lambda \subseteq \Gamma_{F_{tr}}$.

Proof. Since $\overline{F_{sep}}$ contains the separable closure of $\overline{F}$ by (A.8) and $\Gamma_{F_{sep}} = \mathbb{H}(\Gamma_F)$ by (A.9), (i) and (iii)–(vii) follow from Th. A.9.

(ii) Since E is tamely ramified over F if and only if every finite-degree subextension of F in E is tamely ramified over F, we may assume that $[E{:}F] < \infty$. If $E \subseteq F_{tr}$, then E is tamely ramified over F since F_{tr} is tamely ramified over F.

For the converse, suppose E is tamely ramified over F. Since $\overline{E}$ is separable over $\overline{F}$, there is an $s \in \overline{E}$ with $\overline{E} = \overline{F}(s)$. Let $g \in \overline{F}[X]$ be the minimal polynomial of s over $\overline{F}$, and let f be any monic inverse image of g in $\mathcal{O}_F[X]$. Since E is Henselian and s is a simple root of g as s is separable over $\overline{F}$, there is a root a of f in E with $\overline{a} = s$ in $\overline{E}$. Let $L = F(a) \subseteq E$. Then, as $\overline{F}(s) \subseteq \overline{L}$ and

$$[L{:}F] \le deg\, g = deg\, f = [\overline{F}(s){:}\overline{F}] \le [\overline{L}{:}\overline{F}] \le [L{:}F],$$

we have $\overline{L} = \overline{F}(s) = \overline{E}$ and $[L{:}F] = [\overline{L}{:}\overline{F}]$. Hence, L is inertial over F, so $L \subseteq F_{in}$ by Th. A.23(ii).

Now, E is tamely ramified over F, so it is also tamely ramified over L. In addition, E is totally ramified over L, as $\overline{E} = \overline{L}$. Therefore, Lemma A.21 applies for E over L. Take any $\gamma \in \Gamma_E$, and let m be the order of $\gamma + \Gamma_F$ in Γ_E/Γ_F. By the lemma, there is an $a \in E$ with $v(a) = \gamma$ and $a^m \in L$. We will show that $a \in F_{tr}$. Note that $\overline{p} \nmid m$ as E is tamely ramified over L. Hence, $\gamma \in \Gamma_{F_{tr}}$ by (iii) or (iv). Since F_{tr} is totally and tamely ramified over F_{in} by (vi), Lemma A.21 applies for F_{tr} over F_{in}, showing that there is a $b \in F_{tr}$ with $v(b) = \gamma$ and $b^m \in F_{in}$. Let $u = a/b$; so $u^m = a^m/b^m \in F_{in}$ and $v(u) = 0$. Note that F_{in} contains a primitive m-th root of unity ω by the argument in the proof of Prop. A.22(ii), since the separably closed field $\overline{F_{in}}$ contains a primitive m-th root of unity, as $\overline{p} \nmid m$. Our u is a root of the polynomial $X^m - u^m \in F_{in}[X]$, whose roots are $u, u\omega, \ldots, u\omega^{m-1}$. Let h be the minimal polynomial of u over F_{in}, and let $\ell = deg\, h$. Since $h \mid (X^m - u^m)$, the roots of h each have the form $u\omega^i$, so the constant term of h is $(-1)^\ell u^\ell \omega^j$, for some j. Therefore, $u^\ell \in F_{in}$, and hence $h = X^\ell - u^\ell$. The exponent ℓ must then be the minimal power of u lying in F_{in}; hence, $\ell \mid m$, so $\overline{p} \nmid \ell$. Since $\overline{F_{in}}$ is separably closed by Th. A.23(i), the image $\overline{h} = X^\ell - \overline{u^\ell}$ of h in $\overline{F_{in}}[X]$ has a root t in $\overline{F_{in}}$, and t is a simple root as $\overline{p} \nmid \ell$. Therefore, as F_{in} is Henselian, h has a root u' in F_{in}. Then $(u/u')^\ell = 1$. Since $u' \in F_{in}$ and F_{in} contains all ℓ-th roots of unity, u must lie in F_{in}. Hence, $a = bu \in F_{tr}$.

Because $L \subseteq F_{in} \subseteq F_{tr}$ and the proof of Prop. A.22(i) shows that E is generated over L by elements such as a, it follows that $E \subseteq F_{tr}$, as desired. □

Corollary A.25. *Let F be a Henselian field, and let K be any algebraic extension of F. Then, $K \cap F_{in}$ is the unique maximal inertial extension of F in K, and $K \cap F_{tr}$ is the unique maximal tamely ramified extension of F in K.*

Proof. $K \cap F_{in}$ is inertial over F by Th. A.23(ii) since it lies in F_{in}. Likewise, $K \cap F_{tr}$ is tamely ramified over F by Th. A.24(ii). Let L be a field with $F \subseteq L \subseteq K$. If L is inertial over F, then $L \subseteq F_{in}$ by Th. A.23(ii), so $L \subseteq K \cap F_{in}$. Thus, $K \cap F_{in}$ is the unique maximal inertial extension of F in K. Likewise, if L is tamely ramified over F, then $L \subseteq K \cap F_{tr}$ by Th. A.24(ii), so $K \cap F_{tr}$ is the maximal tamely ramified extension of F in K. □

A.3 Properties of the Henselization

Definition A.26. Let F be a field with valuation v. A *Henselization* of (F, v) is a valued field extension (F_h, v_h) of (F, v) such that v_h is Henselian and for any extension (K, w) of (F, v) with w Henselian there is a unique F-homomorphism $\eta \colon F_h \to K$ such that $v_h = w \circ \eta$. We thus refer to $(\eta(F_h), w|_{\eta(F_h)})$ as the Henselization of (F, v) within (K, w).

It is clear from the definition that a Henselization of (F, v) is unique up to unique isomorphism. Thus, we sometimes say that (F_h, v_h) is "the Henselization" of (F, v). The question of existence is settled by the following theorem:

Theorem A.27. *Let (F, v) be a valued field, and let v_{sep} be any extension of v to a separable closure F_{sep} of F. Let F_h be the decomposition field for the extension v_{sep} of v. Then, $(F_h, v_{sep}|_{F_h})$ is a Henselization of (F, v).*

Proof. See Engler–Prestel [73, Th. 5.2.2, p. 121]. □

Corollary A.28. *Let (F_h, v_h) be a Henselization of (F, v). Then, F_h is separable algebraic over F, and v_h is an immediate extension of v.*

Proof. This follows from Th. A.27, since the decomposition field is separable over F and the valuation on a decomposition field is an immediate extension of the valuation on the ground field by Th. A.7(i). □

Remark A.29. Suppose (K, w) is an extension of (F, v) and (L, y) is an extension of (K, w), with w and y each Henselian. Let (F_h, v_h) be the Henselization of (F, v) within (L, y). Then, $F_h \subseteq K$, and (F_h, v_h) is also the Henselization of (F, v) within (K, w). For if (F_h', v_h') is the Henselization of (F, v) within (K, w), then (F_h', v_h') is also a Henselization of (F, v) within (L, y); hence $F_h = F_h' \subseteq K$ and $v_h = v_h'$ by the uniqueness in the universal property for the Henselization.

Proposition A.30. *Let (F_h, v_h) be a Henselization of the valued field (F, v). Let K be any extension field of F lying in the algebraic closure of F_h, and let w be the unique extension of v_h to the compositum $K \cdot F_h$. Then, $(K \cdot F_h, w)$ is a Henselization of $(K, w|_K)$.*

Proof. The valuation w on $K \cdot F_h$ is Henselian since v_h is Henselian. Now that we know that $(K, w|_K)$ has a Henselization (K_h, w_h), the universal property shows that (K_h, w_h) embeds in $(K \cdot F_h, w)$. Thus, we may assume that $K \subseteq K_h \subseteq K \cdot F_h$ and $w_h = w|_{K_h}$. Since (F_h, v_h) is the Henselization of (F, v) within $(K \cdot F_h, w)$, it is also the Henselization of (F, v) within (K_h, w_h), by Remark A.29; so, $F_h \subseteq K_h$. Since also $K \subseteq K_h$, we have $K \cdot F_h \subseteq K_h \subseteq K \cdot F_h$. Hence, $K_h = K \cdot F_h$ and $w = w_h$, showing that $(K \cdot F_h, w)$ is a Henselization of $(K, w|_K)$. □

Proposition A.31. *Let (F, v) be a field with valuation, let w be a coarsening of v on F, and let u be the valuation on $\overline{F}^{w}$ induced by v; so $v = u * w$. Let $(F_{h,v}, v_h)$ be a Henselization of (F, v), let w' be the coarsening of v_h with $w'|_F = w$, and let u' be the valuation on $\overline{F_{h,v}}^{w'}$ induced by v_h. Then, w' and u' are Henselian. Let $(F_{h,w}, w_h)$ be the Henselization of (F, w) within $(F_{h,v}, w')$. Then, $F_{h,w} \subseteq F_{h,v}$ and $F_{h,v}$ is an inertial extension of $F_{h,w}$ with respect to w_h, and $(\overline{F_{h,v}}^{w'}, u')$ is a Henselization of $(\overline{F}^{w}, u)$.*

Proof. For the various algebraic extensions L of F in the proof, we write $\overline{L}$ for the residue field of L with respect to the relevant extension of w to L.

Since v_h is Henselian, Prop. A.15 shows that w' and u' are Henselian. Therefore, the Henselization $(F_{h,w}, w_h)$ exists within $(F_{h,v}, w')$. To simplify notation, let $F_1 = F_{h,w}$, $w_1 = w_h$, $u_1 = u'|_{\overline{F_1}}$, and $v_1 = v_h|_{F_1} = u_1 * w_1$. We identify $\overline{F_1}$ with $\overline{F}$, since w_1 is an immediate extension of w by Cor. A.28. Let $(\overline{F}_{h,u}, u_h)$ be a Henselization of $(\overline{F}, u)$. Since $\overline{F}_{h,u}$ is separable over $\overline{F}$, by Th. A.23(v), there is an inertial extension (F_2, w_2) of (F_1, w_1) with $\overline{F_2} = \overline{F}_{h,u}$. Let $v_2 = u_h * w_2$, which is a valuation on F_2 with $v_2|_F = v$. Then, v_2 is Henselian by Prop. A.15, since w_2 and u_h are Henselian. By the universal property for Henselizations of (F, v), there is an F-homomorphism $\eta\colon F_{h,v} \to F_2$ with $v_h = v_2 \circ \eta$. Let $F_3 = \eta(F_{h,v})$, $v_3 = v_2|_{F_3}$, $w_3 = w_2|_{F_3}$, and u_3 the valuation on $\overline{F_3}$ induced by v_3; so, $v_3 = u_3 * w_3$. Then, v_3 is Henselian, as $(F_3, v_3) \cong (F_{h,v}, v_h)$, so w_3 and u_3 are Henselian by Prop. A.15. Since (F_1, w_1) is the Henselization of (F, w) lying within (F_2, w_2) and w_3 is Henselian on $F_3 \subseteq F_2$, we have $F_1 \subseteq F_3$ and $w_3|_{F_1} = w_1$ by Remark A.29. At the w-residue level, we have inclusions of valued fields,

$$(\overline{F}, u) = (\overline{F_1}, u_1) \subseteq (\overline{F_3}, u_3) \subseteq (\overline{F_2}, u_h).$$

Since u_3 is Henselian and $(\overline{F_2}, u_h)$ is a Henselization of $(\overline{F}, u)$, the Henselization of (F, u) within $(\overline{F_3}, u_3)$ must be $(\overline{F_2}, u_h)$ by Remark A.29, i.e., $\overline{F_3} = \overline{F_2}$. Therefore, $F_3 = F_2$ as (F_2, w_2) is inertial over (F_1, w_1), by the correspondence between inertial extensions and their residue fields, see Th. A.23(v).

From the isomorphism $\eta\colon F_{h,v} \xrightarrow{\sim} F_3$ which maps v_h to v_3, and w' to w_3, it follows that w' on $F_{h,v}$ is inertial over $w_1 = w_h$ on $F_{h,w}$, and

$$(\overline{F_{h,v}}\,, u') \cong (\overline{F_3}, u_3) = (\overline{F_2}, u_2) = (\overline{F}_{h,u}, u_h),$$

which is a Henselization of $(\overline{F}, u)$. □

Theorem A.32. *Let F be a field with a valuation v. Let K be a finite-degree field extension of F, and let $v_1, \ldots, v_r$ be all the extensions of v to K. Let (F_h, v_h) be a Henselization of (F, v), and $(K_{h,i}, v_{i,h})$ a Henselization of (K_i, v_i) for $i = 1, \ldots, r$. Then,*

$$K \otimes_F F_h \cong K_{h,1} \times \ldots \times K_{h,r}.$$

The proof of the theorem follows the proof of the next lemma.

Lemma A.33. *Let $F \subseteq N$ be fields with N Galois over F (possibly of infinite degree), and let $G = \mathcal{G}(N/F)$. Let K and E be subfields of N containing F, with $[K{:}F] < \infty$. Let $H = \mathcal{G}(N/K) \subseteq G$ and $Z = \mathcal{G}(N/E) \subseteq G$. Let $\tau_1, \ldots, \tau_r$ be representatives of the distinct Z–H double cosets of G. (So, $G = \bigcup_{i=1}^{r} Z\tau_i H$, a disjoint union.) Then,*

$$K \otimes_F E \cong \tau_1(K){\cdot}E \times \ldots \times \tau_r(K){\cdot}E.$$

Proof. Since K is separable over F, we have $K = F(a)$ for some a. Let f be the minimal polynomial of a over F. Then f splits over N, as N is normal over F, say $f = (X - a_1)\ldots(X - a_n) \in N[X]$ where the a_i are distinct and $a_1 = a$. Let $\mathcal{A} = \{a_1, \ldots, a_n\}$. Let $f = g_1 \ldots g_r$ be the irreducible factorization of f in $E[X]$, and fix a root b_i of g_i for $i = 1, \ldots, r$. The Galois group G acts transitively on $\mathcal{A}$, but $\mathcal{A}$ decomposes into r disjoint Z-orbits, $\mathcal{A} = \mathcal{B}_1 \cup \ldots \cup \mathcal{B}_r$, where

$$\mathcal{B}_i = Z \cdot b_i = \{\text{roots of } g_i \text{ in } N\}.$$

For each i, choose $\tau_i' \in G$ with $\tau_i'(a) = b_i$. Then, as $H = \{\sigma \in G \mid \sigma(a) = a\}$, we have

$$Z\tau_i'H = \{\sigma \in G \mid \sigma(a) \in \mathcal{B}_i\} \qquad \text{for } i = 1, 2, \ldots, r.$$

Thus, $Z\tau_1'H, \ldots, Z\tau_r'H$ are all the distinct Z–H double cosets in G. Moreover, we may assume that the double coset representatives τ_i' coincide with the τ_i of the lemma, by replacing b_i by $\tau_i(a)$. Since $gcd(g_i, g_j) = 1$ for $i \neq j$, the Chinese Remainder Theorem yields

$$\begin{aligned} K \otimes_F E &\cong \big(F[X]/(f)\big) \otimes_F E \cong E[X]/fE[X] \\ &\cong E[X]/\big[(g_1)\ldots(g_r)\big] \cong E(b_1) \times \ldots \times E(b_r) \\ &\cong \tau_1(K){\cdot}E \times \ldots \times \tau_r(K){\cdot}E. \end{aligned}$$ □

Proof of Th. A.32. The central case when K is separable over F is proved in Endler [72, Th. (17.17), p. 135], and a similar result is given in Efrat [70,

Lemma 17.1.4, p. 152]. But since the theorem is so essential for the results in Ch. 4 above, we give a proof here.

Assume first that K is separable over F. Let F_{sep} be a separable closure of F containing F_h, and let v_{sep} be the unique valuation on F_{sep} extending the Henselian valuation v_h. Let $G = \mathcal{G}(F_{sep}/F)$, $H = \mathcal{G}(F_{sep}/K) \subseteq G$, and $Z = \mathcal{G}(F_{sep}/F_h) \subseteq G$. By Th. A.27 and the universal property of the Henselization, F_h is the decomposition field, so Z is the decomposition subgroup G^{dec} for v_{sep} over v, i.e.,

$$Z = \{\sigma \in G \mid v_{sep} \circ \sigma = v_{sep}\}.$$

Let $\mathcal{E}$ be the set of all valuations on F_{sep} extending v on F. Then, G acts transitively on $\mathcal{E}$ (see Th. A.1), while the distinct H-orbits of $\mathcal{E}$ are $\mathcal{O}_1, \ldots, \mathcal{O}_r$, where $\mathcal{O}_i = \{w \in \mathcal{E} \mid w|_K = v_i\}$. For $i = 1, \ldots, r$, choose $\tau_i \in G$ with $v_{sep} \circ \tau_i|_K = v_i$. Then,

$$\{\sigma \in G \mid v_{sep} \circ \sigma|_K = v_i\} = \{\sigma \in G \mid v_{sep} \circ \sigma \in \mathcal{O}_i\} = Z\tau_i H.$$

So, $G = \bigcup_{i=1}^r Z\tau_i H$ is the disjoint Z–H double coset decomposition of G. We now apply the preceding lemma with $N = F_{sep}$ and $E = F_h$. (So, the K, H and Z of the lemma are the K, H and Z here.) By the lemma,

$$K \otimes_F F_h \cong \prod_{i=1}^{r} \tau_i(K)\cdot F_h \cong \prod_{i=1}^{r} K\cdot\tau_i^{-1}(F_h), \tag{A.10}$$

where the second isomorphism follows by applying τ_i^{-1} to the i-th factor. Note that the F-isomorphism τ_i^{-1} maps (F_h, v_h) to $(\tau_i^{-1}(F_h), v_h \circ \tau_i)$. Hence, $(\tau_i^{-1}(F_h), v_h \circ \tau_i)$ is a Henselization of (F, v). The unique extension of the Henselian valuation $v_h \circ \tau_i$ to $K\cdot\tau_i^{-1}(F_h)$ must be $v_{sep} \circ \tau_i|_{K\cdot\tau_i^{-1}(F_h)}$, whose restriction to K is v_i by the choice of τ_i. Therefore, $K\cdot\tau_i^{-1}(F_h) \cong K_{h,i}$ by Prop. A.30. The theorem then follows from (A.10), for K separable over F.

If K is not separable over F, let S be the separable closure of F in K, and let $y_i = v_i|_S$ for $i = 1, 2, \ldots, r$. Then, $y_1, \ldots, y_r$ are all the extensions of v to S, and since valuations extend uniquely from S to its purely inseparable extension K, $y_i \neq y_j$ for $i \neq j$. Hence, as we just proved, $S \otimes_F F_h \cong \prod_{i=1}^r S_{h,i}$, where $(S_{h,i}, y_{i,h})$ is a Henselization of (S, y_i) in F_{alg}. Therefore,

$$K \otimes_F F_h \cong K \otimes_S (S \otimes_F F_h) \cong \prod_{i=1}^{r} K \otimes_S S_{h,i}. \tag{A.11}$$

Because K is purely inseparable over S while $S_{h,i}$ is separable over S, these fields are linearly disjoint over S; so, $K \otimes_S S_{h,i}$ is a field, which is isomorphic to the compositum $K\cdot S_{h,i}$ in F_{alg}. The Henselian valuation $y_{i,h}$ on $S_{h,i}$ has a unique extension to the field $K \otimes_S S_{h,i}$, whose restriction to K is the unique extension of y_i to K, which is v_i. By Prop. A.30, $K \otimes_S S_{h,i}$ is a Henselization of K with respect to v_i. The theorem thus follows from (A.11). □

A.4 Ranks of value groups and their factor groups

There are various notions of rank for abelian groups. We recall the different kinds of rank arising in connection with the value groups of valuations and gauges, and some significant inequalities relating these ranks.

Rank. Let Γ be a totally ordered abelian group. A subgroup Δ of Γ is said to be *convex* if whenever $\gamma \in \Gamma$ and $\delta \in \Delta$ and $0 \le \gamma \le \delta$, then $\gamma \in \Delta$. (Bourbaki calls such a Δ an *isolated* subgroup of Γ.) It is easy to check that if Δ is convex, then Γ/Δ is torsion-free and the ordering on Γ induces a well-defined ordering on Γ/Δ making it into a totally ordered abelian group. Moreover, if there is a group homomorphism $\psi\colon \Gamma/\Delta \to \Gamma$ splitting the canonical projection $\pi\colon \Gamma \to \Gamma/\Delta$ (i.e., $\pi \circ \psi = id_{\Gamma/\Delta}$), then

$$\Gamma \cong \Delta \times (\Gamma/\Delta), \tag{A.12}$$

and this is an ordered group isomorphism when the direct product is given the right-to-left lexicographic ordering. Note also that the convex subgroups of Γ are linearly ordered by inclusion.

The *rank of* Γ is defined to be

$$rk(\Gamma) \;=\; \text{the number of convex subgroups } \Delta \text{ of } \Gamma \text{ with } \Delta \neq \Gamma. \tag{A.13}$$

Note that if $\mathbb{H}(\Gamma)$ denotes the divisible hull $\Gamma \otimes_{\mathbb{Z}} \mathbb{Q}$ of Γ, and we identify Γ with its canonical image in $\mathbb{H}(\Gamma)$, then the ordering on Γ extends uniquely to an ordering on $\mathbb{H}(\Gamma)$ making $\mathbb{H}(\Gamma)$ into a totally ordered abelian group. Since the map $\Lambda \mapsto \Lambda \cap \Gamma$ gives a one-to-one correspondence between the convex subgroups of $\mathbb{H}(\Gamma)$ and those of Γ, we have

$$rk(\mathbb{H}(\Gamma)) \;=\; rk(\Gamma). \tag{A.14}$$

Recall (see Engler–Prestel [73, Prop. 2.1.1, p. 26]) that $rk(\Gamma) = 1$ if and only if Γ is isomorphic as an ordered group to a nontrivial subgroup of the additive group of $\mathbb{R}$.

If v is a valuation on a field F, then the *rank of* v is defined to be the rank of its value group:

$$rk(v) \;=\; rk(\Gamma_{F,v}). \tag{A.15}$$

Equivalently (see Bourbaki [30, Ch. VI, §4.4, Prop. 5]),

$$rk(v) \;=\; \text{Krull dimension of the valuation ring } \mathcal{O}_{F,v} \text{ of } v. \tag{A.16}$$

Let Δ be a convex subgroup of the totally ordered abelian group Γ. The convex subgroups of Δ are the convex subgroups of Γ lying in Δ. Also, the convex subgroups of Γ/Δ are the groups Λ/Δ where Λ is a convex subgroup of Γ with $\Lambda \supseteq \Delta$. Hence,

$$rk(\Gamma) \;=\; rk(\Delta) + rk(\Gamma/\Delta). \tag{A.17}$$

Rational rank. Let A be an abelian group. The *rational rank of* A is defined to be

$$rr(A) \;=\; dim_{\mathbb{Q}}(A \otimes_{\mathbb{Z}} \mathbb{Q}). \tag{A.18}$$

For example, $rr(A) = 0$ if and only if A is a torsion group. Since $\mathbb{Q}$ is a localization of the commutative ring $\mathbb{Z}$, the functor $- \otimes_{\mathbb{Z}} \mathbb{Q}$ is exact. Hence, for any subgroup B of A,

$$rr(A) \;=\; rr(B) + rr(A/B). \tag{A.19}$$

Now, let Γ be a totally ordered abelian group, and let Λ be any subgroup of Γ. Then, it is known (see Engler–Prestel [73, Prop. 3.4.1, p. 79]) that

$$rk(\Gamma) \;\leq\; rk(\Lambda) + rr(\Gamma/\Lambda). \tag{A.20}$$

Hence in particular, setting $\Lambda = \{0\}$,

$$rk(\Gamma) \;\leq\; rr(\Gamma). \tag{A.21}$$

Remark A.34 (Discrete valuations). If an ordered abelian group Γ has finite rank n, then there is an order-preserving group monomorphism $\iota\colon \Gamma \hookrightarrow \mathbb{R}^n$ where $\mathbb{R}^n$ is given the right-to-left lexicographic ordering. This was noted above if $n = 1$. If $n > 1$, it follows by induction on n. For we can then take a nontrivial convex subgroup Δ of $\mathbb{H}(\Gamma)$, say of rank m with $1 \leq m < n$. So,

$$rk(\mathbb{H}(\Gamma)/\Delta) \;=\; rk(\mathbb{H}(\Gamma)) - rk(\Delta) \;=\; n - m \;<\; n.$$

The projection $\pi\colon \mathbb{H}(\Gamma) \to \mathbb{H}(\Gamma)/\Delta$ has a $\mathbb{Q}$-vector space splitting map $\psi\colon \mathbb{H}(\Gamma)/\Delta \to \mathbb{H}(\Gamma)$. Hence, by (A.12) and induction there are order-preserving monomorphisms

$$\Gamma \;\hookrightarrow\; \mathbb{H}(\Gamma) \;\hookrightarrow\; \Delta \times (\mathbb{H}(\Gamma)/\Delta) \;\hookrightarrow\; \mathbb{R}^m \times \mathbb{R}^{n-m},$$

where $\mathbb{R}^m$, $\mathbb{R}^{n-m}$ and the direct products above are given the right-to-left lexicographic ordering. Take ι to be the composition of these monomorphisms. We say that Γ is a *discrete* ordered group if $im\,\iota$ is a discrete subgroup of $\mathbb{R}^n$. It is easy to check that discreteness is independent of the choice of ι and that the following conditions are equivalent:

(a) Γ is discrete.
(b) Γ is a finitely-generated abelian group and $rr(\Gamma) = rk(\Gamma)$.
(c) Γ is isomorphic as an ordered group to $\mathbb{Z}^n$ with the right-to-left lexicographic ordering.

In particular, Γ is discrete of rank 1 if and only if $\Gamma \cong \mathbb{Z}$.

We say that a valuation v on a field F is *discrete* if $\Gamma_{F,v}$ is a discrete ordered abelian group. Thus, v is discrete of rank 1 if and only if $\Gamma_{F,v} \cong \mathbb{Z}$. Such a valuation has traditionally been called a discrete valuation in commutative algebra, and its ring $\mathcal{O}_{F,v}$ a discrete valuation ring. To avoid ambiguity, we refer to such a v as a discrete rank 1 valuation. Note that a valuation is discrete of rank n if and only if it is the composite of n discrete rank 1 valuations.

The rational rank appears in an important inequality for extensions of valuations. For $F \subseteq K$ fields, let *trdeg* K/F denote the transcendence degree of K over F.

Theorem A.35 (Transcendence Degree Inequality). *Let $F \subseteq K$ be fields. Let v be a valuation on F, and let w be any valuation on K extending v. Then,*

$$trdeg\, \overline{K}/\overline{F} \;+\; rr\left(\Gamma_{K,w}/\Gamma_{F,v}\right) \;\leq\; trdeg\, K/F.$$

Proof. See Engler–Prestel [73, Th. 3.4.3, p. 80] or Bourbaki [30, Ch. 6, §10.3, Cor. 1]. □

Torsion rank. For finite groups there is still another kind of rank. Define the *torsion rank* of a finite abelian group A to be

$$trk(A) \;=\; \text{the number of invariant factors of } A. \tag{A.22}$$

Equivalently,

$$\begin{aligned} trk(A) \;&=\; \text{minimal cardinality of generating sets of } A \\ &=\; \max_{p \text{ prime}} \{dim_{\mathbb{F}_p}(A/pA)\}. \end{aligned} \tag{A.23}$$

Note that for any subgroup B of A, since A is isomorphic to its dual group $A^\vee = Hom(A, \mathbb{Q}/\mathbb{Z})$ and $A^\vee$ maps onto $B^\vee$, we have

$$trk(B) \;=\; trk(B^\vee) \;\leq\; trk(A^\vee) \;=\; trk(A). \tag{A.24}$$

Proposition A.36. *Let Γ be any torsion-free abelian group, and let Λ be a subgroup of Γ with $|\Gamma\!:\!\Lambda| < \infty$. Then,*

$$trk(\Gamma/\Lambda) \;\leq\; rr(\Lambda) \;=\; rr(\Gamma).$$

Proof. By (A.19),

$$rr(\Gamma) - rr(\Lambda) \;=\; rr(\Gamma/\Lambda) \;=\; 0,$$

proving the equality of the proposition.

To prove the rest of the proposition, assume first that the abelian group Γ is finitely-generated. Its torsion-free subgroup Λ is then also finitely-generated, hence a free abelian group with $rr(\Lambda) = rk_{\mathbb{Z}}\,\Lambda$, the finite rank of Λ as a free $\mathbb{Z}$-module. Since Γ/Λ is finite there is some $n \in \mathbb{N}$ with $n\Gamma \subseteq \Lambda$. Hence, $n\Lambda \subseteq n\Gamma \subseteq \Lambda$ and

$$\Gamma/\Lambda \;\cong\; n\Gamma/n\Lambda \;\subseteq\; \Lambda/\,n\Lambda.$$

The factor group $\Lambda/\,n\Lambda$ is a free $\mathbb{Z}/n\mathbb{Z}$-module; so, using (A.24),

$$\begin{aligned} rr(\Lambda) \;=\; rk_{\mathbb{Z}}\,\Lambda \;&=\; rk_{\mathbb{Z}/n\mathbb{Z}}(\Lambda/\,n\Lambda) \\ &=\; trk(\Lambda/\,n\Lambda) \;\geq\; trk(n\Gamma/\,n\Lambda) \;=\; trk(\Gamma/\Lambda). \end{aligned}$$

This proves the proposition if Γ is finitely-generated.

In general, for arbitrary Γ, since $|\Gamma{:}\Lambda| < \infty$ there is a finitely-generated subgroup Γ_0 of Γ with $\Gamma_0 + \Lambda = \Gamma$. Then,

$$\Gamma/\Lambda \;=\; (\Gamma_0 + \Lambda)/\Lambda \cong \Gamma_0/(\Gamma_0 \cap \Lambda).$$

Hence, by the finitely-generated case,

$$trk(\Gamma/\Lambda) \;=\; trk\left(\Gamma_0/(\Gamma_0 \cap \Lambda)\right) \;\le\; rr(\Gamma_0) \;\le rr(\Gamma) \;=\; rr(\Lambda). \qquad \square$$

Appendix B
List of Examples

Chapter 1

J.-P. Tignol, A.R. Wadsworth, *Value Functions on Simple Algebras, and Associated Graded Rings*, Springer Monographs in Mathematics,
DOI 10.1007/978-3-319-16360-4

Chapter 2

Chapter 3

Chapter 4

Chapter 5

Chapter 6

Chapter 7

Chapter 8

Chapter 9

Chapter 10

Chapter 11

Chapter 12

References

Page references at the end of each entry indicate the places in the text where the entry is quoted.

1. Albert, A.A.: Cyclic fields of degree p^n over F of characteristic p. Bull. Amer. Math. Soc. **40**(8), 625–631 (1934) ⟶ *p. 494.*
2. Albert, A.A.: On cyclic fields. Trans. Amer. Math. Soc. **37**(3), 454–462 (1935) ⟶ *p. 487.*
3. Albert, A.A.: Non-cyclic algebras with pure maximal subfields. Bull. Amer. Math. Soc. **44**(8), 576–579 (1938) ⟶ *p. 487.*
4. Albert, A.A.: Structure of Algebras. Revised printing. American Mathematical Society Colloquium Publications, vol. XXIV. American Mathematical Society, Providence, RI (1961) ⟶ *pp. 400, 455, 473, 487, and 492.*
5. Albert, A.A.: Collected Mathematical Papers. Part 1. Associative Algebras and Riemann Matrices. American Mathematical Society, Providence, RI (1993). Edited by Richard E. Block, Nathan Jacobson, J. Marshall Osborn, David J. Saltman and Daniel Zelinsky ⟶ *p. 487.*
6. Amitsur, S.A.: On central division algebras. Israel J. Math. **12**, 408–420 (1972) ⟶ *pp. 476 and 488.*
7. Amitsur, S.A.: The generic division rings. Israel J. Math. **17**, 241–247 (1974) ⟶ *p. 488.*
8. Amitsur, S.A.: Division algebras. A survey. In: Amitsur, S.A., Saltman, D.J., Seligman, G.B. (eds.) Algebraists' Homage: Papers in Ring Theory and Related Topics, New Haven, Conn., 1981. Contemp. Math., vol. 13, pp. 3–26. Amer. Math. Soc., Providence, RI (1982) ⟶ *p. 489.*
9. Amitsur, S.A.: Galois splitting fields of a universal division algebra. J. Algebra **143**(1), 236–245 (1991) ⟶ *p. 489.*
10. Amitsur, S.A.: Selected Papers of S.A. Amitsur with Commentary. Part 2. American Mathematical Society, Providence, RI (2001). Edited by Avinoam Mann, Amitai Regev, Louis Rowen, David J. Saltman and Lance W. Small ⟶ *pp. 487, 488, 489, and 490.*
11. Amitsur, S.A., Rowen, L.H., Tignol, J.P.: Division algebras of degree 4 and 8 with involution. Israel J. Math. **33**(2), 133–148 (1979) ⟶ *pp. 374, 532, and 533.*
12. Amitsur, S.A., Saltman, D.: Generic abelian crossed products and p-algebras. J. Algebra **51**(1), 76–87 (1978) ⟶ *pp. 452, 472, 487, and 577.*
13. Arason, J.K.: Cohomologische Invarianten quadratischer Formen. J. Algebra **36**(3), 448–491 (1975) ⟶ *p. 508.*

J.-P. Tignol, A.R. Wadsworth, *Value Functions on Simple Algebras, and Associated Graded Rings*, Springer Monographs in Mathematics, DOI 10.1007/978-3-319-16360-4

14. Aravire, R., Jacob, B.: p-algebras over maximally complete fields. In: K-Theory and Algebraic Geometry: Connections with Quadratic Forms and Division Algebras, Santa Barbara, CA, 1992. Proc. Sympos. Pure Math., vol. 58, pp. 27–49. Amer. Math. Soc., Providence, RI (1995). With an appendix by Jean-Pierre Tignol ⟶ *p. 454.*
15. Aravire, R., Jacob, B.: Special forms of two symbols are division algebras. Manuscripta Math. **108**(2), 139–162 (2002) ⟶ *p. 533.*
16. Artin, E., Tate, J.: Class Field Theory. W. A. Benjamin, Inc., New York–Amsterdam (1968) ⟶ *p. 481.*
17. Auel, A., Brussel, E.S., Garibaldi, S., Vishne, U.: Open problems on central simple algebras. Transform. Groups **16**(1), 219–264 (2011) ⟶ *p. 490.*
18. Auslander, M., Goldman, O.: The Brauer group of a commutative ring. Trans. Amer. Math. Soc. **97**, 367–409 (1960) ⟶ *pp. 289 and 453.*
19. Baek, S.: Essential dimension of simple algebras in positive characteristic. C. R. Math. Acad. Sci. Paris **349**(7–8), 375–378 (2011) ⟶ *p. 584.*
20. Baek, S.: Essential dimension of simple algebras with involutions. Bull. Lond. Math. Soc. **44**(3), 578–590 (2012) ⟶ *p. 584.*
21. Baek, S., Merkurjev, A.S.: Essential dimension of central simple algebras. Acta Math. **209**(1), 1–27 (2012) ⟶ *pp. 561, 565, 577, 578, 580, 581, and 584.*
22. Barry, D.: Power-central elements in tensor products of symbol algebras (2013). Preprint, available at arXiv:1310.5423 ⟶ *p. 534.*
23. Berhuy, G., Favi, G.: Essential dimension: a functorial point of view (after A. Merkurjev). Doc. Math. **8**, 279–330 (2003) ⟶ *pp. 562 and 583.*
24. Boulagouaz, M.: The graded and tame extensions. In: Cahen, P.J., Costa, D.L., Fontana, M., Kabbaj, S.E. (eds.) Commutative Ring Theory, Fès, 1992. Lecture Notes in Pure and Appl. Math., vol. 153, pp. 27–40. Dekker, New York (1994) ⟶ *p. 236.*
25. Boulagouaz, M.: Le gradué d'une algèbre à division valuée. Comm. Algebra **23**(11), 4275–4300 (1995) ⟶ *pp. 80, 293, 294, and 449.*
26. Boulagouaz, M.: Algèbre à division graduée centrale. Comm. Algebra **26**(9), 2933–2947 (1998) ⟶ *p. 453.*
27. Boulagouaz, M.: Une généralisation du lemme de Hensel. Bull. Belg. Math. Soc. Simon Stevin **5**(5), 665–673 (1998) ⟶ *p. 237.*
28. Boulagouaz, M.: An introduction to the Galois theory for graded fields. In: Boulagouaz, M., Tignol, J.P. (eds.) Algebra and Number Theory, Fez. Lecture Notes in Pure and Appl. Math., vol. 208, pp. 21–31. Dekker, New York (2000) ⟶ *p. 236.*
29. Boulagouaz, M., Mounirh, K.: Generic abelian crossed products and graded division algebras. In: Boulagouaz, M., Tignol, J.P. (eds.) Algebra and number theory, Fez. Lecture Notes in Pure and Appl. Math., vol. 208, pp. 33–47. Dekker, New York (2000) ⟶ *p. 487.*
30. Bourbaki, N.: Commutative algebra. Chapters 1–7. Elements of Mathematics. Springer, Berlin (1989) ⟶ *pp. 21, 585, 586, 608, and 610.*
31. Bourbaki, N.: Algebra I. Chapters 1–3. Elements of Mathematics. Springer, Berlin (1998) ⟶ *pp. 80 and 538.*
32. Bourbaki, N.: General topology. Chapters 1–4. Elements of Mathematics. Springer, Berlin (1998) ⟶ *p. 22.*
33. Bruhat, F., Tits, J.: Schémas en groupes et immeubles des groupes classiques sur un corps local. Bull. Soc. Math. France **112**(2), 259–301 (1984) ⟶ *p. 132.*
34. Brungs, H.H., Gräter, J.: Valuation rings in finite-dimensional division algebras. J. Algebra **120**(1), 90–99 (1989) ⟶ *p. 31.*
35. Brungs, H.H., Gräter, J.: Extensions of valuation rings in central simple algebras. Trans. Amer. Math. Soc. **317**(1), 287–302 (1990) ⟶ *pp. 31 and 132.*
36. Brungs, H.H., Gräter, J.: Noncommutative Prüfer and valuation rings. In: Bokut, L.A., Ershov, Y.L., Kostrikin, A.I. (eds.) Proceedings of the International Conference on Algebra, Part 2, Novosibirsk, 1989. Contemp. Math., vol. 131, pp. 253–269. Amer. Math. Soc., Providence, RI (1992) ⟶ *p. 132.*

37. Brussel, E.S.: Noncrossed products and nonabelian crossed products over $\mathbb{Q}(t)$ and $\mathbb{Q}((t))$. Amer. J. Math. **117**(2), 377–393 (1995) $\longrightarrow$ *pp. 476 and 488.*
38. Brussel, E.S.: Decomposability and embeddability of discretely Henselian division algebras. Israel J. Math. **96**(part A), 141–183 (1996) $\longrightarrow$ *p. 532.*
39. Brussel, E.S.: Division algebras not embeddable in crossed products. J. Algebra **179**(2), 631–655 (1996) $\longrightarrow$ *p. 488.*
40. Brussel, E.S.: Division algebra subfields introduced by an indeterminate. J. Algebra **188**(1), 216–255 (1997) $\longrightarrow$ *p. 487.*
41. Brussel, E.S.: An arithmetic obstruction to division algebra decomposability. Proc. Amer. Math. Soc. **128**(8), 2281–2285 (2000) $\longrightarrow$ *p. 496.*
42. Brussel, E.S.: The division algebras and Brauer group of a strictly Henselian field. J. Algebra **239**(1), 391–411 (2001) $\longrightarrow$ *pp. 374 and 375.*
43. Brussel, E.S.: Noncrossed products over $k_{\mathfrak{p}}(t)$. Trans. Amer. Math. Soc. **353**(5), 2115–2129 (2001) $\longrightarrow$ *p. 489.*
44. Brussel, E.S.: Alternating forms and the Brauer group of a geometric field. Trans. Amer. Math. Soc. **359**(7), 3025–3069 (2007) $\longrightarrow$ *pp. 374 and 375.*
45. Brussel, E.S., McKinnie, K., Tengan, E.: Indecomposable and noncrossed product division algebras over function fields of smooth p-adic curves. Adv. Math. **226**(5), 4316–4337 (2011) $\longrightarrow$ *p. 489.*
46. Buhler, J., Reichstein, Z.: On the essential dimension of a finite group. Compositio Math. **106**(2), 159–179 (1997) $\longrightarrow$ *p. 583.*
47. Caenepeel, S., Van Oystaeyen, F.: Brauer groups and the cohomology of graded rings. Monographs and Textbooks in Pure and Applied Mathematics, vol. 121. Marcel Dekker Inc., New York (1988) $\longrightarrow$ *pp. 293, 294, and 453.*
48. Cauchon, G.: Les T-anneaux et les anneaux à identités polynômiales noethériens. Ph.D. thesis, Orsay (1977) $\longrightarrow$ *p. 452.*
49. Cecioni, F.: Sopra un tipo di algebre prive di divisori dello zero. Rend. Circ. Mat. Palermo **47**(2), 209–254 (1923) $\longrightarrow$ *p. 452.*
50. Chase, S.U., Harrison, D.K., Rosenberg, A.: Galois theory and Galois cohomology of commutative rings. Mem. Amer. Math. Soc. **52**, 15–33 (1965) $\longrightarrow$ *p. 213.*
51. Chen, F.: Indecomposable and noncrossed product division algebras over curves over complete discrete valuation rings. J. Algebra **376**, 89–100 (2013) $\longrightarrow$ *p. 489.*
52. Chernousov, V., Merkurjev, A.S.: R-equivalence and special unitary groups. J. Algebra **209**(1), 175–198 (1998) $\longrightarrow$ *p. 559.*
53. Cohen, I.S.: On non-Archimedean normed spaces. Nederl. Akad. Wetensch., Proc. **51**, 693–698 (1948) = Indagationes Math. **10**, 244–249 $\longrightarrow$ *p. 132.*
54. Cohn, P.M.: On extending valuations in division algebras. Studia Sci. Math. Hungar. **16**(1–2), 65–70 (1981) $\longrightarrow$ *pp. 32 and 190.*
55. Cohn, P.M.: Skew fields. Encyclopedia of Mathematics and its Applications, vol. 57. Cambridge University Press, Cambridge (1995) $\longrightarrow$ *pp. 7, 8, and 32.*
56. Colliot-Thélène, J.L.: Quelques résultats de finitude pour le groupe SK_1 d'une algèbre de biquaternions. K-Theory **10**(1), 31–48 (1996) $\longrightarrow$ *p. 557.*
57. Coyette, C.: Mal'cev-Neumann rings and noncrossed product division algebras. J. Algebra Appl. **11**(3), 125052 (2012) (12 pp.) $\longrightarrow$ *pp. 476 and 489.*
58. Dedecker, P.: Une théorie algébrique des équations approchées. Bull. Soc. Math. France **83**, 331–364 (1955) $\longrightarrow$ *p. 32.*
59. DeMeyer, F., Ingraham, E.: Separable algebras over commutative rings. Lecture Notes in Mathematics, vol. 181. Springer, Berlin (1971) $\longrightarrow$ *pp. 242 and 289.*
60. Dickson, L.E.: Construction of division algebras. Trans. Amer. Math. Soc. **32**(2), 319–334 (1930) $\longrightarrow$ *p. 452.*
61. Draxl, P.K.: Corps gauches dont le groupe des commutateurs n'est pas égal au noyau de la norme réduite. In: Séminaire Delange-Pisot-Poitou, 17e année: 1975/76. Théorie des nombres: Fasc. 2, Exp. No. 26. Secrétariat Math., Paris (1977), 5 pp., available from Numdam $\longrightarrow$ *pp. 554 and 556.*

62. Draxl, P.K.: SK_1 von Algebren über vollständig diskret bewerteten Körpern und Galoiskohomologie abelscher Körpererweiterungen. J. Reine Angew. Math. **293/294**, 116–142 (1977) ⟶ *p. 556.*
63. Draxl, P.K.: Skew fields. London Mathematical Society Lecture Note Series, vol. 81. Cambridge University Press, Cambridge (1983) ⟶ *pp. 18, 48, 49, 58, 78, 349, 351, 356, 402, 536, 538, 545, 549, and 555.*
64. Draxl, P.K.: Ostrowski's theorem for Henselian valued skew fields. J. Reine Angew. Math. **354**, 213–218 (1984) ⟶ *pp. 190 and 375.*
65. Draxl, P.K., Kneser, M. (eds.): SK_1 von Schiefkörpern. Lecture Notes in Mathematics, vol. 778. Springer, Berlin (1980). Seminar held at Bielefeld and Göttingen, 1976 ⟶ *pp. 555 and 556.*
66. Dubisch, R.: Non-cyclic algebras of degree four and exponent two with pure maximal subfields. Bull. Amer. Math. Soc. **47**, 131–133 (1941) ⟶ *p. 487.*
67. Dubrovin, N.I.: Valuation rings in a simple central finite-dimensional algebra over a field. Uspehi Mat. Nauk **33**(3(201)), 167 (1978) ⟶ *p. 132.*
68. Dubrovin, N.I.: Noncommutative valuation rings. Trudy Moskov. Mat. Obshch. **45**, 265–280 (1982). English trans.: Trans. Moscow Math. Soc. **45**, 273–287 (1984) ⟶ *pp. 31 and 132.*
69. Dubrovin, N.I.: Noncommutative valuation rings in simple finite-dimensional algebras over a field. Mat. Sb. (N. S.) **123(165)**(4), 496–509 (1984). English trans.: Math. USSR Sb. **51**, 493–505 (1985) ⟶ *pp. 31 and 132.*
70. Efrat, I.: Valuations, orderings, and Milnor K-theory. Mathematical Surveys and Monographs, vol. 124. American Mathematical Society, Providence, RI (2006) ⟶ *pp. 190, 585, 589, 590, 591, 596, and 606.*
71. Elman, R., Lam, T.Y., Tignol, J.P., Wadsworth, A.R.: Witt rings and Brauer groups under multiquadratic extensions. I. Amer. J. Math. **105**(5), 1119–1170 (1983) ⟶ *pp. 532 and 533.*
72. Endler, O.: Valuation theory. Springer, New York (1972) ⟶ *pp. 139, 181, 585, 586, 590, and 606.*
73. Engler, A.J., Prestel, A.: Valued fields. Springer Monographs in Mathematics. Springer, Berlin (2005) ⟶ *pp. 12, 13, 20, 22, 25, 124, 136, 153, 164, 165, 166, 226, 585, 586, 588, 589, 594, 595, 596, 597, 601, 602, 604, 608, 609, and 610.*
74. Ershov, Y.L.: Valuations of division algebras, and the group SK_1. Dokl. Akad. Nauk SSSR **239**(4), 768–771 (1978). English Trans.: Soviet Math. Dokl. **19**(2), 395–399 (1978) ⟶ *p. 556.*
75. Ershov, Y.L.: Henselian valuations of division rings and the group SK_1. Mat. Sb. (N. S.) **117(159)**(1), 60–68 (1982). English Trans.: Math. USSR-Sb. **45**, 63–71 (1983) ⟶ *pp. 541 and 556.*
76. Ershov, Y.L.: Valued division rings. In: Fifth All-Union Symposium, Theory of Rings, Algebras, and Modules, pp. 53–55. Akad. Nauk SSSR Sibirsk. Otdel. Inst, Mat., Novosibirsk (1982) ⟶ *pp. 32 and 189.*
77. Ershov, Y.L.: Co-Henselian extensions and Henselization of division algebras. Algebra i Logika **27**(6), 649–658 (1988), 736. English Trans.: Algebra and Logic **27**, 401–407 (1988) ⟶ *p. 190.*
78. Fein, B., Schacher, M.M.: Galois groups and division algebras. J. Algebra **38**(1), 182–191 (1976) ⟶ *p. 488.*
79. Ferreira, M.A., Wadsworth, A.R.: Value functions and Dubrovin valuation rings on simple algebras. To appear in Trans. Amer. Math. Soc. (2013), preprint, available at arxiv:1307.5501 ⟶ *pp. 191 and 192.*
80. Ford, T.J.: On the Brauer group of a Laurent polynomial ring. J. Pure Appl. Algebra **51**(1–2), 111–117 (1988) ⟶ *p. 294.*
81. Garibaldi, S., Merkurjev, A., Serre, J.P.: Cohomological invariants in Galois cohomology. University Lecture Series, vol. 28. American Mathematical Society, Providence, RI (2003) ⟶ *p. 577.*

82. Gille, P.: Le problème de Kneser–Tits. Astérisque **326**, Exp. No. 983, vii, 39–81 (2009). Séminaire Bourbaki. Vol. 2007/2008 ⟶ *pp. 555 and 558.*
83. Gille, P., Pianzola, A.: Isotriviality and étale cohomology of Laurent polynomial rings. J. Pure Appl. Algebra **212**(4), 780–800 (2008) ⟶ *p. 375.*
84. Gille, P., Szamuely, T.: Central simple algebras and Galois cohomology. Cambridge Studies in Advanced Mathematics, vol. 101. Cambridge University Press, Cambridge (2006) ⟶ *pp. 49, 283, 450, 451, 468, 494, 508, and 518.*
85. Gille, S., Semenov, N.: On the Brauer group of the product of a torus and a semisimple algebraic group. Israel J. Math. **202**(1), 375–403 (2014) ⟶ *p. 375.*
86. Gordon, L.I.: Normal division algebras of degree four. Master's thesis, University of Chicago (1940) ⟶ *p. 472.*
87. Gräbe, H.G.: Über die Umkehraufgabe der reduzierten K-Theorie. In: Seminar D. Eisenbud/B. Singh/W. Vogel, Vol. 1. Teubner-Texte zur Math., vol. 29, pp. 94–107. Teubner, Leipzig (1980) ⟶ *p. 556.*
88. Gräter, J.: The "Defektsatz" for central simple algebras. Trans. Amer. Math. Soc. **330**(2), 823–843 (1992) ⟶ *pp. 132 and 191.*
89. Gräter, J.: Prime PI-rings in which finitely generated right ideals are principal. Forum Math. **4**(5), 447–463 (1992) ⟶ *p. 132.*
90. Gräter, J.: Dubrovin valuation rings and orders in central simple algebras. In: Kuhlmann, F.V., Kuhlmann, S., Marshall, M. (eds.) Valuation theory and its applications, Vol. I, Saskatoon, SK, 1999. Fields Inst. Commun., vol. 32, pp. 151–162. Amer. Math. Soc., Providence, RI (2002) ⟶ *p. 132.*
91. Greither, C.: Cyclic Galois extensions of commutative rings. Lecture Notes in Mathematics, vol. 1534. Springer, Berlin (1992) ⟶ *p. 213.*
92. Hanke, T.: An explicit example of a noncrossed product division algebra. Math. Nachr. **271**, 51–68 (2004) ⟶ *pp. 476 and 489.*
93. Hanke, T.: A twisted Laurent series ring that is a noncrossed product. Israel J. Math. **150**, 199–203 (2005) ⟶ *p. 489.*
94. Hanke, T.: Galois subfields of inertially split division algebras. J. Algebra **346**, 147–151 (2011) ⟶ *p. 490.*
95. Hanke, T., Neftin, D., Sonn, J.: Noncrossed product bounds over Henselian fields. Algebra Number Theory **8**(4), 837–855 (2014) ⟶ *p. 489.*
96. Hanke, T., Neftin, D., Wadsworth, A.R.: Galois subfields of tame division algebras. To appear in Israel J. Math. (2013), preprint, available at arXiv:1310.4436 ⟶ *pp. 489 and 490.*
97. Hanke, T., Sonn, J.: The location of noncrossed products in Brauer groups of Laurent series fields over global fields. Math. Ann. **350**(2), 313–337 (2011) ⟶ *p. 489.*
98. Hasse, H.: Über $\wp$-adische Schiefkörper und ihre Bedeutung für die Arithmetik hyperkomplexer Zahlsysteme. Math. Ann. **104**(1), 495–534 (1931) ⟶ *p. 30.*
99. Hazrat, R., Wadsworth, A.R.: SK_1 of graded division algebras. Israel J. Math. **183**, 117–163 (2011) ⟶ *pp. 536, 540, 556, and 557.*
100. Hazrat, R., Wadsworth, A.R.: Unitary SK_1 of graded and valued division algebras. Proc. Lond. Math. Soc. (3) **103**(3), 508–534 (2011) ⟶ *p. 559.*
101. Hoobler, R.T.: Functors of graded rings. In: Van Oystaeyen, F. (ed.) Methods in ring theory, Antwerp, 1983. NATO Adv. Sci. Inst. Ser. C Math. Phys. Sci., vol. 129, pp. 161–170. Reidel, Dordrecht (1984) ⟶ *p. 294.*
102. Hwang, Y.S., Wadsworth, A.R.: Algebraic extensions of graded and valued fields. Comm. Algebra **27**(2), 821–840 (1999) ⟶ *pp. 236 and 237.*
103. Hwang, Y.S., Wadsworth, A.R.: Correspondences between valued division algebras and graded division algebras. J. Algebra **220**(1), 73–114 (1999) ⟶ *pp. 80, 294, 295, and 449.*
104. Jacob, B.: Indecomposible division algebras of prime exponent. J. Reine Angew. Math. **413**, 181–197 (1991) ⟶ *pp. 501, 532, and 533.*

105. Jacob, B., Wadsworth, A.R.: A new construction of noncrossed product algebras. Trans. Amer. Math. Soc. **293**(2), 693–721 (1986) ⟶ *pp. 476 and 488.*
106. Jacob, B., Wadsworth, A.R.: Division algebras over Henselian fields. J. Algebra **128**(1), 126–179 (1990) ⟶ *pp. 294, 296, 449, 452, 453, and 490.*
107. Jacobson, N.: PI-algebras, An introduction. Lecture Notes in Mathematics, vol. 441. Springer, Berlin (1975) ⟶ *pp. 476 and 486.*
108. Jacobson, N.: Finite-dimensional division algebras over fields. Springer, Berlin (1996) ⟶ *pp. 283, 473, 475, and 490.*
109. Janusz, G.J.: Algebraic number fields. Pure and Applied Mathematics, vol. 55. Academic Press [A Subsidiary of Harcourt Brace Jovanovich, Publishers], New York–London (1973) ⟶ *p. 496.*
110. de Jong, A.J.: The period-index problem for the Brauer group of an algebraic surface. Duke Math. J. **123**(1), 71–94 (2004) ⟶ *p. 583.*
111. Kahn, B.: Cohomological approaches to SK_1 and SK_2 of central simple algebras. Doc. Math. Extra volume: Andrei A. Suslin sixtieth birthday, 317–369 (2010) ⟶ *pp. 557 and 558.*
112. Karpenko, N.A.: Torsion in CH^2 of Severi–Brauer varieties and indecomposability of generic algebras. Manuscripta Math. **88**(1), 109–117 (1995) ⟶ *p. 534.*
113. Karpenko, N.A.: Codimension 2 cycles on Severi–Brauer varieties. K-Theory **13**(4), 305–330 (1998) ⟶ *p. 533.*
114. Kelley, J.L.: General topology. D. Van Nostrand Company, Inc., Toronto–New York–London (1955) ⟶ *p. 218.*
115. Knus, M.A., Merkurjev, A., Rost, M., Tignol, J.P.: The book of involutions. American Mathematical Society Colloquium Publications, vol. 44. American Mathematical Society, Providence, RI (1998) ⟶ *pp. 71, 191, 325, 453, 507, 508, 557, 558, 565, 577, and 582.*
116. Knus, M.A., Ojanguren, M.: Théorie de la descente et algèbres d'Azumaya. Lecture Notes in Mathematics, vol. 389. Springer, Berlin (1974) ⟶ *pp. 207, 208, and 400.*
117. Krasner, M.: Une généralisation de la notion de corps—corpoïde. Un corpoïde remarquable de la théorie des corps valués. C. R. Acad. Sci. Paris **219**, 345–347 (1944) ⟶ *p. 32.*
118. Kuhlmann, F.V.: Elimination of ramification I: the generalized stability theorem. Trans. Amer. Math. Soc. **362**(11), 5697–5727 (2010) ⟶ *p. 190.*
119. Kuhlmann, F.V.: The defect. In: Fontana, M., Kabbaj, S.E., Olberding, B., Swanson, I. (eds.) Commutative algebra—Noetherian and non-Noetherian perspectives, pp. 277–318. Springer, New York (2011) ⟶ *p. 190.*
120. Kursov, V.V., Yanchevskiĭ, V.I.: Crossed products of simple algebras and their groups of automorphisms. Dokl. Akad. Nauk BSSR **32**(9), 777–780 (1988), 859. English trans.: Selected papers in K-theory, Amer. Math. Soc. Transl., Ser. 2, **154**, 75–80 (1992) ⟶ *p. 400.*
121. Lam, T.Y.: A first course in noncommutative rings. Graduate Texts in Mathematics, vol. 131. Springer, New York (1991) ⟶ *pp. 7, 8, 11, 161, and 550.*
122. Lam, T.Y.: Introduction to quadratic forms over fields. Graduate Studies in Mathematics, vol. 67. American Mathematical Society, Providence, RI (2005) ⟶ *pp. 21, 112, 508, and 527.*
123. Linnell, P.A.: Zero divisors and idempotents in group rings. Math. Proc. Cambridge Philos. Soc. **81**(3), 365–368 (1977) ⟶ *p. 493.*
124. Lipnitskiĭ, V.A.: On the Tannaka–Artin problem over special fields. Dokl. Akad. Nauk SSSR **228**(1), 26–29 (1976). English trans.: Soviet Math. Dokl. **17**, 639–642 (1976) ⟶ *p. 556.*
125. Lipnitskiĭ, V.A.: Triviality of the reduced Whitehead group over certain fields. Mat. Zametki **24**(5), 629–640 (1978), 733. English trans.: Math. Notes **24** (1978), 836–841 (1979) ⟶ *p. 557.*

126. Lipnitskiĭ, V.A.: Norms in fields of real algebraic functions and the reduced Whitehead group. Dokl. Akad. Nauk BSSR **26**(7), 585–588 (1982), 668. English trans.: Selected papers in K-theory, American Math. Soc. Translations, Ser. 2 **154**, 47–52 (1992) ⟶ *p. 556.*
127. Lipnitskiĭ, V.A.: The reduced Whitehead group for weakly and totally ramified division algebras. VestsīAkad. Navuk BSSR Ser. Fīz.-Mat. Navuk **2**, 28–30 (1986), 124. English trans.: Selected papers in K-theory, American Math. Soc. Translations, Ser. 2 **154**, 53–56 (1992) ⟶ *p. 556.*
128. Lorenz, F., Roquette, P.: The theorem of Grunwald–Wang in the setting of valuation theory. In: Kuhlmann, F.V., Kuhlmann, S., Marshall, M. (eds.) Valuation theory and its applications, Vol. II, Saskatoon, SK, 1999. Fields Inst. Commun., vol. 33, pp. 175–212. Amer. Math. Soc., Providence, RI (2003) ⟶ *pp. 481 and 494.*
129. Magid, A.R.: Brauer groups of linear algebraic groups with characters. Proc. Amer. Math. Soc. **71**(2), 164–168 (1978) ⟶ *p. 375.*
130. Mahdavi-Hezavehi, M.: Extending valuations to algebraic division algebras. Comm. Algebra **22**(11), 4373–4378 (1994) ⟶ *p. 32.*
131. Mal'cev, A.I.: On the embedding of group algebras in division algebras. Doklady Akad. Nauk SSSR (N. S.) **60**, 1499–1501 (1948) ⟶ *p. 32.*
132. Marubayashi, H., Miyamoto, H., Ueda, A.: Non-commutative valuation rings and semi-hereditary orders. K-Monographs in Mathematics, vol. 3. Kluwer Academic Publishers, Dordrecht (1997) ⟶ *pp. 31, 132, 190, 192, 265, and 596.*
133. Marubayashi, H., Van Oystaeyen, F.: Prime divisors and noncommutative valuation theory. Lecture Notes in Mathematics, vol. 2059. Springer, Berlin (2012) ⟶ *p. 132.*
134. Mathiak, K.: Bewertungen nicht kommutativer Körper. J. Algebra **48**(2), 217–235 (1977) ⟶ *p. 31.*
135. Mathiak, K.: Zur Bewertungstheorie nicht kommutativer Körper. J. Algebra **73**(2), 586–600 (1981) ⟶ *p. 31.*
136. Mathiak, K.: Valuations of skew fields and projective Hjelmslev spaces. Lecture Notes in Mathematics, vol. 1175. Springer, Berlin (1986) ⟶ *p. 31.*
137. Matsumura, H.: Commutative ring theory, 2nd edn. Cambridge Studies in Advanced Mathematics, vol. 8. Cambridge University Press, Cambridge (1989) ⟶ *pp. 289 and 511.*
138. Matzri, E., Rowen, L.H., Vishne, U.: Non-cyclic algebras with n-central elements. Proc. Amer. Math. Soc. **140**(2), 513–518 (2012) ⟶ *pp. 455, 463, and 487.*
139. McKinnie, K.: Prime to p extensions of the generic abelian crossed product. J. Algebra **317**(2), 813–832 (2007) ⟶ *pp. 487 and 534.*
140. McKinnie, K.: Indecomposable p-algebras and Galois subfields in generic abelian crossed products. J. Algebra **320**(5), 1887–1907 (2008) ⟶ *pp. 455, 472, 487, and 533.*
141. McKinnie, K.: Degeneracy and decomposability in abelian crossed products. J. Algebra **328**, 443–460 (2011) ⟶ *p. 533.*
142. Merkurjev, A.S.: Structure of the Brauer group of fields. Izv. Akad. Nauk SSSR Ser. Mat. **49**(4), 828–846 (1985), 895. English Trans.: Math. USSR Izv. **27**, 141–157 (1986) ⟶ *p. 296.*
143. Merkurjev, A.S.: Generic element in SK_1 for simple algebras. K-Theory **7**(1), 1–3 (1993) ⟶ *p. 558.*
144. Merkurjev, A.S.: Certain K-cohomology groups of Severi–Brauer varieties. In: Jacob, B., Rosenberg, A. (eds.) K-theory and algebraic geometry: connections with quadratic forms and division algebras, Santa Barbara, CA, 1992. Proc. Sympos. Pure Math., vol. 58, pp. 319–331. Amer. Math. Soc., Providence, RI (1995) ⟶ *p. 557.*
145. Merkurjev, A.S.: Invariants of algebraic groups. J. Reine Angew. Math. **508**, 127–156 (1999) ⟶ *p. 557.*
146. Merkurjev, A.S.: Cohomological invariants of simply connected groups of rank 3. J. Algebra **227**(2), 614–632 (2000) ⟶ *p. 559.*

147. Merkurjev, A.S.: The group SK_1 for simple algebras. K-Theory **37**(3), 311–319 (2006) ⟶ *p. 558.*
148. Merkurjev, A.S.: Essential dimension. In: Baeza, R., Chan, W.K., Hoffmann, D.W., Schulze-Pillot, R. (eds.) Quadratic forms—algebra, arithmetic, and geometry. Contemp. Math., vol. 493, pp. 299–325. Amer. Math. Soc., Providence, RI (2009) ⟶ *pp. 577 and 583.*
149. Merkurjev, A.S.: Essential p-dimension of $\mathrm{PGL}(p^2)$. J. Amer. Math. Soc. **23**(3), 693–712 (2010) ⟶ *p. 584.*
150. Merkurjev, A.S.: A lower bound on the essential dimension of simple algebras. Algebra Number Theory **4**(8), 1055–1076 (2010) ⟶ *pp. 561, 565, 575, 577, and 578.*
151. Merkurjev, A.S.: Essential dimension: a survey. Transform. Groups **18**(2), 415–481 (2013) ⟶ *p. 583.*
152. Merkurjev, A.S.: Suslin's conjecture on the reduced Whitehead group of a simple algebra (2014). Preprint, available at `http://www.math.uni-bielefeld.de/LAG/`: No. 544 ⟶ *p. 558.*
153. Meyer, A., Reichstein, Z.: An upper bound on the essential dimension of a central simple algebra. J. Algebra **329**, 213–221 (2011) ⟶ *p. 584.*
154. Monastyrnyĭ, A.P., Yanchevskiĭ, V.I.: Whitehead groups of algebraic groups of classical type, and R-equivalence. Dokl. Akad. Nauk SSSR **314**(1), 110–114 (1990). English trans.: Soviet Math. Dokl. **42**, 351–355 (1991) ⟶ *p. 559.*
155. Monna, A.F.: Sur les espaces linéaires normés. I, II, III, IV. Nederl. Akad. Wetensch., Proc. **49**, 1045–1055, 1056–1062, 1134–1141, 1142–1152 = Indagationes Math. **8**, 643–653, 654–660, 682–689, 690–700 (1946) ⟶ *p. 132.*
156. Morandi, P.: The Henselization of a valued division algebra. J. Algebra **122**(1), 232–243 (1989) ⟶ *pp. 189 and 190.*
157. Morandi, P.: Value functions on central simple algebras. Trans. Amer. Math. Soc. **315**(2), 605–622 (1989) ⟶ *pp. 132 and 133.*
158. Morandi, P.: An approximation theorem for Dubrovin valuation rings. Math. Z. **207**(1), 71–81 (1991) ⟶ *pp. 132 and 191.*
159. Morandi, P.: On defective division algebras. In: Jacob, B., Rosenberg, A. (eds.) K-theory and algebraic geometry: connections with quadratic forms and division algebras, Santa Barbara, CA, 1992. Proc. Sympos. Pure Math., vol. 58, pp. 359–367. Amer. Math. Soc., Providence, RI (1995) ⟶ *p. 190.*
160. Morandi, P.: Field and Galois theory. Graduate Texts in Mathematics, vol. 167. Springer, New York (1996) ⟶ *p. 320.*
161. Morandi, P., Sethuraman, B.A.: Kummer subfields of tame division algebras. J. Algebra **172**(2), 554–583 (1995) ⟶ *p. 490.*
162. Morandi, P., Sethuraman, B.A.: Generalized cocycles with values in one-units of Henselian valued division algebras. J. Algebra **224**(1), 123–139 (2000) ⟶ *p. 490.*
163. Morandi, P., Wadsworth, A.R.: Integral Dubrovin valuation rings. Trans. Amer. Math. Soc. **315**(2), 623–640 (1989) ⟶ *pp. 80 and 132.*
164. Mounirh, K.: Nicely semiramified division algebras over Henselian fields. Int. J. Math. Math. Sci. **2005**(4), 571–577 (2005) ⟶ *p. 452.*
165. Mounirh, K.: Nondegenerate semiramified valued and graded division algebras. Comm. Algebra **36**(12), 4386–4406 (2008) ⟶ *pp. 453, 455, 487, and 533.*
166. Mounirh, K.: Kummer subfields of tame division algebras over Henselian fields. J. Pure Appl. Algebra **214**(4), 440–448 (2010) ⟶ *p. 490.*
167. Mounirh, K., Wadsworth, A.R.: Subfields of nondegenerate tame semiramified division algebras. Comm. Algebra **39**(2), 462–485 (2011) ⟶ *pp. 236 and 237.*
168. Nagata, M.: Local rings. Interscience Tracts in Pure and Applied Mathematics, vol. 13. Interscience Publishers a division of John Wiley & Sons, New York–London (1962) ⟶ *pp. 156 and 596.*
169. Năstăsescu, C.: Anneaux et modules gradués. Rev. Roumaine Math. Pures Appl. **21**(7), 911–931 (1976) ⟶ *p. 80.*

170. Năstăsescu, C., Van Oystaeyen, F.: Graded and filtered rings and modules. Lecture Notes in Mathematics, vol. 758. Springer, Berlin (1979) ⟶ *p. 80.*
171. Neukirch, J., Schmidt, A., Wingberg, K.: Cohomology of number fields. Grundlehren der Mathematischen Wissenschaften [Fundamental Principles of Mathematical Sciences], vol. 323. Springer, Berlin (2000) ⟶ *p. 494.*
172. Neumann, B.H.: On ordered division rings. Trans. Amer. Math. Soc. **66**, 202–252 (1949) ⟶ *p. 32.*
173. Nganou, J.B.: On the Dec group of finite abelian Galois extensions over global fields. J. Algebra **322**(2), 453–462 (2009) ⟶ *p. 453.*
174. Orzech, M., Small, C.: The Brauer group of commutative rings. Lecture Notes in Pure and Applied Mathematics, vol. 11. Marcel Dekker Inc., New York (1975) ⟶ *p. 289.*
175. Ostrowski, A.: Untersuchungen zur arithmetischen Theorie der Körper. (Die Theorie der Teilbarkeit in allgemeinen Körpern.). Math. Z. **39**, I 269–320; II, III 321–404 (1935) ⟶ *p. 190.*
176. Ousmane, C.: Algèbres cycliques génériques. Bull. Soc. Math. Belg. Sér. B **40**(1), 43–59 (1988) ⟶ *p. 493.*
177. Passman, D.S.: The algebraic structure of group rings. Robert E. Krieger Publishing Co. Inc., Melbourne, FL (1985). Reprint of the 1977 original ⟶ *p. 493.*
178. Pierce, R.S.: Associative algebras. Graduate Texts in Mathematics, vol. 88. Springer, New York (1982) ⟶ *pp. 18, 48, 58, 79, 136, 243, 245, 250, 251, 254, 402, 472, 481, 482, 524, 545, and 564.*
179. Platonov, V.P.: On the Tannaka–Artin problem. Dokl. Akad. Nauk SSSR **221**(5), 1038–1041 (1975). English trans.: Soviet Math. Dokl. **16**, 468–473 (1975) ⟶ *pp. 535 and 555.*
180. Platonov, V.P.: The Tannaka–Artin problem, and groups of projective conorms. Dokl. Akad. Nauk SSSR **222**(6), 1299–1302 (1975). English trans.: Soviet Math. Dokl. **16**, 782–786 (1975) ⟶ *pp. 535 and 555.*
181. Platonov, V.P.: Infiniteness of the reduced Whitehead group in the Tannaka–Artin problem. Mat. Sb. (N. S.) **100(142)**(2), 191–200 (1976), 335. English trans.: Math. USSR-Sb. **29**, 167–176 (1976) ⟶ *p. 556.*
182. Platonov, V.P.: Reduced K-theory and approximation in algebraic groups. Trudy Mat. Inst. Steklov. **142**, 198–207 (1976), 270. English trans.: Proc. Steklov Inst. Math. **142**, 213–224 (1979) ⟶ *p. 555.*
183. Platonov, V.P.: The Tannaka–Artin problem, and reduced K-theory. Izv. Akad. Nauk SSSR Ser. Mat. **40**(2), 227–261 (1976), 469. English trans.: Math. USSR Izv. **10**, 211–243 (1976) ⟶ *pp. 535, 555, and 556.*
184. Platonov, V.P.: Birational properties of the reduced Whitehead group. Dokl. Akad. Nauk BSSR **21**(3), 197–198 (1977), 283. English trans.: Selected papers in K-theory, American Math. Soc. Translations, Ser. 2, **154**, 7–9 (1992) ⟶ *p. 556.*
185. Platonov, V.P.: Algebraic groups and reduced K-theory. In: Lehto, O. (ed.) Proceedings of the International Congress of Mathematicians, Helsinki, 1978, pp. 311–317. Academia Scientiarum Fennica, Helsinki (1980) ⟶ *p. 555.*
186. Platonov, V.P., Yanchevskiĭ, V.I.: The structure of unitary groups and of the commutants of a simple algebra over global fields. Dokl. Akad. Nauk SSSR **208**, 541–544 (1973). English trans.: Soviet Math. Dokl. **14**, 132–136 (1973) ⟶ *p. 558.*
187. Platonov, V.P., Yanchevskiĭ, V.I.: The Kneser–Tits conjecture for the unitary groups. Dokl. Akad. Nauk SSSR **225**(1), 48–51 (1975). English trans.: Soviet Math. Dokl. **16**(1975), 1456–1460 (1976) ⟶ *p. 559.*
188. Platonov, V.P., Yanchevskiĭ, V.I.: Stability in reduced K-theory. Dokl. Akad. Nauk SSSR **242**(4), 769–772 (1978). English trans.: Soviet Math. Dokl. **19**, 1200–1203 (1978) ⟶ *pp. 555 and 557.*
189. Platonov, V.P., Yanchevskiĭ, V.I.: SK_1 for division rings of noncommutative rational functions. Dokl. Akad. Nauk SSSR **249**(5), 1064–1068 (1979). English trans.: Soviet Math. Doklady **20**, 1393–1397 (1979) ⟶ *p. 557.*

190. Platonov, V.P., Yanchevskiĭ, V.I.: Dieudonné's conjecture on the structure of unitary groups over a skew-field and Hermitian K-theory. Izv. Akad. Nauk SSSR Ser. Mat. **48**(6), 1266–1294 (1984). English trans.: Math. USSR Izv. **25**, 573–599 (1985) ⟶ *p. 453.*
191. Platonov, V.P., Yanchevskiĭ, V.I.: Finite-dimensional Henselian division algebras. Dokl. Akad. Nauk SSSR **297**(3), 542–547 (1987). English trans.: Soviet Math. Dokl. **36**, 502–506 (1988) ⟶ *p. 452.*
192. Platonov, V.P., Yanchevskiĭ, V.I.: On the theory of Henselian division algebras. Dokl. Akad. Nauk SSSR **297**(2), 294–298 (1987). English trans.: Soviet Math. Dokl. **36**, 468–472 (1988) ⟶ *pp. 449 and 450.*
193. Platonov, V.P., Yanchevskiĭ, V.I.: Finite-dimensional skew fields. In: Vavilov, N.A. (ed.) Algebra, 9 (Russian). Itogi Nauki i Tekhniki, pp. 144–262, 268. Vsesoyuz. Inst. Nauchn. i Tekhn. Inform. (VINITI), Moscow (1992). English trans.: Finite-dimensional division algebras, in Algebra IX, Encyclopaedia Math. Sci., vol. 77, pp. 121–239, Springer, Berlin (1995) ⟶ *p. 555.*
194. Raynaud, M.: Anneaux locaux henséliens. Lecture Notes in Mathematics, vol. 169. Springer, Berlin (1970) ⟶ *pp. 156 and 596.*
195. Reichstein, Z.: On the notion of essential dimension for algebraic groups. Transform. Groups **5**(3), 265–304 (2000) ⟶ *p. 583.*
196. Reichstein, Z.: Essential dimension. In: Proceedings of the International Congress of Mathematicians. Volume II, pp. 162–188. Hindustan Book Agency, New Delhi (2010) ⟶ *p. 583.*
197. Reichstein, Z., Youssin, B.: Essential dimensions of algebraic groups and a resolution theorem for G-varieties. Canad. J. Math. **52**(5), 1018–1056 (2000). With an appendix by János Kollár and Endre Szabó ⟶ *pp. 373 and 375.*
198. Reichstein, Z., Youssin, B.: Splitting fields of G-varieties. Pacific J. Math. **200**(1), 207–249 (2001) ⟶ *p. 476.*
199. Reiner, I.: Maximal orders. London Mathematical Society Monographs, vol. 5. Academic Press, London–New York (1975) ⟶ *pp. 253, 397, 472, 474, 481, 482, 493, 494, 495, 524, 527, 528, 537, 538, and 539.*
200. Renard, J.F., Tignol, J.P., Wadsworth, A.R.: Graded Hermitian forms and Springer's theorem. Indag. Math. (N. S.) **18**(1), 97–134 (2007) ⟶ *p. 132.*
201. de Rham, G.: Sur l'analysis situs des variétés à n dimensions. J. Math. Pures Appl. (9) **10**, 115–200 (1931) ⟶ *p. 373.*
202. Ribenboim, P.: Le théorème d'approximation pour les valuations de Krull. Math. Z. **68**, 1–18 (1957) ⟶ *p. 164.*
203. Ribenboim, P.: Théorie des valuations. Deuxième édition multigraphiée. Séminaire de Mathématiques Supérieures, No. 9 (Été 1964). Les Presses de l'Université de Montréal, Montréal, Québec (1968) ⟶ *p. 190.*
204. Ribenboim, P.: Equivalent forms of Hensel's lemma. Exposition. Math. **3**(1), 3–24 (1985) ⟶ *pp. 595 and 596.*
205. Ribes, L., Zalesskii, P.: Profinite groups, 2nd edn. Ergebnisse der Mathematik und ihrer Grenzgebiete. 3. Folge., vol. 40. Springer, Berlin (2010) ⟶ *p. 373.*
206. Risman, L.J.: Cyclic algebras, complete fields, and crossed products. Israel J. Math. **28**(1–2), 113–128 (1977) ⟶ *p. 488.*
207. Roquette, P.: History of valuation theory. I. In: Kuhlmann, F.V., Kuhlmann, S., Marshall, M. (eds.) Valuation theory and its applications, Vol. I, Saskatoon, SK, 1999. Fields Inst. Commun., vol. 32, pp. 291–355. Amer. Math. Soc., Providence, RI (2002) ⟶ *p. 30.*
208. Rosen, M.: Number theory in function fields. Graduate Texts in Mathematics, vol. 210. Springer, New York (2002) ⟶ *p. 496.*
209. Rosset, S.: Group extensions and division algebras. J. Algebra **53**(2), 297–303 (1978) ⟶ *p. 493.*

210. Rost, M.: Computation of some essential dimensions (2000). Unpublished, available at http://www.math.uni-bielefeld.de/~rost/ed.html ⟶ *p. 584.*
211. Rotman, J.J.: An introduction to the theory of groups, 3rd edn. Allyn and Bacon Inc., Boston, MA (1984) ⟶ *p. 461.*
212. Rowen, L.H.: Correction to: "Cyclic division algebras". Israel J. Math. **43**(3), 277–280 (1982) ⟶ *p. 533.*
213. Rowen, L.H.: Cyclic division algebras. Israel J. Math. **41**(3), 213–234 (1982) ⟶ *pp. 374 and 533.*
214. Rowen, L.H.: Division algebras of exponent 2 and characteristic 2. J. Algebra **90**(1), 71–83 (1984) ⟶ *p. 533.*
215. Rowen, L.H.: Ring theory. Vol. II. Pure and Applied Mathematics, vol. 128. Academic Press Inc., Boston, MA (1988) ⟶ *p. 533.*
216. Ruozzi, A.: Essential p-dimension of PGL_n. J. Algebra **328**, 488–494 (2011) ⟶ *p. 584.*
217. Saltman, D.J.: Splittings of cyclic p-algebras. Proc. Amer. Math. Soc. **62**(2), 223–228 (1977) ⟶ *pp. 475 and 490.*
218. Saltman, D.J.: Noncrossed product p-algebras and Galois p-extensions. J. Algebra **52**(2), 302–314 (1978) ⟶ *pp. 472, 487, and 488.*
219. Saltman, D.J.: Noncrossed products of small exponent. Proc. Amer. Math. Soc. **68**(2), 165–168 (1978) ⟶ *pp. 476, 484, 485, and 490.*
220. Saltman, D.J.: Indecomposable division algebras. Comm. Algebra **7**(8), 791–817 (1979) ⟶ *pp. 375, 492, and 531.*
221. Saltman, D.J.: Division algebras over discrete valued fields. Comm. Algebra **8**(18), 1749–1774 (1980) ⟶ *p. 453.*
222. Saltman, D.J.: The Brauer group and the center of generic matrices. J. Algebra **97**(1), 53–67 (1985) ⟶ *p. 296.*
223. Saltman, D.J.: Lectures on division algebras. CBMS Regional Conference Series in Mathematics, vol. 94. American Mathematical Society, Providence, RI (1999) ⟶ *pp. 476 and 533.*
224. Schacher, M.M., Small, L.W.: Noncrossed products in characteristic p. J. Algebra **24**, 100–103 (1973) ⟶ *p. 488.*
225. Scharlau, W.: Über die Brauer-Gruppe eines Hensel–Körpers. Abh. Math. Sem. Univ. Hamburg **33**, 243–249 (1969) ⟶ *pp. 294 and 375.*
226. Schilling, O.F.G.: Noncommutative valuations. Bull. Amer. Math. Soc. **51**, 297–304 (1945) ⟶ *p. 30.*
227. Schilling, O.F.G.: The Theory of Valuations. Mathematical Surveys, vol. 4. American Mathematical Society, New York, NY (1950) ⟶ *pp. 30, 85, and 136.*
228. Schofield, A., Van den Bergh, M.: The index of a Brauer class on a Brauer–Severi variety. Trans. Amer. Math. Soc. **333**(2), 729–739 (1992) ⟶ *p. 532.*
229. Serre, J.-P.: Local fields. Graduate Texts in Mathematics, vol. 67. Springer, New York (1979). English trans. of *Corps Locaux*, Hermann, Paris (1968) ⟶ *pp. 23, 395, 401, 467, and 499.*
230. Sethuraman, B.A.: Indecomposable division algebras. Proc. Amer. Math. Soc. **114**(3), 661–665 (1992) ⟶ *pp. 493 and 534.*
231. Sivatski, A.S.: On indecomposable algebras of exponent 2. Israel J. Math. **164**, 365–379 (2008) ⟶ *pp. 533 and 534.*
232. Srinivas, V.: Algebraic K-theory, 2nd edn. Progress in Mathematics, vol. 90. Birkhäuser Boston Inc., Boston, MA (1996) ⟶ *p. 518.*
233. Suslin, A.A.: SK_1 of division algebras and Galois cohomology. In: Suslin, A.A. (ed.) Algebraic K-theory. Adv. Soviet Math., vol. 4, pp. 75–99. Amer. Math. Soc., Providence, RI (1991) ⟶ *p. 558.*
234. Suslin, A.A.: SK_1 of division algebras and Galois cohomology revisited. In: Proceedings of the St. Petersburg Mathematical Society. Vol. XII. Amer. Math. Soc. Transl. Ser. 2, vol. 219, pp. 125–147. Amer. Math. Soc., Providence, RI (2006) ⟶ *p. 558.*

235. Tignol, J.P.: Produits croisés abéliens. J. Algebra **70**(2), 420–436 (1981) ⟶ *pp. 450 and 451.*
236. Tignol, J.P.: Sur les décompositions des algèbres à division en produit tensoriel d'algèbres cycliques. In: Verschoren, A.H.M.J. (ed.) Brauer groups in ring theory and algebraic geometry, Wilrijk, 1981. Lecture Notes in Math., vol. 917, pp. 126–145. Springer, Berlin (1982) ⟶ *p. 374.*
237. Tignol, J.P.: Cyclic and elementary abelian subfields of Malcev-Neumann division algebras. J. Pure Appl. Algebra **42**(2), 199–220 (1986) ⟶ *p. 490.*
238. Tignol, J.P.: Algèbres indécomposables d'exposant premier. Adv. Math. **65**(3), 205–228 (1987) ⟶ *pp. 530, 533, and 534.*
239. Tignol, J.P.: Generalized crossed products. Sem. Math. (nouvelle série), vol. 106. Université Catholique de Louvain, Louvain-La-Neuve, Belgium (1987); available at `http://perso.uclouvain.be/jean-pierre.tignol/Textes.html` ⟶ *p. 400.*
240. Tignol, J.P.: Algèbres à division et extensions de corps sauvagement ramifiées de degré premier. J. Reine Angew. Math. **404**, 1–38 (1990) ⟶ *pp. 453 and 454.*
241. Tignol, J.P.: Classification of wild cyclic field extensions and division algebras of prime degree over a Henselian field. In: Proceedings of the International Conference on Algebra, Part 2, Novosibirsk, 1989. Contemp. Math., vol. 131, pp. 491–508. Amer. Math. Soc., Providence, RI (1992) ⟶ *p. 454.*
242. Tignol, J.P., Amitsur, S.A.: Kummer subfields of Mal'cev–Neumann division algebras. Israel J. Math. **50**(1–2), 114–144 (1985) ⟶ *pp. 449 and 490.*
243. Tignol, J.P., Amitsur, S.A.: Symplectic modules. Israel J. Math. **54**(3), 266–290 (1986) ⟶ *pp. 373 and 490.*
244. Tignol, J.P., Amitsur, S.A.: Totally ramified splitting fields of central simple algebras over Henselian fields. J. Algebra **98**(1), 95–101 (1986) ⟶ *p. 487.*
245. Tignol, J.P., Wadsworth, A.R.: Totally ramified valuations on finite-dimensional division algebras. Trans. Amer. Math. Soc. **302**(1), 223–250 (1987) ⟶ *pp. 374, 375, 449, and 487.*
246. Tignol, J.P., Wadsworth, A.R.: Value functions and associated graded rings for semisimple algebras. Trans. Amer. Math. Soc. **362**(2), 687–726 (2010) ⟶ *pp. 80, 133, 190, 265, 294, and 295.*
247. Tignol, J.P., Wadsworth, A.R.: Valuations on algebras with involution. Math. Ann. **351**(1), 109–148 (2011) ⟶ *pp. 132 and 191.*
248. Tits, J.: Groupes de Whitehead de groupes algébriques simples sur un corps (d'après V.P. Platonov et al.). In: Séminaire Bourbaki, 29e année (1976/77), Exp. No. 505. Lecture Notes in Math., vol. 677, pp. 218–236. Springer, Berlin (1978) ⟶ *p. 555.*
249. Tomchin, I.L., Yanchevskiĭ, V.I.: On defects of valued division algebras. Algebra i Analiz **3**(3), 147–164 (1991). English trans.: St. Petersburg Math. J. **3**, 631–647 (1992) ⟶ *p. 190.*
250. Van Geel, J., Van Oystaeyen, F.: About graded fields. Nederl. Akad. Wetensch. Indag. Math. **43**(3), 273–286 (1981) ⟶ *pp. 236 and 237.*
251. Voskresenskiĭ, V.E.: The reduced Whitehead group of a simple algebra. Uspehi Mat. Nauk **32**(6(198)), 247–248 (1977) ⟶ *p. 555.*
252. Voskresenskiĭ, V.E.: Algebraic groups and their birational invariants. Translations of Mathematical Monographs, vol. 179. American Mathematical Society, Providence, RI (1998) ⟶ *p. 555.*
253. Wadsworth, A.R.: Extending valuations to finite-dimensional division algebras. Proc. Amer. Math. Soc. **98**(1), 20–22 (1986) ⟶ *p. 32.*
254. Wadsworth, A.R.: The residue division ring of a valued division algebra. Bull. Soc. Math. Belg. Sér. A **40**(2), 307–322 (1988). Deuxième Contact Franco-Belge en Algèbre (Faulx-les-Tombes, 1987) ⟶ *p. 453.*
255. Wadsworth, A.R.: Dubrovin valuation rings and Henselization. Math. Ann. **283**(2), 301–328 (1989) ⟶ *pp. 31, 132, and 192.*

256. Wadsworth, A.R.: Valuation theory on finite dimensional division algebras. In: Kuhlmann, F.V., Kuhlmann, S., Marshall, M. (eds.) Valuation theory and its applications, Vol. I, Saskatoon, SK, 1999. Fields Inst. Commun., vol. 32, pp. 385–449. Amer. Math. Soc., Providence, RI (2002) ⟶ *pp. 132 and 555.*
257. Wadsworth, A.R.: Unitary SK_1 of semiramified graded and valued division algebras. Manuscripta Math. **139**(3–4), 343–389 (2012) ⟶ *pp. 556 and 559.*
258. Wadsworth, A.R., Yanchevskiĭ, V.I.: Unitary SK_1 for a graded division ring and its quotient division ring. J. Algebra **352**, 62–78 (2012) ⟶ *p. 559.*
259. Wall, C.T.C.: Quadratic forms on finite groups, and related topics. Topology **2**, 281–298 (1963) ⟶ *p. 373.*
260. Witt, E.: Schiefkörper über diskret bewerteten Körpern. J. Reine Angew. Math. **176**, 153–156 (1937) ⟶ *p. 296.*
261. Yanchevskiĭ, V.I.: Commutants of simple algebras with a surjective reduced norm. Dokl. Akad. Nauk SSSR **221**(5), 1056–1058 (1975). English trans.: Soviet Math. Dokl. **16**, 492–495 (1975) ⟶ *p. 557.*
262. Yanchevskiĭ, V.I.: Division algebras over Henselian discretely valued fields, and the Tannaka–Artin problem. Dokl. Akad. Nauk SSSR **226**(2), 281–283 (1976). English trans.: Soviet Math. Dokl. **17**, 113–116 (1976) ⟶ *p. 556.*
263. Yanchevskiĭ, V.I.: Reduced unitary K-theory and division algebras over Henselian discretely valued fields. Izv. Akad. Nauk SSSR Ser. Mat. **42**(4), 879–918 (1978). English trans.: Math. USSR-Izv. **13**, 175–213 (1979) ⟶ *p. 559.*
264. Yanchevskiĭ, V.I.: The inverse problem of reduced K-theory. Mat. Zametki **26**(3), 475–482 (1979), 495. English trans.: Math. Notes **26**(1979), 728–731 (1980) ⟶ *p. 559.*
265. Yanchevskiĭ, V.I.: Reduced unitary K-theory. Applications to algebraic groups. Mat. Sb. (N. S.) **110(152)**(4), 579–596 (1979). English trans.: Math. USSR Sbornik **38**, 533–548 (1981) ⟶ *p. 559.*
266. Yanchevskiĭ, V.I.: Reduced Whitehead unitary groups of skew-fields of noncommutative rational functions. Zap. Nauchn. Sem. Leningrad. Otdel. Mat. Inst. Steklov. (LOMI) **94**, 142–148 (1979), 152. English trans.: J. Soviet Math. **19**, 1067–1071 (1982) ⟶ *p. 559.*
267. Yanchevskiĭ, V.I.: Reduced unitary Whitehead groups and noncommutative rational functions. Dokl. Akad. Nauk BSSR **24**(7), 588–591 (1980), 667. English trans.: Amer. Math. Soc. Transl., Ser. 2, **154**, 63–67 (1992) ⟶ *p. 559.*

Subject Index

J.-P. Tignol, A.R. Wadsworth, *Value Functions on Simple Algebras, and Associated Graded Rings*, Springer Monographs in Mathematics,
DOI 10.1007/978-3-319-16360-4

Index of Notation

J.-P. Tignol, A.R. Wadsworth, *Value Functions on Simple Algebras, and Associated Graded Rings*, Springer Monographs in Mathematics, DOI 10.1007/978-3-319-16360-4

MIX
Papier aus verantwortungsvollen Quellen
Paper from responsible sources
FSC® C105338
FSC
www.fsc.org

If you have any concerns about our products,
you can contact us on
ProductSafety@springernature.com

In case Publisher is established outside the EU,
the EU authorized representative is:
Springer Nature Customer Service Center GmbH
Europaplatz 3, 69115 Heidelberg, Germany

Printed by Libri Plureos GmbH
in Hamburg, Germany